College Algebra and Trigonometry

College Algebra and Trigonometry

MARGARET L. LIAL ▼▼▼
American River College

E. JOHN HORNSBY, JR. ▼▼▼
University of New Orleans

DAVID I. SCHNEIDER ▼▼▼
University of Maryland

 **ADDISON-WESLEY**

An imprint of Addison Wesley Longman, Inc.

Reading, Massachusetts • Menlo Park, California • New York • Harlow, England
Don Mills, Ontario • Sydney • Mexico City • Madrid • Amsterdam

Sponsoring Editor: Anne Kelly
Developmental Editor: Lynn Mooney
Project Editor: Lisa A. De Mol
Design Administrator: Jess Schaal
Text Design: Lesiak/Crampton Design Inc.: Lucy Lesiak
Cover Design: Lesiak/Crampton Design Inc.: Lucy Lesiak
Cover Illustration: Precision Graphics
Production Administrator: Randee Wire
Compositor: Interactive Composition Corporation
Printer and Binder: R.R. Donnelley & Sons
Cover Printer: Phoenix Color Corporation

College Algebra and Trigonometry
Copyright © 1997 by Addison-Wesley Educational Publishers Inc.

Library of Congress Cataloging-in-Publication Data
Lial, Margaret L.
    College algebra and trigonometry / Margaret L. Lial, E. John Hornsby, Jr., David I. Schneider.
        p.   cm.
    Includes index.
    ISBN 0–673–98046–4
    1. Algebra.   2. Trigonometry   I. Hornsby, E. John.   II. Schneider, David I.,
    1942–   .   III. Title
    QA154.2.L522   1996
    512′.13—dc20   96–653
                                                                                    CIP

1 2 3 4 5 6 7 8 9 10—DOC—99 98 97 96

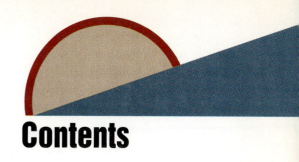

Contents

Preface

College Algebra and Trigonometry is written for students in a traditional college algebra and trigonometry course. This text combines material from the successful Lial, Hornsby, Schneider texts, *College Algebra,* Seventh Edition, and *Trigonometry,* Sixth Edition. We assume students have had at least one earlier course in algebra, but we include a thorough review in Chapter 1. For most students, the topics in Chapter 2 will also be review. We also assume students have some background in geometry.

Although this book is intended for a traditional course, we have acknowledged the growing interest in using graphing calculators to augment and deepen the concepts typically presented in college algebra and trigonometry by including optional graphing calculator material.

Content Highlights ▼▼▼

- The binomial theorem is introduced, using Pascal's Triangle, in the section that reviews polynomial operations. A complete presentation of the binomial theorem is given in the final chapter.
- The content in Chapter 3 includes relations and general function concepts, including symmetry, translation, and reflection. The conic sections are presented in a separate chapter on analytic geometry, which includes a discussion of eccentricity.
- In the chapter on polynomial and rational functions, we emphasize general graphing techniques for polynomial functions, considering end behavior, turning points, and the expected number of zeros.
- Chapter 5, on exponential and logarithmic functions, begins with a section on inverse functions, because this is the first real need for inverses.
- Systems of equations are combined with matrices, because matrix solutions of linear systems have become more important with the increasing use of technology.
- Although we introduce the trigonometric functions using triangles in Chapter 6, an alternative Chapter 6 with the unit circle approach is included as an appendix.
- A section on parametric equations is included in Chapter 8, because this topic is assuming more importance with the increase in the use of technology.

5

Exponential and Logarithmic Functions

5.1 Inverse Functions
5.2 Exponential Functions
5.3 Logarithmic Functions
5.4 Evaluating Logarithms;
   Change of Base
5.5 Exponential and
   Logarithmic Equations
5.6 Exponential Growth or
   Decay

In 1896 Swedish scientist Svante Arrhenius first predicted the greenhouse effect resulting from emissions of carbon dioxide by industrialized countries. In his classic calculation, he was able to estimate that a doubling of the carbon dioxide level in the atmosphere would raise the average global temperature by 7°F to 11°F. Since global warming would not be uniform, changes as small as 4.5°F in the average temperature could have drastic climatic effects, particularly on the central plains of North America. Sea levels could rise dramatically as a result of both thermal expansion and the melting of ice caps. The annual cost to the United States economy could reach $60 billion.

The burning of fossil fuels, deforestation, and changes in land use from 1850 to 1986 put approximately 312 billion tons of carbon into the atmosphere, mostly in the form of carbon dioxide. Burning of fossil fuels produces 5.4 billion tons of carbon each year which is absorbed by both the atmosphere and the oceans. A critical aspect of the accumulation of carbon dioxide in the atmosphere is that it is irreversible and its effect requires hundreds of years to disappear. In 1990 the International Panel of Climate Change (IPCC) reported that if current trends of burning of fossil fuel and deforestation

Sources: Clime, W., *The Economics of Global Warming.* Institute for International Economics, Washington, D.C., 1992.
Kraljic, M. (Editor), *The Greenhouse Effect*, The H. W. Wilson Company, New York, 1992.
International Panel on Climate Change (IPCC), 1990.
Wuebbles, D. and J. Edmonds, *Primer of Greenhouse Gases*. Lewis Publishers, Inc., Chelsea, Michigan, 1991.

Chapter Openers present a genuine application of the material to be presented.

Exercises corresponding to the Chapter Openers are marked with a special symbol.

deer population in using the equation 0,000 is the initial e of growth. *T* is the e passed.
fter 4 years.
as 30,000 and the ately how many years?
can we expect in n is 45,000 and the

world population in al function defined by

$$A(t) = 2600e^{.018t},$$

where *t* is the number of years since 1950.
(a) The world population was about 3700 million in 1970. How closely does the function approximate this value?
(b) Use the function to approximate the population in 1990. (The actual 1990 population was about 5320 million.)
(c) Estimate the population in the year 2000.

57. A sample of 500 g of lead 210 decays to polonium 210 according to the function given by

$$A(t) = 500e^{-.032t},$$

where *t* is time in years. Find the amount of the sample after each of the following times.
(a) 4 years   (b) 8 years
(c) 20 years   (d) Graph $y = A(t)$.

58. Vehicle theft in the United States has been rising exponentially since 1972. The number of stolen vehicles, in millions, is given by

$$f(x) = .88(1.03)^x,$$

where $x = 0$ represents the year 1972. Find the number of vehicles stolen in the following years.
(a) 1975   (b) 1980   (c) 1985   (d) 1990

59. 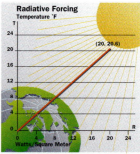 (Refer to Example 8.) Carbon dioxide in the atmosphere traps heat from the sun. Presently, the net incoming solar radiation reaching the earth's surface is 240 watts per square meter (w/m²). The relationship between additional watts per square meter of heat trapped by the increased carbon dioxide *R* and the average rise in global temperature *T* (in °F) is shown in the graph. This additional solar radiation trapped by carbon dioxide is called **radiative forcing.** It is measured in watts per square meter.
(a) Is *T* a linear or exponential function of *R*?
(b) Let *T* represent the temperature increase resulting from an additional radiative forcing of *R* w/m². Use the graph to write *T* as a function of *R*.
(c) Find the global temperature increase when $R = 5$ w/m².

Radiative Forcing
Temperature °F

(20, 20.6)

Watts/Square Meter

Source: Clime, W. *The Economics of Global Warming.* Institute for International Economics, Washington, D.C., 1992.

Titled Examples include detailed, step-by-step solutions and descriptive side comments. Examples relating to the Chapter Openers are also marked with a symbol.

Boxes highlight words, definitions, rules, and procedures.

Notes and Cautions highlight common student errors and address concepts that students often find difficult or confusing.

(b) The ratio A/K for a sample of granite from New Hampshire is .212. How old is the sample?
Since A/K is .212, we have

$$t = (1.26 \times 10^9)\frac{\ln[1 + 8.33(.212)]}{\ln 2} \approx 1.85 \times 10^9.$$

The granite is about 1.85 billion years old. ▶

EXAMPLE 6
Analyzing global temperature increase

Carbon dioxide in the atmosphere traps heat from the sun. The additional solar radiation trapped by carbon dioxide is called *radiative forcing*. It is measured in watts per square meter. In 1896 the Swedish scientist Svante Arrhenius estimated the radiative forcing R caused by additional atmospheric carbon dioxide using the logarithmic equation $R = k \ln(C/C_0)$, where C_0 is the preindustrial amount of carbon dioxide, C is the current carbon dioxide level, and k is a constant. Arrhenius determined that $10 \le k \le 16$ when $C = 2C_0.*$

(a) Let $C = 2C_0$. Is the relationship between R and k linear or logarithmic?
If $C = 2C_0$, $C/C_0 = 2$, so $R = k \ln 2$ is a linear relation, because $\ln 2$ is a constant.

(b) The average global temperature increase T (in °F) is given by $T(R) = 1.03R$. (See Section 5.2, Exercise 59.) Write T as a function of k.
Use the expression for R given in the introduction above.

$$T(R) = 1.03R$$
$$T(k) = 1.03k \ln(C/C_0) \blacktriangleright$$

LOGARITHMS TO OTHER BASES   A calculator can be used to find the values of either natural logarithms (base e) or common logarithms (base 10). However, sometimes it is convenient to use logarithms to other bases. For example, base 2 logarithms are important in computer science. The following theorem can be used to convert logarithms from one base to another.

Change-of-Base Theorem

For any positive real numbers x, a, and b, where $a \ne 1$ and $b \ne 1$:

$$\log_a x = \frac{\log_b x}{\log_b a}.$$

NOTE   As an aid in remembering the change-of-base theorem, notice that x is above a on both sides of the equation.

Source: Cline, W., *The Economics of Global Warming*. Institute for International Economics, Washington, D.C., 1992.

Inverse Function

Let f be a one-to-one function. Then g

$$(f \circ g)(x) = x \quad \text{for every }$$
and $\quad (g \circ f)(x) = x \quad \text{for every }$

A special notation is often used for inv a function f, then g is written as f^{-1} (read " $f(x) = 8x + 5$, and $g(x) = f^{-1}(x) = (x - $

EXAMPLE 3
Deciding whether two functions are inverses

Let functions f and g be defined by $f(x) = x$ tively. Is g the inverse function of f?
A graph indicates that f is one-to-one, is one-to-one, now find $(f \circ g)(x)$ and $(g \circ f)(x)$.

$$(f \circ g)(x) = f(g(x)) = (\sqrt[3]{x + 1})^3 - 1$$
$$= x + 1 - 1$$
$$= x$$
$$(g \circ f)(x) = g(f(x)) = \sqrt[3]{(x^3 - 1) + 1}$$
$$= \sqrt[3]{x^3}$$
$$= x$$

Since both $(f \circ g)(x) = x$ and $(g \circ f)(x) = x$, function g is indeed the inverse of function f, so that f^{-1} is given by

$$f^{-1}(x) = \sqrt[3]{x + 1}. \blacktriangleright$$

CAUTION   Do not confuse the -1 in f^{-1} with a negative exponent. The symbol $f^{-1}(x)$ does not represent $1/f(x)$; it represents the inverse function of f. Keep in mind that a function f can have an inverse function f^{-1} if and only if f is one-to-one.

The definition of inverse function can be used to show that the domain of f equals the range of f^{-1}, and the range of f equals the domain of f^{-1}. See Figure 4.

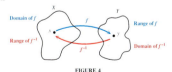

FIGURE 4

CONNECTIONS Inverse functions are used by government agencies and other businesses to send and receive coded information. The functions they use are usually very complicated. A simplified example involves the function $f(x) = 2x + 5$. If each letter of the alphabet is assigned a numerical value according to its position ($a = 1, \ldots, z = 26$), the word ALGEBRA would be encoded as 7 29 19 15 9 41 7. The "message" can be decoded using the inverse function $f^{-1}(x) = \dfrac{x - 5}{2}$.

FOR DISCUSSION OR WRITING
Use the alphabet assignment given above.

1. The function $f(x) = 3x - 2$ was used to encode the following message:

    37 25 19 61 13 34 22 1 55 1 52 52 25 64 13 10.

    Find the inverse function and decode the message.
2. Encode the message SEND HELP using the one-to-one function $f(x) = x^3 - 1$. Give the inverse function that the decoder would need when the message is received.

For the inverse functions f and g discussed earlier $f(10) = 85$ and $g(85) = 10$; that is, (10, 85) belongs to f and (85, 10) belongs to g. The ordered pairs of the inverse of any one-to-one function f can be found by exchanging the components of the ordered pairs of f. The equation of the inverse of a function defined by $y = f(x)$ also is found by exchanging x and y. For example, if $f(x) = 7x - 2$, then $y = 7x - 2$. The function f is one-to-one, so that f^{-1} exists. The ordered pairs in f^{-1} have the form (y, x), so y can be used to produce x, since $x = f^{-1}(y)$. Therefore, the equation for f^{-1} can be found by solving $y = f(x)$ for x. Finally, x and y can be interchanged to conform to our convention of using x for the independent variable and y for the dependent variable.

$$y = 7x - 2$$
$$7x = y + 2 \qquad \text{Add 2.}$$
$$x = \frac{y + 2}{7} = f^{-1}(y) \qquad \text{Divide by 7.}$$
$$y = \frac{x + 2}{7} = f^{-1}(x) \qquad \text{Exchange } x \text{ and } y.$$
$$f^{-1}(x) = \frac{x + 2}{7}.$$

As a check, verify that $(f \circ f^{-1})(x) = x$ and $(f^{-1} \circ f)(x) = x$.

Connections Boxes point out the many connections between mathematics and the "real world" or other mathematical concepts.

Optional Graphing Calculator Boxes offer guidance for students using graphing calculators.

Many examples include optional graphing calculator coverage.

This theorem is proved by using the definition of logarithm to write $= \log_a x$ in exponential form.

Proof
Let
$$y = \log_a x.$$
$$a^y = x \qquad \text{Change to exponential form.}$$
$$\log_b a^y = \log_b x \qquad \text{Take logarithms on both sides.}$$
$$y \log_b a = \log_b x \qquad \text{Property (c) of logarithms}$$
$$y = \frac{\log_b x}{\log_b a} \qquad \text{Divide both sides by } \log_b a.$$
$$\log_a x = \frac{\log_b x}{\log_b a} \qquad \text{Substitute } \log_a x \text{ for } y. \quad \blacktriangleright$$

Any positive number other than 1 can be used for base b in the change of base theorem, but usually the only practical bases are e and 10, since calculators give logarithms only for these two bases. The change-of-base theorem is used to find logarithms for other bases.

The change-of-base theorem is needed to graph logarithmic functions with bases other than 10 and e (and sometimes with one of those bases). For instance,

to graph $y = \log_3(x - 1)$, graph $y = \dfrac{\log(x - 1)}{\log 3}$ or $y = \dfrac{\ln(x - 1)}{\ln 3}$.

The next example shows how the change-of-base theorem is used to find logarithms to bases other than 10 or e with a calculator.

EXAMPLE 7
Using the change-of-base theorem

The result of Example 7(a) is valid for *either* natural or common logarithms.

Use natural logarithms to find each of the following. Round to the nearest hundredth.

(a) $\log_5 17$
Use natural logarithms and the change-of-base theorem.

$$\log_5 17 = \frac{\log_e 17}{\log_e 5}$$
$$= \frac{\ln 17}{\ln 5}$$
$$\approx \frac{2.8332}{1.6094}$$
$$\approx 1.76$$

To check, use a calculator along with the definition of logarithm, to verify that $5^{1.76} \approx 17$.

Discovering Connections

exercises tie together different topics and highlight the relationships among various concepts and skills.

DISCOVERING CONNECTIONS (Exercises 39–44)

The solution set of $f(x) = 0$ consists of all x-values for which the graph of $y = f(x)$ intersects the x-axis (i.e., the x-intercepts). The solution set of $f(x) < 0$ consists of all x-values for which the graph lies below *the x-axis, while the solution set of $f(x) > 0$ consists of all x-values for which the graph lies* above *the x-axis.*

In Chapter 2 we saw how a sign graph can be used to solve a quadratic inequality. Graphical analysis allows us to solve such inequalities as well. Work the following exercises in order. They demonstrate why we must reverse the direction of the inequality sign when multiplying or dividing an inequality by a negative number.

39. Graph $f(x) = x^2 + 2x - 8$. This function has a graph with two x-intercepts. What are they?

40. Based on the graph from Exercise 39, what is the solution set of $x^2 + 2x - 8 < 0$?

41. Now graph $g(x) = -f(x) = -x^2 - 2x + 8$. Using the terminology of Chapter 3, how is the graph of g obtained by a transformation of the graph of f?

42. Based on the graph from Exercise 41, what is the solution set of $-x^2 - 2x + 8 > 0$?

43. How do the two solution sets of the inequalities in Exercises 40 and 42 compare?

44. Write a short paragraph explaining how Exercises 39–43 illustrate the property involving multiplying an inequality by a negative number.

In Exercises 45 and 46, find a polynomial function f whose graph matches the one in the figure. Then use a graphing calculator to graph the function and verify your result.

45.

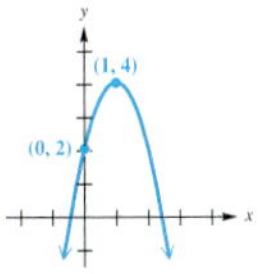

46.

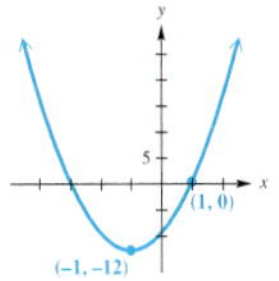

t dollars from 1985
uadratic function
3x + 3954
5 and $f(x)$ is in bil-
pply, what would be
Note: There are pit-
ar into the future.)
c Analysis.)
by the U.S. Court of
4 and 1990 can be
odel
x + 31,676
84. Based on this
ber of cases com-

menced in 1996? (*Note:* There are pitfalls in using models to predict far into the future.)

49. Between 1985 and 1989, the number of female suicides by firearms in the United States each year can be modeled by

$$f(x) = -17x^2 + 44.6x + 2572$$

where $x = 0$ represents 1985. Based on this model, in what year did the number of such suicides reach its peak?

50. The number of infant deaths during the past decade has been decreasing. Between 1980 and 1989, the number of infant deaths per 1000 live births each year can be approximated by the function

$$f(x) = .0234x^2 - .5029x + 12.5$$

where $x = 0$ corresponds to 1980.

Solve the problem involving a polynomial function model. See Example 8.

65. From 1930 to 1990 the rate of breast cancer was nearly constant at 30 cases per 100,000 females whereas the rate of lung cancer in females over the same period increased. The number of lung cancer cases per 100,000 females in the year t (where $t = 0$ corresponds to 1930) can be modeled using the function defined by $f(t) = 2.8 \times 10^{-4}t^3 - .011t^2 + .23t + .93$. (*Source:* Valanis, B., *Epidemiology in Nursing and Health Care,* Appleton & Lange, Norwalk, Connecticut, 1992.)

(a) Use a graphing calculator to graph the rates of breast and lung cancer for $0 \le t \le 60$. Use the window [0, 60] by [0, 40].

(b) Determine the year when rates for lung cancer first exceeded those for breast cancer.

(c) Discuss reasons for the rapid increase of lung cancer in females.

66. The number of military personnel on active duty in the United States during the period 1985 to 1990 can be determined by the cubic model $f(x) = -7.66x^3 + 52.71x^2 - 93.43x + 2151$, where $x = 0$ corresponds to 1985, and $f(x)$ is in thousands. Based on this model, how many military personnel were on active duty in 1990? (*Source:* U.S. Department of Defense.)

67. A survey team measures the concentration (in parts per million) of a particular toxin in a local river. On a normal day, the concentration of the toxin at time x (in hours) after the factory upstream dumps its waste is given by $g(x) = -.006x^4 + .14x^3 - .05x^2 + .02x$, where $0 \le x \le 24$.

(a) Graph $y = g(x)$ in the window [0, 24] by [0, 200].

(b) Estimate the time at which the concentration is greatest.

(c) A concentration greater than 100 parts per million is considered pollution. Using the graph from part (a), estimate the period during which the river is polluted.

68. During the early part of the twentieth century, the deer population of the Kaibab Plateau in Arizona experienced a rapid increase because hunters had reduced the number of natural predators and because the deer were protected from hunters. The increase in population depleted the food resources and eventually caused the population to decline. For the period from 1905 to 1930, the deer population was approximated by $D(x) = -.125x^5 + 3.125x^4 + 4000$, where x is time in years from 1905.

(a) Graph $y = D(x)$ in the window [0, 50] by [0, 120,000].

(b) From the graph, over what period of time (from 1905 to 1930) was the deer population increasing? Relatively stable? Decreasing?

69. The table lists the total annual amount (in millions of dollars) of government-guaranteed student loans from 1986 to 1994. (*Source: USA TODAY.*)

Year	*Amount*
1986	8.6
1987	9.8
1988	11.8
1989	12.5
1990	12.3
1991	13.5
1992	14.7
1993	16.5
1994	18.2

(a) Graph the data with the following three function definitions, where x represents the year.

(i) $f(x) = .4(x - 1986)^2 + 8.6$

(ii) $f(x) = 1.088(x - 1986) + 8.6$

(iii) $f(x) = 1.455\sqrt{x - 1986} + 8.6$

(b) Discuss which function definition models the data best.

70. The table lists the number of Americans (in thousands) who are expected to be over 100 years old for selected years. (*Source:* U.S. Census Bureau.)

Year	*Number*
1994	50
1996	56
1998	65
2000	75
2002	94
2004	110

(a) Use graphing to determine which polynomial best models the number of Americans over 100 years old where $x = 0$ corresponds to 1994.

(i) $f(x) = 6.057x + 44.714$

Many exercises and examples are based on **Real Data**, and many require reading graphs and charts.

Writing and **Conceptual Exercises** are included to aid students in applying the concepts presented.

Features ▼▼▼

The design has been developed to enhance the pedagogical features mentioned below and to ensure their accessibility.

- Each chapter opens with a genuine application of the material to be presented. Corresponding examples and exercises, identified with a special icon, are located throughout the chapter.
- We have made an effort to point out the many connections between mathematical topics in this course and those studied earlier, as well as connections between mathematics and the "real world." Optional Connections boxes presenting such topics are included in many sections throughout the book. Most of them include thought-provoking questions for writing or class discussion. In addition, we have included a feature in many exercise sets called Discovering Connections. These groups of exercises tie together different topics and highlight the relationships among various concepts and skills.
- Graphing calculator comments and screens are given throughout the book as appropriate. These are identified with an icon, so that an instructor may choose whether or not to use them. We know that many students have graphing calculators and may need guidance for using them, even if they are not a required part of the course.
- Many examples and exercises are based on real data, and many require reading charts and graphs.
- The exercise sets contain many conceptual and writing exercises, as well as graphing calculator exercises. Those exercises that require applying the topics in a section to ideas beyond the examples are marked as challenging in the instructor's edition.
- Cautions and notes are included to highlight common student errors and misconceptions. Some of these address concepts that students often find difficult or confusing.

Supplements ▼▼▼

For the Instructor

Instructor's Annotated Exercises   With this volume, instructors have immediate access to the answers to every exercise in the text, excluding proofs and writing exercises. Each answer is printed next to or below the corresponding text exercise. In addition, challenging exercises, which will require most students to stretch beyond the concepts discussed in the text, are marked with the symbol ▲. The conceptual (◉) and writing (✐) exercises are also marked in this edition so instructors may assign these problems at their discretion. (Graphing calculator exercises will be marked by 📟 in both the student's and instructor's editions.)

Instructor's Resource Manual Included here are two forms of a pretest; four versions of a chapter test for each chapter, additional test items for each chapter, and two forms of a final examination. Answers to all tests and additional exercises also are provided. Answers to most of the textbook exercises are included as well.

Instructor's Solution Manual This manual includes complete, worked-out solutions to every even exercise in the textbook (excluding most writing exercises).

Test Generator/Editor for Mathematics with QuizMaster is a computerized test generator that lets instructors select test questions by objective or section or use a ready-made test for each chapter. The software is algorithm driven so that regenerated number values maintain problem types and provide a large number of test items in both multiple-choice and open-response formats for one or more test forms. The **Editor** lets instructors modify existing questions or create their own including graphics and accurate math symbols. Tests created with the **Test Generator** can be used with **QuizMaster,** which records student scores as they take tests on a single computer or network, and prints reports for students, classes, or courses. CLAST and TASP versions of this package are also available. (IBM, DOS/Windows, and Macintosh)

For the Student

Student's Solution Manual Complete, worked-out solutions are given for odd-numbered exercises and chapter review exercises and all chapter test exercises in a volume available for purchase by students. In addition, all new cumulative review exercises with worked-out solutions are provided.

Videotapes A new videotape series has been developed to accompany *College Algebra and Trigonometry*. In a separate lesson for each section of the book, the series covers all objectives, topics, and problem-solving techniques within the text.

Interactive Mathematics Tutorial Software with Management System is an innovative software package that is objective-based, self-paced, and algorithm driven to provide unlimited opportunity for review and practice. Tutorial lessons provide examples, progress-check questions, and access to an on-line glossary. Practice problems include hints for the first incorrect responses, solutions, textbook page references, and on-line tools to aid in computation and understanding. Quick Reviews for each section focus on major concepts. The optional **Management System** records student scores on disk and lets instructors print diagnostic reports for individual students or classes. Student versions, which include record-keeping and practice tests, may be purchased by students for home use.

Acknowledgments ▼▼▼

We are grateful to the many users of our previous editions and to our reviewers for their insightful comments and suggestions. It is because they take the time to write thoughtful reviews that our textbooks continue to meet the needs of students and their instructors.

Reviewers

Bill Ardis, Collin County Community College

Michael Blackwell, Jones County Junior College

Kathleen Burk, Pensacola Junior College

Patrick Cassens, Missouri Southern State College

Oiyin Pauline Chow, Harrisburg Area Community College

Lawrence Clar, Monroe Community College

Randall Crist, Creighton University

Walter Daum, City College of New York

Wanda S. Dixon, Meridian Community College

John Formsma, Los Angeles City College

Odene Forsythe, Westark Community College

Barbara Glass, Sussex County Community College

Heather J. Goodling, University of North Florida

Glenda R. Haynie, North Carolina State University

Joe Howe, St. Charles County Community College

Lloyd R. Jaisingh, Morehead State University

Dr. Sarah Percy Janes, San Jacinto College

Patricia H. Jones, Methodist College

William R. Livingston, Missouri Southern State College

Wanda J. Long, St. Charles County Community College

Andrew D. Martin, Morehead State University

James Lamar Middleton, Polk Community College

Feridoon Moinian, Cameron University

Sandy Morris, College of DuPage

Arumugam Muhundan, Manatee Community College

Lynne Nation, Young Harris College

Frank Neckel, Community College of Aurora

Smruti Patel, Hutchinson Community College

Kathy Rogotzke, North Iowa Area Community College

Cynthia Floyd Sikes, Georgia Southern University

Laurence Small, Los Angeles Pierce College

Jeff Solheim, Emporia State University

Mary Jane Sterling, Bradley University

Martha J. A. Turner, Winthrop University

Dr. Jan Vandever, South Dakota State University

Patrick Ward, Illinois Central College

Mary E. Wilson, Austin Community College

Gail Wiltse, St. Johns River Community College

Dr. Kenneth J. Word, Central Texas College

Adil Yaqub, University of California/Santa Barbara

Accuracy Checkers

John Armon, Illinois Central
   College
Frank Neckel, Community College
   of Aurora

Henry M. Smith, Delgado
   Community College

We are thankful for the assistance given by Gary Rockswold, Mankato State University, who researched and provided the chapter opener applications and exercises. As always, we are grateful to Paul Eldersveld, College of DuPage, for an outstanding job of coordinating the print supplements. We also thank Kitty Pellissier, who checked the answers to all the exercises in her usual careful and thorough manner, and Paul Van Erden, American River College, who created an accurate, complete index. As always, Ed Moura and Anne Kelly were there to lend support and guidance. Special thanks go out to Lisa De Mol and Andrea Coens, who coordinated the production of an extremely complex project. It is only through the cooperation of these and many other individuals that we are able to produce texts that successfully serve both instructors and students.

Margaret L. Lial
E. John Hornsby, Jr.
David I. Schneider

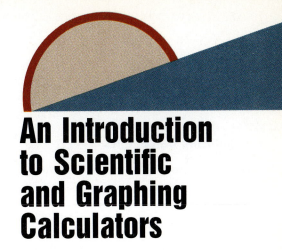

An Introduction to Scientific and Graphing Calculators

In the past, some of the most brilliant minds in mathematics and science spent long, laborious hours calculating values for logarithmic and trigonometric tables. These tables were essential to solve equations in real applications. Because they could not predict what values would be needed, the tables were sometimes incomplete. During the second half of the twentieth century, computers and sophisticated calculators appeared. These computing devices are able to evaluate mathematical expressions and generate tables in a fraction of a second. As a result, the study of mathematics is changing dramatically.

Although computers and calculators have made a profound difference, they have *not* replaced mathematical thought. Calculators cannot decide whether to add or subtract two numbers in order to solve a problem—only you can do that. Once you have made this decision, calculators can efficiently determine the solution to the problem. In addition, graphing calculators also provide important graphical and numerical support to the validity of a mathematical solution. They are capable of exposing errors in logic and pointing to patterns. These patterns can lead to conjectures and theorems about mathematics. The human mind is capable of mathematical insight and decision making, but is not particularly proficient at performing long arithmetic calculations. On the other hand, calculators are incapable of possessing mathematical insight, but are excellent at performing arithmetic and other routine computations. In this way, calculators complement the human mind.

If this is your first experience with a scientific or graphing calculator, the numerous keys and strange symbols that appear on the keyboard may be intimidating. Like any learning experience, take it a step at a time. You do not have to understand every key before you begin using your calculator. Some keys may not even be needed in this course. The following explanations and suggestions are intended to give you a brief overview of scientific and graphing calculators.

It is not intended to be complete or specific toward any particular type of calculator. You may find that some things are different on your calculator. *Remember, a calculator comes with an owner's manual.* This manual is essential in learning how to use your calculator.

Scientific Calculators ▼▼▼

Two basic parts of any calculator are the keyboard and the display. The keyboard is used to input data—the display is used to output data. Without correct input, the displayed output is meaningless. Most scientific calculators do not display the entire arithmetic expression that is entered, but only display the most recent number inputted or outputted. If an arithmetic expression is entered incorrectly, it is not possible to edit it. The entire expression must be entered again.

Order of Operation

The order in which expressions are entered into a calculator is essential to obtaining correct answers. Operations on a calculator can usually be divided into two basic types: unary and binary. Unary operations require that only one number be entered. Examples of unary operations are $\sqrt{x}$, x^2, $x!$, $\sqrt[3]{x}$, and $\log x$. When entering a unary operation on a scientific calculator, the number is usually entered first, followed by the unary operation. For example, to find the square root of 4, press the key $\boxed{4}$, followed by the square root key. Binary operations require that two numbers are entered. Examples of binary operations are $+$, $-$, $\times$, $\div$, and x^y. When evaluating a binary operation on a scientific calculator, the operation symbol is usually entered between the numbers. Thus, to add the two numbers 4 and 5, enter $\boxed{4}$ $\boxed{+}$ $\boxed{5}$ $\boxed{=}$. However, on some calculators, such as those made by Hewlett Packard, it is necessary to use *Reverse Polish Notation* (RPN). In RPN the operation is entered last, after the operands. One advantage of RPN is that parentheses are usually not necessary.

Every calculator has a set of built-in precedence rules that can be found in the owner's manual. For example, suppose that the expression $3 + 4 \times 2 = $ is entered, from left to right, into a scientific calculator. The output will usually be 11 and not 14. This is because multiplication is performed before addition in the absence of parentheses. Parentheses can always be used to override existing precedence rules. *When in doubt, use parentheses.* Try evaluating $\frac{24}{4-2}$. It should be entered as $24 \div (4 - 2) = $ in order to obtain the correct answer of 12. This is because division has precedence over subtraction.

Scientific Notation

Numbers that are either large or small in absolute value are often displayed using scientific notation. The numeric expression 2.46 E12 refers to the large number 2.46×10^{12}, while the expression 2.46 E$-$12 refers to the small positive number 2.46×10^{-12}. Try multiplying one billion times ten million. Observe the output on your calculator.

Precision and Accuracy

Precision refers to the number of digits a calculator will display. When $\frac{1}{3}$ is evaluated, a calculator may display 0.333333333. This answer is approximate. The displayed precision of most calculators is between 8 and 12 digits. Accuracy is different from precision. It refers to the number of correct digits that an answer contains, compared to the true value. If a scale is misread as 129.6 pounds, when the actual answer is 145.8 pounds, then the number 129.6 has four digits of precision, but only one digit of accuracy. Many times when using a calculator to solve a real application, it will display ten digits of precision, but only a few digits will be accurate or meaningful. For example, suppose you drive 100 miles on 3 gallons of gas. A calculator would say that your mileage is $100 \div 3 \approx 33.33333333$. The precision of this answer is ten digits. The accuracy is probably not ten digits unless both the mileage and amount of gasoline were measured in an exceedingly accurate manner. It would be more reasonable or accurate to say that the mileage is about 33 miles per gallon, rather than 33.33333333 miles per gallon.

Second and Inverse Keys

Because the size of the keyboard is limited, there is often a 2nd or INV key. This key can be used to access additional features. These additional features are usually labeled above the key in a different color.

Graphing Calculators ▼▼▼

Graphing calculators provide several features beyond those found on scientific calculators. The bottom rows of keys on a graphing calculator are often similar to those found on scientific calculators. Graphing calculators have additional keys that can be used to create graphs, make tables, analyze data, and change settings. One of the major differences between graphing and scientific calculators is that a graphing calculator has a larger viewing screen with graphing capabilities.

Editing Input

The screen of a graphing calculator can display several lines of text at a time. This feature allows the user to view both previous and current expressions. If an incorrect expression is entered, a brief error message is displayed. It can be viewed and corrected by using various editing keys—much like a word-processing program. You do not need to enter the entire expression again. Many graphing calculators can also recall past expressions for editing or updating.

Order of Operation

Arithmetic expressions on graphing calculators are usually entered as they are written in mathematical equations. As a result, unary operations like $\sqrt{x}$, $\sqrt[3]{x}$, and $\log x$ are entered first, followed by the number. Unary operations like x^2 and $x!$ are entered after the number. Binary operations are entered in a manner similar to most scientific calculators. The order of operation on graphing calculators is also important. For example, try evaluating the expression $\sqrt{2 \times 8}$. If this expression is entered as it is written, without any parentheses, a graphing calculator may display 11.3137085 and not 4. This is because a square root is performed before multiplication. To prevent this error, use parentheses around 2×8.

Calculator Screen

If you look closely at the screen of a graphing calculator, you will notice that the screen is composed of many tiny rectangles or points called pixels. The calculator can darken these rectangles so that output can be displayed. Many graphing calculator screens are approximately 96 pixels across and 64 pixels high. Computer screens are usually 640 by 480 pixels or more. For this reason, you will notice that the resolution on a graphing calculator screen is not as clear as on most computer terminals. With a graphing calculator, a straight line will not always appear to be exactly straight and a circle will not be precisely circular. Because of the screen's low resolution, graphs generated by graphing calculators may require mathematical understanding to interpret them correctly.

Viewing Window

The viewing window for a graphing calculator is similar to the viewfinder in a camera. A camera cannot take a picture of an entire view in a single picture. The camera must be centered on some object and can only photograph a subset of the available scenery. A person may want to photograph a close-up of a face or a person standing in front of a mountain. A camera with a zoom lens can capture different views of the same scene by zooming in and out. Graphing calculators have similar capabilities. The xy-coordinate plane is infinite. The calculator screen can show only a finite, rectangular region in the xy-coordinate plane. This rectangular region must be specified before a graph can be drawn. This is done by setting minimum and maximum values for both the x- and y-axes. Determining an appropriate viewing window is often one of the most difficult things to do. Many times it will take a few attempts before a satisfactory window size is found. Like many cameras, the graphing calculator can also zoom in and out. Zooming in shows more detail in a small region of a graph, whereas zooming out gives a better overall picture of the graph.

Graphing and the Free-moving Cursor

Once a viewing window has been determined, an equation in the form of $y = f(x)$ can be graphed. A simple example of this form is $y = 3x$. Four or more equations of this type can be graphed at once in the same viewing window. A graphing calculator has a free-moving cursor. By using the arrow keys, a small cross-hair can be made to move about on the screen. Its x- and y-coordinates are usually displayed on the screen. The cursor can be used to approximate the locations of features on the graph, such as x-intercepts and points of intersection. Using the trace key, the free-moving cursor can also be made to trace over the graph, displaying the corresponding x- and y-coordinates located on the graph.

Tables

Some graphing calculators have the ability to display tables. For example, if $y = x^2$, then a vertical table like the following can be generated automatically. This is an efficient way to evaluate an equation at selected values of x.

X	Y
0	0
1	1
2	4
3	9
4	16
5	25
6	36

Programming

Graphing calculators can be programmed, much like computers. Complex problems can be solved with the aid of programs. In this course, it will not be necessary for you to program your calculator. However, the capability is there, if you choose to use it.

Additional Features

Graphing calculators have additional features too numerous to list completely. They may be able to generate sequences, find maximums and minimums on graphs, do arithmetic with complex numbers, solve systems of equations using matrices, and analyze data with statistics. The most advanced calculators are capable of performing *symbolic manipulation*. Using these calculators, one can factor $x^2 - 1$ into $(x - 1)(x + 1)$ and simplify $\dfrac{x^2 y^3}{x^{-2} y}$ to $x^4 y^2$ automatically. If the solution to a problem is π, symbolic manipulation routines will display π rather than 3.141592654.

Final Comments ▼▼▼

Mathematicians from the past would have been amazed by today's calculators. Calculators are powerful computing devices that can perform difficult computational tasks. The solutions to many important equations in mathematics cannot be determined by hand. However, the solutions to these equations often can be approximated using a calculator. Calculators also provide the capability to ask questions like "What if . . . ?" more easily. Values in algebraic expressions can be altered and conjectures tested quickly.

At the heart of all mathematics is deductive thought and proof. No robot or artificial intelligence program has been effective at this task. Only the human mind is capable of this. Calculators are an important tool in mathematics. Like any tool, they must be used *appropriately* in order to enhance our ability to understand mathematics. Mathematical insight may often be the quickest and easiest way to solve a problem; a calculator may neither be needed nor appropriate. By using mathematical concepts, you can decide when to use or not to use a calculator.

1

Algebraic Expressions

Since ancient times the need for measurement has played a central role in the development of mathematics. For measurement to occur, numbers and algebraic expressions are often necessary. The Egyptians made sophisticated measurements and calculations in the design of their pyramids. They discovered an amazing formula for their time that could be used to determine the volume of a partially completed pyramid. It is common in applications to be unable to measure a value directly. Even today, scientists must rely on indirect measurement to determine many quantities. The following is an example of this type of situation.

> If the global climate were to warm up significantly as a result of the greenhouse effect or other climatic change, the Greenland ice cap could melt. It is estimated that this ice cap contains 3 million cubic kilometers of ice. Over 200 million people currently live on soil that is less than 1 meter (3.28 feet) above sea level. In the United States there are several large cities that have low average elevations. Three examples are New Orleans (4 feet), Boston (14 feet), and San Diego (13 feet). Would the melting of the ice cap affect these people and cities?

Clearly sea level will rise if the ice cap melts, but by how much? As is often the case, this measurement cannot be done directly as part of an experiment. Therefore, people must use known data and mathematical expressions to approximate the rise in sea level without actually having the event occur. Much of the success of technology has resulted from its ability to make indirect measurements and then use mathematics to determine quantities that could not be measured directly. This requires ingenuity and an understanding of numbers, variables, and formulas. In this chapter you will learn about these concepts and in the exercises for Section 1.7 you will be asked to calculate this potential rise in sea level.

1.1 The Real Numbers ▼▼▼

SETS OF NUMBERS The idea of counting goes back into the mists of antiquity. When people first counted they used only the **natural numbers,** written in set notation as

$$\{1, 2, 3, 4, 5, \ldots\}.$$

More recent is the idea of counting *no* object—that is, the idea of the number 0. As early as A.D. 150 the Greeks used the symbol o or $\bar{o}$ to represent zero.

Including 0 with the set of natural numbers gives the set of **whole numbers.**

$$\{0, 1, 2, 3, 4, 5, \ldots\}$$

(These and other sets of numbers are summarized later in this section.)

About 500 years ago, people came up with the idea of counting backward, from 4 to 3 to 2 to 1 to 0. There seemed no reason not to continue this process, calling the new numbers $-1, -2, -3$, and so on. Including these numbers with the set of whole numbers gives the very useful set of **integers,**

$$\{\ldots, -4, -3, -2, -1, 0, 1, 2, 3, \ldots\}.$$

Integers can be shown pictorially with a **number line.** (A number line is similar to a thermometer on its side.) As an example, the elements of the set $\{-3, -1, 0, 1, 3, 5\}$ are located on the number line in Figure 1.

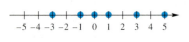

FIGURE 1

The result of dividing two integers, with a nonzero divisor, is called a *rational number*. By definition, the **rational numbers** are the elements of the set

$$\left\{\frac{p}{q} \,\middle|\, p \text{ and } q \text{ are integers and } q \neq 0\right\}.$$

This definition, which is given in *set-builder notation,* is read "the set of all elements p/q such that p and q are integers and $q \neq 0$."* Examples of rational numbers include $3/4, -5/8, 7/2$, and $-14/9$. All integers are rational numbers, since any integer can be written as the quotient of itself and 1.

Rational numbers can be located on a number line by a process of subdivision. For example, $5/8$ can be located by dividing the interval from 0 to 1 into 8 equal parts, then labeling the fifth part $5/8$. Several rational numbers are located on the number line in Figure 2.

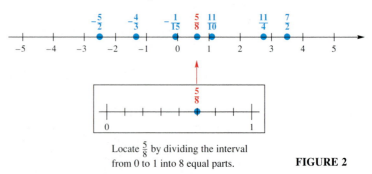

Locate $\frac{5}{8}$ by dividing the interval from 0 to 1 into 8 equal parts.

FIGURE 2

---

*See Appendix A for further discussion of sets.

The set of all numbers that correspond to points on a number line is called the set of **real numbers.** The set of real numbers is shown in Figure 3.

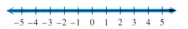

FIGURE 3

A real number that is not rational is called an **irrational number.** The set of irrational numbers includes $\sqrt{3}$ and $\sqrt{5}$ but not $\sqrt{1}$, $\sqrt{4}$, $\sqrt{9}$, . . . , which equal 1, 2, 3, . . . , and hence are rational numbers. Another irrational number is π, which is approximately equal to 3.14159. A calculator shows that $\sqrt{2} \approx$ 1.414 (where $\approx$ is read "is approximately equal to") and $\sqrt{5} \approx 2.236$. Using these approximations, the numbers in the set $\{-2/3, 0, \sqrt{2}, \sqrt{5}, \pi, 4\}$ can be located on a number line as shown in Figure 4.

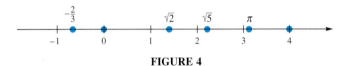

FIGURE 4

Real numbers can also be defined in another way, in terms of decimals. Using repeated subdivisions, any real number can be located (at least in theory) as a point on a number line. By this process, the set of real numbers can be defined as the set of all decimals. Further work would show that the set of rational numbers is the set of all decimals that repeat or terminate. For example,

$$.25 = 1/4,$$
$$.833333 \ldots = 5/6,$$
$$.076923076923 \ldots = 1/13,$$

and so on. Repeating decimals are often written with a bar to indicate the digits that repeat endlessly. With this notation, $5/6 = .8\overline{3}$, and $1/7 = .\overline{142857}$. The set of irrational numbers is the set of decimals that neither repeat nor terminate. For example,

$$\sqrt{2} = 1.414213562373 \ldots \quad \text{and} \quad \pi = 3.14159265358 \ldots .$$

The sets of numbers discussed so far are summarized as follows.

Sets of Numbers

Real Numbers $\{x \mid x \text{ corresponds to a point on a number line}\}$

Integers $\{ \ldots , -3, -2, -1, 0, 1, 2, 3, \ldots \}$

Rational Numbers $\left\{ \dfrac{p}{q} \, \middle| \, p \text{ and } q \text{ are integers and } q \neq 0 \right\}$

Irrational Numbers $\{x \mid x \text{ is real but not rational}\}$

Whole Numbers $\{0, 1, 2, 3, 4, \ldots \}$

Natural Numbers $\{1, 2, 3, 4, \ldots \}$

EXAMPLE 1
Identifying elements of subsets
of the real numbers

Let set $A = \{-8, -6, -3/4, 0, 3/8, 1/2, 1, \sqrt{2}, \sqrt{5}, 6, 9/0\}$. List the elements from set A that belong to each of the sets of numbers just discussed.

(a) All elements of A are real numbers except $9/0$. Division by 0 is not defined, so $9/0$ is not a number.

(b) The irrational numbers are $\sqrt{2}$ and $\sqrt{5}$.

(c) The rational numbers are -8, -6, $-\dfrac{3}{4}$, 0, $\dfrac{3}{8}$, $\dfrac{1}{2}$, 1, and 6.

(d) The integers are -8, -6, 0, 1, and 6.

(e) The whole numbers are 0, 1, and 6.

(f) The natural numbers in set A are 1 and 6. ▶

The relationships among the various subsets of the real numbers are shown in Figure 5.

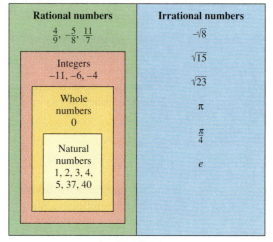

The Real Numbers

FIGURE 5

EXPONENTS   Exponential notation is used to write the products of repeated factors. For example, the product $2 \cdot 2 \cdot 2$ can be written as 2^3, where the 3 shows that three factors of 2 appear in the product. The notation a^n is defined as follows.

Definition of a^n

If n is any positive integer and a is any real number, then

$$a^n = a \cdot a \cdot a \cdots a,$$

where a appears n times.

The integer n is the **exponent,** and a is the **base.** (Read a^n as "a to the nth power," or just "a to the nth.")

Evaluate each exponential expression, or power, and identify the base and the exponent.

EXAMPLE 2
Evaluating exponential expressions

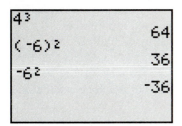

This screen shows how the TI-82 calculates the exponential expressions in Examples 2(a), (b), and (c).

(a) $4^3 = 4 \cdot 4 \cdot 4 = 64$; the base is 4 and the exponent is 3

(b) $(-6)^2 = (-6)(-6) = 36$; the base is -6 and the exponent is 2

(c) $-6^2 = -(6 \cdot 6) = -36$; the base is 6 and the exponent is 2

(d) $4 \cdot 3^2 = 4 \cdot 3 \cdot 3 = 36$; the base is 3 and the exponent is 2

(e) $(4 \cdot 3)^2 = 12^2 = 144$; the base is $4 \cdot 3$ or 12 and the exponent is 2. ▶

CAUTION In Example 2, notice that $4 \cdot 3^2 \neq (4 \cdot 3)^2$.

ORDER OF OPERATIONS When a problem involves more than one operation symbol, we use the following *order of operations.*

Order of Operations

If grouping symbols such as parentheses, square brackets, or fraction bars are present:

1. Work separately above and below each fraction bar.
2. Use the rules below within each set of parentheses or square brackets. Start with the innermost and work outward.

If no grouping symbols are present:

1. Simplify all powers and roots, working from left to right.
2. Do any multiplications or divisions in the order in which they occur, working from left to right.
3. Do any negations, additions, or subtractions in the order in which they occur, working from left to right.

EXAMPLE 3
Using order of operations

The TI-82 uses the standard rules for order of operations. The third display shows how the rational value is converted to a fraction, –22/31. See Example 3.

Use the order of operations given above to evaluate each of the following.

(a) $6 \div 3 + 2^3 \cdot 5 = 6 \div 3 + 8 \cdot 5 = 2 + 8 \cdot 5 = 2 + 40 = 42$

(b) $(8 + 6) \div 7 \cdot 3 - 6 = 14 \div 7 \cdot 3 - 6$
$$= 2 \cdot 3 - 6$$
$$= 6 - 6 = 0$$

(c) $\dfrac{-(-3)^3 + (-5)}{2(-8) - 5(3)} = \dfrac{-(-27) + (-5)}{2(-8) - 5(3)}$ Evaluate the exponential.

$$= \frac{27 + (-5)}{-16 - 15}$$ Multiply.

$$= \frac{22}{-31} = -\frac{22}{31}$$ Add and subtract. ▶

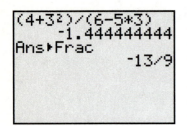

```
(4+3²)/(6-5*3)
          -1.44444444
Ans▶Frac
               -13/9
```

Notice the careful use of parentheses in entering the expression into the calculator. *Ans* is the memory location reserved for the latest "answer." The TI-82 will convert $-1.\overline{4}$ to $-13/9$ using the conversion capability.

▦ Graphing calculators and computers use this same order of operations. It is important to be careful to use parentheses as necessary to get the desired results. For example, to evaluate

$$\frac{4 + 3^2}{6 - 5 \cdot 3}$$

the numerator and denominator must both be enclosed in parentheses. Otherwise the calculator will find 3^2, then divide that by 6, multiply 5 times 3, and then perform the additions and subtractions to get -9.5. Verify that the correct value of the expression is $-13/9 = -1.\overline{4}$. You may want to use the problems given in Example 3 to experiment with your graphing calculator.

◀ **EXAMPLE 4**
Using order of operations

Use the order of operations to evaluate each expression if $x = -2$, $y = 5$, and $z = -3$.

```
-2→X:5→Y: -3→Z: -4
X²-7Y+4Z
                 -63
(2(X-5)²+4Y)/(Z+
4)
                118
```

This screen supports the results in Example 4.

(a) $-4x^2 - 7y + 4z$
Replace x with -2, y with 5, and z with -3.

$$
\begin{aligned}
-4x^2 - 7y + 4z &= -4(-2)^2 - 7(5) + 4(-3) \\
&= -4(4) - 7(5) + 4(-3) \\
&= -16 - 35 - 12 = -63
\end{aligned}
$$

(b) $\dfrac{2(x - 5)^2 + 4y}{z + 4} = \dfrac{2(-2 - 5)^2 + 4(5)}{-3 + 4}$

$$= \frac{2(-7)^2 + 20}{1} \qquad \text{Work inside parentheses; multiply; add.}$$

$$= 2(49) + 20 \qquad \text{Evaluate the exponential.}$$

$$= 118 \ ▶$$

CAUTION Notice the use of parentheses when numbers are substituted for the variables. This is especially important when substituting negative numbers for the variables in a product.

◀ **EXAMPLE 5**
Calculating the "magic number" in baseball

Near the end of a major league baseball season, fans are often interested in the current first-place team's "magic number." The magic number is the sum of the required number of wins of the first-place team and the number of losses of the second-place team (for the remaining games) necessary to clinch the pennant. (In a regulation major league season, each team plays 162 games.) To calculate the magic number M for a first-place team prior to the end of a season, the formula is

$$M = W_2 + N_2 - W_1 + 1,$$

where W_2 = the current number of wins of the second-place team;

N_2 = the number of remaining games of the second-place team;

W_1 = the current number of wins of the first-place team.

On Wednesday, September 6, 1995, baseball fans woke up to the following National League East Division standings in their local newspapers.

	W	L	Pct.
Atlanta	76	44	.633
Philadelphia	61	60	.504
Montreal	58	62	.483
Florida	54	64	.458
New York	52	67	.437

In 1995, due to the strike that shortened the season, each team played 144 games rather than the usual 162 games. To calculate Atlanta's magic number, we note that $W_2 = 61$ (the number of wins for Philadelphia), $N_2 = 144 - (61 + 60) = 23$ (the number of games Philadelphia had remaining), and $W_1 = 76$ (the number of wins Atlanta had). Therefore, the magic number for Atlanta was

$$M = 61 + 23 - 76 + 1 = 9.$$

Later in the season, when the total of Atlanta's wins and Philadelphia's losses became 9, Atlanta clinched the pennant. ▶

PROPERTIES OF REAL NUMBERS There are several properties that describe how real numbers behave. We will discuss each property and then summarize them at the end of the section. The *commutative properties* state that two numbers may be added or multiplied in any order:

$$4 + (-12) = -12 + 4, \qquad 8(-5) = -5(8),$$
$$-9 + (-1) = -1 + (-9), \qquad (-6)(-3) = (-3)(-6),$$

and so on. Generalizing, the **commutative properties** say that for all real numbers a and b,

$$a + b = b + a \qquad \text{and} \qquad ab = ba.$$

◀ **EXAMPLE 6**
Illustrating the commutative properties

The following statements illustrate the commutative properties. Notice that the *order* of the numbers changes from one side of the equals sign to the other.

(a) $(6 + x) + 9 = (x + 6) + 9$ **(b)** $(6 + x) + 9 = 9 + (6 + x)$

(c) $5 \cdot (9 \cdot 8) = (9 \cdot 8) \cdot 5$ **(d)** $5 \cdot (9 \cdot 8) = 5 \cdot (8 \cdot 9)$ ▶

By the *associative properties,* if three numbers are to be added or multiplied, either the first two numbers or the last two may be "associated." For example, the sum of the three numbers -9, 8, and 7 may be found in either of two ways:

$$-9 + (8 + 7) = -9 + 15 = 6,$$

or

$$(-9 + 8) + 7 = -1 + 7 = 6.$$

Also,

$$5(-3 \cdot 2) = 5(-6) = -30$$

or

$$(5 \cdot -3)2 = (-15)2 = -30.$$

In summary, the **associative properties** say that for all real numbers a, b, and c,

$$(a + b) + c = a + (b + c) \qquad \text{and} \qquad (ab)c = a(bc).$$

CAUTION It is a common error to confuse the associative and commutative properties. To avoid this error, check the order of the terms: with the commutative properties the order changes from one side of the equals sign to the other; with the associative properties the order does not change, but the grouping does.

◖EXAMPLE 7
Distinguishing between the commutative and the associative properties

This example shows a list of statements using the same symbols. Notice the difference between the commutative and associative properties.

Commutative Properties	Associative Properties
$(x + 4) + 9 = (4 + x) + 9$	$(x + 4) + 9 = x + (4 + 9)$
$7 \cdot (5 \cdot 2) = (5 \cdot 2) \cdot 7$	$7 \cdot (5 \cdot 2) = (7 \cdot 5) \cdot 2$

◖EXAMPLE 8
Using the commutative and associative properties to simplify expressions

Simplify each expression using the commutative and associative properties as needed.

(a) $6 + (9 + x) = (6 + 9) + x = 15 + x$ Associative property

(b) $\dfrac{5}{8}(16y) = \left(\dfrac{5}{8} \cdot 16\right)y = 10y$ Associative property

(c) $(-10p)\left(\dfrac{6}{5}\right) = \dfrac{6}{5}(-10p)$ Commutative property

$\qquad\qquad = \left[\dfrac{6}{5}(-10)\right]p$ Associative property

$\qquad\qquad = -12p$

The *identity properties* show special properties of the numbers 0 and 1. The sum of 0 and any real number a is a itself. For example,

$$0 + 4 = 4, \qquad -5 + 0 = -5.$$

The number 0 preserves the identity of a number under addition, making 0 the **identity element for addition** (or the **additive identity**).

The number 1, the **identity element for multiplication** (or **multiplicative identity**), preserves the identity of a number under multiplication, since the product of 1 and any number a is a. For example,

$$5 \cdot 1 = 5, \qquad 1\left(-\frac{2}{3}\right) = -\frac{2}{3}.$$

In summary, the **identity properties** say that for every real number a, there exist unique real numbers 0 and 1 such that

$$a + 0 = a \qquad \text{and} \qquad 0 + a = a$$

$$a \cdot 1 = a \qquad \text{and} \qquad 1 \cdot a = a.$$

The sum of the numbers 5 and -5 is the identity element for addition, 0, just as the sum of $-2/3$ and $2/3$ is 0. In fact, for any real number a there is a real number, written $-a$, such that the sum of a and $-a$ is the identity element 0. The number $-a$ is the **additive inverse** or **negative** of a.

CAUTION   Do not confuse the *negative of a number with a negative number*. Since a is a variable, it can represent either a positive or a negative number. The negative of a, written $-a$, can also be either a negative or a positive number (or zero). Do not make the common mistake of thinking that $-a$ *must* be a negative number. For example, if a is -3, then $-a$ is $-(-3) = 3$.

For each real number a (except 0) there is a real number $1/a$ such that the product of a and $1/a$ is the identity element for multiplication, 1. The number $1/a$ is called the **multiplicative inverse** or **reciprocal** of the number a. Every real number except 0 has a reciprocal.

The existence of $-a$ and of $1/a$ comes from the **inverse properties,** which say that for every real number a, there exists a unique real number $-a$ such that

$$a + (-a) = 0 \qquad \text{and} \qquad -a + a = 0,$$

and for every nonzero real number a, there exists a unique real number $1/a$ such that

$$a \cdot \frac{1}{a} = 1 \qquad \text{and} \qquad \frac{1}{a} \cdot a = 1.$$

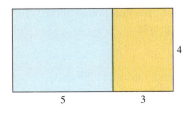

FIGURE 6

The area of the entire region shown in Figure 6 can be found in two ways. One way is to multiply the length of the base of the entire region, or $5 + 3 = 8$, by the width of the region:

$$4(5 + 3) = 4(8) = 32.$$

Another way to find the area of the region is to add the areas of the smaller rectangles on the left and right,

$$4(5) = 20 \quad \text{and} \quad 4(3) = 12,$$

to get a total area of

$$4(5) + 4(3) = 20 + 12 = 32,$$

the same result. The equal results from finding the area in two different ways,

$$4(5 + 3) = 4(5) + 4(3),$$

illustrate the **distributive property of multiplication over addition,** which says that for all real numbers a, b, and c,

$$a(b + c) = ab + ac.$$

We will refer to this as simply "the distributive property" from now on.

Using a commutative property, the distributive property can be rewritten as $(b + c)a = ba + ca$. Another form of the distributive property, $a(b - c) = ab - ac$, comes from the definition of subtraction as "adding the negative." Also, the distributive property can be extended to include more than two numbers in the sum. For example,

$$9(5x + y + 4z) = 9(5x) + 9y + 9(4z)$$
$$= 45x + 9y + 36z.$$

NOTE The distributive property is one of the key properties of the real numbers, because it is used to change products to sums and sums to products.

EXAMPLE 9
Illustrating the distributive property

The following statements illustrate the distributive property.

(a) $3(x + y) = 3x + 3y$

(b) $-(m - 4n) = -1 \cdot (m - 4n) = -m + 4n$

(c) $7p + 21 = 7p + 7 \cdot 3 = 7(p + 3)$

(d) $\dfrac{1}{3}\left(\dfrac{4}{5}m - \dfrac{3}{2}n - 27\right) = \dfrac{1}{3}\left(\dfrac{4}{5}m\right) - \dfrac{1}{3}\left(\dfrac{3}{2}n\right) - \dfrac{1}{3}(27)$

$$= \dfrac{4}{15}m - \dfrac{1}{2}n - 9 \; \blacktriangleright$$

A summary of the properties of the real numbers follows.

Properties of the Real Numbers

For all real numbers a, b, and c:

Commutative Properties $a + b = b + a$
$ab = ba$

Associative Properties $(a + b) + c = a + (b + c)$
$(ab)c = a(bc)$

Identity Properties   There exists a unique real number 0 such that

$$a + 0 = a \qquad \text{and} \qquad 0 + a = a.$$

There exists a unique real number 1 such that

$$a \cdot 1 = a \qquad \text{and} \qquad 1 \cdot a = a.$$

Inverse Properties   There exists a unique real number $-a$ such that

$$a + (-a) = 0 \quad \text{and} \quad (-a) + a = 0.$$

If $a \neq 0$, there exists a unique real number $1/a$ such that

$$a \cdot \frac{1}{a} = 1 \qquad \text{and} \qquad \frac{1}{a} \cdot a = 1.$$

Distributive Property $a(b + c) = ab + ac$

1.1   Exercises ▼▼▼▼▼▼▼▼▼▼▼▼▼▼▼▼▼▼▼▼▼▼▼▼▼▼▼▼▼▼▼▼▼▼▼▼

Let set $B = \{-6, -12/4, -5/8, -\sqrt{3}, 0, 1/4, 1, 2\pi, 3, \sqrt{12}\}$. List all the elements of B that belong to the set. See Example 1.

1. Natural numbers

2. Whole numbers

3. Integers

4. Rational numbers

5. Irrational numbers

6. Real numbers

For Exercises 7–12, choose all words from the following list that apply: (a) natural number, (b) whole number, (c) integer, (d) rational number, (e) irrational number, (f) real number, (g) not defined. See Example 1.

7. 29

8. 0

9. $-\dfrac{5}{6}$

10. $\dfrac{3}{8}$

11. $\sqrt{13}$

12. $-\sqrt{3}$

13. Explain why division by zero is not defined.

Evaluate the exponential expression. See Example 2.

14. 3^4 **15.** -3^5 **16.** -2^6 **17.** $(-3)^4$ **18.** $(-2)^5$ **19.** $(-3)^5$ **20.** $(-3)^6$

21. Based on your answers to Exercises 14–20, complete the following statements. A negative base raised to an odd exponent is _____ . A negative base raised to an even
positive/negative
exponent is _____ .
positive/negative

22. The accompanying graphing calculator screen indicates that $-5^2 = -25$. Use the concepts of this section to explain why this is correct. Why does the calculator *not* give 25 as the answer? What would it give for $(-5)^2$? for $-(-5)^2$?

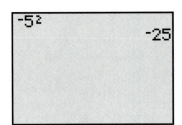

Use the order of operations to evaluate the expression. See Examples 3 and 4.

23. $8^2 - (-4) + 11$

24. $16(-9) - 4$

25. $-2 \cdot 5 + 12 \div 3$

26. $9 \cdot 3 - 16 \div 4$

27. $-4(9 - 8) + (-7)(2)^3$

28. $6(-5) - (-3)(2)^4$

29. $(4 - 2^3)(-2 + \sqrt{25})$

30. $[-3^2 - (-2)][\sqrt{16} - 2^3]$

31. $\left(-\dfrac{2}{9} - \dfrac{1}{4}\right) - \left[-\dfrac{5}{18} - \left(-\dfrac{1}{2}\right)\right]$

32. $\left[-\dfrac{5}{8} - \left(-\dfrac{2}{5}\right)\right] - \left(\dfrac{3}{2} - \dfrac{11}{10}\right)$

33. $\dfrac{-8 + (-4)(-6) \div 12}{4 - (-3)}$

34. $\dfrac{15 \div 5 \cdot 4 \div 6 - 8}{-6 - (-5) - 8 \div 2}$

The graphing calculator screen shown here indicates that when $x = 4$, $y = 3$, and $z = 1$, the value of the expression $x^2 + 2y - 3z$ is 19. Determine what the value of this expression would be for the given values of x, y, and z. Then use your own calculator to support your answer.

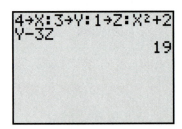

35.

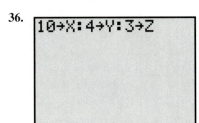

36.

Evaluate the expression if p = −4, q = 8, and r = −10. See Example 4.

37. $2(q - r)$

38. $\dfrac{p}{q} + \dfrac{3}{r}$

39. $\dfrac{q + r}{q + p}$

40. $\dfrac{3q}{3p - 2r}$

41. $\dfrac{3q}{r} - \dfrac{5}{p}$

42. $\dfrac{\dfrac{q}{4} - \dfrac{r}{5}}{\dfrac{p}{2} + \dfrac{q}{2}}$

Solve the problem. See Example 5.

43. On September 6, 1995, the Cleveland Indians led the American League Central Division with a record of 82 wins and 37 losses. The Kansas City Royals were in second place with a record of 61 wins and 58 losses. What was the Indians' magic number on that day?

(See Example 5, and remember that 144 games were to be played that season.)

44. Repeat Exercise 43 for the Cincinnati Reds, who led the Central Division of the National League with a record of 74 wins and 45 losses over the second-place Chicago Cubs, who had a record of 61 wins and 59 losses.

Identify the properties illustrated in each statement. Some will require more than one property. Assume that all variables represent real numbers. See Examples 6–9.

45. $6 \cdot 12 + 6 \cdot 15 = 6(12 + 15)$

46. $8(m + 4) = 8m + 8 \cdot 4$

47. $(x + 6) \cdot \left(\dfrac{1}{x + 6}\right) = 1, \quad \text{if } x + 6 \neq 0$

48. $\dfrac{2 + m}{2 - m} \cdot \dfrac{2 - m}{2 + m} = 1, \quad \text{if } m \neq 2 \text{ or } -2$

49. $(7 + y) + 0 = 7 + y$

50. $[9 + (-9)] \cdot 5 = 5 \cdot 0$

51. Is there a commutative property for subtraction? That is, does $a - b = b - a$? Support your answer with examples.

52. Is there an associative property for subtraction? Does $(a - b) - c = a - (b - c)$? Support your answer with examples.

Use the distributive property to rewrite sums as products and products as sums. See Example 9.

53. $8p - 14p$

54. $15 - 10x$

55. $18y + 6$

56. $9(r - s)$

57. $-3(z - y)$

58. $-2(m + n)$

59. $a(r + s - t)$

60. $p(q - w + x)$

Use the various properties of real numbers to simplify each expression. See Examples 8 and 9.

61. $\dfrac{10}{11}(22z)$

62. $\left(\dfrac{3}{4}r\right)(-12)$

63. $\left(-\dfrac{5}{8}p\right)(-24)$

64. $\dfrac{2}{3}(12y - 6z + 18q)$

65. $-\dfrac{1}{4}(20m + 8y - 32z)$

66. $\dfrac{3}{8}\left(\dfrac{16}{9}y + \dfrac{32}{27}z - \dfrac{40}{9}\right)$

The stated problem involves operations with real numbers. Solve the problem.

67. To find the average of n real numbers, we add the numbers and then divide the sum by n. The lowest PGA championship golf score for 4 rounds of golf was 64, 71, 69, 67 by Bobby Nicholls at Columbus Country Club, Ohio, in 1964. What was his average score per round? (*Source:* PGA Tour.)

68. The average distance from the center of the Earth to the center of the sun is 92,960,000 miles. There are approximately 365.26 days per year. Estimate the average speed (in miles per hour) that the Earth is moving around the Sun if it is assumed that the Earth's orbit is circular. Use $\pi \approx 3.14$ and speed = distance/time. (*Source:* Wright, J. (editor), *The Universal Almanac,* Universal Press Syndicate Company (1994).)

69. The deepest place in the ocean is the Mariana Trench near Japan with a depth of 35,840 feet. If a person dropped a 2.2-pound steel ball there, it would take 1 hour and 4 minutes for it to reach the bottom. What would be the ball's average velocity in miles per hour during this time period? (*Source: The Guinness Book of Records 1995.*)

According to the Census Bureau, home owners' yearly maintenance expenditures can be broken down into ranges according to the pie chart shown here. To find a percent of a real number, we change the percent to a decimal and then multiply the number by this decimal. For example, in a sample of 10,000 home owners, we would expect approximately 45% of them, or .45(10,000) = 4500, to have under $250 worth of expenditures. Use this procedure to find the approximate number of homeowners in the group described.

Yearly Home Owners' Maintenance Expenditures

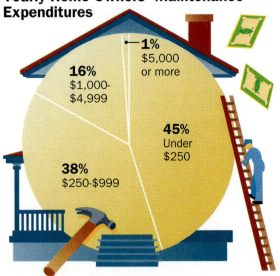

(*Source*: "Yearly homeowner's maintenance expenditures" reprinted with permission. Copyright 1994, Gannett Co., Inc.)

70. A sample of 25,000 homeowners, having expenditures between $250 and $999

71. A sample of 15,000 homeowners, having expenditures between $1000 and $4999

72. A sample of 20,000 homeowners, having expenditures of $5000 or more

73. In a later chapter we will study how one set of real numbers can relate to another set of real numbers. As an example, the table below lists the wave heights produced in the ocean for various wind speeds and wind durations.

(a) What is the expected wave height if a 46-mile-per-hour wind blows for 30 hours?

(b) If the wave height is 5 feet, can you determine the speed of the wind and its duration?

(c) Describe the relationships between wave height, wind speed, and duration of the wind.

Wind Speed	Duration of the Wind			
mph	*10 hr*	*20 hr*	*30 hr*	*40 hr*
11.5	2 ft	2 ft	2 ft	2 ft
17.3	4 ft	5 ft	5 ft	5 ft
23.0	7 ft	8 ft	9 ft	9 ft
34.5	13 ft	17 ft	18 ft	19 ft
46.0	21 ft	28 ft	31 ft	33 ft
57.5	29 ft	40 ft	45 ft	48 ft

(*Source:* Navarra, J., *Atmosphere, Weather and Climate*, W. B. Saunders Company, 1979.)

74. This exercise is designed to show that $\sqrt{2}$ is irrational. Give a reason for each of steps (a) through (h).

There are two possibilities:

(1) A rational number a/b exists such that $(a/b)^2 = 2$.

(2) There is no such rational number.

We work with assumption (1). If it leads to a contradiction, then we will know that (2) must be correct. Start by assuming that a rational number a/b exists with $(a/b)^2 = 2$. Assume also that a/b is written in lowest terms.

(a) Since $(a/b)^2 = 2$, we must have $a^2/b^2 = 2$, or $a^2 = 2b^2$.

(b) $2b^2$ is an even number.

(c) Therefore, a^2, and a itself, must be even numbers.

(d) Since a is an even number, it must be a multiple of 2. That is, we can find a natural number c such that $a = 2c$. This changes $a^2 = 2b^2$ into $(2c)^2 = 2b^2$.

(e) Therefore, $4c^2 = 2b^2$ or $2c^2 = b^2$.

(f) $2c^2$ is an even number.

(g) This makes b^2 an even number, so that b must be even.

(h) We have reached a contradiction. Show where the contradiction occurs.

(i) Since assumption (1) leads to a contradiction, we are forced to accept assumption (2), which says that $\sqrt{2}$ is irrational.

1.2 Order and Absolute Value ▼▼▼

ORDER Figure 7 shows a number line with the points corresponding to several different numbers marked on the line. A number that corresponds to a particular point on a line is called the **coordinate** of the point. For example, the leftmost marked point in Figure 7 has coordinate -4. The correspondence between points on a line and the real numbers is called a **coordinate system** for the line. (From now on, the phrase "the point on a number line with coordinate a" will be abbreviated as "the point with coordinate a," or simply "the point a.")

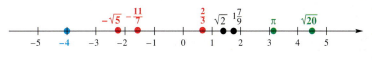

FIGURE 7

If the real number a is to the left of the real number b on a number line, then *a* **is less than** *b,* written $a < b$. If a is to the right of b, then *a* **is greater than** *b,* written $a > b$. For example, in Figure 7, $-\sqrt{5}$ is to the left of $-11/7$ on the number line, so $-\sqrt{5} < -11/7$, while $\sqrt{20}$ is to the right of π, indicating $\sqrt{20} > \pi$.

NOTE Remember that the "point" of the symbol goes toward the smaller number.

As an alternative to this geometric definition of "is less than" or "is greater than," there is an algebraic definition: if a and b are two real numbers and if the difference $a - b$ is positive, then $a > b$. If $a - b$ is negative, then $a < b$. The geometric and algebraic statements of order are summarized as follows.

Statement	*Geometric Form*	*Algebraic Form*
a > *b*	a is to the right of b	$a - b$ is positive
a < *b*	a is to the left of b	$a - b$ is negative

EXAMPLE 1
Identifying the smaller of
two numbers

Part (a) of this example shows how to identify the smaller of two numbers with the geometric approach, and part (b) uses the algebraic approach.

(a) In Figure 7, $-\sqrt{5}$ is to the left of $2/3$, so

$$-\sqrt{5} < \frac{2}{3}.$$

Since $2/3$ is to the right of $-\sqrt{5}$,

$$\frac{2}{3} > -\sqrt{5}.$$

(b) The difference $2/3 - (-11/7) = 2/3 + 11/7$ is positive, showing that

$$\frac{2}{3} > -\frac{11}{7}.$$

The difference $-11/7 - 2/3 = -(11/7 + 2/3)$ is negative, showing that

$$-\frac{11}{7} < \frac{2}{3}. \quad \blacktriangleright$$

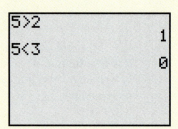

Graphing calculators have a logic function that tests inequality statements. If the inequality is true, the calculator returns a 1; if it is false, it returns a 0. See your manual for instructions for your calculator. The figure shows tests for two inequality statements on a TI-82 calculator.

```
5>2
                    1
5<3
                    0
```

5 > 2 is true, while 5 < 3 is false.

The following variations on $<$ and $>$ are often used.

Symbol	Meaning (Reading Left to Right)
$\leq$	is less than or equal to
$\geq$	is greater than or equal to
$\nless$	is not less than
$\ngtr$	is not greater than

Statements involving these symbols, as well as $<$ and $>$, are called **inequalities.**

EXAMPLE 2
Showing why inequality
statements are true

The list below shows several statements and the reason why each is true.

Statement	Reason
$8 \leq 10$	$8 < 10$
$8 \leq 8$	$8 = 8$
$-9 \geq -14$	$-9 > -14$
$-8 \not> -2$	$-8 < -2$
$4 \not< 2$	$4 > 2$

The inequality $a < b < c$ says that b is *between* a and c, since

$$a < b < c$$

means $\qquad a < b \qquad$ and $\qquad b < c.$

In the same way, $\qquad\qquad a \leq b \leq c$

means $\qquad a \leq b \qquad$ and $\qquad b \leq c.$

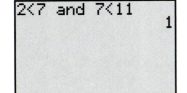

$2 < 7$ and $7 < 11$ is a true statement.

CAUTION When writing these "between" statements, make sure that both inequality symbols point in the same direction, toward the smallest number. For example,

$$\text{both} \quad 2 < 7 < 11 \qquad \text{and} \qquad 5 > 4 > -1$$

are true statements, but $3 < 5 > 2$ is meaningless. Generally, it is best to rewrite statements such as $5 > 4 > -1$ as $-1 < 4 < 5$, which is the order of these numbers on the number line.

The following *properties of order* give the basic properties of $<$ and $>$.

Properties of Order

For all real numbers a, b, and c:

Transitive Property $\qquad$ If $a < b$ and $b < c$, then $a < c$.
Addition Property $\qquad$ If $a < b$, then $a + c < b + c$.
Multiplication Property $\qquad$ If $a < b$, and if $c > 0$, then $ac < bc$.
$\qquad\qquad\qquad\qquad\qquad\qquad$ If $a < b$, and if $c < 0$, then $ac > bc$.

In these properties, replacing $<$ with $>$ and $>$ with $<$ results in equivalent properties.

EXAMPLE 3
Illustrating the properties of order

(a) By the transitive property, if $3z < k$ and $k < p$, then $3z < p$.

(b) By the addition property, any real number can be added to both sides of an inequality. For example, adding -2 to both sides of

$$x + 2 < 5$$

gives
$$x + 2 + (-2) < 5 + (-2)$$
$$x < 3.$$

This process is explained in more detail in Chapter 2.

(c) While any number may be *added* to both sides of an inequality, more care must be used when *multiplying* both sides by a number. For example, multiplying both sides of

$$\frac{1}{2}x < 5$$

by 2 gives
$$2 \cdot \frac{1}{2}x < 2 \cdot 5$$
$$x < 10,$$

by the first part of the multiplication property. On the other hand, multiplying both sides of

$$-\frac{3}{5}r \geq 9$$

by $-\frac{5}{3}$ gives
$$-\frac{5}{3}\left(-\frac{3}{5}r\right) \leq -\frac{5}{3} \cdot 9$$
$$r \leq -15.$$

Here the $\geq$ symbol has changed to $\leq$, using the second part of the multiplication property. ▶

CAUTION   Don't forget to reverse the inequality symbol when multiplying (or dividing) both sides of an equation by a negative quantity.

ABSOLUTE VALUE   The distance on the number line from a number to 0 is called the **absolute value** of that number. The absolute value of the number a is written $|a|$. For example, the distance on the number line from 9 to 0 is 9, as is the distance from -9 to 0. (See Figure 8.) Therefore,

$$|9| = 9 \quad \text{and} \quad |-9| = 9.$$

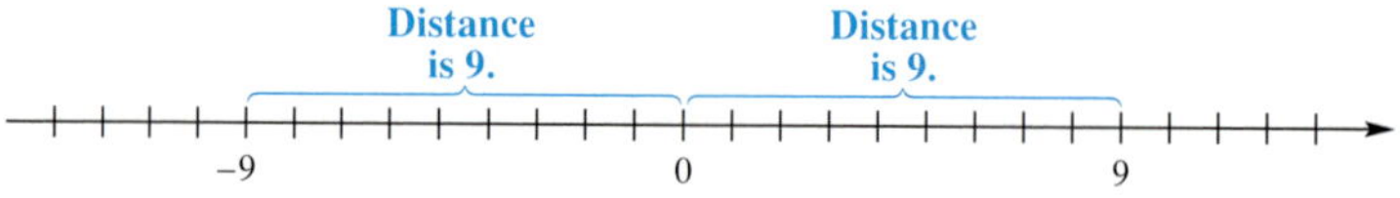

FIGURE 8

NOTE Since distance cannot be negative, the absolute value of a number is always nonnegative.

EXAMPLE 4
Evaluating absolute value

(a) $|2\pi| = 2\pi$

(b) $\left| -\dfrac{5}{8} \right| = \dfrac{5}{8}$

(c) $-|8| = -(8) = -8$

(d) $-|-2| = -(2) = -2$ ▶

The algebraic definition of absolute value can be stated as follows.

Absolute Value

For all real numbers a, $\quad |a| = \begin{cases} a & \text{if } a \geq 0 \\ -a & \text{if } a < 0. \end{cases}$

The second part of this definition requires some thought. If a is a negative number, that is, if $a < 0$, then $-a$ is positive. Thus, for a *negative* number a,

$$|a| = -a,$$

or the negative of a. For example, if $a = -5$, then $|a| = |-5| = -(-5) = 5$. Think of $-a$ as the "opposite" of a.

EXAMPLE 5
Finding absolute value of a sum or difference

Write each of the following without absolute value bars.

(a) $|-8 + 2|$
   Work inside the absolute value bars first. Since $-8 + 2 = -6$, $|-8 + 2| = |-6| = 6$.

(b) $|\sqrt{5} - 2|$
   Since $\sqrt{5} > 2$ ($\sqrt{5} \approx 2.24$), $\sqrt{5} - 2 > 0$, and so $|\sqrt{5} - 2| = \sqrt{5} - 2$.

(c) $|\pi - 4|$
   Here, $\pi < 4$, so that $\pi - 4 < 0$, and

$$|\pi - 4| = -(\pi - 4)$$
$$= -\pi + 4 \qquad \text{or} \qquad 4 - \pi.$$

(d) $|m - 2|$ if $m < 2$
   If $m < 2$, then $m - 2 < 0$, so

$$|m - 2| = -(m - 2)$$
$$= -m + 2 \qquad \text{or} \qquad 2 - m. ▶$$

Graphing calculators have an absolute value function, labeled abs, that returns the absolute value of a numerical expression. For example, if you enter

$$\text{abs}(4 - 10)$$

the calculator returns 6.

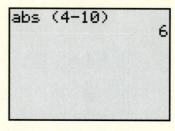

Absolute value is useful in applications where only the *size,* not the *sign,* of the difference between two numbers is important.

EXAMPLE 6
Measuring blood pressure difference

Systolic blood pressure is the maximum pressure produced by each heartbeat. Both low blood pressure and high blood pressure are cause for medical concern. Therefore, health-care professionals are interested in a patient's "pressure difference from normal," or P_d. If 120 is considered a normal systolic pressure, $P_d = |P - 120|$ where P is the patient's recorded systolic pressure. For example, a patient with a systolic pressure, P, of 113 would have a pressure difference from normal of

$$
\begin{aligned}
P_d &= |P - 120| \\
&= |113 - 120| \\
&= |-7| \\
&= 7. \; \blacktriangleright
\end{aligned}
$$

The definition of absolute value can be used to prove the following properties of absolute value.

Properties of Absolute Value

For all real numbers a and b:

$$|a| \geq 0 \qquad\qquad |-a| = |a| \qquad\qquad |a| \cdot |b| = |ab|$$

$$\left|\frac{a}{b}\right| = \frac{|a|}{|b|} \;\; (b \neq 0) \quad |a + b| \leq |a| + |b| \quad \textbf{(the triangle inequality).}$$

EXAMPLE 7
Illustrating the properties of absolute value

(a) $|-15| = 15 \geq 0$

(b) $|-10| = 10$ and $|10| = 10$, so $|-10| = |10|$.

(c) $|5x| = |5| \cdot |x| = 5|x|$ since 5 is positive.

(d) $\left|\dfrac{2}{y}\right| = \dfrac{|2|}{|y|} = \dfrac{2}{|y|}$ for $y \neq 0$

(e) For $a = 3$ and $b = -7$:
$$|a + b| = |3 + (-7)| = |-4| = 4$$
$$|a| + |b| = |3| + |-7| = 3 + 7 = 10$$
$$|a + b| < |a| + |b|$$

(f) For $a = 2$ and $b = 12$:
$$|a + b| = |2 + 12| = |14| = 14$$
$$|a| + |b| = |2| + |12| = 2 + 12 = 14$$
$$|a + b| = |a| + |b| \quad \blacktriangleright$$

Sometimes it is necessary to evaluate expressions involving absolute value. The next example shows how this is done.

EXAMPLE 8
Evaluating absolute value expressions

Let $x = -6$ and $y = 10$. Evaluate each expression.

(a) $|2x - 3y| = |2(-6) - 3(10)|$ Substitute.
$\qquad\qquad\quad = |-12 - 30|$ Work inside the bars; multiply.
$\qquad\qquad\quad = |-42|$ Subtract.
$\qquad\qquad\quad = 42$

(b) $\dfrac{2|x| - |3y|}{|xy|} = \dfrac{2|-6| - |3(10)|}{|-6(10)|}$ Substitute.

$\qquad\qquad\quad = \dfrac{2 \cdot 6 - |30|}{|-60|}$ $|-6| = 6$; multiply.

$\qquad\qquad\quad = \dfrac{12 - 30}{60}$ Multiply; $|30| = 30$; $|-60| = 60$.

$\qquad\qquad\quad = \dfrac{-18}{60} = -\dfrac{3}{10} \quad \blacktriangleright$

1.2 Exercises ▼▼▼▼▼▼▼▼▼▼▼▼▼▼▼▼▼▼▼▼▼▼▼▼▼▼▼▼▼▼▼▼▼▼

Decide whether the statement is true or false.

1. If $a > b$, then the absolute value of $b - a$ is $a - b$.
2. For all real numbers x, $|x| > 0$.
3. If a and b are both positive, $|a + b| = a + b$.
4. If a and b are both negative, $|a + b| = -(a + b)$.

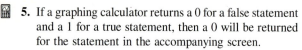 **5.** If a graphing calculator returns a 0 for a false statement and a 1 for a true statement, then a 0 will be returned for the statement in the accompanying screen.

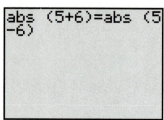

 **6.** (Refer to Exercise 5.) A 1 will be returned for the statement in the accompanying screen.

```
abs (52+635)=abs
  (-52-635)
```

Write the numbers in each list in numerical order, from smallest to largest. Use a calculator as necessary. See Example 1.

7. $|-8|, -|9|, -|-6|$

8. $-|-9|, -|7|, -|-2|$

9. $\sqrt{8}, -4, -\sqrt{3}, -2, -5, \sqrt{6}, 3$

10. $\sqrt{2}, -1, 4, 3, \sqrt{8}, -\sqrt{6}, \sqrt{7}$

11. $\dfrac{3}{4}, \sqrt{2}, \dfrac{7}{5}, \dfrac{8}{5}, \dfrac{22}{15}$

12. $-\dfrac{9}{8}, -3, -\sqrt{3}, -\sqrt{5}, -\dfrac{9}{5}, -\dfrac{8}{5}$

13. Explain why it is necessary to reverse the direction of the inequality symbol when multiplying an inequality by a negative number. Give examples to illustrate your explanation.

14. What is wrong with writing the statement "$x < 2$ or $x > 5$" as $5 < x < 2$?

For the given inequality, multiply both sides by the indicated number to get an equivalent inequality. See Example 3(c).

15. $-2 < 5, 3$

16. $-7 < 1, 2$

17. $-11 \geq -22, -\dfrac{1}{11}$

18. $-2 \geq -14, -\dfrac{1}{2}$

19. $3x < 9, \dfrac{1}{3}$

20. $5x > 45, \dfrac{1}{5}$

21. $-9k > 63, -\dfrac{1}{9}$

22. $-3p \leq 24, -\dfrac{1}{3}$

Let $x = -4$ and $y = 2$. Evaluate each expression. See Examples 4, 5, and 8.

23. $|2x|$

24. $|-3y|$

25. $|x - y|$

26. $|2x + 5y|$

27. $|3x + 4y|$

28. $|-5y + x|$

29. $\dfrac{|-8y + x|}{-|x|}$

30. $\dfrac{|x| + 2|y|}{5 + x}$

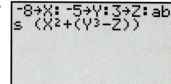 *Use the accompanying graphing calculator screen to predict the value of the expression given the values of x, y, and z shown. Then use your calculator to verify your answer.*

31.
```
-2→X:3→Y:-6→Z:ab
s ((2X-3)/(2Z))
```

32.
```
-8→X:-5→Y:3→Z:ab
s (X²+(Y³-Z))
```

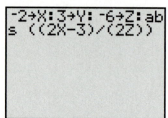

Write the expression without absolute value bars. See Example 5.

33. $|\pi - 3|$

34. $|\pi - 5|$

35. $|x - 4|$,   if $x > 4$

36. $|y - 3|$,   if $y < 3$

37. $|2k - 8|$,   if $k < 4$

38. $|3r - 15|$,   if $r > 5$

39. $|-8 - 4m|$,   if $m > -2$

40. $|6 - 5r|$,   if $r < -2$

41. $|x - y|$,   if $x < y$

42. $|x - y|$,   if $x > y$

43. $|3 + x^2|$

44. $|x^2 + 4|$

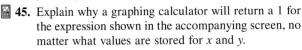 **45.** Explain why a graphing calculator will return a 1 for the expression shown in the accompanying screen, no matter what values are stored for x and y.

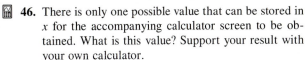 **46.** There is only one possible value that can be stored in x for the accompanying calculator screen to be obtained. What is this value? Support your result with your own calculator.

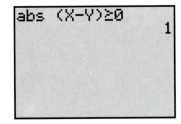

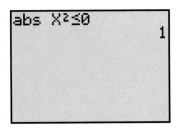

Justify each statement by giving the correct property from this section. Assume that all variables represent real numbers. See Examples 3 and 7.

47. If $x + 8 < 15$, then $x < 7$.

48. If $-4x < 24$, then $x > -6$.

49. If $x < 5$ and $5 < m$, then $x < m$.

50. If $m > 0$, then $9m > 0$.

51. If $k > 0$, then $8 + k > 8$.

52. $|8 + m| \le |8| + |m|$

53. $|k - m| \le |k| + |-m|$

54. $|8| \cdot |-4| = |-32|$

55. $|12 + 11r| \ge 0$

56. $\left| \dfrac{-12}{5} \right| = \dfrac{|-12|}{|5|}$

Refer to Example 6 to work Exercises 57 and 58.

57. Calculate the P_d value for a woman whose actual systolic pressure is 116 and whose normal value should be 125.

58. If a patient's P_d value is 17 and the normal pressure for his sex and age should be 130, what are the two possible values for his systolic blood pressure?

The wind-chill factor is a measure of the cooling effect that the wind has on a person's skin. It calculates the equivalent cooling temperature if there were no wind. The table gives the wind-chill factor for various wind speeds and temperatures.

Wind/°F	40°	30°	20°	10°	0°	−10°	−20°	−30°	−40°	−50°
5 mph	37	27	16	6	−5	−15	−26	−36	−47	−57
10 mph	28	16	4	−9	−21	−33	−46	−58	−70	−83
15 mph	22	9	−5	−18	−36	−45	−58	−72	−85	−99
20 mph	18	4	−10	−25	−39	−53	−67	−82	−96	−110
25 mph	16	0	−15	−29	−44	−59	−74	−88	−104	−118
30 mph	13	−2	−18	−33	−48	−63	−79	−94	−109	−125
35 mph	11	−4	−20	−35	−49	−67	−82	−98	−113	−129
40 mph	10	−6	−21	−37	−53	−69	−85	−100	−116	−132

(*Source:* Miller, A. and J. Thompson, *Elements of Meteorology,* Second Edition, Charles E. Merrill Publishing Co., 1975.)

Suppose that we wish to determine the difference between two of these entries, and are interested only in the magnitude, or absolute value, of this difference. Then we subtract the two entries and find the absolute value. For example, the difference in wind-chill factors for wind at 20 miles per hour with a 20°F temperature and wind at 30 miles per hour with a 40°F temperature is $|-10° - 13°| = 23°F$, or equivalently, $|13° - (-10°)| = 23°F$.

Find the absolute value of the difference of the two indicated wind-chill factors.

59. wind at 15 miles per hour with a 30°F temperature and wind at 10 miles per hour with a −10°F temperature

60. wind at 20 miles per hour with a −20°F temperature and wind at 5 miles per hour with a 30°F temperature

61. wind at 30 miles per hour with a −30°F temperature and wind at 15 miles per hour with a −20°F temperature

62. wind at 40 miles per hour with a 40°F temperature and wind at 25 miles per hour with a −30°F temperature

The accompanying graphs depict the amounts of revenue and expenditures for state and local governments in the United States during the years 1990, 1991, and 1992, in billions of dollars. Determine the absolute value of the difference between the revenue and the expenditure for each year, and tell whether the governments were "in the red" or "in the black." (Note: These descriptions go back to the days when bookkeepers used red ink to indicate losses and black ink to indicate gains.)

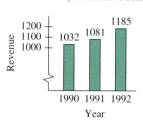

State and Local Government Revenue (in billions of dollars)

Source: U.S. Bureau of the Census, *Government Finances,* Series GF, No. 5, annual.

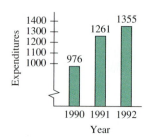

State and Local Government Expenditures (in billions of dollars)

Source: U.S. Bureau of the Census, *Government Finances,* Series GF, No. 5, annual.

63. 1990   **64.** 199   **65.** 1992

1.3 Polynomials; The Binomial Theorem ▼▼▼

RULES FOR EXPONENTS   Positive integer exponents were introduced in Section 1.1. Work with exponents can be simplified by using the rules for exponents. By definition, the notation a^m (where m is a positive integer and a is a real number) means that a appears as a factor m times. In the same way, a^n (where n is a positive integer) means that a appears as a factor n times. In the product $a^m \cdot a^n$, the number a would appear $m + n$ times.

Product Rule

For all positive integers m and n and every real number a,
$$a^m \cdot a^n = a^{m+n}.$$

◀ **EXAMPLE 1**
Using the product rule

Find the following products.

(a) $y^4 \cdot y^7 = y^{4+7} = y^{11}$

(b) $(6z^5)(9z^3)(2z^2)$

$$
\begin{aligned}
(6z^5)(9z^3)(2z^2) &= (6 \cdot 9 \cdot 2) \cdot (z^5 z^3 z^2) \quad &&\text{Commutative and associative properties} \\
&= 108z^{5+3+2} &&\text{Product rule} \\
&= 108z^{10}
\end{aligned}
$$

(c) $(2k^m)(k^{1+m})$

$$
\begin{aligned}
(2k^m)(k^{1+m}) &= 2k^{m+(1+m)} \quad &&\text{Product rule (if } m \text{ is a positive integer)} \\
&= 2k^{1+2m} \quad ▶
\end{aligned}
$$

An exponent of zero is defined as follows.

Definition of a^0

For any nonzero real number a,
$$a^0 = 1.$$

We will show why a^0 is defined this way in Section 1.6. The symbol 0^0 is not defined.

◀ **EXAMPLE 2**
Using the definition of a^0

(a) $3^0 = 1$

(b) $(-4)^0 = 1$
Replace a with -4 in the definition.

(c) $-4^0 = -1$
As shown in Section 1.1, $-4^0 = -(4^0) = -1$.

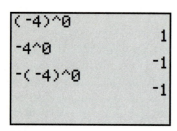

The results in Example 2, parts (b), (c), and (d), are supported here. Notice the use of the caret, ∧, to indicate exponentiation.

(d) $-(-4)^0 = -(1) = -1$

(e) $(7r)^0 = 1$, if $r \neq 0$ ▶

The expression $(2^5)^3$ can be written as

$$(2^5)^3 = 2^5 \cdot 2^5 \cdot 2^5.$$

By a generalization of the product rule for exponents, this product is

$$(2^5)^3 = 2^{5+5+5} = 2^{15}.$$

The same exponent could have been obtained by multiplying 3 and 5. This example suggests the first of the **power rules** given below. The others are found in a similar way.

Power Rules

For all positive integers m and n and all real numbers a and b,

$$(a^m)^n = a^{mn} \qquad (ab)^m = a^m b^m \qquad \left(\frac{a}{b}\right)^m = \frac{a^m}{b^m} \quad (b \neq 0).$$

◀ EXAMPLE 3
Using the power rules

(a) $(5^3)^2 = 5^{3(2)} = 5^6$

(b) $(3^4 x^2)^3 = (3^4)^3 (x^2)^3 = 3^{4(3)} x^{2(3)} = 3^{12} x^6$

(c) $\left(\dfrac{2^5}{b^4}\right)^3 = \dfrac{(2^5)^3}{(b^4)^3} = \dfrac{2^{15}}{b^{12}}$, if $b \neq 0$ ▶

CAUTION Be careful not to confuse examples like mn^2 and $(mn)^2$. The two expressions are *not* equal. The second power rule given above can be used only with the second expression: $(mn)^2 = m^2 n^2$.

POLYNOMIALS An **algebraic expression** is any combination of variables or constants joined by the basic operations of addition, subtraction, multiplication, division (except by zero), or extraction of roots. Here are some examples of algebraic expressions.

$$2x^2 - 3x, \qquad \frac{5y}{2y - 3}, \qquad \sqrt{m^3 - 8}, \qquad (3a + b)^4$$

The simplest algebraic expressions, *polynomials,* are discussed in this section.

The product of a real number and one or more variables raised to powers is called a **term.** The real number is called the **numerical coefficient,** or just the **coefficient.** The coefficient in $-3m^4$ is -3, while the coefficient in $-p^2$ is -1. **Like terms** are terms with the same variables each raised to the same powers. For example, $-13x^3$, $4x^3$, and $-x^3$ are like terms, while $6y$ and $6y^2$ are not.

A **polynomial** is defined as a term or a finite sum of terms, with only nonnegative integer exponents permitted on the variables. If the terms of a polynomial contain only the variable x, then the polynomial is called a **polynomial in x.** (Polynomials in other variables are defined similarly.) Examples of polynomials include

$$5x^3 - 8x^2 + 7x - 4, \qquad 9p^5 - 3, \qquad 8r^2, \qquad \text{and} \qquad 6.$$

The expression $9x^2 - 4x - 6/x$ is not a polynomial because of $-6/x$. The terms of a polynomial cannot have variables in a denominator.

The greatest exponent in a polynomial in one variable is the **degree** of the polynomial. A nonzero constant is said to have degree 0. (The polynomial 0 has no degree.) For example, $3x^6 - 5x^2 + 2x + 3$ is a polynomial of degree 6.

A polynomial can have more than one variable. A term containing more than one variable has degree equal to the sum of all the exponents appearing on the variables in the term. For example, $-3x^4y^3z^5$ is of degree $4 + 3 + 5 = 12$. The degree of a polynomial in more than one variable is equal to the greatest degree of any term appearing in the polynomial. By this definition, the polynomial

$$2x^4y^3 - 3x^5y + x^6y^2$$

is of degree 8 because of the x^6y^2 term.

A polynomial containing exactly three terms is called a **trinomial;** one containing exactly two terms is a **binomial;** and a single-term polynomial is called a **monomial.** For example, $7x^9 - 8x^4 + 1$ is a trinomial of degree 9.

EXAMPLE 4
Determining the degree and type of a polynomial

The list below shows several polynomials, gives the degree of each, and identifies each as a monomial, binomial, trinomial, or none of these.

Polynomial	Degree	Type
$9p^7 - 4p^3 + 8p^2$	7	Trinomial
$29x^{11} + 8x^{15}$	15	Binomial
$-10r^6s^8$	14	Monomial
$5a^3b^7 - 3a^5b^5 + 4a^2b^9 - a^{10}$	11	None of these

ADDITION AND SUBTRACTION   Since the variables used in polynomials represent real numbers, a polynomial represents a real number. This means that all the properties of the real numbers mentioned in this chapter hold for polynomials. In particular, the distributive property holds, so

$$3m^5 - 7m^5 = (3 - 7)m^5 = -4m^5.$$

Thus, polynomials are added by adding coefficients of like terms; polynomials are subtracted by subtracting coefficients of like terms.

◀EXAMPLE 5
Adding and subtracting polynomials

Add or subtract, as indicated.

(a) $(2y^4 - 3y^2 + y) + (4y^4 + 7y^2 + 6y)$
$$= (2 + 4)y^4 + (-3 + 7)y^2 + (1 + 6)y$$
$$= 6y^4 + 4y^2 + 7y$$

(b) $(-3m^3 - 8m^2 + 4) - (m^3 + 7m^2 - 3)$
$$= (-3 - 1)m^3 + (-8 - 7)m^2 + [4 - (-3)]$$
$$= -4m^3 - 15m^2 + 7$$

(c) $8m^4p^5 - 9m^3p^5 + (11m^4p^5 + 15m^3p^5) = 19m^4p^5 + 6m^3p^5$

(d) $4(x^2 - 3x + 7) - 5(2x^2 - 8x - 4)$
$$= 4x^2 - 4(3x) + 4(7) - 5(2x^2)$$
$$\quad - 5(-8x) - 5(-4) \qquad \text{Distributive property}$$
$$= 4x^2 - 12x + 28 - 10x^2 + 40x + 20 \qquad \text{Associative property}$$
$$= -6x^2 + 28x + 48 \qquad \text{Add like terms.} \ ▶$$

As shown in parts (a), (b), and (d) of Example 5, polynomials in one variable are often written with their terms in *descending order,* so the term of greatest degree is first, the one with the next greatest degree is next, and so on.

MULTIPLICATION The associative and distributive properties, together with the properties of exponents, can also be used to find the product of two polynomials. For example, to find the product of $3x - 4$ and $2x^2 - 3x + 5$, we treat $3x - 4$ as a single expression and use the distributive property as follows.

$$(3x - 4)(2x^2 - 3x + 5) = (3x - 4)(2x^2) - (3x - 4)(3x) + (3x - 4)(5)$$

Now use the distributive property three separate times on the right of the equals sign to get

$$(3x - 4)(2x^2 - 3x + 5)$$
$$= (3x)(2x^2) - 4(2x^2) - (3x)(3x) - (-4)(3x) + (3x)5 - 4(5)$$
$$= 6x^3 - 8x^2 - 9x^2 + 12x + 15x - 20$$
$$= 6x^3 - 17x^2 + 27x - 20.$$

It is sometimes more convenient to write such a product vertically, as follows.

$$
\begin{array}{r}
2x^2 - \ \ 3x + \ \ 5 \\
3x - \ \ 4 \\
\hline
- \ \ 8x^2 + 12x - 20 \\
6x^3 - \ \ 9x^2 + 15x \\
\hline
6x^3 - 17x^2 + 27x - 20 \qquad \text{Add in columns.}
\end{array}
$$

◖EXAMPLE 6
Multiplying polynomials

Multiply $(3p^2 - 4p + 1)(p^3 + 2p - 8)$.

    Multiply each term of the second polynomial by each term of the first and add these products. It is most efficient to work vertically with polynomials of more than two terms, so that like terms can be placed in columns.

$$
\begin{array}{r}
3p^2 - 4p + 1 \\
p^3 + 2p - 8 \\
\hline
-24p^2 + 32p - 8 \\
6p^3 - 8p^2 + 2p \\
3p^5 - 4p^4 + p^3 \\
\hline
3p^5 - 4p^4 + 7p^3 - 32p^2 + 34p - 8
\end{array}
$$

Multiply $3p^2 - 4p + 1$ by -8.
Multiply $3p^2 - 4p + 1$ by $2p$.
Multiply $3p^2 - 4p + 1$ by p^3.
Add in columns. ◗

    The FOIL method is a convenient way to find the product of two binomials. The memory aid FOIL (for First, Outside, Inside, Last) gives the pairs of terms to be multiplied to get the product, as shown in the next examples.

◖EXAMPLE 7
Using FOIL to multiply two binomials

Find each product.

        F   **O**   **I**   **L**

(a) $(6m + 1)(4m - 3) = (6m)(4m) + (6m)(-3) + 1(4m) + 1(-3)$
$$= 24m^2 - 14m - 3$$

(b) $(2x + 7)(2x - 7) = 4x^2 - 14x + 14x - 49$
$$= 4x^2 - 49$$

(c) $(2k^n - 5)(k^n + 3) = 2k^{2n} + 6k^n - 5k^n - 15$
$$= 2k^{2n} + k^n - 15 \quad \text{(if } n \text{ is a positive integer)} ◗$$

    In parts (a) and (c) of Example 7, the product of two binomials was a trinomial, while in part (b) the product of two binomials was a binomial. The product of two binomials of the forms $x + y$ and $x - y$ is always a binomial. The squares of binomials are also special products. The products $(x + y)^2$ and $(x - y)^2$ are shown below.

Special Products

Product of the Sum and Difference of Two Terms
Square of a Binomial

$$(x + y)(x - y) = x^2 - y^2$$
$$(x + y)^2 = x^2 + 2xy + y^2$$
$$(x - y)^2 = x^2 - 2xy + y^2$$

It is useful to memorize and be able to apply these special products. In Section 1.4 on factoring polynomials, you will need to recognize them.

EXAMPLE 8
Using the special products

Find each product.

(a) $(3p + 11)(3p - 11)$

Using the pattern discussed above, replace x with $3p$ and y with 11.

$$(3p + 11)(3p - 11) = (3p)^2 - 11^2 = 9p^2 - 121$$

(b) $(5m^3 - 3)(5m^3 + 3) = (5m^3)^2 - 3^2 = 25m^6 - 9$

(c) $(9k - 11r^3)(9k + 11r^3) = (9k)^2 - (11r^3)^2 = 81k^2 - 121r^6$

(d) $(2m + 5)^2 = (2m)^2 + 2(2m)(5) + 5^2$
$$= 4m^2 + 20m + 25$$

(e) $(3x - 7y^4)^2 = (3x)^2 - 2(3x)(7y^4) + (7y^4)^2$
$$= 9x^2 - 42xy^4 + 49y^8$$

CAUTION   As shown in Examples 8(d) and (e), the square of a binomial has three terms. Students often mistakenly give $a^2 + b^2$ as the result of expanding $(a + b)^2$. Be careful to avoid that error.

CONNECTIONS The special product $(a + b)(a - b) = a^2 - b^2$ can be used to perform some multiplication problems. For example,

$$51 \times 49 = (50 + 1)(50 - 1)$$
$$= 50^2 - 1^2$$
$$= 2500 - 1 = 2499.$$

Similarly, the perfect square pattern gives

$$47^2 = (50 - 3)^2$$
$$= 50^2 - 2(50)(3) + 3^2$$
$$= 2500 - 300 + 9$$
$$= 2209.$$

FOR DISCUSSION OR WRITING
Use the special products to evaluate the following expressions.

1. 99×101 **2.** 63×57 **3.** 102^2 **4.** 71^2
5. Discuss how FOIL might be used to evaluate products like 45×87.

BINOMIAL THEOREM   The square of a binomial is a special case of the *binomial theorem,* which gives a pattern for finding any positive integer power of a binomial. The coefficients in the formula can be found from the following array of numbers, known as **Pascal's triangle.**

Pascal's Triangle

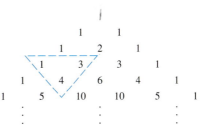

The triangle can be extended by observing the pattern. Each number is the sum of the two numbers above it, one to the right and one to the left. For example, in row 4, 1 is the sum of 1, 4 is the sum of 3 and 1, 6 is the sum of 3 and 3, and so on. Now notice the behavior of the exponents on x and y for powers of the binomial $x + y$.

$$(x + y)^2 = x^2 + 2xy + y^2$$
$$(x + y)^3 = x^3 + 3x^2y + 3xy^2 + y^3$$
$$(x + y)^4 = x^4 + 4x^3y + 6x^2y^2 + 4xy^3 + y^4$$

The last two results can be verified by multiplying out $(x + y)^3$ and $(x + y)^4$. These examples suggest that the variables in the expansion of $(x + y)^n$ should have the following pattern:

$$x^n, \quad x^{n-1}y, \quad x^{n-2}y^2, \quad x^{n-3}y^3, \quad \ldots, \quad xy^{n-1}, \quad y^n,$$

and the coefficients should come from the nth row of Pascal's triangle. Notice that the sum of the exponents in each term is n. The next example puts all this together.

◀**EXAMPLE 9**
Finding the 5th power of a binomial

Write out the binomial expansion of $(m - 2)^5$.

Let $n = 5$, $x = m$, and $y = -2$, since $(m - 2)^5 = [m + (-2)]^5$. Use the coefficients from row 5 of Pascal's triangle and the pattern shown above for the exponents.

$$(m - 2)^5 = m^5 + 5m^4(-2) + 10m^3(-2)^2 + 10m^2(-2)^3 + 5m(-2)^4 + (-2)^5$$
$$= m^5 - 10m^4 + 40m^3 - 80m^2 + 80m - 32 \quad ▶$$

An alternative method for finding the coefficients of $(x + y)^n$ is to multiply the exponent on x in any term by the coefficient of the term and divide the product by the number of the term to get the coefficient of the next term. For instance, in Example 9, the first term is $1m^5$, so the number of the term is 1, the exponent is 5, and the coefficient is 1. The coefficient of the second term thus is $(5 \cdot 1)/1 = 5$. Similarly, the second term is $5m^4(-2)$, so the coefficient of the third term is $(4 \cdot 5)/2 = 10$, and so on.

A complete discussion of the binomial theorem and a proof are given in the last chapter in this book.

CONNECTIONS Over the years many interesting patterns have been discovered in Pascal's triangle.* By forming the sums of the indicated terms of the triangle as shown below, an important sequence of numbers called the Fibonacci sequence is formed. The presence of this sequence in the triangle apparently was not recognized by Pascal.

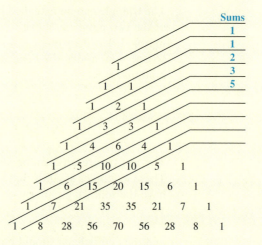

FOR DISCUSSION OR WRITING

1. Complete the sequence of sums for the remaining marked rows. How is the next term of the sequence determined? (*Hint:* Look at the previous two terms.)
2. Find out more about the Fibonacci sequence, including its appearance in nature.

DIVISION   The quotient of two polynomials can be found with a **division algorithm** very similar to that used for dividing whole numbers. (An *algorithm* is a step-by-step procedure for working a problem.) This algorithm requires that both polynomials be written in descending order.

---

* From *Mathematical Ideas,* Seventh Edition, by Charles D. Miller, Vern E. Heeren, and E. John Hornsby, Jr., HarperCollins, 1994, page 624.

EXAMPLE 10
Using the division algorithm

Divide $4m^3 - 8m^2 + 4m + 6$ by $2m - 1$.
Work as follows.

$4m^3$ divided by $2m$ is $2m^2$.
$-6m^2$ divided by $2m$ is $-3m$.
m divided by $2m$ is $\frac{1}{2}$.

$$
\begin{array}{r}
2m^2 - 3m + 1/2 \\
2m - 1 \overline{)\,4m^3 - 8m^2 + 4m + 6} \\
\end{array}
$$

$4m^3 - 2m^2$ $2m^2(2m - 1) = 4m^3 - 2m^2$

$-6m^2 + 4m$ Subtract; bring down the next term.

$-6m^2 + 3m$ $-3m(2m - 1) = -6m^2 + 3m$

$m + 6$ ← Subtract; bring down the next term.

$m - 1/2$ ← $(\frac{1}{2})(2m - 1) = m - (\frac{1}{2})$

$13/2$ ← Subtract. The remainder is $\frac{13}{2}$.

In dividing these polynomials, $4m^3 - 2m^2$ is subtracted from $4m^3 - 8m^2 + 4m + 6$. The complete result, $-6m^2 + 4m + 6$, should be written under the line. However, it is customary to save work and "bring down" just the $4m$, the only term needed for the next step. By this work,

$$\frac{4m^3 - 8m^2 + 4m + 6}{2m - 1} = 2m^2 - 3m + \frac{1}{2} + \frac{13/2}{2m - 1}.$$

The polynomial $3x^3 - 2x^2 - 150$ has a missing term, the term in which the power of x is 1. When a polynomial with a missing term is divided, it is useful to allow for that term by inserting a zero coefficient for the missing term, as shown in the next example.

EXAMPLE 11
Dividing polynomials with missing terms

Divide $3x^3 - 2x^2 - 150$ by $x^2 - 4$.
Both polynomials have missing terms. Insert each missing term with a 0 coefficient.

$$
\begin{array}{r}
3x - 2 \\
x^2 + 0x - 4 \overline{)\,3x^3 - 2x^2 + 0x - 150} \\
3x^3 + 0x^2 - 12x \\
\hline
-2x^2 + 12x - 150 \\
-2x^2 + 0x + 8 \\
\hline
12x - 158
\end{array}
$$

Since $12x - 158$ has lower degree than the divisor, it is the remainder, and the result of the division is written

$$\frac{3x^3 - 2x^2 - 150}{x^2 - 4} = 3x - 2 + \frac{12x - 158}{x^2 - 4}.$$

1.3 Exercises ▼▼▼▼▼▼▼▼▼▼▼▼▼▼▼▼▼▼▼▼▼▼▼▼▼▼▼▼▼▼▼▼▼▼▼▼

Decide whether the statement is true or false.

1. $(-4)^5(-4)^2 = 16^7$

2. $(5^3)^9 = 5^{12}$

3. $k^r \cdot k^{2r+1} = k^{3r+1}$

4. $-3^0 = -1$

5. $(-3)^0 = 1$

6. $4^{-1} + 3^{-1} = 7^{-1}$

Use the properties of exponents to simplify the expression, leaving exponents in your answer. See Examples 1–3.

7. $(2^2)^5$

8. $(6^4)^3$

9. $(2x^5y^4)^3$

10. $(-4m^3n^9)^2$

11. $-\left(\dfrac{p^4}{q}\right)^2$

12. $\left(\dfrac{r^8}{s^2}\right)^3$

Identify each expression as a polynomial *or* not a polynomial. *For each polynomial, give the degree and identify as a* monomial, binomial, trinomial, *or* none of these. *See Example 4.*

13. $-5x^{11}$

14. $9y^{12} + y^2$

15. $18p^5q + 6pq$

16. $2a^6 + 5a^2 + 4a$

17. $\sqrt{2}x^2 + \sqrt{3}x^6$

18. $-\sqrt{7}m^5n^2 + 2\sqrt{3}m^3n^2$

19. $\dfrac{1}{3}r^2s^2 - \dfrac{3}{5}r^4s^2 + rs^3$

20. $\dfrac{13}{10}p^7 - \dfrac{2}{7}p^5$

21. $\dfrac{5}{p} + \dfrac{2}{p^2} + \dfrac{5}{p^3}$

22. $-5\sqrt{z} + 2\sqrt{z^3} - 5\sqrt{z^5}$

Find the sum or difference. See Example 5.

23. $(3x^2 - 4x + 5) + (-2x^2 + 3x - 2)$

24. $(4m^3 - 3m^2 + 5) + (-3m^3 - m^2 + 5)$

25. $(12y^2 - 8y + 6) - (3y^2 - 4y + 2)$

26. $(8p^2 - 5p) - (3p^2 - 2p + 4)$

27. $(6m^4 - 3m^2 + m) - (2m^3 + 5m^2 + 4m) + (m^2 - m)$

28. $-(8x^3 + x - 3) + (2x^3 + x^2) - (4x^2 + 3x - 1)$

Find the product. See Examples 6 and 7.

29. $(4r - 1)(7r + 2)$

30. $(5m - 6)(3m + 4)$

31. $\left(3x - \dfrac{2}{3}\right)\left(5x + \dfrac{1}{3}\right)$

32. $\left(2m - \dfrac{1}{4}\right)\left(3m + \dfrac{1}{2}\right)$

33. $4x^2(3x^3 + 2x^2 - 5x + 1)$

34. $2b^3(b^2 - 4b + 3)$

35. $(2z - 1)(-z^2 + 3z - 4)$

36. $(k + 2)(12k^3 - 3k^2 + k + 1)$

37. $(m - n + k)(m + 2n - 3k)$

38. $(r - 3s + t)(2r - s + t)$

39. Explain why, for *any* values of x and y, the graphing calculator screen will indicate that the equation shown is true (as indicated by the 1).

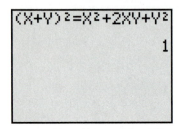

40. For particular stored values of x and y, the graphing calculator screen is obtained.

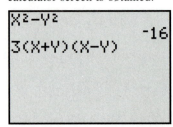

What will the screen display for the value of the polynomial in the final line of the display?

41. Consider the following figure, which is a square divided into two squares and two rectangles.

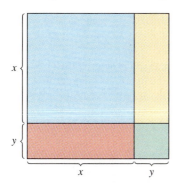

(a) The length of each side of the largest square is $x + y$. Use the formula for the area of a square to write the area of the largest square as a power.

(b) Use the formulas for the area of a square and the area of a rectangle to write the area of the largest square as a trinomial that represents the sum of the areas of the four figures that compose it.

(c) Explain why the expressions in parts (a) and (b) must be equivalent.

(d) What special product from this section does this exercise reinforce geometrically?

42. Use the reasoning process of Exercise 41 and the accompanying figure to geometrically support the distributive property. Write a short paragraph explaining this process.

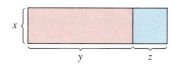

Use the special product formula that applies for each product. See Example 8.

43. $(2m + 3)(2m - 3)$

44. $(8s - 3t)(8s + 3t)$

45. $(4m + 2n)^2$

46. $(a - 6b)^2$

47. $(5r + 3t^2)^2$

48. $(2z^4 - 3y)^2$

49. $[(2p - 3) + q]^2$

50. $[(4y - 1) + z]^2$

51. $[(3q + 5) - p][(3q + 5) + p]$

52. $[(9r - s) + 2][(9r - s) - 2]$

53. $[(3a + b) - 1]^2$

54. $[(2m + 7) - n]^2$

Use the various procedures described in this section to perform the operations.

55. $(p^3 - 4p^2 + p) - (3p^2 + 2p + 7)$

56. $(2z + y)(3z - 4y)$

57. $(7m + 2n)(7m - 2n)$

58. $(3p + 5)^2$

59. $-3(4q^2 - 3q + 2) + 2(-q^2 + q - 4)$

60. $2(3r^2 + 4r + 2) - 3(-r^2 + 4r - 5)$

61. $p(4p - 6) + 2(3p - 8)$

62. $m(5m - 2) + 9(5 - m)$

63. $-y(y^2 - 4) + 6y^2(2y - 3)$

64. $-z^3(9 - z) + 4z(2 + 3z)$

Find the product. Assume that all variables used in exponents represent nonnegative integers. See Examples 7(c) and 8.

65. $(k^m + 2)(k^m - 2)$

66. $(y^x - 4)(y^x + 4)$

67. $(3p^x + 1)(p^x - 2)$

68. $(2^a + 5)(2^a + 3)$

69. $(q^p - 5p^q)^2$

70. $(3y^x - 2x^y)^2$

Suppose one polynomial has degree m and another has degree n, where m and n are natural numbers with $n < m$. Find the degree of the following for the polynomials.

71. Sum

72. Difference

73. Product

74. What would be the degree of the square of the polynomial of degree m?

Write out the binomial expansion. See Example 9.

75. $(x + y)^6$

76. $(m + n)^4$

77. $(p - q)^5$

78. $(a - b)^7$

79. $(r^2 + s)^5$

80. $(m + n^2)^4$

81. $(3r - s)^6$

82. $(7p + 2q)^4$

83. $(4a - 5b)^5$

84. For the expansion of $(x - y)^8$,
 (a) how many terms are there?
 (b) what is true of the signs (positive/negative) of the terms?
 (c) what is the numerical coefficient of the term with the variable factor y^7?

Solve the problem.

85. One of the most amazing formulas in all of ancient mathematics is the formula discovered by the Egyptians to find the volume of the frustum of a square pyramid shown in the figure. Its volume is given by $(1/3)h(a^2 + ab + b^2)$ where b is the length of the base, a is the length of the top, and h is the height. (*Source:* Freebury, H. A., *A History of Mathematics.* MacMillan Company, New York, 1968.)

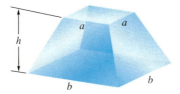

 (a) When the Great Pyramid in Egypt was partially completed to a height h of 200 feet, b was 756 feet, and a was 314 feet. Calculate its volume at this stage of construction.
 (b) Try to visualize the figure if $a = b$. What is the resulting shape? Find its volume.
 (c) Let $a = b$ in the Egyptian formula and simplify. Are the results the same?

86. Refer to the formula and the discussion in Exercise 85.
 (a) Use the expression $(1/3) h(a^2 + ab + b^2)$ to determine a formula for the volume of a pyramid with a square base b and height h by letting $a = 0$.
 (b) The Great Pyramid in Egypt had a square base of 756 feet and a height of 481 feet. Find the volume of the Great Pyramid. Compare it with the 273-foot-tall Superdome in New Orleans, which has an approximate volume of 100 million cubic feet. (*Source: The Guinness Book of Records 1995.*)
 (c) The Superdome covers an area of 13 acres. How many acres does the Great Pyramid cover? (*Hint:* 1 acre $=$ 43,560 ft².)

▼▼▼▼▼▼▼▼▼▼▼▼ **DISCOVERING CONNECTIONS** (Exercises 87–92)*▼▼▼▼▼▼▼▼▼▼▼▼▼

Use the binomial theorem and a calculator and work these exercises in order.

87. Expand $(x + 1)^0$; evaluate $(10 + 1)^0$ or 11^0.

88. Expand $(x + 1)^1$; evaluate $(10 + 1)^1$ or 11^1.

89. Expand $(x + 1)^2$; evaluate $(10 + 1)^2$ or 11^2.

90. Expand $(x + 1)^3$; evaluate $(10 + 1)^3$ or 11^3.

91. Describe the similarities in the two results for each part above.

92. Expand $(x + 1)^4$, and use the result to predict the value of 11^4. Then support your answer using your calculator.

———————

*Many exercise sets will contain groups of exercises under the heading Discovering Connections. These exercises are provided to illustrate how the concepts currently being studied relate to previously learned concepts. In general, they should be worked sequentially. We provide the answers to all such exercises, both even- and odd-numbered, in the Answer Section in the back of the book.

*In Chapters 3 and 4 we will study how polynomials can be graphed in a coordinate plane using ordered pairs of numbers. The **bar graph** shown here illustrates the number of farms in the United States since 1850, in selected years, in millions. Using a technique from statistics, it can be shown that the polynomial*

$$-.00102834793874x^2 + 3.9526021723669x$$
$$- 3791.976211763$$

will give a reasonably good approximation of the number of farms for these specified years by substituting the given year for x and then evaluating the polynomial. For example, if we let x = 1910, the value of the polynomial is approximately 6.0, which is just .4 "off" from the information in the bar graph. Use a calculator as necessary to evaluate the polynomial for the given year, and then compare it to the figure shown in the bar graph.

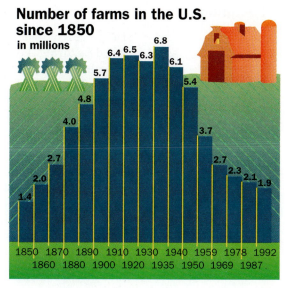

Number of farms in the U.S. since 1850
in millions

Source: U.S. Census

93. 1900   **94.** 1935

95. 1959   **96.** 1992

Perform each division. See Examples 10 and 11.

97. $\dfrac{-4x^7 - 14x^6 + 10x^4 - 14x^2}{-2x^2}$

98. $\dfrac{-8r^3s - 12r^2s^2 + 20rs^3}{4rs}$

99. $\dfrac{10x^8 - 16x^6 - 4x^4}{-2x^6}$

100. $\dfrac{6m^3 + 7m^2 - 4m + 2}{3m + 2}$

101. $\dfrac{2x^3 + 6x^2 - 8x + 10}{2x - 1}$

102. $\dfrac{3x^4 - 6x^2 + 9x - 5}{3x + 3}$

103. $\dfrac{3x^4 + 2x^2 + 6x - 1}{3x^2 - x}$

104. $\dfrac{k^4 - 4k^2 + 2k + 5}{k^2 + 1}$

1.4 Factoring ▼▼▼

The process of finding polynomials whose product equals a given polynomial is called **factoring.** For example, since $4x + 12 = 4(x + 3)$, both 4 and $x + 3$ are called **factors** of $4x + 12$. Also, $4(x + 3)$ is called the **factored form** of $4x + 12$. A nonconstant polynomial that cannot be written as a product of two polynomials of lower degree is a **prime** or **irreducible polynomial.** A polynomial is **factored completely** when it is written as a product of prime polynomials. Factoring polynomials is an important first step in working with quotients of polynomials and in solving polynomial equations.

FACTORING OUT THE GREATEST COMMON FACTOR   Polynomials are factored by using the distributive property. For example, to factor $6x^2y^3 + 9xy^4 + 18y^5$, we look for a monomial that is the greatest common factor of each term. The terms of this polynomial have $3y^3$ as the greatest common factor. By the distributive property,

$$6x^2y^3 + 9xy^4 + 18y^5 = (3y^3)(2x^2) + (3y^3)(3xy) + (3y^3)(6y^2)$$
$$= 3y^3(2x^2 + 3xy + 6y^2).$$

EXAMPLE 1
Factoring out the greatest common factor

Factor out the greatest common factor from each polynomial.

(a) $9y^5 + y^2$

The greatest common factor is y^2.

$$9y^5 + y^2 = y^2 \cdot 9y^3 + y^2 \cdot 1$$
$$= y^2(9y^3 + 1)$$

(b) $6x^2t + 8xt + 12t = 2t(3x^2 + 4x + 6)$

(c) $14m^4(m + 1) - 28m^3(m + 1) - 7m^2(m + 1)$

The greatest common factor is $7m^2(m + 1)$. Use the distributive property as follows.

$$14m^4(m + 1) - 28m^3(m + 1) - 7m^2(m + 1)$$
$$= [7m^2(m + 1)](2m^2 - 4m - 1)$$
$$= 7m^2(m + 1)(2m^2 - 4m - 1) \; ▶$$

CAUTION   In Example 1(a), avoid the common error of forgetting to include the 1. Since $y^2(9y^3) \neq 9y^5 + y^2$, the 1 is essential in the answer. Remember that factoring can always be checked by multiplication.

FACTORING BY GROUPING   When a polynomial has more than three terms, it can sometimes be factored by a method called **factoring by grouping.** For example, to factor

$$ax + ay + 6x + 6y,$$

collect the terms into two groups so that each group has a common factor.

$$ax + ay + 6x + 6y = (ax + ay) + (6x + 6y)$$

Factor each group, getting

$$ax + ay + 6x + 6y = a(x + y) + 6(x + y).$$

The quantity $(x + y)$ is now a common factor, which can be factored out, producing

$$ax + ay + 6x + 6y = (x + y)(a + 6).$$

It is not always obvious which terms should be grouped. Experience and repeated trials are the most reliable tools for factoring by grouping.

EXAMPLE 2
Factoring by grouping

Factor by grouping.

(a) $mp^2 + 7m + 3p^2 + 21$
Group the terms as follows.

$$mp^2 + 7m + 3p^2 + 21 = (mp^2 + 7m) + (3p^2 + 21)$$

Factor out the greatest common factor from each group.

$$(mp^2 + 7m) + (3p^2 + 21) = m(p^2 + 7) + 3(p^2 + 7)$$
$$= (p^2 + 7)(m + 3) \qquad p^2 + 7 \text{ is a common factor.}$$

(b) $2y^2 + az - 2z - ay^2$
One way to regroup the terms gives

$$2y^2 - 2z - ay^2 + az = (2y^2 - 2z) + (-ay^2 + az)$$
$$= 2(y^2 - z) + a(-y^2 + z).$$

The expression $-y^2 + z$ is the negative of $y^2 - z$, so the terms should be grouped as follows.

$$2y^2 - 2z - ay^2 + az = (2y^2 - 2z) - (ay^2 - az)$$
$$= 2(y^2 - z) - a(y^2 - z) \qquad \text{Factor each group.}$$
$$= (y^2 - z)(2 - a). \qquad \text{Factor out } y^2 - z. \ \blacktriangleright$$

Later in this section we show another way to factor by grouping three of the four terms.

FACTORING TRINOMIALS   Factoring is the opposite of multiplication. Since the product of two binomials is usually a trinomial, we can expect factorable trinomials (that have terms with no common factor) to have two binomial factors. Thus, factoring trinomials requires using FOIL backward.

◖**EXAMPLE 3**
Factoring trinomials

Factor each trinomial.

(a) $4y^2 - 11y + 6$

To factor this polynomial, we must find integers a, b, c, and d such that

$$4y^2 - 11y + 6 = (ay + b)(cy + d).$$

By using FOIL, we see that $ac = 4$ and $bd = 6$. The positive factors of 4 are 4 and 1 or 2 and 2. Since the middle term is negative, we consider only negative factors of 6. The possibilities are -2 and -3 or -1 and -6. Now we try various arrangements of these factors until we find one that gives the correct coefficient of y.

$$(2y - 1)(2y - 6) = 4y^2 - \mathbf{14y} + 6 \qquad \text{Incorrect}$$
$$(2y - 2)(2y - 3) = 4y^2 - \mathbf{10y} + 6 \qquad \text{Incorrect}$$
$$(y - 2)(4y - 3) = 4y^2 - \mathbf{11y} + 6 \qquad \text{Correct}$$

The last trial gives the correct factorization.

(b) $6p^2 - 7p - 5$

Again, we try various possibilities. The positive factors of 6 could be 2 and 3 or 1 and 6. As factors of -5 we have only -1 and 5 or -5 and 1. Try different combinations of these factors until the correct one is found.

$$(2p - 5)(3p + 1) = 6p^2 - \mathbf{13p} - 5 \qquad \text{Incorrect}$$
$$(3p - 5)(2p + 1) = 6p^2 - \mathbf{7p} - 5 \qquad \text{Correct}$$

Finally, $6p^2 - 7p - 5$ factors as $(3p - 5)(2p + 1)$. ◗

NOTE   In Example 3, we chose positive factors of the positive first term. Of course, we could have used two negative factors, but the work is easier if positive factors are used.

Each of the special patterns of multiplication given earlier can be used in reverse to get a pattern for factoring. Perfect square trinomials can be factored as follows.

Perfect Square Trinomials

$$x^2 + 2xy + y^2 = (x + y)^2$$
$$x^2 - 2xy + y^2 = (x - y)^2$$

These formulas should be memorized.

EXAMPLE 4
Factoring perfect square trinomials

Factor each polynomial.

(a) $16p^2 - 40pq + 25q^2$

Since $16p^2 = (4p)^2$ and $25q^2 = (5q)^2$, we use the second pattern shown above with $4p$ replacing x and $5q$ replacing y to get

$$16p^2 - 40pq + 25q^2 = (4p)^2 - 2(4p)(5q) + (5q)^2$$
$$= (4p - 5q)^2.$$

Make sure that the middle term of the trinomial being factored, $-40pq$ here, is twice the product of the two terms in the binomial $4p - 5q$.

$$-40pq = 2(4p)(-5q)$$

(b) $169x^2 + 104xy^2 + 16y^4 = (13x + 4y^2)^2$, since $2(13x)(4y^2) = 104xy^2$. ▶

FACTORING BINOMIALS   The pattern for the product of the sum and difference of two terms gives the following factorization.

Difference of Two Squares

$$x^2 - y^2 = (x + y)(x - y)$$

EXAMPLE 5
Factoring a difference of squares

Factor each of the following polynomials.

(a) $4m^2 - 9$

First we recognize that $4m^2 - 9$ is the difference of two squares, since $4m^2 = (2m)^2$ and $9 = 3^2$. Thus, we can use the pattern for the difference of two squares with $2m$ replacing x and 3 replacing y. Doing this gives

$$4m^2 - 9 = (2m)^2 - 3^2$$
$$= (2m + 3)(2m - 3).$$

(b) $256k^4 - 625m^4$

Use the difference of two squares pattern twice, as follows:

$$256k^4 - 625m^4 = (16k^2)^2 - (25m^2)^2$$
$$= (16k^2 + 25m^2)(16k^2 - 25m^2)$$
$$= (16k^2 + 25m^2)(4k + 5m)(4k - 5m).$$

(c) $(a + 2b)^2 - 4c^2 = (a + 2b)^2 - (2c)^2$
$$= [(a + 2b) + 2c][(a + 2b) - 2c]$$
$$= (a + 2b + 2c)(a + 2b - 2c)$$

(d) $y^{4q} - z^{2q} = (y^{2q} + z^q)(y^{2q} - z^q)$ (if q is a positive integer)

(e) $x^2 - 6x + 9 - y^4$

Group the first three terms to get a perfect square trinomial. Then use the difference of squares pattern.

$$
\begin{aligned}
x^2 - 6x + 9 - y^4 &= (x^2 - 6x + 9) - y^4 \\
&= (x - 3)^2 - (y^2)^2 \\
&= [(x - 3) + y^2][(x - 3) - y^2] \\
&= (x - 3 + y^2)(x - 3 - y^2) \;\blacktriangleright
\end{aligned}
$$

Two other special results of factoring are listed below. Each can be verified by multiplying on the right side of the equation.

Difference and Sum of Two Cubes

Difference of Two Cubes $x^3 - y^3 = (x - y)(x^2 + xy + y^2)$

Sum of Two Cubes $x^3 + y^3 = (x + y)(x^2 - xy + y^2)$

EXAMPLE 6
Factoring the sum or difference of cubes

Factor each polynomial.

(a) $x^3 + 27$

Notice that $27 = 3^3$, so the expression is a sum of two cubes. Use the second pattern given above.

$$x^3 + 27 = x^3 + 3^3 = (x + 3)(x^2 - 3x + 9)$$

(b) $m^3 - 64n^3$

Since $64n^3 = (4n)^3$, the given polynomial is a difference of two cubes. To factor, use the first pattern in the box above, replacing x with m and y with $4n$.

$$
\begin{aligned}
m^3 - 64n^3 &= m^3 - (4n)^3 \\
&= (m - 4n)[m^2 + m(4n) + (4n)^2] \\
&= (m - 4n)(m^2 + 4mn + 16n^2)
\end{aligned}
$$

(c) $8q^6 + 125p^9$

Write $8q^6$ as $(2q^2)^3$ and $125p^9$ as $(5p^3)^3$, so that the given polynomial is a sum of two cubes.

$$
\begin{aligned}
8q^6 + 125p^9 &= (2q^2)^3 + (5p^3)^3 \\
&= (2q^2 + 5p^3)[(2q^2)^2 - (2q^2)(5p^3) + (5p^3)^2] \\
&= (2q^2 + 5p^3)(4q^4 - 10q^2p^3 + 25p^6) \;\blacktriangleright
\end{aligned}
$$

METHOD OF SUBSTITUTION   Sometimes a polynomial can be factored by substituting one expression for another. The next example shows this **method of substitution.**

EXAMPLE 7
Factoring by substitution

Factor each polynomial.

(a) $6z^4 - 13z^2 - 5$

Replace z^2 with y, so that $y^2 = (z^2)^2 = z^4$. This replacement gives

$$6z^4 - 13z^2 - 5 = 6y^2 - 13y - 5.$$

Factor $6y^2 - 13y - 5$ as

$$6y^2 - 13y - 5 = (2y - 5)(3y + 1).$$

Replacing y with z^2 gives

$$6z^4 - 13z^2 - 5 = (2z^2 - 5)(3z^2 + 1).$$

(Some students prefer to factor this type of trinomial directly using trial and error with FOIL.)

(b) $10(2a - 1)^2 - 19(2a - 1) - 15$

Replacing $2a - 1$ with m gives

$$10m^2 - 19m - 15 = (5m + 3)(2m - 5).$$

Now replace m with $2a - 1$ in the factored form and simplify.

$$
\begin{aligned}
10(2a - 1)^2 - 19(2a - 1) - 15 \\
= [5(2a - 1) + 3][2(2a - 1) - 5] \quad &\text{Let } m = 2a - 1. \\
= (10a - 5 + 3)(4a - 2 - 5) \quad &\text{Multiply.} \\
= (10a - 2)(4a - 7) \quad &\text{Add.} \\
= 2(5a - 1)(4a - 7) \quad &\text{Factor out the common factor.}
\end{aligned}
$$

(c) $(2a - 1)^3 + 8$

Let $2a - 1 = K$ to get

$$
\begin{aligned}
(2a - 1)^3 + 8 &= K^3 + 8 \\
&= K^3 + 2^3 \\
&= (K + 2)(K^2 - 2K + 2^2).
\end{aligned}
$$

Replacing K with $2a - 1$ gives

$$
\begin{aligned}
(2a - 1)^3 + 8 \\
= (2a - 1 + 2)[(2a - 1)^2 - 2(2a - 1) + 4] \quad &\text{Let } K = 2a - 1. \\
= (2a + 1)(4a^2 - 4a + 1 - 4a + 2 + 4) \quad &\text{Multiply.} \\
= (2a + 1)(4a^2 - 8a + 7). \quad &\text{Combine terms.} \blacktriangleright
\end{aligned}
$$

This idea of replacing a variable expression with a single variable is used in reverse in Chapter 3 when we study functions.

1.4 Exercises ▼▼▼▼▼▼▼▼▼▼▼▼▼▼▼▼▼▼▼▼▼▼▼▼▼▼▼▼▼▼▼▼▼▼▼▼▼

1. When a student was directed to factor the polynomial $4x^3y^5 - 8x^2y^4$ completely on a test, she responded with $2x^2y^3(2xy^2 - 4y)$. When her teacher did not give her full credit she complained, indicating that when her answer is multiplied out, the original polynomial is obtained. Was her teacher justified in doing this? Why or why not?

2. Why does the binomial $x^6 - 64$ fall into *two* special categories discussed in this section?

Factor the greatest common factor from the polynomial. See Example 1.

3. $4k^2m^3 + 8k^4m^3 - 12k^2m^4$

4. $28r^4s^2 + 7r^3s - 35r^4s^3$

5. $2(a + b) + 4m(a + b)$

6. $4(y - 2)^2 + 3(y - 2)$

7. $(5r - 6)(r + 3) - (2r - 1)(r + 3)$

8. $(3z + 2)(z + 4) - (z + 6)(z + 4)$

9. $2(m - 1) - 3(m - 1)^2 + 2(m - 1)^3$

10. $5(a + 3)^3 - 2(a + 3) + (a + 3)^2$

Factor the polynomial by grouping. See Example 2.

11. $6st + 9t - 10s - 15$

12. $10ab - 6b + 35a - 21$

13. $2m^4 + 6 - am^4 - 3a$

14. $15 - 5m^2 - 3r^2 + m^2r^2$

15. $20z^2 + 18x^2 - 8zx - 45zx$

16. Shalita factored $16a^2 - 40a - 6a + 15$ by grouping and obtained $(8a - 3)(2a - 5)$. Jamal factored the same polynomial and gave an answer of $(3 - 8a)(5 - 2a)$. Are both of these answers correct? If not, why not?

Factor the trinomial. See Example 3.

17. $6a^2 - 48a - 120$

18. $8h^2 - 24h - 320$

19. $3m^3 + 12m^2 + 9m$

20. $9y^4 - 54y^3 + 45y^2$

21. $6k^2 + 5kp - 6p^2$

22. $14m^2 + 11mr - 15r^2$

23. $5a^2 - 7ab - 6b^2$

24. $12s^2 + 11st - 5t^2$

25. $9x^2 - 6x^3 + x^4$

26. $30a^2 + am - m^2$

27. $24a^4 + 10a^3b - 4a^2b^2$

28. $18x^5 + 15x^4z - 75x^3z^2$

Factor the perfect square trinomial. See Example 4.

29. $9m^2 - 12m + 4$

30. $16p^2 - 40p + 25$

31. $32a^2 - 48ab + 18b^2$

32. $20p^2 - 100pq + 125q^2$

33. $4x^2y^2 + 28xy + 49$

34. $9m^2n^2 - 12mn + 4$

35. $(a - 3b)^2 - 6(a - 3b) + 9$

36. $(2p + q)^2 - 10(2p + q) + 25$

Factor the difference of two squares. See Example 5.

37. $9a^2 - 16$

38. $16q^2 - 25$

39. $25s^4 - 9t^2$

40. $36z^2 - 81y^4$

41. $(a + b)^2 - 16$

42. $(p - 2q)^2 - 100$

43. $p^4 - 625$

44. $m^4 - 81$

45. Which of the following is the correct complete factorization of $x^4 - 1$?
    (a) $(x^2 - 1)(x^2 + 1)$ (b) $(x^2 + 1)(x + 1)(x - 1)$
    (c) $(x^2 - 1)^2$ (d) $(x - 1)^2(x + 1)^2$

46. Which of the following is the correct factorization of $x^3 + 8$?
    (a) $(x + 2)^3$ (b) $(x + 2)(x^2 + 2x + 4)$
    (c) $(x + 2)(x^2 - 2x + 4)$ (d) $(x + 2)(x^2 - 4x + 4)$

Factor the sum or difference of cubes. See Example 6.

47. $8 - a^3$ **48.** $r^3 + 27$ **49.** $125x^3 - 27$ **50.** $8m^3 - 27n^3$

51. $27y^9 + 125z^6$ **52.** $27z^3 + 729y^3$ **53.** $(r + 6)^3 - 216$ **54.** $(b + 3)^3 - 27$

55. $27 - (m + 2n)^3$ **56.** $125 - (4a - b)^3$

▼▼▼▼▼▼▼▼▼▼▼▼▼ **DISCOVERING CONNECTIONS** (Exercises 57–62) ▼▼▼▼▼▼▼▼▼▼▼▼▼

The polynomial $x^6 - 1$ can be considered either a difference of squares or a difference of cubes. Work Exercises 57–62 in order, and see some interesting connections between the results obtained when two different methods of factoring are used.

57. Factor $x^6 - 1$ by first factoring as the difference of two squares, and then factor further by using the patterns for the sum of two cubes and the difference of two cubes.

58. Factor $x^6 - 1$ by first factoring as the difference of two cubes, and then factor further by using the pattern for the difference of two squares.

59. Compare your answers in Exercises 57 and 58. Based on these results, what is the factorization of $x^4 + x^2 + 1$?

60. The polynomial $x^4 + x^2 + 1$ cannot be factored using the methods described in this section. However, there is a technique that allows us to factor it. This technique is shown below. Supply the reason that each step is valid.

$$x^4 + x^2 + 1 = x^4 + 2x^2 + 1 - x^2$$
$$= (x^4 + 2x^2 + 1) - x^2$$
$$= (x^2 + 1)^2 - x^2$$
$$= (x^2 + 1 - x)(x^2 + 1 + x)$$
$$= (x^2 - x + 1)(x^2 + x + 1)$$

61. Compare your answer in Exercise 59 with the final line in Exercise 60. What do you notice?

62. Factor $x^8 + x^4 + 1$ using the technique outlined in Exercise 60.

Factor each polynomial, using the method of substitution. See Example 7.

63. $m^4 - 3m^2 - 10$ **64.** $a^4 - 2a^2 - 48$ **65.** $7(3k - 1)^2 + 26(3k - 1) - 8$

66. $6(4z - 3)^2 + 7(4z - 3) - 3$ **67.** $9(a - 4)^2 + 30(a - 4) + 25$ **68.** $20(4 - p)^2 - 3(4 - p) - 2$

Factor by any method.

69. $4b^2 + 4bc + c^2 - 16$ **70.** $(2y - 1)^2 - 4(2y - 1) + 4$ **71.** $x^2 + xy - 5x - 5y$

72. $8r^2 - 3rs + 10s^2$ **73.** $p^4(m - 2n) + q(m - 2n)$ **74.** $36a^2 + 60a + 25$

75. $4z^2 + 28z + 49$ **76.** $6p^4 + 7p^2 - 3$ **77.** $1000x^3 + 343y^3$

78. $b^2 + 8b + 16 - a^2$ **79.** $125m^6 - 216$ **80.** $q^2 + 6q + 9 - p^2$

81. $12m^2 + 16mn - 35n^2$ **82.** $216p^3 + 125q^3$ **83.** $4p^2 + 3p - 1$

84. $100r^2 - 169s^2$ **85.** $144z^2 + 121$ **86.** $(3a + 5)^2 - 18(3a + 5) + 81$

87. $(x + y)^2 - (x - y)^2$ **88.** $4z^4 - 7z^2 - 15$

Factor each polynomial. Assume that all variables used in exponents represent positive integers.
See Example 5(d).

89. $r^2 + rs^q - 6s^{2q}$

90. $6z^{2a} - z^a x^b - x^{2b}$

91. $9a^{4k} - b^{8k}$

92. $16y^{2c} - 25x^{4c}$

93. $4y^{2a} - 12y^a + 9$

94. $25x^{4c} - 20x^{2c} + 4$

95. Explain how the accompanying figures give geometric interpretation to the formula

$$x^2 - y^2 = (x + y)(x - y).$$

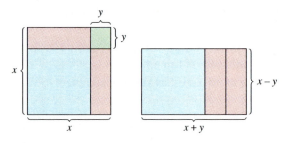

96. Explain how the accompanying figures give geometric interpretation to the formula

$$x^2 + 2xy + y^2 = (x + y)^2.$$

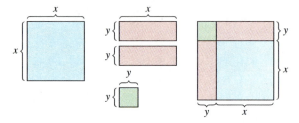

Find a value of b or c that will make the polynomial a perfect square trinomial.

97. $4z^2 + bz + 81$

98. $9p^2 + bp + 25$

99. $100r^2 - 60r + c$

100. $49x^2 + 70x + c$

1.5 Rational Expressions ▼▼▼

An expression that is the quotient of two algebraic expressions (with denominator not 0) is called a **fractional expression.** The most common fractional expressions are the quotients of two polynomials; these are called **rational expressions.** Since fractional expressions involve quotients, it is important to keep track of values of the variable that satisfy the requirement that no denominator be 0. For example, $x \neq -2$ in the rational expression

$$\frac{x + 6}{x + 2}$$

Notice that the expression defined for Y_1, $\frac{x+6}{x+2}$, is not defined for $x = -2$, while the expression defined for Y_2, $\frac{(x+6)(x+4)}{(x+2)(x+4)}$ is not defined for $x = -2$ *and* for $x = -4$. The calculator returns an ERROR message.

because replacing x with -2 makes the denominator equal 0. Similarly, in

$$\frac{(x + 6)(x + 4)}{(x + 2)(x + 4)},$$

$x \neq -2$ and $x \neq -4$.

The restrictions on the variable are found by determining the values that make the denominator equal to zero. In the second example above, finding the values of x that make $(x + 2)(x + 4) = 0$ requires using the property that $ab = 0$ if and only if $a = 0$ or $b = 0$, as follows.

$$(x + 2)(x + 4) = 0$$
$$x + 2 = 0 \qquad \text{or} \qquad x + 4 = 0$$
$$x = -2 \qquad \text{or} \qquad x = -4$$

CONNECTIONS Many relationships can be expressed as rational expressions. One such relationship exists for the Nile River in Africa, which is about 4000 miles long.* The Nile begins as an outlet of Lake Victoria at an altitude of 7000 feet above sea level and empties into the Mediterranean Sea at sea level (0 feet). The distance from its origin in thousands of miles is related to its height above sea level in thousands of feet (x) by the rational expression

$$\frac{7 - x}{.639x + 1.75}.$$

For example, when the river is at an altitude of 600 feet, $x = .6$ (thousand), and the distance from the origin is

$$\frac{7 - .6}{.639(.6) + 1.75} = 3,$$

which represents 3000 miles.

FOR DISCUSSION OR WRITING
1. What is the distance from the origin of the Nile when the river has an altitude of 7000 feet?
2. Find the distance from the origin of the Nile when the river is 1200 feet high.
3. For what value of x is the rational expression undefined? What happens to the rational expression as x gets closer and closer to that value?

* *The Universal Almanac,* 1993, John W. Wright, General Editor, Andrews and McMeel, p. 305.

Just as the fraction 6/8 is written in lowest terms as 3/4, rational expressions can also be written in lowest terms. This is done with the fundamental principle.

Fundamental Principle of Fractions

$$\frac{ac}{bc} = \frac{a}{b} \quad (b \neq 0, c \neq 0)$$

EXAMPLE 1
Writing in lowest terms

Write each expression in lowest terms.

(a) $\dfrac{2p^2 + 7p - 4}{5p^2 + 20p}$

Factor the numerator and denominator to get

$$\frac{2p^2 + 7p - 4}{5p^2 + 20p} = \frac{(2p - 1)(p + 4)}{5p(p + 4)}.$$

By the fundamental principle,

$$\frac{2p^2 + 7p - 4}{5p^2 + 20p} = \frac{2p - 1}{5p}.$$

In the original expression p cannot be 0 or -4, because $5p^2 + 20p \neq 0$, so this result is valid only for values of p other than 0 and -4. From now on, we shall always assume such restrictions when reducing rational expressions.

(b) $\dfrac{6 - 3k}{k^2 - 4}$

Factor to get
$$\frac{6 - 3k}{k^2 - 4} = \frac{3(2 - k)}{(k + 2)(k - 2)}.$$

The factors $2 - k$ and $k - 2$ have opposite signs. Because of this, we multiply numerator and denominator by -1, as follows.

$$\frac{6 - 3k}{k^2 - 4} = \frac{3(2 - k)(-1)}{(k + 2)(k - 2)(-1)}$$

Since $(k - 2)(-1) = -k + 2$, or $2 - k$,

$$\frac{6 - 3k}{k^2 - 4} = \frac{3(2 - k)(-1)}{(k + 2)(2 - k)},$$

giving
$$\frac{6 - 3k}{k^2 - 4} = \frac{-3}{k + 2}.$$

Working in an alternative way would lead to the equivalent result

$$\frac{3}{-k - 2}.$$

CAUTION   Probably the most common error made in algebra is the incorrect use of the fundamental principle to write a fraction in lowest terms. Remember, the fundamental principle requires a pair of common *factors,* one in the numerator and one in the denominator. For example,

$$\frac{2x + 4}{6} = \frac{2(x + 2)}{6} = \frac{2(x + 2)}{2 \cdot 3} = \frac{x + 2}{3}.$$

On the other hand, $\dfrac{2x + 5}{6}$ cannot be simplified further by the fundamental principle, because the numerator cannot be factored.

MULTIPLICATION AND DIVISION   Rational expressions are multiplied and divided using definitions from earlier work with fractions.

Multiplication and Division

For fractions $\dfrac{a}{b}$ and $\dfrac{c}{d}$ $(b \neq 0, d \neq 0)$,

$$\frac{a}{b} \cdot \frac{c}{d} = \frac{ac}{bd}$$

$$\frac{a}{b} \div \frac{c}{d} = \frac{a}{b} \cdot \frac{d}{c} \quad \left(\text{if } \frac{c}{d} \neq 0 \right).$$

EXAMPLE 2
Multiplying or dividing rational expressions

Multiply or divide, as indicated.

(a) $\dfrac{2y^2}{9} \cdot \dfrac{27}{8y^5} = \dfrac{2y^2 \cdot 27}{9 \cdot 8y^5}$

$\phantom{(a) \dfrac{2y^2}{9} \cdot \dfrac{27}{8y^5}} = \dfrac{2 \cdot 9 \cdot 3 \cdot y^2}{2 \cdot 9 \cdot 4 \cdot y^2 \cdot y^3}$ Factor.

$\phantom{(a) \dfrac{2y^2}{9} \cdot \dfrac{27}{8y^5}} = \dfrac{3}{4y^3}$ Fundamental principle

The product was written in lowest terms in the last step.

(b) $\dfrac{3m^2 - 2m - 8}{3m^2 + 14m + 8} \cdot \dfrac{3m + 2}{3m + 4} = \dfrac{(m - 2)(3m + 4)}{(m + 4)(3m + 2)} \cdot \dfrac{3m + 2}{3m + 4}$ Factor.

$ = \dfrac{(m - 2)(3m + 4)(3m + 2)}{(m + 4)(3m + 2)(3m + 4)}$ Multiply fractions.

$ = \dfrac{m - 2}{m + 4}$ Fundamental principle

(c) $\dfrac{5}{8m + 16} \div \dfrac{7}{12m + 24} = \dfrac{5}{8(m + 2)} \div \dfrac{7}{12(m + 2)}$ Factor.

$= \dfrac{5}{8(m + 2)} \cdot \dfrac{12(m + 2)}{7}$ Definition of division

$= \dfrac{5 \cdot 12(m + 2)}{8 \cdot 7(m + 2)}$ Multiply.

$= \dfrac{15}{14}$ Fundamental principle

(d) $\dfrac{3p^2 + 11p - 4}{24p^3 - 8p^2} \div \dfrac{9p + 36}{24p^4 - 36p^3} = \dfrac{(p + 4)(3p - 1)}{8p^2(3p - 1)} \div \dfrac{9(p + 4)}{12p^3(2p - 3)}$

$= \dfrac{(p + 4)(3p - 1)(12p^3)(2p - 3)}{8p^2(3p - 1)(9)(p + 4)}$

$= \dfrac{12p^3(2p - 3)}{9 \cdot 8p^2} = \dfrac{p(2p - 3)}{6}$ ▶

ADDITION AND SUBTRACTION   Adding and subtracting rational expressions also depends on definitions from earlier work with fractions.

Addition and Subtraction

For fractions $\dfrac{a}{b}$ and $\dfrac{c}{d}$ $(b \neq 0, d \neq 0)$,

$$\frac{a}{b} + \frac{c}{d} = \frac{ad + bc}{bd}$$

$$\frac{a}{b} - \frac{c}{d} = \frac{ad - bc}{bd}.$$

In practice, rational expressions are normally added or subtracted after rewriting all the rational expressions with a common denominator found with the steps given below.

Finding a Common Denominator

1. Write each denominator as a product of prime factors.
2. Form a product of all the different prime factors. Each factor should have as exponent the *greatest* exponent that appears on that factor.

EXAMPLE 3
Adding or subtracting rational expressions

Add or subtract, as indicated.

(a) $\dfrac{5}{9x^2} + \dfrac{1}{6x}$

Write each denominator as a product of prime factors, as follows.

$$9x^2 = 3^2 \cdot x^2$$
$$6x = 2^1 \cdot 3^1 \cdot x^1$$

For the common denominator, form the product of all the prime factors, with each factor having the greatest exponent that appears on it. Here the greatest exponent on 2 is 1, while both 3 and x have a greatest exponent of 2. The common denominator is

$$2^1 \cdot 3^2 \cdot x^2 = 18x^2.$$

Now use the fundamental principle to write both of the given expressions with this denominator, then add.

$$\frac{5}{9x^2} + \frac{1}{6x} = \frac{5 \cdot 2}{9x^2 \cdot 2} + \frac{1 \cdot 3x}{6x \cdot 3x}$$

$$= \frac{10}{18x^2} + \frac{3x}{18x^2}$$

$$= \frac{10 + 3x}{18x^2}$$

Always check at this point to see that the answer is in lowest terms.

(b) $\dfrac{y+2}{y^2-y} - \dfrac{3y}{2y^2-4y+2}$

Factor each denominator, giving

$$\frac{y+2}{y^2-y} - \frac{3y}{2y^2-4y+2} = \frac{y+2}{y(y-1)} - \frac{3y}{2(y-1)^2}.$$

The common denominator, by the method above, is $2y(y-1)^2$. Write each rational expression with this denominator and subtract, as follows.

$$\frac{y+2}{y(y-1)} - \frac{3y}{2(y-1)^2} = \frac{(y+2) \cdot 2(y-1)}{y(y-1) \cdot 2(y-1)} - \frac{3y \cdot y}{2(y-1)^2 \cdot y}$$

$$= \frac{2(y^2+y-2)}{2y(y-1)^2} - \frac{3y^2}{2y(y-1)^2}$$

$$= \frac{2y^2+2y-4-3y^2}{2y(y-1)^2} \qquad \text{Subtract.}$$

$$= \frac{-y^2+2y-4}{2y(y-1)^2} \qquad \text{Combine terms.}$$

(c) $\dfrac{3}{(x-1)(x+2)} - \dfrac{1}{(x+3)(x-4)}$

The common denominator here is $(x-1)(x+2)(x+3)(x-4)$. Write each fraction with this common denominator, then perform the subtraction.

$$\dfrac{3}{(x-1)(x+2)} - \dfrac{1}{(x+3)(x-4)}$$

$$= \dfrac{3(x+3)(x-4)}{(x-1)(x+2)(x+3)(x-4)} - \dfrac{(x-1)(x+2)}{(x+3)(x-4)(x-1)(x+2)}$$

$$= \dfrac{3(x^2-x-12) - (x^2+x-2)}{(x-1)(x+2)(x+3)(x-4)}$$

$$= \dfrac{3x^2-3x-36-x^2-x+2}{(x-1)(x+2)(x+3)(x-4)}$$

$$= \dfrac{2x^2-4x-34}{(x-1)(x+2)(x+3)(x-4)} \quad \blacktriangleright$$

CAUTION When subtracting fractions where the second fraction has more than one term in the numerator, as in Example 3(c), be sure to distribute the negative sign to each term. Notice in Example 3(c) how parentheses were used in the second step to avoid an error in the subtraction step.

COMPLEX FRACTIONS Any quotient of two rational expressions is called a **complex fraction.** Complex fractions often can be simplified by the methods shown in the following example.

◀**EXAMPLE 4**
Simplifying complex fractions

Simplify each complex fraction.

(a) $\dfrac{6 - \dfrac{5}{k}}{1 + \dfrac{5}{k}}$

Multiply both numerator and denominator by the least common denominator of all the fractions, k.

$$\dfrac{k\left(6 - \dfrac{5}{k}\right)}{k\left(1 + \dfrac{5}{k}\right)} = \dfrac{6k - k\left(\dfrac{5}{k}\right)}{k + k\left(\dfrac{5}{k}\right)} = \dfrac{6k - 5}{k + 5}$$

(b) $\dfrac{\dfrac{a}{a+1}+\dfrac{1}{a}}{\dfrac{1}{a}+\dfrac{1}{a+1}}$

Multiply both numerator and denominator by the least common denominator of all the fractions, in this case $a(a+1)$.

$$\dfrac{\dfrac{a}{a+1}+\dfrac{1}{a}}{\dfrac{1}{a}+\dfrac{1}{a+1}} = \dfrac{\left(\dfrac{a}{a+1}+\dfrac{1}{a}\right)a(a+1)}{\left(\dfrac{1}{a}+\dfrac{1}{a+1}\right)a(a+1)}$$

$$= \dfrac{\dfrac{a}{a+1}(a)(a+1)+\dfrac{1}{a}(a)(a+1)}{\dfrac{1}{a}(a)(a+1)+\dfrac{1}{a+1}(a)(a+1)} \qquad \text{\color{blue}Distributive property}$$

$$= \dfrac{a^2+(a+1)}{(a+1)+a}$$

$$= \dfrac{a^2+a+1}{2a+1}$$

As an alternative method of solution, first perform the indicated additions in the numerator and denominator, and then divide.

$$\dfrac{\dfrac{a}{a+1}+\dfrac{1}{a}}{\dfrac{1}{a}+\dfrac{1}{a+1}} = \dfrac{\dfrac{a^2+1(a+1)}{a(a+1)}}{\dfrac{1(a+1)+1(a)}{a(a+1)}} \qquad \text{\color{blue}Get a common denominator; add terms in numerator and denominator.}$$

$$= \dfrac{\dfrac{a^2+a+1}{a(a+1)}}{\dfrac{2a+1}{a(a+1)}} \qquad \text{\color{blue}Combine terms in numerator and denominator.}$$

$$= \dfrac{a^2+a+1}{a(a+1)} \cdot \dfrac{a(a+1)}{2a+1} \qquad \text{\color{blue}Definition of division}$$

$$= \dfrac{a^2+a+1}{2a+1} \qquad \text{\color{blue}Multiply fractions and write in lowest terms.} \quad \blacktriangleright$$

1.5 Exercises ▼▼▼▼▼▼▼▼▼▼▼▼▼▼▼▼▼▼▼▼▼▼▼▼▼▼▼▼▼

For each rational expression, give the restrictions on the variable.

1. $\dfrac{x-2}{x+6}$

2. $\dfrac{x+5}{x+3}$

3. $\dfrac{2x}{5x-3}$

4. $\dfrac{6x}{2x-1}$

5. $\dfrac{-8}{x^2+1}$

6. $\dfrac{3x}{3x^2+7}$

7. Which one of the following expressions is equivalent to $\dfrac{x^2 + 4x + 3}{x + 1}$ $(x \neq -1)$?

(a) $x + 3$ (b) $x + 7$ (c) $5x + 3$ (d) $x^2 + 7$

8. Explain why $\dfrac{3}{5}$ is not the simplified form of $\dfrac{2x + 3}{2x + 5}$.

The expression Y_1 is a rational expression of the form $\dfrac{x + 5}{x - k}$ for some real number k. Use the calculator-generated table to determine the value of k.

9.

X	Y₁	
-8	.6	
-7	.5	
-6	.33333	
-5	0	
-4	-1	
-3	ERROR	
-2	3	

X=-8

10.

X	Y₁	
4	ERROR	
5	10	
6	5.5	
7	4	
8	3.25	
9	2.8	
10	2.5	

X=10

Write each of the following in lowest terms. See Example 1.

11. $\dfrac{8k + 16}{9k + 18}$

12. $\dfrac{20r + 10}{30r + 15}$

13. $\dfrac{3(t + 5)}{(t + 5)(t - 3)}$

14. $\dfrac{-8(y + 4)}{(y + 2)(y + 4)}$

15. $\dfrac{8x^2 + 16x}{4x^2}$

16. $\dfrac{36y^2 + 72y}{9y}$

17. $\dfrac{m^2 - 4m + 4}{m^2 + m - 6}$

18. $\dfrac{r^2 - r - 6}{r^2 + r - 12}$

19. $\dfrac{8m^2 + 6m - 9}{16m^2 - 9}$

20. $\dfrac{6y^2 + 11y + 4}{3y^2 + 7y + 4}$

Find the product or quotient. See Example 2.

21. $\dfrac{15p^3}{9p^2} \div \dfrac{6p}{10p^2}$

22. $\dfrac{3r^2}{9r^3} \div \dfrac{8r^3}{6r}$

23. $\dfrac{2k + 8}{6} \div \dfrac{3k + 12}{2}$

24. $\dfrac{5m + 25}{10} \cdot \dfrac{12}{6m + 30}$

25. $\dfrac{x^2 + x}{5} \cdot \dfrac{25}{xy + y}$

26. $\dfrac{3m - 15}{4m - 20} \cdot \dfrac{m^2 - 10m + 25}{12m - 60}$

27. $\dfrac{4a + 12}{2a - 10} \div \dfrac{a^2 - 9}{a^2 - a - 20}$

28. $\dfrac{6r - 18}{9r^2 + 6r - 24} \cdot \dfrac{12r - 16}{4r - 12}$

29. $\dfrac{p^2 - p - 12}{p^2 - 2p - 15} \cdot \dfrac{p^2 - 9p + 20}{p^2 - 8p + 16}$

30. $\dfrac{x^2 + 2x - 15}{x^2 + 11x + 30} \cdot \dfrac{x^2 + 2x - 24}{x^2 - 8x + 15}$

31. $\dfrac{m^2 + 3m + 2}{m^2 + 5m + 4} \div \dfrac{m^2 + 5m + 6}{m^2 + 10m + 24}$

32. $\dfrac{y^2 + y - 2}{y^2 + 3y - 4} \div \dfrac{y^2 + 3y + 2}{y^2 + 4y + 3}$

33. $\dfrac{2m^2 - 5m - 12}{m^2 - 10m + 24} \div \dfrac{4m^2 - 9}{m^2 - 9m + 18}$

34. $\dfrac{6n^2 - 5n - 6}{6n^2 + 5n - 6} \cdot \dfrac{12n^2 - 17n + 6}{12n^2 - n - 6}$

35. $\dfrac{x^3 + y^3}{x^3 - y^3} \cdot \dfrac{x^2 - y^2}{x^2 + 2xy + y^2}$

36. $\dfrac{x^2 - y^2}{(x - y)^2} \cdot \dfrac{x^2 - xy + y^2}{x^2 - 2xy + y^2} \div \dfrac{x^3 + y^3}{(x - y)^4}$

37. Which of the following rational expressions equals -1? (In parts (a), (b), and (d), $x \neq -4$, and in part (c), $x \neq 4$.)

(a) $\dfrac{x - 4}{x + 4}$ (b) $\dfrac{-x - 4}{x + 4}$ (c) $\dfrac{x - 4}{4 - x}$ (d) $\dfrac{x - 4}{-x - 4}$

38. In your own words, explain how to find the least common denominator of two fractions.

Perform each addition or subtraction. See Example 3.

39. $\dfrac{3}{2k} + \dfrac{5}{3k}$

40. $\dfrac{8}{5p} + \dfrac{3}{4p}$

41. $\dfrac{a+1}{2} - \dfrac{a-1}{2}$

42. $\dfrac{y+6}{5} - \dfrac{y-6}{5}$

43. $\dfrac{3}{p} + \dfrac{1}{2}$

44. $\dfrac{9}{r} - \dfrac{2}{3}$

45. $\dfrac{1}{6m} + \dfrac{2}{5m} + \dfrac{4}{m}$

46. $\dfrac{8}{3p} + \dfrac{5}{4p} + \dfrac{9}{2p}$

47. $\dfrac{1}{a} - \dfrac{b}{a^2}$

48. $\dfrac{3}{z} + \dfrac{x}{z^2}$

49. $\dfrac{1}{x+z} + \dfrac{1}{x-z}$

50. $\dfrac{m+1}{m-1} + \dfrac{m-1}{m+1}$

51. $\dfrac{3}{a-2} - \dfrac{1}{2-a}$

52. $\dfrac{q}{p-q} - \dfrac{q}{q-p}$

53. $\dfrac{x+y}{2x-y} - \dfrac{2x}{y-2x}$

54. $\dfrac{m-4}{3m-4} + \dfrac{3m+2}{4-3m}$

55. $\dfrac{1}{x^2+x-12} - \dfrac{1}{x^2-7x+12} + \dfrac{1}{x^2-16}$

56. $\dfrac{2}{2p^2-9p-5} + \dfrac{p}{3p^2-17p+10} - \dfrac{2p}{6p^2-p-2}$

Refer to the Connections in this section to work Exercises 57–60, which illustrate rational expressions used in a "real-world" application. Use a calculator as necessary.

57. In situations involving environmental pollution, a cost-benefit model expresses cost in terms of the percentage of pollutant removed from the environment. Suppose a cost-benefit model is expressed as

$$y = \frac{6.7x}{100-x},$$

where y is the cost in thousands of dollars of removing x percent of a certain pollutant. Find the value of y for each given value of x.
(a) $x = 75$ (75%)
(b) $x = 95$ (95%)
(c) $x = 98.5$ (98.5%)

58. In a recent year, the cost in thousands of dollars per ton, y, to build an oil tanker of x thousand deadweight tons was approximated by

$$y = \frac{110{,}000}{x+225}.$$

Find the value of y for each given value of x.
(a) $x = 25$ **(b)** $x = 100$ **(c)** $x = 400$

*In recent years the economist Arthur Laffer has been a center of controversy because of his **Laffer curve**, an idealized version of which is shown here. According to this curve, increasing a tax rate, say from x_1 percent to x_2 percent on the graph, can actually lead to a decrease in government revenue. All economists agree on the endpoints, 0 revenue at tax rates of both 0% and 100%,* but there is much disagreement on the location of the rate x_1 that produces maximum revenue.

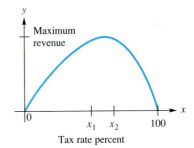

Tax rate percent

59. Suppose an economist studying the Laffer curve produces the rational expression

$$y = \frac{80x-8000}{x-110},$$

with y giving government revenue in tens of millions of dollars for a tax rate of x percent, with the relationship valid for $55 \le x \le 100$. Find the revenue for the following tax rates.
(a) 55%   **(b)** 60%   **(c)** 70%
(d) 90%   **(e)** 100%

60. Suppose an economist studies a different tax, this time producing

$$y = \frac{60x-6000}{x-120},$$

where y is government revenue in millions of dollars from a tax rate of x percent and is valid for $50 \le x \le 100$. Find the revenue from the following tax rates.
(a) 50%   **(b)** 60%   **(c)** 80%   **(d)** 100%

Perform the indicated operations. See Example 4.

61. $\dfrac{1 + \dfrac{1}{x}}{1 - \dfrac{1}{x}}$

62. $\dfrac{2 - \dfrac{2}{y}}{2 + \dfrac{2}{y}}$

63. $\dfrac{\dfrac{1}{x+1} - \dfrac{1}{x}}{\dfrac{1}{x}}$

64. $\dfrac{\dfrac{1}{y+3} - \dfrac{1}{y}}{\dfrac{1}{y}}$

65. $\dfrac{1 + \dfrac{1}{1-b}}{1 - \dfrac{1}{1+b}}$

66. $m - \dfrac{m}{m + \dfrac{1}{2}}$

67. $\dfrac{m - \dfrac{1}{m^2 - 4}}{\dfrac{1}{m+2}}$

68. $\dfrac{\dfrac{3}{p^2 - 16} + p}{\dfrac{1}{p-4}}$

69. $\left(\dfrac{3}{p-1} - \dfrac{2}{p+1}\right)\left(\dfrac{p-1}{p}\right)$

70. $\left(\dfrac{y}{y^2-1} - \dfrac{y}{y^2-2y+1}\right)\left(\dfrac{y-1}{y+1}\right)$

71. $\dfrac{\dfrac{1}{x+h} - \dfrac{1}{x}}{h}$

72. $\dfrac{1}{h}\left(\dfrac{1}{(x+h)^2 + 9} - \dfrac{1}{x^2 + 9}\right)$

 The following two exercises are warm-up exercises for the solution of the Greenland ice cap problem stated at the beginning of this chapter.

73. A square sheet of aluminum foil 30 centimeters on a side has a mass of 47.25 grams. If one cubic centimeter of aluminum has a mass of 2.7 grams, find the thickness of the aluminum foil. (*Hint:* The thickness of the foil will be equal to its volume divided by its area.) (*Source:* Haber-Schaim, U., J. Cross, G. Abegg, J. Dodge, and J. Walter, *Introductory Physical Science,* Prentice Hall, Inc., 1972.)

74. (Refer to Exercise 73.) A drop of oil containing .2 cubic centimeter is spilled onto a lake. The oil spreads out in a circular shape having a diameter of 46 centimeters. Approximate the thickness of the oil film.

1.6 Rational Exponents ▼▼▼

In Section 1.3 we introduced some rules for exponents: the product rule and the power rules. In this section we complete our review of exponential expressions, beginning with a rule for division.

NEGATIVE EXPONENTS AND THE QUOTIENT RULE In the product rule, $a^m \cdot a^n = a^{m+n}$, the exponents are *added.* By the definition of exponent in Section 1.3, if $a \neq 0$,

$$\frac{a^3}{a^7} = \frac{a \cdot a \cdot a}{a \cdot a \cdot a \cdot a \cdot a \cdot a \cdot a} = \frac{1}{a \cdot a \cdot a \cdot a} = \frac{1}{a^4}.$$

This suggests that we should *subtract* exponents when dividing. Subtracting exponents gives

$$\frac{a^3}{a^7} = a^{3-7} = a^{-4}.$$

The only way to keep these results consistent is to define a^{-4} as $1/a^4$. This example suggests the following definition.

Negative Exponents

If a is a nonzero real number and n is any integer, then

$$a^{-n} = \frac{1}{a^n}.$$

EXAMPLE 1
Using the definition of a negative exponent

```
4^-2►Frac
          1/16
(2/5)^-3►Frac
          125/8
-4^-2►Frac
         -1/16
```

A calculator supports the results in Example 1(a), (b), and (c).

Evaluate each expression in parts (a)–(c). In parts (d) and (e), write the expression without negative exponents.

(a) $4^{-2} = \dfrac{1}{4^2} = \dfrac{1}{16}$

(b) $\left(\dfrac{2}{5}\right)^{-3} = \dfrac{1}{\left(\dfrac{2}{5}\right)^3} = \dfrac{1}{\dfrac{8}{125}} = \dfrac{125}{8}$

(c) $-4^{-2} = -\dfrac{1}{4^2} = -\dfrac{1}{16}$

(d) $x^{-4} = \dfrac{1}{x^4} \quad (x \neq 0)$

(e) $xy^{-3} = x \cdot \dfrac{1}{y^3} = \dfrac{x}{y^3} \quad (y \neq 0)$ ▶

CAUTION   A negative exponent indicates a reciprocal, *not* a negative expression.

Part (b) of Example 1 showed that

$$\left(\frac{2}{5}\right)^{-3} = \frac{125}{8} = \left(\frac{5}{2}\right)^3.$$

This result can be generalized. If $a \neq 0$ and $b \neq 0$, then

$$\left(\frac{a}{b}\right)^{-n} = \left(\frac{b}{a}\right)^{n}$$

for any integer n.

The quotient rule for exponents follows from the definition of exponents, as shown above.

Quotient Rule

For all integers m and n and all nonzero real numbers a,

$$\frac{a^m}{a^n} = a^{m-n}.$$

By the quotient rule, if $a \neq 0$,

$$\frac{a^m}{a^m} = a^{m-m} = a^0.$$

On the other hand, any nonzero quantity divided by itself equals 1. This is why we defined $a^0 = 1$ in Section 1.3.

◀ **EXAMPLE 2**
Using the quotient rule

Use the quotient rule to simplify each expression. Assume that all variables represent nonzero real numbers.

(a) $\dfrac{12^5}{12^2} = 12^{5-2} = 12^3$ 　　　　　　 **(b)** $\dfrac{a^5}{a^{-8}} = a^{5-(-8)} = a^{13}$

(c) $\dfrac{16m^{-9}}{12m^{11}} = \dfrac{16}{12} \cdot m^{-9-11} = \dfrac{4}{3}m^{-20} = \dfrac{4}{3} \cdot \dfrac{1}{m^{20}} = \dfrac{4}{3m^{20}}$

(d) $\dfrac{25r^7z^5}{10r^9z} = \dfrac{25}{10} \cdot \dfrac{r^7}{r^9} \cdot \dfrac{z^5}{z^1} = \dfrac{5}{2}r^{-2}z^4 = \dfrac{5z^4}{2r^2}$

(e) $\dfrac{x^{5y}}{x^{3y}} = x^{5y-3y} = x^{2y}$, 　if y is an integer ▶

The rules for exponents from Section 1.3 also apply to negative exponents.

◀ **EXAMPLE 3**
Using the rules for exponents

Use the rules for exponents to simplify each expression. Write answers without negative exponents. Assume that all variables represent nonzero real numbers.

(a) $3x^{-2}(4^{-1}x^{-5})^2 = 3x^{-2}(4^{-2}x^{-10})$ 　　　　Power rule

$= 3 \cdot 4^{-2} \cdot x^{-2+(-10)}$ 　　　Rearrange factors: product rule

$= 3 \cdot 4^{-2} \cdot x^{-12}$

$= \dfrac{3}{16x^{12}}$ 　　　　　Write with positive exponents.

(b) $\dfrac{5m^{-3}}{10m^{-5}} = \dfrac{5}{10}m^{-3-(-5)}$ 　　　　Quotient rule

$= \dfrac{1}{2}m^2$ 　　or　　 $\dfrac{m^2}{2}$

(c) $\dfrac{12p^3q^{-1}}{8p^{-2}q} = \dfrac{12}{8} \cdot \dfrac{p^3}{p^{-2}} \cdot \dfrac{q^{-1}}{q^1}$

$= \dfrac{3}{2} \cdot p^{3-(-2)}q^{-1-1}$ 　　　Quotient rule

$= \dfrac{3}{2}p^5q^{-2}$

$= \dfrac{3p^5}{2q^2}$ 　　　　Write with positive exponents.

(d) $\dfrac{(3x^2)^{-1}(3x^5)^{-2}}{(3^{-1}x^{-2})^2} = \dfrac{3^{-1}x^{-2}3^{-2}x^{-10}}{3^{-2}x^{-4}}$ Power rule

$$= \dfrac{3^{-1+(-2)}x^{-2+(-10)}}{3^{-2}x^{-4}} = \dfrac{3^{-3}x^{-12}}{3^{-2}x^{-4}}$$ Product rule

$$= 3^{-3-(-2)}x^{-12-(-4)} = 3^{-1}x^{-8}$$ Quotient rule

$$= \dfrac{1}{3x^8}$$ Write with positive exponents. ▶

CAUTION   Notice the use of the power rule $(ab)^n = a^n b^n$ in Example 3(d): $(3x^2)^{-1} = 3^{-1}(x^2)^{-1} = 3^{-1}x^{-2}$. It is a common error to forget to apply the exponent to a numerical coefficient.

RATIONAL EXPONENTS   The definition of a^n can be extended to rational values of n by defining $a^{1/n}$ to be the nth root of a. By one of the power rules of exponents (extended to a rational exponent)

$$(a^{1/n})^n = a^{(1/n)n} = a^1 = a,$$

suggesting that $a^{1/n}$ is a number whose nth power is a.

$a^{1/n}$, n **Even**   **(i)** If n is an *even* positive integer, and if $a > 0$, then $a^{1/n}$ is the positive real number whose nth power is a. That is, $(a^{1/n})^n = a$. In this case, $a^{1/n}$ is the principal nth root of a.

$a^{1/n}$, n **Odd**   **(ii)** If n is an *odd* positive integer, and a *is any real number*, then $a^{1/n}$ is the positive or negative real number whose nth power is a. That is, $(a^{1/n})^n = a$.

◀ EXAMPLE 4
Using the definition of $a^{1/n}$

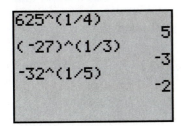

A calculator supports the results of Example 4(d), (f), and (g).

Evaluate each expression.

(a) $36^{1/2} = 6$ because $6^2 = 36$.   **(b)** $-100^{1/2} = -10$

(c) $-(225)^{1/2} = -15$ **(d)** $625^{1/4} = 5$

(e) $(-1296)^{1/4}$ is not a real number, but $-1296^{1/4} = -6$.

(f) $(-27)^{1/3} = -3$ **(g)** $-32^{1/5} = -2$ ▶

What about more general rational exponents? The notation $a^{m/n}$ should be defined so that all the past rules for exponents still hold. For the power rule to hold, $(a^{1/n})^m$ must equal $a^{m/n}$. Therefore, $a^{m/n}$ is defined as follows.

An error message is returned for $(-8)^{2/3}$.

Rational Exponents

For all integers m, all positive integers n, and all real numbers a for which $a^{1/n}$ is a real number:

$$a^{m/n} = (a^{1/n})^m.$$

When a graphing calculator is used to evaluate fractional powers of negative numbers, the theorem on $a^{m/n}$ may not hold true. For example, the calculator may evaluate $[(-8)^{1/3}]^2$ or $[(-8)^2]^{1/3}$, but not $(-8)^{2/3}$. The simplest way to handle this is to evaluate $8^{2/3}$ and determine the appropriate sign mentally.

EXAMPLE 5
Using the definition of $a^{m/n}$

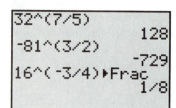

A calculator supports the results of Example 5(b), (c), and (f).

Evaluate each expression.

(a) $125^{2/3} = (125^{1/3})^2 = 5^2 = 25$

(b) $32^{7/5} = (32^{1/5})^7 = 2^7 = 128$

(c) $-81^{3/2} = -(81^{1/2})^3 = -9^3 = -729$

(d) $(-4)^{5/2}$ is not a real number because $(-4)^{1/2}$ is not a real number.

(e) $(-27)^{2/3} = [(-27)^{1/3}]^2 = (-3)^2 = 9$

(f) $16^{-3/4} = \dfrac{1}{16^{3/4}} = \dfrac{1}{(16^{1/4})^3} = \dfrac{1}{2^3} = \dfrac{1}{8}$ ▶

NOTE By starting with $(a^{1/n})^m$ and $(a^m)^{1/n}$, and raising each expression to the nth power, it can be shown that $(a^{1/n})^m$ is equal to $(a^m)^{1/n}$. This means that $a^{m/n}$ could be defined in either of the following ways.

For all real numbers a, integers m, and positive integers n for which $a^{1/n}$ is a real number:

$$a^{m/n} = (a^{1/n})^m \qquad \text{or} \qquad a^{m/n} = (a^m)^{1/n}.$$

Now $a^{m/n}$ can be evaluated in either of two ways: as $(a^{1/n})^m$ or as $(a^m)^{1/n}$. It is usually easier to find $(a^{1/n})^m$. For example, $27^{4/3}$ can be evaluated in either of two ways:

$$27^{4/3} = (27^{1/3})^4 = 3^4 = 81$$
$$27^{4/3} = (27^4)^{1/3} = 531{,}441^{1/3} = 81.$$

The form $(27^{1/3})^4$ is easier to evaluate.

CONNECTIONS Many useful formulas involve fractional exponents. One such formula is used in weather forecasting. Meteorologists can determine the duration of a storm by using the formula $.07D^{3/2} = T$, where D is the diameter of the storm in miles and T is the time in hours. For example, if radar shows that the diameter of a storm is 16 miles, we can expect the storm to last

$$.07(16)^{3/2} = .07(16^{1/2})^3$$
$$= .07(4)^3$$
$$= .07(64)$$
$$= 4.48 \text{ hours.}$$

FOR DISCUSSION OR WRITING

1. The National Weather Service reports that a storm 4 miles in diameter is headed toward New Haven. How long can the residents expect the storm to last?
2. After weeks of dry weather, a thunderstorm is predicted for the farming community of Apple Valley. The crops need at least 1.5 hours of rain. Local radar shows that the storm is 7 miles in diameter. Will it rain long enough to meet the farmers' need?

It can be shown that all the earlier results concerning integer exponents also apply to rational exponents. These definitions and rules are summarized here.

Definitions and Rules for Exponents

Let r and s be rational numbers. The results below are valid for all positive numbers a and b.

$$a^r \cdot a^s = a^{r+s} \qquad (ab)^r = a^r \cdot b^r \qquad (a^r)^s = a^{rs}$$

$$\frac{a^r}{a^s} = a^{r-s} \qquad \left(\frac{a}{b}\right)^r = \frac{a^r}{b^r} \qquad a^{-r} = \frac{1}{a^r}$$

EXAMPLE 6
Using the definitions and rules for exponents

```
((27^(1/3)*27^(5
/3))/27³▶Frac
              1/27
81^(5/4)*4^(-3/2
)▶Frac
            243/8
```

A calculator supports the results of Example 6(a) and (b).

Use the definitions and rules for exponents to simplify each expression.

(a) $\dfrac{27^{1/3} \cdot 27^{5/3}}{27^3} = \dfrac{27^{1/3+5/3}}{27^3}$ Product rule

$$= \frac{27^2}{27^3} = 27^{2-3} \qquad \text{Quotient rule}$$

$$= 27^{-1} = \frac{1}{27}$$

(b) $81^{5/4} \cdot 4^{-3/2} = (81^{1/4})^5 (4^{1/2})^{-3} = 3^5 \cdot 2^{-3} = \dfrac{3^5}{2^3}$ or $\dfrac{243}{8}$.

(c) $6y^{2/3} \cdot 2y^{1/2} = 12y^{2/3+1/2} = 12y^{7/6}$, where $y \geq 0$

(d) $\left(\dfrac{3m^{5/6}}{y^{3/4}}\right)^2 \cdot \left(\dfrac{8y^3}{m^6}\right)^{2/3} = \dfrac{9m^{5/3}}{y^{3/2}} \cdot \dfrac{4y^2}{m^4} = 36m^{5/3-4}y^{2-3/2}$

$\qquad\qquad = \dfrac{36y^{1/2}}{m^{7/3}}$ $(m > 0, y > 0)$

(e) $m^{2/3}(m^{7/3} + 2m^{1/3}) = (m^{2/3+7/3} + 2m^{2/3+1/3}) = m^3 + 2m$

(f) $\dfrac{(x^{2/p})^p(x^{p-1})}{x^{-1/4}} = \dfrac{x^{(2/p)p} \cdot x^{p-1}}{x^{-1/4}}$　　　Power rule

$\qquad = \dfrac{x^2 x^{p-1}}{x^{-1/4}}$

$\qquad = \dfrac{x^{2+p-1}}{x^{-1/4}}$　　　Product rule

$\qquad = \dfrac{x^{1+p}}{x^{-1/4}}$

$\qquad = x^{1+p-(-1/4)}$　　　Quotient rule

$\qquad = x^{(5/4)+p}$　or　$x^{(5+4p)/4}$ ▶

The next example shows how to factor with negative or rational exponents.

EXAMPLE 7
Factoring an expression with negative or rational exponents

Factor out the smallest power of the variable. Assume that all variables represent positive real numbers.

(a) $9x^{-2} - 6x^{-3}$

The smallest exponent here is -3. Since 3 is a common numerical factor, factor out $3x^{-3}$.

$\qquad 9x^{-2} - 6x^{-3} = 3x^{-3}(3x^{-2-(-3)} - 2x^{-3-(-3)}) = 3x^{-3}(3x - 2)$

Check by multiplying on the right. The factored form can now be written without negative exponents as $\dfrac{3(3x-2)}{x^3}$.

(b) $4m^{1/2} + 3m^{3/2} = m^{1/2}(4 + 3m)$

To check this result, multiply $m^{1/2}$ by $4 + 3m$.

(c) $(y-2)^{-1/3} + (y-2)^{2/3} = (y-2)^{-1/3}[1 + (y-2)]$

$\qquad = (y-2)^{-1/3}(y-1)$

The factored form can be written with only positive exponents as

$$\dfrac{y-1}{(y-2)^{1/3}}.$$ ▶

Negative exponents are sometimes used to write complex fractions. Recall, complex fractions are simplified either by first multiplying the numerator and denominator by the least common multiple of all the denominators, or by performing any indicated operations in the numerator and the denominator and then using the definition of division for fractions.

◀ **EXAMPLE 8**
Simplifying a fraction with negative exponents

Simplify $\dfrac{(x + y)^{-1}}{x^{-1} + y^{-1}}$. Write the result with only positive exponents.

Begin by using the definition of a negative integer exponent. Then perform the indicated operations.

$$\frac{(x + y)^{-1}}{x^{-1} + y^{-1}} = \frac{\dfrac{1}{x + y}}{\dfrac{1}{x} + \dfrac{1}{y}}$$

$$= \frac{\dfrac{1}{x + y}}{\dfrac{y + x}{xy}}$$

$$= \frac{1}{x + y} \cdot \frac{xy}{x + y}$$

$$= \frac{xy}{(x + y)^2} \quad \blacktriangleright$$

```
(3+4)-1=3-1+4-1
                      0
(3+4)-1≠3-1+4-1
                      1
```

This screen supports the statement for $x = 3$ and $y = 4$.

CAUTION   Remember that if $r \neq 1$, $(x + y)^r \neq x^r + y^r$. In particular, this means that $(x + y)^{-1} \neq x^{-1} + y^{-1}$.

1.6 Exercises ▼▼▼▼▼▼▼▼▼▼▼▼▼▼▼▼▼▼▼▼▼▼▼▼▼▼▼▼▼▼▼▼▼▼▼

Simplify the expression. Do not use a calculator. See Examples 1, 4, and 5.

1. $(-4)^{-3}$ **2.** $(-5)^{-2}$ **3.** $\left(\dfrac{1}{2}\right)^{-3}$ **4.** $\left(\dfrac{2}{3}\right)^{-2}$

5. $-4^{1/2}$ **6.** $25^{1/2}$ **7.** $8^{2/3}$ **8.** $-81^{3/4}$

9. $27^{-2/3}$ **10.** $(-32)^{-4/5}$ **11.** $\left(-\dfrac{4}{9}\right)^{-3/2}$ **12.** $\left(\dfrac{1}{8}\right)^{-5/3}$

13. $\left(\dfrac{27}{64}\right)^{-4/3}$ **14.** $\left(\dfrac{121}{100}\right)^{-3/2}$

The screen displayed here shows how a graphing calculator can evaluate a power involving a rational exponent. Assuming that the result is rational it can also express the result in a/b form.

🖩 *Use a graphing calculator to find each of the following, and express the result in a/b form.*

15. $64^{3/2}$ **16.** $16^{5/4}$ **17.** $8^{-5/3}$

18. $4^{-7/2}$ **19.** $\left(\dfrac{8}{27}\right)^{-5/3}$ **20.** $\left(\dfrac{27}{8}\right)^{-4/3}$

21. $100^{-2.5}$ **22.** $729^{-5/6}$

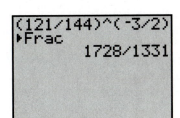

```
(121/144)^(-3/2)
▶Frac
          1728/1331
```

$\left(\dfrac{121}{144}\right)^{-3/2} = \dfrac{1728}{1331}$

23. Why is $a^{1/n}$ defined to be the nth root of a (with appropriate restrictions)?

24. Explain why a must be positive if n is even for $a^{1/n}$ to be a real number.

25. Which of the following expressions is equivalent to $(2x^{-3/2})^2$?

(a) $2x^{-3}$ **(b)** 2^{-3} **(c)** $2^2x^{-3/4}$ **(d)** $\dfrac{2^2}{x^3}$

26. Explain why the following input will lead to an error message on a graphing calculator programmed to work with real numbers only.

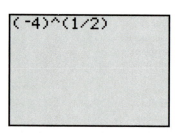

$$(-4)^{\wedge}(1/2)$$

Perform the indicated operations. Write the answer using only positive exponents. Assume that all variables represent positive real numbers and that variables used as exponents represent rational numbers. See Examples 2, 3, and 6.

27. $27^{-2} \cdot 27^{-1}$

28. $9^{-4} \cdot 9^{-1}$

29. $\dfrac{4^{-2} \cdot 4^{-1}}{4^{-3}}$

30. $\dfrac{3^{-1} \cdot 3^{-4}}{3^2 \cdot 3^{-2}}$

31. $(m^{2/3})(m^{5/3})$

32. $(x^{4/5})(x^{2/5})$

33. $(1 + n)^{1/2}(1 + n)^{3/4}$

34. $(m + 7)^{-1/6}(m + 7)^{-2/3}$

35. $(2y^{3/4}z)(3y^{-2}z^{-1/3})$

36. $(4a^{-1}b^{2/3})(a^{3/2}b^{-3})$

37. $(4a^{-2}b^7)^{1/2} \cdot (2a^{1/4}b^3)^5$

38. $(x^{-2}y^{1/3})^5 \cdot (8x^2y^{-2})^{-1/3}$

39. $\left(\dfrac{r^{-2}}{s^{-5}}\right)^{-3}$

40. $\left(\dfrac{p^{-1}}{q^{-5}}\right)^{-2}$

41. $\left(\dfrac{-a}{b^{-3}}\right)^{-1}$

42. $\dfrac{7^{-1/3}7r^{-3}}{7^{2/3}r^{-2}}$

43. $\dfrac{12^{5/4}y^{-2}}{12^{-1}y^{-3}}$

44. $\dfrac{6k^{-4}(3k^{-1})^{-2}}{2^3k^{1/2}}$

45. $\dfrac{8p^{-3}(4p^2)^{-2}}{p^{-5}}$

46. $\dfrac{k^{-3/5}h^{-1/3}t^{2/5}}{k^{-1/5}h^{-2/3}t^{1/5}}$

47. $\dfrac{m^{7/3}n^{-2/5}p^{3/8}}{m^{-2/3}n^{3/5}p^{-5/8}}$

48. $\dfrac{m^{2/5}m^{3/5}m^{-4/5}}{m^{1/5}m^{-6/5}}$

49. $\dfrac{-4a^{-1}a^{2/3}}{a^{-2}}$

50. $\dfrac{8y^{2/3}y^{-1}}{2^{-1}y^{3/4}y^{-1/6}}$

51. $\dfrac{(k + 5)^{1/2}(k + 5)^{-1/4}}{(k + 5)^{3/4}}$

52. $\dfrac{(x + y)^{-5/8}(x + y)^{3/8}}{(x + y)^{1/8}(x + y)^{-1/8}}$

One important application of mathematics to business and management concerns supply and demand. Usually, as the price of an item increases, the supply increases and the demand decreases. By studying past records of supply and demand at different prices, economists can construct an equation that describes (approximately) supply and demand for a given item. The next four exercises show examples of this.

53. The price (in dollars) of a certain type of solar heater is approximated by p, where

$$p = 2x^{1/2} + 3x^{2/3},$$

and x is the number of units supplied. Find the price when the supply is 64 units. Do not use a calculator.

54. Repeat Exercise 53 for a supply of 100 units. Use a calculator.

55. For a certain commodity the demand and the price (in dollars) are related by the equation
$$p = 1000 - 200x^{-2/3} \quad (x > 0),$$
where x is the number of units of the product demanded. Find the price when the demand is 27. Do not use a calculator.

56. Repeat Exercise 55 for a demand of 50. Use a calculator.

In our system of government, the president is elected by the electoral college, and not by individual voters. Because of this, smaller states have a greater voice in the selection of a president than they otherwise would have. Two political scientists have studied the problems of campaigning for president under the current system and have concluded that candidates should allot their money according to the formula

$$\frac{\text{Amount for}}{\text{large state}} = \left(\frac{E_{\text{large}}}{E_{\text{small}}}\right)^{3/2} \times \frac{\text{amount for}}{\text{small state.}}$$

Here E_{large} represents the electoral vote of the large state, and E_{small} represents the electoral vote of the small state. Find the amount that should be spent in each of the following larger states if $1,000,000$ is spent in the small state and the following statements are true.

57. The large state has 48 electoral votes, and the small state has 3.

58. The large state has 36 electoral votes, and the small state has 4.

59. 6 votes in a small state; 28 in a large

60. 9 votes in a small state; 32 in a large

▼▼▼▼▼▼▼▼▼▼▼▼▼ **DISCOVERING CONNECTIONS** (Exercises 61–70) ▼▼▼▼▼▼▼▼▼▼▼▼▼

Using a technique from statistics called **exponential regression**, *it can be shown that the equation $y = 386(1.18)^x$ provides a fairly good model for product sales generated by infomercials, where y is in millions of dollars and $x = 0$ corresponds to the year 1988, $x = 1$ corresponds to 1989, and so on through the year 1993. (Source: National Infomercial Marketing Association.)*

Use a calculator to estimate, to the nearest million dollars, the amount generated during the indicated year.

61. 1988   **62.** 1989

63. 1990   **64.** 1991

65. 1992

*The bar graph indicates another (more accurate) way of expressing the data described in Exercises 61–65. Use it to **(a)** determine the absolute value of the difference between the amount shown on the graph and the value provided by the model given earlier, and **(b)** determine whether the amount indicated by the model is less than, equal to, or greater than the amount indicated on the graph.*

66. 1988   **67.** 1989

68. 1990   **69.** 1991

70. 1992

Product Sales Generated by Infomercials

In millions

$350 ('88) $500 ('89) $600 ('90) $600 ('91) $750 ('92) $900 ('93*)

'88 '89 '90 '91 '92 '93*

Source: National Infomercial Marketing Association   *Estimate

The Galapagos Islands are a chain of islands ranging in size from 2 to 2249 square miles. A biologist has shown that the number of different land-plant species on an island in this chain is related to the size of the island by approximately

$$S = 28.6A^{.32},$$

where A is the area of the island in square miles and S is the number of different plant species on that island. Estimate S (rounding to the nearest whole number) for islands of the following areas.

71. 10 square miles   **72.** 25 square miles   **73.** 300 square miles   **74.** 2000 square miles

Perform the indicated operations. Write the answers without a denominator. Assume that all variables used in denominators are not zero and all variables used as exponents represent rational numbers. See Examples 2(e) and 6(f).

75. $(r^{3/p})^{2p}(r^{1/p})^{p^2}$ **76.** $(m^{2/x})^{x/3}(m^{x/4})^{2/x}$ **77.** $\dfrac{m^{1-a}m^a}{m^{-1/2}}$ **78.** $\dfrac{(y^{3-b})(y^{2b-1})}{y^{1/2}}$

79. $\dfrac{(x^{n/2})(x^{3n})^{1/2}}{x^{1/n}}$ **80.** $\dfrac{(a^{2/3})(a^{1/x})}{(a^{x/3})^{-2}}$ **81.** $\dfrac{(p^{1/n})(p^{1/m})}{p^{-m/n}}$ **82.** $\dfrac{(q^{2r/3})(q^r)^{-1/3}}{(q^{4/3})^{1/r}}$

Find each product. Assume that all variables represent positive real numbers. See Example 6(e). (Hint: Use the special binomial product formulas in Exercises 87, 89, and 90.)

83. $y^{5/8}(y^{3/8} - 10y^{11/8})$ **84.** $p^{11/5}(3p^{4/5} + 9p^{19/5})$

85. $-4k(k^{7/3} - 6k^{1/3})$ **86.** $-5y(3y^{9/10} + 4y^{3/10})$

87. $(x + x^{1/2})(x - x^{1/2})$ **88.** $(2z^{1/2} + z)(z^{1/2} - z)$

89. $(r^{1/2} - r^{-1/2})^2$ **90.** $(p^{1/2} - p^{-1/2})(p^{1/2} + p^{-1/2})$

Factor, using the given common factor. Assume all variables represent positive real numbers. See Example 7.

91. $4k^{-1} + k^{-2};\quad k^{-2}$ **92.** $y^{-5} - 3y^{-3};\quad y^{-5}$ **93.** $9z^{-1/2} + 2z^{1/2};\quad z^{-1/2}$

94. $3m^{2/3} - 4m^{-1/3};\quad m^{-1/3}$ **95.** $p^{-3/4} - 2p^{-7/4};\quad p^{-7/4}$ **96.** $6r^{-2/3} - 5r^{-5/3};\quad r^{-5/3}$

97. $(p + 4)^{-3/2} + (p + 4)^{-1/2} + (p + 4)^{1/2};\quad (p + 4)^{-3/2}$

98. $(3r + 1)^{-2/3} + (3r + 1)^{1/3} + (3r + 1)^{4/3};\quad (3r + 1)^{-2/3}$

Perform all indicated operations and write the answer with positive integer exponents. See Example 8.

99. $\dfrac{a^{-1} + b^{-1}}{(ab)^{-1}}$ **100.** $\dfrac{p^{-1} - q^{-1}}{(pq)^{-1}}$ **101.** $\dfrac{r^{-1} + q^{-1}}{r^{-1} - q^{-1}} \cdot \dfrac{r - q}{r + q}$

102. $\dfrac{xy^{-1} + yx^{-1}}{x^2 + y^2}$ **103.** $\dfrac{x - 9y^{-1}}{(x - 3y^{-1})(x + 3y^{-1})}$ **104.** $\dfrac{(m + n)^{-1}}{m^{-2} - n^{-2}}$

1.7   Radicals ▼▼▼

RADICAL NOTATION   In the last section the notation $a^{1/n}$ was used for the nth root of a for appropriate values of a and n. An alternative (and more familiar) notation for $a^{1/n}$ is *radical notation*.

> ### Radical Notation for $a^{1/n}$
>
> If a is a real number, n is a positive integer, and $a^{1/n}$ is a real number, then
>
> $$\sqrt[n]{a} = a^{1/n}.$$

The symbol $\sqrt[n]{}$ is a **radical sign,** the number a is the **radicand,** and n is the **index** of the radical $\sqrt[n]{a}$. It is customary to use the familiar notation $\sqrt{a}$ instead of $\sqrt[2]{a}$ for the square root.

For even values of n (square roots, fourth roots, and so on), when a is positive, there are two nth roots, one positive and one negative. In such cases, the notation $\sqrt[n]{a}$ represents the positive root, the **principal nth root.** The negative root is written $-\sqrt[n]{a}$.

◀ EXAMPLE 1
Evaluating roots

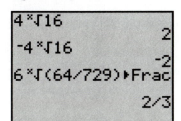

The calculator screen supports the results of Example 1(a), (b), and (f).

Evaluate each root.

(a) $\sqrt[4]{16} = 16^{1/4} = 2$ **(b)** $-\sqrt[4]{16} = -16^{1/4} = -2$

(c) $\sqrt[4]{-16}$ is not a real number.   **(d)** $\sqrt[5]{-32} = -2$

(e) $\sqrt[3]{1000} = 10$ **(f)** $\sqrt[6]{\dfrac{64}{729}} = \dfrac{2}{3}$ ▶

With $a^{1/n}$ written as $\sqrt[n]{a}$, $a^{m/n}$ also can be written using radicals.

> ### Radical Notation for $a^{m/n}$
>
> If a is a real number, m is an integer, n is a positive integer, and $\sqrt[n]{a}$ is a real number, then
>
> $$a^{m/n} = (\sqrt[n]{a})^m = \sqrt[n]{a^m}.$$

◀ EXAMPLE 2
Converting from rational
exponents to radicals

The calculator screen supports the
results in Example 2(a), (b), and (c),
using radical notation.

Write in radical form and simplify.

(a) $8^{2/3} = (\sqrt[3]{8})^2 = 2^2 = 4$

(b) $(-32)^{4/5} = (\sqrt[5]{-32})^4 = (-2)^4 = 16$

(c) $-16^{3/4} = -(\sqrt[4]{16})^3 = -(2)^3 = -8$

(d) $x^{5/6} = \sqrt[6]{x^5} \quad (x \geq 0)$

(e) $3x^{2/3} = 3\sqrt[3]{x^2}$

(f) $2p^{-1/2} = \dfrac{2}{p^{1/2}} = \dfrac{2}{\sqrt{p}} \quad (p > 0)$

(g) $(3a + b)^{1/4} = \sqrt[4]{3a + b} \quad (3a + b \geq 0)$ ▶

CAUTION   It is not possible to "distribute" exponents over a sum, so in Example 2(g), $(3a + b)^{1/4}$ *cannot be written as* $(3a)^{1/4} + b^{1/4}$. More generally,

$$\sqrt[n]{x^n + y^n} \text{ is } not \text{ equal to } x + y.$$

Be alert for this common error.

EXAMPLE 3
Converting from radicals to rational exponents

Write in exponential form.

(a) $\sqrt[4]{x^5} = x^{5/4}$ $(x \geq 0)$ **(b)** $\sqrt{3y} = (3y)^{1/2}$ $(y \geq 0)$

(c) $10(\sqrt[5]{z})^2 = 10z^{2/5}$ **(d)** $5\sqrt[3]{(2x^4)^7} = 5(2x^4)^{7/3} = 5 \cdot 2^{7/3} x^{28/3}$

(e) $\sqrt{p^2 + q} = (p^2 + q)^{1/2}$ $(p^2 + q \geq 0)$ ▶

By the definition of $\sqrt[n]{a}$, for any positive integer n, if $\sqrt[n]{a}$ is a real number, then

$$(\sqrt[n]{a})^n = a.$$

If a is positive, or if a is negative and n is an odd positive integer,

$$\sqrt[n]{a^n} = a.$$

Because of the conditions just given, we *cannot* simply write $\sqrt{x^2} = x$. For example, if $x = -5$,

$$\sqrt{x^2} = \sqrt{(-5)^2} = \sqrt{25} = 5 \neq x.$$

To take care of the fact that a negative value of x can produce a positive result, we use absolute value. For any real number a,

$$\sqrt{a^2} = |a|.$$

For example,

$$\sqrt{(-9)^2} = |-9| = 9, \quad \text{and} \quad \sqrt{13^2} = |13| = 13.$$

This result can be generalized to any even nth root.

$\sqrt[n]{a^n}$ If n is an even positive integer, $\sqrt[n]{a^n} = |a|$, and if n is an odd positive integer, $\sqrt[n]{a^n} = a$.

EXAMPLE 4
Using absolute value to simplify roots

Use absolute value as applicable to simplify the following expressions.

(a) $\sqrt{p^4} = |p^2| = p^2$

(b) $\sqrt[4]{p^4} = |p|$

(c) $\sqrt{16m^8 r^6} = |4m^4 r^3| = 4m^4|r^3|$

(d) $\sqrt[6]{(-2)^6} = |-2| = 2$

(e) $\sqrt[5]{m^5} = m$

(f) $\sqrt{(2k + 3)^2} = |2k + 3|$

(g) $\sqrt{x^2 - 4x + 4} = \sqrt{(x - 2)^2} = |x - 2|$ ▷

NOTE   To avoid difficulties when working with variable radicands, we usually will assume that all variables in radicands represent only nonnegative real numbers.

Three key rules for working with radicals are given below. These rules are just the power rules for exponents written in radical notation.

Rules for Radicals

For all real numbers a and b, and positive integers m and n for which the indicated roots are real numbers,

$$\sqrt[n]{a} \cdot \sqrt[n]{b} = \sqrt[n]{ab} \qquad \sqrt[n]{\frac{a}{b}} = \frac{\sqrt[n]{a}}{\sqrt[n]{b}} \quad (b \neq 0) \qquad \sqrt[m]{\sqrt[n]{a}} = \sqrt[mn]{a}.$$

◀ EXAMPLE 5
Using the rules for radicals to simplify radical expressions

(a) $\sqrt{6} \cdot \sqrt{54} = \sqrt{6 \cdot 54} = \sqrt{324} = 18$

(b) $\sqrt[3]{m} \cdot \sqrt[3]{m^2} = \sqrt[3]{m^3} = m$

(c) $\sqrt{\frac{7}{64}} = \frac{\sqrt{7}}{\sqrt{64}} = \frac{\sqrt{7}}{8}$

(d) $\sqrt[4]{\frac{a}{b^4}} = \frac{\sqrt[4]{a}}{\sqrt[4]{b^4}} = \frac{\sqrt[4]{a}}{b} \quad (a \geq 0, b > 0)$

(e) $\sqrt[7]{\sqrt[3]{2}} = \sqrt[21]{2}$ Use the third rule given above.

(f) $\sqrt[4]{\sqrt{3}} = \sqrt[8]{3}$ ▷

NOTE   In Example 5, converting to fractional exponents would show why these rules work. For example, in part (e)

$$\sqrt[7]{\sqrt[3]{2}} = (2^{1/3})^{1/7} = 2^{(1/3)(1/7)} = 2^{1/21} = \sqrt[21]{2}.$$

SIMPLIFYING RADICALS   In working with numbers, it is customary to write a number in its simplest form. For example, $10/2$ is written as 5, $-9/6$ is written as $-3/2$, and $4/16$ is written as $1/4$. Similarly, expressions with radicals should be written in their simplest forms.

> ### Simplified Radicals
>
> An expression with radicals is simplified when all of the following conditions are satisfied.
>
> **1.** The radicand has no factor raised to a power greater than or equal to the index.
> **2.** The radicand has no fractions.
> **3.** No denominator contains a radical.
> **4.** Exponents in the radicand and the index of the radical have no common factor.
> **5.** All indicated operations have been performed (if possible).

EXAMPLE 6
Simplifying radicals

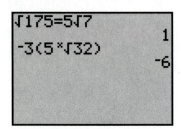

The first display indicates that the result in Example 6(a) is true. The second display supports the result in Example 6(b).

Simplify each of the following. Assume that all variables represent nonnegative real numbers.

(a) $\sqrt{175} = \sqrt{25 \cdot 7} = \sqrt{25} \cdot \sqrt{7} = 5\sqrt{7}$

(b) $-3\sqrt[5]{32} = -3\sqrt[5]{2^5} = -3 \cdot 2 = -6$

(c) $\sqrt[3]{81x^5y^7z^6} = \sqrt[3]{27 \cdot 3 \cdot x^3 \cdot x^2 \cdot y^6 \cdot y \cdot z^6}$ Factor.

$\qquad\qquad = \sqrt[3]{(27x^3y^6z^6)(3x^2y)}$ Group all perfect cubes.

$\qquad\qquad = 3xy^2z^2\sqrt[3]{3x^2y}$ Remove all perfect cubes from the radical. ◗

Radicals with the same radicand and the same index, such as $3\sqrt[4]{11pq}$ and $-7\sqrt[4]{11pq}$, are called **like radicals.** Like radicals are added or subtracted by using the distributive property. Only like radicals can be combined. As shown in parts (b) and (c) of the next example, it is sometimes necessary to simplify radicals before adding or subtracting.

EXAMPLE 7
Adding and subtracting like radicals

Add or subtract as indicated. Assume all variables are positive real numbers.

(a) $3\sqrt[4]{11pq} + (-7\sqrt[4]{11pq}) = -4\sqrt[4]{11pq}$

(b) $\sqrt{98x^3y} + 3x\sqrt{32xy}$

First remove all perfect square factors from under the radical. Then use the distributive property, as follows.

$$\sqrt{98x^3y} + 3x\sqrt{32xy} = \sqrt{49 \cdot 2 \cdot x^2 \cdot x \cdot y} + 3x\sqrt{16 \cdot 2 \cdot x \cdot y}$$
$$= 7x\sqrt{2xy} + (3x)(4)\sqrt{2xy}$$
$$= 7x\sqrt{2xy} + 12x\sqrt{2xy}$$
$$= 19x\sqrt{2xy} \qquad \text{\color{blue}Distributive property}$$

(c) $\sqrt[3]{64m^4n^5} - \sqrt[3]{-27m^{10}n^{14}} = \sqrt[3]{(64m^3n^3)(mn^2)} - \sqrt[3]{(-27m^9n^{12})(mn^2)}$

$\qquad\qquad\qquad\qquad\qquad = 4mn\sqrt[3]{mn^2} - (-3)m^3n^4\sqrt[3]{mn^2}$

$\qquad\qquad\qquad\qquad\qquad = 4mn\sqrt[3]{mn^2} + 3m^3n^4\sqrt[3]{mn^2}$

$\qquad\qquad\qquad\qquad\qquad = (4 + 3m^2n^3)mn\sqrt[3]{mn^2}$ Distributive property ▶

If the index of the radical and an exponent in the radicand have a common factor, the radical can be simplified by writing it in exponential form, simplifying the rational exponent, then writing the result as a radical again.

EXAMPLE 8
Simplifying radicals by writing them with rational exponents

Simplify the following radicals by first rewriting with rational exponents.

(a) $\sqrt[6]{3^2} = 3^{2/6} = 3^{1/3} = \sqrt[3]{3}$

(b) $\sqrt[6]{x^{12}y^3} = (x^{12}y^3)^{1/6} = x^2y^{3/6} = x^2y^{1/2} = x^2\sqrt{y}$ $(y \geq 0)$

(c) $\sqrt[9]{\sqrt{6^3}} = \sqrt[9]{6^{3/2}} = (6^{3/2})^{1/9} = 6^{1/6} = \sqrt[6]{6}$ ▶

In Example 8(a), we simplified $\sqrt[6]{3^2}$ as $\sqrt[3]{3}$. However, to simplify $(\sqrt[6]{x})^2$, the variable x must be nonnegative. For example, consider the statement

$$(-8)^{2/6} = [(-8)^{1/6}]^2.$$

This result is not a real number, since $(-8)^{1/6}$ is not defined. On the other hand,

$$(-8)^{1/3} = -2.$$

Here, even though $2/6 = 1/3$,

$$(\sqrt[6]{x})^2 \neq \sqrt[3]{x}.$$

If a is nonnegative, then it is always true that $a^{m/n} = a^{mp/(np)}$. Reducing rational exponents on negative bases must be considered case by case.

Multiplying radical expressions is much like multiplying polynomials.

EXAMPLE 9
Multiplying radical expressions

Find the product.

(a) $(\sqrt{2} + 3)(\sqrt{8} - 5) = \sqrt{2}(\sqrt{8}) - \sqrt{2}(5) + 3\sqrt{8} - 3(5)$ FOIL

$\qquad\qquad\qquad\qquad = \sqrt{16} - 5\sqrt{2} + 3(2\sqrt{2}) - 15$ Multiply.

$\qquad\qquad\qquad\qquad = 4 - 5\sqrt{2} + 6\sqrt{2} - 15$

$\qquad\qquad\qquad\qquad = -11 + \sqrt{2}$ Combine terms.

(b) $(\sqrt{7} - \sqrt{10})(\sqrt{7} + \sqrt{10}) = (\sqrt{7})^2 - (\sqrt{10})^2$ Product of the sum and difference of two terms

$\qquad\qquad\qquad\qquad\qquad = 7 - 10$

$\qquad\qquad\qquad\qquad\qquad = -3$ ▶

RATIONALIZING THE DENOMINATOR   Condition 3 of the rules for simplifying radicals described above requires that no denominator contain a radical. The process of achieving this is called **rationalizing the denominator.** It is accomplished by multiplying by a form of 1, as explained in Example 10.

EXAMPLE 10
Rationalizing denominators

Rationalize each denominator.

(a) $\dfrac{4}{\sqrt{3}}$

To rationalize the denominator, multiply by $\sqrt{3}/\sqrt{3}$ (which equals 1) so that the denominator of the product is a rational number.

$$\frac{4}{\sqrt{3}} \cdot \frac{\sqrt{3}}{\sqrt{3}} = \frac{4\sqrt{3}}{3}$$

(b) $\sqrt[4]{\dfrac{3}{5}}$

Start by using the fact that the radical of a quotient can be written as the quotient of radicals.

$$\sqrt[4]{\frac{3}{5}} = \frac{\sqrt[4]{3}}{\sqrt[4]{5}}$$

The denominator will be a rational number if it equals $\sqrt[4]{5^4}$. That is, four factors of 5 are needed under the radical. Since $\sqrt[4]{5}$ has just one factor of 5, three additional factors are needed, so multiply by $\sqrt[4]{5^3}/\sqrt[4]{5^3}$.

$$\frac{\sqrt[4]{3}}{\sqrt[4]{5}} = \frac{\sqrt[4]{3} \cdot \sqrt[4]{5^3}}{\sqrt[4]{5} \cdot \sqrt[4]{5^3}} = \frac{\sqrt[4]{3 \cdot 5^3}}{\sqrt[4]{5^4}} = \frac{\sqrt[4]{375}}{5}$$

📉 **What's going on here?**

It is always a temptation to become too dependent on technology. To illustrate the pitfalls of not knowing the limitations of your calculator or computer, consider the results of Example 9(a) and Example 10(a). The screen on the left indicates that the result of Example 9(a) is *false,* as the calculator returns a 0 for the equation that we know to be true! This is due to the fact that we cannot always rely on this feature if we are dealing with irrational numbers. On the other hand, the screen on the right correctly supports our result in Example 10(a).

Just as in the consumer world where the motto *Caveat emptor* ("Let the buyer beware") is so very true, the user of technology must also beware of its limitations as well as its capabilities.

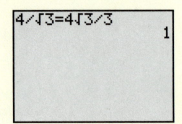

EXAMPLE 11
Simplifying rational
expressions

Simplify. Assume all variables represent positive real numbers.

(a) $\dfrac{\sqrt[4]{ab^3} \cdot \sqrt[4]{ab}}{\sqrt[4]{a^3b^3}}$

To begin, use the product and quotient rules to write all radicals under one radical sign.

$$\frac{\sqrt[4]{ab^3} \cdot \sqrt[4]{ab}}{\sqrt[4]{a^3b^3}} = \sqrt[4]{\frac{ab^3 \cdot ab}{a^3b^3}}$$

$$= \sqrt[4]{\frac{a^2b^4}{a^3b^3}} \qquad \text{Multiply.}$$

$$= \sqrt[4]{\frac{b}{a}} \qquad \text{Write in lowest terms.}$$

$$= \frac{\sqrt[4]{b}}{\sqrt[4]{a}} \cdot \frac{\sqrt[4]{a^3}}{\sqrt[4]{a^3}} \qquad \sqrt[4]{a} \cdot \sqrt[4]{a^3} = \sqrt[4]{a^4} = a.$$

$$= \frac{\sqrt[4]{a^3b}}{a}$$

(b) $\dfrac{-1}{\sqrt[3]{16}} - \dfrac{5}{\sqrt[3]{128}} + \dfrac{4}{\sqrt[3]{2}}$

Begin by simplifying all radicals:

$$\sqrt[3]{16} = \sqrt[3]{8 \cdot 2} = 2\sqrt[3]{2}$$

and
$$\sqrt[3]{128} = \sqrt[3]{64 \cdot 2} = 4\sqrt[3]{2}.$$

Now proceed as follows.

$$\frac{-1}{\sqrt[3]{16}} - \frac{5}{\sqrt[3]{128}} + \frac{4}{\sqrt[3]{2}} = \frac{-1}{2\sqrt[3]{2}} - \frac{5}{4\sqrt[3]{2}} + \frac{4}{\sqrt[3]{2}}$$

$$= \frac{-2}{4\sqrt[3]{2}} - \frac{5}{4\sqrt[3]{2}} + \frac{16}{4\sqrt[3]{2}} \qquad \text{Write with a common denominator.}$$

$$= \frac{9}{4\sqrt[3]{2}} \qquad \text{Combine terms.}$$

$$= \frac{9\sqrt[3]{2^2}}{4\sqrt[3]{2} \cdot \sqrt[3]{2^2}} \qquad \text{Rationalize the denominator.}$$

$$= \frac{9\sqrt[3]{4}}{4 \cdot \sqrt[3]{8}}$$

$$= \frac{9\sqrt[3]{4}}{4 \cdot 2}$$

$$= \frac{9\sqrt[3]{4}}{8}$$

NOTE In Example 11(a), $\sqrt[4]{a^4} = a$ (not $|a|$) because of the assumption that a is positive.

In Example 9(b), we saw that

$$(\sqrt{7} - \sqrt{10})(\sqrt{7} + \sqrt{10}) = -3,$$

a rational number. This suggests a way to rationalize a denominator that is a binomial in which one or both terms is a radical. The expressions $a\sqrt{m} + b\sqrt{n}$ and $a\sqrt{m} - b\sqrt{n}$ are called **conjugates.**

EXAMPLE 12
Rationalizing a binomial denominator

Rationalize the denominator of $\dfrac{1}{1 - \sqrt{2}}$.

As mentioned above, the best approach here is to multiply both numerator and denominator by the conjugate of the denominator, in this case $1 + \sqrt{2}$.

$$\frac{1}{1 - \sqrt{2}} = \frac{1(1 + \sqrt{2})}{(1 - \sqrt{2})(1 + \sqrt{2})} = \frac{1 + \sqrt{2}}{1 - 2} = -1 - \sqrt{2} \quad \blacktriangleright$$

```
1/(1-√2)
        -2.414213562
-1-√2
        -2.414213562
```

The calculator reports that the rational decimal approximations of $\frac{1}{1-\sqrt{2}}$ and $-1 - \sqrt{2}$ agree to nine decimal places, supporting (but not actually proving) the result in Example 12.

CONNECTIONS In calculus, sometimes it is useful to change a radical expression by rationalizing the *numerator*. This is done to avoid a zero denominator. For example, the expression

$$\frac{\sqrt{x} - 2}{x - 4} \text{ is rewritten as } \frac{1}{\sqrt{x} + 2} \text{ as follows.}$$

$$\frac{\sqrt{x} - 2}{x - 4} = \frac{\sqrt{x} - 2}{x - 4} \cdot \frac{\sqrt{x} + 2}{\sqrt{x} + 2} = \frac{x - 4}{(x - 4)(\sqrt{x} + 2)} = \frac{1}{\sqrt{x} + 2}$$

FOR DISCUSSION OR WRITING
1. Verify that if $x = 4$,

$$\frac{\sqrt{x} - 2}{x - 4} = \frac{0}{0}, \text{ which is undefined, but } \frac{1}{\sqrt{x} + 2} = \frac{1}{4}.$$

2. What number makes the following expression undefined? Find an equivalent expression that is defined for that number.

$$\frac{\sqrt{x} - \sqrt{6}}{x - 6}$$

1.7 Exercises ▼▼▼▼▼▼▼▼▼▼▼▼▼▼▼▼▼▼▼▼▼▼▼▼▼▼▼▼▼▼▼▼▼▼▼▼

Match the rational exponent expression in Column I with the equivalent radical expression in Column II. Assume that x is not zero. See Examples 2 and 3.

I

1. $(-3x)^{1/3}$

2. $-3x^{1/3}$

3. $(-3x)^{-1/3}$

4. $-3x^{-1/3}$

5. $(3x)^{1/3}$

6. $3x^{-1/3}$

7. $(3x)^{-1/3}$

8. $3x^{1/3}$

II

(a) $\dfrac{3}{\sqrt[3]{x}}$

(b) $-3\sqrt[3]{x}$

(c) $\dfrac{1}{\sqrt[3]{3x}}$

(d) $\dfrac{-3}{\sqrt[3]{x}}$

(e) $3\sqrt[3]{x}$

(f) $\sqrt[3]{-3x}$

(g) $\sqrt[3]{3x}$

(h) $\dfrac{1}{\sqrt[3]{-3x}}$

Write in radical form. Assume all variables represent positive real numbers. See Example 2.

9. $(-m)^{2/3}$ **10.** $p^{5/4}$ **11.** $(2m + p)^{2/3}$ **12.** $(5r + 3t)^{4/7}$

Write in exponential form. Assume all variables represent nonnegative real numbers. See Example 3.

13. $\sqrt[5]{k^2}$ **14.** $-\sqrt[4]{z^5}$ **15.** $-3\sqrt{5p^3}$ **16.** $m\sqrt{2y^5}$

17. What is wrong with the statement $\sqrt[3]{4} \cdot \sqrt[3]{4} = 4$?

18. Which of the following expressions is *not* simplified? Give the simplified form.

     (a) $\sqrt[3]{2y}$ **(b)** $\dfrac{\sqrt{5}}{2}$ **(c)** $\sqrt[4]{m^3}$ **(d)** $\sqrt{\dfrac{3}{4}}$

19. Explain how to rationalize the denominator of $\sqrt[3]{\dfrac{3}{2}}$.

20. How can we multiply $\sqrt{2}$ and $\sqrt[3]{2}$?

Simplify each radical expression. Assume that all variables represent positive real numbers. See Examples 1, 5, 6, 8, 10, and 11(a).

21. $\sqrt[3]{125}$ **22.** $\sqrt[4]{81}$ **23.** $\sqrt[5]{-3125}$ **24.** $\sqrt[3]{343}$ **25.** $\sqrt{50}$

26. $\sqrt{45}$ **27.** $\sqrt[3]{81}$ **28.** $\sqrt[3]{250}$ **29.** $-\sqrt[4]{32}$ **30.** $-\sqrt[4]{243}$

31. $-\sqrt{\dfrac{9}{5}}$ **32.** $-\sqrt[3]{\dfrac{3}{2}}$ **33.** $-\sqrt[3]{\dfrac{4}{5}}$ **34.** $\sqrt[4]{\dfrac{3}{2}}$ **35.** $\sqrt[3]{16(-2)^4(2)^8}$

36. $\sqrt[3]{25(3)^4(5)^3}$ **37.** $\sqrt{8x^5z^8}$ **38.** $\sqrt{24m^6n^5}$ **39.** $\sqrt[3]{16z^5x^8y^4}$ **40.** $-\sqrt[6]{64a^{12}b^8}$

41. $\sqrt[4]{m^2n^7p^8}$ **42.** $\sqrt[4]{x^8y^7z^9}$ **43.** $\sqrt[4]{x^4 + y^4}$ **44.** $\sqrt[3]{27 + a^3}$ **45.** $\sqrt{\dfrac{2}{3x}}$

46. $\sqrt{\dfrac{5}{3p}}$ **47.** $\sqrt{\dfrac{x^5 y^3}{z^2}}$ **48.** $\sqrt{\dfrac{g^3 h^5}{r^3}}$ **49.** $\sqrt[3]{\dfrac{8}{x^2}}$ **50.** $\sqrt[3]{\dfrac{9}{16p^4}}$

51. $\sqrt[4]{\dfrac{g^3 h^5}{9r^6}}$ **52.** $\sqrt[4]{\dfrac{32x^5}{y^5}}$ **53.** $\dfrac{\sqrt[3]{mn} \cdot \sqrt[3]{m^2}}{\sqrt[3]{n^2}}$ **54.** $\dfrac{\sqrt[3]{8m^2 n^3} \cdot \sqrt[3]{2m^2}}{\sqrt[3]{32m^4 n^3}}$ **55.** $\dfrac{\sqrt[4]{32x^5 y} \cdot \sqrt[4]{2xy^4}}{\sqrt[4]{4x^3 y^2}}$

56. $\dfrac{\sqrt[4]{rs^2 t^3} \cdot \sqrt[4]{r^3 s^2 t}}{\sqrt[4]{r^2 t^3}}$ **57.** $\sqrt[3]{\sqrt{4}}$ **58.** $\sqrt[4]{\sqrt[3]{2}}$

Would a calculator indicate that the display on the screen is true or false, assuming that the calculator is capable of returning an accurate answer? Support your answer, if you wish, by using your own calculator.

59.

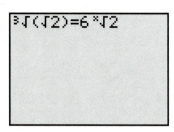

60.

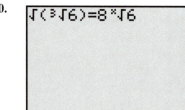

Simplify the expression, assuming that all variables represent nonnegative numbers. See Examples 7, 9, and 11(b).

61. $9\sqrt{8k} + 3\sqrt{18k} - \sqrt{32k}$ **62.** $2\sqrt[3]{3} + 4\sqrt[3]{24} - \sqrt[3]{81}$ **63.** $\sqrt[3]{32} - 5\sqrt[3]{4} + 2\sqrt[3]{108}$

64. $\dfrac{1}{\sqrt{3}} - \dfrac{2}{\sqrt{12}} + 2\sqrt{3}$ **65.** $\dfrac{1}{\sqrt{2}} + \dfrac{3}{\sqrt{8}} + \dfrac{1}{\sqrt{32}}$ **66.** $\dfrac{5}{\sqrt[3]{2}} - \dfrac{2}{\sqrt[3]{16}} + \dfrac{1}{\sqrt[3]{54}}$

67. $\dfrac{-4}{\sqrt[3]{3}} + \dfrac{1}{\sqrt[3]{24}} - \dfrac{2}{\sqrt[3]{81}}$ **68.** $(\sqrt{2} + 3)(\sqrt{2} - 3)$ **69.** $(\sqrt{5} + \sqrt{2})(\sqrt{5} - \sqrt{2})$

70. $(\sqrt[3]{11} - 1)(\sqrt[3]{11^2} + \sqrt[3]{11} + 1)$ **71.** $(\sqrt[3]{7} + 3)(\sqrt[3]{7^2} - 3\sqrt[3]{7} + 9)$ **72.** $(\sqrt{3} + \sqrt{8})^2$

73. $(\sqrt{2} - 1)^2$ **74.** $(3\sqrt{2} + \sqrt{3})(2\sqrt{3} - \sqrt{2})$ **75.** $(4\sqrt{5} - 1)(3\sqrt{5} + 2)$

76. Does the graphing calculator screen shown indicate that the number π is exactly equal to $\sqrt[4]{\dfrac{2143}{22}}$? Explain.

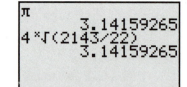

Rationalize the denominator of the radical expression. Assume that all variables represent nonnegative numbers and that no denominators are zero. See Example 12.

77. $\dfrac{\sqrt{3}}{\sqrt{5} + \sqrt{3}}$ **78.** $\dfrac{\sqrt{7}}{\sqrt{3} - \sqrt{7}}$ **79.** $\dfrac{1 + \sqrt{3}}{3\sqrt{5} + 2\sqrt{3}}$ **80.** $\dfrac{\sqrt{7} - 1}{2\sqrt{7} + 4\sqrt{2}}$

81. $\dfrac{p}{\sqrt{p} + 2}$ **82.** $\dfrac{\sqrt{r}}{3 - \sqrt{r}}$ **83.** $\dfrac{a}{\sqrt{a + b} - 1}$ **84.** $\dfrac{3m}{2 + \sqrt{m + n}}$

▼▼▼▼▼▼▼▼▼▼▼▼▼ **DISCOVERING CONNECTIONS** (Exercises 85–88) ▼▼▼▼▼▼▼▼▼▼▼▼▼

The chart below was first seen in Section 1.2. It describes the wind-chill factor for various wind speeds and temperatures.

Wind/°F	40°	30°	20°	10°	0°	−10°	−20°	−30°	−40°	−50°
5 mph	37	27	16	6	−5	−15	−26	−36	−47	−57
10 mph	28	16	4	−9	−21	−33	−46	−58	−70	−83
15 mph	22	9	−5	−18	−36	−45	−58	−72	−85	−99
20 mph	18	4	−10	−25	−39	−53	−67	−82	−96	−110
25 mph	16	0	−15	−29	−44	−59	−74	−88	−104	−118
30 mph	13	−2	−18	−33	−48	−63	−79	−94	−109	−125
35 mph	11	−4	−20	−35	−49	−67	−82	−98	−113	−129
40 mph	10	−6	−21	−37	−53	−69	−85	−100	−116	−132

Work Exercises 85–88 in order, so that you can see how an equation involving radicals can model these data. Use a calculator in Exercises 85 and 86.

85. Consider the expression $T - \left(\dfrac{v}{4} + 7\sqrt{v}\right)\left(1 - \dfrac{T}{90}\right)$ as a model, where T represents the temperature and v represents the wind velocity. Evaluate it for **(a)** $T = -10$ and $v = 30$, and **(b)** $T = -40$ and $v = 5$.

86. Consider the expression $91.4 - (91.4 - T)(.478 + .301\sqrt{v} - .02v)$ as a model, where once again T represents the temperature and v represents the wind velocity. Repeat parts (a) and (b) of Exercise 85.

87. Use the chart to find the wind-chill factors for the information in parts (a) and (b) of Exercise 85.

88. Based on your results in Exercises 85–87, make a conjecture about which formula models the wind-chill factor better.

Write the expression without radicals. Use absolute value if necessary. See Example 4.

89. $\sqrt{(m + n)^2}$

90. $\sqrt[4]{(a + 2b)^4}$

91. $\sqrt{z^2 - 6zx + 9x^2}$

92. $\sqrt[3]{(r + 2s)(r^2 + 4rs + 4s^2)}$

Refer to the Connections box following Example 12. Rationalize the numerator of each expression. Assume that all variables represent positive real numbers and that no denominator is equal to 0.

93. $\dfrac{1 + \sqrt{2}}{2}$

94. $\dfrac{1 - \sqrt{3}}{3}$

95. $\dfrac{\sqrt{x}}{1 + \sqrt{x}}$

96. $\dfrac{\sqrt{p}}{1 - \sqrt{p}}$

97. $\dfrac{\sqrt{x} + \sqrt{x + 1}}{\sqrt{x} - \sqrt{x + 1}}$

98. $\dfrac{\sqrt{p} + \sqrt{p^2 - 1}}{\sqrt{p} - \sqrt{p^2 - 1}}$

Now that we have discussed various topics involving real numbers and their properties, we investigate further the Greenland ice cap problem, first mentioned in the chapter opener.

99. The Greenland ice cap contains approximately 3 million cubic kilometers of ice. If the average global temperature increased significantly, this ice cap could melt and sea level would rise. Since over 200 million people live on land that is less than 1 meter above sea level, an increase in sea level could cause deaths, property damage, and major displacement of people. In this exercise you will estimate the rise in sea level if this cap were to melt and determine whether or not this event would have a significant impact on people.

(a) The surface area of a sphere is given by the expression $4\pi r^2$ where r is its radius. Although the shape of the Earth is not exactly spherical, it has an average diameter of approximately 12,742 kilometers. Estimate the surface area of the Earth.

(b) Water covers approximately 71% of the total surface area of the Earth. Assuming that the oceans account for essentially all of this area, how many square kilometers of the Earth are covered by oceans?

(c) When ice melts it does not turn into an equivalent volume of water. Instead, its volume shrinks by approximately 8%. (This is why ice floats.) Calculate the equivalent volume of water that is contained in the Greenland ice cap.

(d) Approximate the potential rise in sea level by dividing the total volume of the water from the ice cap by the surface area of the oceans.

(e) Discuss the implications of your calculation. How would cities like Boston, New Orleans, and San Diego be affected?

(f) Inland seas and fresh-water lakes were not taken into account in calculating the area of the oceans. If they had been accounted for, how would this have affected your final answer? Try to determine whether or not their effect would be significant to your result.

1.8   Complex Numbers ▼▼▼

So far, we have worked only with real numbers in this book. The set of real numbers, however, does not include all the numbers needed in algebra. For example, with real numbers alone, it is not possible to find a number whose square is -1.

To extend the real number system to include numbers such as $\sqrt{-1}$, the new number i is defined to have the property

$$i^2 = -1.$$

Thus, $i =$ the square root of -1. The number i is called the **imaginary unit.**

> **CONNECTIONS** Although as early as 50 B.C. square roots of negative numbers were known, they were not incorporated into an integrated number system until much later. Italian mathematician Girolamo Cardano (1501–1576) and others computed with them. As algebra became necessary in many applications, mathematicians distinguished between two types of solutions. Eventually these were named "real" and "imaginary" by René Descartes (1596–1650).
>
> The German mathematician Gottfried Leibniz, one of the founders of calculus, wrote to his Dutch colleague Christiaan Huygens in 1679 of the need for an expanded number system: "I have no hope that we can get very

far in physics until we have found some method of abridgment." The void was filled by the complex numbers, which combine real and imaginary numbers. It was the renowned Leonhard Euler in 1748 who first wrote $\sqrt{-1} = i$, and in 1832 Carl Friedrich Gauss stated that numbers of the form $a + bi$ are "complex." The first application of the imaginary numbers to geometry came in 1796 when the Norwegian Caspar Wessel used them in surveying. Charles Steinmetz (1865–1923), an electrical engineer, is said to have "generated electricity with the square root of minus one" when he used complex numbers to develop a theory of alternating currents. Today complex numbers are used extensively in science and engineering.

Numbers of the form $a + bi$, where a and b are real numbers, are called **complex numbers.** In the complex number $a + bi$, a is called the **real part** and b is called the **imaginary part.*** Since a complex number involves two real numbers, a and b, we can think of complex numbers as two dimensional, where real numbers are one dimensional.

Each real number is a complex number, since a real number a may be thought of as the complex number $a + 0i$. A complex number of the form $a + bi$, where b is nonzero, is called an **imaginary number.** Both the set of real numbers and the set of imaginary numbers are subsets of the set of complex numbers. (See Figure 9, which is an extension of Figure 5 in Section 1.1.) A complex number that is written in the form $a + bi$ or $a + ib$ is in **standard form.** (The form $a + ib$ is used to simplify symbols such as $i\sqrt{5}$, since $\sqrt{5}i$ could be too easily mistaken for $\sqrt{5i}$.)

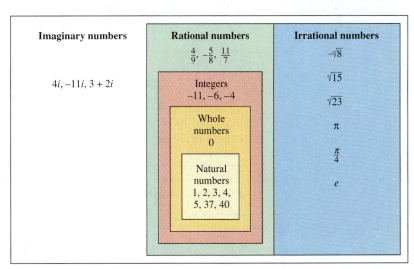

Complex Numbers (Real numbers are shaded.)

FIGURE 9

*In some texts, the term bi is defined to be the imaginary part.

EXAMPLE 1
Identifying kinds of complex numbers

The following statements identify different kinds of complex numbers.

(a) -8, $\sqrt{7}$, and π are real numbers and complex numbers.

(b) $3i$, $-11i$, $i\sqrt{14}$, and $5 + i$ are imaginary numbers and complex numbers. ▶

EXAMPLE 2
Writing complex numbers in standard form

The list below shows several numbers, along with the standard form of each number.

Number	Standard Form
$6i$	$6i$ or $0 + 6i$
-9	-9 or $-9 + 0i$
0	0 or $0 + 0i$
$-i + 2$	$2 - i$
$8 + i\sqrt{3}$	$8 + i\sqrt{3}$

▶

Many of the solutions to quadratic equations in a later chapter will involve expressions such as $\sqrt{-a}$, for a positive real number a, defined as follows.

Definition of $\sqrt{-a}$

If $a > 0$, then
$$\sqrt{-a} = i\sqrt{a}.$$

EXAMPLE 3
Writing $\sqrt{-a}$ as $i\sqrt{a}$

Write each expression as the product of i and a real number.

(a) $\sqrt{-16} = i\sqrt{16} = 4i$

(b) $\sqrt{-70} = i\sqrt{70}$ ▶

Products or quotients with negative radicands are simplified by first rewriting $\sqrt{-a}$ as $i\sqrt{a}$ for positive numbers a. Then the properties of real numbers can be applied, together with the fact that $i^2 = -1$.

CAUTION When working with negative radicands, *use the definition $\sqrt{-a} = i\sqrt{a}$ before using any of the other rules for radicals.* In particular, the rule $\sqrt{c} \cdot \sqrt{d} = \sqrt{cd}$ is valid only when c and d are *not* both negative. For example,

$$\sqrt{(-4)(-9)} = \sqrt{36} = 6,$$

while
$$\sqrt{-4} \cdot \sqrt{-9} = 2i(3i) = 6i^2 = -6,$$

so
$$\sqrt{-4} \cdot \sqrt{-9} \neq \sqrt{(-4)(-9)}.$$

EXAMPLE 4
Finding products and quotients
involving negative radicands

Multiply or divide as indicated.

(a) $\sqrt{-7} \cdot \sqrt{-7} = i\sqrt{7} \cdot i\sqrt{7}$
$\qquad = i^2 \cdot (\sqrt{7})^2$
$\qquad = (-1) \cdot 7 \qquad i^2 = -1$
$\qquad = -7$

(b) $\sqrt{-6} \cdot \sqrt{-10} = i\sqrt{6} \cdot i\sqrt{10}$
$\qquad = i^2 \cdot \sqrt{60}$
$\qquad = -1 \cdot 2\sqrt{15}$
$\qquad = -2\sqrt{15}$

(c) $\dfrac{\sqrt{-20}}{\sqrt{-2}} = \dfrac{i\sqrt{20}}{i\sqrt{2}} = \sqrt{\dfrac{20}{2}} = \sqrt{10}$

(d) $\dfrac{\sqrt{-48}}{\sqrt{24}} = \dfrac{i\sqrt{48}}{\sqrt{24}} = i\sqrt{2}$ ▶

OPERATIONS ON COMPLEX NUMBERS With the definitions $i^2 = -1$ and $\sqrt{-a} = i\sqrt{a}$, if $a > 0$, all the properties of real numbers can be extended to the complex numbers. As a result, complex numbers are added, subtracted, multiplied, and divided as shown by the following definitions and examples.

The *sum* and *difference* of two complex numbers $a + bi$ and $c + di$ are defined as follows.

Definition of Addition and Subtraction

For complex numbers $a + bi$ and $c + di$,

$$(a + bi) + (c + di) = (a + c) + (b + d)i$$

and $\quad (a + bi) - (c + di) = (a - c) + (b - d)i.$

EXAMPLE 5
Adding and subtracting
complex numbers

Add or subtract as indicated.

(a) $(3 - 4i) + (-2 + 6i) = [3 + (-2)] + [-4 + 6]i = 1 + 2i$

(b) $(-9 + 7i) + (3 - 15i) = -6 - 8i$

(c) $(-4 + 3i) - (6 - 7i) = (-4 - 6) + [3 - (-7)]i = -10 + 10i$

(d) $(12 - 5i) - (8 - 3i) = 4 - 2i$ ▶

The product of two complex numbers can be found by multiplying as if the numbers were binomials and using the fact that $i^2 = -1$, as follows.

$$(a + bi)(c + di) = ac + adi + bic + bidi$$
$$= ac + adi + bci + bdi^2$$
$$= ac + (ad + bc)i + bd(-1)$$
$$(a + bi)(c + di) = (ac - bd) + (ad + bc)i$$

Based on this result, the *product* of the complex numbers $a + bi$ and $c + di$ is defined in the following way.

> **Multiplication of Complex Numbers**
> $$(a + bi)(c + di) = (ac - bd) + (ad + bc)i$$

This definition is not practical to use. To find a given product, it is easier just to multiply as with binomials.

EXAMPLE 6
Multiplying complex numbers

Find each of the following products.

(a) $(2 - 3i)(3 + 4i) = 2(3) + 2(4i) - 3i(3) - 3i(4i)$
$\qquad\qquad\qquad\qquad = 6 + 8i - 9i - 12i^2$
$\qquad\qquad\qquad\qquad = 6 - i - 12(-1)$ $\qquad i^2 = -1$
$\qquad\qquad\qquad\qquad = 18 - i$

(b) $(5 - 4i)(7 - 2i) = 5(7) + 5(-2i) - 4i(7) - 4i(-2i)$
$\qquad\qquad\qquad\qquad = 35 - 10i - 28i + 8i^2$
$\qquad\qquad\qquad\qquad = 35 - 38i + 8(-1)$
$\qquad\qquad\qquad\qquad = 27 - 38i$

(c) $(6 + 5i)(6 - 5i) = 6^2 - 25i^2$ Product of the sum and difference of two terms

$\qquad\qquad\qquad\qquad = 36 - 25(-1)$ $\qquad i^2 = -1$
$\qquad\qquad\qquad\qquad = 36 + 25$
$\qquad\qquad\qquad\qquad = 61 \quad$ or $\quad 61 + 0i$ Standard form

(d) $(4 + 3i)^2 = 4^2 + 2(4)(3i) + (3i)^2$ Square of a binomial
$\qquad\qquad\quad = 16 + 24i + (-9)$
$\qquad\qquad\quad = 7 + 24i$ ▶

Powers of i can be simplified using the facts that $i^2 = -1$ and $i^4 = 1$. The next example shows how this is done.

EXAMPLE 7
Simplifying powers of i

Simplify each power of i.

(a) i^{15}

Since $i^2 = -1$, the value of a power of i is found by writing the given power as a product involving i^2 or i^4. For example, $i^3 = i^2 \cdot i = (-1) \cdot i = -i$. Also, $i^4 = i^2 \cdot i^2 = (-1)(-1) = 1$. Using i^4 and i^3 to rewrite i^{15} gives

$$i^{15} = i^{12} \cdot i^3 = (i^4)^3 \cdot i^3 = (1)^3(-i) = -i.$$

(b) $i^{-3} = i^{-4} \cdot i = (i^4)^{-1} \cdot i = (1)^{-1} \cdot i = i$ ▶

We can use the method of Example 7 to construct the following table of powers of i.

Powers of i

$$i^1 = i \qquad i^5 = i \qquad i^9 = i$$
$$i^2 = -1 \qquad i^6 = -1 \qquad i^{10} = -1$$
$$i^3 = -i \qquad i^7 = -i \qquad i^{11} = -i$$
$$i^4 = 1 \qquad i^8 = 1 \qquad i^{12} = 1, \qquad \text{and so on.}$$

Example 6(c) showed that $(6 + 5i)(6 - 5i) = 61$. The numbers $6 + 5i$ and $6 - 5i$ differ only in their middle signs; for this reason these numbers are called **conjugates** of each other. The product of a complex number and its conjugate is always a real number.

Property of Complex Conjugates

For real numbers a and b:

$$(a + bi)(a - bi) = a^2 + b^2.$$

EXAMPLE 8
Examining conjugates and their products

The following list shows several pairs of conjugates, together with their products.

Number	Conjugate	Product
$3 - i$	$3 + i$	$(3 - i)(3 + i) = 9 + 1 = 10$
$2 + 7i$	$2 - 7i$	$(2 + 7i)(2 - 7i) = 53$
$-6i$	$6i$	$(-6i)(6i) = 36$

The conjugate of the divisor is used to find the *quotient* of two complex numbers. The quotient is found by multiplying both the numerator and the denominator by the conjugate of the denominator. The result should be written in standard form.

EXAMPLE 9
Dividing complex numbers

(a) Find $\dfrac{3 + 2i}{5 - i}$.

Multiply numerator and denominator by the conjugate of $5 - i$.

$$\frac{3 + 2i}{5 - i} = \frac{(3 + 2i)(\mathbf{5 + i})}{(5 - i)(\mathbf{5 + i})}$$

$$= \frac{15 + 3i + 10i + 2i^2}{25 - i^2} \qquad \text{Multiply.}$$

$$= \frac{13 + 13i}{26} \qquad i^2 = -1$$

$$= \frac{13}{26} + \frac{13i}{26} \qquad \frac{a + bi}{c} = \frac{a}{c} + \frac{bi}{c}$$

$$= \frac{1}{2} + \frac{1}{2}i \qquad \text{Lowest terms; standard form}$$

To check this answer, show that

$$(5 - i)\left(\frac{1}{2} + \frac{1}{2}i\right) = 3 + 2i.$$

(b) $\dfrac{3}{i} = \dfrac{3(-i)}{i(-i)}$ *−i is the conjugate of i.*

$\quad\quad = \dfrac{-3i}{-i^2}$

$\quad\quad = \dfrac{-3i}{1}$ $-i^2 = -(-1) = 1$

$\quad\quad = -3i$ or $0 - 3i$ Standard form ▶

> 🖩 The more advanced graphing calculators can work with complex numbers. The TI-83 represents them in the form $a + bi$, and the TI-85 represents them as ordered pairs. If your calculator has this ability, refer to your manual for details.

1.8 Exercises ▼▼

1. Discuss the following true statement: A complex number may be a real number.

2. Discuss the following true statement: A complex number might not be an imaginary number.

Identify the number as real or imaginary. See Example 1.

3. -5 **4.** π **5.** $i\sqrt{6}$

6. $-3i$ **7.** $2 + 5i$ **8.** $-7 - 6i$

Write the number without a negative radicand. See Examples 3 and 4.

9. $\sqrt{-100}$ **10.** $\sqrt{-169}$ **11.** $-\sqrt{-400}$ **12.** $-\sqrt{-225}$

13. $-\sqrt{-39}$ **14.** $-\sqrt{-95}$ **15.** $5 + \sqrt{-4}$ **16.** $-7 + \sqrt{-100}$

17. $9 - \sqrt{-50}$ **18.** $-11 - \sqrt{-24}$ **19.** $\sqrt{-5} \cdot \sqrt{-5}$ **20.** $\sqrt{-20} \cdot \sqrt{-20}$

21. $\dfrac{\sqrt{-40}}{\sqrt{-10}}$ **22.** $\dfrac{\sqrt{-190}}{\sqrt{-19}}$

Add or subtract. Write the result in standard form. See Example 5.

23. $(3 + 2i) + (4 - 3i)$ **24.** $(4 - i) + (2 + 5i)$

25. $(-2 + 3i) - (-4 + 3i)$ **26.** $(-3 + 5i) - (-4 + 3i)$

27. $(2 - 5i) - (3 + 4i) - (-2 + i)$ **28.** $(-4 - i) - (2 + 3i) + (-4 + 5i)$

Multiply. Write the result in standard form. See Example 6.

29. $(2 + 4i)(-1 + 3i)$ **30.** $(1 + 3i)(2 - 5i)$ **31.** $(-3 + 2i)^2$ **32.** $(2 + i)^2$

33. $(2 + 3i)(2 - 3i)$ **34.** $(6 - 4i)(6 + 4i)$ **35.** $(\sqrt{6} + i)(\sqrt{6} - i)$ **36.** $(\sqrt{2} - 4i)(\sqrt{2} + 4i)$

37. $i(3 - 4i)(3 + 4i)$ **38.** $i(2 + 7i)(2 - 7i)$

Find each power of i. See Example 7.

39. i^5 **40.** i^8 **41.** i^9 **42.** i^{11} **43.** i^{12} **44.** i^{25}

45. i^{43} **46.** $\dfrac{1}{i^9}$ **47.** $\dfrac{1}{i^{12}}$ **48.** i^{-6} **49.** i^{-15} **50.** i^{-49}

51. Suppose that your friend, Anne Kelly, tells you that she has discovered a method of simplifying a positive power of i. "Just divide the exponent by 4," she says, "and then look at the remainder. Then refer to the table of powers of i in this section. The large power of i is equal to i to the power indicated by the remainder. And if the remainder is 0, the result is $i^0 = 1$." Explain why Anne's method works.

52. Explain why the following method of simplifying i^{-46} works.

$$i^{-46} = i^{-46} \cdot i^{48} = i^{-46+48} = i^2 = -1$$

Divide. Write the result in standard form. See Example 9.

53. $\dfrac{1+i}{1-i}$ **54.** $\dfrac{2-i}{2+i}$ **55.** $\dfrac{4-3i}{4+3i}$ **56.** $\dfrac{5-2i}{6-i}$ **57.** $\dfrac{3-4i}{2-5i}$

58. $\dfrac{1-3i}{1+i}$ **59.** $\dfrac{-3+4i}{2-i}$ **60.** $\dfrac{5+6i}{5-6i}$ **61.** $\dfrac{2}{i}$ **62.** $\dfrac{-7}{3i}$

63. Show that $\dfrac{\sqrt{2}}{2} + \dfrac{\sqrt{2}}{2}i$ is a square root of i.

64. Show that $\dfrac{\sqrt{3}}{2} + \dfrac{1}{2}i$ is a cube root of i.

65. Evaluate $3z - z^2$ if $z = 3 - 2i$.

66. Evaluate $-2z + z^3$ if $z = -6i$.

▼▼▼▼▼▼▼▼▼▼▼▼▼ **DISCOVERING CONNECTIONS** (Exercises 67–70) ▼▼▼▼▼▼▼▼▼▼▼▼▼

In Section 1.3 we saw how to expand a binomial using Pascal's triangle. The pattern also applies to raising a complex number of the form $a + bi$ to a positive integer power n. For example, since

$$(x + y)^3 = x^3 + 3x^2y + 3xy^2 + y^3,$$

replacing x with 2 and y with i allows us to expand $(2 + i)^3$:

$$(2 + i)^3 = 2^3 + 3(2)^2i + 3(2)i^2 + i^3.$$

Knowing how to simplify large powers of i allows us to simplify the right side of the equation above. We obtain

$$8 + 3(4)i + 6(-1) + (-i)$$
$$= 8 - 6 + 12i - i$$
$$= 2 + 11i.$$

Work Exercises 67–70 in order, so that you can see how this method yields the same result as direct multiplication.

67. Why is $(2 + i)^3 = (2 + i)^2(2 + i)$ a true statement?

68. Square $2 + i$ by using the method described in this section.

69. Multiply your result in Exercise 68 by $2 + i$. What is the product? Does it agree with the result obtained in the discussion above?

70. Use Pascal's triangle to expand and then simplify $(1 + i)^6$.

Chapter 1 Summary ▼▼▼▼▼▼▼▼▼▼▼▼▼▼▼▼▼▼▼▼▼▼▼▼▼▼▼▼▼▼▼

KEY TERMS	KEY IDEAS
1.1 The Real Numbers	

KEY TERMS	KEY IDEAS

exponent
base
identity element for addition
   (additive identity)
identity element for multiplication
   (multiplicative identity)
additive inverse (negative)
multiplicative inverse (reciprocal)
	Sets of Numbers
	
	Real Numbers
	$$\{x \mid x \text{ corresponds to a point on a number line}\}$$
	Integers
	$$\{\ldots, -3, -2, -1, 0, 1, 2, 3, \ldots\}$$
	Rational Numbers
	$$\left\{\frac{p}{q} \;\middle|\; p \text{ and } q \text{ are integers and } q \neq 0\right\}$$
	Irrational Numbers
	$$\{x \mid x \text{ is real but not rational}\}$$
	Whole Numbers
	$$\{0, 1, 2, 3, 4, \ldots\}$$
	Natural Numbers
	$$\{1, 2, 3, 4, \ldots\}$$
	Properties of the Real Numbers
	For all real numbers a, b, and c:
	
	Commutative Properties
	$$a + b = b + a$$
	$$ab = ba$$
	Associative Properties
	$$(a + b) + c = a + (b + c)$$
	$$(ab)c = a(bc)$$
	Identity Properties
	There exists a unique real number 0 such that
	$$a + 0 = a \quad \text{and} \quad 0 + a = a.$$
	There exists a unique real number 1 such that
	$$a \cdot 1 = a \quad \text{and} \quad 1 \cdot a = a.$$
	Inverse Properties
	There exists a unique real number $-a$ such that
	$$a + (-a) = 0 \quad \text{and} \quad (-a) + a = 0.$$
	If $a \neq 0$, there exists a unique real number $1/a$ such that
	$$a \cdot \frac{1}{a} = 1 \quad \text{and} \quad \frac{1}{a} \cdot a = 1.$$
	Distributive Property
	$$a(b + c) = ab + ac$$

KEY TERMS	KEY IDEAS
1.2 Order and Absolute Value	
coordinate inequalities absolute value	$a > b$ if a is to the right of b or if $a - b$ is positive. $a < b$ if a is to the left of b or if $a - b$ is negative. $$\lvert a \rvert = \begin{cases} a & \text{if } a \geq 0 \\ -a & \text{if } a < 0 \end{cases}$$
1.3 Polynomials; The Binomial Theorem	
algebraic expressions trinomial term binomial coefficient monomial like terms Pascal's triangle polynomial n-factorial degree binomial coefficient	**Special Products** $$(x + y)(x - y) = x^2 - y^2$$ $$(x + y)^2 = x^2 + 2xy + y^2$$ $$(x - y)^2 = x^2 - 2xy + y^2$$
1.4 Factoring	
factoring factor factored form prime (irreducible) polynomial	**Factoring Patterns** $$x^2 - y^2 = (x + y)(x - y)$$ $$x^2 + 2xy + y^2 = (x + y)^2$$ $$x^2 - 2xy + y^2 = (x - y)^2$$ $$x^3 - y^3 = (x - y)(x^2 + xy + y^2)$$ $$x^3 + y^3 = (x + y)(x^2 - xy + y^2)$$
1.5 Rational Expressions	
rational expression complex fraction	
1.6 Rational Exponents	
	Rules for Exponents Let r and s be rational numbers. The results below are valid for all positive numbers a and b. $a^r \cdot a^s = a^{r+s}$ $(ab)^r = a^r \cdot b^r$ $(a^r)^s = a^{rs}$ $\dfrac{a^r}{a^s} = a^{r-s}$ $\left(\dfrac{a}{b}\right)^r = \dfrac{a^r}{b^r}$ $a^{-r} = \dfrac{1}{a^r}$

KEY TERMS	KEY IDEAS
1.7 Radicals	

radical sign
radicand
index
rationalizing the denominator
conjugate
	Radical Notation
	If a is a real number, n is a positive integer, and $a^{1/n}$ is defined, then
	$$\sqrt[n]{a} = a^{1/n}.$$
	If m is an integer, n is a positive integer, and a is a real number for which $\sqrt[n]{a}$ is defined, then
	$$a^{m/n} = (\sqrt[n]{a})^m = \sqrt[n]{a^m}.$$

1.8 Complex Numbers	
imaginary unit complex number real part imaginary part imaginary number conjugates	**Definition of i** $$i^2 = -1 \qquad \text{or} \qquad i = \sqrt{-1}$$ **Definition of $\sqrt{-a}$** For $a > 0$, $\sqrt{-a} = i\sqrt{a}$.

Chapter 1   Review Exercises ▼▼▼▼▼▼▼▼▼▼▼▼▼▼▼▼▼▼▼▼▼▼▼▼▼▼

Let set $K = \{-12, -6, -.9, -\sqrt{7}, -\sqrt{4}, 0, 1/8, \pi/4, 6, \sqrt{11}\}$. List the elements of K that are members of the set named.

1. Integers

2. Rational numbers

3. Irrational numbers

For Exercises 4–7 choose all words from the following list that apply.
(a) *natural number*   **(b)** *whole number*   **(c)** *integer*   **(d)** *rational number*
(e) *irrational number*   **(f)** *real number*

4. 0

5. $-\sqrt{36}$

6. $\dfrac{16}{17}$

7. $\dfrac{4\pi}{5}$

Use the order of operations to simplify the expression.

8. $-4 - [2 - (-3^2)]$

9. $[2^3 - (-5)] - 2^2$

10. $(-4 - 1)(-3 - 5) - 2^3$

11. $(6 - 9)(-2 - 7) - (-4)$

12. $\left(-\dfrac{5}{9} - \dfrac{2}{3}\right) - \dfrac{5}{6}$

13. $\left(-\dfrac{2^3}{5} - \dfrac{3}{4}\right) - \left(-\dfrac{1}{2}\right)$

14. $\dfrac{6(-4) - 3^2(-2)^3}{-5[-2 - (-6)]}$

15. $\dfrac{(-7)(-3) - (-2^3)(-5)}{(-2^2 - 2)(-1 - 6)}$

Evaluate the expression if $a = -1$, $b = -2$, and $c = 4$.

16. $9a - 5b + 4c$

17. $-4(2a - 5b)$

18. $\dfrac{9a + 2b}{a + b + c}$

19. Suppose your friend missed class when the order of operations was discussed. Explain to her how to work the following problem. Evaluate $\dfrac{4a - 5c}{2a + 3b}$ if $a = -1$, $b = -2$, and $c = 4$.

20. The table lists the wave heights produced in the ocean for various wind speeds and fetches. (The *fetch* is the distance or stretch of water that the wind blows over.)

Wind Speed	*Fetch*			
mph	*11.5 mi*	*115 mi*	*575 mi*	*1150 mi*
11.5	2 ft	2 ft	2 ft	2 ft
17.3	3 ft	5 ft	5 ft	5 ft
23.0	4 ft	8 ft	9 ft	9 ft
34.5	6 ft	16 ft	19 ft	20 ft
46.0	8 ft	23 ft	33 ft	34 ft
57.5	10 ft	30 ft	47 ft	51 ft

(*Source:* Navarra, J., *Atmosphere, Weather and Climate*. W. B. Saunders Company, 1979.)

(a) What is the expected wave height if a 34.5-mile-per-hour wind blows across a fetch of 115 miles?

(b) If the wave height is 8 feet can you determine the speed of the wind and the fetch?

(c) Describe the relationship between wave height, wind speed, and fetch.

Identify the properties illustrated in the equation.

21. $8(5 + 9) = (5 + 9)8$

22. $4 \cdot 6 + 4 \cdot 12 = 4(6 + 12)$

23. $3 \cdot (4 \cdot 2) = (3 \cdot 4) \cdot 2$

24. $-8 + 8 = 0$

25. $(9 + p) + 0 = 9 + p$

Use the distributive property to rewrite sums as products and products as sums.

26. $-11(m - n)$

27. $k(r + s - t)$

28. $mn - 9m$

Write the numbers in numerical order from smallest to largest.

29. $|6 - 4|, -|-2|, |8 + 1|, -|3 - (-2)|$

30. $\sqrt{7}, -\sqrt{8}, -|\sqrt{16}|, |-\sqrt{12}|$

Write without absolute value bars.

31. $-|-6| + |3|$

32. $7 - |-8|$

33. $|\sqrt{8} - 3|$

34. $|\sqrt{97} - 8|$

35. $|m - 3|$ if $m > 3$

36. $|-6 - x^2|$

37. $|\pi - 4|$

38. $|3 + 5k|$ if $k < -3/5$

Perform the indicated operations.

39. $(3q^3 - 9q^2 + 6) + (4q^3 - 8q + 3)$

40. $2(3y^6 - 9y^2 + 2y) - (5y^6 - 10y^2 - 4y)$

41. $(8y - 7)(2y + 7)$

42. $(2r + 11s)(4r - 9s)$

43. $(3k - 5m)^2$

44. $(4a - 3b)^2$

The accompanying bar graph depicts the number of North American users, in millions, of on-line services, as reported by Forrester Research, Inc. Using these figures, it can be determined that the polynomial

$$.035x^4 - .266x^3 + 1.005x^2 + .509x + 2.986$$

*gives a good approximation of the number of users in the year x, where x = 0 corresponds to 1992, x = 1 corresponds to 1993, and so on. For the given year (**a**) use the bar graph to determine the number of users and then (**b**) use the polynomial to determine the number of users.*

45. 1992   **46.** 1993

47. 1994   **48.** 1995

Number of Users of On-Line Services in North America

In millions

19.5
13.2
9.1
6.5
3.0 4.2

'92 '93 '94 '95 '96 '97

Source: Forrester Research Inc.

Use Pascal's triangle to expand.

49. $(x + 2y)^4$

50. $\left(\dfrac{k}{2} - g\right)^5$

Perform each division.

51. $\dfrac{72r^2 + 59r + 12}{8r + 3}$

52. $\dfrac{30m^3 - 9m^2 + 22m + 5}{5m + 1}$

53. $\dfrac{5m^3 - 7m^2 + 14}{m^2 - 2}$

54. $\dfrac{3b^3 - 8b^2 + 12b - 30}{b^2 + 4}$

Factor as completely as possible.

55. $7z^2 - 9z^3 + z$

56. $3(z - 4)^2 + 9(z - 4)^3$

57. $r^2 + rp - 42p^2$

58. $z^2 - 6zk - 16k^2$

59. $6m^2 - 13m - 5$

60. $48a^8 - 12a^7b - 90a^6b^2$

61. $169y^4 - 1$

62. $49m^8 - 9n^2$

63. $8y^3 - 1000z^6$

64. $6(3r - 1)^2 + (3r - 1) - 35$

65. $ar - 3as + 5rb - 15sb$

66. $15mp + 9mq - 10np - 6nq$

67. $(16m^2 - 56m + 49) - 25a^2$

68. A student factored $64a^3 + 8b^3$ as $(4a + 2b)(16a^2 - 8ab + 4b^2)$. Is the polynomial factored completely? Explain.

69. Which one of the following is equal to $\dfrac{2a + b}{4a^2 - b^2}$?

(a) $\dfrac{1}{2a - b}$ (b) $\dfrac{2}{2a - b}$ (c) $\dfrac{1}{2ab}$ (d) $-\dfrac{1}{2ab}$

Perform each operation.

70. $\dfrac{k^2 + k}{8k^3} \cdot \dfrac{4}{k^2 - 1}$

71. $\dfrac{3r^3 - 9r^2}{r^2 - 9} \div \dfrac{8r^3}{r + 3}$

72. $\dfrac{x^2 + x - 2}{x^2 + 5x + 6} \div \dfrac{x^2 + 3x - 4}{x^2 + 4x + 3}$

73. $\dfrac{27m^3 - n^3}{3m - n} \div \dfrac{9m^2 + 3mn + n^2}{9m^2 - n^2}$

74. $\dfrac{p^2 - 36q^2}{(p - 6q)^2} \cdot \dfrac{p^2 - 5pq - 6q^2}{p^2 - 6pq + 36q^2} \div \dfrac{5p}{p^3 + 216q^3}$

75. $\dfrac{1}{4y} + \dfrac{8}{5y}$

76. $\dfrac{m}{4 - m} + \dfrac{3m}{m - 4}$

77. $\dfrac{3}{x^2 - 4x + 3} - \dfrac{2}{x^2 - 1}$

78. $\dfrac{\dfrac{1}{p} + \dfrac{1}{q}}{1 - \dfrac{1}{pq}}$

79. $\dfrac{3 + \dfrac{2m}{m^2 - 4}}{\dfrac{5}{m - 2}}$

80. The largest known star is Alpha Orionis, or Betelgeuse, which is the top left star in the constellation Orion. Its diameter is 400 million miles compared to our sun with a diameter of .8 million miles. If the volume of a sphere is given by $(4/3)\pi r^3$ where r is the radius, how many times greater is the volume of Alpha Orionis than our sun? (*Hint:* Find the quotient of the volumes.) (*Source: The Guinness Book of Records 1995.*)

Simplify the expression. Write results with only positive exponents. Assume that all variables represent positive real numbers.

81. 2^{-6}

82. -3^{-2}

83. $\left(-\dfrac{5}{4}\right)^{-2}$

84. $3^{-1} - 4^{-1}$

85. $(5z^3)(-2z^5)$

86. $(8p^2q^3)(-2p^5q^{-4})$

87. $(-6p^5w^4m^{12})^0$

88. $(-6x^2y^{-3}z^2)^{-2}$

89. $\dfrac{-8y^7p^{-2}}{y^{-4}p^{-3}}$

90. $\dfrac{a^{-6}(a^{-8})}{a^{-2}(a^{11})}$

91. $\dfrac{(p + q)^4(p + q)^{-3}}{(p + q)^6}$

92. $\dfrac{[p^2(m + n)^3]^{-2}}{p^{-2}(m + n)^{-5}}$

93. $(7r^{1/2})(2r^{3/4})(-r^{1/6})$

94. $(a^{3/4}b^{2/3})(a^{5/8}b^{-5/6})$

95. $\dfrac{y^{5/3} \cdot y^{-2}}{y^{-5/6}}$

96. $\left(\dfrac{25m^3n^5}{m^{-2}n^6}\right)^{-1/2}$

Find each product. Assume that all variables represent positive real numbers.

97. $2z^{1/3}(5z^2 - 2)$

98. $-m^{3/4}(8m^{1/2} + 4m^{-3/2})$

99. $(p + p^{1/2})(3p - 5)$

100. $(m^{1/2} - 4m^{-1/2})^2$

Simplify. Assume that all variables represent positive numbers.

101. $\sqrt{200}$

102. $\sqrt[3]{16}$

103. $\sqrt[4]{1250}$

104. $-\sqrt{\dfrac{16}{3}}$

105. $-\sqrt[3]{\dfrac{2}{5p^2}}$

106. $\sqrt{\dfrac{2^7y^8}{m^3}}$

107. $\sqrt[4]{\sqrt[3]{m}}$

108. $\dfrac{\sqrt[4]{8p^2q^5} \cdot \sqrt[4]{2p^3q}}{\sqrt[4]{p^5q^2}}$

109. $(\sqrt[3]{2} + 4)(\sqrt[3]{2^2} - 4\sqrt[3]{2} + 16)$

110. $\dfrac{3}{\sqrt{5}} - \dfrac{2}{\sqrt{45}} + \dfrac{6}{\sqrt{80}}$

111. $\sqrt{18m^3} - 3m\sqrt{32m} + 5\sqrt{m^3}$

112. $\dfrac{2}{7 - \sqrt{3}}$

113. $\dfrac{6}{3 - \sqrt{2}}$

114. $\dfrac{k}{\sqrt{k} - 3}$

115. $\dfrac{\sqrt{x} - \sqrt{x - 2}}{\sqrt{x} + \sqrt{x - 2}}$

Write as a product of i and a real number.

116. $\sqrt{-3}$

117. $\sqrt{-49}$

118. $-\sqrt{-100}$

Perform each operation. Write the result in standard form.

119. $(6 - i) + (4 - 2i)$ **120.** $(-11 + 2i) - (5 - 7i)$ **121.** $15i - (3 + 2i) - 5$ **122.** $-6 + 4i - (5i - 2)$

123. $(5 - i)(3 + 4i)$ **124.** $(8 - 2i)(1 - i)$ **125.** $(5 - 11i)(5 + 11i)$ **126.** $(7 + 6i)(7 - 6i)$

127. $(4 - 3i)^2$ **128.** $(1 + 7i)^2$ **129.** $\dfrac{6 + i}{1 - i}$ **130.** $\dfrac{5 + 2i}{5 - 2i}$

131. The product of a complex number and its conjugate is always a _____ number.

132. True or false: A real number is a complex number.

Find the power of i.

133. i^7

134. i^{150}

135. i^{-35}

136. i^{-20}

Chapter 1   Test ▼▼▼

1. Let $A = \{-13, -12/4, 0, 3/5, \pi/4, 5.9, \sqrt{49}\}$. List the elements of A that belong to the given set.
   (a) Integers   **(b)** Rational numbers   **(c)** Real numbers

2. Evaluate the expression if $x = -2$, $y = -4$, and $z = 5$: $\left| \dfrac{x^2 + 2yz}{3(x + z)} \right|$.

3. Identify the property illustrated. Let a, b, and c represent any real numbers.
   (a) $a + (b + c) = (a + b) + c$ **(b)** $a + (c + b) = a + (b + c)$
   (c) $a(b + c) = ab + ac$ **(d)** $a + [b + (-b)] = a + 0$

Perform the indicated operations.

4. $(x^2 - 3x + 2) - (x - 4x^2) + 3x(2x + 1)$

5. $(6r - 5)^2$

6. $(t + 2)(3t^2 - t + 4)$

7. $\dfrac{2x^3 - 11x^2 + 28}{x - 5}$

8. According to data supplied by the Congressional Budget Office, the adjusted poverty threshold for a single person between the years 1984 and 1990 can be approximated by the polynomial $18.7x^2 + 105.3x + 4814.1$, where $x = 0$ corresponds to 1984, $x = 1$ corresponds to 1985, and so on, and the amount is in dollars. According to this model, what was the adjusted poverty threshold in 1987?

9. Explain how the accompanying figures geometrically support the equation $(a + 1)^2 = a^2 + 2a + 1$.

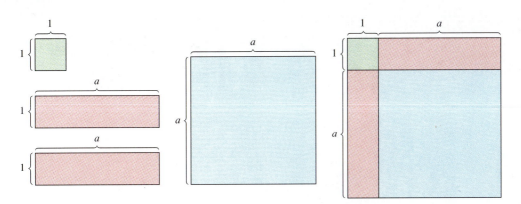

10. Use Pascal's triangle to expand $(2x - 3y)^4$.

Factor completely.

11. $x^4 - 16$

12. $24m^3 - 14m^2 - 24m$

13. $x^3y^2 - 9x^3 - 8y^2 + 72$

Perform the indicated operations.

14. $\dfrac{5x^2 - 9x - 2}{30x^3 + 6x^2} \cdot \dfrac{2x^8 + 6x^7 + 4x^6}{x^4 - 3x^2 - 4}$

15. $\dfrac{x}{x^2 + 3x + 2} + \dfrac{2x}{2x^2 - x - 3}$

16. $\dfrac{a + b}{2a - 3} - \dfrac{a - b}{3 - 2a}$

17. $\dfrac{y - 2}{y - \dfrac{4}{y}}$

18. Simplify $\left(\dfrac{x^{-2}y^{-1/3}z}{x^{-5/3}y^{-2/3}z^{2/3}} \right)^3$ so that there are no negative exponents. Assume that all variables represent positive real numbers.

Simplify. Assume that all variables represent positive real numbers.

19. $\sqrt{18x^5y^8}$

20. $\sqrt{32x} + \sqrt{2x} - \sqrt{18x}$

21. $(\sqrt{x} - \sqrt{y})(\sqrt{x} + \sqrt{y})$

22. The period t in seconds of the swing of a pendulum is given by the equation $t = 2\pi\sqrt{L/32}$, where L is the length of the pendulum in feet. Find the period of a pendulum 3.5 feet long. Use a calculator.

Perform the operation. Write the result in standard form.

23. $(7 - 3i) - (2 + 5i)$

24. $(4 + 3i)(-5 + 3i)$

25. $\dfrac{5 - 5i}{1 - 3i}$

26. Is i^{301} equal to i, $-i$, 1, or -1?

2

Equations and Inequalities

In A.D. 61 the Roman philosopher Seneca wrote about the poor air quality in Rome caused by fireplaces. Although indoor air quality has been a problem ever since people first started building fires inside their houses, it has become a major health concern during the past decade. Today, indoor air pollution has become more hazardous as people spend 80 to 90 percent of their time in tightly sealed, energy-efficient buildings. These buildings often lack proper ventilation. Many contaminants, such as tobacco smoke, formaldehyde, radon, lead, and carbon monoxide, are allowed to increase to unsafe levels. Some indoor air pollutants occur in such low levels that until recently scientists were unable to detect or measure them. Radon levels as low as 8 atoms per cubic centimeter are believed to be a health risk by the EPA. Although modern technology has succeeded in significantly increasing our life expectancy, it has also introduced chemicals and by-products into our air that can cause cancer, respiratory problems, skin irritations, headaches and a number of other health ailments.

A major objective of scientists and mathematicians has been to determine the amount of risk that each contaminant poses. Formaldehyde is one of over 700 organic contaminants that occur indoors. It is found in building products such as plywood, carpeting, paneling, and certain types of insulation. It readily diffuses into the air and has an unpleasant odor causing eye and skin irritation. It has also been classified as a probable carcinogen for humans. Some mobile homes with urea formaldehyde foam insulation have average indoor concentrations that exceed .4 part per million.

Mathematics plays a central role in the determination of risk assessment. In order to determine the risk associated with exposure to contaminants in the air, mathematicians use equations and inequalities. These equations and inequalities can be used to describe the effect that scientists believe pollutants have on human health. In this chapter we will be introduced to many of the fundamental equations needed to answer health questions like these. Using this knowledge we will be able to calculate cancer risks, ventilation requirements for rooms, approximate lethal carbon monoxide levels, and estimate the annual number of cancer cases that can be attributed to passive smoking. A world without risk is impossible, but, with the aid of mathematics, we can make informed decisions that will help minimize our health risks.*

2.1 Linear Equations ▼▼▼

An **equation** is a statement that two expressions are equal. Examples of equations include

$$x + 2 = 9, \qquad 11y = 5y + 6y, \qquad \text{and} \qquad x^2 - 2x - 1 = 0.$$

To **solve** an equation means to find all numbers that make the equation a true statement. Such numbers are called **solutions** or **roots** of the equation. A number that is a solution of an equation is said to **satisfy** the equation, and the solutions of an equation make up its **solution set.**

An equation satisfied by every number that is a meaningful replacement for the variable is called an **identity.** Examples of identities are

$$3x + 4x = 7x \qquad \text{and} \qquad x^2 - 3x + 2 = (x - 2)(x - 1).$$

Equations satisfied by some numbers but not by others are called **conditional equations.** Examples of conditional equations are

$$2m + 3 = 7 \qquad \text{and} \qquad \frac{5r}{r - 1} = 7.$$

The equation

$$3(x + 1) = 5 + 3x$$

is neither an identity nor a conditional equation. Multiplying on the left side gives

$$3x + 3 = 5 + 3x,$$

which is false for every value of x. Such an equation is called a **contradiction.**

Sources: Boubel, R., D. Fox, D. Turner, and A. Stern, *Fundamentals of Air Pollution,* Academic Press, 1994.
Godish, T., *Indoor Air Pollution Control,* Lewis Publishers, 1989.

EXAMPLE 1
Identifying an equation as conditional, an identity, or a contradiction

Decide whether each of the following equations is an identity, a conditional equation, or a contradiction.

(a) $9p^2 - 25 = (3p + 5)(3p - 5)$

Since the product of $3p + 5$ and $3p - 5$ is $9p^2 - 25$, the given equation is true for *every* value of p and is an identity. Its solution set is {all real numbers}.

(b) $5y - 4 = 11$

Replacing y with 3 gives

$$5 \cdot 3 - 4 = 11$$
$$11 = 11,$$

a true statement. On the other hand, $y = 4$ leads to

$$5 \cdot 4 - 4 = 11$$
$$16 = 11,$$

a false statement. The equation $5y - 4 = 11$ is true for some values of y, but not all, and thus is a conditional equation. (The word *some* in mathematics means "at least one." We can therefore say that the statement $5y - 4 = 11$ is true for *some* replacements of y, even though it turns out to be true only for $y = 3$.) Since 3 is the only number that is a solution (as can be shown using methods discussed later), the solution set is {3}.

(c) $(a - 2)^2 + 3 = a^2 - 4a + 2$

The left side can be rewritten as follows.

$$a^2 - 4a + 4 + 3 = a^2 - 4a + 2$$
$$a^2 - 4a + 7 = a^2 - 4a + 2$$

The final equation is false for every value of a, so this equation is a contradiction. The solution set contains no elements. It is called the *empty* or *null* set, and is symbolized ∅. ▶

Equations with the same solution set are called **equivalent equations.** For example, $x + 1 = 5$ and $6x + 3 = 27$ are equivalent equations because they have the same solution set, {4}.

EXAMPLE 2
Determining whether two equations are equivalent

Decide which of the following pairs of equations are equivalent.

(a) $2x - 1 = 3$ and $12x + 7 = 31$

Each of these equations has solution set {2}. Since the solution sets are equal, the equations are equivalent.

(b) $x = 3$ and $x^2 = 9$

The solution set for $x = 3$ is {3}, while the solution set for the equation $x^2 = 9$ is {3, −3}. Since the solution sets are not equal, the equations are not equivalent. ▶

CONNECTIONS* When we solve an equation, we must make sure that it remains "balanced"—that is, any operation performed on one side of an equation must also be performed on the other side in order to assure that the set of solutions remains the same.

Underlying the rules for solving equations are four axioms of equality, listed below. For all real numbers a, b, and c,

1. **Reflexive axiom** $a = a$
2. **Symmetric axiom**   If $a = b$, then $b = a$.
3. **Transitive axiom**   If $a = b$ and $b = c$, then $a = c$.
4. **Substitution axiom**   If $a = b$, then a may replace b in any statement without affecting the truth or falsity of the statement.

FOR DISCUSSION OR WRITING
1. Does $<$ satisfy any of these axioms? If so, which ones?
2. Does the transitive axiom hold in sports competition, with the relation "defeats"?
3. Give an example of a relation that does not satisfy the transitive axiom.

One way to solve an equation is to rewrite it as a series of simpler equivalent equations. These simpler equations often can be obtained with the *addition and multiplication properties of equality.*

Addition and Multiplication Properties of Equality

For real numbers a, b, and c:

$a = b$ and $a + c = b + c$ **are equivalent.** *(The same number may be added to both sides of an equation without changing the solution set.)*

If $c \neq 0$, then $a = b$ and $ac = bc$ **are equivalent.** *(Both sides of an equation may be multiplied by the same nonzero number without changing the solution set.)*

---

* From Charles Miller, Vern Heeren, and E. John Hornsby, Jr., *Mathematical Ideas,* Seventh Ed., HarperCollins, 1994.

EXAMPLE 3
Solving a linear equation

Solve $3(2x - 4) = 7 - (x + 5)$.

Use the distributive property and then collect like terms to get the following sequence of simpler equivalent equations.

$$3(2x - 4) = 7 - (x + 5)$$

$$6x - 12 = 7 - x - 5 \qquad \text{Distributive property}$$

$$6x - 12 = 2 - x$$

$$x + 6x - 12 = x + 2 - x \qquad \text{Add } x \text{ to each side.}$$

$$7x - 12 = 2 \qquad \text{Combine terms.}$$

$$12 + 7x - 12 = 12 + 2 \qquad \text{Add 12 to each side.}$$

$$7x = 14 \qquad \text{Combine terms.}$$

$$\frac{1}{7} \cdot 7x = \frac{1}{7} \cdot 14 \qquad \text{Multiply both sides by } \tfrac{1}{7}.$$

$$x = 2$$

To check, replace x with 2 in the original equation, getting

$$3(2x - 4) = 7 - (x + 5) \qquad \text{Original equation}$$

$$3(2 \cdot 2 - 4) = 7 - (2 + 5) \qquad ? \quad \text{Let } x = 2.$$

$$3(4 - 4) = 7 - (7) \qquad ?$$

$$0 = 0. \qquad \text{True}$$

Since replacing x with 2 results in a true statement, 2 is the solution of the given equation. The solution set is therefore $\{2\}$. ▶

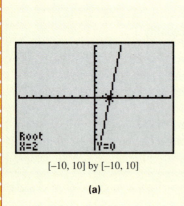

Root
X=2   Y=0

[−10, 10] by [−10, 10]

(a)

You will recall from your previous studies in algebra that if ordered pairs (x, y) that satisfy $y = mx + b$ are graphed in a two-dimensional rectangular coordinate system, the points representing the ordered pairs lie in a straight line. In Chapter 3 we will study such graphs in detail. For now, we can use this idea to show how a calculator is used to solve a linear equation.

A graphing calculator can be used to solve equations in two ways using graphs. One way is to rewrite the equation with one side equal to 0 and then to let y equal the expression on the other side. For the equation in Example 3, we would have $y = 3(2x - 4) - 7 + (x + 5)$. To solve $y = 0$, we graph $y = 3(2x - 4) - 7 + (x + 5)$ and use the capability of the calculator to find the value of x where the graph intersects the x-axis. As Figure (a) shows, $y = 0$ at $x = 2$. (Recall that *root* is another name for "solution.") It is important to choose an appropriate viewing rectangle. This may require some trial and error. Graphing calculators have a standard setting, which is often a good way to start. We will indicate our choice of window by using,

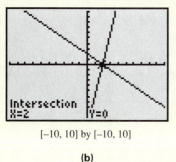

[−10, 10] by [−10, 10]

(b)

for example, the notation $[-10, 10]$ by $[-10, 10]$ to represent $-10 \le x \le 10$ and $-10 \le y \le 10$.

Another way to solve an equation is to graph the expressions on each side of the equation simultaneously, letting one equal Y_1 and the other Y_2. In Figure (b), $Y_1 = 3(2x - 4)$ and $Y_2 = 7 - (x + 5)$. Now, we use the capability of the calculator to find the x-coordinate of the point where the two graphs intersect. The result is the same value of x, 2. (*Note:* The point of intersection will not always be on the x-axis, as it is in Figure (b).

The equation in Example 3 is a *linear equation,* as are most of the equations in this section.

Linear Equation in One Variable

A **linear equation** in one variable is an equation that can be written in the form

$$ax + b = 0,$$

where $a \ne 0$.

The next examples will show how some equations can be simplified before solving.

EXAMPLE 4
Simplifying an equation before solving

Solve each equation.

(a) $\dfrac{3p - 1}{3} - \dfrac{2p}{p - 1} = p$

This equation does not satisfy the definition of a linear equation. However, the equation can be written as a linear equation and solved as one if the restrictions on the denominator are noted. Multiply both sides of the equation by the common denominator, $3(p - 1)$, assuming $p \ne 1$. Doing this gives

$$3(p - 1)\left(\frac{3p - 1}{3}\right) - 3(p - 1)\left(\frac{2p}{p - 1}\right) = 3(p - 1)p$$

$$(p - 1)(3p - 1) - 3(2p) = 3p(p - 1)$$

$$3p^2 - 4p + 1 - 6p = 3p^2 - 3p.$$

A simpler equivalent equation comes from combining terms and adding $-3p^2$ to both sides, producing

$$-10p + 1 = -3p.$$

Now add $10p$ to both sides to get

$$1 = 7p.$$

Finally, multiplying both sides by $1/7$ gives

$$\frac{1}{7} = p.$$

Check $1/7$ in the given equation to verify that the solution set is $\{1/7\}$. The restriction $p \neq 1$ does not affect the solution set here, since $1/7 \neq 1$. Since the original equation and the linear equation found by multiplying both sides by $3(p - 1)$ have the same solution set, $\{1/7\}$, they are equivalent equations.

(b) $\dfrac{x}{x - 2} = \dfrac{2}{x - 2} + 2$

Multiply both sides of the equation by $x - 2$, assuming that $x \neq 2$.

$$x = 2 + 2(x - 2)$$
$$x = 2 + 2x - 4$$
$$x = 2$$

It is necessary to assume $x \neq 2$ in order to be able to multiply both sides of the equation by $x - 2$. Since $x = 2$, however, the multiplication property of equality does not apply. The solution set is $\emptyset$. (Substituting 2 for x in the original equation would result in a denominator of 0.) The original equation and the linear equation $x = 2 + 2(x - 2)$ are *not* equivalent. ▶

CAUTION   It is essential to check proposed solutions (such as 2 in Example 4(b)) whenever each side of an equation is multiplied by a variable expression. Do not forget this important step.

◀**EXAMPLE 5**
Simplifying an equation before solving

Solve $\dfrac{2}{x - 1} - \dfrac{4}{3x} = \dfrac{1}{x^2 - x}$.

Note that $x \neq 1$ and $x \neq 0$ are restrictions on the solution. To begin, we find a common denominator. Since $x^2 - x$ can be factored as $x(x - 1)$, the least common denominator is $3x(x - 1)$. Multiply both sides of the equation by $3x(x - 1)$.

$$3x(x - 1)\frac{2}{x - 1} - 3x(x - 1)\frac{4}{3x} = 3x(x - 1)\frac{1}{x(x - 1)} \quad (x \neq 0, 1)$$
$$6x - 4(x - 1) = 3$$
$$6x - 4x + 4 = 3$$
$$2x = -1$$
$$x = -\frac{1}{2}$$

Since $-1/2 \neq 1$ or 0, the solution set is $\{-1/2\}$. ▶

2.1 Exercises ▼▼▼▼▼▼▼▼▼▼▼▼▼▼▼▼▼▼▼▼▼▼▼▼▼▼▼▼▼▼▼▼▼▼▼▼▼▼

In Exercises 1–4, decide whether the statement is true or false.

1. The solution set of $2x + 3 = x - 5$ is $\{-8\}$.

2. The equation $5(x - 9) = 5x - 45$ is an example of an identity.

3. The equations $x^2 = 9$ and $x = 3$ are equivalent equations.

4. It is possible for a linear equation to have exactly two solutions.

5. Explain the difference between an identity and a conditional equation.

6. Make a complete list of the steps needed to solve a linear equation. (Some equations will not require every step.)

Decide whether the equation is an identity, a conditional equation, or a contradiction. Give the solution set. See Example 1.

7. $x^2 + 6x = x(x + 6)$

8. $3(x + 2) - 5(x + 2) = -2x - 4$

9. $3t + 4 = 5(t - 2)$

10. $2(x - 7) + x = 5x + 3$

11. $2x - 4 = 2(x + 2)$

12. $x^2 + 2 = x^2 + 4$

Decide whether the pair of equations is an example of equivalent equations. See Example 2.

13. $\dfrac{5x}{x - 2} = \dfrac{20}{x - 2}$ and $5x = 20$

14. $\dfrac{x + 3}{13} = \dfrac{6}{13}$ and $x = 3$

15. $\dfrac{x + 3}{x + 1} = \dfrac{2}{x + 1}$ and $x = -1$

16. $\dfrac{x}{x + 7} = \dfrac{-7}{x + 7}$ and $x = -7$

17. Which one of the following is not a linear equation?
    (a) $5x + 7(x - 1) = -3x$ (b) $8x^2 - 4x + 3 = 0$ (c) $7y + 8y = 13y$ (d) $.04t - .08t = .40$

18. In solving the equation $3(2t - 4) = 6t - 12$, a student obtains the result $0 = 0$ and gives the solution set $\{0\}$. Is this correct? Explain.

Solve the equation. See Examples 3–5.

19. $2m - 5 = m + 7$

20. $.01p + 3.1 = 2.03p - 2.96$

21. $\dfrac{5}{6}k - 2k + \dfrac{1}{3} = \dfrac{2}{3}$

22. $\dfrac{3}{4} + \dfrac{1}{5}r - \dfrac{1}{2} = \dfrac{4}{5}r$

23. $3r + 2 - 5(r + 1) = 6r + 4$

24. $5(a + 3) + 4a - 5 = -(2a - 4)$

25. $2[m - (4 + 2m) + 3] = 2m + 2$

26. $4[2p - (3 - p) + 5] = -7p - 2$

27. $\dfrac{3x - 2}{7} = \dfrac{x + 2}{5}$

28. $\dfrac{2p + 5}{5} = \dfrac{p + 2}{3}$

29. $\dfrac{1}{4p} + \dfrac{2}{p} = 3$

30. $\dfrac{2}{t} + 6 = \dfrac{5}{2t}$

31. $\dfrac{m}{2} - \dfrac{1}{m} = \dfrac{6m + 5}{12}$

32. $\dfrac{-3k}{2} + \dfrac{9k - 5}{6} = \dfrac{11k + 8}{k}$

33. $\dfrac{2r}{r - 1} = 5 + \dfrac{2}{r - 1}$

34. $\dfrac{3x}{x + 2} = \dfrac{1}{x + 2} - 4$

35. $\dfrac{5}{2a + 3} + \dfrac{1}{a - 6} = 0$

36. $\dfrac{2}{x+1} = \dfrac{3}{5x+5}$

37. $\dfrac{4}{x-3} - \dfrac{8}{2x+5} + \dfrac{3}{x-3} = 0$

38. $\dfrac{5}{2p+3} - \dfrac{3}{p-2} = \dfrac{4}{2p+3}$

39. $\dfrac{2p}{p-2} = 3 + \dfrac{4}{p-2}$

40. $\dfrac{5k}{k+4} = 3 - \dfrac{20}{k+4}$

41. $\dfrac{3}{y-2} + \dfrac{1}{y+1} = \dfrac{1}{y^2-y-2}$

42. $\dfrac{2}{p+3} - \dfrac{5}{p-1} = \dfrac{1}{3-2p-p^2}$

43. $.08w + .06(w+12) = 7.72$

44. $.04(x-12) + .06x = 1.52$

45. Recall that a graphing calculator will return a 1 for a true statement and a 0 for a false statement. What value must be stored in X for the following screen to come up on a graphing calculator?

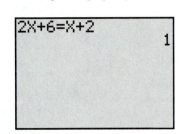

46. (See Exercise 45.) What is the only value that *cannot* be stored in T for the following screen to come up on a graphing calculator?

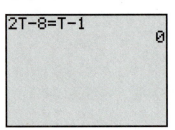

▼▼▼▼▼▼▼▼▼▼▼▼▼ **DISCOVERING CONNECTIONS** (Exercises 47–50) ▼▼▼▼▼▼▼▼▼▼▼▼▼

In Chapter 1 we learned how to evaluate an expression for a given value of the variable. We can use this knowledge to create a linear equation with a predetermined solution of our choice. (This is how mathematics teachers are able to write equations with "nice" solutions for their students). Work Exercises 47–50 in order, so that you can seen how this is done.

47. Suppose we wish to write an equation of the form

$$3(2x-5) + 4x = \underline{\quad\quad},$$

where the solution is 5 and the blank is filled in with a constant. Evaluate the polynomial on the left for $x = 5$. What value must be substituted for the blank so that the equation has solution set $\{5\}$?

48. Suppose we want an equation of the same form as the one in Exercise 47, but with solution set $\{-4\}$. What number must be substituted for the blank to accomplish this?

49. Consider the equation

$$-5x + 2(4-2x) = x + \underline{\quad\quad}.$$

Suppose we want it to have solution set $\{1\}$. Evaluate the left side for 1, and then substitute 1 for x on the right side. What number must be substituted for the blank to accomplish our goal?

50. Repeat Exercise 49 to obtain a solution set of $\{3.5\}$.

In the metric system of weights and measures, temperature is measured in degrees Celsius (°C) instead of degrees Fahrenheit (°F). To convert back and forth between the two systems, we use the equations

$$C = \dfrac{5(F-32)}{9} \quad \text{and} \quad F = \dfrac{9}{5}C + 32.$$

In each of the following exercises, convert to the other system. Round answers to the nearest tenth of a degree if necessary.

51. 20°C   **52.** 100°C   **53.** 59°F   **54.** 86°F   **55.** 100°F   **56.** 350°F

When a consumer borrows money, the lender must tell the consumer the true annual interest rate of the loan. The method of finding the exact true annual interest rate requires special tables available from the government, but a quick approximate rate can be found by using the equation

$$A = \frac{2pf}{b(q+1)},$$

where p is the number of payments made in one year, f is the finance charge, b is the balance owed on the loan, and q is the total number of payments. Find the value of the variables not given in Exercises 57–60. Round A to the nearest percent and round other variables to the nearest whole number. (This formula is not accurate enough for the requirements of federal law.)

57. $p = 12, f = \$800, b = \$4000, q = 36;$ find A

58. $p = 12, f = \$60, b = \$740, q = 12;$ find A

59. $A = 14\%$ (or .14), $p = 12, b = \$2000, q = 36;$ find f

60. $A = 11\%, p = 12, b = \$1500, q = 24;$ find f

When a loan is paid off early, a portion of the finance charge must be returned to the borrower. By one method of calculating finance charge (called the rule of 78), the amount of unearned interest (finance charge to be returned) is given by

$$u = f \cdot \frac{n(n+1)}{q(q+1)},$$

where u represents unearned interest, f is the original finance charge, n is the number of payments remaining when the loan is paid off, and q is the original number of payments. Find the amount of the unearned interest in each of the following.

61. Original finance charge = $800, loan scheduled to run 36 months, paid off with 18 payments remaining

62. Original finance charge = $1400, loan scheduled to run 48 months, paid off with 12 payments remaining

63. Original finance charge = $950, loan scheduled to run 24 months, paid off with 6 payments remaining

64. Original finance charge = $175, loan scheduled to run 12 months, paid off with 3 payments remaining

---

65. Kitchen gas ranges are a source of indoor pollutants such as carbon monoxide and nitrogen dioxide. One of the most effective ways of removing contaminants from the air while cooking is to use a *vented* range hood. If a range hood removes *F* liters of air per second, then the percent *P* of contaminants that are also removed from the surrounding air can be expressed by the linear equation $P = 1.06F + 7.18$ where $10 \le F \le 75$. What flow rate must a range hood have to remove 50% of the contaminants from the air? (*Source:* Rezvan, R. L., "Effectiveness of Local Ventilation in Removing Simulated Pollutants from Point Sources," 65–75. In *Proceedings of the Third International Conference on Indoor Air Quality and Climate*, 1984.)

66. Venus is the hottest planet with a surface temperature of 864°F. What is this temperature in Celsius? (See the formulas preceding Exercise 51.) (*Source: The Guinness Book of World Records 1995.*)

67. The accompanying graph illustrates how the percent of alcohol-related traffic deaths in the United States over the period from 1982 to 1993 has declined. These data can be modeled by the equation $y = -1.18x + 57.03$, where $x = 0$ corresponds to 1982, $x = 1$ corresponds to 1983, and so on, and y is the percent of alcohol-related traffic deaths.

Percent of Alcohol-related Traffic Deaths

57%

44%

Year

1982 1984 1986 1988 1990   1993

Source: National Highway Traffic Safety Administration.

(a) Use the model equation to find the year in which the percent was 50%. (*Hint*: Let $y = 50$ and solve.)

(b) Use the graph to answer the item in part (a). How closely do the answers correspond?

(c) If this model were used to calculate the percent in 1997, what would the percent be?

(d) Discuss any pitfalls that might be associated with the procedure used in part (c).

68. The changes in the minimum hourly wage in the United States for selected years during the period from 1938 to 1991 are shown in the accompanying table. Using these data, a technique from statistics allows us to determine the linear model $y = .0769x - 149.156$ to describe the data, where x is the year and y is the minimum wage in dollars.

Changes in the Minimum Hourly Wage	
1938	$.25
1950	.75
1956	1.00
1963	1.25
1974	2.00
1980	3.10
1981	3.35
1990	3.80
1991	4.25

(*Source*: U.S. Labor Department.)

(a) Use the model to approximate the minimum wage in 1980. How close is it to the information in the table?

(b) Use the model to approximate the year in which the minimum wage is $4.25. How close is it to the information in the table?

Refer to the graphing calculator box in this section that explains how to solve an equation graphically. Use the first method described (*finding the value of x where the line crosses the x-axis*) to solve the equation. (*Hint*: In Exercises 73 and 74, write an equivalent equation with 0 on one side first.)

69. $2(x - 5) + 3x - x - 6 = 0$

70. $6x - 3(5x + 2) - 4 + 5x = 0$

71. $4x - 3(4 - 3x) - 2(x - 3) - 6x - 2 = 0$

72. $-9x - (4 + 3x) + 2x - 1 + 5 = 0$

73. $\dfrac{x}{2} + \dfrac{x}{3} = 5$

74. $\dfrac{x - 2}{3} + \dfrac{x}{4} = \dfrac{1}{2}$

Refer to the graphing calculator box in this section, and use the second method described (*finding the x-value of the point of intersection of the two lines*) to solve the equation.

75. $3(2x + 1) - 2(x - 2) = 5$

76. $-3x - 2 = 6 + 4x - 1$

77. $-(8 + 3x) + 5 = 2x + 3$

78. $4(x + 3) - 5 = 8x + 7$

79. $\dfrac{x - 2}{4} + \dfrac{x + 1}{2} = 1$

80. $\dfrac{2x + 1}{3} + \dfrac{x - 1}{4} = \dfrac{13}{2}$

Most graphing calculators have a built-in equation solver. Various makes and models differ in the syntax in which they accept the data, so you should read your instruction manual to see how your calculator handles equation-solving. One popular model requires that the equation be written in the form

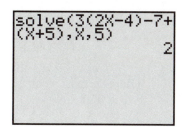

$$\text{expression in } x = 0.$$

Then we direct the calculator to solve the equation by inputting the expression in x, the variable for which we are solving (x), and a guess at the solution. For example, if we want to solve $3(2x - 4) - 7 + (x + 5) = 0$, we input the information shown on the screen (our guess is 5) and the calculator returns the solution, 2. Compare to the graphing calculator box in this section.

Use the built-in equation-solving feature of a graphing calculator to solve the equation.

81. $2x + 7 = 3(x + 3) - 8$ **82.** $4x - 2(x - 1) = 12$

83. $5x - 2(x + 4) = 6x + 3$ **84.** $x - (2x + 1) - 7 = -3x - 3$

85. $-2(x - 1) - 10 = 2(2 + x)$ **86.** $6 - 4(x - 1) = 3 + 2x$

Use any technology-based method to solve the linear equation. Give solutions to the nearest hundredth.

87. $4(.23x + \sqrt{5}) = \sqrt{2}x + 1$ **88.** $9(-.84x + \sqrt{17}) = \sqrt{6}x - 4$

89. $2\pi x + \sqrt[3]{4} = .5\pi x - \sqrt{28}$ **90.** $3\pi x - \sqrt[4]{3} = .75\pi x + \sqrt{19}$

91. $.23(\sqrt{3} + 4x) - .82(\pi x + 2.3) = 5$ **92.** $-.15(6 + \sqrt{2}x) + 1.4(2\pi x - 6.1) = 10$

2.2 Applications of Linear Equations ▼▼▼

Mathematics is an important problem-solving tool. Many times the solution of a problem depends on the use of a formula that expresses a relationship among several variables. For example, the formula

$$A = \frac{24f}{b(p + 1)} \qquad (*)$$

gives the approximate annual interest rate for a consumer loan paid off with monthly payments. Here f is the finance charge on the loan, p is the number of payments, and b is the original amount of the loan.

Suppose the number of payments, p, must be found when the other quantities are known. To do this, we might first solve the equation for p by treating p as the variable and the other letters as constants. Begin by multiplying both sides of formula $(*)$ by $p + 1$.

$$(p + 1)A = \frac{24f}{b}$$

Multiplying both sides by $1/A$ gives

$$p + 1 = \frac{24f}{Ab}.$$

(Here we must assume $A \neq 0$. Why is this a very safe assumption?) Finally, add -1 to both sides.

$$p = \frac{24f}{Ab} - 1$$

This process is called **solving for a specified variable.**

◖**EXAMPLE 1**
Solving a formula for a specified variable

Solve $J\left(\dfrac{x}{k} + a\right) = x$ for x.

To get all terms with x on one side of the equation and all terms without x on the other, first use the distributive property.

$$J\left(\frac{x}{k}\right) + Ja = x$$

Eliminate the denominator, k, by assuming $k \neq 0$ and multiplying both sides by k.

$$kJ\left(\frac{x}{k}\right) + kJa = kx$$

$$Jx + kJa = kx$$

Then add $-Jx$ to both sides to get the two terms with x together.

$$kJa = kx - Jx$$

$$kJa = x(k - J) \qquad \text{Factor the right side.}$$

We factored on the right, because we need to get x by itself on one side. Now, assuming $k \neq J$, we multiply both sides by $1/(k - J)$ to find the solution.

$$x = \frac{kJa}{k - J} \qquad ▶$$

CAUTION   Errors often occur in problems like Example 1 because students forget to factor out the variable for which they are solving. It is necessary to have that variable as a factor so that in the next step we can multiply by the reciprocal of its coefficient.

■ **PROBLEM SOLVING**   One of the main reasons for learning mathematics is to be able to use it in solving practical problems. For most students, however, learning how to apply mathematical skills to real situations is the most difficult task they face. In the rest of this section a few hints are given that may help with applications.

A common difficulty with applied problems is trying to do everything at once. It is usually best to attack the problem in stages.

Solving Applied Problems

1. **Decide on an unknown, and name it with some variable that you *write down*.** Do not try to skip this step. This is an important step. If you don't know what "x" represents, how can you write a meaningful equation or interpret a result?
2. **Draw a sketch or make a chart, if appropriate,** showing the information given in the problem.
3. **Decide on a variable expression to represent any other unknowns** in the problem. For example, if W represents the width of a rectangle, L represents the length, and you know that the length is one more than twice the width, *write down $L = 2W + 1$.*
4. **Use the results of Steps 1 and 3 to write an equation with one variable.**
5. **Solve the equation.**
6. **Check the solution in the words of the original problem.** Be sure that the answer makes sense. ■

Notice how each of the steps listed above is carried out in the following examples.

◀EXAMPLE 2
Solving a geometry problem

If the length of a side of a square is increased by 3 centimeters, the perimeter of the new square is 40 centimeters more than twice the length of a side of the original square. Find the dimensions of the original square.

First, decide what the variable should represent (Step 1). Since the length of a side of the original square is to be found, let the variable represent it.

$$x = \text{length of side of the original square in centimeters}$$

Now draw a figure using the given information, as in Figure 1 (Step 2).

The length of a side of the new square is 3 centimeters more than the length of a side of the old square. Write a variable expression for that relationship (Step 3).

$$x + 3 = \text{length of side of the new square}$$

Now write a variable expression for the perimeter of the new square. Since the perimeter of a square is 4 times the length of a side,

$$4(x + 3) = \text{perimeter of the new square.}$$

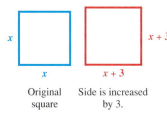

| Original square | Side is increased by 3. |

x and $x + 3$ are in centimeters.

FIGURE 1

Now you can use the information given in the problem to write an equation (Step 4). The perimeter of the new square is 40 centimeters more than twice the length of a side of the original square, so the equation is written as follows.

The new perimeter	is	40	more than	twice the side of the original square.
$4(x + 3)$	$=$	40	$+$	$2x$

Solve the equation (Step 5).

$$4(x + 3) = 40 + 2x$$
$$4x + 12 = 40 + 2x$$
$$2x = 28$$
$$x = 14$$

This solution should be checked using the words of the original problem (Step 6). The length of a side of the new square would be $14 + 3 = 17$ centimeters; its perimeter would be $4(17) = 68$ centimeters. Twice the length of a side of the original square is $2(14) = 28$ centimeters. Since $40 + 28 = 68$, the solution satisfies the problem. Each side of the original square measures 14 centimeters. ▶

In the remaining examples, the steps are not numbered. See if you can identify them. (Some steps may not apply in some problems.)

■ **PROBLEM SOLVING**   The next example is a constant velocity problem. The components *distance, rate,* and *time* are denoted by the letters *d, r,* and *t,* respectively. (The *rate* is also called the *speed* or *velocity*.) These variables are related by the equation

$$d = rt.$$

This equation is easily solved for *r* and *t*; $r = d/t$ and $t = d/r$. ■

EXAMPLE 3
Solving a constant velocity problem

Maria and Eduardo are traveling to a business conference. Maria travels 110 miles in the same time that Eduardo travels 140 miles. Eduardo travels 15 miles per hour faster than Maria. Find the average rate of each person.

Let x = Maria's rate,
$x + 15$ = Eduardo's rate.

Summarize the given information in a table.

	d	r	t
Maria	110	x	$\dfrac{110}{x}$
Eduardo	140	$x + 15$	$\dfrac{140}{x + 15}$

Use $t = d/r$.

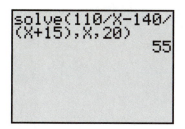

A built-in equation solver in a graphing calculator can solve the equation given in Example 3. Here we see that the solution is indeed 55.

The unused information from the problem is "in the same time that."

$$\text{Maria's time} = \text{Eduardo's time}$$

$$\frac{110}{x} = \frac{140}{x + 15}$$

$$x(x + 15)\frac{110}{x} = x(x + 15)\frac{140}{x + 15} \qquad \text{Multiply both sides by } x(x + 15).$$

$$110(x + 15) = 140x$$

$$110x + 1650 = 140x$$

$$1650 = 30x$$

$$55 = x \qquad \text{Maria's rate}$$

$$x + 15 = 55 + 15 = 70 \qquad \text{Eduardo's rate}$$

Therefore, Maria's rate is 55 miles per hour and Eduardo's rate is 70 miles per hour.

   Check:

$$\text{Time traveled by Maria:} \quad 110/55 = 2 \text{ hr}$$
$$\text{Time traveled by Eduardo:} \quad 140/70 = 2 \text{ hr}$$

Same ▶

■ PROBLEM SOLVING   In Example 3 (a constant velocity problem), we used the formula relating rate, time, and distance. In problems involving rate of work (as in Example 4, which follows), we use a similar idea. If a person or a machine can do a job in t units of time, then the rate of work is $1/t$ job per time unit. Therefore,

$$\textbf{rate} \times \textbf{time} = \textbf{portion of the job completed}.$$

   If the letters r, t, and A represent the rate at which work is done, the time, and the amount of work accomplished, respectively, then

$$A = rt.$$

Amounts of work are often measured in terms of the number of jobs accomplished. For instance, if one job is accomplished in t hours, then $A = 1$ and $r = 1/t$. ■

◀**EXAMPLE 4**
Solving a problem about work

One computer can do a job twice as fast as another. Working together, both computers can do the job in 2 hours. How long would it take each computer, working alone, to do the job?

   Let x represent the number of hours it would take the faster computer, working alone, to do the job. The time for the slower computer to do the job alone is then $2x$ hours. Therefore, the rates for the two computers are as follows:

$$\frac{1}{x} = \text{rate of faster computer (job per hour)}$$

$$\frac{1}{2x} = \text{rate of slower computer (job per hour)}.$$

The time for the computers to do the job together is 2 hours. Multiplying each rate by the time will give the fractional part of the job accomplished by each. This is summarized in the chart that follows.

	Rate	*Time*	*Part of the Job Accomplished*
Faster computer	$\dfrac{1}{x}$	2	$2\left(\dfrac{1}{x}\right) = \dfrac{2}{x}$ ⟵ $A = rt$
Slower computer	$\dfrac{1}{2x}$	2	$2\left(\dfrac{1}{2x}\right) = \dfrac{1}{x}$

The sum of the two parts of the job accomplished is 1, since one whole job is done. The equation can now be written and solved.

$$\frac{2}{x} + \frac{1}{x} = 1 \qquad \text{The sum of the two parts is 1.}$$

$$2 + 1 = x \qquad \text{Multiply by } x.$$

$$x = 3$$

The faster computer could do the entire job, working alone, in 3 hours. The slower computer would need $2(3) = 6$ hours. ▶

NOTE In problems involving rates of work, the formula given above and used in Example 4 assumes a uniform rate. In other words, the work does not speed up or slow down as the job is carried out.

◖EXAMPLE 5
Solving a mixture problem

Constance Morganstern is a chemist. She needs a 20% solution of alcohol. She has a 15% solution on hand, as well as a 30% solution. How many liters of the 15% solution should she add to 3 liters of the 30% solution to get her 20% solution?

Let x be the number of liters of the 15% solution to be added. See Figure 2.

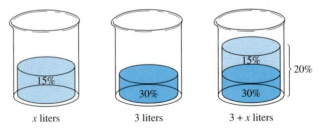

x liters   3 liters $3 + x$ liters

FIGURE 2

Arrange the information of the problem in a chart.

Strength	Liters of Solution	Liters of Pure Alcohol
15%	x	**.15x**
30%	3	**.30(3)**
20%	3 + x	**.20(3 + x)**

Since the number of liters of pure alcohol in the 15% solution plus the number of liters in the 30% solution must equal the number of liters in the final 20% solution,

Liters in 15%   Liters in 30%   Liters in 20%

$$.15x \quad + \quad .30(3) \quad = \quad .20(3 + x).$$

Solve this equation as follows.

$$.15x + .90 = .60 + .20x \qquad \text{Distributive property}$$
$$.30 = .05x$$
$$6 = x$$

By this result, 6 liters of the 15% solution should be mixed with 3 liters of the 30% solution, giving 6 + 3 = 9 liters of 20% solution. ▶

■ **PROBLEM SOLVING**   In Example 5 (a mixture problem), we multiplied rate of concentration by the number of liters to get the amount of pure chemical present. Similarly, in Example 6 (an investment problem) which follows, we will multiply interest rate by principal to find the amount of interest earned. ■

EXAMPLE 6
Solving an investment problem

A financial manager has $14,000 to invest for her company. She plans to invest part of the money in tax-free bonds at 6% interest and the remainder at 9%. She wants to earn $1005 per year in interest from the investments. Find the amount she should invest at each rate.

Let x represent the dollar amount to be invested at 6%, so that $14,000 - x$ is the amount to be invested at 9%. Interest is given by the product of principal, rate, and time in years ($i = prt$). Summarize this information in a chart.

Amount Invested	Interest Rate	Interest Earned in 1 Yr
x	6% = .06	**.06x**
14,000 − x	9% = .09	**.09(14,000 − x)**

Since the total interest is to be $1005,

$$.06x + .09(14,000 - x) = 1005.$$

To clear decimal points, we first multiply both sides of the equation by 100.

$$6x + 9(14,000 - x) = 100,500$$
$$6x + 126,000 - 9x = 100,500$$
$$126,000 - 3x = 100,500$$
$$-3x = -25,500$$
$$x = 8500$$

The manager should invest $8500 at 6%, and $14,000 - $8500 = $5500 at 9%. ▶

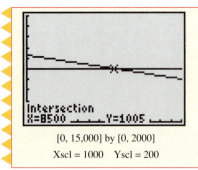

[0, 15,000] by [0, 2000]
Xscl = 1000   Yscl = 200

We can use a graphing calculator to picture the situation in Example 6. Let $Y_1 = .06x + .09(14,000 - x)$, the left side of the equation. This represents the amount of interest earned as x changes. Let $Y_2 = 1005$, the amount of interest required. The figure shows the graphs of Y_1 and Y_2. Notice that as x increases the value of Y_1 decreases. This makes sense, because x represents the amount invested at the smaller interest rate. The x-value at the point where Y_1 and Y_2 intersect gives the amount to invest at 6% to get total interest of $1005. This value of x is 8500 and the corresponding y-value is 1005, which agrees with our analytic approach.

NOTE The interest formula used in Example 6 ($i = prt$) applies to simple interest. In most real-life applications, compound interest is used. We will study the formula for compound interest and its applications in Chapter 5.

The final example illustrates an application of linear equations using data obtained from A. Hines, T. Ghosh, S. Layalka, and R. Warder, *Indoor Air Quality & Control*, Prentice Hall, 1993.

EXAMPLE 7
Examining the effect of formaldehyde on the eyes

Formaldehyde is a volatile organic compound that has come to be recognized as a highly toxic indoor air pollutant. Its source is found in building materials such as fiberboard, plywood, foam insulation, and carpeting. When concentrations of formaldehyde in the air exceed 33 $\mu g/ft^3$ (1 μg = 1 microgram = 10^{-6} gram) a strong odor and irritation to the eyes often occurs. One square foot of hardwood plywood paneling can emit 3365 μg of formaldehyde per day. A 4- by 8-foot sheet of this paneling is attached to an 8-foot wall in a room having dimensions of 10 by 10 feet.

(a) How many cubic feet of air are there in the room?
The volume of the room is $8 \times 10 \times 10 = 800$ cubic feet.

(b) Find the total number of micrograms of formaldehyde that are released into the air by the paneling each day.

The paneling releases 3365 μg for each square foot of area. The area of the sheet is 32 square feet, so it will release $32 \times 3365 = 107,680\ \mu$g of formaldehyde into the air each day.

(c) If there is no ventilation in the room, write a linear equation that gives the amount of formaldehyde F in the room after x days.

The paneling emits formaldehyde at a constant rate of 107,680 μg per day. Thus, $F = 107,680x$.

(d) How long will it take before a person's eyes become irritated in the room?

We must determine when concentration exceeds 33 μg/ft^3. Since the room has 800 cubic feet this will occur when the total amount reaches $33 \times 800 = 26,400\ \mu$g.

$$F = 107,680x = 26,400$$

$$x = \frac{26,400}{107,680}$$

$$x \approx .25$$

It will take approximately 1/4 day, or 6 hours. ▶

2.2 Exercises ▼▼▼▼▼▼▼▼▼▼▼▼▼▼▼▼▼▼▼▼▼▼▼▼▼▼▼▼▼▼▼▼▼▼▼▼▼

The following exercises should be done mentally, without the use of pencil and paper or a calculator. They will prepare you for some of the applications found in this exercise set.

1. If a train travels at 80 miles per hour for 15 minutes, what is the distance traveled?

2. If 40 liters of an acid solution is 75% acid, how much pure acid is there in the mixture?

3. If a person invests $100 at 4% simple interest for two years, how much interest is earned?

4. If a jar of coins contains 30 half-dollars and 100 quarters, what is the monetary value of the coins?

Solve the formula for the indicated variable. Assume that the denominator is not zero if variables appear in the denominator. See Example 1.

5. $V = lwh$ for l (volume of a rectangular box)

6. $i = prt$ for p (simple interest)

7. $P = a + b + c$ for c (perimeter of a triangle)

8. $P = 2l + 2w$ for w (perimeter of a rectangle)

9. $A = \frac{1}{2}(B + b)h$ for B (area of a trapezoid)

10. $A = \frac{1}{2}(B + b)h$ for h (area of a trapezoid)

11. $S = 2\pi rh + 2\pi r^2$ for h (surface area of a right circular cylinder)

12. $S = 2lw + 2wh + 2hl$ for h (surface area of a rectangular box)

13. $C = \frac{5}{9}(F - 32)$ for F (Fahrenheit to Celsius)

14. $s = \frac{1}{2}gt^2$ for g (distance traveled by a falling object)

15. $u = f \cdot \frac{k(k + 1)}{n(n + 1)}$ for f (unearned interest)

16. $\frac{1}{R} = \frac{1}{r_1} + \frac{1}{r_2}$ for R (electricity)

17. Refer to Example 1. Suppose someone tells you there is no reason to solve for x, since the right side of the equation is already equal to x. How would you respond?

18. Suppose two acid solutions are mixed. One is 26% acid and the other is 32% acid. Which one of the following concentrations cannot possibly be the concentration of the mixture? Explain.

  (a) 36%   **(b)** 28%   **(c)** 30%   **(d)** 31%

19. Suppose that a computer that originally sells for x dollars has been discounted 30%. Which one of the following expressions does not represent its sale price?

  (a) $x - .30x$ **(b)** $.70x$

  (c) $\dfrac{7}{10}x$ **(d)** $x - .30$

20. Refer to Exercise 11. Why is it not possible to solve this formula for r using the methods of this section?

Solve the problem. See Example 2.

21. The length of a rectangular label is 3 cm less than twice the width. The perimeter is 54 cm. Find the width.

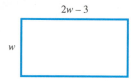

2w − 3

w

Side lengths are in
centimeters.

22. A puzzle piece in the shape of a triangle has a perimeter of 30 cm. Two sides of the triangle are each twice as long as the shortest side. Find the length of the shortest side.

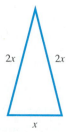

2x   2x

x

Side lengths are
in centimeters.

23. The world's largest tablecloth has a perimeter of 23,803.2 inches. It is made of damask and was manufactured by Tonrose Limited of Manchester, England, in June 1988. Its length is 11,757.6 inches more than its width. What is its length in yards? (*Source: Guinness Book of World Records 1995.*)

24. The smallest ticket ever produced was for admission to the Asian-Pacific Exposition Fukuoka '89 in Japan. It was rectangular, with length 4 mm more than its width. If the width had been increased by 1 mm and the length had been increased by 4 mm, the ticket would have had perimeter equal to 30 mm. What were the actual dimensions of the ticket? (*Source: Guinness Book of World Records 1995.*)

25. A right circular cylinder has radius 6 in and volume 144π cu in. What is its height? (In the figure, let $h =$ height.)

h

6 in

h is in inches.

26. A recycling bin is in the shape of a rectangular box. Find the height of the box if its length is 18 ft, its width is 8 ft, and its surface area is 496 sq ft. (In the figure, let $h =$ height. Assume that the given surface area includes that of the top lid of the box.)

h

8 ft   18 ft

h is in feet.

27. Which of the following cannot be a correct equation to solve a geometry problem, if x represents the length of a rectangle? (*Hint:* Solve each equation and consider the solutions.)

  (a) $2x + 2(x - 1) = 14$ **(b)** $-2x + 7(5 - x) = 62$
  (c) $4(x + 2) + 4x = 8$ **(d)** $2x + 2(x - 3) = 22$

To find the average of n numbers, find their sum and divide by n. Use this idea to solve Exercises 28 and 29.

28. Hien has grades of 84, 88, and 92 on his first three calculus tests. What grade on his fourth test will give him an average of 90?

29. Jamal scored 78, 94, and 60 on his three trigonometry tests. If his final exam score is to be counted as two test grades in determining his course average, what grade must he make on his final exam to give him an average of 80?

Exercises 30 and 31 depend on the idea of the octane rating of gasoline, a measure of its antiknock qualities. In one measure of octane, a standard fuel is made with only two ingredients: heptane and isooctane. For this fuel, the octane rating is the percent of isooctane. An actual gasoline blend is then compared to a standard fuel. For example, a gasoline with an octane rating of 98 has the same antiknock properties as a standard fuel that is 98% isooctane.

30. How many liters of 94-octane gasoline should be mixed with 200 liters of 99-octane gasoline to get a mixture that is 97-octane?

31. A service station has 92-octane and 98-octane gasoline. How many liters of each should be mixed to provide 12 liters of 96-octane gasoline needed for chemical research?

A person's intelligence quotient (IQ) is found by multiplying the mental age by 100 and dividing by the chronological age. Use this to solve Exercises 32 and 33.

32. Jack is 7 years old. His IQ is 130. Find his mental age.

33. If a person is 16 years old with a mental age of 20, what is the person's IQ?

Solve the problem. See Example 3.

34. In 1994 Leroy Burrell (USA) set a world record in the 100-meter dash with a time of 9.85 seconds. If this pace could be maintained for an entire 26-mile marathon, how would this time compare to the fastest time for a marathon of 2 hours, 6 minutes, and 50 seconds? (*Hint:* 1 meter ≈ 3.281 feet.) (*Source:* International Amateur Athletic Association.)

35. Johnny gets to work in 20 min when he drives his car. Riding his bike (by the same route) takes him 45 min. His average driving speed is 4.5 mph greater than his average speed on his bike. How far does he travel to work?

36. In the morning, Marge drove to a business appointment at 50 mph. Her average speed on the return trip in the afternoon was 40 mph. The return trip took 1/4 hr longer because of heavy traffic. How far did she travel to the appointment?

37. On a vacation trip, Alvaro averaged 50 mph traveling from Denver to Minneapolis. Returning by a different route that covered the same number of miles, he averaged 55 mph. What is the distance between the two cities if his total traveling time was 32 hr?

38. Tanika left by plane to visit her mother in Hartford, 420 km away. Fifteen minutes later, her mother left to meet her at the airport. She drove the 20 km to the airport at 40 km/hr, arriving just as the plane taxied in. What was the speed of the plane?

39. Russ and Janet are running in the Apple Hill Fun Run. Russ runs at 7 mph, Janet at 5 mph. If they start at the same time, how long will it be before they are 1/2 mi apart?

40. If the run in Exercise 39 has a staggered start, and Janet starts first, with Russ starting 10 min later, how long will it be before he catches up with her?

Solve the problem. See Example 4.

41. Two chemical plants are polluting a river. If plant A produces a predetermined maximum amount of pollutant twice as fast as plant B, and together they produce the maximum pollutant in 26 hr, how long will it take plant B alone?

42. A sewage treatment plant has two inlet pipes to its settling pond. One can fill the pond in 10 hours, the other in 12 hours. If the first pipe is open for 5 hours and then the second pipe is opened, how long will it take to fill the pond?

43. An inlet pipe can fill Reynaldo's pool in 5 hr, while an outlet pipe can empty it in 8 hr. In his haste to watch television, Reynaldo left both pipes open. How long did it take to fill the pool?

44. Suppose Reynaldo discovered his error (see Exercise 43) after an hour-long program. If he then closed the outlet pipe, how much more time would be needed to fill the pool?

Solve the problem. See Example 5.

45. A pharmacist wishes to strengthen a mixture from 10% alcohol to 30% alcohol. How much pure alcohol should be added to 7 liters of the 10% mixture?

46. A student needs 10% hydrochloric acid for a chemistry experiment. How much 5% acid should be mixed with 60 ml of 20% acid to get a 10% solution?

47. How much pure acid should be added to 6 liters of 30% acid to increase the concentration to 50% acid?

48. How much water should be added to 8 ml of 6% saline solution to reduce the concentration to 4%?

Solve the problem. See Example 6.

49. Adam Bryer wishes to sell a piece of property for $125,000. He wishes the money to be paid off in two ways—a short-term note at 12% interest and a long-term note at 10%. Find the amount of each note if the total annual interest paid is $13,700.

50. In planning her retirement, Shirley Cicero deposits some money at 4.5% interest with twice as much deposited at 5%. Find the amount deposited at each rate if the total annual interest income is $2900.

51. A church building fund has invested some money in two ways: part of the money at 7% interest and four times as much at 11%. Find the amount invested at each rate if the total annual income from interest is $7650.

52. Orlando won $200,000 in a state lottery. He first paid income tax of 30% on the winnings. Of the rest, he invested some at 8.5% and some at 7%, making $10,700 interest per year. How much is invested at each rate?

53. Marjorie Williams earned $48,000 from royalties on her cookbook. She paid a 28% income tax on these royalties. The balance was invested in two ways, some of it at 6.5% interest and some at 6.25%. The investments produced $2210 interest per year. Find the amount invested at each rate.

54. Anoa bought two plots of land for a total of $120,000. When she sold the first plot, she made a profit of 15%. When she sold the second, she lost 10%. Her total profit was $5500. How much did she pay for each piece of land?

Solve the problem. See Example 7.

55. Ventilation is an effective method for removing indoor air pollutants. According to the American Society of Heating, Refrigerating and Air-Conditioning Engineers, Inc. (ASHRAE), a nonsmoking classroom should have a ventilation rate of 15 cubic feet per minute for each person in the classroom. (*Source:* ASHRAE, 1989.)

(a) Write an equation that describes the total ventilation V (in cubic feet per hour) necessary for a classroom with x students.

(b) A common unit of ventilation is an air change per hour (ach). 1 ach is equivalent to exchanging all of the air in a room every hour. If x students are in a classroom having a volume of 15,000 cubic feet, determine how many air exchanges per hour are necessary to keep the room properly ventilated.

(c) Find the necessary number of ach A if the classroom has 40 students in it.

(d) In areas like bars and lounges that allow smoking, the ventilation rate should be increased to 50 cubic feet per minute per person. Compared to classrooms, ventilation should be increased by what factor in heavy smoking areas?

56. Ski resorts require large amounts of water in order to make snow. Snowmass Ski Area in Colorado plans to pump between 1120 and 1900 gallons of water per minute at least 12 hours per day from Snowmass Creek between mid-October and late December. Environmentalists are concerned about the effects on the ecosystem. (*Source:* York Snow Incorporated.)

(a) Determine an equation that will calculate the *minimum* amount of water A (in gallons) pumped after x days during mid-October to late December.

(b) Find the minimum amount of water pumped in 30 days.

(c) Suppose the water being pumped from Snowmass Creek was used to fill swimming pools. The average backyard swimming pool holds 20,000 gallons of water. Determine an equation that will give the minimum number of pools P that could be filled after x days. How many pools could be filled each day?

(d) In how many days could a minimum of 1000 pools be filled?

57. The excess lifetime cancer risk R is a measure of the likelihood that an individual will develop cancer from a particular pollutant. For example, if $R = .01$ then a person has a 1% increased chance of developing cancer during a lifetime. This would translate into 1 case of cancer for every 100 people during an average lifetime. For nonsmokers exposed to environmental tobacco smoke (passive smokers) $R = 1.5 \times 10^{-3}$. (*Source:* Hines, A., T. Ghosh, S. Layalka, and R. Warder, *Indoor Air Quality & Control*, Prentice Hall, 1993.)

(a) If the average life expectancy is 72 years, what is the excess lifetime cancer risk per year?

(b) Write a linear equation that will give the expected number of cancer cases C per year if there are x passive smokers.

(c) Estimate the number of cancer cases per 100,000 passive smokers.

(d) The excess lifetime risk of death from smoking is $R = .44$. Currently 26% of the U.S. population smoke. If the U.S. population is 260 million, approximate the excess number of deaths caused by smoking each year.

58. 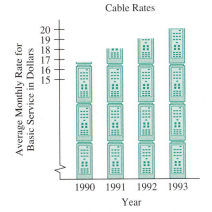 (See Exercise 57.) The excess lifetime cancer risk R for formaldehyde can be calculated using the linear equation $R = kd$ where k is a constant and d is the daily dose in parts per million. The constant k for formaldehyde can be calculated using the formula $k = .132(B/W)$ where B is the total number of cubic meters of air a person breathes in one day and W is a person's weight in kilograms. (*Sources:* A. Hines, T. Ghosh, S. Layalka, and R. Warder, *Indoor Air Quality & Control,* Prentice Hall, 1993; I. Ritchie, and R. Lehnen, "An Analysis of Formaldehyde Concentration in Mobile and Conventional Homes." *J. Env. Health* 47: 300–305.)

(a) Find k for a person that breathes in 20 cubic meters of air per day and weighs 75 kg.

(b) Mobile homes in Minnesota were found to have a mean daily dose d of .42 part per million. Calculate R using the value of k found in part (a).

(c) For every 5000 people, how many cases of cancer could be expected each year from these levels of formaldehyde? Assume an average life expectancy of 72 years.

Cable Rates

Source: National Cable Television Association, Nations Bank, Paul Kagan Associates.

59. The linear model $y = 1.082x + 16.882$ provides the approximate average monthly rate for basic cable television subscribers between the years 1990 and 1993, where $x = 0$ corresponds to 1990, $x = 1$ corresponds to 1991, and so on, and y is in dollars. Use this model to answer the following questions.

(a) What was the approximate average monthly rate in 1991?

(b) What was the approximate average monthly rate in 1993?

(c) During 1994, the rates dropped substantially to $18.86. The model above is based on data from 1990 through 1993. If you were to use the model for 1994, what would the rate be?

(d) Why do you think there is such a discrepancy between the actual rate and the rate based on the model in part (c)? Discuss the pitfalls of using the model to predict for years following 1993.

60. In a survey of 1014 households, ICR Survey Research Group asked the following question: "Which do you trust more to send a message: computer e-mail or the U.S. Postal Service?" The pie chart shows how the respondents answered. Determine the number of people responding in each of the four groups.

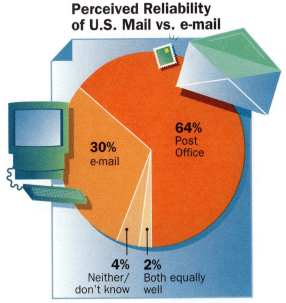

Perceived Reliability of U.S. Mail vs. e-mail

64% Post Office

30% e-mail

4% Neither/ don't know

2% Both equally well

Source: USA Today, 7/6/'95, ICR Survey Research Group, survey of 1,014 households

2.3 Quadratic Equations ▼▼▼

As mentioned earlier, an equation of the form $ax + b = 0$ is a linear equation. A *quadratic equation* is defined as follows.

Quadratic Equation in One Variable

An equation that can be written in the form

$$ax^2 + bx + c = 0,$$

where a, b, and c are real numbers with $a \neq 0$, is a **quadratic equation.**

(Why is the restriction $a \neq 0$ necessary?) A quadratic equation written in the form $ax^2 + bx + c = 0$ is in *standard form*.

The simplest method of solving a quadratic equation, but one that is not always easily applied, is by factoring. This method depends on the following property.

Zero-Factor Property

If a and b are complex numbers, with $ab = 0$, then $a = 0$ or $b = 0$ or both.

The next example shows how the zero-factor property is used to solve a quadratic equation.

◀ **EXAMPLE 1**
Using the zero-factor property

Solve $6r^2 + 7r = 3$.

First write the equation in standard form as

$$6r^2 + 7r - 3 = 0.$$

Now factor $6r^2 + 7r - 3$ to get

$$(3r - 1)(2r + 3) = 0.$$

By the zero-factor property, the product $(3r - 1)(2r + 3)$ can equal 0 only if

$$3r - 1 = 0 \qquad \text{or} \qquad 2r + 3 = 0.$$

Solve each of these linear equations separately to find that the solutions of the original equation are $1/3$ and $-3/2$. Check these solutions by substituting in the original equation. The solution set is $\{1/3, -3/2\}$. ▶

📉 We can use a graphing calculator to solve the equation in Example 1. Recall from your earlier studies that the graph of an equation of the form $y = ax^2 + bx + c$ is a *parabola*. Let y equal the standard form of the quadratic expression, so $y = 6x^2 + 7x - 3$. (We use the variable x instead of r, because graphing calculators typically use the variable x.) Graph the polynomial and use the capabilities of your calculator to find the values of x where y is 0—that is, where the graph crosses the x-axis. The solutions indicated in the windows shown below agree with the solutions found in Example 1.

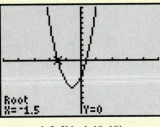

[−5, 5] by [−10, 10]

−1.5 is the decimal form of $-\frac{3}{2}$.

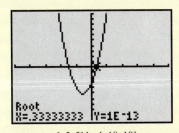

[−5, 5] by [−10, 10]

The expression shown on the screen for y, 1E −13, means 1×10^{-13}. It is as close as the calculator can get to 0 using its internal routines. This will occur often, and the user should be aware of this *limitation of technology*.

A quadratic equation of the form $x^2 = k$ can be solved by factoring with the following sequence of equivalent equations.

$$x^2 = k$$
$$x^2 - k = 0$$
$$(x - \sqrt{k})(x + \sqrt{k}) = 0$$
$$x - \sqrt{k} = 0 \quad \text{or} \quad x + \sqrt{k} = 0$$
$$x = \sqrt{k} \quad \text{or} \quad x = -\sqrt{k}$$

This proves the following statement, which we call the **square root property.**

Square Root Property

The solution set of $x^2 = k$ is $\{\sqrt{k}, -\sqrt{k}\}$.

This solution set is often abbreviated as $\{\pm\sqrt{k}\}$. Both solutions are real if $k > 0$ and imaginary if $k < 0$. If $k < 0$, we write the solution set as $\{\pm i\sqrt{k}\}$. (If $k = 0$, there is only one distinct solution, sometimes called a *double* solution.)

EXAMPLE 2
Using the square root property

Solve each quadratic equation.

(a) $z^2 = 17$
The solution set is $\{\pm\sqrt{17}\}$.

(b) $m^2 = -25$
Since $\sqrt{-25} = 5i$, the solution set of $m^2 = -25$ is $\{\pm 5i\}$.

(c) $(y - 4)^2 = 12$

Use a generalization of the square root property, working as follows.

$$(y - 4)^2 = 12$$
$$y - 4 = \pm\sqrt{12}$$
$$y = 4 \pm \sqrt{12}$$
$$y = 4 \pm 2\sqrt{3}$$

The solution set is $\{4 \pm 2\sqrt{3}\}$. ▶

COMPLETING THE SQUARE   As suggested by Example 2(c), any quadratic equation can be solved using the square root property if it is first written in the form $(x + n)^2 = k$ for suitable numbers n and k. The next example shows how to write a quadratic equation in this form.

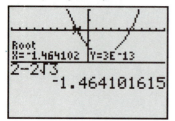

◀ **EXAMPLE 3**
Using the method of completing the square

[−10, 10] by [−15, 10]
Xscl = 1   Yscl = 5

The equation in Example 3, $x^2 - 4x = 8$, is equivalent to $x^2 - 4x - 8 = 0$. We can graph $y = x^2 - 4x - 8$ to support our solutions. The *exact* value of the negative root in Example 3 is $2 - 2\sqrt{3}$. This is supported by the graph and the calculator approximation for $2 - 2\sqrt{3}$. A similar support can be given for $2 + 2\sqrt{3}$.

Solve $x^2 - 4x = 8$.

To write $x^2 - 4x = 8$ in the form $(x + n)^2 = k$, we must find a number that can be added to the left side of the equation to get a perfect square. The equation $(x + n)^2 = k$ can be written as $x^2 + 2xn + n^2 = k$. Comparing this equation with $x^2 - 4x = 8$ shows that

$$2xn = -4x$$
$$n = -2.$$

If $n = -2$, then $n^2 = 4$. Adding 4 to both sides of $x^2 - 4x = 8$ and factoring on the left gives

$$x^2 - 4x + 4 = 8 + 4$$
$$(x - 2)^2 = 12.$$

Now the square root property can be used as follows.

$$x - 2 = \pm\sqrt{12}$$
$$x = 2 \pm 2\sqrt{3}$$

The solution set is $\{2 \pm 2\sqrt{3}\}$. ▶

The steps used in solving a quadratic equation by completing the square follow.

Solving by Completing the Square

To solve $ax^2 + bx + c = 0$, $a \neq 0$, by completing the square:

1. If $a \neq 1$, multiply both sides of the equation by $1/a$.
2. Rewrite the equation so that the constant term is alone on one side of the equals sign.
3. Square half the coefficient of x, and add this square to both sides of the equation.
4. Factor the resulting trinomial as a perfect square and combine terms on the other side.
5. Use the square root property to complete the solution.

EXAMPLE 4
Using the method of
completing the square

Solve $9z^2 - 12z - 1 = 0$.

The coefficient of z^2 must be 1. Multiply both sides by $1/9$.

$$z^2 - \frac{4}{3}z - \frac{1}{9} = 0$$

Now add $1/9$ to both sides of the equation.

$$z^2 - \frac{4}{3}z = \frac{1}{9}$$

Half the coefficient of z is $-2/3$, and $(-2/3)^2 = 4/9$. Add $4/9$ to both sides, getting

$$z^2 - \frac{4}{3}z + \frac{4}{9} = \frac{1}{9} + \frac{4}{9}.$$

Factoring on the left and combining terms on the right gives

$$\left(z - \frac{2}{3}\right)^2 = \frac{5}{9}.$$

Now use the square root property and the quotient property for radicals to get

$$z - \frac{2}{3} = \pm\sqrt{\frac{5}{9}}$$

$$z - \frac{2}{3} = \pm\frac{\sqrt{5}}{3}$$

$$z = \frac{2}{3} \pm \frac{\sqrt{5}}{3}.$$

These two solutions can be written as

$$\frac{2 \pm \sqrt{5}}{3},$$

with the solution set abbreviated as $\left\{\dfrac{2 \pm \sqrt{5}}{3}\right\}$. ▶

QUADRATIC FORMULA   The method of completing the square can be used to solve any quadratic equation. However, in the long run it is better to start with the general quadratic equation,

$$ax^2 + bx + c = 0, \qquad a \neq 0,$$

and use the method of completing the square to solve this equation for x in terms of the constants a, b, and c. The result will be a general formula for solving any quadratic equation. For now, assume that $a > 0$ and multiply both sides by $1/a$ to get

$$x^2 + \frac{b}{a}x + \frac{c}{a} = 0.$$

Add $-c/a$ to both sides.

$$x^2 + \frac{b}{a}x = -\frac{c}{a}$$

Now take half of b/a, and square the result:

$$\frac{1}{2} \cdot \frac{b}{a} = \frac{b}{2a} \qquad \text{and} \qquad \left(\frac{b}{2a}\right)^2 = \frac{b^2}{4a^2}.$$

Add the square to both sides, producing

$$x^2 + \frac{b}{a}x + \frac{b^2}{4a^2} = \frac{b^2}{4a^2} - \frac{c}{a}.$$

The expression on the left side of the equals sign can be written as the square of a binomial, while the expression on the right can be simplified.

$$\left(x + \frac{b}{2a}\right)^2 = \frac{b^2 - 4ac}{4a^2}$$

By the square root property, this last statement leads to

$$x + \frac{b}{2a} = \sqrt{\frac{b^2 - 4ac}{4a^2}} \qquad \text{or} \qquad x + \frac{b}{2a} = -\sqrt{\frac{b^2 - 4ac}{4a^2}}.$$

Since $4a^2 = (2a)^2$, or $4a^2 = (-2a)^2$,

$$x + \frac{b}{2a} = \frac{\sqrt{b^2 - 4ac}}{2a} \qquad \text{or} \qquad x + \frac{b}{2a} = \frac{-\sqrt{b^2 - 4ac}}{2a}.$$

Adding $-b/(2a)$ to both sides of each result gives

$$x = \frac{-b + \sqrt{b^2 - 4ac}}{2a} \qquad \text{or} \qquad x = \frac{-b - \sqrt{b^2 - 4ac}}{2a}.$$

It can be shown that these two results are also valid if $a < 0$. A compact form of these two equations, called the *quadratic formula*, follows.

Quadratic Formula

The solutions of the quadratic equation $ax^2 + bx + c = 0$, where $a \neq 0$, are

$$\frac{-b \pm \sqrt{b^2 - 4ac}}{2a}.$$

CAUTION Notice that the fraction bar in the quadratic formula extends under the $-b$ term in the numerator.

◖**EXAMPLE 5**
Using the quadratic formula
(real solutions)

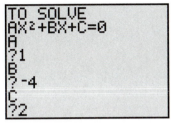

If the program for the TI-82 supplied in this section is used for the equation in Example 5, the screen above is obtained.

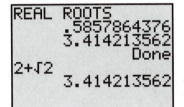

The program displays approximations for the two irrational roots. Notice that the approximation for $2 + \sqrt{2}$ agrees with the second root displayed in the output.

Solve $x^2 - 4x + 2 = 0$.

Here $a = 1$, $b = -4$, and $c = 2$. Substitute these values into the quadratic formula to get

$$x = \frac{-b \pm \sqrt{b^2 - 4ac}}{2a}$$

$$= \frac{-(-4) \pm \sqrt{(-4)^2 - 4(1)2}}{2(1)} \qquad a = 1, b = -4, c = 2$$

$$= \frac{4 \pm \sqrt{16 - 8}}{2}$$

$$= \frac{4 \pm 2\sqrt{2}}{2} \qquad \sqrt{16 - 8} = \sqrt{8} = 2\sqrt{2}$$

$$= \frac{2(2 \pm \sqrt{2})}{2} \qquad \text{Factor out a 2 in the numerator.}$$

$$= 2 \pm \sqrt{2}. \qquad \text{Lowest terms}$$

The solution set is $\{2 + \sqrt{2}, 2 - \sqrt{2}\}$, abbreviated as $\{2 \pm \sqrt{2}\}$. ◗

▱ This excellent quadratic formula program for the TI-81 calculator was
⊞ written by Mr. Charles W. Gantner of Miami-Dade Community College. The authors thank Mr. Gantner and Texas Instruments for permission to include this program here. Similar programs exist for other makes and models of graphing calculators.

: Prgm#: QUADFORM	: Disp R
: ClrHome	: Stop
: Disp "TO SOLVE"	: Lbl 1
: Disp "AX² + BX + C = 0"	: √D/2A → S
: Disp "A"	: R − S → T
: Input A	: R + S → U
: Disp "B"	: Disp "REAL ROOTS"
: Input B	: Disp T
: Disp "C"	: Disp U
: Input C	: Stop
: ClrHome	: Lbl 2
: B² − 4AC → D	: √−D/2A → I
: −B/2A → R	: Disp "COMPLEX ROOTS"
: If D > 0	: Disp "REAL PART"
: Goto 1	: Disp R
: If D < 0	: Disp "+ − IMAGINARY"
: Goto 2	: Disp I
: Disp "DOUBLE ROOT"	

◖EXAMPLE 6
Using the quadratic formula
(imaginary solutions)

The program for the quadratic formula displays the real and imaginary parts of the roots for the equation in Example 6. Note that $\frac{\sqrt{31}}{4} \approx 1.391941091$.

Solve $2x^2 = x - 4$.

To find the values of a, b, and c, first rewrite the equation in standard form as $2x^2 - x + 4 = 0$. Then $a = 2$, $b = -1$, and $c = 4$. By the quadratic formula,

$$x = \frac{-(-1) \pm \sqrt{(-1)^2 - 4(2)(4)}}{2(2)}$$

$$= \frac{1 \pm \sqrt{1 - 32}}{4}$$

$$= \frac{1 \pm \sqrt{-31}}{4}$$

$$= \frac{1 \pm i\sqrt{31}}{4}.$$

The solution set is $\left\{ \dfrac{1}{4} \pm \dfrac{i\sqrt{31}}{4} \right\}$. ◗

CONNECTIONS We know that the quadratic formula gives the solutions of the quadratic equation $ax^2 + bx + c = 0$. Let r_1 and r_2 be the solutions with

$$r_1 = \frac{-b + \sqrt{b^2 - 4ac}}{2a} \quad \text{and} \quad r_2 = \frac{-b - \sqrt{b^2 - 4ac}}{2a}.$$

Then we have

$$r_1 + r_2 = \frac{-b + \sqrt{b^2 - 4ac} + (-b - \sqrt{b^2 - 4ac})}{2a}$$

$$= \frac{-2b}{2a} = -\frac{b}{a}$$

and

$$r_1 \cdot r_2 = \frac{(-b + \sqrt{b^2 - 4ac})(-b - \sqrt{b^2 - 4ac})}{2a \cdot 2a}$$

$$= \frac{(-b)^2 - (\sqrt{b^2 - 4ac})^2}{4a^2}$$

$$= \frac{b^2 - (b^2 - 4ac)}{4a^2} = \frac{4ac}{4a^2} = \frac{c}{a}.$$

These formulas, called the **sum and product of the solutions,** are useful for checking the solutions of a quadratic equation without having to substitute a rational expression involving radicals into the original equation.

> **FOR DISCUSSION OR WRITING**
> 1. Solve each equation and use the formulas for the sum and product of the solutions to check your answers.
>   **(a)** $3x^2 - 5x = 2$ **(b)** $2x^2 = 3x + 1$ **(c)** $3y^2 + 4y + 2 = 0$
> 2. Check the solutions to the equations in item 1(a)–(c) by substitution into the original equation.
> 3. Compare the two methods of checking solutions used in items 1 and 2. Which do you prefer? Why?

The equation in Example 7 is called a *cubic* equation, because of the term of degree 3. In Chapter 4 we will discuss such higher degree equations in more detail. However, the equation $x^3 + 8 = 0$, for example, can be solved using factoring and the quadratic formula.

EXAMPLE 7
Using the quadratic formula in solving a particular cubic equation

Solve $x^3 + 8 = 0$.

Factor on the left side, and then set each factor equal to zero.

$$x^3 + 8 = 0$$
$$(x + 2)(x^2 - 2x + 4) = 0$$
$$x + 2 = 0 \quad \text{or} \quad x^2 - 2x + 4 = 0$$

The solution of $x + 2 = 0$ is $x = -2$. Now use the quadratic formula to solve $x^2 - 2x + 4 = 0$.

$$x^2 - 2x + 4 = 0$$
$$x = \frac{2 \pm \sqrt{4 - 16}}{2} \qquad a = 1, b = -2, c = 4$$
$$x = \frac{2 \pm \sqrt{-12}}{2}$$
$$x = \frac{2 \pm 2i\sqrt{3}}{2}$$
$$x = 1 \pm i\sqrt{3} \qquad \text{Factor out a 2 in the numerator and reduce to lowest terms.}$$

The solution set is $\{-2, 1 \pm i\sqrt{3}\}$. ▶

Sometimes it is necessary to solve a literal equation for a variable that is squared. In such cases, we usually apply the square root property of equations or the quadratic formula.

EXAMPLE 8
Solving for a variable that is squared

(a) Solve for d: $A = \dfrac{\pi d^2}{4}$.

Start by multiplying both sides by 4 to get

$$4A = \pi d^2.$$

Now divide by π.

$$d^2 = \frac{4A}{\pi}$$

Use the square root property and rationalize the denominator on the right.

$$d = \pm\sqrt{\frac{4A}{\pi}}$$

$$d = \frac{\pm 2\sqrt{A}}{\sqrt{\pi}}$$

$$d = \frac{\pm 2\sqrt{A\pi}}{\pi}$$

(b) Solve for t: $rt^2 - st = k$ $(r \neq 0)$.

Because this equation has a term with t as well as t^2, we use the quadratic formula. Subtract k from both sides to get

$$rt^2 - st - k = 0.$$

Now use the quadratic formula to find t, with $a = r$, $b = -s$, and $c = -k$.

$$t = \frac{-b \pm \sqrt{b^2 - 4ac}}{2a}$$

$$t = \frac{-(-s) \pm \sqrt{(-s)^2 - 4(r)(-k)}}{2(r)}$$

$$t = \frac{s \pm \sqrt{s^2 + 4rk}}{2r} \quad \blacktriangleright$$

THE DISCRIMINANT   The quantity under the radical in the quadratic formula, $b^2 - 4ac$, is called the **discriminant.** When the numbers a, b, and c are *integers* (but not necessarily otherwise), the value of the discriminant can be used to determine whether the solutions will be rational, irrational, or imaginary numbers. If the discriminant is 0, there will be only one distinct solution. (Why?)

The discriminant of a quadratic equation gives the following information about the solutions of the equation.

Discriminant

Discriminant	Number of Solutions	Kind of Solutions
Positive, perfect square	Two	Rational
Positive, but not a perfect square	Two	Irrational
Zero	One (a double solution)	Rational
Negative	Two	Imaginary

CAUTION The restriction that a, b, and c be integers is important. For example, for the equation

$$x^2 - \sqrt{5}x - 1 = 0,$$

the discriminant is $b^2 - 4ac = 5 + 4 = 9$, which would otherwise indicate two rational solutions. By the quadratic formula, however, the two solutions

$$x = \frac{\sqrt{5} \pm 3}{2}$$

are *irrational* numbers.

EXAMPLE 9
Using the discriminant

Use the discriminant to determine whether the solutions of $5x^2 + 2x - 4 = 0$ are rational, irrational, or imaginary.

The discriminant is

$$b^2 - 4ac = (2)^2 - (4)(5)(-4) = 84.$$

Because the discriminant is positive and a, b, and c are integers, there are two real number solutions. Since 84 is not a perfect square, the solutions will be irrational numbers. ▶

Recall from Section 1.5 that a rational expression is not defined when its denominator is 0. Restrictions on the variable are found by determining the value or values that cause the expression in the denominator to equal 0.

If the denominator is a quadratic polynomial, we can use the methods of this section to find the restrictions.

EXAMPLE 10
Determining restrictions on the variable

For each of the following, give the real number restrictions on the variable.

(a) $\dfrac{2x - 5}{2x^2 - 9x - 5}$

Set the denominator equal to 0 and solve.

$$2x^2 - 9x - 5 = 0$$
$$(2x + 1)(x - 5) = 0$$
$$2x + 1 = 0 \qquad \text{or} \qquad x - 5 = 0$$
$$x = -\frac{1}{2} \qquad \text{or} \qquad x = 5$$

The restrictions on the variable are $x \neq -1/2$ and $x \neq 5$.

(b) $\dfrac{1}{3x^2 - x + 4}$

Solve $3x^2 - x + 4 = 0$. Since the polynomial does not factor, use the quadratic formula.

$$x = \frac{-(-1) \pm \sqrt{(-1)^2 - 4(3)(4)}}{2(3)} = \frac{1 \pm \sqrt{-47}}{6}$$

Both solutions are imaginary numbers, so there are no real numbers that make the denominator equal to zero. Thus there are no real number restrictions on x. ▶

2.3 Exercises ▼▼▼▼▼▼▼▼▼▼▼▼▼▼▼▼▼▼▼▼▼▼▼▼▼▼▼▼▼▼▼▼▼▼▼▼▼▼

1. Which one of the following equations is set up for direct use of the zero-factor property? Solve it.
   (a) $3x^2 - 17x - 6 = 0$ (b) $(2x + 5)^2 = 7$ (c) $x^2 + x = 12$ (d) $(3x + 1)(x - 7) = 0$

2. Which one of the following equations is set up for direct use of the square root property? Solve it.
   (a) $3x^2 - 17x - 6 = 0$ (b) $(2x + 5)^2 = 7$ (c) $x^2 + x = 12$ (d) $(3x + 1)(x - 7) = 0$

3. Only one of the following equations does not require Step 1 of the method of completing the square described in this section. Which one is it? Solve it.
   (a) $3x^2 - 17x - 6 = 0$ (b) $(2x + 5)^2 = 7$ (c) $x^2 + x = 12$ (d) $(3x + 1)(x - 7) = 0$

4. Only one of the following equations is set up so that the values of a, b, and c can be determined immediately. Which one is it? Solve it.
   (a) $3x^2 - 17x - 6 = 0$ (b) $(2x + 5)^2 = 7$ (c) $x^2 + x = 12$ (d) $(3x + 1)(x - 7) = 0$

Solve the equation by factoring or by the square root property. See Examples 1 and 2.

5. $p^2 = 16$	**6.** $k^2 = 25$	**7.** $x^2 = 27$	**8.** $r^2 = 48$
9. $t^2 = -16$	**10.** $y^2 = -100$	**11.** $x^2 = -18$	**12.** $(p + 2)^2 = 7$
13. $(3k - 1)^2 = 12$	**14.** $(4t + 1)^2 = 20$	**15.** $p^2 - 5p + 6 = 0$	**16.** $q^2 + 2q - 8 = 0$
17. $(5r - 3)^2 = -3$	**18.** $(-2w + 5)^2 = -8$		

Solve the equation by completing the square. See Examples 3 and 4.

19. $p^2 - 8p + 15 = 0$	**20.** $m^2 + 5m = 6$	**21.** $x^2 - 2x - 4 = 0$
22. $r^2 + 8r + 13 = 0$	**23.** $2p^2 + 2p + 1 = 0$	**24.** $9z^2 - 12z + 8 = 0$

25. Francisco claimed that the equation $x^2 - 4x = 0$ cannot be solved by the quadratic formula since there is no value for c. Is he correct?

26. Francesca, Francisco's twin sister, claimed that the equation $x^2 - 17 = 0$ cannot be solved by the quadratic formula since there is no value for b. Is she correct?

Solve the equation by using the quadratic formula. See Examples 5 and 6.

27. $m^2 - m - 1 = 0$	**28.** $y^2 - 3y - 2 = 0$	**29.** $x^2 - 6x + 7 = 0$
30. $11p^2 - 7p + 1 = 0$	**31.** $4z^2 - 12z + 11 = 0$	**32.** $x^2 = 2x - 5$
33. $\dfrac{1}{2}t^2 + \dfrac{1}{4}t - 3 = 0$	**34.** $\dfrac{2}{3}x^2 + \dfrac{1}{4}x = 3$	
35. $4 + \dfrac{3}{x} - \dfrac{2}{x^2} = 0$	**36.** $4 - \dfrac{11}{x} - \dfrac{3}{x^2} = 0$	

Solve the cubic equation by first factoring and then using the quadratic formula. See Example 7.

37. $x^3 - 8 = 0$	**38.** $x^3 - 1 = 0$	**39.** $x^3 - 27 = 0$
40. $x^3 + 27 = 0$	**41.** $x^3 + 64 = 0$	**42.** $x^3 - 64 = 0$

Use any method to solve the equation.

43. $8p^3 + 125 = 0$

44. $2 - \dfrac{5}{k} + \dfrac{2}{k^2} = 0$

45. $(m - 3)^2 = 5$

46. $t^2 - t = 3$

47. $x^2 + x = -1$

48. $64r^3 - 343 = 0$

49. $(3y + 1)^2 = -7$

50. $\dfrac{1}{3}x^2 + \dfrac{1}{6}x + \dfrac{1}{9} = 0$

In this section we explain how a quadratic equation can be solved by a graphing calculator, using a graph or using a program for the quadratic formula. Use one of these methods to find the solutions of the equation. (These equations have only real solutions.) Give as many decimal places as the calculator shows.

51. $x^2 - \sqrt{2}x - 1 = 0$

52. $x^2 - \sqrt{3}x - 2 = 0$

53. $\sqrt{2}x^2 - 3x = -\sqrt{2}$

54. $-\sqrt{6}x^2 - 2x = -\sqrt{6}$

55. $x(x + \sqrt{5}) = -1$

56. $x(3\sqrt{5}x - 2) = \sqrt{5}$

Use a quadratic formula program with complex number capability to find the complex solutions of the equation. Give as many decimal places as the calculator shows.

57. $2x^2 + \sqrt{6}x + 7 = 0$

58. $3x^2 + \sqrt{3}x + 4 = 0$

59. $\sqrt{6}x^2 + 5x = -\sqrt{10}$

60. $-\sqrt{5}x^2 + 3x = \sqrt{13}$

61. $8.4x(x - 1) = -8$

62. $-12.5x(x - 1) = \sqrt{13}$

The built-in equation-solving feature of graphing calculators, first discussed in the exercises for Section 2.1, can be used to solve quadratic equations. For example, if we wish to solve $2x^2 - 11x - 40 = 0$, we obtain the solutions -2.5 and 8 as shown in the accompanying screen. (We used -5 and 5 as our guesses.) Solve the quadratic equation using the equation-solving feature of a calculator. If your model requires guesses, use those given. (Some advanced models do not require such guesses.)

```
solve(2X²-11X-40
,X,-5)
               -2.5
solve(2X²-11X-40
,X,5)
                  8
```

63. $2x^2 - 13x - 7 = 0$ (guesses: -5 and 5)

64. $4x^2 - 11x = 3$ (guesses: 0 and 5)

65. $\sqrt{6}x^2 - 2x - 1.4 = 0$ (guesses: -1 and 2)

66. $\sqrt{10}x(x - 1) = 8.6$ (guesses: -2 and 2)

Solve the equation for the indicated variable. Assume that no denominators are zero. See Example 8.

67. $s = \dfrac{1}{2}gt^2$ for t

68. $A = \pi r^2$ for r

69. $F = \dfrac{kMv^4}{r}$ for v

70. $s = s_0 + gt^2 + k$ for t

71. $P = \dfrac{E^2R}{(r + R)^2}$ for R

72. $S = 2\pi rh + 2\pi r^2$ for r

For the given equation, (a) solve for x in terms of y, and (b) solve for y in terms of x. See Example 8.

73. $4x^2 - 2xy + 3y^2 = 2$

74. $3y^2 + 4xy - 9x^2 = -1$

Identify the values of a, b, and c for the equation and then evaluate the discriminant $b^2 - 4ac$. Use it to predict the type of solutions. Do not solve the equation. See Example 9.

75. $x^2 + 8x + 16 = 0$

76. $x^2 - 5x + 4 = 0$

77. $3m^2 - 5m + 2 = 0$

78. $8y^2 = 14y - 3$

79. $4p^2 = 6p + 3$

80. $2r^2 - 4r + 1 = 0$

81. $9k^2 + 11k + 4 = 0$

82. $3z^2 = 4z - 5$

83. $8x^2 - 72 = 0$

84. Show that the discriminant for the equation $\sqrt{2}m^2 + 5m - 3\sqrt{2} = 0$ is 49. If this equation is completely solved, it can be shown that the solution set is $\{-3\sqrt{2}, \sqrt{2}/2\}$. Here we have a discriminant that is positive and a perfect square, yet the two solutions are irrational. Does this contradict the discussion in this section? Explain.

▼▼▼▼▼▼▼▼▼▼▼▼ **DISCOVERING CONNECTIONS** (Exercises 85–92) ▼▼▼▼▼▼▼▼▼▼▼▼

In the discussion of the discriminant in this section, we saw that the value of the discriminant determines the number of real solutions of a quadratic equation. Suppose we want to determine the value of k for which the equation

$$x^2 + 11x + k = 0$$

has only one real solution. Work Exercises 85–92 in order, and see how this can be done.

85. What must the value of the discriminant be for a quadratic equation to have only one solution?

86. Give an expression for the discriminant of the equation above.

87. Determine an equation that must be solved so that the given equation has only one solution.

88. Solve the equation involving k found in Exercise 87.

89. Write the original equation with the value of k you found in Exercise 88.

90. Solve the equation you wrote in Exercise 89.

91. Use the procedure of this group of exercises to determine the value of k for which $25x^2 - 10x + k = 0$ has only one solution. Then solve the equation.

92. Use the procedure of this group of exercises to determine the *two* values of k for which $16x^2 + kx + 25 = 0$ has only one solution. Then solve the two resulting equations.

Give the real number restrictions on the variable for the rational expression. See Example 10.

93. $\dfrac{3 + 2x}{3x^2 - 19x - 14}$

94. $\dfrac{-7 + 3x}{2x^2 + 7x - 15}$

95. $\dfrac{y - 3}{16y^2 + 8y + 1}$

96. $\dfrac{-18}{25y^2 - 30y + 9}$

97. $\dfrac{-8 + 7x}{x^2 + x + 1}$

98. $\dfrac{-9x + 4}{7x^2 + 2x + 2}$

2.4 Applications of Quadratic Equations ▼▼▼

Many applied problems lead to quadratic equations. In this section we give examples of several kinds of such problems.

CAUTION   When solving problems that lead to quadratic equations, we may get a solution that does not satisfy the physical constraints of the problem. For example, if x represents a width and the two solutions of the quadratic equation are -9 and 1, the value -9 must be rejected, since a width must be a positive number.

EXAMPLE 1
Solving a geometry problem

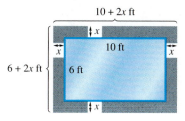

FIGURE 3

A landscape contractor wants to make an exposed gravel border of uniform width around a rectangular pool in a garden. The pool is 10 feet long and 6 feet wide. There is enough material to cover 36 square feet. How wide should the border be?

A diagram of the pool with the border is shown in Figure 3. Since we are asked to find the width of the border, let

$$x = \text{the width of the border in feet.}$$

Then $6 + 2x = $ the width of the larger rectangle in feet,

and $10 + 2x = $ the length of the larger rectangle in feet.

The area of the larger rectangle is $(6 + 2x)(10 + 2x)$ square feet, and the area of the pool is $6 \cdot 10 = 60$ square feet. The area of the border is found by subtracting the area of the pool from the area of the larger rectangle. This difference should be 36 square feet.

Area of the larger rectangle	minus	area of the pool	is	36 square feet.
$(6 + 2x)(10 + 2x)$	$-$	60	$=$	36

Now solve this equation.

$$60 + 32x + 4x^2 - 60 = 36$$
$$4x^2 + 32x - 36 = 0$$
$$x^2 + 8x - 9 = 0$$
$$(x + 9)(x - 1) = 0$$

The solutions are -9 and 1. The width of the border cannot be negative, so the border should be 1 foot wide. ▶

Problems involving rate of work were first introduced in Section 2.2. Recall that if a job can be done in x units of time, the rate of work is $1/x$ job per unit of time.

EXAMPLE 2
Solving a work problem

Pat and Mike clean the offices in a downtown building each night. Working alone, Pat takes 1 hour less time than Mike to complete the job. Working together, they can finish the job in 6 hours. One night Pat calls in sick. How long should it take Mike to do the job alone?

Let

$$x = \text{the time for Mike to do the job alone}$$

and

$$x - 1 = \text{the time for Pat to do the job alone.}$$

The rates for Mike and Pat are, respectively, $1/x$ and $1/(x-1)$ job per hour. If we multiply the time worked together, 6 hours, by each rate, we get the fractional part of the job done by each person. This is summarized in the following chart.

	Rate	Time	Part of the Job Accomplished
Mike	$\dfrac{1}{x}$	6	$6\left(\dfrac{1}{x}\right) = \dfrac{6}{x}$
Pat	$\dfrac{1}{x-1}$	6	$6\left(\dfrac{1}{x-1}\right) = \dfrac{6}{x-1}$

Since one whole job can be done by the two people, the sum of the parts must equal 1, as indicated by the equation

$$\frac{6}{x} + \frac{6}{x-1} = 1.$$

To clear fractions, multiply both sides of the equation by the least common denominator, $x(x-1)$.

$$x(x-1)\frac{6}{x} + x(x-1)\frac{6}{x-1} = 1x(x-1)$$

$$6(x-1) + 6x = x(x-1)$$

$$6x - 6 + 6x = x^2 - x$$

$$0 = x^2 - 13x + 6$$

$$x = \frac{13 \pm \sqrt{(-13)^2 - 4(1)(6)}}{2(1)} \qquad \begin{aligned} a &= 1, \\ b &= -13, \\ c &= 6 \end{aligned}$$

$$x = \frac{13 \pm \sqrt{169 - 24}}{2}$$

$$x = \frac{13 \pm \sqrt{145}}{2}$$

$$x \approx \frac{13 \pm 12.04}{2}$$

Use a calculator to find that to the nearest tenth, $x = 12.5$ or $x = .5$. The solution $x = .5$ does not satisfy the conditions of the problem, since then Pat takes $x - 1 = -.5$ hour to complete the work. It will take Mike 12.5 hours to do the job alone. ▶

EXAMPLE 3
Solving a motion problem

A river excursion boat traveled upstream from Galt to Isleton, a distance of 12 miles. On the return trip downstream, the boat traveled 3 miles per hour faster. If the return trip took 8 minutes less time, how fast did the boat travel upstream?

The chart below summarizes the information in the problem, where x represents the rate upstream.

	d	r	t
Upstream	12	x	$\dfrac{12}{x}$
Downstream	12	$x + 3$	$\dfrac{12}{x + 3}$

$t = \frac{d}{r}$

The entries in the column for time are found by solving the distance formula, $d = rt$, for t in each case. Since rates are given in miles per hour, convert 8 minutes to hours as follows, letting H represent the equivalent number of hours.

$$\frac{H \text{ hr}}{1 \text{ hr}} = \frac{8 \text{ min}}{60 \text{ min}}$$

$$H = \frac{8}{60} = \frac{2}{15}$$

Now write an equation using the fact that the time for the return trip (downstream) was 8 minutes or 2/15 hour less than the time upstream.

Time downstream   is   time upstream   less $\frac{2}{15}$ hour.

$$\frac{12}{x + 3} = \frac{12}{x} - \frac{2}{15}$$

Solve the equation, first multiplying on both sides by the common denominator, $15x(x + 3)$, to get

$$12(15x) = 12(15)(x + 3) - 2x(x + 3)$$
$$180x = 180x + 540 - 2x^2 - 6x$$
$$2x^2 + 6x - 540 = 0 \qquad \text{Standard form}$$
$$x^2 + 3x - 270 = 0 \qquad \text{Divide by 2.}$$
$$(x + 18)(x - 15) = 0$$
$$x = -18 \quad \text{or} \quad x = 15.$$

Reject the negative solution. The boat traveled 15 miles per hour upstream. ▶

CAUTION   When problems involve different units of time (as in Example 3, where rate was given in miles per hour and time was given in minutes), it is necessary to convert to the same unit before setting up the equation.

Example 4 requires the use of the **Pythagorean theorem** from geometry.

> **Pythagorean Theorem**
>
> In a right triangle, the sum of the squares of the lengths of the legs is equal to the square of the length of the hypotenuse.
>
>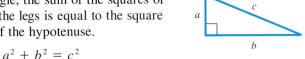
>
> $$a^2 + b^2 = c^2$$

◀EXAMPLE 4
Solving a problem requiring the Pythagorean theorem

A lot is in the shape of a right triangle. The longer leg of the triangle is 20 meters longer than twice the length of the shorter leg. The hypotenuse is 10 meters longer than the longer leg. Find the lengths of the three sides of the lot.

Let s = length of the shorter leg in meters. Then $2s + 20$ meters represents the length of the longer leg, and $(2s + 20) + 10 = 2s + 30$ meters represents the length of the hypotenuse. See Figure 4.

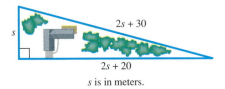

s is in meters.

FIGURE 4

Application of the Pythagorean theorem gives the equation

$$s^2 + (2s + 20)^2 = (2s + 30)^2$$
$$s^2 + 4s^2 + 80s + 400 = 4s^2 + 120s + 900$$
$$s^2 - 40s - 500 = 0$$
$$(s - 50)(s + 10) = 0$$
$$s = 50 \quad \text{or} \quad s = -10.$$

Since s represents a length, the value -10 is not reasonable. The shorter leg is 50 meters long, the longer leg 120 meters long, and the hypotenuse 130 meters long. ▸

CONNECTIONS Three numbers $a, b,$ and c that satisfy the Pythagorean theorem $a^2 + b^2 = c^2$ are called **Pythagorean triples.** There are infinitely many such triples. Various formulas will generate Pythagorean triples. For example, if we choose positive integers r and s, with $r > s$, then the set of equations

$$a = r^2 - s^2, b = 2rs, c = r^2 + s^2$$

generates a Pythagorean triple (a, b, c).

EXAMPLE 5
Solving a problem involving motion of a projectile

If a projectile is shot vertically upward with an initial velocity of 100 feet per second, neglecting air resistance, its height s (in feet) above the ground t seconds after projection is given by

$$s = -16t^2 + 100t.$$

(a) After how many seconds will it be 50 feet above the ground?

We must find the value of t so that $s = 50$. Let $s = 50$ in the equation, and use the quadratic formula.

$$50 = -16t^2 + 100t$$

$$16t^2 - 100t + 50 = 0 \qquad \text{Standard form}$$

$$8t^2 - 50t + 25 = 0 \qquad \text{Divide by 2.}$$

$$t = \frac{-(-50) \pm \sqrt{(-50)^2 - 4(8)(25)}}{2(8)}$$

$$t = \frac{50 \pm \sqrt{1700}}{16}$$

$$t \approx .55 \qquad \text{or} \qquad t \approx 5.70 \qquad \text{Use a calculator.}$$

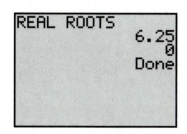

REAL ROOTS
.548058984
5.701941016
Done

The quadratic formula program given in Section 2.3 provides these solutions to the equation in Example 5(a). *Both* are acceptable, due to the interpretation of the variable.

Here, both solutions are acceptable, since the projectile reaches 50 feet twice: once on its way up (after .55 second) and once on its way down (after 5.70 seconds).

(b) How long will it take for the projectile to return to the ground?

When it returns to the ground, its height s will be 0 feet, so let $s = 0$ in the equation.

$$0 = -16t^2 + 100t$$

This can be solved by factoring.

$$0 = -4t(4t - 25)$$

$$-4t = 0 \qquad \text{or} \qquad 4t - 25 = 0$$

$$t = 0 \qquad\qquad\qquad 4t = 25$$

$$t = 6.25$$

REAL ROOTS
6.25
0
Done

While the quadratic formula program gives the two solutions in Example 5(b), we see the need to understand the *concepts.* Technology cannot tell us that 0 is not a valid response here.

The first solution, 0, represents the time at which the projectile was on the ground prior to being launched, so it does not answer the question. The projectile will return to the ground 6.25 seconds after it is launched. ▸

2.4 Exercises ▼▼▼▼▼▼▼▼▼▼▼▼▼▼▼▼▼▼▼▼▼▼▼▼▼▼▼▼▼▼▼▼▼▼▼▼▼▼

Use the concepts introduced in this section to answer the question.

1. For the rectangular parking area of the shopping center shown, which one of the following equations says that the area is 40,000 square yards?

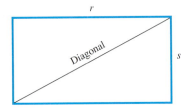

2x + 200

x is in yards.

   (a) $x(2x + 200) = 40,000$
   (b) $2x + 2(2x + 200) = 40,000$
   (c) $x + (2x + 200) = 40,000$
   (d) none of the above

2. If a rectangle is *r* feet long and *s* feet wide, which one of the following is the length of its diagonal in terms of *r* and *s*?

   r

   Diagonal   *s*

   (a) $\sqrt{rs}$ (b) $r + s$
   (c) $\sqrt{r^2 + s^2}$ (d) $r^2 + s^2$

3. To solve for the lengths of the right triangle sides, which equation is correct?

   x   *x* + 4

   2*x* − 2

   (a) $x^2 = (2x - 2)^2 + (x + 4)^2$
   (b) $x^2 + (x + 4)^2 = (2x - 2)^2$
   (c) $x^2 = (2x - 2)^2 - (x + 4)^2$
   (d) $x^2 + (2x - 2)^2 = (x + 4)^2$

4. The mat around the picture shown measures *x* inches across. Which one of the following equations says that the area of the picture itself is 600 square inches?

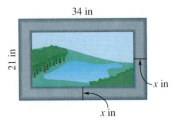

34 in

21 in

x in

x in

   (a) $2(34 - 2x) + 2(21 - 2x) = 600$
   (b) $(34 - 2x)(21 - 2x) = 600$
   (c) $(34 - x)(21 - x) = 600$
   (d) $x(34)(21) = 600$

Solve the problem. See Example 1.

5. A shopping center has a rectangular area of 40,000 sq yd enclosed on three sides for a parking lot. The length is 200 yd more than twice the width. What are the dimensions of the lot? (See Exercise 1.)

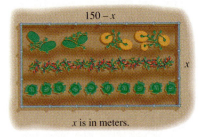

150 − *x*

x

x is in meters.

6. An ecology center wants to set up an experimental garden using 300 m of fencing to enclose a rectangular area of 5000 sq m. Find the dimensions of the rectangle.

7. Greg McRill went into a frame-it-yourself shop. He wanted a frame 3 inches longer than it was wide. The frame he chose extended 1.5 inches beyond the picture on each side. Find the outside dimensions of the frame if the area of the unframed picture is 70 sq inches.

8. Cynthia Herring wants to buy a rug for a room that is 12 ft wide and 15 ft long. She wants to leave a uniform strip of floor around the rug. She can afford to buy 108 sq ft of carpeting. What dimensions should the rug have?

9. A rectangular page in a book is to have an 18 cm by 23 cm illustration in the center with equal margins on all four sides. How wide should the margins be if the page has an area of 594 sq cm?

10. A landscape architect has included a rectangular flower bed measuring 9 ft by 5 ft in her plans for a new building. She wants to use two colors of flowers in the bed, one in the center and the other for a border of the same width on all four sides. If she has enough plants to cover 24 sq ft for the border, how wide can the border be?

Solve the problem. See Example 2.

11. Felipe Martinez can clean a garage in 9 hr less time than his brother Felix. Working together, they can do the job in 20 hr. How long would it take each one to do the job alone?

12. An experienced roofer can do a complete roof in a housing development in half the time it takes an inexperienced roofer. If the two work together on a roof, they complete the job in 2 2/3 hr. How long would it take the experienced roofer to do a roof by himself?

13. Two data processors are working on a special project. The experienced employee could complete the project in 2 hr less time than the new employee. Together they complete the project in 2.4 hr. How long would it have taken the experienced data processor working alone?

14. It takes two copy machines 1.2 hr to make the copies for a company newsletter. One copy machine would take 1 hr longer than the other to do the job alone. How long would it take the faster machine to complete the job alone?

Solve the problem. See Example 3.

15. Paula drives 10 mph faster than Steve. Both start at the same time for Atlanta from Chattanooga, a distance of about 100 mi. It takes Steve 1/3 hr longer than Paula to make the trip. What is Steve's average speed?

16. Marjorie Seachrist walks 1 mph faster than her daughter Sibyl. In a walk for charity, both walked the full distance of 24 mi. Sibyl took 2 hr longer than Marjorie. What was Sibyl's average speed?

17. A plane flew 1000 mi. It later took off and flew 2025 mi at an average speed of 50 mph faster than the first trip. The second trip took 2 hr more time than the first. Find the time for the first trip.

18. The Branson family traveled 100 mi to a lake for their vacation. On the return trip their average speed was 50 mph faster. The total time for the round trip was 11/3 hr. What was the family's average speed on their trip to the lake?

Solve the problem. See Example 4.

19. A kite is flying on 50 ft of string. How high is it above the ground if its height is 10 ft more than the horizontal distance from the person flying it? Assume the string is being released at ground level.

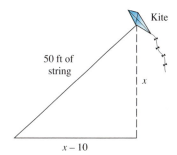

20. A boat is being pulled into a dock with a rope attached at water level. When the boat is 12 ft from the dock, the length of the rope from the boat to the dock is 3 ft longer than twice the height of the dock above the water. Find the height of the dock.

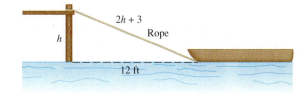

21. Chris and Josh have received walkie-talkies for Christmas. If they leave from the same point at the same time, Chris walking north at 2.5 mph and Josh walking east at 3 mph, how long will they be able to talk to each other if the range of the walkie-talkies is 4 miles? Round your answer to the nearest minute.

22. There is a bamboo 10 ft high, the upper end of which, being broken, reaches the ground 3 ft from the stem. Find the height of the break. (Adapted from an ancient Chinese work, *Arithmetic in Nine Sections.*)

Solve the problem. See Example 5.

23. A toy rocket is launched from ground level. After t seconds, its height h in feet is given by the formula $h = -16t^2 + 128t$. After how many seconds will it reach a height of 80 feet?

24. How long will it take for the rocket in Exercise 23 to return to the ground?

25. An astronaut on the moon throws a baseball upward. The astronaut is 6 feet, 6 inches tall, and the initial velocity of the ball is 30 feet per second. The height of the ball is approximated by the equation $h = -2.7t^2 + 30t + 6.5$, where t is the number of seconds after the ball was thrown. After how many seconds is the ball 12 feet above the moon's surface? How many seconds will it take for the ball to return to the surface?

26. The ball in Exercise 25 will never reach a height of 100 feet. How can this be determined mathematically?

Solve the problem.

 Carbon monoxide (CO) is a dangerous combustion product. It combines with the hemoglobin of the blood to form carboxyhemoglobin (COHb), which reduces transport of oxygen to tissues. A person's health is affected by both the concentration of carbon monoxide in the air and the exposure time. Smokers routinely have a 4% to 6% COHb level in their blood, which can cause symptoms such as blood flow alterations, visual impairment, and poorer vigilance ability. The quadratic equation $T = .00787x^2 - 1.528x + 75.89$ approximates the exposure time in hours necessary to reach this 4% to 6% level where $50 \leq x \leq 100$ is the amount of carbon monoxide present in the air in parts per million (ppm). (Source: Indoor Air Quality Environmental Information Handbook: Combustion Sources, Report No. DOE/EV/10450-1, U.S. Department of Energy, 1985.)

27. A kerosene heater or a room full of smokers is capable of producing 50 ppm of carbon monoxide. How long would it take for a nonsmoking person to start feeling the above symptoms?

28. Find the carbon monoxide concentration necessary for a person to reach the 4% to 6% COHb level in 3 hours.

 High concentrations of carbon monoxide can cause coma and possible death. The time required for a person to reach a COHb level capable of causing a coma can be approximated by the quadratic equation $T = .0002x^2 - .316x + 127.9$ where T is the exposure time in hours necessary to reach this level and $500 \leq x \leq 800$ is the amount of carbon monoxide in parts per million (ppm). (Source: Indoor Air Quality Environmental Information Handbook: Combustion Sources, Report No. DOE/EV/10450-1, U.S. Department of Energy, 1985.)

29. What is the exposure time when $x = 600$ ppm?

30. Estimate the concentration of CO necessary to produce a coma in 4 hours.

31. Based on information provided by the National Park Service, the number y of visitor hours in millions to Federal Recreation Areas between 1985 and 1992 can be approximated by the model $y = -21.99x^2 + 353.44x + 6507.5$, where $x = 0$ corresponds to 1985. Based on this model, in what year was the number of visitor hours about 7500 million?

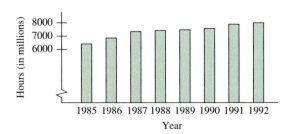

Visitor Hours to Federal Recreation Areas

32. To determine the appropriate landing speed of a small airplane, the formula $.1s^2 - 3s + 22 = D$ is used, where s is the initial landing speed in feet per second and D is the distance needed in feet. If the landing speed is too fast, the pilot may run out of runway; if the speed is too slow, the plane may stall. If the runway is 800 feet long, what is the appropriate landing speed?

A basic fact of economics is this: if we charge x dollars per unit and sell p units, the revenue we obtain is given by the product px. Now use this idea to analyze the following problem, working Exercises 33–38 in order.

The manager of an 80-unit apartment complex knows from experience that at a rent of $300, all the units will be full. On the average, one additional unit will remain vacant for each $20 increase in rent over $300. Furthermore, the manager must keep at least 30 units rented due to other financial considerations. Currently, the revenue from the complex is $35,000. How many apartments are rented?

33. Suppose that x represents the number of $20 increases over $300. How can we represent the number of apartment units that will be rented in terms of x?

34. How can we represent the rent per unit in terms of x?

35. Use the answers in Exercises 33 and 34 to write an expression that defines the revenue generated when there are x $20 increases over $300.

36. According to the problem, the revenue currently generated is $35,000. Write a quadratic equation in standard form using your expression from Exercise 35.

37. Solve the equation from Exercise 36 and answer the question in the problem.

38. Use the concepts from Exercises 33–37 to solve the following problem: A local club is arranging a charter flight to Miami. The cost of the trip is $225 each for 75 passengers, with a refund of $5 per passenger for each passenger in excess of 75. How many passengers must take the flight to produce a revenue of $16,000?

2.5   Other Types of Equations ▼▼▼

Many equations that are not actually quadratic equations can be solved by the methods discussed earlier in this chapter.

EQUATIONS QUADRATIC IN FORM   The equation $12m^4 - 11m^2 + 2 = 0$ is not a quadratic equation because of the m^4 term. However, with the substitutions

$$x = m^2 \quad \text{and} \quad x^2 = m^4$$

the given equation becomes

$$12x^2 - 11x + 2 = 0,$$

which is a quadratic equation. This quadratic equation can be solved to find x, and then $x = m^2$ can be used to find the values of m, the solutions to the original equation.

> ### Quadratic in Form
> An equation is said to be **quadratic in form** if it can be written as
> $$au^2 + bu + c = 0$$
> where $a \neq 0$ and u is some algebraic expression.

EXAMPLE 1
Solving an equation quadratic in form

Solve $12m^4 - 11m^2 + 2 = 0$.

As mentioned above, this equation is quadratic in form. By making the substitution $x = m^2$, the equation becomes

$$12x^2 - 11x + 2 = 0,$$

which can be solved by factoring in the following way.

$$12x^2 - 11x + 2 = 0$$
$$(3x - 2)(4x - 1) = 0$$
$$x = \frac{2}{3} \quad \text{or} \quad x = \frac{1}{4}$$

The original equation contains the variable m. To find m, use the fact that $x = m^2$ and replace x with m^2, getting

$$m^2 = \frac{2}{3} \qquad \text{or} \qquad m^2 = \frac{1}{4}$$

$$m = \pm\sqrt{\frac{2}{3}} \qquad\qquad m = \pm\sqrt{\frac{1}{4}}$$

$$m = \pm\frac{\sqrt{2}}{\sqrt{3}} \cdot \frac{\sqrt{3}}{\sqrt{3}}$$

$$m = \frac{\pm\sqrt{6}}{3} \qquad \text{or} \qquad m = \pm\frac{1}{2}.$$

These four solutions of the given equation $12m^4 - 11m^2 + 2 = 0$ make up the solution set $\{\sqrt{6}/3, -\sqrt{6}/3, 1/2, -1/2\}$, abbreviated as $\{\pm\sqrt{6}/3, \pm1/2\}$. ▶

The equation solved in Example 1 involved a fourth-degree polynomial, and had four solutions. In Chapter 4 we shall see that an nth-degree polynomial equation has at most n solutions.

NOTE Some equations that are quadratic in form, such as the one in Example 1, can be solved quite easily by direct factorization. The polynomial there can be factored as $(3m^2 - 2)(4m^2 - 1)$, and by setting each factor equal to zero, the same solution set is obtained.

To graph the equation in Example 1 using a graphing calculator, we must first consider a suitable window. The solutions found in Example 1 suggest using $[-2, 2]$ for the Xmin and Xmax. It is a good idea to start with the standard window (each graphing calculator has one that can easily be set), and then adjust the Xmin, Xmax, Ymin, and Ymax until the graph looks reasonable. The figures below show the four solutions. Since $\sqrt{6}/3 \approx$ .81649658, the first and last windows show approximations for the solutions $-\sqrt{6}/3$ and $\sqrt{6}/3$, respectively.

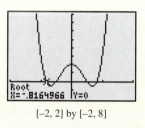

$[-2, 2]$ by $[-2, 8]$ $[-2, 2]$ by $[-2, 8]$

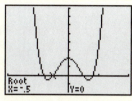

$[-2, 2]$ by $[-2, 8]$ $[-2, 2]$ by $[-2, 8]$

EXAMPLE 2
Solving an equation quadratic in form

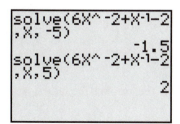

The built-in equation solver supports the solutions in Example 2.

Solve $6p^{-2} + p^{-1} = 2$.

Let $u = p^{-1}$ so that $u^2 = p^{-2}$. Then substitute and rearrange terms to get

$$6u^2 + u - 2 = 0.$$

Factor on the left, and then place each factor equal to 0, giving

$$(3u + 2)(2u - 1) = 0$$

$$3u + 2 = 0 \qquad \text{or} \qquad 2u - 1 = 0$$

$$u = -\frac{2}{3} \qquad \text{or} \qquad u = \frac{1}{2}.$$

Since $u = p^{-1}$, $\qquad p^{-1} = -\dfrac{2}{3} \qquad$ or $\qquad p^{-1} = \dfrac{1}{2}$,

from which $\qquad p = -\dfrac{3}{2} \qquad$ or $\qquad p = 2.$

The solution set of $6p^{-2} + p^{-1} = 2$ is $\{-3/2, 2\}$. ▶

CAUTION When solving an equation that is quadratic in form, if a substitution variable is used, do not forget the step that gives the solution in terms of the original variable.

EQUATIONS WITH RADICALS OR RATIONAL EXPONENTS To solve equations containing radicals or rational exponents, such as $x = \sqrt{15 - 2x}$, or $(x + 1)^{1/2} = x$, use the following property.

> If P and Q are algebraic expressions, then every solution of the equation $P = Q$ is also a solution of the equation $(P)^n = (Q)^n$, for any positive integer n.

CAUTION Be very careful when using this result. It does *not* say that the equations $P = Q$ and $(P)^n = (Q)^n$ are equivalent; it says only that each solution of the original equation $P = Q$ is also a solution of the new equation $(P)^n = (Q)^n$.

When using this property to solve equations, we must be aware that the new equation may have *more* solutions than the original equation. For example, the solution set of the equation $x = -2$ is $\{-2\}$. If we square both sides of the equation $x = -2$, we get the new equation $x^2 = 4$, which has solution set $\{-2, 2\}$. Since the solution sets are not equal, the equations are not equivalent. Because of this, when an equation contains radicals or rational exponents, **it is *essential* to check all proposed solutions in the original equation.**

◀EXAMPLE 3
Solving an equation containing a radical

Solve $x = \sqrt{15 - 2x}$.

The equation $x = \sqrt{15 - 2x}$ can be solved by squaring both sides as follows.

$$x^2 = (\sqrt{15 - 2x})^2$$
$$x^2 = 15 - 2x$$
$$x^2 + 2x - 15 = 0$$
$$(x + 5)(x - 3) = 0$$
$$x = -5 \quad \text{or} \quad x = 3$$

Now the proposed solutions *must* be checked in the original equation, $x = \sqrt{15 - 2x}$.

If $x = -5$,		If $x = 3$,	
$x = \sqrt{15 - 2x}$		$x = \sqrt{15 - 2x}$	
$-5 = \sqrt{15 - 2(-5)}$	?	$3 = \sqrt{15 - 2(3)}$	?
$-5 = \sqrt{15 + 10}$	?	$3 = \sqrt{15 - 6}$	?
$-5 = \sqrt{25}$	?	$3 = \sqrt{9}$	?
$-5 = 5.$	False	$3 = 3.$	True

As this check shows, only 3 is a solution, giving the solution set $\{3\}$. ▶

📊 A graphing calculator-generated graph of the equation in Example 3, $x = \sqrt{15 - 2x}$, represented by $Y_1 = x - \sqrt{15 - 2x}$, is shown on the left.

The graph shows just the one solution, 3. When $x = -5$, the graph shows $y = -10$. The corresponding equation with both sides squared, $x^2 = 15 - 2x$, represented by $Y_2 = x^2 - 15 + 2x$, however, has the two solutions indicated in the second and third calculator-generated graphs.

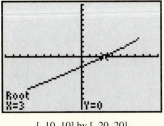

[−10, 10] by [−20, 20]
Xscl = 1   Yscl = 2

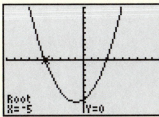

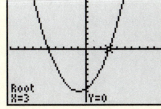

[−10, 10] by [−20, 20]

FOR DISCUSSION OR WRITING
The graph of $Y_1 = x - \sqrt{15 - 2x}$ in the first figure looks like a straight line. Do you think it is? Explain why or why not.

To solve an equation containing radicals, we follow these steps.

Solving an Equation Involving Radicals

1. Isolate the radical on one side of the equation.
2. Raise each side of the equation to a power that is the same as the index of the radical so that the radical is eliminated.
3. Solve the resulting equation. If it still contains a radical, repeat Steps 1 and 2.
4. Check each proposed solution in the *original* equation.

◀ **EXAMPLE 4**
Solving an equation containing two radicals

Solve $\sqrt{2x + 3} - \sqrt{x + 1} = 1$.

When an equation contains two radicals, begin by isolating one of the radicals on one side of the equation. For this one, let us isolate $\sqrt{2x + 3}$ (Step 1).

$$\sqrt{2x + 3} = 1 + \sqrt{x + 1}$$

Now square both sides (Step 2). Be very careful when squaring on the right side of this equation. Recall that $(a + b)^2 = a^2 + 2ab + b^2$; replace a with 1 and b with $\sqrt{x + 1}$ to get the next equation, the result of squaring both sides of $\sqrt{2x + 3} = 1 + \sqrt{x + 1}$.

$$2x + 3 = 1 + 2\sqrt{x + 1} + x + 1$$
$$x + 1 = 2\sqrt{x + 1}$$

One side of the equation still contains a radical; to eliminate it, square both sides again (Step 3).

$$x^2 + 2x + 1 = 4(x + 1)$$
$$x^2 - 2x - 3 = 0$$
$$(x - 3)(x + 1) = 0$$
$$x = 3 \qquad \text{or} \qquad x = -1$$

Check these proposed solutions in the original equation (Step 4).

Let $x = 3$.

$$\sqrt{2x + 3} - \sqrt{x + 1} = 1$$
$$\sqrt{2(3) + 3} - \sqrt{3 + 1} = 1 \quad ?$$
$$\sqrt{9} - \sqrt{4} = 1 \quad ?$$
$$3 - 2 = 1 \quad ?$$
$$1 = 1 \quad \text{True}$$

Let $x = -1$.

$$\sqrt{2x + 3} - \sqrt{x + 1} = 1$$
$$\sqrt{2(-1) + 3} - \sqrt{-1 + 1} = 1 \quad ?$$
$$\sqrt{1} - \sqrt{0} = 1 \quad ?$$
$$1 - 0 = 1 \quad ?$$
$$1 = 1 \quad \text{True}$$

Both proposed solutions 3 and -1 are solutions of the original equation, giving $\{3, -1\}$ as the solution set. ▶

EXAMPLE 5
Solving an equation containing a rational exponent

Solve $(5x^2 - 6)^{1/4} = x$.

Since the equation involves a fourth root, begin by raising both sides to the fourth power.

$$[(5x^2 - 6)^{1/4}]^4 = x^4$$
$$5x^2 - 6 = x^4$$
$$x^4 - 5x^2 + 6 = 0$$

Now substitute y for x^2.

$$y^2 - 5y + 6 = 0$$
$$(y - 3)(y - 2) = 0$$
$$y = 3 \qquad \text{or} \qquad y = 2$$

Since $y = x^2$,

$$x^2 = 3 \qquad \text{or} \qquad x^2 = 2$$
$$x = \pm\sqrt{3} \qquad \text{or} \qquad x = \pm\sqrt{2}.$$

Checking the four proposed solutions, $\sqrt{3}, -\sqrt{3}, \sqrt{2}$, and $-\sqrt{2}$ in the original equation shows that only $\sqrt{3}$ and $\sqrt{2}$ are solutions, so the solution set is $\{\sqrt{3}, \sqrt{2}\}$. ▶

NOTE In the equation of Example 5, we can use the fact that $b^{1/4} = \sqrt[4]{b}$ is a principal fourth root, and thus the right side, x, cannot be negative. Therefore, the two negative proposed solutions must be rejected.

2.5 Exercises ▼▼▼▼▼▼▼▼▼▼▼▼▼▼▼▼▼▼▼▼▼▼▼▼▼▼▼▼▼▼▼▼▼▼▼

1. Only one of the following equations can be solved by the method of substitution described in this section. Which one is it?

(a) $x^9 + x^3 - 2 = 0$ **(b)** $x^6 + 2x^2 - 4 = 0$ **(c)** $x^8 + x^6 + x^4 = 0$ **(d)** $x^8 - 4x^4 - 5 = 0$

2. What is wrong with the following solution?

$$\text{Solve } 4x^4 - 11x^2 - 3 = 0.$$

$$\text{Let } t = x^2.$$

$$4t^2 - 11t - 3 = 0$$

$$(4t + 1)(t - 3) = 0$$

$$4t + 1 = 0 \qquad \text{or} \qquad t - 3 = 0$$

$$t = -\frac{1}{4} \qquad \text{or} \qquad t = 3$$

The solution set is $\{-1/4, 3\}$.

3. What is wrong with the following solution?

$$\text{Solve } x = \sqrt{3x + 4}.$$

Square both sides to get

$$x^2 = 3x + 4$$

$$x^2 - 3x - 4 = 0$$

$$(x - 4)(x + 1) = 0$$

$$x - 4 = 0 \qquad \text{or} \qquad x + 1 = 0$$

$$x = 4 \qquad\qquad\qquad x = -1.$$

The solution set is $\{4, -1\}$.

4. Without solving the equation, how can we tell that $x^{1/4} = -4$ has no real solutions?

Solve the equation by using the method of substitution to rewrite it as a quadratic equation. See Examples 1 and 2.

5. $m^4 + 2m^2 - 15 = 0$

6. $3k^4 + 10k^2 - 25 = 0$

7. $2r^4 - 7r^2 + 5 = 0$

8. $4x^4 - 8x^2 + 3 = 0$

9. $(g - 2)^2 - 6(g - 2) + 8 = 0$

10. $(p + 2)^2 - 2(p + 2) - 15 = 0$

11. $-(r + 1)^2 - 3(r + 1) + 3 = 0$

12. $-2(z - 4)^2 + 2(z - 4) + 3 = 0$

13. $6(k + 2)^4 - 11(k + 2)^2 + 4 = 0$

14. $8(m - 4)^4 - 10(m - 4)^2 + 3 = 0$

15. $7p^{-2} + 19p^{-1} = 6$

16. $5k^{-2} - 43k^{-1} = 18$

An equation such as

$$(r - 1)^{2/3} + (r - 1)^{1/3} - 12 = 0$$

can be solved using the substitution method. Notice that the larger rational exponent, $2/3$, is twice the smaller one, $1/3$. Thus we can let $u = (r - 1)^{1/3}$ and proceed as usual. Solve the equation using this method.

17. $(r - 1)^{2/3} + (r - 1)^{1/3} - 12 = 0$

18. $(y + 3)^{2/3} - 2(y + 3)^{1/3} - 3 = 0$

Solve the equation. See Example 3.

19. $\sqrt{3z + 7} = 3z + 5$ **20.** $\sqrt{4r + 13} = 2r - 1$ **21.** $\sqrt{4k + 5} - 2 = 2k - 7$

22. $\sqrt{6m + 7} - 1 = m + 1$ **23.** $\sqrt{4x} - x + 3 = 0$ **24.** $\sqrt{2t} - t + 4 = 0$

25. Refer to the equation in Exercise 19. A student attempted to solve the equation by "squaring both sides" to get $3z + 7 = 9z^2 + 25$. What was incorrect about the student's method?

26. Refer to the equation in Exercise 32. What should be the first step in solving the equation using the method described in this section?

Solve the equation. See Example 4.

27. $\sqrt{y} = \sqrt{y - 5} + 1$ **28.** $\sqrt{2m} = \sqrt{m + 7} - 1$ **29.** $\sqrt{m + 7} + 3 = \sqrt{m - 4}$

30. $\sqrt{r + 5} - 2 = \sqrt{r - 1}$ **31.** $\sqrt{2z} = \sqrt{3z + 12} - 2$ **32.** $\sqrt{5k + 1} - \sqrt{3k} = 1$

33. $\sqrt{r + 2} = 1 - \sqrt{3r + 7}$ **34.** $\sqrt{2p - 5} - 2 = \sqrt{p - 2}$

Solve the equation. See Example 5.

35. $\sqrt[3]{4n + 3} = \sqrt[3]{2n - 1}$ **36.** $\sqrt[3]{2z} = \sqrt[3]{5z + 2}$ **37.** $\sqrt[3]{t^2 + 2t - 1} = \sqrt[3]{t^2 + 3}$

38. $\sqrt[3]{2x^2 - 5x + 4} = \sqrt[3]{2x^2}$ **39.** $(2r + 5)^{1/3} = (6r - 1)^{1/3}$ **40.** $(3m + 7)^{1/3} = (4m + 2)^{1/3}$

41. $(z^2 + 24z)^{1/4} = 3$ **42.** $(3t^2 + 52t)^{1/4} = 4$ **43.** $(2r - 1)^{2/3} = r^{1/3}$

44. $(z - 3)^{2/5} = (4z)^{1/5}$

▼▼▼▼▼▼▼▼▼▼▼▼▼ **DISCOVERING CONNECTIONS** (Exercises 45–48) ▼▼▼▼▼▼▼▼▼▼▼▼▼

In this section we introduced methods of solving equations quadratic in form by substitution and solving equations involving radicals by raising both sides of the equation to a power. Suppose we wish to solve

$$x - \sqrt{x} - 12 = 0.$$

We can solve this equation using either of the two methods. Work Exercises 45–48 in order, so that you can see how both methods apply.

45. Let $u = \sqrt{x}$ and solve the equation by substitution. What is the value of u that does not lead to a solution of the equation?

46. Solve the equation by isolating $\sqrt{x}$ on one side and then squaring. What is the value of x that does not satisfy the equation?

47. Which one of the methods used in Exercises 45 and 46 do you prefer? Why?

48. Solve $3x - 2\sqrt{x} - 8 = 0$ using one of the two methods described.

The graphing calculator feature in this section illustrates how a graph can support the solution of the equation in Example 3. The screen to the right shows how the equation-solving feature of a graphing calculator can also provide the solution. Use a graphing calculator to find the single real solution of the equation. Use either method. If the equation-solving feature of the calculator requires a guess, use the guess suggested.

49. $2\sqrt{x} - \sqrt{3x + 4} = 0$ (Guess: 0)   **50.** $2\sqrt{x} - \sqrt{5x - 16} = 0$ (Guess: 10)

51. $\sqrt{x^2} - \sqrt{5x + 6} = x + \pi$ (Guess: 0)   **52.** $\sqrt{x^2 - \pi x + \pi} = x + \sqrt{2}$ (Guess: 0)

53. What is wrong with the following solution?

Solve $x^4 - x^2 = 0$.

Since x^2 is a common factor, divide both sides by x^2 to get

$$x^2 - 1 = 0$$
$$(x - 1)(x + 1) = 0$$
$$x = 1 \quad \text{or} \quad x = -1.$$

The solution set is $\{-1, 1\}$.

Solve the equation for the indicated variable. Assume that all denominators are nonzero.

54. $d = k\sqrt{h}$ for h

55. $x^{2/3} + y^{2/3} = a^{2/3}$ for y

56. $m^{3/4} + n^{3/4} = 1$ for m

2.6 Variation ▼▼▼

In many applications of mathematics, it is necessary to express relationships between quantities. For example, in chemistry, the ideal gas law shows how temperature, pressure, and volume are related. In physics, various formulas in optics show how the focal length of a lens and the size of an image are related.

DIRECT VARIATION   When a change in one quantity causes a proportional change in another quantity, the two quantities are said to *vary directly*. For example, if you work for an hourly wage of $6, then [pay] = 6[hours worked]. Doubling the hours worked doubles the pay. Tripling the hours worked triples the pay, and so on. This is stated more precisely as follows.

Direct Variation

y **varies directly** as x, or y is **directly proportional** to x, if a nonzero real number k, called the **constant of variation,** exists such that

$$y = kx.$$

The phrase "directly proportional" is sometimes abbreviated to just "proportional."

◀ EXAMPLE 1
Solving a direct variation problem

Suppose the area of a certain rectangle varies directly as the length. If the area is 50 square meters when the length is 10 meters, find the area when the length is 25 meters.

Since the area varies directly as the length,

$$A = kl,$$

where A represents the area of the rectangle, l is the length, and k is a nonzero constant that must be found. Since $A = 50$ when $l = 10$, the equation $A = kl$ becomes

$$50 = 10k \quad \text{or} \quad k = 5.$$

Using this value of k, the relationship between the area and the length can be expressed as

$$A = 5l.$$

To find the area when the length is 25, replace l with 25 to get

$$A = 5l = 5(25) = 125.$$

The area of the rectangle is 125 square meters when the length is 25 meters. ▶

Sometimes y varies as a power of x. In this case y is a polynomial function of x.

Direct Variation as *n*th Power

Let n be a positive real number. Then y **varies directly as the *n*th power** of x, or y is **directly proportional to the *n*th power** of x, if a nonzero real number k exists such that

$$y = kx^n.$$

For example, the area of a square of side x is given by the formula $A = x^2$, so that the area varies directly as the square of the length of a side. Here $k = 1$.

INVERSE VARIATION   The case where y increases as x decreases is an example of inverse variation. In this case, the product of the variables is constant.

Inverse Variation

Let n be a positive real number. Then y **varies inversely as the *n*th power** of x, or y is **inversely proportional to the *n*th power** of x, if a nonzero real number k exists such that

$$y = \frac{k}{x^n}.$$

If $n = 1$, then $y = k/x$, and y **varies inversely** as x.

◀**EXAMPLE 2**
Solving an inverse variation problem

In a certain manufacturing process, the cost of producing a single item varies inversely as the square of the number of items produced. If 100 items are produced, each costs $2. Find the cost per item if 400 items are produced.

Let x represent the number of items produced and y the cost per item, and write

$$y = \frac{k}{x^2}$$

for some nonzero constant k. Since $y = 2$ when $x = 100$,

$$2 = \frac{k}{100^2} \qquad \text{or} \qquad k = 20{,}000.$$

Thus, the relationship between x and y is given by

$$y = \frac{20{,}000}{x^2}.$$

When 400 items are produced, the cost per item is

$$y = \frac{20{,}000}{400^2} = .125, \text{ or } 12.5\text{¢}. \quad \blacktriangleright$$

The steps involved in solving a variation problem are summarized here.

Solving Variation Problems

1. Write the general relationship among the variables as an equation. Use the constant k.
2. Substitute given values of the variables and find the value of k.
3. Substitute this value of k into the equation from Step 1, obtaining a specific formula.
4. Solve for the required unknown.

COMBINED AND JOINT VARIATION   One variable may depend on more than one other variable. Such variation is called *combined variation*. More specifically, when a variable depends on the *product* of two or more other variables, it is referred to as *joint variation*.

Joint Variation

Let m and n be real numbers. Then y **varies jointly** as the nth power of x and the mth power of z if a nonzero real number k exists such that

$$y = kx^n z^m.$$

◀ **EXAMPLE 3**
Solving a joint variation problem

The area of a triangle varies jointly as the lengths of the base and the height. A triangle with a base of 10 feet and a height of 4 feet has an area of 20 square feet. Find the area of a triangle with a base of 3 centimeters and a height of 8 centimeters.

Let A represent the area, b the base, and h the height of the triangle. Then

$$A = kbh$$

for some number k. Since A is 20 when b is 10 and h is 4,

$$20 = k(10)(4)$$

$$\frac{1}{2} = k.$$

Then

$$A = \frac{1}{2}bh,$$

which is the familiar formula for the area of a triangle. When $b = 3$ centimeters and $h = 8$ centimeters,

$$A = \frac{1}{2}(3)(8) = 12 \text{ square centimeters.} \quad \blacktriangleright$$

The final example in this section shows combined variation.

EXAMPLE 4
Solving a combined variation problem

The number of vibrations per second (the pitch) of a steel guitar string varies directly as the square root of the tension and inversely as the length of the string. If the number of vibrations per second is 5 when the tension is 225 kilograms and the length is .60 meter, find the number of vibrations per second when the tension is 196 kilograms and the length is .65 meter.

Let n represent the number of vibrations per second, T represent the tension, and L represent the length of the string. Then, from the information in the problem,

$$n = \frac{k\sqrt{T}}{L}.$$

Substitute the given values for n, T, and L to find k.

$$5 = \frac{k\sqrt{225}}{.60} \qquad \text{Let } n = 5, T = 225, L = .60.$$

$$3 = k\sqrt{225} \qquad \text{Multiply by .60.}$$

$$3 = 15k \qquad \sqrt{225} = 15$$

$$k = \frac{1}{5} = .2 \qquad \text{Divide by 15.}$$

Now substitute for k and use the second set of values for T and L to find n.

$$n = \frac{.2\sqrt{196}}{.65} \qquad \text{Let } k = .2, T = 196, L = .65.$$

$$n \approx 4.3$$

The number of vibrations per second is 4.3. $\quad \blacktriangleright$

2.6 Exercises ▼▼▼▼▼▼▼▼▼▼▼▼▼▼▼▼▼▼▼▼▼▼▼▼▼▼▼▼▼▼▼

A formula from geometry or physics is given. Fill in the blank or blanks with the appropriate response.

1. $A = 5w$; Area of a rectangle with length 5
The area of this rectangle varies directly as its _____ .
The constant of variation is _____ .

2. $A = \frac{1}{2}(4)h$; Area of a triangle with base 4
The area of this triangle varies directly as its _____ .
The constant of variation is _____ .

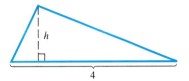

3. $A = \pi r^2$; Area of a circle with radius r
The area of a circle varies _____ as the _____ of its _____ . The constant of variation is _____ .

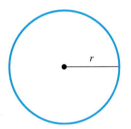

4. $d = 16t^2$; Distance a falling object has traveled after t seconds
The distance a falling object has traveled varies _____ as the _____ of _____ . The constant of variation is _____ .

5. $C = 2\pi r$; Circumference of a circle with radius r
The circumference of a circle varies _____ as its _____ . The constant of variation is _____ .

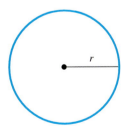

6. $V = 36h$; Volume of a rectangular solid with a square base measuring 6 by 6
The volume of this solid varies _____ as its _____ .
The constant of variation is _____ .

7. $b = \dfrac{24}{h}$; Base of a parallelogram with area 24

The base of this parallelogram varies _____ as its _____. The constant of variation is _____.

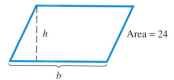

h Area = 24

b

8. $h = \dfrac{300/\pi}{r^2}$; Height of a right circular cone with volume 100

The height of this cone varies _____ as the _____ of the _____ of its base. The constant of variation is _____.

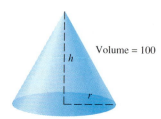

h Volume = 100

r

Solve the variation problem. See Examples 1–4.

9. If m varies directly as x and y, and $m = 10$ when $x = 4$ and $y = 7$, find m when $x = 11$ and $y = 8$.

10. Suppose m varies directly as z and p. If $m = 10$ when $z = 3$ and $p = 5$, find m when $z = 5$ and $p = 7$.

11. Suppose r varies directly as the square of m, and inversely as s. If $r = 12$ when $m = 6$ and $s = 4$, find r when $m = 4$ and $s = 10$.

12. Suppose p varies directly as the square of z, and inversely as r. If $p = 32/5$ when $z = 4$ and $r = 10$, find p when $z = 2$ and $r = 16$.

13. Let a be proportional to m and n^2, and inversely proportional to y^3. If $a = 9$ when $m = 4, n = 9$, and $y = 3$, find a if $m = 6$, $n = 2$, and $y = 5$.

14. If y varies directly as x, and inversely as m^2 and r^2, and $y = 5/3$ when $x = 1$, $m = 2$, and $r = 3$, find y if $x = 3$, $m = 1$, and $r = 8$.

▼▼▼▼▼▼▼▼▼▼▼▼▼ **DISCOVERING CONNECTIONS** (Exercises 15–18) ▼▼▼▼▼▼▼▼▼▼▼▼▼

In Section 2.1 we introduced a method for solving equations involving fractions: multiply both sides by the least common denominator of the fractions. We can use this method to solve an equation that biologists use to estimate the number of fish in a lake.

Here is the procedure biologists use: they first catch a sample of fish and mark each specimen with a harmless tag. Some weeks later, they catch a similar sample of fish from the same areas of the lake and determine the proportion of previously tagged fish in the new sample. The total fish population is estimated by assuming that the proportion of tagged fish in the new sample is the same as the proportion of tagged fish in the entire lake.

Now let us consider the following problem.

Suppose a group of biologists tag 300 fish in a lake known as False River on May 1. When they return and take a new sample of 400 fish on June 1, 5 of the 400 were tagged previously.

Work Exercises 15–18 in order, so that you can see how the biologists estimate the number of fish in the lake.

15. Let x represent the total fish population in the lake. Set up two fractions of the form

$$\frac{\text{number tagged}}{\text{fish population}},$$

and set the two fractions equal to each other.

16. What is the least common denominator of the fractions in the equation?

17. Multiply both sides of the equation to obtain a linear equation in x.

18. Solve the equation. About how many fish are there in False River?

19. For $k > 0$, if y varies directly as x, when x increases, y _____ , and when x decreases, y

_____ .

20. For $k > 0$, if y varies inversely as x, when x increases, y _____ , and when x decreases,

y _____ .

21. What happens to y if y varies inversely as x, and x is doubled?

22. If y varies directly as x, and x is halved, how is y changed?

23. Suppose y is directly proportional to x, and x is replaced by $(1/3)x$. What happens to y?

24. What happens to y if y is inversely proportional to x, and x is tripled?

25. Suppose p varies directly as r^3 and inversely as t^2. If r is halved and t is doubled, what happens to p?

26. If m varies directly as p^2 and q^4, and p doubles while q triples, what happens to m?

Solve the problem. A calculator will be helpful in performing the arithmetic. See Examples 1–4.

27. Hooke's law for an elastic spring states that the distance a spring stretches varies directly as the force applied. If a force of 15 lb stretches a certain spring 8 in, how much will a force of 30 lb stretch the spring?

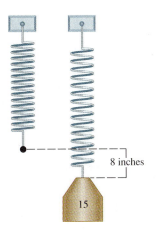

8 inches

15

28. In electric current flow, it is found that the resistance (measured in units called ohms) offered by a fixed length of wire of a given material varies inversely as the square of the diameter of the wire. If a wire .01 inch in diameter has a resistance of .4 ohm, what is the resistance of a wire of the same length and material with a diameter of .03 inch?

29. The illumination produced by a light source varies inversely as the square of the distance from the source. The illumination of a light source at 5 m is 70 candela. What is the illumination 12 m from the source?

30. The pressure exerted by a certain liquid at a given point is proportional to the depth of the point below the surface of the liquid. If the pressure 20 m below the surface is 70 kg per cm^2, what pressure is exerted 40 m below the surface?

31. The distance that a person can see to the horizon from a point above the surface of the earth varies directly as the square root of the height of the point above the earth. A person on a hill 121 m high can see for 15 km to the horizon. How far can a person see to the horizon from a hill 900 m high?

32. Simple interest varies jointly as principal and time. If $1000 left at interest for 2 yr earned $110, find the amount of interest earned by $5000 for 5 yr.

33. The volume of a right circular cylinder is jointly proportional to the square of the radius of the circular base and to the height. If the volume is 300 cu cm when the height is 10.62 cm and the radius is 3 cm, find the volume of a cylinder with a radius of 4 cm and a height of 15.92 cm.

34. The roof of a new sports arena rests on round concrete pillars. The maximum load a cylindrical column of circular cross section can hold varies directly as the fourth power of the diameter and inversely as the square of the height. The arena has 9 m tall columns that are 1 m in diameter and will support a load of 8 metric tons. How many metric tons will be supported by a column 12 m high and 2/3 m in diameter?

35. The sports arena in Exercise 34 requires a beam 16 m long, 24 cm wide, and 8 cm high. The maximum load of a horizontal beam that is supported at both ends varies directly as the width and square of the height and inversely as the length between supports. If a beam of the same material 8 m long, 12 cm wide, and 15 cm high can support a maximum of 400 kg, what is the maximum load the beam in the arena will support?

36. The period of a pendulum varies directly as the square root of the length of the pendulum and inversely as the square root of the acceleration due to gravity. Find the period when the length is 121 cm and the acceleration due to gravity is 980 cm per sec squared, if the period is 6π sec when the length is 289 cm and the acceleration due to gravity is 980 cm per sec squared.

37. The maximum speed possible on a length of railroad track is directly proportional to the cube root of the amount of money spent on maintaining the track. Suppose that a maximum speed of 25 km/hr is possible on a stretch of track for which $450,000 was spent on maintenance. Find the maximum speed if the amount spent on maintenance is increased to $1,750,000.

38. The number of long-distance phone calls between two cities in a certain time period varies directly as the populations p_1 and p_2 of the cities, and inversely as the distance between them. If 10,000 calls are made between two cities 500 mi apart, having populations of 50,000 and 125,000, find the number of calls between two cities 800 mi apart having populations of 20,000 and 80,000.

39. According to Poiseuille's law, the resistance to flow of a blood vessel, R, is directly proportional to the length, l, and inversely proportional to the fourth power of the radius, r. If $R = 25$ when $l = 12$ and $r = .2$, find R as r increases to .3, while l is unchanged.

40. The Stefan-Boltzmann law says that the radiation of heat R from an object is directly proportional to the fourth power of the Kelvin temperature of the object. For a certain object, $R = 213.73$ at room temperature (293° Kelvin). Find R if the temperature increases to 335° Kelvin.

41. Suppose a nuclear bomb is detonated at a certain site. The effects of the bomb will be felt over a distance from the point of detonation that is directly proportional to the cube root of the yield of the bomb. Suppose a 100-kiloton bomb has certain effects to a radius of 3 km from the point of detonation. Find the distance that the effects would be felt for a 1500-kiloton bomb.

42. A measure of malnutrition, called the *pelidisi,* varies directly as the cube root of a person's weight in grams and inversely as the person's sitting height in centimeters. A person with a pelidisi below 100 is considered to be undernourished, while a pelidisi greater than 100 indicates overfeeding. A person who weighs 48,820 g with a sitting height of 78.7 cm has a pelidisi of 100. Find the pelidisi (to the nearest whole number) of a person whose weight is 54,430 g and whose sitting height is 88.9 cm. Is this individual undernourished or overfed?

Variation can be seen extensively in the field of photography. The formula $L = \dfrac{25F^2}{st}$ represents a combined variation. The luminance, L, varies directly as the square of the F-stop, F. It also varies inversely as the product of the film ASA number, s, and the shutter speed, t. The constant of variation is 25. Use this information to work the problems in Exercises 43 and 44.

43. Suppose we want to use 200 ASA film and a shutter speed of 1/250 when 500 footcandles of light are available. What would be an appropriate F-stop?

44. If 125 footcandles of light are available and an F-stop of 2 is used with 200 ASA film, what shutter speed should be used?

Refer to the Discovering Connections exercises done earlier to work the problems in Exercises 45 and 46.

45. Biologists tagged 250 fish in Willow Lake on October 5. On a later date they found 7 tagged fish in a sample of 350. Estimate the total number of fish in Willow Lake.

46. On May 13, researchers at Argyle Lake tagged 420 fish. When they returned a few weeks later, their sample of 500 fish contained 9 that were tagged. Give a good approximation of the fish population in Argyle Lake.

2.7 Inequalities ▼▼▼

An equation says that two expressions are equal, while an **inequality** says that one expression is greater than, greater than or equal to, less than, or less than or equal to, another. As with equations, a value of the variable for which the inequality is true is a solution of the inequality, and the set of all such solutions is the solution set of the inequality. Two inequalities with the same solution set are **equivalent inequalities.**

Inequalities are solved with the following properties of inequality. (These were first introduced in Chapter 1.)

Properties of Inequality

For real numbers, a, b, and c:

(a) $a < b$ **and** $a + c < b + c$ **are equivalent.**
(The same number may be added to both sides of an inequality without changing the solution set.)

(b) **If** $c > 0$, **then** $a < b$ **and** $ac < bc$ **are equivalent.**
(Both sides of an inequality may be multiplied by the same positive number without changing the solution set.)

(c) **If** $c < 0$, **then** $a < b$ **and** $ac > bc$ **are equivalent.**
(Both sides of an inequality may be multiplied by the same negative number without changing the solution set, as long as the direction of the inequality symbol is reversed.)

Replacing $<$ with $>$, $\leq$, or $\geq$ results in equivalent properties.

NOTE Because division is defined in terms of multiplication, the word "multiplied" may be replaced by "divided" in parts (b) and (c) of the properties of inequality. Always remember to reverse the direction of the inequality sign when multiplying or dividing by a negative number.

LINEAR INEQUALITIES A linear inequality is defined in a way similar to a linear equation.

> ### Linear Inequality
>
> A **linear inequality** in one variable is an inequality that can be written in the form
>
> $$ax + b > 0,$$
>
> where $a \neq 0$. (Any of the symbols $\geq$, $<$, or $\leq$ may also be used.)

EXAMPLE 1
Solving a linear inequality

Solve the inequality $-3x + 5 > -7$.

Use the properties of inequality. Adding -5 on both sides gives

$$-3x + 5 + (-5) > -7 + (-5)$$
$$-3x > -12.$$

Now multiply both sides by $-1/3$. (We could also divide by -3.) Since $-1/3 < 0$, reverse the direction of the inequality symbol.

$$-\frac{1}{3}(-3x) < -\frac{1}{3}(-12)$$

$$x < 4$$

The original inequality is satisfied by any real number less than 4. The solution set can be written $\{x \mid x < 4\}$. A graph of the solution set is shown in Figure 5, where the parenthesis is used to show that 4 itself does not belong to the solution set.

FIGURE 5

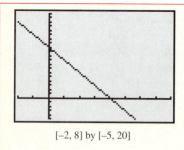

[−2, 8] by [−5, 20]

A graphing calculator can be used to solve the inequality in Example 1 as follows. Write the inequality with 0 on one side: $-3x + 12 > 0$. Graph $y = -3x + 12$. See the figure. Since y equals $-3x + 12$, the y-values of points on the graph show where $-3x + 12$ is positive or negative. As the graph shows, the x-value where $y = 0$ is the dividing point. From the graph we see that $-3x + 12 > 0$ when $x < 4$. This is the same as the solution in Example 1. Note that the graph does not tell us whether the endpoint is included in the solution. We must decide that analytically.

The analytic method is simpler for linear inequalities, but the graphical method is useful for inequalities of higher degree.

The set $\{x \mid x < 4\}$, the solution set for the inequality in Example 1, is an example of an **interval**. A simplified notation, called **interval notation,** is used for writing intervals. With this notation, the interval in Example 1 can be written as $(-\infty, 4)$. The symbol $-\infty$ is not a real number; it is used to show that the interval includes all real numbers less than 4. The interval $(-\infty, 4)$ is an example of an **open interval,** since the endpoint, 4, is not part of the interval. Examples of other sets written in interval notation are shown below. A square bracket is used to show that a number *is* part of the graph, and a parenthesis is used to indicate that a number *is not* part of the graph. Whenever two real numbers *a* and *b* are used to write an interval in the chart that follows, it is assumed that $a < b$.

Type of Interval	Set	Interval Notation	Graph
Open interval	$\{x \mid x > a\}$	(a, ∞)	 *a*
	$\{x \mid a < x < b\}$	(a, b)	 *a* *b*
	$\{x \mid x < b\}$	$(-\infty, b)$	 *b*
Half-open interval	$\{x \mid x \geq a\}$	$[a, \infty)$	 *a*
	$\{x \mid a < x \leq b\}$	$(a, b]$	 *a* *b*
	$\{x \mid a \leq x < b\}$	$[a, b)$	 *a* *b*
	$\{x \mid x \leq b\}$	$(-\infty, b]$	 *b*
Closed interval	$\{x \mid a \leq x \leq b\}$	$[a, b]$	 *a* *b*
All real numbers	$\{x \mid x \text{ is real}\}$	$(-\infty, \infty)$	

EXAMPLE 2
Solving a linear inequality

Solve $4 - 3y \leq 7 + 2y$. Write the solution in interval notation and graph the solution on a number line.

Write the following series of equivalent inequalities.

$$4 - 3y \leq 7 + 2y$$

$$-4 - 2y + 4 - 3y \leq -4 - 2y + 7 + 2y \qquad \text{Subtract 4 and } 2y.$$

$$-5y \leq 3$$

$$\left(-\frac{1}{5}\right)(-5y) \geq \left(-\frac{1}{5}\right)(3) \qquad \text{Multiply by } -\tfrac{1}{5}; \text{ reverse the inequality symbol.}$$

$$y \geq -\frac{3}{5}$$

In set-builder notation, the solution set is $\{y \mid y \geq -3/5\}$, while in interval notation the solution set is $[-3/5, \infty)$. See Figure 6 for the graph of the solution set. ◗

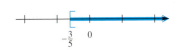

$-\frac{3}{5}$ 0

FIGURE 6

From now on, the solutions of all inequalities will be written with interval notation.

THREE-PART INEQUALITIES The inequality $-2 < 5 + 3x < 20$ in the next example says that $5 + 3x$ is between -2 and 20. This inequality can be solved using an extension of the properties of inequality given above, working with all three expressions at the same time.

EXAMPLE 3
Solving a three-part inequality

Solve $-2 < 5 + 3x < 20$.
 Write equivalent inequalities as follows.

$$-2 < 5 + 3x < 20$$
$$-7 < 3x < 15. \qquad \text{Add } -5.$$
$$-\frac{7}{3} < x < 5 \qquad \text{Multiply by } \tfrac{1}{3}.$$

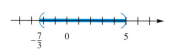

FIGURE 7

The solution, graphed in Figure 7, is the interval $(-7/3, 5)$. ▶

An inequality like $-2 < 5 + 3x < 20$ can be solved with a graphing calculator by slightly altering the graphical method given earlier. Here, it is not possible to get 0 on one side, since there are three "sides." Instead, we must graph three expressions, -2, $5 + 3x$, and 20, simultaneously. We can use the capabilities of the calculator to find (or approximate) the x-values where the graph of $5 + 3x$ intersects the graphs of -2 and 20. Then we must decide on what interval the graph of $5 + 3x$ lies between the graphs of $y = -2$ and $y = 20$. See the figure. (Don't forget to decide about the endpoints separately.) We used an x-scale of 1 and a y-scale of 5 here.

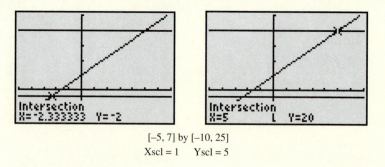

[−5, 7] by [−10, 25]
Xscl = 1   Yscl = 5

A product will break even, or begin to produce a profit, only if the revenue from selling the product at least equals the cost of producing it. If R represents revenue and C is cost, then the **break-even point** is the point where $R = C$.

EXAMPLE 4
Finding the break-even point

If the revenue and cost of a certain product are given by $R = 4x$ and $C = 2x + 1000$, where x is the number of units produced, where does R at least equal C?
Set $R \geq C$ and solve for x.

$$R \geq C$$
$$4x \geq 2x + 1000$$
$$2x \geq 1000$$
$$x \geq 500$$

The break-even point is at $x = 500$. This product will at least break even only if the number of units produced is in the interval $[500, \infty)$. ▶

QUADRATIC INEQUALITIES The solution of *quadratic inequalities* depends on the solution of quadratic equations, which were introduced in Section 2.3.

Quadratic Inequality

A **quadratic inequality** is an inequality that can be written in the form
$$ax^2 + bx + c < 0$$
for real numbers $a \neq 0$, b, and c. (The symbol $<$ can be replaced with $>$, $\leq$, or $\geq$.)

EXAMPLE 5
Solving a quadratic inequality

Solve the quadratic inequality $x^2 - x - 12 < 0$.
Begin by finding the values of x that satisfy $x^2 - x - 12 = 0$.

$$x^2 - x - 12 = 0$$
$$(x + 3)(x - 4) = 0 \qquad \text{Factor.}$$
$$x = -3 \quad \text{or} \quad x = 4 \qquad \text{Use the zero-factor property.}$$

The two points, -3 and 4, divide a number line into the three regions shown in Figure 8. If a point in region A, for example, makes the polynomial $x^2 - x - 12$ negative, then all points in region A will make that polynomial negative.

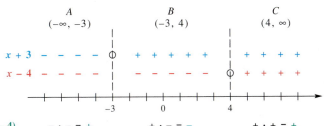

FIGURE 8

To find the regions that make $x^2 - x - 12$ negative (< 0), draw a number line that shows where factors are positive or negative, as in Figure 8. First decide on the sign of the factor $x + 3$ in each of the three regions; then do the same thing for the factor $x - 4$. The results are shown in Figure 8.

Now consider the sign of the product of the two factors in each region. As Figure 8 shows, both factors are negative in the interval $(-\infty, -3)$; therefore their product is positive in that interval. For the interval $(-3, 4)$, one factor is positive and the other is negative, giving a negative product. In the last region, $(4, \infty)$, both factors are positive, so their product is positive. The polynomial $x^2 - x - 12$ is negative (what the original inequality calls for) when the product of its factors is negative, that is, for the interval $(-3, 4)$. The graph of this solution set is shown in Figure 9. ◗

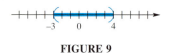

FIGURE 9

Figure 8 is an example of a **sign graph,** a graph that shows the values of the variable that make the factors of a quadratic inequality positive or negative. The steps used in solving a quadratic inequality are summarized below.

Solving a Quadratic Inequality

1. Solve the corresponding quadratic equation.
2. Identify the intervals determined by the solutions of the equation.
3. Use a sign graph to determine which intervals are in the solution set.

A number line graph of an inequality can be simulated by a graphing calculator. If we enter $Y_1 = (x^2 - x - 12) < 0$, the calculator will return a 1 when the inequality is true and a 0 when false. Thus, a horizontal line segment will appear when x is between -3 and 4. See Example 5 for the analytic justification. (We use dot mode in the accompanying screen.)

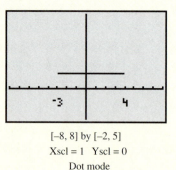

$[-8, 8]$ by $[-2, 5]$
Xscl = 1   Yscl = 0
Dot mode

◀EXAMPLE 6
Solving a quadratic inequality

Solve the inequality $2x^2 + 5x - 12 \geq 0$.

Begin by finding the values of x that satisfy $2x^2 + 5x - 12 = 0$.

$$2x^2 + 5x - 12 = 0$$

$$(2x - 3)(x + 4) = 0$$

$$x = \frac{3}{2} \quad \text{or} \quad x = -4$$

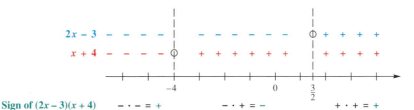

FIGURE 10

These two points divide the number line into the three regions shown in the sign graph in Figure 10. Since both factors are negative in the first interval, their product, $2x^2 + 5x - 12$, is positive there. In the second interval, the factors have opposite signs, and therefore their product is negative. Both factors are positive in the third interval, and their product also is positive there. Thus, the polynomial $2x^2 + 5x - 12$ is positive or zero in the interval $(-\infty, -4]$ and also in the interval $[3/2, \infty)$. Since both of the intervals belong to the solution set, the result can be written as the *union** of the two intervals,

$$(-\infty, -4] \cup \left[\frac{3}{2}, \infty\right).$$

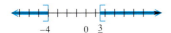

FIGURE 11

The graph of the solution set is shown in Figure 11. ◗

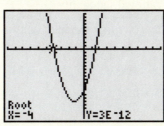 If we graph $y = 2x^2 + 5x - 12$ with a graphing calculator, we can support the result found analytically in Example 6. The graph should look like the one below. From the graph we see that $2x^2 + 5x - 12$ is nonnegative (above or on the x-axis) for $x \le -4$ and again for $x \ge 3/2$.

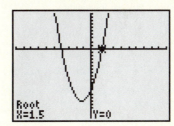

[−10, 10] by [−20, 10]

FOR DISCUSSION OR WRITING

1. Solve each of the following using the analytic method.
   (a) $2x^2 + x - 10 = 0$ (b) $2x^2 + x - 10 > 0$
   (c) $2x^2 + x - 10 < 0$
2. Use your calculator to graph $2x^2 + x - 10$ as Y_1. For the values of x in each of the solution sets in item 1, decide if Y_1 is below the x-axis, on the x-axis, or above the x-axis.
3. Draw a conclusion by comparing the answers to items 1 and 2.

---

*The **union** of sets A and B, written $A \cup B$, is defined as $A \cup B = \{x \mid x$ is an element of A or x is an element of $B\}$.

RATIONAL INEQUALITIES The inequalities discussed in the remainder of this section involve quotients of algebraic expressions, and for this reason they are called **rational inequalities.** These inequalities can be solved with a sign graph in much the same way as quadratic inequalities.

Solving a Rational Inequality

1. Rewrite the inequality, if necessary, so that 0 is on one side.
2. Make a sign graph with intervals determined by the numbers that cause either the numerator or the denominator of the rational expression to equal zero.
3. Determine the appropriate interval(s) of the solution set.

CAUTION Solving a rational inequality such as

$$\frac{5}{x + 4} \geq 1$$

by multiplying both sides by $x + 4$ to get $5 \geq x + 4$, would require considering two cases since the sign of $x + 4$ depends on the value of x. If $x + 4$ were negative, then we would have to reverse the inequality sign. The procedure described in the box above and used in the next two examples eliminates the need for considering separate cases.

◖EXAMPLE 7
Solving a rational inequality

Solve the rational inequality $\dfrac{5}{x + 4} \geq 1$.

Start by rewriting the inequality so that 0 is on one side, getting

$$\frac{5}{x + 4} - 1 \geq 0.$$

Writing the left side as a single fraction gives

$$\frac{5 - (x + 4)}{x + 4} \geq 0$$

or

$$\frac{1 - x}{x + 4} \geq 0.$$

The quotient can change sign only when the denominator is 0 or when the numerator is 0. This occurs at

$$1 - x = 0 \quad \text{or} \quad x + 4 = 0$$
$$x = 1 \quad \text{or} \quad x = -4.$$

Make a sign graph as before. This time, consider the sign of the quotient of the two quantities rather than the sign of their product. See Figure 12.

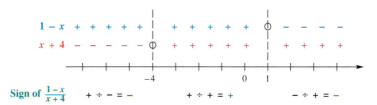

FIGURE 12

The quotient of two numbers is positive if both numbers are positive or if both numbers are negative. On the other hand, the quotient is negative if the two numbers have opposite signs. The sign graph in Figure 12 shows that values in the interval $(-4, 1)$ give a positive quotient and are part of the solution. With a quotient, the endpoints must be considered separately to make sure that no denominator is 0. Here, -4 gives a 0 denominator but 1 satisfies the given inequality. In interval notation, the solution set is $(-4, 1]$. ▶

CAUTION   As suggested by Example 7, be very careful with the endpoints of the intervals in the solution of rational inequalities.

◀EXAMPLE 8
Solving a rational inequality

Solve $\dfrac{2x - 1}{3x + 4} < 5$.

Begin by subtracting 5 on both sides and combining the terms on the left into a single fraction.

$$\frac{2x - 1}{3x + 4} < 5$$

$$\frac{2x - 1}{3x + 4} - 5 < 0 \qquad \text{Subtract 5.}$$

$$\frac{2x - 1 - 5(3x + 4)}{3x + 4} < 0 \qquad \text{Common denominator is } 3x + 4.$$

$$\frac{-13x - 21}{3x + 4} < 0 \qquad \text{Combine terms.}$$

To draw a sign graph, first solve the equations

$$-13x - 21 = 0 \quad \text{and} \quad 3x + 4 = 0,$$

getting the solutions

$$x = -\frac{21}{13} \quad \text{and} \quad x = -\frac{4}{3}.$$

Use the values $-21/13$ and $-4/3$ to divide the number line into three intervals. Now complete a sign graph and find the intervals where the quotient is negative. See Figure 13.

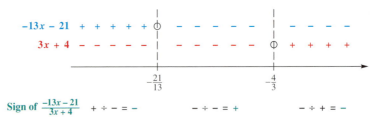

FIGURE 13

From the sign graph, values of x in the two intervals $(-\infty, -21/13)$ and $(-4/3, \infty)$ make the quotient negative, as required. Neither endpoint satisfies the given inequality, so the solution set should be written $(-\infty, -21/13) \cup (-4/3, \infty)$. ◗

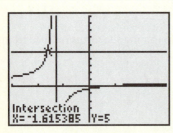

Intersection
X=-1.615385 Y=5

[−3, 3] by [−5, 10]

⊡ The inequality in Example 8 can be solved in the same way as the inequality in Example 6, by first getting one side equal to 0, or by using the intersection of two graphs method, as we did for Example 3. We discuss the intersection-of-graphs method here. The figure shows the graphs

$$Y_1 = \frac{2x - 1}{3x + 4} \quad \text{and} \quad Y_2 = 5.$$

To solve the inequality we must decide in which x-intervals the graph of the rational expression is less than the graph of $Y_2 = 5$ (the horizontal line). This occurs in the intervals $(-\infty, -1.615385)$ and $(-4/3, \infty)$. (Verify that $-1.615385 \approx -21/13$.)

Using the graphical method with this inequality presents some problems. It is necessary to use an x-interval small enough to show clearly where the two graphs intersect. To determine the interval $(-4/3, \infty)$, we can use the vertical line that the graph approaches from each side. Note that this line is not part of the graph. We can use the trace key to approximate this x-value, but we cannot obtain this value exactly. (We will have more to say about this line in Section 4.5.)

2.7 Exercises ▼▼▼▼▼▼▼▼▼▼▼▼▼▼▼▼▼▼▼▼▼▼▼▼▼▼▼▼▼▼▼▼▼▼▼▼

Match the inequality in the exercise with its equivalent interval notation in the column on the right.

1. $x < -4$

2. $x \leq 4$

3. $-2 < x \leq 6$

4. $0 \leq x \leq 8$

5. $x \geq -3$

6. $4 \leq x$

7.

8.

9.

10.

A. $(-2, 6]$

B. $[-2, 6)$

C. $(-\infty, -4]$

D. $[4, \infty)$

E. $(3, \infty)$

F. $(-\infty, -4)$

G. $(0, 8)$

H. $[0, 8]$

I. $[-3, \infty)$

J. $(-\infty, 4]$

11. Explain how to determine whether a parenthesis or a square bracket is used when graphing the solution set of a linear inequality.

12. The three-part inequality $a < x < b$ means "a is less than x and x is less than b." Which one of the following inequalities is not satisfied by some real number x?
   (a) $-3 < x < 5$ **(b)** $0 < x < 4$ **(c)** $-3 < x < -2$ **(d)** $-7 < x < -10$

Solve the following inequalities. Write the solutions in interval notation. Graph each solution. See Examples 1–3.

13. $-3p - 2 \leq 1$

14. $-5r + 3 \geq -2$

15. $2(m + 5) - 3m + 1 \geq 5$

16. $6m - (2m + 3) \geq 4m - 5$

17. $8k - 3k + 2 < 2(k + 7)$

18. $2 - 4x + 5(x - 1) < -6(x - 2)$

19. $\dfrac{4x + 7}{-3} \leq 2x + 5$

20. $\dfrac{2z - 5}{-8} \geq 1 - z$

21. $2 \leq y + 1 \leq 5$

22. $-3 \leq 2t \leq 6$

23. $-10 > 3r + 2 > -16$

24. $4 > 6a + 5 > -1$

25. $-3 \leq \dfrac{x - 4}{-5} < 4$

26. $1 < \dfrac{4m - 5}{-2} < 9$

27. The capital outlay for education by state and local governments during the period from 1985 to 1992 can be approximated by the linear model $y = 2480x + 12,726$, where $x = 0$ corresponds to 1985 and y is in millions of dollars. Based on this model, when did this amount first exceed $17,600 million?

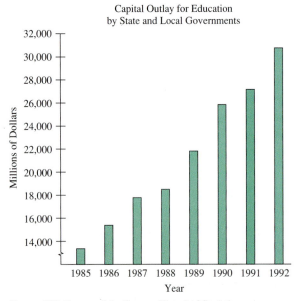

Capital Outlay for Education
by State and Local Governments

Source: U.S. Bureau of the Census, *Historical Statistics on Governmental Finances and Employment*, and *Government Finances*, series GF.

28. The number of farms with milk cows in the United States has steadily declined since 1985. The linear model $y = -13.5x + 261.5$ gives a good approximation of the number of such farms in each year from 1985 to 1992, with $x = 0$ representing 1985 and with y in thousands. If these trends continue, in what year would we expect the number of such farms to first fall below 150,000?

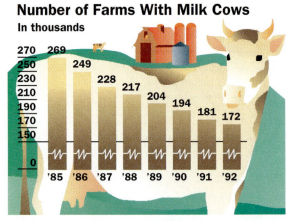

Number of Farms With Milk Cows

In thousands

Source: U.S. Dept. of Agriculture, National Agriculture Statistics Service, *Dairy Products*, annual; and *Milk: Production, Disposition, and Income*, annual.

Find all intervals where the following products will at least break even. See Example 4.

29. The cost to produce x units of wire is $C = 50x + 5000$, while the revenue is $R = 60x$.

30. The cost to produce x units of squash is $C = 100x + 6000$, while the revenue is $R = 500x$.

31. $C = 85x + 900$; $R = 105x$

32. $C = 70x + 500$; $R = 60x$

Solve the following quadratic inequalities. Write the solutions in interval notation. Graph each solution. (Hint: In Exercises 43–46, use the quadratic formula.) See Examples 5 and 6.

33. $x^2 \leq 9$

34. $p^2 > 16$

35. $r^2 + 4r + 6 \geq 3$

36. $z^2 + 6z + 16 < 8$

37. $x^2 - x \leq 6$

38. $r^2 + r < 12$

39. $2k^2 - 9k > -4$

40. $3n^2 \leq -10 - 13n$

41. $x^2 > 0$

42. $p^2 < -1$

43. $x^2 + 5x - 2 < 0$

44. $4x^2 + 3x + 1 \leq 0$

45. $m^2 - 2m \leq 1$

46. $p^2 + 4p > -1$

Solve the following rational inequalities. Give the answers in interval notation. See Examples 7 and 8.

47. $\dfrac{m - 3}{m + 5} \leq 0$

48. $\dfrac{r + 1}{r - 4} > 0$

49. $\dfrac{k - 1}{k + 2} > 1$

50. $\dfrac{a - 6}{a + 2} < -1$

51. $\dfrac{3}{x - 6} \leq 2$

52. $\dfrac{1}{k - 2} < \dfrac{1}{3}$

53. $\dfrac{1}{m - 1} < \dfrac{5}{4}$

54. $\dfrac{6}{5 - 3x} \leq 2$

55. $\dfrac{10}{3 + 2x} \leq 5$

56. $\dfrac{1}{x + 2} \geq 3$

57. $\dfrac{7}{k + 2} \geq \dfrac{1}{k + 2}$

58. $\dfrac{5}{p + 1} > \dfrac{12}{p + 1}$

59. $\dfrac{3}{2r - 1} > -\dfrac{4}{r}$

60. $-\dfrac{5}{3h + 2} \geq \dfrac{5}{h}$

61. $\dfrac{4}{y - 2} \leq \dfrac{3}{y - 1}$

62. $\dfrac{4}{n + 1} < \dfrac{2}{n + 3}$

63. $\dfrac{y + 3}{y - 5} \leq 1$

64. $\dfrac{a + 2}{3 + 2a} \leq 5$

▼▼▼▼▼▼▼▼▼▼▼▼▼ **DISCOVERING CONNECTIONS** (Exercises 65–72) ▼▼▼▼▼▼▼▼▼▼▼▼▼

Inequalities that involve more than two factors, such as

$$(3x - 4)(x + 2)(x + 6) \leq 0,$$

can be solved using an extension of the method shown in Examples 5 and 6. Work Exercises 65–72 in order, to obtain the solution set of this inequality.

65. Consider the associated *equation* $(3x - 4)(x + 2)(x + 6) = 0$. Use the zero-factor property to solve this equation.

66. Construct a number line and plot the three solutions in Exercise 65.

67. The number line from Exercise 66 should show four regions formed by the three points. Choose a number from the region farthest to the left. Does it satisfy the original inequality? If it does, graph the region on the number line. (Include the endpoint.)

68. Repeat Exercise 67, but use the second region from the left.

69. Repeat Exercise 67, but use the third region from the left.

70. Repeat Exercise 67, using the region farthest to the right.

71. On a single number line, graph all the regions of solution. This is the solution set of the inequality.

72. Use the technique just described to graph the solution set of the inequality $x^3 + 4x^2 - 9x - 36 > 0$. (*Hint:* Start by factoring by grouping.)

73. Which of the following inequalities have solution set $(-\infty, \infty)$?
  (a) $(x + 3)^2 \geq 0$ **(b)** $(5x - 6)^2 \leq 0$ **(c)** $(6y + 4)^2 > 0$
  (d) $(8p - 7)^2 < 0$ **(e)** $\dfrac{x^2 + 7}{2x^2 + 4} \geq 0$ **(f)** $\dfrac{2x^2 + 8}{x^2 + 9} < 0$

74. Which of the inequalities in Exercise 73 have solution set $\emptyset$?

⊞ *Use one of the graphing calculator methods described in this section to solve the inequality. See the graphing calculator boxes in this section.*

75. $3(x + 2) - 5x < x$

76. $-2(x + 4) \leq 6x + 8$

77. $x^2 - 4x - 5 \leq 0$

78. $x^2 - x - 12 > 0$

79. $\dfrac{x - 2}{x + 2} \leq 2$

80. $\dfrac{x}{x + 2} \geq 2$

81. A student attempted to solve the inequality

$$\frac{2x - 1}{x + 2} \leq 0$$

by multiplying both sides by $x + 2$ to get

$$2x - 1 \leq 0$$

$$x \leq \frac{1}{2}.$$

He wrote the solution set as $(-\infty, 1/2]$. Is his solution correct? Explain.

82. A student solved the inequality $p^2 \leq 16$ by taking the square root of both sides to get $p \leq 4$. She wrote the solution set as $(-\infty, 4]$. Is her solution correct? Explain.

83. Radon gas occurs naturally in homes and is produced when uranium radioactively decays into lead. Uranium occurs in small traces in most soils throughout the world.

As a result, radon is present in the soil. It enters buildings through basements, water supplies, and natural gas, with soil being the major contributor. Exposure to radon gas is a known lung cancer risk. According to the Environmental Protection Agency (EPA) the individual lifetime excess cancer risk R for radon exposure is between 1.5×10^{-3} and 6.0×10^{-3} where $R = .01$ represents a 1% increase in risk of developing lung cancer.

(a) Calculate the range of individual annual risk by dividing R by an average life expectancy of 72 years.

(b) Approximate the range of new cases of lung cancer each year (to the nearest hundred) caused by radon if the population of the United States is 260 million.

(*Source: Indoor-Air-Assessment: A Review of Indoor Air Quality Risk Characterization Studies.* Report No. EPA/600/8-90/044, Environmental Protection Agency, 1991.)

84. (a) If $a > b$, is it always true that $1/a < 1/b$? Explain.

(b) If $a > b$, is it always true that $a^2 > b^2$? Explain.

2.8 Absolute Value Equations and Inequalities ▼▼▼

In this section we describe methods of solving equations and inequalities involving absolute value. Recall from Chapter 1 that the absolute value of a number a, written $|a|$, gives the distance from a to 0 on a number line. By this definition, the absolute value equation $|x| = 3$ can be solved by finding all real numbers at a distance of 3 units from 0. As shown in Figure 14, there are two numbers satisfying this condition, 3 and -3, so that the solution set of the equation $|x| = 3$ is the set $\{3, -3\}$.

ABSOLUTE VALUE EQUATIONS If a and b represent two real numbers, then the absolute value of their difference, $|a - b|$ or $|b - a|$, represents the distance between the points on the number line whose coordinates are a and b. (Verify this for 3 and -3 in Figure 14.) This concept is used in simple equations involving absolute value.

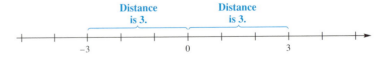

FIGURE 14

EXAMPLE 1
Using the distance definition to solve an absolute value equation

Solve $|p - 4| = 5$.

The expression $|p - 4|$ represents the distance between p and 4. The equation $|p - 4| = 5$ can be solved by finding all real numbers that are 5 units from 4. As shown in Figure 15, these numbers are -1 and 9. The solution set is $\{-1, 9\}$.

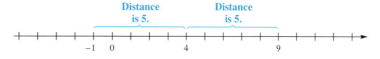

Distance is 5.   Distance is 5.

FIGURE 15

The abs function on a graphing calculator is used to enter the absolute value of an expression. This may be a second function or may be located in a special menu of functions. See the manual for your calculator. For example the absolute value of $P - 4$ might be entered as abs $(P - 4)$. Remember to use parentheses around the expression or it will be interpreted differently than you intend.

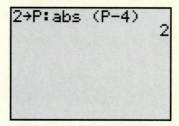

When $P = 2$, $|P - 4| = |2 - 4| = 2$.

The definition of absolute value leads to the following properties of absolute value that can be used to solve absolute value equations algebraically.

Solving Absolute Value Equations

1. For $b > 0$, $|a| = b$ if and only if $a = b$ or $a = -b$
2. $|a| = |b|$ if and only if $a = b$ or $a = -b$.

EXAMPLE 2
Solving absolute value
equations

Solve each equation.

(a) $|5 - 3m| = 12$

Use property (1) above, with $a = 5 - 3m$, to write

$$5 - 3m = 12 \quad \text{or} \quad 5 - 3m = -12.$$

Solve each equation.

$$
\begin{array}{ll}
5 - 3m = 12 \qquad \text{or} & 5 - 3m = -12 \\
\quad -3m = 7 & \quad -3m = -17 \\
\quad\quad m = -\dfrac{7}{3} & \quad\quad m = \dfrac{17}{3}
\end{array}
$$

The solution set is $\{-7/3,\ 17/3\}$.

(b) $|4m - 3| = |m + 6|$

By property (2) above, this equation will be true if

$$4m - 3 = m + 6 \quad \text{or} \quad 4m - 3 = -(m + 6).$$

Solve each equation.

$$
\begin{array}{ll}
4m - 3 = m + 6 \qquad \text{or} & 4m - 3 = -(m + 6) \\
\quad\quad 3m = 9 & \quad 4m - 3 = -m - 6 \\
\quad\quad\ m = 3 & \quad\quad\quad 5m = -3 \\
& \quad\quad\quad\ m = -\dfrac{3}{5}
\end{array}
$$

The solution set of $|4m - 3| = |m + 6|$ is thus $\{3,\ -3/5\}$. ▶

ABSOLUTE VALUE INEQUALITIES   The method used to solve absolute value equations can be extended to solve inequalities with absolute value.

EXAMPLE 3
Using the distance definition
for absolute value inequalities

(a) Solve $|x| < 5$.

Since absolute value gives the distance between a number and 0, the inequality $|x| < 5$ is satisfied by all real numbers whose distance from 0 is less than 5. As shown in Figure 16, the solution includes all numbers from -5 to 5, or $\{x | -5 < x < 5\}$. In interval notation, the solution is written as the open interval $(-5, 5)$. A graph of the solution set is shown in Figure 16.

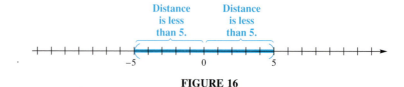

FIGURE 16

(b) Solve $|x| > 5$.

In a manner similar to part (a), we see that the solution of $|x| > 5$ consists of all real numbers whose distance from 0 is greater than 5. This includes those numbers greater than 5 or those less than -5: $x < -5$ or $x > 5$.

In interval notation, the solution is written $(-\infty, -5) \cup (5, \infty)$. The solution set is shown in Figure 17. ▶

Distance is greater than 5.   Distance is greater than 5.

FIGURE 17

The following properties of absolute value, which can be obtained from the definition of absolute value, are used to solve absolute value inequalities.

Solving Absolute Value Inequalities

For any positive number b:

1. $|a| < b$ if and only if $-b < a < b$;

2. $|a| > b$ if and only if $a < -b$ or $a > b$.

◀ **EXAMPLE 4**
Solving an absolute value inequality

Solve $|x - 2| < 5$.

This inequality is satisfied by all real numbers whose distance from 2 is less than 5. As shown in Figure 18, the solution set is the interval $(-3, 7)$. Property (1) above can be used to solve the inequality as follows. Let $a = x - 2$ and $b = 5$, so that $|x - 2| < 5$ if and only if

$$-5 < x - 2 < 5.$$

Adding 2 to each part of this three-part inequality produces

$$-3 < x < 7,$$

giving the interval solution $(-3, 7)$. ▶

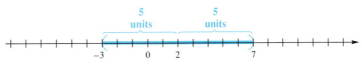

5 units   5 units

FIGURE 18

EXAMPLE 5
Solving an absolute value inequality

Solve $|x - 8| \geq 1$.

All numbers whose distance from 8 is greater than or equal to 1 are solutions. To find the solution using property (2) above, let $a = x - 8$ and $b = 1$ so that $|x - 8| \geq 1$ if and only if

$$x - 8 \leq -1 \qquad \text{or} \qquad x - 8 \geq 1$$
$$x \leq 7 \qquad \text{or} \qquad x \geq 9.$$

The solution set, $(-\infty, 7] \cup [9, \infty)$, is shown in Figure 19. ▶

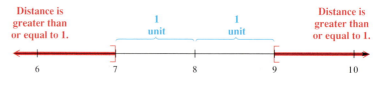

FIGURE 19

The properties given above for solving absolute value inequalities require that the absolute value expression be *alone* on one side of the inequality. Example 6 shows how to meet this requirement when it is not the case at first.

EXAMPLE 6
Solving an absolute value inequality requiring a transformation

Solve $|2 - 7m| - 1 > 4$.

In order to use the properties of absolute value given above, first add 1 to both sides; this gives

$$|2 - 7m| > 5.$$

Now use property (2) above. By this property, $|2 - 7m| > 5$ if and only if

$$2 - 7m < -5 \qquad \text{or} \qquad 2 - 7m > 5.$$

Solve each of these inequalities separately to get the solution set $(-\infty, -3/7) \cup (1, \infty)$. ▶

If an absolute value equation or inequality is written with 0 or a negative number on one side, such as $|2 - 5x| \geq -4$, we do not solve by applying the methods of the earlier examples. Use the fact that the absolute value of any expression must be a nonnegative number to solve the equation or inequality.

EXAMPLE 7
Solving special cases of absolute value equations and inequalities

Use the fact that absolute value is always nonnegative to solve each equation or inequality.

(a) $|2 - 5x| \geq -4$

Since the absolute value of a number is always nonnegative, $|2 - 5x| \geq -4$ is always true. The solution set includes all real numbers, written $(-\infty, \infty)$.

(b) $|4x - 7| < -3$

The absolute value of any number will never be less than -3 (or less than *any* negative number). For this reason, the solution set of this inequality is ∅.

(c) $|5x + 15| = 0$

The absolute value of a number will be zero only if that number is 0. Therefore, this equation is equivalent to $5x + 15 = 0$, which has solution set $\{-3\}$. ▶

We end this section with an example showing how certain statements involving distance can be described using absolute value inequalities.

EXAMPLE 8
Using absolute value inequalities to describe distances

Write each statement using an absolute value inequality.

(a) k is not less than 5 units from 8.

Since the distance from k to 8, written $|k - 8|$ or $|8 - k|$, is not less than 5, the distance is greater than or equal to 5. Write this as

$$|k - 8| \geq 5.$$

(b) n is within .001 of 6.

This statement indicates that n may be .001 more than 6 or .001 less than 6. That is, the distance of n from 6 is no more than .001, written

$$|n - 6| \leq .001. \quad ▶$$

CONNECTIONS In quality control and other applications, as well as in more advanced mathematics, we often wish to keep the difference between two quantities within some predetermined amount. For example, suppose $y = 2x + 1$ and we want y to be within .01 unit of 4. This can be written using absolute value as $|y - 4| < .01$. To find the values of x that will satisfy this condition on y, we use properties of absolute value as follows.

$|y - 4| < .01$	
$|2x + 1 - 4| < .01$	Substitute $2x + 1$ for y.
$|2x - 3| < .01$	
$-.01 < 2x - 3 < .01$	Property 1
$2.99 < 2x < 3.01$	Add 3 to each part.
$1.495 < x < 1.505$	Divide each part by 2.

By reversing these steps, we can show that keeping x between 1.495 and 1.505 will ensure that the difference between y and 4 is less than .01.

FOR DISCUSSION OR WRITING

1. Follow the steps used above to find the values of x that will make the difference between y and 1 no more than .1. (Use the same equation relating x and y.)

2. What are some real world applications of this idea?

2.8 Exercises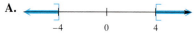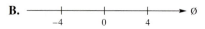

Match the graph in the columns on the right having the solution set of the given absolute value equation or inequality. Do this without actually solving the equation or inequality algebraically.

1. $|x| = 4$

2. $|x| = -4$

3. $|x| > -4$

4. $|x| > 4$

5. $|x| < 4$

6. $|x| \geq 4$

7. $|x| \leq 4$

8. $|x| \neq 4$

A. (number line with brackets at -4 and 4, shaded outside)

B. (number line, marks at -4, 0, 4) $\varnothing$

C. (number line with brackets at -4 and 4, shaded between)

D. (number line fully shaded) $(-\infty, \infty)$

E. (number line, shaded outside -4 and 4)

F. (number line with dots at -4 and 4)

G. (number line shaded between -4 and 4)

H. (number line shaded with open circles at -4 and 4)

Solve the equation. See Examples 1, 2, and 7(c).

9. $|3m - 1| = 2$

10. $|4p + 2| = 5$

11. $|5 - 3x| = 3$

12. $|-3a + 7| = 3$

13. $\left|\dfrac{z - 4}{2}\right| = 5$

14. $\left|\dfrac{m + 2}{2}\right| = 7$

15. $\left|\dfrac{5}{r - 3}\right| = 10$

16. $\left|\dfrac{3}{2h - 1}\right| = 4$

17. $|4w + 3| - 2 = 7$

18. $|8 - 3t| - 3 = -2$

19. $|6x + 9| = 0$

20. $|12t - 3| = 0$

21. $\left|\dfrac{6y + 1}{y - 1}\right| = 3$

22. $\left|\dfrac{3a - 4}{2a + 3}\right| = 1$

23. $|2k - 3| = |5k + 4|$

24. $|p + 1| = |3p - 1|$

25. $|4 - 3y| = |2 - 3y|$

26. $|3 - 2x| = |5 - 2x|$

27. Without actually going through the solution process, we can say that the equation $|5x - 6| = 6x$ cannot have a negative solution. Why is this true?

28. Determine by inspection the solution set of each of the following absolute value equations.
   (a) $-|x| = |x|$ **(b)** $|-x| = |x|$ **(c)** $|x^2| = |x|$ **(d)** $-|x| = 3$

Using x rather than p as the variable in Example 1, the equation solved there can be solved graphically with a graphing calculator by graphing $y = |x - 4| - 5$ and finding the points at which the graph intersects the x-axis. As seen in the accompanying screens, these two values are -1 and 9, supporting the work done in Example 1.

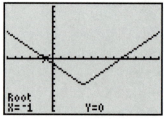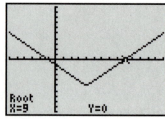

[−6, 14] by [−10, 10]

Use a graphing calculator to solve the absolute value equation.

29. $|x - 2| = 1$

30. $|x - 3| = 2$

31. $|2x + 7| = 5$

32. $|3x + 1| = 10$

33. $|2 + 5x| = |4 - 6x|$

34. $|x + 2| = |3x + 8|$

Solve the absolute value inequality. Give the solution set using interval notation. See Examples 3–7.

35. $|m| > 1$

36. $|z| \geq 5$

37. $|2x + 5| < 3$

38. $\left| x - \dfrac{1}{2} \right| < 2$

39. $4|x - 3| > 12$

40. $3|x + 1| > 6$

41. $|3z + 1| \geq 7$

42. $|8b + 5| \geq 7$

43. $\left| \dfrac{2}{3}t + \dfrac{1}{2} \right| \leq \dfrac{1}{6}$

44. $\left| \dfrac{5}{3} - \dfrac{1}{2}x \right| > \dfrac{2}{9}$

45. $\left| 5x + \dfrac{1}{2} \right| - 2 < 5$

46. $\left| x + \dfrac{2}{3} \right| + 1 < 4$

47. $|6x + 3| \geq -2$

48. $|7 - 8x| < -4$

49. $\left| \dfrac{1}{2}x + 6 \right| > 0$

50. $\left| \dfrac{2}{3}r - 4 \right| \leq 0$

51. Write an equation involving absolute value that says the distance between p and q is 5 units.

52. Write an inequality involving absolute value that says the distance between r and s is less than 9 units.

53. Suppose you hear someone say, "The absolute value of a number is always positive." How might you politely correct this person's misconception?

54. The following triangle inequality holds for all real numbers x and y:

$$|x| + |y| \geq |x + y|.$$

Illustrate cases for this inequality by choosing values of x and y that satisfy the following.
(a) x and y are both positive.   (b) x and y are both negative.
(c) x is positive and y is negative.   (d) x is negative and y is positive.

Write the statement as an absolute value equation or inequality. See Example 8.

55. m is no more than 8 units from 9.   **56.** z is no less than 2 units from 12.   **57.** p is at least 5 units from 9.

58. k is 6 units from 1.   **59.** r is 5 units from 3.

60. If x is within .0004 unit of 2, then y is within .00001 unit of 7.

▼▼▼▼▼▼▼▼▼▼▼▼▼ **DISCOVERING CONNECTIONS** (Exercises 61–66) ▼▼▼▼▼▼▼▼▼▼▼▼▼

In Section 2.3 we introduced several methods of solving quadratic equations, and in this section we presented a method for solving absolute value equations. Suppose we want to solve an equation that involves the absolute value of a quadratic polynomial, such as

$$|x^2 - x| = 6.$$

By combining the methods studied in Section 2.3 and this section, we can solve this equation. Work Exercises 61–66 in order, to see how this can be done.

61. For $x^2 - x$ to have an absolute value equal to 6, what are the two possible values that it may be? (*Hint:* One is positive and the other is negative.)

62. Write an equation stating that $x^2 - x$ is equal to the positive value you found in Exercise 61.

63. Solve the equation from Exercise 62 using the method of factoring.

64. Write the equation stating that $x^2 - x$ is equal to the negative value you found in Exercise 61.

65. Solve the equation from Exercise 64 using the quadratic formula. (*Hint:* The solutions are not real numbers.)

66. Give the complete solution set of $|x^2 - x| = 6$, using the results from Exercises 63 and 65.

Solve the problem.

When humans breathe, carbon dioxide is emitted. In one study, the emission rates of carbon dioxide by college students were measured during both lectures and exams. The average individual rate R_L (in grams per hour) during a lecture class satisfied the inequality $|R_L - 26.75| \le 1.42$ whereas during an exam the rate R_E satisfied the inequality $|R_E - 38.75| \le 2.17$. (Source: Wang, T. C., ASHRAE Trans., 81 (Part 1), 32 (1975).)

Use this information in Exercises 67–69.

67. Find the range of values for R_L and R_E.

68. The class had 225 students. If T_L and T_E represent the total amounts of carbon dioxide in grams emitted during a one-hour lecture and exam, respectively, write inequalities that describe the ranges for T_L and T_E.

69. Discuss any reasons that might account for these differences between the rates during lectures and the rates during exams.

---

70. The temperatures on the surface of Mars in degrees Celsius approximately satisfy the inequality

$$|C + 84| \le 56.$$

What range of temperatures corresponds to this inequality?

71. Dr. Tydings has found that, over the years, 95% of the babies he has delivered weighed y pounds, where $|y - 8.0| \le 1.5$. What range of weights corresponds to this inequality?

72. The industrial process that is used to convert methanol to gasoline is carried out at a temperature range of 680°F to 780°F. Using F as the variable, write an absolute value inequality that corresponds to this range.

73. When a model kite was flown in crosswinds in tests to determine its limits of power extraction, it attained speeds of 98 to 148 feet per second in winds of 16 to 26 feet per second. Using x as the variable in each case, write absolute value inequalities that correspond to these ranges.

74. Is $|a - b|^2$ always equal to $(b - a)^2$? Why?

Chapter 2   Summary ▼▼▼▼▼▼▼▼▼▼▼▼▼▼▼▼▼▼▼▼▼▼▼▼▼▼▼▼▼▼▼▼

KEY TERMS		KEY IDEAS
2.1 Linear Equations		

equation
solution
solution set
identity
conditional
  equation
	contradiction
	equivalent
	  equations
	linear equation
		Addition and Multiplication Properties of Equality
		For real numbers a, b, and c:
		
		$\quad a = b$ and $a + c = b + c$ are equivalent.
		$\quad$ If $c \ne 0$, then $a = b$ and $ac = bc$ are equivalent.

KEY TERMS	KEY IDEAS

2.2 Applications of Linear Equations

solving for a specified variable	**Steps in Problem Solving** **1.** Decide what the variable is to represent. **2.** Make a sketch or chart, if appropriate. **3.** Decide on a variable expression for any other unknown quantity. **4.** Write an equation. **5.** Solve the equation. **6.** Check.

2.3 Quadratic Equations

quadratic equation completing the square quadratic formula discriminant	**Zero-Factor Property** If a and b are complex numbers, with $ab = 0$, then $a = 0$ or $b = 0$ or both. **Square Root Property** The solution set of $x^2 = k$ is $\{\sqrt{k}, -\sqrt{k}\}$. **Quadratic Formula** The solutions of the quadratic equation $ax^2 + bx + c = 0$, where $a \neq 0$, are given by $$x = \frac{-b \pm \sqrt{b^2 - 4ac}}{2a}.$$

2.4 Applications of Quadratic Equations

	Pythagorean Theorem In a right triangle, the sum of the squares of the legs a and b is equal to the square of the hypotenuse c: $$a^2 + b^2 = c^2.$$

2.5 Other Types of Equations

quadratic in form	

2.6 Variation

constant of variation	**Direct Variation** y varies directly as the nth power of x if a nonzero real number k exists such that $y = kx^n$. **Inverse Variation** y varies inversely as the nth power of x if a nonzero real number k exists such that $y = \dfrac{k}{x^n}$. **Joint Variation** For real numbers m and n, y varies jointly as the nth power of x and the mth power of z if a nonzero real number k exists such that $y = kx^n z^m$.

KEY TERMS	KEY IDEAS
2.7 Inequalities	
inequality equivalent inequalities linear inequality interval interval notation open interval half-open interval closed interval three-part inequality quadratic inequality sign graph rational inequality	**Properties of Inequality** For real numbers a, b, and c: **(a)** $a < b$ and $a + c < b + c$ are equivalent. **(b)** If $c > 0$, then $a < b$ and $ac < bc$ are equivalent. **(c)** If $c < 0$, then $a < b$ and $ac > bc$ are equivalent. **For Quadratic or Rational Inequalities:** **1.** Be sure that 0 is on one side. **2.** Make a sign graph. **3.** Determine the intervals of the solution set.
2.8 Absolute Value Equations and Inequalities	
	Properties of Absolute Value If b is a positive number, then $\|a\| = b$ if and only if $a = b$ or $a = -b$; $\|a\| < b$ if and only if $-b < a < b$; $\|a\| > b$ if and only if $a < -b$ or $a > b$. For any real numbers a and b, $\|a\| = \|b\|$ if and only if $a = b$ or $a = -b$.

Chapter 2   Review Exercises ▼▼▼▼▼▼▼▼▼▼▼▼▼▼▼▼▼▼▼▼▼▼▼▼▼▼▼▼▼

Solve the equation.

1. $2m + 7 = 3m + 1$

2. $4k - 2(k - 1) = 12$

3. $5y - 2(y + 4) = 3(2y + 1)$

4. $\dfrac{2}{x - 3} = \dfrac{3}{x + 3}$

5. $\dfrac{10}{4z - 4} = \dfrac{1}{1 - z}$

6. $\dfrac{2}{p} - \dfrac{4}{3p} = 8 + \dfrac{3}{p}$

7. $\dfrac{5}{3r} - 10 = \dfrac{3}{2r}$

8. $\dfrac{p}{p - 2} - \dfrac{3}{5} = \dfrac{2}{p - 2}$

Solve for x.

9. $3(x + 2b) + a = 2x - 6$

10. $9x - 11(k + p) = x(a - 1)$

Solve the formula for the specified variable.

11. $A = \dfrac{24f}{B(p + 1)}$ for f (approximate annual interest rate)

12. $A = \dfrac{24f}{B(p + 1)}$ for B (approximate annual interest rate)

13. $A = P\left(1 + \dfrac{i}{m}\right)$ for m (compound interest)

14. $V = \dfrac{1}{3}\pi r^2 h$ for h (volume of a right circular cone)

A baseball pitcher's earned run average (E.R.A.) *gives the average number of earned runs given up per* 9 *innings. An equation that gives this average is*

$$\text{E.R.A.} = \frac{9(\text{number of earned runs allowed})}{\text{number of innings pitched}}.$$

In each of the following exercises, two of the three values in this equation are given for a particular major league pitcher during a recent season. Find the remaining value.

15. Roger Clemens;   E.R.A.: 3.13;   innings pitched: $253\dfrac{1}{3}$

16. Dwight Gooden;   E.R.A.: 2.89;   innings pitched: $118\dfrac{1}{3}$

17. Greg Swindell;   innings pitched: $184\dfrac{1}{3}$;   earned runs allowed: 69

18. Frank Wills;   innings pitched: $71\dfrac{1}{3}$;   earned runs allowed: 29

19. Bert Blyleven;   E.R.A.: 2.73;   earned runs allowed: 73
20. Tim Belcher;   E.R.A.: 2.82;   earned runs allowed: 72

Solve the problem.

21. To make a special candy mix for Valentine's Day, the owner of a candy store wants to combine chocolate hearts that sell for $5 per lb with candy kisses that sell for $3.50 per lb. How many pounds of each kind should be used to get 30 lb of a mix that can be sold for $4.50 per lb?

22. Alison can ride her bike to the university library in 20 min. The trip home, which is all uphill, takes her 30 min. If her rate is 8 mph slower on the return trip, how far does she live from the library?

23. A chemist wishes to strengthen a mixture that is 10% alcohol to one that is 30% alcohol. How much pure alcohol should be added to 12 liters of the 10% mixture?

24. A student needs 10% hydrochloric acid for a chemistry experiment. How much 5% acid should be mixed with 120 ml of 20% acid to get a 10% solution?

25. Lynn Mooney earns take-home pay of $592 a week. If her deductions for taxes, retirement, union dues, and medical plan amount to 26% of her wages, what is her weekly pay before deductions?

26. A realtor borrowed $90,000 to develop some property. He was able to borrow part of the money at 11.5% interest and the rest at 12%. The annual interest on the two loans amounts to $10,525. How much was borrowed at each rate?

27. An excursion boat travels upriver to a landing and then returns to its starting point. The trip upriver takes 1.2 hr, and the trip back takes .9 hr. If the average speed on the return trip is 5 mph faster than on the trip upriver, what is the boat's speed upriver?

28. If x represents the number of pennies in a jar in an applied problem, which of the following equations cannot be a correct equation for finding x? (*Hint:* Solve each equation and consider the solutions.)
(a) $5x + 3 = 9$ (b) $12x + 3 = -4$
(c) $100x = 50(x + 3)$ (d) $6(x + 4) = x + 24$

29.  Lead is a neurotoxin found in drinking water, old paint, and air. It is particularly hazardous to people because it is not easily eliminated from the body. As directed by the "Safe Drinking Water Act" of December 1974 the EPA proposed a maximum lead level in public drinking water of .05 milligram per liter. This standard assumed an individual consumption of two liters of water per day. (*Source:* Nemerow, N. and A. Dasgupta, *Industrial and Hazardous Waste Treatment,* Van Nostrand Reinhold, New York, 1991.)

(a) If EPA guidelines are followed, write an equation that gives the maximum amount of lead ingested in x years. Assume that there are 365.25 days in a year.

(b) If the average life expectancy is 72 years, find the EPA maximum lead intake from water over a lifetime.

30. The Pro Billiards Tour has grown substantially over the past few years. Based on data provided by the Tour, the total prize money offered from 1991 through 1994 can be modeled by the equation $y = 220,000x + 320,000$ where $x = 0$ corresponds to 1991 and y is in dollars. Based on this model, what would we expect the 1995 prize money to be?

Pro Billiards Tour Prize Money

In dollars
$1,000,000
$750,000
$500,000
$350,000

'91 '92 '93 '94

Source: The Tour

Solve the equation.

31. $(b + 7)^2 = 5$

32. $(3y - 2)^2 = 8$

33. $2a^2 + a - 15 = 0$

34. $12x^2 = 8x - 1$

35. $2q^2 - 11q = 21$

36. $3x^2 + 2x = 16$

37. $2 - \dfrac{5}{p} = \dfrac{3}{p^2}$

38. $\dfrac{4}{m^2} = 2 + \dfrac{7}{m}$

39. $\sqrt{2}x^2 - 4x + \sqrt{2} = 0$

40. Which one of the following equations has two real, distinct solutions? Do not actually solve.
  (a) $(3x - 4)^2 = -4$ **(b)** $(4 + 7x)^2 = 0$ **(c)** $(5x + 9)(5x + 9) = 0$ **(d)** $(7x + 4)^2 = 11$

41. Which equations in Exercise 40 have only one distinct, real solution?

42. Which one of the equations in Exercise 40 has two imaginary solutions?

Evaluate the discriminant for the equation, and then use it to predict the number and type of solutions.

43. $8y^2 = 2y - 6$

44. $6k^2 - 2k = 3$

45. $16r^2 + 3 = 26r$

46. $25x^2 - 110x = -121$

Solve the problem.

47. A projectile is fired from ground level. After t sec its height above the ground is $220t - 16t^2$ ft. At what times is the projectile 624 ft above the ground?

48. Paula Story plans to replace the vinyl floor covering in her 10- by 12-ft kitchen. She wants to have a border of even width of a special material. She can afford only 21 sq ft of this material. How wide a border can she have?

49. Mohammed wants to fence off a rectangular playground beside an apartment building. The building forms one boundary, so he needs to fence only the other three sides. The area of the playground is to be 11,250 sq m. He has enough material to build 325 m of fence. Find the length and width of the playground.

50. It takes two gardeners 3 hr (working together) to mow the lawns in a city park. One gardener could do the entire job in 1 hr less time than the other. How long would it take the slower gardener to complete the work alone? Give the answer to the nearest tenth.

51. The lengths of the sides of a right triangle are such that the shortest side is 7 inches shorter than the middle side, while the longest side (the hypotenuse) is 1 inch longer than the middle side. Find the lengths of the sides.

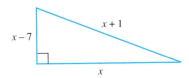

52. Based on information from the U.S. Bureau of Labor Statistics, the Consumer Price Index for college tuition can be modeled by the equation $y = 1.20x^2 + 8.55x + 139.7$ for the period beginning in 1987 and ending in 1993, where $x = 0$ corresponds to 1987. Based on this model, for what year in the period was the Consumer Price Index equal to 150.0?

Solve the problem involving variation.

53. Suppose r varies directly as x and inversely as the square of y. If r is 10 when x is 5 and y is 3, find r when x is 12 and y is 4.

54. Suppose m varies jointly as n and the square of p, and inversely as q. If m is 20 when n is 5, p is 6, and q is 18, find m when n is 7, p is 11, and q is 2.

55. The pressure on a point in a liquid is directly proportional to the distance from the surface to the point. In a certain liquid the pressure at a depth of 4 m is 60 kg per m^2. Find the pressure at a depth of 10 m.

56. The force needed to keep a car from skidding on a curve varies inversely as the radius of the curve and jointly as the weight of the car and the square of the speed. It takes 3000 lb of force to keep a 2000-lb car from skidding on a curve of radius 500 ft at 30 mph. What force is needed to keep the same car from skidding on a curve of radius 800 ft at 60 mph?

57. The power a windmill obtains from the wind varies directly as the cube of the wind velocity. If a 10 km/hr wind produces 10,000 units of power, how much power is produced by a wind of 15 km/hr?

58. The weight w of an object varies inversely as the square of the distance d between the object and the center of the earth. If a man weighs 90 kg on the surface of the earth, how much would he weigh 800 km above the surface? (The radius of the earth is about 6400 km.)

Solve the equation.

59. $4a^4 + 3a^2 - 1 = 0$

60. $2x^4 - x^2 = 0$

61. $(2z + 3)^{2/3} + (2z + 3)^{1/3} = 6$

62. $5\sqrt{m} = \sqrt{3m + 2}$

63. $\sqrt{4y - 2} = \sqrt{3y + 1}$

64. $\sqrt{2x + 3} = x + 2$

65. $\sqrt{p + 2} = 2 + p$

66. $\sqrt{k} = \sqrt{k + 3} - 1$

67. $\sqrt{x + 3} - \sqrt{3x + 10} = 1$

68. $\sqrt{5x - 15} - \sqrt{x + 1} = 2$

69. $\sqrt[3]{6y + 2} = \sqrt[3]{4y}$

70. $(x - 2)^{2/3} = x^{1/3}$

Solve the following inequalities. Write answers in interval notation.

71. $-9x < 4x + 7$

72. $11y \geq 2y - 8$

73. $-5z - 4 \geq 3(2z - 5)$

74. $-(4a + 6) < 3a - 2$

75. $3r - 4 + r > 2(r - 1)$

76. $7p - 2(p - 3) \leq 5(2 - p)$

77. $5 \leq 2x - 3 \leq 7$

78. $-8 > 3a - 5 > -12$

79. $x^2 + 3x - 4 \leq 0$

80. $p^2 + 4p > 21$

81. $6m^2 - 11m - 10 < 0$

82. $k^2 - 3k - 5 \geq 0$

83. $x^2 - 6x + 9 \le 0$

84. $\dfrac{3a - 2}{a} > 4$

85. $\dfrac{5p + 2}{p} < -1$

86. $\dfrac{3}{r - 1} \le \dfrac{5}{r + 3}$

87. $\dfrac{3}{x + 2} > \dfrac{2}{x - 4}$

88. If $0 < a < b$, on what interval is $(x - a)(x - b)$ positive? negative? zero?

89. Without actually solving the inequality, explain why 3 cannot be in the solution set of $\dfrac{2x + 5}{x - 3} < 0$.

90. Without actually solving the inequality, explain why -4 must be in the solution set of $\dfrac{x + 4}{x - 3} \ge 0$.

Solve the problem.

 Tropospheric ozone (ground level ozone) is a gas toxic to both plants and animals. Ozone causes respiratory problems and eye irritation in humans. Automobiles are a major source of this type of harmful ozone. It is often present when smog levels are significant. Ozone in outdoor air can enter buildings through ventilation systems. Guideline levels for indoor ozone are less than 50 parts per billion (ppb). Ozone can be removed from the air using filters. In a scientific study a purafil air filter was used to reduce an initial ozone concentration of 140 ppb. The filter removed 43% of the ozone. (Source: Parmar and Grosjean, "Removal of Air Pollutants from Museum Display Cases," Getty Conservation Institute, Marina del Rey, CA, 1989.)

91. Determine if this type of filter reduced the ozone concentration to acceptable levels.

92. What is the maximum initial concentration of ozone that this filter will reduce to an acceptable level?

93. A company produces videotapes. The revenue from the sale of x units of tapes is $R = 8x$. The cost to produce x units of tapes is $C = 3x + 1500$. In what interval will the company at least break even?

94. A projectile is launched upward. Its height in feet above the ground after t seconds is $320t - 16t^2$.
(a) After how many seconds in the air will it hit the ground?
(b) During what time interval is the projectile more than 576 ft above the ground?

Solve the equation.

95. $|a + 4| = 7$

96. $|-y + 2| = -4$

97. $\left| \dfrac{7}{2 - 3a} \right| = 9$

98. $|5 - 8x| + 1 = 3$

99. $|5r - 1| = |2r + 3|$

100. $|k + 7| = 2k$

Solve the inequality. Write solutions with interval notation.

101. $|m| \le 7$

102. $|r| < 2$

103. $|p| > 3$

104. $|z| > -1$

105. $|2z + 9| \le 3$

106. $|5m - 8| \le 2$

107. $|7k - 3| < 5$

108. $|2p - 1| > 2$

109. $|3r + 7| - 5 > 0$

110. Write as an absolute value equation or inequality.
(a) k is 12 units from 3 on the number line.
(b) k is at least 5 units from 1 on the number line.

Chapter 2   Test ▼▼▼▼▼▼▼▼▼▼▼▼▼▼▼▼▼▼▼▼▼▼▼▼▼▼▼▼▼▼▼▼▼▼

Solve the equation.

1. $3(x - 4) - 5(x + 2) = 2 - (x + 24)$

2. $\dfrac{2}{t - 3} - \dfrac{3}{t + 3} = \dfrac{12}{t^2 - 9}$

3. The formula for the surface area of a rectangular solid is

$$S = 2HW + 2LW + 2LH,$$

where S, H, W, and L represent surface area, height, width, and length, respectively. Solve this formula for W.

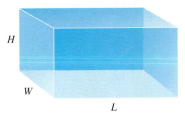

Solve the problem.

Radon comes from the soil and often enters a building through cracks and pores in its foundation. The amount of radon gas present in a house can be measured using a home detection kit. These kits often use activated charcoal as an absorbent. The equation to determine the concentration C of radon in picocuries per liter of air (pCi/L) is

$$C = \frac{5.48 \times 10^3 q^{.571}(T + 273)^{-1}}{[2.10 \times 10^{-11} - 6.58 \times 10^{-14}(T + 273)]^{.571}},$$

where T is the temperature of the room in Celsius and q is the cm³ of radon absorbed in each gram of charcoal. A concentration below 4 pCi/L is considered safe by the EPA while readings as high as 4000 have been recorded in some homes. (One pCi/L is equivalent to 1700 atoms of radon in one liter of air whereas there are about 2.5 × 10²² molecules of oxygen and nitrogen in the same liter of air.) (Source: A. Hines, T. Ghosh, S. Layalka, and R. Warder, Indoor Air Quality & Control, Prentice Hall, 1993.)

4. Use a calculator to show why the radon level is unsafe if the temperature of a basement is $20°$ and q is measured by a laboratory to be 3.1×10^{-13}.

5. What type of action might be taken to make the radon level safe in the previous problem?

6. How many quarts of a 60% alcohol solution must be added to 40 quarts of a 20% alcohol solution to obtain a mixture that is 30% alcohol?

7. Fred and Wilma start from the same point and travel on a straight road. Fred travels at 30 mph, while Wilma travels at 50 mph. If Wilma starts 3 hr after Fred, find the distance they travel before Wilma catches up with Fred.

Solve the equation.

8. $3x^2 - 5x = -2$

9. $(5t - 3)^2 = 17$

10. $6s(2 - s) = 7$

11. What must be the value of k for the equation $4x^2 - 5x - k = 0$ to have a single real solution?

12. More Americans are flying on commuter airlines. The number of fliers on 10- to 30-seat commuter aircraft was 1.4 million in 1975 and 3.1 million in 1994. It is estimated that there will be 9.3 million fliers in the year 2006. Here are three possible models for these data, where $x = 0$ corresponds to the year 1975.

(a) $y = .24x + .6$ **(b)** $y = .0138x^2 - .172x + 1.4$ **(c)** $y = .0125x^2 - .193x + 1.4$

The table shows each equation evaluated at the years 1975, 1994, and 2006. Decide which equation best models the data.

Year	(a)	(b)	(c)
0	.60	1.40	1.40
19	5.16	3.11	2.25
31	8.04	9.33	7.43

(*Source: Federal Aviation Administration (USA Today 3/27/95 p. 1B*))

13. Use each equation given in the previous problem to estimate the number of commuter fliers in the year 2000.

14. A projectile is launched from ground level with an initial velocity of 96 feet per second. Its height in feet, h, after t seconds is given by the equation $h = -16t^2 + 96t$.
   (a) At what time(s) will it reach a height of 80 feet?
   (b) After how many seconds will it return to the ground?

Solve the equation.

15. $\sqrt{5 + 2x} - x = 1$

16. $x^4 + 6x^2 - 40 = 0$

17. $\sqrt[3]{3x - 8} = \sqrt[3]{9x + 4}$

18. $6 = \dfrac{7}{2y - 3} + \dfrac{3}{(2y - 3)^2}$

Solve the problem.

19. Stratospheric ozone occurs in the atmosphere between altitudes of 20 to 30 kilometers and is an important filter of ultraviolet light from the sun. Ozone is frequently measured in Dobson units, and these units vary directly with the thickness of the ozone layer. If 300 Dobson units corresponds to an ozone layer of 3 millimeters, what was the thickness at the Antarctic ozone hole in 1991, which measured 110 Dobson units? (*Source:* Robert E. Huffman, *Atmospheric Ultraviolet Remote Sensing,* Academic Press, 1992.)

20. Under certain conditions, the length of time that it takes for fruit to ripen during the growing season varies inversely as the average maximum temperature during the season. If it takes 25 days for fruit to ripen with an average maximum temperature of 80°, find the number of days it would take at 75°.

Solve the inequality. Give the answer using interval notation.

21. $-2(x - 1) - 10 < 2(x + 2)$

22. $-2 \leq \dfrac{1}{2}x + 3 \leq 4$

23. $2x^2 - x - 3 \geq 0$

24. $\dfrac{x + 1}{x - 3} < 5$

Solve the absolute value equation or inequality.

25. $|3x + 5| = 4$

26. $|2x - 5| < 9$

27. $|2x + 1| \geq 11$

Relations and Functions

3

Approximating data with linear relations and functions is one of the most important and fundamental mathematical techniques we use today. Although most real world applications are nonlinear, we can often use linear approximations to give accurate estimations. For example, the shape of Earth is round, not flat. Yet, when a building is constructed, the curvature of Earth's surface is seldom taken into account. Instead, it is assumed that the surface is level over the relatively small distance covered by the building. In this case, we use a linear approximation to accurately solve a nonlinear problem. However, when freeways were built across the United States, the curvature of Earth's surface had to be taken into account. If the distance or interval is small, linear approximations can lead to accurate estimations. Their advantage is that they are simple and easy to compute. On the other hand, if the distance or interval is large, then a linear approximation may lead to incorrect results.

In the late 1920s the famous observational astronomer Edwin P. Hubble (1889–1953) determined by careful measurement both the distances to several galaxies and the velocities at which they were moving away from Earth. Four galaxies with their distances in megaparsecs and velocities in kilometers per second are listed in the table. (One megaparsec is approximately 1.9×10^{19} miles.)

Galaxy	Distance	Velocity
Virgo	15	1600
Ursa Minor	200	15,000
Corona Borealis	290	24,000
Bootes	520	40,000
Hydra	?	60,000

Is there any relationship between the data that could be used to predict the distance from Earth to the galaxy Hydra? Could the age of the universe be estimated using these data? Edwin Hubble made one of the most important discoveries in astronomy when he determined that a linear relationship existed between the distance and velocity of a galaxy. His important finding resulted in Hubble's Law. Because of this significant contribution to the understanding of our expanding universe, the Hubble Space Telescope was named after him. For galaxies relatively close to Earth, Hubble's linear relationship has been shown to be accurate. How far into deep space this linear relationship holds remains uncertain. Before we can calculate the distance to the galaxy Hydra or approximate the age of the universe, we must first understand relations and functions. Using relations and functions to approximate real data, we will be able to answer these and other important questions. The answers to the questions posed here are given in Section 3.4.*

A major goal of this text is to make a careful study of *relations* and *functions*. In this chapter we discuss several important first- and second-degree relations.

3.1 Relations and the Rectangular Coordinate System ▼▼▼

Many things in daily life are related. For example, a student's grade in a course usually is related to the amount of time spent studying, while the number of miles per gallon of gas used on a car trip depends on the speed of the car. Also, the total amount (in millions of dollars) of government-guaranteed student loans is related to the year as shown in the table.

Year	Amount
1988	11.8
1991	13.5
1994	18.2

Pairs of related numbers, such as 1988 and 11.8 or 1991 and 13.5 in the table, can be written as *ordered pairs*. An **ordered pair** of numbers consists of two numbers, written inside parentheses, in which the sequence of the numbers is important. For example, (4, 2) and (2, 4) are different ordered pairs because

---

*Sources: Acker, A., and C. Jaschek, *Astronomical Methods and Calculations,* John Wiley & Sons, 1986.
Sharov, A., and I. Novikov, *Edwin Hubble, The Discoverer of the Big Bang Universe,* Cambridge University Press, 1993.

the order of the numbers is different. Notation such as (3, 4) has already been used in this book to show an interval on the number line. Now the same notation is used to indicate an ordered pair of numbers. In virtually every case, the intended use will be clear from the context of the discussion.

RELATIONS   A set of ordered pairs is called a **relation.** The **domain** of a relation is the set of first elements in the ordered pairs, and the **range** of the relation is the set of all possible second elements. In the student loan example above, the domain is the set of years, and the range is the set of resulting student loan amounts. In this text, we usually confine domains and ranges to real number values.

Ordered pairs are used to express the solutions of equations in two variables. For example, we say that (1, 2) is a solution of $2x - y = 0$, since substituting 1 for x and 2 for y in the equation gives

$$2(1) - 2 = 0$$
$$0 = 0,$$

a true statement. When an ordered pair represents the solution of an equation with the variables x and y, the x-value is written first.

Although any set of ordered pairs is a relation, in mathematics we are most interested in those relations that are solution sets of equations. We may say that an equation *defines a relation,* or that it is the *equation of the relation.* For simplicity, we often refer to equations such as

$$y = 3x + 5 \qquad \text{or} \qquad x^2 + y^2 = 16$$

as relations, although technically the solution set of the equation is the relation.

To determine the domain of a relation defined by an equation, think of any reason a number should be *excluded* as an input value. For example, 0 cannot replace x in the equation $y = 1/x$, because 0 makes the expression undefined. Thus, the domain is $(-\infty, 0) \cup (0, \infty)$. Similarly, in the equation $y = \sqrt{x}$, x cannot be negative, so the domain is all nonnegative numbers, written in interval notation as $[0, \infty)$.

EXAMPLE 1
Finding ordered pairs, domains, and ranges

For each relation defined below, give three ordered pairs that belong to the relation, and state the domain and the range of the relation.

(a) $\{(2, 5), (7, -1), (10, 3), (-4, 0), (0, 5)\}$

Three ordered pairs from the relation are any three of the five ordered pairs in the set. The domain is the set of first elements.

$$\{2, 7, 10, -4, 0\},$$

and the range is the set of second elements,

$$\{5, -1, 3, 0\}.$$

(b) $y = 4x - 1$

To find an ordered pair of the relation, we can choose any number for x or y and substitute in the equation to get the corresponding value of the other variable. For example, let $x = -2$. Then

$$y = 4(-2) - 1 = -9,$$

giving the ordered pair $(-2, -9)$. If $y = 3$, then

$$3 = 4x - 1$$
$$4 = 4x$$
$$1 = x,$$

and the ordered pair is $(1, 3)$. Verify that $(0, -1)$ also belongs to the relation. Since x and y can take any real-number values, both the domain and range include all real numbers, symbolized $(-\infty, \infty)$.

(c) $x = \sqrt{y - 1}$

Verify that the ordered pairs $(1, 2)$, $(0, 1)$, and $(2, 5)$ belong to the relation. Since x equals the principal square root of $y - 1$, the domain is restricted to $[0, \infty)$. Also, only nonnegative numbers have a real square root, so the range is determined by the inequality

$$y - 1 \geq 0$$
$$y \geq 1,$$

giving $[1, \infty)$ as the range. ▶

THE RECTANGULAR COORDINATE SYSTEM Since the study of relations often involves looking at their *graphs,* this section includes a brief review of the coordinate plane. As mentioned in Chapter 1, each real number corresponds to a point on a number line. This correspondence is set up by establishing a coordinate system for the line. This idea is extended to the two dimensions of a plane by drawing two perpendicular lines, one horizontal and one vertical. These lines intersect at a point O called the **origin.** The horizontal line is called the **x-axis,** and the vertical line is called the **y-axis.**

Starting at the origin, the x-axis can be made into a number line by placing positive numbers to the right and negative numbers to the left. The y-axis can be made into a number line with positive numbers going up and negative numbers going down.

The x-axis and y-axis together make up a **rectangular coordinate system,** or **Cartesian coordinate system** (named for one of its co-inventors, René Descartes; the other co-inventor was Pierre de Fermat). The plane into which the coordinate system is introduced is the **coordinate plane,** or **xy-plane.** The x-axis and y-axis divide the plane into four regions, or **quadrants,** labeled as shown in Figure 1. The points on the x-axis and y-axis belong to no quadrant.

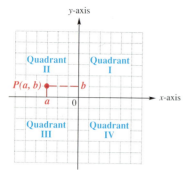

FIGURE 1

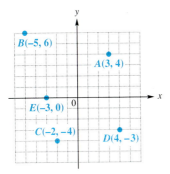

FIGURE 2

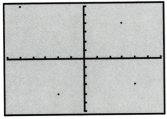

[−6, 6] by [−6, 6]

The points in Figure 2 are plotted on this graphing calculator screen. Why do you think that E (−3, 0) is not visible?

Each point P in the xy-plane corresponds to a unique ordered pair (a, b) of real numbers. The numbers a and b are the **coordinates** of point P. To locate on the xy-plane the point corresponding to the ordered pair (3, 4), for example, draw a vertical line through 3 on the x-axis and a horizontal line through 4 on the y-axis. These two lines cross at point A in Figure 2. Point A corresponds to the ordered pair (3, 4). Also in Figure 2, B corresponds to the ordered pair (−5, 6), C to (−2, −4), D to (4, −3), and E to (−3, 0). The point P corresponding to the ordered pair (a, b) often is written as $P(a, b)$ as in Figure 1 and referred to as "the point (a, b)."

As we shall see later in this chapter the **graph** of a relation is the set of points in the plane that correspond to the ordered pairs of the relation.

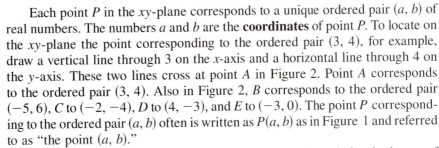

CONNECTIONS In an earlier chapter, we discussed *equations in one variable*. We saw there that in most cases the solution sets were finite. For example, a linear equation in one variable, such as $2x + 1 = 7$, has just one solution, and a quadratic equation in one variable, such as $2x^2 + 3x = −1$, has two solutions. In this chapter and in much of the rest of the book, we will be interested in *equations in two variables*. As we saw in Example 1(b), equations in two variables have an infinite number of solutions that are ordered pairs of numbers, rather than a finite number of solutions that are numbers. We graphed the solutions of equations in one variable as one or more points on a number line. Now we see that the graphs of solutions of equations in two variables require two number lines (that are placed perpendicular to each other and intersect at their origins).

FOR DISCUSSION OR WRITING

1. What kind of solutions do you think an equation in three variables would have?
2. How might we graph the solutions of an equation in three variables?
3. Would it be possible to graph solutions of an equation in four variables? If so, how? If not, why not?

In the rest of this section, we will use coordinates of ordered pairs to derive two useful formulas, the *distance formula* and the *midpoint formula,* and to determine the equations of circles.

THE DISTANCE FORMULA By using the Pythagorean theorem, we can develop a formula to find the distance between any two points in a plane. For example, Figure 3 shows the points $P(−4, 3)$ and $R(8, −2)$.

To find the distance between these two points, we complete a right triangle as shown in the figure. This right triangle has its 90° angle at (8, 3). The horizontal side of the triangle has length

$$|8 − (−4)| = 12,$$

where absolute value is used to make sure that the distance is not negative. The vertical side of the triangle has length

$$|3 − (−2)| = 5.$$

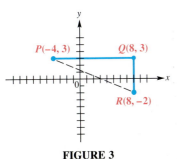

FIGURE 3

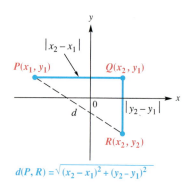

$$d(P, R) = \sqrt{(x_2 - x_1)^2 + (y_2 - y_1)^2}$$

FIGURE 4

By the Pythagorean theorem, the length of the remaining side of the triangle is

$$\sqrt{12^2 + 5^2} = \sqrt{144 + 25} = \sqrt{169} = 13.$$

Thus, the distance between $(-4, 3)$ and $(8, -2)$ is 13.

To obtain a general formula for the distance between two points on a coordinate plane, we let $P(x_1, y_1)$ and $R(x_2, y_2)$ be any two distinct points in a plane, as shown in Figure 4. We complete a triangle by locating point Q with coordinates (x_2, y_1). Using the Pythagorean theorem gives the distance between P and R, written $d(P, R)$, as

$$d(P, R) = \sqrt{(x_2 - x_1)^2 + (y_2 - y_1)^2}.$$

NOTE   The use of absolute value bars is not necessary in this formula, since for all real numbers a and b, $|a - b|^2 = (a - b)^2$.

The distance formula can be summarized as follows.

> **Distance Formula**
>
> Suppose that $P(x_1, y_1)$ and $R(x_2, y_2)$ are two points in a coordinate plane. Then the distance between P and R, written $d(P, R)$, is given by the **distance formula,**
>
> $$d(P, R) = \sqrt{(x_2 - x_1)^2 + (y_2 - y_1)^2}.$$

Although the proof of the distance formula assumes that P and R are not on a horizontal or vertical line, the result is true for any two points.

◀ EXAMPLE 2
Using the distance formula

Find the distance between $P(-8, 4)$ and $Q(3, -2)$.
According to the distance formula,

$$d(P, Q) = \sqrt{[3 - (-8)]^2 + (-2 - 4)^2} \qquad x_1 = -8, y_1 = 4, x_2 = 3, y_2 = -2$$
$$= \sqrt{11^2 + (-6)^2}$$
$$= \sqrt{121 + 36} = \sqrt{157}. \quad \blacktriangleright$$

NOTE   As shown in Example 2, it is customary to leave the distance between two points in radical form rather than approximating the radical with a calculator (unless, of course, it is otherwise specified).

A statement of the form "If p, then q" is called a *conditional* statement. The related statement "If q, then p" is called its *converse.* In Chapter 2 we studied the Pythagorean theorem. The converse of the Pythagorean theorem is also a true statement: If the sides a, b, and c of a triangle satisfy $a^2 + b^2 = c^2$, then the triangle is a right triangle with legs having lengths a and b and hypotenuse having length c. This can be used to determine whether three points are the vertices of a right triangle, as shown in the next example.

EXAMPLE 3
Determining whether three points are the vertices of a right triangle

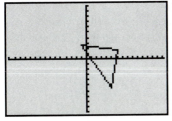

[–30, 30] by [–20, 20]
Xscl = 2   Yscl = 2

The triangle in Figure 5 has been drawn on this screen. Notice that due to the limited resolution, the straight lines *appear* to be jagged. These "jaggies" actually represent straight line segments.

Are the three points $M(-2, 5)$, $N(12, 3)$, and $Q(10, -11)$ the vertices of a right triangle?

A triangle with the three given points as vertices is shown in Figure 5. This triangle is a right triangle if the square of the length of the longest side equals the sum of the squares of the lengths of the other two sides. Use the distance formula to find the length of each side of the triangle.

$$d(M, N) = \sqrt{[12 - (-2)]^2 + (3 - 5)^2} = \sqrt{196 + 4} = \sqrt{200}$$
$$d(M, Q) = \sqrt{[10 - (-2)]^2 + (-11 - 5)^2} = \sqrt{144 + 256} = \sqrt{400} = 20$$
$$d(N, Q) = \sqrt{(10 - 12)^2 + (-11 - 3)^2} = \sqrt{4 + 196} = \sqrt{200}$$

The longest side has a length of 20 units. Since

$$(\sqrt{200})^2 + (\sqrt{200})^2 = 400 = 20^2,$$

the triangle is a right triangle with hypotenuse connecting M and Q. ▶

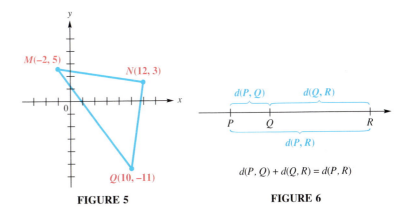

FIGURE 5   FIGURE 6

Using a procedure similar to that of Example 3, we can tell whether three points lie on a straight line. Points that lie on a line are called collinear. Three points are **collinear** if the sum of the distances between two pairs of the points is equal to the distance between the remaining pair of points. See Figure 6.

EXAMPLE 4
Determining whether three points are collinear

Are the points $(-1, 5)$, $(2, -4)$, and $(4, -10)$ collinear?

The distance between $(-1, 5)$ and $(2, -4)$ is

$$\sqrt{(-1 - 2)^2 + [5 - (-4)]^2} = \sqrt{9 + 81} = \sqrt{90} = 3\sqrt{10}.$$

The distance between $(2, -4)$ and $(4, -10)$ is

$$\sqrt{(2 - 4)^2 + [-4 - (-10)]^2} = \sqrt{4 + 36} = \sqrt{40} = 2\sqrt{10}.$$

Finally, the distance between the remaining pair of points, $(-1, 5)$ and $(4, -10)$ is

$$\sqrt{(-1 - 4)^2 + [5 - (-10)]^2} = \sqrt{25 + 225} = \sqrt{250} = 5\sqrt{10}.$$

Because $3\sqrt{10} + 2\sqrt{10} = 5\sqrt{10}$, the three points are collinear. ▶

O is the center and *r* is
the radius of this circle.

FIGURE 7

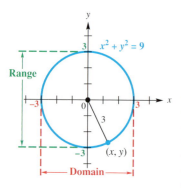

FIGURE 8

CIRCLES By definition, a **circle** is the set of all points in a plane that lie a given distance from a given point. The given distance is the **radius** of the circle and the given point is the **center.** See Figure 7. Since a circle is a set of points, it corresponds to a relation. The equation of a circle can be found from its definition by using the distance formula.

Figure 8 shows a circle of radius 3 with center at the origin. To find the equation of this circle, let (x, y) be any point on the circle. The distance between (x, y) and the center of the circle, $(0, 0)$, is given by

$$\sqrt{(x - 0)^2 + (y - 0)^2}.$$

Since this distance equals the radius, 3,

$$\sqrt{(x - \mathbf{0})^2 + (y - \mathbf{0})^2} = \mathbf{3}$$
$$\sqrt{x^2 + y^2} = 3$$
$$x^2 + y^2 = 9.$$

As suggested by Figure 8, the domain of the relation is $[-3, 3]$, and the range of the relation is $[-3, 3]$.

To use a graphing calculator in *function* mode to graph the circle $x^2 + y^2 = 9$, for example, we must first solve the equation for y. We then graph the two resulting equations, $Y_1 = \sqrt{9 - x^2}$ and $Y_2 = -\sqrt{9 - x^2}$, in the same window. The graph is shown in the figures below. The graph appears elliptical unless a square window is used. Note that the calculator may not completely connect the two graphs. This is due to limitations of the calculator.

Some calculators can draw a circle given the center and radius. Consult your owner's manual to see whether your calculator can do this.

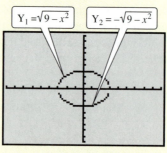

$[-9, 9]$ by $[-9, 9]$

Because the screen ratio of a graphing calculator is usually about 3 to 2, the circle appears elliptical. This could lead to false conclusions.

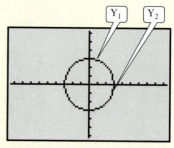

$[-9.4, 9.4]$ by $[-6.2, 6.2]$

We use a *square viewing window* to gain a true perspective for this circle.

EXAMPLE 5
Finding the equation of a circle

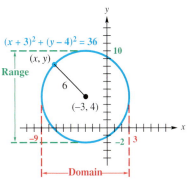

FIGURE 9

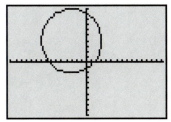

[−15, 15] by [−10, 10]

This circle was created by using a DRAW command on the TI-82 calculator. Compare to Figure 9.

Find an equation for the circle having radius 6 and center at $(-3, 4)$. Graph the circle.

This circle is shown in Figure 9. We can find its equation by using the distance formula. Let (x, y) be any point on the circle. The distance from (x, y) to $(-3, 4)$ is given by

$$\sqrt{[x - (-3)]^2 + (y - 4)^2} = \sqrt{(x + 3)^2 + (y - 4)^2}.$$

This distance equals the radius, 6. Therefore,

$$\sqrt{(x + 3)^2 + (y - 4)^2} = 6$$

or

$$(x + 3)^2 + (y - 4)^2 = 36.$$

The domain is $[-9, 3]$, and the range is $[-2, 10]$, as seen in Figure 9. ▶

Generalizing from the work in Example 5 gives the following result.

> **Center-Radius Form of the Equation of a Circle**
>
> The circle with center (h, k) and radius r has equation
>
> $$(x - h)^2 + (y - k)^2 = r^2,$$
>
> the **center-radius form** of the equation of a circle. A circle with center $(0, 0)$ and radius r has equation
>
> $$x^2 + y^2 = r^2.$$

Starting with the center-radius form of the equation of a circle, $(x - h)^2 + (y - k)^2 = r^2$, and squaring $x - h$ and $y - k$ gives an equation of the form

$$x^2 + y^2 + cx + dy + e = 0, \qquad (*)$$

where c, d, and e are real numbers. This form is called the **general form of the equation of a circle.** Also, starting with an equation in the form of $(*)$, the process of completing the square can be used to get an equation of the form

$$(x - h)^2 + (y - k)^2 = m$$

for some number m. If $m > 0$, then $r^2 = m$, and the equation represents a circle with radius $\sqrt{m}$. If $m = 0$, then the equation represents the single point (h, k). If $m < 0$, no points satisfy the equation.

EXAMPLE 6
Finding the center and radius by completing the square

Decide whether or not each equation has a circle as its graph.

(a) $x^2 - 6x + y^2 + 10y + 25 = 0$

Since this equation has the form of equation $(*)$, it either represents a circle, a single point, or no points at all. To decide which, complete the square on x and y separately, as explained in Section 2.3. Start with

$$(x^2 - 6x \quad) + (y^2 + 10y \quad) = -25.$$

Half of -6 is -3, and $(-3)^2 = 9$. Also, half of 10 is 5, and $5^2 = 25$. Add 9 and 25 on the left, and to compensate, add 9 and 25 on the right.

$$(x^2 - 6x + 9) + (y^2 + 10y + 25) = -25 + 9 + 25$$
$$(x - 3)^2 + (y + 5)^2 = 9$$

Since $9 > 0$, the equation represents a circle that has its center at $(3, -5)$ and radius 3.

(b) $x^2 + 10x + y^2 - 4y + 33 = 0$

Complete the square as above.

$$(x^2 + 10x + 25) + (y^2 - 4y + 4) = -33 + 25 + 4$$
$$(x + 5)^2 + (y - 2)^2 = -4$$

Since $-4 < 0$, there are no ordered pairs (x, y), with x and y both real numbers, satisfying the equation. The graph of the given equation contains no points.

(c) $2x^2 + 2y^2 - 6x + 10y = 1$

After grouping and completing the square, divide both sides of the equation by 2, so the coefficients of x^2 and y^2 are 1.

$$2x^2 + 2y^2 - 6x + 10y = 1$$
$$2\left(x^2 - 3x + \frac{9}{4}\right) + 2\left(y^2 + 5y + \frac{25}{4}\right) = 2\left(\frac{9}{4}\right) + 2\left(\frac{25}{4}\right) + 1 \quad \text{Complete the square.}$$
$$2\left(x - \frac{3}{2}\right)^2 + 2\left(y + \frac{5}{2}\right)^2 = 18$$
$$\left(x - \frac{3}{2}\right)^2 + \left(y + \frac{5}{2}\right)^2 = 9 \quad \text{Divide both sides by 2.}$$

The equation has a circle with center at $(3/2, -5/2)$ and radius 3 as its graph. ▶

EXAMPLE 7
Finding the equation of a circle given the center and the diameter

Find the equation of a circle with center $(-2, 3)$ that has a diameter with one endpoint at $(1, 0)$.

The diameter of a circle goes through the center of the circle and has endpoints on the circle, so the point $(1, 0)$ is on the circle and satisfies its equation. Thus, the radius is the distance between $(-2, 3)$ and $(1, 0)$.

$$r = \sqrt{(-3)^2 + 3^2} = \sqrt{18} = 3\sqrt{2}$$

Since the center is at $(-2, 3)$, the equation is

$$(x + 2)^2 + (y - 3)^2 = 18. ▶$$

THE MIDPOINT FORMULA   The midpoint formula is used to find the coordinates of the midpoint of a line segment. (Recall that the midpoint of a line segment is equidistant from the endpoints of the segment.) To develop the midpoint formula, let (x_1, y_1) and (x_2, y_2) be any two distinct points in a plane. (Although

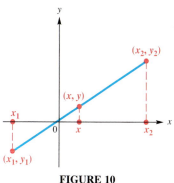

FIGURE 10

Figure 10 shows $x_1 < x_2$, no particular order is required.) Assume that the two points are not on a horizontal or vertical line. Let (x, y) be the midpoint of the segment connecting (x_1, y_1) and (x_2, y_2). Draw vertical lines from each of the three points to the x-axis, as shown in Figure 10.

Since (x, y) is the midpoint of the line segment connecting (x_1, y_1) and (x_2, y_2), the distance between x and x_1 equals the distance between x and x_2, so that

$$x_2 - x = x - x_1$$
$$x_2 + x_1 = 2x$$
$$x = \frac{x_1 + x_2}{2}.$$

By this result, the x-coordinate of the midpoint is the average of the x-coordinates of the endpoints of the segment. In a similar manner, the y-coordinate of the midpoint is $(y_1 + y_2)/2$, proving the following statement.

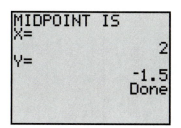

Explore the programming features of your calculator to see whether you can write a program to find the midpoint. This display was obtained using the endpoints in Example 8.

Midpoint Formula

The midpoint of the line segment with endpoints (x_1, y_1) and (x_2, y_2) is

$$\left(\frac{x_1 + x_2}{2}, \frac{y_1 + y_2}{2} \right).$$

In other words, the midpoint formula says that the coordinates of the midpoint of a segment are found by calculating the *average* of the x-coordinates and the average of the y-coordinates of the endpoints of the segment. In Exercise 37, you are asked to verify that the coordinates above satisfy the definition of midpoint.

◀ **EXAMPLE 8**
Using the midpoint formula

Find the midpoint M of the segment with endpoints $(8, -4)$ and $(-4, 1)$.

Use the midpoint formula to find that the coordinates of M are

$$\left(\frac{8 + (-4)}{2}, \frac{-4 + 1}{2} \right) = \left(2, -\frac{3}{2} \right). \quad ▶$$

◀ **EXAMPLE 9**
Using the midpoint formula

A line segment has an endpoint at $(2, -8)$ and a midpoint at $(-1, -3)$. Find the other endpoint of the segment.

The formula for the x-coordinate of the midpoint is $(x_1 + x_2)/2$. Here the x-coordinate of the midpoint is -1. Letting $x_1 = 2$ gives

$$-1 = \frac{2 + x_2}{2}$$
$$-2 = 2 + x_2$$
$$-4 = x_2.$$

In the same way, $y_2 = 2$ and the endpoint is $(-4, 2)$. ▶

3.1 Exercises ▼▼▼▼▼▼▼▼▼▼▼▼▼▼▼▼▼▼▼▼▼▼▼▼▼▼▼▼▼▼

Decide whether the statement is true or false.

1. For a particular point (x, y), if $xy > 0$, then the point will lie in either quadrant I or quadrant III.

2. The midpoint of the segment joining (a, b) and $(-a, -b)$ is the origin.

3. The point $(a, 0)$ lies on the x-axis and its distance from the origin is $|a|$.

4. The circle with equation $x^2 + y^2 = 16$ has center at the origin and radius 16.

For the given relation, give three ordered pairs that belong to the relation and state its domain and its range. See Example 1.

5. $\{(-4, 6), (3, 2), (5, 7)\}$

6. $\{(\pi, -3), (\sqrt{2}, 5), (7, \sqrt[3]{12})\}$

7. $y = 9x - 3$

8. $y = -6x + 4$

9. $y = -\sqrt{x}$

10. $y = \sqrt[3]{x}$

11. $y = |x + 2|$

12. $y = -|x - 4|$

 13.

X	Y₁
0	3
1	5.1
2	7.2
3	9.3
4	11.4
5	13.5
6	15.6

X=0

14.

X	Y₁
0	4
1	.9
2	-2.2
3	-5.3
4	-8.4
5	-11.5
6	-14.6

X=0

*For the points P and Q, **(a)** find the distance d(P, Q) and **(b)** find the midpoint of the segment PQ. See Examples 2 and 8.*

15. $P(-5, -7), Q(-13, 1)$

16. $P(-4, 3), Q(2, -5)$

17. $P(8, 2), Q(3, 5)$

18. $P(-6, -5), Q(6, 10)$

19. $P(3\sqrt{2}, 4\sqrt{5}), Q(\sqrt{2}, -\sqrt{5})$

20. $P(-\sqrt{7}, 8\sqrt{3}), Q(5\sqrt{7}, -\sqrt{3})$

21.

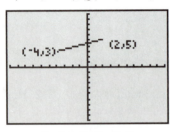

$P = (-4, 3), Q = (2, 5)$

22.

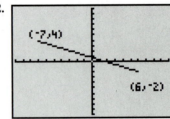

$P = (-7, 4), Q = (6, -2)$

23. The table lists how poverty level income cutoffs (in dollars) for a family of four have changed over time.

Year	Income
1960	3022
1970	3968
1975	5500
1980	8414
1985	10,989
1990	13,359

(*Source:* U.S. Census Bureau.)

(a) Use the midpoint formula to approximate the poverty level cutoff in 1965.
(b) The midpoint formula will give an exact answer if the data have what type of relationship?
(c) Do the income cutoffs have a linear relationship with time? (*Hint:* Consider the data for 1970 and 1980.)

24. The accompanying graph shows an idealized linear relationship for the average monthly family payment to families with dependent children in 1994 dollars. Based on this information, what was the average payment in 1987?

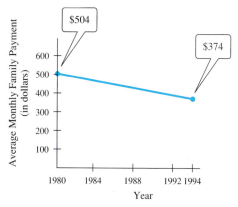

Source: Office of Financial Management, Administration for Children and Families.

Determine whether the three points are the vertices of a right triangle. See Example 3.

25. $(-6, -4)$, $(0, -2)$, $(-10, 8)$

26. $(-2, -8)$, $(0, -4)$, $(-4, -7)$

27. $(-4, 1)$, $(1, 4)$, $(-6, -1)$

28. $(-2, -5)$, $(1, 7)$, $(3, 15)$

Determine whether the three points are collinear. See Example 4.

29. $(0, -7)$, $(-3, 5)$, $(2, -15)$

30. $(-1, 4)$, $(-2, -1)$, $(1, 14)$

31. $(0, 9)$, $(-3, -7)$, $(2, 19)$

32. $(-1, -3)$, $(-5, 12)$, $(1, -11)$

Find the coordinates of the other endpoint of the segment, given its midpoint and one endpoint. See Example 9.

33. midpoint $(5, 8)$, endpoint $(13, 10)$

34. midpoint $(-7, 6)$, endpoint $(-9, 9)$

35. midpoint $(12, 6)$, endpoint $(19, 16)$

36. midpoint $(-9, 8)$, endpoint $(-16, 9)$

37. Show that if M is the midpoint of the segment with endpoints $P(x_1, y_1)$ and $Q(x_2, y_2)$, then $d(P, M) + d(M, Q) = d(P, Q)$ and $d(P, M) = d(M, Q)$.

38. The distance formula as given in the text involves a square root radical. Write the distance formula using rational exponents.

Find the center-radius form of the equation of the circle described. Then graph the circle and give the domain and the range. See Example 5.

39. center $(0, 0)$, radius 6

40. center $(0, 0)$, radius 9

41. center $(2, 0)$, radius 6

42. center $(0, -3)$, radius 7

43. center $(-2, 5)$, radius 4

44. center $(4, 3)$, radius 5

45. center $(5, -4)$, radius 7

46. center $(-3, -2)$, radius 6

47. Find the center-radius form of the equation of a circle with center $(3, 2)$ and tangent to the x-axis. (*Hint: Tangent to* a circle means touching it at exactly one point.)

48. Find the equation of a circle with center at $(-4, 3)$, passing through the point $(5, 8)$. Give it in center-radius form.

49. Use a graphing calculator to graph the circle $(x + 4)^2 + (y - 2)^2 = 25$. Use a square viewing window. Give the equations used for Y_1 and Y_2.

50. A graphing calculator was used to *draw* the circle with center $(-3, 5)$ and radius 4. Which one of the two choices is the correct one?

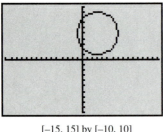

           $[-15, 15]$ by $[-10, 10]$ $[-15, 15]$ by $[-10, 10]$

          (a)   **(b)**

Find the center and the radius of the circle. See Example 6.

51. $x^2 + 6x + y^2 + 8y + 9 = 0$ **52.** $x^2 + 8x + y^2 - 6y + 16 = 0$

53. $x^2 - 4x + y^2 + 12y = -4$ **54.** $x^2 - 12x + y^2 + 10y = -25$

55. $x^2 + y^2 - 2y - 48 = 0$ **56.** $x^2 + 4x + y^2 - 21 = 0$

57. Describe the graph of the equation $x^2 - 6x + y^2 - 6y + 18 = 0$.

58. Describe the graph of the equation $(x - 4)^2 + (y + 4)^2 = -1$.

59. Without actually graphing, state whether or not the graphs of $x^2 + y^2 = 16$ and $x^2 + y^2 = 49$ will intersect. Explain your answer.

60. Can a circle have its center at $(4, 2)$ and be tangent to both axes? Explain.

▼▼▼▼▼▼▼▼▼▼▼▼▼ **DISCOVERING CONNECTIONS** (Exercises 61–66) ▼▼▼▼▼▼▼▼▼▼▼▼▼

In this section we introduced the distance formula, the midpoint formula, and the center-radius form of the equation of a circle. These three ideas are connected quite closely in the following problem.

   A circle has a diameter with endpoints $(-1, 3)$ and $(5, -9)$. Find the center-radius form of the equation of this circle.

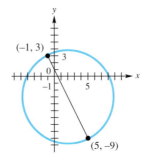

Work Exercises 61–66 in order, so that you can see the relationships among these concepts in solving the problem.

61. To find the center-radius form, we must find both the radius and the coordinates of the center. Find the coordinates of the center using the midpoint formula. (The center of the circle must be the midpoint of the diameter.)

62. There are several ways to find the radius of the circle. One way is to find the distance between the center and the point $(-1, 3)$. Use your result from Exercise 61 and the distance formula to find the radius.

63. Another way to find the radius is to repeat Exercise 62, but use the point $(5, -9)$ rather than $(-1, 3)$. Do this to obtain the same answer you found in Exercise 62.

64. There is yet another way to find the radius. Because the radius is half the diameter, it can be found by finding half the length of the diameter. Using the endpoints of the diameter given in the problem, find the radius in this manner. You should once again obtain the same answer you found in Exercise 62.

65. Using the center found in Exercise 61 and the radius found in Exercises 62–64, give the center-radius form of the circle.

66. Use the method described in Exercises 61–65 to find the center-radius form of the equation of the circle having a diameter whose endpoints are $(3, -5)$ and $(-7, 3)$.

Seismologists can locate the epicenter of an earthquake by determining the intersection of three circles. The radii of these circles represent the distances from the epicenter to each of three receiving stations. The centers of the circles represent the receiving stations.

Suppose receiving stations A, B, and C are located on a coordinate plane at the points $(1, 4)$, $(-3, -1)$, and $(5, 2)$. Let the distance from the earthquake epicenter to each station be 2 units, 5 units, and 4 units, respectively. Where on the coordinate plane is the epicenter located?

Graphically, it appears that the epicenter is located at $(1, 2)$. To check this algebraically, determine the equation for each circle and substitute $x = 1$ and $y = 2$.

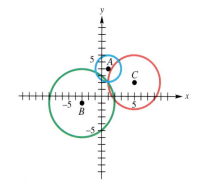

Station A:
$$(x - 1)^2 + (y - 4)^2 = 4$$
$$(1 - 1)^2 + (2 - 4)^2 = 4$$
$$0 + 4 = 4$$
$$4 = 4$$

Station B:
$$(x + 3)^2 + (y + 1)^2 = 25$$
$$(1 + 3)^2 + (2 + 1)^2 = 25$$
$$16 + 9 = 25$$
$$25 = 25$$

Station C:
$$(x - 5)^2 + (y - 2)^2 = 16$$
$$(1 - 5)^2 + (2 - 2)^2 = 16$$
$$16 + 0 = 16$$
$$16 = 16$$

Thus, we can be sure that the epicenter lies at $(1, 2)$.

Use the ideas explained in this section to solve the problems in Exercises 67 and 68.

67. If three receiving stations at $(1, 4)$, $(-6, 0)$, and $(5, -2)$ record distances to an earthquake epicenter of 4 units, 5 units, and 10 units, respectively, show algebraically that the epicenter would lie at $(-3, 4)$.

68. Three receiving stations record the presence of an earthquake. The location of the receiving station and the distance to the epicenter are contained in the following three equations: $(x - 2)^2 + (y - 1)^2 = 25$, $(x + 2)^2 + (y - 2)^2 = 16$, and $(x - 1)^2 + (y + 2)^2 = 9$. Determine the location of the earthquake epicenter.

69. Show that the points $(-2, 2)$, $(13, 10)$, $(21, -5)$, and $(6, -13)$ are the vertices of a rhombus (all sides equal in length).

70. Are the points $A(1, 1)$, $B(5, 2)$, $C(3, 4)$, $D(-1, 3)$ the vertices of a parallelogram (opposite sides equal in length)? Of a rhombus (all sides equal in length)?

71. Suppose that a circle is tangent to both axes, is in the third quadrant, and has radius $\sqrt{2}$. Find the center-radius form of its equation.

72. One circle has center at $(3, 4)$ and radius 5. A second circle has center at $(-1, -3)$ and radius 4. Do these circles intersect?

73. Find all points (x, y) with $x = y$ that are 4 units from $(1, 3)$.

74. Find all points satisfying $x + y = 0$ that are 8 units from $(-2, 3)$.

75. Find the coordinates of a point whose distance from $(1, 0)$ is $\sqrt{10}$ and whose distance from $(5, 4)$ is $\sqrt{10}$.

76. Find the equation of the circle of smallest radius that contains the points $(1, 4)$ and $(-3, 2)$ within or on its boundary.

77. Find all values of y such that the distance between $(3, y)$ and $(-2, 9)$ is 12.

78. Find the coordinates of the points that divide the line segment joining $(4, 5)$ and $(10, 14)$ into three equal parts.

3.2 Functions ▼▼▼

In business, the price of an item often is directly related to the cost of producing the item. The relationship may be expressed as an equation. In such a situation it would be undesirable to have a particular cost lead to more than one price. A special type of relation, which assigns exactly one range value to each value in the domain, is most suitable for applications and is so useful it is given a special name.

> **Function**
>
> A **function** is a relation in which for each element in the domain there corresponds exactly one element in the range.*

In the example above, we say price is a function of production cost. If x represents any element in the domain (the set of possible production costs) x is called the **independent variable.** If y represents an element in the range (the set of prices) y is called the **dependent variable,** because the value of y *depends* on the value of x.

---

*A function may also be defined as a *mapping* from one set, the domain, into another set, the range.

In most mathematical applications of functions, the correspondence between the domain and range elements is defined with an equation, like $y = 5x - 11$. The equation is usually solved for y, as it is here, because y is the dependent variable. As we choose values from the domain for x, we can easily determine the corresponding y-values of the ordered pairs of the function. (These equations need not use only x and y as variables; any appropriate letters may be used. In physics, for example, t is often used to represent the independent variable *time*.)

> Now we see why we had to express the equation of a circle, $x^2 + y^2 = 9$, by using two equations. In the function mode, graphing calculators will graph only functions of x, and the relation defined by $x^2 + y^2 = 9$ is not a function. This is not a problem if the DRAW command is used.

We can think of a function as an input-output machine. If we input an element from the domain, the function (machine) outputs an element belonging to the range. See Figure 11.

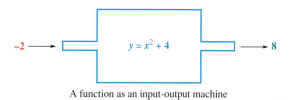

A function as an input-output machine

FIGURE 11

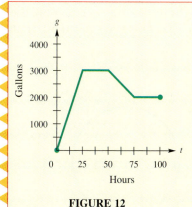

FIGURE 12

CONNECTIONS Figure 12 shows the relationship between the number of gallons of water in a small swimming pool and time in hours. By looking at this graph of the function, we can answer many questions about the water level in the pool at various times. For example, we can describe what is happening to the water level. At time zero, the pool is empty. The water level then increases, stays constant for a while, decreases, then remains constant again.

FOR DISCUSSION OR WRITING
1. What is the maximum number of gallons of water in the pool?
2. When is the maximum water level first reached?
3. For how long is the water level increasing? decreasing?
4. How many gallons are in the pool after 90 hours?
5. Describe a series of events that could account for the water level changes shown in the graph.

EXAMPLE 1
Deciding whether a relation is a function

Decide whether the following represent functions.

(a) $\{(1, 2), (3, 4), (5, 6), (7, 8), (9, 10)\}$

Since each element in the domain corresponds to exactly one element in the range, this set is a function. The correspondence is shown below using D for the domain and R for the range.

$$D = \{1, 3, 5, 7, 9\}$$
$$\downarrow \ \downarrow \ \downarrow \ \downarrow \ \downarrow$$
$$R = \{2, 4, 6, 8, 10\}$$

(b) $\{(1, 1), (1, 2), (1, 3), (2, 4)\}$

As shown in the correspondence below, one element in the domain, 1, has been assigned three different elements from the range, so this relation is not a function.

$$D = \{1, 2\}$$
$$R = \{1, 2, 3, 4\}$$

(c) The x^2 key on a calculator

Each real number input for x produces exactly one square, so this is an example of a function.

(d) $\{(x, y) \mid x = |y|\}$

Most values of x are assigned *two* y-values by the equation $x = |y|$. For example, both $(3, 3)$ and $(3, -3)$ belong to this relation, so it is not a function.

(e) $\{(x, y) \mid y = x\}$

Typical ordered pairs are $(1, 1)$, $(-2, -2)$, $(3, 3)$, and so on. Each x-value is assigned just one y-value (which equals that x-value), so this relation is a function, sometimes called the **identity function.** ▶

VERTICAL LINE TEST There is a quick way to tell whether a given graph is the graph of a function. Figure 13 shows two graphs. In the graph for part (a), each value of x leads to only one value of y, so this is the graph of a function. On the

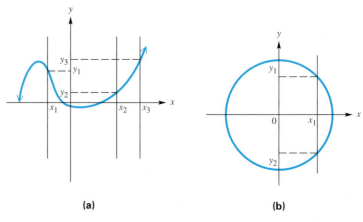

(a)   **(b)**

FIGURE 13

other hand, the graph in part (b) is not the graph of a function. For example, if $x = x_1$, the vertical line through x_1 intersects the graph at two points, showing that two values of y correspond to this x-value. This idea leads to the *vertical line test* for a function.

Vertical Line Test

If each vertical line intersects a graph at no more than one point, the graph is the graph of a function.

FUNCTION NOTATION   It is common to use the letters f, g, and h to name functions. Function names are helpful when we wish to discuss more than one function. If f is a function and x is an element in the domain of f, then $f(x)$, read "f of x," is the corresponding element in the range. The notation $f(x)$ is an abbreviation for "(the function) f evaluated at x." For example, if f is the function in which a value of x is squared to give the corresponding value in the range, f could be defined as

$$f(x) = x^2.$$

If $x = -5$ is an element from the domain of f, the corresponding element from the range is found by replacing x with -5. This is written

$$f(-5) = (-5)^2 = 25.$$

The number -5 in the domain corresponds to 25 in the range, and the ordered pair $(-5, 25)$ belongs to the function f.

This *function notation* can be summarized as follows.

Name of the function

Name of the independent variable

$$y = f(x) = 3x + 2$$

Value of the function at x Defining expression

NOTE   Keep in mind that $f(x)$ is just another (more meaningful) notation for y. You can always replace the notation $f(x)$ with y or y with $f(x)$.

EXAMPLE 2
Using function notation

Let $g(x) = 3\sqrt{x}$, $h(x) = 1 + 4x$, and $k(x) = x^2 + 3$. Find each of the following.

(a) $g(16)$
     To find $g(16)$, replace x in $g(x) = 3\sqrt{x}$ with 16, getting

$$g(16) = 3\sqrt{16} = 3 \cdot 4 = 12.$$

(b) $h(-3) = 1 + 4(-3) = -11$

(c) $k(-2) = (-2)^2 + 3 = 4 + 3 = 7$

(d) $g(-4)$ is not a real number; -4 is not in the domain of g since $\sqrt{-4}$ is not real.*

(e) $h(\pi) = 1 + 4\pi$

(f) $g(m) = 3\sqrt{m}$, if m represents a nonnegative real number.

(g) $k(a + b)$
Let $x = a + b$ in $k(x) = x^2 + 3$.

$$k(\boldsymbol{a + b}) = (\boldsymbol{a + b})^2 + 3 = a^2 + 2ab + b^2 + 3. \quad \blacktriangleright$$

Notation such as $k(a + b)$ will be useful in the last two sections of this chapter.

A graphing calculator can be used to evaluate $f(x)$ for specific values of x. Instead of $f(x)$, the calculator uses the notation Y_1. Some calculators will give values for $Y_1(x)$. For example, by entering $Y_1 = 1 + 4x$ in the calculator, we can then find $Y_1(-3)$, and so on. See the figure.

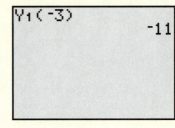

When $Y_1 = 1 + 4x$, and $x = -3$,
$Y_1 = -11$.

Sometimes we need to find the value of the dependent variable from a graph.

◀ EXAMPLE 3
Finding function values from a graph

The graph in Figure 12 on page 201 shows the number of gallons of water $g = f(t)$ in a small swimming pool at time t in hours. Use the graph to find and describe the following function values.

(a) $f(30)$
From the graph, when $t = 30$, $g = f(30) = 3000$, so there are 3000 gallons of water in the pool after 30 hours.

(b) $f(75)$
The graph shows that $f(75) = 2000$, which means that after 75 hours, there are 2000 gallons of water in the pool. $\quad \blacktriangleright$

---

*In our work with functions, we restrict all variables to *real numbers*.

DOMAIN AND RANGE   Throughout this book, if the domain for a function specified by an algebraic formula is not given, it will be assumed to be the largest possible set of real numbers for which the formula is meaningful. For example, if

$$f(x) = \frac{-4}{2x - 3},$$

then any real number can be used for x except $x = 3/2$, which makes the denominator equal to 0. Based on our assumption, the domain of this function must be $\{x \mid x \neq 3/2\}$ or $(-\infty, 3/2) \cup (3/2, \infty)$ in interval form.

EXAMPLE 4
Finding domain and range

Give the domain and range of each of the following functions.

(a) $f(x) = 4 - 3x$

Here x can be any real number. Multiplying by -3 and adding 4 will give one real number $f(x)$ for each value of x. Thus, both the domain and range are the set of all real numbers. In interval notation, both the domain and range are $(-\infty, \infty)$.

(b) $f(x) = x^2 + 4$

Since any real number can be squared, the domain is the set of all real numbers, $(-\infty, \infty)$. The square of any number is nonnegative, so that $x^2 \geq 0$ and $x^2 + 4 \geq 4$. The range is the interval $[4, \infty)$.

(c) $g(x) = \sqrt{x - 2}$

If $\sqrt{x - 2}$ is to be a real number, then

$$x - 2 \geq 0 \qquad \text{or} \qquad x \geq 2,$$

so the domain of the function is given by the interval $[2, \infty)$. Since $\sqrt{x - 2}$ is nonnegative, the range is $[0, \infty)$.

(d) $h(x) = -\sqrt{100 - x^2}$

For $-\sqrt{100 - x^2}$ to be a real number,

$$100 - x^2 \geq 0.$$

Since $100 - x^2$ factors as $(10 - x)(10 + x)$, use a sign graph to verify that

$$-10 \leq x \leq 10,$$

making the domain of the function $[-10, 10]$. As x takes values from -10 to 10, $h(x)$ (or y) goes from 0 to -10 and back to 0. (Verify this by substituting -10, 0, 10, and some values in between for x.) Thus, the range is $[-10, 0]$.

(e) $k(x) = \dfrac{2}{x - 5}$

The quotient is undefined if the denominator is 0, so x cannot equal 5. Therefore, the domain is written in interval notation as $(-\infty, 5) \cup (5, \infty)$. As x takes on all real-number values except 5, y will take on all real-number values except 0. The range is $(-\infty, 0) \cup (0, \infty)$. ◗

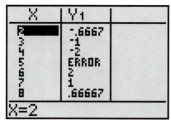

For $Y_1 = \sqrt{x-2}$, when $x \geq 2$, the function is defined. But for $x < 2$, the calculator gives an error message. Compare to Example 4(c).

X	Y₁
2	-.6667
3	-1
4	-2
5	ERROR
6	2
7	1
8	.66667

X=2

For $Y_1 = \frac{2}{x-5}$, an error message is returned at $x = 5$. Compare to Example 4(e).

Example 4 suggests some generalizations about the types of functions where the domain is restricted. As shown in parts (c) and (d), for functions involving even roots, the domain includes only numbers that make a radicand nonnegative, and in part (e) the domain includes only numbers that make a denominator nonzero.

For most of the functions discussed in this book, the domain can be found with the methods already presented. As Example 4 suggests, one reason for studying inequalities is to find domains. The range, however, often must be found by using graphing, more involved algebra, or calculus. For example, the graph in Figure 14 suggests that the function has domain $[-3, 3]$ and range $[-1, 3]$.

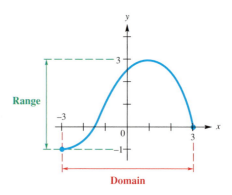

FIGURE 14

Informally speaking, a function **increases** on an interval of its domain if its graph rises from left to right. It **decreases** on an interval if its graph falls from left to right. It is **constant** on an interval if its graph is horizontal on the interval.

The formal definitions of these concepts follow.

Increasing, Decreasing, and Constant Functions

Suppose that a function f is defined over an interval I. If x_1 and x_2 are in I,

(a) f increases on I if, whenever $x_1 < x_2, f(x_1) < f(x_2)$;
(b) f decreases on I if, whenever $x_1 < x_2, f(x_1) > f(x_2)$;
(c) f is constant on I if, for every x_1 and $x_2, f(x_1) = f(x_2)$.

Figure 15 illustrates these ideas.

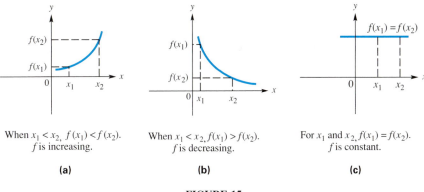

When $x_1 < x_2$, $f(x_1) < f(x_2)$.
f is increasing.

(a)

When $x_1 < x_2$, $f(x_1) > f(x_2)$.
f is decreasing.

(b)

For x_1 and x_2, $f(x_1) = f(x_2)$.
f is constant.

(c)

FIGURE 15

EXAMPLE 5
Determining intervals over which a function is increasing, decreasing, or constant

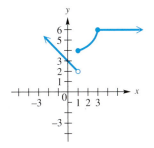

FIGURE 16

Figure 16 shows the graph of a function. Determine the intervals over which the function is increasing, decreasing, or constant.

In making our determination, we must always ask, "What is the y-value doing as x is getting larger?" For this graph, we see that on the interval $(-\infty, 1)$, the y-values are *decreasing;* on the interval $[1, 3]$, the y-values are *increasing;* and on the interval $[3, \infty)$, the y-values are *constant* (all are 6). Therefore, the function is

decreasing on $(-\infty, 1)$,

increasing on $[1, 3]$,

constant on $[3, \infty)$. ▶

CAUTION   When specifying intervals where a function is increasing, decreasing, or constant, we use *domain* values. Range values are not involved when writing the intervals.

APPLYING FUNCTIONS   In manufacturing, the cost of making a product usually consists of two parts. One part is a *fixed cost* for designing the product, setting up a factory, training workers, and so on. Usually the fixed cost is constant for a particular product and does not change as more items are made. The other part of the cost is a *variable cost* per item for labor, materials, packaging, shipping, and so on. The variable cost is often the same per item, so that the total amount of variable cost increases as more items are produced. A *linear cost function* has the form $C(x) = mx + b$, where m represents the variable cost per item and b represents the fixed cost. The revenue from selling a product depends on the price per item and the number of items sold, as given by the *revenue function,* $R(x) = px$, where p is the price per item and $R(x)$ is the revenue from the sale of x items. The profit is described by the *profit function* given by $P(x) = R(x) - C(x)$.

EXAMPLE 6
Writing linear cost and profit functions

(a) Assume that the cost to produce an item is given by a linear function. If the fixed cost is $1500 and the variable cost is $100, write a cost function for the product.

Since the cost function is linear it will have the form $C(x) = mx + b$, with $m = 100$ and $b = 1500$. That is,

$$C(x) = 100x + 1500.$$

(b) Find the revenue function if the item in part (a) sells for $125.

The revenue function is

$$R(x) = px = 125x. \qquad \text{Let } p = 125.$$

(c) Give the profit function for the item in part (a).

The profit function is given by

$$P(x) = R(x) - C(x)$$
$$= 125x - (100x + 1500)$$
$$= 125x - 100x - 1500$$
$$= 25x - 1500.$$

(d) How many items must be produced and sold before the company makes a profit?

To make a profit, $P(x)$ must be positive. Set $P(x) = 25x - 1500 > 0$ and solve for x.

$$25x - 1500 > 0$$
$$25x > 1500 \qquad \text{Add 1500 to each side.}$$
$$x > 60 \qquad \text{Divide by 25.}$$

At least 61 items must be sold for the company to make any profit. ▶

3.2 Exercises ▼▼▼▼▼▼▼▼▼▼▼▼▼▼▼▼▼▼▼▼▼▼▼▼▼▼▼▼▼▼▼▼▼▼▼▼▼▼▼

Decide whether the set represents y as a function of x. See Example 1.

1. $\{(1, 3), (2, 4), (3, 5)\}$

2. $\{(-1, 6), (-2, 9), (-3, 12)\}$

3. $\{(x, y) \,|\, x = y^2\}$

4. $\{(x, y) \,|\, x = y^4\}$

5. $\{(x, y) \,|\, y = 3x - 7\}$

6. $\{(x, y) \,|\, y = -.4x + 2\}$

Use the vertical line test to determine whether the graph is that of a function.

7.

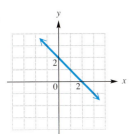

8.

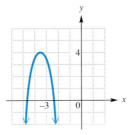

9.

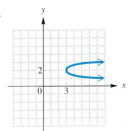

10.

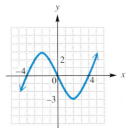

11.

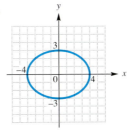

12.

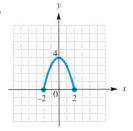

A calculator-generated graph of a relation is shown. Determine whether the graph is that of a function by using the vertical line test. Assume the graph extends infinitely where it reaches the edge of the screen, following the pattern established.

13.

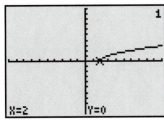

[−10, 10] by [−10, 10]

14.

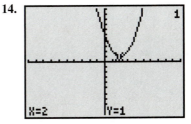

[−10, 10] by [−10, 10]

15.

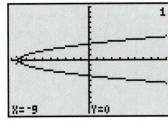

[−10, 10] by [−10, 10]

16.

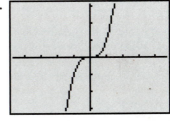

[−4.7, 4.7] by [−3.1, 3.1]

Given $f(x) = 4x - 2$ and $g(x) = x^2 + 3$, find each function value. See Example 2.

17. $f(-3)$ **18.** $f(5)$ **19.** $g(-8)$ **20.** $g(7)$ **21.** $f(k)$

22. $g(h)$ **23.** $f(-1) + g(9)$ **24.** $f(3) \cdot g(6)$ **25.** $g(x + k)$

26. Complete the following sentence: If the ordered pair (2, 7) lies on the graph of the function g, then $g(\underline{\hspace{0.5cm}}) = \underline{\hspace{0.7cm}}$.

27. The graph of $Y_1 = f(x)$ is shown with a display at the bottom. What is $f(3)$?

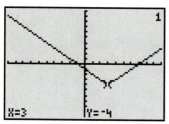

X=3   Y=⁻4

[−10, 10] by [−10, 10]

28. The graph of $Y_1 = f(x)$ is shown with a display at the bottom. What is $f(-2)$?

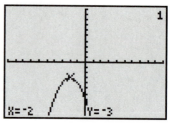

X=⁻2   Y=⁻3

[−10, 10] by [−10, 10]

Use the graph to find the function values: **(a)** $f(-2)$, **(b)** $f(0)$, **(c)** $f(1)$, and **(d)** $f(4)$. *See Example 3.*

29.

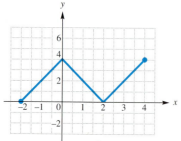

30.

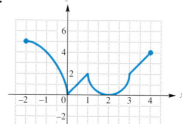

31.

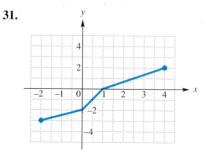

32.

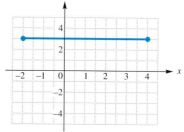

Two functions, $Y_1 = f(x)$ and $Y_2 = g(x)$, are entered into a graphing calculator with table-forming capability. The screen here shows values for Y_1 and Y_2 for $x = -1, 0, 1, 2, 3, 4,$ and 5. Use the screen to evaluate the expressions in Exercises 33–36 and to answer the questions in Exercises 37 and 38.

X	Y₁	Y₂
-1	-23	2.5
0	-15	2
1	-7	1.5
2	1	1
3	9	.5
4	17	0
5	25	-.5

X=2

$Y_1 = f(x), Y_2 = g(x)$

33. $f(0) + g(2)$

34. $f(4) - g(-1)$

35. $g[f(2)]$

36. $f[g(2)]$

37. For what value of x is $f(x) = g(x)$ true?

38. What would be the result if you attempted to evaluate $\dfrac{f(4)}{g(4)}$?

Solve each problem involving functions.

39. The table contains incidence ratios by age for deaths from coronary heart disease (CHD) and lung cancer (LC) when comparing smokers (21–39 cigarettes per day) to nonsmokers.

Age	CHD	LC
55–64	1.9	10
65–74	1.7	9

For example, the incidence ratio of 10 means that smokers are 10 times more likely than nonsmokers to die of lung cancer between the ages of 55 and 64. If the incidence ratio is x, then the percent P (in decimal form) of deaths caused by smoking can be calculated using the function $P(x) = \dfrac{x-1}{x}$. (*Source:* Walker, A., *Observation and Inference: An Introduction to the Methods of Epidemiology*, Epidemiology Resources Inc., Newton Lower Falls, MA, 1991.)

(a) Calculate the percent of lung cancer deaths between the ages of 65 and 74 that can be attributed to smoking.

(b) Calculate the percent of coronary heart disease deaths between the ages of 55 and 64 that can be attributed to smoking.

40. Insulation workers who were exposed to asbestos and employed before 1960 experienced an increased likelihood of lung cancer. If a group of insulation workers have a cumulative total of 100,000 years of work experience with their first date of employment t years ago, then the number of lung cancer cases occurring within the group can be modeled using the function $N(t) = .00437t^{3.2}$. (*Source:* Walker, A., *Observation and Inference: An Introduction to the Methods of Epidemiology*, Epidemiology Resources Inc., Newton Lower Falls, MA, 1991.)

(a) Calculate $N(t)$ when $t = 5$, 10, and 20.

(b) If the years of employment are doubled, does the number of cancer cases also double?

According to an article in the December 1994 issue of Scientific American, the coast-down time for a typical 1993 car as it drops 10 miles per hour from an initial speed depends on variations from the standard condition (automobile in neutral, average drag, and tire pressure). The accompanying graph illustrates some of these conditions with coast-down time in seconds graphed as a function of initial speed in miles per hour.

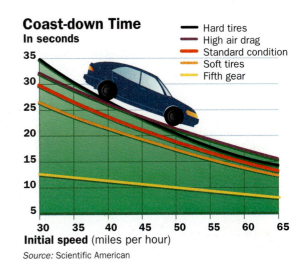

Coast-down Time
In seconds

Legend: — Hard tires — High air drag — Standard condition — Soft tires — Fifth gear

Initial speed (miles per hour)

Source: Scientific American

Use the graph to answer the questions in Exercises 41 and 42.

41. What is the approximate coast-down time in fifth gear if the initial speed is 40 miles per hour?

42. For what speed is the coast-down time the same for the conditions of high air drag and hard tires?

Give the domain and the range of the function. See Example 4.

43.

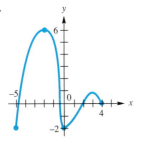

44.

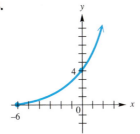

45.

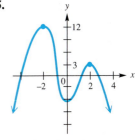

46.
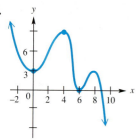

47.
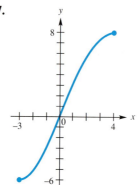

48.
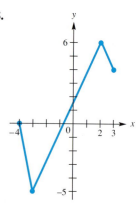

49. the function graphed in Exercise 27

50. the function graphed in Exercise 28

51. $f(x) = 3x - 9$

52. $g(x) = -6x + 2$

53. $f(x) = x^6$

54. $h(x) = (x + 3)^2$

55. $h(x) = \sqrt{9 + x}$

56. $k(x) = \sqrt{3 - x}$

57. $f(x) = -\sqrt{4 - x^2}$

58. $F(x) = -\sqrt{121 - x^2}$

59. $g(x) = \dfrac{3}{7 + x}$

60. $G(x) = \dfrac{-3}{12 + 2x}$

61. $f(x) = \sqrt[3]{x + 2}$

62. $F(x) = \sqrt[3]{7 - 3x}$

Determine the intervals of the domain for which the function is **(a)** *increasing,* **(b)** *decreasing,* **(c)** *constant. See Example 5.*

63.
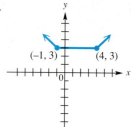

64.
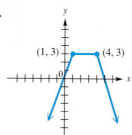

65.
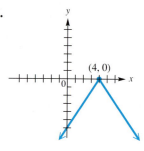

66.
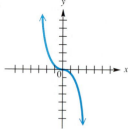

67.
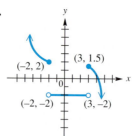

68.
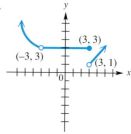

69. Refer to Figure 12 in this section. Give the intervals over which the function is **(a)** increasing, **(b)** decreasing, **(c)** constant.

70. Give an example of a function from everyday life. (*Hint:* Fill in the blanks: _____ depends on _____ , so _____ is a function of _____ .)

A firm will break even (no profit and no loss) as long as revenue just equals cost. The value of x (the number of items produced and sold) where $C(x) = R(x)$ is called the break-even point. Assume that each of the following can be expressed as a linear cost function. Find (a) the cost function, (b) the revenue function, and (c) the profit function. (d) Then find the break-even point and decide whether the product should be produced based on the restrictions on sales. See Example 6.

	Fixed Cost	Variable Cost	Price of Item	
71.	$500	$10	$35	No more than 18 units can be sold.
72.	$180	$11	$20	No more than 30 units can be sold.
73.	$2700	$150	$280	No more than 25 units can be sold.
74.	$1650	$400	$305	All units produced can be sold.

75. The manager of a small company that produces roof tile has determined that the total cost in dollars, $C(x)$, of producing x units of tile is given by $C(x) = 200x + 1000$, while the revenue in dollars, $R(x)$, from the sale of x units of tile is given by $R(x) = 240x$.
(a) Find the break-even point.
(b) Graph functions R and C on the same axes. Identify the break-even point. Identify any regions of profit or loss.
(c) What is the cost/revenue at the break-even point?

76. Suppose the manager of the company in Exercise 75 finds he has miscalculated his variable cost, and that it is actually $220 per unit, instead of $200. How does this affect the break-even point? Is he better off or not?

3.3 Linear Functions ▼▼▼

In this section, we begin the study of specific functions by looking at *linear functions*. The name "linear" comes from the fact that the graph of every linear function is a straight line.

> **Definition of Linear Function**
>
> A function f is a **linear function** if
> $$f(x) = ax + b,$$
> for real numbers a and b.

NOTE Linear functions are sometimes written in the form $Ax + By = C$, where A, B, and C are real, and B is not 0. This is called the **standard form.**

In the equation $Ax + By = C$, any number can be used for x or y, so both the domain and range of a linear function in which neither A nor B is 0 are the set of real numbers $(-\infty, \infty)$.

CAUTION The definition of "standard form" is not standard from one text to another. Any linear equation can be written in many different (all equally correct) forms. For example, the equation $2x + 3y = 8$ can be written as $2x = 8 - 3y$, $3y = 8 - 2x$, $x + (3/2)y = 4$, $4x + 6y = 16$, and so on. In addition to writing it in the form $Ax + By = C$ (with A, B, and C integers and $A \geq 0$), let us agree that the form $2x + 3y = 8$ is preferred over any multiples of both sides, such as $4x + 6y = 16$.

GRAPHING LINEAR FUNCTIONS The graph of a linear function can be found by plotting at least two points. Two points that are especially useful for sketching the graph of a line are found with the *intercepts*. An **x-intercept** is an x-value at which a graph crosses the x-axis. A **y-intercept** is a y-value at which a graph crosses the y-axis.* Since $y = 0$ on the x-axis, an x-intercept is found by setting y equal to 0 in the equation and solving for x. Similarly, a y-intercept is found by setting $x = 0$ in the equation and solving for y.

EXAMPLE 1
Graphing a linear function using intercepts

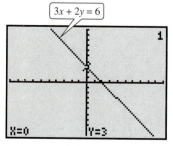

[−10, 10] by [−10, 10]

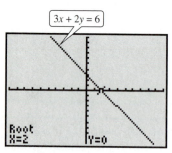

[−10, 10] by [−10, 10]

We graph $3x + 2y = 6$ on a graphing calculator by solving for y to get $y = -\frac{3}{2}x + 3$. The screens show the x- and y-intercepts.

Graph $3x + 2y = 6$. Give the domain and the range.

Use the intercepts. The y-intercept is found by letting $x = 0$.

$$3 \cdot \mathbf{0} + 2y = 6$$
$$2y = 6$$
$$y = 3$$

For the x-intercept, let $y = 0$, getting

$$3x + 2 \cdot \mathbf{0} = 6$$
$$3x = 6$$
$$x = 2.$$

Plotting $(0, 3)$ and $(2, 0)$ gives the graph in Figure 17. A third point could be found as a check if desired. The domain and the range are both $(-\infty, \infty)$. ▶

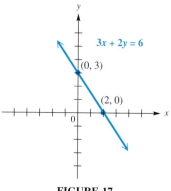

FIGURE 17

---

*The intercepts are sometimes defined as ordered pairs instead of numbers. At this level, however, they are usually defined as numbers.

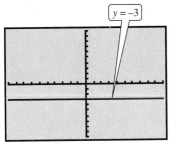

[−10, 10] by [−10, 10]

Compare with Figure 18(a).

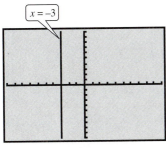

[−10, 10] by [−10, 10]

Because $x = -3$ does not define a function, we must use the DRAW capability of the TI-82 to obtain the graph.

EXAMPLE 2
Graphing horizontal and vertical lines

(a) Graph $y = -3$.

Since y always equals -3, the value of y can never be 0. This means that the graph has no x-intercept. The only way a straight line can have no x-intercept is for it to be parallel to the x-axis, as shown in Figure 18(a). Notice that the domain of this linear function is $(-\infty, \infty)$, but the range is $\{-3\}$.

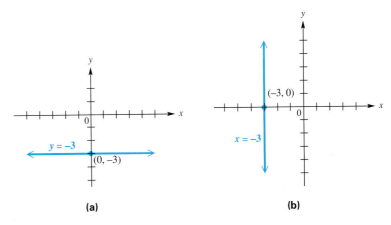

(a)   **(b)**

FIGURE 18

(b) Graph $x = -3$.

Here, since x always equals -3, the value of x can never be 0, and the graph has no y-intercept. Using reasoning similar to that of part (a), we find that this graph is parallel to the y-axis, as shown in Figure 18(b). The domain of this relation, which is *not* a function, is $\{-3\}$, while the range is $(-\infty, \infty)$.

From this example we may conclude that a linear function of the form $y = k$ has as its graph a horizontal line through $(0, k)$, and a relation of the form $x = k$ has as its graph a vertical line through $(k, 0)$. ▶

EXAMPLE 3
Graphing a line through the origin

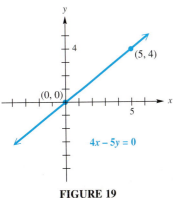

FIGURE 19

Graph $4x - 5y = 0$.

Find the intercepts. If $x = 0$, then

$$4(0) - 5y = 0$$
$$-5y = 0$$
$$y = 0.$$

Letting $y = 0$ leads to the same ordered pair, $(0, 0)$. The graph of this function has just one intercept—at the origin. Find another point by choosing a different value for x (or y). Choosing $x = 5$ gives

$$4(5) - 5y = 0$$
$$20 - 5y = 0$$
$$20 = 5y$$
$$4 = y,$$

which leads to the ordered pair $(5, 4)$. Complete the graph using the two points $(0, 0)$ and $(5, 4)$, with a third point as a check. See Figure 19. ▶

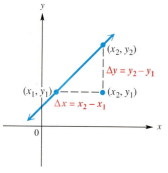

FIGURE 20

SLOPE   An important characteristic of a straight line is its *slope,* a numerical measure of the steepness of the line. (Geometrically, this may be interpreted as the ratio of *rise* to *run.*) To find this measure, start with the line through the two distinct points (x_1, y_1) and (x_2, y_2), as shown in Figure 20, where $x_1 \neq x_2$. The difference

$$x_2 - x_1$$

is called the **change in x** and denoted by Δx (read "delta x"), where Δ is the Greek letter *delta.* In the same way, the **change in y** can be written

$$\Delta y = y_2 - y_1.$$

The *slope* of a nonvertical line is defined as the quotient of the change in y and the change in x, as follows.

Slope

The **slope** m of the line through the points (x_1, y_1) and (x_2, y_2) is

$$m = \frac{\text{rise}}{\text{run}} = \frac{\Delta y}{\Delta x} = \frac{y_2 - y_1}{x_2 - x_1},$$

where $\Delta x \neq 0$.

CAUTION   When using the slope formula, be sure it is applied correctly. It makes no difference which point is (x_1, y_1) or (x_2, y_2); however, it is important to be consistent. Start with the x- and y-values of *one* point (either one) and subtract the corresponding values of the *other* point. Be sure to put the difference of the y-values in the numerator and the difference of the x-values in the denominator.

The slope of a line can be found only if the line is nonvertical. This guarantees that $x_2 \neq x_1$ so that the denominator $x_2 - x_1 \neq 0$. It is not possible to define the slope of a vertical line.

The slope of a vertical line is undefined.

◀EXAMPLE 4
Finding slopes with the slope formula

Find the slope of the line through each of the following pairs of points.

(a) $(-4, 8), (2, -3)$

Let $x_1 = -4$, $y_1 = 8$ and $x_2 = 2$, $y_2 = -3$. Then

$$\Delta y = -3 - 8 = -11$$

and

$$\Delta x = 2 - (-4) = 6.$$

The slope is

$$m = \frac{\Delta y}{\Delta x} = -\frac{11}{6}.$$

(b) $(2, 7)$, $(2, -4)$

If we use the formula, we get

$$m = \frac{-4 - 7}{2 - 2} = \frac{-11}{0},$$

which is undefined. The formula is not valid here because $\Delta x = x_2 - x_1 = 2 - 2 = 0$. A sketch would show that the line through $(2, 7)$ and $(2, -4)$ is vertical. As mentioned above, the slope of a vertical line is not defined.

(c) $(5, -3)$ and $(-2, -3)$

By the definition of slope,

$$m = \frac{-3 - (-3)}{-2 - 5} = \frac{0}{-7} = 0. \ \blacktriangleright$$

Drawing a graph through the points in Example 4(c) would produce a line that is horizontal, which suggests the following generalization.

> The slope of a horizontal line is 0.

Figure 21 shows lines with various slopes. As the figure shows, a line with a positive slope goes up from left to right, but a line with a negative slope goes down from left to right. When the slope is positive, the function is increasing, and when the slope is negative, the function is decreasing.

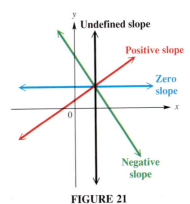

FIGURE 21

It can be shown, using theorems for similar triangles, that the slope is independent of the choice of points on the line. That is, the slope of a line is the same no matter which pair of distinct points on the line are used to find it.

Since the slope of a line is the ratio of vertical change to horizontal change, if we know the slope of a line and the coordinates of a point on the line, the graph of the line can be drawn. The next example illustrates this.

EXAMPLE 5
Graphing a line using a point and the slope

Graph the line passing through $(-1, 5)$ and having slope $-5/3$.

First locate the point $(-1, 5)$ as shown in Figure 22. Since the slope of this line is $-5/3$, a change of -5 units vertically (that is, 5 units down) produces a change of 3 units horizontally (3 units to the right). This gives a second point, $(2, 0)$, which can then be used to complete the graph.

Because $-5/3 = 5/(-3)$, another point could be obtained by starting at $(-1, 5)$ and moving 5 units *up* and 3 units to the *left*. We would reach a different second point, but the graph would be the same. ▶

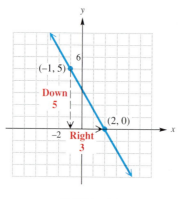

FIGURE 22

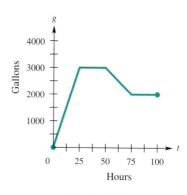

FIGURE 23

EXAMPLE 6
Interpreting slope from a graph

The graph in Figure 23 was shown in the Connections Box in Section 3.2. It gives the number of gallons of water g in a small swimming pool after t hours. How is the slope of the graph related to the flow of water into and out of the pool?

The slope is positive for the first 25 hours as the amount of water in the pool increases. From 25 hours to 50 hours the slope is 0, indicating no change in the amount of water. The slope is negative from 50 hours to 75 hours, as the amount of water decreases. Finally, the slope is 0 again from 75 hours to 100 hours, because the amount of water is unchanged in that time period. ▶

EXAMPLE 7
Finding the difference quotient

Let $f(x) = 2x^2 - 3x$. If h represents any nonzero number, then the quotient

$$\frac{f(x + h) - f(x)}{h}, \quad h \neq 0,$$

represents the slope of the line through $(x, f(x))$ and $(x + h, f(x + h))$. This expression is called a *difference quotient* and is used in calculus to determine the steepness of a curve at a point. Find and simplify the quotient.

To find $f(x + h)$, replace x in $f(x)$ with $x + h$, to get

$$f(x + h) = 2(x + h)^2 - 3(x + h).$$

Then

$$\frac{f(x + h) - f(x)}{h} = \frac{2(x + h)^2 - 3(x + h) - (2x^2 - 3x)}{h}$$

$$= \frac{2(x^2 + 2xh + h^2) - 3x - 3h - 2x^2 + 3x}{h}$$ Square $x + h$; use the distributive property.

$$= \frac{2x^2 + 4xh + 2h^2 - 3x - 3h - 2x^2 + 3x}{h}$$

$$= \frac{4xh + 2h^2 - 3h}{h}$$ Combine terms.

$$= \frac{h(4x + 2h - 3)}{h}$$ Factor out h.

$$= 4x + 2h - 3.$$ Divide. ▶

CAUTION   Notice that $f(x + h)$ is not the same as $f(x) + f(h)$. For $f(x) = 2x^2 - 3x$, as shown in Example 7,

$$f(x + h) = 2(x + h)^2 - 3(x + h) = 2x^2 + 4xh + 2h^2 - 3x - 3h$$

but

$$f(x) + f(h) = (2x^2 - 3x) + (2h^2 - 3h) = 2x^2 - 3x + 2h^2 - 3h.$$

These expressions differ by $4xh$.

3.3 Exercises ▼▼▼▼▼▼▼▼▼▼▼▼▼▼▼▼▼▼▼▼▼▼▼▼▼▼▼▼▼▼▼▼▼▼▼

Decide whether the statement is true or false.

1. To find the x-intercept of the graph of a linear function $f(x) = ax + b$, we solve $f(x) = 0$, and to find the y-intercept, we evaluate $f(0)$.

2. The graph of $f(x) = -3$ is a horizontal line.

3. The graph of $f(x) = ax$ is a straight line that passes through the origin.

4. The slope of the graph of a linear function cannot be undefined.

5. The slope of the line in Figure 17 is $3/2$.

6. The slope of the line in Figure 18(a) is undefined.

Graph the function. Give the domain and the range. See Examples 1–3.

7. $-x + y = -4$ **8.** $x + y = 4$ **9.** $y = 3x - 6$ **10.** $y = \frac{2}{3}x + 2$

11. $2x + 5y = 10$ **12.** $-4x + 3y = 9$ **13.** $y = -4$ **14.** $y = 3$

15. $y = 3x$ **16.** $y = -2x$

Match the equation with the sketch that most closely resembles its graph.

17. $y = 2$

18. $y = -2$

19. $x = 2$

20. $x = -2$

A.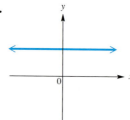

B.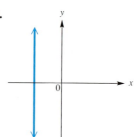

C.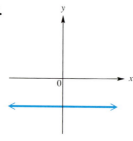

D.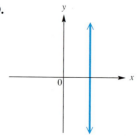

Use a graphing calculator to graph the linear function in the standard window. If the equation is in the form Ax + By = C, solve for y first.

21. $y = 3x + 4$ **22.** $y = -2x + 3$ **23.** $3x + 4y = 6$ **24.** $-2x + 5y = 10$

25. $y = -3x$ **26.** $y = 2.5x$

Find the slope of the line satisfying the given conditions. See Example 4.

27. through $(2, -1)$ and $(-3, -3)$ **28.** through $(5, -3)$ and $(1, -7)$

29. through $(5, 9)$ and $(-2, 9)$ **30.** through $(-2, 4)$ and $(6, 4)$

31. horizontal, through $(3, -7)$ **32.** horizontal, through $(-6, 5)$

33. vertical, through $(3, -7)$ **34.** vertical, through $(-6, 5)$

35. A linear function is graphed in a window, and tracing yields the *x*- and *y*-values displayed. What is the slope of the line?

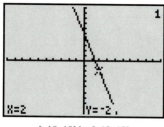

[−10, 10] by [−10, 10]

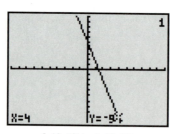

[−10, 10] by [−10, 10]

36. A linear function is designated as Y_1, and a table of points is generated. See the figure. What is the slope of the line?

X	Y₁	
-3	-10	
-2	-6	
-1	-2	
0	2	
1	6	
2	10	
3	14	

X=-3

37. Explain in your own words what is meant by the slope of a line.

38. Explain how to graph a line using a point on the line and the slope of the line.

Graph the line passing through the given point and having the indicated slope. Indicate two points on the line. See Example 5.

39. Through $(-1, 3)$, $m = 3/2$

40. Through $(-2, 8)$, $m = -1$

41. Through $(3, -4)$, $m = -1/3$

42. Through $(-2, -3)$, $m = -3/4$

43. Through $(-1, 4)$, $m = 0$

44. Through $(9/4, 2)$, undefined slope

Solve the problem. See Example 6.

45. The graph shows the winning times (in minutes) at the Olympic Games for the 5000 meter run together with a linear approximation of the data.

(a) The equation for the linear approximation is $y = -.0221x + 57.14$. What does the slope of this line represent? Why is the slope negative?

(b) Can you think of any reason why there are no data points for the years 1940 and 1944?

(c) Do you see a limitation of this linear model?

Olympic Time for 5000 Meter Run
In minutes

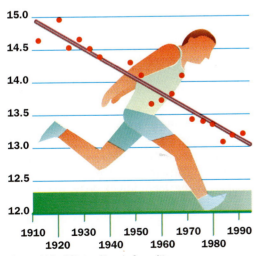

Source: United States Olympic Committee

46. The graph shows the number of U.S. radio stations on the air along with a linear function that models the data.

(a) Discuss the accuracy of the linear function.

(b) Use the two data points $(1950, 2773)$ and $(1994, 11,600)$ to find the approximate slope of the line shown. Interpret this number.

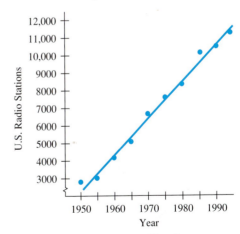

Source: National Association of Broadcasters.

▼▼▼▼▼▼▼▼▼▼▼▼ **DISCOVERING CONNECTIONS** (Exercises 47–56) ▼▼▼▼▼▼▼▼▼▼▼▼

The accompanying table was generated for a linear function by using a graphing calculator. It identifies seven points on the graph of the function. Work Exercises 47–56 in order, to see connections between the slope formula, the distance formula, the midpoint formula, and linear functions.

X	Y1
0	-6
1	-3
2	0
3	3
4	6
5	9
6	12

X=0

47. Use the first two points in the table to find the slope of the line.

48. Use the second and third points in the table to find the slope of the line.

49. Make a conjecture by filling in the blanks: If we use any two points on a line to find its slope, we find that the slope is _____ in all cases.

50. Use the distance formula to find the distance between the first two points in the table.

51. Use the distance formula to find the distance between the second and fourth points in the table.

52. Use the distance formula to find the distance between the first and fourth points in the table.

53. Add the results in Exercises 50 and 51, and compare the sum to the answer you found in Exercise 52. What do you notice?

54. Fill in the blanks, basing your answers on your observations in Exercises 50–53: If points A, B, and C lie on a line in that order, then the distance between A and B added to the distance between _____ and _____ is equal to the distance between _____ and _____.

55. Use the midpoint formula to find the midpoint of the segment joining $(0, -6)$ and $(6, 12)$. Compare your answer to the middle entry in the table. What do you notice?

56. If the table were set up to show an x-value of 4.5, what would be the corresponding y-value?

For each of the functions defined as follows, find **(a)** $f(x + h)$, **(b)** $f(x + h) - f(x)$, *and* **(c)** $\dfrac{f(x + h) - f(x)}{h}$. *See Example 7.*

57. $f(x) = 6x + 2$ **58.** $f(x) = 4x + 11$ **59.** $f(x) = -2x + 5$
60. $f(x) = 1 - x^2$ **61.** $f(x) = x^2 - 4$ **62.** $f(x) = 8 - 3x^2$

3.4 Equations of a Line ▼▼▼

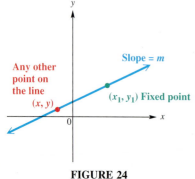

FIGURE 24

In the previous section we assumed that the graph of a linear function is a straight line. In this section we develop various forms for the equation of a line. Figure 24 shows the line passing through the fixed point (x_1, y_1) and having slope m. (Assuming that the line has a slope guarantees that it is not vertical.) Let (x, y) be any other point on the line. By the definition of slope, the slope of the line is

$$\frac{y - y_1}{x - x_1}.$$

Since the slope of the line is m,

$$\frac{y - y_1}{x - x_1} = m.$$

Multiplying both sides by $x - x_1$ gives

$$y - y_1 = m(x - x_1).$$

This result, called the *point-slope form* of the equation of a line, identifies points on a given line: a point (x, y) lies on the line through (x_1, y_1) with slope m if and only if

$$y - y_1 = m(x - x_1).$$

Point-Slope Form

The line with slope m passing through the point (x_1, y_1) has an equation
$$y - y_1 = m(x - x_1),$$
the **point-slope form** of the equation of a line.

◀**EXAMPLE 1**
Using the point-slope form (given a point and the slope)

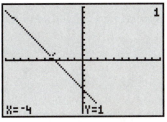

[–10, 10] by [–20, 20]
Xscl = 1   Yscl = 2

When $Y_1 = -3x - 11$ is graphed, the calculator supports the fact that the point $(-4, 1)$ is on the line.

Write an equation of the line through $(-4, 1)$ with slope -3.

Here $x_1 = -4$, $y_1 = 1$, and $m = -3$. Use the point-slope form of the equation of a line to get

$$y - 1 = -3[x - (-4)] \qquad x_1 = -4,\ y_1 = 1,\ m = -3$$
$$y - 1 = -3(x + 4)$$
$$y - 1 = -3x - 12 \qquad \text{Distributive property}$$
$$y = -3x - 11. \quad ▶$$

◀**EXAMPLE 2**
Using the point-slope form (given two points)

Find an equation of the line through $(-3, 2)$ and $(2, -4)$.

Find the slope first. By the definition of slope,

$$m = \frac{-4 - 2}{2 - (-3)} = -\frac{6}{5}.$$

Either $(-3, 2)$ or $(2, -4)$ can be used for (x_1, y_1). Using $x_1 = -3$ and $y_1 = 2$ in the point-slope form gives

$$y - 2 = -\frac{6}{5}[x - (-3)]$$

$$5(y - 2) = -6(x + 3) \qquad \text{Multiply by 5.}$$
$$5y - 10 = -6x - 18 \qquad \text{Distributive property}$$

$$y = -\frac{6}{5}x - \frac{8}{5}.$$

Verify that the same equation results if $(2, -4)$ is used instead of $(-3, 2)$ in the point-slope form. ◗

As a special case of the point-slope form of the equation of a line, suppose that a line passes through the point $(0, b)$, so the line has y-intercept b. If the line has slope m, then using the point-slope form with $x_1 = 0$ and $y_1 = b$ gives

$$y - y_1 = m(x - x_1)$$
$$y - b = m(x - 0)$$
$$y = mx + b$$

as an equation of the line. Since this result shows the slope of the line and the y-intercept, it is called the *slope-intercept form* of the equation of the line.

Slope-Intercept Form

The line with slope m and y-intercept b has an equation

$$y = mx + b,$$

the **slope-intercept** form of the equation of a line.

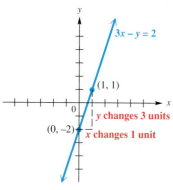

FIGURE 25

EXAMPLE 3
Using the slope-intercept form to graph a line

Find the slope and y-intercept of $3x - y = 2$. Graph the line using this information.

First write $3x - y = 2$ in the slope-intercept form, $y = mx + b$, by solving for y, getting $y = 3x - 2$. This result shows that the slope is $m = 3$ and the y-intercept is $b = -2$. To draw the graph, first locate the y-intercept. See Figure 25. Then, as in Section 3.3, use the slope of 3, or $3/1$, to get a second point on the graph. The line through these two points is the graph of $3x - y = 2$. ◗

In the preceding discussion, it was assumed that the given line had a slope. The only lines having undefined slope are vertical lines. The vertical line through the point (a, b) passes through all the points of the form (a, y), for any value of y. This fact determines the equation of a vertical line.

Equation of a Vertical Line

An equation of the vertical line through the point (a, b) is $x = a$.

For example, the vertical line through $(-4, 9)$ has equation $x = -4$, while the vertical line through $(0, 1/4)$ has equation $x = 0$. (This is the y-axis).

The horizontal line through the point (a, b) passes through all points of the form (x, b), for any value of x. Therefore, the equation of a horizontal line involves only the variable y.

> ### Equation of a Horizontal Line
> An equation of the horizontal line through the point (a, b) is $y = b$.

For example, the horizontal line through $(1, -3)$ has the equation $y = -3$. See Figure 18(a) in the previous section for the graph of this equation. Since each point on the x-axis has y-coordinate 0, the equation of the x-axis is $y = 0$.

PARALLEL AND PERPENDICULAR LINES   Slopes can be used to decide whether or not two lines are parallel. Since two parallel lines are equally "steep," they should have the same slope. Also, two distinct lines with the same "steepness" are parallel. The following result summarizes this discussion.

> ### Parallel Lines
> Two distinct nonvertical lines are parallel if and only if they have the same slope.

Slopes are also used to determine if two lines are perpendicular. Whenever two lines have slopes with a product of -1, the lines are perpendicular.

> ### Perpendicular Lines
> Two lines, neither of which is vertical, are perpendicular if and only if their slopes have a product of -1. Thus the slopes of perpendicular lines are negative reciprocals.

For example, if the slope of a line is $-3/4$, the slope of any line perpendicular to it is $4/3$, since $(-3/4)(4/3) = -1$. (Numbers like $-3/4$ and $4/3$ are called "negative reciprocals.") A proof of this result is outlined in Exercises 35–42.

◀ **EXAMPLE 4**
Using the slope relationships for parallel and perpendicular lines

Find the equation in standard form of the line that passes through the point $(3, 5)$ and satisfies the given condition.

(a) parallel to the line $2x + 5y = 4$

Since it is given that the point $(3, 5)$ is on the line, we need only find the slope to use the point-slope form. Find the slope by writing the equation of the given line in slope-intercept form. (That is, solve for y.)

$$2x + 5y = 4$$

$$y = -\frac{2}{5}x + \frac{4}{5}$$

The slope is $-2/5$. Since the lines are parallel, $-2/5$ is also the slope of the line whose equation is to be found. Substituting $m = -2/5$, $x_1 = 3$, and $y_1 = 5$ into

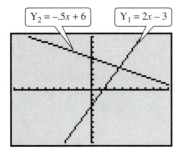

Standard viewing window

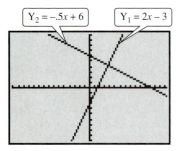

[−15, −15] by [−10, 10]
Square viewing window

To gain the correct perspective when graphing perpendicular lines with a calculator (such as $Y_1 = 2x - 3$ and $Y_2 = -.5x + 6$), we use a *square viewing window*.

the point-slope form gives

$$y - y_1 = m(x - x_1)$$

$$y - 5 = -\frac{2}{5}(x - 3)$$

$$5(y - 5) = -2(x - 3)$$

$$5y - 25 = -2x + 6$$

$$2x + 5y = 31.$$

(b) perpendicular to the line $2x + 5y = 4$

In part (a) it was found that the slope of this line is $-2/5$, so the slope of any line perpendicular to it is $5/2$. Therefore, use $m = 5/2$, $x_1 = 3$, and $y_1 = 5$ in the point-slope form.

$$y - 5 = \frac{5}{2}(x - 3)$$

$$2(y - 5) = 5(x - 3)$$

$$2y - 10 = 5x - 15$$

$$-5x + 2y = -5$$

or

$$5x - 2y = 5 \quad \blacktriangleright$$

All the lines discussed above have equations that could be written in the form $Ax + By = C$ for real numbers A, B, and C. As mentioned earlier, the equation $Ax + By = C$ is the standard form of the equation of a line.

Linear Equations

General Equation	Type of Equation
$Ax + By = C$	*Standard form* (if $A \neq 0$ and $B \neq 0$), x-intercept C/A, y-intercept C/B, slope $-A/B$
$x = a$	*Vertical line*, x-intercept a, no y-intercept, undefined slope
$y = b$	*Horizontal line*, y-intercept b, no x-intercept, slope 0
$y = mx + b$	*Slope-intercept form*, y-intercept b, slope m
$y - y_1 = m(x - x_1)$	*Point-slope form*, slope m, through (x_1, y_1)

■ **PROBLEM SOLVING** A straight line is often the best approximation of a set of data points that result from a real situation. If the equation is known, it can be used to predict the value of one variable, given a value of the other. For this reason, the equation is written as a linear relation in slope-intercept form. One way to find the equation of such a straight line is to use two typical data points and the point-slope form of the equation of a line. ■

EXAMPLE 5
Finding an equation from data points

Estimates for Medicare costs (in billions of dollars) are shown in the table below. The data are graphed in Figure 26. (*Source:* U.S. Office of Management and Budget.)

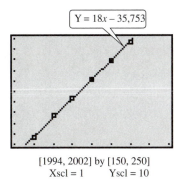

[1994, 2002] by [150, 250]
Xscl = 1   Yscl = 10

The data in Example 5 can be analyzed using the statistics mode of a graphing calculator. The line Y = 18x − 35,753 is a "close fit" for the points as determined in the example.

Year	Cost
1995	157
1996	178
1997	194
1998	211
1999	229
2000	247

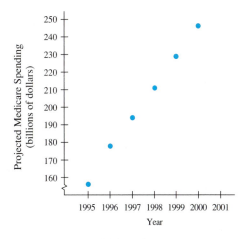

FIGURE 26

(a) Find a linear function where $f(x)$ approximates the cost in year x.

Start by choosing two data points that the line should go through. Suppose we choose the points (1995, 157) and (2000, 247). The slope of the line through these points is

$$\frac{247 - 157}{2000 - 1995} = 18.$$

Now we use the point-slope form of the equation of a line with $(x_1, y_1) = (2000, 247)$ to get the equation.

$$y - 247 = 18(x - 2000)$$
$$y - 247 = 18x - 36,000$$
$$y = 18x - 35,753$$

Thus $f(x) = 18x - 35,753$.

(b) Use $f(x)$ to predict the cost of Medicare in 2002.

$$f(2002) = 18(2002) - 35,753 = 283 \text{ billion dollars.}$$

3.4 Exercises ▼▼▼▼▼▼▼▼▼▼▼▼▼▼▼▼▼▼▼▼▼▼▼▼▼▼▼▼▼▼▼▼▼▼▼▼▼▼

Decide whether the statement is true or false.

1. An equation of the line through (3, 4) with slope 6 is $y - 4 = 6(x - 3)$.
2. An equation of the line through (0, −5) with slope .5 is $y = .5x - 5$.
3. The graph of $y = x + 4$ is parallel to the graph of $y = -x + 4$.
4. The graph of $y = (-1/2)x + 3$ is perpendicular to the graph of $y = 2x - 6$.

Write an equation for the line described. Give answers in standard form for Exercises 5–10 and in slope-intercept form (if possible) for Exercises 11–16. See Examples 1 and 2.

5. Through $(1, 3)$, $m = -2$

6. Through $(2, 4)$, $m = -1$

7. Through $(-5, 4)$, $m = -3/2$

8. Through $(-4, 3)$, $m = 3/4$

9. Through $(-8, 4)$, undefined slope

10. Through $(5, 1)$, $m = 0$

11. Through $(-1, 3)$ and $(3, 4)$

12. Through $(8, -1)$ and $(4, 3)$

13. x-intercept 3, y-intercept -2

14. x-intercept -2, y-intercept 4

15. Vertical, through $(-6, 4)$

16. Horizontal, through $(2, 7)$

17. Fill in each blank with the appropriate response: The line $x + 2 = 0$ has x-intercept _____ . It _____ have a y-intercept. The slope of this line is _____ .
  (does/does not)   (zero/undefined)

The line $4y = 2$ has y-intercept _____ . It _____ have an x-intercept. The
  (does/does not)

slope of this line is _____ .
  (zero/undefined)

18. Match the equation with the line that would most closely resemble its graph. (*Hint:* Consider the signs of m and b in the slope-intercept form.)
  (a) $y = 3x + 2$ **(b)** $y = -3x + 2$ **(c)** $y = 3x - 2$ **(d)** $y = -3x - 2$

A. 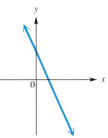   **B.** 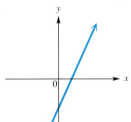   **C.** 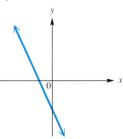   **D.**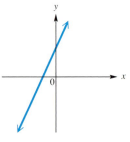

19. Match the equation with its calculator-generated graph. The standard window is used in each case, but no tick marks are shown.
  (a) $y = 2x + 3$
  (b) $y = -2x + 3$
  (c) $y = 2x - 3$
  (d) $y = -2x - 3$

A. 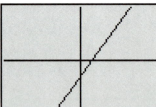   **B.**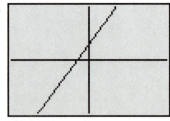

C. 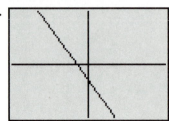   **D.**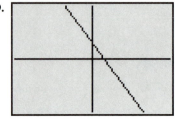

20. The accompanying table was generated by a graphing calculator for a linear function.

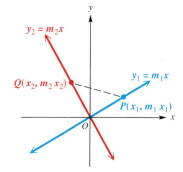

(a) Find the slope of the line defined by the equation in Y_1.

(b) Find the y-intercept of the line.

(c) Find the equation for this line in slope-intercept form.

Give the slope and y-intercept of each line. See Example 3.

21. $y = 3x - 1$

22. $y = -2x + 7$

23. $4x - y = 7$

24. $2x + 3y = 16$

25. $4y = -3x$

26. $2y - x = 0$

Write an equation in standard form for the line described. See Example 4.

27. Through $(-1, 4)$, parallel to $x + 3y = 5$

28. Through $(3, -2)$, parallel to $2x - y = 5$

29. Through $(1, 6)$, perpendicular to $3x + 5y = 1$

30. Through $(-2, 0)$, perpendicular to $8x - 3y = 7$

31. Through $(-5, 6)$, perpendicular to $y = -2$

32. Through $(4, -4)$, perpendicular to $x = 4$

33. Find k so that the line through $(4, -1)$ and $(k, 2)$ is
(a) parallel to $3y + 2x = 6$;
(b) perpendicular to $2y - 5x = 1$.

34. Find r so that the line through $(2, 6)$ and $(-4, r)$ is
(a) parallel to $2x - 3y = 4$;
(b) perpendicular to $x + 2y = 1$.

▼▼▼▼▼▼▼▼▼▼▼▼▼ **DISCOVERING CONNECTIONS** (Exercises 35–42) ▼▼▼▼▼▼▼▼▼▼▼▼▼

In this section we state that two lines, neither of which is vertical, are perpendicular if and only if their slopes have a product of -1. In Exercises 35–42, we outline a proof of this for the case where the two lines intersect at the origin. Work these exercises in order, and refer to the figure as needed.

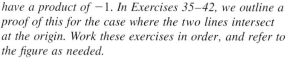

35. In triangle OPQ, angle POQ is a right angle if and only if

$$[d(O, P)]^2 + [d(O, Q)]^2 = [d(P, Q)]^2.$$

What theorem from geometry assures us of this?

36. Find an expression for the distance $d(O, P)$.

37. Find an expression for the distance $d(O, Q)$.

38. Find an expression for the distance $d(P, Q)$.

39. Use your results from Exercises 36–38, and substitute into the equation in Exercise 35. Simplify to show that this leads to the equation $-2m_1m_2x_1x_2 - 2x_1x_2 = 0$.

40. Factor $-2x_1x_2$ from the final form of the equation in Exercise 39.

41. Use the zero-factor property to solve the equation in Exercise 40 to show that $m_1m_2 = -1$.

42. State your conclusion based on Exercises 35–41.

Solve the problem. See Example 5.

43. The table lists the distances (in megaparsecs) and velocities (in kilometers per second) of four galaxies moving rapidly away from Earth.

Galaxy	Distance	Velocity
Virgo	15	1600
Ursa Minor	200	15,000
Corona Borealis	290	24,000
Bootes	520	40,000

(*Sources:* Acker, A., and C. Jaschek, *Astronomical Methods and Calculations.* John Wiley & Sons, 1986. Karttunen, H. (editor), *Fundamental Astronomy,* Springer-Verlag, 1994.)

(a) Plot the data using distance for the *x*-values and velocity for the *y*-values. (You may want to use a graphing calculator.) What type of relationship seems to hold between the data?

(b) Find a linear equation in the form $y = mx$ that models these data using the points $(520, 40,000)$ and $(0, 0)$. Graph your equation with the data on the same coordinate axes.

(c) The galaxy Hydra has a velocity of 60,000 km/sec. How far away is it?

(d) The value of *m* is called the **Hubble constant.** The Hubble constant can be used to estimate the age of the universe *A* (in years) using the formula $A = \dfrac{9.5 \times 10^{11}}{m}$. Approximate *A* using your value of *m*.

(e) Astronomers currently place the value of the Hubble constant between 50 and 100. What is the range for the age of the universe *A*?

44. The table lists the average annual cost (in dollars) of tuition and fees at private four-year colleges for selected years.

(a) Determine a linear function defined by $f(x) = mx + b$ that models the data, using the points $(1, 4113)$ and $(13, 11,025)$. Graph *f* and the data on the same coordinate axes. (You may wish to use a graphing calculator.) What does the slope of the graph of *f* indicate?

Year	Tuition & Fees
1981	4113
1983	5093
1985	6121
1987	7116
1989	8446
1991	10,017
1993	11,025

(*Source:* The College Board.)

(b) Use this function to approximate the tuition and fees in the year 1990. Compare it with the true value of $9340.

(c) Discuss the accuracy of using *f* to estimate the cost of private colleges in the years 1970 and 2010.

45. The table lists the total federal debt (in billions of dollars) from 1985 to 1989.

Year	Federal Debt
1985	1828
1986	2130
1987	2354
1988	2615
1989	2881

(*Source:* U.S. Office of Management and Budget.)

(a) Plot the data by letting $x = 0$ correspond to 1985. Discuss any trends of the federal debt over this time period.

(b) Find a linear function defined by $f(x) = mx + b$ that approximates the data, using the points $(0, 1828)$ and $(4, 2881)$. What does the slope of the graph of *f* represent? Graph *f* and the data on the same coordinate axes.

(c) Use *f* to predict the federal debt in the years 1984 and 1990. Compare your results to the true values of 1577 and 3191 billion dollars.

(d) Now use f to predict the federal debt in the years 1980 and 1994. Compare your results to the true values of 914 and 4690 billion dollars.

(e) Discuss the accuracy of using a linear approximation to model the federal debt over different domains.

46. The table lists the average annual cost (in dollars) of tuition and fees at public four-year colleges for selected years.

Year	Tuition & Fees
1981	909
1983	1148
1985	1318
1987	1537
1989	1781
1991	2137
1993	2527

(*Source:* The College Board.)

(a) Plot the cost of public colleges by letting $x = 0$ correspond to 1980. Is the data *exactly* linear? Could the data be *approximated* by a linear function?

(b) Determine a linear function f defined by $f(x) = mx + b$ that models the data, using the points $(1, 909)$ and $(11, 2137)$. Graph f and the data on the same coordinate axes. What does the slope of the graph of f indicate?

(c) Use this function to approximate the tuition and fees in the year 1984. Compare it with the true value of $1228.

(d) Discuss the accuracy of using f to estimate the cost of public colleges in the years 1970 and 2010.

47. The time interval between a person's initial infection with HIV and that person's eventual development of AIDS symptoms is an important issue. The method of infection with HIV affects the time interval before AIDS develops. One study of HIV patients who were infected by intravenous drug use found that 17% of the patients had AIDS after 4 years and 33% had developed the disease after 7 years. The relationship between the time interval and the percentage of patients with AIDS can accurately be modeled using a linear function. (*Source:* Alcabes, P., A. Munoz, D. Vlahov, and G. Friedland, "Incubation Period of Human Immunodeficiency Virus," *Epidemiologic Review,* Vol. 15, No. 2, The Johns Hopkins University School of Hygiene and Public Health, 1993.)

(a) Write a linear function defined by $f(x) = mx + b$ that models this data, using the points $(4, .17)$ and $(7, .33)$.

(b) Find the slope of the linear equation in part (a). What does the slope tell us about the percent of HIV patients who will develop AIDS?

48. (a) When the Celsius temperature is $0°$, the corresponding Fahrenheit temperature is $32°$. When the Celsius temperature is $100°$, the corresponding Fahrenheit temperature is $212°$. Let C represent the Celsius temperature and F the Fahrenheit temperature. Express F as an exact linear function of C.

(b) Solve the equation in part (a) for C, thus expressing C as a function of F.

(c) For what temperature is $F = C$?

49. During the period from 1988 to 1994, U.S. automakers decreased their number of employees as depicted in the line graph. Using the information given for 1988 and 1994 on the graph, find a linear model of the form $f(x) = mx + b$ that approximates the number of employees in thousands as a function of x, where $x = 0$ corresponds to 1988.

U.S. Auto Industry Employees
In thousands

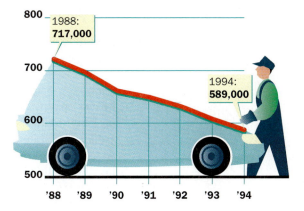

Source: USA Today, Sec. B, p. 1, 6/29/94

50. Despite the decrease in employees (see Exercise 49), the number of problems per 100 cars in the first three months declined between 1989 and 1993 as depicted in the bar graph. Using the data for 1989 and 1993, find a linear model of the form $f(x) = mx + b$ that approximates the number of problems as a function of x, where $x = 0$ corresponds to 1989. (*Source:* J. D. Power and Associates, General Motors, Ford Motor, Chrysler, *USA TODAY* research.)

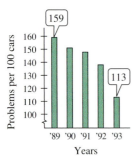

51. Suppose a baseball is thrown at 85 mph. The ball will travel 320 ft when hit by a bat swung at 50 mph and will travel 440 ft when hit by a bat swung at 80 mph. Let y be the number of feet traveled by the ball when hit by a bat swung at x mph. (*Note:* This function is valid for $50 \le x \le 90$, where the bat is 35 inches long, weighs 32 oz, and is swung slightly upward to drive the ball at an angle of 35°.) Assume the relationship is linear, and find an equation of the line. How much farther will a ball travel for each one-mile-per-hour increase in the speed of the bat? (*Source:* Adair, Robert K., *The Physics of Baseball,* New York, Harper & Row, 1990.)

52. In the snake *Lampropeltis polyzona,* total length y is related to tail length x in the domain $30 \text{ mm} \le x \le 200 \text{ mm}$ by a linear function. Find such a linear function if a snake 455 mm long has a 60-mm tail, and a 1050-mm snake has a 140-mm tail.

---

53. Use your own words to describe how to find the equation of a line through two given points.

54. Discuss the advantages and disadvantages of each of the three forms of the equation of a line. When would you use each of them and why? Which form is written in function notation?

55. The product of the slopes of two perpendicular lines is -1. Is this true for *any* two perpendicular lines? Explain. (*Hint:* Is slope defined for every line?)

56. Show that the line $y = x$ is the perpendicular bisector of the segment with endpoints (a, b) and (b, a), where $a \ne b$. (*Hint:* Use the midpoint formula and the slope formula.)

3.5 Graphing Relations and Functions ▼▼▼

In this section we begin discussing a major theme of this book—graphing relations and functions. We begin here with the most basic tool, point plotting.

The set of all points in the plane corresponding to the ordered pairs of a relation is the **graph of the relation.** For now we will find graphs of relations by determining a reasonable number of ordered pairs, locating the corresponding points, and then connecting the points, guessing at the shape of the entire graph. A graphing calculator provides another option that is becoming more widely available. Later in this text, we will develop methods for quickly producing and identifying the graphs of many relations.

EXAMPLE 1
Graphing an absolute value function by point plotting

Graph $y = |x|$.

Start with a table. Use this table to get the points in Figure 27. The graph drawn through these points is made up of portions of two straight lines. The domain is $(-\infty, \infty)$ and the range is $[0, \infty)$. By the vertical line test, $y = |x|$ defines a function.

x	y
-4	4
-3	3
-2	2
-1	1
0	0
1	1
2	2
3	3
4	4

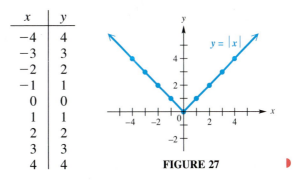

FIGURE 27

CAUTION There is a danger in the method used in Example 1—we might choose a few values for x, find the corresponding values of y, begin to sketch a graph through these few points, and then make a completely wrong guess as to the shape of the graph. For example, choosing only $-1, 0$, and 1 as values of x in Example 1 would produce only the three points $(-1, 1)$, $(0, 0)$, and $(1, 1)$. These three points alone would not give enough information to determine the proper graph for $y = |x|$. However, this section involves only elementary graphs; when more complicated graphs are presented later, we will develop more accurate methods of working with them.

Some graphing calculators will plot points and draw lines connecting two points. To graph equations that can be written in the form $y = $ (an expression in x), it is easier to use the graph function of the calculator.

The absolute value function is often a built-in function in a graphing calculator. It is designated by abs(x) or $|x|$. If neither of these forms is available, use the fact that $|x| = \sqrt{x^2}$.

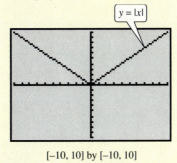

$y = |x|$

$[-10, 10]$ by $[-10, 10]$

EXAMPLE 2
Graphing a second-degree relation by point plotting

Graph $x = y^2 - 4$.

Since $y^2 \geq 0$, the domain is $[-4, \infty)$. It is easier here to choose values of y, then find the corresponding x-values. Choosing 1 for y, for example, gives $x = (1)^2 - 4 = -3$. Choosing -1 for y gives the same result. The table shown with Figure 28 gives values of x corresponding to various values of y. The ordered pairs from this table were used to get the points plotted in Figure 28. (Don't forget that x always goes first in the ordered pair.) A smooth curve was then drawn through the resulting points. Here, y can take on any value, so the range is $(-\infty, \infty)$. As mentioned earlier, the domain is $[-4, \infty)$. Note that this is not the graph of a function, because it fails the vertical line test.

x	y
5	3
0	2
-3	1
-4	0
-3	-1
0	-2
5	-3

FIGURE 28

As mentioned earlier, a graphing calculator requires an equation to be in the form $y = $ (an expression in x). Thus, to graph $x = y^2 - 4$ we must first solve the equation for y. This gives the two equations $Y_1 = \sqrt{(x + 4)}$ and $Y_2 = -\sqrt{(x + 4)}$. (Remember to use parentheses around $x + 4$ or the calculator will graph $y = \sqrt{x} + 4$.) Then we use the graphing calculator to graph both equations in the same window.

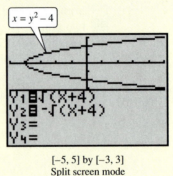

$[-5, 5]$ by $[-3, 3]$
Split screen mode

EXAMPLE 3
Graphing a square root function by point plotting

Graph $y = \sqrt{x + 4}$.

The domain is determined by the fact that $x + 4 \geq 0$ or $x \geq -4$, giving $[-4, \infty)$. The range is $[0, \infty)$. Selecting some values of x in the domain and calculating the corresponding y-values leads to the ordered pairs shown in the table.

x	-4	-3	0	5
y	0	1	2	3

Plotting these points and drawing a curve through them gives the graph in Figure 29. ▶

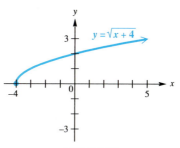

FIGURE 29

Some applications require a relation whose graph includes parts of two or more different lines or curves. For instance, the absolute value function, graphed in Example 1, is composed of parts of the lines with equations $y = x$ and $y = -x$. The next examples show other such functions, called **piecewise-defined functions.**

EXAMPLE 4
Graphing a piecewise-defined function

Graph the functions defined as follows.

(a) $f(x) = \begin{cases} x + 1 & \text{if } x > 2 \\ -2x + 5 & \text{if } x \le 2 \end{cases}$

We must graph the function on each portion of the domain separately. If $x \le 2$, the graph has an endpoint at at $x = 2$. Find the y-value by substituting 2 for x in $-2x + 5$ to get $y = 1$. To get another point on this part of the graph, let's choose $x = 0$, so $y = 5$. Draw the graph through $(2, 1)$ and $(0, 5)$ as a ray with endpoint at $(2, 1)$. Graph the ray for $x > 2$ similarly. This ray has an open endpoint at $(2, 3)$. Use $y = x + 1$ to find another point with x-value greater than 2 to complete the graph. See Figure 30.

(b) $f(x) = \begin{cases} 2x + 3 & \text{if } x \le 1 \\ -x + 6 & \text{if } x > 1 \end{cases}$

Graph $f(x) = 2x + 3$ for $x \le 1$. For $x > 1$, graph $f(x) = -x + 6$. The graph consists of the two pieces shown in Figure 31. The two lines meet at the point $(1, 5)$, which is an endpoint of the first graph. ▶

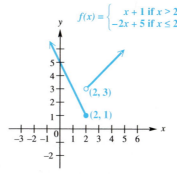

FIGURE 30

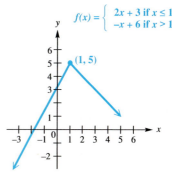

FIGURE 31

Piecewise-defined functions can be graphed with some graphing calculators. You may need to use dot mode. See the instruction manual for your calculator. The graph will not distinguish between open and closed endpoints, so it is still important to understand what the graph should look like.

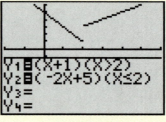

[−2, 6] by [−1, 6]
Split screen mode

This is the graph of

$$f(x) = \begin{cases} x + 1 \text{ if } x > 2 \\ -2x + 5 \text{ if } x \leq 2. \end{cases}$$

Compare with Figure 30.

◀EXAMPLE 5
Graphing the greatest-integer function

The symbol $[\![x]\!]$ is used to represent the greatest integer less than or equal to x. For example, $[\![8.4]\!] = 8$, $[\![-5]\!] = -5$, $[\![\pi]\!] = 3$, $[\![-6.9]\!] = -7$, and so on. In general, if $f(x) = [\![x]\!]$,

$$\text{for } 0 \leq x < 1, \qquad f(x) = 0,$$
$$\text{for } 1 \leq x < 2, \qquad f(x) = 1,$$
$$\text{for } 2 \leq x < 3, \qquad f(x) = 2,$$

and so on. A graph of $f(x)$ is shown in Figure 32. ▶

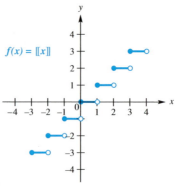

FIGURE 32

The function defined by $f(x) = [\![x]\!]$, discussed in Example 5, is called the **greatest-integer function.** The greatest-integer function is an example of a **step function,** a function with a graph that looks like a series of steps. Some applications of step functions are included in the exercises.

📉🖩 Graphing calculators have built-in functions, such as LOG, SIN, ABS, and so on. Some have INT for the greatest-integer function. If your calculator does not have this built-in function, you can rewrite it as a piecewise-defined function for the desired domain.

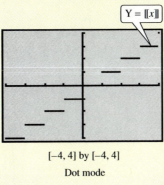

$Y = [\![x]\!]$

$[-4, 4]$ by $[-4, 4]$

Dot mode

Piecewise-defined functions can be used to describe many everyday situations. The next example gives one instance and others are included in the exercises.

EXAMPLE 6
Applying a piecewise-defined function

Professional basketball player Shaquille O'Neal is 7-foot 1-inch tall and weighs 300 pounds. The table lists his age and shoe size.

Age	Size
20	19
21	20
22	21
23	22

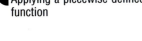

$$Y = \begin{cases} x-1 \text{ if } 20 \le x \le 24 \\ 23 \text{ if } 24 < x \le 30 \end{cases}$$

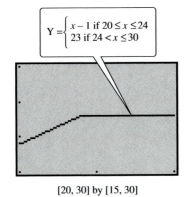

$[20, 30]$ by $[15, 30]$
Xscl = 5   Yscl = 5

FIGURE 33

(a) Determine a linear function defined by $f(x)$ that models the data where x is Shaquille O'Neal's age and $f(x)$ computes his shoe size. Interpret the slope of the graph of f.

The shoe size is one less than his age, so we let $f(x) = x - 1$. The slope of 1 indicates that his shoe size is increasing at a rate of 1 size per year.

(b) Could $f(x)$ be used to predict O'Neal's shoe size at any age?

No, most people's feet eventually stop growing.

(c) Suppose his feet continue to grow at the present rate and then stop at age 24. Graph a piecewise-defined function that describes his shoe size between the ages of 20 and 30.

On the interval $[20, 24]$, $y = x - 1$, and on $(24, 30]$, $y = 23$. A calculator-generated graph is shown in Figure 33. ▸

3.5 Exercises ▼▼▼▼▼▼▼▼▼▼▼▼▼▼▼▼▼▼▼▼▼▼▼▼▼▼▼▼▼▼▼▼▼▼▼▼▼

Exercises 1–10 list ten of the most basic graphs studied in elementary algebra. You should be able to identify the graphs immediately, with a minimum of point plotting, as you progress through this text. Match the equation of the relation with its graph.

1. $y = x$

2. $y = x^2$

3. $x = |y|$

4. $y = |x|$

5. $y = x^3$

6. $y = \sqrt{x}$

7. $y = \sqrt[3]{x}$

8. $x = y^2$

9. $y = [\![x]\!]$

10. $y = \dfrac{1}{x}$

A.

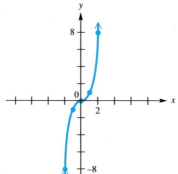

B.

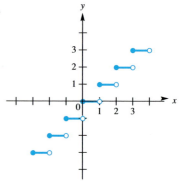

C.

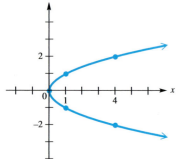

D.

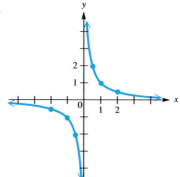

E.

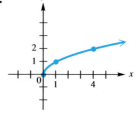

F.

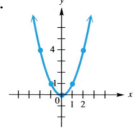

G.

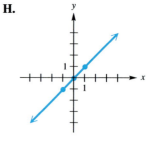

H.

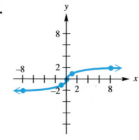

I.

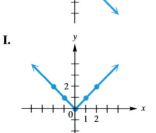

J.

11. One of the graphs in choices A–J is also the graph of $x = y^3$. Which one is it?

12. What one point is common to all the graphs in choices A–J? What one point is common to all but one of the graphs?

Use point plotting to sketch the graph of the relation. Give the domain and the range. See Examples 1–3.

13. $y = 3x - 2$ **14.** $y = -2x + 4$ **15.** $3x = y^2$ **16.** $4x = y^2$

17. $16x^2 = -y$ **18.** $4x^2 = -y$ **19.** $y = |x| + 4$ **20.** $y = |x| - 3$

21. $x = |y| + 1$ **22.** $x = |y| - 1$ **23.** $y = -|x + 1|$ **24.** $y = -|x - 2|$

25. $x = \sqrt{y} - 2$ **26.** $x = -\sqrt{y} + 1$ **27.** $x = -\sqrt{y} - 2$ **28.** $x = \sqrt{y} - 4$

29. $y = \sqrt{2x + 4}$ **30.** $y = \sqrt{3x + 9}$ **31.** $y = -2\sqrt{x}$ **32.** $y = -\sqrt{x}$

33. Which of the relations in Exercises 13–32 express y as a function of x?

34. Which of the relations in Exercises 13–32 do not express y as a function of x?

*For the given piecewise-defined function, find (**a**) $f(-5)$, (**b**) $f(-1)$, (**c**) $f(0)$, and (**d**) $f(3)$.*

35. $f(x) = \begin{cases} 2x & \text{if } x \le -1 \\ x - 1 & \text{if } x > -1 \end{cases}$ **36.** $f(x) = \begin{cases} x - 2 & \text{if } x < 3 \\ 5 - x & \text{if } x \ge 3 \end{cases}$

Graph the piecewise-defined function. See Example 4.

37. $f(x) = \begin{cases} x - 1 & \text{if } x \le 3 \\ 2 & \text{if } x > 3 \end{cases}$ **38.** $f(x) = \begin{cases} 6 - x & \text{if } x \le 3 \\ 3x - 6 & \text{if } x > 3 \end{cases}$

39. $f(x) = \begin{cases} 4 - x & \text{if } x < 2 \\ 1 + 2x & \text{if } x \ge 2 \end{cases}$ **40.** $f(x) = \begin{cases} 2x + 1 & \text{if } x \ge 0 \\ x & \text{if } x < 0 \end{cases}$

41. $f(x) = \begin{cases} 2 + x & \text{if } x < -4 \\ -x & \text{if } -4 \le x \le 5 \\ 3x & \text{if } x > 5 \end{cases}$ **42.** $f(x) = \begin{cases} -2x & \text{if } x < -3 \\ 3x - 1 & \text{if } -3 \le x \le 2 \\ -4x & \text{if } x > 2 \end{cases}$

Give a rule for the piecewise-defined function. Also give the domain and the range. See Example 4.

43. 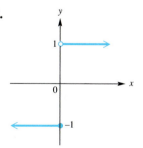   **44.** 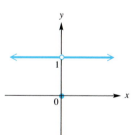   **45.** 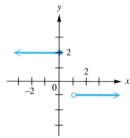   **46.**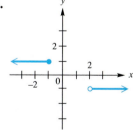

Graph the function. Give the domain and the range. See Example 5.

47. $f(x) = [\![-x]\!]$ **48.** $f(x) = [\![2x]\!]$ **49.** $g(x) = [\![2x - 1]\!]$

50. Describe how the y-values of the greatest-integer function are determined for negative x-values.

Graph the function *f* in the standard window of a graphing calculator.

51. $f(x) = x$ **52.** $f(x) = x^2$ **53.** $f(x) = x^3$

54. $f(x) = |x|$ **55.** $f(x) = \sqrt[3]{x}$ **56.** $f(x) = \sqrt{x}$

Graph the relation in the standard window of a graphing calculator. First solve for Y_1 and Y_2, and give the equations for them.

57. $x = y^2$ **58.** $x = |y|$

Solve the problem. See Example 6.

59. The table lists the federal minimum hourly wage from 1978 to 1995. Sketch a graph of the data as a piece-wise-defined function. (*Source:* U.S. Department of Labor.)

Year(s)	Wage
1978	$2.65
1979	$2.90
1980	$3.10
1981–90	$3.35
1991–95	$4.50

60. The following table lists the approximate number of animal rabies cases in the United States from 1988 to 1992. (*Source: USA TODAY* research.)

Year	Cases
1988	4800
1989	4900
1990	5000
1991	6700
1992	8400

   (a) Describe the change in the data from one year to the next.
   (b) Determine a piecewise-defined function $f(x)$ that approximates the data. Let $x = 0$ correspond to the year 1988.

61. When a diabetic takes long-acting insulin, the insulin reaches its peak effect on the blood sugar level in about 3 hr. This effect remains fairly constant for 5 hr, then declines, and is very low until the next injection. In a typical patient, the level of insulin might be given by the following function.

$$i(t) = \begin{cases} 40t + 100 & \text{if } 0 \le t \le 3 \\ 220 & \text{if } 3 < t \le 8 \\ -80t + 860 & \text{if } 8 < t \le 10 \\ 60 & \text{if } 10 < t \le 24 \end{cases}$$

Here $i(t)$ is the blood sugar level, in appropriate units, at time t measured in hours from the time of the injection. Suppose a patient takes insulin at 6 A.M. Find the blood sugar level at each of the following times:
(a) 7 A.M.   **(b)** 9 A.M.   **(c)** 10 A.M.   **(d)** noon
(e) 2 P.M.   **(f)** 5 P.M.   **(g)** midnight.
(h) Graph $y = i(t)$.

62. The snow depth in Michigan's Isle Royale National Park varies throughout the winter. In a typical winter, the snow depth in inches is approximated by the following function.

$$f(x) = \begin{cases} 6.5x & \text{if } 0 \le x \le 4 \\ -5.5x + 48 & \text{if } 4 < x \le 6 \\ -30x + 195 & \text{if } 6 < x \le 6.5 \end{cases}$$

Here, x represents the time in months with $x = 0$ representing the beginning of October, $x = 1$ representing the beginning of November, and so on.
(a) Graph $f(x)$.
(b) In what month is the snow deepest? What is the deepest snow depth?
(c) In what months does the snow begin and end?

Some models of graphing calculators are capable of graphing piecewise-defined functions. Match the piecewise-defined function with its calculator-generated graph.

63. $f(x) = \begin{cases} x^2 - 4 & \text{if } x \geq 0 \\ -x + 5 & \text{if } x < 0 \end{cases}$

64. $g(x) = \begin{cases} |x - 4| & \text{if } x \geq -1 \\ -x^2 & \text{if } x < -1 \end{cases}$

65. $h(x) = \begin{cases} 6 & \text{if } x \geq 0 \\ -6 & \text{if } x < 0 \end{cases}$

66. $k(x) = \begin{cases} \sqrt{x} & \text{if } x \geq 0 \\ -x^2 & \text{if } x < 0 \end{cases}$

A.
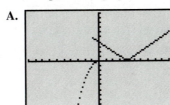
[−10, 10] by [−10, 10]
Dot mode

B.
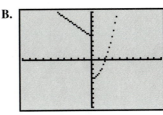
[−10, 10] by [−10, 10]
Dot mode

C.
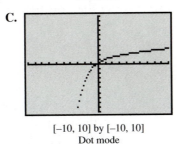
[−10, 10] by [−10, 10]
Dot mode

D.
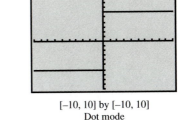
[−10, 10] by [−10, 10]
Dot mode

3.6   General Graphing Techniques ▼▼▼

One of the main objectives of this course is to recognize and learn to graph various functions. Several graphing techniques presented in this section show how to graph functions that are defined by altering the equation of a basic function in certain ways.

STRETCHING AND SHRINKING   We begin by considering how the graph of $y = a \cdot f(x)$ compares to the graph of $y = f(x)$.

◀**EXAMPLE 1**
Graphing functions of the form $y = a \cdot f(x)$

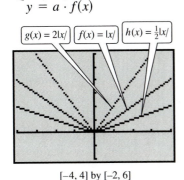

[−4, 4] by [−2, 6]
Stretching and shrinking as illustrated on a graphing calculator screen

Graph each of the following.

(a) $g(x) = 2|x|$

Use a calculator or plot a few points to get the graph of $g(x)$, shown in blue in Figure 34. The graph of $f(x) = |x|$ is shown in red for comparison. Since each y-value in $g(x)$ is twice the corresponding y-value in $f(x)$, the graph of $g(x)$ is narrower than that of $f(x)$.

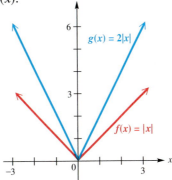

FIGURE 34

(b) $h(x) = \dfrac{1}{2}|x|$

The graph of $h(x)$ is again the same general shape as that of $f(x)$, but here the coefficient $1/2$ causes the graph of $h(x)$ to be broader than the graph of $f(x)$. See Figure 35. ▶

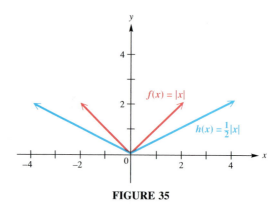

FIGURE 35

The graphs in Figures 34 and 35 suggest the following generalizations.

Stretching and Shrinking

The graph of $g(x) = a \cdot f(x)$ has the same general shape as the graph of $f(x)$. It is

stretched vertically compared to the graph of $f(x)$ if $|a| > 1$;

shrunken vertically compared to the graph of $f(x)$ if $0 < |a| < 1$.

◀EXAMPLE 2
Graphing the negative of a function

Graph $g(x) = -|x|$.

Use a calculator or plot enough points to sketch the graph of $g(x) = -|x|$. The result is shown in blue in Figure 36. The graph of $f(x) = |x|$ is shown in red for comparison. As the equation suggests, every y-value of the graph of $g(x) = -|x|$ is the negative of the corresponding y-value of $f(x) = |x|$. This has the effect of reflecting the graph about the x-axis. ▶

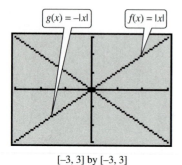

[−3, 3] by [−3, 3]

Reflecting across the x-axis as illustrated on a graphing calculator screen

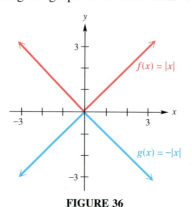

FIGURE 36

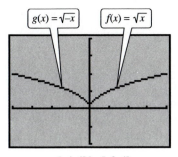

[−4, 4] by [−2, 4]

Reflecting across the y-axis as illustrated on a graphing calculator screen

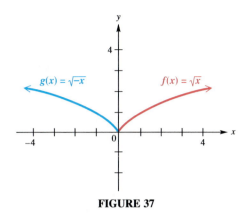

FIGURE 37

The graph of the function defined by $f(x) = \sqrt{x}\ (x \geq 0)$ is shown in red in Figure 37. The figure also shows the graph of $g(x) = \sqrt{-x}\ (x \leq 0)$. The graphs in Figures 36 and 37 suggest the following generalization.

Reflection About an Axis

The graph of $y = -f(x)$ is the same as the graph of $y = f(x)$ reflected about the x-axis.

The graph of $y = f(-x)$ is the same as the graph of $y = f(x)$ reflected about the y-axis.

Notice that if Figure 36 were folded on the x-axis, the graph of $f(x)$ would exactly match the graph of $g(x)$. The same match occurs if Figure 37 is folded on the y-axis.

SYMMETRY   The graph shown in Figure 38 (a) is cut in half by the y-axis with each half the mirror image of the other half. A graph with this property is said to be *symmetric with respect to the y-axis*. As this graph suggests, a graph is symmetric with respect to the y-axis if the point $(-x, y)$ is on the graph whenever (x, y) is on the graph.

If the graph in Figure 38(b) were folded in half along the x-axis, the portion at the top would exactly match the portion at the bottom. Such a graph is *symmetric with respect to the x-axis*: the point $(x, -y)$ is on the graph whenever the point (x, y) is on the graph.

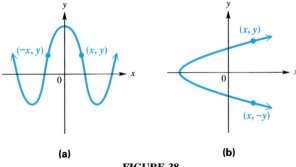

(a)   (b)

FIGURE 38

The following test tells when a graph is symmetric with respect to the *x*-axis or *y*-axis.

Symmetry with Respect to an Axis

The graph of an equation is **symmetric with respect to the *y*-axis** if the replacement of *x* with $-x$ results in an equivalent equation.

The graph of an equation is **symmetric with respect to the *x*-axis** if the replacement of *y* with $-y$ results in an equivalent equation.

EXAMPLE 3
Testing for symmetry with respect to an axis

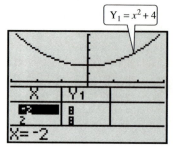

$[-3, 3]$ by $[-1, 10]$
Xscl = 1   Yscl = 2

The table shows that when $x = -2$ or $x = 2$, $Y_1 = 8$. This is a specific case resulting from the *y*-axis symmetry.

Test for symmetry with respect to the *x*-axis or the *y*-axis.

(a) $y = x^2 + 4$
   Replace *x* with $-x$.

$$y = x^2 + 4 \qquad \text{becomes} \qquad y = (-x)^2 + 4 = x^2 + 4$$

The result is the same as the original equation, so the graph, shown in Figure 39, is symmetric with respect to the *y*-axis. Check that the graph is *not* symmetric with respect to the *x*-axis.

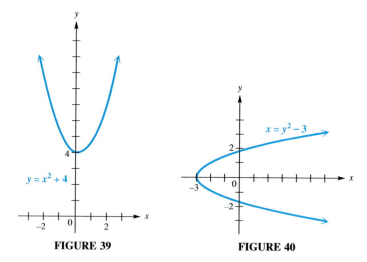

FIGURE 39   FIGURE 40

(b) $x = y^2 - 3$
   Replace *y* with $-y$ to get $x = (-y)^2 - 3 = y^2 - 3$, the same as the original equation. The graph is symmetric with respect to the *x*-axis, as shown in Figure 40. Is the graph symmetric with respect to the *y*-axis?

(c) $x^2 + y^2 = 16$.

Here,

$$(-x)^2 + y^2 = 16 \qquad \text{and} \qquad x^2 + (-y)^2 = 16$$

both become $\qquad\qquad\qquad x^2 + y^2 = 16.$

Thus, the graph, a circle of radius 4 centered at the origin, is symmetric with respect to both axes.

(d) $2x + y = 4$

Replace x with $-x$, and then replace y with $-y$; in neither case does an equivalent equation result. This graph is symmetric with respect to neither the x-axis nor the y-axis. ▶

Another kind of symmetry is found when a graph can be rotated 180° about the origin, with the result coinciding exactly with the original graph. Symmetry of this type is called *symmetry with respect to the origin.* It turns out that rotating a graph 180° is equivalent to saying that whenever the point (x, y) is on the graph, the point $(-x, -y)$ is also on the graph. Figure 41 shows two graphs that are symmetric with respect to the origin.

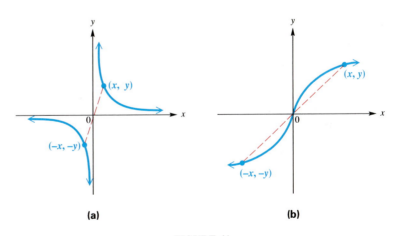

(a)   (b)

FIGURE 41

Symmetry with Respect to the Origin

The graph of an equation is **symmetric with respect to the origin** if the replacement of both x with $-x$ and y with $-y$ results in an equivalent equation.

EXAMPLE 4
Testing for symmetry with respect to the origin

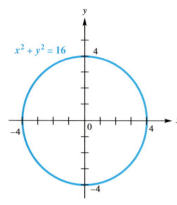

$x^2 + y^2 = 16$

FIGURE 42

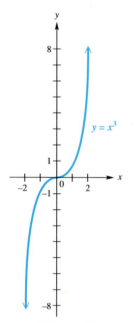

$y = x^3$

FIGURE 43

Are the following graphs symmetric with respect to the origin?

(a) $x^2 + y^2 = 16$

Replace x with $-x$ and y with $-y$ to get

$$(-x)^2 + (-y)^2 = 16 \quad \text{or} \quad x^2 + y^2 = 16,$$

an equivalent equation. This graph, shown in Figure 42, is symmetric with respect to the origin.

(b) $y = x^3$

Replace x with $-x$ and y with $-y$ to get

$$-y = (-x)^3, \quad \text{or} \quad -y = -x^3, \quad \text{or} \quad y = x^3,$$

an equivalent equation. The graph, symmetric with respect to the origin, is shown in Figure 43. ▶

A graph symmetric with respect to both the x-and y-axes is automatically symmetric with respect to the origin. However, a graph symmetric with respect to the origin need not be symmetric with respect to either axis. (See Figure 43.) Of the three types of symmetry—with respect to the x-axis, the y-axis, and the origin—a graph possessing any two must have the third type also.

The various tests for symmetry are summarized below.

Tests for Symmetry

	Symmetric with Respect to:		
	x-axis	*y-axis*	*Origin*
Equation is unchanged if:	y is replaced with $-y$	x is replaced with $-x$	x is replaced with $-x$ and y is replaced with $-y$
Example:			

A function whose graph is symmetric with respect to the y-axis is an **even function,** while a function whose graph is symmetric with respect to the origin is an **odd function.**

CONNECTIONS A figure has *rotational symmetry* around an axis *l* if it coincides with itself by all rotations about *l*. Because of their complete rotational symmetry, the circle in the plane and the sphere in space were considered by the early Greeks to be the most perfect geometric figures. Aristotle assumed a spherical shape for the celestial bodies because any other would detract from their heavenly perfection.

Symmetry has been an important characteristic of art from the earliest times. The art of M. C. Escher (1898–1972)* is composed of symmetries and translations, and Leonardo da Vinci's sketches indicate a superior understanding of symmetry. Almost all nature exhibits symmetry—from the hexagons of snowflakes to the diatom, a microscopic sea plant. Perhaps the most striking examples of symmetry in nature are crystals.

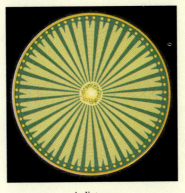

A diatom

A cross-section of tourmaline

Source: Mathematics, Life Science Library, Time Inc., New York, 1963.

FOR DISCUSSION AND WRITING
Discuss other examples of symmetry in art and nature.

TRANSLATIONS The next examples show the results of horizontal and vertical shifts, called **translations,** of the graph of $f(x) = |x|$.

---

* See World of Escher on the Internet at http://www.texas.net/escher/.

EXAMPLE 5
Graphing a function of the form $y = f(x) - c$

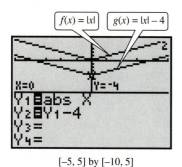

[-5, 5] by [-10, 5]

A vertical translation as illustrated on a graphing calculator screen

Graph $g(x) = |x| - 4$.

By comparing the tables of values for $g(x) = |x| - 4$ and $f(x) = |x|$ shown with Figure 44, we see that for corresponding x-values, the y-values of g are each 4 less than those for f. Thus, the graph of $g(x) = |x| - 4$ is the same as that of $f(x) = |x|$, but translated 4 units down. See Figure 44. The "vertex" (here the lowest point) is at $(0, -4)$. The graph is symmetric with respect to the y-axis, which has equation $x = 0$. ▶

$g(x) = |x| - 4$

x	y
-4	0
-1	-3
0	-4
1	-3
4	0

$f(x) = |x|$

x	y
-4	4
-1	1
0	0
1	1
4	4

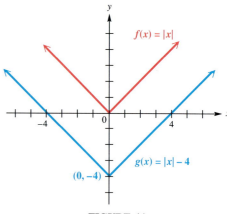

FIGURE 44

EXAMPLE 6
Graphing a function of the form $y = f(x - c)$

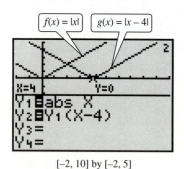

[-2, 10] by [-2, 5]

A horizontal translation as illustrated on a graphing calculator screen

Graph $g(x) = |x - 4|$.

Comparing the tables of values shown with Figure 45 shows that the graph of $g(x)$ is the same as that of $f(x) = |x|$, but it has been translated 4 units to the right. The "vertex" is at $(4, 0)$. As suggested by Figure 45, this graph is symmetric with respect to the line $x = 4$. ▶

$g(x) = |x - 4|$

x	y
0	4
3	1
4	0
5	1
8	4

$f(x) = |x|$

x	y
-4	4
-1	1
0	0
1	1
4	4

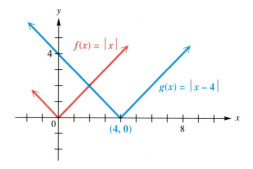

FIGURE 45

EXAMPLE 7
Graphing a function of the form $y = f(x + h) + c$

Graph $f(x) = -|x + 3| + 1$.

The "vertex" of this graph is translated 3 units to the left and 1 unit up, as shown in Figure 46. The graph opens downward because of the negative sign in front of the absolute value symbol, making the "vertex" the *highest* point on the graph. The graph is symmetric with respect to the line $x = -3$. ▶

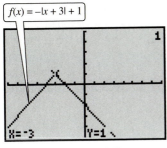

[-8, 8] by [-5, 5]

The graph of $y = |x|$ is translated *and* reflected on this screen. Compare with Figure 46.

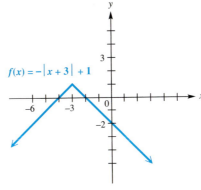

FIGURE 46

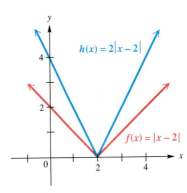

FIGURE 47

EXAMPLE 8
Graphing a function of the form $y = a \cdot f(x - h)$

Graph $h(x) = |2x - 4|$.
    Factor out 2 as follows.

$$
\begin{aligned}
h(x) &= |2x - 4| \\
&= |2(x - 2)| \\
&= |2| \cdot |x - 2| \\
&= 2|x - 2|
\end{aligned}
$$

The graph of h is the graph of $f(x) = |x|$ translated 2 units to the right, and stretched by a factor of 2. See Figure 47. ▶

[-8, 8] by [-5, 5]

Vertical translations of $Y_1 = -\sqrt[3]{x}$

In general, the graph of a function g, defined by $g(x) = f(x) + c$, where c is a real number, can be found from the graph of the function f as follows. For every point (x, y) on the graph of f, there will be a corresponding point $(x, y + c)$ on the graph of g. The new graph will be the same as the graph of f, but translated c units upward if c is positive, or $|c|$ units downward if c is negative. The graph of g is called a **vertical translation** of the graph of f. Figure 48 shows a graph of a function f and two different vertical translations of f.

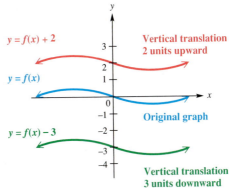

FIGURE 48

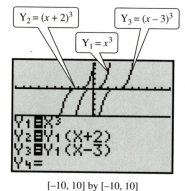

[−10, 10] by [−10, 10]

Xscl = 1   Yscl = 2

Horizontal translations of $Y_1 = x^3$
(Note the use of function notation
for Y_2 and Y_3.)

If a function g is defined by $g(x) = f(x - c)$, for each ordered pair (x, y) of f, there will be a corresponding ordered pair $(x - c, y)$ on the graph of g. This has the effect of translating the graph of f horizontally; c units to the right if c is positive and $|c|$ units to the left if c is negative. Figure 49 shows the graph of $y = f(x)$ along with the graphs of $y = f(x - 3)$ and $y = f(x + 2)$; each of these graphs is obtained from that of $y = f(x)$ by a **horizontal translation.**

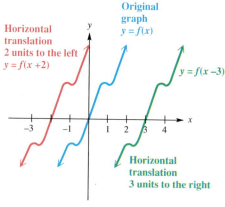

FIGURE 49

Translations of the Graph of a Function

Let f be a function, and let c be a positive number.

To graph:	Shift the graph of $y = f(x)$ by c units:
$y = f(x) + c$	upward
$y = f(x) - c$	downward
$y = f(x + c)$	left
$y = f(x - c)$	right

CAUTION Errors frequently occur when horizontal translations are involved. In order to determine the direction and magnitude of horizontal translations, find the value that would cause the expression in parentheses to equal 0. For example, the graph of $y = (x - 5)^2$ would be shifted 5 units to the *right* of $y = x^2$, because +5 would cause $x - 5$ to equal 0. On the other hand, the graph of $y = (x + 4)^2$ would be shifted 4 units to the *left* of $y = x^2$, because −4 would cause $x + 4$ to equal 0.

EXAMPLE 9
Graphing translations of
$y = f(x)$

A graph of a function defined by $y = f(x)$ is shown in Figure 50. Use this graph to find each of the following graphs.

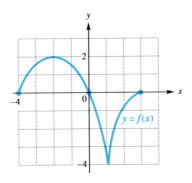

FIGURE 50

(a) $g(x) = f(x) + 3$

This graph is the same as the graph in Figure 50, translated 3 units upward. See Figure 51(a).

(b) $h(x) = f(x + 3)$

To get the graph of $y = f(x + 3)$, translate the graph of $y = f(x)$ three units to the left. See Figure 51(b).

(c) $k(x) = f(x - 2) + 3$

This graph will look like the graph of $f(x)$ translated 2 units to the right and 3 units up, as shown in Figure 51(c). ▶

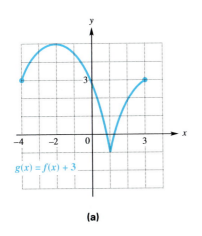

(a)

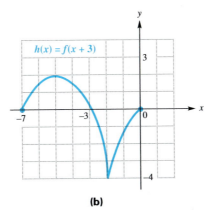

(b)

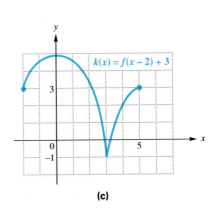

(c)

FIGURE 51

3.6 Exercises ▼▼▼▼▼▼▼▼▼▼▼▼▼▼▼▼▼▼▼▼▼▼▼▼▼▼▼▼▼▼▼▼▼▼▼▼

Use the concepts of this section to decide whether the statement is true or false.

1. The graph of a nonzero function cannot be symmetric with respect to the *x*-axis.

2. The graph of an even function is symmetric with respect to the *y*-axis.

3. The graph of an odd function is symmetric with respect to the origin.

4. If (a, b) is on the graph of an even function, so is $(a, -b)$.

5. If (a, b) is on the graph of an odd function, so is $(-a, b)$.

6. A nonzero function cannot be both even and odd.

Plot the point, and then plot the points that are symmetric to the given point with respect to the
(a) x-axis,   (b) y-axis,   (c) origin.

7. $(5, -3)$ **8.** $(-6, 1)$ **9.** $(-4, -2)$ **10.** $(-8, 0)$

For Exercises 11 and 12, see Examples 1, 2, 5, 6, 7, 8, and 9.

11. Given the graph of $y = g(x)$ in the figure, sketch the graph of each of the following and explain how it is obtained from the graph of $y = g(x)$.
  (a) $y = g(-x) + 1$ **(b)** $y = g(x - 2)$
  (c) $y = g(x + 1) - 2$ **(d)** $y = -g(x) + 2$

12. Use the graph of $y = f(x)$ in the figure to obtain the graph of each of the following. Explain how each graph is related to the graph of $y = f(x)$.
  (a) $y = -f(x)$ **(b)** $y = 2f(x)$
  (c) $y = f(x - 1) + 3$ **(d)** $y = f(-x)$

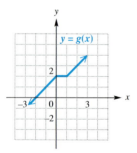

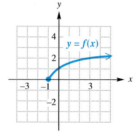

Without graphing, determine whether the equation has a graph that is symmetric with respect to the x-axis, the y-axis, the origin, or none of these. See Examples 3 and 4.

13. $y = x^2 + 2$

14. $y = 2x^4 - 1$

15. $x^2 + y^2 = 10$

16. $y^2 = \dfrac{-5}{x^2}$

17. $y = -3x^3$

18. $y = x^3 - x$

19. $y = x^2 - x + 7$

20. $y = x + 12$

Use the techniques of this section to help graph the relation. See Examples 1–9.

21. $y = |x| - 1$

22. $y = |x + 3| + 2$

23. $y = \dfrac{1}{x}$

24. $y = \dfrac{1}{x^2}$

25. $y = -(x + 1)^3$

26. $y = (-x + 1)^3$

27. $y = 2x^2 - 1$

28. $y = \dfrac{2}{3}(x - 2)^2$

Suppose $f(3) = 6$. For the given assumptions in Exercises 29–34, find another function value.

29. The graph of $y = f(x)$ is symmetric with respect to the origin.

30. The graph of $y = f(x)$ is symmetric with respect to the y-axis.

31. The graph of $y = f(x)$ is symmetric with respect to the line $x = 6$.

32. For all x, $f(-x) = f(x)$.

33. For all x, $f(-x) = -f(x)$.

34. f is an odd function.

35. Find the function g whose graph can be obtained by translating the graph of $f(x) = 2x + 5$ up 2 units and left 3 units.

36. Find the function g whose graph can be obtained by translating the graph of $f(x) = 3 - x$ down 2 units and right 3 units.

37. Complete the left half of the graph of $y = f(x)$ in the figure for the each of the following conditions:
(a) $f(-x) = f(x)$ **(b)** $f(-x) = -f(x)$.

38. Complete the right half of the graph of $y = f(x)$ in the figure for each of the following conditions:
(a) $f(x)$ is odd **(b)** $f(x)$ is even.

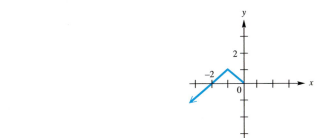

39. Suppose the equation $y = F$ is changed to $y = c \cdot F$, for some constant c. What is the effect on the graph of $y = F$? Discuss the effect depending on whether $c > 0$ or $c < 0$, and $|c| > 1$ or $|c| < 1$.

40. Suppose $y = F(x)$ is changed to $y = F(x + h)$. How are the graphs of these equations related? Is the graph of $y = F(x) + h$ the same as the graph of $y = F(x + h)$? If not, how do they differ?

Sketch an example of a graph having the given characteristics.

41. Symmetric with respect to the x-axis but not to the y-axis

42. Symmetric with respect to the y-axis but not to the x-axis

Shown on the left is the graph of $Y_1 = (x - 2)^2 + 1$ in the standard viewing window of a graphing calculator. Six other functions, Y_2 through Y_7, are graphed according to the rules shown in the screen on the right.

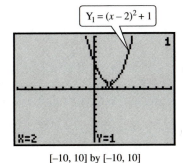

$Y_1 = (x - 2)^2 + 1$

```
Y2=-Y1
Y3=Y1(-X)
Y4=2Y1
Y5=-2Y1
Y6=Y1-2
Y7=Y1+2
Y8=
Y9=
```

X=2   Y=1

[−10, 10] by [−10, 10]

Match the function with its calculator-generated graph from choices A–F first without using a calculator, by applying the techniques of this section. Then confirm your answer by graphing the function on your calculator.

43. Y_2

44. Y_3

45. Y_4

46. Y_5

47. Y_6

48. Y_7

A.

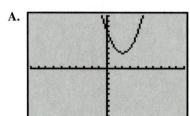

[–10, 10] by [–10, 10]

B.

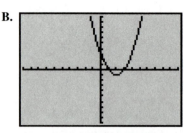

[–10, 10] by [–10, 10]

C.

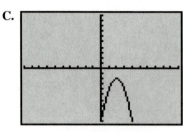

[–10, 10] by [–10, 10]

D.

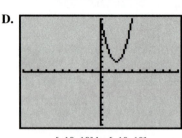

[–10, 10] by [–10, 10]

E.

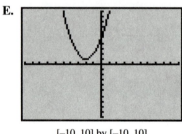

[–10, 10] by [–10, 10]

F.

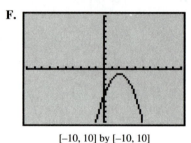

[–10, 10] by [–10, 10]

▼▼▼▼▼▼▼▼▼▼▼▼▼ **DISCOVERING CONNECTIONS** (Exercises 49–56) ▼▼▼▼▼▼▼▼▼▼▼▼▼

Recall from Section 3 of this chapter that a unique line is determined by two different points on the line, and that the values of m and b can then be determined for the general form of the linear function $f(x) = mx + b$. Make the connections between the concepts introduced there and in this section to answer the following.

49. Sketch by hand the line that passes through the points $(1, -2)$ and $(3, 2)$.

50. Use the slope formula to find the slope of this line.

51. Find the equation of this line and write it in the form $y_1 = mx + b$.

52. Keeping the same two x-values as indicated in Exercise 49, add 6 to each y-value. What are the coordinates of the two new points?

53. Find the slope of the line through the points determined in Exercise 52.

54. Find the equation of this new line and write it in the form $y_2 = mx + b$.

55. Graph both Y_1 and Y_2 by hand or in the standard viewing window of a graphing calculator, and describe how the graph of Y_2 can be obtained by vertically translating the graph of Y_1. What is the value of the constant by which this vertical translation occurs? Where do you think this comes from?

56. Fill in the blanks with the correct responses, based on your work in Exercises 49–55.

If the points (x_1, y_1) and (x_2, y_2) lie on a line, then when we add the positive constant c to each y-value, we obtain the points $(x_1, y_1 + $ _____ $)$ and $(x_2, y_2 + $ _____ $)$. The slope of the new line is _____ the slope of the original line. The graph

(the same as/different from)

of the new line can be obtained by shifting the graph of the original line _____ units in the _____ direction.

The 8-hour maximum carbon monoxide levels (in parts per million) for the United States from 1982 to 1992 can be modeled by the function $f(x) = -.012053x^2 - .046607x + 9.125$ where $x = 0$ corresponds to 1982. (Source: U.S. Environmental Protection Agency, 1992.)

57. Discuss the general trend in these carbon monoxide levels.

58. Find a function $g(x)$ that models the same carbon monoxide levels except that x is the actual year between 1982 and 1992. For example, $g(1985) = f(3)$ and $g(1990) = f(8)$. (*Hint:* Use a horizontal translation.)

3.7 Algebra of Functions; Composite Functions ▼▼▼

As mentioned near the end of Section 3.2, economists frequently use the equation "profit equals revenue minus cost," or $P(x) = R(x) - C(x)$, where x is the number of items produced and sold. That is, the profit function is found by subtracting the cost function from the revenue function. New functions can be formed by using other operations as well.

The various operations on functions are defined below.

Definition of Operations on Functions

Given two functions f and g, then for all values of x for which both $f(x)$ and $g(x)$ are defined, the functions $f + g, f - g, fg,$ and f/g are defined as follows.

Sum	$(f + g)(x) = f(x) + g(x)$
Difference	$(f - g)(x) = f(x) - g(x)$
Product	$(fg)(x) = f(x) \cdot g(x)$
Quotient	$\left(\dfrac{f}{g}\right)(x) = \dfrac{f(x)}{g(x)}, \quad g(x) \neq 0$

NOTE   The condition $g(x) \neq 0$ in the definition of the quotient means that the domain of $\left(\dfrac{f}{g}\right)(x)$ consists of all values of x for which $g(x)$ is not zero. The condition does not mean that $g(x)$ is a function that is never zero.

EXAMPLE 1
Using the operations on functions

Let $f(x) = x^2 + 1$ and $g(x) = 3x + 5$. Find each of the following.

(a) $(f + g)(1)$

Since $f(1) = 2$ and $g(1) = 8$, use the definition above to get

$$(f + g)(1) = f(1) + g(1) = 2 + 8 = 10.$$

(b) $(f - g)(-3) = f(-3) - g(-3) = 10 - (-4) = 14$

(c) $(fg)(5) = f(5) \cdot g(5) = 26 \cdot 20 = 520$

(d) $\left(\dfrac{f}{g}\right)(0) = \dfrac{f(0)}{g(0)} = \dfrac{1}{5}$ ▶

EXAMPLE 2
Using the operations on functions

Let $f(x) = 8x - 9$ and $g(x) = \sqrt{2x - 1}$.

(a) $(f + g)(x) = f(x) + g(x) = 8x - 9 + \sqrt{2x - 1}$

(b) $(f - g)(x) = f(x) - g(x) = 8x - 9 - \sqrt{2x - 1}$

(c) $(fg)(x) = f(x) \cdot g(x) = (8x - 9)\sqrt{2x - 1}$

(d) $\left(\dfrac{f}{g}\right)(x) = \dfrac{f(x)}{g(x)} = \dfrac{8x - 9}{\sqrt{2x - 1}}$ ▶

```
Y1◻X²+1
Y2◻3X+5
Y3=
Y4=
Y5=
Y6=
Y7=
Y8=
```

```
Y1(1)+Y2(1)
                10
Y1(-3)-Y2(-3)
                14
Y1(5)*Y2(5)
               520
```

By letting $Y_1 = f(x)$ and $Y_2 = g(x)$ as defined in Example 1, this screen supports the results of parts (a), (b), and (c).

In Example 2, the domain of f is the set of all real numbers, while the domain of g, where $g(x) = \sqrt{2x - 1}$, includes just those real numbers that make $2x - 1 \geq 0$; the domain of g is the interval $[1/2, \infty)$. The domains of $f + g, f - g$, and fg are thus $[1/2, \infty)$. With f/g, the denominator cannot be zero, so the value $1/2$ is excluded from the domain. The domain of f/g is $(1/2, \infty)$.

The domains of $f + g, f - g, fg$, and f/g are summarized below. (Recall that the intersection of two sets is the set of all elements belonging to *both sets*.)

Domains

For functions f and g, the domains of $f + g, f - g$, and fg include all real numbers in the intersection of the domains of f and g, while the domain of f/g includes those real numbers in the intersection of the domains of f and g for which $g(x) \neq 0$.

COMPOSITION OF FUNCTIONS   The sketch in Figure 52 shows a function f that assigns to each element x of set X some element y of set Y. Suppose also that a function g takes each element of set Y and assigns a value z of set Z. Using both f and g, then, we can assign to any element x in X exactly one element z in Z. The result of this process is a new function h, that takes an element x in X and assigns to it an element z in Z. This function h is called the *composition* of functions g and f, written $g \circ f$, and is defined as follows.

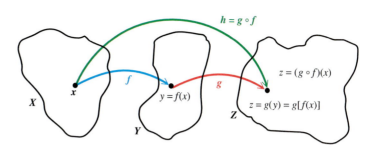

FIGURE 52

> ## Composition of Functions
>
> If f and g are functions, then the **composite function,** or **composition,** of g and f is
>
> $$(g \circ f)(x) = g[f(x)]$$
>
> for all x in the domain of f such that $f(x)$ is in the domain of g.

As a real-life example of function composition, suppose an oil well off the California coast is leaking, with the leak spreading oil in a circular layer over the surface. At any time, t, in minutes, after the beginning of the leak, the radius of the circular oil slick is $r(t) = 5t$ feet. Since $A(r) = \pi r^2$ gives the area of a circle of radius r, the area can be expressed as a function of time by substituting $5t$ for r in $A(r) = \pi r^2$ to get

$$A(r) = \pi r^2$$
$$A[r(t)] = \pi (5t)^2 = 25\pi t^2.$$

The function $A[r(t)]$ is a composite function of the functions A and r.

◀ EXAMPLE 3
Evaluating composite functions

Given $f(x) = 2x - 1$ and $g(x) = \dfrac{4}{x - 1}$, find each of the following.

(a) $f(g(2))$

First find $g(2)$. Since $g(x) = \dfrac{4}{x - 1}$,

$$g(2) = \frac{4}{2 - 1} = \frac{4}{1} = 4.$$

Now find $f(g(2)) = f(4)$:

$$f(x) = 2x - 1$$
$$f(g(2)) = f(4) = 2(4) - 1 = 7.$$

(b) $g(f(-3))$

$$f(-3) = 2(-3) - 1 = -7$$

$$g(f(-3)) = g(-7) = \frac{4}{-7 - 1} = \frac{4}{-8} = -\frac{1}{2} \quad ▶$$

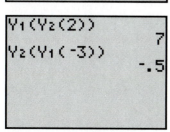

By letting $Y_1 = f(x)$ and $Y_2 = g(x)$ as defined in Example 3, this screen supports the results of parts (a) and (b).

EXAMPLE 4
Finding composite functions

Let $f(x) = 4x + 1$ and $g(x) = 2x^2 + 5x$. Find each of the following.

(a) $(g \circ f)(x)$

By definition, $(g \circ f)(x) = g[f(x)]$. Using the given functions,

$$
\begin{aligned}
(g \circ f)(x) = g[f(x)] &= g(4x + 1) & & f(x) = 4x + 1 \\
&= 2(4x + 1)^2 + 5(4x + 1) & & g(x) = 2x^2 + 5x \\
&= 2(16x^2 + 8x + 1) + 20x + 5 & & \text{Square } 4x + 1. \\
&= 32x^2 + 16x + 2 + 20x + 5 & & \text{Multiply.} \\
&= 32x^2 + 36x + 7. & & \text{Combine terms.}
\end{aligned}
$$

(b) $(f \circ g)(x)$

If we use the definition above with f and g interchanged, $(f \circ g)(x)$ becomes $f[g(x)]$, with

$$
\begin{aligned}
(f \circ g)(x) = f[g(x)] & \\
&= f(2x^2 + 5x) & & g(x) = 2x^2 + 5x \\
&= 4(2x^2 + 5x) + 1 & & f(x) = 4x + 1 \\
&= 8x^2 + 20x + 1. & & \text{Multiply.} \blacktriangleright
\end{aligned}
$$

As this example shows, it is not always true that $f \circ g = g \circ f$. In fact, two composite functions are equal only for a special class of functions, discussed in Section 5.1. In Example 4, the domain of both composite functions is the set of all real numbers.

CAUTION   In general, the composite function $f \circ g$ is not the same as the product fg. For example, with f and g defined as in Example 4,

$$(f \circ g)(x) = 8x^2 + 20x + 1$$

but $(fg)(x) = (4x + 1)(2x^2 + 5x) = 8x^3 + 22x^2 + 5x.$

EXAMPLE 5
Finding composite functions and their domains

Let $f(x) = 1/x$ and $g(x) = \sqrt{3 - x}$. Find $f \circ g$ and $g \circ f$. Give the domain of each.

First find $f \circ g$.

$$
\begin{aligned}
(f \circ g)(x) = f[g(x)] & \\
&= f(\sqrt{3 - x}) & & g(x) = \sqrt{3 - x} \\
&= \frac{1}{\sqrt{3 - x}} & & f(x) = \tfrac{1}{x}
\end{aligned}
$$

The radical $\sqrt{3 - x}$ is a nonzero real number only when $3 - x > 0$ or $x < 3$, so the domain of $f \circ g$ is the interval $(-\infty, 3)$.

Use the same functions to find $g \circ f$, as follows.

$$(g \circ f)(x) = g[f(x)]$$

$$= g\left(\frac{1}{x}\right) \qquad f(x) = \frac{1}{x}$$

$$= \sqrt{3 - \frac{1}{x}} \qquad g(x) = \sqrt{3 - x}$$

$$= \sqrt{\frac{3}{1} \cdot \frac{x}{x} - \frac{1}{x}} \qquad \text{Get a common denominator.}$$

$$= \sqrt{\frac{3x - 1}{x}} \qquad \text{Combine terms.}$$

The domain of $g \circ f$ is the set of all real numbers x such that $x \neq 0$ and $3 - f(x) \geq 0$. As shown above,

$$3 - f(x) = \frac{3x - 1}{x}.$$

We need to solve the inequality

$$\frac{3x - 1}{x} \geq 0.$$

By the methods of Section 2.7, first find values of x that make the numerator or denominator zero. The required numbers are 0 and 1/3. Then use a sign graph to verify that the domain of $g \circ f$ is the set $(-\infty, 0) \cup [1/3, \infty)$. ▶

◀**EXAMPLE 6**
Finding composite functions
and their domains

Given $f(x) = \sqrt{x - 2}$ and $g(x) = x^2 + 2$, find $f \circ g$ and $g \circ f$ and their domains.

$$(f \circ g)(x) = \sqrt{(x^2 + 2) - 2} \qquad \text{Substitute } g(x) \text{ for } x \text{ in } f(x).$$
$$= \sqrt{x^2} \qquad \text{Combine terms.}$$
$$= |x| \qquad \sqrt{x^2} \text{ is the principal square root.}$$
$$(g \circ f)(x) = (\sqrt{x - 2})^2 + 2 \qquad \text{Substitute } f(x) \text{ for } x \text{ in } g(x).$$
$$= (x - 2) + 2 \qquad x - 2 \text{ must be } \geq 0.$$

The domain of g is $(-\infty, \infty)$ and, since $g(x) \geq 2$ for all x, the domain of $f \circ g$ is $(-\infty, \infty)$. The domain of f is $[2, \infty)$ and $f(x) \geq 0$ for x in $[2, \infty)$. Thus, the domain of $g \circ f$ is $[2, \infty)$, and

$$(g \circ f)(x) = x - 2 + 2 = x. \text{ ▶}$$

NOTE Comparing the domains of $f \circ g$ and $g \circ f$ in Examples 5 and 6 shows that we cannot always look just at the equation of a composite function to determine the domain.

In calculus it is sometimes useful to treat a function as a composition of two functions. The next example shows how this can be done.

◀ EXAMPLE 7
Finding functions that form a given composite

Suppose $h(x) = \sqrt{2x + 3}$. Find functions f and g, so that $(f \circ g)(x) = h(x)$.

Since there is a quantity, $2x + 3$, under a radical, one possibility is to choose $f(x) = \sqrt{x}$ and $g(x) = 2x + 3$. Then $(f \circ g)(x) = \sqrt{2x + 3}$, as required. Other combinations are possible. For example, we could choose $f(x) = \sqrt{x + 3}$ and $g(x) = 2x.$ ▶

CONNECTIONS Earlier in this book, we introduced several basic functions: $f(x) = x$, $f(x) = x^2$, $f(x) = \sqrt{x}$, $f(x) = 1/x$, $f(x) = |x|$, and so on. In this section we have shown how these basic functions can be combined using the operations on functions, and formed into composite functions. With these processes we can build all the algebraic functions from just a few basic ones.

Without realizing it, we have been using composite functions. In Chapter 1, we used substitution to factor expressions like $2(3m + 1)^2 - (3m + 1) + 1$, by substituting $x = 3m + 1$ before factoring. If we choose $f(x) = 2x^2 - x + 1$ and $g(m) = 3m + 1$, then $f(g(m)) = 2(3m + 1)^2 - (3m + 1) + 1$. We also used substitution to rewrite an expression such as

$$5 + \frac{7}{p^2 - 4} - \frac{6}{(p^2 - 4)^2}$$

as

$$5 + \frac{7}{x} - \frac{6}{x^2}.$$

Here, we choose

$$f(x) = 5 + \frac{7}{x} - \frac{6}{x^2} \quad \text{and} \quad g(p) = p^2 - 4$$

to get the composite function

$$f(g(p)) = 5 + \frac{7}{p^2 - 4} - \frac{6}{(p^2 - 4)^2}.$$

FOR DISCUSSION OR WRITING
Give at least one other example where we have used composite functions in this book before this section. (*Hint:* There is an example in Section 3.2.)

3.7 Exercises ▼▼▼▼▼▼▼▼▼▼▼▼▼▼▼▼▼▼▼▼▼▼▼▼▼▼▼▼▼▼▼▼

Decide whether the statement is true or false.

1. If $f(x) = x$ and $g(x) = x^2$, then $(f + g)(2) = 6$.

2. If $f(x) = x + 1$ and $g(x) = x + 2$, then $(f + g)(x) = x + 3$.

3. If $f(x) = x$ and $g(x) = \dfrac{1}{x}$, then $(fg)(x) = 1$ $(x \neq 0)$.

4. If $f(x) = x$ and $g(x) = 2$, then $(f \circ g)(x) = 2$.

For the pair of functions defined, find $f + g$, $f - g$, fg, and f/g. Give the domain of each. See Example 2.

5. $f(x) = 3x + 4$, $g(x) = 2x - 5$ **6.** $f(x) = 6 - 3x$, $g(x) = -4x + 1$

7. $f(x) = 2x^2 - 3x$, $g(x) = x^2 - x + 3$ **8.** $f(x) = 4x^2 + 2x - 3$, $g(x) = x^2 - 3x + 2$

9. $f(x) = \sqrt{4x - 1}$, $g(x) = \sqrt{x + 3}$ **10.** $f(x) = \sqrt{5x - 4}$, $g(x) = \sqrt{3x - 1}$

Let $f(x) = 5x^2 - 2x$ and let $g(x) = 6x + 4$. Find each of the following. See Examples 1 and 3.

11. $(f + g)(3)$ **12.** $(f - g)(-5)$ **13.** $(fg)(4)$ **14.** $(fg)(-3)$

15. $\left(\dfrac{f}{g}\right)(-1)$ **16.** $\left(\dfrac{f}{g}\right)(4)$ **17.** $(f - g)(m)$ **18.** $(f + g)(2k)$

19. $(f \circ g)(2)$ **20.** $(f \circ g)(-5)$ **21.** $(g \circ f)(2)$ **22.** $(g \circ f)(-5)$

The graphs of functions f and g are shown. Use these graphs to find the value.

23. $f(1) + g(1)$ **24.** $f(4) - g(3)$

25. $f(-2) \cdot g(4)$ **26.** $\dfrac{f(4)}{g(2)}$

27. $(f \circ g)(2)$ **28.** $(g \circ f)(2)$

29. $(g \circ f)(-4)$ **30.** $(f \circ g)(-2)$

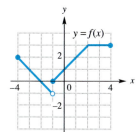 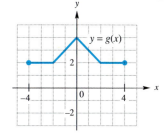

The tables below give some selected ordered pairs for functions f and g.

x	3	4	6
$f(x)$	1	3	9

x	2	7	1	9
$g(x)$	3	6	9	12

Find each of the following.

31. $(f \circ g)(2)$ **32.** $(f \circ g)(7)$ **33.** $(g \circ f)(3)$

34. $(g \circ f)(6)$ **35.** $(f \circ f)(4)$ **36.** $(g \circ g)(1)$

37. Why can you not determine $(f \circ g)(1)$ given the information in the tables for Exercises 31–36?

38. Extend the concept of composition of functions to evaluate $(g \circ (f \circ g))(7)$ using the tables for Exercises 31–36.

Find $f \circ g$ and $g \circ f$ for each pair of functions. See Examples 4–6.

39. $f(x) = -6x + 9$, $g(x) = 5x + 7$

40. $f(x) = 8x + 12$, $g(x) = 3x - 1$

41. $f(x) = 4x^2 + 2x + 8$, $g(x) = x + 5$

42. $f(x) = 5x + 3$, $g(x) = -x^2 + 4x + 3$

43. $f(x) = \dfrac{2}{x^4}$, $g(x) = 2 - x$

44. $f(x) = \dfrac{1}{x}$, $g(x) = x^2$

45. $f(x) = 9x^2 - 11x$, $g(x) = 2\sqrt{x + 2}$

46. $f(x) = \sqrt{x + 2}$, $g(x) = 8x^2 - 6$

For each pair of functions defined below, show that $(f \circ g)(x) = x$ and $(g \circ f)(x) = x$.

47. $f(x) = 2x^3 - 1$, $g(x) = \sqrt[3]{(x + 1)/2}$

48. $f(x) = x^3 + 4$, $g(x) = \sqrt[3]{x - 4}$

49. Describe the steps required to find the composite function $f \circ g$, given $f(x) = 2x - 5$ and $g(x) = x^2 + 3$.

50. Composition is an operation that is unique to functions. Is composition of functions commutative? That is, does $f \circ g = g \circ f$ for all functions f and g? Explain.

51. The graphs of functions f and g are shown. Draw the graph of $f \circ g$.

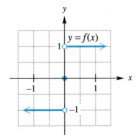 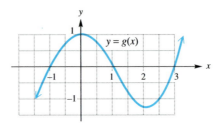

52. Suppose $g(x) = x - 5$.
   (a) For any function f, the graph of $f \circ g$ is a translation of the graph of f. Describe the translation.
   (b) For any function f, the graph of $g \circ f$ is a translation of the graph of f. Describe the translation.

In Chapter 5 we will investigate inverse functions. Two functions f and g are inverses provided certain conditions are met. One of these conditions is

$$(f \circ g)(x) = x \qquad and \qquad (g \circ f)(x) = x.$$

Show that this condition is true for the functions f and g as defined.

53. $f(x) = 2x + 3$, $g(x) = \dfrac{x - 3}{2}$

54. $f(x) = \dfrac{x + 5}{3}$, $g(x) = 3x - 5$

55. $f(x) = \sqrt[3]{\dfrac{x + 1}{2}}$, $g(x) = 2x^3 - 1$

56. $f(x) = \sqrt[3]{x - 4}$, $g(x) = x^3 + 4$

Find functions f and g such that $(f \circ g)(x) = h(x)$. (There are many possible ways to do this.) See Example 7.

57. $h(x) = (6x - 2)^2$

58. $h(x) = (11x^2 + 12x)^2$

59. $h(x) = \sqrt{x^2 - 1}$

60. $h(x) = (2x - 3)^3$

61. $h(x) = \sqrt{6x + 12}$

62. $h(x) = \sqrt[3]{2x + 3} - 4$

63. The graphing calculator screen on the left shows three functions, Y_1, Y_2, and Y_3. The last of these, Y_3, is defined as $Y_1 \circ Y_2$, indicated by the notation $Y_3 = Y_1(Y_2)$. The table on the right shows selected values of X, along with the calculated values of Y_3. Verify by direct calculation that the result is true for the given value of X.

```
Y1=2X-5
Y2=X²
Y3 ⊟Y1(Y2)
Y4=
Y5=
Y6=
Y7=
Y8=
```

X	Y3
0	-5
1	-3
2	3
3	13
4	27
5	45
6	67

X=0

(a) X = 0   **(b)** X = 1   **(c)** X = 2   **(d)** X = 3

64. Use a square window of a graphing calculator to graph $Y_1 = 2x + 5$ and $Y_2 = .5(x - 5)$. Then graph $Y_3 = x$ in the same window. What do you notice?

Solve the problem.

65. The function $f(x) = 12x$ computes the number of inches in x feet and the function $g(x) = 5280x$ computes the number of feet in x miles. What does $(f \circ g)(x)$ compute?

66. The perimeter x of a square with side of length s is given by the formula $x = 4s$.

(a) Solve for s in terms of x.
(b) If y represents the area of this square, write y as a function of the perimeter x.
(c) Use the composite function of part (b) to analytically find the area of a square with perimeter 6.

67. The area of an equilateral triangle with sides of length x is given by the function $A(x) = \dfrac{\sqrt{3}}{4}x^2$.

(a) Find $A(2x)$, the function representing the area of an equilateral triangle with sides of length twice the original length.
(b) Find analytically the area of an equilateral triangle with side length 16. Use the formula for $A(2x)$ found in part (a).

68. A textbook author invests his royalties in two accounts for 1 year.
(a) The first account pays 4% simple interest. If he invests x dollars in this account, write an expression for y_1 in terms of x, where y_1 represents the amount of interest earned.
(b) He invests in a second account $500 more than he invested in the first account. This second account pays 2.5% simple interest. Write an expression for y_2, where y_2 represents the amount of interest earned.
(c) What does $y_1 + y_2$ represent?
(d) How much interest will he receive if $250 is invested in the first account?

69. An oil well off the Gulf Coast is leaking, with the leak spreading oil over the surface as a circle. At any time t, in minutes, after the beginning of the leak, the radius of the circular oil slick on the surface is $r(t) = 4t$ ft. Let $A(r) = \pi r^2$ represent the area of a circle of radius r.

(a) Find $(A \circ r)(t)$.
(b) Interpret $(A \circ r)(t)$.
(c) What is the area of the oil slick after 3 minutes?

70. When a thermal inversion layer is over a city, pollutants cannot rise vertically but are trapped below the layer and must disperse horizontally. Assume that a factory smokestack begins emitting a pollutant at 8 A.M. Assume that the pollutant disperses horizontally over a circular area. If t represents the time, in hours, since the factory began emitting pollutants ($t = 0$ represents 8 A.M.), assume that the radius of the circle of pollution is $r(t) = 2t$ mi. Let $A(r) = \pi r^2$ represent the area of a circle of radius r.

(a) Find $(A \circ r)(t)$.

(b) Interpret $(A \circ r)(t)$.

(c) What is the area of the circular region covered by the layer at noon?

71. A couple planning their wedding has found that the cost to hire a caterer for the reception depends on the number of guests attending. If 100 people attend, the cost per person will be $2. For each person less than 100, the cost will increase by $.20. Assume that no more than 100 people will attend. Let x represent the number less than 100 who do not attend. For example, if 95 attend, $x = 5$.

(a) Write a function $N(x)$ for the possible number of guests.

(b) Write a function $G(x)$ for the possible cost per guest.

(c) Write the function $N(x) \cdot G(x)$ for the total cost, $C(x)$.

(d) What is the total cost if 40 people attend?

72. The area of a square is x^2 square inches. If 3 inches is added to one dimension and 1 inch is subtracted from the other dimension, express the area $A(x)$ of the resulting rectangle as a product of two functions.

Chapter 3 Summary ▼▼▼▼▼▼▼▼▼▼▼▼▼▼▼▼▼▼▼▼▼▼▼▼▼▼▼▼▼▼▼▼

KEY TERMS		KEY IDEAS
3.1 Relations and the Rectangular Coordinate System		
ordered pair	coordinate plane	**Distance Formula**
relation	(xy-plane)	Suppose $P(x_1, y_1)$ and $R(x_2, y_2)$ are two points in a coordinate plane.
domain	quadrants	Then the distance between P and R, written $d(P, R)$, is
range	coordinates	$$d(P, R) = \sqrt{(x_2 - x_1)^2 + (y_2 - y_1)^2}.$$
origin	graph	
x-axis	collinear	**Midpoint Formula**
y-axis	circle	The midpoint of the line segment with endpoints (x_1, y_1) and (x_2, y_2) is
rectangular	radius	$$\left(\frac{x_1 + x_2}{2}, \frac{y_1 + y_2}{2}\right).$$
(Cartesian)	center of a circle	
coordinate		**Center-Radius Form of the Equation of a Circle**
system		$$(x - h)^2 + (y - k)^2 = r^2$$
		General Form of the Equation of a Circle
		$$x^2 + y^2 + cx + dy + e = 0$$
3.2 Functions		
function	$f(x)$ notation	A function is a relation in which for each element in the domain there corresponds exactly one element in the range.
independent	increasing	
variable	decreasing	**Vertical Line Test**
dependent	constant	If each vertical line intersects a graph at no more than one point, the graph is the graph of a function.
variable		
function		
identity function		

KEY TERMS	KEY IDEAS

3.3 Linear Functions

linear function x-intercept y-intercept slope change in x change in y	**Definition of Slope** The slope m of the line through the points (x_1, y_1) and (x_2, y_2) is $$m = \frac{\Delta y}{\Delta x} = \frac{y_2 - y_1}{x_2 - x_1},$$ where $\Delta x \neq 0$.

3.4 Equations of a Line

point-slope form
slope-intercept form

Linear Equations

General Equation	Type of Equation
$Ax + By = C$	*Standard form* (if $A \neq 0$ and $B \neq 0$), x-intercept C/A, y-intercept C/B, slope $-A/B$
$x = k$	*Vertical line*, x-intercept k, no y-intercept, undefined slope
$y = k$	*Horizontal line*, y-intercept k, no x-intercept, slope 0
$y = mx + b$	*Slope-intercept form*, y-intercept b, slope m
$y - y_1 = m(x - x_1)$	*Point-slope form*, slope m, through (x_1, y_1)

3.5 Graphing Relations and Functions

graph of a relation piecewise-defined function step function	**Basic Functions** **Absolute Value Function $y = \lvert x \rvert$** **Square Root Function $y = \sqrt{x}$** **Greatest-Integer Function $y = [\![x]\!]$**

3.6 General Graphing Techniques

Stretching and Shrinking
The graph of $g(x) = a \cdot f(x)$ has the same shape as the graph of $f(x)$, and it is

narrower if $\lvert a \rvert > 1$;
broader if $0 < \lvert a \rvert < 1$.

Reflection About an Axis
The graph of $y = -f(x)$ is the same as the graph of $y = f(x)$ reflected about the x-axis.

The graph of $y = f(-x)$ is the same as the graph of $y = f(x)$ reflected about the y-axis.

KEY TERMS	KEY IDEAS
even function odd function	**Symmetry** The graph of an equation is **symmetric with respect to the y-axis** if the replacement of x with $-x$ results in an equivalent equation. The graph of an equation is **symmetric with respect to the x-axis** if the replacement of y with $-y$ results in an equivalent equation. The graph of an equation is **symmetric with respect to the origin** if the replacement of both x with $-x$ and y with $-y$ results in an equivalent equation. **Translations** Let f be a function and c be a positive number. | To graph:	Shift the graph of $y = f(x)$ by c units:
$y = f(x) + c$	upward
$y = f(x) - c$	downward
$y = f(x + c)$	left
$y = f(x - c)$	right	

3.7 Algebra of Functions; Composite Functions

Operations on Functions

Given two functions f and g, then for all values of x for which both $f(x)$ and $g(x)$ are defined, the following operations are defined.

Sum $\quad (f + g)(x) = f(x) + g(x)$

Difference $\quad (f - g)(x) = f(x) - g(x)$

Product $\quad (fg)(x) = f(x) \cdot g(x)$

Quotient $\quad \left(\dfrac{f}{g}\right)(x) = \dfrac{f(x)}{g(x)}, \quad g(x) \neq 0$

Composition of Functions

If f and g are functions, then the composite function, or composition, of g and f is

$$(g \circ f)(x) = g[f(x)]$$

for all x in the domain of f such that $f(x)$ is in the domain of g.

Chapter 3   Review Exercises ▼▼▼▼▼▼▼▼▼▼▼▼▼▼▼▼▼▼▼▼▼▼▼▼▼▼▼

Give the domain and the range of each relation.

1. $\{(-3, 6), (-1, 4), (8, 5)\}$

2. $y = \sqrt{-x}$

Find the distance between the pair of points, and state the coordinates of the midpoint of the segment joining them.

3. $P(3, -1)$, $Q(-4, 5)$

4. $M(-8, 2)$, $N(3, -7)$

5. $A(-6, 3)$, $B(-6, 8)$

6. Are the points $(5, 7)$, $(3, 9)$, $(6, 8)$ the vertices of a right triangle?

7. Find all possible values of k so that $(-1, 2)$, $(-10, 5)$, and $(-4, k)$ are the vertices of a right triangle.

8. Use the distance formula to determine whether the points $(-2, -5)$, $(1, 7)$, and $(3, 15)$ are collinear.

Find an equation for the circle satisfying the given conditions.

9. Center $(-2, 3)$, radius 15

10. Center $(\sqrt{5}, -\sqrt{7})$, radius $\sqrt{3}$

11. Center $(-8, 1)$, passing through $(0, 16)$

12. Center $(3, -6)$, tangent to the x-axis

Find the center and radius of the circle.

13. $x^2 - 4x + y^2 + 6y + 12 = 0$

14. $x^2 - 6x + y^2 - 10y + 30 = 0$

15. $2x^2 + 14x + 2y^2 + 6y + 2 = 0$

16. $3x^2 + 33x + 3y^2 - 15y = 0$

17. Find all possible values of x so that the distance between $(x, -9)$ and $(3, -5)$ is 6.

18. Find all points (x, y) with $x = 6$ so that (x, y) is 4 units from $(1, 3)$.

19. Find all points (x, y) with $x + y = 0$ so that (x, y) is 6 units from $(-2, 3)$.

20. Describe the graph of $(x - 4)^2 + (y + 5)^2 = 0$.

Decide whether the curve is the graph of a function of x. Give the domain and range of each relation.

21.
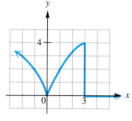

22.
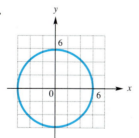

23.
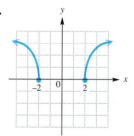

24.
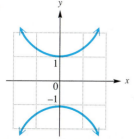

25.
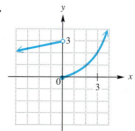

26.
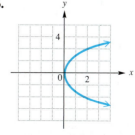

Determine whether the equation defines y as a function of x.

27. $x = \dfrac{1}{2}y^2$ **28.** $y = 3 - x^2$ **29.** $y = \dfrac{-8}{x}$ **30.** $y = \sqrt{x - 7}$

Give the domain of the function defined.

31. $y = -4 + |x|$ **32.** $y = \dfrac{8 + x}{8 - x}$ **33.** $y = -\sqrt{\dfrac{5}{x^2 + 9}}$ **34.** $y = \sqrt{49 - x^2}$

35. For the function graphed in Exercise 23, give the interval over which it is **(a)** increasing and **(b)** decreasing.

36. The screen shows the graph of $x = y^2 - 4$. Give the two functions Y_1 and Y_2 that must be used to graph this relation if the calculator is in function mode.

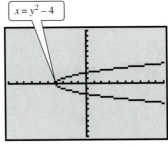

$x = y^2 - 4$

[−10, 10] by [−10, 10]

For the given function, find and simplify $\dfrac{f(x + h) - f(x)}{h}$.

37. $f(x) = 2x + 9$

38. $f(x) = x^2 - 5x + 3$

39. The figure shows average prices for domestic crude oil from 1980 to mid-1992.
  (a) Is this the graph of a function?
  (b) In what year were oil prices lowest? Highest?
  (c) What was the lowest price? The highest price?
  (d) What is the general trend of prices over the given period?
  (e) What do the horizontal portions of the graph indicate?

40. The figure shows the number of jobs gained or lost in the Sacramento area in a recent period from September to May.
  (a) Is this the graph of a function?
  (b) In what month were the most jobs lost? The most gained?
  (c) What was the largest number of lost jobs? The most gained?
  (d) Do these data show an upward or downward trend? If so, which is it?

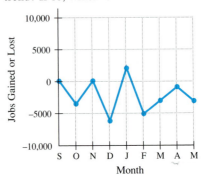

Find the slope for the line provided that it has a slope.

41. Through $(8, 7)$ and $(1/2, -2)$ **42.** Through $(2, -2)$ and $(3, -4)$ **43.** Through $(5, 6)$ and $(5, -2)$
44. Through $(0, -7)$ and $(3, -7)$ **45.** $9x - 4y = 2$ **46.** $11x + 2y = 3$
47. $x - 5y = 0$ **48.** $x - 2 = 0$

49.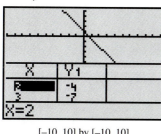

[−10, 10] by [−10, 10]
Xscl = 1   Yscl = 2

50.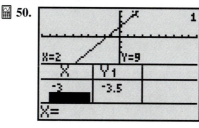

[−10, 10] by [−10, 10]
Xscl = 1   Yscl = 2

Graph each of the following. Give the domain and range.

51. $3x + 7y = 14$ **52.** $2x - 5y = 5$ **53.** $3y = x$
54. $f(x) = 3$ **55.** $x = -5$ **56.** $f(x) = x$

For the line described, write the equation in standard form.

57. Through $(-2, 4)$ and $(1, 3)$ **58.** Through $(3, -5)$ with slope -2
59. x-intercept -3, y-intercept 5 **60.** Through $(2, -1)$, parallel to $3x - y = 1$
61. Through $(0, 5)$, perpendicular to $8x + 5y = 3$ **62.** Through $(2, -10)$, perpendicular to a line with undefined slope
63. Through $(3, -5)$, parallel to $y = 4$
64. Through $(-7, 4)$, perpendicular to $y = 8$

Graph the line satisfying the given conditions.

65. Through $(2, -4)$, $m = 3/4$ **66.** Through $(0, 5)$, $m = -2/3$

The following table estimates the growth in airline passengers (in millions) at some of the fastest-growing airports in the United States between 1992 and 2005.

Airport	1992	2005
Harrisburg Intl.	.7	1.4
Dayton Intl.	1.1	2.4
Austin Robert Mueller	2.2	4.7
Milwaukee Gen. Mitchell Intl.	2.2	4.4
Sacramento Metropolitan	2.6	5.0
Fort Lauderdale-Hollywood	4.1	8.1
Washington Dulles Intl.	5.3	10.9
Greater Cincinnati Airport	5.8	12.3

(*Source:* FAA.)

67. Determine a linear function $y = f(x)$ that approximates the data using the two points $(.7, 1.4)$ and $(5.3, 10.9)$.

68. How does the slope of the graph of f relate to growth in airline passengers at these airports?

69. 4.9 million passengers used Raleigh-Durham International Airport in 1992. Approximate the number of passengers using this airport in 2005 and compare it with the Federal Aviation Administration's estimation of 10.3 million passengers.

70. Lynn Mooney, while on vacation in Canada, found that the value (or purchasing power) of her U.S. dollars had increased by 16%. On her return, she expected a 16% decrease when converting her Canadian money back into U.S. dollars. Write an equation for each of these conversion functions.

Graph each function.

71. $f(x) = -|x|$

72. $f(x) = |x| - 3$

73. $f(x) = -|x| - 2$

74. $f(x) = -|x + 1| + 3$

75. $f(x) = 2|x - 3| - 4$

76. $f(x) = [\![x - 3]\!]$

77. $f(x) = \left[\!\!\left[\dfrac{1}{2}x - 2\right]\!\!\right]$

78. $f(x) = \begin{cases} -4x + 2 & \text{if } x \le 1 \\ 3x - 5 & \text{if } x > 1 \end{cases}$

79. $f(x) = \begin{cases} 3x + 1 & \text{if } x < 2 \\ -x + 4 & \text{if } x \ge 2 \end{cases}$

80. $f(x) = \begin{cases} |x| & \text{if } x < 3 \\ 6 - x & \text{if } x \ge 3 \end{cases}$

The percentages of babies delivered by cesarean birth are listed in the table.

Year	Percentage
1970	5%
1975	11%
1980	17%
1985	23%
1990	23%

(*Source:* Teutsch S., R. Churchill, *Principles and Practice of Public Health Surveillance,* Oxford University Press, New York, 1994.)

81. Determine a piecewise linear function defined by $f(x)$ that models these data where $x = 0$ corresponds to 1970.

82. Use f to estimate the percentage of cesarean deliveries in 1973 and 1982.

Decide whether the equation has a graph that is symmetric with respect to the x-axis, the y-axis, the origin, or none of these.

83. $3y^2 - 5x^2 = 15$

84. $x + y^2 = 8$

85. $y^3 = x + 1$

86. $x^2 = y^3$

87. $|y| = -x$

88. $|x + 2| = |y - 3|$

89. $|x| = |y|$

Describe how the graph of the function can be obtained from the graph of $f(x) = |x|$.

90. $g(x) = -|x|$

91. $h(x) = |x| - 2$

92. $k(x) = 2|x - 4|$

Let $f(x) = 3x - 4$. Find an equation for each reflection of the graph of $f(x)$.

93. About the x-axis

94. About the y-axis

95. About the origin

96. The graph of a function f is shown in the figure. Sketch the graph of each function defined as follows.

(a) $y = f(x) + 3$

(b) $y = f(x - 2)$

(c) $y = f(x + 3) - 2$

(d) $y = |f(x)|$

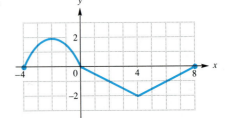

Let $f(x) = 3x^2 - 4$ *and* $g(x) = x^2 - 3x - 4$. *Find each of the following.*

97. $(f + g)(x)$ **98.** $(fg)(x)$ **99.** $(f - g)(4)$

100. $(f + g)(-4)$ **101.** $(f + g)(2k)$ **102.** $\left(\dfrac{f}{g}\right)(3)$

103. $\left(\dfrac{f}{g}\right)(-1)$ **104.** Give the domain of $(fg)(x)$.   **105.** Give the domain of $\left(\dfrac{f}{g}\right)(x)$.

106. Which of the following is *not* equal to $(f \circ g)(x)$ for $f(x) = 1/x$ and $g(x) = x^2 + 1$? (*Hint:* There may be more than one.)

     (a) $f[g(x)]$ **(b)** $\dfrac{1}{x^2 + 1}$ **(c)** $\dfrac{1}{x^2}$ **(d)** $(g \circ f)(x)$

Let $f(x) = \sqrt{x - 2}$ *and* $g(x) = x^2$. *Find each of the following.*

107. $(f \circ g)(x)$ **108.** $(g \circ f)(x)$ **109.** $(f \circ g)(-6)$ **110.** $(g \circ f)(3)$

The graphs of two functions f and g are shown in the figures here.

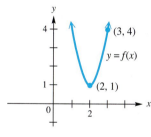 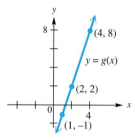

111. Find $(f \circ g)(2)$.   **112.** Find $(g \circ f)(3)$.

Solve the problem.

113. The population P of a certain mammal depends on the number x (in hundreds) of a smaller mammal that serves as its primary food supply. The number x (in hundreds) of the smaller mammal depends upon the amount (in appropriate units) of its food supply, a type of plant. Suppose $P(x) = 2x^2 + 1$ and $x = f(a) = 3a + 2$. Find $(P \circ f)(a)$, the relationship between the population P of the larger mammal and the amount a of plants available to serve as food for the smaller mammal.

114. The formula for the volume of a sphere is $V(r) = \dfrac{4}{3}\pi r^3$, where r represents the radius of the sphere. Construct a model representing the amount of volume gained when a sphere of radius r inches is increased by 3 inches.

115. Suppose the length of a rectangle is twice its width. Let x represent the width of the rectangle.

Write a formula for the perimeter P of the rectangle in terms of x alone. Then use $P(x)$ notation to describe it as a function. What type of function is this?

116. Cylindrical cans make the most efficient use of materials when their height is the same as the diameter of their top.

    (a) Express the volume V of such a can as a function of the diameter d of its top.

    (b) Express the surface area S of such a can as a function of the diameter d of its top. (*Hint:* The curved side is made from a rectangle whose length is the circumference of the top of the can.)

Chapter 3 Test ▼▼▼▼▼▼▼▼▼▼▼▼▼▼▼▼▼▼▼▼▼▼▼▼▼▼▼▼▼

The graph shows the line that passes through the points
$(-2, 1)$ *and* $(3, 4)$. *Refer to it to answer test items 1–6.*

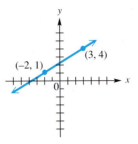

1. Without actually computing the slope, tell whether the slope is positive or negative based on a visual observation of the graph.

2. What is the slope of the line?

3. What is the distance between the two points shown?

4. What are the coordinates of the midpoint of the *segment* joining the two points?

5. Find the standard form of the equation of the line.

6. Write the linear function $f(x) = ax + b$ that has this line as its graph.

7. Suppose point A has coordinates $(5, -3)$.
   (a) What is the equation of the vertical line through A?
   (b) What is the equation of the horizontal line through A?

8. Find the slope-intercept form of the equation of the line passing through $(2, 3)$
   (a) parallel to the graph of $y = -3x + 2$ (b) perpendicular to the graph of $y = -3x + 2$.

9. The calculator-generated table shows several points that lie on the graph of a linear function. Find the equation that defines this function.

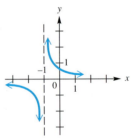

Tell whether the graph is that of a function. Give the domain and the range. If it is a function, give the intervals where it is increasing, decreasing, or constant.

10.

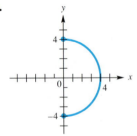

11.

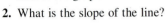

Graph the relation.

12. $y = |x - 2| - 1$

13. $f(x) = [\![x + 1]\!]$

14. $f(x) = \begin{cases} 3 & \text{if } x < -2 \\ 2 - \dfrac{1}{2}x & \text{if } x \geq -2 \end{cases}$

15. Explain how the graph of $y = -2\sqrt{x + 2} - 3$ can be obtained from the graph of $y = \sqrt{x}$.

16. Determine whether the graph of $3x^2 - 2y^2 = 3$ is symmetric with respect to
    (a) the *x*-axis,   **(b)** the *y*-axis,   **(c)** the origin.

Given $f(x) = 2x^2 - 3x + 2$ *and* $g(x) = -2x + 1$, *find the following.*

17. $f(-3)$

18. $\dfrac{f(x + h) - f(x)}{h}$

19. $(f \circ g)(x)$

Solve the problem.

20. The graph shows the total federal debt from 1982 to 1994 (in billions of dollars).
    (a) Find a linear function that models these data, letting $x = 0$ represent 1982 and $x = 12$ represent 1994.
    (b) Based on your model, what was the debt in 1990? How does this compare to the actual debt of $3190 billion?

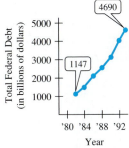

Source: Office of Management and Budget.

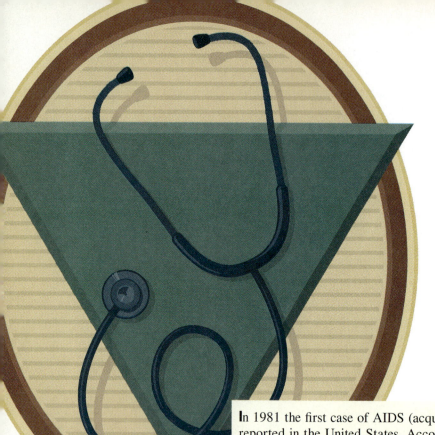

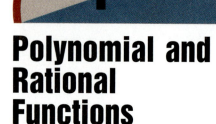

4

Polynomial and Rational Functions

In 1981 the first case of AIDS (acquired immune deficiency syndrome) was reported in the United States. According to the Centers of Disease Control and Prevention over 360,000 individuals have been diagnosed with AIDS and of them over 220,000 have died. AIDS is one of the most devastating diseases of our time. The World Health Organization estimates that over 17 million people have been infected with HIV (human immune deficiency virus) and by the year 2000 this number will increase to 30–40 million.

The emergence of new diseases and drug-resistant strains of old ones have eliminated modern medicine's hope of eradicating infectious diseases. AIDS, toxic shock, Lyme disease, and Legionnaires' disease were unknown 30 years ago. In order to understand how diseases spread, mathematicians and scientists analyze data that have been reported to health officials. This information can be used to create a mathematical model. Mathematical models help officials forecast future needs for health care and determine risk factors for different populations of people. The following table lists the total (cumulative) number of AIDS cases diagnosed in the United States through 1993.

Sources: Teutsch, S. and R. Churchill, *Principles and Practice of Public Health Surveillance,* Oxford University Press, New York, 1994.

U.S. Dept. of Health and Human Services, Centers for Disease Control and Prevention, *HIV/AIDS Surveillance,* March 1994.

World Health Organization, Global Programme on AIDS, *The Current Situation of HIV/AIDS Pandemic,* July 1994.

Wright, J. (editor), *The Universal Almanac,* Universal Press Syndicate Company, 1994.

Year	AIDS Cases
1982	1563
1983	4647
1984	10,845
1985	22,620
1986	41,662
1987	70,222
1988	105,489
1989	147,170
1990	193,245
1991	248,023
1992	315,329
1993	361,509

How can the number of new AIDS cases in the year 2000 be predicted? This question is answered in Section 4.1. First, mathematicians and scientists must create a model. A model not only explains present data but also makes predictions about future phenomena. Polynomial and rational functions are the most common functions used to model data. Using these functions and their graphs, predictions regarding future trends can be made. In this chapter you will see how mathematics can be used to model the number of new cases of AIDS, cancer, rabies, and coronary heart disease.

Functions such as $f(x) = 5x - 1$, $f(x) = x^4 + \sqrt{2}x^3 - 4x^2$, and $f(x) = 4x^2 - x + 2$ are examples of *polynomial functions*. Polynomial functions are the simplest type of function, because a polynomial involves only the operations of addition, subtraction, and multiplication. In calculus it is shown how polynomial functions can be used to approximate more complicated functions.

Polynomial Function

A **polynomial function of degree n,** where n is a nonnegative integer, is a function defined by an expression of the form

$$f(x) = a_n x^n + a_{n-1}x^{n-1} + \ldots + a_1 x + a_0,$$

where $a_n, a_{n-1}, \ldots, a_1$, and a_0 are real numbers, with $a_n \neq 0$.

For the polynomial $f(x) = 2x^3 - \frac{1}{2}x + 5$, n is 3 and the polynomial has the form $a_3 x^3 + a_2 x^2 + a_1 x + a_0$, where a_3 is 2, a_2 is 0, a_1 is $-1/2$, and a_0 is 5. The polynomial functions defined by $f(x) = x^4 + \sqrt{2}x^3 - 4x^2$ and $f(x) = 4x^2 - x + 2$ have degrees 4 and 2, respectively. The number a_n is the **leading coefficient** of $f(x)$. The function defined by $f(x) = 0$ is called the **zero polynomial.** The zero polynomial has no degree. However, a polynomial $f(x) = a_0$ for a nonzero number a_0 has degree 0.

In this chapter we discuss the graphs of polynomial functions of degree 2 or higher and methods of finding, or at least approximating, the values of x that satisfy $f(x) = 0$, called the **zeros** of $f(x)$. The chapter ends with a section on *rational functions,* which are defined as quotients of polynomials.

4.1 Quadratic Functions ▼▼▼

In Sections 3.3 and 3.4, we discussed first-degree (*linear*) polynomial functions, where the highest power of the variable is 1. In this section we look at polynomial functions of degree 2, called *quadratic functions.*

Quadratic Function

A function f is a **quadratic function** if

$$f(x) = ax^2 + bx + c,$$

where a, b, and c are real numbers, with $a \neq 0$.

The simplest quadratic function is given by $f(x) = x^2$ with $a = 1$, $b = 0$, and $c = 0$. To find some points on the graph of this function, choose some values for x and find the corresponding values for $f(x)$, as in the chart with Figure 1. Then plot these points, and draw a smooth curve through them. This graph is called a **parabola.** Every quadratic function has a graph that is a parabola.

x	$f(x)$
-2	4
-1	1
0	0
1	1
2	4

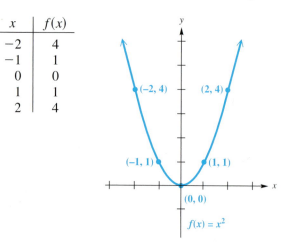

FIGURE 1

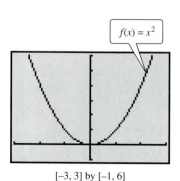

$[-3, 3]$ by $[-1, 6]$

A calculator graph of $f(x) = x^2$

$y = 3x^2 - 6x - 1$

NOTE The domain and the range of a parabola with a vertical axis, such as the one in Figure 1, can be determined by looking at the graph. Since the graph extends indefinitely to the right and to the left, we see that the domain is $(-\infty, \infty)$. Since the lowest point on the graph is $(0, 0)$, the minimum range value (y-value) is 0. The graph extends upward indefinitely, indicating that there is no maximum y-value, and so the range is $[0, \infty)$. (Domains and ranges of other types of relations can also be determined by observing their graphs.)

Parabolas are symmetric with respect to a line (the y-axis in Figure 1). The line of symmetry for a parabola is called the **axis** of the parabola. The point where the axis intersects the parabola is the **vertex** of the parabola. As Figure 2 shows, the vertex of a parabola that opens downward is the highest point of the graph and the vertex of a parabola that opens upward is the lowest point of the graph.

The methods of Section 3.6 can be applied to the graph of $f(x) = x^2$ to obtain the graph of *any* quadratic function. The graph of $g(x) = ax^2$ is a parabola with vertex at the origin that opens upward if a is positive and downward if a is negative. The width of $g(x)$ is determined by the magnitude of a. That is, $g(x)$ is narrower than $f(x) = x^2$ if $|a| > 1$ and is broader than $f(x) = x^2$ if $|a| < 1$. By completing the square, a technique discussed in Chapter 2, any quadratic function can be written in the form

$$h(x) = a(x - h)^2 + k.$$

The graph of $h(x)$ is the same as the graph of $g(x) = ax^2$ translated h units horizontally (to the right if h is positive and to the left if h is negative) and translated k units vertically (up if k is positive and down if k is negative).

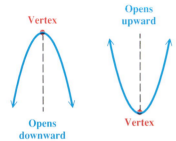

Vertex

Opens upward

Opens downward

Vertex

FIGURE 2

EXAMPLE 1
Graphing functions of the form $f(x) = a(x - h)^2 + k$

Graph the functions defined as follows.

(a) $g(x) = -\frac{1}{2}x^2$

The function $g(x)$ can be thought of as $-\left(\frac{1}{2}x^2\right)$. The graph of $\frac{1}{2}x^2$ is a broad version of the graph of x^2 and the graph of $g(x) = -\left(\frac{1}{2}x^2\right)$ is a reflection of the graph of $y = \frac{1}{2}x^2$ about the x-axis. See Figure 3. The vertex is $(0, 0)$ and the axis of the parabola is the line $x = 0$ (the y-axis).

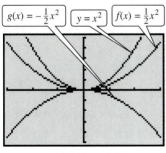

$[-3, 3]$ by $[-5, 5]$

A calculator graph of $y = x^2$, $f(x) = \frac{1}{2}x^2$, and $g(x) = -\frac{1}{2}x^2$

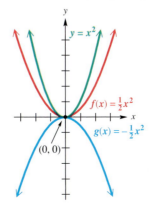

$y = x^2$

$f(x) = \frac{1}{2}x^2$

$g(x) = -\frac{1}{2}x^2$

$(0, 0)$

FIGURE 3

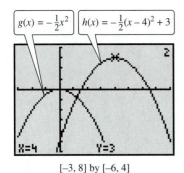

$g(x) = -\frac{1}{2}x^2$ $h(x) = -\frac{1}{2}(x-4)^2 + 3$

[−3, 8] by [−6, 4]

The vertex of the graph of h is (4, 3).

(b) $h(x) = -\frac{1}{2}(x - 4)^2 + 3$

The function $h(x)$ is $g(x - h) + k$, where $g(x)$ is the function of part (a), h is 4, and k is 3. Therefore, $h(x)$ is obtained by translating the graph of $g(x)$ 4 units to the right and 3 units up. See Figure 4. The vertex is (4, 3) and the axis of the parabola is the line $x = 4$. ▶

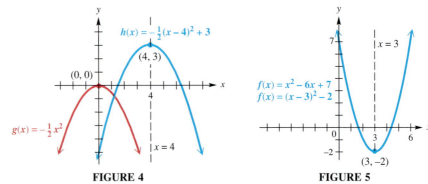

FIGURE 4   FIGURE 5

In general, the graph of the quadratic function

$$f(x) = a(x - h)^2 + k$$

is a parabola with vertex (h, k) and axis $x = h$. The parabola opens upward if a is positive and opens downward if a is negative. With these facts in mind, we will apply *completing the square* to the graphing of a quadratic function.

◀ **EXAMPLE 2**
Graphing a function of the form $f(x) = x^2 + bx + c$

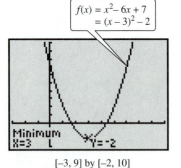

$f(x) = x^2 - 6x + 7$
$= (x - 3)^2 - 2$

[−3, 9] by [−2, 10]

This screen shows that the vertex of the graph of f is (3, −2). Because it is the lowest point on the graph, the calculator is directed to find the *minimum*.

Graph $f(x) = x^2 - 6x + 7$.

To graph this parabola, $x^2 - 6x + 7$ must be rewritten in the form $(x - h)^2 + k$. Start as follows.

$$f(x) = (x^2 - 6x \quad) + 7$$

As shown earlier, a number must be added inside the parentheses to get a perfect square trinomial. To find this number, take half the coefficient of x and then square the result. Half of -6 is -3, and $(-3)^2$ is 9. Now add and subtract 9 inside the parentheses. (This is the same as adding 0.)

$$f(x) = (x^2 - 6x \ + 9 - 9) + 7$$
$$f(x) = (x^2 - 6x + 9) - 9 + 7 \qquad \text{Group terms.}$$
$$f(x) = (x - 3)^2 - 2 \qquad \text{Factor.}$$

This result shows that the vertex of the parabola is (3, −2) and the axis is the line $x = 3$.

The y-intercept is 7. Verify that (1, 2) also satisfies the equation. Plotting these points and using symmetry about the axis of the parabola gives the graph shown in Figure 5. ▶

NOTE In Example 2 we added and subtracted 9 *on the same side* of the equation to complete the square. This differs from adding the same number to *each side of the equation,* as when we completed the square in Chapter 2. Here, since we want just y on one side of the equation, we chose to slightly change that step in the process of completing the square.

◄ **EXAMPLE 3**
Graphing a function of the form $f(x) = ax^2 + bx + c$

Graph $f(x) = -3x^2 - 2x + 1$.

To complete the square, first factor out -3 from the x-terms.

$$f(x) = -3\left(x^2 + \frac{2}{3}x \qquad\right) + 1$$

(This is necessary to make the coefficient of x^2 equal to 1.) Half the coefficient of x is 1/3, and $(1/3)^2 = 1/9$. Add and subtract 1/9 inside the parentheses as follows.

$$f(x) = -3\left(x^2 + \frac{2}{3}x + \frac{1}{9} - \frac{1}{9}\right) + 1$$

Use the distributive property and simplify.

$$f(x) = -3\left(x^2 + \frac{2}{3}x + \frac{1}{9}\right) - 3\left(-\frac{1}{9}\right) + 1$$

$$f(x) = -3\left(x^2 + \frac{2}{3}x + \frac{1}{9}\right) + \frac{1}{3} + 1$$

$$f(x) = -3\left(x^2 + \frac{2}{3}x + \frac{1}{9}\right) + \frac{4}{3}$$

$$f(x) = -3\left(x + \frac{1}{3}\right)^2 + \frac{4}{3} \qquad \text{Factor.}$$

Now the equation of the parabola is written in the form $f(x) = a(x - h)^2 + k$. In this form, the equation shows that the axis of the parabola is the vertical line

$$x + \frac{1}{3} = 0 \qquad \text{or} \qquad x = -\frac{1}{3}$$

and that the vertex is $(-1/3, 4/3)$. Additional points can be found by substituting x-values near the vertex into the original equation. For example, $(1/2, -3/4)$ is on the graph shown in Figure 6. The intercepts are often good additional points to find. Here, the y-intercept is

$$y = -3(0)^2 - 2(0) + 1 = 1,$$

giving the point $(0, 1)$. The x-intercepts are found by setting $f(x)$ equal to zero in the original equation.

$$0 = -3x^2 - 2x + 1$$

$$3x^2 + 2x - 1 = 0 \qquad \text{Multiply by } -1.$$

$$(3x - 1)(x + 1) = 0 \qquad \text{Factor.}$$

Therefore, the x-intercepts are 1/3 and -1. ▶

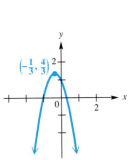

$f(x) = -3x^2 - 2x + 1$
$= -3(x + \frac{1}{3})^2 + \frac{4}{3}$

Maximum
X=-.333333 Y=1.3333333

[−4.7, 4.7] by [−3.1, 3.1]

This screen shows that the vertex of the graph of f is $(-.\overline{3}, 1.\overline{3}) = \left(-\frac{1}{3}, \frac{4}{3}\right)$. Because it is the highest point on the graph, the calculator is directed to find the *maximum.*

$\left(-\frac{1}{3}, \frac{4}{3}\right)$

$f(x) = -3x^2 - 2x + 1$
$f(x) = -3\left(x + \frac{1}{3}\right)^2 + \frac{4}{3}$

FIGURE 6

We can now generalize the work above to get a formula for the vertex of a parabola. Starting with the general quadratic function $f(x) = ax^2 + bx + c$ and completing the square will change the function to the form $f(x) = a(x - h)^2 + k$.

$$f(x) = ax^2 + bx + c$$

$$= a\left(x^2 + \frac{b}{a}x \quad\right) + c \qquad\qquad \text{Factor } a \text{ from the first two terms.}$$

$$= a\left(x^2 + \frac{b}{a}x + \frac{b^2}{4a^2}\right) + c - a\left(\frac{b^2}{4a^2}\right) \qquad \text{Add } (\frac{1}{2} \cdot \frac{b}{a})^2 = \frac{b^2}{4a^2} \text{ in the parentheses; subtract } a(\frac{b^2}{4a^2}) \text{ from } c.$$

$$= a\left(x + \frac{b}{2a}\right)^2 + c - \frac{b^2}{4a} \qquad\qquad \text{Factor the trinomial.}$$

Comparing the last result with $f(x) = a(x - h)^2 + k$ shows that

$$h = -\frac{b}{2a} \qquad \text{and} \qquad k = c - \frac{b^2}{4a}.$$

Letting $x = h$ in $f(x) = a(x - h)^2 + k$ gives $f(h) = a(h - h)^2 + k = k$, so $k = f(h)$, or $k = f(-b/(2a))$.

The following statement summarizes this discussion.

Graph of a Quadratic Function

The quadratic function defined by $f(x) = ax^2 + bx + c$ can be written in the form

$$y = f(x) = a(x - h)^2 + k, \quad a \neq 0,$$

where

$$h = -\frac{b}{2a} \quad \text{and} \quad k = f(h).$$

The graph of f has the following characteristics:

1. It is a parabola with vertex (h, k), and the vertical line $x = h$ as axis.
2. It opens upward if $a > 0$ and downward if $a < 0$.
3. It is broader than $y = x^2$ if $|a| < 1$ and narrower than $y = x^2$ if $|a| > 1$.
4. The y-intercept is $f(0) = c$.
5. The x-intercepts are $\dfrac{-b \pm \sqrt{b^2 - 4ac}}{2a}$, if $b^2 - 4ac \geq 0$.

The vertex and axis of a parabola can be found from its equation either by completing the square or by memorizing the formula $h = -b/2a$ and letting $k = f(h)$.

EXAMPLE 4
Finding the axis and the vertex of a parabola using the formula

Find the axis and the vertex of the parabola having equation $f(x) = 2x^2 + 4x + 5$ using the formula given above.

Here $a = 2$, $b = 4$, and $c = 5$. The axis of the parabola is the vertical line

$$x = h = -\frac{b}{2a} = -\frac{4}{2(2)} = -1.$$

The vertex is the point

$$(-1, f(-1)) = (-1, 3). \quad \blacktriangleright$$

CONNECTIONS Parabolas have many applications due to their geometric definition, which is discussed in a later chapter. For example, the reflectors of solar ovens and flashlights are made by revolving a parabola about its axis, and the path of a projectile takes the shape of a parabola. An interesting and unusual application is found in the shape of the Arctic poppy.* Its parabolic-shaped flower follows the sun, concentrating the sun's rays at the focus (a point in the interior) of the parabola. Because of this, the temperature within the flower may be several degrees above ambient air temperature.

FOR DISCUSSION OR WRITING
Look up some other applications of the parabola.

■ **PROBLEM SOLVING**   The fact that the vertex of a vertical parabola is the highest or lowest point on the graph makes equations of the form $y = ax^2 + bx + c$ important in problems where the maximum or minimum value of some quantity is to be found. When $a < 0$, the y-value of the vertex gives the maximum value of y and the x-value tells where it occurs. Similarly, when $a > 0$, the y-value of the vertex gives the minimum y-value.   ■

EXAMPLE 5
Finding the maximum spending on "smart" highways

Since 1990 Congress has authorized more than $700 million for research and development of "smart" highways.** The spending per year is approximated by the function

$$f(x) = -19.321x^2 + 3608.7x - 168,310,$$

---

*From "Looking Down in Alaska," Wayne Merry, *Motorland/CSAA*, July/August 1995.
**Source: IVHS America.

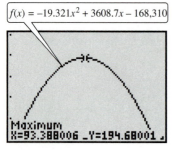

$$f(x) = -19.321x^2 + 3608.7x - 168{,}310$$

Maximum
X=93.388006 _Y=194.68001

[90, 97] by [0, 250]
Xscl = 1   Yscl = 50

FIGURE 7

where x is the number of years since 1990 and $y = f(x)$ is in millions of dollars. Use a graphing calculator to graph the function and to find the maximum amount spent in one year. In what year was the maximum amount spent?

The graph is shown in Figure 7. Since $a < 0$, it opens downward (as expected) and thus has a maximum. Using the capability of the calculator, we find the coordinates of the maximum point are approximately (93.4, 194.68). This means that the maximum spending of $194.68 million occurred in 1993. ▶

4.1 Exercises ▼▼

In Exercises 1–4, you are given an equation and the graph of a quadratic function. Do each of the following. See Examples 1(b), and 2–4.
(a) Give the domain and the range.   *(b) Give the coordinates of the vertex.*
(c) Give the equation of its axis.   *(d) Find the y-intercept.*
(e) Find the x-intercepts.

1. $f(x) = (x + 3)^2 - 4$

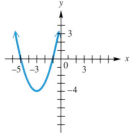

$f(x) = (x + 3)^2 - 4$

2. $f(x) = (x - 5)^2 - 4$

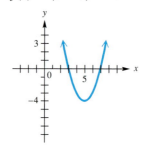

$f(x) = (x - 5)^2 - 4$

3. $f(x) = -2(x + 3)^2 + 2$

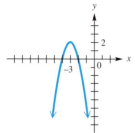

$f(x) = -2(x + 3)^2 + 2$

4. $f(x) = -3(x - 2)^2 + 1$

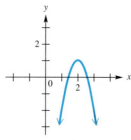

$f(x) = -3(x - 2)^2 + 1$

The graphs of the functions in Exercises 5–8 are shown in Figures A–D, as generated by a graphing calculator. Match each function with its graph using the concepts of this section without actually entering it into your calculator. Then, after you have completed the exercises, check your answers with your calculator. Use the standard viewing window.

5. $f(x) = (x - 4)^2 - 3$

6. $f(x) = -(x - 4)^2 + 3$

7. $f(x) = (x + 4)^2 - 3$

8. $f(x) = -(x + 4)^2 + 3$

A.

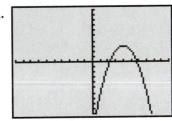

[–10, 10] by [–10, 10]

B.

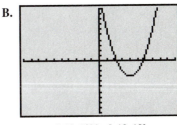

[–10, 10] by [–10, 10]

C.

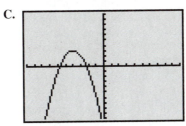

[–10, 10] by [–10, 10]

D.

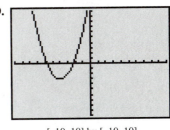

[–10, 10] by [–10, 10]

9. Graph the following on the same coordinate system. See Example 1(a).

  (a) $y = 2x^2$ **(b)** $y = 3x^2$ **(c)** $y = \dfrac{1}{2}x^2$ **(d)** $y = \dfrac{1}{3}x^2$

  (e) How does the coefficient of x^2 affect the shape of the graph?

10. Graph the following on the same coordinate system.

  (a) $y = x^2 + 2$ **(b)** $y = x^2 - 1$ **(c)** $y = x^2 + 1$ **(d)** $y = x^2 - 2$

  (e) How do these graphs differ from the graph of $y = x^2$?

11. Graph the following on the same coordinate system.

  (a) $y = (x - 2)^2$ **(b)** $y = (x + 1)^2$ **(c)** $y = (x + 3)^2$ **(d)** $y = (x - 4)^2$

  (e) How do these graphs differ from the graph of $y = x^2$?

12. Match each equation with the description of the parabola that is its graph.

  (a) $y = (x - 4)^2 - 2$ **A.** vertex $(2, -4)$, opens down

  (b) $y = (x - 2)^2 - 4$ **B.** vertex $(2, -4)$, opens up

  (c) $y = -(x - 4)^2 - 2$ **C.** vertex $(4, -2)$, opens down

  (d) $y = -(x - 2)^2 - 4$ **D.** vertex $(4, -2)$, opens up

13. For the graph of $y = a(x - h)^2 + k$, in what quadrant is the vertex if:

  (a) $h < 0, k < 0$;   **(b)** $h < 0, k > 0$;   **(c)** $h > 0, k < 0$;   **(d)** $h > 0, k > 0$?

14. Explain what causes the graph of a parabola with equation of the form $y = ax^2$ to be wider or narrower than the graph of $y = x^2$.

Graph the parabola. Give the vertex, axis, domain, and range. See Examples 1(b) and 2–4.

15. $f(x) = (x - 2)^2$ **16.** $f(x) = (x + 4)^2$ **17.** $f(x) = (x + 3)^2 - 4$

18. $f(x) = (x - 5)^2 - 4$ **19.** $f(x) = -2(x + 3)^2 + 2$ **20.** $f(x) = -3(x - 2)^2 + 1$

21. $f(x) = -\dfrac{1}{2}(x + 1)^2 - 3$ **22.** $f(x) = \dfrac{2}{3}(x - 2)^2 - 1$ **23.** $f(x) = x^2 - 2x + 3$

24. $f(x) = x^2 + 6x + 5$ **25.** $f(x) = 2x^2 - 4x + 5$ **26.** $f(x) = -3x^2 + 24x - 46$

The figure shows the graph of a quadratic function y = f(x).

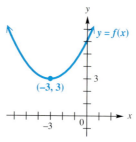

27. What is the minimum value of $f(x)$?

28. For what value of x is $f(x)$ as small as possible?

29. How many real solutions are there to the equation $f(x) = 1$?

30. How many real solutions are there to the equation $f(x) = 4$?

31. In Chapter 3 we saw how certain changes to an equation cause the graph of the equation to be stretched, shrunk, reflected about an axis, or translated vertically or horizontally. It is important to notice that the order in which these changes are done affects the final graph. For example, stretching and then shifting vertically produces a graph that differs from the one produced by shifting vertically, then stretching. To see this, use a graphing calculator to graph $y = 3x^2 - 2$ and $y = 3(x^2 - 2)$, and then compare the results. Are the two expressions equivalent algebraically?

32. Suppose that a quadratic function with $a > 0$ is written in the form $f(x) = a(x - h)^2 + k$. Match each of the items (a), (b), and (c) with one of the items (A), (B), or (C).

    (a) k is positive.   **(A)** f intersects the x-axis at only one point.

    (b) k is negative.   **(B)** f does not intersect the x-axis.

    (c) k is zero.   **(C)** f intersects the x-axis twice.

The figures below show several possible graphs of $f(x) = ax^2 + bx + c$. For the restrictions on a, b, and c given in Exercises 33–38, select the corresponding graph from (A) through (F) below.

33. $a < 0, b^2 - 4ac = 0$

34. $a > 0, b^2 - 4ac < 0$

35. $a < 0, b^2 - 4ac < 0$

36. $a < 0, b^2 - 4ac > 0$

37. $a > 0, b^2 - 4ac > 0$

38. $a > 0, b^2 - 4ac = 0$

A.
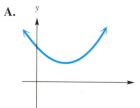

B.
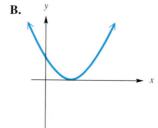

C.
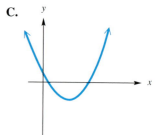

D.
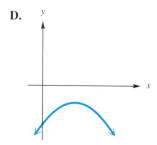

E.
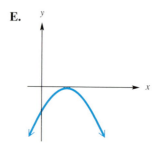

F.
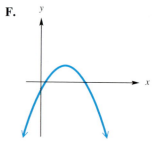

▼▼▼▼▼▼▼▼▼▼▼▼ **DISCOVERING CONNECTIONS** (Exercises 39–44) ▼▼▼▼▼▼▼▼▼▼▼▼

The solution set of $f(x) = 0$ consists of all x-values for which the graph of $y = f(x)$ intersects the x-axis (i.e., the x-intercepts). The solution set of $f(x) < 0$ consists of all x-values for which the graph lies below *the x-axis, while the solution set of $f(x) > 0$ consists of all x-values for which the graph lies* above *the x-axis.*

In Chapter 2 we saw how a sign graph can be used to solve a quadratic inequality. Graphical analysis allows us to solve such inequalities as well. Work the following exercises in order. They demonstrate why we must reverse the direction of the inequality sign when multiplying or dividing an inequality by a negative number.

39. Graph $f(x) = x^2 + 2x - 8$. This function has a graph with two x-intercepts. What are they?

40. Based on the graph from Exercise 39, what is the solution set of $x^2 + 2x - 8 < 0$?

41. Now graph $g(x) = -f(x) = -x^2 - 2x + 8$. Using the terminology of Chapter 3, how is the graph of g obtained by a transformation of the graph of f?

42. Based on the graph from Exercise 41, what is the solution set of $-x^2 - 2x + 8 > 0$?

43. How do the two solution sets of the inequalities in Exercises 40 and 42 compare?

44. Write a short paragraph explaining how Exercises 39–43 illustrate the property involving multiplying an inequality by a negative number.

In Exercises 45 and 46, find a polynomial function f whose graph matches the one in the figure. Then use a graphing calculator to graph the function and verify your result.

45.

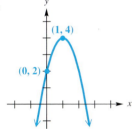

46.

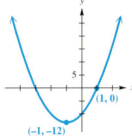

Solve the problem. See Example 5.

47. The Gross State Product in current dollars from 1985 through 1989 is modeled by the quadratic function

$$f(x) = 18.14x^2 + 234.03x + 3954$$

where $x = 0$ corresponds to 1985 and $f(x)$ is in billions. If this model continues to apply, what would be the gross state product in 1996? (*Note:* There are pitfalls in using models to predict far into the future.) (*Source:* U.S. Bureau of Economic Analysis.)

48. The number of cases commenced by the U.S. Court of Appeals each year between 1984 and 1990 can be approximated by the quadratic model

$$f(x) = 68.90x^2 + 1165.29x + 31,676$$

where $x = 0$ corresponds to 1984. Based on this model, what would be the number of cases com-

menced in 1996? (*Note:* There are pitfalls in using models to predict far into the future.)

49. Between 1985 and 1989, the number of female suicides by firearms in the United States each year can be modeled by

$$f(x) = -17x^2 + 44.6x + 2572$$

where $x = 0$ represents 1985. Based on this model, in what year did the number of such suicides reach its peak?

50. The number of infant deaths during the past decade has been decreasing. Between 1980 and 1989, the number of infant deaths per 1000 live births each year can be approximated by the function

$$f(x) = .0234x^2 - .5029x + 12.5$$

where $x = 0$ corresponds to 1980.

(a) Based on this model, how many deaths per 1000 live births occurred during 1985?

(b) If the trend continued, how many deaths per 1000 live births would we have expected during the year 1990?

51. The following table lists the total (cumulative) number of AIDS cases diagnosed in the United States up to 1993. For example, a total of 22,620 AIDS cases were diagnosed between 1981 and 1985. (*Source:* U.S. Dept. of Health and Human Services, Centers for Disease Control and Prevention, *HIV/AIDS Surveillance,* March 1994.)

Year	AIDS Cases
1982	1563
1983	4647
1984	10,845
1985	22,620
1986	41,662
1987	70,222
1988	105,489
1989	147,170
1990	193,245
1991	248,023
1992	315,329
1993	361,509

(a) Plot the data. Let $x = 0$ correspond to the year 1980.

(b) Would a linear or quadratic function model this data best? Explain.

(c) Find a quadratic function $f(x) = a(x - h)^2 + k$ that models the data. (*Hint:* Use (2, 1563) as the vertex and then choose a second point such as (13, 361,509) to determine a.)

(d) Plot the data together with f on the same coordinate plane. How well does f model the number of AIDS cases?

(e) Use f to predict the total number of AIDS cases that will be diagnosed by the years 1999 and 2000.

(f) How many new cases will be diagnosed in the year 2000?

(g) Discuss factors that could cause this trend to change.

52. The following table lists the total (cumulative) number of known deaths caused by AIDS in the United States up to 1993. (*Source:* U.S. Dept. of Health and Human Services, Centers for Disease Control and Prevention, *HIV/AIDS Surveillance,* March 1994.)

Year	Deaths
1982	620
1983	2122
1984	5600
1985	12,529
1986	24,550
1987	40,820
1988	61,723
1989	89,172
1990	119,821
1991	154,567
1992	191,508
1993	220,592

(a) Plot the data. Let $x = 0$ correspond to the year 1980.

(b) Would a linear or quadratic function model these data best? Explain.

(c) Find a quadratic function $g(x) = a(x - h)^2 + k$ that models the data. Use (2, 620) as the vertex and (13, 220,592) as the other point to determine a.

(d) Plot the data together with g on the same coordinate plane. How well does g model the number of deaths caused by AIDS?

(e) Use g to predict the number of new deaths caused by AIDS during the year 2000.

(f) Both f (from the previous exercise) and g are quadratic functions. Discuss why it is reasonable to expect that the number of AIDS cases and AIDS-related deaths could be modeled using the same type of function.

53. A frog leaps from a stump 3 feet high and lands 4 feet from the base of the stump. We can consider the initial position of the frog to be at $(0, 3)$ and its landing position to be at $(4, 0)$. See the figure. It is determined that the height of the frog as a function of its distance x from the base of the stump is given by the function

$$h(x) = -.5x^2 + 1.25x + 3,$$

where x and $h(x)$ are both in feet.

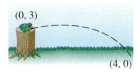

(0, 3)

(4, 0)

   (a) How high was the frog when its horizontal distance from the base of the stump was 2 feet?
   (b) At what two times after it jumped was the frog 3.25 feet above the ground?
   (c) At what distance from the base of the stump did the frog reach its highest point?
   (d) What was the maximum height reached by the frog?

54. Refer to Exercise 53. Suppose that the initial position of the frog is $(0, 4)$ and its landing position is $(6, 0)$. The height of the frog is given by $h(x) = -\frac{1}{3}x^2 + \frac{4}{3}x + 4$. After how many seconds did it reach its maximum height? What was the maximum height?

55. Glenview College wants to construct a rectangular parking lot on land bordered on one side by a highway. It has 320 ft of fencing which it will use to fence off the other three sides. What should be the dimensions of the lot if the enclosed area is to be a maximum? See the figure. (*Hint:* Let x represent the width of the lot and let $320 - 2x$ represent the length. Graph the area parabola, $A = x(320 - 2x)$, and investigate the vertex.)

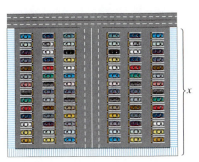

x

56. A charter flight charges a fare of $200 per person plus $4 per person for each unsold seat on the plane. If the plane holds 100 passengers, and if x represents the number of unsold seats, find the following.
   (a) a quadratic function that models the total revenue R, in dollars, received for the flight (*Hint:* Multiply the number of people flying, $100 - x$, by the price per ticket.)
   (b) the graph for the expression of part (a)
   (c) the number of unsold seats that will produce the maximum revenue
   (d) the maximum revenue

An important application of quadratic functions deals with the height of a propelled object as a function of the time elapsed after it is propelled.

Formula for the Height of a Propelled Object

If air resistance is neglected, the height s (in feet) of an object propelled directly upward from an initial height s_0 feet with initial velocity v_0 feet per second is described by the function

$$s(x) = -16x^2 + v_0x + s_0,$$

where x is the number of seconds after the object is propelled.

Use this function in Exercises 57–60.

57. A ball is thrown directly upward from an initial height of 100 feet with an initial velocity of 80 feet per second.
   (a) After how many seconds does the ball reach its maximum height?
   (b) What is the maximum height reached by the ball?

58. A rock is thrown directly upward from ground level with an initial velocity of 90 feet per second.
   (a) After how many seconds does the rock reach its maximum height?
   (b) What is the maximum height reached by the rock?

59. A toy rocket is launched from the top of a building 50 feet tall at an initial velocity of 200 feet per second.
   (a) Determine the time at which the rocket reaches its maximum height, and the maximum height in feet.
   (b) For what time interval will the rocket be more than 300 feet above ground level?
   (c) After how many seconds will it hit the ground?

60. Use a graphing calculator to determine whether a ball thrown from ground level with an initial velocity of 150 feet per second will reach a height of 400 feet. If it will, determine the time(s) at which this happens. If it will not, explain why using a graphical interpretation.

---

61. Find a value of c so that $y = x^2 - 10x + c$ has exactly one x-intercept.

62. For what values of a does $y = ax^2 - 8x + 4$ have no x-intercepts?

63. Find b so that $y = x^2 + bx + 9$ has exactly one x-intercept.

64. Define the quadratic function f having x-intercepts 2 and 5, and y-intercept 5.

65. Define the quadratic function f having x-intercepts 1 and -2, and y-intercept 4.

66. Find the largest possible value of y if $y = -(x - 2)^2 + 9$. Then find the following.
  (a) The largest possible value of $\sqrt{-(x - 2)^2 + 9}$
  (b) The smallest possible positive value of $\dfrac{1}{-(x - 2)^2 + 9}$

67. Find the smallest possible value of y if $y = 3 + (x + 5)^2$. Then find the following.
  (a) The smallest possible value of $\sqrt{3 + (x + 5)^2}$
  (b) The largest possible value of $\dfrac{1}{3 + (x + 5)^2}$

68. From the distance formula in Section 3.1, the distance between the two points $P(x_1, y_1)$ and $R(x_2, y_2)$ is $d(P, R) = \sqrt{(x_1 - x_2)^2 + (y_1 - y_2)^2}$. Using the results of Exercises 66 and 67, find the closest point on the line $y = 2x$ to the point $(1, 7)$. (*Hint:* Every point on $y = 2x$ has the form $(x, 2x)$, and the closest point has the minimum distance.)

4.2 Synthetic Division ▼▼▼

The quotient of two polynomials was found in Chapter 1 with a division algorithm similar to that used to divide with whole numbers.

> **Division Algorithm**
>
> Let $f(x)$ and $g(x)$ be polynomials with $g(x)$ of lower degree than $f(x)$ and $g(x)$ of degree one or more. There exist unique polynomials $q(x)$ and $r(x)$ such that
>
> $$f(x) = g(x) \cdot q(x) + r(x),$$
>
> where either $r(x) = 0$ or the degree of $r(x)$ is less than the degree of $g(x)$.

Recall that $q(x)$ is the quotient polynomial and $r(x)$ is the remainder polynomial or the remainder.

A shortcut method of performing long division with certain polynomials, called **synthetic division,** will be useful in applying the theorems presented in this and the next two sections. The method is used only when a polynomial is divided by a first-degree binomial of the form $x - k$, where the coefficient of x is 1. To illustrate, notice the example worked on the left below. On the right the division process is simplified by omitting all variables and writing only coefficients, with 0 used to represent the coefficient of any missing terms. Since the coefficient of x in the divisor is always 1 in these divisions, it too can be omitted. These omissions simplify the problem as shown on the right below.

$$
\begin{array}{r}
3x^2 + 10x + 40 \\
x - 4\overline{)3x^3 - 2x^2 \qquad - 150} \\
3x^3 - 12x^2 \\
\overline{10x^2} \\
10x^2 - 40x \\
\overline{40x - 150} \\
40x - 160 \\
\overline{10}
\end{array}
\qquad
\begin{array}{r}
3 \quad 10 \quad 40 \\
-4\overline{)3 - 2 + 0 - 150} \\
3 - 12 \\
\overline{10} \\
10 - 40 \\
\overline{40 - 150} \\
40 - 160 \\
\overline{10}
\end{array}
$$

The numbers in color that are repetitions of the numbers directly above them can also be omitted.

$$
\begin{array}{r}
3 \quad 10 \quad 40 \\
-4\overline{)3 - 2 + 0 - 150} \\
- 12 \\
\overline{10} \\
- 40 \\
\overline{40 - 150} \\
- 160 \\
\overline{10}
\end{array}
$$

The entire problem can now be condensed vertically, and the top row of numbers can be omitted since it duplicates the bottom row if the 3 is brought down.

$$
\begin{array}{r}
-4\overline{\smash{\underline{}}}3 \quad - 2 \quad\quad 0 \quad -150 \\
-12 \quad -40 \quad -160 \\
\hline
3 \quad\quad 10 \quad\quad 40 \quad\quad 10
\end{array}
$$

The rest of the bottom row is obtained by subtracting -12, -40, and -160 from the corresponding terms above.

With synthetic division it is useful to change the sign of the divisor, so the -4 at the left is changed to 4, which also changes the sign of the numbers in the second row. To compensate for this change, subtraction is changed to addition. Doing this gives the following result.

$$
\begin{array}{r}
4\overline{\smash{\underline{}}}3 \quad - 2 \quad\quad 0 \quad -150 \\
12 \quad\quad 40 \quad\quad 160 \\
\hline
3 \quad\quad 10 \quad\quad 40 \quad\quad 10
\end{array}
$$

In summary, to use synthetic division to divide a polynomial by a binomial of the form $x - k$, begin by writing the coefficients of the polynomial in decreasing powers of the variable, using 0 as the coefficient of any missing powers. The number k is written to the left in the same row. In the example above, $x - k$ is $x - 4$, so k is 4. Next bring down the leading coefficient of the polynomial, 3 in the previous example, as the first number in the last row. Multiply the 3 by 4 to get the first number in the second row, 12. Add 12 to -2; this gives 10, the second number in the third row. Multiply 10 by 4 to get 40, the next number in the second row. Add 40 to 0 to get the third number in the third row, and so on. This process of multiplying each result in the third row by k and adding the product to the number in the next column is repeated until there is a number in the last row for each coefficient in the first row.

CAUTION   To avoid incorrect results, it is essential to use a 0 for any missing terms, including a missing constant, when you set up the division.

EXAMPLE 1
Using synthetic division

Use synthetic division to divide $5x^3 - 6x^2 - 28x - 2$ by $x + 2$.
   Begin by writing

$$-2 \,\vert\, 5 \quad -6 \quad -28 \quad -2.$$

The value of k is -2, since k is found by writing $x + 2$ as $x - (-2)$. Next, bring down the 5.

$$
\begin{array}{r|rrrr}
-2 & 5 & -6 & -28 & -2 \\
\hline
& 5 & & &
\end{array}
$$

Now, multiply -2 by 5 to get -10, and add it to the -6 in the first row. The result is -16.

$$
\begin{array}{r|rrrr}
-2 & 5 & -6 & -28 & -2 \\
& & -10 & & \\
\hline
& 5 & -16 & &
\end{array}
$$

Next, $(-2)(-16) = 32$. Add this to the -28 in the first row.

$$
\begin{array}{r|rrrr}
-2 & 5 & -6 & -28 & -2 \\
& & -10 & 32 & \\
\hline
& 5 & -16 & 4 &
\end{array}
$$

Finally, $(-2)(4) = -8$, which is added to the -2 to get -10.

$$
\begin{array}{r|rrrr}
-2 & 5 & -6 & -28 & -2 \\
& & -10 & 32 & -8 \\
\hline
& 5 & -16 & 4 & -10
\end{array}
$$

The coefficients of the quotient polynomial and the remainder are read directly from the bottom row. Since the degree of the quotient will always be one less than the degree of the polynomial to be divided,

$$\frac{5x^3 - 6x^2 - 28x - 2}{x + 2} = 5x^2 - 16x + 4 + \frac{-10}{x + 2}.$$

The result of the division in Example 1 can be written as

$$5x^3 - 6x^2 - 28x - 2 = (x + 2)(5x^2 - 16x + 4) + (-10)$$

by multiplying both sides by the denominator $x + 2$. The following theorem is a generalization of the division process illustrated above.

> For any polynomial $f(x)$ and any complex number k, there exists a unique polynomial $q(x)$ and number r such that
>
> $$f(x) = (x - k)q(x) + r.$$

For example, in the synthetic division above,

$$5x^3 - 6x^2 - 28x - 2 = (x + 2)\ (5x^2 - 16x + 4) + (-10).$$

$$f(x) \qquad = \quad (x - k) \cdot \quad q(x) \qquad + \quad r$$

This theorem is a special case of the division algorithm given earlier. Here $g(x)$ is the first-degree polynomial $x - k$.

Evaluating $f(k)$ By the division algorithm, $f(x) = (x - k)q(x) + r$. This equality is true for all complex values of x, so it is true for $x = k$. Replacing x with k gives

$$f(k) = (k - k)q(k) + r$$
$$f(k) = r.$$

This proves the following **remainder theorem,** which gives a new method of evaluating polynomial functions.

> **Remainder Theorem**
>
> If the polynomial $f(x)$ is divided by $x - k$, the remainder is $f(k)$.

As an illustration of this theorem, we have seen that when the polynomial $f(x) = 5x^3 - 6x^2 - 28x - 2$ is divided by $x + 2$ or $x - (-2)$, the remainder is -10. Now substituting -2 for x in $f(x)$ gives

$$f(-2) = 5(-2)^3 - 6(-2)^2 - 28(-2) - 2$$
$$= -40 - 24 + 56 - 2$$
$$= -10.$$

As shown here, the simpler way to find the value of a polynomial is often by using synthetic division. By the remainder theorem, instead of replacing x by -2 to find $f(-2)$, divide $f(x)$ by $x + 2$ using synthetic division as in Example 1. Then $f(-2)$ is the remainder, -10.

$$\begin{array}{r|rrrr} -2 & 5 & -6 & -28 & -2 \\ & & -10 & 32 & -8 \\ \hline & 5 & -16 & 4 & -10 \end{array} \quad \leftarrow f(-2)$$

EXAMPLE 2
Applying the remainder theorem

Let $f(x) = -x^4 + 3x^2 - 4x - 5$. Find $f(-3)$.
Use the remainder theorem and synthetic division.

$$
\begin{array}{r|rrrrr}
-3 & -1 & 0 & 3 & -4 & -5 \\
& & 3 & -9 & 18 & -42 \\
\hline
& -1 & 3 & -6 & 14 & -47
\end{array}
$$

The remainder when $f(x)$ is divided by $x - (-3) = x + 3$ is -47, so $f(-3) = -47$. ▶

The remainder theorem gives a quick way to decide if a number k is a zero of a polynomial $f(x)$. Use synthetic division to find $f(k)$; if the remainder is zero, then $f(k) = 0$ and k is a zero of $f(x)$. A zero of $f(x)$ is called a **root** or **solution** of the equation $f(x) = 0$.

EXAMPLE 3
Deciding whether a number is a zero

Decide whether or not the given number is a zero of the given polynomial.

(a) 1; $f(x) = x^3 - 4x^2 + 9x - 6$
Use synthetic division.

$$
\begin{array}{r|rrrr}
1 & 1 & -4 & 9 & -6 \\
& & 1 & -3 & 6 \\
\hline
& 1 & -3 & 6 & 0
\end{array}
$$

Since the remainder is 0, $f(1) = 0$, and 1 is a zero of the polynomial $f(x) = x^3 - 4x^2 + 9x - 6$.

```
Y1日-X^4+3X²-4X-5
Y2=
Y3=
Y4=
Y5=
Y6=
Y7=
```

```
Y1(-3)
                    -47
```

For $Y_1 = f(x) = -x^4 + 3x^2 - 4x - 5$, $Y_1(-3) = f(-3) = -47$. Compare to the result in Example 2.

(b) -4; $f(x) = x^4 + x^2 - 3x + 1$
Remember to use a coefficient of 0 for the missing x^3 term in the synthetic division.

$$
\begin{array}{r|rrrrr}
-4 & 1 & 0 & 1 & -3 & 1 \\
& & -4 & 16 & -68 & 284 \\
\hline
& 1 & -4 & 17 & -71 & 285
\end{array}
$$

The remainder is not 0, so -4 is not a zero of $f(x) = x^4 + x^2 - 3x + 1$. In fact, $f(-4) = 285$.

(c) $1 + 2i$; $f(x) = x^4 - 2x^3 + 4x^2 + 2x - 5$
Use synthetic division and operations with complex numbers.

$$
\begin{array}{r|rrrrr}
1 + 2i & 1 & -2 & 4 & 2 & -5 \\
& & 1 + 2i & -5 & -1 - 2i & 5 \\
\hline
& 1 & -1 + 2i & -1 & 1 - 2i & 0
\end{array}
$$

Since the remainder is zero, $1 + 2i$ is a zero of the given polynomial. ▶

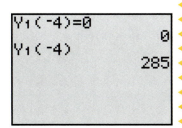

For $Y_1 = f(x) = x^4 + x^2 - 3x + 1$, $Y_1(-4) = 0$ is false, as indicated by the 0 (rather than 1). Evaluating $Y_1(-4) = f(-4)$ gives 285. Compare to Example 3(b).

CONNECTIONS In Example 3(c) we found that $f(x) = x^4 - 2x^3 + 4x^2 + 2x - 5$ has an imaginary zero $1 + 2i$. At the beginning of this chapter, we defined a polynomial with *real* coefficients. The theorems and definitions of this chapter apply also to polynomials with *imaginary* coefficients. For example, we can show that $f(x) = 3x^3 + (-1 + 3i)x^2 + (-12 + 5i)x + 4 - 2i$ has $2 - i$ as a zero. We must show that $f(2 - i) = 0$.

$$\begin{array}{r|rrrr} 2 - i & 3 & -1 + 3i & -12 + 5i & 4 - 2i \\ & & 6 - 3i & 10 - 5i & -4 + 2i \\ \hline & 3 & 5 & -2 & 0 \end{array}$$

By the division algorithm, $f(x) = (x - 2 + i)(3x^2 + 5x - 2)$. We can factor $q(x) = 3x^2 + 5x - 2$ as $(3x - 1)(x + 2)$. Thus $q(x)$ has zeros of $1/3$ and -2, which are also zeros of $f(x)$. This shows that even though $f(x)$ has imaginary coefficients, it has two real zeros in addition to the one imaginary zero we tested for above.

FOR DISCUSSION OR WRITING

1. Find $f(-2 + i)$ if $f(x) = x^3 - 4x^2 + 2x - 29i$.
2. Is i a zero of $f(x) = x^3 + 2ix^2 + 2x + i$? Is $-i$?
3. Give a simple function with real coefficients that has at least one imaginary zero.

4.2 Exercises ▼▼▼▼▼▼▼▼▼▼▼▼▼▼▼▼▼▼▼▼▼▼▼▼▼▼▼▼▼▼▼▼

Decide whether the statement is true or false.

1. When the polynomial $f(x)$ is divided by $x - r$, the remainder is $f(r)$.
2. If $f(c) = 0$, then $x - c$ is a factor of the polynomial $f(x)$.
3. If $x^3 - 1$ is divided by $x + 1$, the remainder is 0.
4. If $x^3 - 1$ is divided by $x - 1$, the remainder is 0.

Use synthetic division to perform the indicated division. See Example 1.

5. $\dfrac{x^3 + 4x^2 - 5x + 42}{x + 6}$

6. $\dfrac{x^3 + 2x^2 - 8x - 17}{x - 3}$

7. $\dfrac{4x^3 - 3x - 2}{x + 1}$

8. $\dfrac{3x^3 - 4x + 2}{x - 1}$

9. $\dfrac{x^4 - 3x^3 - 4x^2 + 12x}{x - 3}$

10. $\dfrac{x^4 - 3x^3 - 5x^2 + 2x - 16}{x + 2}$

11. $\dfrac{x^5 + 3x^4 + 2x^3 + 2x^2 + 3x + 1}{x + 2}$

12. $\dfrac{\dfrac{1}{3}x^3 - \dfrac{2}{9}x^2 + \dfrac{1}{27}x + 1}{x - \dfrac{1}{3}}$

Express the polynomial in the form $f(x) = (x - k)q(x) + r$ for the given value of k.

13. $f(x) = 2x^3 + x^2 + x - 8; \quad k = -1$ **14.** $f(x) = 2x^3 + 3x^2 - 16x + 10; \quad k = -4$

15. $f(x) = -x^3 + 2x^2 + 4; \quad k = -2$ **16.** $f(x) = -4x^3 + 2x^2 - 3x - 10; \quad k = 2$

17. $f(x) = 4x^4 - 3x^3 - 20x^2 - x; \quad k = 3$ **18.** $f(x) = 2x^4 + x^3 - 15x^2 + 3x; \quad k = -3$

For the given polynomial, use the remainder theorem and synthetic division to find $f(k)$. See Example 2.

19. $k = 3; \quad f(x) = x^2 - 4x + 5$ **20.** $k = -2; \quad f(x) = x^2 + 5x + 6$

21. $k = 2; \quad f(x) = 2x^2 - 3x - 3$ **22.** $k = 4; \quad f(x) = -x^3 + 8x^2 + 63$

23. $k = -1; \quad f(x) = x^3 - 4x^2 + 2x + 1$ **24.** $k = 2; \quad f(x) = 2x^3 - 3x^2 - 5x + 4$

25. $k = 3; \quad f(x) = 2x^5 - 10x^3 - 19x^2 - 45$ **26.** $k = 4; \quad f(x) = x^4 + 6x^3 + 9x^2 + 3x - 3$

27. $k = -8; \quad f(x) = x^6 + 7x^5 - 5x^4 + 22x^3 - 16x^2 + x + 19$

28. $k = -\dfrac{1}{2}; \quad f(x) = 6x^3 - 31x^2 - 15x$

29. $k = 2 + i; \quad f(x) = x^2 - 5x + 1$ **30.** $k = 3 - 2i; \quad f(x) = x^2 - x + 3$

Use synthetic division to decide whether the given number is a zero of the given polynomial. See Example 3.

31. $3; \quad f(x) = 2x^3 - 6x^2 - 9x + 4$ **32.** $-6; \quad f(x) = 2x^3 + 9x^2 - 16x + 12$

33. $-5; \quad f(x) = x^3 + 7x^2 + 10x$ **34.** $-2; \quad f(x) = 2x^3 - 3x^2 - 5x$

35. $\dfrac{2}{5}; \quad f(x) = 5x^4 + 2x^3 - x + 15$ **36.** $\dfrac{1}{2}; \quad f(x) = 2x^4 - 3x^2 + 4$

37. $2 - i; \quad f(x) = x^2 + 3x + 4$ **38.** $1 - 2i; \quad f(x) = x^2 - 3x + 5$

▼▼▼▼▼▼▼▼▼▼▼▼▼ **DISCOVERING CONNECTIONS** (Exercises 39–44) ▼▼▼▼▼▼▼▼▼▼▼▼▼

In Example 3(a) we used synthetic division to evaluate $f(1)$ for $f(x) = x^3 - 4x^2 + 9x - 6$. There is an interesting quick method for evaluating a polynomial $f(x)$ for $x = 1$. Work Exercises 39–44 in order, so that you can discover this method and another for $x = -1$.

39. If 1 is raised to *any* power, what is the result?

40. If we multiply the result found in Exercise 39 by a real number, what is the value of the product? That is, how does it compare with the real number?

41. Based on your answer to Exercise 40, how can we evaluate $f(1)$ for $f(x) = x^3 - 4x^2 + 9x - 6$ without using direct substitution or synthetic division?

42. Support your answer in Exercise 41 by actually applying it. Does your answer agree with the one found in Example 3(a)?

43. Find $f(-x)$ and $f(-1)$.

44. Add the coefficients of $f(-x)$ in Exercise 43, and compare to $f(-1)$. Make a conjecture about how to find $f(-1)$.

4.3 Zeros of Polynomial Functions ▼▼▼

In this section we will build upon some of the ideas presented in the previous section to learn more about finding zeros of polynomial functions.

By the remainder theorem, if $f(k) = 0$, then the remainder when $f(x)$ is divided by $x - k$ is zero. This means that $x - k$ is a factor of $f(x)$. Conversely, if $x - k$ is a factor of $f(x)$, then $f(k)$ must equal 0. This is summarized in the following **factor theorem.**

> **Factor Theorem**
>
> The polynomial $x - k$ is a factor of the polynomial $f(x)$ if and only if $f(k) = 0$.

EXAMPLE 1
Deciding whether $x - k$ is a factor of $f(x)$

Is $x - 1$ a factor of $f(x) = 2x^4 + 3x^2 - 5x + 7$?

By the factor theorem, $x - 1$ will be a factor of $f(x)$ only if $f(1) = 0$. Use synthetic division and the remainder theorem to decide.

$$
\begin{array}{r|rrrrr}
1 & 2 & 0 & 3 & -5 & 7 \\
  &   & 2 & 2 & 5 & 0 \\
\hline
  & 2 & 2 & 5 & 0 & 7
\end{array}
$$

Since the remainder is 7, $f(1) = 7$ and not 0, so $x - 1$ is not a factor of the polynomial $f(x)$. ▶

The factor theorem can be used to factor a polynomial of higher degree into linear factors. Linear factors are factors of the form $ax - b$.

EXAMPLE 2
Factoring a polynomial given a zero

Factor $f(x) = 6x^3 + 19x^2 + 2x - 3$ into linear factors given that -3 is a zero of f.

Since -3 is a zero of f, $x - (-3) = x + 3$ is a factor. Use synthetic division to divide $f(x)$ by $x + 3$.

$$
\begin{array}{r|rrrr}
-3 & 6 & 19 & 2 & -3 \\
   &   & -18 & -3 & 3 \\
\hline
   & 6 & 1 & -1 & 0
\end{array}
$$

The quotient is $6x^2 + x - 1$, so

$$f(x) = (x + 3)(6x^2 + x - 1).$$

Factor $6x^2 + x - 1$ as $(2x + 1)(3x - 1)$ to get

$$f(x) = (x + 3)(2x + 1)(3x - 1),$$

where all factors are linear. ▶

CONNECTIONS Example 2 showed how to factor a polynomial if its zeros are known. The following rational zeros theorem is a useful method for finding a set of possible zeros of a polynomial with integer coefficients. The numbers in the set can then be tested by synthetic division to find the actual zeros.

> Let $f(x) = a_n x^n + a_{n-1} x^{n-1} + \ldots + a_1 x + a_0$, $a_n \neq 0$, be a polynomial with only integer coefficients. If p/q is a rational number written in lowest terms and if p/q is a zero of $f(x)$, then p is a factor of the constant term a_0 and q is a factor of the leading coefficient a_n.

For example, since $f(x) = 2x^3 - x^2 - 2x + 1$ has only integer coefficients, the theorem applies. For this polynomial, $a_0 = 1$ and $a_n = a_3 = 2$. By the rational zeros theorem, p must be a factor of $a_0 = 1$ so p is either 1 or -1. Also, q must be a factor of $a_3 = 2$, so q is ± 1 or ± 2. Any rational zeros are in the form p/q, and so must be either ± 1 or $\pm 1/2$. We use synthetic division to try 1 first.

$$
\begin{array}{r|rrrr}
1 & 2 & -1 & -2 & 1 \\
  &   & 2 & 1 & -1 \\
\hline
  & 2 & 1 & -1 & 0
\end{array}
$$

One zero is 1. To find any other zeros, we set the quotient polynomial $2x^2 + x - 1$ equal to zero and solve the quadratic equation. Doing this produces the remaining zeros, -1 and $1/2$.

FOR DISCUSSION OR WRITING
Use the rational zeros theorem to find the possible zeros of the following functions.

1. $f(x) = 15x^3 + 59x^2 + 4x - 1$
2. $f(x) = x^4 - 2x^3 + x^2 + 18$
3. Use the theorem to show that $f(x) = x^2 - 7$ has no rational zeros, so $\sqrt{7}$ must be irrational.

The next theorem says that every polynomial of degree 1 or more has a zero, which means that every such polynomial can be factored. The theorem was first proved by Carl Friedrich Gauss (1777-1855) as part of his doctoral dissertation completed in 1799. This theorem, which Gauss named the "fundamental theorem of algebra," had challenged the world's finest mathematicians for at least

200 years. Peter Rothe had stated it in 1608, followed by Albert Girard in 1629 and René Descartes in 1637. Jean LeRond D'Alembert (1717-1783) thought he had a proof in 1746; consequently it is known in France today as D'Alembert's theorem. Two of the world's greatest mathematicians, Euler and Lagrange, had attempted unsuccessfully to prove it in 1749 and 1772, respectively. Their errors were noted in Gauss's dissertation. Gauss's proof used advanced mathematical concepts outside the field of algebra. To this day, no purely algebraic proof has been discovered.

Fundamental Theorem of Algebra

Every polynomial of degree 1 or more has at least one complex zero.

From the fundamental theorem, if $f(x)$ is of degree 1 or more then there is some number k_1 such that $f(k_1) = 0$. By the factor theorem, then

$$f(x) = (x - k_1) \cdot q_1(x)$$

for some polynomial $q_1(x)$. If $q_1(x)$ is of degree 1 or more, the fundamental theorem and the factor theorem can be used to factor $q_1(x)$ in the same way. There is some number k_2 such that $q_1(k_2) = 0$, so that

$$q_1(x) = (x - k_2)q_2(x)$$

and
$$f(x) = (x - k_1)(x - k_2)q_2(x).$$

Assuming that $f(x)$ has degree n and repeating this process n times gives

$$f(x) = a(x - k_1)(x - k_2) \ldots (x - k_n),$$

where a is the leading coefficient of $f(x)$. Each of these factors leads to a zero of $f(x)$, so $f(x)$ has the n zeros $k_1, k_2, k_3, \ldots, k_n$. This result suggests the next theorem.

Number of Zeros Theorem

A polynomial of degree n has at most n distinct zeros.

NOTE The theorem says that there exist *at most* n distinct zeros. For example, the polynomial $f(x) = x^3 + 3x^2 + 3x + 1 = (x + 1)^3$ is of degree 3 but has only one zero, -1. Actually, the zero -1 occurs three times, since there are three factors of $x + 1$; this zero is called a **zero of multiplicity 3.**

A graphing calculator is useful for finding the number of distinct real zeros of a polynomial function. If a number x is a zero of a function f, then $y = f(x) = 0$. By definition, the real values of x where $y = 0$ are the x-intercepts of the graph of $y = f(x)$. Therefore, we can use a graphing calculator to determine the real zeros of a function. If the zeros are rational numbers, the calculator will give exact values (perhaps as repeating decimals). Otherwise, the calculator will give an approximation of the zero. It is important, however, to be sure that the x-values in the window include all the real zeros. The accompanying figures show the three zeros of $f(x) = .5x^3 - 2.5x^2 + x + 4$, which is discussed in Example 3(a) below. Notice also that $f(1) = 3$, as required in the example.

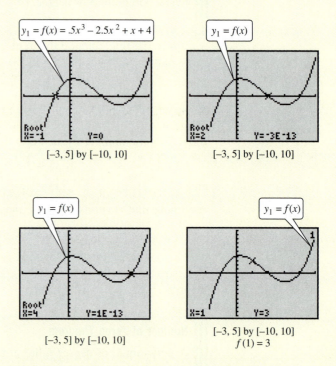

$y_1 = f(x) = .5x^3 - 2.5x^2 + x + 4$

$[-3, 5]$ by $[-10, 10]$

$y_1 = f(x)$

$[-3, 5]$ by $[-10, 10]$

$y_1 = f(x)$

$[-3, 5]$ by $[-10, 10]$

$y_1 = f(x)$

$[-3, 5]$ by $[-10, 10]$
$f(1) = 3$

The fundamental theorem says that every polynomial has at least one complex zero, and the number of zeros theorem tells us that a polynomial of degree n has at most n zeros. If a graph showing all real zeros shows fewer than n real zeros, either the missing zeros are imaginary or there are one or more zeros of multiplicity greater than 1.

EXAMPLE 3
Finding a polynomial that satisfies given conditions (real zeros)

Find a polynomial $f(x)$ of degree 3 that satisfies the following conditions.

(a) Zeros of -1, 2, and 4; $f(1) = 3$

These three zeros give $x - (-1) = x + 1, x - 2$, and $x - 4$ as factors of $f(x)$. Since $f(x)$ is to be of degree 3, these are the only possible factors by the theorem just stated. Therefore, $f(x)$ has the form

$$f(x) = a(x + 1)(x - 2)(x - 4)$$

for some real number a. To find a, use the fact that $f(1) = 3$.

$$f(1) = a(1 + 1)(1 - 2)(1 - 4) = 3$$
$$a(2)(-1)(-3) = 3$$
$$6a = 3$$
$$a = \frac{1}{2}$$

Thus, $$f(x) = \frac{1}{2}(x + 1)(x - 2)(x - 4),$$

or, by multiplication,

$$f(x) = \frac{1}{2}x^3 - \frac{5}{2}x^2 + x + 4.$$

(b) -2 is a zero of multiplicity 3; $f(-1) = 4$

The polynomial $f(x)$ has the form

$$f(x) = a(x + 2)(x + 2)(x + 2)$$
$$= a(x + 2)^3.$$

Since $f(-1) = 4$,

$$f(-1) = a(-1 + 2)^3 = 4,$$
$$a(1)^3 = 4$$

or

$$a = 4,$$

and $f(x) = 4(x + 2)^3 = 4x^3 + 24x^2 + 48x + 32.$ ▶

NOTE   In Example 3(a), we cannot clear the denominators in $f(x)$ by multiplying both sides by 2, because the result would equal $2 \cdot f(x)$, not $f(x)$ itself.

The remainder theorem can be used to show that both $2 + i$ and $2 - i$ are zeros of $f(x) = x^3 - x^2 - 7x + 15$. In general, if $a + bi$ is a zero of a polynomial function with *real* coefficients, then so is $a - bi$. To prove this requires the following properties of complex conjugates. Let $z = a + bi$, and write $\bar{z}$ for the conjugate of z, so that $\bar{z} = a - bi$. For example, if $z = -5 + 2i$, then $\bar{z} = -5 - 2i$. The proofs of the following equalities are left for the exercises (see Exercises 49–52).

Properties of Conjugates

For any complex numbers c and d,

$$\overline{c + d} = \overline{c} + \overline{d}$$
$$\overline{c \cdot d} = \overline{c} \cdot \overline{d}$$
$$\overline{c^n} = (\overline{c})^n.$$

If the complex number z is a zero of $f(x)$, then the conjugate of z, written as $\overline{z}$, is also a zero of $f(x)$. This is shown by starting with the polynomial

$$f(x) = a_n x^n + a_{n-1} x^{n-1} + \ldots + a_1 x + a_0,$$

where all coefficients are real numbers. If $z = a + bi$ is a zero of $f(x)$, then

$$f(z) = a_n z^n + a_{n-1} z^{n-1} + \ldots + a_1 z + a_0 = 0.$$

Taking the conjugate of both sides of this last equation gives

$$\overline{a_n z^n + a_{n-1} z^{n-1} + \ldots + a_1 z + a_0} = \overline{0}.$$

Using generalizations of the properties $\overline{c + d} = \overline{c} + \overline{d}$ and $\overline{c \cdot d} = \overline{c} \cdot \overline{d}$ gives

$$\overline{a_n z^n} + \overline{a_{n-1} z^{n-1}} + \ldots + \overline{a_1 z} + \overline{a_0} = \overline{0}$$

or

$$\overline{a_n}\, \overline{z^n} + \overline{a_{n-1}}\, \overline{z^{n-1}} + \ldots + \overline{a_1}\, \overline{z} + \overline{a_0} = \overline{0}.$$

Now use the third property from above and the fact that for any real number a, $\overline{a} = a$, to get

$$a_n (\overline{z})^n + a_{n-1} (\overline{z})^{n-1} + \ldots + a_1 (\overline{z}) + a_0 = 0.$$

Hence $\overline{z}$ is also a zero of $f(x)$, which completes the proof of the **conjugate zeros theorem.**

Conjugate Zeros Theorem

If $f(x)$ is a polynomial *having only real coefficients* and if $a + bi$ is a zero of $f(x)$, where a and b are real numbers, then $a - bi$ is also a zero of $f(x)$.

CAUTION The requirement that the polynomial have only real coefficients is *essential.* For example, $f(x) = x - (1 + i)$ has $1 + i$ as a zero, but the conjugate $1 - i$ is not a zero.

EXAMPLE 4
Finding a polynomial that satisfies given conditions (complex zeros)

Find a polynomial of lowest degree having only real coefficients and zeros 3 and $2 + i$.

The complex number $2 - i$ also must be a zero, so the polynomial has at least three zeros, 3, $2 + i$, and $2 - i$. For the polynomial to be of lowest degree these must be the only zeros. By the factor theorem there must be three factors, $x - 3$, $x - (2 + i)$, and $x - (2 - i)$. A polynomial of lowest degree is

$$f(x) = (x - 3)[x - (2 + i)][x - (2 - i)]$$
$$= (x - 3)(x - 2 - i)(x - 2 + i)$$
$$= x^3 - 7x^2 + 17x - 15.$$

Other polynomials, such as $2(x^3 - 7x^2 + 17x - 15)$ or $\sqrt{5}(x^3 - 7x^2 + 17x - 15)$, for example, also satisfy the given conditions on zeros. The information on zeros given in the problem is not enough to give a specific value for the leading coefficient. ▶

The theorem on conjugate zeros is important in helping predict the number of real zeros of polynomials with real coefficients. A polynomial with real coefficients of odd degree n, where $n \geq 1$, must have at least one real zero (since zeros of the form $a + bi$, where $b \neq 0$, occur in conjugate pairs). On the other hand, a polynomial with real coefficients of even degree n may have no real zeros.

◀**EXAMPLE 5**
Finding all zeros of a
polynomial given one zero

Find all zeros of $f(x) = x^4 - 7x^3 + 18x^2 - 22x + 12$, given that $1 - i$ is a zero.

Since the polynomial has only real coefficients and since $1 - i$ is a zero, by the conjugate zeros theorem $1 + i$ is also a zero. To find the remaining zeros, first divide the original polynomial by $x - (1 - i)$.

$$
\begin{array}{r|rrrrr}
1 - i & 1 & -7 & 18 & -22 & 12 \\
      &   & 1 - i & -7 + 5i & 16 - 6i & -12 \\
\hline
      & 1 & -6 - i & 11 + 5i & -6 - 6i & 0
\end{array}
$$

By the factor theorem, since $x = 1 - i$ is a zero of $f(x)$, $x - (1 - i)$ is a factor, and $f(x)$ can be written as

$$f(x) = [x - (1 - i)][x^3 + (-6 - i)x^2 + (11 + 5i)x + (-6 - 6i)].$$

We know that $x = 1 + i$ is also a zero of $f(x)$, so that

$$f(x) = [x - (1 - i)][x - (1 + i)]q(x).$$

Thus, $x^3 + (-6 - i)x^2 + (11 + 5i)x + (-6 - 6i) = (x - (1 + i))q(x)$. Use synthetic division to find $q(x)$.

$$
\begin{array}{r|rrrr}
1 + i & 1 & -6 - i & 11 + 5i & -6 - 6i \\
      &   & 1 + i & -5 - 5i & 6 + 6i \\
\hline
      & 1 & -5 & 6 & 0
\end{array}
$$

Since $q(x) = x^2 - 5x + 6$, $f(x)$ can be written as

$$f(x) = [x - (1 - i)][x - (1 + i)](x^2 - 5x + 6).$$

Now find the zeros of the quadratic polynomial $x^2 - 5x + 6$. Factoring the polynomial shows that the zeros are 2 and 3, so the four zeros of $f(x)$ are $1 - i$, $1 + i$, 2, and 3. ▶

In Example 5 we saw how to find additional zeros when at least one zero is known. A graphing calculator can be used to locate a real zero. If the exact value of the zero can be determined, then the polynomial can be factored, and the techniques of this section could be used to find other zeros.

4.3 Exercises ▼▼▼▼▼▼▼▼▼▼▼▼▼▼▼▼▼▼▼▼▼▼▼▼▼▼▼▼▼▼▼▼▼▼

Decide whether the statement is true or false.

1. Given that $x - 1$ is a factor of $f(x) = x^6 - x^4 + 2x^2 - 2$, we are assured that $f(1) = 0$.

2. Given that $f(1) = 0$ for $f(x) = x^6 - x^4 + 2x^2 - 2$, we are assured that $x - 1$ is a factor of $f(x)$.

3. For the function $f(x) = (x + 2)^4(x - 3)$, 2 is a zero of multiplicity 4.

4. Given that $2 + 3i$ is a zero of $f(x) = x^2 - 4x + 13$, we are assured that $2 - 3i$ is also a zero.

Use the factor theorem to decide whether the second polynomial is a factor of the first. See Example 1.

5. $4x^2 + 2x + 54$; $x - 4$

6. $5x^2 - 14x + 10$; $x + 2$

7. $x^3 + 2x^2 - 3$; $x - 1$

8. $2x^3 + x + 2$; $x + 1$

9. $2x^4 + 5x^3 - 2x^2 + 5x + 6$; $x + 3$

10. $5x^4 + 16x^3 - 15x^2 + 8x + 16$; $x + 4$

For each polynomial, one zero is given. Find all others. See Examples 2 and 5.

11. $f(x) = x^3 - x^2 - 4x - 6$;   3

12. $f(x) = x^3 + 4x^2 - 5$;   1

13. $f(x) = 4x^3 + 6x^2 - 2x - 1$; $\dfrac{1}{2}$

14. $f(x) = x^3 - 7x^2 + 17x - 15$; $2 - i$

15. $f(x) = x^4 + 5x^2 + 4$; $-i$

16. $f(x) = x^4 + 10x^3 + 27x^2 + 10x + 26$; i

For the given information, find a polynomial of degree 3 with only real coefficients that satisfies the given conditions. See Examples 3 and 4.

17. Zeros of -3, 1, and 4; $f(2) = 30$

18. Zeros of 1, -1, and 0; $f(2) = 3$

19. Zeros of -2, 1, and 0; $f(-1) = -1$

20. Zeros of 2, -3, and 5; $f(3) = 6$

21. Zeros of 5, i, and $-i$; $f(2) = 5$

22. Zeros of -2, i, and $-i$; $f(-3) = 30$

▼▼▼▼▼▼▼▼▼▼▼▼▼ **DISCOVERING CONNECTIONS** (Exercises 23–26) ▼▼▼▼▼▼▼▼▼▼▼▼▼

A calculator-generated graph of the polynomial function $f(x) = x^3 - 21x - 20$ is shown here. Notice that there are three x-intercepts and thus three real zeros of the function. We will investigate polynomial functions in general in the next section. For now, let us see how synthetic division allows us to make connections between the zeros of f and those of the functions defined by the other factors of $f(x)$ obtained when using the methods discussed in this section. Work Exercises 23–26 in order.

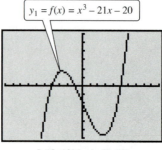

$y_1 = f(x) = x^3 - 21x - 20$

[−10, 10] by [−60, 60]
Xscl = 1   Yscl = 10

23. Given that $x + 4$ is a factor of $f(x)$, find the other factor and call it $g(x)$. Use synthetic division and the factor theorem.

24. Using the same window as shown above, graph both f and g on the same screen. What kind of function is g? What do you notice about the x-intercepts of the two functions in comparison to each other?

25. Given that $x - 5$ is a factor of $g(x)$, find the other factor and call it $h(x)$. Use either synthetic division and the factor theorem, or direct factorization.

26. Using the same window as shown above, graph both g and h on the same screen. What kind of function is h? What do you notice about the x-intercepts in comparison to each other?

Factor $f(x)$ into linear factors given that k is a zero of $f(x)$. See Example 2.

27. $f(x) = 2x^3 - 3x^2 - 17x + 30;$ $k = 2$

28. $f(x) = 2x^3 - 3x^2 - 5x + 6;$ $k = 1$

29. $f(x) = 6x^3 + 13x^2 - 14x + 3;$ $k = -3$

30. $f(x) = 6x^3 + 17x^2 - 63x + 10;$ $k = -5$

For the polynomial function, find all zeros and their multiplicities.

31. $f(x) = 7x^3 + x$

32. $f(x) = (x + 1)^2(x - 1)^3(x^2 - 10)$

33. $f(x) = 3(x - 2)(x + 3)(x^2 - 1)$

34. $f(x) = 5x^2(x + 1 - \sqrt{2})(2x + 5)$

35. $f(x) = (x^2 + x - 2)^5(x - 1 + \sqrt{3})^2$

36. $f(x) = (7x - 2)^3(x^2 + 9)^2$

For each of the following, find a polynomial of lowest degree with only real coefficients and having the given zeros. See Example 4.

37. $3 + i$ and $3 - i$

38. $7 - 2i$ and $7 + 2i$

39. $1 + \sqrt{2}, 1 - \sqrt{2}$, and 3

40. $1 - \sqrt{3}, 1 + \sqrt{3}$, and 1

41. $-2 + i, -2 - i, 3$, and -3

42. $3 + 2i, -1$, and 2

43. 2 and $3i$

44. -1 and $6 - 3i$

45. $1 + 2i, 2$ (multiplicity 2)

46. $2 + i, -3$ (multiplicity 2)

47. Show that -2 is a zero of multiplicity 2 of $f(x) = x^4 + 2x^3 - 7x^2 - 20x - 12$ and find all other complex zeros. Then write $f(x)$ in factored form.

48. Show that -1 is a zero of multiplicity 3 of $f(x) = x^5 - 4x^3 - 2x^2 + 3x + 2$ and find all other complex zeros. Then write $f(x)$ in factored form.

If c and d are complex numbers, prove each statement. (Hint: Let $c = a + bi$ and $d = m + ni$ and form all the conjugates, the sums, and the products.)

49. $\overline{c + d} = \overline{c} + \overline{d}$

50. $\overline{cd} = \overline{c} \cdot \overline{d}$

51. $\overline{a} = a$ for any real number a

52. $\overline{c^n} = (\overline{c})^n$

In the screens accompanying the Connections box in this section, we showed how a graphing calculator can find real zeros of a polynomial function if the graph is generated. Modern graphing calculators also have the capability of finding (or approximating) real zeros using a "solver" feature. For example, the screen shown here indicates the three zeros of $Y_1 = f(x) = .5x^3 - 2.5x^2 + x + 4$, which were found from the graph. Since different makes and models of calculators require different keystrokes, consult your owner's manual to see how your particular model accomplishes equation solving.

```
solve(Y₁,X,-5)
               -1
solve(Y₁,X,1)
               2
solve(Y₁,X,6)
               4
```

The TI-82 requires "guesses." Shown here are –5, 1, and 6 as guesses that yield the three zeros, –1, 2, and 4.

Use a graphing calculator to find the real zeros of the given function defined by $f(x)$. Use the "solver" feature. Express decimal approximations to the nearest hundredth.

53. $f(x) = .86x^3 - 5.24x^2 + 3.55x + 7.84$

54. $f(x) = -2.47x^3 - 6.58x^2 - 3.33x + .14$

55. $f(x) = 2.45x^4 - 3.22x^3 + .47x^2 - 6.54x + 3$

56. $f(x) = 4x^4 + 8x^3 - 4x^2 + 4x + 1$

57. $f(x) = -\sqrt{7}\,x^3 + \sqrt{5}\,x + \sqrt{17}$

58. $f(x) = \sqrt{10}\,x^3 - \sqrt{11}\,x - \sqrt{8}$

4.4 Graphs of Polynomial Functions ▼▼▼

We have already discussed the graphs of polynomial functions of degree 0 to 2. In this section we show how to graph polynomial functions of degree 3 or more. The domains will be restricted to real numbers, since we will be graphing on the real number plane.

The recent strides made in computer technology and graphing calculators have made some of the material in this section of limited value. However, the concepts presented here allow students to understand the ideas of finding real zeros of polynomial functions. Once these ideas are mastered, students may then wish to investigate the use of computers and graphing calculators to find real zeros of polynomial functions. Learning the methods of this section first will help students realize the power of today's technology.

We begin the discussion of graphing polynomial functions with the graphs of polynomial functions defined by equations of the form $f(x) = ax^n$.

◀ **EXAMPLE 1**
Graphing functions of the form $f(x) = ax^n$

Graph each function f defined as follows.
(a) $f(x) = x^3$

Choose several values for x, and find the corresponding values of $f(x)$, or y, as shown in the left table in Figure 8. Plot the resulting ordered pairs and connect the points with a smooth curve. The graph of $f(x) = x^3$ is shown in blue in Figure 8.

$f(x) = x^3$	
x	$f(x)$
-2	-8
-1	-1
0	0
1	1
2	8

$f(x) = x^5$	
x	$f(x)$
-1.5	-7.6
-1	-1
0	0
1	1
1.5	7.6

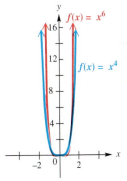

FIGURE 8

(b) $f(x) = x^5$

Work as in part (a) of this example to get the graph shown in red in Figure 8. Notice that the graphs of $f(x) = x^3$ and $f(x) = x^5$ are both symmetric with respect to the origin.

(c) $f(x) = x^4, \ f(x) = x^6$

Some typical ordered pairs for the graphs of $f(x) = x^4$ and $f(x) = x^6$ are given in the tables in Figure 9. These graphs are symmetric with respect to the y-axis, as is the graph of $f(x) = ax^2$ for a nonzero real number a. ▶

$f(x) = x^4$	
x	$f(x)$
-2	16
-1	1
0	0
1	1
2	16

$f(x) = x^6$	
x	$f(x)$
-1.5	11.4
-1	1
0	0
1	1
1.5	11.4

FIGURE 9

As with the graph of $y = ax^2$ in Section 1 of this chapter, the value of a in $f(x) = ax^n$ determines how the graph is affected. When $|a| > 1$ the graph is stretched vertically, making it narrower, while when $0 < |a| < 1$, the graph is shrunk or compressed vertically, so the graph is broader. The graph of $f(x) = -ax^n$ is reflected across the x-axis as compared to the graph of $f(x) = ax^n$.

The ZOOM feature of a graphing calculator is useful in graphs like those below to show the difference between the graphs of $y = x^3$ and $y = x^5$ and between $y = x^4$ and $y = x^6$ for values of x in $[-1.5, 1.5]$. See the figures below. In each case the first window shows x in $[-10, 10]$ and the second window shows x in $[-1.5, 1.5]$.

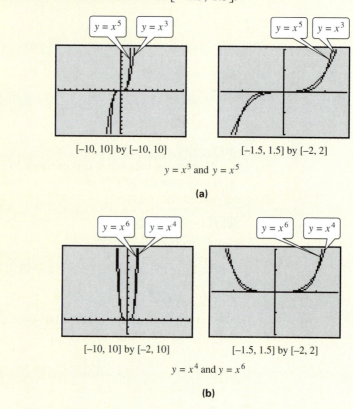

$[-10, 10]$ by $[-10, 10]$ $[-1.5, 1.5]$ by $[-2, 2]$

$y = x^3$ and $y = x^5$

(a)

$[-10, 10]$ by $[-2, 10]$ $[-1.5, 1.5]$ by $[-2, 2]$

$y = x^4$ and $y = x^6$

(b)

Compared with the graph of $f(x) = ax^n$, the graph of $f(x) = ax^n + k$ is translated (shifted) k units up if $k > 0$ and $|k|$ units down if $k < 0$. Also, the graph of $f(x) = a(x - h)^n$ is translated h units to the right if $h > 0$ and $|h|$ units to the left if $h < 0$, when compared with the graph of $f(x) = ax^n$.

The graph of $f(x) = a(x - h)^n + k$ shows a combination of these translations. The effects here are the same as those we saw earlier with quadratic functions.

◀EXAMPLE 2
Examining vertical and horizontal translations

Graph each of the following.

(a) $f(x) = x^5 - 2$

The graph will be the same as that of $f(x) = x^5$, but translated down 2 units. See Figure 10.

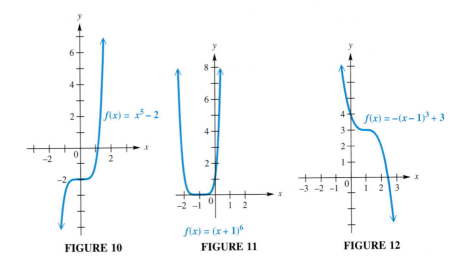

FIGURE 10   FIGURE 11   FIGURE 12

(b) $f(x) = (x + 1)^6$

This function f has a graph like that of $f(x) = x^6$, but since $x + 1 = x - (-1)$, it is translated one unit to the left as shown in Figure 11.

(c) $f(x) = -(x - 1)^3 + 3$

The negative sign causes the graph to be reflected about the x-axis when compared with the graph of $f(x) = x^3$. As shown in Figure 12, the graph is also translated 1 unit to the right and 3 units up. ▶

The domain of every polynomial function is the set of all real numbers. The range of a polynomial function of odd degree is also the set of all real numbers. Some typical graphs of polynomial functions of odd degree are shown in Figure 13. These graphs suggest that for every polynomial function f of odd degree there is at least one real value of x that makes $f(x) = 0$. The zeros are the x-intercepts of the graph.

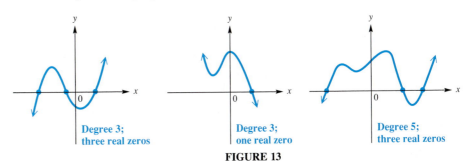

Degree 3;   Degree 3;   Degree 5;
three real zeros   one real zero   three real zeros

FIGURE 13

A polynomial function of even degree will have a range of the form $(-\infty, k]$ or else $[k, \infty)$ for some real number k. Figure 14 shows two typical graphs of polynomial functions of even degree.

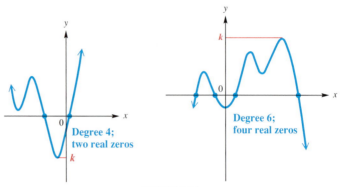

FIGURE 14

The graphs in Figures 13 and 14 show that polynomial functions often have *turning points* where the function changes from increasing to decreasing or from decreasing to increasing.

Turning Points

A polynomial function of degree n has at most $n - 1$ turning points with at least one turning point between each pair of successive zeros.

The graphs shown in Figures 13 and 14 illustrate this.

The end behavior of a polynomial graph is determined by the term of highest degree. That is, a polynomial of the form $f(x) = a_n x^n + a_{n-1} x^{n-1} + \cdots + a_0$ has the same end behavior as the polynomial $f(x) = a_n x^n$. For instance, the polynomial $f(x) = 2x^3 - 8x^2 + 9$ has the same end behavior as $f(x) = 2x^3$. It is large and positive for large positive values of x and large and negative for negative values of x with large absolute value. The arrows at the ends of the graph look like those of the first graph in Figure 13; the right arrow points up and the left arrow points down. The end behavior of polynomials is summarized in the following table.

End Behavior of $f(x) = a_n x^n + a_{n-1} x^{n-1} + \cdots + a_0$

Degree	Sign of a_n	Left arrow	Right arrow	Example
Odd	Positive	Down	Up	First graph of Figure 13
Odd	Negative	Up	Down	Second graph of Figure 13
Even	Positive	Up	Up	First graph of Figure 14
Even	Negative	Down	Down	Second graph of Figure 14

We have discussed several characteristics of the graphs of polynomial functions that are useful in graphing the function. We now define what we mean by a complete graph of a polynomial function.

Complete Graph of a Polynomial Function

A complete graph of a polynomial function will exhibit the following characteristics.

1. all x-intercepts (zeros)
2. the y-intercept
3. all turning points
4. enough of the domain to show the end behavior

If the zeros of a polynomial function are known, its graph can be approximated without plotting very many points; this method is shown in the next example.

EXAMPLE 3
Graphing a polynomial function in factored form

Graph $f(x) = (x - 1)(2x + 3)(x + 2)$.
 The three zeros of f are $x = 1$, $x = -3/2$, and $x = -2$. These three zeros divide the x-axis into four regions, shown in Figure 15.

Regions   I   II   III   IV

-2 $-\dfrac{3}{2}$ 0 1 x

FIGURE 15

In any of these regions, the values of $f(x)$ are either always positive or always negative. To find the sign of $f(x)$ in each region, select an x-value in each region and substitute it into the equation for $f(x)$ to determine if the values of the function are positive or negative in that region. A typical selection of test points and the results of the tests are shown below.

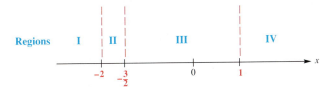

$f(x) = 2x^3 + 5x^2 - x - 6$ $y = x^3$

$[-3, 3]$ by $[-6, 6]$

Notice that the end behavior of f is the same as that of $y = x^3$.

Region	Test point	Value of $f(x)$	Sign of $f(x)$
I $(-\infty, -2)$	-3	-12	**Negative**
II $\left(-2, -\dfrac{3}{2}\right)$	$-\dfrac{7}{4}$	$\dfrac{11}{32}$	**Positive**
III $\left(-\dfrac{3}{2}, 1\right)$	0	-6	**Negative**
IV $(1, \infty)$	2	28	**Positive**

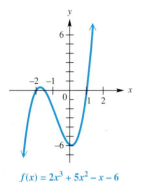

$f(x) = 2x^3 + 5x^2 - x - 6$

FIGURE 16

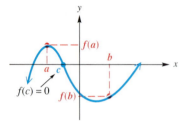

FIGURE 17

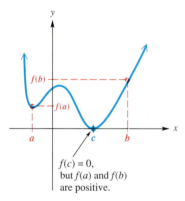

$f(c) = 0$,
but $f(a)$ and $f(b)$
are positive.

FIGURE 18

Plot the three zeros and the test points and connect them with a smooth curve to get the graph. The graph in Figure 16 shows that this function has 2 turning points, the maximum number for a third-degree polynomial function. When the values of $f(x)$ are negative, the graph is below the x-axis, and when $f(x)$ takes on positive values, the graph is above the x-axis, as shown in Figure 16. The sketch could be improved by plotting additional points in each region. Notice that the left arrow points down and the right arrow points up. This end behavior is correct since when the linear factors are multiplied out, the highest term of the polynomial is $2x^3$. ▶

As Example 3 shows, the key to graphing a polynomial function is locating its zeros. In the special case where the zeros are rational numbers, the zeros can be found by a technique presented in the Connections box in the previous section. Occasionally, irrational zeros can be found by inspection. For instance, $f(x) = x^3 - 2$ has the irrational zero $\sqrt[3]{2}$. Two theorems presented in this section apply to the zeros of every polynomial function with real coefficients. The first theorem uses the fact that graphs of polynomial functions are unbroken curves, with no gaps or sudden jumps. The proof requires advanced methods, so it is not given here. Figure 17 illustrates the theorem.

Intermediate Value Theorem for Polynomials

If $f(x)$ is a polynomial with only real coefficients, and if for real numbers a and b, the values $f(a)$ and $f(b)$ are opposite in sign, then there exists at least one real zero between a and b.

This theorem helps to identify intervals where zeros of polynomials are located. If $f(a)$ and $f(b)$ are opposite in sign, then 0 is between $f(a)$ and $f(b)$, and there must be a number c between a and b where $f(c) = 0$.

CAUTION Be careful how you interpret the intermediate value theorem. If $f(a)$ and $f(b)$ are *not* opposite in sign, it does not necessarily mean that there is no zero between a and b. For example, the fact that $f(c) = 0$, for c between a and b, does not imply that 0 is between $f(a)$ and $f(b)$. See Figure 18.

◀ **EXAMPLE 4**
Using the intermediate value
theorem to locate a zero

Show that $f(x) = x^3 - 2x^2 - x + 1$ has a real zero between 2 and 3.
Use synthetic division to find $f(2)$ and $f(3)$.

$$
\begin{array}{r}
2\,|\,1 \quad -2 \quad -1 \quad 1 \\
\underline{\quad 2 \quad\ 0 \quad -2} \\
1 \quad\ 0 \quad -1 \quad -1
\end{array}
\qquad
\begin{array}{r}
3\,|\,1 \quad -2 \quad -1 \quad 1 \\
\underline{\quad 3 \quad\ 3 \quad\ 6} \\
1 \quad\ 1 \quad\ 2 \quad\ 7
\end{array}
$$

Since $f(2)$ is negative but $f(3)$ is positive, there must be a real zero between 2 and 3. ▶

The intermediate value theorem for polynomials is helpful in limiting the search for real zeros to smaller and smaller intervals. In Example 4 the theorem was used to verify that there is a real zero between 2 and 3. The theorem could then be used repeatedly to locate the zero more accurately. The next theorem, the *boundedness theorem,* shows how the bottom row of a synthetic division can be used to place upper and lower bounds on the possible real zeros of a polynomial.

Boundedness Theorem

Let $f(x)$ be a polynomial of degree $n \geq 1$ with real coefficients and with a positive leading coefficient. If $f(x)$ is divided synthetically by $x - c$ and

(a) if $c > 0$ and all numbers in the bottom row of the synthetic division are nonnegative, then $f(x)$ has no zero greater than c;

(b) if $c < 0$ and the numbers in the bottom row of the synthetic division alternate in sign (with 0 considered positive or negative, as needed), then $f(x)$ has no zero less than c.

An outline of the proof of part (a) is given here. The proof for part (b) is similar. By the division algorithm, if $f(x)$ is divided by $x - c$, then

$$f(x) = (x - c)q(x) + r,$$

where all coefficients of $q(x)$ are nonnegative, $r \geq 0$, and $c > 0$. If $x > c$, then $x - c > 0$. Since $q(x) > 0$ and $r \geq 0$,

$$f(x) = (x - c)q(x) + r > 0.$$

This means that $f(x)$ will never be 0 for $x > c$.

EXAMPLE 5
Using the boundedness theorem

Show that the real zeros of $f(x) = 2x^4 - 5x^3 + 3x + 1$ satisfy the following conditions.

(a) No real zero is greater than 3.

Since $f(x)$ has real coefficients and the leading coefficient, 2, is positive, the boundedness theorem can be used. Divide $f(x)$ synthetically by $x - 3$.

$$
\begin{array}{r|rrrrr}
3 & 2 & -5 & 0 & 3 & 1 \\
  &   & 6 & 3 & 9 & 36 \\
\hline
  & 2 & 1 & 3 & 12 & 37
\end{array}
$$

Since $3 > 0$ and all numbers in the last row of the synthetic division are nonnegative, $f(x)$ has no real zero greater than 3.

(b) No real zero is less than -1.

Divide $f(x)$ by $x + 1$.

$$
\begin{array}{r|rrrrr}
-1 & 2 & -5 & 0 & 3 & 1 \\
   &   & -2 & 7 & -7 & 4 \\
\hline
   & 2 & -7 & 7 & -4 & 5
\end{array}
$$

Here $-1 < 0$ and the numbers in the last row alternate in sign, so $f(x)$ has no zero less than -1. ▶

In the next example we use the information from this section and the previous one to approximate the irrational real zeros of a polynomial function.

Approximating real zeros of a polynomial

Approximate the real zeros of $f(x) = x^4 - 6x^3 + 8x^2 + 2x - 1$.

The highest degree term is x^4, so the graph will have end behavior similar to the graph of $f(x) = x^4$, which is positive for all values of x with large absolute values. That is, the end behavior is upward at the left and the right. There are at most four real zeros, since the polynomial is fourth-degree.

By substitution, $f(0)$ is the constant term, -1. Because the end behavior is positive on the left and the right, by the intermediate value theorem f has at least one zero on either side of $x = 0$. To approximate the zeros, we will use synthetic division to find $f(-1), f(-2), f(-3)$, and so on, until there is a change of sign, indicating the location of a negative zero. The boundedness theorem can be used to determine when we need to look no further. We will then use the same process to locate positive zeros, finding $f(1), f(2), f(3)$, and so on. It is helpful to use the shortened form of synthetic division in the table below. Only the last row of the synthetic division is shown for each division. The first row of the table is used for each division and the work in the second row of the division is done mentally.

A table generated by a graphing calculator gives the same function values shown in the chart. Here, $Y_1 = x^4 - 6x^3 + 8x^2 + 2x - 1$.

x					$f(x) \approx$	
	1	-6	8	2	-1	
-1	1	-7	15	-13	12	Alternating signs; no zero < -1
0	1	-6	8	2	-1	← Zero between -1 and 0
1	1	-5	3	5	4	← Zero between 0 and 1
2	1	-4	0	2	3	
3	1	-3	-1	-1	-4	← Zero between 2 and 3
4	1	-2	0	2	7	← Zero between 3 and 4

Since the polynomial is of degree 4, there are no more than 4 zeros. Expand the table to approximate the real zeros to the nearest tenth. For example, for the zero between 0 and 1, the work might go as follows. Start halfway between 0 and 1 with $x = .5$. Since $f(.5) > 0$ and $f(0) < 0$, try $x = .4$ next.

Compare to the chart. A zero must be between $x = .3$ and $x = .2$.

x					$f(x) \approx$	
	1	-6	8	2	-1	
.5	1	-5.5	5.25	4.63	1.31	
.4	1	-5.6	5.76	4.30	.72	
.3	1	-5.7	6.29	3.89	.17	← Zero between .3 and .2
.2	1	-5.8	6.84	3.37	$-.33$	

The value $f(.3) \approx .17$ is closer to 0 than $f(.2) \approx -.33$, so to the nearest tenth, the zero is .3. Use synthetic division to verify that the remaining zeros are approximately 2.4, 3.7, and $-.4$. ▶

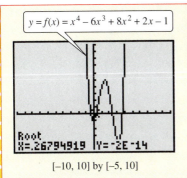

$y = f(x) = x^4 - 6x^3 + 8x^2 + 2x - 1$

Root
X=.26794919 Y=-2E-14

$[-10, 10]$ by $[-5, 10]$

Alternatively, a graphing calculator can be used to search for the real zeros of a polynomial. First choose a window that shows a complete graph. Use the boundedness theorem to determine suitable ranges for x and y. Then use the capabilities of the calculator to approximate the real zeros. It asks you for upper and lower bounds on the zero, and then operates in much the same way as we illustrate in Example 6. (The calculator does this much faster than we can!)

◀ **EXAMPLE 7**
Graphing a polynomial by first approximating the zeros

Graph $f(x) = 3x^4 - 14x^3 + 24x - 3$.

Because the highest power term is $3x^4$, the end behavior is positive at both ends. This fourth-degree polynomial has at most four zeros. Again $f(0) = -3$, a negative number, so there will be at least one positive and one negative zero by the intermediate value theorem.

To find the zeros and other points to plot, use synthetic division to make a table like the one shown below. Start with $x = 0$ and work up through the positive integers until a row with all positive numbers is found. Then work down through the negative integers until a row with alternating signs is found.

x					$f(x)$	Ordered pair	
	3	-14	0	24	-3		
5	**3**	**1**	**5**	**49**	242	(5, 242)	All positive / ← Zero
4	3	-2	-8	-8	-35	$(4, -35)$	
3	3	-5	-15	-21	-66	$(3, -66)$	
2	3	-8	-16	-8	-19	$(2, -19)$	← Zero
1	3	-11	-11	13	10	(1, 10)	← Zero
0	3	-14	0	54	-3	$(0, -3)$	
-1	3	-17	17	7	-10	$(-1, -10)$	← Zero
-2	**3**	**-20**	**40**	**-56**	109	$(-2, 109)$	Alternating signs

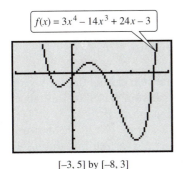

$f(x) = 3x^4 - 14x^3 + 24x - 3$

$[-3, 5]$ by $[-8, 3]$

A calculator graph of the function in Example 7

By the changes in sign of $f(x)$, the polynomial has zeros between -2 and -1, 0 and 1, 1 and 2, and 4 and 5. Each of these zeros is an x-intercept of the graph. Plotting the points from the table and drawing a continuous curve through them gives the graph shown in Figure 19 (on the following page). Notice that the graph has the maximum number of turning points, 3. ▶

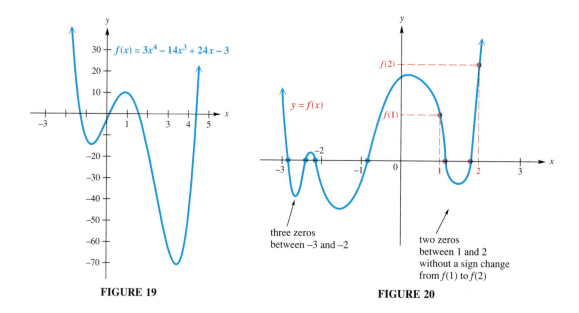

FIGURE 19　　　　　　　**FIGURE 20**

CAUTION　As the graph in Figure 20 suggests, it is possible to have several zeros between a pair of consecutive integers or two or more zeros without a sign change. Knowing the number of zeros to expect will help to avoid missing any zeros in such cases.

We saw earlier that data can sometimes be fit to linear models, while in this chapter we saw how quadratic functions can be used to model data. Higher degree polynomial functions may be used as well, and modern graphing calculators provide in their statistical functions the capability for determining many types of models. The final example shows how a cubic function can be used to model data.

◀**EXAMPLE 8**
Analyzing a cubic polynomial model

Based on information provided by the U.S. National Institute on Drug Abuse, the percent y of 18- to 25-year-olds who had used hallucinogens between 1974 and 1991 can be modeled by the function

$$y = .025x^3 - .70x^2 + 4.43x + 16.77$$

where $x = 0$ corresponds to the year 1974. In what year during the early years of this period did this type of drug use reach its maximum? Based on this model, what percent of 18- to 25-year-olds had used hallucinogens?

The graph of this function in the window [0, 20] by [0, 30] is shown in Figure 21. We can use the capabilities of the calculator to find that during the early years, the "highest point" is approximately (4.04, 24.89), meaning that during the fourth year after 1974 (that is, 1978), such drug use reached its maximum, with almost 25% of persons in this age group having reported use of hallucinogens. ◗

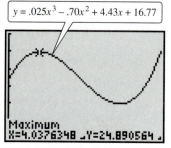

[0, 20] by [0, 30]
Xscl = 1　Yscl = 5

FIGURE 21

4.4 Exercises ▼▼▼▼▼▼▼▼▼▼▼▼▼▼▼▼▼▼▼▼▼▼▼▼▼▼▼▼▼▼▼▼▼▼▼▼

The graphs of four polynomial functions are shown here.

A.

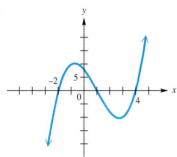

B.

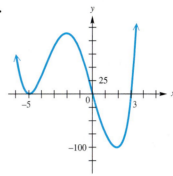

C.

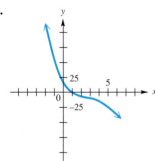

D.

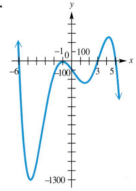

They represent the graphs of functions defined by these four equations, but not necessarily in the order listed:

$$y = x^3 - 3x^2 - 6x + 8$$
$$y = x^4 + 7x^3 - 5x^2 - 75x$$
$$y = -x^3 + 9x^2 - 27x + 17$$
$$y = -x^5 + 36x^3 - 22x^2 - 147x - 90.$$

Apply the concepts of this section to answer the question.

1. Which one of the graphs is that of $y = x^3 - 3x^2 - 6x + 8$?

2. Which one of the graphs is that of $y = x^4 + 7x^3 - 5x^2 - 75x$?

3. How many real zeros does the graph in (C) have?

4. Which one of (C) and (D) is the graph of $y = -x^3 + 9x^2 - 27x + 17$? (*Hint:* Look at the y-intercept.)

5. Which of the graphs cannot be that of a cubic polynomial function?

6. How many positive real zeros does the function graphed in (D) have?

7. How many negative real zeros does the function graphed in (A) have?

8. Which one of the graphs is that of a function whose range is *not* $(-\infty, \infty)$?

Sketch the graph of the polynomial function. See Examples 1 and 2.

9. $f(x) = 2x^4$

10. $f(x) = \dfrac{1}{4}x^6$

11. $f(x) = -\dfrac{2}{3}x^5$

12. $f(x) = -\dfrac{5}{4}x^5$

13. $f(x) = \dfrac{1}{2}x^3 + 1$

14. $f(x) = -x^4 + 2$

15. $f(x) = -(x + 1)^3$

16. $f(x) = (x + 2)^3 - 1$

17. $f(x) = (x - 1)^4 + 2$

18. $f(x) = \dfrac{1}{3}(x + 3)^4$

19. Which one of the following does not define a polynomial function?

(a) $f(x) = x^2$ (b) $f(x) = (x + 1)^3$ (c) $f(x) = \dfrac{1}{x}$ (d) $f(x) = 2x^5$

20. Write a short explanation of how the values of a, h, and k affect the graph of $f(x) = a(x - h)^n + k$ in comparison to the graph of $f(x) = x^n$.

Graph the polynomial function. Factor first if the expression is not in factored form. See Example 3.

21. $f(x) = 2x(x - 3)(x + 2)$

22. $f(x) = x^2(x + 1)(x - 1)$

23. $f(x) = x^2(x - 2)(x + 3)^2$

24. $f(x) = x^2(x - 5)(x + 3)(x - 1)$

25. $f(x) = x^3 - x^2 - 2x$

26. $f(x) = -x^3 - 4x^2 - 3x$

27. $f(x) = (x + 2)(x - 1)(x + 1)$

28. $f(x) = (x - 4)(x + 2)(x - 1)$

29. $f(x) = (3x - 1)(x + 2)^2$

30. $f(x) = (4x + 3)(x + 2)^2$

31. $f(x) = x^3 + 5x^2 - x - 5$

32. $f(x) = x^3 + x^2 - 36x - 36$

At the beginning of this exercise set, four polynomial functions were graphed by an artist using a computer. The screens below show these same four functions graphed by a graphing calculator. Compare the earlier graphs with these screens to prepare you for Exercises 33–36.

A.

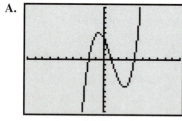

[–10, 10] by [–20, 20]
Xscl = 1 Yscl = 2

B.

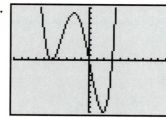

[–10, 10] by [–100, 100]
Xscl = 1   Yscl = 10

C.

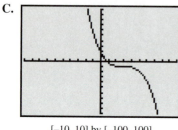

[–10, 10] by [–100, 100]
Xscl = 1 Yscl = 10

D.

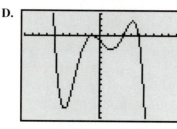

[–10, 10] by [–1500, 400]
Xscl = 1   Yscl = 100

Use a graphing calculator to graph the function defined by $f(x)$ in the window specified. Then compare the graph to the one shown in the answer section of this text.

33. $f(x) = 2x(x - 3)(x + 2)$ window: $[-3, 4]$ by $[-20, 12]$ Compare to Exercise 21.
34. $f(x) = x^2(x - 2)(x + 3)^2$ window: $[-4, 3]$ by $[-24, 4]$ Compare to Exercise 23.
35. $f(x) = (3x - 1)(x + 2)^2$ window: $[-4, 2]$ by $[-15, 15]$ Compare to Exercise 29.
36. $f(x) = x^3 + 5x^2 - x - 5$ window: $[-6, 2]$ by $[-30, 30]$ Compare to Exercise 31.

Use the intermediate value theorem for polynomials to show that the polynomial has a real zero between the numbers given. See Example 4.

37. $f(x) = 2x^2 - 7x + 4$;   2 and 3
38. $f(x) = 3x^2 - x - 4$;   1 and 2
39. $f(x) = 2x^3 - 5x^2 - 5x + 7$;   0 and 1
40. $f(x) = 2x^3 - 9x^2 + x + 20$;   2 and 2.5
41. $f(x) = 2x^4 - 4x^2 + 4x - 8$;   1 and 2
42. $f(x) = x^4 - 4x^3 - x + 3$;   1 and .5

Use a graphing calculator to approximate the real zero discussed in the specified exercise.

43. Exercise 37   **44.** Exercise 39   **45.** Exercise 40   **46.** Exercise 41

In Exercises 47 and 48, find a cubic polynomial having the graph shown.

47.
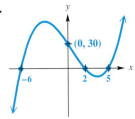

48.
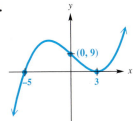

Show that the real zeros of each polynomial satisfy the given conditions. See Example 5.

49. $f(x) = 4x^3 - 3x^2 + 4x + 7$;   no real zero greater than 1
50. $f(x) = x^4 - x^3 + 2x^2 - 3x - 5$;   no real zero greater than 2
51. $f(x) = x^4 + x^3 - x^2 + 3$;   no real zero less than -2
52. $f(x) = x^5 + 2x^3 - 2x^2 + 5x + 5$;   no real zero less than -1

For the given polynomial, approximate each zero as a decimal to the nearest tenth. See Example 6.

53. $f(x) = x^3 + 3x^2 - 2x - 6$
54. $f(x) = x^3 - 3x + 3$
55. $f(x) = -2x^4 - x^2 + x + 5$
56. $f(x) = -x^4 + 2x^3 + 3x^2 + 6$

57. Explain why the graph of a polynomial having a as a zero of odd multiplicity crosses the x-axis at $x = a$.

58. Explain why the graph of a polynomial having a as a zero of even multiplicity touches, but does not cross, the x-axis at $x = a$.

Graph the polynomial function defined by $f(x)$. See Example 7.

59. $f(x) = x^3 - 2x^2 - x + 1$
60. $f(x) = -x^3 - x^2 + 2x + 1$
61. $f(x) = -4x^3 + 7x^2 - 2$
62. $f(x) = 5x^3 - 9x^2 + 1$
63. $f(x) = x^4 - 5x^2 + 2$
64. $f(x) = 2x^4 - 6x^3 + 7x - 2$

Solve the problem involving a polynomial function model.
See Example 8.

65. From 1930 to 1990 the rate of breast cancer was nearly constant at 30 cases per 100,000 females whereas the rate of lung cancer in females over the same period increased. The number of lung cancer cases per 100,000 females in the year t (where $t = 0$ corresponds to 1930) can be modeled using the function defined by $f(t) = 2.8 \times 10^{-4}t^3 - .011t^2 + .23t + .93$. (*Source: Valanis, B., Epidemiology in Nursing and Health Care, Appleton & Lange, Norwalk, Connecticut, 1992.*)

(a) Use a graphing calculator to graph the rates of breast and lung cancer for $0 \le t \le 60$. Use the window $[0, 60]$ by $[0, 40]$.

(b) Determine the year when rates for lung cancer first exceeded those for breast cancer.

(c) Discuss reasons for the rapid increase of lung cancer in females.

66. The number of military personnel on active duty in the United States during the period 1985 to 1990 can be determined by the cubic model $f(x) = -7.66x^3 + 52.71x^2 - 93.43x + 2151$, where $x = 0$ corresponds to 1985, and $f(x)$ is in thousands. Based on this model, how many military personnel were on active duty in 1990? (*Source: U.S. Department of Defense.*)

67. A survey team measures the concentration (in parts per million) of a particular toxin in a local river. On a normal day, the concentration of the toxin at time x (in hours) after the factory upstream dumps its waste is given by $g(x) = -.006x^4 + .14x^3 - .05x^2 + .02x$, where $0 \le x \le 24$.

(a) Graph $y = g(x)$ in the window $[0, 24]$ by $[0, 200]$.

(b) Estimate the time at which the concentration is greatest.

(c) A concentration greater than 100 parts per million is considered pollution. Using the graph from part (a), estimate the period during which the river is polluted.

68. During the early part of the twentieth century, the deer population of the Kaibab Plateau in Arizona experienced a rapid increase because hunters had reduced the number of natural predators and because the deer were protected from hunters. The increase in population depleted the food resources and eventually caused the population to decline. For the period from 1905 to 1930, the deer population was approximated by $D(x) = -.125x^5 + 3.125x^4 + 4000$, where x is time in years from 1905.

(a) Graph $y = D(x)$ in the window $[0, 50]$ by $[0, 120,000]$.

(b) From the graph, over what period of time (from 1905 to 1930) was the deer population increasing? Relatively stable? Decreasing?

69. The table lists the total annual amount (in millions of dollars) of government-guaranteed student loans from 1986 to 1994. (*Source: USA TODAY.*)

Year	Amount
1986	8.6
1987	9.8
1988	11.8
1989	12.5
1990	12.3
1991	13.5
1992	14.7
1993	16.5
1994	18.2

(a) Graph the data with the following three function definitions, where x represents the year.

(i) $f(x) = .4(x - 1986)^2 + 8.6$

(ii) $f(x) = 1.088(x - 1986) + 8.6$

(iii) $f(x) = 1.455\sqrt{x - 1986} + 8.6$

(b) Discuss which function definition models the data best.

70. The table lists the number of Americans (in thousands) who are expected to be over 100 years old for selected years. (*Source: U.S. Census Bureau.*)

Year	Number
1994	50
1996	56
1998	65
2000	75
2002	94
2004	110

(a) Use graphing to determine which polynomial best models the number of Americans over 100 years old where $x = 0$ corresponds to 1994.

(i) $f(x) = 6.057x + 44.714$

(ii) $g(x) = .4018x^2 + 2.039x + 50.071$
(iii) $h(x) = -.06x^3 + .506x^2 + 1.659x + 50.238$

(b) Use your choice from part (a) to predict the number of Americans who will be over 100 years old in the year 2008.

71. The time delay between an individual's initial infection with HIV and when that individual develops symptoms of AIDS is an important issue. The method of infection with HIV affects the time interval before AIDS develops. In one study of HIV patients who were infected by intravenous drug use, it was found that after 4 years 17% of the patients had AIDS and after 7 years 33% had developed the disease. The relationship between the time interval and the percent of patients with AIDS can accurately be modeled with a linear function defined by $f(x) = .05\overline{3}x - .04\overline{3}$, where x represents the time interval in years. (See Section 3.4, Exercise 47.) (*Source:* Alcabes, P., A. Munoz, D. Vlahov, and G. Friedland, "Incubation Period of Human Immunodeficiency Virus," *Epidemiologic Review,* Vol. 15, No. 2, The Johns Hopkins University School of Hygiene and Public Health, 1993.)

(a) Assuming the function continues to model the situation, determine the percent of patients with AIDS after 10 years.

(b) Predict the number of years before half of these patients will have AIDS.

72. The linear function defined by $f(x) = 1.5457x - 3067.7$ where x is the year can be used to estimate the percentage of gonorrhea cases with antibiotic resistance diagnosed from 1985 to 1990. (*Source:* Teutsch, S. and R. Churchill, *Principles and Practice of Public Health Surveillance,* Oxford University Press, New York, 1994.)

(a) Determine this percentage in 1988.

(b) Interpret the slope of the graph of f.

73. A simple pendulum will swing back and forth in regular time intervals. Grandfather clocks use pendulums to keep accurate time. The relationship between the length of a pendulum L and the time T for one complete oscillation can be expressed by the equation $L = kT^n$ where k is a constant and n is a positive integer to be determined. The following data were taken for different lengths of pendulums.

(a) If the length of the pendulum increases, what happens to T?

L (ft)	T (sec)
1.0	1.11
1.5	1.36
2.0	1.57
2.5	1.76
3.0	1.92
3.5	2.08
4.0	2.22

(b) Discuss how n and k could be found.

(c) Use the data to approximate k and determine the best value for n.

(d) Using the values of k and n from part (c), predict T for a pendulum having a length of 5 feet.

(e) If the length L of a pendulum doubles, what happens to the period T?

74. A technique for measuring cardiac output depends on the concentration of a dye in the bloodstream after a known amount is injected into a vein near the heart. For a normal heart, the concentration of dye in the bloodstream at time x (in seconds) is given by the function defined as follows.

$$g(x) = -.006x^4 + .140x^3 - .053x^2 + 1.79x$$

What is the concentration after 20 seconds?

Exercises 75–78 are problems that are geometric in nature, and lead to polynomial models. Solve each problem.

75. A piece of rectangular sheet metal is 20 inches wide. It is to be made into a rain gutter by turning up the edges to form parallel sides. Let x represent the length of each of the parallel sides.

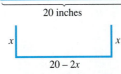

20 inches

x x

$20 - 2x$

(a) Give the restrictions on x.

(b) Describe a function A that gives the area of a cross-section of the gutter.

(c) For what value of x will A be a maximum (and thus maximize the amount of water that the gutter will hold)? What is this maximum area?

(d) For what values of x will the area of a cross-section be less then 40 square inches?

76. Find the value of x in the figure that will maximize the area of rectangle $ABCD$.

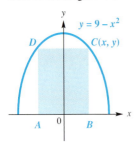

77. A certain right triangle has an area of 84 square inches. One leg of the triangle measures 1 inch less than the hypotenuse. Let x represent the length of the hypotenuse.

(a) Express the length of the leg mentioned above in terms of x.
(b) Express the length of the other leg in terms of x.
(c) Write an equation based on the information determined thus far. Square both sides and then write the equation with one side as a polynomial with integer coefficients, in descending powers, and the other side equal to 0.
(d) Solve the equation in part (c) graphically. Find the lengths of the three sides of the triangle.

78. A storage tank for butane gas is to be built in the shape of a right circular cylinder of altitude 12 feet, with a half sphere attached to each end. If x represents the radius of each half sphere, what radius should be used to cause the volume of the tank to be 144π cubic feet?

Exercises 79 and 80 provide a bar graph and a line graph, depicting information concerning data from California community colleges. Data such as these can be modeled by graphing calculators using built-in **polynomial fitting** *capability. Refer to your owner's manual as you work the problems.*

79. Use the points $(0, 118{,}000)$, $(2, 122{,}000)$, $(3, 63{,}000)$, and $(4, 70{,}000)$ to determine a cubic (third-degree) polynomial model for these data, where $x = 0$ corresponds to 1990 and $x = 4$ corresponds to 1994. Graph the function in the window $[0, 4]$ by $[0, 180{,}000]$.

Students Attending Community Colleges After Earning a B.A. or B.S. degree

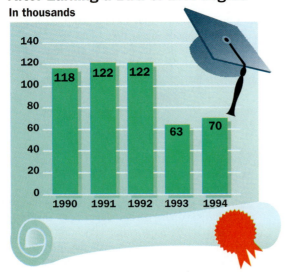

Source: Chancellor's Office, California Community Colleges

(a) Based on the cubic model, what was the statewide total of students with bachelor's degrees in 1991?
(b) How does your answer in (a) compare with the bar graph?

80. Use the points $(0, 1.24)$, $(2, 1.18)$, $(7, 1.5)$, $(9, 1.5)$, and $(11, 1.36)$ to determine a quartic (fourth-degree) model for these data, where $x = 0$ corresponds to 1983 and $x = 11$ corresponds to 1994. Graph the function in the window $[0, 11]$ by $[1.0, 1.6]$.

(a) Based on the quartic model, what was the statewide fall enrollment in millions in 1988?
(b) How does your answer in (a) compare with the line graph?

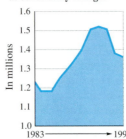

Statewide fall enrollment at community colleges

Source: Chancellor's Office, California Community Colleges

4.5 Rational Functions ▼▼▼

A rational expression is a fraction that is the quotient of two polynomials. A function defined by a rational expression is called a *rational function.*

Definition of Rational Function

If $p(x)$ and $q(x)$ are polynomials with $q(x) \neq 0$, then

$$f(x) = \frac{p(x)}{q(x)}$$

defines a **rational function.**

Since any values of x such that $q(x) = 0$ are excluded from the domain, a rational function usually has a graph with one or more breaks in it.

The simplest rational function with a variable denominator is defined by

$$f(x) = \frac{1}{x}.$$

The domain of this function is the set of all real numbers except 0. The number 0 cannot be used as a value of x, but it is helpful to find the values of $f(x)$ for several values of x close to 0. The following table shows what happens to $f(x)$ as x gets closer and closer to 0 from either side.

x approaches 0.

x	-1	$-.1$	$-.01$	$-.001$	$.001$	$.01$	$.1$	1
$f(x)$	-1	-10	-100	-1000	1000	100	10	1

$|f(x)|$ gets larger and larger.

The table suggests that $|f(x)|$ gets larger and larger as x gets closer and closer to 0, written in symbols as

$$|f(x)| \to \infty \quad \text{as} \quad x \to 0.$$

(The symbol $x \to 0$ means that x approaches as close as desired to 0, without ever being equal to 0.) Since x cannot equal 0, the graph of $f(x) = 1/x$ will never intersect the vertical line $x = 0$. The line $x = 0$, the y-axis, is called a *vertical asymptote* for the graph. The graph gets closer and closer to the y-axis as x gets closer and closer to 0.

On the other hand, as $|x|$ gets larger and larger, the values of $f(x) = 1/x$ get closer and closer to 0. (See the table.)

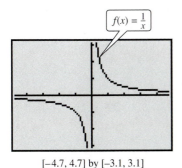

$f(x) = \frac{1}{x}$

[−4.7, 4.7] by [−3.1, 3.1]

A calculator graph of the simplest rational function

x	$-10{,}000$	-1000	-100	-10	10	100	1000	$10{,}000$
$f(x)$	$-.0001$	$-.001$	$-.01$	$-.1$	$.1$	$.01$	$.001$	$.0001$

Letting $|x|$ get larger and larger without bound (written $|x| \to \infty$) causes the graph of $y = 1/x$ to move closer and closer to the horizontal line $y = 0$, the x-axis. That is,

$$f(x) \to 0 \quad \text{as} \quad |x| \to \infty,$$

read "$f(x)$ approaches zero as the absolute value of x approaches infinity." The line $y = 0$ is called a *horizontal asymptote*.

To graph $f(x)$, first replace x with $-x$, getting $f(-x) = 1/(-x) = -1/x = -f(x)$, showing that $f(x) = 1/x$ is an odd function, symmetric with respect to the origin. Choosing some positive values of x and finding the corresponding values of $f(x)$ gives the first-quadrant part of the graph shown in Figure 22. The other part of the graph (in the third quadrant) can be found by symmetry.

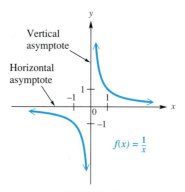

FIGURE 22

◀ EXAMPLE 1
Graphing a rational function using reflection

Graph $f(x) = -\dfrac{2}{x}$.

Rewrite $f(x)$ as

$$f(x) = -2 \cdot \frac{1}{x}.$$

Compared to $f(x) = 1/x$, the graph will be reflected about the x-axis (because of the negative sign) and each point will be twice as far from the x-axis. See the graph in Figure 23. The y-axis is the vertical asymptote and the horizontal asymptote is $y = 0$, or the x-axis. ▶

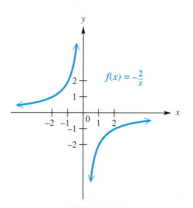

FIGURE 23

EXAMPLE 2
Graphing a rational function using translations

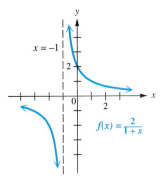

FIGURE 24

Graph $f(x) = \dfrac{2}{1+x}$.

The domain of this function is the set of all real numbers except -1. Since $x \neq -1$, the graph cannot cross the line $x = -1$, which is thus a vertical asymptote. The horizontal asymptote is $y = 0$. As shown in Figure 24, the graph is similar to that of $f(x) = 1/x$, translated 1 unit to the left. The y-intercept is 2. ▶

Special care must be taken in interpreting rational function graphs generated by a graphing calculator. If the calculator is in the connected mode, it may show a vertical line for x-values that produce vertical asymptotes. While this may be interpreted as a graph of the asymptote, a more realistic graph can be obtained by using the dot mode. For example, in the figures that follow we show graphs of the function defined by

$$f(x) = \frac{1}{x+3}$$

in both connected mode and dot mode.

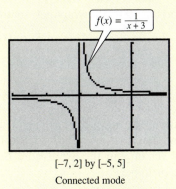

[−7, 2] by [−5, 5]

Connected mode

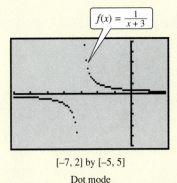

[−7, 2] by [−5, 5]

Dot mode

The window used may affect whether or not the vertical lines appear in a calculator graph in connected mode. To see this on the TI−82 calculator, change the window used above slightly to $[-7, 1]$ by $[-5, 5]$. The connected graph will not show the vertical line.

The examples above suggest the following definitions of vertical and horizontal asymptotes.

Definition of Asymptotes

For the rational function defined by $y = f(x)$, and for real numbers a and b,

if $|f(x)| \to \infty$ as $x \to a$, then the line $x = a$ is a **vertical asymptote;**
if $|f(x)| \to b$ as $|x| \to \infty$, then the line $y = b$ is a **horizontal asymptote.**

Locating asymptotes is an important part of sketching the graphs of rational functions. Vertical asymptotes are found by determining the values of x that make the denominator equal to 0 but do not make the numerator equal to 0. Horizontal asymptotes (and, in some cases, oblique asymptotes) are found by considering what happens to $f(x)$ as $|x| \to \infty$. The next example shows how to find asymptotes.

EXAMPLE 3
Finding asymptotes of graphs of rational functions

For each rational function f, find all asymptotes.

(a) $f(x) = \dfrac{x + 1}{(2x - 1)(x + 3)}$

To find the vertical asymptotes, set the denominator equal to zero and solve.

$$(2x - 1)(x + 3) = 0$$

$$2x - 1 = 0 \quad \text{or} \quad x + 3 = 0 \qquad \textcolor{blue}{\text{Zero-factor property}}$$

$$x = \frac{1}{2} \quad \text{or} \quad x = -3$$

The equations of the vertical asymptotes are $x = 1/2$ and $x = -3$.

To find the equation of the horizontal asymptote, we divide each term by the largest power of x in the expression. Begin by multiplying the factors in the denominator to get

$$f(x) = \frac{x + 1}{(2x - 1)(x + 3)} = \frac{x + 1}{2x^2 + 5x - 3}.$$

Now divide each term in the numerator and denominator by x^2, since 2 is the largest exponent on x. This gives

$$f(x) = \frac{\dfrac{x}{x^2} + \dfrac{1}{x^2}}{\dfrac{2x^2}{x^2} + \dfrac{5x}{x^2} - \dfrac{3}{x^2}} = \frac{\dfrac{1}{x} + \dfrac{1}{x^2}}{2 + \dfrac{5}{x} - \dfrac{3}{x^2}}.$$

As $|x|$ gets larger and larger, the quotients $1/x$, $1/x^2$, $5/x$, and $3/x^2$ all approach 0, and the value of $f(x)$ approaches

$$\frac{0 + 0}{2 + 0 - 0} = \frac{0}{2} = 0.$$

The line $y = 0$ (that is, the x-axis) is therefore the horizontal asymptote.

(b) $f(x) = \dfrac{2x + 1}{x - 3}$

Set the denominator equal to zero to find that the vertical asymptote has the equation $x = 3$. To find the horizontal asymptote, divide each term in the rational expression by x, since the greatest power of x in the expression is 1.

$$f(x) = \frac{2x + 1}{x - 3} = \frac{\dfrac{2x}{x} + \dfrac{1}{x}}{\dfrac{x}{x} - \dfrac{3}{x}} = \frac{2 + \dfrac{1}{x}}{1 - \dfrac{3}{x}}$$

As $|x|$ gets larger and larger, both $1/x$ and $3/x$ approach 0, and $f(x)$ approaches

$$\frac{2 + 0}{1 - 0} = \frac{2}{1} = 2,$$

so the line $y = 2$ is the horizontal asymptote.

(c) $f(x) = \dfrac{x^2 + 1}{x - 2}$

Setting the denominator equal to zero shows that the vertical asymptote has the equation $x = 2$. If we divide by the largest power of x as before (x^2 in this case), we see that there is no horizontal asymptote because

$$f(x) = \frac{\dfrac{x^2}{x^2} + \dfrac{1}{x^2}}{\dfrac{x}{x^2} - \dfrac{2}{x^2}} = \frac{1 + \dfrac{1}{x^2}}{\dfrac{1}{x} - \dfrac{2}{x^2}}$$

does not approach any real number as $|x| \to \infty$, since $1/0$ is undefined. This will happen whenever the degree of the numerator is greater than the degree of the denominator. In such cases, divide the denominator into the numerator to write the expression in another form. Using synthetic division gives

$$
\begin{array}{r|rrr}
2 & 1 & 0 & 1 \\
  &   & 2 & 4 \\
\hline
  & 1 & 2 & 5.
\end{array}
$$

The function can now be written as

$$f(x) = \frac{x^2 + 1}{x - 2} = x + 2 + \frac{5}{x - 2}.$$

For very large values of $|x|$, $5/(x - 2)$ is close to 0, and the graph approaches the line $y = x + 2$. This line is an **oblique asymptote** (neither vertical nor horizontal) for the graph of the function.

In general, if the degree of the numerator is exactly one more than the degree of the denominator, a rational function may have an oblique asymptote. The equation of this asymptote is found by dividing the numerator by the denominator and disregarding the remainder. ▶

The results of Example 3 can be summarized as follows.

Determining Asymptotes

In order to find asymptotes of a rational function defined by a rational expression *in lowest terms,* use the following procedures.

1. **Vertical Asymptotes**
   Find any vertical asymptotes by setting the denominator equal to 0 and solving for x. If a is a zero of the denominator, then the line $x = a$ is a vertical asymptote.
2. **Other Asymptotes**
   Determine any other asymptotes. We consider three possibilities:
   (a) If the numerator has lower degree than the denominator, there is a horizontal asymptote, $y = 0$ (the x-axis).
   (b) If the numerator and denominator have the same degree, and the function is of the form

   $$f(x) = \frac{a_n x^n + \cdots + a_0}{b_n x^n + \cdots + b_0}, \quad \text{where } b_n \neq 0,$$

   dividing by x^n in the numerator and denominator produces the horizontal asymptote

   $$y = \frac{a_n}{b_n}.$$

   (c) If the numerator is of degree exactly one more than the denominator, there will be an oblique asymptote. To find it, divide the numerator by the denominator and disregard any remainder. Set the rest of the quotient equal to y to get the equation of the asymptote.

NOTE The graph of a rational function may have more than one vertical asymptote, or it may have none at all. The graph cannot intersect any vertical asymptote. There can be only one other (non-vertical) asymptote, and the graph *may* intersect that asymptote. This will be seen in Example 6. The method of graphing a rational function having common variable factors in the numerator and denominator of the defining expression will be covered in Example 8.

The following procedure can be used to graph functions defined by rational expressions reduced to lowest terms.

Graphing Rational Functions

Let $f(x) = \dfrac{p(x)}{q(x)}$ define a function where the rational expression is written in lowest terms. To sketch its graph, follow the steps below.

1. Find any vertical asymptotes.
2. Find any horizontal or oblique asymptotes.
3. Find the y-intercept by evaluating $f(0)$.
4. Find the x-intercepts, if any, by solving $f(x) = 0$. (These will be the zeros of the numerator, $p(x)$.)
5. Determine whether the graph will intersect its non-vertical asymptote by solving $f(x) = b$, or $f(x) = mx + b$, where b (or $mx + b$) is the y-value of the non-vertical asymptote.
6. Plot a few selected points, as necessary. Choose an x-value in each interval of the domain as determined by the vertical asymptotes and x-intercepts.
7. Complete the sketch.

The next example shows how the above guidelines can be used to graph a rational function.

EXAMPLE 4
Graphing a rational function with the x-axis as the horizontal asymptote

Graph $f(x) = \dfrac{x + 1}{(2x - 1)(x + 3)}$.

Step 1 As shown in Example 3(a), the vertical asymptotes have equations $x = 1/2$ and $x = -3$.

Step 2 Again, as shown in Example 3(a), the horizontal asymptote is the x-axis.

Step 3 Since $f(0) = \dfrac{0 + 1}{(2(0) - 1)((0) + 3)} = -\dfrac{1}{3}$, the y-intercept is $-1/3$.

Step 4 The x-intercept is found by solving $f(x) = 0$.

$$\frac{x + 1}{(2x - 1)(x + 3)} = 0$$

$$x + 1 = 0 \qquad \text{Multiply by } (2x - 1)(x + 3).$$

$$x = -1$$

The x-intercept is -1.

Step 5 To determine whether the graph intersects its horizontal asymptote, solve

$$f(x) = \underset{\underset{\textstyle y\text{-value of horizontal asymptote}}{\uparrow}}{0}.$$

Since the horizontal asymptote is the x-axis, the solution of this equation was found in Step 4. The graph intersects its horizontal asymptote at $(-1, 0)$.

Step 6   Since this graph has an *x*-intercept, consider the sign of $f(x)$ in each region determined by a vertical asymptote or an *x*-intercept. Here there are four regions to be considered, as shown in the chart below.

Region	Test point	Value of $f(x)$	Sign of $f(x)$
$(-\infty, -3)$	-4	$-1/3$	$-$
$(-3, -1)$	-2	$1/5$	$+$
$(-1, 1/2)$	0	$-1/3$	$-$
$(1/2, \infty)$	2	$1/5$	$+$

Step 7   Using the asymptotes and intercepts and plotting a few points (shown in the table of values) gives the graph in Figure 25. ▶

x	y
-4	$-1/3$
-2	$1/5$
-1	0
0	$-1/3$
1	$1/2$
2	$1/5$

$$f(x) = \frac{x+1}{(2x-1)(x+3)}$$

FIGURE 25

In the remaining examples, we will not specifically number the steps.

◀**EXAMPLE 5**
Graphing a rational function that does not intersect its horizontal asymptote

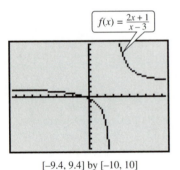

$$f(x) = \frac{2x+1}{x-3}$$

[–9.4, 9.4] by [–10, 10]

A calculator graph of the function in Example 5

Graph $f(x) = \dfrac{2x + 1}{x - 3}$.

As shown in Example 3(b), the equation of the vertical asymptote is $x = 3$ and the equation of the horizontal asymptote is $y = 2$. Since $f(0) = -1/3$, the *y*-intercept is $-1/3$. The solution of $f(x) = 0$ is $-1/2$, so the only *x*-intercept is $-1/2$. The graph does not intersect its horizontal asymptote, since $f(x) = 2$ has no solution. (Verify this.) The points $(-4, 1)$ and $(6, 13/3)$ are on the graph and can be used to complete the sketch, as shown in Figure 26. ▶

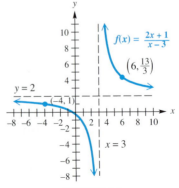

$$f(x) = \frac{2x+1}{x-3}$$

FIGURE 26

EXAMPLE 6
Graphing a rational function that intersects its horizontal asymptote

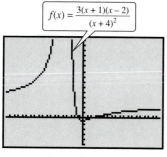

[−18.8, 18.8] by [−5, 15]

A calculator graph of the function in Example 6

Graph $f(x) = \dfrac{3(x + 1)(x - 2)}{(x + 4)^2}$.

The only vertical asymptote is the line $x = -4$. To find any horizontal asymptotes, multiply the factors in the numerator and denominator.

$$f(x) = \frac{3x^2 - 3x - 6}{x^2 + 8x + 16}$$

As explained in the guidelines above, the equation of the horizontal asymptote can be shown to be

$$y = \frac{3}{1} \begin{array}{l} \leftarrow \text{Leading coefficient of numerator} \\ \leftarrow \text{Leading coefficient of denominator} \end{array}$$

or $y = 3$. The y-intercept is $-3/8$ and the x-intercepts are -1 and 2. By setting $f(x) = 3$ and solving, we can find the point where the graph intersects the horizontal asymptote.

$$f(x) = \frac{3x^2 - 3x - 6}{x^2 + 8x + 16}$$

$$3 = \frac{3x^2 - 3x - 6}{x^2 + 8x + 16}$$

$3x^2 - 3x - 6 = 3x^2 + 24x + 48 \qquad \text{Multiply by } x^2 + 8x + 16.$

$-3x - 6 = 24x + 48 \qquad \text{Subtract } 3x^2.$

$-27x = 54$

$x = -2$

The graph intersects its horizontal asymptote at $(-2, 3)$.

Some other points that lie on the graph are $(-10, 9), (-3, 30),$ and $(5, 2/3)$. These can be used to complete the graph, shown in Figure 27.

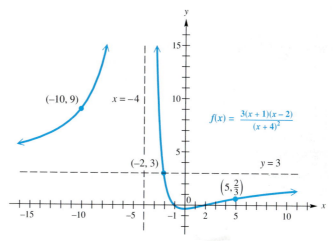

FIGURE 27

The next example discusses a rational function defined by an expression having the degree of its numerator greater than the degree of its denominator.

Graph $f(x) = \dfrac{x^2 + 1}{x - 2}$.

EXAMPLE 7
Graphing a rational function with an oblique asymptote

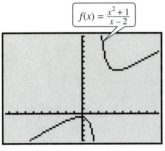

[−9.4, 9.4] by [−5, 15]

A calculator graph of the function in Example 7

As shown in Example 3(c), the vertical asymptote has the equation $x = 2$, and the graph has an oblique asymptote with the equation $y = x + 2$. The y-intercept is $-1/2$, and the graph has no x-intercepts, since the numerator, $x^2 + 1$, has no real zeros. It can be shown that the graph does not intersect its oblique asymptote. Using the intercepts, asymptotes, the points $(4, 17/2)$ and $(-1, -2/3)$, and the general behavior of the graph near its asymptotes, leads to the graph shown in Figure 28. ▶

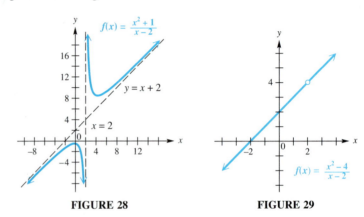

FIGURE 28   **FIGURE 29**

As mentioned earlier, a rational function must be defined by an expression in lowest terms before we can use the methods discussed in this section to determine the graph. The next example shows a typical rational function defined by an expression that is not in lowest terms.

EXAMPLE 8
Graphing a rational function defined by an expression that is not in lowest terms

X	Y₁	Y₂
-1	1	1
0	2	2
1	3	3
2	ERROR	4
3	5	5
4	6	6
5	7	7

Y₁=ERROR

The table supports the fact that if $Y_1 = \frac{x^2 - 4}{x - 2}$ and if $Y_2 = x + 2$, then $Y_1 = Y_2$ except when $x = 2$, where Y_1 is undefined.

Graph $f(x) = \dfrac{x^2 - 4}{x - 2}$.

Start by noticing that the domain of this function cannot contain 2. The rational expression $(x^2 - 4)/(x - 2)$ can be reduced to lowest terms by factoring the numerator, and using the fundamental principle.

$$f(x) = \frac{x^2 - 4}{x - 2} = \frac{(x + 2)(x - 2)}{x - 2} = x + 2 \quad (x \neq 2)$$

Therefore, the graph of this function will be the same as the graph of $y = x + 2$ (a straight line), with the exception of the point with x-value 2. A "hole" appears in the graph at $(2, 4)$. See Figure 29. ▶

In Example 8, if the window of a graphing calculator is set so that 2 is an x-value for the location of the cursor, then the display will show an unlit pixel at 2. To see this, look carefully at the figure below at the point on the screen where $x = 2$. However, such points of discontinuity will often *not* be evident from calculator-generated graphs—once again showing us a reason for studying the concepts along with the technology.

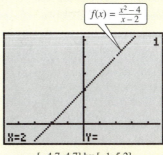

$$f(x) = \frac{x^2 - 4}{x - 2}$$

[−4.7, 4.7] by [−1, 5.2]

There is a tiny gap in the graph at $x = 2$.

Rational functions have a variety of applications. The next example discusses one, and several others are given in the exercises for this section.

EXAMPLE 9
Applying rational functions to traffic intensity*

Vehicles arrive randomly at a parking ramp with an average rate of 2.6 vehicles per minute. The parking attendant can admit 3.2 vehicles per minute. However, since arrivals are random, lines form at various times.

(a) The traffic intensity x is defined as the ratio of the average arrival rate to the average admittance rate. Determine x for this parking ramp.

The average arrival rate is 2.6 vehicles and the average admittance rate is 3.2 vehicles, so

$$x = \frac{2.6}{3.2} = .8125.$$

(b) The average number of vehicles waiting in line to enter the ramp is given by the rational function $f(x) = \dfrac{x^2}{2(1 - x)}$ where $0 \le x < 1$ is the traffic intensity. Compute $f(x)$ for this parking ramp.

In part (a) we found that $x = .8125$. Thus,

$$f(x) = \frac{.8125^2}{2(1 - .8125)} \approx 1.76 \text{ vehicles.}$$

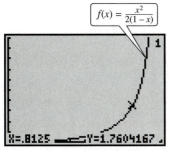

$$f(x) = \frac{x^2}{2(1 - x)}$$

X=.8125   Y=1.7604167

[0, 1] by [0, 5]
Xscl = .1   Yscl = .5

The display supports the result in Example 9(b).

FIGURE 30

(c) Graph $f(x)$. What happens to the number of vehicles waiting as the traffic intensity approaches 1?

From the graph, shown in Figure 30, we see that as x approaches 1, $y = f(x)$ gets very large. This is not surprising; it is what we would expect.

*Source: Mannering, F. and W. Kilareski, *Principles of Highway Engineering and Traffic Control*, John Wiley & Sons, 1990.

4.5 Exercises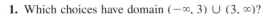

Use the graphs of the rational functions in (A) through (D) to answer the questions in Exercises 1–8. Give all possible answers, as there may be more than one correct choice.

1. Which choices have domain $(-\infty, 3) \cup (3, \infty)$?

2. Which choices have range $(-\infty, 3) \cup (3, \infty)$?

3. Which choices have range $(-\infty, 0) \cup (0, \infty)$?

4. Which choices have range $(0, \infty)$?

5. If f represents the function, only one choice has a single solution to the equation $f(x) = 3$. Which one is it?

6. What is the range of the function in (D)?

7. Which choices have the x-axis as a horizontal asymptote?

8. Which choices are symmetric with respect to a vertical line?

A.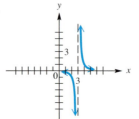

B.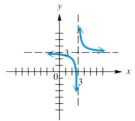

C.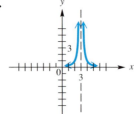

D.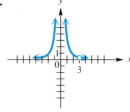

Use reflections, symmetry, and translations to graph the rational function. See Examples 1 and 2.

9. $f(x) = \dfrac{2}{x}$

10. $f(x) = -\dfrac{3}{x}$

11. $f(x) = \dfrac{1}{x + 2}$

12. $f(x) = \dfrac{1}{x - 3}$

13. $f(x) = \dfrac{1}{x} + 1$

14. $f(x) = \dfrac{1}{x} - 2$

Give the equations of the vertical, horizontal, and/or oblique asymptotes of the rational function. See Example 3.

15. $f(x) = \dfrac{3}{x - 5}$

16. $f(x) = \dfrac{-6}{x + 9}$

17. $f(x) = \dfrac{4 - 3x}{2x + 1}$

18. $f(x) = \dfrac{2x + 6}{x - 4}$

19. $f(x) = \dfrac{x^2 - 1}{x + 3}$

20. $f(x) = \dfrac{x^2 + 4}{x - 1}$

21. $f(x) = \dfrac{(x - 3)(x + 1)}{(x + 2)(2x - 5)}$

22. $f(x) = \dfrac{3(x + 2)(x - 4)}{(5x - 1)(x - 5)}$

23. Which one of the following has a graph that does not have a vertical asymptote?

   (a) $f(x) = \dfrac{1}{x^2 + 2}$ **(b)** $f(x) = \dfrac{1}{x^2 - 2}$ **(c)** $f(x) = \dfrac{3}{x^2}$ **(d)** $f(x) = \dfrac{2x + 1}{x - 8}$

24. Which one of the following has a graph that does not have a horizontal asymptote?

   (a) $f(x) = \dfrac{2x - 7}{x + 3}$ **(b)** $f(x) = \dfrac{3x}{x^2 - 9}$ **(c)** $f(x) = \dfrac{x^2 - 9}{x + 3}$ **(d)** $f(x) = \dfrac{x + 5}{(x + 2)(x - 3)}$

25. The figures below show the four ways that the graph of a rational function can approach the vertical line $x = 2$ as an asymptote. Identify the graph of each rational function defined as follows.

(a) $f(x) = \dfrac{1}{(x - 2)^2}$

(b) $f(x) = \dfrac{1}{x - 2}$

(c) $f(x) = \dfrac{-1}{x - 2}$

(d) $f(x) = \dfrac{-1}{(x - 2)^2}$

A.
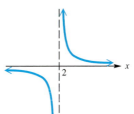

B.
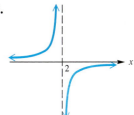

C.
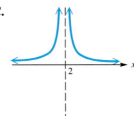

D.
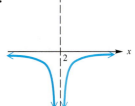

26. Describe in your own words what is meant by an *asymptote* of the graph of a rational function.

Sketch the graph of the rational function. See Examples 4–8.

27. $f(x) = \dfrac{x + 1}{x - 4}$

28. $f(x) = \dfrac{x - 5}{x + 3}$

29. $f(x) = \dfrac{3x}{(x + 1)(x - 2)}$

30. $f(x) = \dfrac{2x + 1}{(x + 2)(x + 4)}$

31. $f(x) = \dfrac{5x}{x^2 - 1}$

32. $f(x) = \dfrac{x}{4 - x^2}$

33. $f(x) = \dfrac{(x - 3)(x + 1)}{(x - 1)^2}$

34. $f(x) = \dfrac{x(x - 2)}{(x + 3)^2}$

35. $f(x) = \dfrac{x}{x^2 - 9}$

36. $f(x) = \dfrac{4}{5 + 3x}$

37. $f(x) = \dfrac{1}{x^2 + 1}$

38. $f(x) = \dfrac{(x - 5)(x - 2)}{x^2 + 9}$

39. $f(x) = \dfrac{x^2 + 1}{x + 3}$

40. $f(x) = \dfrac{2x^2 + 3}{x - 4}$

41. $f(x) = \dfrac{x^2 + 2x}{2x - 1}$

42. $f(x) = \dfrac{x^2 - x}{x + 2}$

43. $f(x) = \dfrac{x^2 - 9}{x + 3}$

44. $f(x) = \dfrac{x^2 - 16}{x + 4}$

Find an equation for the rational function graph.

45.
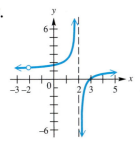

46.
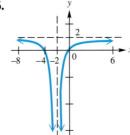

47.

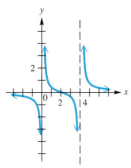

48.

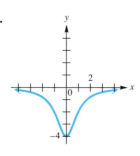

Because rational functions may have gaps or breaks in their graphs, graphing them with a graphing calculator sometimes leads to a view that does not accurately depict the true behavior of the graph. Choosing an appropriate window, along with the dot graphing mode (rather than connected), is an important consideration when using a graphing calculator for rational functions. Occasionally a "well-chosen" window will not yield the vertical lines when the calculator is in connected mode, but it takes a lot of experience and a knowledge of how your particular calculator is designed in order to accomplish this.

In the two screens below, we graph the function $f(x) = \dfrac{x}{x^2 - 16}$ in two different windows.

The first is in the standard window, and the second is in the decimal window $[-4.7, 4.7]$ by $[-3.1, 3.1]$, as generated by a TI-82 calculator. Both are in connected mode.

You can see how different the two screens appear. Which one do you think gives a more accurate depiction of the behavior of the graph?

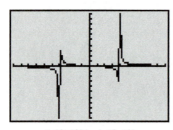

$[-10, 10]$ by $[-10, 10]$

Standard viewing window

$f(x) = \dfrac{x}{x^2 - 16}$

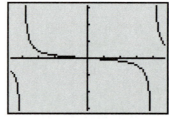

$[-4.7, 4.7]$ by $[-3.1, 3.1]$

Decimal viewing window

$f(x) = \dfrac{x}{x^2 - 16}$

Use a graphing calculator to graph the rational function in the window specified. Then compare the graph to the one shown in the answer section of this text.

49. $f(x) = \dfrac{x + 1}{x - 4}$ window: $[-6, 10]$ by $[-8, 8]$ Use dot mode and compare to Exercise 27.

50. $f(x) = \dfrac{5x}{x^2 - 1}$ window: $[-4, 4]$ by $[-4, 4]$ Use dot mode and compare to Exercise 31.

51. $f(x) = \dfrac{(x - 3)(x + 1)}{(x - 1)^2}$ window: $[-3, 5]$ by $[-4, 2]$ Use connected mode and compare to Exercise 33.

52. $f(x) = \dfrac{x^2 + 1}{x + 3}$ window: $[-10, 6]$ by $[-20, 5]$ Use dot mode and compare to Exercise 39.

53. Let $f(x) = p(x)/q(x)$ define a rational function where the expression is reduced to lowest terms. Suppose the degree of $p(x)$ is m and the degree of $q(x)$ is n. Write an explanation of how you would determine the nonvertical asymptote in each of the following situations.
   (a) $m < n$ **(b)** $m = n$ **(c)** $m > n$

54. Suppose a friend tells you that the graph of
$$f(x) = \frac{x^2 - 25}{x + 5}$$
has a vertical asymptote with equation $x = -5$. Is this correct? If not, describe the behavior of the graph at $x = -5$.

Solve the problem. See Example 9.

55. The table contains incidence ratios by age for deaths due to coronary heart disease (CHD) and lung cancer (LC) when comparing smokers (21–39 cigarettes per day) to nonsmokers.

Age	CHD	LC
55–64	1.9	10
65–74	1.7	9

The incidence ratio of 10 means that smokers are 10 times more likely than nonsmokers to die of lung cancer between the ages of 55 and 64. If the incidence ratio is x, then the percent P (in decimal form) of deaths caused by smoking can be calculated using the rational function $P(x) = \dfrac{x - 1}{x}$. (*Source:* Walker, A., *Observation and Inference: An Introduction to the Methods of Epidemiology*, Epidemiology Resources Inc., Newton Lower Falls, MA, 1991.)
   (a) As x increases, what value does $P(x)$ approach?
   (b) Why do you suppose the incidence ratios are slightly smaller for ages 65–74 than for ages 55–64?

56. Refer to Example 9. Let the average number of vehicles arriving at the gate of an amusement park per minute be equal to k, and let the average number of vehicles admitted by the park attendants be equal to r. Then, the average waiting time T (in minutes) for each vehicle arriving at the park is given by the rational function $T(r) = \dfrac{2r - k}{2r^2 - 2kr}$ where $r > k$. (*Source:* Mannering, F. and W. Kilareski, *Principles of Highway Engineering and Traffic Control*, John Wiley & Sons, Inc., 1990.)

(a) It is known from experience that on Saturday afternoon k is equal to 25 vehicles per minute. Use graphing to estimate the admittance rate r that is necessary to keep the average waiting time T for each vehicle to 30 seconds.
   (b) If one park attendant can serve 5.3 vehicles per minute, how many park attendants will be needed to keep the average wait to 30 seconds?

57. The rational function defined by $d(x) = \dfrac{(8.71 \times 10^3)x^2 - (6.94 \times 10^4)x + (4.70 \times 10^5)}{(1.08)x^2 - (3.24 \times 10^2)x + (8.22 \times 10^4)}$ can be used to accurately model the braking distance for automobiles traveling at x miles per hour where $20 \le x \le 70$. (*Source:* Mannering, F. and W. Kilareski, *Principles of Highway Engineering and Traffic Control*, John Wiley & Sons, Inc., 1990.)
   (a) Use graphing to estimate x when $d(x) = 300$.
   (b) Complete the table for each value of x.

x	$d(x)$
20	
25	
30	
35	
40	
45·	
50	
55	
60	
65	
70	

(c) If a car doubles its speed, does the stopping distance double or more than double? Explain.

(d) Suppose the stopping distance doubled whenever the speed doubled. What type of relationship would exist between the stopping distance and the speed?

58. Refer to Exercises 51 and 52 in Section 4.1.

(a) Make a table listing the ratio of total deaths caused by AIDS to total cases of AIDS in the United States for each year from 1982 to 1993. (For example, in 1982 there were 620 deaths and 1563 cases so the ratio is $\dfrac{620}{1563} \approx .397$.)

(b) As time progresses what happens to the values of the ratio?

(c) 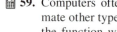 Using the polynomial functions f and g that were found in the exercises cited above, define the rational function h, where $h(x) = \dfrac{g(x)}{f(x)}$. Graph $h(x)$ on the interval $[2, 20]$. Compare $h(x)$ to the values for the ratio found in your table.

(d) Use $h(x)$ to write an equation that approximates the relationship between the functions $f(x)$ and $g(x)$ as x increases.

(e) The ratio of AIDS deaths to AIDS cases can be used to estimate the total number of AIDS deaths. According to the World Health Organization, in 1994 there had been 4 million AIDS cases diagnosed worldwide since the disease began. Predict the total number of deaths caused by AIDS.

59. Computers often use rational functions to approximate other types of functions. Use graphing to match the function with its rational approximation on the interval $[1, 15]$.

(a) $f_1(x) = \sqrt{x}$ **(b)** $f_2(x) = \sqrt{4x + 1}$

(c) $f_3(x) = \sqrt[3]{x}$ **(d)** $f_4(x) = \dfrac{1 - \sqrt{x}}{1 + \sqrt{x}}$

(i) $r_1(x) = \dfrac{2 - 2x^2}{3x^2 + 10x + 3}$

(ii) $r_2(x) = \dfrac{15x^2 + 75x + 33}{x^2 + 23x + 31}$

(iii) $r_3(x) = \dfrac{10x^2 + 80x + 32}{x^2 + 40x + 80}$

(iv) $r_4(x) = \dfrac{7x^3 + 42x^2 + 30x + 2}{2x^3 + 30x^2 + 42x + 7}$

60. In recent years, economist Arthur Laffer has been a center of controversy because of his **Laffer curve,** an idealized version of which is shown here.

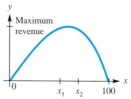

Tax rate percent

According to this curve, increasing a tax rate, say from x_1 percent to x_2 percent on the graph above, can actually lead to a decrease in government revenue. All economists agree on the endpoints, 0 revenue at tax rates of both 0% and 100%, but there is much disagreement on the location of the rate x_1 that produces maximum revenue.

(a) Suppose an economist studying the Laffer curve produces the rational function defined by

$$R(x) = \dfrac{80x - 8000}{x - 110},$$

with $R(x)$ giving government revenue in tens of millions of dollars for a tax rate of x percent, with the function valid for $55 \leq x \leq 100$. Find the revenue for the following tax rates.

(i) 55% **(ii)** 60% **(iii)** 70% **(iv)** 90% **(v)** 100%

(vi) Graph R in the window $[0, 100]$ by $[0, 80]$.

(b) Suppose an economist studies a different tax, this time producing

$$R(x) = \dfrac{60x - 6000}{x - 120}$$

where $y = R(x)$ is government revenue in millions of dollars from a tax rate of x percent, with $y = R(x)$ valid for $50 \leq x \leq 100$. Find the revenue for the following tax rates.

(i) 50% **(ii)** 60% **(iii)** 80% **(iv)** 100%

(v) Graph R in the window $[0, 100]$ by $[0, 50]$.

Chapter 4   Summary ▼▼▼▼▼▼▼▼▼▼▼▼▼▼▼▼▼▼▼▼▼▼▼▼▼▼▼▼▼▼▼▼▼

KEY TERMS	EXAMPLES
polynomial function leading coefficient zero polynomial zero of a polynomial function	

4.1 Quadratic Functions

quadratic function parabola axis vertex	**Equations of Vertical Parabolas** $y = a(x - h)^2 + k;$ vertex at $(h, k);$ axis $x = h$ $y = f(x) = ax^2 + bx + c$ vertex at $\left(-\dfrac{b}{2a}, f\left(-\dfrac{b}{2a}\right)\right);$ axis $x = -\dfrac{b}{2a}$

4.2 Synthetic Division

	Synthetic division is a shortcut method for dividing a polynomial by a binomial of the form $x - k$.

4.3 Zeros of Polynomial Functions

zero of multiplicity n	**Factor Theorem** $x - k$ is a factor of the polynomial $f(x)$ if and only if $f(k) = 0$. **Fundamental Theorem of Algebra** Every polynomial function has at least one complex zero. **Number of Zeros Theorem** A polynomial function has at most n distinct zeros. **Conjugate Zeros Theorem** See page 300. A polynomial function of degree n has at most $n - 1$ turning points.

4.4 Graphs of Polynomial Functions

end behavior turning points complete graph	**Intermediate Value Theorem** See page 310. **Boundedness Theorem** See page 311.

4.5 Rational Functions

rational function vertical asymptote horizontal asymptote oblique asymptote	**Graphing Rational Functions** To graph a rational function defined by an expression in lowest terms, find asymptotes and intercepts. Determine whether the graph intersects the nonvertical asymptote. Plot a few points, as necessary, to complete the sketch. If a rational function is defined by an expression not reduced to lowest terms, there may be a "hole" in the graph.

Chapter 4   Review Exercises ▼▼▼▼▼▼▼▼▼▼▼▼▼▼▼▼▼▼▼▼▼▼▼▼▼▼▼▼▼▼▼▼▼▼▼

Graph the quadratic function. Give the vertex, axis, x-intercepts, and y-intercept of each graph.

1. $f(x) = 3(x + 4)^2 - 5$

2. $f(x) = -\dfrac{2}{3}(x - 6)^2 + 7$

3. $f(x) = -3x^2 - 12x - 1$

4. $f(x) = 4x^2 - 4x + 3$

In Exercises 5–8, consider the graph of $f(x) = a(x - h)^2 + k$, with $a > 0$.

5. What is the y-coordinate of the lowest point of the graph?

6. What is the x-coordinate of the lowest point of the graph?

7. What is the y-intercept of the graph?

8. Under what conditions, involving the letters a, h, or k, will the graph have one or more x-intercepts? For these conditions, express the x-intercept(s) in terms of a, h, and k.

9. If a is positive, what is the smallest value of $ax^2 + bx + c$?

10. Use a parabola to find the dimensions of the rectangular region of maximum area that can be enclosed with 180 meters of fencing if no fencing is needed along one side of the region.

11. The daily measurement (in particles) of a certain type of pollen during the first 10 days of June is approximated by the function $G(x) = 15 + 24x - 2x^2$, where x is the day in June, with $x = 1$ representing June 1.
(a) Sketch the graph of G.
(b) Find the maximum pollen measurement, and determine when it occurs.

12. During the course of a year, the number of volunteers available to run a food bank each month is approximated by $V(x)$, where $V(x) = 2x^2 - 32x + 150$ between the months of January and August. Here x is time in months, with $x = 1$ representing January. From August to December, $V(x)$ is approximated by $V(x) = 31x - 226$. Find the number of volunteers in each of the following months:
(a) January   (b) May   (c) August   (d) October   (e) December.
(f) Sketch a graph of $y = V(x)$ for January through December. In what month are the fewest volunteers available?

Consider the function $f(x) = -2.64x^2 + 5.47x + 3.54$ for Exercises 13–17.

13. Use the discriminant to explain how you can determine the number of x-intercepts the graph of f will have even before graphing it on your calculator.

14. Graph the function in the standard window of your calculator, and use the root-finding capabilities to solve the equation $f(x) = 0$. Express solutions as approximations to the nearest hundredth.

15. Use your answer to Exercise 14 and the graph of f to solve
(a) $f(x) > 0$ and (b) $f(x) < 0$.

16. Use the capabilities of your calculator to find the coordinates of the vertex of the graph. Express coordinates to the nearest hundredth.

17. Verify *analytically* that your answer in Exercise 16 is correct.

Use synthetic division to find the quotient $q(x)$ and the remainder r.

18. $\dfrac{x^3 + x^2 - 11x - 10}{x - 3}$

19. $\dfrac{3x^3 + 8x^2 + 5x + 10}{x + 2}$

Use synthetic division to find $f(2)$.

20. $f(x) = -x^3 + 5x^2 - 7x + 1$

21. $f(x) = 2x^3 - 3x^2 + 7x - 12$

22. $f(x) = 5x^4 - 12x^2 + 2x - 8$

23. $f(x) = x^5 + 4x^2 - 2x - 4$

24. If $f(x)$ is defined by a polynomial with real coefficients, and $7 + 2i$ is a zero of the function, what other complex number must also be a zero?

Find a polynomial function with real coefficients and of lowest degree having the given zeros.

25. $-1, 4, 7$

26. $8, 2, 3$

27. $\sqrt{3}, -\sqrt{3}, 2, 3$

28. $-2 + \sqrt{5}, -2 - \sqrt{5}, -2, 1$

29. Is -1 a zero of $f(x) = 2x^4 + x^3 - 4x^2 + 3x + 1$?

30. Is $x + 1$ a factor of $f(x) = x^3 + 2x^2 + 3x + 2$?

31. Find a polynomial function with real coefficients of degree 4 with 3, 1, and $-1 - 3i$ as zeros, and $f(2) = -36$.

32. Find a polynomial function of degree 3 with -2, 1, and 4 as zeros, and $f(2) = 16$.

33. Give an example of a fourth-degree polynomial function having exactly two distinct real zeros, and then sketch its graph.

34. Give an example of a cubic polynomial function having exactly one real zero, and then sketch its graph.

35. Find all zeros of $f(x) = x^4 - 3x^3 - 8x^2 + 22x - 24$, given that $1 - i$ is a zero.

36. Find all zeros of $f(x) = 2x^4 - x^3 + 7x^2 - 4x - 4$, given that 1 and $2i$ are zeros.

37. Find a value of s such that $x - 4$ is a factor of $f(x) = x^3 - 2x^2 + sx + 4$.

38. Find a value of s such that when the polynomial $x^3 - 3x^2 + sx - 4$ is divided by $x - 2$, the remainder is 5.

Use a graphing calculator to solve the problem.

39. After a 2-inch slice is cut off the top of a cube, the resulting solid has a volume of 32 cubic inches. Find the dimensions of the original cube.

40. The width of a rectangular box is three times its height, and its length is 11 inches more than its height. Find the dimensions of the box if its volume is 720 cubic inches.

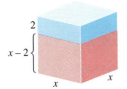

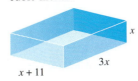

Show that the polynomials in Exercises 41 and 42 have real zeros satisfying the given conditions.

41. $f(x) = 3x^3 - 8x^2 + x + 2$;   zero in $[-1, 0]$ and $[2, 3]$

42. $f(x) = 6x^4 + 13x^3 - 11x^2 - 3x + 5$;   no real zero greater than 1 or less than -3

43. The function $f(x) = 1/x$ is negative at $x = -1$ and positive at $x = 1$, but has no zero between -1 and 1. Explain why this does not contradict the intermediate value theorem.

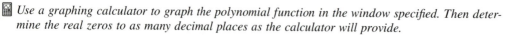 *Use a graphing calculator to graph the polynomial function in the window specified. Then determine the real zeros to as many decimal places as the calculator will provide.*

44. $f(x) = x^3 - 8x^2 + 2x + 5$ window: $[-10, 10]$ by $[-60, 60]$

45. $f(x) = x^4 - 4x^3 - 5x^2 + 14x - 15$ window: $[-10, 10]$ by $[-60, 60]$

After factoring the polynomial and locating its zeros, sketch the graph of the function.

46. $f(x) = 2x^3 + x^2 - x$ **47.** $f(x) = x^4 - 3x^2 + 2$

48. The polynomial function defined by

$$A(x) = -.015x^3 + 1.058x$$

gives the approximate alcohol concentration (in tenths of a percent) in an average person's bloodstream x hours after drinking about 8 ounces of 100-proof whiskey. The function is approximately valid for x in the interval $[0, 8]$.

(a) Graph A in the window $[0, 8]$ by $[0, 5]$.

(b) Using the graph from part (a), approximate the time of maximum alcohol concentration.

(c) In one state, a person is legally drunk if the blood alcohol concentration exceeds .08 percent. Use the graph from part (a) to approximate the time period in which the average person is legally drunk.

Graph the rational function.

49. $f(x) = \dfrac{4}{x - 1}$ **50.** $f(x) = \dfrac{4x - 2}{3x + 1}$ **51.** $f(x) = \dfrac{6x}{(x - 1)(x + 2)}$ **52.** $f(x) = \dfrac{2x}{x^2 - 1}$

53. $f(x) = \dfrac{x^2 + 4}{x + 2}$ **54.** $f(x) = \dfrac{x^2 - 1}{x}$ **55.** $f(x) = \dfrac{-2}{x^2 + 1}$ **56.** $f(x) = \dfrac{4x^2 - 9}{2x + 3}$

---

57. Antique-car owners often enter their cars in a *concours d'elegance* in which a maximum of 100 points can be awarded to a particular car. Points are awarded for the general attractiveness of the car. The function defined by

$$C(x) = \frac{10x}{49(101 - x)}$$

expresses the cost, in thousands of dollars, of restoring a car so that it will win x points.

(a) Use a graphing calculator to graph the function in the window $[0, 101]$ by $[0, 10]$.

(b) How much would an owner expect to pay to restore a car in order to earn 95 points?

58. In situations involving environmental pollution, a cost-benefit model expresses cost as a function of the percentage of pollutant removed from the environment. Suppose a cost-benefit model is expressed as

$$C(x) = \frac{6.7x}{100 - x},$$

where $C(x)$ is the cost in thousands of dollars of removing x percent of a certain pollutant.

(a) Use a graphing calculator to graph the function in the window $[0, 100]$ by $[0, 100]$.

(b) How much would it cost to remove 95% of the pollutant?

59. **(a)** Sketch the graph of a function that has the line $x = 3$ as a vertical asymptote, the line $y = 1$ as a horizontal asymptote, and x-intercepts 2 and 4.
(b) Find an equation for a possible corresponding rational function.

60. **(a)** Sketch the graph of a function that is never negative and has the lines $x = -1$ and $x = 1$ as vertical asymptotes, the x-axis as a horizontal asymptote, and 0 as an x-intercept.
(b) Find an equation for a possible corresponding rational function.

61. Suppose the degree of the numerator of a rational function expressed in lowest terms is one more than the degree of the denominator. Explain in your own words how you would find the equation of the oblique asymptote.

62. Find an equation for the rational function.

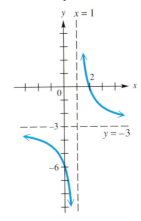

Chapter 4   Test ▼▼▼▼▼▼▼▼▼▼▼▼▼▼▼▼▼▼▼▼▼▼▼▼▼▼▼▼▼▼▼▼▼▼▼▼▼

1. Sketch the graph of the quadratic function $f(x) = -2x^2 + 6x - 3$. Give the intercepts, the vertex, the axis, the domain, and the range.

2. In the United States, the number of post-secondary degrees below the bachelor degree earned during the years 1985 to 1990 can be approximated by the model

$$f(x) = 5334.6x^2 - 23{,}040x + 617{,}519.1$$

where $x = 0$ corresponds to 1985.
(a) Based on this model, how many such degrees were earned in 1986?
(b) In what year during the period did the number of degrees reach a minimum? What was this minimum value?

Use synthetic division to find the quotient $q(x)$ and the remainder r.

3. $\dfrac{3x^3 + 4x^2 - 9x + 6}{x + 2}$

4. $\dfrac{2x^3 - 11x^2 + 28}{x - 5}$

5. Use synthetic division to determine $f(5)$, if $f(x) = 2x^3 - 9x^2 + 4x + 8$.

6. Use the factor theorem to determine whether $x - 3$ is a factor of $6x^4 - 11x^3 - 35x^2 + 34x + 24$. If it is, what is the other factor? If it is not, explain why.

7. Find all zeros of $f(x)$, given that $f(x) = 2x^3 - x^2 - 13x - 6$, and -2 is one zero of $f(x)$.

8. The accompanying screen shows the graph of a fourth-degree polynomial $f(x)$ having only real coefficients. It has -1, 2, and i as zeros, and the point $(3, 80)$ lies on the graph as indicated at the bottom of the screen. Find $f(x)$. (You may wish to graph it to confirm your answer.)

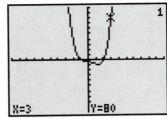

[−10, 10] by [−100, 100]
Xscl = 1   Yscl = 10

9. Explain why the polynomial function $f(x) = x^4 + 8x^2 + 12$ cannot have any real zeros.

10. Consider the function $f(x) = x^3 - 5x^2 + 2x + 7$.
   (a) Use the intermediate value theorem to show that f has a zero between 1 and 2.
   (b) Use a graphing calculator to find all real zeros to as many decimal places as the calculator will give.

11. Graph the functions $f_1(x) = x^4$ and $f_2(x) = -2(x + 5)^4 + 3$ on the same axes. Explain how the graph of f_2 can be obtained by a translation of the graph of f_1.

12. Use end behavior to determine which one of the following graphs is that of $f(x) = -x^7 + x - 4$.

A.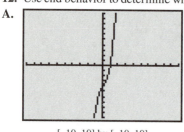

B.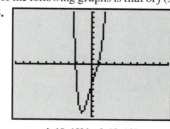

C.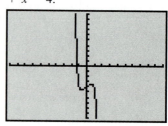

| [−10, 10] by [−10, 10] | [−10, 10] by [−10, 10] | [−10, 10] by [−10, 10] |

Graph the polynomial function. Factor first if the expression is not in factored form.

13. $f(x) = (3 - x)(x + 2)(x + 5)$

14. $f(x) = 2x^4 - 8x^3 + 8x^2$

15. Find a cubic polynomial having the graph shown.

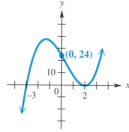

16. The pressure of the oil in a reservoir tends to drop with time. By taking sample pressure readings for a particular reservoir, petroleum engineers found that the change in pressure is modeled by the cubic function defined by $f(x) = 1.06x^3 - 24.6x^2 + 180x$, for x in the interval $[0, 15]$.
   (a) What was the change after 2 years?
   (b) For what time periods is the amount of change in pressure increasing? Decreasing? Use a graph to decide.

Graph the rational function.

17. $f(x) = \dfrac{3x - 1}{x - 2}$

18. $f(x) = \dfrac{x^2 - 1}{x^2 - 9}$

19. $f(x) = \dfrac{x^2 - 16}{x + 4}$

20. For the rational function defined by $f(x) = \dfrac{2x^2 + x - 6}{x - 1}$,
   (a) determine the equation of the oblique asymptote.   (b) determine the x-intercepts.
   (c) determine the y-intercept.   (d) determine the equation of the vertical asymptote.
   (e) sketch the graph.

Exponential and Logarithmic Functions

In 1896 Swedish scientist Svante Arrhenius first predicted the greenhouse effect resulting from emissions of carbon dioxide by industrialized countries. In his classic calculation, he was able to estimate that a doubling of the carbon dioxide level in the atmosphere would raise the average global temperature by 7°F to 11°F. Since global warming would not be uniform, changes as small as 4.5°F in the average temperature could have drastic climatic effects, particularly on the central plains of North America. Sea levels could rise dramatically as a result of both thermal expansion and the melting of ice caps. The annual cost to the United States economy could reach $60 billion.

The burning of fossil fuels, deforestation, and changes in land use from 1850 to 1986 put approximately 312 billion tons of carbon into the atmosphere, mostly in the form of carbon dioxide. Burning of fossil fuels produces 5.4 billion tons of carbon each year which is absorbed by both the atmosphere and the oceans. A critical aspect of the accumulation of carbon dioxide in the atmosphere is that it is irreversible and its effect requires hundreds of years to disappear. In 1990 the International Panel of Climate Change (IPCC) reported that if current trends of burning of fossil fuel and deforestation

Sources: Clime, W., *The Economics of Global Warming,* Institute for International Economics, Washington, D.C., 1992.

Kraljic, M. (Editor), *The Greenhouse Effect,* The H. W. Wilson Company, New York, 1992.

International Panel on Climate Change (IPCC), 1990.

Wuebbles, D. and J. Edmonds, *Primer of Greenhouse Gases,* Lewis Publishers, Inc., Chelsea, Michigan, 1991.

continue, then future amounts of atmospheric carbon dioxide in parts per million (ppm) will increase as shown in the table.

Year	Carbon Dioxide
1990	353
2000	375
2075	590
2175	1090
2275	2000

How can these data be used to predict when the amount of carbon dioxide will double? What will be the resulting global warming? How are carbon dioxide levels and global temperature increases related? These questions will be addressed in several sections of this chapter.

Since hard data on the greenhouse effect are lacking, mathematical models play a central role in analyzing the reality of the greenhouse effect and answering questions like these. In order for Svante Arrhenius to make his first calculation about global warming, he needed both logarithmic and exponential functions. These functions are central to many real applications found throughout science, business, and environmental forecasting. Using these functions we will be able to analyze the greenhouse effect, model population growth, and predict the time it takes for the planet Pluto to orbit the sun.

As we show later, exponential and logarithmic functions are inverses of each other, so we begin by discussing inverse functions.

5.1 Inverse Functions ▼▼▼

Addition and subtraction are inverse operations: starting with a number x, adding 5, and subtracting 5 gives x back as a result. Similarly, some functions are inverses of each other. For example, the functions

$$f(x) = 8x \quad \text{and} \quad g(x) = \frac{1}{8}x$$

are inverses of each other with respect to function composition. This means that if a value of x such as $x = 12$ is chosen, so that

$$f(12) = 8 \cdot 12 = 96,$$

calculating $g(96)$ gives

$$g(96) = \frac{1}{8} \cdot 96 = 12.$$

Thus,

$$g[f(12)] = 12.$$

Also, $f[g(12)] = 12$. For these functions f and g, it can be shown that

$$f[g(x)] = x \quad \text{and} \quad g[f(x)] = x$$

for any value of x.

This section will show how to start with a function such as $f(x) = 8x$, and obtain the inverse function $g(x) = (1/8)x$. Not all functions have inverse functions. Only functions that are *one-to-one functions* have inverse functions.

ONE-TO-ONE FUNCTIONS For the function $y = 5x - 8$, any two different values of x produce two different values of y. On the other hand, for the function $y = x^2$, two different values of x can lead to the *same* value of y; for example, both $x = 4$ and $x = -4$ give $y = 4^2 = (-4)^2 = 16$. A function such as $y = 5x - 8$, where different elements from the domain always lead to different elements from the range, is called a *one-to-one function*.

> **One-to-One Function**
>
> A function f is a **one-to-one function** if, for elements a and b from the domain of f,
> $$a \neq b \quad \text{implies} \quad f(a) \neq f(b).$$

◀**EXAMPLE 1**
Deciding whether a function is one-to-one

Decide whether the following functions are one-to-one.

(a) $f(x) = -4x + 12$

For this function, two different x-values will always generate two different y-values. To see this, suppose that $a \neq b$. Then $-4a \neq -4b$, and $-4a + 12 \neq -4b + 12$. Thus, the fact that $a \neq b$ implies that $f(a) \neq f(b)$ shows that f is one-to-one.

(b) $f(x) = \sqrt{25 - x^2}$

If $a = 3$ and $b = -3$, then $3 \neq -3$, but

$$f(3) = \sqrt{25 - 3^2} = \sqrt{25 - 9} = \sqrt{16} = 4$$

and $$f(-3) = \sqrt{25 - (-3)^2} = \sqrt{25 - 9} = 4.$$

Here, even though $3 \neq -3$, $f(3) = f(-3)$. By definition, this is not a one-to-one function. ▶

As shown in Example 1(b), a way to show that a function is *not* one-to-one is to produce a pair of unequal numbers that lead to the same function value. There is also a useful graphical test that tells whether or not a function is one-to-one. This *horizontal line test* for one-to-one functions can be summarized as follows.

> **Horizontal Line Test**
>
> If each horizontal line intersects the graph of a function in no more than one point, then the function is one-to-one.

NOTE In Example 1(b), the graph of the function is a semicircle. There are infinitely many horizontal lines that cut the graph of a semicircle in two points, so the horizontal line test shows that the function is not one-to-one.

EXAMPLE 2
Using the horizontal line test

Use the horizontal line test to determine whether the graphs in Figures 1 and 2 are graphs of one-to-one functions.

Each point in Figure 1 where the horizontal line intersects the graph has the same value of y but a different value of x. Since more than one (here three) different values of x lead to the same value of y, the function is not one-to-one.

Every horizontal line will intersect the graph in Figure 2 in exactly one point. This function is one-to-one. ▶

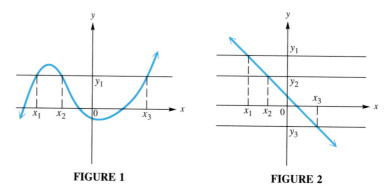

FIGURE 1   **FIGURE 2**

Using a graphing calculator to graph a function and then applying the horizontal line test provides a quick way to decide whether a function is one-to-one.

Some graphing calculators have the capability of "drawing" the inverse of a function. The feature does not require that the function be one-to-one, however, so the resulting figure may not be the graph of a function. Again, it is necessary to understand the mathematics to make a correct interpretation of results.

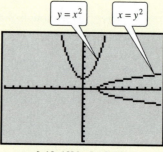

$y = x^2$ $x = y^2$

[−10, 10] by [−10, 10]

Despite the fact that $y = x^2$ is not one-to-one, the calculator will draw its "inverse," $x = y^2$.

INVERSE FUNCTIONS   As mentioned earlier, certain pairs of one-to-one functions "undo" one another. For example, if

$$f(x) = 8x + 5 \qquad \text{and} \qquad g(x) = \frac{x - 5}{8}$$

$$f(10) = 8 \cdot 10 + 5 = 85 \qquad \text{and} \qquad g(85) = \frac{85 - 5}{8} = 10.$$

Starting with 10, we "applied" function f and then "applied" function g to the result, which gave back the number 10. See Figure 3.

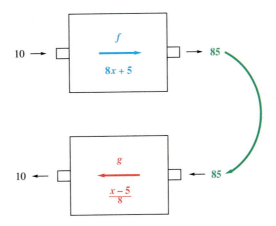

FIGURE 3

Similarly, for these same functions, check that

$$f(3) = 29 \qquad \text{and} \qquad g(29) = 3,$$
$$f(-5) = -35 \qquad \text{and} \qquad g(-35) = -5,$$
$$g(2) = -\frac{3}{8} \qquad \text{and} \qquad f\left(-\frac{3}{8}\right) = 2.$$

In particular, for these functions,

$$f[g(2)] = 2 \qquad \text{and} \qquad g[f(2)] = 2.$$

In fact for *any* value of x,

$$f[g(x)] = x \qquad \text{and} \qquad g[f(x)] = x,$$

or

$$(f \circ g)(x) = x \qquad \text{and} \qquad (g \circ f)(x) = x.$$

Because of this property, g is called the *inverse* of f.

> **Inverse Function**
>
> Let f be a one-to-one function. Then g is the **inverse function** of f if
>
> $$(f \circ g)(x) = x \quad \text{for every } x \text{ in the domain of } g,$$
> and $\quad(g \circ f)(x) = x \quad \text{for every } x \text{ in the domain of } f.$

A special notation is often used for inverse functions: if g is the inverse of a function f, then g is written as f^{-1} (read "f-inverse"). In the example above, $f(x) = 8x + 5$, and $g(x) = f^{-1}(x) = (x - 5)/8$.

◖EXAMPLE 3
Deciding whether two functions are inverses

Let functions f and g be defined by $f(x) = x^3 - 1$ and $g(x) = \sqrt[3]{x + 1}$, respectively. Is g the inverse function of f?

A graph indicates that f is one-to-one, so it does have an inverse. Since it is one-to-one, now find $(f \circ g)(x)$ and $(g \circ f)(x)$.

$$(f \circ g)(x) = f(g(x)) = (\sqrt[3]{x + 1})^3 - 1$$
$$= x + 1 - 1$$
$$= x$$
$$(g \circ f)(x) = g(f(x)) = \sqrt[3]{(x^3 - 1) + 1}$$
$$= \sqrt[3]{x^3}$$
$$= x$$

Since both $(f \circ g)(x) = x$ and $(g \circ f)(x) = x$, function g is indeed the inverse of function f, so that f^{-1} is given by

$$f^{-1}(x) = \sqrt[3]{x + 1}. \quad ◗$$

CAUTION Do not confuse the -1 in f^{-1} with a negative exponent. The symbol $f^{-1}(x)$ does not represent $1/f(x)$; it represents the inverse function of f. Keep in mind that a function f can have an inverse function f^{-1} if and only if f is one-to-one.

The definition of inverse function can be used to show that the domain of f equals the range of f^{-1}, and the range of f equals the domain of f^{-1}. See Figure 4.

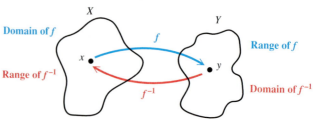

FIGURE 4

CONNECTIONS Inverse functions are used by government agencies and other businesses to send and receive coded information. The functions they use are usually very complicated. A simplified example involves the function $f(x) = 2x + 5$. If each letter of the alphabet is assigned a numerical value according to its position ($a = 1, \ldots, z = 26$), the word ALGEBRA would be encoded as 7   29   19   15   9   41   7. The "message" can be decoded using the inverse function $f^{-1}(x) = \dfrac{x - 5}{2}$.

FOR DISCUSSION OR WRITING
Use the alphabet assignment given above.

1. The function $f(x) = 3x - 2$ was used to encode the following message:

   37   25   19   61   13   34   22   1   55   1   52   52   25   64   13   10.

   Find the inverse function and decode the message.
2. Encode the message SEND HELP using the one-to-one function $f(x) = x^3 - 1$. Give the inverse function that the decoder would need when the message is received.

For the inverse functions f and g discussed earlier $f(10) = 85$ and $g(85) = 10$; that is, $(10, 85)$ belongs to f and $(85, 10)$ belongs to g. The ordered pairs of the inverse of any one-to-one function f can be found by exchanging the components of the ordered pairs of f. The equation of the inverse of a function defined by $y = f(x)$ also is found by exchanging x and y. For example, if $f(x) = 7x - 2$, then $y = 7x - 2$. The function f is one-to-one, so that f^{-1} exists. The ordered pairs in f^{-1} have the form (y, x), so y can be used to produce x, since $x = f^{-1}(y)$. Therefore, the equation for f^{-1} can be found by solving $y = f(x)$ for x. Finally, x and y can be interchanged to conform to our convention of using x for the independent variable and y for the dependent variable.

$$y = 7x - 2$$
$$7x = y + 2 \qquad \text{Add 2.}$$
$$x = \frac{y + 2}{7} = f^{-1}(y) \qquad \text{Divide by 7.}$$
$$y = \frac{x + 2}{7} = f^{-1}(x) \qquad \text{Exchange } x \text{ and } y.$$
$$f^{-1}(x) = \frac{x + 2}{7}.$$

As a check, verify that $(f \circ f^{-1})(x) = x$ and $(f^{-1} \circ f)(x) = x$.

In summary, the equation of an inverse function can be found with the following steps.

Finding an Equation for f^{-1}

1. Check that the function f defined by $y = f(x)$ is a one-to-one function.
2. Solve for x. Let $x = f^{-1}(y)$.
3. Exchange x and y to get $y = f^{-1}(x)$.
4. Check that $(f \circ f^{-1})(x) = x$ and $(f^{-1} \circ f)(x) = x$.

Any restrictions on x or y should be considered.

NOTE   There may be cases where it is very difficult or even impossible to solve for x in Step 2. In such cases, go on to Step 3, exchanging x and y. This will give an expression that defines the inverse function implicitly.

◀ **EXAMPLE 4**
Finding the inverse of a function

X	Y1
1	2
6	6
11	10
-4	-2
-9	-6
51	42
101	146

Y1☐(4X+6)/5

X	Y2
2	1
6	6
10	11
-2	-4
-6	-9
42	51
146	101

Y2☐(5X-6)/4

The inverses discussed in Example 4(a) are defined in Y_1 and Y_2. Notice that the x- and y-values in their ordered pairs are reversed.

For each of the following functions, find the inverse function, if it exists.

(a) $f(x) = \dfrac{4x + 6}{5}$

This function is one-to-one and thus has an inverse. Let $f(x) = y$, and solve for x, getting

$$y = \frac{4x + 6}{5}$$

$$5y = 4x + 6 \qquad \text{Multiply by 5.}$$

$$5y - 6 = 4x \qquad \text{Subtract 6.}$$

$$x = \frac{5y - 6}{4} = f^{-1}(y). \qquad \text{Divide by 4.}$$

Finally, exchange x and y to get

$$y = \frac{5x - 6}{4} = f^{-1}(x)$$

or

$$f^{-1}(x) = \frac{5x - 6}{4}.$$

The domain and range of both f and f^{-1} are the set of real numbers. In function f, the value of y is found by multiplying x by 4, adding 6 to the product, then dividing that sum by 5. In the equation for the inverse, x is *multiplied* by 5, then 6 is *subtracted*, and the result is *divided* by 4. This shows how an inverse function is used to "undo" what a function does to the variable x.

(b) $f(x) = x^3 - 1$

Two different values of x will produce two different values of $x^3 - 1$, so the function is one-to-one and has an inverse. To find the inverse, first solve $y = x^3 - 1$ for x, as follows.

$$y = x^3 - 1$$

$$y + 1 = x^3 \qquad \text{Add 1.}$$

$$\sqrt[3]{y + 1} = x \qquad \text{Take cube roots.}$$

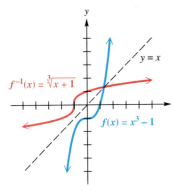

$f^{-1}(x) = \sqrt[3]{x+1}$

$y = x$

$f(x) = x^3 - 1$

FIGURE 5

Exchange x and y, giving

$$\sqrt[3]{x + 1} = y,$$

or

$$f^{-1}(x) = \sqrt[3]{x + 1}.$$

In Example 3, we verified that $\sqrt[3]{x + 1}$ is the inverse of $x^3 - 1$.

(c) $f(x) = x^2$

We can find two different values of x that give the same value of y. For example, both $x = 4$ and $x = -4$ give $y = 16$. Therefore, the function is not one-to-one and thus has no inverse function. ▶

We have seen that $f(x) = x^3 - 1$ and $f^{-1}(x) = \sqrt[3]{x + 1}$ define a pair of inverse functions. Figure 5 shows the graph of $f(x) = x^3 - 1$ in blue and the graph of $f^{-1}(x) = \sqrt[3]{x + 1}$ in red. The graphs are symmetric with respect to the line $y = x$. For instance, $(1, 0)$ is on the graph of f and $(0, 1)$ is on the graph of f^{-1}. This is true in general. For inverse functions f and f^{-1}, if $f(a) = b$, then $f^{-1}(b) = a$. This shows that if a point (a, b) is on the graph of f, then (b, a) will belong to the graph of f^{-1}. As shown in Figure 6, the points (a, b) and (b, a) are symmetric with respect to the line $y = x$. Thus, the graph of f^{-1} can be obtained from the graph of f by reflecting the graph of f about the line $y = x$.

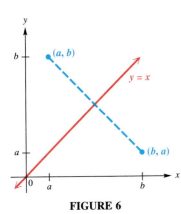

(a, b)

$y = x$

(b, a)

FIGURE 6

◀EXAMPLE 5

Finding the inverse of a function with a restricted domain

Let $f(x) = \sqrt{x + 5}$ with domain $[-5, \infty)$. Find $f^{-1}(x)$.

The function f is one-to-one and has an inverse function. To find this inverse function, start with

$$y = \sqrt{x + 5}$$

and solve for x, to get

$$y^2 = x + 5 \qquad \text{Square both sides.}$$

$$y^2 - 5 = x. \qquad \text{Subtract 5.}$$

Exchanging x and y gives

$$x^2 - 5 = y.$$

We cannot give just $x^2 - 5$ as $f^{-1}(x)$. In the definition of f above, the domain was given as $[-5, \infty)$. The range of f is $[0, \infty)$. As mentioned above, the range of f equals the domain of f^{-1}, so f^{-1} must be defined

$$f^{-1}(x) = x^2 - 5, \qquad \text{domain } [0, \infty).$$

As a check, the range of f^{-1}, $[-5, \infty)$, equals the domain of f. Graphs of f and f^{-1} are shown in Figure 7. The line $y = x$ is included on the graph to show that the graphs of f and f^{-1} are mirror images with respect to this line. ▶

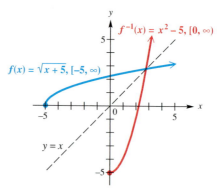

FIGURE 7

5.1 Exercises ▼▼

Based on your reading of the examples and exposition in this section, answer each of the following.

1. In order for a function to have an inverse, it must be _____ .

2. For a function f to be of the type mentioned in Exercise 1, if $a \neq b$, then _____ .

3. If f and g are inverses, then $(f \circ g)(x) = $ _____ , and _____ $= x$.

4. The domain of f is equal to the _____ of f^{-1}, and the range of f is equal to the _____ of f^{-1}.

5. If the point (a, b) lies on the graph of f, and f has an inverse, then the point _____ lies on the graph of f^{-1}.

6. If the graphs of f and f^{-1} intersect, they do so at a point that satisfies what condition?

7. If a function f has an inverse, then the graph of f^{-1} may be obtained by reflecting the graph of f across the line with equation _____ .

8. If a function f has an inverse and $f(-3) = 6$, then $f^{-1}(6) = $ _____ .

Decide whether each function defined or graphed as follows is one-to-one. See Examples 1 and 2.

9.

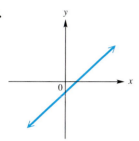

10.

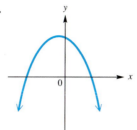

11.

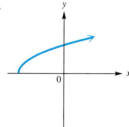

12.

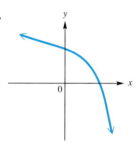

13.

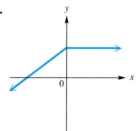

14.

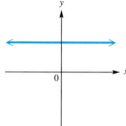

15. $y = (x - 2)^2$

16. $y = -(x + 3)^2 - 8$

17. $y = \sqrt{36 - x^2}$

18. $y = -\sqrt{100 - x^2}$

19. $y = 2x^3 + 1$

20. $y = -\sqrt[3]{x + 5}$

21. $y = \dfrac{1}{x + 2}$

22. $y = \dfrac{-4}{x - 8}$

23. Calculator-generated tables of values for two functions are shown below.

X	Y₁
.7	-1.5
.8	-1
.9	-.5
1	0
1.1	.5
1.2	1
1.3	1.5

X=.7

X	Y₁
-1.5	.7
-1	.8
-.5	.9
0	1
.5	1.1
1	1.2
1.5	1.3

X=-1.5

Based on the pairs of numbers shown, are the functions inverses of one another? Explain your answer.

▼▼▼▼▼▼▼▼▼▼▼▼ **DISCOVERING CONNECTIONS** (Exercises 24–32) ▼▼▼▼▼▼▼▼▼▼▼▼

The idea of an inverse is not unique to functions. Most mathematical operations have inverses as do many everyday activities. Work through the following exercises in order.

24. If $f(g(x)) =$ _____ and $g(f(x)) =$ _____, then f and g are _____ functions.

25. **(a)** $6 + (x -$ _____ $) = x$, so addition and _____ are inverse operations.
(b) $4 \cdot (x \div$ _____ $) = x$, so multiplication and _____ are inverse operations.

26. When the appropriate operation is applied to an inverse, the result is the _____ for that operation.

In Exercises 27–32, an everyday activity is described. Keeping in mind that an inverse operation "undoes" what an operation does, describe the inverse activity.

27. Tying your shoelaces

28. Starting a car

29. Entering a room

30. Climbing the stairs

31. Taking off in an airplane

32. Filling a cup

Which pairs of functions graphed or defined as follows are inverses of each other? See Example 3.

33.

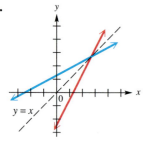

34.

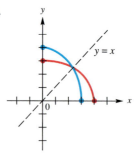

35.

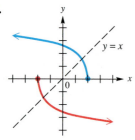

36.

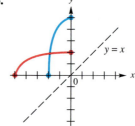

37. $f(x) = 2x + 4$, $g(x) = \dfrac{1}{2}x - 2$

38. $f(x) = 8x - 7$, $g(x) = \dfrac{x + 8}{7}$

39. $f(x) = \dfrac{2}{x + 6}$, $g(x) = \dfrac{6x + 2}{x}$

40. $f(x) = \dfrac{1}{x + 1}$, $g(x) = \dfrac{1 - x}{x}$

41. $f(x) = x^2 + 3$, domain $[0, \infty)$, and $g(x) = \sqrt{x - 3}$, domain $[3, \infty)$

42. $f(x) = \sqrt{x + 8}$, domain $[-8, \infty)$, and $g(x) = x^2 - 8$, domain $[0, \infty)$

43. $f(x) = -|x + 5|$, domain $[-5, \infty)$, and $g(x) = |x - 5|$, domain $[5, \infty)$

44. $f(x) = |x - 1|$, domain $[-1, \infty)$, and $g(x) = |x + 1|$, domain $[1, \infty)$

Graph the inverse of each one-to-one function.

45.

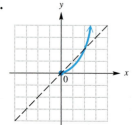

46.

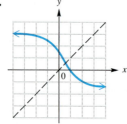

47.

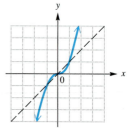

48.

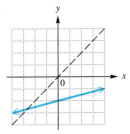

49.

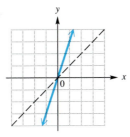

50.

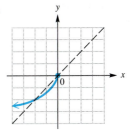

The graph of a function f is shown in the figure. Use the graph to find each value.

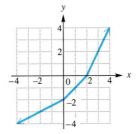

51. $f^{-1}(4)$ **52.** $f^{-1}(2)$ **53.** $f^{-1}(0)$

54. $f^{-1}(-2)$ **55.** $f^{-1}(-3)$ **56.** $f^{-1}(-4)$

For each function defined as follows that is one-to-one, write an equation for the inverse function in the form of $y = f^{-1}(x)$, and then graph f and f^{-1} on the same axes. See Examples 4 and 5.

57. $y = 3x - 4$ **58.** $y = 4x - 5$ **59.** $y = x^3 + 1$ **60.** $y = -x^3 - 2$

61. $y = x^2$ **62.** $y = -x^2 + 2$ **63.** $y = \dfrac{1}{x}$ **64.** $y = \dfrac{4}{x}$

65. $f(x) = \sqrt{6 + x}$ **66.** $f(x) = -\sqrt{x^2 - 16}, \; x \geq 4$

Let $f(x) = x^2 + 5x$ for $x \geq -5/2$. Find the value of the expression in Exercises 67 and 68, rounding to the nearest hundredth.

67. $f^{-1}(7)$ **68.** $f^{-1}(-3)$

69. Suppose $f(x)$ is the number of cars that can be built for x dollars. What does $f^{-1}(1000)$ represent?

70. Suppose $f(r)$ is the volume (in cu inches) of a sphere of radius r inches. What does $f^{-1}(5)$ represent?

71. The brightness B of a star is related to its distance from Earth, d, by the equation $B = k/d$, where k is a constant. Write an equation that expresses d as a function of B.

72. Young's rule for determining the correct medicine dosage c for a five-year-old child from the adult dosage d is $c = f(d) = (5/17)d$. Find an expression in terms of c for f^{-1}.

73. If a line has slope a, what is the slope of its reflection in the line $y = x$?

74. Find $f^{-1}(f(2))$, where $f(2) = 3$.

 Use a graphing calculator to graph each of the following using the given window. Use the graph to decide which functions are one-to-one. If a function is one-to-one, give the equation of its inverse function and graph the inverse function on the same coordinate system.

75. $f(x) = 6x^3 + 11x^2 - x - 6;$ $[-3, 2]$ by $[-10, 10]$ **76.** $f(x) = x^4 - 5x^2 + 6;$ $[-3, 3]$ by $[-1, 8]$

77. $f(x) = \dfrac{x - 5}{x + 3};$ $[-8, 8]$ by $[-6, 8]$ **78.** $f(x) = \dfrac{-x}{x - 4};$ $[-1, 8]$ by $[-6, 6]$

5.2   Exponential Functions ▼▼▼

Recall from Chapter 1 the definition of a^r, where r is a rational number: if $r = m/n$, then for appropriate values of m and n,

$$a^{m/n} = (\sqrt[n]{a})^m.$$

For example,

$$16^{3/4} = (\sqrt[4]{16})^3 = 2^3 = 8,$$

$$27^{-1/3} = \frac{1}{27^{1/3}} = \frac{1}{\sqrt[3]{27}} = \frac{1}{3},$$

and

$$64^{-1/2} = \frac{1}{64^{1/2}} = \frac{1}{\sqrt{64}} = \frac{1}{8}.$$

In this section the definition of a^r is extended to include all real (not just rational) values of the exponent r. For example, the new symbol $2^{\sqrt{3}}$ might be evaluated by approximating the exponent $\sqrt{3}$ by the numbers 1.7, 1.73, 1.732, and so on. Since these decimals approach the value of $\sqrt{3}$ more and more closely, it seems reasonable that $2^{\sqrt{3}}$ should be approximated more and more closely by the numbers $2^{1.7}$, $2^{1.73}$, $2^{1.732}$, and so on. (Recall, for example, that $2^{1.7} = 2^{17/10} = \sqrt[10]{2^{17}}$.) In fact, this is exactly how $2^{\sqrt{3}}$ is defined (in a more advanced course). To show that this assumption is reasonable, Figure 8 gives the graphs of the function $f(x) = 2^x$ with three different domains.

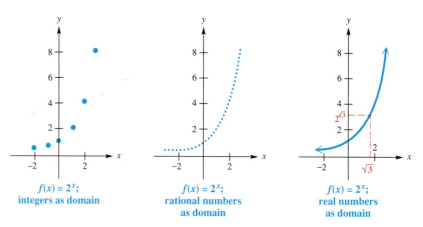

$f(x) = 2^x$;	$f(x) = 2^x$;	$f(x) = 2^x$;
integers as domain	rational numbers as domain	real numbers as domain

FIGURE 8

With this interpretation of real exponents, all rules and theorems for exponents are valid for real-number exponents as well as rational ones. In addition to the rules for exponents presented earlier, several new properties are used in this chapter. For example, if $y = 2^x$, then each real value of x leads to exactly one value of y, and therefore, $y = 2^x$ defines a function. Furthermore,

$$\text{if } 3^x = 3^4, \quad \text{then} \quad x = 4,$$

and for $p > 0$,

$$\text{if } p^2 = 3^2, \quad \text{then} \quad p = 3.$$

Also,

$$4^2 < 4^3 \quad \text{but} \quad \left(\frac{1}{2}\right)^2 > \left(\frac{1}{2}\right)^3,$$

so that when $a > 1$, increasing the exponent on a leads to a *larger* number, but if $0 < a < 1$, increasing the exponent on a leads to a *smaller* number.

These properties are generalized below. Proofs of the properties are not given here, as they require more advanced mathematics.

Additional Properties of Exponents

For any real number $a > 0$, $a \neq 1$, and any real number x, the following statements are true:

(a) a^x **is a unique real number.**
(b) $a^b = a^c$ **if and only if** $b = c$.
(c) **If** $a > 1$ **and** $m < n$, **then** $a^m < a^n$.
(d) **If** $0 < a < 1$ **and** $m < n$, **then** $a^m > a^n$.

Properties (a) and (b) require $a > 0$ so that a^x is always defined. For example, $(-6)^x$ is not a real number if $x = 1/2$. This means that a^x will always be positive, since a must be positive. In part (a), $a \neq 1$ because $1^x = 1$ for every real-number value of x, so that each value of x leads to the *same* real number, 1. For Property (b) to hold, a must not equal 1 since, for example, $1^4 = 1^5$, even though $4 \neq 5$.

With most calculators, values of a^x are computed with either a key labeled x^y (or y^x or a^b) or with the key marked ^. In each case enter the base, then the appropriate exponentiation key, then the exponent.

GRAPHING EXPONENTIAL FUNCTIONS   As mentioned, the expression a^x satisfies all the properties of exponents from Chapter 1. We can now define a function $f(x) = a^x$ whose domain is the set of all real numbers (and not just the rationals).

Exponential Function

If $a > 0$ and $a \neq 1$, then

$$f(x) = a^x$$

defines the **exponential function** with base a.

NOTE   If $a = 1$, the function is the constant function $f(x) = 1$, and not an exponential function.

EXAMPLE 1
Evaluating an exponential expression

If $f(x) = 2^x$, find each of the following.

(a) $f(-1)$

Replace x with -1.

$$f(-1) = 2^{-1} = \frac{1}{2}$$

(b) $f(3) = 2^3 = 8$

(c) $f(5/2) = 2^{5/2} = (2^5)^{1/2} = 32^{1/2} = \sqrt{32} = 4\sqrt{2}$

(d) $f(4.92) \approx 30.2738447$ ▶

Figure 9 shows the graph of $f(x) = 2^x$. The base of this exponential function is 2. The y-intercept is

$$y = 2^0 = 1.$$

Since $2^x > 0$ for all x and $2^x \to 0$ as $x \to -\infty$, the x-axis is a horizontal asymptote. The table to the left of Figure 9 gives several points on the graph of the function. Plotting these points and then drawing a smooth curve through them gives the graph in Figure 9. As the graph suggests, the domain of the function is $(-\infty, \infty)$ and the range is $(0, \infty)$. The function is increasing on its entire domain, and it is one-to-one by the horizontal line test.

x	$f(x)$
-2	$\frac{1}{4}$
-1	$\frac{1}{2}$
0	1
1	2
2	4
3	8

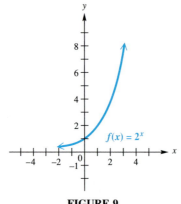

$f(x) = 2^x$

FIGURE 9

x	$f(x)$
-3	8
-2	4
-1	2
0	1
1	$\frac{1}{2}$
2	$\frac{1}{4}$

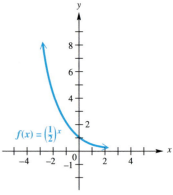

$f(x) = \left(\frac{1}{2}\right)^x$

FIGURE 10

EXAMPLE 2
Graphing an exponential function

Graph $f(x) = (1/2)^x$.

Again the y-intercept is 1 and the x-axis is a horizontal asymptote. Plot a few ordered pairs and draw a smooth curve through them. For example, several points are shown in the table to the left of Figure 10. Like the function $f(x) = 2^x$, this function also has domain $(-\infty, \infty)$ and range $(0, \infty)$ and is one-to-one. The graph is decreasing on the entire domain. ▶

Starting with $f(x) = 2^x$ and replacing x with $-x$ gives $f(-x) = 2^{-x} = (2^{-1})^x = (1/2)^x$. For this reason, the graphs of $f(x) = 2^x$ and $f(x) = (1/2)^x$ are reflections of each other about the y-axis. This is suggested by the graphs in Figures 9 and 10.

The graph of $f(x) = 2^x$ is typical of graphs of $f(x) = a^x$ where $a > 1$. For larger values of a, the graphs rise more steeply, but the general shape is similar to the graph in Figure 9. When $0 < a < 1$ the graph decreases in a manner similar to the graph of $f(x) = (1/2)^x$. In Figure 11, the graphs of several typical exponential functions illustrate these facts.

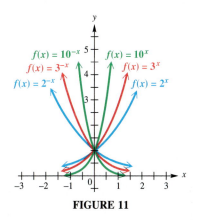

FIGURE 11

In summary, the graph of a function of the form $f(x) = a^x$ has the following features.

Graph of $f(x) = a^x$

1. The points $(0, 1)$ and $(1, a)$ are on the graph.
2. If $a > 1$, f is an increasing function; if $0 < a < 1$, f is a decreasing function.
3. The x-axis is a horizontal asymptote.
4. The domain is $(-\infty, \infty)$ and the range is $(0, \infty)$.

CONNECTIONS We can use function composition to produce more general exponential functions. Let $h(x) = ka^x$, where k is a constant, and let $g(x)$ be any function. Then

$$f(x) = h(g(x)) = ka^{g(x)}.$$

For example, if $a = 5$, $g(x) = 2x + 3$, and $k = 4$, then

$$f(x) = 4 \cdot 5^{2x+3}.$$

FOR DISCUSSION OR WRITING
1. If $h(g(x)) = 2^{x+3}$, how could $h(x)$ and $g(x)$ be defined?
2. If $h(x) = 2^x$ and $g(x) = -x^2$, how is $h(g(x))$ defined?
3. Give possible expressions for $f(x)$ and $g(x)$ if $f(g(x)) = k(1 + x)^{nt}$, where k, n, and t are constants.

◀ EXAMPLE 3
Graphing reflections and translations

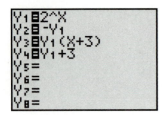

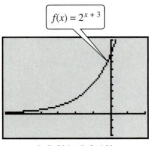

[−2, 2] by [−6, 6]

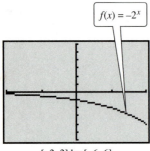

[−5, 2] by [−3, 10]

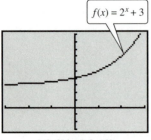

[−3, 3] by [−3, 10]

The top screen shows how Y_2, Y_3, and Y_4 are defined in terms of Y_1, using function notation. Compare the graphs to those in Figure 12.

Graph each of the following.

(a) $f(x) = -2^x$

The graph is that of $f(x) = 2^x$, reflected about the x-axis. The domain is $(-\infty, \infty)$, and the range is $(-\infty, 0)$. See Figure 12(a).

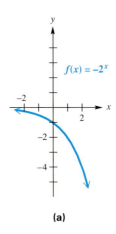

(a)

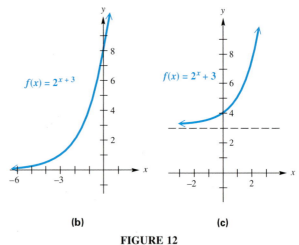

(b)   (c)

FIGURE 12

(b) $f(x) = 2^{x+3}$

The graph will be the graph of $f(x) = 2^x$ translated 3 units to the left, as shown in Figure 12(b).

(c) $f(x) = 2^x + 3$

This graph is that of $f(x) = 2^x$ translated 3 units upward. See Figure 12(c). ▶

EXAMPLE 4
Graphing a composite exponential function

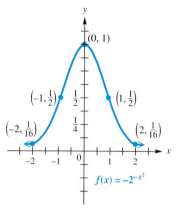

FIGURE 13

Graph $f(x) = 2^{-x^2}$.

Write $f(x) = 2^{-x^2}$ as $f(x) = 1/(2^{x^2})$ to find ordered pairs that belong to the function. Some ordered pairs are shown in the table.

x	-2	-1	0	1	2
y	$\dfrac{1}{16}$	$\dfrac{1}{2}$	1	$\dfrac{1}{2}$	$\dfrac{1}{16}$

As the table suggests, $0 < y \le 1$ for all values of x. The y-intercept is 1. The x-axis is a horizontal asymptote. Replacing x with $-x$ shows that the graph is symmetric with respect to the y-axis. Plotting the y-intercept and the points in the table, and drawing a smooth curve through them gives the graph in Figure 13. It is necessary to plot several points close to $(0, 1)$, to determine the correct shape of the graph there. This type of "bell-shaped" curve is important in statistics. ▶

EXPONENTIAL EQUATIONS Property (b) given at the beginning of this section is useful in solving equations, as shown by the next examples.

EXAMPLE 5
Using a property of exponents to solve an equation

Solve $\left(\dfrac{1}{3}\right)^x = 81$.

First, write $1/3$ as 3^{-1}, so that $(1/3)^x = (3^{-1})^x = 3^{-x}$. Since $81 = 3^4$,

$$\left(\frac{1}{3}\right)^x = 81$$

becomes

$$3^{-x} = 3^4.$$

By Property (b),

$$-x = 4, \qquad \text{or} \qquad x = -4.$$

The solution set of the given equation is $\{-4\}$. ▶

Later in this chapter, we describe a more general method for solving exponential equations where the approach used in Example 5 is not possible. For instance, this method could not be used to solve an equation like $7^x = 12$, since it is not easy to express both sides as exponential expressions with the same base.

EXAMPLE 6
Using a property of exponents to solve an equation

Solve $81 = b^{4/3}$.

Begin by writing $b^{4/3}$ as $(\sqrt[3]{b})^4$.

$$81 = b^{4/3}$$
$$81 = (\sqrt[3]{b})^4 \qquad \text{Definition of rational exponent}$$
$$\pm 3 = \sqrt[3]{b} \qquad \text{Take fourth roots on both sides.}$$
$$\pm 27 = b \qquad \text{Cube both sides.}$$

Check both solutions in the original equation. Since both solutions check, the solution set is $\{-27, 27\}$. ▶

COMPOUND INTEREST   The formula for *compound interest* (interest paid on both principal and interest) is an important application of exponential functions. You may recall the formula for simple interest, $I = Prt$, where P is the principal (amount left at interest), r is the rate of interest expressed as a decimal, and t is time in years that the principal earns interest. Suppose $t = 1$ year. Then at the end of the year the amount has grown to

$$P + Pr = P(1 + r),$$

the original principal plus the interest. If this amount is left at the same interest rate for another year, the total amount becomes

$$[P(1 + r)] + [P(1 + r)]r = [P(1 + r)](1 + r)$$
$$= P(1 + r)^2.$$

After the third year, this will grow to

$$[P(1 + r)^2] + [P(1 + r)^2]r = [P(1 + r)^2](1 + r)$$
$$= P(1 + r)^3.$$

Continuing in this way produces the following formula for compound interest.

Compound Interest

If P dollars is deposited in an account paying an annual rate of interest r compounded (paid) m times per year, then after t years the account will contain A dollars, where

$$A = P\left(1 + \frac{r}{m}\right)^{tm}.$$

For example, let $1000 be deposited in an account paying 8% per year compounded quarterly, or four times per year. After 10 years the account will contain

$$P\left(1 + \frac{r}{m}\right)^{tm} = 1000\left(1 + \frac{.08}{4}\right)^{10(4)}$$
$$= 1000(1 + .02)^{40}$$
$$= 1000(1.02)^{40}$$

dollars. The number $(1.02)^{40}$ can be found using a calculator. To five decimal places, $(1.02)^{40} = 2.20804$. The amount on deposit after 10 years is

$$1000(1.02)^{40} = 1000(2.20804) = 2208.04,$$

or $2208.04.

In the formula for compound interest, A is sometimes called the **future value** and P the **present value.**

EXAMPLE 7
Finding present value

An accountant wants to buy a new computer in three years that will cost $20,000.

(a) How much should be deposited now, at 6% interest compounded annually, to give the required $20,000 in three years?

Since the money deposited should amount to $20,000 in three years, $20,000 is the future value of the money. To find the present value P of $20,000 (the amount to deposit now), use the compound interest formula with $A = 20,000$, $r = .06$, $m = 1$, and $t = 3$.

$$A = P\left(1 + \frac{r}{m}\right)^{tm}$$

$$20,000 = P\left(1 + \frac{.06}{1}\right)^{3(1)} = P(1.06)^3$$

$$\frac{20,000}{(1.06)^3} = P$$

$$P = 16,792.39$$

The accountant must deposit $16,792.39.

(b) If only $15,000 is available to deposit now, what annual interest rate is required for it to increase to $20,000 in three years?

Here $P = 15,000$, $A = 20,000$, $m = 1$, $t = 3$, and r is unknown. Substitute the known values into the compound interest formula and solve for r.

$$A = P\left(1 + \frac{r}{m}\right)^{tm}$$

$$20,000 = 15,000\left(1 + \frac{r}{1}\right)^3$$

$$\frac{4}{3} = (1 + r)^3 \qquad \text{Divide both sides by 15,000.}$$

$$\left(\frac{4}{3}\right)^{1/3} = 1 + r \qquad \text{Take the cube root on both sides.}$$

$$\left(\frac{4}{3}\right)^{1/3} - 1 = r \qquad \text{Subtract 1 on both sides.}$$

$$r \approx .10 \qquad \text{Use a calculator.}$$

An interest rate of 10% will produce enough interest to increase the $15,000 deposit to the $20,000 needed at the end of three years. ▶

Perhaps the single most useful base for an exponential function is the irrational number e. Base e exponential functions provide a good model for many natural, as well as economic, phenomena. The letter e was chosen to represent this number in honor of the Swiss mathematician Leonhard Euler (pronounced "oiler") (1707–1783). Applications of the exponential function with base e are given later in this chapter.

The number e comes up in a natural way when using the formula for compound interest. Suppose a lucky investment produces an annual interest rate of 100%, so that $r = 1.00$, or $r = 1$. Suppose also that only $1 can be deposited at this rate, and for only one year. Then $P = 1$ and $t = 1$. Substitute into the formula for compound interest:

$$P\left(1 + \frac{r}{m}\right)^{tm} = 1\left(1 + \frac{1}{m}\right)^{1(m)} = \left(1 + \frac{1}{m}\right)^{m}.$$

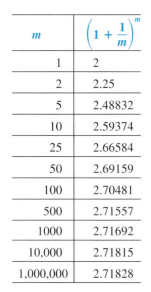

m	$\left(1 + \dfrac{1}{m}\right)^{m}$
1	2
2	2.25
5	2.48832
10	2.59374
25	2.66584
50	2.69159
100	2.70481
500	2.71557
1000	2.71692
10,000	2.71815
1,000,000	2.71828

As interest is compounded more and more often, the value of this expression will increase. If interest is compounded annually, making $m = 1$, the total amount on deposit is

$$\left(1 + \frac{1}{m}\right)^{m} = \left(1 + \frac{1}{1}\right)^{1} = 2^{1} = 2,$$

so an investment of $1 becomes $2 in one year. As interest is compounded more and more often, the value of this expression will increase.

A calculator with a y^x key was used to get the results in the table at the left. These results have been rounded when necessary to five decimal places. The table suggests that, as m increases, the value of $(1 + 1/m)^m$ gets closer and closer to some fixed number. It turns out that this is indeed the case. This fixed number is called e.

Value of e

To nine decimal places,

$$e \approx 2.718281828.$$

NOTE   Values of e^x can be found with a calculator that has a key marked e^x or by using a pair of keys marked INV and ln. See your instruction booklet for details or ask your instructor for assistance. The reason the second of these methods works will be apparent in the next section.

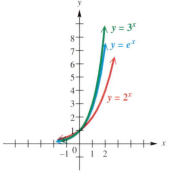

FIGURE 14

In Figure 14 the functions $y = 2^x$, $y = e^x$, and $y = 3^x$ are graphed for comparison.

CONNECTIONS In calculus, it is shown that

$$e^x = 1 + x + \frac{x^2}{2 \cdot 1} + \frac{x^3}{3 \cdot 2 \cdot 1} + \frac{x^4}{4 \cdot 3 \cdot 2 \cdot 1} + \frac{x^5}{5 \cdot 4 \cdot 3 \cdot 2 \cdot 1} + \ldots.$$

By using more and more terms, a more and more accurate approximation may be obtained for e^x.

FOR DISCUSSION OR WRITING
1. Use the terms shown here and replace x with 1 to approximate $e^1 = e$ to three decimal places. Check your results with a calculator.
2. Use the terms shown here and replace x with $-.05$ to approximate $e^{-.05}$ to four decimal places. Check your results with a calculator.
3. Give the next term in the sum for e^x.

EXPONENTIAL GROWTH OR DECAY   As mentioned above, the number e is important as the base of an exponential function because many practical applications require an exponential function with base e. For example, it can be shown that in situations involving growth or decay of a quantity, the amount or number present at time t often can be closely approximated by a function defined by

$$y = y_0 e^{kt},$$

where y_0 is the amount or number present at time $t = 0$ and k is a constant.

The next example, which refers to the problem stated at the beginning of this chapter, illustrates exponential growth.

 **EXAMPLE 8**
Using data to determine an exponential growth function

 The International Panel on Climate Change (IPCC) in 1990 published its finding that if current trends of burning fossil fuel and deforestation continue, then future amounts of atmospheric carbon dioxide in parts per million (ppm) will increase as shown in the table.*

Year	Carbon Dioxide
1990	353
2000	375
2075	590
2175	1090
2275	2000

*Source: International Panel on Climate Change (IPCC), 1990.

(a) Plot the data. Do the carbon dioxide levels appear to grow exponentially?

We show a calculator-generated graph for the data in Figure 15(a). The data do appear to have the shape of the graph of an increasing exponential function.

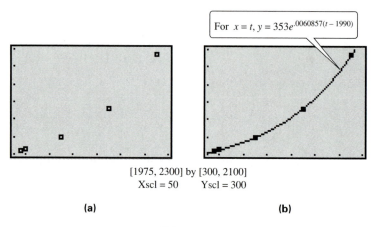

For $x = t, y = 353e^{.0060857(t-1990)}$

[1975, 2300] by [300, 2100]
Xscl = 50   Yscl = 300

(a)   (b)

FIGURE 15

(b) The function defined by $y = 353e^{.0060857(t-1990)}$ is a good model for the data.

(Later in this chapter we will show how this expression for y was obtained.) A graph of the function in Figure 15(b) shows that it is very close to the data points. From the graph, estimate when future levels of carbon dioxide will double and triple over the preindustrial level of 280 ppm.

We graph $y = 2 \cdot 280 = 560$ and $y = 3 \cdot 280 = 840$ on the same coordinate axes as the function in Figure 16 and use the calculator to find the intersec-

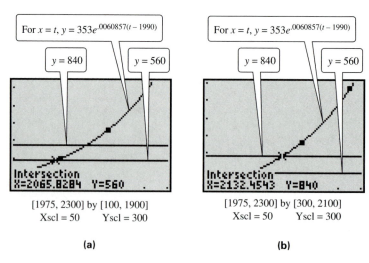

For $x = t, y = 353e^{.0060857(t-1990)}$ For $x = t, y = 353e^{.0060857(t-1990)}$

$y = 840$ $y = 560$ $y = 840$ $y = 560$

Intersection
X=2065.8284 Y=560   Intersection
X=2132.4543 Y=840

[1975, 2300] by [100, 1900]   [1975, 2300] by [300, 2100]
Xscl = 50   Yscl = 300   Xscl = 50   Yscl = 300

(a)   (b)

FIGURE 16

tion points. The graph of the function intersects the horizontal lines at approximately 2065.8 and 2132.5. Carbon dioxide levels will double by 2065 and triple by 2132. ▶

Further examples of exponential growth and decay are given in Section 5.6.

5.2 Exercises ▼▼▼

▼▼▼▼▼▼▼▼▼▼▼▼ **DISCOVERING CONNECTIONS** (Exercises 1–5) ▼▼▼▼▼▼▼▼▼▼▼▼▼

In Exercises 1–5, assume $f(x) = a^x$, where $a > 1$.

1. Is f a one-to-one function? If so, based on Section 5.1, what kind of related function exists for f?

2. If f has an inverse function f^{-1}, sketch f and f^{-1} on the same set of axes.

3. If f^{-1} exists, find an equation for $y = f^{-1}(x)$ using the method described in Section 5.1. You need not solve for y.

4. If $a = 10$, what is the equation for $f^{-1}(x)$?

5. If $a = e$, what is the equation for $f^{-1}(x)$?

6. Graph f for each of the following. Compare the graphs to that of $f(x) = 2^x$. See Examples 2–4.
   (a) $f(x) = 2^x + 1$ (b) $f(x) = 2^x - 4$ (c) $f(x) = 2^{x+1}$ (d) $f(x) = 2^{x-4}$

7. Graph f for each of the following. See Examples 2–4.
   (a) $f(x) = 3^{-x} - 2$ (b) $f(x) = 3^{-x} + 4$ (c) $f(x) = 3^{-x-2}$ (d) $f(x) = 3^{-x+4}$

8. Explain how you could use the graph of $y = 4^x$ to graph $y = -4^x$.

Graph f for each of the following. See Examples 1–4.

9. $f(x) = 3^x$

10. $f(x) = 4^x$

11. $f(x) = \left(\dfrac{3}{2}\right)^x$

12. $f(x) = e^{-x}$

13. $f(x) = 2^{|x|}$

14. $f(x) = 2^{-|x|}$

 In the given figure, the graphs of $y = a^x$ for $a = 1.8$, 2.3, 3.2, .4, .75, and .31 are given. They are identified by letter, but not necessarily in the same order as the values of a just given. Use your knowledge of how the exponential function behaves for various values of a to identify each lettered graph.

[–10, 10] by [–5, 10]

15. A   16. B   17. C

18. D   19. E   20. F

21. For $a > 1$, how does the value of $f(x) = a^x$ change as x increases? What if $0 < a < 1$?

22. What two points on the graph of $f(x) = a^x$ can be found with no computation?

23. A function of the form $f(x) = x^r$, where r is a constant, is called a *power function*. Discuss the difference between an exponential function and a power function.

24. If $f(x) = a^x$ and $f(3) = 27$, find the following values of $f(x)$.
   (a) $f(1)$ **(b)** $f(-1)$ **(c)** $f(2)$ **(d)** $f(0)$

Give an equation of the form $f(x) = a^x$ to define the exponential function whose graph contains the given point.

25. $(3, 8)$

26. $(-3, 64)$

Use properties of exponents to write each of the following in the form $f(t) = ka^t$, where k is a constant. (Hint: Recall $4^{x+y} = 4^x \cdot 4^y$.)

27. $f(t) = 3^{2t+3}$

28. $f(t) = \left(\dfrac{1}{3}\right)^{1-2t}$

29. Explain why the exponential equation $3^x = 12$ cannot be solved by using the properties of exponentials given in this section.

Solve the equation. See Examples 5 and 6.

30. $4^x = 2$

31. $125^r = 5$

32. $\left(\dfrac{1}{2}\right)^k = 4$

33. $\left(\dfrac{2}{3}\right)^x = \dfrac{9}{4}$

34. $2^{3-y} = 8$

35. $5^{2p+1} = 25$

36. $\dfrac{1}{27} = b^{-3}$

37. $\dfrac{1}{81} = k^{-4}$

38. $4 = r^{2/3}$

39. $z^{5/2} = 32$

40. $27^{4z} = 9^{z+1}$

41. $32^t = 16^{1-t}$

42. $\left(\dfrac{1}{2}\right)^{-x} = \left(\dfrac{1}{4}\right)^{x+1}$

43. $\left(\dfrac{2}{3}\right)^{k-1} = \left(\dfrac{81}{16}\right)^{k+1}$

Use a graphing calculator to graph f in Exercises 44–47.

44. $f(x) = \dfrac{e^x - e^{-x}}{2}$

45. $f(x) = \dfrac{e^x + e^{-x}}{2}$

46. $f(x) = x \cdot 2^x$

47. $f(x) = x^2 \cdot 2^{-x}$

Use the formula for compound interest to find each future value.

48. $8906.54 at 5% compounded semiannually for 9 years

49. $56,780 at 5.3% compounded quarterly for 23 quarters

Find the present value for each future value. See Example 7(a).

50. $25,000, if interest is 6% compounded quarterly for 11 quarters

51. $45,678.93, if interest is 9.6% compounded monthly for 11 months

Find the required annual interest rate to the nearest tenth. See Example 7(b).

52. $65,000 compounded monthly for 6 months to yield $65,325

53. $1200 compounded quarterly for 5 years to yield $1780

54. Suppose you can borrow money at 8% compounded daily or 8.3% compounded annually. Which will cost you more? (*Hint:* Calculate a loan of $1 for 1 year both ways.) Use the TABLE feature of a graphing calculator to find the difference after 1 year, after 2 years, and after 5 years.

Solve each applied problem.

55. The exponential growth of the deer population in Massachusetts can be calculated using the equation $T = 50{,}000(1 + .06)^n$, where 50,000 is the initial deer population and .06 is the rate of growth. T is the total population after n years have passed.

   (a) Predict the total population after 4 years.

   (b) If the initial population was 30,000 and the growth rate was .12, approximately how many deer would be present after 3 years?

   (c) How many additional deer can we expect in 5 years if the initial population is 45,000 and the current growth rate is .08?

56. Since 1950, the growth in the world population in millions closely fits the exponential function defined by

$$A(t) = 2600e^{.018t},$$

where t is the number of years since 1950.

   (a) The world population was about 3700 million in 1970. How closely does the function approximate this value?

   (b) Use the function to approximate the population in 1990. (The actual 1990 population was about 5320 million.)

   (c) Estimate the population in the year 2000.

57. A sample of 500 g of lead 210 decays to polonium 210 according to the function given by

$$A(t) = 500e^{-.032t},$$

where t is time in years. Find the amount of the sample after each of the following times.

   (a) 4 years   **(b)** 8 years

   (c) 20 years   **(d)** Graph $y = A(t)$.

58. Vehicle theft in the United States has been rising exponentially since 1972. The number of stolen vehicles, in millions, is given by

$$f(x) = .88(1.03)^x,$$

where $x = 0$ represents the year 1972. Find the number of vehicles stolen in the following years.

   (a) 1975   **(b)** 1980   **(c)** 1985   **(d)** 1990

59. (Refer to Example 8.) Carbon dioxide in the atmosphere traps heat from the sun. Presently, the net incoming solar radiation reaching the earth's surface is 240 watts per square meter (w/m²). The relationship between additional watts per square meter of heat trapped by the increased carbon dioxide R and the average rise in global temperature T (in °F) is shown in the graph. This additional solar radiation trapped by carbon dioxide is called **radiative forcing**. It is measured in watts per square meter.

   (a) Is T a linear or exponential function of R?

   (b) Let T represent the temperature increase resulting from an additional radiative forcing of R w/m². Use the graph to write T as a function of R.

   (c) Find the global temperature increase when $R = 5$ w/m².

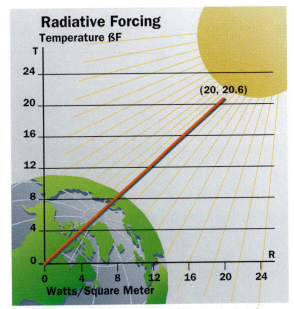

Source: Clime, W. *The Economics of Global Warming,* Institute for International Economics, Washington, D.C., 1992.

60. 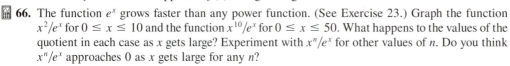 The atmospheric pressure (in millibars) at a given altitude (in meters) is shown in the table. (*Source:* Miller, A. and J. Thompson, *Elements of Meteorology,* Charles E. Merrill Publishing Company, Columbus, Ohio, 1975.)

Altitude	Pressure
0	1013
1000	899
2000	795
3000	701
4000	617
5000	541
6000	472
7000	411
8000	357
9000	308
10,000	265

(a) Use a graphing calculator to plot the data for atmospheric pressure P at altitude x.

(b) Would a linear or exponential function fit the data better?

(c) The function defined by $P(x) = 1013e^{-.0001341x}$ approximates the data. Use a graphing calculator to graph P and the data on the same coordinate axes.

(d) Use P to predict the pressure at 1500 m and 11,000 m and compare it to the actual values of 846 millibars and 227 millibars, respectively.

Any points where the graphs of functions f and g intersect give solutions of the equation $f(x) = g(x)$. Use a graphing calculator and this idea to estimate the solution(s) of the equation.

61. $x = 2^x$ **62.** $5e^{3x} = 75$ **63.** $6^{-x} = 1 - x$ **64.** $3x + 2 = 4^x$

65. Graph the function $f(x) = (1 + (1/x))^x$ and the horizontal line $y = 2.71828$ with $1 \le x \le 25$ and $0 \le y \le 3$. What happens to $f(x)$ as x gets large?

66. The function e^x grows faster than any power function. (See Exercise 23.) Graph the function x^2/e^x for $0 \le x \le 10$ and the function x^{10}/e^x for $0 \le x \le 50$. What happens to the values of the quotient in each case as x gets large? Experiment with x^n/e^x for other values of n. Do you think x^n/e^x approaches 0 as x gets large for any n?

5.3   Logarithmic Functions ▼▼▼

The previous section dealt with exponential functions of the form $y = a^x$ for all positive values of a, where $a \ne 1$. As mentioned there, the horizontal line test shows that exponential functions are one-to-one, and thus have inverse functions. In this section we discuss the inverses of exponential functions. The equation defining the inverse of a function is found by exchanging x and y in the equation that defines the function. Doing so with $y = a^x$ gives

$$x = a^y$$

as the equation of the inverse function of the exponential function defined by $y = a^x$. This equation can be solved for y by using the following definition.

> **Logarithm**
>
> For all real numbers y, and all positive numbers a and x, where $a \neq 1$:
>
> $$y = \log_a x \qquad \text{if and only if} \qquad x = a^y.$$

The "log" in the definition above is an abbreviation for *logarithm.* Read $\log_a x$ as "the logarithm to the base a of x."

Consider the following simple fill-in-the-box problems.

$$4^3 = \boxed{} \qquad 5^{\boxed{}} = 25$$

The answers, of course, are

$$4^3 = \boxed{64} \qquad 5^{\boxed{2}} = 25.$$

When we solve the problem on the left, we are "doing" exponents. When we solve the right problem, we are "doing" logarithms. That is, we are finding the power to which 5 must be raised in order to get 25. Therefore, $2 = \log_5 25$. In a certain sense, logarithms are just exponents.

By the definition of logarithm, if $s = \log_a r$, then the power to which a *must* be raised to get r is s, or $r = a^s$.

$$s = \log_a r \quad \text{if and only if} \quad r = a^s.$$

This key statement should be memorized. It is important to remember the location of the base and exponent in each part.

$$\text{Logarithmic form:} \quad s = \log_a r$$

(Exponent → s; Base ↑ a)

$$\text{Exponential form:} \quad a^s = r$$

(Exponent → s; Base ↑ a)

CAUTION The "log" in $y = \log_a x$ is the notation for a particular function and there must be a replacement for x following it, as in $\log_a 3$, $\log_a(2x - 1)$, or $\log_a x^2$. Avoid writing meaningless notation such as $y = \log$ or $y = \log_a$.

EXAMPLE 1
Converting between exponential and logarithmic statements

The chart below shows several pairs of equivalent statements. The same statement is written in both exponential and logarithmic forms.

Exponential Form	Logarithmic Form
$2^3 = 8$	$\log_2 8 = 3$
$\left(\dfrac{1}{2}\right)^{-4} = 16$	$\log_{1/2} 16 = -4$
$10^5 = 100{,}000$	$\log_{10} 100{,}000 = 5$
$3^{-4} = \dfrac{1}{81}$	$\log_3\left(\dfrac{1}{81}\right) = -4$
$5^1 = 5$	$\log_5 5 = 1$
$\left(\dfrac{3}{4}\right)^0 = 1$	$\log_{3/4} 1 = 0$

LOGARITHMIC EQUATIONS   The definition of logarithm can be used to solve logarithmic equations, as shown in the next example.

EXAMPLE 2
Solving logarithmic equations

Solve each equation.

(a) $\log_x \dfrac{8}{27} = 3$

First, write the expression in exponential form.

$$x^3 = \frac{8}{27}$$

$$x^3 = \left(\frac{2}{3}\right)^3 \qquad \tfrac{8}{27} = \left(\tfrac{2}{3}\right)^3$$

$$x = \frac{2}{3} \qquad \text{Property (b) of exponents}$$

The solution set is {2/3}.

(b) $\log_4 x = 5/2$

In exponential form, the given statement becomes

$$4^{5/2} = x$$
$$(4^{1/2})^5 = x$$
$$2^5 = x$$
$$32 = x.$$

The solution set is {32}.

LOGARITHMIC FUNCTIONS   The logarithmic function with base a is defined as follows.

> ## Logarithmic Function
>
> If $a > 0$, $a \neq 1$, and $x > 0$, then
>
> $$f(x) = \log_a x$$
>
> defines the **logarithmic function** with base a.

Exponential and logarithmic functions are inverses of each other. Since the domain of an exponential function is the set of all real numbers, the range of a logarithmic function also will be the set of all real numbers. In the same way, both the range of an exponential function and the domain of a logarithmic function are the set of all positive real numbers, so logarithms can be found for positive numbers only.

The graph of $y = 2^x$ is shown in red in Figure 17. The graph of its inverse is found by reflecting the graph of $y = 2^x$ about the line $y = x$. The graph of the inverse function, defined by $y = \log_2 x$, shown in blue, has the y-axis as a vertical asymptote.

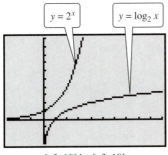

[-3, 10] by [-3, 10]

The graph of $y = \log_2 x$ can be obtained by drawing the inverse of $y = 2^x$. (It can also be graphed using the change-of-base theorem, introduced in Section 5.4.)

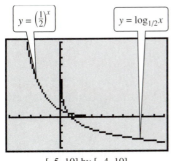

[-5, 10] by [-4, 10]

Compare to Figure 18.

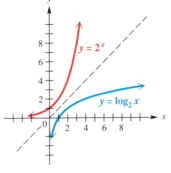

FIGURE 17

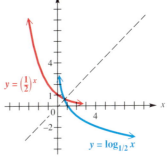

FIGURE 18

The graph of $y = (1/2)^x$ is shown in red in Figure 18. The graph of its inverse, defined by $y = \log_{1/2} x$, in blue, is found by reflecting the graph of $y = (1/2)^x$ about the line $y = x$. As Figure 18 suggests, the graph of $y = \log_{1/2} x$ also has the y-axis for a vertical asymptote.

The graphs of $y = \log_2 x$ in Figure 17 and $y = \log_{1/2} x$ in Figure 18 suggest the following generalizations about the graphs of logarithmic functions of the form $f(x) = \log_a x$.

> ## Graph of $f(x) = \log_a x$
>
> 1. The points $(1, 0)$ and $(a, 1)$ are on the graph.
> 2. If $a > 1$, f is an increasing function; if $0 < a < 1$, f is a decreasing function.
> 3. The y-axis is a vertical asymptote.
> 4. The domain is $(0, \infty)$ and the range is $(-\infty, \infty)$.

Compare these generalizations to those for exponential functions discussed in Section 5.2.

>  Calculator-generated graphs of logarithmic functions do not, in general, give an accurate picture of the behavior of the graphs near the vertical asymptotes. While it may seem as if the graph has an endpoint, this is not the case. The resolution of the calculator screen is not precise enough to indicate that the graph approaches the vertical asymptote as the value of x gets closer to it. Do not draw incorrect conclusions just because the calculator does not show this behavior.

More general logarithmic functions can be obtained by forming the composition of $h(x) = \log_a x$ with a function $g(x)$ to get

$$f(x) = h[g(x)] = \log_a[g(x)].$$

In writing composite logarithmic functions, it is important to use parentheses to make the intent clear. Just as we put parentheses around $x - 2$ in $f(x - 2)$, we put parentheses around $x - 2$ in $\log_a(x - 2)$. Similarly, we write $\log_a(xy)$ to avoid the misinterpretation $(\log_a x)y$. Also, we write $\log_a(x^2)$ to distinguish it from $(\log_a x)^2$. However, we will continue to write $\log_a x$ without parentheses, because the meaning is clear in that case.

The next examples illustrate some composite logarithmic functions.

EXAMPLE 3
Graphing a translated logarithmic function

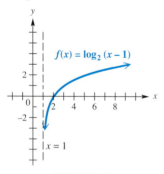

FIGURE 19

Graph each function.

(a) $f(x) = \log_2(x - 1)$

The graph of $f(x) = \log_2(x - 1)$ will be the graph of $f(x) = \log_2 x$, translated one unit to the right. The vertical asymptote is $x = 1$. The domain of the function defined by $f(x) = \log_2(x - 1)$ is $(1, \infty)$, since logarithms can be found only for positive numbers. To find some ordered pairs to plot, use the equivalent equations

$$x - 1 = 2^y \quad \text{or} \quad x = 2^y + 1,$$

choosing values for y and then calculating each of the corresponding x-values. See Figure 19.

(b) $f(x) = (\log_3 x) - 1$

This function has the same graph as $g(x) = \log_3 x$ translated down one unit. Ordered pairs to plot can be found by writing $y = (\log_3 x) - 1$ as follows.

$$y = (\log_3 x) - 1$$
$$y + 1 = \log_3 x$$
$$x = 3^{y+1}$$

Again, it is easier to choose y-values and calculate the corresponding x-values. The graph is shown in Figure 20. ▶

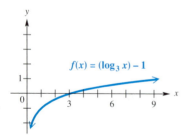

FIGURE 20

CAUTION   If you write a logarithmic function in exponential form, choosing y-values to calculate x-values as we did in Example 3, be careful to get the ordered pairs in the correct order.

Since a logarithmic statement can be written as an exponential statement, it is not surprising that there are properties of logarithms based on the properties of exponents. The properties of logarithms allow us to change the form of logarithmic statements so that products can be converted to sums, quotients can be converted to differences, and powers can be converted to products. These properties will be used to solve logarithmic and exponential equations later in this chapter.

Properties of Logarithms

If x and y are any positive real numbers, r is any real number, and a is any positive real number, $a \neq 1$, then the following properties are true.

(a) $\log_a xy = \log_a x + \log_a y$ **(b)** $\log_a \dfrac{x}{y} = \log_a x - \log_a y$

(c) $\log_a x^r = r \log_a x$ **(d)** $\log_a a = 1$

(e) $\log_a 1 = 0$

Proof   To prove Property (a), let

$$m = \log_a x \qquad \text{and} \qquad n = \log_a y.$$

Change to exponential form.

$$a^m = x \qquad \text{and} \qquad a^n = y$$

Multiplication gives

$$a^m \cdot a^n = xy.$$

By a property of exponents,

$$a^{m+n} = xy.$$

Now use the definition of logarithm to write

$$\log_a xy = m + n.$$

Since $m = \log_a x$ and $n = \log_a y$,

$$\log_a xy = \log_a x + \log_a y. \ \blacktriangleright$$

Properties (b) and (c) are proven in a similar way. (See Exercises 73 and 74.) Properties (d) and (e) follow directly from the definition of logarithm since $a^1 = a$ and $a^0 = 1$.

The properties of logarithms are useful for rewriting expressions with logarithms in different forms, as shown in the next examples.

EXAMPLE 4
Using the properties of logarithms

Assuming that all variables represent positive real numbers, use the properties of logarithms to rewrite each of the following expressions.

(a) $\log_6(7 \cdot 9)$

$$\log_6(7 \cdot 9) = \log_6 7 + \log_6 9$$

(b) $\log_9\left(\dfrac{15}{7}\right)$

$$\log_9\left(\dfrac{15}{7}\right) = \log_9 15 - \log_9 7$$

(c) $\log_5 \sqrt{8}$

$$\log_5 \sqrt{8} = \log_5(8^{1/2}) = \frac{1}{2} \log_5 8$$

(d) $\log_a\left(\dfrac{mnq}{p^2}\right) = \log_a m + \log_a n + \log_a q - 2 \log_a p$

(e) $\log_a \sqrt[3]{m^2} = \dfrac{2}{3} \log_a m$

(f) $\log_b \sqrt[n]{\dfrac{x^3 y^5}{z^m}} = \dfrac{1}{n} \log_b\left(\dfrac{x^3 y^5}{z^m}\right)$

$$= \frac{1}{n}(\log_b(x^3) + \log_b(y^5) - \log_b(z^m))$$

$$= \frac{1}{n}(3 \log_b x + 5 \log_b y - m \log_b z)$$

$$= \frac{3}{n} \log_b x + \frac{5}{n} \log_b y - \frac{m}{n} \log_b z$$

Notice the use of parentheses in the second step. The factor $1/n$ applies to each term. ▶

EXAMPLE 5
Using the properties of logarithms

Use the properties of logarithms to write each of the following as a single logarithm with a coefficient of 1. Assume that all variables represent positive real numbers.

(a) $\log_3(x + 2) + \log_3 x - \log_3 2$
Using Properties (a) and (b),

$$\log_3(x + 2) + \log_3 x - \log_3 2 = \log_3\left[\frac{(x + 2)x}{2}\right].$$

(b) $2 \log_a m - 3 \log_a n = \log_a(m^2) - \log_a(n^3) = \log_a\left(\dfrac{m^2}{n^3}\right)$

Here we used Property (c), then Property (b).

(c) $\dfrac{1}{2} \log_b m + \dfrac{3}{2} \log_b(2n) - \log_b(m^2 n)$

$$= \log_b(m^{1/2}) + \log_b[(2n)^{3/2}] - \log_b(m^2 n) \qquad \text{Property (c)}$$

$$= \log_b\left(\frac{m^{1/2}(2n)^{3/2}}{m^2 n}\right) \qquad \text{Properties (a) and (b)}$$

$$= \log_b\left(\frac{2^{3/2} n^{1/2}}{m^{3/2}}\right) \qquad \text{Rules for exponents}$$

$$= \log_b\left[\left(\frac{2^3 n}{m^3}\right)^{1/2}\right] \qquad \text{Rules for exponents}$$

$$= \log_b \sqrt{\frac{8n}{m^3}} \qquad \text{Definition of } a^{1/n} \; \blacktriangleright$$

CAUTION   There is no property of logarithms to rewrite a logarithm of a *sum* or *difference*. That is why, in Example 5(a), $\log_3(x + 2)$ was not written as $\log_3 x + \log_3 2$. Remember, $\log_3 x + \log_3 2 = \log_3(x \cdot 2)$.

The distributive property does not apply here, because $\log(x + y)$ is one term; "log" is not a factor.

◖ EXAMPLE 6
Using the properties of logarithms with numerical values

Assume that $\log_{10} 2 = .3010$. Find the base 10 logarithms of 4 and 5.

By the properties of logarithms,

$$\log_{10} 4 = \log_{10}(2^2) = 2 \log_{10} 2 = 2(.3010) = .6020$$

$$\log_{10} 5 = \log_{10}\left(\frac{10}{2}\right) = \log_{10} 10 - \log_{10} 2 = 1 - .3010 = .6990.$$

We used Property (d) to replace $\log_{10} 10$ with 1.   ▶

Compositions of the exponential and logarithmic functions can be used to get two more useful properties. If $f(x) = a^x$ and $g(x) = \log_a x$, then

$$f[g(x)] = a^{\log_a x}$$

and
$$g[f(x)] = \log_a(a^x).$$

Theorem on Inverses

For $a > 0$, $a \neq 1$:

$$a^{\log_a x} = x \qquad \text{and} \qquad \log_a(a^x) = x.$$

Proof   Exponential and logarithmic functions are inverses of each other, so $f[g(x)] = x$ and $g[f(x)] = x$. Letting $f(x) = a^x$ and $g(x) = \log_a x$ gives both results.   ▶

By the results of the last theorem,

$$\log_5 5^3 = 3, \qquad 7^{\log_7 10} = 10, \qquad \text{and} \quad \log_r r^{k+1} = k + 1.$$

The second statement in the theorem will be useful in Sections 5.5 and 5.6 when solving logarithmic or exponential equations.

5.3 Exercises ▼▼▼▼▼▼▼▼▼▼▼▼▼▼▼▼▼▼▼▼▼▼▼▼▼▼▼▼▼▼▼▼▼▼▼▼

Complete the statement or answer the question in Exercises 1–4.

1. $y = \log_a x$ if and only if _____ .
2. What is wrong with the expression $y = \log_b$?
3. The statement $\log_5 125 = 3$ tells us that _____ is the power of _____ that equals _____ .
4. Let $f(x) = \log_a x$. If $a > 1$, f is a(n) _____ function; if $0 < a < 1$, f is a(n) _____ function.

For each statement, write an equivalent statement in logarithmic form. See Example 1.

5. $3^4 = 81$
6. $2^5 = 32$
7. $(2/3)^{-3} = 27/8$
8. $10^{-4} = .0001$

For each statement, write an equivalent statement in exponential form. See Example 1.

9. $\log_6 36 = 2$
10. $\log_5 5 = 1$
11. $\log_{\sqrt{3}} 81 = 8$
12. $\log_4\left(\dfrac{1}{64}\right) = -3$

13. Explain why logarithms of negative numbers are not defined.
14. Why does $\log_a 1$ always equal 0 for any valid base a?

Find the value of each expression. (Hint: In Exercises 15–20, let the expression equal y, and write in exponential form.)

15. $\log_5 25$
16. $\log_3 81$
17. $\log_{10} .001$
18. $\log_6\left(\dfrac{1}{216}\right)$
19. $\log_4\left(\dfrac{\sqrt[3]{4}}{2}\right)$
20. $\log_9\left(\dfrac{\sqrt[4]{27}}{3}\right)$
21. $2^{\log_2 9}$
22. $8^{\log_8 11}$

Solve the equation. See Example 2.

23. $x = \log_2 32$
24. $x = \log_2 128$
25. $\log_x 25 = -2$
26. $\log_x\left(\dfrac{1}{16}\right) = -2$

27. Compare the summary of facts about the graph of $f(x) = \log_a x$ with the similar summary about the graph of $f(x) = a^x$ in Section 5.2. Make a list of the facts that reinforce the idea that these are inverse functions.

28. Graph each function. Compare the graphs to that of $f(x) = \log_2 x$. See Example 3.
    (a) $f(x) = (\log_2 x) + 3$
    (b) $f(x) = \log_2(x + 3)$
    (c) $f(x) = |\log_2(x + 3)|$

29. Graph each function. Compare the graphs to that of $f(x) = \log_{1/2} x$. See Example 3.
    (a) $f(x) = (\log_{1/2} x) - 2$
    (b) $f(x) = \log_{1/2}(x - 2)$
    (c) $f(x) = |\log_{1/2}(x - 2)|$

30. A calculator-generated graph of $y = \log_2 x$ is shown with the values of the ordered pair with $x = 5$. What does the value of y represent?

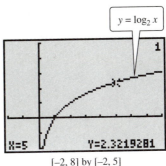

[-2, 8] by [-2, 5]

Graph each function. See Example 3.

31. $f(x) = \log_3 x$

32. $f(x) = \log_{10} x$

33. $f(x) = \log_{1/2}(1 - x)$

34. $f(x) = \log_{1/3}(3 - x)$

35. $f(x) = \log_3(x - 1)$

36. $f(x) = \log_2(x^2)$

37. Graph $y = \log_{10} x^2$ and $y = 2 \log_{10} x$ on separate viewing screens. (Use the log key on your calculator; base ten is understood.) It would seem at first glance that by applying the power rule for logarithms, these graphs should be the same. Are they? If not, why not? (*Hint:* Consider the domain in each case.)

For each function, identify the corresponding graph below.

38. $f(x) = \log_2 x$

39. $f(x) = \log_2(2x)$

40. $f(x) = \log_2\left(\dfrac{1}{x}\right)$

41. $f(x) = \log_2\left(\dfrac{x}{2}\right)$

42. $f(x) = \log_2(x - 1)$

43. $f(x) = \log_2(-x)$

A.

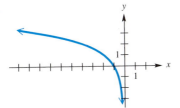

B.

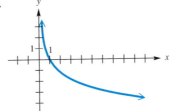

C.

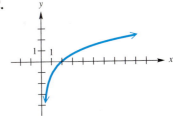

D.

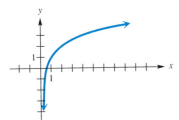

E.

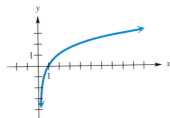

F.

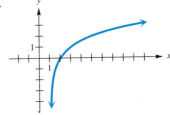

Use the log *key on your graphing calculator (for* $\log_{10} x$*) to graph each function.*

44. $f(x) = x \log_{10} x$

45. $f(x) = x^2 \log_{10} x$

Use a graphing calculator to estimate the solution(s) of each equation to the nearest hundredth.

46. $\log_{10} x = x - 2$

47. $2^{-x} = \log_{10} x$

▼▼▼▼▼▼▼▼▼▼▼▼ **DISCOVERING CONNECTIONS** (Exercises 48–51) ▼▼▼▼▼▼▼▼▼▼▼▼

Do Exercises 48–51 in order.

48. Complete the following statement of the quotient rule for logarithms: If x and y are positive numbers, $\log_a \dfrac{x}{y} = $ _____ .

49. Use the quotient rule to explain how the graph of $f(x) = \log_2\left(\dfrac{x}{4}\right)$ can be obtained from the graph of $g(x) = \log_2 x$ by a vertical shift.

50. Graph f and g on the same axes and explain how these graphs support your answer in Exercise 49.

51. If $x = 4$, $\log_2\left(\dfrac{x}{4}\right) = $ _____ ; since $\log_2 x = $ _____ and $\log_2 4 = $ _____ , $\log_2 x - \log_2 4 = $ _____ . How does this support the quotient rule stated in Exercise 48?

Write each expression as a sum, difference, or product of logarithms. Simplify the result if possible. Assume that all variables represent positive real numbers. See Example 4.

52. $\log_2\left(\dfrac{6x}{y}\right)$

53. $\log_3\left(\dfrac{4p}{q}\right)$

54. $\log_5\left(\dfrac{5\sqrt{7}}{3}\right)$

55. $\log_2\left(\dfrac{2\sqrt{3}}{5}\right)$

56. $\log_4(2x + 5y)$

57. $\log_6(7m + 3q)$

58. $\log_m\sqrt{\dfrac{5r^3}{z^5}}$

59. $\log_p\sqrt[3]{\dfrac{m^5 n^4}{t^2}}$

Write each expression as a single logarithm with a coefficient of 1. Assume that all variables represent positive real numbers. See Example 5.

60. $\log_a x + \log_a y - \log_a m$

61. $(\log_b k - \log_b m) - \log_b a$

62. $2 \log_m a - 3 \log_m(b^2)$

63. $\dfrac{1}{2}\log_y(p^3 q^4) - \dfrac{2}{3}\log_y(p^4 q^3)$

64. $2 \log_a(z - 1) + \log_a(3z + 2), \quad z > 1$

65. $\log_b(2y + 5) - \dfrac{1}{2}\log_b(y + 3)$

Given $\log_{10} 2 = .3010$ and $\log_{10} 3 = .4771$, find each logarithm without using a calculator. See Example 6.

66. $\log_{10} 6$

67. $\log_{10} 12$

68. $\log_{10}(9/4)$

69. $\log_{10}(20/27)$

Suppose f is a logarithmic function and $f(3) = 2$. Determine the function values in Exercises 70–71.

70. $f(1/9)$

71. $f(27)$

72. The following table lists the interest rates for various U.S. Treasury Securities in January 1996. (*Source*: Reuters.)

Time	Yield
3-month	5.71%
6-month	6.37%
1-year	6.87%
2-year	7.34%
3-year	7.52%
5-year	7.63%
10-year	7.68%
30-year	7.79%

(a) Plot the data.

(b) Discuss which type of function will model these data best: linear, exponential, or logarithmic.

73. Prove Property (b): $\log_a \dfrac{x}{y} = \log_a x - \log_a y$.

74. Prove Property (c): $\log_a x^r = r \log_a x$.

5.4 Evaluating Logarithms; Change of Base ▼▼▼

COMMON LOGARITHMS The bases 10 and e are so important for logarithms that scientific and graphing calculators have keys for these bases. Base 10 logarithms are called **common logarithms.** The common logarithm of the number x, or $\log_{10} x$, is often abbreviated as just $\log x$, and we will use that convention from now on. A calculator with a log key can be used to find base 10 logarithms of any positive number. (If your calculator has an ln key, but not a log key, you will need to use the *change-of-base theorem* discussed later in this section.)

EXAMPLE 1
Evaluating common logarithms

```
log 142
        2.152288344
log .005832
        -2.234182485
```

Compare these results to those in Example 1.

Use a calculator to evaluate the following logarithms.

(a) log 142

Enter 142 and press the log key. This may be a second function key on some calculators. With other calculators, these steps may be reversed. Consult your owner's manual if you have any problem using this key. The result should be 2.152 to the nearest thousandth. (This means that $10^{2.152} \approx 142$.)

(b) log .005832
A calculator gives

$$\log .005832 \approx -2.234.$$

(Thus, $10^{-2.234} \approx .005832$.) ▶

NOTE   Base a, $a > 1$, logarithms of numbers less than 1 are always negative, as suggested by the graphs in Section 5.3.

In chemistry, the pH of a solution is defined as

$$pH = -\log[H_3O^+],$$

where $[H_3O^+]$ is the hydronium ion concentration in moles* per liter. The pH value is a measure of the acidity or alkalinity of solutions. Pure water has a pH of 7.0, substances with pH values greater than 7.0 are alkaline, and substances with pH values less than 7.0 are acidic.

EXAMPLE 2
Finding pH

(a) Find the pH of a solution with $[H_3O^+] = 2.5 \times 10^{-4}$.

$$pH = -\log[\mathbf{H_3O^+}]$$
$$pH = -\log(\mathbf{2.5 \times 10^{-4}}) \qquad \text{Substitute.}$$
$$= -(\log 2.5 + \log 10^{-4}) \qquad \text{Property (a) of logarithms}$$
$$= -(.3979 - 4)$$
$$= -.3979 + 4$$
$$\approx 3.6$$

It is customary to round pH values to the nearest tenth.

(b) Find the hydronium ion concentration of a solution with pH $= 7.1$.

$$\mathbf{pH} = -\log[H_3O^+]$$
$$\mathbf{7.1} = -\log[H_3O^+] \qquad \text{Substitute.}$$
$$-7.1 = \log[H_3O^+] \qquad \text{Multiply by } -1.$$
$$[H_3O^+] = 10^{-7.1} \qquad \text{Write in exponential form.}$$

Evaluate $10^{-7.1}$ with a calculator to get

$$[H_3O^+] \approx 7.9 \times 10^{-8}. \quad \blacktriangleright$$

EXAMPLE 3
Measuring the loudness of sound

The loudness of sounds is measured in a unit called a *decibel*. To measure with this unit, we first assign an intensity of I_0 to a very faint sound, called the *threshold sound*. If a particular sound has intensity I, then the decibel rating of this louder sound is

$$d = 10 \log \frac{I}{I_0}.$$

Find the decibel rating of a sound with intensity $10{,}000I_0$.

---

*A *mole* is the amount of a substance that contains the same number of molecules as the number of atoms in exactly 12 grams of carbon 12.

Let $I = 10,000I_0$ and find d.

$$d = 10 \log \frac{10,000I_0}{I_0}$$

$$= 10 \log 10,000$$

$$= 10(4) \qquad \qquad \text{log } 10,000 = 4$$

$$= 40$$

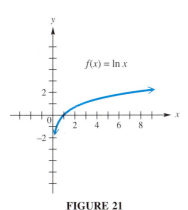

FIGURE 21

The sound has a decibel rating of 40. ▶

NATURAL LOGARITHMS In most practical applications of logarithms, the number $e \approx 2.718281828$ is used as base. The number e is irrational, like π. Logarithms to base e are called **natural logarithms,** since they occur in the life sciences and economics in natural situations that involve growth and decay. The base e logarithm of x is written ln x (read "el-en x"). A graph of the natural logarithm function defined by $f(x) = \ln x$ is given in Figure 21.

Natural logarithms can be found with a calculator that has an ln key.

EXAMPLE 4
Evaluating natural logarithms

```
ln 85
      4.442651256
ln 127.8
      4.850466542
ln .049
     -3.015934981
```

Compare these results to those in Example 4.

Use a calculator to find the following logarithms.

(a) ln 85
With a calculator, enter 85, press the ln key, and read the result, 4.4427. The steps may be reversed with some calculators. If your calculator has an e^x key, but not a key labeled ln x, natural logarithms can be found by entering the number, pressing the INV key and then the e^x key. This works because $y = e^x$ is the inverse function of $y = \ln x$ (or $y = \log_e x$). Because ln 85 $\approx$ 4.4427, $e^{4.4427} \approx 85$.

(b) ln 127.8 $\approx$ 4.850

(c) ln .049 $\approx$ −3.02
As with common logarithms, natural logarithms of numbers between 0 and 1 are negative. ▶

EXAMPLE 5
Measuring the age of rocks

Geologists sometimes measure the age of rocks by using "atomic clocks." By measuring the amounts of potassium 40 and argon 40 in a rock, the age t of the specimen in years is found with the formula

$$t = (1.26 \times 10^9) \frac{\ln[1 + 8.33 \, (A/K)]}{\ln 2}.$$

A and K are respectively the numbers of atoms of argon 40 and potassium 40 in the specimen.

(a) How old is a rock in which $A = 0$ and $K > 0$?
If $A = 0$, $A/K = 0$ and the equation becomes

$$t = (1.26 \times 10^9) \frac{\ln 1}{\ln 2} = (1.26 \times 10^9)(0) = 0.$$

The rock is 0 years old or new.

(b) The ratio A/K for a sample of granite from New Hampshire is .212. How old is the sample?

Since A/K is .212, we have

$$t = (1.26 \times 10^9) \frac{\ln[1 + 8.33(.212)]}{\ln 2} \approx 1.85 \times 10^9.$$

The granite is about 1.85 billion years old. ▶

EXAMPLE 6
Analyzing global
temperature increase

Carbon dioxide in the atmosphere traps heat from the sun. The additional solar radiation trapped by carbon dioxide is called *radiative forcing*. It is measured in watts per square meter. In 1896 the Swedish scientist Svante Arrhenius estimated the radiative forcing R caused by additional atmospheric carbon dioxide using the logarithmic equation $R = k \ln(C/C_0)$, where C_0 is the preindustrial amount of carbon dioxide, C is the current carbon dioxide level, and k is a constant. Arrhenius determined that $10 \leq k \leq 16$ when $C = 2C_0$.*

(a) Let $C = 2C_0$. Is the relationship between R and k linear or logarithmic?

If $C = 2C_0$, $C/C_0 = 2$, so $R = k \ln 2$ is a linear relation, because $\ln 2$ is a constant.

(b) The average global temperature increase T (in °F) is given by $T(R) = 1.03R$. (See Section 5.2, Exercise 59.) Write T as a function of k.

Use the expression for R given in the introduction above.

$$T(R) = 1.03R$$
$$T(k) = 1.03k \ln(C/C_0) ▶$$

LOGARITHMS TO OTHER BASES A calculator can be used to find the values of either natural logarithms (base e) or common logarithms (base 10). However, sometimes it is convenient to use logarithms to other bases. For example, base 2 logarithms are important in computer science. The following theorem can be used to convert logarithms from one base to another.

Change-of-Base Theorem

For any positive real numbers x, a, and b, where $a \neq 1$ and $b \neq 1$:

$$\log_a x = \frac{\log_b x}{\log_b a}.$$

NOTE As an aid is remembering the change-of-base theorem, notice that x is above a on both sides of the equation.

Source: Clime, W., *The Economics of Global Warming*, Institute for International Economics, Washington, D.C., 1992.

This theorem is proved by using the definition of logarithm to write $y = \log_a x$ in exponential form.

Proof

Let $\qquad\qquad\qquad\qquad y = \log_a x.$

$$a^y = x \qquad \text{Change to exponential form.}$$

$$\log_b a^y = \log_b x \qquad \text{Take logarithms on both sides.}$$

$$y \log_b a = \log_b x \qquad \text{Property (c) of logarithms}$$

$$y = \frac{\log_b x}{\log_b a} \qquad \text{Divide both sides by } \log_b a.$$

$$\log_a x = \frac{\log_b x}{\log_b a} \qquad \text{Substitute } \log_a x \text{ for } y. \;\; \blacktriangleright$$

Any positive number other than 1 can be used for base b in the change of base theorem, but usually the only practical bases are e and 10, since calculators give logarithms only for these two bases. The change-of-base theorem is used to find logarithms for other bases.

The change-of-base theorem is needed to graph logarithmic functions with bases other than 10 and e (and sometimes with one of those bases). For instance,

$$\text{to graph } y = \log_3(x - 1), \text{ graph } y = \frac{\log(x - 1)}{\log 3} \text{ or } y = \frac{\ln(x - 1)}{\ln 3}.$$

The next example shows how the change-of-base theorem is used to find logarithms to bases other than 10 or e with a calculator.

EXAMPLE 7
Using the change-of-base theorem

```
log 17/log 5
        1.760374428
ln 17/ln 5
        1.760374428
```

The result of Example 7(a) is valid for *either* natural or common logarithms.

Use natural logarithms to find each of the following. Round to the nearest hundredth.

(a) $\log_5 17$

Use natural logarithms and the change-of-base theorem.

$$\log_5 17 = \frac{\log_e 17}{\log_e 5}$$

$$= \frac{\ln 17}{\ln 5}$$

$$\approx \frac{2.8332}{1.6094}$$

$$\approx 1.76$$

To check, use a calculator along with the definition of logarithm, to verify that $5^{1.76} \approx 17$.

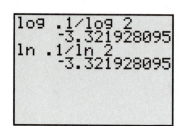

The result of Example 7(b) is valid for *either* common or natural logarithms.

(b) $\log_2 .1$

$$\log_2 .1 = \frac{\ln .1}{\ln 2} \approx \frac{-2.3026}{.6931} \approx -3.32 \quad \blacktriangleright$$

NOTE   In Example 7, logarithms evaluated in the intermediate steps, such as ln 17 and ln 5, were shown to four decimal places. However, the final answers were obtained *without* rounding off these intermediate values, using all the digits obtained with the calculator. In general, it is best to wait until the final step to round off the answer; otherwise, a build-up of round-off error may cause the final answer to have an incorrect final decimal place digit.

EXAMPLE 8
Solving an application with base 2 logarithms

One measure of the diversity of the species in an ecological community is given by the formula

$$H = -[P_1 \log_2 P_1 + P_2 \log_2 P_2 + \cdots + P_n \log_2 P_n],$$

where $P_1, P_2, \ldots, P_n$ are the proportions of a sample belonging to each of n species found in the sample. For example, in a community with two species, where there are 90 of one species and 10 of the other, $P_1 = 90/100 = .9$ and $P_2 = 10/100 = .1$. Thus,

$$H = -[.9 \log_2 .9 + .1 \log_2 .1].$$

In Example 7(b), $\log_2 .1$ was found to be -3.32. Now find $\log_2 .9$.

$$\log_2 .9 = \frac{\ln .9}{\ln 2}$$

$$\approx \frac{-.1054}{.6931}$$

$$\approx -.152$$

Therefore,

$$H \approx -[(.9)(-.152) + (.1)(-3.32)] \approx .469.$$

If the number in each species is the same, the measure of diversity is 1, representing "perfect" diversity. In a community with little diversity, H is close to 0. In this example, since $H \approx .5$, there is neither great nor little diversity. $\quad \blacktriangleright$

5.4 Exercises ▼▼▼▼▼▼▼▼▼▼▼▼▼▼▼▼▼▼▼▼▼▼▼▼▼▼▼▼▼▼▼▼▼▼▼

Use a calculator to evaluate each logarithm to four decimal places. See Examples 1 and 4.

1. log 43   **2.** log 1247   **3.** log .014   **4.** log .0069

5. ln 580   **6.** ln .08   **7.** ln .7   **8.** ln 81,000

9. The graph of $y = \log x$ is shown with the coordinates of a point displayed at the bottom of the screen. Write the logarithmic equation associated with the display.

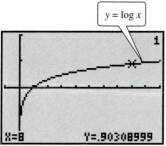

[−1, 10] by [−2, 2]

10. The graph of $y = \ln x$ is shown with the coordinates of a point displayed at the bottom of the screen. Write the logarithmic equation associated with the display.

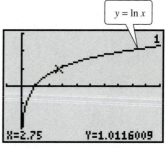

[−1, 10] by [−3, 3]

11. Is the logarithm to the base 3 of 4 written as $\log_4 3$ or $\log_3 4$?

▼▼▼▼▼▼▼▼▼▼▼▼▼ **DISCOVERING CONNECTIONS** (Exercises 12–17) ▼▼▼▼▼▼▼▼▼▼▼▼▼

Work Exercises 12–17 in order.

12. What is the exact value of $\log_3 9$?

13. What is the exact value of $\log_3 27$?

14. Between what two consecutive integers must $\log_3 16$ lie? Explain your answer.

15. Use the change-of-base theorem to support your answer for Exercise 14.

16. Repeat Exercises 12 and 13 for $\log_5(1/5)$ and $\log_5 1$.

17. Repeat Exercises 14 and 15 for $\log_5 .68$.

Use the change-of-base theorem to find each logarithm to the nearest hundredth. See Example 7.

18. $\log_5 10$ **19.** $\log_9 12$ **20.** $\log_{15} 5$ **21.** $\log_{1/2} 3$

22. $\log_{100} 83$ **23.** $\log_{200} 175$ **24.** $\log_{2.9} 7.5$ **25.** $\log_{5.8} 12.7$

26. Consider the function defined by $f(x) = \log_3 |x|$.
   (a) What is the domain of this function?
   (b) Use a graphing calculator to graph $f(x) = \log_3 |x|$ in the window $[-4, 4]$ by $[-4, 4]$.
   (c) How might one easily misinterpret the domain of the function simply by observing the calculator-generated graph?

27. The table is for $Y_1 = \log_3(4 - x)$. Why do the values of Y_1 show ERROR for $X \geq 4$?

X	Y₁
1	1
2	.63093
3	0
4	ERROR
5	ERROR
6	ERROR
7	ERROR

X=1

Graph each function.

28. $f(x) = \log_5 x$

29. $f(x) = \log_x 5$ in *connected* mode. What does the vertical line in the graph simulate?

30. Explain the error in the following "proof" that $2 < 1$.

$$\frac{1}{9} < \frac{1}{3}$$

$$\left(\frac{1}{3}\right)^2 < \frac{1}{3} \qquad \text{Rewrite the left side.}$$

$$\log\left(\frac{1}{3}\right)^2 < \log\left(\frac{1}{3}\right) \qquad \text{Take the log on each side.}$$

$$2 \log \frac{1}{3} < 1 \log\left(\frac{1}{3}\right) \qquad \text{Property of logarithms}$$

$$2 < 1 \qquad \text{Divide both sides by } \log\left(\frac{1}{3}\right).$$

For each substance, find the pH from the given hydronium ion concentration. See Example 2(a).

31. Grapefruit, 6.3×10^{-4}

32. Crackers, 3.9×10^{-9}

33. Limes, 1.6×10^{-2}

34. Sodium hydroxide (lye), 3.2×10^{-14}

Find the $[H_3O^+]$ for each substance from the given pH. See Example 2(b).

35. Soda pop, 2.7

36. Wine, 3.4

37. Beer, 4.8

38. Drinking water, 6.5

Solve each problem. See Example 3.

39. Find the decibel ratings of sounds having the following intensities:
   (a) $100I_0$ (b) $1000I_0$
   (c) $100,000I_0$ (d) $1,000,000I_0$

40. Find the decibel ratings of the following sounds, having intensities as given. Round each answer to the nearest whole number.
   (a) whisper, $115I_0$
   (b) busy street, $9,500,000I_0$
   (c) heavy truck, 20 m away, $1,200,000,000I_0$
   (d) rock music, $895,000,000,000I_0$
   (e) jetliner at takeoff, $109,000,000,000,000I_0$

41. The magnitude of an earthquake, measured on the Richter scale, is $\log_{10}(I/I_0)$, where I_0 is the magnitude of an earthquake of a certain (small) size. Find the Richter scale ratings for earthquakes having the following magnitudes.
   (a) $1000I_0$ (b) $1,000,000I_0$ (c) $100,000,000I_0$

42. On July 14, 1991, Peshawar, Pakistan, was shaken by an earthquake that measured 6.6 on the Richter scale.
   (a) Express this reading in terms of I_0. See Exercise 41.

(b) In February of the same year a quake measuring 6.5 on the Richter scale killed about 900 people in the mountains of Pakistan and Afghanistan. Express the magnitude of a 6.5 reading in terms of I_0.

(c) How much greater was the force of the earthquake with a measure of 6.6?

43. (a) The San Francisco earthquake of 1906 had a Richter scale rating of 8.3. Express the magnitude of this earthquake as a multiple of I_0.

(b) In 1989, the San Francisco region experienced an earthquake with a Richter scale rating of 7.1. Express the magnitude of this earthquake as a multiple of I_0.

(c) Compare the magnitudes of the two San Francisco earthquakes discussed above.

Solve each problem. See Example 5.

44. The number of years, n, since two independently evolving languages split off from a common ancestral language is approximated by $n \approx -7600 \log r$, where r is the proportion of words from the ancestral language common to both languages.

(a) Find n if $r = .9$.
(b) Find n if $r = .3$.
(c) How many years have elapsed since the split if half of the words of the ancestral language are common to both languages?

45. The number of species in a sample is given by

$$S(n) = a \ln\left(1 + \frac{n}{a}\right).$$

Here n is the number of individuals in the sample and a is a constant that indicates the diversity of species in the community. If $a = .36$, find $S(n)$ for the following values of n. (*Hint:* n must be a whole number.)
(a) 100   (b) 200   (c) 150   (d) 10

46. In Exercise 45, find $S(n)$ if a changes to .88. Use the following values of n. (*Hint:* $S(n)$ must be a whole number.)
(a) 50   (b) 100   (c) 250

In Exercises 47 and 48, refer to Example 8.

47. Suppose a sample of a small community shows two species with 50 individuals each. Find the index of diversity H.

48. A virgin forest in northwestern Pennsylvania has 4 species of large trees with the following proportions of each: hemlock, .521; beech, .324; birch, .081; maple, .074. Find the index of diversity H.

49. Given $g(x) = e^x$, evaluate the following.
(a) $g(\ln 3)$ (b) $g[\ln(5^2)]$ (c) $g\left[\ln\left(\frac{1}{e}\right)\right]$

50. Given $f(x) = 3^x$, evaluate the following.
(a) $f(\log_3 7)$ (b) $f[\log_3(\ln 3)]$
(c) $f[\log_3(2 \ln 3)]$

51. Given $f(x) = \ln x$, evaluate the following.
(a) $f(e^5)$ (b) $f(e^{\ln 3})$ (c) $f(e^{2\ln 3})$

52. Given $f(x) = \log_2 x$, evaluate the following.
(a) $f(2^3)$ (b) $f(2^{\log_2 2})$ (c) $f(2^{2\log_2 2})$

53. The heights in the bar graph represent the number of visitors (in millions) to U.S. National Parks from 1950 to 1994. Suppose x represents the number of years since 1900—thus, 1950 is represented by 50, 1960 is represented by 60, and so on. The logarithmic function defined by $f(x) = -266 + 72 \ln x$ closely approximates the data. Use this function to estimate the number of visitors in the year 2000. What assumption must we make to estimate the number of visitors in years beyond 1993?

54. The growth of outpatient surgery as a percent of total surgeries at hospitals is shown in the accompanying graph. Connecting the tops of the bars with a continuous curve would give a graph that indicates logarithmic growth. The function with $f(x) = -1317 + 304 \ln x$, where x represents the number of years since 1900 and $f(x)$ is the percent, approximates the curve reasonably well. What does this function predict for the percent of outpatient surgeries in 1998?

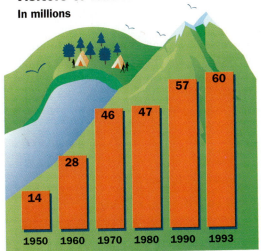

Visitors to National Parks

In millions

14   28   46   47   57   60
1950 1960 1970 1980 1990 1993

Source: Statistical Abstract of the United States 1995

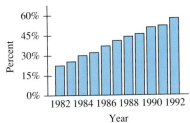

Outpatient Surgery Grows at Hospitals...
Outpatient surgery as a percentage of total surgeries

Percent
60%
45%
30%
15%
0%
1982 1984 1986 1988 1990 1992
Year

Source: American Hospital Association, Chicago

55. In Example 6, we expressed the average global temperature increase T (in °F) as $T(k) = 1.03k \ln(C/C_0)$ where C_0 is the preindustrial amount of carbon dioxide, C is the current carbon dioxide level, and k is a constant. Arrhenius determined that $10 \leq k \leq 16$ when C was double the value C_0. Use $T(k)$ to find the range of the rise in global temperature T (rounded to the nearest degree) that Arrhenius predicted. (*Source:* Clime, W., *The Economics of Global Warming,* Institute for International Economics, Washington, D.C., 1992.)

 56. (Refer to Exercise 55.) According to the IPCC if present trends continue, future increases in average global temperatures (in °F) can be modeled by $T = 6.489 \cdot \ln(C/280)$ where C is the concentration of atmospheric carbon dioxide (in ppm). C can be modeled by the function with $C(x) = 353(1.006)^{x-1990}$ where x is the year. (*Source:* International Panel on Climate Change (IPCC), 1990.)

(a) Write T as a function of x.

(b) With a graphing calculator graph $C(x)$ and $T(x)$ on the interval $[1990, 2275]$ using different coordinate axes. Describe each function's graph. How are C and T related?

(c) Approximate the slope of the graph of T. What does this slope represent?

(d) Use graphing to estimate x and C when $T = 10°F$.

57. The following table contains the planets' average distances D from the sun and their periods P of revolution

around the sun in years. The distances have been normalized so that Earth is one unit away from the sun. For example, since Jupiter's distance is 5.2, its distance from the sun is 5.2 times farther than Earth's. (*Source:* Ronan, C., *The Natural History of the Universe,* MacMillan Publishing Co., New York, 1991.)

Planet	D	P
Mercury	.39	.24
Venus	.72	.62
Earth	1	1
Mars	1.52	1.89
Jupiter	5.2	11.9
Saturn	9.54	29.5
Uranus	19.2	84.0
Neptune	30.1	164.8

(a) Plot the point $(\ln D, \ln P)$ for each planet on the xy-coordinate axes using a graphing calculator. Do the data points appear to be linear?

(b) Determine a linear equation that approximates the data points. Graph your line and the data on the same coordinate axes.

(c) Use this linear equation to predict the period of the planet Pluto if its distance is 39.5. Compare your answer to the true value of 248.5 years.

5.5 Exponential and Logarithmic Equations ▼▼▼

Exponential and logarithmic functions are important in many useful applications of mathematics. Using these functions in applications often requires solving exponential and logarithmic equations. Some simple equations were solved in earlier sections of this chapter. More general methods for solving these equations depend on the property below. This property follows from the fact that logarithmic functions are one-to-one.

> **Property of Logarithms**
>
> **(f)** If $x > 0$, $y > 0$, $a > 0$, and $a \neq 1$, then
>
> $$x = y \quad \text{if and only if} \quad \log_a x = \log_a y.$$

EXPONENTIAL EQUATIONS The first examples illustrate a general method, using the new property, for solving exponential equations.

EXAMPLE 1
Solving an exponential equation

Solve the equation $7^x = 12$.

The properties given in Section 5.2 cannot be used to solve this equation, so we apply Property (f). While any appropriate base b can be used, the best practical base is base 10 or base e. Taking base e (natural) logarithms of both sides gives

$$7^x = 12$$

$$\ln 7^x = \ln 12$$

$$x \ln 7 = \ln 12 \qquad \text{Property (c) of logarithms}$$

$$x = \frac{\ln 12}{\ln 7} \qquad \text{Divide by ln 7.}$$

$$\approx \frac{2.4849}{1.9459} \approx 1.277.$$

A calculator can be used to check this answer. Evaluate $7^{1.277}$; the result should be approximately 12. This step verifies that, to the nearest thousandth, the solution set is $\{1.277\}$. ▶

CAUTION Be careful when evaluating a quotient like $\dfrac{\ln 12}{\ln 7}$ in Example 1. Do not confuse this quotient with $\ln\left(\dfrac{12}{7}\right)$ which can be written as $\ln 12 - \ln 7$. You *cannot* change the quotient of two *logarithms* to a difference of logarithms.

$$\frac{\ln 12}{\ln 7} \neq \ln\left(\frac{12}{7}\right)$$

EXAMPLE 2
Solving an exponential equation

Solve $3^{2x-1} = 4^{x+2}$.

Taking natural logarithms on both sides gives

$$\ln 3^{2x-1} = \ln 4^{x+2}.$$

Now use a property of logarithms.

$$(2x - 1)\ln 3 = (x + 2)\ln 4$$

$$2x \ln 3 - \ln 3 = x \ln 4 + 2 \ln 4 \qquad \text{Distributive property}$$

$$2x \ln 3 - x \ln 4 = 2 \ln 4 + \ln 3$$

$$x(2 \ln 3 - \ln 4) = 2 \ln 4 + \ln 3 \qquad \text{Factor out } x.$$

$$x = \frac{2 \ln 4 + \ln 3}{2 \ln 3 - \ln 4}$$

$$x = \frac{\ln 16 + \ln 3}{\ln 9 - \ln 4} \qquad \text{Properties of logarithms}$$

$$x = \frac{\ln 48}{\ln\left(\dfrac{9}{4}\right)}$$

This quotient could be approximated by a decimal if desired.

$$x = \frac{\ln 48}{\ln 2.25} \approx \frac{3.8712}{.8109} \approx 4.774$$

To the nearest thousandth, the solution set is {4.774}. ▶

EXAMPLE 3
Solving a base *e* exponential equation

Solve $e^{x^2} = 200$.

Take natural logarithms on both sides; then use properties of logarithms.

$$e^{x^2} = 200$$
$$\ln e^{x^2} = \ln 200$$
$$x^2 = \ln 200 \qquad \ln e^{x^2} = x^2$$
$$x = \pm\sqrt{\ln 200}$$
$$x \approx \pm 2.302$$

To the nearest thousandth, the solution set is {−2.302, 2.302}. ▶

CONNECTIONS Sometimes it is necessary to solve more complicated exponential equations. (A graphing calculator is very handy for these!) A typical equation that arises in some situations is

$$\frac{e^x - 1}{e^{-x} - 1} = -3.$$

To solve this equation, begin by multiplying both sides by the denominator.

$$e^x - 1 = -3(e^{-x} - 1)$$
$$e^x - 1 = -3e^{-x} + 3 \qquad \text{Distributive property}$$
$$e^x - 4 + 3e^{-x} = 0 \qquad \text{Get 0 alone on one side.}$$
$$e^{2x} - 4e^x + 3 = 0 \qquad \text{Multiply both sides by } e^x.$$

Rewrite this equation as

$$(e^x)^2 - 4e^x + 3 = 0,$$

a quadratic equation in e^x. Let $u = e^x$ and solve first for u getting the solutions 1 and 3. Then solve for x.

$$u = 1 \quad \text{or} \quad u = 3$$
$$e^x = 1 \qquad e^x = 3 \qquad \text{Replace } u \text{ with } e^x.$$
$$x = 0 \qquad x = \ln 3$$

FOR DISCUSSION OR WRITING

1. Carry out the steps of the solution of the quadratic equation to find u.
2. Check the solutions for x in the original equation. One is extraneous. Explain why one solution does not satisfy the equation. Give the solution set.
3. In solving the equation here, we used a two-step process. What are the two steps?

LOGARITHMIC EQUATIONS   The next examples show some ways to solve logarithmic equations. The properties of logarithms given in Section 5.3 are useful here, as is Property (f).

EXAMPLE 4
Solving a logarithmic equation

Solve $\log_a(x + 6) - \log_a(x + 2) = \log_a x$.
   Using a property of logarithms, rewrite the equation as

$$\log_a \frac{x + 6}{x + 2} = \log_a x. \qquad \text{Property (b) of logarithms}$$

Now the equation is in the proper form to use Property (f).

$$\frac{x + 6}{x + 2} = x \qquad \text{Property (f)}$$
$$x + 6 = x(x + 2) \qquad \text{Multiply by } x + 2.$$
$$x + 6 = x^2 + 2x \qquad \text{Distributive property}$$
$$x^2 + x - 6 = 0 \qquad \text{Get 0 on one side.}$$
$$(x + 3)(x - 2) = 0 \qquad \text{Use the zero-factor property.}$$
$$x = -3 \qquad \text{or} \qquad x = 2$$

The negative solution ($x = -3$) cannot be used since it is not in the domain of $\log_a x$ in the original equation. For this reason, the only valid solution is the positive number 2, giving the solution set {2}. ◗

CAUTION   Recall that the domain of $y = \log_b x$ is $(0, \infty)$. For this reason, it is always necessary to check that the apparent solution of a logarithmic equation results in the logarithms of positive numbers in the original equation.

EXAMPLE 5
Solving a logarithmic equation

Solve $\log(3x + 2) + \log(x - 1) = 1$.
   Since $\log x$ is an abbreviation for $\log_{10} x$, and $1 = \log_{10} 10$, the properties of logarithms give

$$\log(3x + 2)(x - 1) = \log 10 \qquad \text{Property (a) of logarithms}$$
$$(3x + 2)(x - 1) = 10 \qquad \text{Property (f)}$$
$$3x^2 - x - 2 = 10$$
$$3x^2 - x - 12 = 0.$$

Now use the quadratic formula to get

$$x = \frac{1 \pm \sqrt{1 + 144}}{6}.$$

The number $(1 - \sqrt{145})/6$ is negative, so $x - 1$ is negative. Therefore, $\log(x - 1)$ is not defined and this proposed solution must be discarded. Since

```
solve(log (3X+2)
+log (X-1)-1,X,5
)
        2.173599096
(1+√145)/6
        2.173599096
```

The equation-solving feature of a
graphing calculator supports the result
of Example 5.

$(1 + \sqrt{145})/6 > 1$, both $3x + 2$ and $x - 1$ are positive and the solution set is

$$\left\{ \frac{1 + \sqrt{145}}{6} \right\}.$$ ▶

The definition of logarithm could have been used in Example 5 by first writing

$$\log(3x + 2) + \log(x - 1) = 1$$
$$\log_{10}(3x + 2)(x - 1) = 1 \qquad \text{Property (a)}$$
$$(3x + 2)(x - 1) = 10^1, \qquad \text{Definition of logarithm}$$

then continuing as shown above.

◀ **EXAMPLE 6**
Solving a logarithmic equation

Solve $\ln e^{\ln x} - \ln(x - 3) = \ln 2$.

On the left, $\ln e^{\ln x}$ can be written as $\ln x$ using the theorem on inverses at the end of Section 5.3. The equation becomes

$$\ln x - \ln(x - 3) = \ln 2$$

$$\ln \frac{x}{x - 3} = \ln 2 \qquad \text{Property (b)}$$

$$\frac{x}{x - 3} = 2 \qquad \text{Property (f)}$$

$$x = 2x - 6 \qquad \text{Multiply by } x - 3.$$

$$6 = x.$$

Verify that the solution set is $\{6\}$. ▶

🔲 A graphing calculator is helpful for determining the number of solutions and approximating the solutions. For instance, to examine the equation in Example 6, we could graph $Y_1 = \ln e^{\ln x} - \ln(x - 3)$ and $Y_2 = \ln 2$, and find the x-values of any intersection points. The result, in the figure below, supports our analytic solution.

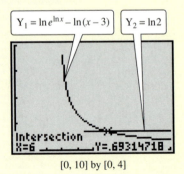

$[0, 10]$ by $[0, 4]$

A summary of the methods used for solving equations in this section follows.

> ## Solving Exponential or Logarithmic Equations
>
> In summary, to solve an exponential or logarithmic equation, first use the properties of algebra to change the given equation into one of the following forms, where a and b are real numbers with appropriate restrictions.
>
> 1. $a^{f(x)} = b$
>   To solve, take logarithms on both sides.
> 2. $\log_a f(x) = b$
>   Solve by changing to the exponential form $a^b = f(x)$.
> 3. $\log_a f(x) = \log_a g(x)$
>   From the given equation, obtain the equation $f(x) = g(x)$, then solve algebraically.
> 4. In a more complicated equation, such as the one in the Connections box, it may be necessary to first solve for $e^{f(x)}$ or $\log_a f(x)$ and then solve the resulting equation using one of the methods given above.

The next examples show applications of exponential and logarithmic equations.

EXAMPLE 7
Solving a composite exponential equation

The strength of a habit is a function of the number of times the habit is repeated. If N is the number of repetitions and H is the strength of the habit, then, according to psychologist C. L. Hull,

$$H = 1000(1 - e^{-kN}),$$

where k is a constant. Solve this equation for k.

We must first solve the equation for e^{-kN}.

$$\frac{H}{1000} = 1 - e^{-kN} \qquad \text{Divide by 1000.}$$

$$\frac{H}{1000} - 1 = -e^{-kN} \qquad \text{Subtract 1.}$$

$$e^{-kN} = 1 - \frac{H}{1000} \qquad \text{Multiply by } -1.$$

Now solve for k. As shown earlier, we take logarithms on each side of the equation and use the fact that $\ln e^x = x$.

$$\ln e^{-kN} = \ln\left(1 - \frac{H}{1000}\right)$$

$$-kN = \ln\left(1 - \frac{H}{1000}\right) \qquad \text{\textcolor{blue}{$\ln e^x = x$}}$$

$$k = -\frac{1}{N}\ln\left(1 - \frac{H}{1000}\right) \qquad \text{\textcolor{blue}{Multiply by $-\frac{1}{N}$.}}$$

With the last equation, if one pair of values for H and N is known, k can be found, and the equation can then be used to find either H or N, for given values of the other variable. ◗

EXAMPLE 8
Solving a composite logarithmic equation

In the exercises for Section 5.4, we saw that the number of species in a sample is given by $S(n)$ or S, where

$$S = a\ln\left(1 + \frac{n}{a}\right),$$

n is the number of individuals in the sample, and a is a constant. Solve this equation for n.

We begin by solving for $\ln\left(1 + \frac{n}{a}\right)$. Then we can change to exponential form and solve the resulting equation for n.

$$\frac{S}{a} = \ln\left(1 + \frac{n}{a}\right) \qquad \text{\textcolor{blue}{Divide by a.}}$$

$$e^{S/a} = 1 + \frac{n}{a} \qquad \text{\textcolor{blue}{Write in exponential form.}}$$

$$e^{S/a} - 1 = \frac{n}{a} \qquad \text{\textcolor{blue}{Subtract 1.}}$$

$$n = a(e^{S/a} - 1) \qquad \text{\textcolor{blue}{Multiply by a.}}$$

Using this equation and given values of S and a, the number of individuals in a sample can be found. ◗

5.5 Exercises ▼▼▼▼▼▼▼▼▼▼▼▼▼▼▼▼▼▼▼▼▼▼▼▼▼▼▼▼▼▼▼▼▼▼▼▼

1. Between what two consecutive integers must x be if $5^x = 28$? Why is this so?

2. Is the statement $\dfrac{\log 16}{\log 3} = \dfrac{\ln 16}{\ln 3}$ true? If so, how can you support the statement?

3. Explain why an equation of the form $2^x = a$ does not always have a solution.

4. Without solving, explain why an equation of the form $\log_2 x = a$ must always have a solution for x.

Solve the equation. When necessary, give answers as decimals rounded to the nearest thousandth. See Examples 1–6.

5. $3^x = 6$

6. $4^x = 12$

7. $6^{1-2k} = 8$

8. $3^{2m-5} = 13$

9. $e^{k-1} = 4$

10. $e^{2-y} = 12$

11. $2e^{5a+2} = 8$

12. $10e^{3z-7} = 5$

13. $2^x = -3$

14. $\left(\dfrac{1}{4}\right)^p = -4$

15. $e^{2x} \cdot e^{5x} = e^{14}$

16. $e^3 \cdot e^{\ln x} = \dfrac{1}{8}$

17. $100(1 + .02)^{3+n} = 150$

18. $500(1 + .05)^{p/4} = 200$

19. $\log(t - 1) = 1$

20. $\log q^2 = 1$

21. $\ln(y + 2) = \ln(y - 7) + \ln 4$

22. $\ln p - \ln(p + 1) = \ln 5$

23. $\ln(5 + 4y) - \ln(3 + y) = \ln 3$

24. $\ln m + \ln(2m + 5) = \ln 7$

25. $2 \ln(x - 3) = \ln(x + 5) + \ln 4$

26. $\ln(k + 5) + \ln(k + 2) = \ln(14k)$

27. $\log_3(a - 3) = 1 + \log_3(a + 1)$

28. $\log w + \log(3w - 13) = 1$

29. $\ln e^x - \ln e^3 = \ln e^5$

30. $\ln e^x - 2 \ln e = \ln e^4$

31. $\log_2 \sqrt{2y^2} - 1 = \dfrac{1}{2}$

32. $\log_2(\log_2 x) = 1$

33. $\log z = \sqrt{\log z}$

34. $\log x^2 = (\log x)^2$

35. Suppose you overhear the following statement: "I must reject any negative answer when I solve an equation involving logarithms." Is this correct? Write an explanation of why it is or is not correct.

36. What values of x could not possibly be solutions of the following equation?

$$\log_a(4x - 7) + \log_a(x^2 + 4) = 0$$

Solve each equation for the indicated variable. Use logarithms to the appropriate bases. See Examples 7 and 8.

37. $I = \dfrac{E}{R}(1 - e^{-Rt/2})$ for t

38. $r = p - k \ln t$ for t

39. $p = a + \dfrac{k}{\ln x}$ for x

40. $T = T_0 + (T_1 - T_0)10^{-kt}$ for t

▼▼▼▼▼▼▼▼▼▼▼▼▼▼ **DISCOVERING CONNECTIONS** (Exercises 41–46) ▼▼▼▼▼▼▼▼▼▼▼▼▼▼

Earlier, we introduced methods of solving quadratic equations and showed how they can be applied to equations that are not actually quadratic, but are quadratic in form. Consider the equation

$$e^{2x} - 4e^x + 3 = 0,$$

and work Exercises 41–46 in order.

41. The expression e^{2x} is equivalent to $(e^x)^2$. Explain why this is so.

42. The given equation is equivalent to $(e^x)^2 - 4e^x + 3 = 0$. Factor the left side of this equation.

43. Solve the equation in Exercise 42 by using the zero-factor property. Give exact values.

44. Support your solution(s) in Exercise 43 using a calculator-generated graph of $y = e^{2x} - 4e^x + 3$.

45. Use the graph from Exercise 44 to identify the x-intervals where $y > 0$. These intervals give the solutions of $e^{2x} - 4e^x + 3 > 0$.

46. Use the graph from Exercise 44 and your answer to Exercise 45 to give the intervals where $e^{2x} - 4e^x + 3 < 0$.

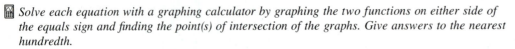 *Solve each equation with a graphing calculator by graphing the two functions on either side of the equals sign and finding the point(s) of intersection of the graphs. Give answers to the nearest hundredth.*

47. $e^x + \ln x = 5$ **48.** $e^x - \ln(x + 1) = 3$ **49.** $2e^x + 1 = 3e^{-x}$

50. $e^x + 6e^{-x} = 5$ **51.** $\log x = x^2 - 8x + 14$ **52.** $\ln x = -\sqrt[3]{x + 3}$

In Exercises 53 and 54 find the equation, domain, and range of $f^{-1}(x)$.

53. $f(x) = e^{3x+1}$ **54.** $f(x) = 3 \cdot 10^x$

Solve each inequality with a graphing calculator by rearranging terms so that one side is zero, then graphing the expression Y on the other side and observing from the graph where Y is positive or negative as applicable. These inequalities are studied in calculus to determine where certain functions are increasing.

55. $\log_3 x > 3$ **56.** $\log_x .2 < -1$

---

57. Recall from Section 5.4 that the formula for the decibel rating of the loudness of a sound is

$$d = 10 \log \frac{I}{I_0}.$$

A few years ago, there was a controversy about a proposed government limit on factory noise. One group wanted a maximum of 89 decibels, while another group wanted 86. This difference seemed very small to many people. Find the percent by which the 89-decibel intensity exceeds that for 86 decibels.

For Exercises 58–61, refer to the formula for compound interest,

$$A = P\left(1 + \frac{r}{m}\right)^{tm},$$

given in Section 5.2.

58. George Tom wants to buy a $30,000 car. He has saved $27,000. Find the number of years (to the nearest tenth) it will take for his $27,000 to grow to $30,000 at 6% interest compounded quarterly.

59. Find t to the nearest hundredth if $1786 becomes $2063.40 at 11.6%, with interest compounded monthly.

60. Find the interest rate that will produce $2500 if $2000 is left at interest compounded semiannually for 3.5 years.

61. At what interest rate will $16,000 grow to $20,000 if invested for 5.25 years, and interest is compounded quarterly?

62. The population of an animal species introduced into a certain area may grow rapidly at first but then increase more slowly as time goes on. A logarithmic function can provide an excellent description of such growth. Suppose the population of foxes in an area t months after the foxes were first introduced there is

$$F = 500 \log(2t + 3).$$

Solve the equation for t. Then find t to the nearest tenth for the following values of F.
(a) 600   **(b)** 1000

63. India has become an important exporter of software to the United States. The chart shows India's software exports (in millions of U.S. dollars) in selected years since 1985. (*Sources:* NIIT, NASSCOM. From *Scientific American*, September, 1994, page 95.)

Year (x)	1985	1987	1989	1991	1993	1995	1997
Million $ (y)	6	39	67	128	225	483	1000

The figure for 1997 is an estimate. Letting y represent software (in millions of dollars) and x represent the number of years since 1900, we find that the function with

$$f(x) = 6.2(10)^{-12}(1.4)^x$$

approximates the data reasonably well. According to this function, when will software exports double their 1997 value?

64. The graph shows that the percent y of U.S. children growing up without a father has increased rapidly since 1950. If x represents the number of years since 1900, the function defined by

$$f(x) = \frac{25}{1 + 1364.3e^{-x/9.316}}$$

models the data fairly well.
(a) From the graph, in what year were 20% of U.S. children living without a father?
(b) If the percent continues to increase in the same way, according to $f(x)$, in what year will 30% of U.S. children live in a home without a father?

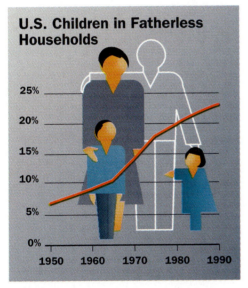

U.S. Children in Fatherless Households

Sources: National Longitudinal Survey of Youth; U.S. Department of Commerce; Bureau of the Census

65. One action that government could take to reduce carbon emissions into the atmosphere is to place a tax on fossil fuel. This tax would be based on the amount of carbon dioxide emitted into the air when the fuel is burned. The **cost-benefit** equation $\ln(1 - P) = -.0034 - .0053T$ describes the approximate relationship between a tax of T dollars per ton of carbon and the corresponding percent reduction P (in decimal form) of emissions of carbon dioxide. (*Source:* Nordhause, W., "To Slow or Not to Slow: The Economics of the Greenhouse Effect." Yale University, New Haven, Connecticut.)
(a) Write P as a function of T.
(b) Graph P for $0 \le T \le 1000$. Discuss the benefit of continuing to raise taxes on carbon.
(c) Determine P when $T = \$60$, and interpret this result.
(d) What value of T will give a 50% reduction in carbon emissions?

66. (Refer to Example 6 in Section 5.4 and Exercise 59 in Section 5.2.) Using computer models the International Panel on Climate Change (IPCC) in 1990 estimated k to be 6.3 in the radiative forcing equation $R = k \ln(C/C_0)$, where C_0 is the preindustrial amount of carbon dioxide and C is the current level. (*Source:* Clime, W., *The Economics of Global Warming,* Institute for International Economics, Washington, D.C., 1992.)
(a) What radiative forcing R (in w/m^2) is expected by the IPCC if the carbon dioxide level in the atmosphere doubles from its preindustrial level?
(b) Determine the global temperature increase predicted by the IPCC if the carbon dioxide levels were to double.

5.6 Exponential Growth or Decay ▼▼▼

In many situations that occur in biology, economics, and the social sciences, a quantity changes at a rate proportional to the amount present. In such cases the amount present at time t is a function of t called the **exponential growth or decay function.**

> **Exponential Growth or Decay Function**
>
> Let y_0 be the amount or number present at time $t = 0$. Then, under certain conditions, the amount present at any time t is given by
>
> $$y = y_0 e^{kt},$$
>
> where k is a constant.

In this section we see how to determine the constant k of the function from given data. When $k > 0$, the function describes growth; when $k < 0$, the function describes decay. Radioactive decay is an important application; it has been shown that radioactive substances decay exponentially; that is, in the function

$$y = y_0 e^{kt},$$

k is a negative number.

◖EXAMPLE 1
Determining an exponential decay function

If 600 grams of a radioactive substance are present initially and three years later only 300 grams remain, how much of the substance will be present after six years?

To express the situation as an exponential equation,

$$y = y_0 e^{kt},$$

we first find y_0 and then find k. From the statement of the problem, $y = 600$ when $t = 0$ (that is, initially), so

$$600 = y_0 e^{k(0)}$$
$$600 = y_0,$$

giving the exponential decay equation

$$y = 600 e^{kt}.$$

Since 300 grams of the substance remain after three years, use the fact that $y = 300$ when $t = 3$ to find k.

$$300 = 600 e^{3k}$$
$$\frac{1}{2} = e^{3k}$$

Take natural logarithms on both sides, then solve for k.

$$\ln \frac{1}{2} = \ln e^{3k}$$
$$\ln .5 = 3k \qquad\qquad \ln e^x = x$$
$$\frac{\ln .5}{3} = k$$
$$k \approx -.231,$$

giving

$$y = 600 e^{-.231t}$$

as the exponential decay equation. To find the amount present after six years, let $t = 6$.

$$y = 600 e^{-.231(6)} \approx 600 e^{-1.386} \approx 150$$

After six years, about 150 grams of the substance remain. ▶

EXAMPLE 2
Solving an exponential decay problem

Nuclear energy derived from radioactive isotopes can be used to supply power to space vehicles. The output of the radioactive power supply for a certain satellite is given by the function

$$y = 40e^{-.004t},$$

where y is in watts and t is the time in days.

(a) How much power will be available at the end of 180 days?
Let $t = 180$ in the formula.

$$y = 40e^{-.004(180)}$$
$$y \approx 19.5 \qquad \text{Use a calculator.}$$

About 19.5 watts will be left.

(b) How long will it take for the amount of power to be half of its original strength?
The original amount of power is 40 watts. (Why?) Since half of 40 is 20, replace y with 20 in the formula, and solve for t.

$$20 = 40e^{-.004t}$$
$$.5 = e^{-.004t} \qquad \text{Divide by 40.}$$
$$\ln .5 = \ln e^{-.004t}$$
$$\ln .5 = -.004t \qquad \ln e^x = x$$
$$t = \frac{\ln .5}{-.004}$$
$$t \approx 173 \qquad \text{Use a calculator.}$$

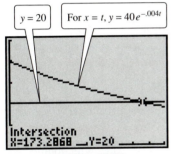

[0, 200] by [0, 50]
Xscl = 20 Yscl = 10

This screen provides graphical support for the result in Example 2.

After about 173 days, the amount of available power will be half of its original amount. ▶

The **half-life** of a quantity that decays exponentially is the time it takes for half of a given amount to decay. Thus, in Example 2(b), we found the half-life of the power supply. In the next example we find the **doubling time,** the time it takes for a given substance that grows exponentially to double.

EXAMPLE 3
Determining an exponential function to model the increase of carbon dioxide

In Example 8, Section 5.2, we discussed the growth of atmospheric carbon dioxide over time. The function was given in that example. Now we can see how to determine it from the data.*

(a) Find an exponential function that gives the amount of carbon dioxide y in year t.
Recall, the graph of the data points showed exponential growth, so the equation will take the form $y = y_0 e^{kt}$. We must find the values of y_0 and k. The data begin with the year 1990, so to simplify the work, we let 1990 correspond

*Source: International Panel on Climate Change (IPCC), 1990.

to $t = 0$, 1991 correspond to $t = 1$, and so on. Since y_0 is the initial amount, $y_0 = 353$ in 1990, that is, when $t = 0$. Thus the equation is

$$y = 353e^{kt}.$$

From the last pair of values in the table in Example 8, we know that in 2275 the carbon dioxide level is expected to be 2000 ppm. The year 2275 corresponds to $2275 - 1990 = 285$. Substituting 2000 for y and 285 for t gives

$$2000 = 353e^{k(285)}$$

$$\frac{2000}{353} = e^{285k} \qquad \text{Divide by 353.}$$

$$\ln\left(\frac{2000}{353}\right) = \ln(e^{285k}) \qquad \text{Take logarithms on both sides.}$$

$$\ln\left(\frac{2000}{353}\right) = 285k \qquad \ln e^x = x$$

$$k = \frac{1}{285} \cdot \ln\left(\frac{2000}{353}\right) \approx .00609.$$

The function that models the data is

$$y = 353e^{.00609t}.$$

(Note that this is a different function than the one given in Example 8, because we used 0 to represent 1990 here.)

(b) Estimate the year when future levels of carbon dioxide will double the preindustrial level of 280 ppm.

Let $y = 2(280) = 560$ and find t.

$$560 = 353e^{.00609t}$$

$$\frac{560}{353} = e^{.00609t}$$

$$\ln\left(\frac{560}{353}\right) = \ln e^{.00609t} = .00609t$$

$$t = \frac{1}{.00609} \cdot \ln\left(\frac{560}{353}\right) \approx 75.8$$

Since $t = 0$ corresponds to 1990, the carbon dioxide level will double in about 76 years after 1990 or in about 2066. ◗

◖EXAMPLE 4
Solving a carbon dating problem

Carbon 14, also known as radiocarbon, is a radioactive form of carbon that is found in all living plants and animals. After a plant or animal dies, the radiocarbon disintegrates. Scientists can determine the age of the remains by comparing the amount of radiocarbon with the amounts present in living plants and animals. This technique is called *carbon dating*. The amount of radiocarbon present after t years is given by

$$y = y_0 e^{-(\ln 2)(1/5700)t},$$

where y_0 is the amount present in living plants and animals.

(a) Find the half-life.
Let $y = (1/2)y_0$.

$$\frac{1}{2}y_0 = y_0 e^{-(\ln 2)(1/5700)t}$$

$$\frac{1}{2} = e^{-(\ln 2)(1/5700)t} \qquad \text{Divide by } y_0.$$

$$\ln\frac{1}{2} = \ln e^{-(\ln 2)(1/5700)t} \qquad \text{Take logarithms on both sides.}$$

$$\ln\frac{1}{2} = -\frac{\ln 2}{5700}t \qquad \ln e^x = x$$

$$-\frac{5700}{\ln 2}\ln\frac{1}{2} = t \qquad \text{Multiply by } -\frac{5700}{\ln 2}.$$

$$-\frac{5700}{\ln 2}(\ln 1 - \ln 2) = t \qquad \text{Property (b)}$$

$$-\frac{5700}{\ln 2}(-\ln 2) = t \qquad \ln 1 = 0$$

$$5700 = t$$

The half-life is 5700 years.

(b) Charcoal from an ancient fire pit on Java contained $1/4$ the carbon 14 of a living sample of the same size. Estimate the age of the charcoal.
Let $y = \frac{1}{4}y_0$.

$$\frac{1}{4}y_0 = y_0 e^{-(\ln 2)(1/5700)t}$$

$$\frac{1}{4} = e^{-(\ln 2)(1/5700)t}$$

$$\ln\frac{1}{4} = \ln e^{-(\ln 2)(1/5700)t}$$

$$\ln\frac{1}{4} = -\frac{\ln 2}{5700}t$$

$$-\frac{5700}{\ln 2}\ln\frac{1}{4} = t$$

$$t = 11{,}400$$

The charcoal is about 11,400 years old. ▶

The compound interest formula

$$A = P\left(1 + \frac{r}{m}\right)^{mt}$$

was discussed in Section 2 of this chapter. The table presented there shows that increasing the frequency of compounding makes smaller and smaller differences in the amount of interest earned. In fact, it can be shown that even if interest is compounded at intervals of time as small as one chooses (such as each hour, each minute, or each second), the total amount of interest earned will be only slightly more than the interest earned with daily compounding. This is true even for a process called **continuous compounding**, which can be described loosely as compounding every instant. As suggested in Section 2, the value of the expression $(1 + 1/m)^m$ approaches e as m gets larger. Because of this, the formula for continuous compounding involves the number e.

Continuous Compounding

If P dollars is deposited at a rate of interest r compounded continuously for t years, the final amount on deposit is

$$A = Pe^{rt}$$

dollars.

EXAMPLE 5
Solving a continuous compounding problem

Suppose $5000 is deposited in an account paying 3% interest compounded continuously for five years. Find the total amount on deposit at the end of five years.

Let $P = 5000$, $t = 5$, and $r = .03$. Then

$$A = 5000e^{.03(5)} = 5000e^{.15}$$
$$\approx 5000(1.161834)$$
$$= 5809.17$$

or $5809.17. Check that daily compounding would have produced a compound amount about 3¢ less. ▶

EXAMPLE 6
Finding doubling time for money

How long will it take for the money in an account that is compounded continuously at 3% interest to double?

Use the formula for continuous compounding, $A = Pe^{rt}$, to find the time t that makes $A = 2P$. Substitute $2P$ for A and .03 for r; then solve for t.

$$A = Pe^{rt}$$
$$2P = Pe^{.03t}$$
$$2 = e^{.03t}$$

Taking natural logarithms on both sides gives

$$\ln 2 = \ln e^{.03t}$$

$$\ln 2 = .03t \qquad \ln e^x = x$$

$$\frac{\ln 2}{.03} = t$$

$$23.10 \approx t.$$

It will take about 23 years for the amount to double. ◗

5.6 Exercises ▼▼▼▼▼▼▼▼▼▼▼▼▼▼▼▼▼▼▼▼▼▼▼▼▼▼▼▼▼▼▼▼▼▼

Solve the following applied problems from the physical sciences. See Examples 1–3.

1. A sample of 500 grams of radioactive lead 210 decays to polonium 210 according to the function $A(t) = 500e^{-.032t}$, where t is time in years. Find the amount of the sample remaining after **(a)** 4 years **(b)** 8 years **(c)** 20 years. Then **(d)** find the half-life.

2. Repeat Exercise 1 for 500 grams of plutonium 241, which decays according to the function

$$A(t) = A_0 e^{-.053t},$$

where t is time in years.

3. Find the half-life of radium 226, which decays according to the function $A(t) = A_0 e^{-.00043t}$, where t is in years.

4. How long will it take any quantity of iodine 131 to decay to 25% of its initial amount, knowing that it decays according to the function $A(t) = A_0 e^{-.087t}$, where t is in days?

5. Explain why the half-life of a radioactive material does not depend on the amount of material present initially.

6. Suppose an 80 g sample of radioactive material has decay constant $-.02$. On the same axes, graph the three functions $y = 80e^{-.02t}$, $y = 40$, and $y = 20$. Determine the t-coordinates of the intersection points of the decay curve and the horizontal lines and observe that the time required for the material to decay from 80 g to 40 g is the same as the time required to decay from 40 g to 20 g. Explain why this is true.

7. In the central Sierra Nevada mountains of California, the percent of moisture that falls as snow rather than rain is approximated reasonably well by the function $p(h) = 86.3 \ln h - 680$, where h is the altitude in feet, and $p(h)$ is the percent of snow. (This model is valid for $h \geq 3000$.) Find the percent of snow that falls at the following altitudes.
   (a) 3000 feet **(b)** 4000 feet **(c)** 7000 feet

8. The magnitude M of a star is defined by the equation $M = 6 - \frac{5}{2} \log \frac{I}{I_0}$, where I_0 is the measure of a just-visible star and I is the actual intensity of the star being measured. The dimmest stars are of magnitude 6, and the brightest are of magnitude 1. Determine the ratio of light intensities between a star of magnitude 1 and a star of magnitude 3.

For Exercises 9–11, refer to Example 4.

9. Suppose an Egyptian mummy is discovered in which the amount of carbon 14 present is only about one-third the amount found in living human beings. About how long ago did the Egyptian die?

10. A sample from a refuse deposit near the Strait of Magellan had 60% of the carbon 14 of a contemporary living sample. How old was the sample?

11. Paint from the Lascaux caves of France contains 15% of the normal amount of carbon 14. Estimate the age of the paintings.

12. The amount of a chemical that will dissolve in a solution increases exponentially as the (Celsius) temperature t is increased according to the equation $A(t) = 10e^{.0095t}$. At what temperature will 15 g dissolve?

13. By Newton's law of cooling, the temperature of a body at time t after being introduced into an environment having constant temperature T_0 is

$$A(t) = T_0 + Ce^{-kt},$$

where C and k are constants. If $C = 100$, $k = .1$, and t is time measured in minutes, how long will it take a hot cup of coffee to cool to a temperature of 25°C in a room at 20°C?

14. Use the function defined by

$$t = T\frac{\ln[1 + 8.33(A/K)]}{\ln 2}$$

to estimate the age of a rock sample, if tests show that A/K is .103 for the sample. Let $T = 1.26 \times 10^9$.

Solve the following problems from finance. Use the formula for compound interest given in Section 5.2, and referred to in this section. See Examples 5 and 6.

15. How long will it take for $1000 to grow to $5000 at an interest rate of 3.5% if interest is compounded **(a)** quarterly **(b)** continuously?

16. How long will it take for $5000 to grow to $8400 at an interest rate of 6% if interest is compounded **(a)** semiannually **(b)** continuously?

17. Find the doubling time of an investment earning 2.5% interest if interest is compounded **(a)** quarterly **(b)** continuously.

18. Linda Youngman, who is self-employed, wants to invest $60,000 in a pension plan. One investment offers 7% compounded quarterly. Another offers 6.75% compounded continuously. Which investment will earn more interest in 5 years? How much more will the better plan earn?

19. If Ms. Youngman (see Exercise 18) chooses the plan with continuous compounding, how long will it take for her $60,000 to grow to $80,000?

20. If interest is compounded continuously and the interest rate is tripled, what effect will this have on the time required for an investment to double?

21. On the same axes, graph $Y_1 = 5\left(1 + \dfrac{.08}{2}\right)^{2x}$ and $Y_2 = 5e^{.08x}$, for $0 \le x \le 25, 0 \le Y \le 40$. Compare the graphs and determine the two Y-values when $x = 25$. Repeat with the 2s in the first function replaced by 12s.

22. Use the TABLE feature of your graphing calculator to find how long it will take $1500 invested at 5.75% compounded daily to triple in value. Zoom in on the solution by systematically decreasing ΔTbl. Find the answer to the nearest day. (Find your answer to the nearest day by eventually letting ΔTbl = 1/365. The decimal part of the solution can be multiplied by 365 to determine the number of days greater than the nearest year. For example, if the solution is determined to be 16.2027 years, then multiply .2027 by 365 to get 73.9855. The solution is then, to the nearest day, 16 years and 74 days.) Confirm your answer analytically.

Solve the following applied problems from the biological sciences and medicine.

23. The number of cesarean section deliveries in the United States has increased over the years. Between the years 1980 and 1989, the number of such births, in thousands, can be approximated by the function with $f(t) = 625e^{.0516t}$, where $t = 1$ corresponds to the year 1980. Based on this function, what would be the approximate number of cesarean section deliveries in 1996? (*Source:* U.S. National Center for Health Statistics.)

24. When physicians prescribe medication they must consider how the drug's effectiveness decreases over time. If, each hour, a drug is only 90% as effective as the previous hour, at some point the patient will not be receiving enough medication and must receive another dose. This situation can be modeled with a geometric sequence. (See the section on Geometric Sequences and Series.) If the initial dose was 200 mg and the drug was administered 3 hours ago, the expression $200(.90)^2$ represents the amount of effective medication still available. Thus, $200(.90)^2 \approx 162$ mg are still in the system. (The exponent is equal to the number of hours since the drug was administered, less one.) How long will it take for this initial dose to reach the dangerously low level of 50 mg?

Many environmental situations place effective limits on the growth of the number of an organism in an area. Many such limited growth situations are described by the logistic function, *defined by*

$$G(t) = \frac{MG_0}{G_0 + (M - G_0)e^{-kMt}},$$

where G_0 is the initial number present, M is the maximum possible size of the population, and k is a positive constant. Assume $G_0 = 100$, $M = 2500$, $k = .0004$, and t is time in decades (10-yr periods). Use this information in Exercises 25–27.

25. Use a calculator to graph the S-shaped limited growth function using $0 \le t \le 8$, $0 \le y \le 2500$.

26. Estimate the value of $G(2)$ from the graph. Then evaluate $G(2)$ analytically to find the population after 20 yr.

27. Find the t-coordinate of the intersection of the curve with the horizontal line $Y = 1000$ to estimate the number of decades required for the population to reach 1000. Then solve $G(t) = 1000$ analytically to obtain the exact value of t.

28. Chlorofluorocarbons (CFCs) are greenhouse gases that were produced by humans after 1930. CFCs are found in refrigeration units, foaming agents, and aerosols. They have great potential for destroying the ozone layer. As a result, governments have agreed to phase out their production by the year 2000. CFC-11 is an example of a CFC that has increased faster than any other greenhouse gas. The following graph displays approximate concentrations of atmospheric CFC-11 in parts per billion (ppb) since 1950.

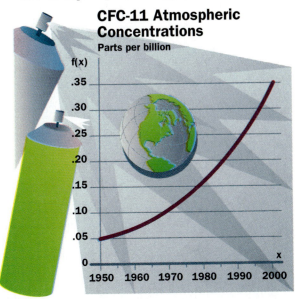

CFC-11 Atmospheric Concentrations
Parts per billion

Source: Nilsson, A., *Greenhouse Earth*, John Wiley & Sons, New York, 1992

(a) Use the graph to find an exponential function with $f(x) = A_0 a^{x-1950}$ that models the concentration of CFC-11 in the atmosphere from 1950 to 2000 where x is the year.

(b) Approximate the average annual percent increase of CFC-11 during this time period.

Solve the following applied problems from economics.

29. The number of books, in millions, sold per year in the United States between 1985 and 1990, can be approximated by the function $A(t) = 1757e^{.0264t}$, where $t = 0$ corresponds to the year 1985. Based on this model, how many books would be sold in 1996? (*Source:* Book Industry Study Group.)

30. Personal consumption expenditures for recreation in billions of dollars in the United States during the years 1984 through 1990 can be approximated by the function $A(t) = 185.4e^{.0587t}$, where $t = 0$ corresponds to the year 1984. Based on this model, how much would personal consumption expenditures be in 1996? (*Source:* U.S. Bureau of Economic Analysis.)

31. The U.S. Consumer Price Index is approximated by $A(t) = 34e^{.04t}$, where t represents the number of years after 1960. Assuming the same equation continues to apply, find the year in which costs will be 50% higher than in 1987, when the CPI was set at 100.

32. Experiments have shown that the sales of a product, under relatively stable market conditions, but in the absence of promotional activities such as advertising, tend to decline at a constant yearly rate. This rate of sales decline varies considerably from product to product, but seems to remain the same for any particular product. The sales decline can be expressed by a function of the form

$$S(t) = S_0 e^{-at},$$

where $S(t)$ is the rate of sales at time t measured in years, S_0 is the rate of sales at time $t = 0$, and a is the sales decay constant.

(a) Suppose the sales decay constant for a particular product is $a = .10$. Let $S_0 = 50,000$ and find $S(1)$ and $S(3)$.

(b) Find $S(2)$ and $S(10)$ if $S_0 = 80,000$ and $a = .05$.

33. Use the sales decline function given in Exercise 32. If $a = .1$, $S_0 = 50,000$, and t is time measured in years, find the number of years it will take for sales to fall to half the initial sales.

34. Assume the cost of a loaf of bread is $1. With continuous compounding, find the time it would take to triple at an annual inflation rate of 6%.

35. Historically, the consumption of electricity has increased at a continuous rate of 6% per year. If it continued to increase at this rate, find the number of years before twice as much electricity would be needed.

36. Suppose a conservation campaign together with higher rates caused demand for electricity to increase at only 2% per year. (See Exercise 35.) Find the number of years before twice as much electricity would be needed.

Solve the following problems from the social sciences.

37. One measure of living standards in the United States is given by $L = 9 + 2e^{.15t}$, where t is the number of years since 1982. Find L for the following years.
   (a) 1982   **(b)** 1986   **(c)** 1992
   ▦ **(d)** Graph L in the window $[0, 10]$ by $[0, 30]$.
   (e) What can be said about the growth of living standards in the U.S. according to this equation?

38. A midwestern city finds its residents moving to the suburbs. Its population is declining according to the relationship $P = P_0 e^{-.04t}$, where t is time measured in years and P_0 is the population at time $t = 0$. Assume that $P_0 = 1,000,000$.
   (a) Find the population at time $t = 1$.
   (b) Estimate the time it will take for the population to be reduced to 750,000.
   (c) How long will it take for the population to decline to half the initial number?

39. Recall from Section 5.4 Exercises that the number of years, n, since two independently evolving languages split off from a common language is approximated by $n = -7600 \log r$, where r is the proportion of words from the ancestral language still common to the two languages. Solve the formula for r. Then find r if the languages split
   (a) 1000 years ago,   **(b)** 2500 years ago.

40. The number of bacteria in a jar doubles every minute. If the jar is full at 2:00 P.M., when was it half full?

41. The maximum Social Security tax in 1985 was $2,791.80 and in 1995 it was $4,681.80. Use an exponential growth function to approximate the annual percentage increase. Then compare it to consumer prices which rose at an annual rate of 3.56%. (*Source:* Social Security Administration.)

42. In 1995, the U.S. racial mix was 75.3% white, 9.0% Hispanic, 12.0% African American, 2.9% Asian/Pacific Islanders, and .8% Native American. Their respective percentages of executives, managers, and administrators in private-industry communication firms were 84.0%, 9.5%, 3.4%, 2.0%, and 0%, respectively. Assume these percents remain constant. If the number of African American executives, managers, and administrators increases at a rate of 5% per year, determine the year when their representation will reach 12.0% of the total number of executives, managers, and administrators in communications. (*Source:* U.S. Labor Department's Glass Ceiling Commission.)

43. (Refer to Exercise 42.) In private-industry retail trade the percentages in 1995 were 80.8% white, 4.8% Hispanic, 4.9% African American, 5.2% Asian/Pacific Islanders, and 0% Native American. Assume these percents remain constant. If the number of Hispanic executives, managers, and administrators increases at a rate of 3% per year, determine the year when their representation will reach 9.0% of the total number in retail trade. (*Source:* U.S. Labor Department's Glass Ceiling Commission.)

44. The number of Internet users was estimated to be 1.6 million in October of 1989 and 39 million in October of 1994. This number has grown exponentially. The function $f(x) = a(b^x)$ can be used to model the number of users, where a and b are constants and $x = 0$ (in months) corresponds to October of 1989. (*Source:* Genesis Corp.)
   (a) Write the information in the first sentence as two ordered pairs of the form $(x, f(x))$, and substitute these values into the function to get a system of two equations with two variables. (See Section 6.1.)
   (b) Solve the system from part (a) to approximate the values of a and b to four decimal places, and determine $f(x)$.
   (c) Use $f(x)$ to estimate the number of users in October of 1992. Compare your estimate with the actual value of 10.3 million.
   ▦ **(d)** Graphically determine the month and year when there were 30 million users.

Chapter 5   Summary ▼▼▼▼▼▼▼▼▼▼▼▼▼▼▼▼▼▼▼▼▼▼▼▼▼▼▼▼▼▼▼

KEY TERMS	KEY IDEAS
5.1 Inverse Functions	
	One-to-One Function
	A function f is a one-to-one function if, for any elements a and b from the domain of f,
	$$a \neq b \quad \text{implies} \quad f(a) \neq f(b).$$
	Horizontal Line Test
	If each horizontal line intersects the graph of a function in no more than one point, then the function is one-to-one.
	Inverse Functions
	Let f be a one-to-one function. Then g is the inverse function of f if
	$$(f \circ g)(x) = x \text{ for every } x \text{ in the domain of } g,$$
	and $\quad (g \circ f)(x) = x$ for every x in the domain of f.
5.2 Exponential Functions	
exponential function	**Additional Properties of Exponents**
compound interest	**(a)** If $a > 0$ and $a \neq 1$, then a^x is a unique real number for all real
future value	numbers x.
present value	**(b)** If $a > 0$ and $a \neq 1$, then $a^b = a^c$ if and only if $b = c$.
exponential growth or decay	**(c)** If $a > 1$ and $m < n$, then $a^m < a^n$.
	(d) If $0 < a < 1$ and $m < n$, then $a^m > a^n$.
5.3 Logarithmic Functions	
logarithm	**Properties of Logarithms**
logarithmic function	For any positive real numbers x and y, real number r, and positive real number a, $a \neq 1$:
	(a) $\log_a xy = \log_a x + \log_a y$
	(b) $\log_a \dfrac{x}{y} = \log_a x - \log_a y$
	(c) $\log_a x^r = r \log_a x$
	(d) $\log_a a = 1$
	(e) $\log_a 1 = 0$.
5.4 Evaluating Logarithms; Change of Base	
common logarithm	**Change-of-Base Theorem**
natural logarithm	For any positive real numbers x, a, and b, where $a \neq 1$ and $b \neq 1$:
	$$\log_a x = \frac{\log_b x}{\log_b a}.$$

KEY TERMS	KEY IDEAS
5.5 Exponential and Logarithmic Equations	
	Property of Logarithms **(f)** If $x > 0$, $y > 0$, $b > 0$ and $b \neq 1$, then $\qquad \log_b x = \log_b y$ if and only if $x = y$.
5.6 Exponential Growth or Decay	
half-life doubling time continuous compounding	**Growth or Decay Function** $y = y_0 e^{kt}$, for constants y_0 and k.

Chapter 5   Review Exercises ▼▼▼▼▼▼▼▼▼▼▼▼▼▼▼▼▼▼▼▼▼▼▼▼▼▼▼▼▼

Write an equation for the inverse function in the form $y = f^{-1}(x)$.

1. $f(x) = x^3 - 3$ $\qquad\qquad\qquad\qquad$ **2.** $f(x) = \sqrt{25 - x^2}$

3. Suppose $f(t)$ is the amount an investment will grow to t years after 1992. What does $f^{-1}(\$50,000)$ represent?

4. Alice, on vacation in Canada, found that her U.S. dollars were increased by 16%. On her return, she expected a 16% decrease when converting her Canadian money back into U.S. dollars. Write an equation for each of these conversion functions. Show that one is not the inverse of the other. What should the conversion factor be for Canadian to U.S. dollars?

5. The graphs of two functions are shown below. Based on their graphs, are these functions inverses?

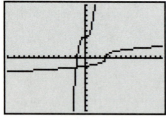

[−15, 15] by [−10, 10]

6. $f(x) = \left(\dfrac{5}{4}\right)^x$ defines a(n) _____ function.
$\qquad\qquad\qquad\qquad\qquad$ increasing/decreasing

7. $f(x) = \log_{2/3} x$ defines a(n) _____ function.
$\qquad\qquad\qquad\qquad$ increasing/decreasing

Match each equation with one of the graphs below.

8. $y = \log_{.3} x$ **9.** $y = e^x$ **10.** $y = \ln x$ **11.** $y = (.3)^x$

A. 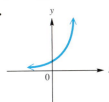   **B.** 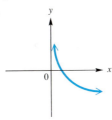   **C.** 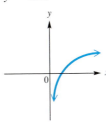   **D.**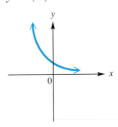

Write each equation in logarithmic form.

12. $2^5 = 32$ **13.** $100^{1/2} = 10$ **14.** $(1/16)^{1/4} = 1/2$ **15.** $(3/4)^{-1} = 4/3$ **16.** $10^{.4771} = 3$ **17.** $e^{2.4849} = 12$

18. Explain how the graph of $f(x) = 8 - 2^{x-1}$ can be obtained from the graph of $y = 2^x$ using translations and reflection.

Consider the exponential function $y = f(x) = a^x$ graphed here.

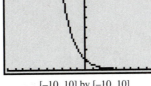
[–10, 10] by [–10, 10]

19. What is true about the value of a in comparison to 1?

20. What is the domain of f?

21. What is the range of f?

22. What is the value of $f(0)$?

23. Sketch the graph of $y = f^{-1}(x)$ by hand.

24. What is the expression that defines $f^{-1}(x)$?

Write each equation in exponential form.

25. $\log_{10} .001 = -3$ **26.** $\log_2 \sqrt{32} = \dfrac{5}{2}$ **27.** $\log 3.45 = .537819$

28. $\ln 45 = 3.806662$ **29.** $\log_9 27 = \dfrac{3}{2}$

30. One of your friends is taking a mathematics course and tells you, "I have no idea what an expression like $\log_5 27$ really means." Write a clear explanation of what it means and how you can find an approximation for it using a calculator.

31. What is the base of the logarithmic function whose graph contains the point $(81, 4)$?

32. What is the base of the exponential function whose graph contains the point $(-4, 1/16)$?

Use properties of logarithms to write each of the following logarithms as a sum, difference, or product of logarithms.

33. $\log_3\left(\dfrac{mn}{5r}\right)$ **34.** $\log_2\left(\dfrac{\sqrt{7}}{15}\right)$ **35.** $\log_5(x^2y^4 \sqrt[5]{m^3p})$ **36.** $\log_7(7k + 5r^2)$

37. Correct the mistakes in the following equation.

$$\log_5 125 - \log_5 25 = \frac{\log_5 125}{\log_5 25} = \log_5\left(\frac{125}{25}\right) = \log_5 5 = 1$$

Find each logarithm. Round to the nearest thousandth.

38. log 45.6

39. log .0411

40. ln 470

41. ln 144,000

42. $\log_3 769$

43. $\log_{2/3}(5/8)$

44. A population is increasing according to the growth law $y = 2e^{.02t}$, where y is in millions and t is in years. Match each question with one of the solutions (A), (B), (C), or (D).
  (a) How long will it take for the population to triple?　　(A) Evaluate $2e^{.02(1/3)}$.
  (b) When will the population reach 3 million?　　(B) Solve $2e^{.02t} = 3 \cdot 2$ for t.
  (c) How large will the population be in 3 years?　　(C) Evaluate $2e^{.02(3)}$.
  (d) How large will the population be in 4 months?　　(D) Solve $2e^{.02t} = 3$ for t.

45. The population of the world is expected to double in the next 44 years. Without solving for the growth constant k, determine how much the population will increase in 22 years.

46. If the world population continues to grow at the current rate, by what factor will the population grow in the next 220 years? See Exercise 45. (*Hint:* $220 = 5 \cdot 44$.)

Solve each equation. Round to the nearest thousandth if necessary.

47. $8^k = 32$

48. $\dfrac{8}{27} = b^{-3}$

49. $10^{2r-3} = 17$

50. $e^{p+1} = 10$

51. $\log_{64} y = \dfrac{1}{3}$

52. $\ln(6x) - \ln(x + 1) = \ln 4$

53. $\log_{16}\sqrt{x + 1} = \dfrac{1}{4}$

54. $\ln x + 3 \ln 2 = \ln \dfrac{2}{x}$

55. $\ln[\ln(e^{-x})] = \ln 3$

---

56. What annual interest rate, to the nearest tenth, will produce $8780 if $3500 is left at interest for 10 years?

57. Find the number of years (to the nearest tenth) needed for $48,000 to become $58,344 at 5% interest compounded semiannually.

58. Manuel deposits $10,000 for 12 years in an account paying 12% compounded annually. He then puts this total amount on deposit in another account paying 10% compounded semiannually for another 9 years. Find the total amount on deposit after the entire 21-year period.

59. Anne Kelly deposits $12,000 for 8 years in an account paying 5% compounded annually. She then leaves the money alone with no further deposits at 6% compounded annually for an additional 6 years. Find the total amount on deposit after the entire 14-year period.

60. If the inflation rate were 10%, use the formula for continuous compounding to find the number of years for a $1 item to cost $2.

61. The function defined by

$$A(t) = (5 \times 10^{12})e^{-.04t}$$

gives the known coal reserves in the world in year t (in tons), where $t = 0$ corresponds to 1970, and $-.04$ indicates the rate of consumption.
  (a) Find the amount of coal available in 1990.
  (b) When were the coal reserves half of what they were in 1970?

62. Give the property that justifies each step of the following derivation. Let a be any number.
  (a) (a, e^a) is on the graph of $f(x) = e^x$.
  (b) (e^a, a) is on the graph of $g(x) = \ln x$.
  (c) $\ln e^a = a$.

63. The graphs of $y = x^2$ and $y = 2^x$ have the points $(2, 4)$ and $(4, 16)$ in common. There is a third point in common to the graphs whose coordinates can be approximated by using a graphing calculator. Find the coordinates, giving as many decimal places as your calculator displays.

64. Let $Y_1 = 3^{x+4}$ and $Y_2 = 27^{x+1}$. Graph Y_1 and Y_2 in the same window with a graphing calculator and find the coordinates of the point of intersection. (Use the window [0, 3] by [0, 300].)

65. Use the same functions for Y_1 and Y_2 as in Exercise 64. Graph $Y_3 = Y_1 - Y_2$, and explain how the graph of Y_3 supports your answer in Exercise 64.

66. Solve $\log_2 x + \log_2(x + 2) = 3$.

67. To support the solution in Exercise 66, we could graph $Y_1 = \log_2 x + \log_2(x + 2) - 3$ and find the x-intercept. Write an expression for Y_1 using the change-of-base theorem, with base 10.

68. Consider $f(x) = \log_4(2x^2 - x)$.
  (a) Use the change-of-base theorem with base e to write $\log_4(2x^2 - x)$ in a suitable form to graph with a calculator.
  (b) Graph the function using a graphing calculator. Use the window $[-2.5, 2.5]$ by $[-5, 2.5]$.
  (c) What are the x-intercepts?
  (d) Give the equations of the vertical asymptotes.
  (e) Explain why there is no y-intercept.

69. (Refer to Exercise 60 in Section 5.2.) The atmospheric pressure (in millibars) at a given altitude (in meters) is listed in the table.

Altitude (x)	Pressure (P)
0	1013
1000	899
2000	795
3000	701
4000	617
5000	541
6000	472
7000	411
8000	357
9000	308
10,000	265

  (a) Plot the points $(x, \ln P)$ on the coordinate axes. Is there a linear relationship between x and $\ln P$?
  (b) If $P = Ce^{kx}$ with constants C and k, explain why there is a linear relationship between x and $\ln P$.

70. After a medical drug is injected directly into the bloodstream it is gradually eliminated from the body. Graph each of the following functions on the interval [0, 10]. Use [0, 500] for the range of $A(t)$. Determine the function that best describes the amount $A(t)$ (in milligrams) of a drug remaining in the body after t hours if 350 milligrams were initially injected.
  (a) $A(t) = t^2 - t + 350$
  (b) $A(t) = 350 \log(t + 1)$
  (c) $A(t) = 350(.75)^t$
  (d) $A(t) = 100(.95)^t$

71. Computing power of personal computers has increased dramatically as a result of the ability to place an increasing number of transistors on a single processor chip. The table lists the number of transistors on some popular computer chips made by Intel. (*Source:* Intel.)

Year	Chip	Transistors
1971	4004	2300
1986	386DX	275,000
1989	486DX	1,200,000
1993	Pentium	3,300,000
1995	P6	5,500,000

  (a) Plot the data. Let the x-axis represent the year, where $x = 0$ corresponds to 1971, and let the y-axis represent the number of transistors.
  (b) Discuss which type of function $y = f(x)$ describes the data best, where a and b are constants.
  (i) $f(x) = ax + b$ (linear)
  (ii) $f(x) = a \ln b(x + 1)$ (logarithmic)
  (iii) $f(x) = ae^{bx}$ (exponential)
  (c) Determine a function f that approximates these data. Plot f and the data on the same coordinate system.
  (d) Assuming that the present trend continues, use f to predict the number of transistors on a chip in the year 2000.

Chapter 5   Test ▼▼▼▼▼▼▼▼▼▼▼▼▼▼▼▼▼▼▼▼▼▼▼▼▼▼▼▼▼▼▼▼▼▼▼▼

1. Solve $25^{2x-1} = 125^{x+1}$.

Write in logarithmic form.

2. $a^2 = b$

3. $e^c = 4.82$

Write in exponential form.

4. $\log_3 \sqrt{27} = \dfrac{3}{2}$

5. $\ln 5 = a$

6. What two points on the graph of $y = \log_a x$ can be found without computation?

Graph each function.

7. $y = (1.5)^{x+2}$

8. $y = \log_{1/2} x$

9. Use properties of logarithms to write the following as a sum, difference, or product of logarithms.

$$\log_7\left(\frac{x^2 \sqrt[4]{y}}{z^3}\right)$$

Find each logarithm. Round to the nearest thousandth.

10. $\ln 2300$

11. $\log_{2.7} 94.6$

12. What values of x cannot possibly be solutions of the equation $\log_a(2x - 3) = -1$?

13. The equation $3^x = 243$ can be solved analytically in two ways, but the equation $3^x = 17$ can be solved in only one way. Explain the difference between these two types of exponential equations.

Solve the equation. Round answers to the nearest thousandth.

14. $8^{2w-4} = 100$

15. $\log_3(m + 2) = 2$

16. $\ln x - 4 \ln 3 = \ln\left(\dfrac{5}{x}\right)$

---

17. The amount of radioactive material, in grams, present after t days is given by $A(t) = 600e^{-.05t}$.
(a) Find the amount present after 12 days.
(b) Find the half-life of the material.

18. A skydiver in free fall travels at the speed of $v(t) = 176(1 - e^{-.18t})$ ft per sec after t sec. How long will it take for the skydiver to attain the speed of 147 ft per sec (100 mph)?

19. How many years, to the nearest tenth, will be needed for $5000 to increase to $18,000 at 6.8% compounded monthly?

20. In July of 1994 the population of New York state was 18.2 million and increasing at a rate of .1% per year. The population of Florida was 14.0 million and increasing at a rate of 1.7% per year. (*Source:* U.S. Census Bureau.)
(a) Write an exponential equation expressing the growth of the population for each state. (Let $x = 0$ correspond to July 1994.)
(b) If this trend continues, estimate graphically when the population of Florida will exceed the population of New York.

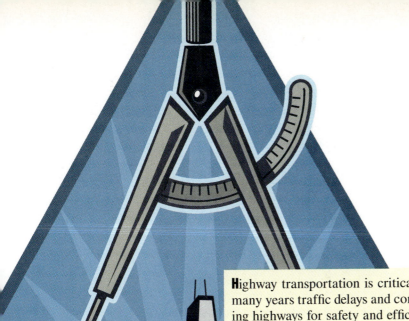

6

Trigonometric Functions

Highway transportation is critical to the economy of the United States. For many years traffic delays and congestion have continued to increase. Designing highways for safety and efficiency has become a critical issue that saves both lives and time. Many design problems can be solved with trigonometry. One such problem involves travel around curves. While it is essential to maintain visibility on curves for safe stopping distances, it is expensive to clear land and move buildings that obstruct visibility. If the highway curve shown in the figure has a radius of 600 feet with a speed limit of 55 miles per hour, how far to the inside of the curve should the highway department clear the land? The solution to this problem is found in Section 6.4. We also see in this chapter how trigonometry is used to compute grade resistance and calculate safe stopping distances.

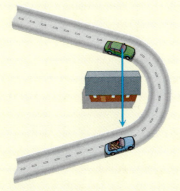

Finding energy sources has been important since the beginning of civilization. During the last hundred years fossil fuels have been the main source of energy, but they are finite and limited. Heavy use of fossil fuels has caused increasing environmental damage. Nuclear energy can provide almost unlimited amounts of energy. Unfortunately it creates health risks and dangerous nuclear waste. Solar energy has recently evolved from a mere kilowatt of electricity to hundreds of megawatts. It has many advantages: it does not create pollution; it is unlimited and potentially cheap; and it is readily available worldwide. Understanding the movement and knowing the position of the sun at any time on any date are fundamental concepts in solar energy collection. How high in the sky will the sun be in Sacramento or New Orleans on February 29, 2000, at 3 P.M.? Answers to such questions are necessary to generate electricity from sunlight. The circular functions are essential for answering this kind of question, as we see later in this chapter.

Cities throughout the world experience unique seasons. Temperature changes are a primary cause of seasons. The table lists the average monthly temperatures in Vancouver, Canada.

Month	Jan.	Feb.	Mar.	Apr.	May	June	July	Aug.	Sept.	Oct.	Nov.	Dec.
Temp	36	39	43	48	55	59	64	63	57	50	43	39

The average temperatures in Vancouver are coldest in January and warmest in July. These temperatures cycle yearly and may change only slightly over many years. Seasonal temperature changes occur periodically because Earth's axis is tilted and its orbit around the sun is nearly circular. Thus, the circular functions may be used to mathematically model the data and allow us to answer questions about temperatures. We will look at some examples in Sections 7 and 8.*

6.1 Angles ▼▼▼

A line may be drawn through the two distinct points A and B. This line is called **line AB**. The portion of the line between A and B, including points A and B themselves, is **segment AB**. The portion of line AB that starts at A and continues through B, and on past B, is called **ray AB**. Point A is the endpoint of the ray. (See Figure 1.)

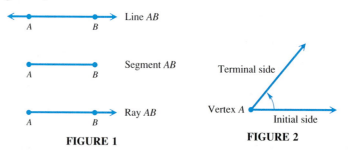

FIGURE 1

FIGURE 2

An **angle** is formed by rotating a ray around its endpoint. The ray in its initial position is called the **initial side** of the angle, while the ray in its location after the rotation is the **terminal side** of the angle. The endpoint of the ray is the **vertex** of the angle. Figure 2 shows the initial and terminal sides of an angle with vertex A.

*Sources: Mannering, F., and W. Kilareski, *Principles of Highway Engineering and Traffic Control,* John Wiley & Sons, Inc., 1990.

Miller, A., and J. Thompson, *Elements of Meteorology,* Charles E. Merrill Publishing Company, Columbus, Ohio, 1975.

Winter, C., R. Sizmann, and Vant-Hunt (Editors), *Solar Power Plants,* Springer-Verlag, 1991.

If the rotation of the terminal side is counterclockwise, the angle is **positive.** If the rotation is clockwise, the angle is **negative.** Figure 3 shows two angles, one positive and one negative.

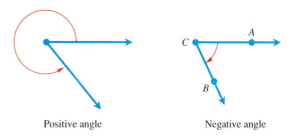

Positive angle   Negative angle

FIGURE 3

An angle can be named by using the name of its vertex. For example, the angle on the right in Figure 3 can be called angle *C*. Alternatively, an angle can be named using three letters, with the vertex letter in the middle. For example, the angle on the right also could be named angle *ACB* or angle *BCA*.

There are two systems in common use for measuring the size of angles. The most common unit of measure is the **degree.** (The other common unit of measure, called the *radian,* is discussed in Section 6.5.) Degree measure was developed by the Babylonians, four thousand years ago. To use degree measure, we assign 360 degrees to a complete rotation of a ray. In Figure 4, notice that the terminal side of the angle corresponds to its initial side when it makes a complete rotation. One degree, written 1°, represents 1/360 of a rotation. For example, 90° represents $90/360 = 1/4$ of a complete rotation, and 180° represents $180/360 = 1/2$ of a complete rotation.

An angle having a measure between 0° and 90° is called an **acute angle.** An angle whose measure is exactly 90° is a **right angle.** An angle measuring more than 90° but less than 180° is an **obtuse angle,** and an angle of exactly 180° is a **straight angle**. See Figure 5.

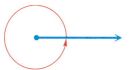

A complete rotation of a ray gives an angle whose measure is 360°.

FIGURE 4

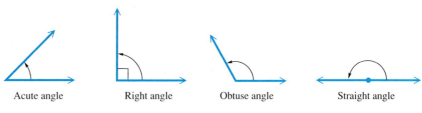

Acute angle   Right angle   Obtuse angle   Straight angle

FIGURE 5

If the sum of the measures of two angles is 90°, the angles are called **complementary.** Two angles with measures whose sum is 180° are **supplementary.**

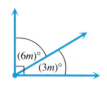

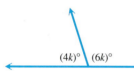

FIGURE 6

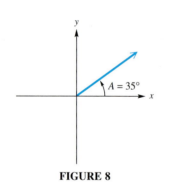

FIGURE 7

Find the measure of each angle in Figures 6 and 7.

(a) In Figure 6, since the two angles form a right angle (as indicated by the symbol),

$$6m + 3m = 90$$
$$9m = 90$$
$$m = 10.$$

The two angles have measures of $6 \cdot 10° = 60°$ and $3 \cdot 10° = 30°$.

(b) The angles in Figure 7 are supplementary, so

$$4k + 6k = 180$$
$$10k = 180$$
$$k = 18.$$

These angle measures are $4(18°) = 72°$ and $6(18°) = 108°$. ▶

Do not confuse an angle with its measure. The angle in Figure 8 is a rotation; the measure of the rotation is 35°. This measure is often expressed by saying that $m(\text{angle } A)$ is 35°, where $m(\text{angle } A)$ is read "the measure of angle A." It saves a lot of work, however, to abbreviate $m(\text{angle } A) = 35°$ as simply angle $A = 35°$.

Traditionally, portions of a degree have been measured with minutes and seconds. One **minute,** written 1′, is 1/60 of a degree.

$$1' = \frac{1}{60}^{\circ} \qquad \text{or} \qquad 60' = 1°$$

One **second,** 1″, is 1/60 of a minute.

$$1'' = \frac{1}{60}^{'} = \frac{1}{3600}^{\circ} \qquad \text{or} \qquad 60'' = 1'$$

The measure 12° 42′ 38″ represents 12 degrees, 42 minutes, 38 seconds.

The next example shows how to perform calculations with degrees, minutes, and seconds.

FIGURE 8

Perform each calculation.

(a) $51° 29' + 32° 46'$

Add the degrees and the minutes separately.

$$\begin{array}{r} 51° \ 29' \\ + \ 32° \ 46' \\ \hline 83° \ 75' \end{array}$$

Since $75' = 60' + 15' = 1° \, 15'$, the sum is written

$$\begin{array}{r} 83° \\ + \quad 1° \; 15' \\ \hline 84° \; 15'. \end{array}$$

(b) $90° - 73° \, 12'$

Write $90°$ as $89° \, 60'$. Then

$$\begin{array}{r} 89° \; 60' \\ - \; 73° \; 12' \\ \hline 16° \; 48'. \end{array}$$

The calculations explained in Example 2 can be done with a graphing calculator capable of working with degrees, minutes, and seconds.

☑ Many graphing calculators have the capability to add and subtract angles given in degrees, minutes, and seconds. They are also able to convert from degrees, minutes, and seconds to decimal degrees (see the next paragraph) and vice versa. Read your owner's manual for details on these capabilities. Also find out whether your calculator has a function to convert between degree measure and radian measure.

The real number π is used extensively to express angle measure in trigonometry. Learn where the π key is located on your calculator keyboard. Be aware that when *exact* values involving π, such as $\pi/3$ and $\pi/4$, are required, decimal approximations given by the calculator are not acceptable.

A graphing calculator will express the first few digits of the irrational number π, as shown in the first line of the display. If a statement is *false*, such as those in the second and third displays, the logic feature returns a zero. If a statement is *true*, a one is returned.

The second and third lines display false statements because 22/7 and 3.14 are rational *approximations* of π; they are not equal to π.

Because calculators are an integral part of our world today, it is now common to measure angles in **decimal degrees.** For example, $12.4238°$ represents

$$12.\mathbf{4238}° = 12 \, \frac{\mathbf{4238}}{\mathbf{10,000}}°.$$

The next example shows how to change between decimal degrees and degrees, minutes, and seconds.

EXAMPLE 3
Converting between decimal degrees and degrees, minutes, seconds

The conversions in Example 3 can be done on some graphing calculators. The second displayed result was obtained by setting the calculator to show only three places after the decimal point.

(a) Convert $74° \, 8' \, 14''$ to decimal degrees. Round to the nearest thousandth of a degree.

Since $1' = \dfrac{1}{60}°$ and $1'' = \dfrac{1}{3600}°$,

$$74° \, 8' \, 14'' = 74° + \frac{8}{60}° + \frac{14}{3600}°$$
$$= 74° + .1333° + .0039°$$
$$= 74.137° \text{ (rounded)}.$$

(b) Convert 34.817° to degrees, minutes, and seconds.

$$34.817° = 34° + .817°$$
$$= 34° + (.817)(60')$$ 1 degree = 60 minutes
$$= 34° + 49.02'$$
$$= 34° + 49' + .02'$$
$$= 34° + 49' + (.02)(60'')$$ 1 minute = 60 seconds
$$= 34° + 49' + 1'' \text{ (rounded)}$$
$$= 34° \, 49' \, 1'' \quad \blacktriangleright$$

An angle is in **standard position** if its vertex is at the origin and its initial side is along the positive x-axis. The two angles in Figure 9 are in standard position. An angle in standard position is said to lie in the quadrant in which its terminal side lies. For example, an acute angle is in quadrant I and an obtuse angle is in quadrant II. Angles in standard position having their terminal sides along the x-axis or y-axis, such as angles with measures 90°, 180°, 270°, and so on, are called **quadrantal angles.**

A complete rotation of a ray results in an angle of measure 360°. But there is no reason why the rotation need stop at 360°. By continuing the rotation, angles of measure larger than 360° can be produced. The angles in Figure 10(a) have measures 60° and 420°. These two angles have the same initial side and the same terminal side, but different amounts of rotation. Angles that have the same initial side and the same terminal side are called **coterminal angles.** As shown in Figure 10(b), angles with measures 110° and 830° are coterminal.

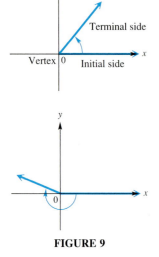

FIGURE 9

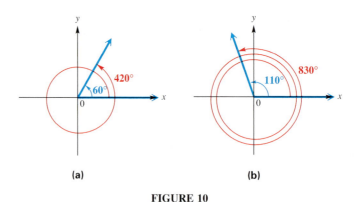

(a)   (b)

FIGURE 10

◀ **EXAMPLE 4**
Finding measures of coterminal angles

Find the angles of smallest possible positive measure coterminal with the following angles.

(a) 908°

Add or subtract 360° as many times as needed to get an angle with measure greater than 0° but less than 360°. Since $908° - 2 \cdot 360° = 908° - 720° = 188°$, an angle of 188° is coterminal with an angle of 908°. See Figure 11.

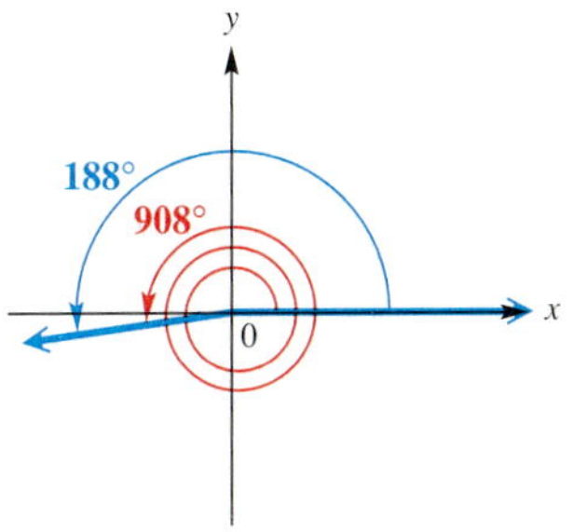

FIGURE 11

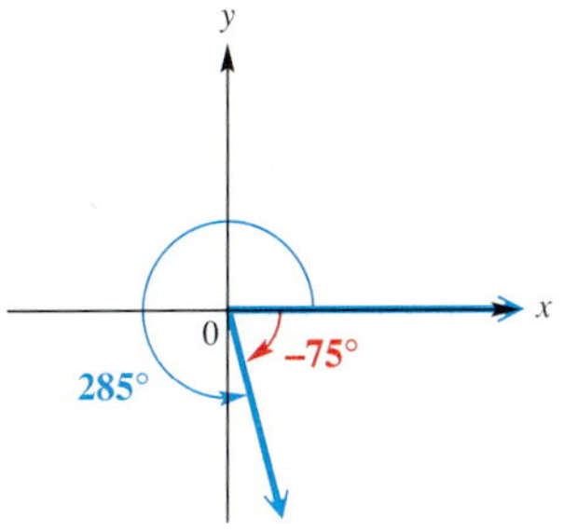

FIGURE 12

(b) $-75°$

Use a rotation of $360° + (-75°) = 285°$. See Figure 12.

Sometimes it is necessary to find an expression that will generate all angles coterminal with a given angle. For example, suppose we wish to do this for a $60°$ angle. The table below shows a few of the possibilities.

Value of n	Angle Coterminal with 60°
2	$60° + 2 \cdot 360° = $ **780°**
1	$60° + 1 \cdot 360° = $ **420°**
0	$60° + 0 \cdot 360° = $ **60°** (the angle itself)
−1	$60° + (-1) \cdot 360° = $ **−300°**

Since any angle coterminal with $60°$ can be obtained by adding an appropriate integer multiple of $360°$ to $60°$, we can let n represent any integer, and the expression

$$60° + n \cdot 360°$$

will represent all such coterminal angles.

EXAMPLE 5
Analyzing the revolution of a phonograph record

A phonograph record makes 45 revolutions per minute. Through how many degrees will a point on the edge of the record move in 2 seconds?

The record revolves 45 times per minute or $45/60 = 3/4$ times per second (since there are 60 seconds in a minute). In 2 seconds, the record will revolve $2 \cdot (3/4) = 3/2$ times. Each revolution is $360°$, so a point on the edge will revolve $(3/2) \cdot 360° = 540°$ in 2 seconds.

6.1 Exercises ▼▼▼▼▼▼▼▼▼▼▼▼▼▼▼▼▼▼▼▼▼▼▼▼▼▼▼▼▼▼▼▼▼▼▼▼

1. Explain the difference between a segment and a ray.

2. What part of a complete revolution is an angle of 45°?

3. What angle is its own complement?

4. What angle is its own supplement?

5. Does a merry-go-round turn in a clockwise or counterclockwise direction?

6. Does the shadow of a sundial move in a clockwise or counterclockwise direction?

Find the measure of each angle in Exercises 7–12. See Example 1.

7.

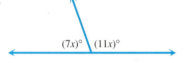

8.

9.

10. Supplementary angles with measures $10m + 7$ and $7m + 3$ degrees

11. Supplementary angles with measures $6x - 4$ and $8x - 12$ degrees

12. Complementary angles with measures $9z + 6$ and $3z$ degrees

13. If an angle measures x degrees, how can we represent its complement?

14. If an angle measures x degrees, how can we represent its supplement?

15. If a positive angle x has a measure between $0°$ and $60°$, how can we represent the first negative angle coterminal with it?

16. If a negative angle x has a measure between $0°$ and $-60°$, how can we represent the first positive angle coterminal with it?

Perform the calculation. See Example 2.

17. $62° \ 18' + 21° \ 41'$ **18.** $75° \ 15' + 83° \ 32'$ **19.** $71° \ 58' + 47° \ 29'$ **20.** $90° - 73° \ 48'$

21. $90° - 51° \ 28'$ **22.** $180° - 124° \ 51'$ **23.** $90° - 72° \ 58' \ 11''$ **24.** $90° - 36° \ 18' \ 47''$

Convert the angle measure to decimal degrees. Use a calculator, and round to the nearest thousandth of a degree. See Example 3.

25. $20° \ 54'$ **26.** $38° \ 42'$ **27.** $91° \ 35' \ 54''$

28. $34° \ 51' \ 35''$ **29.** $274° \ 18' \ 59''$ **30.** $165° \ 51' \ 9''$

Convert the angle measure to degrees, minutes, and seconds. Use a calculator as necessary. See Example 3.

31. $31.4296°$ **32.** $59.0854°$ **33.** $89.9004°$

34. $102.3771°$ **35.** $178.5994°$ **36.** $122.6853°$

37. Read about the degree symbol ($°$) in the manual for your graphing calculator. How is it used?

38. Show that 1.21 hours is the same as 1 hour, 12 minutes, and 36 seconds. Discuss the similarity between converting hours, minutes, and seconds to decimal hours and converting degrees, minutes, and seconds to decimal degrees.

Find the angles of smallest positive measure coterminal with the following angles. See Example 4.

39. $-40°$ **40.** $-98°$ **41.** $-125°$ **42.** $-203°$

43. $539°$ **44.** $699°$ **45.** $850°$ **46.** $1000°$

Give an expression that generates all angles coterminal with the given angle. Let n represent any integer.

47. 30°   **48.** 45°   **49.** 60°   **50.** 90°

51. 135°   **52.** 270°   **53.** −90°   **54.** −135°

55. Explain why the answers to Exercises 52 and 53 give the same set of angles.

56. Which two of the following are not coterminal with $r°$?
   (a) $360° + r°$ **(b)** $r° − 360°$ **(c)** $360° − r°$ **(d)** $r° + 180°$

Consider the function $Y_1 = 360((X/360) − \text{int}(X/360))$ *specified on a graphing calculator. (Note: The value of* int(x) *is the largest integer less than or equal to x. With some calculators,* int *is found in the MATH menu.) The following screen shows that for* X = 908 *and* X = −75, *the function returns the smallest possible positive measure coterminal with the angle. See Example 4.*

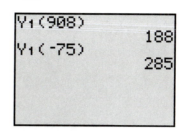

57. Rework Exercise 39 with a graphing calculator.

58. Rework Exercise 40 with a graphing calculator.

Sketch the angle in standard position. Draw an arrow representing the correct amount of rotation. Find the measure of two other angles, one positive and one negative, that are coterminal with the given angle. Give the quadrant of each angle.

59. 75°   **60.** 89°   **61.** 122°   **62.** 174°   **63.** 234°   **64.** 250°

65. 300°   **66.** 512°   **67.** 624°   **68.** −52°   **69.** −61°   **70.** −159°

Locate the following points in a coordinate system. Draw a ray from the origin through the given point. Indicate with an arrow the angle in standard position having smallest positive measure. Then find the distance r from the origin to the point, using the distance formula.

71. $(−3, −3)$ **72.** $(−5, 2)$ **73.** $(−3, −5)$ **74.** $(\sqrt{3}, 1)$ **75.** $(−2, 2\sqrt{3})$ **76.** $(4\sqrt{3}, −4)$

77. A windmill makes 90 revolutions per minute. How many revolutions does it make per second?

78. A turntable makes 45 revolutions per minute. How many revolutions does it make per second?

Solve each problem. See Example 5.

79. A tire rotates 600 times per minute. Through how many degrees does a point on the edge of the tire move in 1/2 second?

80. An airplane propeller rotates 1000 times per minute. Find the number of degrees that a point on the edge of the propeller will rotate through in 1 second.

81. A certain pulley rotates through 75° in one minute. How many rotations, then, does the pulley make in an hour?

82. One student in a surveying class measures an angle as 74.25°, while another student measures the same angle as 74° 20′. Find the difference between these measurements, both to the nearest minute and to the nearest hundredth of a degree.

83. Due to Earth's rotation, celestial objects like the moon and the stars appear to move across the sky, rising in the east and setting in the west. As a result, if a telescope on Earth remains stationary while viewing a celestial object, the object will slowly move outside of the viewing field of the telescope. For this reason a motor is often attached to telescopes so that the telescope rotates at the same rate as Earth. Determine how long it should take for the motor to turn the telescope through an angle of 1 minute in a direction perpendicular to Earth's axis.

6.2   Definitions of the Trigonometric Functions ▼▼▼

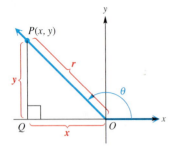

FIGURE 13

The study of trigonometry covers the six trigonometric functions defined in this section. To define these six basic functions, start with an angle θ (the Greek letter *theta*) in standard position. Choose any point P having coordinates (x, y) on the terminal side of angle θ. (The point P must not be the vertex of the angle.) See Figure 13.

A perpendicular from P to the x-axis at point Q determines a triangle having vertices at O, P, and Q. The distance r from $P(x, y)$ to the origin, $(0, 0)$, can be found from the distance formula.

$$r = \sqrt{(x - 0)^2 + (y - 0)^2}$$
$$r = \sqrt{x^2 + y^2}$$

Notice that $r > 0$, since distance is never negative.

FINDING TRIGONOMETRIC FUNCTION VALUES   The six trigonometric functions of angle θ are called **sine, cosine, tangent, cotangent, secant,** and **cosecant.** In the following definitions, we use the customary abbreviations for the names of the functions.

Trigonometric Functions

Let (x, y) be a point other than the origin on the terminal side of an angle θ in standard position. The distance from the point to the origin is $r = \sqrt{x^2 + y^2}$. The six trigonometric functions of θ are:

$$\sin \theta = \frac{y}{r} \qquad\qquad \csc \theta = \frac{r}{y} \ (y \neq 0)$$

$$\cos \theta = \frac{x}{r} \qquad\qquad \sec \theta = \frac{r}{x} \ (x \neq 0)$$

$$\tan \theta = \frac{y}{x} \ (x \neq 0) \qquad \cot \theta = \frac{x}{y} \ (y \neq 0).$$

NOTE   Although Figure 13 shows a second quadrant angle, these definitions apply to any angle θ. Because of the restrictions on the denominators in the definitions of tangent, cotangent, secant, and cosecant, some angles will have undefined function values.

◀**EXAMPLE 1**
Finding the function values of an angle

The terminal side of an angle α in standard position goes through the point $(8, 15)$. Find the values of the six trigonometric functions of angle α.

Figure 14 shows angle α and the triangle formed by dropping a perpendicular from the point $(8, 15)$ to the x-axis. The point $(8, 15)$ is 8 units to the right of the y-axis and 15 units above the x-axis, so that $x = 8$ and $y = 15$.

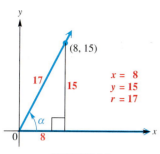

FIGURE 14

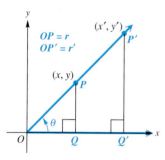

FIGURE 15

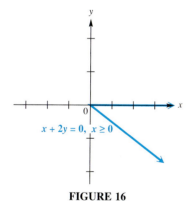

FIGURE 16

Since $r = \sqrt{x^2 + y^2}$,

$$r = \sqrt{8^2 + 15^2}$$
$$= \sqrt{64 + 225}$$
$$= \sqrt{289}$$
$$= 17.$$

The values of the six trigonometric functions of angle α can now be found with the definitions given above.

$$\sin \alpha = \frac{y}{r} = \frac{15}{17} \qquad \csc \alpha = \frac{r}{y} = \frac{17}{15}$$

$$\cos \alpha = \frac{x}{r} = \frac{8}{17} \qquad \sec \alpha = \frac{r}{x} = \frac{17}{8}$$

$$\tan \alpha = \frac{y}{x} = \frac{15}{8} \qquad \cot \alpha = \frac{x}{y} = \frac{8}{15} \quad \blacktriangleright$$

The six trigonometric functions can be found from *any* point on the terminal side of the angle other than the origin. To see why any point may be used, refer to Figure 15, which shows an angle θ and two distinct points on its terminal side. Point P has coordinates (x, y) and point P' (read "P-prime") has coordinates (x', y'). Let r be the length of the hypotenuse of triangle OPQ, and let r' be the length of the hypotenuse of triangle $OP'Q'$. Since corresponding sides of similar triangles are in proportion,

$$\frac{y}{r} = \frac{y'}{r'},$$

so that $\sin \theta = y/r$ is the same no matter which point is used to find it. A similar result holds for the other five functions.

We can also find the trigonometric function values of an angle if we know the equation of the line coinciding with the terminal ray. The graph of the equation

$$Ax + By = 0$$

is a line that passes through the origin. If we restrict x to have only nonpositive or only nonnegative values, we obtain as the graph a ray with endpoint at the origin. For example, the graph of $x + 2y = 0$, $x \geq 0$, is shown in Figure 16. A ray such as the one described above can serve as the terminal side of an angle in standard position. By finding a point on the ray, the trigonometric function values of the angle can be found.

◀EXAMPLE 2
Finding the function values of an angle

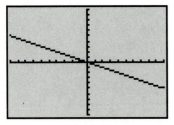

[−10, 10] by [−10, 10]

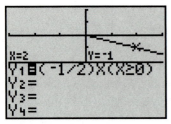

[−3, 3] by [−2, 2]

The line $x + 2y = 0$ is graphed in a standard viewing window in the first screen above. It is done by entering the equation as $Y_1 = (−1/2)x$ or $Y_1 = −.5x$.

The second screen is in split-screen mode. With the restriction $x \geq 0$ we are able to simulate the angle θ in standard position as shown in Figure 17.

Find the six trigonometric function values of the angle θ in standard position, if the terminal side of θ is defined by $x + 2y = 0$, $x \geq 0$.

The angle is shown in Figure 17. We can use *any* point except $(0, 0)$ on the terminal side of θ to find the trigonometric function values, so if we let $x = 2$, we can find the corresponding value of y.

$$x + 2y = 0, \ x \geq 0$$
$$\mathbf{2} + 2y = 0 \qquad \text{Arbitrarily choose } x = 2.$$
$$2y = -2$$
$$y = -1$$

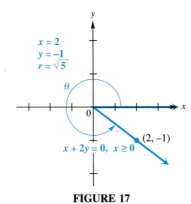

FIGURE 17

The point $(2, -1)$ lies on the terminal side, and the corresponding value of r is $r = \sqrt{2^2 + (-1)^2} = \sqrt{5}$. Now use the definitions of the trigonometric functions.

$$\sin \theta = \frac{y}{r} = \frac{-1}{\sqrt{5}} = \frac{-1}{\sqrt{5}} \cdot \frac{\sqrt{5}}{\sqrt{5}} = -\frac{\sqrt{5}}{5} \qquad \csc \theta = \frac{r}{y} = \frac{\sqrt{5}}{-1} = -\sqrt{5}$$

$$\cos \theta = \frac{x}{r} = \frac{2}{\sqrt{5}} = \frac{2}{\sqrt{5}} \cdot \frac{\sqrt{5}}{\sqrt{5}} = \frac{2\sqrt{5}}{5} \qquad \sec \theta = \frac{r}{x} = \frac{\sqrt{5}}{2}$$

$$\tan \theta = \frac{y}{x} = \frac{-1}{2} = -\frac{1}{2} \qquad\qquad \cot \theta = \frac{x}{y} = \frac{2}{-1} = -2 \ \ \textcolor{red}{\blacktriangleright}$$

Recall that when the equation of a line is written in the form $y = mx + b$, the coefficient of x is the slope of the line. In Example 2, $x + 2y = 0$ can be written as $y = (-1/2)x$, so the slope is $-1/2$. Notice that $\tan \theta = -1/2$. In general, it is true that $m = \tan \theta$.

NOTE The trigonometric function values we found in Examples 1 and 2 are *exact*. If we were to use a calculator to approximate these values, the decimal results would not be acceptable if exact values were required.

CONNECTIONS A convenient way to see the three basic trigonometric ratios geometrically is shown in the two figures below at left for θ in quadrants I and II. The circle, which has a radius of 1, is called a *unit circle*. We will see the unit circle again later in this chapter. By memorizing this figure and the segments that represent the sine, cosine, and tangent functions, you can quickly recall the properties of the trigonometric functions. Horizontal line segments to the left of the origin and vertical line segments below the x-axis represent negative values. Note that the tangent line that contains the tangent segment must be tangent to the circle at (1, 0), no matter which quadrant θ lies in.

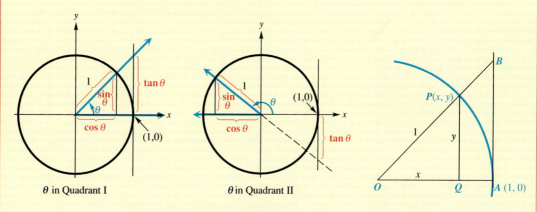

FOR DISCUSSION OR WRITING

1. Label the triangles as shown in the figure above at right. Use the definition of the trigonometric functions and similar triangles to show that $PQ = \sin\theta$, $OQ = \cos\theta$, and $AB = \tan\theta$.
2. Sketch similar figures for θ in quadrants III and IV.

   If the terminal side of an angle in standard position lies along the y-axis, any point on this terminal side has x-coordinate 0. Similarly, an angle with terminal side on the x-axis has y-coordinate 0 for any point on the terminal side. Since the values of x and y appear in the denominators of some of the trigonometric functions, and since a fraction is undefined if its denominator is 0, some of the trigonometric function values of quadrantal angles (i.e., those with the terminal side on an axis) will be undefined.

EXAMPLE 3
Finding trigonometric function values of a quadrantal angle

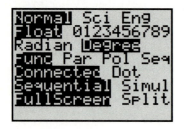

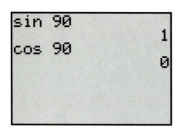

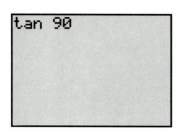

When the calculator is set to degree mode, it returns the correct values for sin 90° and cos 90°. Notice that it returns an ERROR message for tan 90°, since 90° is not in the domain of the tangent function. Compare these results to those found in Example 3.

Find the values of the six trigonometric functions for the following angles.

(a) an angle of 90°

First, select any point on the terminal side of a 90° angle. Let us select the point $(0, 1)$, as shown in Figure 18(a). Here $x = 0$ and $y = 1$. Verify that $r = 1$. Then, by the definition of the trigonometric functions,

$$\sin 90° = \frac{1}{1} = 1 \qquad\qquad \csc 90° = \frac{1}{1} = 1$$

$$\cos 90° = \frac{0}{1} = 0 \qquad\qquad \sec 90° = \frac{1}{0} \text{ (undefined)}$$

$$\tan 90° = \frac{1}{0} \text{ (undefined)} \qquad \cot 90° = \frac{0}{1} = 0.$$

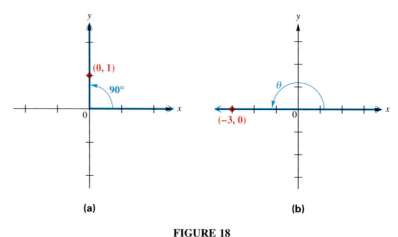

FIGURE 18

(b) an angle in standard position with terminal side through $(-3, 0)$

Figure 18(b) shows the angle. Here, $x = -3$, $y = 0$, and $r = 3$, so the trigonometric functions have the following values.

$$\sin \theta = \frac{0}{3} = 0 \qquad\qquad \csc \theta = \frac{3}{0} \text{ (undefined)}$$

$$\cos \theta = \frac{-3}{3} = -1 \qquad \sec \theta = \frac{3}{-3} = -1$$

$$\tan \theta = \frac{0}{-3} = 0 \qquad\qquad \cot \theta = \frac{-3}{0} \text{ (undefined)} \; \blacktriangleright$$

Since the most commonly used quadrantal angles are 0°, 90°, 180°, and 270°, the values of the functions of these angles are summarized in the following table. This table is for reference only; you should be able to reproduce it quickly.

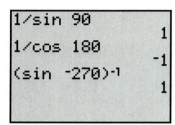

This screen shows how csc 90°, sec 180° and csc (−270°) can be found, using the appropriate reciprocal identities. Compare these results with the ones found in the chart of quadrantal angle function values. Be sure not to use the *inverse trigonometric function* keys to find the reciprocal function values. Consult your owner's manual if further information is needed.

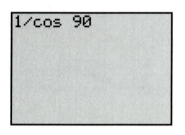

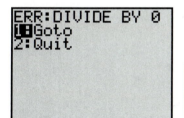

Attempting to find sec 90° by entering 1/cos 90° leads to an ERROR message, indicating that division by 0 is not allowed. Compare to the chart of quadrantal angle function values, where we indicate that sec 90° is undefined.

Quadrantal Angles

θ	$\sin\theta$	$\cos\theta$	$\tan\theta$	$\cot\theta$	$\sec\theta$	$\csc\theta$
0°	0	1	0	Undefined	1	Undefined
90°	1	0	Undefined	0	Undefined	1
180°	0	−1	0	Undefined	−1	Undefined
270°	−1	0	Undefined	0	Undefined	−1

The values given in this table can also be found with a calculator that has trigonometric function keys. First, make sure the calculator is set for *degree mode*.

CAUTION   One of the most common errors involving calculators in trigonometry occurs when the calculator is set for *radian measure*, rather than *degree measure*. (Radian measure of angles is studied in Section 6.5.) For this reason, be sure that you know how to set your calculator to *degree mode*.

THE RECIPROCAL IDENTITIES   The definitions of the trigonometric functions given earlier were written so that functions on the same line are reciprocals of each other. Since $\sin\theta = y/r$ and $\csc\theta = r/y$,

$$\sin\theta = \frac{1}{\csc\theta} \qquad \text{and} \qquad \csc\theta = \frac{1}{\sin\theta}.$$

Also, $\cos\theta$ and $\sec\theta$ are reciprocals, as are $\tan\theta$ and $\cot\theta$. In summary, we have the **reciprocal identities** that hold for any angle θ that does not lead to a zero denominator.

Reciprocal Identities

$$\sin\theta = \frac{1}{\csc\theta} \qquad \csc\theta = \frac{1}{\sin\theta}$$

$$\cos\theta = \frac{1}{\sec\theta} \qquad \sec\theta = \frac{1}{\cos\theta}$$

$$\tan\theta = \frac{1}{\cot\theta} \qquad \cot\theta = \frac{1}{\tan\theta}$$

Identities are equations that are true for all meaningful values of the variable. For example, both $(x + y)^2 = x^2 + 2xy + y^2$ and $2(x + 3) = 2x + 6$ are identities. Identities are studied in more detail in the next chapter.

NOTE   When studying identities, be aware that various forms exist. For example,

$$\sin \theta = \frac{1}{\csc \theta}$$

can also be written

$$\csc \theta = \frac{1}{\sin \theta}$$

and

$$(\sin \theta)(\csc \theta) = 1.$$

You should become familiar with all forms of these identities.

EXAMPLE 4
Using the reciprocal identities

Find each function value.

(a) $\cos \theta$, if $\sec \theta = 5/3$
Since $\cos \theta$ is the reciprocal of $\sec \theta$,

$$\cos \theta = \frac{1}{\sec \theta} = \frac{1}{5/3} = \frac{3}{5}.$$

(b) $\sin \theta$, if $\csc \theta = -\sqrt{12}/2$

$$\sin \theta = \frac{1}{-\sqrt{12}/2}$$

$$= \frac{-2}{\sqrt{12}}$$

$$= \frac{-2}{2\sqrt{3}} \qquad \sqrt{12} = \sqrt{4 \cdot 3} = 2\sqrt{3}$$

$$= \frac{-1}{\sqrt{3}}$$

$$= \frac{-\sqrt{3}}{3} \qquad \text{Multiply by } \tfrac{\sqrt{3}}{\sqrt{3}} \text{ to rationalize the denominator.} \quad \blacktriangleright$$

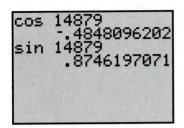

This screen was obtained with the calculator in degree mode. How can we use it to justify that an angle of 14,879° is a second quadrant angle?

In the definition of the trigonometric functions, r is the distance from the origin to the point (x, y), so $r > 0$. If we choose a point (x, y) in quadrant I, then both x and y will be positive. Since $r > 0$, all six of the fractions used in the definitions of the trigonometric functions will be positive, so that the values of all six functions will be positive in quadrant I.

A point (x, y) in quadrant II has $x < 0$ and $y > 0$. This makes the value of sine and cosecant positive for quadrant II angles, while the other four functions take on negative values. Similar results can be obtained for the other quadrants, as summarized below.

Signs of Function Values

θ in Quadrant	$\sin \theta$	$\cos \theta$	$\tan \theta$	$\cot \theta$	$\sec \theta$	$\csc \theta$
I	+	+	+	+	+	+
II	+	−	−	−	−	+
III	−	−	+	+	−	−
IV	−	+	−	−	+	−

II
$x < 0$
$y > 0$
$r > 0$
Sine and cosecant positive

I
$x > 0$
$y > 0$
$r > 0$
All functions positive

III
$x < 0$
$y < 0$
$r > 0$
Tangent and cotangent positive

IV
$x > 0$
$y < 0$
$r > 0$
Cosine and secant positive

EXAMPLE 5
Identifying the quadrant of an angle

Identify the quadrant (or quadrants) of any angle θ that satisfies $\sin \theta > 0$, $\tan \theta < 0$.

Since $\sin \theta > 0$ in quadrants I and II, while $\tan \theta < 0$ in quadrants II and IV, both conditions are met only in quadrant II. ▶

Figure 19 shows an angle θ as it increases in measure from near $0°$ toward $90°$. In each case, the value of r is the same. As the measure of the angle increases, y increases but never exceeds r, so that $y \leq r$. Dividing both sides by the positive number r gives

$$y \leq r$$

$$\frac{y}{r} \leq 1.$$

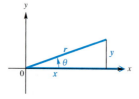
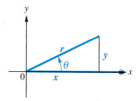
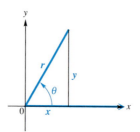
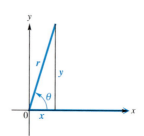

FIGURE 19

In a similar way, angles in the fourth quadrant suggest that

$$-1 \leq \frac{y}{r},$$

so

$$-1 \leq \frac{y}{r} \leq 1.$$

Since $y/r = \sin \theta$,

$$-1 \leq \sin \theta \leq 1$$

for any angle θ. In the same way,

$$-1 \leq \cos \theta \leq 1.$$

The tangent of an angle is defined as y/x. It is possible that $x < y$, that $x = y$, or that $x > y$. For this reason y/x can take on any value at all, so $\tan \theta$ can be any real number, as can $\cot \theta$.

The functions $\sec \theta$ and $\csc \theta$ are reciprocals of the functions $\cos \theta$ and $\sin \theta$, respectively, making

$$\sec \theta \leq -1 \quad \text{or} \quad \sec \theta \geq 1,$$
$$\csc \theta \leq -1 \quad \text{or} \quad \csc \theta \geq 1.$$

In summary, the ranges of the trigonometric functions are as follows.

Ranges of Trigonometric Functions

For any angle θ for which the indicated functions exist:

1. $-1 \leq \sin \theta \leq 1$ and $-1 \leq \cos \theta \leq 1$;
2. $\tan \theta$ and $\cot \theta$ may be equal to any real number;
3. $\sec \theta \leq -1$ or $\sec \theta \geq 1$ and $\csc \theta \leq -1$ or $\csc \theta \geq 1$.

(Notice that $\sec \theta$ and $\csc \theta$ are *never* between -1 and 1.)

EXAMPLE 6
Deciding whether a trigonometric function value is in the range

Decide whether the following statements are *possible* or *impossible*.

(a) $\sin \theta = \sqrt{8}$
For any value of θ, $-1 \leq \sin \theta \leq 1$. Since $\sqrt{8} > 1$, there is no value of θ with $\sin \theta = \sqrt{8}$.

(b) $\tan \theta = 110.47$
Tangent can take on any value. Thus, $\tan \theta = 110.47$ is possible.

(c) $\sec \theta = .6$
Since $\sec \theta \leq -1$ or $\sec \theta \geq 1$, the statement $\sec \theta = .6$ is impossible.

THE PYTHAGOREAN IDENTITIES   We can derive three very useful new identities from the relationship $x^2 + y^2 = r^2$. Dividing both sides by r^2 gives

$$\frac{x^2}{r^2} + \frac{y^2}{r^2} = \frac{r^2}{r^2},$$

or

$$\left(\frac{x}{r}\right)^2 + \left(\frac{y}{r}\right)^2 = 1.$$

Since $\sin \theta = y/r$ and $\cos \theta = x/r$, this result becomes

$$(\sin \theta)^2 + (\cos \theta)^2 = 1,$$

or, as it is usually written,

$$\sin^2 \theta + \cos^2 \theta = 1.$$

Similarly, starting with $x^2 + y^2 = r^2$ and dividing through by x^2 gives

$$\tan^2 \theta + 1 = \sec^2 \theta.$$

On the other hand, dividing through by y^2 leads to

$$1 + \cot^2 \theta = \csc^2 \theta.$$

These three identities are called the **Pythagorean identities** since the original equation that led to them, $x^2 + y^2 = r^2$, comes from the Pythagorean theorem.

> **Pythagorean Identities**
>
> $$\sin^2 \theta + \cos^2 \theta = 1 \qquad \tan^2 \theta + 1 = \sec^2 \theta$$
> $$1 + \cot^2 \theta = \csc^2 \theta$$

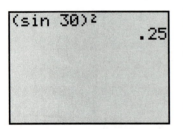

This screen supports the statement found in the Note.

NOTE Although we usually write $\sin^2 \theta$, for example, it should be entered as $(\sin \theta)^2$ in your calculator. To test yourself, verify that $\sin^2 30° = .25$.

As before, we have given only one form of each identity. However, algebraic transformations can be made to get equivalent identities. For example, by subtracting $\sin^2 \theta$ from both sides of $\sin^2 \theta + \cos^2 \theta = 1$, we get the equivalent identity

$$\cos^2 \theta = 1 - \sin^2 \theta.$$

You should be able to transform these identities quickly, and also recognize their equivalent forms.

THE QUOTIENT IDENTITIES   Recall that $\sin \theta = y/r$ and $\cos \theta = x/r$. Consider the quotient of $\sin \theta$ and $\cos \theta$, where $\cos \theta \neq 0$.

$$\frac{\sin \theta}{\cos \theta} = \frac{y/r}{x/r} = \frac{y}{r} \div \frac{x}{r} = \frac{y}{r} \cdot \frac{r}{x} = \frac{y}{x} = \tan \theta$$

Similarly, it can be shown that $(\cos \theta)/(\sin \theta) = \cot \theta$, for $\sin \theta \neq 0$. Thus we have two more identities, called the **quotient identities.**

> **Quotient Identities**
>
> $$\frac{\sin \theta}{\cos \theta} = \tan \theta \qquad \frac{\cos \theta}{\sin \theta} = \cot \theta$$

◀ **EXAMPLE 7**
Finding other function values, given one function value and the quadrant of θ

Find $\sin \theta$ and $\cos \theta$, if $\tan \theta = 4/3$ and θ is in quadrant III.

   Since θ is in quadrant III, $\sin \theta$ and $\cos \theta$ will both be negative. It is tempting to say that since $\tan \theta = (\sin \theta)/(\cos \theta)$ and $\tan \theta = 4/3$, then $\sin \theta = -4$ and $\cos \theta = -3$. This is *incorrect*, however, since both $\sin \theta$ and $\cos \theta$ must be in the interval $[-1, 1]$.

Use the Pythagorean identity $\tan^2 \theta + 1 = \sec^2 \theta$ to find $\sec \theta$, and then the reciprocal identity $\cos \theta = 1/\sec \theta$.

$$\tan^2 \theta + 1 = \sec^2 \theta$$

$$\left(\frac{4}{3}\right)^2 + 1 = \sec^2 \theta \qquad \tan \theta = \tfrac{4}{3}$$

$$\frac{16}{9} + 1 = \sec^2 \theta$$

$$\frac{25}{9} = \sec^2 \theta$$

$$-\frac{5}{3} = \sec \theta \qquad \text{Choose the negative square root since } \theta \text{ is in quadrant III.}$$

$$-\frac{3}{5} = \cos \theta \qquad \text{Secant and cosine are reciprocals.}$$

Since $\sin^2 \theta = 1 - \cos^2 \theta$,

$$\sin^2 \theta = 1 - \left(-\frac{3}{5}\right)^2 \qquad \cos \theta = -\tfrac{3}{5}$$

$$= 1 - \frac{9}{25}$$

$$= \frac{16}{25}$$

$$\sin \theta = -\frac{4}{5}. \qquad \text{Choose the negative square root.}$$

Therefore, we have $\sin \theta = -4/5$ and $\cos \theta = -3/5$. ▶

NOTE   Example 7 can also be worked by drawing θ in standard position in quadrant III, choosing $x = -3$, $y = -4$, and finding r to be 5. See Figure 20. Then the definitions of $\sin \theta$ and $\cos \theta$ in terms of x, y, and r can be used.

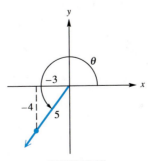

FIGURE 20

6.2 Exercises ▼▼▼▼▼▼▼▼▼▼▼▼▼▼▼▼▼▼▼▼▼▼▼▼▼▼▼▼▼▼▼▼▼▼▼▼▼▼

Sketch an angle θ in standard position such that θ has the smallest possible positive measure, and the given point is on the terminal side of θ.

1. $(5, -12)$ **2.** $(-12, -5)$

Find the values of the six trigonometric functions for the angles in standard position having the following points on their terminal sides. Rationalize denominators when applicable. See Examples 1 and 3.

3. $(-3, 4)$ **4.** $(-4, -3)$ **5.** $(0, 2)$ **6.** $(-4, 0)$ **7.** $(1, \sqrt{3})$ **8.** $(-2\sqrt{3}, -2)$

9. For any nonquadrantal angle θ, $\sin \theta$ and $\csc \theta$ will have the same sign. Explain why this is so.

10. If the terminal side of an angle β is in quadrant III, what is the sign of each of the trigonometric function values of β?

Suppose the point (x, y) is in the indicated quadrant. Decide whether the given ratio is positive or negative. (Hint: It may be helpful to draw a sketch.)

11. II, $\dfrac{x}{r}$ **12.** III, $\dfrac{y}{r}$ **13.** IV, $\dfrac{y}{x}$ **14.** IV, $\dfrac{x}{y}$

In Exercises 15–18, an equation with a restriction on x is given. This is an equation of the terminal side of an angle θ in standard position. Sketch the smallest positive such angle θ, and find the values of the six trigonometric functions of θ. See Example 2.

15. $2x + y = 0, \quad x \geq 0$ **16.** $3x + 5y = 0, \quad x \geq 0$

17. $-6x - y = 0, \quad x \leq 0$ **18.** $-5x - 3y = 0, \quad x \leq 0$

19. Rework Example 2 using a different value for x. Find the corresponding y-value, and then show that the six trigonometric function values you obtain are the same as the ones obtained in Example 2.

Use the trigonometric function values of quadrantal angles in this section to evaluate each of the following. An expression such as $\cot^2 90°$ means $(\cot 90°)^2$ which is equal to $0^2 = 0$.

20. $3 \sec 180° - 5 \tan 360°$ **21.** $4 \csc 270° + 3 \cos 180°$

22. $\tan 360° + 4 \sin 180° + 5 \cos^2 180°$ **23.** $2 \sec 0° + 4 \cot^2 90° + \cos 360°$

24. $\sin^2 180° + \cos^2 180°$ **25.** $\sin^2 360° + \cos^2 360°$

If n is an integer, $n \cdot 180°$ represents an integer multiple of $180°$, and $(2n + 1) \cdot 90°$ represents an odd integer multiple of $90°$. Decide whether each of the following is equal to 0, 1, -1, or is undefined.

26. $\cos[(2n + 1) \cdot 90°]$ **27.** $\tan[n \cdot 180°]$

28. The angles $15°$ and $75°$ are complementary. With your calculator determine $\sin 15°$ and $\cos 75°$. Make a conjecture about the sines and cosines of complementary angles and test your hypothesis with other pairs of complementary angles. (This relationship will be discussed in detail in Section 7.3.)

29. With your calculator determine $\sin 10°$ and $\sin(-10°)$. Make a conjecture about the sine of an angle and its negative and test your hypothesis with other angles. Also, use a geometry argument with the definition of $\sin \theta$ to justify your hypothesis. (This relationship will be discussed in detail in Section 7.1.)

30. With your calculator determine cos 20° and cos(−20°). Make a conjecture about the cosine of an angle and its negative and test your hypothesis with other angles. Also, use a geometry argument with the definition of cos θ to justify your hypothesis. (This relationship will be discussed in detail in Section 7.1.)

The figure at left in the Connections box shows that the intersection of the unit circle and the terminal side of an angle is the point (cos θ, sin θ). *Use this fact for the following exercises.*

31. Define the cosine function in terms of the *x*-coordinate of a point on the unit circle.

32. Define the sine function in terms of the *y*-coordinate of a point on the unit circle.

In Exercises 33–38, place your graphing calculator in parametric and degree modes. Set the window and functions as shown, and graph. A circle of radius 1 will appear on the screen. Use the trace feature to move a short distance around the circle. In the accompanying graphics screen the point on the circle corresponds to an angle of T = 25°; *cos 25° is .90630779, and sin 25° is .42261826.*

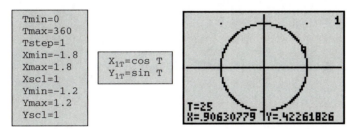

33. Use the right- and left-arrow keys to move to the point corresponding to 20°. What are cos 20° and sin 20°?

In Exercises 34–36, assume 0° ≤ T ≤ 90°.

34. For what angle *T* is cos *T* ≈ .766?

35. For what angle *T* is sin *T* ≈ .574?

36. For what angle *T* does cos *T* = sin *T*?

37. As *T* increases from 0° to 90° does the cosine increase or decrease? How about the sine?

38. As *T* increases from 90° to 180° does the cosine increase or decrease? How about the sine?

39. What positive number *a* is its own reciprocal? Find a value of θ for which sin θ = csc θ = a.

40. What negative number *a* is its own reciprocal? Find a value of θ for which cos θ = sec θ = a.

Use the appropriate reciprocal identity to find each function value. Rationalize denominators when applicable. In Exercises 45 and 46, use a calculator. See Example 4.

41. cos α, if sec α = −2.5

42. cot β, if tan β = $-\dfrac{1}{5}$

43. csc α, if sin α = $\dfrac{\sqrt{2}}{4}$

44. tan θ, if cot θ = $-\dfrac{\sqrt{5}}{3}$

45. sin θ, if csc θ = 1.42716321

46. cos α, if sec α = 9.80425133

47. Can a given angle γ satisfy both sin γ > 0 and csc γ < 0? Explain.

48. Suppose that the following item appears on a trigonometry test:

$$\text{Find sec } \theta, \text{ given that cos } \theta = \frac{3}{2}.$$

What is wrong with this test item?

Find the tangent of each angle. See Example 4.

49. ϕ, if cot $\phi = -3$

50. ω, if cot $\omega = \dfrac{\sqrt{3}}{3}$

Find a value of the variable in each of the following.

51. $\tan(3B - 4°) = \dfrac{1}{\cot(5B - 8°)}$

52. $\sec(2\alpha + 6°)\cos(5\alpha + 3°) = 1$

Identify the quadrant or quadrants for the angle satisfying the given conditions. See Example 5.

53. sin $\alpha > 0$, cos $\alpha < 0$

54. cos $\beta > 0$, tan $\beta > 0$

55. tan $\gamma > 0$, cot $\gamma > 0$

56. csc $\theta < 0$, cos $\theta < 0$

Give the signs of the six trigonometric functions for each angle.

57. 183°   **58.** 298°   **59.** 302°   **60.** 412°   **61.** −82°   **62.** −121°

Without using a calculator, decide which is greater.

63. sin 30° or tan 30°

64. sin 33° or sec 33°

Decide whether each statement is possible *or* impossible. *See Example 6.*

65. sin $\theta = 2$ **66.** cos $\alpha = -1.001$ **67.** tan $\beta = .92$

68. cot $\omega = -12.1$ **69.** csc $\alpha = \dfrac{1}{2}$ **70.** sec $\alpha = 1$

71. sin $\alpha = \dfrac{1}{2}$ and csc $\alpha = 2$

72. tan $\beta = 2$ and cot $\beta = -2$

73. Draw the graph of $f(x) = \sin^2 x + \cos^2 x$ with a graphing calculator using the window $[0, 360]$ by $[-2, 2]$ as a further verification that $\sin^2 \theta + \cos^2 \theta = 1$.

74. Draw the graph of $f(x) = \tan^2 x - \sec^2 x$ with a graphing calculator using the window $[0, 360]$ by $[-2, 2]$ as a further verification that $\tan^2 \theta + 1 = \sec^2 \theta$.

Use identities to find the indicated function value. Use a calculator in Exercises 81 and 82. See Example 7.

75. tan α, if sec $\alpha = 3$, with α in quadrant IV

76. sin α, if cos $\alpha = -1/4$, with α in quadrant II

77. tan θ, if cos $\theta = 1/3$, with θ in quadrant IV

78. sec θ, if tan $\theta = \sqrt{7}/3$, with θ in quadrant III

79. cos β, if csc $\beta = -4$, with β in quadrant III

80. sin θ, if sec $\theta = 2$, with θ in quadrant IV

81. sin β, if cot $\beta = 2.40129813$, with β in quadrant I

82. cot α, if csc $\alpha = -3.5891420$, with α in quadrant III

In Exercises 83 and 84, the given graphing calculator screen is obtained for a particular stored value of X. *What will the screen display for the value of the expression in the final line of the display?*

83.

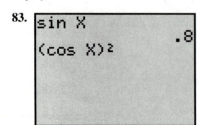

```
sin X
          .8
(cos X)²
```

84.

```
tan X
          2
(1/cos X)²
```

85. Does there exist an angle θ with cos $\theta = -.6$ and sin $\theta = .8$?

Find all the trigonometric function values for each angle. Use a calculator in Exercises 90 and 91. See Example 7.

86. $\tan \alpha = -15/8$, with α in quadrant II

87. $\cos \alpha = -3/5$, with α in quadrant III

88. $\tan \beta = \sqrt{3}$, with β in quadrant III

89. $\sin \beta = \sqrt{5}/7$, with $\tan \beta > 0$

90. $\cot \theta = -1.49586$, with θ in quadrant IV

91. $\sin \alpha = .164215$, with α in quadrant II

92. Derive the identity $1 + \cot^2 \theta = \csc^2 \theta$ by dividing $x^2 + y^2 = r^2$ by y^2.

93. Using a method similar to the one given in this section showing that $(\sin \theta)/(\cos \theta) = \tan \theta$, show that

$$\frac{\cos \theta}{\sin \theta} = \cot \theta.$$

94. True or false: For all angles θ, $\sin \theta + \cos \theta = 1$. If false, give an example showing why it is false.

95. True or false: Since $\cot \theta = \cos \theta/\sin \theta$, if $\cot \theta = 1/2$ with θ in quadrant I, then $\cos \theta = 1$ and $\sin \theta = 2$. If false, explain why.

96. The straight line in the figure below determines both the angle α and the angle β with the positive x-axis. Explain why $\tan \alpha = \tan \beta$.

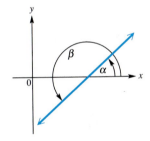

6.3 Finding Trigonometric Function Values ▼▼▼

Figure 21 shows an acute angle A in standard position. The definitions of the trigonometric function values of angle A require x, y, and r. As drawn in Figure 21, x and y are the lengths of the two legs of right triangle ABC, and r is the length of the hypotenuse.

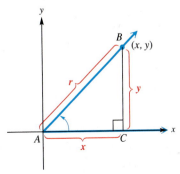

FIGURE 21

The side of length y is called the **side opposite** angle A, and the side of length x is called the **side adjacent** to angle A. The lengths of these sides can be used to replace x and y in the definition of the trigonometric functions, with r replaced by the length of the hypotenuse, to get the following right triangle-based definitions.

Right Triangle-Based Definitions of Trigonometric Functions

For any acute angle A in standard position,

$$\sin A = \frac{y}{r} = \frac{\text{side opposite}}{\text{hypotenuse}} \qquad \csc A = \frac{r}{y} = \frac{\text{hypotenuse}}{\text{side opposite}}$$

$$\cos A = \frac{x}{r} = \frac{\text{side adjacent}}{\text{hypotenuse}} \qquad \sec A = \frac{r}{x} = \frac{\text{hypotenuse}}{\text{side adjacent}}$$

$$\tan A = \frac{y}{x} = \frac{\text{side opposite}}{\text{side adjacent}} \qquad \cot A = \frac{x}{y} = \frac{\text{side adjacent}}{\text{side opposite}}.$$

EXAMPLE 1
Finding trigonometric function values of an acute angle in a right triangle

Find the values of $\sin A$, $\cos A$, and $\tan A$ in the right triangle in Figure 22.

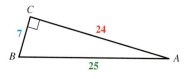

FIGURE 22

The length of the side opposite angle A is 7. The length of the side adjacent to angle A is 24, and the length of the hypotenuse is 25. Using the relationships given above,

$$\sin A = \frac{\text{side opposite}}{\text{hypotenuse}} = \frac{7}{25}$$

$$\cos A = \frac{\text{side adjacent}}{\text{hypotenuse}} = \frac{24}{25}$$

$$\tan A = \frac{\text{side opposite}}{\text{side adjacent}} = \frac{7}{24}. \quad \blacktriangleright$$

NOTE Because the cosecant, secant, and cotangent ratios are the reciprocals of the sine, cosine, and tangent values, in Example 1 we can conclude that $\csc A = 25/7$, $\sec A = 25/24$, and $\cot A = 24/7$.

EXACT TRIGONOMETRIC FUNCTION VALUES OF SPECIAL ANGLES Certain special angles, such as 30°, 45°, and 60°, occur so often in trigonometry and in more advanced mathematics that they deserve special study. We can find the exact trigonometric function values of these angles by using properties of geometry and the Pythagorean theorem.

To find the trigonometric function values for 30° and 60°, we start with an equilateral triangle, a triangle with all sides of equal length. Each angle of such a triangle has a measure of 60°. While the results we will obtain are independent of the length, for convenience, we choose the length of each side to be 2 units. See Figure 23(a).

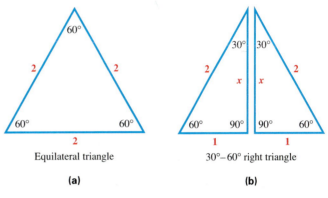

Equilateral triangle

(a)

30°–60° right triangle

(b)

FIGURE 23

Bisecting one angle of this equilateral triangle leads to two right triangles, each of which has angles of 30°, 60°, and 90°, as shown in Figure 23(b). Since the hypotenuse of one of these right triangles has a length of 2, the shortest side will have a length of 1. (Why?) If x represents the length of the medium side, then, by the Pythagorean theorem,

$$2^2 = 1^2 + x^2$$
$$4 = 1 + x^2$$
$$3 = x^2$$
$$\sqrt{3} = x.$$

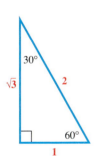

FIGURE 24

Figure 24 summarizes our results, showing a 30°–60° right triangle.

As shown in the figure, the side opposite the 30° angle has length 1; that is, for the 30° angle,

$$\text{hypotenuse} = 2, \quad \text{side opposite} = 1, \quad \text{side adjacent} = \sqrt{3}.$$

Using the definitions of the trigonometric functions,

$$\sin 30° = \frac{\text{side opposite}}{\text{hypotenuse}} = \frac{1}{2} \qquad \csc 30° = \frac{2}{1} = 2$$

$$\cos 30° = \frac{\text{side adjacent}}{\text{hypotenuse}} = \frac{\sqrt{3}}{2} \qquad \sec 30° = \frac{2}{\sqrt{3}} = \frac{2\sqrt{3}}{3}$$

$$\tan 30° = \frac{\text{side opposite}}{\text{side adjacent}} = \frac{1}{\sqrt{3}} = \frac{\sqrt{3}}{3} \qquad \cot 30° = \frac{\sqrt{3}}{1} = \sqrt{3}.$$

The denominator was rationalized for tan 30° and sec 30°.

EXAMPLE 2
Finding the function values
for 60°

Find the six trigonometric function values for a 60° angle.
From Figure 24 on page 441,

$$\sin 60° = \frac{\sqrt{3}}{2} \qquad \tan 60° = \sqrt{3} \qquad \sec 60° = 2$$

$$\cos 60° = \frac{1}{2} \qquad \cot 60° = \frac{\sqrt{3}}{3} \qquad \csc 60° = \frac{2\sqrt{3}}{3}.$$ ▶

EXAMPLE 3
Finding trigonometric function
values for a non-acute angle

Find the sine, cosine, and tangent values for 210°.
 Draw an angle of 210° in standard position as shown in Figure 25. Choose point P on the terminal side of the angle so that the distance from the origin to P is 2. Draw PR to form the right triangle OPR. By the results for 30°–60° right triangles, the coordinates of P become $(-\sqrt{3}, -1)$, with $x = -\sqrt{3}$, $y = -1$, and $r = 2$. Then, by the definitions of the trigonometric functions,

$$\sin 210° = -\frac{1}{2} \qquad \cos 210° = -\frac{\sqrt{3}}{2} \qquad \tan 210° = \frac{\sqrt{3}}{3}.$$ ▶

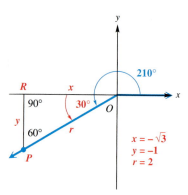

FIGURE 25

 The values of the trigonometric functions for 45° can be found by starting with a 45°–45° right triangle, as shown in Figure 26. This triangle is isosceles, and, for convenience, we choose the lengths of the equal sides to be 1 unit. (As before, the results are independent of the length of the equal sides of the right triangle.) Since the shorter sides each have length 1, if r represents the length of the hypotenuse, then

$$1^2 + 1^2 = r^2$$
$$2 = r^2$$
$$\sqrt{2} = r.$$

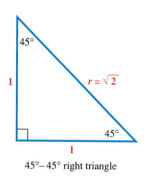

45°–45° right triangle

FIGURE 26

Using the measures indicated on the 45°–45° right triangle in Figure 26, we find

$$\sin 45° = \frac{1}{\sqrt{2}} = \frac{\sqrt{2}}{2} \qquad \tan 45° = \frac{1}{1} = 1 \qquad \sec 45° = \frac{\sqrt{2}}{1} = \sqrt{2}$$

$$\cos 45° = \frac{1}{\sqrt{2}} = \frac{\sqrt{2}}{2} \qquad \cot 45° = \frac{1}{1} = 1 \qquad \csc 45° = \frac{\sqrt{2}}{1} = \sqrt{2}.$$

◀EXAMPLE 4
Using trigonometric values for 45° to find values for 315°

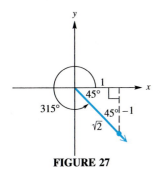

FIGURE 27

Find sin 315°, sec 315°, and cot 315°.

Sketch a 315° angle in standard position as shown in Figure 27. The angle from the terminal ray to the x-axis is 45°. Drawing a line from the terminal ray perpendicular to the x-axis completes a 45°–45° right triangle. We can label the sides of the triangle as $x = 1$, $y = -1$, and $r = \sqrt{2}$. The value for y must be negative because the angle terminates in quadrant IV, where y is negative. Now we can give the required trigonometric function values for an angle of 315°.

$$\sin 315° = \frac{-1}{\sqrt{2}} = -\frac{\sqrt{2}}{2}, \qquad \sec 315° = \frac{\sqrt{2}}{1} = \sqrt{2},$$

and

$$\cot 315° = \frac{1}{-1} = -1 \quad \blacktriangleright$$

CAUTION It is very important to choose the correct signs for the quadrant of the angle. Because r is always positive, if an angle has a negative trigonometric function value in a quadrant, then either x or y must be negative. The choice depends on the quadrant: x is negative in quadrants II and III; y is negative in quadrants III and IV. In quadrant III, tangent θ is positive because both x and y are negative there.

The importance of the exact trigonometric function values of 30°, 60°, and 45° angles cannot be overemphasized. It is essential to memorize them. They are summarized in the chart that follows.

Function Values of Special Angles

θ	$\sin \theta$	$\cos \theta$	$\tan \theta$	$\cot \theta$	$\sec \theta$	$\csc \theta$
30°	$\frac{1}{2}$	$\frac{\sqrt{3}}{2}$	$\frac{\sqrt{3}}{3}$	$\sqrt{3}$	$\frac{2\sqrt{3}}{3}$	2
45°	$\frac{\sqrt{2}}{2}$	$\frac{\sqrt{2}}{2}$	1	1	$\sqrt{2}$	$\sqrt{2}$
60°	$\frac{\sqrt{3}}{2}$	$\frac{1}{2}$	$\sqrt{3}$	$\frac{\sqrt{3}}{3}$	2	$\frac{2\sqrt{3}}{3}$

NOTE You should be able to reproduce the 30°–60° and 45°–45° triangles or this chart quickly. It is not difficult to do if you learn the values of sin 30°, sin 45°, and sin 60°. Then complete the rest of the chart using the reciprocal identities or the quotient identities.

CONNECTIONS A convenient way to quickly produce a chart of the trigonometric values for the special angles is shown in the chart below. Write the angles in the first column. In the second column each numerator is a radical with the numbers 0, 1, 2, 3, and 4, in order, placed under it. Each denominator is 2. In the third column each numerator is a radical with the numbers 4, 3, 2, 1, and 0, in order, placed under it. Each denominator is 2. Simplifying these fractions gives the values shown in the chart above for sin θ and cos θ. *Note that this works only for the degree measures shown below and cannot be extended to other values of θ.* The other trigonometric function values are easily found from these basic ones.

θ	$\sin \theta$	$\cos \theta$
0°	$\dfrac{\sqrt{0}}{2}$	$\dfrac{\sqrt{4}}{2}$
30°	$\dfrac{\sqrt{1}}{2}$	$\dfrac{\sqrt{3}}{2}$
45°	$\dfrac{\sqrt{2}}{2}$	$\dfrac{\sqrt{2}}{2}$
60°	$\dfrac{\sqrt{3}}{2}$	$\dfrac{\sqrt{1}}{2}$
90°	$\dfrac{\sqrt{4}}{2}$	$\dfrac{\sqrt{0}}{2}$

FOR DISCUSSION OR WRITING
Verify that the simplified forms of the fractions in the chart agree with the values shown earlier.

USING A CALCULATOR Until now, exact values of the trigonometric functions have been found only for certain special values of θ. The methods used to find these special values were based primarily on geometry. For angles whose function values cannot easily be found by geometry, approximate values can be found with a scientific calculator. Calculators have keys labeled $\boxed{\text{sin}}$, $\boxed{\text{cos}}$, and $\boxed{\text{tan}}$. The reciprocal identities [such as cot θ = 1/(tan θ)] must be used to obtain values of sec, csc, and cot.

EXAMPLE 5
Finding function values with a calculator

Use a calculator to find approximate values of the following trigonometric functions.

(a) sin 49° 12′

Convert 49° 12′ to decimal degrees, as explained in Section 6.1.

$$49° \, 12′ = 49\frac{12}{60}^{°} = 49.2°$$

To eight decimal places,

$$\sin 49° \, 12′ = \sin 49.2° \approx .75699506.$$

(b) sec 97.977°

Calculators do not have secant keys. However,

$$\sec \theta = \frac{1}{\cos \theta}$$

for all angles θ where $\cos \theta \neq 0$. So find sec 97.977° by first finding cos 97.977° and then taking the reciprocal to get

$$\sec 97.977° \approx -7.205879213.$$

(c) cot 51.4283°

Use the identity $\cot \theta = 1/\tan \theta$.

$$\cot 51.4283° \approx .79748114$$

(d) $\sin(-246°) \approx .91354546$

(e) sin 130° 48′

130° 48′ is equal to 130.8°.

$$\sin 130° \, 48′ = \sin 130.8° \approx .75699506 \quad \blacktriangleright$$

Notice that the values found in parts (a) and (e) of Example 5 are the same.

FINDING ANGLES   So far in this section we have used a calculator to find trigonometric function values of angles. This process can be reversed. For now we restrict our attention to angles in the interval [0°, 90°]. An angle is found from its trigonometric function value as shown in the next example.

EXAMPLE 6
Finding angle measures with a calculator

Use a calculator to find a value of θ in the interval [0°, 90°] satisfying each of the following. Leave answers in decimal degrees.

(a) $\sin \theta = .81815000$

We find θ using a key labeled `arc` or `inv` together with the `sin` key. Some calculators may require a key labeled `sin⁻¹` instead. Check your owner's manual to see how your calculator handles this. Again, make sure the calculator is set for degree measure. You should get

$$\theta \approx 54.900028°.$$

(b) sec $\theta = 1.0545829$

Use the identity $\cos \theta = 1/\sec \theta$. Enter 1.0545829 and find the reciprocal. This gives $\cos \theta \approx .9482421913$. Now find θ as shown in part (a). The result is

$$\theta \approx 18.514704°. \quad \blacktriangleright$$

CAUTION Compare Examples 5(b) and 6(b). Note that the reciprocal is used *before* the inverse cosine key when finding the angle, but *after* the cosine key when finding the trigonometric function value.

EXAMPLE 7
Finding grade resistance

When an automobile travels uphill or downhill on a highway, it experiences a force due to gravity. This force F is called **grade resistance** and is computed using the equation $F = W \sin \theta$ where θ is the grade and W is the weight in lb of the automobile.* If the automobile is moving uphill, $\theta > 0$ and if downhill, $\theta < 0$. See Figure 28.

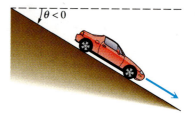

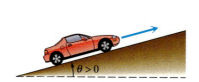

FIGURE 28

(a) Calculate F to two significant digits for a 2500-lb car traveling an uphill grade with $\theta = 2.5°$.

$$F = W \sin \theta = 2500 \sin 2.5° \approx 110 \text{ lb}$$

(b) Calculate F to two significant digits for a 5000-lb truck traveling a downhill grade with $\theta = -6.1°$.

$$F = W \sin \theta = 5000 \sin(-6.1°) \approx -530 \text{ lb}$$

It is negative because the truck is moving downhill.

(c) Calculate F for $\theta = 0°$ and $\theta = 90°$. Do these answers agree with your intuition?

$$F = W \sin \theta = W \sin 0° = W(0) = 0 \text{ lb}$$
$$F = W \sin \theta = W \sin 90° = W(1) = W \text{ lb}$$

This agrees with intuition because if $\theta = 0°$ then there is level ground and gravity does not cause the vehicle to roll. If $\theta = 90°$ the road would be vertical and the full weight of the vehicle would be pulled downward by gravity, so $F = W$. $\blacktriangleright$

---

*Source: Mannering, F., and W. Kilareski, *Principles of Highway Engineering and Traffic Control,* John Wiley & Sons, Inc., 1990.

6.3 Exercises ▼▼▼▼▼▼▼▼▼▼▼▼▼▼▼▼▼▼▼▼▼▼▼▼▼▼▼▼▼▼▼▼

Find expressions for the six trigonometric functions for angle A. See Example 1.

1.

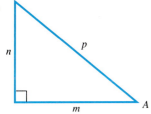

2.

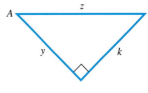

Suppose ABC is a right triangle with sides of lengths a, b, and c with right angle at C. Also, a is opposite angle A and b is opposite angle B. Find the unknown side length using the Pythagorean theorem, and then find the values of the six trigonometric functions for angle B. Rationalize denominators when applicable.

3. $a = 5, b = 12$ **4.** $a = 3, b = 5$ **5.** $a = 6, c = 7$ **6.** $b = 7, c = 12$

Refer to the discussion in this section to give the exact trigonometric function values for each of the following. Do not use a calculator.

7. $\tan 30°$ **8.** $\csc 45°$ **9.** $\sec 45°$ **10.** $\cot 60°$

11. In Example 3, why was 2 a good choice for r? Could any other positive number have been used?

12. In Example 4, is there any reason that $\sqrt{2}$ was an especially good choice for r?

13. Explain why two coterminal angles have the same values for their trigonometric functions.

14. If two angles have the same values for each of the six trigonometric functions, must the angles be coterminal? Explain your reasoning.

Complete the following table with exact trigonometric function values using the methods of this section. See Examples 2–4.

θ	$\sin \theta$	$\cos \theta$	$\tan \theta$	$\cot \theta$	$\sec \theta$	$\csc \theta$
15. 30°	$\dfrac{1}{2}$	$\dfrac{\sqrt{3}}{2}$			$\dfrac{2\sqrt{3}}{3}$	2
16. 45°			1	1		
17. 60°		$\dfrac{1}{2}$	$\sqrt{3}$		2	
18. 120°	$\dfrac{\sqrt{3}}{2}$		$-\sqrt{3}$			$\dfrac{2\sqrt{3}}{3}$
19. 135°	$\dfrac{\sqrt{2}}{2}$	$-\dfrac{\sqrt{2}}{2}$			$-\sqrt{2}$	$\sqrt{2}$
20. 150°		$-\dfrac{\sqrt{3}}{2}$	$-\dfrac{\sqrt{3}}{3}$			2
21. 210°	$-\dfrac{1}{2}$		$\dfrac{\sqrt{3}}{3}$	$\sqrt{3}$		-2
22. 240°	$-\dfrac{\sqrt{3}}{2}$	$-\dfrac{1}{2}$			-2	$-\dfrac{2\sqrt{3}}{3}$

Find the exact values of the six trigonometric functions for each of the following angles. Rationalize denominators when applicable. See Examples 3 and 4.

23. 300°   **24.** 315°   **25.** 330°   **26.** 225°   **27.** 405°

28. 570°   **29.** 390°   **30.** 420°   **31.** −120°   **32.** −30°

33. What trigonometric functions are Y_1 and Y_2?

X	Y₁	Y₂
0	0	0
15	.25882	.26795
30	.5	.57735
45	.70711	1
60	.86603	1.7321
75	.96593	3.7321
90	1	ERROR

X=0

34. What trigonometric functions are Y_1 and Y_2?

X	Y₁	Y₂
0	1	ERROR
15	.96593	3.8637
30	.86603	2
45	.70711	1.4142
60	.5	1.1547
75	.25882	1.0353
90	0	1

X=0

35. What value of A between 0° and 90° will produce the output for the accompanying graphing calculator screen?

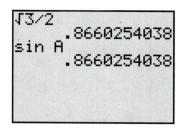

```
√3/2
            .8660254038
sin A
            .8660254038
```

---

36. A student was asked to give the exact value of sin 45°. Using a calculator, he gave the answer .7071067812. The teacher did not give him credit. What was the teacher's reason for this?

37. With a graphing calculator, find the coordinates of the point of intersection of $y = x$ and $y = \sqrt{1 - x^2}$ and give their *exact* forms. (*Hint:* They are irrational.) These coordinates are the cosine and sine of what angle between 0° and 90°?

38. Find the equation of the line passing through the origin and making a 60° angle with the positive x-axis.

39. Find the equation of the line passing through the origin and making a 30° angle with the positive x-axis.

40. Construct an equilateral triangle with each side having length $2k$.
   (a) What is the measure of each angle?
   (b) Label one angle A. Drop a perpendicular from A to the side opposite A. Two 30° angles are formed at A, and two right triangles are formed. What is the length of each side opposite each 30° angle?

(c) What is the length of the perpendicular constructed in part (b)?

(d) From the results of parts (a)–(c), complete the following statement: In a 30°–60° right triangle, the hypotenuse is always _____ times as long as the shorter leg, and the longer leg has a length that is _____ times as long as that of the shorter leg. Also, the shorter leg is opposite the _____ angle, and the longer leg is opposite the _____ angle.

41. Construct a square with each side of length k.
   (a) Draw a diagonal of the square. What is the measure of each angle formed by a side of the square and this diagonal?
   (b) What is the length of the diagonal?
   (c) From the results of parts (a) and (b), complete the following statement: In a 45°–45° right triangle, the hypotenuse has a length that is _____ times as long as either leg.

Use the results of Exercises 40 and 41 to find the exact value of each labeled part in each figure.

42.

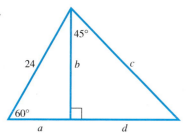

43.

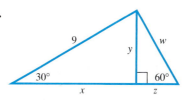

44.

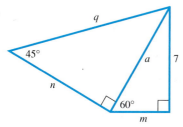

45.

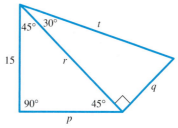

46. Suppose you know the length of one side and one acute angle of a right triangle. Can you determine the measures of all the sides and angles of the triangle?

47. Refer to the table in the Connections box in this section. Explain why this pattern cannot possibly continue past 90°. (*Hint:* What is the maximum value of the sine ratio?)

48. Refer to Figure 23(b). Explain why the length of the side opposite the 30° angle is 1.

Use a calculator to find a decimal approximation for each value. Give as many digits as your calculator displays. In Exercises 61–64, simplify the expression before using the calculator. See Example 5.

49. $\tan 29° \, 30'$

50. $\sin 38° \, 42'$

51. $\cot 41° \, 24'$

52. $\cos 27° \, 10'$

53. $\sec 13.25°$

54. $\csc 44.5°$

55. $\csc 145° \, 45'$

56. $\cot 183° \, 48'$

57. $\cos 421.5°$

58. $\sec 312.2°$

59. $\tan(-80° \, 6')$

60. $\sin(-317° \, 36')$

61. $\dfrac{1}{\sec 14.8°}$

62. $\dfrac{1}{\cot 23.4°}$

63. $\dfrac{\sin 33°}{\cos 33°}$

64. $\dfrac{\cos 77°}{\sin 77°}$

Tell whether each statement is true or false. If false, tell why. Use a calculator in Exercises 69–72.

65. $\sin 30° + \sin 60° = \sin(30° + 60°)$

66. $\sin(30° + 60°) = \sin 30° \cdot \cos 60° + \sin 60° \cdot \cos 30°$

67. $\cos 60° = 2 \cos^2 30° - 1$

68. $\cos 60° = 2 \cos 30°$

69. $\cos 40° = 2 \cos 20°$

70. $\sin 10° + \sin 10° = \sin 20°$

71. $\cos 70° = 2 \cos^2 35° - 1$

72. $\sin 50° = 2 \sin 25° \cdot \cos 25°$

Find a value of θ in [0°, 90°) that satisfies the statement. Leave your answer in decimal degrees. See Example 6.

73. $\sin \theta = .84802194$

74. $\tan \theta = 1.4739716$

75. $\sec \theta = 1.1606249$

76. $\cot \theta = 1.2575516$

77. $\sec \theta = 2.7496222$

78. $\sin \theta = .27843196$

79. What value of A between 0° and 90° will produce the output for the accompanying graphing calculator screen, assuming the calculator is in degree mode?

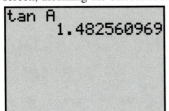

80. What value of A will produce the output for the accompanying graphing calculator screen, assuming the calculator is in degree mode?

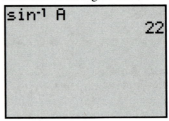

81.

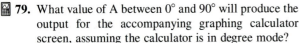

If aerodynamic resistance is ignored, the braking distance D (in feet) for an automobile to change its velocity from V_1 to V_2 (feet per second) can be calculated using the equation

$$D = \frac{1.05(V_1^2 - V_2^2)}{64.4(K_1 + K_2 + \sin \theta)}.$$

K_1 is a constant determined by the efficiency of the brakes and tires, K_2 is a constant determined by the rolling resistance of the automobile, and θ is the grade of the highway. (*Source:* Mannering, F., and W. Kilareski, *Principles of Highway Engineering and Traffic Control,* John Wiley & Sons, Inc., 1990.)

(a) Compute the number of feet required to slow a car from 55 to 30 miles per hour while traveling uphill with a grade of $\theta = 3.5°$. Let $K_1 = .4$ and $K_2 = .02$. (*Hint:* Change miles per hour to feet per second.)

(b) Repeat part (a) with $\theta = -2°$.

(c) How is braking distance affected by the grade θ? Does this agree with your driving experience?

82.

(Refer to Exercise 81.) An automobile is traveling at 90 miles per hour on a highway with a downhill grade of $\theta = -3.5°$. The driver sees a stalled truck in the road 200 feet away and immediately applies the brakes. Assuming that a collision cannot be avoided, how fast (in miles per hour) is the car traveling when it hits the truck? (Use the values for K_1 and K_2 given in Exercise 81.)

When a light ray travels from one medium, such as air, to another medium, such as water or glass, the speed of the light changes, and the direction that the ray is traveling changes. (This is why a fish under water is in a different position than it appears to be.) These changes are given by Snell's law

$$\frac{c_1}{c_2} = \frac{\sin \theta_1}{\sin \theta_2},$$

where c_1 is the speed of light in the first medium, c_2 is the speed of light in the second medium, and θ_1 and θ_2 are the angles shown in the figure. In the following exercises, assume that $c_1 = 3 \times 10^8$ m per sec. Find the speed of light in the second medium.

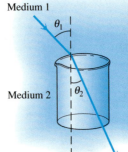

Medium 1 — If this medium is less dense, light travels at a faster speed, c_1.

Medium 2 — If this medium is more dense, light travels at a slower speed, c_2.

83. $\theta_1 = 46°, \theta_2 = 31°$

84. $\theta_1 = 39°, \theta_2 = 28°$

Find θ_2 for the following values of θ_1 and c_2. Round to the nearest degree.

85. $\theta_1 = 40°$, $c_2 = 1.5 \times 10^8$ m per sec

86. $\theta_1 = 62°$, $c_2 = 2.6 \times 10^8$ m per sec

The figure below shows a fish's view of the world above the surface of the water. Suppose that a light ray comes from the horizon, enters the water, and strikes the fish's eye. (Source: Walker, Jearl, "The Amateur Scientist," Scientific American, March 1984. Copyright © 1984 by Scientific American, Inc. Reprinted by permission of Scientific American, Inc. All rights reserved.)

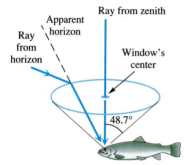

87. Let us assume that this ray gives a value of 90° for angle θ_1 in the formula for Snell's law. (In a practical situation this angle would probably be a little less than 90°.) The speed of light in water is about 2.254×10^8 m per sec. Find angle θ_2.

   (Your result should have been about 48.7°. This means that a fish sees the world above the water as a cone, making an angle of 48.7° with the vertical.)

88. Suppose an object is located at a true angle of 29.6° above the horizon. Find the apparent angle above the horizon to a fish.

   Refer to Example 7 for Exercises 89–94.

89. What is the grade resistance of a 2400-pound car traveling on a $-2.4°$ downhill grade?

90. What is the grade resistance of a 2100-pound car traveling on a 1.8° uphill grade?

91. A car traveling on a $-3°$ downhill grade has a grade resistance of -145 pounds. What is the weight of the car?

92. A 2600-pound car traveling downhill has a grade resistance of 130 pounds. What is the angle of the grade?

93. Which has the greater grade resistance: a 2200-pound car on a 2° uphill grade or a 2000-pound car on a 2.2° uphill grade?

94. Complete the table for the values of $\sin \theta$, $\tan \theta$, and $\pi\theta/180$ to four decimal places.

θ	$\sin \theta$	$\tan \theta$	$\dfrac{\pi\theta}{180}$
0°			
.5°			
1°			
1.5°			
2°			
2.5°			
3°			
3.5°			
4°			

(a) How do $\sin \theta$, $\tan \theta$, and $\pi\theta/180$ compare for small grades of θ?

(b) Highway grades are usually small. Give two approximations to the grade resistance $F = W \sin \theta$ that do not use the sine function.

(c) A stretch of highway has a 4-ft vertical rise for every 100 ft of horizontal run. Use an approximation from part (a) to estimate the grade resistance for a 2000-lb car on this stretch of highway.

(d) A stretch of highway has a 3.75° grade. Without calculating a trigonometric function, estimate the grade resistance for an 1800-lb car on this stretch of highway.

95. When a highway goes downhill and then uphill it is said to have a *sag curve.* Sag curves are designed so that at night, headlights shine sufficiently far down the road to allow for a safe stopping distance. See the figure. The minimum length L of a sag curve is determined by the height h of the car's headlights above the pavement, the downhill grade $\theta_1 < 0°$, the uphill grade $\theta_2 > 0°$, and the safe stopping distance S for a given speed limit. In addition, L is dependent on the vertical alignment of the headlights. Headlights are usually pointed upward at a slight angle α above the horizontal of the car. Using these quantities, L can then be computed using the formula $L = \dfrac{(\theta_2 - \theta_1)S^2}{200(h + S \tan \alpha)}$ where $S < L$. (*Source:* Mannering, F., and W.

Kilareski, *Principles of Highway Engineering and Traffic Control,* John Wiley & Sons, Inc., 1990.)

(a) Compute L for a 55 mile per hour speed limit where $h = 1.9$ ft, $\alpha = .9°$, $\theta_1 = -3°$, $\theta_2 = 4°$, and $S = 336$ feet.

(b) Repeat part (a) with $\alpha = 1.5°$.

(c) How does the alignment of the headlights affect the value of L?

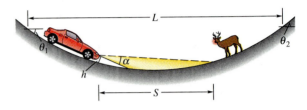

6.4 Solving Right Triangles ▼▼▼

Many applications require finding a measurement that cannot be measured directly, such as the height of a tree or flagpole or the angle formed between the horizontal and the line of sight to the top of a building. These measurements are often found by solving right triangles.

To solve a triangle means to find the measures of all the angles and sides of the triangle. Before we solve triangles, a short discussion concerning accuracy and significant digits is appropriate. If a wall measured to the nearest foot is found to be 18 feet long, actually this means that the wall has a length between 17.5 feet and 18.5 feet. If the wall is measured more accurately and found to be 18.3 feet long, then its length is really between 18.25 feet and 18.35 feet. A measurement of 18.00 feet would indicate that the length of the wall is between 17.995 feet and 18.005 feet. The measurement 18 feet is said to have two **significant digits** of accuracy; 18.0 has three significant digits, and 18.00 has four.

What about the measurement 900 meters? We cannot tell whether this represents a measurement to the nearest meter, ten meters, or hundred meters. To avoid this problem the number can be written in scientific notation as 9.00×10^2 to the nearest meter, 9.0×10^2 to the nearest ten meters, or 9×10^2 to the nearest hundred meters. These three cases have 3, 2, and 1 significant digits, respectively.

A significant digit is a digit obtained by actual measurement. A number that represents the result of counting, or a number that results from theoretical work and is not the result of a measurement, is an **exact number.** There are fifty states in the United States, so 50 is an exact number. Most values of trigonometric functions used in applications are approximations, and virtually all measurements are approximations. To perform calculations on such approximate numbers, follow the rules given below.

Calculation with Significant Digits

For *adding* and *subtracting,* round the answer so that the last digit you keep is in the right-most column in which all the numbers have significant digits.

For *multiplying* or *dividing,* round the answer to the least number of significant digits found in any of the given numbers.

For *powers* and *roots,* round the answer so that it has the same number of significant digits as the number whose power or root you are finding.

When solving triangles, use the following table for deciding on significant digits in angle measure.

Significant Digits for Angles

Number of Significant Digits	Angle Measure to Nearest:
2	Degree
3	Ten minutes, or nearest tenth of a degree
4	Minute, or nearest hundredth of a degree
5	Tenth of a minute, or nearest thousandth of a degree

For example, an angle measuring 52° 30′ has three significant digits (assuming that 30′ is measured to the nearest ten minutes).

In using trigonometry to solve triangles, a labeled sketch is an important aid. It is conventional to use a to represent the length of the side opposite angle A, b for the length of the side opposite angle B, and so on. As mentioned earlier, in a right triangle the letter c is reserved for the hypotenuse. Figure 29 shows the labeling of a typical right triangle.

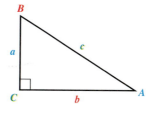

FIGURE 29

EXAMPLE 1
Solving a right triangle

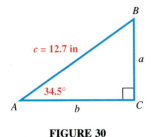

FIGURE 30

Solve right triangle ABC, with $A = 34.5°$ and $c = 12.7$ in. See Figure 30.

To solve the triangle, find the measures of the remaining sides and angles. The value of a can be found with a trigonometric function involving the known values of angle A and side c. Since the sine of angle A is given by the quotient of the side opposite A and the hypotenuse, use $\sin A$.

$$\sin A = \frac{a}{c}$$

Substituting known values gives

$$\sin 34.5° = \frac{a}{12.7},$$

or, upon multiplying both sides by 12.7,

$$a = 12.7 \sin 34.5°$$
$$a = 12.7(.56640624) \qquad \text{Use a calculator.}$$
$$a = 7.19 \text{ in.}$$

The value of b could be found with the Pythagorean theorem. It is better, however, to use the information given in the problem rather than a result just calculated. If a mistake were made in finding a, then b also would be incorrect. Also, rounding more than once may cause the result to be less accurate. Using $\cos A$ gives

$$\cos A = \frac{\text{side adjacent}}{\text{hypotenuse}} = \frac{b}{c}$$

$$\cos 34.5° = \frac{b}{12.7}$$

$$b = 12.7 \cos 34.5°$$

$$b = 10.5 \text{ in.}$$

Once b has been found, the Pythagorean theorem could be used as a check. All that remains to solve triangle ABC is to find the measure of angle B. Since $A + B = 90°$ and $A = 34.5°$,

$$A + B = 90°$$
$$B = 90° - A$$
$$B = 90° - 34.5°$$
$$B = 55.5°. \quad \blacktriangleright$$

NOTE In Example 1 we could have started by finding the measure of angle B and then used the trigonometric function values of B to find the unknown sides. The process of solving a right triangle (like many problems in mathematics) can usually be done in several ways, each resulting in the correct answer. However, in order to retain as much accuracy as can be expected, always use given information as much as possible, and avoid rounding off in intermediate steps.

EXAMPLE 2
Using trigonometry to measure a distance

A method that surveyors use to determine a small distance d between two points P and Q is called the **subtense bar method.** The subtense bar with length b is centered at Q and situated perpendicular to the line of sight between P and Q.* See Figure 31. The angle θ is measured and then the distance d can be determined.

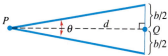

FIGURE 31

(a) Find d when $\theta = 1° \, 23' \, 12''$ and $b = 2$ meters.
From Figure 31 we see that

$$\cot \frac{\theta}{2} = \frac{d}{b/2}$$

$$d = \frac{b}{2} \cot \frac{\theta}{2}.$$

To evaluate $\theta/2$, we change to decimal degrees: $1° \, 23' \, 12'' \approx 1.386667°$, so

$$d = \frac{2}{2} \cot \frac{1.386667°}{2} \approx 82.6341 \text{ m.}$$

(b) The angle θ usually cannot be measured more accurately than to the nearest $1''$. How much change would there be in the value of d if θ were measured $1''$ larger?
Use $\theta = 1° \, 23' \, 13'' \approx 1.386944°$.

$$d = \frac{2}{2} \cot \frac{1.386944°}{2} \approx 82.6176 \text{ m.}$$

The error is $82.6341 - 82.6176 \approx .017$ m. ▶

■ **PROBLEM SOLVING**   The process of solving right triangles is easily adapted to solving applied problems. A crucial step in such applications involves sketching the triangle and labeling the given parts correctly. Then we can use the methods described in the earlier examples to find the unknown value or values. ■

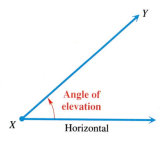

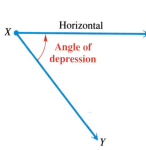

FIGURE 32

Many applications of right triangles involve the angle of elevation or the angle of depression. The **angle of elevation** from point X to point Y (above X) is the acute angle formed by ray XY and a horizontal ray with endpoint at X. The angle of elevation is always measured to or from the horizontal. See Figure 32. The **angle of depression** from point X to point Y (below X) is the angle made by ray XY and a horizontal ray with endpoint X. Again, see Figure 32.

*Source: Mueller, I. and K. Ramsayer, *Introduction to Surveying,* Frederick Ungar Publishing Co., New York, 1979.

CAUTION Errors are often made in interpreting the angle of depression. Remember that both the angle of elevation *and* the angle of depression are measured between the line of sight and the horizontal.

◀EXAMPLE 3
Finding the angle of elevation when lengths are known

The length of the shadow of a building 34.09 meters tall is 37.62 meters. Find the angle of elevation of the sun.

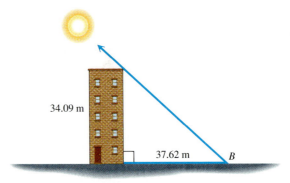

FIGURE 33

As shown in Figure 33, the angle of elevation of the sun is angle *B*. Since the side opposite *B* and the side adjacent to *B* are known, use the tangent ratio to find *B*.

$$\tan B = \frac{34.09}{37.62}$$

$$B = 42.18° \qquad \textcolor{blue}{\text{Use inverse tangent.}}$$

The angle of elevation of the sun is 42.18°. ▶

◀EXAMPLE 4
Finding a height given two angles of elevation

Francisco needs to know the height of a tree. From a given point on the ground he finds that the angle of elevation to the top of the tree is 36.7°. He then moves back 50 feet. From the second point, the angle of elevation to the top of the tree is 22.2°. See Figure 34. Find the height of the tree.

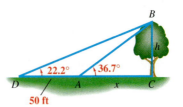

FIGURE 34

The figure shows two unknowns: x, the distance from the center of the trunk of the tree to the point where the first observation was made, and h, the height of the tree. Since nothing is given about the length of the hypotenuse of either triangle ABC or triangle BCD, use a ratio that does not involve the hypotenuse—the tangent.

$$\text{In triangle } ABC, \quad \tan 36.7° = \frac{h}{x} \quad \text{or} \quad h = x \tan 36.7°.$$

$$\text{In triangle } BCD, \quad \tan 22.2° = \frac{h}{50 + x} \quad \text{or} \quad h = (50 + x) \tan 22.2°.$$

Since each expression equals h, these expressions must be equal. Thus,

$$x \tan 36.7° = (50 + x) \tan 22.2°.$$

Now use algebra to solve for x.

$$x \tan 36.7° = 50 \tan 22.2° + x \tan 22.2° \qquad \text{Distributive property}$$

$$x \tan 36.7° - x \tan 22.2° = 50 \tan 22.2° \qquad \text{Get } x \text{ terms on one side.}$$

$$x(\tan 36.7° - \tan 22.2°) = 50 \tan 22.2° \qquad \text{Factor out } x \text{ on the left.}$$

$$x = \frac{50 \tan 22.2°}{\tan 36.7° - \tan 22.2°} \qquad \text{Divide by the coefficient of } x.$$

We saw above that $h = x \tan 36.7°$. Substituting for x,

$$h = \left(\frac{50 \tan 22.2°}{\tan 36.7° - \tan 22.2°}\right)(\tan 36.7°).$$

From a calculator,

$$\tan 36.7° = .74537703$$

$$\tan 22.2° = .40809244$$

so

$$\tan 36.7° - \tan 22.2° = .74537703 - .40809244 = .33728459$$

and

$$h = \left(\frac{50(.40809244)}{.33728459}\right)(.74537703) = 45 \text{ (rounded)}.$$

The height of the tree is approximately 45 feet. ▶

NOTE In practice we usually do not write down the intermediate calculator approximation steps. However, we have done this in Example 4 so that the reader may follow the steps more easily.

CONNECTIONS An alternative approach to solving the problem in Example 4 uses the intersection of graphs capability of the graphing calculator. This approach is based on a similar solution proposed by a student, John Cree, as explained in a letter written by Cree's teacher, Robert Ruzich, in the January 1995 issue of *Mathematics Teacher*.*

In the figure below, we have superimposed Figure 34 on a coordinate axis with the origin at D. Since the tangent of the angle between the x-axis and the graph of a line with equation $y = mx + b$ is the slope of the line, m, the segment BD lies on the graph of $Y_1 = (\tan 22.2°)x$. (Here, the value of b is 0.) By similar reasoning, we can find an equation of the line containing segment AB. Then we can find the point of intersection of the two graphs. The y-coordinate of this point gives the length of segment BC, which is the height of the tree.

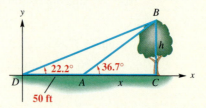

FOR DISCUSSION OR WRITING

1. Find the equation of the line containing segment AB, following the reasoning we used to get Y_1.
2. Use the intersection of graphs method to find the coordinates of the point of intersection. Use a window of [0, 200] by [0, 100]. Does the y-value agree with the solution in Example 4?
3. Explain why the tangent of angle D in the figure gives the slope of the line containing segment BD.

Other applications of right triangles involve **bearing,** an important idea in navigation. There are two common ways to express bearing. *When a single angle is given, such as* 164°, *it is understood that the bearing is measured in a clockwise direction from due north.* Several sample bearings using this first type of system are shown in Figure 35.

NOTE   In the following examples and exercises, the problems all result in right triangles, so the methods of this section apply. Chapter 8 will include problems involving bearing that result in triangles that are *not* right triangles and require other methods to solve.

_____
* Robert Ruzich, letter to the editor, *Mathematics Teacher,* January 1995. Reprinted by permission.

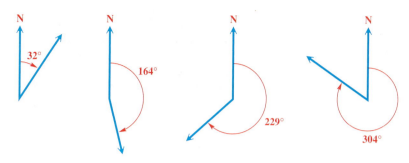

FIGURE 35

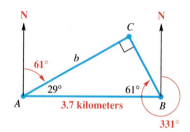

FIGURE 36

> **EXAMPLE 5**
> Solving a problem involving bearing (first type)

Radar stations A and B are on an east-west line, 3.7 kilometers apart. Station A detects a plane at C, on a bearing of 61°. Station B simultaneously detects the same plane, on a bearing of 331°. Find the distance from A to C.

Draw a sketch showing the given information, as in Figure 36. Since a line drawn due north is perpendicular to an east-west line, right angles are formed at A and B, so that angles CAB and CBA can be found. Angle C is a right angle because angles CAB and CBA are complementary. (If C were not a right angle, the methods of Chapter 8 would be needed.) Find distance b by using the cosine function.

$$\cos 29° = \frac{b}{3.7}$$
$$3.7 \cos 29° = b$$
$$b = 3.2 \text{ kilometers}$$

Use a calculator and round to the nearest tenth. ▶

The second common system for expressing bearing starts with a north-south line and uses an acute angle to show the direction, either east or west, from this line. Figure 37 shows several sample bearings using this system. Either N or S always comes first, followed by an acute angle, and then E or W.

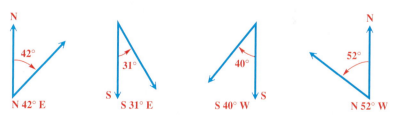

FIGURE 37

EXAMPLE 6
Solving a problem involving bearing (second type)

The bearing from A to C is S 52° E. The bearing from A to B is N 84° E. The bearing from B to C is S 38° W. A plane flying at 250 miles per hour takes 2.4 hours to go from A to B. Find the distance from A to C.

Make a sketch of the situation. First draw the two bearings from point A. Choose a point B on the bearing N 84° E from A and draw the bearing to C. Point C will be located where the bearing lines from A and B intersect as shown in Figure 38.

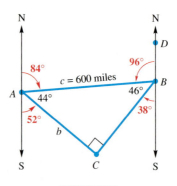

FIGURE 38

Since the bearing from A to B is N 84° E, angle ABD is $180° - 84° = 96°$. Thus, angle ABC is 46°. Also, angle BAC is $180° - (84° + 52°) = 44°$. Angle C is $180° - (44° + 46°) = 90°$. From the statement of the problem, a plane flying at 250 miles per hour takes 2.4 hours to go from A to B. The distance from A to B is the product of rate and time, or

$$c = \text{rate} \times \text{time} = 250(2.4) = 600 \text{ miles.}$$

To find b, the distance from A to C, use the sine. (The cosine could also have been used.)

$$\sin 46° = \frac{b}{c}$$

$$\sin 46° = \frac{b}{600}$$

$$600 \sin 46° = b$$

$$b \approx 430 \text{ miles} \quad \blacktriangleright$$

6.4 Exercises ▼▼▼▼▼▼▼▼▼▼▼▼▼▼▼▼▼▼▼▼▼▼▼▼▼▼▼▼▼▼▼▼▼▼▼▼

Refer to the discussion of accuracy and significant digits in this section to work the first six exercises.

1. What is the difference between a measurement of 23.0 feet and a measurement of 23.00 feet?

2. What number indicates a measurement between 25.95 and 26.05 pounds?

Fill in the blanks in Exercises 3 and 4.

3. If h is the actual height of a building and the height is measured as 58.6 ft, then $|h - 58.6| \leq$ ___.

4. If w is the actual weight of a new small car and the weight is measured as 15.00×10^2 pounds, then $|w - 1500| \leq$ ___.

Find the error in each statement.

5. I have 2 bushel baskets, each containing 65 apples. I know that $2 \times 65 = 130$, but 2 has only one significant digit, so I must write the answer as 1×10^2, or 100. I therefore have 100 apples.

6. The formula for the circumference of a circle is $C = 2\pi r$. My circle has a radius of 54.98 cm, and my calculator has a $\boxed{\pi}$ key, giving fifteen digits of accuracy. Pressing the right buttons gives 345.44953. Because 2 has only one significant digit, however, the answer must be given as 3×10^2, or 300 cm. (What is the correct answer?)

In the remaining exercises in this set, use a calculator as necessary.

Solve each right triangle. See Example 1.

7.

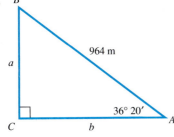

8.

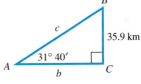

9.

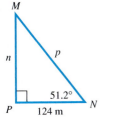

10.

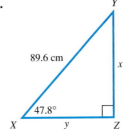

In Exercises 11 and 12, assume the calculator is in degree mode.

11. Make up a right triangle problem whose solution is obtained from the accompanying graphing calculator screen.

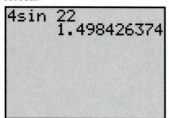

12. Make up a right triangle problem whose solution is obtained from the accompanying graphing calculator screen.

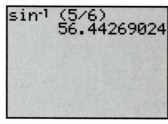

13. Can a right triangle be solved if we are given the measures of its two acute angles and no side lengths? Explain.

14. If we are given an acute angle and a side in a right triangle, what unknown part of the triangle requires the least work to find?

15. Explain why you can always solve a right triangle if you know the measures of one side and one acute angle.

16. Explain why you can always solve a right triangle if you know the lengths of two sides.

Solve each right triangle. In each case, C = 90°. If the angle information is given in degrees and minutes, give the answers in the same way. If given in decimal degrees, do likewise in your answers. When two sides are given, give answers in degrees and minutes. See Example 1.

17. $A = 28.00°$, $c = 17.4$ ft

18. $B = 46.00°$, $c = 29.7$ m

19. $B = 73.00°$, $b = 128$ in

20. $A = 61° 00'$, $b = 39.2$ cm

21. $a = 76.4$ yd, $b = 39.3$ yd

22. $a = 958$ m, $b = 489$ m

Solve each problem.

23. A 13.5-m fire-truck ladder is leaning against a wall. Find the distance the ladder goes up the wall if it makes an angle of $43° 50'$ with the ground.

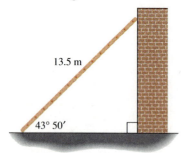

13.5 m

43° 50'

24. A guy wire 77.4 m long is attached to the top of an antenna mast that is 71.3 m high. Find the angle that the wire makes with the ground.

25. To find the distance *RS* across a lake, a surveyor lays off $RT = 53.1$ m, with angle $T = 32° 10'$, and angle $S = 57° 50'$. Find length *RS*.

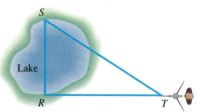

S

Lake

R   T

26. The length of the base of an isosceles triangle is 42.36 in. Each base angle is $38.12°$. Find the length of each of the two equal sides of the triangle. (*Hint:* Divide the triangle into two right triangles.)

27. Find the altitude of an isosceles triangle having a base of 184.2 cm if the angle opposite the base is $68° 44'$.

28. To determine the diameter of the sun, an astronomer might sight with a transit (a device used by surveyors for measuring angles) first to one edge of the sun and then to the other, finding that the included angle equals 1° 4′. Assuming that the distance from Earth to the sun is 92,919,800 mi, calculate the diameter of the sun. (See the figure.)

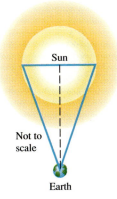

29. Find the minimum height h above the surface of Earth so that a pilot at point A in the figure can see an object on the horizon at C, 125 mi away. Assume that the radius of Earth is 4.00×10^3 mi.

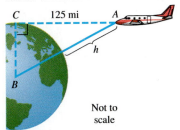

30. A tunnel is to be dug from A to B (see the figure). Both A and B are visible from C. If AC is 1.4923 mi and BC is 1.0837 mi, and if C is 90°, find the measures of angles A and B.

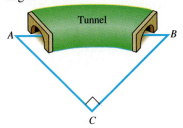

31. A piece of land is shaped as shown in the figure. Find x.

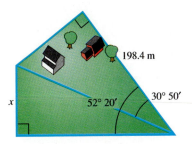

32. The highest mountain peak in the world is Mt. Everest located in the Himalayas. The height of this enormous mountain was determined in 1856 by surveyors using trigonometry long before it was first climbed in 1953. This difficult measurement had to be done at a great distance. At an altitude of 14,545 feet on a different mountain, the straight line distance to the peak of Mt. Everest is 27.0134 miles and its angle of elevation is $\theta = 5.82°$. See the figure below. (*Source:* Dunham, W., *The Mathematical Universe,* John Wiley & Sons, Inc., 1994.)

(a) Approximate the height of Mt. Everest.

(b) In the actual measurement, Mt. Everest was over 100 miles away and the curvature of Earth had to be taken into account. Would the curvature of Earth make the peak appear taller or shorter than it actually is?

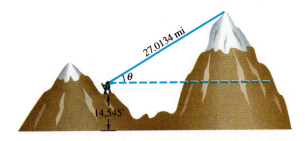

33. When is an angle of elevation 90°?

34. Can an angle of elevation be more than 90°?

35. Explain why the angle of depression *DAB* has the same measure as the angle of elevation *ABC* in the accompanying figure.

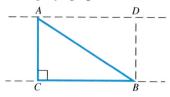

AD is parallel to *BC*.

36. Why is angle *CAB* *not* an angle of depression in the figure for Exercise 35?

Work each problem involving angles of elevation or depression. See Examples 3 and 4.

37. Suppose the angle of elevation of the sun is 23.4°. Find the length of the shadow cast by Cindy Newman, who is 5.75 ft tall.

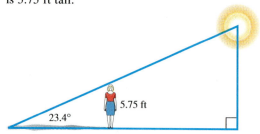

38. The shadow of a vertical tower is 40.6 m long when the angle of elevation of the sun is 34.6°. Find the height of the tower.

39. A company safety committee has recommended that a floodlight be mounted in a parking lot so as to illuminate the employee exit (see the figure). Find the angle of depression of the light.

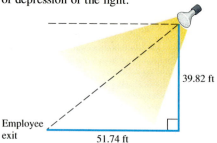

40. In one area, the lowest angle of elevation of the sun in winter is 23° 20′. Find the minimum distance *x* that a plant needing full sun can be placed from a fence 4.65 ft high (see the figure).

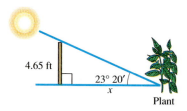

41. An airplane is flying 10,500 feet above level ground. The angle of depression from the plane to the base of a tree is 13° 50′. How far horizontally must the plane fly to be directly over the tree?

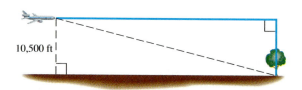

42. The angle of elevation from the top of a small building to the top of a nearby taller building is 46° 40′, while the angle of depression to the bottom is 14° 10′. If the smaller building is 28.0 m high, find the height of the taller building.

An observer for a radar station is located at the origin of a coordinate system. For each of the points in Exercises 43–46, find the bearing for an airplane located at that point. Express the bearing using both methods.

43. $(-4, 0)$ **44.** $(-3, -3)$

45. $(-5, 5)$ **46.** $(0, -2)$

47. The ray $y = x$, $x \geq 0$ contains the origin and all points in the coordinate system whose bearing from the origin is 45°. Determine the equation of a ray consisting of the origin and all points whose bearing from the origin is 240°.

48. Repeat Exercise 47 for a bearing of 150°.

Work each problem. In these exercises, assume the course of a plane or ship is on the indicated bearing. See Examples 5 and 6.

49. A plane flies 1.3 hr at 110 mph on a bearing of 40°. It then turns and flies 1.5 hr at the same speed on a bearing of 130°. How far is the plane from its starting point?

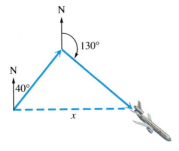

50. A ship travels 50 km on a bearing of 27°, and then travels on a bearing of 117° for 140 km. Find the distance traveled from the starting point to the ending point.

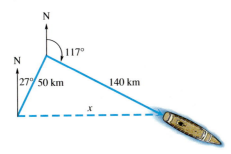

51. Two lighthouses are located on a north-south line. From lighthouse *A* the bearing of a ship 3742 m away is 129° 43′. From lighthouse *B* the bearing of the ship is 39° 43′. Find the distance between the lighthouses.

52. A ship leaves its home port and sails on a bearing of N 28° 10′ E. Another ship leaves the same port at the same time and sails on a bearing of S 61° 50′ E. If the first ship sails at 24.0 mph and the second sails at 28.0 mph, find the distance between the two ships after 4 hr.

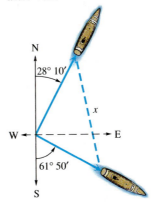

53. Radio direction finders are set up at points *A* and *B*, which are 2.50 mi apart on an east-west line. From *A* it is found that the bearing of the signal from a radio transmitter is N 36° 20′ E, while from *B* the bearing of the same signal is N 53° 40′ W. Find the distance of the transmitter from *B*.

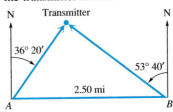

54. The bearing from Atlanta to Macon is S 27° E, and the bearing from Macon to Augusta is N 63° E. A car traveling at 60 mph needs 1 1/4 hr to go from Atlanta to Macon and 1 3/4 hr to go from Macon to Augusta. Find the distance from Atlanta to Augusta.

55. Solve the equation $ax = b + cx$ for x in terms of a, b, and c. (*Note:* The is in essence the calculation carried out in Example 4.)

56. Explain why the line $y = (\tan \theta)(x - a)$ passes through the point $(a, 0)$ and makes an angle of $\theta°$ with the x-axis.

In Exercises 57 and 58 use the method of Example 4 and then check your answer with the graphing method described in the Connections discussion. Drawing a sketch for these problems where one is not given may be helpful.

57. Find *h* as indicated in the figure.

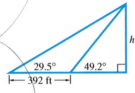

29.5°   49.2°

|← 392 ft →|

58. Find *h* as indicated in the figure.

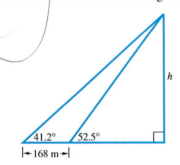

h

41.2°   52.5°

|← 168 m →|

59. The angle of elevation from a point on the ground to the top of a pyramid is 35° 30'. The angle of elevation from a point 135 ft farther back to the top of the pyramid is 21° 10'. Find the height of the pyramid.

60. Debbie Maybury, a whale researcher standing at the top of a tower, is watching a whale approach the tower directly. When she first begins watching the whale, the angle of depression of the whale is 15° 50'. Just as the whale turns away from the tower, the angle of depression is 35° 40'. If the height of the lighthouse is 68.7 m, find the distance traveled by the whale as it approaches the tower.

61. (Refer to Example 2.) A variation of the subtense bar method that surveyors use to determine larger distances *d* between two

points *P* and *Q* is shown in the figure. In this case the subtense bar with length *b* is placed between the points *P* and *Q* so that the bar is centered on and perpendicular to the line of sight connecting *P* and *Q*. The angles α and β are measured from points *P* and *Q*, respectively. (*Source:* Mueller, I., and K. Ramsayer, *Introduction to Surveying,* Frederick Ungar Publishing Co., New York, 1979.)

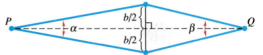

(**a**) Find a formula for *d* involving α, β, and *b*.
(**b**) Use your formula to determine *d* if $\alpha = 37'\ 48''$, $\beta = 42'\ 3''$, and *b* = 2 meters.

62. 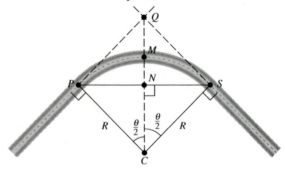 A basic highway curve connecting two straight sections of road is often circular. See the figure. The points *P* and *S* mark the beginning and end of the curve. Let *Q* be the point of intersection where the two straight sections of highway leading into the curve would meet if extended. The radius of the curve is *R* and the central angle θ denotes how many degrees the curve turns. (*Source:* Mannering, F., and W. Kilareski, *Principles of Highway Engineering and Traffic Control,* John Wiley & Sons, Inc., 1990.)

(**a**) If *R* = 965 ft and $\theta = 37.0°$, find the distance *d* between *P* and *Q*.
(**b**) Find an expression in terms of *R* and θ for the distance between points *M* and *N*.

63. 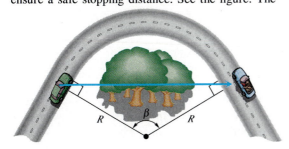 (Refer to Exercise 62.) When an automobile travels along a circular curve, objects like trees and buildings situated on the inside of the curve can obstruct a driver's vision. These obstructions prevent the driver from seeing sufficiently far down the highway to ensure a safe stopping distance. See the figure. The minimum distance d that should be cleared on the inside of the highway is given by the equation $d = R(1 - \cos(\beta/2))$. (*Source:* Mannering, F., and W. Kilareski, *Principles of Highway Engineering and Traffic Control,* John Wiley & Sons, Inc., 1990.)

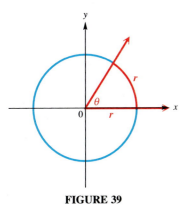

(a) It can be shown that if β is measured in degrees then $\beta \approx 57.3S/R$ where S is the safe stopping distance for the given speed limit. Compute d for a 55 mile per hour speed limit if $S = 336$ feet and $R = 600$ feet.

(b) Compute d for a 65 mile per hour speed limit if $S = 485$ feet.

(c) How does the speed limit affect the amount of land that should be cleared on the inside of the curve?

6.5 Radian Measure ▼▼▼

In most work involving applications of trigonometry, angles are measured in degrees. In more advanced work in mathematics, the use of *radian measure* of angles is preferred. Radian measure also allows us to treat our familiar trigonometric functions as functions with domains of *real numbers,* rather than angles.

Figure 39 shows an angle θ in standard position along with a circle of radius r. The vertex of θ is at the center of the circle. Angle θ intercepts an arc on the circle equal in length to the radius of the circle. Because of this, the measure of angle θ is defined as one radian.

FIGURE 39

> **Radian**
>
> An angle that has its vertex at the center of a circle and that intercepts an arc on the circle equal in length to the radius of the circle has a measure of **one radian.**

The circumference of a circle, the distance around the circle, is given by $C = 2\pi r$, where r is the radius of the circle. The formula $C = 2\pi r$ shows that the radius can be laid off 2π times around a circle. Therefore, an angle of 360°, which corresponds to a complete circle, intercepts an arc equal in length to 2π times the radius of the circle. Because of this, an angle of 360° has a measure of 2π radians:

$$360° = 2\pi \text{ radians.}$$

An angle of 180° is half the size of an angle of 360°, so an angle of 180° has half the radian measure of an angle of 360°.

$$180° = \frac{1}{2}(2\pi) \text{ radians}$$

Degree/Radian Relationship

$$180° = \pi \text{ radians}$$

We can use this simple relationship to convert from degree measure to radian measure or from radians to degrees. It helps to remember that radians are associated with π and degrees with 180°. We convert by using ratios as shown in the next example.

◀ EXAMPLE 1
Converting degrees to radians

Convert each degree measure to radians.

(a) 45°

Let r represent the radian measure. Use the proportion

$$\frac{r}{\pi} = \frac{45}{180}.$$

Solve for r and reduce the fraction.

$$r = \frac{45\pi}{180} = \frac{\pi}{4}$$

(b) 249.8°

$$\frac{r}{\pi} = \frac{249.8}{180}$$

$$r = \frac{249.8\pi}{180} \approx 4.360$$

to the nearest thousandth. Note that in the second step of each example the radian measure r equals the degree measure multiplied by $\pi/180°$. ▶

EXAMPLE 2
Converting radians to degrees

Convert each of the following radian measures to degrees.

(a) $\dfrac{9\pi}{4}$

Let d represent the degree measure.

$$\frac{\frac{9\pi}{4}}{\pi} = \frac{d}{180°}$$

$$\frac{9\pi}{4} \cdot \frac{1}{\pi} = \frac{d}{180°}$$

$$\frac{9}{4} = \frac{d}{180°}$$

$$d = \frac{9}{4} \cdot 180° = 405°$$

As a shortcut method, since π radians $= 180°$, when converting from radian measure given as a multiple of π, simply replace π with $180°$ and simplify.

(b) 4.25 (Give the answer in decimal degrees.)

$$\frac{4.25}{\pi} = \frac{d}{180°}$$

$$d = \frac{(4.25)180°}{\pi} \approx 243.5°$$

In the last step we used the π key on a calculator to complete the computation. ▶

Converting Between Degrees and Radians

Use the proportion

$$\frac{\text{radian measure}}{\pi} = \frac{\text{degree measure}}{180°}$$

or the following shortcuts:

1. If a radian measure involves a multiple of π, replace π with $180°$ and simplify to convert to degrees.

2. Multiply a degree measure by $\pi/180°$ to convert to radians.

Some calculators have a key that automatically converts between decimal degrees and radians. Check your owner's manual to see if your calculator has this feature. If not, you may wish to write a program to accomplish this.

CAUTION   Figure 40 shows angles measuring 30 radians and 30°. These angle measures are not at all close, so be careful not to confuse them.

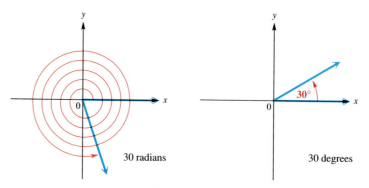

FIGURE 40

The following table and Figure 41 give some equivalent angles measured in degrees and radians. It will be useful to memorize these equivalent values. Keep in mind that $180° = \pi$ radians. Then it will be easy to reproduce the rest of the table.

Equivalent Angle Measures in Degrees and Radians

Degrees	Radians Exact	Radians Approximate	Degrees	Radians Exact	Radians Approximate
0°	0	0	90°	$\dfrac{\pi}{2}$	1.57
30°	$\dfrac{\pi}{6}$	.52	180°	π	3.14
45°	$\dfrac{\pi}{4}$	.79	270°	$\dfrac{3\pi}{2}$	4.71
60°	$\dfrac{\pi}{3}$	1.05	360°	2π	6.28

Radian measure is used to simplify formulas that are more complicated if expressed using degree measure.

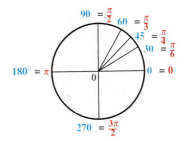

FIGURE 41

ARC LENGTH OF A CIRCLE   The formula for an arc length of a circle depends on the fact (proven in geometry) that the length of an arc is proportional to the measure of its central angle.

In Figure 42, angle QOP has a measure of 1 radian and intercepts an arc of length r on the circle. Angle ROT has a measure of θ radians and intercepts an arc of length s on the circle. Since the lengths of the arcs are proportional to the measure of their central angles,

$$\frac{s}{r} = \frac{\theta}{1}.$$

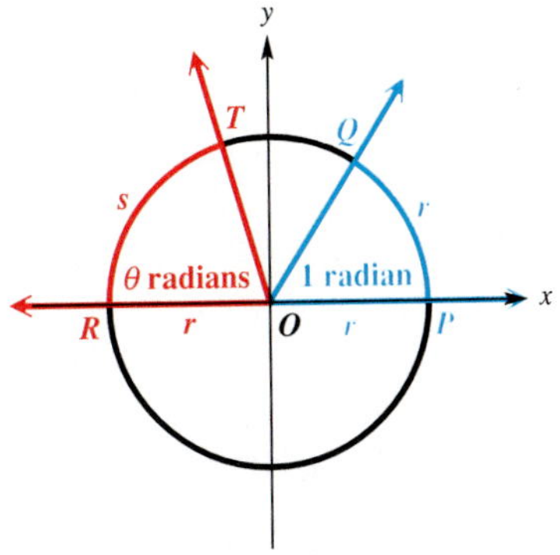

FIGURE 42

Multiplying both sides by r gives the following result.

Length of Arc

The length s of the arc intercepted on a circle of radius r by a central angle of measure θ radians is given by the product of the radius and the radian measure of the angle, or

$$s = r\theta, \quad \theta \text{ in radians.}$$

This formula is a good example of the usefulness of radian measure. To see why, write the equivalent formula for an angle measured in degrees.

CAUTION When applying the formula $s = r\theta$, the value of θ *must be* expressed in *radians*.

EXAMPLE 3
Finding arc length using
$s = r\theta$

A circle has a radius of 18.2 centimeters. Find the length of the arc intercepted by a central angle having each of the following measures.

(a) $\dfrac{3\pi}{8}$ radians

Here $r = 18.2$ cm and $\theta = 3\pi/8$. Since $s = r\theta$,

$$s = 18.2\left(\frac{3\pi}{8}\right) \text{ centimeters}$$

$$s = \frac{54.6\pi}{8} \text{ centimeters} \qquad \text{The exact answer}$$

or $\qquad s \approx 21.4$ centimeters. $\qquad$ Calculator approximation

(b) $144°$

The formula $s = r\theta$ requires that θ be measured in radians. First, convert θ to radians by the methods explained earlier in this section. Using the shortcut, multiply by $\pi/180°$.

$$144° = 144°\left(\frac{\pi}{180°}\right) \text{ radians} \qquad \text{Change from degrees to radians.}$$

$$144° = \frac{4\pi}{5} \text{ radians}$$

Now $\qquad s = 18.2\left(\dfrac{4\pi}{5}\right)$ centimeters $\qquad$ Use $s = r\theta$.

$$s \approx 45.7 \text{ centimeters.} \quad \blacktriangleright$$

EXAMPLE 4
Finding the distance between
two cities using latitudes

Reno, Nevada, is approximately due north of Los Angeles. The latitude of Reno is 40° N, while that of Los Angeles is 34° N. (The N in 34° N means *north* of the equator.) If the radius of Earth is 6400 kilometers, find the north-south distance between the two cities.

Latitude gives the measure of a central angle with vertex at Earth's center whose initial side goes through the equator and whose terminal side goes through the given location. As shown in Figure 43, the central angle for Reno and Los Angeles is 6°. The distance between the two cities can thus be found by the formula $s = r\theta$, after 6° is first converted to radians.

$$6° = 6°\left(\frac{\pi}{180°}\right) = \frac{\pi}{30} \text{ radian}$$

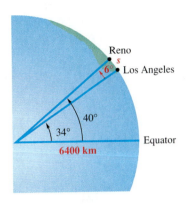

FIGURE 43

The distance between the two cities is

$$s = r\theta$$

$$s = 6400\left(\frac{\pi}{30}\right) \text{ kilometers} \qquad r = 6400, \ s = \frac{\pi}{30}$$

$$\approx 670 \text{ kilometers.} \quad \blacktriangleright$$

EXAMPLE 5
Finding a length using
$s = r\theta$

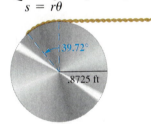

FIGURE 44

A rope is being wound around a drum with radius .8725 foot. (See Figure 44.) How much rope will be wound around the drum if the drum is rotated through an angle of 39.72°?

The length of rope wound around the drum is just the arc length for a circle of radius .8725 feet and a central angle of 39.72°. Use the formula $s = r\theta$, with the angle converted to radian measure.

$$s = r\theta$$

$$s = (.8725)\left[(39.72°)\left(\frac{\pi}{180°}\right)\right] \qquad \text{Convert to radians.}$$

$$\approx .6049$$

The length of the rope wound around the drum is approximately .6049 foot. ▶

EXAMPLE 6
Finding an angle measure
using $s = r\theta$

FIGURE 45

Two gears are adjusted so that the smaller gear drives the larger one as shown in Figure 45. If the smaller gear rotates through 225°, through how many degrees will the larger gear rotate?

First find the radian measure of the angle, which will give the arc length on the smaller gear that determines the motion of the larger gear. Since 225° = $5\pi/4$ radians, for the smaller gear,

$$s = r\theta = 2.5\left(\frac{5\pi}{4}\right) \approx 9.8 \text{ centimeters.}$$

An arc length of 9.8 centimeters on the larger gear corresponds to an angle measure θ, in radians, of

$$s = r\theta$$
$$9.8 = 4.8\theta$$
$$2.0 \approx \theta.$$

Changing back to degrees shows that the larger gear rotates through

$$2.0\left(\frac{180°}{\pi}\right) \approx 110°,$$

to two significant figures. ▶

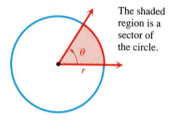

The shaded region is a sector of the circle.

FIGURE 46

SECTOR OF A CIRCLE A **sector of a circle** is the portion of the interior of a circle intercepted by a central angle. See Figure 46. To find the area of a sector, assume that the radius of the circle is r. A complete circle can be thought of as an angle with a measure of 2π radians. If a central angle for the sector has measure θ radians, then the sector makes up a fraction $\dfrac{\theta}{2\pi}$ of a complete circle. The area of a complete circle is $A = \pi r^2$. Therefore, the area of the sector is given by the product of the fraction $\dfrac{\theta}{2\pi}$ and the total area, πr^2, or

$$\text{area of sector} = \frac{\theta}{2\pi}(\pi r^2) = \frac{1}{2}r^2\theta, \qquad \theta \text{ in radians.}$$

This discussion is summarized as follows.

> **Area of a Sector**
>
> The area of a sector of a circle of radius r and central angle θ in radians is given by
>
> $$A = \frac{1}{2}r^2\theta, \quad \theta \text{ in radians}.$$

CAUTION As in the formula for arc length, the measure of θ must be in radians when using this formula for the area of a sector.

◀EXAMPLE 7
Finding area using $A = \frac{1}{2}r^2\theta$

Figure 47 shows a field in the shape of a sector of a circle. The central angle is 15° and the radius of the circle is 321 meters. Find the area of the field.

First, convert 15° to radians.

$$15° = 15°\left(\frac{\pi}{180°}\right) = \frac{\pi}{12} \text{ radian}$$

FIGURE 47

Now use the formula for the area of a sector.

$$A = \frac{1}{2} r^2 \theta$$

$$= \frac{1}{2} (321)^2 \left(\frac{\pi}{12} \right)$$

$$\approx 13{,}500 \text{ m}^2. \quad \blacktriangleright$$

6.5 Exercises ▼▼▼▼▼▼▼▼▼▼▼▼▼▼▼▼▼▼▼▼▼▼▼▼▼▼▼▼▼▼▼▼▼

In Exercises 1–4, each angle is an integer when measured in radians. Give the radian measure of the angle.

1.

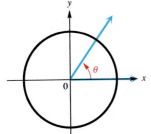

2.

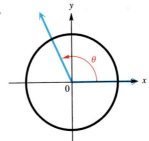

3.

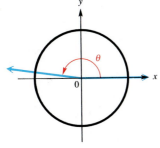

4.

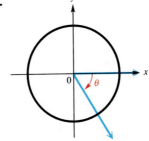

Convert the degree measure to radians. Leave answers as multiples of π. See Example 1(a).

5. $60°$ **6.** $90°$ **7.** $150°$ **8.** $270°$ **9.** $315°$ **10.** $480°$

11. In your own words, explain how to convert degree measure to radian measure.

12. In your own words, explain how to convert radian measure to degree measure.

13. In your own words, explain the meaning of radian measure.

14. Explain the difference between degree measure and radian measure.

Convert the radian measure to degrees. See Example 2(a).

15. $\dfrac{\pi}{3}$ **16.** $\dfrac{8\pi}{3}$ **17.** $\dfrac{7\pi}{4}$ **18.** $-\dfrac{\pi}{6}$ **19.** $\dfrac{8\pi}{5}$ **20.** $\dfrac{11\pi}{15}$ **21.** $\dfrac{4\pi}{15}$ **22.** $\dfrac{7\pi}{20}$

Convert the degree measure to radians. Give answers to three significant digits. See Example 1(b).

23. $39°$ **24.** $74°$ **25.** $139°\ 10'$ **26.** $174°\ 50'$ **27.** $85.04°$ **28.** $122.62°$

Convert the following radian measures to degrees. Write answers in decimal degrees. In Exercises 29 and 30 give answers to the nearest tenth. Use the appropriate number of significant digits in Exercises 31–34. See Example 2(b).

29. 2 **30.** 5 **31.** 1.74 **32.** $.3417$ **33.** 9.847630 **34.** -3.471890

Find the exact value without using a calculator.

35. $\cos \dfrac{\pi}{6}$ **36.** $\tan \dfrac{\pi}{4}$ **37.** $\csc \dfrac{\pi}{4}$ **38.** $\sin \dfrac{\pi}{2}$

39. $\sec \pi$ **40.** $\tan \dfrac{2\pi}{3}$ **41.** $\cot \dfrac{2\pi}{3}$ **42.** $\sin \dfrac{5\pi}{6}$

43. The figure shows the same angles measured in both degrees and radians. Complete the missing measures.

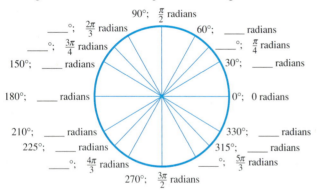

44. Find the measure (in both degrees and radians) of the angle θ formed in the accompanying screen by the line passing through the origin and the positive part of the x-axis. Use the displayed values of X and Y at the bottom of the screen.

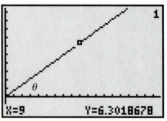

45. In the accompanying screen, was the calculator in degree or radian mode?

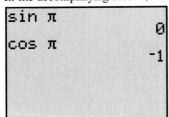

```
sin π
              0
cos π
             -1
```

46. The **solar constant** S is the amount of energy per unit area that reaches Earth's atmosphere from the sun. It is equal to 1367 watts per square meter but varies slightly throughout the seasons. This fluctuation ΔS in S can be calculated using the formula

$$\Delta S = .034S \sin\left[\frac{2\pi(82.5 - N)}{365.25}\right].$$ In this formula, N is the day number covering a

four-year period where $N = 1$ corresponds to January 1 of a leap year and $N = 1461$ corresponds to December 31 of the fourth year. (*Source:* Winter, C., R. Sizmann, and Vant-Hunt (Editors), *Solar Power Plants,* Springer-Verlag, 1991.)

(a) Calculate ΔS for $N = 80$ which is the spring equinox in the first year.
(b) Calculate ΔS for $N = 1268$ which is the summer solstice in the fourth year.
(c) What is the maximum value of ΔS?
(d) Find a value of N where ΔS is equal to zero.

In Exercises 47 and 48, find the length of the arc intercepted by the given angle.

47.

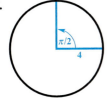

$\pi/2$
4

48.

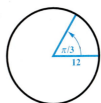

$\pi/3$
12

In Exercises 49 and 50, find the radius of the circle. (The measure of a central angle (in radians) and length of its intercepted arc are given.)

49.

6π

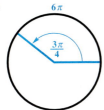

$\frac{3\pi}{4}$

50.

3π

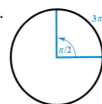

$\pi/2$

Find the length of the arc intercepted by a central angle θ in a circle of radius r. See Example 3.

51. $r = 12.3$ cm, $\theta = \dfrac{2\pi}{3}$ radians

52. $r = .892$ cm, $\theta = \dfrac{11\pi}{10}$ radians

53. $r = 253$ m, $\theta = \dfrac{2\pi}{5}$ radians

54. $r = 120$ mm, $\theta = \dfrac{\pi}{9}$ radian

55. $r = 4.82$ m, $\theta = 60°$

56. $r = 71.9$ cm, $\theta = 135°$

57. If the radius of a circle is doubled, how is the length of the arc intercepted by a fixed central angle changed?

58. Consider the length of arc formula $s = r\theta$. What well-known formula corresponds to the special case $\theta = 2\pi$?

Find the distance in kilometers between each of the following pairs of cities, assuming they lie on the same north-south line. See Example 4.

59. Panama City, Panama, 9° N, and Pittsburgh, Pennsylvania, 40° N

60. Farmersville, California, 36° N, and Penticton, British Columbia, 49° N

61. New York City, New York, 41° N, and Lima, Peru, 12° S

62. Halifax, Nova Scotia, 45° N, and Buenos Aires, Argentina, 34° S

63. Madison, South Dakota, and Dallas, Texas, are 1200 km apart and lie on the same north-south line. The latitude of Dallas is 33° N. What is the latitude of Madison?

64. Charleston, South Carolina, and Toronto, Canada, are 1100 km apart and lie on the same north-south line. The latitude of Charleston is 33° N. What is the latitude of Toronto?

Work each problem. See Examples 5 and 6.

65. (a) How many inches will the weight in the figure rise if the pulley is rotated through an angle of 71° 50′?

(b) Through what angle, to the nearest minute, must the pulley be rotated in order to raise the weight 6 in?

9.27 in

66. Find the radius of the pulley in the figure if a rotation of 51.6° raises the weight 11.4 cm.

r

67. The rotation of the smaller wheel in the figure causes the larger wheel to rotate. Through how many degrees will the larger wheel rotate if the smaller one rotates through 60.0°?

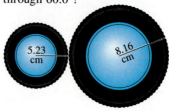

5.23 cm   8.16 cm

68. Find the radius of the larger wheel in the figure if the smaller wheel rotates 80.0° when the larger wheel rotates 50.0°.

69. The figure shows the chain drive of a bicycle. How far will the bicycle move if the pedals are rotated through 180°? Assume that the radius of the bicycle wheel is 13.6 in.

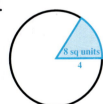

70. The speedometer of a small pickup truck is designed to be accurate with tires of radius 14 in.

   (a) Find the number of rotations of a tire in 1 hr if the truck is driven at 55 mph.

(b) Suppose that oversize tires of radius 16 in are placed on the truck. If the truck is now driven for 1 hr with the speedometer reading 55 mph, how far has the truck gone? If the speed limit is 55 mph, does the driver deserve a speeding ticket?

If a central angle is very small, there is little difference in length between an arc and the inscribed chord. See the figure. Approximate each of the following lengths by finding the necessary arc length. (Note: When a central angle intercepts an arc, the arc is said to subtend the angle.)

Arc length ≈ length of inscribed chord

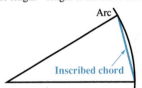

71. An oil tanker 2.3 km at sea subtends a 1° 30′ angle horizontally. Find the length of the ship.

72. The full moon subtends an angle of 1/2°. The moon is 240,000 mi away. Find the diameter of the moon.

In Exercises 73 and 74, find the area of the sector. See Example 7.

73.

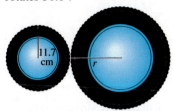

74.

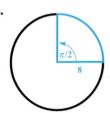

In Exercises 75 and 76, find the measure (in radians) of the central angle. The number inside the sector is the area. See Example 7.

75.

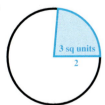

76.

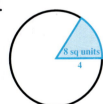

Find the area of a sector of a circle having radius r and central angle θ in each of the following. See Example 7.

77. $r = 52$ cm, $\theta = 3\pi/10$ radian

78. $r = 25$ mm, $\theta = \pi/15$ radian

79. $r = 12.7$ cm, $\theta = 81°$

80. $r = 18.3$ m, $\theta = 125°$

---

81. Find the measure (in radians) of a central angle of a sector of area 16 square inches in a circle of radius 3.0 in.

82. Find the radius of a circle in which a central angle of $\pi/4$ radian determines a sector of area 36 square ft.

83. Consider the area-of-a-sector formula $A = (1/2)r^2\theta$. What well-known formula corresponds to the special case $\theta = 2\pi$?

84. If the radius of a circle is doubled, and the central angle of a sector is unchanged, how is the area of the sector determined by this fixed central angle changed?

85. The sector in the accompanying graphing calculator screen is bounded above by the line $y = (\sqrt{3}/3)x$, below by the *x*-axis, and on the right by the circle $x^2 + y^2 = 4$. What is the area of the sector?

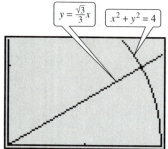

86. The figure shows Medicine Wheel, a Native American structure in northern Wyoming. This circular structure is perhaps 200 years old. There are 32 spokes in the wheel, all equally spaced.

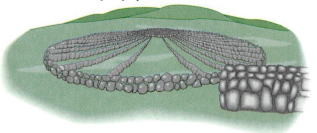

(a) Find the measure of each central angle in degrees and in radians.

(b) If the radius of the wheel is 76 ft, find the circumference.

(c) Find the length of each arc intercepted by consecutive pairs of spokes.

(d) Find the area of each sector formed by consecutive spokes.

87. The unusual corral in the figure is separated into 26 areas, many of which approximate sectors of a circle. Assume that the corral has a diameter of 50 m.

(a) Find the central angle for each region, assuming that the 26 regions are all equal sectors, with the fences meeting at the center.

(b) What is the area of each sector?

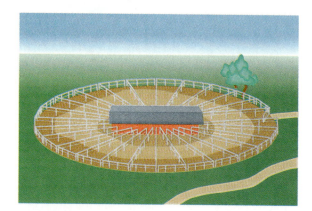

88. Eratosthenes (*ca.* 230 B.C.) made a famous measurement of Earth. He observed at Syene (the modern Aswan) at noon and at the summer solstice that a vertical stick had no shadow, while at Alexandria (on the same meridian as Syene) the sun's rays were inclined 1/50 of a complete circle to the vertical. See the figure. He then calculated the circumference of Earth from the known distance of 5000 stades between Alexandria and Syene. Obtain Eratosthenes' result of 250,000 stades for the circumference of Earth. There is reason to suppose that a stade is about equal to

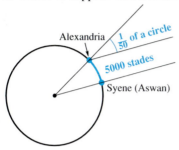

516.7 ft. Assuming this, use Eratosthenes' result to calculate the polar diameter of Earth in miles. (The actual polar diameter of Earth, to the nearest mile, is 7900 mi.) (*Source:* Eves, Howard, *A Survey of Geometry,* Vol 1. Reprinted by permission of the author.)

89. A 300-megawatt solar-power plant requires approximately 950,000 square meters of land area in order to collect the required amount of energy from sunlight.

(a) If this land area is circular, what is its radius?

(b) If this land area is a 35° sector of a circle, what is its radius?

6.6 Circular Functions and Their Applications ▼▼▼

Earlier we defined the six trigonometric functions for *angles.* The angles can be measured either in degrees or in radians. While the domain of the trigonometric functions is a set of angles, the range is a set of real numbers. In advanced work, such as calculus, it is necessary to modify the trigonometric functions so that the domain contains not angles, but real numbers. To do this we use the relationship between an angle θ and an arc of length s on a circle.

Look at Figure 48. In the figure, starting at the point $(1, 0)$, we lay off an arc of length $|s|$ along the circle. We go counterclockwise if s is positive and clockwise if s is negative. The endpoint of the arc is the point (x, y). The circle in Figure 48 is a **unit circle,** a circle with center at the origin and a radius of one unit (hence the name *unit circle*). Recall from algebra that the equation of this circle is

$$x^2 + y^2 = 1.$$

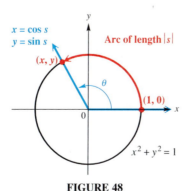

FIGURE 48

We saw earlier that the radian measure of θ is related to the arc length s. In fact, for θ measured in radians, we know that $s = r\theta$. Here, $r = 1$, so s, which is measured in linear units (such as inches or centimeters), is numerically equal to θ, measured in radians. Thus, the trigonometric functions of angle θ in radians found by choosing a point (x, y) on the unit circle can be rewritten as functions of the arc length s, a real number. To distinguish these from the trigonometric functions of angles, they are called **circular functions.**

Circular Functions

$$\sin s = y \qquad \tan s = \frac{y}{x}, x \neq 0 \qquad \sec s = \frac{1}{x}, x \neq 0$$

$$\cos s = x \qquad \cot s = \frac{x}{y}, y \neq 0 \qquad \csc s = \frac{1}{y}, y \neq 0$$

NOTE Since $\sin s = y$ and $\cos s = x$, we can replace x and y in the equation $x^2 + y^2 = 1$ and obtain the familiar Pythagorean identity

$$\cos^2 s + \sin^2 s = 1.$$

Since the ordered pair (x, y) represents a point on the unit circle,

$$-1 \leq x \leq 1 \qquad \text{and} \qquad -1 \leq y \leq 1,$$

making $\qquad\qquad -1 \leq \cos s \leq 1 \qquad \text{and} \qquad -1 \leq \sin s \leq 1.$

For any value of s, both $\sin s$ and $\cos s$ exist, so the domain of these functions is the set of all real numbers. For $\tan s$, defined as y/x, $x \neq 0$. The only way x can equal 0 is when the arc length s is $\pi/2$, $-\pi/2$, $3\pi/2$, $-3\pi/2$, and so on. To avoid a zero denominator, the domain of tangent must be restricted to those values of s satisfying

$$s \neq \frac{\pi}{2} + n\pi, \quad n \text{ any integer.}$$

The definition of secant also has x in the denominator, making the domain of secant the same as the domain of tangent. Both cotangent and cosecant are defined with a denominator of y. To guarantee that $y \neq 0$, the domain of these functions must be the set of all values of s satisfying

$$s \neq n\pi, \quad n \text{ any integer.}$$

The domains of the circular functions are summarized in the following box. Compare these with the domains of the trigonometric functions.

Domains of the Circular Functions

The domains of the circular functions are as follows. Assume that n is any integer, and s is a real number.

Sine and Cosine Functions: $(-\infty, \infty)$

Tangent and Secant Functions: $\left\{ s \mid s \neq \dfrac{\pi}{2} + n\pi \right\}$

Cotangent and Cosecant Functions: $\{ s \mid s \neq n\pi \}$

As mentioned at the beginning of this section, in Figure 48, s is the radian measure of angle θ and the radius r equals 1. Using the definition of the trigonometric functions,

$$\sin \theta = \frac{y}{r} = \frac{y}{1} = y = \sin s$$

$$\cos \theta = \frac{x}{r} = \frac{x}{1} = x = \cos s.$$

Similar results hold for the other four functions.

As shown above, the trigonometric functions and the circular functions lead to the same function values. Because of this, a value such as $\sin \pi/2$ can be found without worrying about whether $\pi/2$ is a real number or the radian measure of an angle. In either case, $\sin \pi/2 = 1$. All the formulas developed in this book are valid for either angles or real numbers. For example, $\sin \theta = 1/\csc \theta$ is equally valid for θ as the measure of an angle in degrees or radians or for θ as a real number.

We find function values of arc lengths s just as we found the function values for angles measured in radians.

EXAMPLE 1
Finding the function value of an arc length

Find the following circular function values.

(a) $\cos \dfrac{2\pi}{3}$

Since

$$\frac{2\pi}{3} = \frac{2}{3} \cdot 180° = 120°,$$

$$\cos \frac{2\pi}{3} = \cos 120° = -\frac{1}{2}.$$

(b) $\sin .5149$

Use a calculator.

$$\sin .5149 \approx .4924478173$$

(c) cot 1.3209

Recall, because calculators do not have keys for cotangent, secant, and cosecant, we must use the appropriate reciprocal function to find these values. To find cot 1.3209, we first find tan 1.3209 and then find the reciprocal.

$$\cot 1.3209 = \frac{1}{\tan 1.3209} \approx .25523149 \quad \blacktriangleright$$

CAUTION One of the most common errors in trigonometry involves using calculators in degree mode when radian mode should be used. Remember that if you are finding a circular function value of a real number, the calculator *must* be in radian mode.

◀**EXAMPLE 2**
Finding arc length
given x or y

Find s in the interval $[0, \pi/2]$ if $x = .96854556$.

Since $x = \cos s$, we know $\cos s = .96854556$. Use the $\cos^{-1}$ key (or other appropriate keys) of a calculator to find

$$s \approx .25147856. \quad \blacktriangleright$$

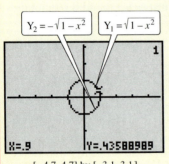

$Y_2 = -\sqrt{1-x^2}$ $Y_1 = \sqrt{1-x^2}$

X=.9   Y=.43588989

$[-4.7, 4.7]$ by $[-3.1, 3.1]$

▱ At the beginning of this section we introduced the concept of circular ▦ functions. We showed how the unit circle $x^2 + y^2 = 1$ can be used to find the cosine and sine of a real number s. To illustrate this concept, we can use a graphing calculator to graph the circle. We must first solve the equation for y, getting the two functions

$$Y_1 = \sqrt{1 - x^2} \qquad \text{and} \qquad Y_2 = -\sqrt{1 - x^2}.$$

With your calculator in function mode, graph these two equations in the same window to get the graph of the circle. To get an undistorted figure, a *square window* must be used. (See your instruction manual for details.)

The TRACE function of the calculator allows us to find coordinates of points on the graph. Experiment with this feature, and notice that the x and y coordinates are displayed below the graph. The x coordinate represents the cosine of the length of the arc from the point $(1, 0)$ to the point indicated by the cursor. For example, one such point is

$$x = .9 \qquad y = .43588989.$$

This point is shown in the figure.

While the calculator does not give the arc length, it can be found by setting the calculator in radian mode and finding either $\cos^{-1} .9$ or $\sin^{-1} .43588989$. By doing this, we find the arc length is approximately .4510268.

LINEAR AND ANGULAR VELOCITY   In many situations we need to know the speed at which a point on a circular disk is moving or how fast the central angle of such a disk is changing. Some examples occur with machinery involving gears or pulleys or the speed of a car around a curved portion of a highway.

Suppose that point P moves at a constant speed along a circle of radius r and center O. See Figure 49. The measure of how fast the position of P is changing is called **linear velocity.** If v represents linear velocity, then

$$\text{velocity} = \frac{\text{distance}}{\text{time}}$$

$$v = \frac{s}{t},$$

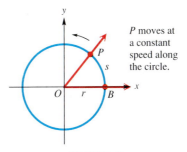

P moves at a constant speed along the circle.

FIGURE 49

where s is the length of the arc traced by point P at time t. (This formula is just a restatement of the familiar result $d = rt$ with s as distance, v as the rate, and t as time.)

Look at Figure 49 again. As point P moves along the circle, ray OP rotates around the origin. Since the ray OP is the terminal side of angle POB, the measure of the angle changes as P moves along the circle. The measure of how fast angle POB is changing is called **angular velocity.** Angular velocity, written ω (the Greek letter *omega*) can be given as

$$\omega = \frac{\theta}{t}, \ \theta \text{ in radians},$$

where θ is the measure of angle POB at time t. As with the earlier formulas in this section, θ must be measured in radians, with ω expressed as radians per unit of time. Angular velocity is used in physics and engineering, among other applications.

The length s of the arc intercepted on a circle of radius r by a central angle of measure θ radians is $s = r\theta$. Using this formula, the formula for linear velocity, $v = s/t$, becomes

$$v = \frac{r\theta}{t}$$

$$v = r \cdot \frac{\theta}{t}$$

or $\qquad\qquad v = r\omega \qquad \omega = \frac{\theta}{t}.$

This last formula relates linear and angular velocity.

A radian is a "pure number," with no units associated with it. This is why the product of the length r, measured in units such as centimeters, and ω, measured in units such as radians per second, is velocity, v, measured in units such as centimeters per second.

The linear and angular velocity formulas are summarized below.

Angular and Linear Velocity

Angular Velocity	Linear Velocity
$\omega = \dfrac{\theta}{t}$	$v = \dfrac{s}{t}$
(ω in radians per unit time, θ in radians)	$v = \dfrac{r\theta}{t}$
	$v = r\omega$

EXAMPLE 3
Using the linear and angular velocity formulas

Suppose that point P is on a circle with a radius of 10 centimeters, and ray OP is rotating with angular velocity of $\pi/18$ radian per second.

(a) Find the angle generated by P in 6 seconds.

The angular velocity of ray OP is $\omega = \pi/18$ radian per second. Since $\omega = \theta/t$, in 6 seconds

$$\frac{\pi}{18} = \frac{\theta}{6},$$

or $\theta = 6(\pi/18) = \pi/3$ radians.

(b) Find the distance traveled by P along the circle in 6 seconds.

In 6 seconds P generates an angle of $\pi/3$ radians. Since $s = r\theta$,

$$s = 10\left(\frac{\pi}{3}\right) = \frac{10\pi}{3} \text{ centimeters.}$$

(c) Find the linear velocity of P.
Since $v = s/t$, in 6 seconds

$$v = \frac{\frac{10\pi}{3}}{6} = \frac{5\pi}{9} \text{ centimeters per second.} \quad \blacktriangleright$$

■ **PROBLEM SOLVING**   In practical applications, angular velocity is often given as revolutions per unit of time, which must be converted to radians per unit of time before using the formulas given in this section. ■

EXAMPLE 4
Using the linear and angular velocity formulas

A belt runs a pulley of radius 6 centimeters at 80 revolutions per minute.

(a) Find the angular velocity of the pulley in radians per second.
In one minute, the pulley makes 80 revolutions. Each revolution is 2π radians, for a total of

$$80(2\pi) = 160\pi \text{ radians per minute.}$$

Since there are 60 seconds in a minute, ω, the angular velocity in radians per second, is found by dividing 160π by 60.

$$\omega = \frac{160\pi}{60} = \frac{8\pi}{3} \text{ radians per second}$$

(b) Find the linear velocity of the belt in centimeters per second.
The linear velocity of the belt will be the same as that of a point on the circumference of the pulley. Thus,

$$v = r\omega$$
$$v = 6\left(\frac{8\pi}{3}\right)$$
$$v = 16\pi \text{ centimeters per second}$$
$$v \approx 50.3 \text{ centimeters per second.} \quad \blacktriangleright$$

EXAMPLE 5
Finding the linear velocity and distance traveled by a satellite

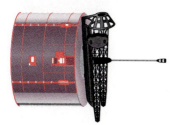

A satellite traveling in a circular orbit 1600 kilometers above the surface of Earth takes two hours to make an orbit. Assume that the radius of Earth is 6400 kilometers.

(a) Find the linear velocity of the satellite.
The distance of the satellite from the center of Earth is

$$r = 1600 + 6400 = 8000 \text{ kilometers.}$$

For one orbit $\theta = 2\pi$, and

$$s = r\theta = 8000(2\pi) \text{ kilometers.}$$

Then the linear velocity is

$$v = \frac{s}{t} = \frac{8000(2\pi)}{2}$$

$$= 8000\pi$$

$$\approx 25{,}000 \text{ kilometers per hour.}$$

(b) Find the distance traveled in 4.5 hours.

$$s = vt = (8000\pi)(4.5)$$

$$= 36{,}000\pi$$

$$\approx 110{,}000 \text{ kilometers.} \quad \blacktriangleright$$

6.6 Exercises ▼▼▼▼▼▼▼▼▼▼▼▼▼▼▼▼▼▼▼▼▼▼▼▼▼▼▼▼▼▼▼▼▼▼▼

1. In your own words, describe what is meant by the unit circle.

2. In your own words, explain how the cosine function associates a value with the real number s.

Find the exact circular function value for each of the following. See Example 1(a).

3. $\sin \dfrac{7\pi}{6}$ **4.** $\cos \dfrac{5\pi}{3}$ **5.** $\tan \dfrac{3\pi}{4}$ **6.** $\sec \dfrac{2\pi}{3}$

7. $\csc \dfrac{11\pi}{6}$ **8.** $\cot \dfrac{5\pi}{6}$ **9.** $\cos\left(-\dfrac{4\pi}{3}\right)$ **10.** $\sin\left(-\dfrac{5\pi}{6}\right)$

11. Suppose a student attempts to work Exercises 3–10 using a calculator. Can the student expect to get the correct results, according to the directions given for those exercises? Explain.

Use a calculator to find an approximation for each circular function value. Be sure your calculator is set in radian mode. See Examples 1(b) and 1(c).

12. $\sin .8203$ **13.** $\cot .6632$ **14.** $\cos .6429$ **15.** $\tan .9047$

16. $\csc 1.3875$ **17.** $\cot 7.4526$ **18.** $\sin (-2.2864)$ **19.** $\cos (-3.0602)$

Find the value of s in the interval $[0, \pi/2]$ that makes each statement true. Use a calculator. See Example 2.

20. $\cos s = .78269876$ **21.** $\sin s = .99184065$ **22.** $\sin s = .98771924$ **23.** $\cos s = .92728460$

24. $\cot s = .62084613$ **25.** $\csc s = 1.0219553$

26. What (exact) value of S between 0 and $\pi/2$ produces the output for the accompanying graphing calculator screen?

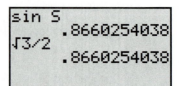

Find the exact value of s in the given interval that has the given circular function value. Do not use a calculator. See Example 2.

27. $\left[\dfrac{\pi}{2}, \pi\right]$; $\sin s = \dfrac{1}{2}$

28. $\left[\dfrac{\pi}{2}, \pi\right]$; $\cos s = -\dfrac{1}{2}$

29. $\left[\pi, \dfrac{3\pi}{2}\right]$; $\tan s = \sqrt{3}$

30. $\left[\pi, \dfrac{3\pi}{2}\right]$; $\sin s = -\dfrac{1}{2}$

31. $\left[\dfrac{3\pi}{2}, 2\pi\right]$; $\tan s = -1$

32. $\left[\dfrac{3\pi}{2}, 2\pi\right]$; $\cos s = \dfrac{\sqrt{3}}{2}$

 The graphing calculator screen shows a point on the unit circle. What is the length of the shortest arc of the circle from (1, 0) to the point?

33.

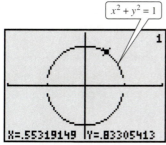

[–2, 2] by [–1.5, 1.5]

34.

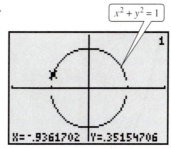

[–2, 2] by [–1.5, 1.5]

Suppose an arc of length s lies on the unit circle $x^2 + y^2 = 1$, starting at the point (1, 0) and terminating at the point (x, y). (See Figure 48.) Use a calculator to find the approximate coordinates for (x, y). (Hint: $x = \cos s$ and $y = \sin s$.)

35. $s = 2.5$

36. $s = 3.4$

37. $s = -7.4$

38. $s = -3.9$

For each value of s, use a calculator to find sin s and cos s and then use the results to decide in which quadrant an angle of s radians lies.

39. $s = 51$

40. $s = 49$

41. $s = 65$

42. $s = 79$

---

43. The values of the circular functions repeat every 2π. For this reason, circular functions are used to describe things that repeat periodically. For example, the maximum afternoon temperature in a given city might be approximated by

$$t = 60 - 30 \cos \dfrac{x\pi}{6},$$

where t represents the maximum afternoon temperature in month x, with $x = 0$ representing January, $x = 1$ representing February, and so on. Find the maximum afternoon temperature for each of the following months.
(a) January   **(b)** April
(c) May   **(d)** June
(e) August   **(f)** October

44. The temperature in Fairbanks is approximated by

$$T(x) = 37 \sin\left[\dfrac{2\pi}{365}(x - 101)\right] + 25,$$

where $T(x)$ is the temperature in degrees Fahrenheit on day x, with $x = 1$ corresponding to January 1 and $x = 365$ corresponding to December 31. Use a calculator to estimate the temperature on the following days. (*Source:* Lando, Barbara, and Clifton Lando, "Is the Graph of Temperature Variation a Sine Curve?" *The Mathematics Teacher,* 70, September, 1977, 534–37.)
(a) March 1 (day 60)   **(b)** April 1 (day 91)
(c) Day 150   **(d)** June 15
(e) September 1   **(f)** October 31

Solve the following equations for $0 \leq x \leq 2\pi$. (Hint: Use the unit circle definition of cosine and sine discussed in this section.)

45. $\sin x = \sin(x + 2)$

46. $\cos x = \cos(x + 1)$

47. If a point moves around the circumference of the unit circle at an angular velocity of 1 radian per second, how long will it take for the point to move around the entire circle?

48. If a point moves around the circumference of the unit circle at the speed of one unit per second, how long will it take for the point to move around the entire circle?

49. What is the difference between linear velocity and angular velocity?

50. Explain why linear velocity is affected by the radius of the circle, whereas angular velocity is not.

Use the formula $\omega = \theta/t$ to find the value of the missing variable. Use a calculator in Exercises 55 and 56. See Example 3.

51. $\omega = 2\pi/3$ radians per sec, $t = 3$ sec

52. $\omega = \pi/4$ radian per min, $t = 5$ min

53. $\theta = 2\pi/9$ radian, $\omega = 5\pi/27$ radian per min

54. $\theta = 3\pi/8$ radians, $\omega = \pi/24$ radian per min

55. $\theta = 3.871142$ radians, $t = 21.4693$ sec

56. $\omega = .90674$ radian per min, $t = 11.876$ min

Use the formula $v = r\omega$ to find the value of the missing variable. Use a calculator in Exercises 59 and 60. See Example 3.

57. $r = 12$ m, $\omega = 2\pi/3$ radians per sec

58. $v = 18$ ft per sec, $r = 3$ ft

59. $v = 107.692$ m per sec, $r = 58.7413$ m

60. $r = 24.93215$ cm, $\omega = .372914$ radian per sec

The formula $\omega = \theta/t$ can be rewritten as $\theta = \omega t$. Using ωt for θ changes $s = r\theta$ to $s = r\omega t$. Use the formula $s = r\omega t$ to find the values of the missing variables.

61. $r = 6$ cm, $\omega = \pi/3$ radians per sec, $t = 9$ sec

62. $r = 9$ yd, $\omega = 2\pi/5$ radians per sec, $t = 12$ sec

63. $s = 6\pi$ cm, $r = 2$ cm, $\omega = \pi/4$ radian per sec

64. $s = 8\pi/9$ m, $r = 4/3$ m, $t = 12$ sec

65. Explain the similarities between the familiar $d = rt$ formula and the formula $s = vt$.

66. Suppose that you must convert k radians per second to degrees per minute. Explain how you would do this.

Find ω for each of the following.

67. The hour hand of a clock

68. The minute hand of a clock

Find v for each of the following.

69. The tip of the second hand of a clock, if the hand is 28 mm long

70. A point on the tread of a tire of radius 18 cm, rotating 35 times per min

71. The tip of an airplane propeller 3 m long, rotating 500 times per min (*Hint: $r = 1.5$ m*)

72. A point on the edge of a gyroscope of radius 83 cm, rotating 680 times per minute

Solve the following problems, which review the ideas of this section. See Examples 1–5.

73. Earth travels about the sun in an orbit that is almost circular. Assume that the orbit is a circle, with a radius of 93,000,000 mi. See the figure.
  (a) Assume that a year is 365 days, and find θ, the angle formed by Earth's movement in one day.
  (b) Give the angular velocity in radians per hour.
  (c) Find the linear velocity of Earth in miles per hour.

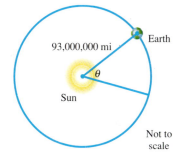

Not to scale

74. The pulley shown has a radius of 12.96 cm. Suppose that it takes 18 sec for 56 cm of belt to go around the pulley. Find the angular velocity of the pulley in radians per second.

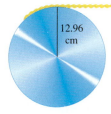

75. The two pulleys in the figure have radii of 15 cm and 8 cm, respectively. The larger pulley rotates 25 times in 36 sec. Find the angular velocity of each pulley in radians per sec.

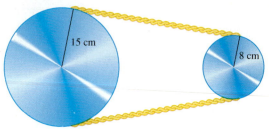

76. A thread is being pulled off a spool at the rate of 59.4 cm per sec. Find the radius of the spool if it makes 152 revolutions per min.

77. Earth revolves on its axis once every 24 hr. Assuming that Earth's radius is 6400 km, find the following.
  (a) Angular velocity of Earth in radians per day and radians per hr
  (b) Linear velocity at the North Pole or South Pole
  (c) Linear velocity at Quito, Ecuador, a city on the equator
  (d) Linear velocity at Salem, Oregon (halfway from the equator to the North Pole)

78. A railroad track is laid along the arc of a circle of radius 1800 ft. The circular part of the track subtends a central angle of 40°. How long (in seconds) will it take a point on the front of a train traveling 30 mph to go around this portion of the track?

79. A 90-horsepower outboard motor at full throttle rotates its propeller at 5000 revolutions per minute. Find the angular velocity of the propeller in radians per second.

6.7 Graphs of the Sine and Cosine Functions ▼▼▼

Many things in daily life repeat with a predictable pattern: in warm areas electricity use goes up in the summer and down in the winter, the price of fresh fruit goes down in the summer and up in the winter, and attendance at amusement parks increases in the summer and declines in autumn. Because the sine and cosine functions repeat their values over and over in a regular pattern, they are examples of *periodic functions*. Figure 50 shows a sine function that represents a normal heartbeat.

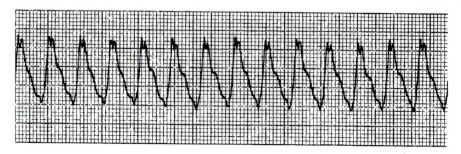

FIGURE 50

The circumference of the unit circle is 2π, so the smallest value of p for which the sine and cosine functions repeat is 2π. Therefore, the sine and cosine functions are periodic functions with period 2π.

Earlier, we saw that if an arc of length s is traced along the unit circle $x^2 + y^2 = 1$, starting at the point $(1, 0)$, the terminal point of the arc has coordinates $(\cos s, \sin s)$. Both functions $f(s) = \sin s$ and $g(s) = \cos s$ have domain $(-\infty, \infty)$ and range $[-1, 1]$. Since their period is 2π, the functions can be graphed using values of s from 0 to 2π.

Look at Figure 51, which shows a unit circle with a point (p, q) on it. Based on the definitions of circular functions, for any arc s or angle s in radians, $p = \cos s$ and $q = \sin s$. As angle s increases from 0 to $\pi/2$ (or 0° to 90°), $q = \sin s$ increases from 0 to 1, while $p = \cos s$ decreases from 1 to 0. As s increases from $\pi/2$ to π (or 90° to 180°), $q = \sin s$ decreases from 1 to 0, while $p = \cos s$ decreases from 0 to -1. Similar results can be found for the other quadrants, as shown in the following table.

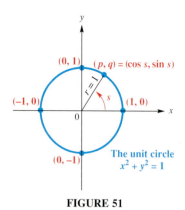

FIGURE 51

As s Increases from	sin s	cos s
0 to $\dfrac{\pi}{2}$	Increases from 0 to 1	Decreases from 1 to 0
$\dfrac{\pi}{2}$ to π	Decreases from 1 to 0	Decreases from 0 to -1
π to $\dfrac{3\pi}{2}$	Decreases from 0 to -1	Increases from -1 to 0
$\dfrac{3\pi}{2}$ to 2π	Increases from -1 to 0	Increases from 0 to 1

Any letter could be used instead of s for the arc length. In the rest of this chapter we will use circular functions of x (rather than s), so that we are graphing on the familiar xy-coordinate system.

Selecting key values of x and finding the corresponding values of sin x give the following results. (Decimals are rounded to the nearest tenth.)

x	0	$\dfrac{\pi}{4}$	$\dfrac{\pi}{2}$	$\dfrac{3\pi}{4}$	π	$\dfrac{5\pi}{4}$	$\dfrac{3\pi}{2}$	$\dfrac{7\pi}{4}$	2π
sin x	0	.7	1	.7	0	$-.7$	-1	$-.7$	0

Plotting the points from the table of values and connecting them with a smooth curve gives the solid portion of the graph in Figure 52. Since $y = \sin x$ is periodic and has $(-\infty, \infty)$ as its domain, the graph continues in both directions indefinitely, as indicated by the arrows. This graph is sometimes called a **sine wave** or **sinusoid.** You should learn the shape of this graph and be able to sketch it quickly. The key points of the graph are $(0, 0)$, $(\pi/2, 1)$, $(\pi, 0)$, $(3\pi/2, -1)$, and $(2\pi, 0)$. By plotting these five points and connecting them with the characteristic sine wave, you can quickly sketch the graph.

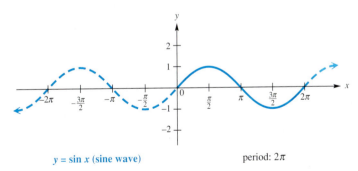

$y = \sin x$ **(sine wave)**   period: 2π

FIGURE 52

The same scales are used on both the x and y axes of Figure 52 so as not to distort the graph. Since the period of $y = \sin x$ is 2π, it is convenient to use subdivisions of 2π on the x-axis. The more familiar x-values, 1, 2, 3, 4, and so on, are still present, but are usually not shown to avoid cluttering the graph. These values are shown in Figure 53.

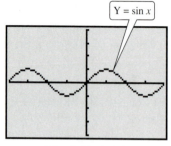

$[-2\pi, 2\pi]$ by $[-4, 4]$

Xscl $= \frac{\pi}{2}$ Yscl $= 1$

The graph of Y = $\sin x$ is shown in the standard *trig window*, having the dimensions and scales indicated.

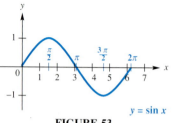

FIGURE 53

Sine graphs occur in many different practical applications. For one application, look back at Figure 51 and assume that the line from the origin to the point (p, q) is part of the pedal of a bicycle wheel, with a foot placed at (p, q). As mentioned earlier, q is equal to $\sin x$, showing that the height of the pedal from the horizontal axis in Figure 51 is given by $\sin x$. By choosing various angles for the pedal and calculating q for each angle, the height of the pedal leads to the sine curve shown in Figure 54. Two sample points are shown in Figure 54.

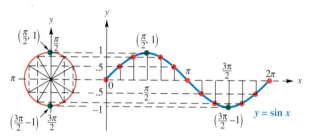

FIGURE 54

The graph of $y = \cos x$ can be found in much the same way as the graph of $y = \sin x$. A table of values is shown below for $y = \cos x$.

x	0	$\frac{\pi}{4}$	$\frac{\pi}{2}$	$\frac{3\pi}{4}$	π	$\frac{5\pi}{4}$	$\frac{3\pi}{2}$	$\frac{7\pi}{4}$	2π
$\cos x$	1	.7	0	$-.7$	-1	$-.7$	0	.7	1

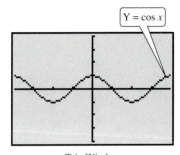

Trig Window
This is a calculator-generated graph of
the cosine function.

Here the key points are $(0, 1)$, $(\pi/2, 0)$, $(\pi, -1)$, $(3\pi/2, 0)$, and $(2\pi, 1)$.

The graph of $y = \cos x$, in Figure 55, has the same shape as the graph of $y = \sin x$. In fact, it is the graph of the sine function, shifted $\pi/2$ units to the left.

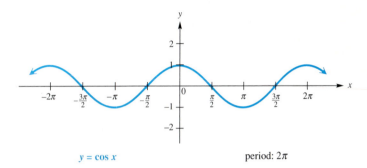

$y = \cos x$ period: 2π

FIGURE 55

STRETCHING AND SHRINKING The next examples show graphs that are "stretched" either vertically, horizontally, or both when compared with the graphs of $y = \sin x$ or $y = \cos x$.

◀ **EXAMPLE 1**
Graphing $y = a \sin x$

Graph $y = 2 \sin x$.

For a given value of x, the value of y is twice as large as it would be for $y = \sin x$, as shown in the table of values. The only change in the graph is the range, which becomes $[-2, 2]$. See Figure 56, which also shows a graph of $y = \sin x$ for comparison.

x	0	$\dfrac{\pi}{2}$	π	$\dfrac{3\pi}{2}$	2π
$\sin x$	0	1	0	-1	0
$2 \sin x$	0	2	0	-2	0

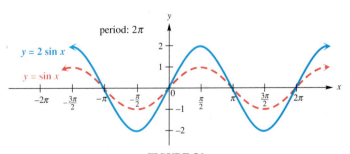

FIGURE 56

Generalizing from Example 1 and assuming $a \neq 0$ gives the following.

Amplitude of the Sine and Cosine Functions

The graph of $y = a \sin x$ or $y = a \cos x$ will have the same shape as the graph of $y = \sin x$ or $y = \cos x$, respectively, except with range $[-|a|, |a|]$. The number $|a|$ is called the **amplitude.** (The amplitude of a periodic function can be interpreted as half the difference between its maximum and minimum values.)

No matter what the value of the amplitude, the period of $y = a \sin x$ and $y = a \cos x$ is still 2π. Now suppose $y = \sin 2x$. We can complete a table of values for the interval $[0, 2\pi]$.

x	0	$\dfrac{\pi}{4}$	$\dfrac{\pi}{2}$	$\dfrac{3\pi}{4}$	π	$\dfrac{5\pi}{4}$	$\dfrac{3\pi}{2}$	$\dfrac{7\pi}{4}$	2π
$\sin 2x$	0	1	0	-1	0	1	0	-1	0

The period here is π, which equals $2\pi/2$. What about $y = \sin 4x$? Look at the table below.

x	0	$\dfrac{\pi}{8}$	$\dfrac{\pi}{4}$	$\dfrac{3\pi}{8}$	$\dfrac{\pi}{2}$	$\dfrac{5\pi}{8}$	$\dfrac{3\pi}{4}$	$\dfrac{7\pi}{8}$	π
$\sin 4x$	0	1	0	-1	0	1	0	-1	0

These values suggest that a complete cycle is achieved in $\pi/2$ units, which is reasonable since

$$\sin\left(4 \cdot \frac{\pi}{2}\right) = \sin 2\pi = 0.$$

In general, since the period of $\sin x$ or $\cos x$ is 2π, to find the period of either $\sin bx$ or $\cos bx$, $b > 0$, we solve the compound inequality

$$0 \leq bx \leq 2\pi$$

for x, getting

$$0 \leq x \leq \frac{2\pi}{b}.$$

Period of the Sine and Cosine Functions

The graph of $y = \sin bx$ will look like that of $y = \sin x$, but with a period of $|2\pi/b|$. Also, the graph of $y = \cos bx$ will look like that of $y = \cos x$, but with a period of $|2\pi/b|$.

EXAMPLE 2
Graphing $y = \cos bx$

Graph $y = \cos \dfrac{2}{3}x.$

For this function the period is

$$\frac{2\pi}{\dfrac{2}{3}} = 3\pi,$$

and the amplitude is 1. Dividing the interval $[0, 3\pi]$ into four equal parts, we get the x-values that will yield minimum points, maximum points, and x-intercepts.

$$0 \quad \frac{3\pi}{4} \quad \frac{3\pi}{2} \quad \frac{9\pi}{4} \quad 3\pi$$

These values are used to get a table of key points for one period.

x	0	$\dfrac{3\pi}{4}$	$\dfrac{3\pi}{2}$	$\dfrac{9\pi}{4}$	3π
$\dfrac{2}{3}x$	0	$\dfrac{\pi}{2}$	π	$\dfrac{3\pi}{2}$	2π
$\cos \dfrac{2}{3}x$	1	0	-1	0	1

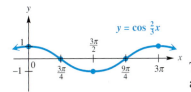

FIGURE 57

The amplitude is 1, because the minimum value is -1, the maximum value is 1, and half of $1 - (-1) = (1/2)(2) = 1$. Now plot these points and join them with a smooth curve. The graph is shown in Figure 57. ▶

The steps used to graph $y = a \sin bx$ or $y = a \cos bx$ are given below.

> **Graphing the Sine and Cosine Functions**
>
> To graph $y = a \sin bx$ or $y = a \cos bx$:
>
> 1. Find the period, $|2\pi/b|$. Start at 0 on the x-axis and lay off a distance of $|2\pi/b|$.
> 2. Divide the interval into four equal parts.
> 3. Evaluate the function for each of the five x-values resulting from Step 2. The points will be maximum points, minimum points, and x-intercepts.
> 4. Plot the points found in Step 3, and join them with a sinusoidal curve with amplitude $|a|$.
> 5. Draw additional cycles of the graph, to the right and to the left, as needed.

The function in the next example has both amplitude and period affected by constants.

EXAMPLE 3
Graphing $y = a \sin bx$

Graph $y = -2 \sin 3x$.

Step 1 For this function, $b = 3$, so the period is $2\pi/3$. We will graph the function over the interval $[0, 2\pi/3]$.

Step 2 Dividing the interval $[0, 2\pi/3]$ into four equal parts gives the x-values 0, $\pi/6$, $\pi/3$, $\pi/2$, and $2\pi/3$.

Step 3 Make a table of points determined by the x-values resulting from Step 2.

x	0	$\dfrac{\pi}{6}$	$\dfrac{\pi}{3}$	$\dfrac{\pi}{2}$	$\dfrac{2\pi}{3}$
$3x$	0	$\dfrac{\pi}{2}$	π	$\dfrac{3\pi}{2}$	2π
$\sin 3x$	0	1	0	-1	0
$-2 \sin 3x$	0	-2	0	2	0

Step 4 Plot the points $(0, 0)$, $(\pi/6, -2)$, $(\pi/3, 0)$, $(\pi/2, 2)$, and $(2\pi/3, 0)$, and join them with a sinusoidal curve with amplitude 2. See Figure 58.

Step 5 If necessary, the graph in Figure 58 can be extended by repeating the cycle over and over.

Notice the effect of the negative value of a. When a is negative, the graph of $y = a \sin bx$ will be the reflection about the x-axis of the graph of $y = a \sin bx$ for positive a. ▶

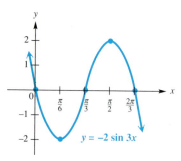

$y = -2 \sin 3x$

FIGURE 58

The graphing calculator can provide an enlightening look at how the constants a and b affect the graph of a function of the form $y = a \sin bx$ or $y = a \cos bx$. Be sure the calculator is set for radians. Some graphing calculators have a built-in "Trig" window with domain $[-2\pi, 2\pi]$ and range $[-4, 4]$. Of course, other settings may be preferable in some cases. By graphing pairs of functions such as $y = \sin x$ and $y = 2 \sin x$, or $y = \sin x$ and $y = -2 \sin x$, we can observe the effect of the coefficient a. See the figures below.

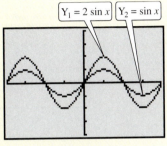

Trig Window

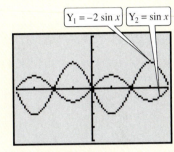

Trig Window

To determine the effect of the positive constant b on the graph of $y = \cos bx$ or $y = \sin bx$, keep the graph of $y = \sin x$ as Y_1 and enter $Y_2 = \sin 2x$, then replace Y_2 with $Y_3 = -\sin 2x$. Notice how the graphs have the same basic shape, but the periods of Y_2 and Y_3 are half that of Y_1. See the figures below.

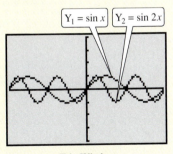

Trig Window

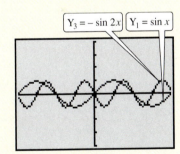

Trig Window

We can also use the graph of $y = \sin x$, for example, to find ordered pairs that satisfy the function, by using the TRACE feature. One such pair is

$$X = 1.1243595, \quad Y = .90199124.$$

This means that $\sin 1.1243595 \approx .90199124$.

◀EXAMPLE 4
Interpreting a sine function
model

The average temperature (in °F) at Mould Bay, Canada, can be approximated by the circular function

$$f(x) = 34 \sin\left[\frac{\pi}{6}(x - 4.3)\right],$$

where x is the month and $x = 1$ corresponds to January.

(a) Graph f over the interval $1 \leq x \leq 25$. Determine the amplitude and period of the graph.

The graph is shown in Figure 59. Its amplitude is 34 and the period is

$$\frac{2\pi}{\dfrac{\pi}{6}} = 12.$$

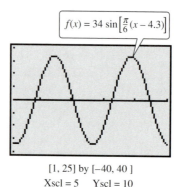

$$f(x) = 34 \sin\left[\frac{\pi}{6}(x - 4.3)\right]$$

[1, 25] by [−40, 40]
Xscl = 5   Yscl = 10

FIGURE 59

The function f has a period of 12 months or 1 year, which agrees with the changing of the seasons.

(b) What is the average temperature during the month of May?

May is the 5th month, so the average temperature during the month of May is

$$f(5) = 34 \sin\left[\frac{\pi}{6}(5 - 4.3)\right] \approx 12°F.$$

(c) What would be an approximation for the average *yearly* temperature in Mould Bay?

From the graph it appears that the average yearly temperature is about 0°F, since the graph is centered vertically about the line $y = 0$. ▶

TRANSLATIONS In Section 3.6, we saw that the graph of the function $y = f(x - d)$ is translated *horizontally* when compared to the graph of $y = f(x)$. The translation is d units to the right if $d > 0$ and $|d|$ units to the left if $d < 0$. In the function $y = f(x - d)$, the expression $x - d$ is called the **argument.** With circular functions, a horizontal translation is called a **phase shift.**

In the next example, we show two methods that can be used to graph a circular function involving a phase shift.

◀EXAMPLE 5
Graphing $y = \sin(x - d)$

Graph $y = \sin\left(x - \dfrac{\pi}{3}\right)$.

Method 1   The argument $x - \pi/3$ indicates that the graph will be translated $\pi/3$ units to the *right* (the phase shift) as compared to the graph of $y = \sin x$. In Figure 60 the graph $y = \sin x$ is shown as a dashed curve, and the graph of $y = \sin(x - \pi/3)$ is shown as a solid curve. Therefore, to graph a function using this method, first graph the basic circular function, and then graph the desired function by using the appropriate translation.

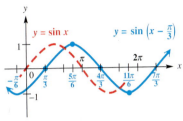

FIGURE 60

Method 2   For the argument $x - \pi/3$ to result in all possible values throughout one period, it must take on all values between 0 and 2π, inclusive. Therefore, to find an interval of one period, we solve the compound inequality

$$0 \le x - \frac{\pi}{3} \le 2\pi.$$

Add $\pi/3$ to each expression to find the interval

$$\frac{\pi}{3} \le x \le \frac{7\pi}{3} \qquad \text{or} \qquad \left[\frac{\pi}{3}, \frac{7\pi}{3}\right].$$

Divide this interval into four equal parts to get the following values.

$$\frac{\pi}{3} \quad \frac{5\pi}{6} \quad \frac{4\pi}{3} \quad \frac{11\pi}{6} \quad \frac{7\pi}{3}$$

Make a table of points using the x-values above.

x	$\dfrac{\pi}{3}$	$\dfrac{5\pi}{6}$	$\dfrac{4\pi}{3}$	$\dfrac{11\pi}{6}$	$\dfrac{7\pi}{3}$
$x - \dfrac{\pi}{3}$	0	$\dfrac{\pi}{2}$	π	$\dfrac{3\pi}{2}$	2π
$\sin\left(x - \dfrac{\pi}{3}\right)$	0	1	0	-1	0

Join these points to get the graph shown in Figure 60. The period is 2π and the amplitude is 1. ▶

As we saw in Section 3.6, the graph of a function of the form $y = c + f(x)$ is shifted *vertically* compared with the graph of $y = f(x)$. The function $y = c + f(x)$ is called a **vertical translation** of $y = f(x)$. The next example illustrates a vertical translation of a circular function.

EXAMPLE 6
Graphing $y = c + a \cos bx$

Graph $y = 3 - 2 \cos 3x$.

The values of y will be 3 greater than the corresponding values of y in $y = -2 \cos 3x$. This means that the graph of $y = 3 - 2 \cos 3x$ is the same as the graph of $y = -2 \cos 3x$, except with a vertical translation of 3 units upward. Since the period of $y = -2 \cos 3x$ is $2\pi/3$, the key points have the following x-values.

$$0 \quad \frac{\pi}{6} \quad \frac{\pi}{3} \quad \frac{\pi}{2} \quad \frac{2\pi}{3}$$

The amplitude is 2. The key points are shown on the graph in Figure 61. ▶

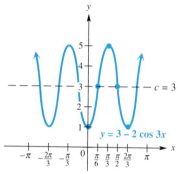

FIGURE 61

Horizontal and Vertical Translations

The graph of $y = \sin(x - d)$ or $y = \cos(x - d)$ has the shape of $y = \sin x$ or $y = \cos x$, but with a translation of $|d|$ units—to the left if $d < 0$ and to the right if $d > 0$. The number d is the **phase shift** of the graph.

The graph of $y = c + \sin x$ or $y = c + \cos x$ has the same shape as the graph of $y = \sin x$ or $y = \cos x$, but is translated $|c|$ units—down if $c < 0$ and up if $c > 0$.

The graphing calculator also helps to reinforce the concepts of horizontal and vertical translations. Again, let Y_1 remain as $\sin x$, and enter Y_2 as $\sin(x - \pi/3)$, taking care to insert parentheses as necessary. The graph of Y_2 should be the same as the graph of Y_1 shifted $\pi/3$ units to the right. See the figure at left.

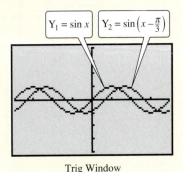

Trig Window

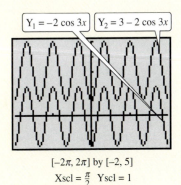

$[-2\pi, 2\pi]$ by $[-2, 5]$
$Xscl = \frac{\pi}{2}$ $Yscl = 1$

To see the effect of a vertical translation, enter $Y_1 = -2 \cos 3x$ and $Y_2 = 3 - 2 \cos 3x$. The constant $c = 3$ in Y_2 will have the effect of shifting the graph of Y_1 three units upward. Adjust the range so that the minimum and maximum values of y are -2 and 5, respectively. Now graph these two functions and observe the results, shown in the figure at right.

We can now graph a trigonometric function that involves all the types of stretching, compressing, and shifting.

EXAMPLE 7
Graphing $y = c + a \sin b(x - d)$

Graph $y = -1 + 2 \sin(4x + \pi)$.

First write the expression in the form $c + a \sin b(x - d)$ by factoring 4 out of the argument as follows.

$$y = -1 + 2 \sin(4x + \pi) = -1 + 2 \sin 4\left(x + \frac{\pi}{4}\right)$$

Now the amplitude is 2, the period is $2\pi/4 = \pi/2$, and the graph is translated down 1 unit and $\pi/4$ unit to the left as compared to the graph of $y = \sin x$. Since the graph is translated $\pi/4$ unit to the left, start at the x-value $0 - \pi/4 = -\pi/4$. The first period will end at $-\pi/4 + \pi/2 = \pi/4$. The maximum y-values will be $2 - 1 = 1$ and the minimum y-values will be $-2 - 1 = -3$. Sketch the graph using the typical sine curve. See Figure 62.

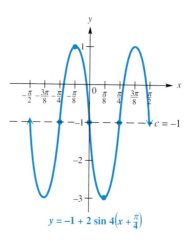

$$y = -1 + 2 \sin 4\left(x + \frac{\pi}{4}\right)$$

FIGURE 62

Alternatively, start by finding an interval over which the graph will complete one period. To do this, use the argument $4(x + \pi/4)$ in a compound inequality, with 0 as one endpoint and 2π as the other. Then solve the inequality for x.

$$0 \le 4\left(x + \frac{\pi}{4}\right) \le 2\pi$$

$$0 \le x + \frac{\pi}{4} < \frac{\pi}{2} \qquad \text{Divide by 4.}$$

$$-\frac{\pi}{4} \le x \le \frac{\pi}{4} \qquad \text{Subtract } \tfrac{\pi}{4}.$$

Divide the interval $[-\pi/4, \pi/4]$ into four equal parts to find the key points on the graph, as shown in the following table.

x	$-\dfrac{\pi}{4}$	$-\dfrac{\pi}{8}$	0	$\dfrac{\pi}{8}$	$\dfrac{\pi}{4}$
$x + \dfrac{\pi}{4}$	0	$\dfrac{\pi}{8}$	$\dfrac{\pi}{4}$	$\dfrac{3\pi}{8}$	$\dfrac{\pi}{2}$
$4\left(x + \dfrac{\pi}{4}\right)$	0	$\dfrac{\pi}{2}$	π	$\dfrac{3\pi}{2}$	2π
$-1 + 2 \sin 4\left(x + \dfrac{\pi}{4}\right)$	-1	1	-1	-3	-1

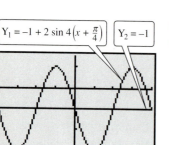

$Y_1 = -1 + 2 \sin 4\left(x + \frac{\pi}{4}\right)$ $Y_2 = -1$

$\left[-\frac{\pi}{2}, \frac{\pi}{2}\right]$ by $[-3.5, 1.5]$
$\text{Xscl} = \frac{\pi}{8}$ $\text{Yscl} = 1$

Compare to the graph in Figure 62.

Join the key points with a smooth curve as shown in Figure 62. This function has period $\pi/2$ and amplitude $[1 - (-3)]/2 = 2$. The phase shift is $\pi/4$ units to the left, and there is a vertical translation of 1 unit down. ▶

Graphing General Sine and Cosine Functions

To graph the general function $y = c + a \sin b(x - d)$ or $y = c + a \cos b(x - d)$, where $b > 0$, follow these steps.

1. Find an interval whose length is one period $(2\pi/b)$ by solving the compound inequality

$$0 \le b(x - d) \le 2\pi.$$

2. Divide the interval into four equal parts.
3. Evaluate the function for each of the five x-values resulting from Step 2. The points will be maximum points, minimum points, and points that intersect the line $y = c$ ("middle" points of the wave).
4. Plot the points found in Step 3, and join them with a sinusoidal curve.
5. Draw the graph over additional periods, to the right and to the left, as needed.

The amplitude of the function is $|a|$. The vertical translation is c units up if $c > 0$, $|c|$ units down if $c < 0$. The horizontal translation (phase shift) is d units to the right if $d > 0$, and $|d|$ units to the left if $d < 0$.

You have probably noticed that the graphs of $\sin x$ and $\cos x$ are the same shape and each is a horizontal translation (or phase shift) of the other. Because of this we can rewrite any cosine function as a sine function or any sine function as a cosine function. The table of values below was produced by a graphing calculator. It shows that the corresponding y-values for $Y_1 = 2 \sin 2x$ and $Y_2 = 2 \cos 2(x - \pi/4)$ are the same in the intervals shown.

X	Y1	Y2
0	0	0
.5	1.6829	1.6829
1	1.8186	1.8186
1.5	.28224	.28224
2	-1.514	-1.514
2.5	-1.918	-1.918
3	-.5588	-.5588

Y1▪2sin 2X

X	Y1	Y2
3.5	1.314	1.314
4	1.9787	1.9787
4.5	.82424	.82424
5	-1.088	-1.088
5.5	-2	-2
6	-1.073	-1.073
6.5	.84033	.84033

Y2▪2cos 2(X-π/4)

FOR DISCUSSION OR WRITING
1. Graph these two functions and compare their graphs. What do you expect to find?
2. Find a sine function that is equivalent to $y = .5 \cos(x - \pi)$.

EXAMPLE 8
Modeling temperature with a sine function

The maximum average monthly temperature in New Orleans is 82°F and the minimum is 54°F. The table shows the average monthly temperatures.

Month	Temperature
Jan	54
Feb	55
Mar	61
Apr	69
May	73
June	79
July	82
Aug	81
Sept	77
Oct	71
Nov	59
Dec	55

(a) Using only the maximum and minimum temperatures, determine a function of the form $f(x) = a \sin b(x - d) + c$, where a, b, c, and d are constants, that models the average monthly temperature in New Orleans. Let x represent the month, with January corresponding to $x = 1$.

We can use the maximum and minimum average monthly temperatures to find the amplitude a.

$$a = \frac{82 - 54}{2} = 14$$

The average of the maximum and minimum temperatures is a good choice for c. The average is

$$\frac{82° + 54°}{2} = 68°F.$$

Since the coldest month is January, when $x = 1$, and the hottest month is July, when $x = 7$, we should choose d to be about 4. The table shows that temperatures are actually a little warmer after July than before, so we try $d = 4.2$. Since temperatures repeat every 12 months, b is $2\pi/12 = \pi/6$. Thus,

$$f(x) = a \sin b(x - d) + c = 14 \sin\left[\frac{\pi}{6}(x - 4.2)\right] + 68.$$

(b) On the same coordinate axes, graph f for a two-year period together with the actual data values found in the table.

We show a graphing calculator graph of f together with the data points in Figure 63. The function models the data quite accurately. ▶

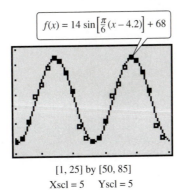

$$f(x) = 14 \sin\left[\frac{\pi}{6}(x - 4.2)\right] + 68$$

[1, 25] by [50, 85]
Xscl = 5   Yscl = 5

FIGURE 63

6.7 Exercises ▼▼▼▼▼▼▼▼▼▼▼▼▼▼▼▼▼▼▼▼▼▼▼▼▼▼▼▼▼▼▼▼▼▼▼▼

1. Describe the graph of $y = \sin x$ as if you were explaining it to a friend on the phone.

2. Describe the graph of $y = \cos x$ as if you were explaining it to a friend on the phone.

3. Explain why a graphing calculator will return a 1 for the expression shown in the accompanying screen, no matter what value is stored for X.

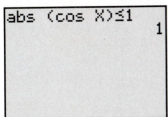

4. Give two possible values in the interval $[0, 2\pi)$ that can be stored in X for the accompanying calculator screen to be obtained. Support your result with your own calculator.

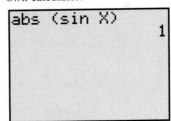

5. What is the minimum value of $y = 2 + \cos 4x$?

6. What is the maximum value of $y = 5 - 4 \sin 3x$?

7. What positive phase shift converts the graph of $y = \sin x$ to the graph of $y = \cos x$?

8. What negative phase shift converts the graph of $y = \sin x$ to the graph of $y = \cos x$?

Graph the defined function over the interval $[-2\pi, 2\pi]$. Do not use a graphing calculator. Identify the amplitude. See Example 1.

9. $y = 3 \sin x$

10. $y = 2 \cos x$

11. $y = \dfrac{3}{4} \cos x$

12. $y = \dfrac{2}{3} \sin x$

13. $y = -\sin x$

14. $y = -\cos x$

15. $y = -3 \cos x$

16. $y = -2 \sin x$

Graph the defined function over a two-period interval. Give the period, the amplitude, and any vertical translations. See Examples 1–3 and 6.

17. $y = \sin \frac{2}{3}x$

18. $y = \cos \frac{3}{4}x$

19. $y = \cos 3x$

20. $y = \sin 3x$

21. $y = -\sin 4x$

22. $y = -\cos 3x$

23. $y = 3 \cos \frac{1}{3}x$

24. $y = \frac{1}{2} \sin 2x$

25. $y = -2 \cos \frac{3}{2}x$

26. $y = 5 \sin 6x$

27. $y = \frac{2}{3} \cos \frac{1}{2}x$

28. $y = \frac{1}{2} \sin 3x$

29. $y = 1 + \sin x$

30. $y = 2 - \cos x$

31. $y = 2 - 3 \cos x$

32. $y = -3 + 2 \sin x$

Find the amplitude, the period, any vertical translation, and any phase shift of each graph. Draw the graph over a one-period interval. See Examples 5 and 7.

33. $y = \sin\left(x + \frac{\pi}{2}\right)$

34. $y = \cos\left(x - \frac{\pi}{4}\right)$

35. $y = 2 \sin\left(x + \frac{\pi}{2}\right)$

36. $y = \frac{3}{2} \sin 2\left(x - \frac{\pi}{4}\right)$

37. $y = 2 \cos\left(x - \frac{\pi}{3}\right)$

38. $y = 3 \cos(4x + \pi)$

39. $y = -2 \cos 4\left(x + \frac{\pi}{3}\right)$

40. $y = -\sin(2x - \pi)$

41. $y = \frac{1}{3} \cos\left(\frac{1}{4}x - \frac{\pi}{3}\right)$

42. $y = -3 + 2 \sin\left(x - \frac{\pi}{2}\right)$

43. $y = 4 - 3 \cos(x + \pi)$

44. $y = -\frac{1}{4} \sin\left(\frac{3}{4}x + \frac{\pi}{8}\right)$

Match each function with its graph in Exercises 45–52.

45. $y = \sin\left(x - \frac{\pi}{4}\right)$

46. $y = \sin\left(x + \frac{\pi}{4}\right)$

47. $y = \sin 2x$

48. $y = \cos 2x$

49. $y = 2 \sin x$

50. $y = 2 \cos x$

51. $y = 1 + \cos x$

52. $y = -1 + \cos x$

A.

B.

C.

D.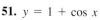

E.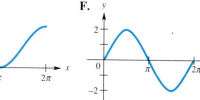

F.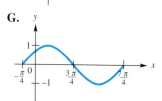

G.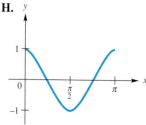

H.

53. Refer to the graph of $y = \sin x$ in Figure 52. The graph completes one cycle between $x = 0$ and $x = 2\pi$. Consider the statement, "The function $y = \sin(bx)$ completes b cycles between 0 and 2π." Use your graphing calculator to confirm the statement for some positive integer values of b, such as 3, 4, and 5. Interpret and confirm the statement for $b = 1/2$ and $b = 3/2$.

54. The graph shown gives the variation in blood pressure for a typical person. Systolic and diastolic pressures are the upper and lower limits of the periodic changes in pressure that produce the pulse. The length of time between peaks is called the period of the pulse.
(a) Find the amplitude of the graph.
(b) Find the pulse rate (the number of pulse beats in one minute) for this person.

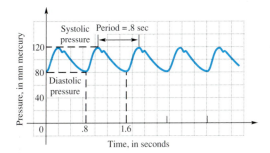

55. Many of the activities of living organisms are periodic. For example, the graph below shows the time that a certain nocturnal animal begins its evening activity.
(a) Find the amplitude of this graph.
(b) Find the period.

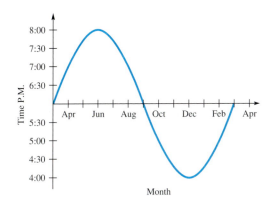

56. The figure shows schematic diagrams of a rhythmically moving arm. The upper arm RO rotates back and forth about the point R; the position of the arm is measured by the angle y between the actual position and the downward vertical position. (*Source*: De Sapio, Rodolfo, *Calculus for the Life Sciences*, W. H. Freeman and Company, Copyright © 1978. Reprinted by permission.)
(a) Find an equation of the form $y = a \sin kt$ for the graph shown.
(b) How long does it take for a complete movement of the arm?

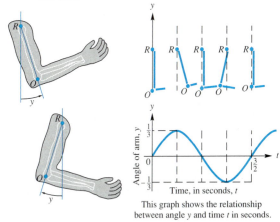

This graph shows the relationship between angle y and time t in seconds.

Pure sounds produce single sine waves on an oscilloscope. Find the amplitude and period of each sine wave in the following figures. On the vertical scale, each square represents .5, and on the horizontal scale each square represents 30° or $\pi/6$.

57.

58.

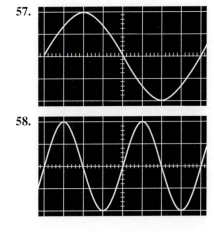

59. The voltage E in an electrical circuit is given by
$$E = 5 \cos 120\pi t,$$
where t is time measured in seconds.
(a) Find the amplitude and the period.
(b) How many cycles are completed in one second? (The number of cycles (periods) completed in one second is the **frequency** of the function.)
(c) Find E when $t = 0, .03, .06, .09, .12$.
(d) Graph E, for $0 \le t \le 1/30$.

60. For another electrical circuit, the voltage E is given by
$$E = 3.8 \cos 40\pi t,$$
where t is time measured in seconds.
(a) Find the amplitude and the period.
(b) Find the frequency. See Exercise 59(b).
(c) Find E when $t = .02, .04, .08, .12, .14$.
(d) Graph one period of E.

 61. 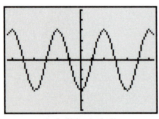 At Mauna Loa, Hawaii, atmospheric carbon dioxide levels in parts per million (ppm) have been measured regularly since 1958. The function defined by
$$L(x) = .022x^2 + .55x + 316 + 3.5 \sin(2\pi x)$$
can be used to model these levels, where x is in years and $x = 0$ corresponds to 1960. (*Source:* Nilsson, A., *Greenhouse Earth,* John Wiley & Sons, New York, 1992.)
(a) Graph L for $15 \le x \le 35$. (*Hint:* Use $325 \le y \le 365$.)

(b) When do the seasonal maximum and minimum carbon dioxide levels occur?
(c) L is the sum of a quadratic function and a sine function. What is the significance of each of these functions? Discuss what physical phenomena may be responsible for each function.

62. (Refer to the previous exercise.) The carbon dioxide content in the atmosphere at Barrow, Alaska, in parts per million (ppm) can be modeled using the function defined by
$$C(x) = .04x^2 + .6x + 330 + 7.5 \sin(2\pi x),$$
where $x = 0$ corresponds to 1970. (*Source:* Zeilik, M., S. Gregory, and E. Smith, *Introductory Astronomy and Astrophysics,* Saunders College Publishing, 1992.)
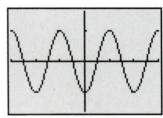 (a) Graph C for $5 \le x \le 25$. (*Hint:* Use $320 \le y \le 380$.)
(b) Discuss possible reasons why the amplitude of the oscillations in the graph of C are larger than the amplitude of the oscillations in the graph of L which models Hawaii.
(c) Define a new function C that is valid if x represents the actual year where $1975 \le x \le 1995$.

63. The graph of $y = .5 \sin x + .866 \cos x$ is the same as that of a function of the form $y = a \sin(x + \alpha)$. Graph the function with a graphing calculator and then estimate the values of a and α.

For each graph in Exercises 64 and 65, give the equation of a sine function having that graph.

64. (*Note:* Xscl is approximately $\pi/4$.)

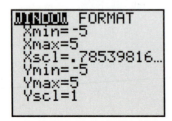

[−5, 5] by [−5, 5]

Xscl = $\frac{\pi}{4}$ Yscl = 1

```
WINDOW FORMAT
Xmin=-5
Xmax=5
Xscl=.78539816…
Ymin=-5
Ymax=5
Yscl=1
```

65.

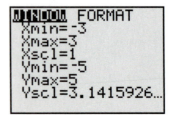

[−3, 3] by [−5, 5]

Xscl = 1   Yscl = π

```
WINDOW FORMAT
Xmin=-3
Xmax=3
Xscl=1
Ymin=-5
Ymax=5
Yscl=3.1415926…
```

66. Give the equation of a cosine function having the graph of the figure in Exercise 64.

67. Give the equation of a cosine function having the graph of the figure in Exercise 65.

Exercises 68 and 69 refer to the table below. The numbers in the X column are approximations of $\frac{\pi}{2}, \frac{3\pi}{4}, \pi, \ldots$ (that is, 90°, 135°, 180°, . . .). The functions Y_1 and Y_2 have the form $y = c + a \sin b(x - d)$.

X	Y₁	Y₂
1.5708	2	-1
2.3562	1.5	1E-13
3.1416	1.134	.73205
3.927	1	1
4.7124	1.134	.73205
5.4978	1.5	0
6.2832	2	-1

X=6.28318530718

X	Y₁	Y₂
7.0686	2.5	-2
7.854	2.866	-2.732
8.6394	3	-3
9.4248	2.866	-2.732
10.21	2.5	-2
10.996	2	-1
11.781	1.5	2E-13

X=11.78097245096

68. Determine a possible expression for the function Y_1.

69. Determine a possible expression for the function Y_2.

---

70. The average temperature (in °F) in Austin, Texas, can be modeled using the circular function

$$f(x) = 17.5 \sin[(\pi/6)(x - 4)] + 67.5$$

where x is the month and $x = 1$ corresponds to January. (*Source:* Miller, A., and J. Thompson, *Elements of Meteorology,* Charles E. Merrill Publishing Company, Columbus, Ohio, 1975.)

(a) Graph f over the interval $1 \le x \le 25$. Determine the amplitude, period, phase shift, and vertical translation of f. See Example 4.

(b) What is the average monthly temperature for the month of December?

(c) Determine the maximum and minimum average monthly temperatures and the months when they occur.

(d) What would be an approximation for the average *yearly* temperature in Austin? How is this related to the vertical translation of the sine function in the formula of f?

Month	Temperature
Jan	36
Feb	39
Mar	43
Apr	48
May	55
June	59
July	64
Aug	63
Sept	57
Oct	50
Nov	43
Dec	39

71. The average monthly temperature (in °F) in Vancouver, Canada is shown in the table. (*Source:* Miller, A., and J. Thompson, *Elements of Meteorology,* Charles E. Merrill Publishing Company, Columbus, Ohio, 1975.)

(a) Plot the average monthly temperature over a two-year period by letting $x = 1$ correspond to the month of January during the first year. Do the data seem to indicate a translated sine graph? See Example 8.

(b) The highest average monthly temperature is 64°F in July and the lowest average monthly temperature is 36°F in January. Their average is 50°F. Graph the data together with the line $y = 50$. What does this line represent with regard to temperature in Vancouver?

(c) Approximate the amplitude, period, and phase shift of the translated sine wave indicated by the data.

(d) Determine a function that has the form $f(x) = a \sin b(x - d) + c$ where a, b, c, and d are constants, which models the data.

(e) Graph f together with the data on the same coordinate axes. Comment on how well f models the given data.

72. The average monthly temperature (in °F) in Phoenix, Arizona, is shown in the table. (*Source:* Miller, A., and J. Thompson, *Elements of Meteorology,* Charles E. Merrill Publishing Company, Columbus, Ohio, 1975.)

(a) Predict the average yearly temperature and compare it to the actual value of 70°F.
(b) Plot the average monthly temperature over a two-year period by letting $x = 1$ correspond to January of the first year.
(c) Determine a function that has the form $f(x) = a \cos b(x - d) + c$ where a, b, c, and d are constants, which models the data.
(d) Graph f together with the data on the same coordinate axes.

Month	Temperature
Jan	51
Feb	55
Mar	63
Apr	67
May	77
June	86
July	90
Aug	90
Sept	84
Oct	71
Nov	59
Dec	52

6.8 Graphs of the Other Circular Functions ▼▼▼

In this section we discuss the graphs of the four remaining circular functions: cosecant, secant, tangent, and cotangent.

GRAPHS OF COSECANT AND SECANT Since cosecant values are reciprocals of the corresponding sine values, the period of the function $y = \csc x$ is 2π, the same as for $y = \sin x$. The following table shows several values for $y = \sin x$ and the corresponding values of $y = \csc x$.

x	0	$\dfrac{\pi}{4}$	$\dfrac{\pi}{2}$	$\dfrac{3\pi}{4}$	π	$\dfrac{5\pi}{4}$	$\dfrac{3\pi}{2}$	2π
$\sin x$	0	$\dfrac{\sqrt{2}}{2}$	1	$\dfrac{\sqrt{2}}{2}$	0	$-\dfrac{\sqrt{2}}{2}$	-1	0
$\csc x$	undefined	$\sqrt{2}$	1	$\sqrt{2}$	undefined	$-\sqrt{2}$	-1	undefined

When $\sin x = 1$, the value of $\csc x$ is also 1, and when $0 < \sin x < 1$, then $\csc x > 1$. Also, if $-1 < \sin x < 0$, then $\csc x < -1$. As x approaches 0, $\sin x$ approaches 0, and $|\csc x|$ gets larger and larger. The lines $x = n\pi$, where n is any integer, are all vertical asymptotes of the graph. Using this information and plotting a few points shows that the graph takes the shape of the solid curve

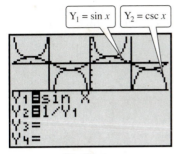

Trig Window

To graph $y = \csc x$, use the fact that $\csc x = \dfrac{1}{\sin x}$. The graphs of both $Y_1 = \sin x$ and $Y_2 = \csc x$ are shown. The calculator is in split-screen and connected modes.

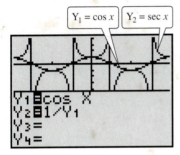

Trig Window

To graph $Y_2 = \sec x$, use the fact that $\sec x = \dfrac{1}{\cos x}$.

shown in Figure 64. To show how the two graphs are related, the graph of $y = \sin x$ is also shown, as a dashed curve. The domain of the function $y = \csc x$ is $\{x \mid x \neq n\pi$, where n is any integer$\}$, and the range is $(-\infty, -1] \cup [1, \infty)$.

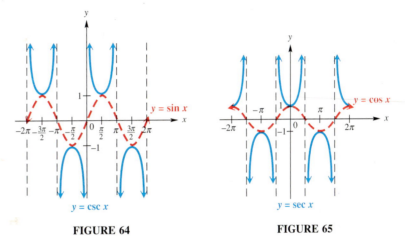

FIGURE 64   **FIGURE 65**

   The graph of $y = \sec x$, shown in Figure 65, is related to the cosine graph in the same way that the graph of $y = \csc x$ is related to the sine graph, because $\sec x = 1/\cos x$. The domain of the function $y = \sec x$ is $\{x \mid x \neq \pi/2 + n\pi$, where n is any integer$\}$, and the range is $(-\infty, -1] \cup [1, \infty)$.
   In order to graph functions based on the cosecant and secant, see the summary that follows.

Graphing the Cosecant and Secant Functions

To graph $y = a \csc bx$ or $y = a \sec bx$, with $b > 0$, follow these steps.

1. Graph the corresponding reciprocal function as a guide, using a dashed curve. That is,

To graph	Use as a guide
$y = a \csc bx$	$y = a \sin bx$
$y = a \sec bx$	$y = a \cos bx.$

2. Sketch the vertical asymptotes. They will have equations of the form $x = k$, where k is an x-intercept of the graph of the guide function.
3. Sketch the graph of the desired function by drawing the typical U-shaped branches between the adjacent asymptotes. The branches will be above the graph of the guide function when the guide function values are positive, and below the graph of the guide function when the guide function values are negative. The graph will resemble the graphs in Figures 64 and 65.

Like the sine and cosine functions, the secant and cosecant function graphs may be translated vertically and horizontally. The period of both functions is 2π.

EXAMPLE 1
Graphing $y = a \sec bx$

Graph $y = 2 \sec \dfrac{1}{2}x$.

Use the guidelines above.

Step 1 This function involves the secant, so the corresponding reciprocal function will involve the cosine. The function that we will graph as a guide is

$$y = 2 \cos \frac{1}{2}x.$$

Using the guidelines of Section 6.7, we find that one period of the graph lies along the interval that satisfies the compound inequality $0 \le (1/2)x \le 2\pi$, or $[0, 4\pi]$. Dividing this interval into four equal parts gives the following key points.

$$(0, 2) \qquad (\pi, 0) \qquad (2\pi, -2) \qquad (3\pi, 0) \qquad (4\pi, 2)$$

These are joined with a smooth curve; it is dashed to indicate that this graph is only a guide. An additional period is graphed as seen in Figure 66(a).

Step 2 Sketch the vertical asymptotes. These occur at x-values for which the guide function equals 0. A few of them have the equations

$$x = -3\pi, \quad x = -\pi, \quad x = \pi, \quad \text{and} \quad x = 3\pi.$$

See Figure 66(a).

Step 3 Sketch the graph of $y = 2 \sec(1/2)x$ by drawing in the typical U-shaped branches, approaching the asymptotes. See Figure 66(b). ▶

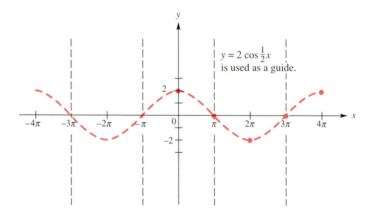

FIGURE 66(a)

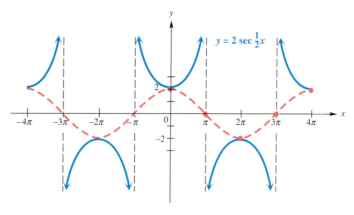

FIGURE 66(b)

◖EXAMPLE 2
Graphing $y = a \csc(x - d)$

Graph $y = \dfrac{3}{2} \csc\left(x - \dfrac{\pi}{2}\right)$.

This function can be graphed by first graphing the corresponding reciprocal function $y = (3/2)\sin(x - \pi/2)$. We can analyze the function as follows. Compared with the graph of $y = \csc x$, this graph has a phase shift of $\pi/2$ units to the right. Thus, the asymptotes are the lines $x = \pi/2$, $3\pi/2$, and so on. Also, there are no values of y between $-3/2$ and $3/2$. As shown in Figure 67, this is related to the increased amplitude of $y = (3/2)\sin x$ compared with $y = \sin x$. (Amplitude does not apply to the secant or cosecant functions; it enters only indirectly from the corresponding cosine or sine graphs.) This means that the graph goes through the points $(\pi, 3/2)$, $(2\pi, -3/2)$, and so on. Two periods are shown in Figure 67. (The graph of the "guide" function, $y = (3/2)\sin(x - \pi/2)$, is shown in red.) ◗

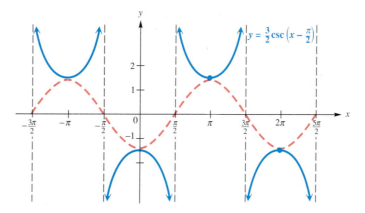

FIGURE 67

To graph the functions discussed in this section with a graphing calculator, we need to use the reciprocal functions of sin x, cos x, and tan x to get csc x, sec x, and cot x, respectively. Because the graphs of all circular functions except sin x and cos x have asymptotes, you may prefer to use dot mode for their graphs. The figures below show the graph of $y = (3/2)\csc(x - (\pi/2))$. As shown in the figures, connected mode will draw vertical appearing lines between the portions of the graph, while dot mode does not. Compare these graphs with Figure 67.

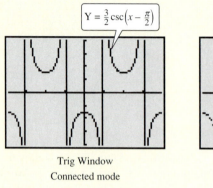

Trig Window
Connected mode

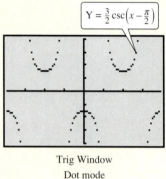

Trig Window
Dot mode

GRAPHS OF TANGENT AND COTANGENT   Earlier we showed that the values of $y = \tan x$ are positive in quadrants I and III, and negative in quadrants II and IV, so

$$\tan(x + \pi) = \tan x,$$

and the period of $y = \tan x$ is π. Thus, the tangent function need be investigated only within an interval of π units. A convenient interval for this purpose is $(-\pi/2, \pi/2)$ because, although the endpoints $-\pi/2$ and $\pi/2$ are not in the domain of $y = \tan x$ (why?), tan x exists for all other values in the interval. In the interval $(0, \pi/2)$, tan x is positive. As x goes from 0 to $\pi/2$, a calculator shows that tan x gets larger and larger without bound. As x goes from $-\pi/2$ to 0, the values of tan x approach 0 through negative values. These results are summarized in the following table.

As x increases from	tan x
0 to $\dfrac{\pi}{2}$	Increases from 0, without bound
$-\dfrac{\pi}{2}$ to 0	Increases to 0

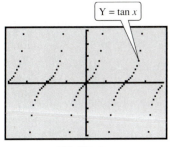

Y = tan x

Trig Window
Dot mode

This is a calculator-generated graph of
Y = tan x.

Based on these results, the graph of $y = \tan x$ will approach the vertical line $x = \pi/2$ but never touch it, so the line $x = \pi/2$ is a vertical asymptote. The lines $x = \pi/2 + n\pi$, where n is any integer, are all vertical asymptotes. These asymptotes are indicated with light dashed lines on the graph in Figure 68. In the interval $(-\pi/2, 0)$, which corresponds to quadrant IV on the unit circle, $\tan x$ is negative, and as x goes from 0 to $-\pi/2$, $\tan x$ gets smaller and smaller. A table of values for $\tan x$, where $-\pi/2 < x < \pi/2$, follows.

x	$-\dfrac{\pi}{3}$	$-\dfrac{\pi}{4}$	$-\dfrac{\pi}{6}$	0	$\dfrac{\pi}{6}$	$\dfrac{\pi}{4}$	$\dfrac{\pi}{3}$
$\tan x$	-1.7	-1	$-.6$	0	$.6$	1	1.7

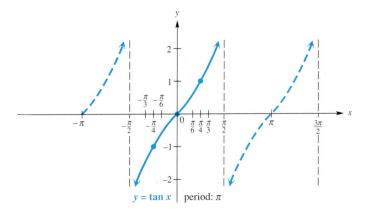

$y = \tan x$ │ period: π

FIGURE 68

Plotting the points from the table and letting the graph approach the asymptotes at $x = \pi/2$ and $x = -\pi/2$ gives the portion of the graph shown with a solid curve in Figure 68. More of the graph can be sketched by repeating the same curve, also as shown in the figure. This graph, like the graphs for the sine and cosine functions, should be remembered well enough so that a quick sketch can easily be made. Convenient key points are $(-\pi/4, -1)$, $(0, 0)$, and $(\pi/4, 1)$. These points are shown in Figure 68. The lines $x = \pi/2$ and $x = -\pi/2$ are vertical asymptotes. The idea of *amplitude,* discussed earlier, applies only to the sine and cosine functions. However, here it indicates that each y-value of $y = a \tan x$ is a times the corresponding y-value of $y = \tan x$. The domain of the tangent function is $\{x \mid x \neq \pi/2 + n\pi$, where n is any integer$\}$. The range is $(-\infty, \infty)$.

The definition $\cot x = 1/(\tan x)$ can be used to find the graph of $y = \cot x$. The period of the cotangent, like that of the tangent, is π. The domain of $y = \cot x$ excludes $0 + n\pi$, where n is any integer, since $1/\tan x$ is undefined for these values of x. Thus, the vertical lines $x = n\pi$ are asymptotes. Values of x that lead to asymptotes for $\tan x$ will make $\cot x = 0$, so $\cot(-\pi/2) = 0$, $\cot(\pi/2) = 0$, $\cot(3\pi/2) = 0$, and so on. The values of $\tan x$ increase as x

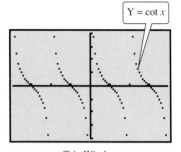

Trig Window
Dot mode

To graph Y = cot x, use the identity
cot $x = \dfrac{1}{\tan x}$ or the identity cot $x = \dfrac{\cos x}{\sin x}$.

goes from $-\pi/2$ to $\pi/2$, so the values of cot x will *decrease* as x goes from $-\pi/2$ to $\pi/2$. A table of values for cot x, where $0 < x < \pi$, is shown below.

x	$\dfrac{\pi}{6}$	$\dfrac{\pi}{4}$	$\dfrac{\pi}{3}$	$\dfrac{\pi}{2}$	$\dfrac{2\pi}{3}$	$\dfrac{3\pi}{4}$	$\dfrac{5\pi}{6}$
cot x	**1.7**	**1**	**.6**	**0**	**−.6**	**−1**	**−1.7**

Plotting these points and using the information discussed above gives the graph of $y = $ cot x shown in Figure 69. (The graph shows two periods.) The domain of the cotangent function is $\{x \mid x \neq n\pi$, where n is any integer$\}$. The range is $(-\infty, \infty)$.

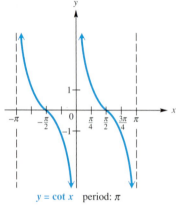

$y = $ cot x period: π

FIGURE 69

Graphing the Tangent and Cotangent Functions

To graph $y = a$ tan bx or $y = a$ cot bx, with $b > 0$, follow these steps.

1. The period is π/b. To locate two adjacent vertical asymptotes, solve the following equations for x:

   For $y = a$ tan bx: $bx = -\dfrac{\pi}{2}$ and $bx = \dfrac{\pi}{2}$

   For $y = a$ cot bx: $bx = 0$ and $bx = \pi$.

2. Sketch the two vertical asymptotes found in Step 1.
3. Divide the interval formed by the vertical asymptotes into four equal parts.
4. Evaluate the function for the first-quarter point, midpoint, and third-quarter point, using the x-values found in Step 3.
5. Join the points with a smooth curve, approaching the vertical asymptotes. Draw additional asymptotes and periods of the graph as necessary.

Like the other circular functions, the graphs of the tangent and cotangent functions may be shifted horizontally as well as vertically.

EXAMPLE 3
Graphing $y = \tan bx$

Graph $y = \tan 2x$.

Step 1 The period of this function is $\pi/2$. To locate two adjacent vertical asymptotes, solve $2x = -\pi/2$ and $2x = \pi/2$ (since this is a tangent function). We find that two asymptotes have equations

$$x = -\frac{\pi}{4} \quad \text{and} \quad x = \frac{\pi}{4}.$$

Step 2 Sketch the two vertical asymptotes $x = \pm\pi/4$, as shown in Figure 70.

Step 3 Divide the interval $(-\pi/4, \pi/4)$ into four equal parts. This gives the following key x-values:

$$\text{first-quarter value: } -\frac{\pi}{8}$$

$$\text{middle value: } \mathbf{0}$$

$$\text{third-quarter value: } \frac{\pi}{8}.$$

Step 4 Evaluate the function for the x-values found in Step 3, as shown in the following table.

x	$-\dfrac{\pi}{8}$	0	$\dfrac{\pi}{8}$
$2x$	$-\dfrac{\pi}{4}$	0	$\dfrac{\pi}{4}$
$\tan 2x$	-1	$\mathbf{0}$	$\mathbf{1}$

Step 5 Join these points with a smooth curve, approaching the vertical asymptotes. See Figure 70. Another period has been graphed as well, one-half period to the left and one-half period to the right. ▶

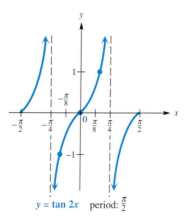

$y = \tan 2x$ period: $\frac{\pi}{2}$

FIGURE 70

EXAMPLE 4
Graphing $y = a \tan bx$

Graph $y = -3 \tan \dfrac{1}{2} x$.

The period is $\pi/(1/2) = 2\pi$. Adjacent asymptotes are at $x = -\pi$ and $x = \pi$. Dividing the interval $-\pi < x < \pi$ into four equal parts gives key x-values of $-\pi/2$, 0, and $\pi/2$. Evaluating the function at these x values gives these key points:

$$\left(-\frac{\pi}{2}, 3\right) \qquad (0, 0) \qquad \left(\frac{\pi}{2}, -3\right).$$

Plotting these points and joining them with a smooth curve, we get the graph shown in Figure 71. Notice that because the coefficient -3 is negative, the graph is reflected about the x-axis compared to the graph of $y = 3 \tan(1/2)x$. ▶

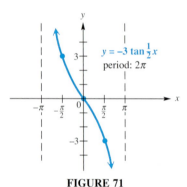

FIGURE 71

NOTE   The function $y = -3 \tan(1/2)x$ of Example 4, graphed in Figure 71, has a graph that compares to the graph of $y = \tan x$ as follows:

1. The period is larger, because $b = 1/2$, and $1/2 < 1$.
2. The graph is "stretched," because $a = -3$, and $|-3| > 1$.
3. Each branch of the graph goes down from left to right (that is, the function decreases) between each pair of adjacent asymptotes, because $a = -3 < 0$. When $a < 0$, the graph is reflected about the x-axis.

We graph a cotangent function in a similar manner.

EXAMPLE 5
Graphing $y = a \cot bx$

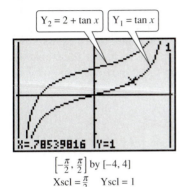

$\left[-\frac{\pi}{2}, \frac{\pi}{2}\right]$ by $[-4, 4]$
Xscl = $\frac{\pi}{2}$ Yscl = 1

For X = $\frac{\pi}{4}$ ≈ .78539816, Y_1 = 1.

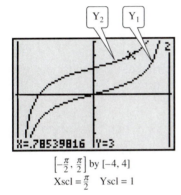

$\left[-\frac{\pi}{2}, \frac{\pi}{2}\right]$ by $[-4, 4]$
Xscl = $\frac{\pi}{2}$ Yscl = 1

For X = $\frac{\pi}{4}$ ≈ .78539816, Y_2 = 2 + Y_1 = 2 + 1 = 3. Notice from the two screens above, Y_2 = 2 + Y_1. The graph of Y_2 is obtained by translating the graph of Y_1 2 units upward.

Graph $y = \frac{1}{2} \cot 2x$.

Because this function involves the cotangent, we can locate two adjacent asymptotes by solving the equations $2x = 0$ and $2x = \pi$. We find that the lines $x = 0$ (the y-axis) and $x = \pi/2$ are two such asymptotes. We divide the interval $0 < x < \pi/2$ into four equal parts, getting key x-values of $\pi/8$, $\pi/4$, and $3\pi/8$. Evaluating the function at these x-values gives the following key points.

$$\left(\frac{\pi}{8}, \frac{1}{2}\right) \qquad \left(\frac{\pi}{4}, 0\right) \qquad \left(\frac{3\pi}{8}, -\frac{1}{2}\right)$$

Joining these points with a smooth curve approaching the asymptotes gives the graph shown in Figure 72. ▶

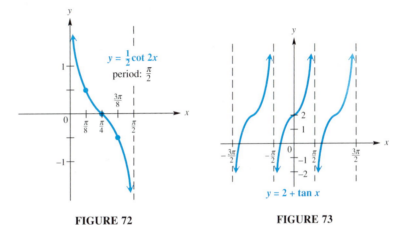

FIGURE 72   **FIGURE 73**

As stated earlier, tangent and cotangent function graphs may be translated as shown in the next two examples.

EXAMPLE 6
Graphing a tangent function with a vertical translation

Graph $y = 2 + \tan x$.

Every value of y for this function will be 2 units more than the corresponding value of y in $y = \tan x$, causing the graph $y = 2 + \tan x$ to be translated 2 units upward as compared with the graph of $y = \tan x$. See Figure 73. ▶

Graph $y = -2 - \cot\left(x - \dfrac{\pi}{4}\right)$.

Here $b = 1$, so the period is π. The graph will be translated down 2 units (because $c = -2$), reflected about the x-axis (because of the negative sign in front of the cotangent) and will have a phase shift (horizontal translation) of $\pi/4$ unit to the right (because of the argument $(x - \pi/4)$). To locate the translated adjacent asymptotes, since this function involves the cotangent, we solve the following equations:

$$x - \frac{\pi}{4} = 0 \qquad x - \frac{\pi}{4} = \pi$$

$$x = \frac{\pi}{4} \qquad x = \frac{5\pi}{4}.$$

Dividing the interval $\pi/4 < x < 5\pi/4$ into four equal parts and evaluating the function at the three key x-values within the interval gives these key points.

$$\left(\frac{\pi}{2}, -3\right) \qquad \left(\frac{3\pi}{4}, -2\right) \qquad (\pi, -1)$$

These points are joined by a smooth curve. This period of the graph, along with one in the interval $-3\pi/4 < x < \pi/4$, is shown in Figure 74. ▶

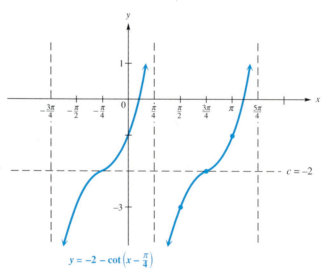

$y = -2 - \cot\left(x - \frac{\pi}{4}\right)$

FIGURE 74

New functions are often formed by adding or subtracting other functions. A function formed by combining two other functions, such as

$$y = \cos x + \sin x,$$

has historically been graphed using a method known as *addition of ordinates*. (The ordinate of a point is its y-coordinate.) To apply this method to this function, we would graph the functions $y = \cos x$ and $y = \sin x$. Then, for selected values of x, we would add $\cos x$ and $\sin x$, and plot the points $(x, \cos x + \sin x)$. Connecting the selected points with a typical circular function-type curve would give the graph of the desired function. While this method illustrates some valuable concepts involving the arithmetic of functions, it is very time-consuming.

With the technology of the graphing calculator, such an exercise can easily be accomplished. Let $Y_1 = \cos x$, $Y_2 = \sin x$, and $Y_3 = Y_1 + Y_2$. Then graph these three function in a "Trig" window and observe carefully as the calculator plots them in order. See the figure below.

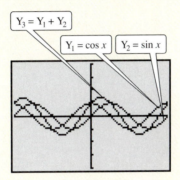

Trig Window

Use TRACE to locate points on each graph with the same x-value. Add the resulting coordinates Y_1 and Y_2. The sum should agree with Y_3. (There may be a slight discrepancy in the last decimal place.) Do you see how the term "addition of ordinates" originated? If your calculator has a TABLE feature, you can use it to see corresponding values of Y_1, Y_2, and Y_3.

FOR DISCUSSION OR WRITING
Use a graphing calculator to graph the following functions directly and then by graphing each term and the sum in one window.

1. $y = \sin x + \sin 2x$
2. $y = \cos x + \sec x$
3. Discuss the effects of subtracting ordinates.

6.8 Exercises ▼▼▼▼▼▼▼▼▼▼▼▼▼▼▼▼▼▼▼▼▼▼▼▼▼▼▼▼▼▼▼▼▼▼

1. Describe the graph of $y = \sec x$ as if you were explaining it to a friend on the phone.

2. Describe the graph of $y = \tan x$ as if you were explaining it to a friend on the phone.

3. True or false: The graph of $y = 3 \csc x$ is the same as the graph of $y = 1/(3 \sin x)$. If false, explain why.

4. If c is any number, how many solutions does the equation $c = \tan x$ have in the interval $(-2\pi, 2\pi]$?

5. True or false: The graph of $y = \tan x$ in Figure 68 suggests that $\tan(-x) = -\tan x$ for all x in the domain of $\tan x$.

6. True or false: The graph of $y = \sec x$ in Figure 65 suggests that $\sec(-x) = \sec x$ for all x in the domain of $\sec x$.

Match each equation with its graph in Exercises 7–12.

7. $y = -\csc x$

8. $y = -\sec x$

9. $y = -\tan x$

10. $y = -\cot x$

11. $y = \tan\left(x - \dfrac{\pi}{4}\right)$

12. $y = \cot\left(x - \dfrac{\pi}{4}\right)$

A.

B.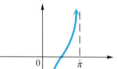

C.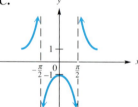

D.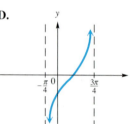

E.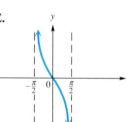

F.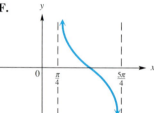

Graph each defined function over a one-period interval. See Examples 1 and 2.

13. $y = \csc\left(x - \dfrac{\pi}{4}\right)$

14. $y = \sec\left(x + \dfrac{3\pi}{4}\right)$

15. $y = \sec\left(x + \dfrac{\pi}{4}\right)$

16. $y = \csc\left(x + \dfrac{\pi}{3}\right)$

17. $y = \sec\left(\dfrac{1}{2}x + \dfrac{\pi}{3}\right)$

18. $y = \csc\left(\dfrac{1}{2}x - \dfrac{\pi}{4}\right)$

19. $y = 2 + 3 \sec(2x - \pi)$

20. $y = 1 - 2 \csc\left(x + \dfrac{\pi}{2}\right)$

21. $y = 1 - \dfrac{1}{2} \csc\left(x - \dfrac{3\pi}{4}\right)$

22. $y = 2 + \dfrac{1}{4} \sec\left(\dfrac{1}{2}x - \pi\right)$

Graph each defined function over a one-period interval. See Examples 3–5.

23. $y = 2 \tan x$

24. $y = 2 \cot x$

25. $y = \dfrac{1}{2} \cot x$

26. $y = 2 \tan \dfrac{1}{4}x$

27. $y = \cot 3x$

28. $y = -\cot \dfrac{1}{2}x$

Graph each defined function over a two-period interval. See Examples 6 and 7.

29. $y = \tan(2x - \pi)$

30. $y = \tan\left(\dfrac{x}{2} + \pi\right)$

31. $y = \cot\left(3x + \dfrac{\pi}{4}\right)$

32. $y = \cot\left(2x - \dfrac{3\pi}{2}\right)$

33. $y = 1 + \tan x$

34. $y = -2 + \tan x$

35. $y = 1 - \cot x$

36. $y = -2 - \cot x$

37. $y = -1 + 2\tan x$

38. $y = 3 + \dfrac{1}{2}\tan x$

39. $y = -1 + \dfrac{1}{2}\cot(2x - 3\pi)$

40. $y = -2 + 3\tan(4x + \pi)$

41. $y = \dfrac{2}{3}\tan\left(\dfrac{3}{4}x - \pi\right) - 2$

42. $y = 1 - 2\cot 2\left(x + \dfrac{\pi}{2}\right)$

For each graph in Exercises 43 and 44, give the equation of a function having that graph.

43.

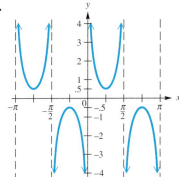

44.

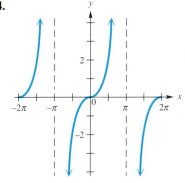

45. A rotating beacon is located at point A next to a long wall. (See the figure.) The beacon is 4 m from the wall. The distance d is given by

$$d = 4\tan 2\pi t,$$

where t is time measured in seconds since the beacon started rotating. (When $t = 0$, the beacon is aimed at point R. When the beacon is aimed to the right of R, the value of d is positive; d is negative if the beacon is aimed to the left of R.) Find d for the following times.

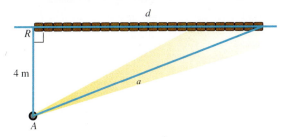

(a) $t = 0$ **(b)** $t = .4$

(c) $t = .8$ **(d)** $t = 1.2$

(e) Why is .25 a meaningless value for t?

46. In the figure for Exercise 45, the distance a is given by

$$a = 4|\sec 2\pi t|.$$

Find a for the following times.

(a) $t = 0$ **(b)** $t = .86$ **(c)** $t = 1.24$

47. Simultaneously sketch the graphs of $y = \tan x$ and $y = x$ for $-1 \le x \le 1$ and $-1 \le y \le 1$ with a graphing calculator. Write a sentence or two describing the relationship of $\tan x$ and x for small x-values.

48. Between each pair of successive asymptotes, a portion of the graph of $y = \sec x$ or $y = \csc x$ resembles a parabola. Can each of these portions actually be a parabola? Explain.

Chapter 6 Summary ▼▼▼▼▼▼▼▼▼▼▼▼▼▼▼▼▼▼▼▼▼▼▼▼▼▼▼▼▼

KEY TERMS	KEY IDEAS

6.1 Angles

angle
initial side
terminal side
vertex
degree measure
complementary angles
supplementary angles
standard position
quadrantal angle
coterminal angles

Types of Angles

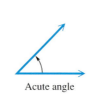
Acute angle

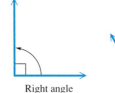
Right angle

Obtuse angle

Straight angle

6.2 Definitions of the Trigonometric Functions

sine
cosine
tangent
cotangent
secant
cosecant
identity

Trigonometric Functions

Let (x, y) be a point other than the origin on the terminal side of an angle θ in standard position. Let $r = \sqrt{x^2 + y^2}$, the distance from the origin to (x, y). Then

$$\sin \theta = \frac{y}{r} \qquad \csc \theta = \frac{r}{y} \quad (y \neq 0)$$

$$\cos \theta = \frac{x}{r} \qquad \sec \theta = \frac{r}{x} \quad (x \neq 0)$$

$$\tan \theta = \frac{y}{x} \quad (x \neq 0) \qquad \cot \theta = \frac{x}{y} \quad (y \neq 0).$$

Reciprocal Identities

$$\sin \theta = \frac{1}{\csc \theta} \qquad \csc \theta = \frac{1}{\sin \theta}$$

$$\cos \theta = \frac{1}{\sec \theta} \qquad \sec \theta = \frac{1}{\cos \theta}$$

$$\tan \theta = \frac{1}{\cot \theta} \qquad \cot \theta = \frac{1}{\tan \theta}$$

Pythagorean Identities

$$\sin^2 \theta + \cos^2 \theta = 1$$
$$\tan^2 \theta + 1 = \sec^2 \theta$$
$$1 + \cot^2 \theta = \csc^2 \theta$$

Quotient Identities

$$\frac{\sin \theta}{\cos \theta} = \tan \theta \qquad \frac{\cos \theta}{\sin \theta} = \cot \theta$$

KEY TERMS	KEY IDEAS
	Signs of Trigonometric Functions

6.3 Finding Trigonometric Function Values

Right Triangle-Based Definitions of the Trigonometric Functions

For any acute angle A in standard position,

$$\sin A = \frac{y}{r} = \frac{\text{side opposite}}{\text{hypotenuse}} \qquad \csc A = \frac{r}{y} = \frac{\text{hypotenuse}}{\text{side opposite}}$$

$$\cos A = \frac{x}{r} = \frac{\text{side adjacent}}{\text{hypotenuse}} \qquad \sec A = \frac{r}{x} = \frac{\text{hypotenuse}}{\text{side adjacent}}$$

$$\tan A = \frac{y}{x} = \frac{\text{side opposite}}{\text{side adjacent}} \qquad \cot A = \frac{x}{y} = \frac{\text{side adjacent}}{\text{side opposite}}.$$

Function Values of Special Angles

θ	$\sin \theta$	$\cos \theta$	$\tan \theta$	$\cot \theta$	$\sec \theta$	$\csc \theta$
30°	$\frac{1}{2}$	$\frac{\sqrt{3}}{2}$	$\frac{\sqrt{3}}{3}$	$\sqrt{3}$	$\frac{2\sqrt{3}}{3}$	2
45°	$\frac{\sqrt{2}}{2}$	$\frac{\sqrt{2}}{2}$	1	1	$\sqrt{2}$	$\sqrt{2}$
60°	$\frac{\sqrt{3}}{2}$	$\frac{1}{2}$	$\sqrt{3}$	$\frac{\sqrt{3}}{3}$	2	$\frac{2\sqrt{3}}{3}$

KEY TERMS	KEY IDEAS
6.4 Solving Right Triangles	
solve a triangle significant digits angle of elevation angle of depression	**Bearing** Type I 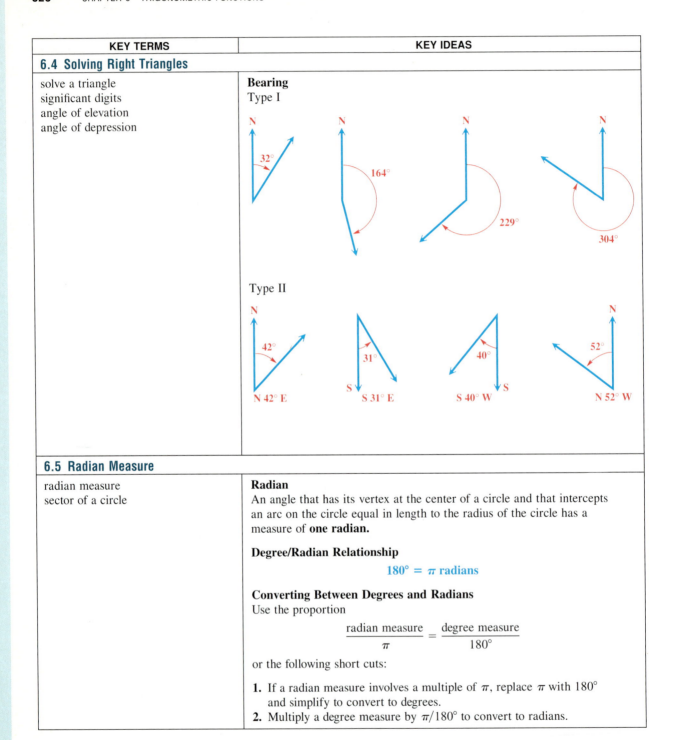

6.5 Radian Measure

radian measure
sector of a circle

Radian

An angle that has its vertex at the center of a circle and that intercepts an arc on the circle equal in length to the radius of the circle has a measure of **one radian.**

Degree/Radian Relationship

$$180° = \pi \text{ radians}$$

Converting Between Degrees and Radians
Use the proportion

$$\frac{\text{radian measure}}{\pi} = \frac{\text{degree measure}}{180°}$$

or the following short cuts:

1. If a radian measure involves a multiple of π, replace π with $180°$ and simplify to convert to degrees.
2. Multiply a degree measure by $\pi/180°$ to convert to radians.

KEY TERMS	KEY IDEAS

6.6 Circular Functions and Their Applications

unit circle

Circular Functions

From $(1, 0)$ on the unit circle lay off an arc of length $|s|$, going counterclockwise if $s > 0$ and clockwise if $s < 0$. Let the endpoint of the arc be (x, y). The circular functions of s are defined as follows. (Assume no denominators are zero.)

$$\sin s = y \qquad \tan s = \frac{y}{x} \qquad \sec s = \frac{1}{x}$$

$$\cos s = x \qquad \cot s = \frac{x}{y} \qquad \csc s = \frac{1}{y}$$

Angular and Linear Velocity

Angular Velocity	Linear Velocity
$$\omega = \frac{\theta}{t}$$	$$v = \frac{s}{t}$$
(ω in radians per unit time, θ in radians)	$$v = \frac{r\theta}{t}$$
	$$v = r\omega$$

6.7 Graphs of the Sine and Cosine Functions

periodic function
period
sine wave (sinusoid)
argument
amplitude
phase shift

Cosine and Sine Functions

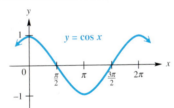

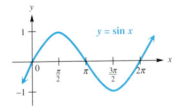

Domain: $(-\infty, \infty)$
Range: $[-1, 1]$
Amplitude: 1
Period: 2π

Domain: $(-\infty, \infty)$
Range: $[-1, 1]$
Amplitude: 1
Period: 2π

The graph of $y = c + a \sin b(x - d)$ or $y = c + a \cos b(x - d)$ has amplitude $|a|$, period $|2\pi/b|$, a vertical translation c units up if $c > 0$ or $|c|$ units down if $c < 0$, and a phase shift d units to the right if $d > 0$ or $|d|$ units to the left if $d < 0$.

KEY TERMS	KEY IDEAS
6.8 Graphs of the Other Circular Functions	

Tangent, Cotangent, Secant, and Cosecant Functions

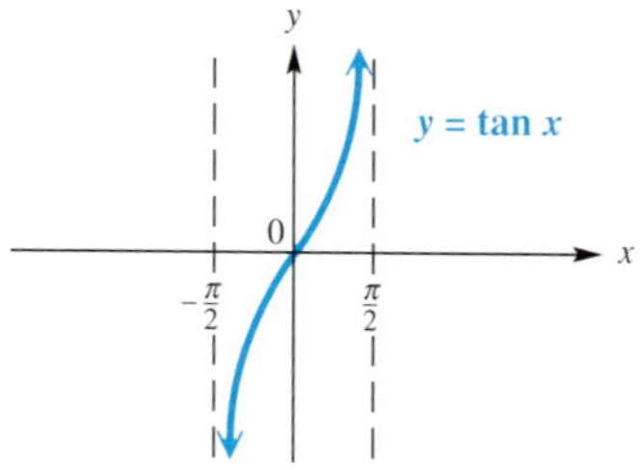

$y = \tan x$

Domain: $\{x \mid x \neq \pi/2 + n\pi, n \text{ any integer}\}$
Range: $(-\infty, \infty)$
Period: π

$y = \cot x$

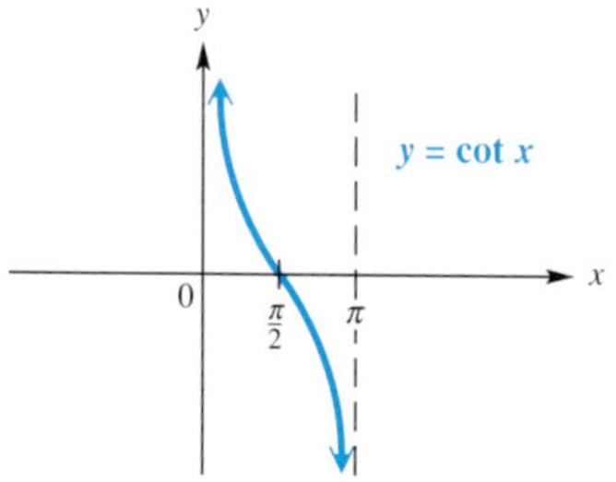

Domain: $\{x \mid x \neq n\pi, n \text{ any integer}\}$
Range: $(-\infty, \infty)$
Period: π

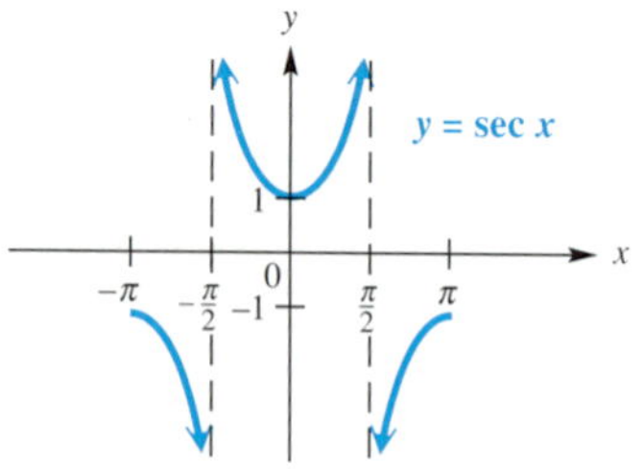

$y = \sec x$

Domain: $\{x \mid x \neq \pi/2 + n\pi, n \text{ any integer}\}$
Range: $(-\infty, -1] \cup [1, \infty)$
Period: 2π

KEY TERMS	KEY IDEAS
	  $y = \csc x$ **Domain:** $\{x \mid x \neq n\pi, n \text{ any integer}\}$ **Range:** $(-\infty, -1] \cup [1, \infty)$ **Period:** 2π

Chapter 6   Review Exercises ▼▼▼▼▼▼▼▼▼▼▼▼▼▼▼▼▼▼▼▼▼▼▼▼▼▼▼▼▼

1. Find the angle of smallest possible positive measure coterminal with an angle of $-51°$.

2. Let n represent any integer, and write an expression for all angles coterminal with an angle of $270°$.

Work each problem.

3. A pulley is rotating 320 times per minute. Through how many degrees does a point on the edge of the pulley move in $2/3$ second?

4. The propeller of a speedboat rotates 650 times per minute. Through how many degrees will a point on the edge of the propeller rotate in 2.4 seconds?

Find the trigonometric function values of each angle. If undefined, say so.

5.

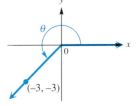

6.

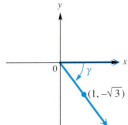

7. $180°$

Find the values of the trigonometric functions for angles in standard position having the following points on their terminal sides.

8. $(-8, 15)$ **9.** $(3, -4)$ **10.** $(6\sqrt{3}, -6)$

11. If the terminal side of a quadrantal angle lies along the y-axis, which of its trigonometric functions are undefined?

In Exercises 12 and 13, consider an angle θ in standard position whose terminal side has the equation $y = -5x$, with $x \leq 0$.

12. Sketch θ and use an arrow to show the rotation if $0° \leq \theta < 360°$.

13. Find the exact values of $\sin \theta$ and $\cos \theta$.

14. Which one of the following statements is possible?
   (a) $\sin \theta = 3/4$ and $\csc \theta = 4/3$ **(b)** $\sec \theta = -2/3$
   (c) $\cos \theta = .25$ and $\sec \theta = -4$

15. If, for some particular angle θ, $\sin \theta < 0$ and $\cos \theta > 0$, in what quadrant must θ lie? What is the sign of $\tan \theta$?

Find the values of the trigonometric functions for each angle A.

16.
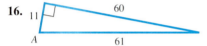

17.
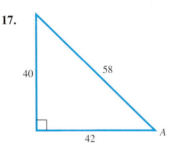

18. Explain why, in the figure, the cosine of angle A is equal to the sine of angle B.

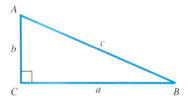

Find the values of the six trigonometric functions for each angle. Give exact values. Do not use a calculator. Rationalize denominators when applicable.

19. $120°$ **20.** $225°$ **21.** $-390°$

22. Give the exact value of $\sec^2 300° - 2 \cos^2 150° + \tan 45°$.

Use a calculator to find the value of each function.

23. $\sin 72° \, 30'$ **24.** $\sec 222° \, 30'$ **25.** $\cos 89.0043°$ **26.** $\cot 1.49783°$

27. Which one of the following cannot be *exactly* determined using the methods of this chapter?
   (a) $\cos 135°$ **(b)** $\cot (-45°)$ **(c)** $\sin 300°$ **(d)** $\tan 140°$

28. A student wants to use a calculator to find the value of $\cot 25°$. However, with the calculator in degree mode, instead of calculating $1/\tan 25$, he calculates $\tan^{-1} 25$. Will this produce the correct answer?

29. For $\theta = 2976°$, use a calculator to find $\cos \theta$ and $\sin \theta$. Use your results to decide what quadrant the angle lies in.

Solve each right triangle. In Exercise 30 give the angles in decimal degrees. In Exercise 31, label the triangle as shown in Figure 29 in Section 6.4. Use a calculator as necessary.

30.

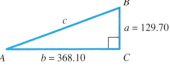

31. $B = 47.88°$, $b = 298.6$ m

Solve the problem.

32. The angle of elevation from a point 93.2 ft from the base of a tower to the top of the tower is 38° 20'. Find the height of the tower.

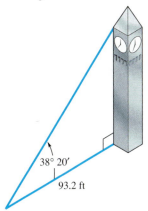

33. The angle of depression of a television tower to a point on the ground 36.0 m from the bottom of the tower is 29.5°. Find the height of the tower.

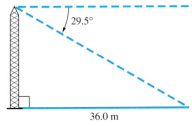

34. The bearing of B from C is 254°. The bearing of A from C is 344°. The bearing of A from B is 32°. The distance from A to C is 780 m. Find the distance from A to B.

35. Two cars leave an intersection at the same time. One heads due south at 55 mph. The other travels due west. After two hours, the bearing of the car headed west from the car headed south is 324°. How far apart are they at that time?

In Exercises 36 and 37, find a line segment in the accompanying figure whose length is equal to the function value. (Hint: Use right triangles.)

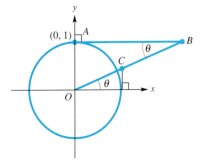

36. cot θ

37. sec θ

38. Artificial satellites that orbit Earth often use VHF signals to communicate with the ground. VHF signals travel in straight lines. The height h of the satellite above Earth and the time T that the satellite can communicate with a fixed location on the ground are related by the equation

$$h = R\left(\frac{1}{\cos(180T/P)} - 1\right)$$

where $R = 3955$ miles is the radius of Earth and P is the period for the satellite to orbit Earth. (*Source:* Schlosser, W., T. Schmidt-Kaler, and E. Milone, *Challenges of Astronomy,* Springer-Verlag, 1991.)
   (a) Find h when $T = 25$ min and $P = 140$ min. (Evaluate the cosine function in degree mode.)
   (b) What is the value of h if T is increased to 30?

39. 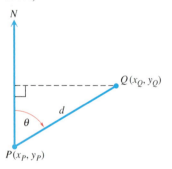 The first fundamental problem of surveying is to determine the coordinates of a point Q given the coordinates of a point P, the distance between P and Q, and the bearing θ from P to Q. See the figure. (*Source:* Mueller, I., and K. Ramsayer, *Introduction to Surveying,* Frederick Ungar Publishing Co., New York, 1979.)

(a) Find a formula for the coordinates (x_Q, y_Q) of the point Q given θ, the coordinates (x_P, y_P) of P, and the distance d between P and Q.
(b) Use your formula to determine (x_Q, y_Q) if $(x_P, y_P) = (123.62, 337.95)$, $\theta = 17°\,19'\,22''$, and $d = 193.86$ ft.

40. At present, the north star Polaris is located very near the celestial north pole. However, because Earth is inclined 23.5°, the moon's gravitational pull on Earth is uneven. As a result, Earth slowly processes (moves in) like a spinning top and the direction of the celestial north pole traces out a circular path once every 26,000 years. See the figure. For example, in approximately A. D. 14,000 the star Vega, not the star Polaris, will be located at the celestial north pole. As viewed from the center C of this circular path, calculate the angle in seconds that the celestial north pole moves each year. (*Source:* Zeilik, M., S. Gregory, and E. Smith, *Introductory Astronomy and Astrophysics,* Saunders College Publishers, 1992.)

41. Which is bigger, an angle of 1° or an angle of 1 radian? Discuss and justify your observations.

42. Give an expression that generates all angles coterminal with an angle of $\pi/6$ radian. Let n represent any integer.

Convert each degree measure to radians. Leave answers as multiples of π.

43. 45°

44. 175°

Convert each radian measure to degrees.

45. $\dfrac{8\pi}{3}$

46. $-\dfrac{11\pi}{18}$

Find the exact function value. Do not use a calculator.

47. $\tan \dfrac{\pi}{3}$

48. $\sin\left(-\dfrac{5\pi}{6}\right)$

49. $\sec \dfrac{3\pi}{4}$

50. Find the length of an arc intercepted by a central angle of .769 radian on a circle with a radius of 11.4 cm.

51. A central angle of $7\pi/4$ radians forms a sector of a circle. Find the area of the sector if the radius of the circle is 28.69 in.

52. Find the measure of the central angle θ (in radians) and the area of the sector.

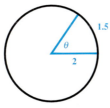

53. The hour hand of a wall clock measures six inches from its tip to the center of the clock.
   (a) Through what angle (in radians) does the hour hand pass between 1 o'clock and 3 o'clock?
   (b) What distance does the tip of the hour hand travel during the time period from 1 o'clock to 3 o'clock?

Use a calculator to find an approximation for each circular function value. Be sure your calculator is set in radian mode.

54. $\cos(-.2443)$

55. $\cot 3.0543$

56. Find the value of s in the interval $[0, \pi/2]$ if $\cos s = .92500448$.

Find the exact value of s in the given interval that has the given circular function value. Do not use a calculator.

57. $\left[\pi, \dfrac{3\pi}{2}\right]$; $\sec s = -\dfrac{2\sqrt{3}}{3}$

58. $\left[\dfrac{3\pi}{2}, 2\pi\right]$; $\sin s = -\dfrac{1}{2}$

59. Find the measure (in both degrees and radians) of the angle θ formed in the accompanying screen by the line passing through the origin and the positive part of the x-axis. Use the displayed values of X and Y at the bottom of the screen.

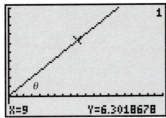

60. The graphing calculator screen shows a point on the unit circle. What is the length of the shortest arc of the circle from $(1, 0)$ to the point?

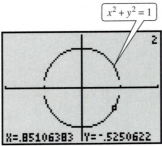

$[-2, 2]$ by $[-1.5, 1.5]$

Solve each problem.

61. Find t if $\theta = 5\pi/12$ radians and $\omega = 8\pi/9$ radians per sec.

62. Find the linear velocity of a point on the edge of a flywheel of radius 7 m if the flywheel is rotating 90 times per sec.

63. Find the distance in kilometers between cities on a north-south line that are on latitudes 28° N and 12° S, respectively.

64. 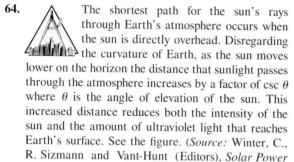 The shortest path for the sun's rays through Earth's atmosphere occurs when the sun is directly overhead. Disregarding the curvature of Earth, as the sun moves lower on the horizon the distance that sunlight passes through the atmosphere increases by a factor of csc θ where θ is the angle of elevation of the sun. This increased distance reduces both the intensity of the sun and the amount of ultraviolet light that reaches Earth's surface. See the figure. (*Source:* Winter, C., R. Sizmann and Vant-Hunt (Editors), *Solar Power Plants*, Springer-Verlag, 1991.)

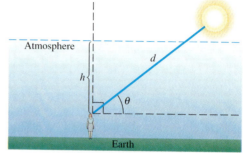

(a) Verify that $d = h$ csc θ.

(b) Determine θ when $d = 2h$.

(c) The atmosphere filters out the ultraviolet light that causes skin to burn. Compare the difference between sunbathing when $\theta = \pi/2$ and $\pi/3$. Which measure gives less ultraviolet light?

65. Give all basic trigonometric functions that satisfy the condition $f(x + \pi) = -f(x)$.

66. Which basic trigonometric functions can have the value $1/2$?

For each defined function, give the amplitude, period, vertical translation, and phase shift, as applicable.

67. $y = 2 \sin x$

68. $y = \tan 3x$

69. $y = -\dfrac{1}{2} \cos 3x$

70. $y = 2 \sin 5x$

71. $y = 1 + 2 \sin \dfrac{1}{4}x$

72. $y = 3 - \dfrac{1}{4} \cos \dfrac{2}{3}x$

73. $y = 3 \cos\left(x + \dfrac{\pi}{2}\right)$

74. $y = \cot\left(\dfrac{x}{2} + \dfrac{3\pi}{4}\right)$

Which one of the six circular functions satisfies the description?

75. Period is π, x-intercepts are of the form $n\pi$, where n is an integer

76. Period is 2π, passes through the origin

77. Period is 2π, passes through the point $(\pi/2, 0)$

78. Suppose f is a sine function with period 10 and $f(5) = 2$. Explain why $f(25) = 2$.

Graph each function defined in Exercises 79–84 over a one-period interval.

79. $y = 3 \cos 2x$

80. $y = \dfrac{1}{2} \cot 3x$

81. $y = \tan\left(x - \dfrac{\pi}{2}\right)$

82. $y = \sec\left(2x + \dfrac{\pi}{3}\right)$

83. $y = 1 + 2 \cos 3x$

84. $y = -1 - 3 \sin 2x$

---

85. Let a person h_1 ft tall stand d ft from an object h_2 ft tall, where $h_2 > h_1$. Let θ be the angle of elevation to the top of the object. (See the figure.)

(**a**) Show that $d = (h_2 - h_1) \cot \theta$.

(**b**) Let $h_2 = 55$ and $h_1 = 5$. Graph d for the interval $0 < \theta < \pi/2$.

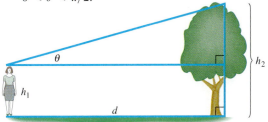

86. 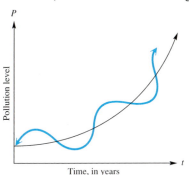 The amount of pollution in the air fluctuates with the seasons. It is lower after heavy spring rains and higher after periods of little rain. In addition to this seasonal fluctuation, the long-term trend is upward. An idealized graph of this situation is shown in the figure. Circular functions can be used to describe the fluctuating part of the pollution levels. Powers of the number e (e is the base of natural logarithms; to six

decimal places, $e = 2.718282$) can be used to show the long-term growth. In fact, the pollution level in a certain area might be given by

$$P(t) = 7(1 - \cos 2\pi t)(t + 10) + 100 e^{.2t},$$

where t is time in years, with $t = 0$ representing January 1 of the base year. Thus, July 1 of the same year would be represented by $t = .5$, and October 1 of the following year would be represented by $t = 1.75$. Find the pollution levels on the following dates.

(**a**) January 1, base year

(**b**) July 1, base year

(**c**) January 1, following year

(**d**) July 1, following year

87. The figure shows the populations of lynx and hares in Canada for the years 1847–1903. The hares are food for the lynx. An increase in hare population causes an increase in lynx population some time later. The increasing lynx population then causes a decline in hare population.

(**a**) Estimate the length of one period.

(**b**) Estimate maximum and minimum hare populations.

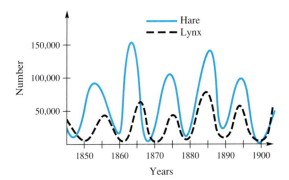

88. The average monthly temperature (in °F) in Chicago, Illinois, is shown in the table. (*Source:* Miller, A., and J. Thompson, *Elements of Meteorology,* Charles E. Merrill Publishing Company, Columbus, Ohio, 1975.)

Month	Temperature
Jan	25
Feb	28
Mar	36
Apr	48
May	61
June	72
July	74
Aug	75
Sept	66
Oct	55
Nov	39
Dec	28

   (a) Plot the average monthly temperature over a two-year period by letting $x = 1$ correspond to January of the first year.
   (b) Determine a function that has the form $f(x) = a \sin b(x - d) + c$ where $a, b, c,$ and d are constants, which models the data.
   (c) Explain the significance of each constant.
   (d) Graph f together with the data on the same coordinate axes.

Chapter 6   Test ▼▼▼▼▼▼▼▼▼▼▼▼▼▼▼▼▼▼▼▼▼▼▼▼▼▼▼▼▼▼▼▼▼▼▼▼▼▼

1. Find the angle of smallest positive measure coterminal with $-157°$.

2. A tire rotates 450 times per minute. Through how many degrees does a point on the edge of the tire move in 1 second?

3. If $(2, -5)$ is on the terminal side of an angle θ in standard position, find $\sin \theta$, $\cos \theta$, and $\tan \theta$.

4. $\cos \theta > 0$ and $\cot \theta < 0$. In what quadrant does θ terminate?

5. $\cos \theta = 4/5$ and θ is in quadrant IV. Find the values of the other trigonometric functions of θ.

6. Explain how you would find $\cot \theta$ if $\tan \theta$ is 1.6778490 using a calculator. Then give $\cot \theta$.

7. A test item read, "Find the exact value of $\cos 45°$." A student gave the answer .7071067812, but did not receive credit. Explain why.

8. Find the exact value of $\cot(-750°)$.

9. Use a calculator to approximate the following.
   (a) $\sin 78° \, 21'$ **(b)** $\tan 11.7689°$ **(c)** $\sec 58.9041°$

10. Solve the triangle in the figure.

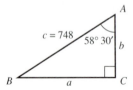

11. A rectangle has adjacent sides measuring 10.93 cm and 15.24 cm. The angle between the diagonal and the longer side is 35.65°. Find the length of the diagonal.

12. A ship leaves a pier on a bearing of S 55° E and travels for 80 km. It then turns and continues on a bearing of N 35° E for 74 km. How far is the ship from the pier?

13. Find the measure of the central angle in radians if $r = 150$ cm and $s = 200$ cm.

14. Convert 120° to radians.

15. Convert $\dfrac{9\pi}{10}$ to degrees.

16. Use a calculator to approximate θ in the interval $[0°, 90°)$ if $\sin \theta = .82584121$.

17. A wheel with a radius of 9 inches rotates 120 times per second.
    (a) Find the angular velocity of the wheel in radians per second.
    (b) Find the linear velocity of a point on the edge of the wheel in inches per second.

18. Give the amplitude, period, vertical translation, and phase shift, as applicable, of the graph of $y = 1 + 2 \sin(1/4)x$.

Graph each function defined as follows.

19. $y = \tan\left(x - \dfrac{\pi}{2}\right)$

20. $y = 1 + 2 \cos 3x$

7

Trigonometric Identities and Equations

The electricity supplied to most homes is produced by electric generators that rotate at 60 cycles per second. Because of this rotation, electric current alternates its direction in electrical wires and can be modeled accurately using either the sine or cosine functions. Household current is often rated at 115 volts. If the current is alternating direction in the wires, is the voltage always 115 volts or does it actually vary with time? Electric companies charge customers according to the wattage that an electrical device uses and how long it is turned on. Given both the voltage and current supplied to a light bulb, how can its wattage be determined? The answers to these questions are found in Sections 7.3 and 7.4. In order to understand phenomena such as electric current, sound waves, or stress on your back muscles when you bend at the waist, we will need to use not only trigonometric functions but also the many identities that relate the trigonometric functions to each other.

When musicians tune instruments, they are able to compare like tones and accurately determine whether their pitches are the same frequency simply by listening, even though these tones vibrate hundreds or thousands of times per second. Some radios and telephones have small speakers that cannot vibrate slower than 200 times per second—yet 35 keys on a piano have frequencies below 200 and all of them can be clearly heard on these speakers. How can we explain these phenomena? What is the advantage of having larger speakers for a stereo if small speakers are capable of reproducing the lower tones? Explanations of musical phenomena like these require a mathematical understanding of sound. Music is made up of sound waves that cause rapid

Sources: Benade, Arthur, *Fundamentals of Musical Acoustics,* Oxford University Press, New York, 1976.

Pierce, John, *The Science of Musical Sound,* Scientific American Books, 1992.

Weidner, R., and R. Sells, *Elementary Classical Physics, Vol. 2,* Allyn and Bacon, 1968.

Wright, J., (editor), *The Universal Almanac 1995,* Universal Press Syndicate Company, Kansas City, 1994.

increases and decreases in air pressure on a person's eardrum. Sound often involves periodic motion through the air. This periodic motion can be modeled using trigonometric functions. Using trigonometric equations and graphs, important aspects of music can be analyzed.

The study of electricity, music, noise control, and biophysics are all fascinating subjects that require knowledge of trigonometric identities and equations.

An **identity** is an equation that is true for *every* value in the domain of its variable. Examples of identities include

$$5(x + 3) = 5x + 15 \qquad \text{and} \qquad (a + b)^2 = a^2 + 2ab + b^2.$$

This chapter discusses identities involving trigonometric and circular functions. The variables in these functions represent either angles or real numbers. The domain of the variable is assumed to be all values for which a given function is defined.

7.1 Fundamental Identities ▼▼▼

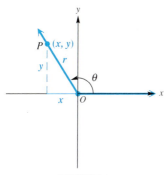

FIGURE 1

In this section we discuss some of the ways we can use the fundamental trigonometric identities first introduced in Chapter 6.* In Chapter 6, the definitions of the trigonometric functions were used to derive the following **reciprocal identities,** which are true for all suitable replacements of the variable.

$$\cot \theta = \frac{1}{\tan \theta} \qquad \csc \theta = \frac{1}{\sin \theta} \qquad \sec \theta = \frac{1}{\cos \theta}$$

Each of these reciprocal identities leads to other forms of the identity. For example, $\csc \theta = 1/\sin \theta$ gives $\sin \theta = 1/\csc \theta$.

From the definitions of the trigonometric functions, with x, y, and r as shown in Figure 1,

$$\frac{\sin \theta}{\cos \theta} = \frac{y/r}{x/r} = \frac{y}{x} = \tan \theta$$

or

$$\tan \theta = \frac{\sin \theta}{\cos \theta}.$$

In a similar manner,

$$\cot \theta = \frac{\cos \theta}{\sin \theta}.$$

---

*All the identities given in this chapter are summarized at the end of the chapter and inside the back cover.

These last two identities, also derived in Chapter 6, are called the **quotient identities**.

Furthermore, we saw in Chapter 6 that the definitions of the trigonometric functions were used to derive the identity

$$\sin^2 \theta + \cos^2 \theta = 1.$$

Dividing both sides by $\cos^2 \theta$ then leads to

$$\tan^2 \theta + 1 = \sec^2 \theta$$

while dividing through by $\sin^2 \theta$ gives

$$1 + \cot^2 \theta = \csc^2 \theta.$$

These last three identities are the **Pythagorean identities.**

A graphing calculator can be used to support an identity—that is, to decide whether two functions are identical. For example, to verify the identity $\sin^2 x + \cos^2 x = 1$, let $Y_1 = \sin^2 x + \cos^2 x$ and let $Y_2 = 1$. (Be sure your calculator is set in radian mode.) Now use the "trig" window and graph the two functions. If the identity is true, you should see no difference in the two graphs. If the equation is not an identity, the graphs of Y_1 and Y_2 will not coincide. As a check, to guard against the possibility that the graphs are different, but one of them is not showing in the window being used, repeat the process, graphing Y_2 first.

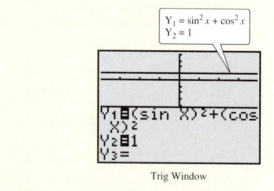

$Y_1 = \sin^2 x + \cos^2 x$
$Y_2 = 1$

Trig Window

As suggested by the circle shown in Figure 2, an angle θ having the point (x, y) on its terminal side has a corresponding angle $-\theta$ with a point $(x, -y)$ on its terminal side. From the definition of sine,

$$\sin(-\theta) = \frac{-y}{r} \qquad \text{and} \qquad \sin \theta = \frac{y}{r},$$

so that $\sin(-\theta)$ and $\sin \theta$ are negatives of each other, or

$$\sin(-\theta) = -\sin \theta.$$

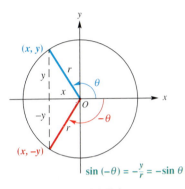

$$\sin(-\theta) = -\frac{y}{r} = -\sin\theta$$

FIGURE 2

Figure 2 shows an angle θ in quadrant II, but the same result holds for θ in any quadrant. Also, by definition,

$$\cos(-\theta) = \frac{x}{r} \quad \text{and} \quad \cos\theta = \frac{x}{r},$$

so that

$$\cos(-\theta) = \cos\theta.$$

These formulas for $\sin(-\theta)$ and $\cos(-\theta)$ can be used to find $\tan(-\theta)$ in terms of $\tan\theta$:

$$\tan(-\theta) = \frac{\sin(-\theta)}{\cos(-\theta)} = \frac{-\sin\theta}{\cos\theta} = -\frac{\sin\theta}{\cos\theta}$$

or

$$\tan(-\theta) = -\tan\theta.$$

The preceding three identities are **negative-angle identities.**

CONNECTIONS We can derive the negative-angle identities by using the properties of even and odd functions. An **even function** has the property that $f(-x) = f(x)$ for all x. Because of this property, the graph of an even function is symmetric with respect to the y-axis—that is, if folded along the y-axis, the two halves would match. The graph of $y = \cos x$ is symmetric about the y-axis, so cosine is an even function and therefore

$$\cos(-x) = \cos x.$$

A function is an **odd function** if $f(-x) = -f(x)$ for all x. The graph of an odd function is symmetric about the origin, which means that if (x, y) belongs to the function, then $(-x, -y)$ also belongs to the function. The sine graph exhibits this property: for example, $(\pi/2, 1)$ and $(-\pi/2, -1)$ are points on the graph of $y = \sin x$. Thus,

$$\sin(-x) = -\sin x.$$

By recalling (from algebra) the type of symmetry shown by the graph of a trigonometric function, we can remember the three negative-angle identities.

FOR DISCUSSION OR WRITING

Is the tangent function even or odd or neither? Does your answer tell you the correct negative-angle identity for tangent?

The identities given in this section are summarized below. As a group, these are called the **fundamental identities.**

Fundamental Identities

Reciprocal Identities

$$\cot \theta = \frac{1}{\tan \theta} \qquad \sec \theta = \frac{1}{\cos \theta} \qquad \csc \theta = \frac{1}{\sin \theta}$$

Quotient Identities

$$\tan \theta = \frac{\sin \theta}{\cos \theta} \qquad \cot \theta = \frac{\cos \theta}{\sin \theta}$$

Pythagorean Identities

$$\sin^2 \theta + \cos^2 \theta = 1 \qquad \tan^2 \theta + 1 = \sec^2 \theta$$

$$1 + \cot^2 \theta = \csc^2 \theta$$

Negative-Angle Identities

$$\sin(-\theta) = -\sin \theta \qquad \cos(-\theta) = \cos \theta \qquad \tan(-\theta) = -\tan \theta$$

NOTE The forms of the identities given above are the most commonly recognized forms. Throughout this chapter it will be necessary to recognize alternative forms of these identities as well. For example, two other forms of $\sin^2 \theta + \cos^2 \theta = 1$ are

$$\sin^2 \theta = 1 - \cos^2 \theta$$

and

$$\cos^2 \theta = 1 - \sin^2 \theta.$$

You should be able to transform the basic identities using algebraic transformations.

The fundamental identities are used extensively in trigonometry and in calculus, so it is important to be very familiar with them. One use for these identities is to find the values of other trigonometric functions from the value of a given trigonometric function. Of course, we could find these values by using a right triangle instead, but this is a good way to practice using the fundamental identities. For example, given a value of $\tan \theta$, the value of $\cot \theta$ can be found from the identity $\cot \theta = 1/\tan \theta$. In fact, given any trigonometric function value and the quadrant in which θ lies, the values of all the other trigonometric functions can be found by using identities, as in the following example.

EXAMPLE 1
Finding all trigonometric function values, given one value and the quadrant

If $\tan \theta = -5/3$ and θ is in quadrant II, find the values of the other trigonometric functions using fundamental identities.

The identity $\cot \theta = 1/\tan \theta$ leads to $\cot \theta = -3/5$. Next, find $\sec \theta$ from the identity $\tan^2 \theta + 1 = \sec^2 \theta$.

$$\left(-\frac{5}{3}\right)^2 + 1 = \sec^2 \theta$$

$$\frac{25}{9} + 1 = \sec^2 \theta$$

$$\frac{34}{9} = \sec^2 \theta$$

$$-\sqrt{\frac{34}{9}} = \sec \theta$$

$$-\frac{\sqrt{34}}{3} = \sec \theta$$

We choose the negative square root since $\sec \theta$ is negative in quadrant II. Now find $\cos \theta$:

$$\cos \theta = \frac{1}{\sec \theta} = \frac{-3}{\sqrt{34}} = -\frac{3\sqrt{34}}{34},$$

after rationalizing the denominator. Find $\sin \theta$ by using the identity $\sin^2 \theta + \cos^2 \theta = 1$, with $\cos \theta = -3/\sqrt{34}$.

$$\sin^2 \theta + \left(-\frac{3}{\sqrt{34}}\right)^2 = 1$$

$$\sin^2 \theta = 1 - \frac{9}{34}$$

$$\sin^2 \theta = \frac{25}{34}$$

$$\sin \theta = \frac{5}{\sqrt{34}}$$

$$\sin \theta = \frac{5\sqrt{34}}{34} \qquad \text{Rationalize.}$$

The positive square root is used since $\sin \theta$ is positive in quadrant II. Finally, since $\csc \theta$ is the reciprocal of $\sin \theta$,

$$\csc \theta = \frac{\sqrt{34}}{5}.$$

CAUTION   Several comments can be made concerning Example 1.

1. We are given $\tan \theta = -5/3$. Although $\tan \theta = (\sin \theta)/(\cos \theta)$, we should *not* assume that $\sin \theta = -5$ and $\cos \theta = 3$. (Why can these values not possibly be correct?)
2. Problems of this type can usually be worked in more than one way. For example, after finding $\cot \theta = -3/5$, we could have then found $\csc \theta$ using the identity $1 + \cot^2 \theta = \csc^2 \theta$. The remaining function values could then be found as well.
3. The most common error made in problems like this is an incorrect sign choice for the functions. When taking the square root, be sure to choose the sign based on the quadrant of θ and the function being found.

Each of $\tan \theta$, $\cot \theta$, $\sec \theta$, and $\csc \theta$ can easily be expressed in terms of $\sin \theta$ and/or $\cos \theta$. For this reason, we often make such substitutions in an expression so that the expression can be simplified as in the next example.

◀EXAMPLE 2
Simplifying an expression by writing it in terms of sine and cosine

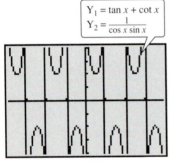

Trig Window
This graph supports the result in Example 2.

Use the fundamental identities to write $\tan \theta + \cot \theta$ in terms of $\sin \theta$ and $\cos \theta$, and then simplify the expression.

From the fundamental identities,

$$\tan \theta + \cot \theta = \frac{\sin \theta}{\cos \theta} + \frac{\cos \theta}{\sin \theta}.$$

Simplify this expression by adding the two fractions on the right side, using the common denominator $\cos \theta \sin \theta$.

$$\tan \theta + \cot \theta = \frac{\sin^2 \theta}{\cos \theta \sin \theta} + \frac{\cos^2 \theta}{\cos \theta \sin \theta}$$

$$= \frac{\sin^2 \theta + \cos^2 \theta}{\cos \theta \sin \theta}$$

Now substitute **1** for $\sin^2 \theta + \cos^2 \theta$.

$$\tan \theta + \cot \theta = \frac{1}{\cos \theta \sin \theta} \quad ▶$$

Every trigonometric function of an angle θ or a number x can be expressed in terms of every other function.

◀EXAMPLE 3
Expressing one function in
terms of another

Express $\cos x$ in terms of $\tan x$.

Since $\sec x$ is related to both $\cos x$ and $\tan x$ by identities, start with $\tan^2 x + 1 = \sec^2 x$. Then take reciprocals to get

$$\frac{1}{\tan^2 x + 1} = \frac{1}{\sec^2 x}$$

or

$$\frac{1}{\tan^2 x + 1} = \cos^2 x$$

$$\pm \sqrt{\frac{1}{\tan^2 x + 1}} = \cos x \qquad \text{Take the square root of both sides.}$$

$$\cos x = \frac{\pm 1}{\sqrt{\tan^2 x + 1}}.$$

Rationalize the denominator to get

$$\cos x = \frac{\pm \sqrt{\tan^2 x + 1}}{\tan^2 x + 1}.$$

We choose the $+$ sign or the $-$ sign, depending on the quadrant of x. ▶

CAUTION   When working with trigonometric expressions and identities, be sure to write the argument of the function. For example, we would *not* write $\sin^2 + \cos^2 = 1$; an argument such as θ is necessary in this identity.

7.1 Exercises ▼▼▼▼▼▼▼▼▼▼▼▼▼▼▼▼▼▼▼▼▼▼▼▼▼▼▼▼▼▼▼▼▼▼▼▼▼▼▼

In Exercises 1–4, the given graphing calculator screen is obtained for a particular stored value of X. What will the screen display for the value of the expression in the final line of the display?

1.
```
tan X
                2.6
tan (-X)
```

2.
```
cos X
               -.65
cos (-X)
```

3.
```
(tan X)²
                1.5
(1/cos X)²
```

4.
```
cos X
                 .8
sin X
                 .6
tan X
```

In Exercises 5–10, find sin s. See Example 1.

5. $\cos s = \dfrac{3}{4}$, s in quadrant I

6. $\cot s = -\dfrac{1}{3}$, s in quadrant IV

7. $\cos s = \dfrac{\sqrt{5}}{5}$, $\tan s < 0$

8. $\tan s = -\dfrac{\sqrt{7}}{2}$, $\sec s > 0$

9. $\sec s = \dfrac{11}{4}$, $\tan s < 0$

10. $\csc s = -\dfrac{8}{5}$

11. Why is it unnecessary to give the quadrant of s in Exercise 10?

12. What is wrong with this problem? "Find sin s if csc $s = -9/5$ and s is in quadrant II."

13. Find $\tan \theta$ if $\cos \theta = -2/5$, and $\sin \theta < 0$.

14. Find $\csc \alpha$ if $\tan \alpha = 6$, and $\cos \alpha > 0$.

Use the fundamental identities to find the remaining five trigonometric functions of θ. See Example 1.

15. $\sin \theta = \dfrac{2}{3}$, θ in quadrant II

16. $\cos \theta = \dfrac{1}{5}$, θ in quadrant I

17. $\tan \theta = -\dfrac{1}{4}$, θ in quadrant IV

18. $\csc \theta = -\dfrac{5}{2}$, θ in quadrant III

19. $\cot \theta = \dfrac{4}{3}$, $\sin \theta > 0$

20. $\sin \theta = -\dfrac{4}{5}$, $\cos \theta < 0$

21. $\sec \theta = \dfrac{4}{3}$, $\sin \theta < 0$

22. $\cos \theta = -\dfrac{1}{4}$, $\sin \theta > 0$

For each expression in Column I, choose the expression from Column II that completes a fundamental identity.

Column I

Column II

23. $\dfrac{\cos x}{\sin x}$ **(a)** $\sin^2 x + \cos^2 x$

24. $\tan x$ **(b)** $\cot x$

25. $\cos(-x)$ **(c)** $\sec^2 x$

26. $\tan^2 x + 1$ **(d)** $\dfrac{\sin x}{\cos x}$

27. 1 **(e)** $\cos x$

For each expression in Column I, choose the expression from Column II that completes an identity. You will have to rewrite one or both expressions, using a fundamental identity, to recognize the matches.

Column I   Column II

28. $-\tan x \cos x$ (a) $\dfrac{\sin^2 x}{\cos^2 x}$

29. $\sec^2 x - 1$ (b) $\dfrac{1}{\sec^2 x}$

30. $\dfrac{\sec x}{\csc x}$ (c) $\sin(-x)$

31. $1 + \sin^2 x$ (d) $\csc^2 x - \cot^2 x + \sin^2 x$

32. $\cos^2 x$ (e) $\tan x$

33. A student writes, "$1 + \cot^2 = \csc^2$." Comment on this student's work.

34. Another student makes the following claim: "Since $\sin^2 \theta + \cos^2 \theta = 1$, I should be able to also say $\sin \theta + \cos \theta = 1$ if I take the square root of both sides." Comment on this student's statement.

35. Suppose that $\cos \theta = \dfrac{x}{x + 1}$. Find $\sin \theta$.   **36.** Find $\tan \alpha$ if $\sec \alpha = \dfrac{p + 4}{p}$.

Use the fundamental identities to get an equivalent expression involving only sines and cosines, and then simplify it. See Example 2.

37. $\cot \theta \sin \theta$ **38.** $\sec \theta \cot \theta \sin \theta$

39. $\cos \theta \csc \theta$ **40.** $\cot^2 \theta(1 + \tan^2 \theta)$

41. $\sin^2 \theta(\csc^2 \theta - 1)$ **42.** $(\sec \theta - 1)(\sec \theta + 1)$

43. $(1 - \cos \theta)(1 + \sec \theta)$ **44.** $\dfrac{\cos \theta + \sin \theta}{\sin \theta}$

45. $\dfrac{\cos^2 \theta - \sin^2 \theta}{\sin \theta \cos \theta}$ **46.** $\dfrac{1 - \sin^2 \theta}{1 + \cot^2 \theta}$

47. $\tan \theta + \cot \theta$ **48.** $(\sec \theta + \csc \theta)(\cos \theta - \sin \theta)$

49. $\sin \theta(\csc \theta - \sin \theta)$ **50.** $\dfrac{1 + \tan^2 \theta}{1 + \cot^2 \theta}$

51. $\sin^2 \theta + \tan^2 \theta + \cos^2 \theta$ **52.** $\dfrac{\tan(-\theta)}{\sec \theta}$

53. Let $\cos x = 1/5$. Find all possible values for $\dfrac{\sec x - \tan x}{\sin x}$.

54. Let $\csc x = -3$. Find all possible values for $\dfrac{\sin x + \cos x}{\sec x}$.

55. Look at the graphs of $\cot x$, $\csc x$, and $\sec x$ and determine which of them are odd and which are even.

56. Write each of the functions $\cot x$, $\csc x$, and $\sec x$ in terms of the tangent, sine, and cosine functions, respectively, and determine which of them are odd and which are even.

▼▼▼▼▼▼▼▼▼▼▼▼▼ **DISCOVERING CONNECTIONS** (Exercises 57–61) ▼▼▼▼▼▼▼▼▼▼▼▼▼

In Chapter 4 we graphed functions of the form $y = c + a \cdot f[b(x - d)]$ with the assumption that $b > 0$. To see what happens when $b < 0$, work Exercises 57–61 in order.

57. Use a negative-angle identity to write $y = \sin(-2x)$ as a function of $2x$.

58. How does your answer to Exercise 57 relate to $y = \sin(2x)$?

59. Use a negative-angle identity to write $y = \cos(-4x)$ as a function of $4x$.

60. How does your answer to Exercise 59 relate to $y = \cos(4x)$?

61. Use your results from Exercises 57–60 to rewrite the following with a positive value of b.
   (a) $y = \sin(-4x)$ **(b)** $y = \cos(-2x)$ **(c)** $y = -5\sin(-3x)$

---

62. How does the graph of $y = \sec(-x)$ compare to that of $y = \sec x$?

63. How does the graph of $y = \csc(-x)$ compare to that of $y = \csc x$?

64. How does the graph of $y = \tan(-x)$ compare to that of $y = \tan x$?

65. How does the graph of $y = \cot(-x)$ compare to that of $y = \cot x$?

Show that the equation is not an identity by replacing the variables with numbers that show the result to be false.

66. $2 \sin s = \sin 2s$ **67.** $\sin x = \sqrt{1 - \cos^2 x}$ **68.** $\sin(x + y) = \sin x + \sin y$

69. Show that $\sin 1° + \sin 2° + \sin 3° + \cdots + \sin 358° + \sin 359° = 0$. (*Hint:* Pair the first term with the last term, the second term with the next to last, and so on. Then use a negative-angle identity.)

70. (Refer to Exercise 69.) Can you determine the sum $\cos 1° + \cos 2° + \cdots + \cos 358° + \cos 359°$ in a similar manner? Explain why or why not.

7.2 Verifying Trigonometric Identities ▼▼▼

One of the skills required for more advanced work in mathematics (and especially in calculus) is the ability to use the trigonometric identities to write trigonometric expressions in alternate forms. This skill is developed by using the fundamental identities to verify that a trigonometric equation is an identity (for those values of the variable for which it is defined). Here are some hints that may help you get started.

Verifying Identities

1. Learn the fundamental identities given in the last section. Whenever you see either side of a fundamental identity, the other side should come to mind. Also, be aware of equivalent forms of the fundamental identities. For example $\sin^2 \theta = 1 - \cos^2 \theta$ is an alternative form of $\sin^2 \theta + \cos^2 \theta = 1$.

2. Try to rewrite the more complicated side of the equation so that it is identical to the simpler side.

3. It is often helpful to express all trigonometric functions in the equation in terms of sine and cosine and then simplify the result.

4. Usually any factoring or indicated algebraic operations should be performed. For example, the expression $\sin^2 x + 2 \sin x + 1$ can be factored as follows: $(\sin x + 1)^2$. The sum or difference of two trigonometric expressions, such as

$$\frac{1}{\sin \theta} + \frac{1}{\cos \theta},$$

can be added or subtracted in the same way as any other rational expressions:

$$\frac{1}{\sin \theta} + \frac{1}{\cos \theta} = \frac{\cos \theta}{\sin \theta \cos \theta} + \frac{\sin \theta}{\sin \theta \cos \theta}$$

$$= \frac{\cos \theta + \sin \theta}{\sin \theta \cos \theta}.$$

5. As you select substitutions, keep in mind the side you are not changing, because it represents your goal. For example, to verify the identity

$$\tan^2 x + 1 = \frac{1}{\cos^2 x},$$

try to think of an identity that relates $\tan x$ to $\cos x$. Here, since $\sec x = 1/\cos x$ and $\sec^2 x = \tan^2 x + 1$, the secant function is the best link between the two sides.

6. If an expression contains $1 + \sin x$, multiplying both numerator and denominator by $1 - \sin x$ would give $1 - \sin^2 x$, which could be replaced with $\cos^2 x$. Similar results for $1 - \sin x$, $1 + \cos x$, and $1 - \cos x$ may be useful.

These hints are used in the examples of this section.

CAUTION Verifying identities is not the same as solving equations. Techniques used in solving equations, such as adding the same terms to both sides, or multiplying both sides by the same term, are not valid when working with identities since you are starting with a statement (to be verified) that may not be true.

EXAMPLE 1
Verifying an identity (working with one side)

Verify that

$$\cot s + 1 = \csc s(\cos s + \sin s)$$

is an identity.

Use the fundamental identities to rewrite one side of the equation so that it is identical to the other side. Since the right side is more complicated, it is probably a good idea to work with it. Here we use the method of changing all the trigonometric functions to sine or cosine.

Steps   **Reasons**

$$\csc s(\cos s + \sin s) = \frac{1}{\sin s}(\cos s + \sin s) \qquad \csc s = \tfrac{1}{\sin s}$$

$$= \frac{\cos s}{\sin s} + \frac{\sin s}{\sin s} \qquad \text{Distributive property}$$

$$= \cot s + 1 \qquad \tfrac{\cos s}{\sin s} = \cot s; \ \tfrac{\sin s}{\sin s} = 1$$

The given equation is an identity since the right side equals the left side. ▶

EXAMPLE 2
Verifying an identity (working with one side)

Verify that

$$\tan^2 \alpha(1 + \cot^2 \alpha) = \frac{1}{1 - \sin^2 \alpha}$$

is an identity.

Working with the left side gives the following.

$$\tan^2 \alpha(1 + \cot^2 \alpha) = \tan^2 \alpha + \tan^2 \alpha \cot^2 \alpha \qquad \text{Distributive property}$$

$$= \tan^2 \alpha + \tan^2 \alpha \cdot \frac{1}{\tan^2 \alpha} \qquad \cot^2 \alpha = \tfrac{1}{\tan^2 \alpha}$$

$$= \tan^2 \alpha + 1 \qquad \tan^2 \alpha \cdot \tfrac{1}{\tan^2 \alpha} = 1$$

$$= \sec^2 \alpha \qquad \tan^2 \alpha + 1 = \sec^2 \alpha$$

$$= \frac{1}{\cos^2 \alpha} \qquad \sec^2 \alpha = \tfrac{1}{\cos^2 \alpha}$$

$$= \frac{1}{1 - \sin^2 \alpha} \qquad \cos^2 \alpha = 1 - \sin^2 \alpha$$

Since the left side equals the right side, the given equation is an identity. ▶

As mentioned earlier, we can use a graphing calculator to support our algebraic verification of an identity. In Example 2, if we let

$$Y_1 = \tan^2 x(1 + \cot^2 x) \quad \text{and} \quad Y_2 = \frac{1}{1 - \sin^2 x},$$

and graph these functions in the same window, the graphs coincide, supporting our analytic work in Example 2. See the figure at left. (Enter $\tan^2 x$ as $(\tan x)^2$.)

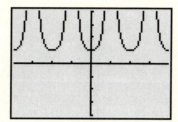

Trig Window

A table of values for Y_1 and Y_2 is shown in the figure at right. The table also supports the identity for selected values in the interval $[1, 4]$.

◀ **EXAMPLE 3**
Verifying an identity (working with one side)

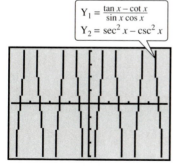

$Y_1 = \frac{\tan x - \cot x}{\sin x \cos x}$
$Y_2 = \sec^2 x - \csc^2 x$

Trig Window
This graph supports the result in Example 3.

Verify that

$$\frac{\tan t - \cot t}{\sin t \cos t} = \sec^2 t - \csc^2 t$$

is an identity.

Since the left side is the more complicated one, transform the left side to equal the right side.

$$\frac{\tan t - \cot t}{\sin t \cos t}$$

$$= \frac{\tan t}{\sin t \cos t} - \frac{\cot t}{\sin t \cos t} \qquad \frac{a-b}{c} = \frac{a}{c} - \frac{b}{c}$$

$$= \tan t \cdot \frac{1}{\sin t \cos t} - \cot t \cdot \frac{1}{\sin t \cos t} \qquad \frac{a}{b} = a \cdot \frac{1}{b}$$

$$= \frac{\sin t}{\cos t} \cdot \frac{1}{\sin t \cos t} - \frac{\cos t}{\sin t} \cdot \frac{1}{\sin t \cos t} \qquad \tan t = \frac{\sin t}{\cos t}; \cot t = \frac{\cos t}{\sin t}$$

$$= \frac{1}{\cos^2 t} - \frac{1}{\sin^2 t}$$

$$= \sec^2 t - \csc^2 t \qquad \frac{1}{\cos^2 t} = \sec^2 t; \frac{1}{\sin^2 t} = \csc^2 t$$

Here, writing in terms of sine and cosine only was used in the third line. ▶

EXAMPLE 4
Verifying an identity (working with one side)

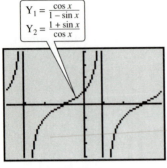

Trig Window
This graph supports the result in Example 4.

Verify that

$$\frac{\cos x}{1 - \sin x} = \frac{1 + \sin x}{\cos x}$$

is an identity.

This time we will work on the right side. Use the suggestion given at the beginning of the section to multiply the numerator and denominator on the right by $1 - \sin x$.

$$\frac{1 + \sin x}{\cos x} = \frac{(1 + \sin x)(1 - \sin x)}{\cos x(1 - \sin x)} \qquad \text{Multiply by 1.}$$

$$= \frac{1 - \sin^2 x}{\cos x(1 - \sin x)}$$

$$= \frac{\cos^2 x}{\cos x(1 - \sin x)} \qquad 1 - \sin^2 x = \cos^2 x$$

$$= \frac{\cos x}{1 - \sin x} \qquad \text{Reduce to lowest terms.} \blacktriangleright$$

If both sides of an identity appear to be equally complex, the identity can be verified by working independently on the left side and on the right side, until each side is changed into some common third result. *Each step, on each side, must be reversible.* With all steps reversible, the procedure is as follows.

left $=$ right

common third
expression

The left side leads to the third expression, which leads back to the right side. This procedure is just a shortcut for the procedure used in the first examples of this section: the left side is changed into the right side, but by going through an intermediate step.

EXAMPLE 5
Verifying an identity (working with both sides)

Verify that

$$\frac{\sec \alpha + \tan \alpha}{\sec \alpha - \tan \alpha} = \frac{1 + 2 \sin \alpha + \sin^2 \alpha}{\cos^2 \alpha}$$

is an identity.

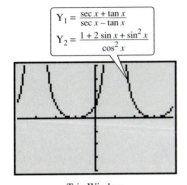

$$Y_1 = \frac{\sec x + \tan x}{\sec x - \tan x}$$

$$Y_2 = \frac{1 + 2\sin x + \sin^2 x}{\cos^2 x}$$

Trig Window
This graph supports the result in Example 5.

Both sides appear equally complex, so verify the identity by changing each side into a common third expression. Work first on the left, multiplying numerator and denominator by $\cos \alpha$.

$$\frac{\sec \alpha + \tan \alpha}{\sec \alpha - \tan \alpha} = \frac{(\sec \alpha + \tan \alpha)\,\cos \alpha}{(\sec \alpha - \tan \alpha)\,\cos \alpha} \qquad \frac{\cos \alpha}{\cos \alpha} = 1; \quad \text{multiplicative identity}$$

$$= \frac{\sec \alpha \cos \alpha + \tan \alpha \cos \alpha}{\sec \alpha \cos \alpha - \tan \alpha \cos \alpha} \qquad \text{Distributive property}$$

$$= \frac{1 + \tan \alpha \cos \alpha}{1 - \tan \alpha \cos \alpha} \qquad \sec \alpha \cos \alpha = 1$$

$$= \frac{1 + \dfrac{\sin \alpha}{\cos \alpha} \cdot \cos \alpha}{1 - \dfrac{\sin \alpha}{\cos \alpha} \cdot \cos \alpha} \qquad \tan \alpha = \frac{\sin \alpha}{\cos \alpha}$$

$$= \frac{1 + \sin \alpha}{1 - \sin \alpha}$$

On the right side of the original statement, begin by factoring.

$$\frac{1 + 2\sin \alpha + \sin^2 \alpha}{\cos^2 \alpha}$$

$$= \frac{(1 + \sin \alpha)^2}{\cos^2 \alpha} \qquad a^2 + 2ab + b^2 = (a + b)^2$$

$$= \frac{(1 + \sin \alpha)^2}{1 - \sin^2 \alpha} \qquad \cos^2 \alpha = 1 - \sin^2 \alpha$$

$$= \frac{(1 + \sin \alpha)^2}{(1 + \sin \alpha)(1 - \sin \alpha)} \qquad 1 - \sin^2 \alpha = (1 + \sin \alpha)(1 - \sin \alpha)$$

$$= \frac{1 + \sin \alpha}{1 - \sin \alpha} \qquad \text{Reduce to lowest terms.}$$

We now have shown that

$$\frac{\sec \alpha + \tan \alpha}{\sec \alpha - \tan \alpha} = \frac{1 + \sin \alpha}{1 - \sin \alpha} = \frac{1 + 2\sin \alpha + \sin^2 \alpha}{\cos^2 \alpha},$$

verifying that the original equation is an identity. ▶

CAUTION This method should be used *only* if the steps are reversible. A good check is to reverse the steps used on the right side to show that you could continue from the last step on the left to get the original expression on the right.

There are usually several ways to verify a given identity. You may wish to go through the examples of this section and verify each using a method different from the one given. For instance, another way to begin verifying the identity in Example 5 is to work on the left as follows.

$$\frac{\sec \alpha + \tan \alpha}{\sec \alpha - \tan \alpha} = \frac{\dfrac{1}{\cos \alpha} + \dfrac{\sin \alpha}{\cos \alpha}}{\dfrac{1}{\cos \alpha} - \dfrac{\sin \alpha}{\cos \alpha}} \qquad \text{\color{blue}Fundamental identities}$$

$$= \frac{\dfrac{1 + \sin \alpha}{\cos \alpha}}{\dfrac{1 - \sin \alpha}{\cos \alpha}} \qquad \text{\color{blue}Add fractions;\quad subtract fractions.}$$

$$= \frac{1 + \sin \alpha}{1 - \sin \alpha} \qquad \text{\color{blue}Divide fractions.}$$

Now we compare this with the result shown in Example 5 for the right side to see that the two sides agree.

CONNECTIONS Much of our work with identities in this chapter is preparation for calculus, where many of the identities we verify here are used. Some calculus problems are simplified by making an appropriate trigonometric substitution. We show one example here where trigonometric substitution and the use of an identity makes it possible to take the square root $\sqrt{9 + x^2}$. To do this, we choose $x = 3 \tan \theta$. The reason for this choice will become clear as we continue.

Letting $x = 3 \tan \theta$ gives

$$\begin{aligned}
\sqrt{9 + x^2} &= \sqrt{9 + (\mathbf{3 \tan \theta})^2} \\
&= \sqrt{9 + 9 \tan^2 \theta} \\
&= \sqrt{9(1 + \tan^2 \theta)} \\
&= 3\sqrt{1 + \tan^2 \theta} \\
&= 3\sqrt{\sec^2 \theta}.
\end{aligned}$$

In the interval $(0, \pi/2)$, the value of $\sec \theta$ is positive, giving

$$\sqrt{9 + x^2} = 3 \sec \theta.$$

FOR DISCUSSION OR WRITING

Substitute $\cos \theta$ for x in $\sqrt{(1 - x^2)^3}$ and simplify. Why is $\cos \theta$ an appropriate choice here?

7.2 Exercises ▼▼▼▼▼▼▼▼▼▼▼▼▼▼▼▼▼▼▼▼▼▼▼▼▼▼▼▼▼▼▼▼▼▼▼▼▼▼

Perform the indicated operation and simplify the result.

1. $\tan \theta + \dfrac{1}{\tan \theta}$

2. $\dfrac{\cos x}{\sin x} + \dfrac{\sin x}{\cos x}$

3. $\cot s(\tan s + \sin s)$

4. $\sec \beta(\cos \beta + \sin \beta)$

5. $\dfrac{1}{\csc^2 \theta} + \dfrac{1}{\sec^2 \theta}$

6. $\dfrac{1}{\sin \alpha - 1} - \dfrac{1}{\sin \alpha + 1}$

7. $\dfrac{\cos x}{\sec x} + \dfrac{\sin x}{\csc x}$

8. $\dfrac{\cos \gamma}{\sin \gamma} + \dfrac{\sin \gamma}{1 + \cos \gamma}$

9. $(1 + \sin t)^2 + \cos^2 t$

10. $(1 + \tan s)^2 - 2 \tan s$

11. $\dfrac{1}{1 + \cos x} - \dfrac{1}{1 - \cos x}$

12. $(\sin \alpha - \cos \alpha)^2$

Factor each trigonometric expression.

13. $\sin^2 \gamma - 1$

14. $\sec^2 \theta - 1$

15. $(\sin x + 1)^2 - (\sin x - 1)^2$

16. $(\tan x + \cot x)^2 - (\tan x - \cot x)^2$

17. $2 \sin^2 x + 3 \sin x + 1$

18. $4 \tan^2 \beta + \tan \beta - 3$

19. $\cos^4 x + 2 \cos^2 x + 1$

20. $\cot^4 x + 3 \cot^2 x + 2$

21. $\sin^3 x - \cos^3 x$

22. $\sin^3 \alpha + \cos^3 \alpha$

Each expression simplifies to a constant, a single circular function, or a power of a circular function. Use the fundamental identities to simplify each expression.

23. $\tan \theta \cos \theta$

24. $\cot \alpha \sin \alpha$

25. $\sec r \cos r$

26. $\cot t \tan t$

27. $\dfrac{\sin \beta \tan \beta}{\cos \beta}$

28. $\dfrac{\csc \theta \sec \theta}{\cot \theta}$

29. $\sec^2 x - 1$

30. $\csc^2 t - 1$

31. $\dfrac{\sin^2 x}{\cos^2 x} + \sin x \csc x$

32. $\dfrac{1}{\tan^2 \alpha} + \cot \alpha \tan \alpha$

Given a trigonometric identity such as those discussed in this section, it is possible to form another identity by replacing each function with its cofunction. Therefore, for example, since the equation

$$\frac{\cot \theta}{\csc \theta} = \cos \theta,$$

in Exercise 33, represents an identity, the equation

$$\frac{\tan \theta}{\sec \theta} = \sin \theta$$

is also an identity. After verifying the identities in Exercises 33–68, you may wish to construct other identities using this method, in order to provide more practice exercises for yourself.

Verify each trigonometric identity. See Examples 1–5.

33. $\dfrac{\cot \theta}{\csc \theta} = \cos \theta$

34. $\dfrac{\tan \alpha}{\sec \alpha} = \sin \alpha$

35. $\dfrac{1 - \sin^2 \beta}{\cos \beta} = \cos \beta$

36. $\dfrac{\tan^2 \gamma + 1}{\sec \gamma} = \sec \gamma$

37. $\cos^2 \theta(\tan^2 \theta + 1) = 1$

38. $\sin^2 \beta(1 + \cot^2 \beta) = 1$

39. $\cot s + \tan s = \sec s \csc s$

40. $\sin^2 \alpha + \tan^2 \alpha + \cos^2 \alpha = \sec^2 \alpha$

41. $\dfrac{\cos \alpha}{\sec \alpha} + \dfrac{\sin \alpha}{\csc \alpha} = \sec^2 \alpha - \tan^2 \alpha$

42. $\dfrac{\sin^2 \gamma}{\cos \gamma} = \sec \gamma - \cos \gamma$

43. $\sin^4 \theta - \cos^4 \theta = 2 \sin^2 \theta - 1$

44. $\dfrac{\cos \theta}{\sin \theta \cot \theta} = 1$

45. $(1 - \cos^2 \alpha)(1 + \cos^2 \alpha) = 2 \sin^2 \alpha - \sin^4 \alpha$

46. $\tan^2 \gamma \sin^2 \gamma = \tan^2 \gamma + \cos^2 \gamma - 1$

47. $\dfrac{\cos \theta + 1}{\tan^2 \theta} = \dfrac{\cos \theta}{\sec \theta - 1}$

48. $\dfrac{(\sec \theta - \tan \theta)^2 + 1}{\sec \theta \csc \theta - \tan \theta \csc \theta} = 2 \tan \theta$

49. $\dfrac{1}{1 - \sin \theta} + \dfrac{1}{1 + \sin \theta} = 2 \sec^2 \theta$

50. $\dfrac{1}{\sec \alpha - \tan \alpha} = \sec \alpha + \tan \alpha$

51. $\dfrac{\tan s}{1 + \cos s} + \dfrac{\sin s}{1 - \cos s} = \cot s + \sec s \csc s$

52. $\dfrac{1 - \cos x}{1 + \cos x} = (\cot x - \csc x)^2$

53. $\dfrac{\cot \alpha + 1}{\cot \alpha - 1} = \dfrac{1 + \tan \alpha}{1 - \tan \alpha}$

54. $\dfrac{1}{\tan \alpha - \sec \alpha} + \dfrac{1}{\tan \alpha + \sec \alpha} = -2 \tan \alpha$

55. $\sin^2 \alpha \sec^2 \alpha + \sin^2 \alpha \csc^2 \alpha = \sec^2 \alpha$

56. $\dfrac{\csc \theta + \cot \theta}{\tan \theta + \sin \theta} = \cot \theta \csc \theta$

57. $\sec^4 x - \sec^2 x = \tan^4 x + \tan^2 x$

58. $\dfrac{1 - \sin \theta}{1 + \sin \theta} = \sec^2 \theta - 2 \sec \theta \tan \theta + \tan^2 \theta$

59. $\sin \theta + \cos \theta = \dfrac{\sin \theta}{1 - \dfrac{\cos \theta}{\sin \theta}} + \dfrac{\cos \theta}{1 - \dfrac{\sin \theta}{\cos \theta}}$

60. $\dfrac{\sin \theta}{1 - \cos \theta} - \dfrac{\sin \theta \cos \theta}{1 + \cos \theta} = \csc \theta(1 + \cos^2 \theta)$

61. $\dfrac{\sec^4 s - \tan^4 s}{\sec^2 s + \tan^2 s} = \sec^2 s - \tan^2 s$

62. $\dfrac{\cot^2 t - 1}{1 + \cot^2 t} = 1 - 2 \sin^2 t$

63. $\dfrac{\tan^2 t - 1}{\sec^2 t} = \dfrac{\tan t - \cot t}{\tan t + \cot t}$

64. $(1 + \sin x + \cos x)^2 = 2(1 + \sin x)(1 + \cos x)$

65. $\dfrac{1 + \cos x}{1 - \cos x} - \dfrac{1 - \cos x}{1 + \cos x} = 4 \cot x \csc x$

66. $(\sec \alpha - \tan \alpha)^2 = \dfrac{1 - \sin \alpha}{1 + \sin \alpha}$

67. $(\sec \alpha + \csc \alpha)(\cos \alpha - \sin \alpha) = \cot \alpha - \tan \alpha$

68. $\dfrac{\sin^4 \alpha - \cos^4 \alpha}{\sin^2 \alpha - \cos^2 \alpha} = 1$

---

69. A student claims that the equation

$$\cos \theta + \sin \theta = 1$$

is an identity, since by letting $\theta = 90°$ (or $\pi/2$ radians) we get $0 + 1 = 1$, a true statement. Comment on this student's reasoning.

70. Explain why the method described in the text involving working on both sides of an identity to show that each side is equal to the same expression is a valid method of verifying an identity. When using this method, what must be true about each step taken? (*Hint:* See the discussion preceding Example 5.)

📈 *In Exercises 71–74, graph the function and conjecture an identity. Then prove your conjecture.*

71. $(\sec \theta + \tan \theta)(1 - \sin \theta)$

72. $(\csc \theta + \cot \theta)(\sec \theta - 1)$

73. $\dfrac{\cos \theta + 1}{\sin \theta + \tan \theta}$

74. $\tan \theta \sin \theta + \cos \theta$

📈 *In Exercises 75–84, graph the functions on each side of the equals sign to determine whether the equation might be an identity. (Note: Use a domain whose length is at least 2π.) If the equation looks like an identity, prove it algebraically.*

75. $\dfrac{2 + 5 \cos s}{\sin s} = 2 \csc s + 5 \cot s$

76. $1 + \cot^2 s = \dfrac{\sec^2 s}{\sec^2 s - 1}$

77. $\dfrac{\tan s - \cot s}{\tan s + \cot s} = 2 \sin^2 s$

78. $\dfrac{1}{1 + \sin s} + \dfrac{1}{1 - \sin s} = \sec^2 s$

79. $\dfrac{1 - \tan^2 s}{1 + \tan^2 s} = \cos^2 s - \sin s$

80. $\dfrac{\sin^3 s - \cos^3 s}{\sin s - \cos s} = \sin^2 s + 2 \sin s \cos s + \cos^2 s$

81. $\sin^2 s + \cos^2 s = \dfrac{1}{2}(1 - \cos 4s)$

82. $\cos 3s = 3 \cos s + 4 \cos^3 s$

83. $\tan^2 x - \sin^2 x = (\tan x \sin x)^2$

84. $\dfrac{\cot \theta}{\csc \theta + 1} = \sec \theta - \tan \theta$

By substituting a number for s or t, show that the equation is not an identity for all real numbers.

85. $\sin(\csc s) = 1$

86. $\sqrt{\cos^2 s} = \cos s$

87. $\csc t = \sqrt{1 + \cot^2 t}$

88. $\cos t = \sqrt{1 - \sin^2 t}$

89. Let $\tan \theta = t$ and show that

$$\sin \theta \cos \theta = \frac{t}{t^2 + 1}.$$

90. When does $\sin x = \sqrt{1 - \cos^2 x}$?

91. An equation that is an identity has an infinite number of solutions. If an equation has an infinite number of solutions, is it necessarily an identity? Discuss.

7.3 Sum and Difference Identities ▼▼▼

Several examples presented throughout this book should have convinced you by now that $\cos(A - B)$ does *not* equal $\cos A - \cos B$. For example, if $A = \pi/2$ and $B = 0$,

$$\cos(A - B) = \cos\left(\frac{\pi}{2} - 0\right) = \cos \frac{\pi}{2} = 0,$$

while

$$\cos A - \cos B = \cos \frac{\pi}{2} - \cos 0 = 0 - 1 = -1.$$

The actual formula for $\cos(A - B)$ is derived in this section. Start by locating angles A and B in standard position on a unit circle, with $B < A$. Let S and Q be the points where angles A and B, respectively, intersect the circle. Locate point R on the unit circle so that angle POR equals the difference $A - B$. See Figure 3.

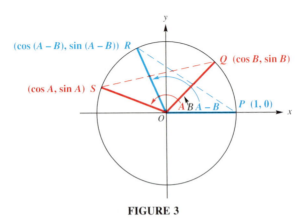

FIGURE 3

Point Q is on the unit circle, so by the work with circular functions in Chapter 6, the x-coordinate of Q is given by the cosine of angle B, while the y-coordinate of Q is given by the sine of angle B:

$$Q \text{ has coordinates } (\cos B, \sin B).$$

In the same way,

$$S \text{ has coordinates } (\cos A, \sin A),$$

and

$$R \text{ has coordinates } (\cos(A - B), \sin(A - B)).$$

Angle SOQ also equals $A - B$. Since the central angles SOQ and POR are equal, chords PR and SQ are equal. By the distance formula, since $PR = SQ$,

$$\sqrt{[\cos(A - B) - 1]^2 + [\sin(A - B) - 0]^2}$$
$$= \sqrt{(\cos A - \cos B)^2 + (\sin A - \sin B)^2}.$$

Squaring both sides and clearing parentheses gives

$$\cos^2(A - B) - 2\cos(A - B) + 1 + \sin^2(A - B)$$
$$= \cos^2 A - 2\cos A \cos B + \cos^2 B + \sin^2 A - 2\sin A \sin B + \sin^2 B.$$

Since $\sin^2 x + \cos^2 x = 1$ for any value of x, rewrite the equation as

$$2 - 2\cos(A - B) = 2 - 2\cos A \cos B - 2\sin A \sin B$$
$$\cos(A - B) = \cos A \cos B + \sin A \sin B.$$

This is the identity for $\cos(A - B)$. Although Figure 3 shows angles A and B in the second and first quadrants, respectively, it can be shown that this result is the same for any values of these angles.

To find a similar expression for $\cos(A + B)$, rewrite $A + B$ as $A - (-B)$ and use the identity for $\cos(A - B)$ found above, along with the fact that $\cos(-B) = \cos B$ and $\sin(-B) = -\sin B$.

$$\cos(A + B) = \cos[A - (-B)]$$
$$= \cos A \cos(-B) + \sin A \sin(-B)$$
$$= \cos A \cos B + \sin A(-\sin B)$$
$$\cos(A + B) = \cos A \cos B - \sin A \sin B$$

The two formulas we have just derived are summarized as follows.

Cosine of Sum or Difference

$$\cos(A - B) = \cos A \cos B + \sin A \sin B$$
$$\cos(A + B) = \cos A \cos B - \sin A \sin B$$

These identities are important in calculus and other areas of mathematics and useful in certain applications. Although a calculator can be used to find an approximation for $\cos 15°$, for example, the method shown below can be applied to give practice using the sum and difference identities, as well as to get an exact value.

EXAMPLE 1
Using the cosine sum and difference identities to find exact values

Find the *exact* value of the following.

(a) $\cos 15°$

To find $\cos 15°$, write $15°$ as the sum or difference of two angles with known function values. Since we know the exact trigonometric function values of both $45°$ and $30°$, write $15°$ as $45° - 30°$. (We could also use $60° - 45°$.) Then use the identity for the cosine of the difference of two angles.

$$\cos 15° = \cos(45° - 30°)$$
$$= \cos 45° \cos 30° + \sin 45° \sin 30° \qquad \text{Use the cosine of the difference identity.}$$
$$= \frac{\sqrt{2}}{2} \cdot \frac{\sqrt{3}}{2} + \frac{\sqrt{2}}{2} \cdot \frac{1}{2}$$
$$= \frac{\sqrt{6} + \sqrt{2}}{4}$$

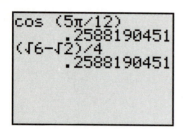

See Example 1(b). The calculator provides support by giving the same approximation for both $\cos\frac{5\pi}{12}$ and $\frac{\sqrt{6}-\sqrt{2}}{4}$.

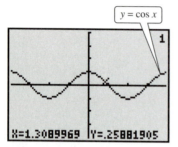

Trig Window

When we input $\frac{5\pi}{12}$ for x, we get the calculator approximation for $\frac{\sqrt{6}-\sqrt{2}}{4}$. See the screen above.

(b) $\cos\dfrac{5\pi}{12} = \cos\left(\dfrac{\pi}{6} + \dfrac{\pi}{4}\right)$

$\qquad = \cos\dfrac{\pi}{6}\cos\dfrac{\pi}{4} - \sin\dfrac{\pi}{6}\sin\dfrac{\pi}{4}$ Use the cosine of the sum identity.

$\qquad = \dfrac{\sqrt{3}}{2}\cdot\dfrac{\sqrt{2}}{2} - \dfrac{1}{2}\cdot\dfrac{\sqrt{2}}{2}$

$\qquad = \dfrac{\sqrt{6} - \sqrt{2}}{4}$ ▸

COFUNCTION IDENTITIES The identities for the cosine of the sum and difference of two angles can be used to derive other identities.

Cofunction Identities

$$\cos(90° - \theta) = \sin\theta \qquad \cot(90° - \theta) = \tan\theta$$
$$\sin(90° - \theta) = \cos\theta \qquad \sec(90° - \theta) = \csc\theta$$
$$\tan(90° - \theta) = \cot\theta \qquad \csc(90° - \theta) = \sec\theta$$

Similar identities can be obtained for a real number domain by replacing 90° by $\pi/2$.

For example, substituting 90° for A and θ for B in the identity given above for $\cos(A - B)$ gives

$$\cos(90° - \theta) = \cos 90° \cos\theta + \sin 90° \sin\theta$$
$$= 0 \cdot \cos\theta + 1 \cdot \sin\theta$$
$$= \sin\theta.$$

This result is true for *any* value of θ since the identity for $\cos(A - B)$ is true for any values of A and B. The other cofunction identities can be applied similarly.

◀ **EXAMPLE 2**
Using the cofunction identities

Find an angle θ that satisfies each of the following.

(a) $\cot\theta = \tan 25°$
  Since tangent and cotangent are cofunctions,

$$\cot\theta = \tan(90° - \theta).$$

This means that

$$\tan(90° - \theta) = \tan 25°,$$
or $\qquad\qquad 90° - \theta = 25°$
$$\theta = 65°.$$

(b) $\sin \theta = \cos(-30°)$

In the same way,

$$\sin \theta = \cos(90° - \theta) = \cos(-30°),$$

giving

$$90° - \theta = -30°$$

$$\theta = 120°.$$

(c) $\csc \dfrac{3\pi}{4} = \sec \theta$

Cosecant and secant are cofunctions, so

$$\csc \frac{3\pi}{4} = \sec\left(\frac{\pi}{2} - \frac{3\pi}{4}\right) = \sec \theta$$

$$\sec\left(-\frac{\pi}{4}\right) = \sec \theta$$

$$-\frac{\pi}{4} = \theta. \quad \blacktriangleright$$

NOTE Because trigonometric (and circular) functions are periodic, the solutions in Example 2 are not unique. In each case, we give only one of infinitely many possibilities.

If one of the angles A or B in the identities for $\cos(A + B)$ and $\cos(A - B)$ is a quadrantal angle, then the identity allows us to write the expression in terms of a single function of A or B. The next example illustrates this.

EXAMPLE 3
Reducing $\cos(A - B)$ to a function of a single variable

Write $\cos(180° - \theta)$ as a trigonometric function of θ.

Use the difference identity. Replace A with $180°$ and B with θ.

$$\cos(180° - \theta) = \cos 180° \cos \theta + \sin 180° \sin \theta$$

$$= (-1) \cos \theta + (0) \sin \theta$$

$$= -\cos \theta \quad \blacktriangleright$$

SINE AND TANGENT OF SUM AND DIFFERENCE Formulas for $\sin(A + B)$ and $\sin(A - B)$ can be developed from the results above. Start with the cofunction relationship

$$\sin \theta = \cos(90° - \theta).$$

Replace θ with $A + B$.

$$\sin(A + B) = \cos[90° - (A + B)]$$

$$= \cos[(90° - A) - B]$$

Using the formula for $\cos(A - B)$ from the previous section gives

$$\sin(A + B) = \cos(90° - A) \cos B + \sin(90° - A) \sin B$$

or

$$\sin(A + B) = \sin A \cos B + \cos A \sin B.$$

(The cofunction relationships were used in the last step.)

Now write $\sin(A - B)$ as $\sin[A + (-B)]$ and use the identity for $\sin(A + B)$ to get

$$\sin(A - B) = \sin[A + (-B)]$$
$$= \sin A \cos(-B) + \cos A \sin(-B)$$
$$= \sin A \cos B - \cos A \sin B$$

since $\cos(-B) = \cos B$ and $\sin(-B) = -\sin B$. In summary,

$$\sin(A - B) = \sin A \cos B - \cos A \sin B.$$

Using the identities for $\sin(A + B)$, $\cos(A + B)$, $\sin(A - B)$, and $\cos(A - B)$, and the identity $\tan \theta = \sin \theta / \cos \theta$, gives the following identities.

$$\tan(A + B) = \frac{\tan A + \tan B}{1 - \tan A \tan B}$$

$$\tan(A - B) = \frac{\tan A - \tan B}{1 + \tan A \tan B}$$

We show the proof for the first of these two identities. The proof for the other is very similar. Start with

$$\tan(A + B) = \frac{\sin(A + B)}{\cos(A + B)}$$

$$= \frac{\sin A \cos B + \cos A \sin B}{\cos A \cos B - \sin A \sin B}.$$

To express this result in terms of the tangent function, multiply both numerator and denominator by $1/(\cos A \cos B)$.

$$\tan(A + B) = \frac{\dfrac{\sin A \cos B + \cos A \sin B}{1}}{\dfrac{\cos A \cos B - \sin A \sin B}{1}} \cdot \frac{\dfrac{1}{\cos A \cos B}}{\dfrac{1}{\cos A \cos B}}$$

$$= \frac{\dfrac{\sin A \cos B}{\cos A \cos B} + \dfrac{\cos A \sin B}{\cos A \cos B}}{\dfrac{\cos A \cos B}{\cos A \cos B} - \dfrac{\sin A \sin B}{\cos A \cos B}}$$

$$= \frac{\dfrac{\sin A}{\cos A} + \dfrac{\sin B}{\cos B}}{1 - \dfrac{\sin A}{\cos A} \cdot \dfrac{\sin B}{\cos B}}$$

Using the identity $\tan \theta = \sin \theta / \cos \theta$,

$$\tan(A + B) = \frac{\tan A + \tan B}{1 - \tan A \tan B}.$$

The identities given in this section are summarized below.

Sine and Tangent of Sum or Difference

$$\sin(A + B) = \sin A \cos B + \cos A \sin B$$

$$\sin(A - B) = \sin A \cos B - \cos A \sin B$$

$$\tan(A + B) = \frac{\tan A + \tan B}{1 - \tan A \tan B}$$

$$\tan(A - B) = \frac{\tan A - \tan B}{1 + \tan A \tan B}$$

Again, the following examples and the corresponding exercises are given primarily to offer practice in using these new identities.

◀ **EXAMPLE 4**
Using the sine and tangent sum and difference identities to find exact values

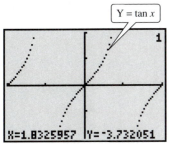

See Example 4(b). The calculator provides support by giving the same approximation for both $\tan \frac{7\pi}{12}$ and $-2 - \sqrt{3}$.

Find the *exact* value of the following.

(a) $\sin 75° = \sin(45° + 30°)$

$$= \sin 45° \cos 30° + \cos 45° \sin 30°$$

$$= \frac{\sqrt{2}}{2} \cdot \frac{\sqrt{3}}{2} + \frac{\sqrt{2}}{2} \cdot \frac{1}{2}$$

$$= \frac{\sqrt{6}}{4} + \frac{\sqrt{2}}{4} = \frac{\sqrt{6} + \sqrt{2}}{4}$$

(b) $\tan \frac{7\pi}{12} = \tan\left(\frac{\pi}{3} + \frac{\pi}{4}\right)$

$$= \frac{\tan \frac{\pi}{3} + \tan \frac{\pi}{4}}{1 - \tan \frac{\pi}{3} \tan \frac{\pi}{4}}$$

$$= \frac{\sqrt{3} + 1}{1 - \sqrt{3} \cdot 1}$$

$$= \frac{\sqrt{3} + 1}{1 - \sqrt{3}} \cdot \frac{1 + \sqrt{3}}{1 + \sqrt{3}}$$

$$= \frac{\sqrt{3} + 3 + 1 + \sqrt{3}}{1 - 3}$$

$$= \frac{4 + 2\sqrt{3}}{-2}$$

$$= -2 - \sqrt{3}$$

Y = tan x

Trig Window
Dot mode

When we input $\frac{7\pi}{12}$ for x, we get the calculator approximation for $-2 - \sqrt{3}$. See the screen above this one.

(c) $\sin 40° \cos 160° - \cos 40° \sin 160° = \sin(40° - 160°)$

$$= \sin(-120°)$$

$$= -\sin 120°$$

$$= -\frac{\sqrt{3}}{2}$$ ▶

◀EXAMPLE 5
Writing a function as an expression involving functions of θ

Write each of the following as an expression involving functions of θ.

(a) $\sin(30° + \theta)$

Using the identity for $\sin(A + B)$,

$$\sin(30° + \theta) = \sin 30° \cos \theta + \cos 30° \sin \theta$$

$$= \frac{1}{2} \cos \theta + \frac{\sqrt{3}}{2} \sin \theta.$$

(b) $\tan(45° - \theta) = \dfrac{\tan 45° - \tan \theta}{1 + \tan 45° \tan \theta} = \dfrac{1 - \tan \theta}{1 + \tan \theta}$

(c) $\sin(180° + \theta) = \sin 180° \cos \theta + \cos 180° \sin \theta$

$$= 0 \cdot \cos \theta + (-1) \sin \theta$$

$$= -\sin \theta \quad ▶$$

◀EXAMPLE 6
Finding functions and the quadrant of $A + B$ given information about A and B

If $\sin A = 4/5$ and $\cos B = -5/13$, where A is in quadrant II and B is in quadrant III, find each of the following.

(a) $\sin(A + B)$

The identity for $\sin(A + B)$ requires $\sin A$, $\cos A$, $\sin B$, and $\cos B$. Two of these values are given. The two missing values, $\cos A$ and $\sin B$, must be found first. These values can be found with the identity $\sin^2 x + \cos^2 x = 1$. To find $\cos A$, use

$$\sin^2 A + \cos^2 A = 1$$

$$\frac{16}{25} + \cos^2 A = 1$$

$$\cos^2 A = \frac{9}{25}$$

$$\cos A = -\frac{3}{5}. \quad \text{\color{teal}Since } A \text{ is in quadrant II, } \cos A < 0.$$

In the same way, $\sin B = -12/13$. Now use the formula for $\sin(A + B)$.

$$\sin(A + B) = \frac{4}{5}\left(-\frac{5}{13}\right) + \left(-\frac{3}{5}\right)\left(-\frac{12}{13}\right)$$

$$= -\frac{20}{65} + \frac{36}{65} = \frac{16}{65}$$

(b) $\tan(A + B)$

Use the values of sine and cosine from part (a) to get $\tan A = -4/3$ and $\tan B = 12/5$. Then

$$\tan(A + B) = \frac{-\dfrac{4}{3} + \dfrac{12}{5}}{1 - \left(-\dfrac{4}{3}\right)\left(\dfrac{12}{5}\right)} = \frac{\dfrac{16}{15}}{1 + \dfrac{48}{15}} = \frac{\dfrac{16}{15}}{\dfrac{63}{15}} = \frac{16}{63}.$$

(c) the quadrant of $A + B$

From the results of parts (a) and (b), we find that $\sin(A + B)$ is positive and $\tan(A + B)$ is also positive. Therefore, $A + B$ must be in quadrant I, since it is the only quadrant in which both sine and tangent are positive. ▶

◀EXAMPLE 7
Applying the cosine
difference identity to
voltage

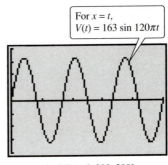

Common household electrical current is called alternating current because the current alternates direction within the wires. The voltage V in a typical 115-volt outlet can be expressed using the equation $V =$ $163 \sin \omega t$, where ω is the angular velocity (in radians per second) of the rotating generator at the electrical plant and t is time measured in seconds.*

(a) It is essential for electrical generators to rotate at precisely 60 cycles per second so that household appliances and computers will function properly. Determine ω for these electrical generators. (Alternating current that cycles 60 times per second is often listed as 60 **hertz**. 1 hertz is equal to 1 cycle per second.)

Since each cycle is 2π radians, at 60 cycles per second, $\omega = 60(2\pi) = 120$ radians per second.

(b) Graph V on the interval $0 \le t \le .05$.

$V = 163 \sin \omega t = 163 \sin 120\pi t$. Because the amplitude is 163 here, we choose $-200 \le V \le 200$ for the range, as shown in Figure 4.

For $x = t$,
$V(t) = 163 \sin 120\pi t$

[0, .05] by [−200, 200]
Xscl = .01   Yscl = 50

FIGURE 4

(c) For what value of ϕ will the graph of $V = 163 \cos(\omega t - \phi)$ be the same as the graph of $V = 163 \sin \omega t$?

Since $\cos(x - \pi/2) = \sin x$, choose $\phi = \pi/2$. ▶

---

*Bell, D., *Fundamentals of Electric Circuits,* Second Edition, Reston Publishing Company, Inc., Reston, Virginia, 1981.

7.3 Exercises ▼▼▼▼▼▼▼▼▼▼▼▼▼▼▼▼▼▼▼▼▼▼▼▼▼▼▼▼▼▼▼▼▼▼▼▼▼

1. Compare the formulas for $\cos(A - B)$ and $\cos(A + B)$. How do they differ? How are they alike?

2. What does the cofunction identity $\cos(\pi/2 - \theta) = \sin\theta$ imply about the graphs of the cosine and sine functions? [*Hint:* First observe that $\cos(\pi/2 - \theta)$ is the same as $\cos(\theta - \pi/2)$.]

3. Compare the formulas for $\sin(A - B)$ and $\sin(A + B)$. How do they differ? How are they alike?

4. Compare the formulas for $\tan(A - B)$ and $\tan(A + B)$. How do they differ? How are they alike?

Use the identities of this section to find the exact value of each expression. See Examples 1 and 4.

5. $\cos(-15°)$ 6. $\sin 105°$ 7. $\tan 15°$ 8. $\tan 105°$ 9. $\sin(-105°)$

10. $\cos\dfrac{5\pi}{12}$ 11. $\cos\dfrac{7\pi}{12}$ 12. $\sin\dfrac{5\pi}{12}$ 13. $\tan\dfrac{\pi}{12}$ 14. $\cos\left(-\dfrac{\pi}{12}\right)$

15. $\sin 76° \cos 31° - \cos 76° \sin 31°$ 16. $\cos(-10°) \cos 35° + \sin(-10°) \sin 35°$

17. $\dfrac{\tan 80° + \tan 55°}{1 - \tan 80° \tan 55°}$ 18. $\dfrac{\tan 80° - \tan(-55°)}{1 + \tan 80° \tan(-55°)}$

19. $\cos\dfrac{2\pi}{5} \cos\dfrac{\pi}{10} - \sin\dfrac{2\pi}{5} \sin\dfrac{\pi}{10}$ 20. $\sin\dfrac{\pi}{5} \cos\dfrac{3\pi}{10} + \cos\dfrac{\pi}{5} \sin\dfrac{3\pi}{10}$

Write each of the following in terms of the cofunction of a complementary angle. See Example 2.

21. $\tan 87°$ 22. $\sin 15°$ 23. $\cot\dfrac{9\pi}{10}$ 24. $\cos\dfrac{\pi}{12}$

Use the cofunction identities to fill in the blanks with the appropriate trigonometric function name. See Example 2.

25. $\cot\dfrac{\pi}{3} = $ _____ $\dfrac{\pi}{6}$ 26. $\sin\dfrac{2\pi}{3} = $ _____ $\left(-\dfrac{\pi}{6}\right)$

27. _____ $33° = \sin 57°$ 28. _____ $72° = \cot 18°$

Use the identities of this section to write each of the following as an expression involving functions of x or θ. See Examples 3 and 5.

29. $\cos(45° - \theta)$ 30. $\cos\left(\dfrac{3\pi}{4} - x\right)$ 31. $\sin\left(\dfrac{\pi}{4} + x\right)$

32. $\sin(270° - \theta)$ 33. $\sin(180° - \theta)$ 34. $\tan(\pi - x)$

35. Why is it not possible to follow Example 2 and find a formula for $\tan(270° - \theta)$?

36. What happens when you try to evaluate $\dfrac{\tan 65.902° + \tan 24.098°}{1 - \tan 65.902° \tan 24.098°}$?

For each of the following, find $\sin(s + t)$, $\sin(s - t)$, $\tan(s + t)$, $\tan(s - t)$, the quadrant of $s + t$, and the quadrant of $s - t$. See Example 6.

37. $\cos s = 3/5$ and $\sin t = 5/13$, s and t in quadrant I

38. $\cos s = -1/5$ and $\sin t = 3/5$, s and t in quadrant II

39. $\sin s = 2/3$ and $\sin t = -1/3$, s in quadrant II and t in quadrant IV

40. $\sin s = 3/5$ and $\sin t = -12/13$, s in quadrant I and t in quadrant III

41. $\cos s = -8/17$ and $\cos t = -3/5$, s and t in quadrant III

📊 *In Exercises 42–45, graph each expression and use the graph to conjecture an identity. Then verify your conjecture using the identities in this section.*

42. $\sin\left(\dfrac{\pi}{2} + x\right)$ **43.** $\sin\left(\dfrac{3\pi}{2} + x\right)$ **44.** $\tan\left(\dfrac{\pi}{2} + x\right)$ **45.** $\dfrac{1 + \tan x}{1 - \tan x}$

Verify that each of the following is an identity.

46. $\sin 2x = 2 \sin x \cos x$ [*Hint:* $\sin 2x = \sin(x + x)$]   **47.** $\sin(x + y) + \sin(x - y) = 2 \sin x \cos y$

48. $\tan(x - y) - \tan(y - x) = \dfrac{2(\tan x - \tan y)}{1 + \tan x \tan y}$ **49.** $\sin(210° + x) - \cos(120° + x) = 0$

50. $\dfrac{\cos(\alpha - \beta)}{\cos \alpha \sin \beta} = \tan \alpha + \cot \beta$ **51.** $\dfrac{\sin(s + t)}{\cos s \cos t} = \tan s + \tan t$

52. $\dfrac{\sin(x - y)}{\sin(x + y)} = \dfrac{\tan x - \tan y}{\tan x + \tan y}$ **53.** $\dfrac{\sin(x + y)}{\cos(x - y)} = \dfrac{\cot x + \cot y}{1 + \cot x \cot y}$

54. $\dfrac{\sin(s - t)}{\sin t} + \dfrac{\cos(s - t)}{\cos t} = \dfrac{\sin s}{\sin t \cos t}$ **55.** $\dfrac{\tan(\alpha + \beta) - \tan \beta}{1 + \tan(\alpha + \beta) \tan \beta} = \tan \alpha$

56. Suppose a fellow student tells you that the cosine of the sum of two angles is the sum of their cosines. Write in your own words how you would correct this student's statement.

57. By a cofunction identity, $\cos 20° = \sin 70°$. What are some values other than 70° that make $\cos 20° = \sin \theta$ a true statement?

▼▼▼▼▼▼▼▼▼▼▼▼▼ **DISCOVERING CONNECTIONS** (Exercises 58–61) ▼▼▼▼▼▼▼▼▼▼▼▼▼

The identities for $\cos(A + B)$ and $\cos(A - B)$ can be used to find exact values of expressions like $\cos 195°$ and $\cos 255°$, where the angle is not in the first quadrant. Work Exercises 58–61 in order to see how this is done.

58. By writing 195° as 180° + 15°, use the identity for $\cos(A + B)$ to express $\cos 195°$ as $-\cos 15°$.

59. Use the identity for $\cos(A - B)$ to find $-\cos 15°$.

60. By the results of Exercises 58 and 59, $\cos 195° = $ _____ .

61. Find the exact value of each of the following using the method shown in Exercises 58–60.

 (a) $\cos 255°$ **(b)** $\sin \dfrac{11\pi}{12}$

---

62. Derive the identity for $\tan(A - B)$ using the identity for $\tan(A + B)$, and the fact that $A - B = A + (-B)$.

63. Derive the identity for $\tan(A - B)$ using the identities for $\sin(A - B)$ and $\cos(A - B)$, and the fact that $\tan(A - B) = \dfrac{\sin(A - B)}{\cos(A - B)}$.

▼▼▼▼▼▼▼▼▼▼▼▼▼ **DISCOVERING CONNECTIONS** (Exercises 64–69) ▼▼▼▼▼▼▼▼▼▼▼▼▼

Refer to the left sketch. It can be shown that m = tan θ, where m is the slope and θ is the angle of inclination of the line. The following exercises, which depend on properties of triangles, refer to the triangle ABC in the right sketch. Assume that all angles are measured in degrees.

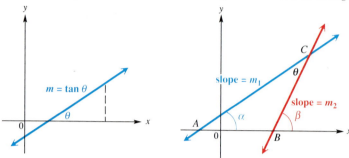

64. In terms of β, what is the measure of angle ABC?

65. Use the fact that the sum of the measures of the angles in a triangle is $180°$ to express θ in terms of α and β.

66. Apply the formula for $\tan(A - B)$ to obtain an expression for $\tan \theta$ in terms of $\tan \alpha$ and $\tan \beta$.

67. Replace $\tan \alpha$ by m_1 and $\tan \beta$ by m_2 to obtain $\tan \theta = \dfrac{m_2 - m_1}{1 + m_1 m_2}$.

In Exercises 68 and 69, use the result from Exercise 67 to find the angle between the pair of lines. Use a calculator and round to the nearest tenth of a degree.

68. $x + y = 9, 2x + y = -1$ **69.** $5x - 2y + 4 = 0, 3x + 5y = 6$

 Exercises 70 and 71 refer to Example 7.

70. How many times does the current oscillate in .05 second?

71. What are the maximum and minimum voltages in this outlet? Is the voltage always equal to 115 volts?

---

72. Sound is a result of waves applying pressure to a person's eardrum. For a pure sound wave radiating outward in a spherical shape, the trigonometric function $P = \dfrac{a}{r} \cos\left[\dfrac{2\pi r}{\lambda} - ct\right]$ can be used to express the sound pressure at a radius of r feet from the source. t is the time in seconds, λ is the length of the sound wave in feet, c is the speed of sound in feet per second, and a is the maximum sound pressure at the source measured in pounds per square foot. (*Source:* Beranek, L., *Noise and Vibration Control,* Institute of Noise Control Engineering, Washington, D.C., 1988.)

(a) Let $a = .4$ lb per ft², $\lambda = 4.9$ ft, and $c = 1026$ ft per sec. Graph the sound pressure at a distance of $r = 10$ feet from its source over the interval $0 \leq t \leq .05$. Describe P at this distance.

(b) Now let $a = 3$ and $t = 10$. Graph the sound pressure for $0 \leq r \leq 20$. What happens to the pressure P as the radius r increases?

(c) Suppose a person stands at a radius r so that $r = n\lambda$ where n is a positive integer. Use the difference identity for cosine to simplify P in this situation.

73. When the two voltages $V_1 = 30 \sin 120\pi t$ and $V_2 = 40 \cos 120\pi t$ are applied to the same circuit, the resulting voltage V will be equal to their sum. (*Source:* Bell, D., *Fundamentals of Electric Circuits,* Second Edition, Reston Publishing Company, Inc., Reston, Virginia, 1981.)

(a) Graph $V = V_1 + V_2$ over the interval $0 \le t \le .05$.

(b) Use the graph to estimate values for a and ϕ so that $V = a \sin(120\pi t + \phi)$.

(c) Use identities to verify that your expression for V is valid.

74. If a person bends at the waist with a straight back making an angle of θ degrees with the horizontal, then the force F exerted on the back muscles can be approximated by the equation $F = \dfrac{.6W \sin(\theta + 90°)}{\sin 12°}$ where W is the weight of the person. (*Source:* Metcalf, H., *Topics in Classical Biophysics,* Prentice-Hall, Inc., Englewood Cliffs, New Jersey, 1980.)

(a) Calculate F when $W = 170$ lb and $\theta = 30°$.

(b) Use an identity to show that F is approximately equal to $2.9W \cos \theta$.

(c) For what value of θ is F maximum?

7.4 Double-Angle Identities and Half-Angle Identities ▼▼▼

DOUBLE-ANGLE IDENTITIES Some special cases of the identities for the sum of two angles are used often enough to be expressed as separate identities. These are the identities that result from the addition identities when $A = B$, so that $A + B = 2A$. These identities are called the **double-angle identities.**

In the identity $\cos(A + B) = \cos A \cos B - \sin A \sin B$, let $B = A$ to derive an expression for $\cos 2A$.

$$\cos 2A = \cos(A + A)$$
$$= \cos A \cos A - \sin A \sin A$$
$$\mathbf{\cos 2A = \cos^2 A - \sin^2 A}$$

Two other useful forms of this identity can be obtained by substituting either $\cos^2 A = 1 - \sin^2 A$ or $\sin^2 A = 1 - \cos^2 A$.

$$\cos 2A = \cos^2 A - \sin^2 A \qquad \text{or} \qquad \cos 2A = \cos^2 A - (1 - \cos^2 A)$$
$$= (1 - \sin^2 A) - \sin^2 A \qquad \qquad = \cos^2 A - 1 + \cos^2 A$$
$$\mathbf{\cos 2A = 1 - 2 \sin^2 A} \qquad\qquad \mathbf{\cos 2A = 2 \cos^2 A - 1}$$

We find $\sin 2A$ with the identity $\sin(A + B) = \sin A \cos B + \cos A \sin B$, letting $B = A$.

$$\sin 2A = \sin(A + A)$$
$$= \sin A \cos A + \cos A \sin A$$
$$\mathbf{\sin 2A = 2 \sin A \cos A}$$

Using the identity for $\tan(A + B)$, we find $\tan 2A$.

$$\tan 2A = \tan(A + A)$$

$$= \frac{\tan A + \tan A}{1 - \tan A \tan A}$$

$$\tan 2A = \frac{2 \tan A}{1 - \tan^2 A}$$

A summary of the double-angle identities follows.

Double-Angle Identities

$$\cos 2A = \cos^2 A - \sin^2 A \qquad \cos 2A = 1 - 2 \sin^2 A$$

$$\cos 2A = 2 \cos^2 A - 1 \qquad \sin 2A = 2 \sin A \cos A$$

$$\tan 2A = \frac{2 \tan A}{1 - \tan^2 A}$$

◀ **EXAMPLE 1**
Using the double-angle
identities

Given $\cos \theta = 3/5$ and $\sin \theta < 0$, use identities to find $\sin 2\theta$, $\cos 2\theta$, and $\tan 2\theta$.

In order to find $\sin 2\theta$, we must first find the value of $\sin \theta$. From the identity $\sin^2 \theta + \cos^2 \theta = 1$, we obtain

$$\sin^2 \theta + \left(\frac{3}{5}\right)^2 = 1$$

$$\sin^2 \theta = \frac{16}{25}$$

$$\sin \theta = -\frac{4}{5}. \qquad \text{Choose the negative square root, since } \sin \theta < 0.$$

Using the double-angle identity for sine, we get

$$\sin 2\theta = 2 \sin \theta \cos \theta = 2\left(-\frac{4}{5}\right)\left(\frac{3}{5}\right) = -\frac{24}{25}.$$

Now find $\cos 2\theta$, using the first form of the identity. (Any form may be used.)

$$\cos 2\theta = \cos^2 \theta - \sin^2 \theta = \frac{9}{25} - \frac{16}{25} = -\frac{7}{25}$$

The value of $\tan 2\theta$ can be found in either of two ways. We can use the double-angle identity, and the fact that $\tan \theta = (\sin \theta)/(\cos \theta) = (-4/5)/(3/5) = -4/3$.

$$\tan 2\theta = \frac{2 \tan \theta}{1 - \tan^2 \theta} = \frac{2\left(-\frac{4}{3}\right)}{1 - \frac{16}{9}} = \frac{-\frac{8}{3}}{-\frac{7}{9}} = \frac{24}{7}$$

As an alternative method, we can find tan 2θ by finding the quotient of sin 2θ and cos 2θ.

$$\tan 2\theta = \frac{\sin 2\theta}{\cos 2\theta} = \frac{-\dfrac{24}{25}}{-\dfrac{7}{25}} = \frac{24}{7} \ \blacktriangleright$$

EXAMPLE 2
Finding functions of θ given information about 2θ

Find the values of the six trigonometric functions of θ if cos $2\theta = 4/5$ and $90° < \theta < 180°$.

Use one of the double-angle identities for cosine to get a trigonometric function of θ.

$$\cos 2\theta = 1 - 2 \sin^2 \theta$$

$$\frac{4}{5} = 1 - 2 \sin^2 \theta$$

$$-\frac{1}{5} = -2 \sin^2 \theta$$

$$\frac{1}{10} = \sin^2 \theta$$

$$\sin \theta = \sqrt{\frac{1}{10}} = \frac{\sqrt{10}}{10}$$

Here, we choose the positive square root since θ terminates in quadrant II. Values of cos θ and tan θ can now be found by using the fundamental identities or by sketching and labeling a right triangle in quadrant II. Using a triangle as in Figure 5, we have

$$\cos \theta = \frac{-3}{\sqrt{10}} = -\frac{3\sqrt{10}}{10}$$

and tan $\theta = 1/(-3) = -1/3$. Find the other three functions using reciprocals.

$$\csc \theta = \frac{1}{\sin \theta} = \sqrt{10}, \qquad \sec \theta = \frac{1}{\cos \theta} = -\frac{\sqrt{10}}{3},$$

$$\cot \theta = \frac{1}{\tan \theta} = -3 \ \blacktriangleright$$

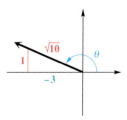

FIGURE 5

EXAMPLE 3
Simplifying expressions using double-angle identities

Simplify each of the following using double-angle identities.

(a) $\cos^2 7x - \sin^2 7x$

This expression suggests one of the identities for $\cos 2A$: $\cos 2A = \cos^2 A - \sin^2 A$. Substituting $7x$ for A gives

$$\cos^2 7x - \sin^2 7x = \cos 2(7x) = \cos 14x.$$

(b) $\sin 15° \cos 15°$

If this expression were $2 \sin 15° \cos 15°$, then we could apply the identity for $\sin 2A$ directly, since $\sin 2A = 2 \sin A \cos A$. We can still apply the identity with $A = 15°$ by writing the multiplicative identity element 1 as $(1/2)(2)$:

$$\sin 15° \cos 15° = \left(\frac{1}{2}\right)(2) \sin 15° \cos 15° \qquad \text{Multiply by 1 in the form } \tfrac{1}{2}(2).$$

$$= \frac{1}{2}(2 \sin 15° \cos 15°) \qquad \text{Associative property}$$

$$= \frac{1}{2}(\sin 2 \cdot 15°) \qquad 2 \sin A \cos A = \sin 2A, \text{ with } A = 15$$

$$= \frac{1}{2} \sin 30°$$

$$= \frac{1}{2} \cdot \frac{1}{2} \qquad \sin 30° = \tfrac{1}{2}$$

$$= \frac{1}{4}. \quad \blacktriangleright$$

Double-angle identities can be used to verify certain identities, as shown in the next example.

EXAMPLE 4
Deriving a multiple-angle identity

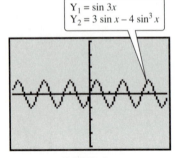

Y₁ = sin 3x
Y₂ = 3 sin x − 4 sin³ x

Trig Window
This graph supports the result in Example 4.

Write $\sin 3s$ in terms of $\sin s$.

$$\sin 3s = \sin(2s + s)$$

$$= \sin 2s \cos s + \cos 2s \sin s$$

$$= (2 \sin s \cos s) \cos s + (\cos^2 s - \sin^2 s) \sin s$$

$$= 2 \sin s \cos^2 s + \cos^2 s \sin s - \sin^3 s$$

$$= 2 \sin s(1 - \sin^2 s) + (1 - \sin^2 s) \sin s - \sin^3 s$$

$$= 2 \sin s - 2 \sin^3 s + \sin s - \sin^3 s - \sin^3 s$$

$$= 3 \sin s - 4 \sin^3 s \quad \blacktriangleright$$

CONNECTIONS The identities for $\cos(A + B)$ and $\cos(A - B)$ can be used to derive a group of identities that are useful in calculus because they make it possible to rewrite a product as a sum.

Adding the identities for $\cos(A + B)$ and $\cos(A - B)$ gives

$$\cos(A + B) = \cos A \cos B - \sin A \sin B$$
$$\underline{\cos(A - B) = \cos A \cos B + \sin A \sin B}$$
$$\cos(A + B) + \cos(A - B) = 2 \cos A \cos B$$

or
$$\cos A \cos B = \frac{1}{2}\left[\cos(A + B) + \cos(A - B)\right].$$

Similarly, subtracting $\cos(A + B)$ from $\cos(A - B)$ gives the following identity.

$$\sin A \sin B = \frac{1}{2}\left[\cos(A - B) - \cos(A + B)\right]$$

Using the identities for $\sin(A + B)$ and $\sin(A - B)$ in the same way, we get two more identities.

$$\sin A \cos B = \frac{1}{2}\left[\sin(A + B) + \sin(A - B)\right]$$

$$\cos A \sin B = \frac{1}{2}\left[\sin(A + B) - \sin(A - B)\right]$$

These last two identities make it possible to rewrite an expression involving both sine and cosine as an expression with just one of these functions. In solving a conditional equation, this conversion can be very useful.

FOR DISCUSSION OR WRITING
1. Show that the double-angle identity $\sin 2A = 2 \sin A \cos A$ is a special case of the identity for $\sin A \cos B$ given above.
2. Show that the double-angle identity $\cos 2A = 2 \cos^2 A - 1$ is a special case of the identity for $\cos A \cos B$ given above.

The next example answers one of the questions posed in the introduction to this chapter.

EXAMPLE 5
Determining wattage consumption

If a toaster is plugged into a common household outlet, the wattage consumed is not constant. Instead, it varies at a high frequency according to the equation $W = \dfrac{V^2}{R}$, where V is the voltage and R is a constant that measures the resistance of the toaster in ohms.* Graph the wattage W

---

*Bell, D., *Fundamentals of Electric Circuits,* Second Edition, Reston Publishing Company, Inc., Reston, Virginia, 1981.

consumed by a typical toaster with $R = 15$ and $V = 163 \sin 120\pi t$ over the interval $0 \le t \le .05$. How many oscillations are there?

By substituting the given values into the wattage equation, we get

$$W = \frac{V^2}{R} = \frac{(163 \sin 120\pi t)^2}{15}.$$

The graph is shown in Figure 6. To determine the range for W, we note that $\sin 120\pi t$ has a maximum value of 1, so the expression for W has a maximum value of $\dfrac{163^2}{15} \approx 1771$. The minimum value is 0. The graph shows that there are 6 oscillations. In Exercise 66 you will be asked to show that the equation given above for W is equivalent to the equation $W = a \cos(\omega t) + c$, for specific values of a, c, and ω.

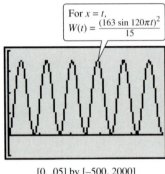

For $x = t$,
$W(t) = \dfrac{(163 \sin 120\pi t)^2}{15}$

[0, .05] by [−500, 2000]
Xscl = .01   Yscl = 500

FIGURE 6

HALF-ANGLE IDENTITIES   From the alternative forms of the identity for $\cos 2A$, we can derive three additional identities for $\sin A/2$, $\cos A/2$, and $\tan A/2$. These are known as **half-angle identities.**

To derive the identity for $\sin A/2$, start with the following double-angle identity for cosine.

$$\cos 2x = 1 - 2 \sin^2 x$$

Then solve for $\sin x$.

$$2 \sin^2 x = 1 - \cos 2x$$

$$\sin x = \pm \sqrt{\frac{1 - \cos 2x}{2}}$$

Now let $2x = A$, so that $x = A/2$, and substitute into this last expression.

$$\sin \frac{A}{2} = \pm \sqrt{\frac{1 - \cos A}{2}}$$

The $\pm$ sign in the identity above indicates that, in practice, the appropriate sign is chosen depending upon the quadrant of $A/2$. For example, if $A/2$ is a third quadrant angle, we choose the negative sign since the sine function is negative there.

The identity for $\cos A/2$ is derived in a very similar way, starting with the double-angle identity $\cos 2x = 2 \cos^2 x - 1$. Solve for $\cos x$.

$$\cos 2x + 1 = 2 \cos^2 x$$

$$\cos x = \pm \sqrt{\frac{1 + \cos 2x}{2}}$$

Replacing x with $A/2$ gives

$$\cos \frac{A}{2} = \pm \sqrt{\frac{1 + \cos A}{2}}.$$

The $\pm$ sign is used as described earlier.

Finally, an identity for $\tan A/2$ comes from the half-angle identities for sine and cosine.

$$\tan \frac{A}{2} = \frac{\pm \sqrt{\dfrac{1 - \cos A}{2}}}{\pm \sqrt{\dfrac{1 + \cos A}{2}}}$$

$$\tan \frac{A}{2} = \pm \sqrt{\frac{1 - \cos A}{1 + \cos A}} \qquad \pm \text{ chosen depending upon quandrant of } \tfrac{A}{2}.$$

Proofs of the other two identities for $\tan A/2$ given below are required in Exercise 65.

Half-Angle Identities

$$\cos \frac{A}{2} = \pm \sqrt{\frac{1 + \cos A}{2}} \qquad \sin \frac{A}{2} = \pm \sqrt{\frac{1 - \cos A}{2}}$$

$$\tan \frac{A}{2} = \pm \sqrt{\frac{1 - \cos A}{1 + \cos A}} \qquad \tan \frac{A}{2} = \frac{\sin A}{1 + \cos A}$$

$$\tan \frac{A}{2} = \frac{1 - \cos A}{\sin A}$$

EXAMPLE 6
Using a half-angle identity to find an exact value

Find the exact value of cos 15° using the half-angle identity for cosine.

$$\cos 15° = \cos \frac{1}{2}(30°)$$

$$= \sqrt{\frac{1 + \cos 30°}{2}} \qquad \text{Choose the positive square root.}$$

$$= \sqrt{\frac{1 + \frac{\sqrt{3}}{2}}{2}}$$

$$= \sqrt{\frac{\left(1 + \frac{\sqrt{3}}{2}\right) \cdot 2}{2 \cdot 2}}$$

$$= \frac{\sqrt{2 + \sqrt{3}}}{2} \quad \blacktriangleright$$

NOTE Compare the value of cos 15° obtained in Example 6 to the value obtained in Example 1 of Section 7.3, where we used the identity for the cosine of the difference of two angles. Although the expressions look completely different, they are indeed equal, as suggested by a calculator approximation for both, .96592583.

EXAMPLE 7
Using a half-angle identity to find an exact value

Find the exact value of tan 22.5° using the identity $\tan \frac{A}{2} = \frac{\sin A}{1 + \cos A}$.
Since 22.5° = (1/2)(45°), replacing A with 45° gives

$$\tan 22.5° = \tan \frac{45°}{2} = \frac{\sin 45°}{1 + \cos 45°} = \frac{\frac{\sqrt{2}}{2}}{1 + \frac{\sqrt{2}}{2}}.$$

Now multiply numerator and denominator by 2. Then rationalize the denominator.

$$\tan 22.5° = \frac{\sqrt{2}}{2 + \sqrt{2}} = \frac{\sqrt{2}}{2 + \sqrt{2}} \cdot \frac{2 - \sqrt{2}}{2 - \sqrt{2}}$$

$$= \frac{2\sqrt{2} - 2}{2} = \sqrt{2} - 1 \quad \blacktriangleright$$

EXAMPLE 8
Finding functions of $A/2$ given information about A

Given $\cos s = 2/3$, with $3\pi/2 < s < 2\pi$, find $\cos s/2$, $\sin s/2$, and $\tan s/2$.

Since

$$\frac{3\pi}{2} < s < 2\pi,$$

dividing through by 2 gives

$$\frac{3\pi}{4} < \frac{s}{2} < \pi,$$

showing that $s/2$ terminates in quadrant II. In this quadrant the value of $\cos s/2$ is negative and the value of $\sin s/2$ is positive. Use the appropriate half-angle identities to get

$$\sin \frac{s}{2} = \sqrt{\frac{1 - \frac{2}{3}}{2}} = \sqrt{\frac{1}{6}} = \frac{\sqrt{6}}{6};$$

and

$$\cos \frac{s}{2} = -\sqrt{\frac{1 + \frac{2}{3}}{2}} = -\sqrt{\frac{5}{6}} = -\frac{\sqrt{30}}{6}.$$

Also,

$$\tan \frac{s}{2} = \frac{\sin \frac{s}{2}}{\cos \frac{s}{2}} = \frac{\frac{\sqrt{6}}{6}}{-\frac{\sqrt{30}}{6}} = -\frac{\sqrt{5}}{5}.$$

Notice that it is not necessary to use a half-angle identity for $\tan s/2$ once we find $\sin s/2$ and $\cos s/2$. However, using this identity would provide an excellent check. ▶

EXAMPLE 9
Simplifying expressions using the half-angle identities

Simplify each of the following using half-angle identities.

(a) $\pm\sqrt{\dfrac{1 + \cos 12x}{2}}$

Start with the identity for $\cos A/2$,

$$\cos \frac{A}{2} = \pm\sqrt{\frac{1 + \cos A}{2}},$$

and replace A with $12x$ to get

$$\pm\sqrt{\frac{1 + \cos 12x}{2}} = \cos \frac{12x}{2} = \cos 6x.$$

(b) $\dfrac{1 - \cos 5\alpha}{\sin 5\alpha}$

Use the third identity for $\tan A/2$ given earlier to get

$$\frac{1 - \cos 5\alpha}{\sin 5\alpha} = \tan \frac{5\alpha}{2}. \text{▶}$$

7.4 Exercises ▼▼▼▼▼▼▼▼▼▼▼▼▼▼▼▼▼▼▼▼▼▼▼▼▼▼▼▼▼▼▼▼

▦ *In Exercises 1–4, the given graphing calculator screen is obtained for a particular stored value of X. What will the screen display for the value of the expression in the final line of the display?*

1.

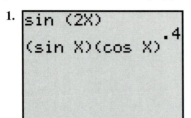

```
sin (2X)
                    .4
(sin X)(cos X)
```

2.

```
cos (2X)
                    .3
2(sin X)²
```

3.

```
cos X
       .9682458366
sin X
              .25
tan (X/2)
```

4.

```
cos X
              -.75
sin X
       .6614378278
tan (X/2)
```

Determine whether the positive or negative square root should be selected.

5. $\sin 195° = \pm \sqrt{\dfrac{1 - \cos 390°}{2}}$

6. $\cos 58° = \pm \sqrt{\dfrac{1 + \cos 116°}{2}}$

7. $\tan 225° = \pm \sqrt{\dfrac{1 - \cos 450°}{1 + \cos 450°}}$

8. $\sin(-10°) = \pm \sqrt{\dfrac{1 - \cos(-20°)}{2}}$

Use the identities in this section to find values of the six trigonometric functions for each angle. See Examples 1 and 2.

9. θ, given $\cos 2\theta = 3/5$ and θ terminates in quadrant I

10. α, given $\cos 2\alpha = 3/4$ and α terminates in quadrant III

11. $2x$, given $\tan x = 2$ and $\cos x > 0$

12. $2x$, given $\tan x = 5/3$ and $\sin x < 0$

Use an identity to write each expression as a single trigonometric function or as a single number. See Example 3.

13. $2 \cos^2 15° - 1$

14. $\cos^2 15° - \sin^2 15°$

15. $\dfrac{2 \tan 15°}{1 - \tan^2 15°}$

16. $1 - 2 \sin^2 15°$

17. $2 \sin \dfrac{\pi}{3} \cos \dfrac{\pi}{3}$

18. $\dfrac{2 \tan \dfrac{\pi}{3}}{1 - \tan^2 \dfrac{\pi}{3}}$

19. $\dfrac{\tan 51°}{1 - \tan^2 51°}$

20. $\dfrac{1}{4} - \dfrac{1}{2} \sin^2 47.1°$

21. $2 \sin 5x \cos 5x$

22. $2 \cos^2 6\alpha - 1$

Use a fundamental identity to simplify each expression.

23. $\sin^2 2x + \cos^2 2x$

24. $1 + \tan^2 4\alpha$

25. $\cot^2 3r + 1$

26. $\sin^2 11\alpha + \cos^2 11\alpha$

Use the half-angle identities of this section to find the exact value. See Examples 6 and 7.

27. $\cos \dfrac{\pi}{8}$

28. $\tan\left(-\dfrac{\pi}{8}\right)$

29. $\cos 195°$

30. $\tan 195°$

31. $\cos 165°$

32. $\sin 165°$

33. Explain how you could use an identity of this section to find the exact value of $\sin 7.5°$. (*Hint:* $7.5 = (1/2)(1/2)(30)$.)

34. The identity

$$\tan \frac{A}{2} = \pm \sqrt{\frac{1 - \cos A}{1 + \cos A}}$$

can be used to find $\tan 22.5° = \sqrt{3 - 2\sqrt{2}}$, and the identity

$$\tan \frac{A}{2} = \frac{\sin A}{1 + \cos A}$$

can be used to get $\tan 22.5° = \sqrt{2} - 1$. Show that these answers are the same, without using a calculator. (*Hint:* If $a > 0$ and $b > 0$ and $a^2 = b^2$, then $a = b$.)

Find each value. See Example 8.

35. $\sin \alpha/2$, given $\tan \alpha = 2$, with $0 < \alpha < \pi/2$

36. $\cos \alpha/2$, given $\cot \alpha = -3$, with $\pi/2 < \alpha < \pi$

37. $\tan \beta/2$, given $\tan \beta = \sqrt{7}/3$, with $180° < \beta < 270°$

38. $\cot \beta/2$, given $\tan \beta = -\sqrt{5}/2$, with $90° < \beta < 180°$

39. $\sin \theta$, given $\cos 2\theta = 3/5$ and θ terminates in quadrant I

40. $\cos \theta$, given $\cos 2\theta = 1/2$ and θ terminates in quadrant II

Use an identity to write each expression as a single trigonometric function. See Example 9.

41. $\sqrt{\dfrac{1 - \cos 147°}{1 + \cos 147°}}$

42. $\sqrt{\dfrac{1 + \cos 165°}{1 - \cos 165°}}$

43. $\pm \sqrt{\dfrac{1 + \cos 18x}{2}}$

44. $\pm \sqrt{\dfrac{1 + \cos 20\alpha}{2}}$

45. $\dfrac{\sin 158.2°}{1 + \cos 158.2°}$

In Exercises 46–49, graph each expression and use the graph to conjecture an identity. Then verify your conjecture.

46. $\cos^4 x - \sin^4 x$

47. $\dfrac{4 \tan x \cos^2 x - 2 \tan x}{1 - \tan^2 x}$

48. $\dfrac{\sin x}{1 + \cos x}$

49. $1 - 8 \sin^2 \dfrac{x}{2} \cos^2 \dfrac{x}{2}$

Verify each identity.

50. $(\sin \gamma + \cos \gamma)^2 = \sin 2\gamma + 1$

51. $\cos 2y = \dfrac{2 - \sec^2 y}{\sec^2 y}$

52. $\sin 2\gamma = \dfrac{2 \tan \gamma}{1 + \tan^2 \gamma}$

53. $\cot s + \tan s = 2 \csc 2s$

54. $\cot^2 \dfrac{x}{2} = \dfrac{(1 + \cos x)^2}{\sin^2 x}$

55. $\sin^2 \dfrac{x}{2} = \dfrac{\tan x - \sin x}{2 \tan x}$

56. $\dfrac{2}{1 + \cos x} - \tan^2 \dfrac{x}{2} = 1$

Use the formulas developed in the Connections box to rewrite each of the following as a sum or difference of trigonometric functions.

57. $\cos 45° \sin 25°$

58. $2 \sin 74° \cos 114°$

59. $3 \cos 5x \cos 3x$

60. $2 \sin 2x \sin 4x$

61. $\sin(-\theta) \sin(-3\theta)$

62. $4 \cos 8\alpha \sin(-4\alpha)$

63. $-8 \cos 4y \cos 5y$

64. $2 \sin 3k \sin 14k$

65. (a) Derive the identity

$$\tan \frac{A}{2} = \frac{\sin A}{1 + \cos A}.$$

(*Hint:* Use the identity $\tan \theta = \sin \theta / \cos \theta$, then multiply the numerator and denominator by $2 \cos(A/2)$.)

(b) Multiply both the numerator and denominator of the right side by $1 - \cos A$ to obtain the equivalent form

$$\tan \frac{A}{2} = \frac{1 - \cos A}{\sin A}.$$

 66. Refer to Example 5. Use an identity to determine values for a, c, and ω so that $W = a \cos(\omega t) + c$. Check your answer by graphing both expressions for W on the same coordinate axes.

An airplane flying faster than sound sends out sound waves that form a cone, as shown in the figure. The cone intersects the ground to form a hyperbola. See Chapter 10. As this hyperbola passes over a particular point on the ground, a sonic boom is heard at that point. If α is the angle at the vertex of the cone, then

$$\sin \frac{\alpha}{2} = \frac{1}{m},$$

where m is the Mach number for the speed of the plane. (We assume $m > 1$.) The Mach number is the ratio of the speed of the plane and the speed of sound. Thus, a speed of Mach 1.4 means that the plane is flying at 1.4 times the speed of sound. Find α or m, as necessary, for each of the following.

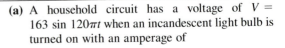

67. $m = 3/2$ **68.** $m = 5/4$ **69.** $\alpha = 30°$ **70.** $\alpha = 60°$

 71. Amperage is a measure of the amount of electricity that is moving through a circuit while voltage is a measure of the force pushing the electricity. The wattage W consumed by an electrical device can be determined by calculating the product of the amperage I and voltage V. (*Source:* Wilcox, G., and C. Hesselberth, *Electricity for Engineering Technology*, Allyn and Bacon, Inc., Boston, 1970.)

(a) A household circuit has a voltage of $V = 163 \sin 120\pi t$ when an incandescent light bulb is turned on with an amperage of

$$I = 1.23 \sin 120\pi t.$$

Graph the wattage $W = VI$ consumed by the light bulb over the interval $0 \leq t \leq .05$.

(b) Determine the maximum and minimum wattages used by the light bulb.

(c) Use identities to determine values for a, c, and ω so that $W = a\cos(\omega t) + c$.

(d) Check your answer by graphing both expressions for W on the same coordinate axes.

(e) Use the graph to estimate the average wattage used by the light. How many watts do you think this incandescent light bulb is rated for?

72. (Refer to Exercise 71.) Suppose that the voltage for an electric heater is given by $V = a\sin(2\pi\omega t)$ and the amperage by $I = b\sin(2\pi\omega t)$ where t is time in seconds.

(a) Find the period of the graph for the voltage.

(b) Show that the graph of the wattage $W = VI$ will have half the period of the voltage. Interpret this result.

▼▼▼▼▼▼▼▼▼▼▼▼▼ **DISCOVERING CONNECTIONS** (Exercises 73–80) ▼▼▼▼▼▼▼▼▼▼▼▼▼

These exercises use results from plane geometry, instead of the half-angle formulas, to obtain exact values of the trigonometric functions of 15°. *Start with a right triangle having a* 60° *angle at A and a* 30° *angle at B. Let the hypotenuse of this triangle have length* 2. *Extend side BC and draw a semicircle with diameter along BC extended, center at B, and radius AB. Draw segment AE. (See the figure.) Since any angle inscribed in a semicircle is a right angle, triangle AED is a right triangle.*

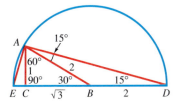

73. Why does $AB = BD$? Conclude that triangle ABD is isosceles.

74. Why does angle ABD have measure 150°?

75. Why do angles DAB and ADB both have measures of 15°?

76. What is the length of DC?

77. Use the Pythagorean theorem to show that the length of AD is $\sqrt{6} + \sqrt{2}$. (*Note:* $(\sqrt{6} + \sqrt{2})^2 = 8 + 4\sqrt{3}$.)

78. Use angle ADB of triangle ADE to find $\cos 15°$.

79. Show that AE has length $\sqrt{6} - \sqrt{2}$ and find $\sin 15°$.

80. Use triangle ACE and find $\tan 15°$.

7.5   Inverse Trigonometric Functions ▼▼▼

In the rest of this chapter we see how to solve trigonometric equations. To do this we will need to "work backwards"—that is, to find an angle or number given its trigonometric value. For example, given $\sin x = .5$, one solution is $x = 30°$.

Section 5.1 introduced the idea of an inverse function and we saw in Chapter 5 that the exponential and logarithmic functions are inverses of each other. Now we want to define inverse functions for the six trigonometric functions.

We begin with $y = \sin x$. From Figure 7 and the horizontal line test, it is clear that $y = \sin x$ is not a one-to-one function. By suitably restricting the domain of the sine function, however, a one-to-one function can be defined. It is common to restrict the domain of $y = \sin x$ to the interval $[-\pi/2, \pi/2]$, which gives the part of the graph shown in color in Figure 7. As Figure 7 shows, the range of $y = \sin x$ is $[-1, 1]$. Reflecting the graph of $y = \sin x$ on the restricted domain about the line $y = x$ gives the graph of the inverse function, shown in Figure 8. Some key points are labeled on the graph.

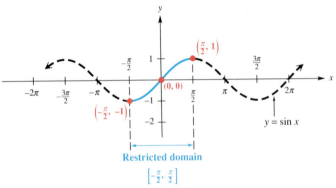

Restricted domain

$\left[-\frac{\pi}{2}, \frac{\pi}{2}\right]$

FIGURE 7

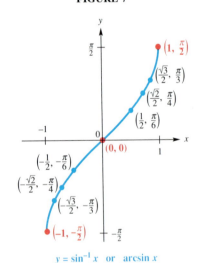

$y = \sin^{-1} x$ or arcsin x

FIGURE 8

The roles of x and y are reversed in a pair of inverse functions. Therefore, the equation of the inverse of $y = \sin x$ is found by exchanging x and y to get $x = \sin y$. This equation then is solved for y by writing $y = \sin^{-1} x$, read "inverse sine of x." (Note that $\sin^{-1} x$ does not mean $1/\sin x$.) As Figure 8 shows, the domain of $y = \sin^{-1} x$ is $[-1, 1]$, while the restricted domain of $y = \sin x$, $[-\pi/2, \pi/2]$, is the range of $y = \sin^{-1} x$. An alternative notation for $\sin^{-1} x$ is **arcsin x.**

> ## $\sin^{-1} x$ or $\arcsin x$
>
> $y = \sin^{-1} x$ or $y = \arcsin x$ means $x = \sin y$, for y in $[-\pi/2,\ \pi/2]$.

Thus, we may think of $y = \sin^{-1} x$ or $y = \arcsin x$ as "y is the number in $[-\pi/2,\ \pi/2]$ whose sine is x." These two types of notation will be used in the rest of this book.

Graphing calculators use the notation $\sin^{-1} x$ (rather than $\arcsin x$) for the inverse sine of x. A calculator-generated graph of $\sin^{-1} x$ is shown in the figure below.

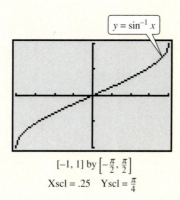

$y = \sin^{-1} x$

$[-1, 1]$ by $\left[-\frac{\pi}{2}, \frac{\pi}{2}\right]$

Xscl = .25   Yscl = $\frac{\pi}{4}$

EXAMPLE 1
Finding inverse sine values

Find y in each of the following.

(a) $y = \arcsin \dfrac{1}{2}$

The graph of the function $y = \arcsin x$ (Figure 8) shows that the point $(1/2,\ \pi/6)$ lies on the graph. Therefore, $\arcsin(1/2) = \pi/6$. Alternatively, we may think of $y = \arcsin(1/2)$ as

$$y \text{ is the number in } \left[-\frac{\pi}{2}, \frac{\pi}{2}\right] \text{ whose sine is } \frac{1}{2}.$$

Thus, we can rewrite the equation as $\sin y = 1/2$. Since $\sin \pi/6 = 1/2$ and $\pi/6$ is in the range of the arcsin function, $y = \pi/6$.

(b) $y = \sin^{-1}(-1)$

Writing the alternative equation, $\sin y = -1$, shows that $y = -\pi/2$. This can be verified by noticing that the point $(-1,\ -\pi/2)$ is on the graph of $y = \sin^{-1} x$. ◗

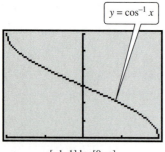

$y = \cos^{-1} x$

$[-1, 1]$ by $[0, \pi]$
Xscl = .5 Yscl = $\frac{\pi}{6}$

This is a calculator graph of the inverse cosine function.

CAUTION In Example 1(b), it is tempting to give the value of $\sin^{-1}(-1)$ as $3\pi/2$, since $\sin(3\pi/2) = -1$. Notice, however, that $3\pi/2$ is not in the range of the inverse sine function. Be certain (in dealing with *all* inverse trigonometric functions) that the number given for an inverse function value is in the range of the particular inverse function being considered.

The function $y = \cos^{-1} x$ (or $y = \arccos x$) is defined by restricting the domain of $y = \cos x$ to $[0, \pi]$. This domain becomes the range of $y = \cos^{-1} x$. The range of $y = \cos x$, the interval $[-1, 1]$, becomes the domain of $y = \cos^{-1} x$. The graphs of $y = \cos x$ with domain $[0, \pi]$ and $y = \cos^{-1} x$ are shown in Figure 9. Compare the key points on the two graphs.

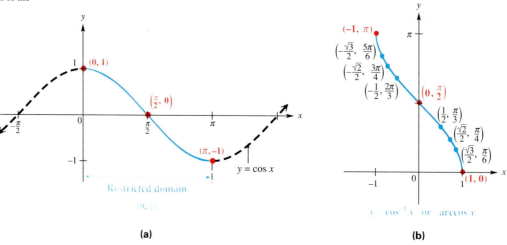

(a)

(b)

FIGURE 9

$\cos^{-1} x$ or arccos x

$y = \cos^{-1} x$ or $y = \arccos x$ means $x = \cos y$, for y in $[0, \pi]$.

EXAMPLE 2
Finding inverse cosine values

Find y in each of the following.

(a) $y = \arccos 1$
   Since the point $(1, 0)$ lies on the graph of $y = \arccos x$, the value of y is 0. Alternatively, we may think of $y = \arccos 1$ as "y is the number in $[0, \pi]$ whose cosine is 1," or $\cos y = 1$. Then $y = 0$, since $\cos 0 = 1$ and 0 is in the range of the arccos function.

(b) $y = \cos^{-1}\left(-\dfrac{\sqrt{2}}{2}\right)$
   We must find the value of y that satisfies $\cos y = -\sqrt{2}/2$, where y is in the interval $[0, \pi]$, the range of the function $y = \cos^{-1} x$. The only value for y that satisfies these conditions is $3\pi/4$. This can be verified from the graph in Figure 9(b). ▶

The inverse tangent function is defined next. Its graph is shown in Figure 10(b).

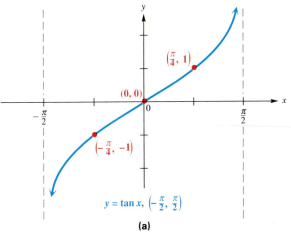

(a)

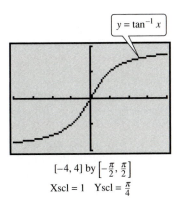

$y = \tan^{-1} x$

$[-4, 4]$ by $\left[-\frac{\pi}{2}, \frac{\pi}{2}\right]$

$\text{Xscl} = 1 \quad \text{Yscl} = \frac{\pi}{4}$

This is a calculator graph of the inverse tangent function.

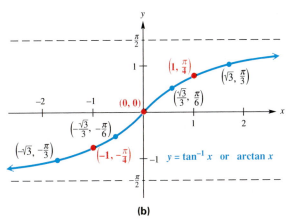

(b)

FIGURE 10

$\tan^{-1} x$ or arctan x

$y = \tan^{-1} x$ or $y = \arctan x$ means $x = \tan y$, for y in $(-\pi/2, \pi/2)$.

From the graph we can see that arctan $0 = 0$, arctan $1 = \pi/4$, $\arctan(-\sqrt{3}) = -\pi/3$, and so on.

We have defined the inverse sine, cosine, and tangent functions with suitable restrictions on the domains. The other three inverse trigonometric functions are similarly defined. The six inverse trigonometric functions with their domains and ranges are given in the table.* This information, particularly the range for each function, should be memorized. (The graphs of the last three inverse trigonometric functions are left for the exercises.)

_____
* $\sec^{-1}$ and $\csc^{-1}$ are sometimes defined differently.

Inverse Trigonometric Functions

Function	Domain	Range	Quadrants of the Unit Circle Range Values Come From
$y = \sin^{-1} x$	$[-1, 1]$	$\left[-\dfrac{\pi}{2}, \dfrac{\pi}{2}\right]$	I and IV
$y = \cos^{-1} x$	$[-1, 1]$	$[0, \pi]$	I and II
$y = \tan^{-1} x$	$(-\infty, \infty)$	$\left(-\dfrac{\pi}{2}, \dfrac{\pi}{2}\right)$	I and IV
$y = \cot^{-1} x$	$(-\infty, \infty)$	$(0, \pi)$	I and II
$y = \sec^{-1} x$	$(-\infty, -1] \cup [1, \infty)$	$[0, \pi], y \neq \dfrac{\pi}{2}$	I and II
$y = \csc^{-1} x$	$(-\infty, -1] \cup [1, \infty)$	$\left[-\dfrac{\pi}{2}, \dfrac{\pi}{2}\right], y \neq 0$	I and IV

CONNECTIONS The sine function with limited domain $[-\pi/2, \pi/2]$ is sometimes differentiated from the sine function with its natural domain of $(-\infty, \infty)$ by the notation Sin x (with a capital S). Its inverse function is then denoted Sin$^{-1} x$. The other trigonometric functions with restricted domains and their inverses are also denoted with capital letters. We have chosen not to use this notation because graphing calculators use lowercase letters to denote the inverse trigonometric functions.

As we mention in a footnote, sometimes the inverse secant and inverse cosecant functions are defined with different ranges than are given here. We have elected to use intervals that match their reciprocal functions, except for one missing point. In calculus, different ranges are considered more convenient: for sec$^{-1} x$, $[0, \pi/2) \cup [\pi, 3\pi/2)$; for csc$^{-1} x$, $[\pi/2, \pi) \cup [3\pi/2, 2\pi)$.

The inverse trigonometric functions are formally defined with real number ranges. However, there are times when it may be convenient to find the degree-measured angles equivalent to these real number values. It is also often convenient to think in terms of the unit circle, and choose the inverse function values based on the quadrants given earlier in the table that summarizes these functions. The next example uses these ideas.

EXAMPLE 3
Finding inverse values
(degree-measured angles)

Find the *degree measure* of θ in each of the following.

(a) θ, if $\theta = \arctan 1$

Here θ must be in $(-90°, 90°)$, but since $1 > 0$, θ must be in quadrant I. The alternative statement, $\tan \theta = 1$, leads to $\theta = 45°$.

(b) θ, if $\theta = \sec^{-1} 2$

Write the equation as $\sec \theta = 2$. For $\sec^{-1} x$, θ is in quadrant I or II. Because 2 is positive, θ is in quadrant I and $\theta = 60°$, since $\sec 60° = 2$. Note that $60°$ (the degree equivalent of $\pi/3$) is in the range of the inverse secant function. ▶

The inverse trignometric function keys on a calculator give results in the proper quadrant for the $\sin^{-1}$, $\cos^{-1}$, and $\tan^{-1}$ functions, according to the definitions of these functions. For example, on a calculator, in degrees,

$$\sin^{-1} .5 = 30°, \qquad\qquad \sin^{-1}(-.5) = -30°,$$
$$\tan^{-1}(-1) = -45°, \qquad \text{and} \qquad \cos^{-1}(-.5) = 120°.$$

Similar results are found when the calculator is set for radian measure. This is not the case for $\cot^{-1}$. For example, since we can take the reciprocal of the inverse tangent to find $\cot^{-1}$, the calculator gives values of $\cot^{-1}$ with the same range as $\tan^{-1}$, $(-\pi/2, \pi/2)$, which is not the correct range for $\cot^{-1}$. For $\cot^{-1}$ the proper range must be considered and the results adjusted accordingly.

EXAMPLE 4
Finding an inverse function
value with a calculator

Find θ in degrees if $\theta = \text{arccot}(-.3541)$.

A calculator gives $\tan^{-1}(1/-.3541) \approx -70.500946°$. The restriction on the range of arccot means that θ must be in quadrant II, and the absolute value of the angle obtained in the display is the first quadrant angle with the same cotangent. Therefore,

$$\theta = 180° - |-70.500946°| = 109.499054°.$$

Use a calculator to verify that $\cot 109.499054° \approx -.3541$. ▶

EXAMPLE 5
Finding function values without
a calculator

Evaluate each of the following without a calculator.

(a) $\sin\left(\tan^{-1} \dfrac{3}{2}\right)$

Let

$$\theta = \tan^{-1} \frac{3}{2}, \text{ so that } \tan \theta = \frac{3}{2}.$$

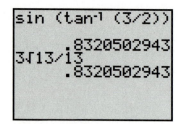

This screen supports the result in Example 5(a).

Since $\tan^{-1}$ is defined only in quadrants I and IV and since $3/2$ is positive, θ is in quadrant I. Sketch θ in quadrant I, and label a triangle as shown in Figure 11. The hypotenuse is $\sqrt{13}$ and the value of sine is the quotient of the side opposite and the hypotenuse, so

$$\sin\left(\tan^{-1}\frac{3}{2}\right) = \sin\theta = \frac{3}{\sqrt{13}} = \frac{3\sqrt{13}}{13}.$$

To check this result on a calculator, enter $3/2$ as 1.5. Then find $\tan^{-1} 1.5$, and finally find $\sin(\tan^{-1} 1.5)$. Store this result and calculate $3\sqrt{13}/13$, which should agree with the result for $\sin(\tan^{-1} 1.5)$. Since the values are only approximations, this check does not *prove* that the result is correct, but it is highly suggestive that it is correct.

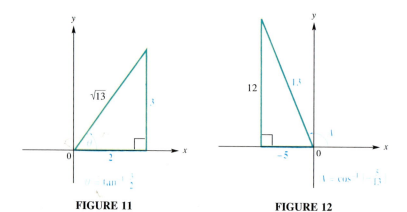

FIGURE 11   **FIGURE 12**

(b) $\tan\left(\cos^{-1}\left(-\frac{5}{13}\right)\right)$

Let $A = \cos^{-1}(-5/13)$. Then $\cos A = -5/13$. Since $\cos^{-1} x$ for a negative value of x is in quadrant II, sketch A in quadrant II, as shown in Figure 12. From the triangle in Figure 12,

$$\tan\left(\cos^{-1}\left(-\frac{5}{13}\right)\right) = \tan A = -\frac{12}{5}.$$

(c) $\cos(\cos^{-1}(-.5))$

Recall, for inverse functions f and g, $f(g(x)) = x$. Since cosine and inverse cosine are inverse functions, $\cos(\cos^{-1}(-.5)) = -.5$.

(d) $\cos^{-1}\left(\cos\frac{5\pi}{4}\right) = \cos^{-1}\left(-\frac{\sqrt{2}}{2}\right) = \frac{3\pi}{4}$ ▶

EXAMPLE 6
Finding function values using sum and double-angle formulas

Evaluate the following without using a calculator.

(a) $\cos\left(\arctan \sqrt{3} + \arcsin \dfrac{1}{3}\right)$

Let $A = \arctan \sqrt{3}$ and $B = \arcsin 1/3$ so that $\tan A = \sqrt{3}$ and $\sin B = 1/3$. Sketch both A and B in quadrant I, as shown in Figure 13.

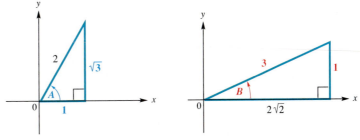

FIGURE 13

Now use the identity for $\cos(A + B)$.

$$\cos(A + B) = \cos A \cos B - \sin A \sin B$$

$$\cos\left(\arctan \sqrt{3} + \arcsin \dfrac{1}{3}\right) = \cos(\arctan \sqrt{3})\cos\left(\arcsin \dfrac{1}{3}\right)$$

$$- \sin(\arctan \sqrt{3}) \sin\left(\arcsin \dfrac{1}{3}\right) \quad \textbf{(1)}$$

From the sketch in Figure 13,

$$\cos(\arctan \sqrt{3}) = \cos A = \dfrac{1}{2}, \qquad \cos\left(\arcsin \dfrac{1}{3}\right) = \cos B = \dfrac{2\sqrt{2}}{3},$$

$$\sin(\arctan \sqrt{3}) = \sin A = \dfrac{\sqrt{3}}{2}, \qquad \sin\left(\arcsin \dfrac{1}{3}\right) = \sin B = \dfrac{1}{3}.$$

Substitute these values into equation (1) to get

$$\cos\left(\arctan \sqrt{3} + \arcsin \dfrac{1}{3}\right) = \dfrac{1}{2} \cdot \dfrac{2\sqrt{2}}{3} - \dfrac{\sqrt{3}}{2} \cdot \dfrac{1}{3}$$

$$= \dfrac{2\sqrt{2}}{6} - \dfrac{\sqrt{3}}{6}$$

$$= \dfrac{2\sqrt{2} - \sqrt{3}}{6}.$$

(b) $\tan\left(2 \arcsin \frac{2}{5}\right)$

Let $\arcsin(2/5) = B$. Then, from the identity for the tangent of the double angle,

$$\tan\left(2 \arcsin \frac{2}{5}\right) = \tan(2B)$$

$$= \frac{2 \tan B}{1 - \tan^2 B}.$$

Since $\arcsin(2/5) = B$, $\sin B = 2/5$. Sketch a triangle in quadrant I, find the length of the third side, and then find $\tan B$. From the triangle in Figure 14, $\tan B = 2/\sqrt{21}$, and

$$\tan\left(2 \arcsin \frac{2}{5}\right) = \frac{2\left(\dfrac{2}{\sqrt{21}}\right)}{1 - \left(\dfrac{2}{\sqrt{21}}\right)^2} = \frac{\dfrac{4}{\sqrt{21}}}{1 - \dfrac{4}{21}}$$

$$= \frac{\dfrac{4}{\sqrt{21}}}{\dfrac{17}{21}} = \frac{4\sqrt{21}}{17}. \quad \blacktriangleright$$

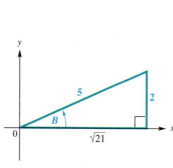

FIGURE 14　　　　　　　　　　**FIGURE 15**

◀ EXAMPLE 7
Writing a function value in terms of u

Write $\sin(\tan^{-1} u)$ as an expression in u.

Let $\theta = \tan^{-1} u$, so that $\tan \theta = u$. Here u may be positive or negative. Since $-\pi/2 < \tan^{-1} u < \pi/2$, sketch θ in quadrants I and IV and label two triangles as shown in Figure 15. Since sine is given by the ratio of the opposite side and the hypotenuse,

$$\sin(\tan^{-1} u) = \sin \theta = \frac{u}{\sqrt{u^2 + 1}}$$

$$= \frac{u\sqrt{u^2 + 1}}{u^2 + 1}.$$

The result is positive when u is positive and negative when u is negative. $\quad \blacktriangleright$

7.5 Exercises ▼▼▼▼▼▼▼▼▼▼▼▼▼▼▼▼▼▼▼▼▼▼▼▼▼▼▼

1. The accompanying screen contains the graph of $Y_1 = \sin^{-1} x$ along with the coordinates of a point on the graph. What is the exact value of Y?

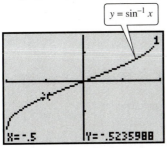

$y = \sin^{-1} x$

X=-.5   Y=-.5235988

[−1, 1] by [−2, 2]
Xscl = .5   Yscl = 1

2. The accompanying screen contains the graph of $Y_1 = \cos^{-1} x$ along with the coordinates of a point on the graph. What is the exact value of Y?

$y = \cos^{-1} x$

X=-.5   Y=2.0943951

[−1, 1] by [−.5, π]
Xscl = .5   Yscl = $\frac{\pi}{4}$

3. Is $\sec^{-1} a$ calculated as $\cos^{-1}\left(\dfrac{1}{a}\right)$ or as $\dfrac{1}{\cos^{-1} a}$?

4. For positive values of a, $\cot^{-1} a$ is calculated as $\tan^{-1}(1/a)$. How is $\cot^{-1} a$ calculated for negative values of a?

Find the exact value of y. Do not use a calculator. See Examples 1 and 2.

5. $y = \arcsin\left(-\dfrac{1}{2}\right)$

6. $y = \arccos\dfrac{\sqrt{3}}{2}$

7. $y = \tan^{-1} 1$

8. $y = \sin^{-1} 0$

9. $y = \cos^{-1}(-1)$

10. $y = \arctan(-1)$

11. $y = \sin^{-1}\left(-\dfrac{\sqrt{3}}{2}\right)$

12. $y = \cos^{-1}\dfrac{1}{2}$

13. $y = \arctan 0$

14. $y = \arcsin\left(-\dfrac{\sqrt{3}}{2}\right)$

15. $y = \arccos 0$

16. $y = \tan^{-1}(-1)$

17. $y = \sin^{-1}\dfrac{\sqrt{2}}{2}$

18. $y = \cos^{-1}\left(-\dfrac{1}{2}\right)$

19. $y = \arccos\left(-\dfrac{\sqrt{3}}{2}\right)$

20. $y = \arcsin\left(-\dfrac{\sqrt{2}}{2}\right)$

21. $y = \cot^{-1}(-1)$

22. $y = \sec^{-1}(-\sqrt{2})$

23. $y = \csc^{-1}(-2)$

24. $y = \text{arccot}(-\sqrt{3})$

25. $y = \text{arcsec}\dfrac{2\sqrt{3}}{3}$

26. $y = \csc^{-1}\sqrt{2}$

27. $y = \text{arccot}\dfrac{\sqrt{3}}{3}$

28. $y = \text{arcsec } 2$

Give the degree measure of θ. Do not use a calculator. See Example 3.

29. $\theta = \arctan(-1)$

30. $\theta = \arccos\left(-\dfrac{1}{2}\right)$

31. $\theta = \arcsin\left(-\dfrac{\sqrt{3}}{2}\right)$

32. $\theta = \arcsin\left(-\dfrac{\sqrt{2}}{2}\right)$

33. $\theta = \cot^{-1}\left(-\dfrac{\sqrt{3}}{3}\right)$

34. $\theta = \sec^{-1}(-2)$

35. $\theta = \csc^{-1}(-2)$

36. $\theta = \csc^{-1}(-1)$

Give the real number value. See Example 4.

37. $\arctan 1.1111111$

38. $\arcsin .81926439$

39. $\cot^{-1}(-.92170128)$

40. $\sec^{-1}(-1.2871684)$

41. $\arcsin .92837781$

42. $\arccos .44624593$

Give the value in decimal degrees. See Example 4.

43. $\sin^{-1}(-.13349122)$

44. $\cos^{-1}(-.13348816)$

45. $\arccos(-.39876459)$

46. $\arcsin .77900016$

47. $\csc^{-1} 1.9422833$

48. $\cot^{-1} 1.7670492$

Graph each function as defined in the text, and give the domain and range.

49. $y = \cot^{-1} x$ **50.** $y = \text{arccsc } x$ **51.** $y = \text{arcsec } x$

52. The following expressions were used by the mathematicians who computed the value of π to 100,000 decimal places. Use a calculator to verify that each is (approximately) correct.

(a) $\pi = 16 \tan^{-1} \dfrac{1}{5} - 4 \tan^{-1} \dfrac{1}{239}$

(b) $\pi = 24 \tan^{-1} \dfrac{1}{8} + 8 \tan^{-1} \dfrac{1}{57} + 4 \tan^{-1} \dfrac{1}{239}$

(c) $\pi = 48 \tan^{-1} \dfrac{1}{18} + 32 \tan^{-1} \dfrac{1}{57} - 20 \tan^{-1} \dfrac{1}{239}$

53. Explain why attempting to find $\sin^{-1} 1.003$ on your calculator will result in an error message.

54. Explain why you are able to find $\tan^{-1} 1.003$ on your calculator. Why is this situation different from the one described in Exercise 53?

55. Find $\sin 1.74$ on your calculator (set for radians) and then find the inverse sine of that number. You get 1.401592654 instead of 1.74. What happened?

56. Recall that a graphing calculator will return a 1 for a true statement and a 0 for a false statement. Determine the possible values stored in X for which the screen will come up on a graphing calculator.

(a)
```
cos-1 (cos X)=X
                 1
```

(b)
```
cos (cos-1 X)=X
                 1
```

(c)
```
sin (sin-1 X)=X
                 1
```

(d)
```
sin-1 (sin X)=X
                 1
```

Give each value without using a calculator. See Examples 5 and 6.

57. $\tan\left(\arccos \dfrac{3}{4}\right)$ **58.** $\sin\left(\arccos \dfrac{1}{4}\right)$ **59.** $\cos(\tan^{-1}(-2))$ **60.** $\sec\left(\sin^{-1}\left(-\dfrac{1}{5}\right)\right)$

61. $\cot\left(\arcsin\left(-\dfrac{2}{3}\right)\right)$ **62.** $\cos\left(\arctan \dfrac{8}{3}\right)$ **63.** $\sec(\sec^{-1} 2)$ **64.** $\csc(\csc^{-1} \sqrt{2})$

65. $\arccos\left(\cos \dfrac{\pi}{4}\right)$ **66.** $\arctan\left(\tan\left(-\dfrac{\pi}{4}\right)\right)$ **67.** $\arcsin\left(\sin \dfrac{\pi}{3}\right)$ **68.** $\arccos(\cos 0)$

69. $\sin\left(2 \tan^{-1} \dfrac{12}{5}\right)$ **70.** $\cos\left(2 \sin^{-1} \dfrac{1}{4}\right)$ **71.** $\cos\left(2 \arctan \dfrac{4}{3}\right)$ **72.** $\tan\left(2 \cos^{-1} \dfrac{1}{4}\right)$

73. $\sin\left(2 \cos^{-1} \dfrac{1}{5}\right)$ **74.** $\cos(2 \arctan(-2))$ **75.** $\tan\left(2 \arcsin\left(-\dfrac{3}{5}\right)\right)$ **76.** $\sin\left(2 \arccos \dfrac{2}{9}\right)$

77. $\sin\left(\sin^{-1}\dfrac{1}{2} + \tan^{-1}(-3)\right)$

78. $\cos\left(\tan^{-1}\dfrac{5}{12} - \cot^{-1}\dfrac{4}{3}\right)$

79. $\cos\left(\arcsin\dfrac{3}{5} + \arccos\dfrac{5}{13}\right)$

80. $\tan\left(\arccos\dfrac{\sqrt{3}}{2} - \arcsin\left(-\dfrac{3}{5}\right)\right)$

Use a calculator to find each value. Give answers as real numbers.

81. $\cos(\tan^{-1}.5)$

82. $\sin(\cos^{-1}.25)$

83. $\tan(\arcsin .12251014)$

84. $\cot(\arccos .58236841)$

Write each of the following as an expression in u. See Example 7.

85. $\sin(\arccos u)$

86. $\tan(\arccos u)$

87. $\cot(\arcsin u)$

88. $\cos(\arcsin u)$

89. $\sin\left(\sec^{-1}\dfrac{u}{2}\right)$

90. $\cos\left(\tan^{-1}\dfrac{3}{u}\right)$

91. $\tan\left(\arcsin\dfrac{u}{\sqrt{u^2+2}}\right)$

92. $\cos\left(\arccos\dfrac{u}{\sqrt{u^2+5}}\right)$

---

93. Explain why $\sin^{-1}\dfrac{1}{2} \neq \dfrac{5\pi}{6}$, despite the fact that $\sin\dfrac{5\pi}{6} = \dfrac{1}{2}$.

94. A student observed, "$\cos^{-1}\dfrac{1}{2}$ has two values: $\dfrac{\pi}{3}$ and $\dfrac{5\pi}{3}$." Comment on this observation.

95. Suppose an airplane flying faster than sound goes directly over you. Assume that the plane is flying level. At the instant you feel the sonic boom from the plane, the angle of elevation to the plane is given by

$$\alpha = 2\arcsin\dfrac{1}{m},$$

where *m* is the Mach number of the plane's speed. (See Exercises 67–70 at the end of Section 7.4.) Find α to the nearest degree for each of the following values of *m*.

(a) $m = 1.2$ **(b)** $m = 1.5$
(c) $m = 2$ **(d)** $m = 2.5$

96. A painting 1 m high and 3 m from the floor will cut off an angle θ to an observer, where

$$\theta = \tan^{-1}\left(\dfrac{x}{x^2+2}\right).$$

Assume that the observer is *x* m from the wall where the painting is displayed and that the eyes of the observer are 2 m above the ground. See the figure. Find the value of θ for the following values of *x*. Round to the nearest degree.

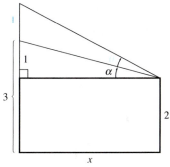

(a) 1   **(b)** 2   **(c)** 3
(d) Derive the formula given above. (*Hint:* Use the identity for $\tan(\theta + \alpha)$. Use right triangles.)
(e) Graph the function for θ with a graphing calculator and determine the distance that maximizes the angle.

97. The following calculator trick will not work on all calculators. However, if you have a Texas Instruments or a Sharp scientific calculator, it will work. The calculator must be in the *degree* mode.
(a) Enter the year of your birth (all four digits).
(b) Subtract the number of years that have elapsed since 1980. For example, if it is 1997, subtract 17.
(c) Find the sine of the display.
(d) Find the inverse sine of the new display. The result should be your age when you celebrate your birthday this year.

98. Explain why the procedure in Exercise 97 works as it does.

7.6   Trigonometric Equations ▼▼▼

Earlier in this chapter we studied trigonometric equations that were identities. We now consider trigonometric equations that are **conditional;** that is, equations that are satisfied by some values but not others.

Conditional equations with trigonometric (or circular) functions can usually be solved by using algebraic methods and trigonometric identities. For example, suppose we wish to find the solutions of the equation

$$2 \sin \theta + 1 = 0$$

for all θ in the interval $[0°, 360°)$. We use the same method here as we would in solving the algebraic equation $2y + 1 = 0$. Subtract 1 from both sides, and divide by 2.

$$2 \sin \theta + 1 = 0$$
$$2 \sin \theta = -1$$
$$\sin \theta = -\frac{1}{2}$$

We know that θ must be in either quadrant III or IV, since the sine function is negative in these two quadrants. Since $\sin 30° = 1/2$, the sketches in Figure 16 show the two possible values of θ, $210°$ and $330°$. The solution set is $\{210°, 330°\}$.

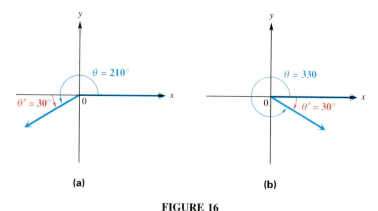

(a)   (b)

FIGURE 16

CAUTION   One value that satisfies $\sin \theta = -1/2$ is $\sin^{-1}(-1/2)$, which in degrees is $-30°$. However, when solving an equation such as this one, we must pay close attention to the domain, $[0°, 360°)$ here.

In some cases we are required to find *all* solutions of conditional trigonometric equations. All solutions of the equation $2 \sin \theta + 1 = 0$ would be written as

$$\theta = 210° + 360° \cdot n \qquad \text{or} \qquad \theta = 330° + 360° \cdot n,$$

where n is any integer. We add integer multiples of 360° to obtain all angles coterminal with 210° or 330°. If we had been required to solve this equation for real numbers (or angles in radians) in the interval $[0, 2\pi)$, the two solutions would be $7\pi/6$ and $11\pi/6$, while all solutions would be written as

$$\theta = \frac{7\pi}{6} + 2n\pi \quad \text{or} \quad \theta = \frac{11\pi}{6} + 2n\pi,$$

where n is any integer.

In the examples in this section, we will find solutions in the intervals $[0°, 360°)$ or $[0, 2\pi)$. Remember that *all* solutions can be found using the methods described above.

EXAMPLE 1
Solving a trigonometric equation

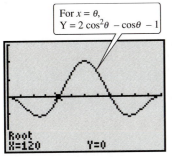

For $x = \theta$,
$Y = 2\cos^2\theta - \cos\theta - 1$

Root
X=120   Y=0

[0, 360] by [−3, 3]
Xscl = 30   Yscl = 1

By graphing $Y = 2\cos^2\theta - \cos\theta - 1$ (where $x = \theta$), we can support the solutions in Example 1. The calculator must be in degree mode. The display shows that 120° is a solution; the others can be supported similarly.

Solve $2\cos^2\theta - \cos\theta - 1 = 0$ in the interval $[0°, 360°)$.

We solve this equation by factoring. (It is quadratic in the term $\cos\theta$.)

$$2\cos^2\theta - \cos\theta - 1 = 0$$
$$(2\cos\theta + 1)(\cos\theta - 1) = 0$$
$$2\cos\theta + 1 = 0 \quad \text{or} \quad \cos\theta - 1 = 0$$
$$\cos\theta = -\frac{1}{2} \qquad\qquad \cos\theta = 1$$

In the first case, we have $\cos\theta = -1/2$, indicating that θ must be in either quadrant II or III, with $\theta = \cos^{-1}(-1/2) = 120°$. Using a sketch similar to those in Figure 16 would indicate that two solutions are 120° and 240°. The second case, $\cos\theta = 1$, has the quadrantal angle 0° as its only solution in the interval. (We do not include 360° since it is not in the stated interval.) Therefore, the solutions of this equation are 0°, 120°, and 240° and the solution set is {0°, 120°, 240°}. Check these solutions by substituting them in the given equation. ▸

EXAMPLE 2
Solving a trigonometric equation

Solve $\sin x \tan x = \sin x$ in the interval $[0°, 360°)$.

Subtract $\sin x$ from both sides, then factor on the left.

$$\sin x \tan x = \sin x$$
$$\sin x \tan x - \sin x = 0$$
$$\sin x(\tan x - 1) = 0$$

Now set each factor equal to 0.

$$\sin x = 0 \quad \text{or} \quad \tan x - 1 = 0$$
$$\tan x = 1$$
$$x = 0° \quad \text{or} \quad x = 180° \qquad x = 45° \quad \text{or} \quad x = 225°$$

The solution set is {0°, 45°, 180°, 225°}. Verify these solutions by substitution. ▸

CAUTION There are four solutions for Example 2. Trying to solve the equation by dividing both sides by $\sin x$ would give just $\tan x = 1$, which would give $x = 45°$ or $x = 225°$. The other two solutions would not appear. The missing solutions are the ones that make the divisor, $\sin x$, equal 0. For this reason, it is best to avoid dividing by a variable expression.

Recall from Chapter 2 that squaring both sides of an equation, such as $\sqrt{x + 4} = x + 2$, will yield all solutions, but may also give extraneous values. (In this equation, 0 is a solution, while -3 is extraneous. Verify this.) The same situation may occur when trigonometric equations are solved in this manner, as shown in the next example.

EXAMPLE 3
Solving a trigonometric equation

Solve $\tan x + \sqrt{3} = \sec x$ in the interval $[0, 2\pi)$.

Since the tangent and secant functions are related by the identity $1 + \tan^2 x = \sec^2 x$, one method of solving this equation is to square both sides, and express $\sec^2 x$ in terms of $\tan^2 x$.

$$\tan x + \sqrt{3} = \sec x$$
$$\tan^2 x + 2\sqrt{3} \tan x + 3 = \sec^2 x$$
$$\tan^2 x + 2\sqrt{3} \tan x + 3 = \mathbf{1 + \tan^2\, x}$$
$$2\sqrt{3} \tan x = -2$$
$$\tan x = -\frac{1}{\sqrt{3}} = -\frac{\sqrt{3}}{3}$$

The possible solutions in the given interval are $\dfrac{5\pi}{6}$ and $\dfrac{11\pi}{6}$. Now check the possible solutions. Try $\dfrac{5\pi}{6}$ first.

Left side: $\tan x + \sqrt{3} = \tan \dfrac{5\pi}{6} + \sqrt{3} = -\dfrac{\sqrt{3}}{3} + \sqrt{3} = \dfrac{2\sqrt{3}}{3}$

Right side: $\sec x = \sec \dfrac{5\pi}{6} = \dfrac{-2\sqrt{3}}{3}$

The check shows that $\dfrac{5\pi}{6}$ is not a solution. Now check $\dfrac{11\pi}{6}$.

Left side: $\tan \dfrac{11\pi}{6} + \sqrt{3} = -\dfrac{\sqrt{3}}{3} + \sqrt{3} = \dfrac{2\sqrt{3}}{3}$

Right side: $\sec \dfrac{11\pi}{6} = \dfrac{2\sqrt{3}}{3}$

This solution satisfies the equation, so the solution set is $\left\{\dfrac{11\pi}{6}\right\}$. ◢

We have suggested checking the solutions of trigonometric equations by substitution. Another way to check solutions is to solve the equation by using a graphing calculator. For instance, in Example 3, let $Y_1 = \tan x + \sqrt{3}$ and $Y_2 = \sec x = \dfrac{1}{\cos x}$. Graph both functions in the same window and use the capabilities of the calculator to find the intersection points. The figure at left below shows the graphs of Y_1 and Y_2 in the same window, with the value of the only intersection point at the bottom of the graph. The graph supports our results in Example 3. It is not clear from the graph whether there are one or two other solutions near $x = \pi/2$. To determine the number of solutions in the interval $[0, 2\pi)$, we graph $Y = \tan x + \sqrt{3} - \sec x$ and find the values of the x-intercepts. See the figure at right. The only x-intercept in the given interval is approximately 5.7595865. Since $11\pi/6 \approx 5.7595865$, this supports our analytic solution.

$\boxed{Y_1 = \tan x + \sqrt{3}}$ $\boxed{Y_2 = \sec x}$ $\boxed{Y = \tan x + \sqrt{3} - \sec x}$

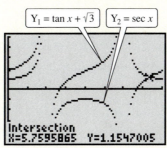
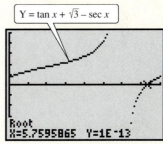

$[0, 2\pi]$ by $[-5, 5]$

Xscl $= \frac{\pi}{4}$ Yscl $= 1$

Both are dot mode.

In some cases trigonometric equations require a calculator to obtain approximate solutions, as in the next example.

EXAMPLE 4
Solving a trigonometric equation using a calculator

Solve $\tan^2 x + \tan x - 2 = 0$ in the interval $[0, 2\pi)$.

Like Example 1, this equation is quadratic in form and may be solved for $\tan x$ by factoring.

$$\tan^2 x + \tan x - 2 = 0$$
$$(\tan x - 1)(\tan x + 2) = 0$$

Set each factor equal to 0.

$$\tan x - 1 = 0 \quad \text{or} \quad \tan x + 2 = 0$$
$$\tan x = 1 \quad \text{or} \quad \tan x = -2$$

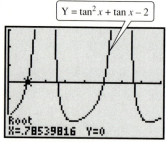

$[0, 2\pi]$ by $[-3, 3]$
Xscl = $\frac{\pi}{4}$ Yscl = 1

This graph supports the results in Example 4. Here, the calculator is in radian mode. The display indicates that $\frac{\pi}{4} \approx .78539816$ is a solution. The other three can be supported similarly.

The solutions for tan $x = 1$ in the interval $[0, 2\pi)$ are $x = \pi/4$ or $5\pi/4$. To solve tan $x = -2$ in that interval, use a calculator set in the *radian* mode. We find that $\tan^{-1}(-2) \approx -1.1071487$. This is a quadrant IV number, based on the range of the inverse tangent function. However, since we want solutions in the interval $[0, 2\pi)$, we must first add π to -1.1071487, and then add 2π:

$$x \approx -1.1071487 + \pi \approx 2.03444394$$
$$x \approx -1.1071487 + 2\pi \approx 5.1760366.$$

The solutions in the required interval are

$$\frac{\pi}{4}, \quad \frac{5\pi}{4}, \qquad 2.0, \quad 5.2.$$

Exact values   Approximate values to the nearest tenth

The solution set is $\{\pi/4, 5\pi/4, 2.0, 5.2\}$. Note that these solutions are the x-intercepts of the graph of $y = \tan^2 x + \tan x - 2$. ▶

When a trigonometric equation that is quadratic in form cannot be factored, the quadratic formula can be used to solve the equation.

EXAMPLE 5
Solving a trigonometric equation with the quadratic formula

Solve $\cot^2 x + 3 \cot x = 1$ in $[0°, 360°)$.
    Write the equation in standard form, with 0 on one side.

$$\cot^2 x + 3 \cot x - 1 = 0$$

Since this quadratic equation cannot be solved by factoring, use the quadratic formula, with $a = 1$, $b = 3$, $c = -1$, and cot x as the variable.

$$\cot x = \frac{-3 \pm \sqrt{9 + 4}}{2} = \frac{-3 \pm \sqrt{13}}{2} \approx \frac{-3 \pm 3.6055513}{2}$$

$$\cot x \approx .30277564 \qquad \text{or} \qquad \cot x \approx -3.3027756$$
$$x \approx 73.2°, 253.2°, 163.2°, 343.2°$$

The final answers were obtained using a calculator set in degree mode. The solution set is $\{73.2°, 163.2°, 253.2°, 343.2°\}$. ▶

Some trigonometric equations involve functions of half angles and multiples of angles. These may require using identities to get functions of the same angle.

EXAMPLE 6
Solving an equation with a double angle

Solve $\cos 2x = \cos x$ in the interval $[0, 2\pi)$.
    First change $\cos 2x$ to a trigonometric function of x. Use the identity $\cos 2x = 2 \cos^2 x - 1$ so that the equation involves only the cosine of x. Then use the methods of the previous examples.

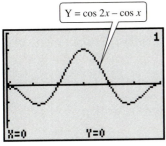

$Y = \cos 2x - \cos x$

$[0, 2\pi]$ by $[-3, 3]$

Xscl $= \frac{\pi}{3}$ Yscl $= 1$

This screen supports the solution 0 found in Example 1. By using Xscl $= \frac{\pi}{3}$ and observing that the graph intersects the *x*-axis at the second and fourth tick marks, we can feel fairly sure that $\frac{2\pi}{3}$ and $\frac{4\pi}{3}$ are also solutions, as found analytically.

$$\cos 2x = \cos x$$
$$2 \cos^2 x - 1 = \cos x$$
$$2 \cos^2 x - \cos x - 1 = 0$$
$$(2 \cos x + 1)(\cos x - 1) = 0$$
$$2 \cos x + 1 = 0 \qquad \text{or} \qquad \cos x - 1 = 0$$
$$\cos x = -\frac{1}{2} \qquad \text{or} \qquad \cos x = 1$$

In the required interval,

$$x = \frac{2\pi}{3} \text{ or } \frac{4\pi}{3} \text{ or } x = 0.$$

The solution set is $\{0, 2\pi/3, 4\pi/3\}$. ▶

CAUTION   In Example 6 it is important to notice that $\cos 2x$ cannot be changed to $\cos x$ by dividing by 2, since 2 is not a factor of $\cos 2x$.

$$\frac{\cos 2x}{2} \neq \cos x$$

The only way to change $\cos 2x$ to a trigonometric function of *x* is by using one of the identities for $\cos 2x$.

The equation in Example 6 could also be solved graphically by finding the points of intersection of the graphs of $Y_1 = \cos 2x$ and $Y_2 = \cos x$. The figures below show two views of the graphs of Y_1 and Y_2, corresponding to the two nonzero solutions we found analytically.

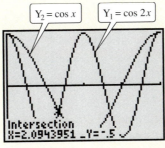

$Y_2 = \cos x$ $Y_1 = \cos 2x$

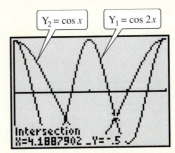

$Y_2 = \cos x$ $Y_1 = \cos 2x$

$[0, 2\pi]$ by $[-1, 1]$

Xscl $= \frac{\pi}{2}$ Yscl $= 1$

$$\frac{2\pi}{3} \approx 2.0943951$$
$$\frac{4\pi}{3} \approx 4.1887902$$

Conditional trigonometric equations in which a half angle or multiple angle is involved often require an additional step to solve. This step involves adjusting the interval of solution to fit the requirements of the half or multiple angle. This is shown in the following examples.

EXAMPLE 7
Solving an equation with a half angle

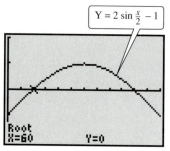

$Y = 2 \sin \frac{x}{2} - 1$

Root
X=60   Y=0

[0, 360] by [−2, 2]
Xscl = 30   Yscl = 1

This screen supports the root 60° found in Example 7. The other root, 300°, is represented by the other x-intercept shown.

Solve $2 \sin \frac{x}{2} = 1$ in the interval [0°, 360°).

As a compound inequality, the interval [0°, 360°) is written

$$0° \le x < 360°.$$

Dividing through by 2 gives

$$0° \le \frac{x}{2} < 180°.$$

To find all values of $x/2$ in the interval 0° to 180°, begin by solving for the trigonometric function.

$$2 \sin \frac{x}{2} = 1$$

$$\sin \frac{x}{2} = \frac{1}{2} \qquad \text{Divide by 2.}$$

Both $\sin 30° = 1/2$ and $\sin 150° = 1/2$ and 30° and 150° are in the given interval for $x/2$, so

$$\frac{x}{2} = 30° \qquad \text{or} \qquad \frac{x}{2} = 150°$$

$$x = 60° \qquad \text{or} \qquad x = 300°. \qquad \text{Multiply by 2.}$$

The solutions in the given interval are 60° and 300°; the solution set is {60°, 300°}. ▶

EXAMPLE 8
Solving an equation with a multiple angle

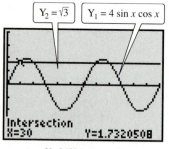

$Y_2 = \sqrt{3}$ $Y_1 = 4 \sin x \cos x$

Intersection
X=30   Y=1.7320508

[0, 360] by [−4, 4]
Xscl = 30   Yscl = 1

The graphs of $Y_1 = 4 \sin x \cos x$ and $Y_2 = \sqrt{3}$ intersect in four points in the given interval. The screen supports the solution 30° found in Example 8; the other solutions 60°, 210°, and 240° are supported similarly.

Solve $4 \sin x \cos x = \sqrt{3}$ in the interval [0°, 360°).
The identity $2 \sin x \cos x = \sin 2x$ is useful here.

$$4 \sin x \cos x = \sqrt{3}$$

$$2(2 \sin x \cos x) = \sqrt{3} \qquad 4 = 2 \cdot 2$$

$$2 \sin 2x = \sqrt{3} \qquad 2 \sin x \cos x = \sin 2x$$

$$\sin 2x = \frac{\sqrt{3}}{2} \qquad \text{Divide by 2.}$$

From the given domain $0° \le x < 360°$, the domain for $2x$ is $0° \le 2x < 720°$. Now list all solutions in this interval:

$$2x = 60°, 120°, 420°, 480°$$

or

$$x = 30°, 60°, 210°, 240°. \qquad \text{Divide by 2.}$$

The last two solutions for $2x$ were found by adding 360° to 60° and 120°, respectively. The solution set is {30°, 60°, 210°, 240°}. ▶

The methods for solving trigonometric equations illustrated in the examples can be summarized as follows.

Solving Trigonometric Equations Analytically

1. If functions of multiple angles are involved, use multiple-angle identities to rewrite the equation.
2. If only one trigonometric function is present, first solve the equation for that function.
3. If more than one trigonometric function is present, rearrange the equation so that one side equals 0. Then try to factor and set each factor equal to 0 to solve.
4. If Step 3 does not work, try using identities to change the form of the equation. It may be helpful to square both sides of the equation first. If this is done, check for extraneous solutions.
5. If the equation is quadratic in form, but not factorable, use the quadratic formula.

The final example in this section is a continuation of the discussion in the chapter opener.

◀**EXAMPLE 9**
Describing a musical tone from a graph

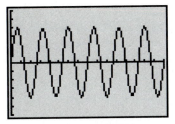

[0, .04] by [−.006, .006]
Xscl = .0025   Yscl = .001

FIGURE 17

A basic component of music is a pure tone. The graph in Figure 17 shows the sinusoidal pressure P in pounds per square foot from a pure tone at time t in seconds.

(a) The frequency of a pure tone is often measured in a unit called **hertz.** One hertz is equal to one cycle per second and is abbreviated as Hz. What is the frequency f in hertz of the pure tone shown in the graph?

From the graph we can see that there are 6 cycles in .04 sec. This is equivalent to $6/.04 = 150$ cycles per second. The pure tone has a frequency of $f = 150$ Hz.

(b) The time for the tone to produce one complete cycle is called the *period.* Approximate the period T in seconds of the pure tone.

Six periods cover a time of .04 second. One period would be equal to $T = .04/6 = 1/150$ or $.00\overline{6}$ sec.

(c) An equation for the graph is $y = .004 \sin(300\pi x)$. Use a calculator to estimate all solutions to the equation $y = .004$ on the interval $[0, .02]$.

If we reproduce the graph in Figure 17 on a calculator and also graph the second function $y = .004$, we can determine that the approximate values of x at the points of intersection of the graphs are .0017, .0083, and .015. Verify these values with your own calculator. ▶

7.6 Exercises ▼▼▼▼▼▼▼▼▼▼▼▼▼▼▼▼▼▼▼▼▼▼▼▼▼▼▼▼▼▼▼

1. An equation of the form $\sin x - b = 0$ has either zero, one, or two solutions in the interval $[0, 2\pi)$. For what values of b will there be two solutions? One solution? No solutions? (*Hint:* Look at the graph of $y = \sin x$.)

2. Suppose an equation of the form $\tan x - b = 0$ has the solution $x = a$ in the interval $(-\pi/2, \pi/2)$. Give an expression for all solutions to the equation.

3. The accompanying screen shows the graphs of $Y_1 = 2 \sin x$ and $Y_2 = \cos \dfrac{x}{2}$ on the interval $[0, 2\pi)$ with $Xscl = .5$ and $Yscl = 1$. Estimate the solutions of the equation $2 \sin x = \cos \dfrac{x}{2}$ in the interval $[0, 2\pi)$.

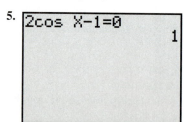

$[0, 2\pi]$ by $[-2.25, 2.25]$
$Xscl = .5$ $Yscl = 1$

4. The accompanying screen shows the graphs of $Y_1 = \sin 2x$ and $Y_2 = \cos^2 x$ on the interval $[0, 2\pi)$ with $Xscl = .5$ and $Yscl = .5$. The equation $\sin 2x = \cos^2 x$ has four solutions in the interval $[0, 2\pi)$. Two of the solutions can easily be found exactly. Find them. What are approximations of the other two solutions?

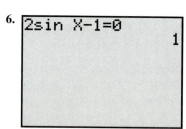

$[0, 2\pi]$ by $[-1.5, 1.5]$
$Xscl = .5$ $Yscl = .5$

Recall that a graphing calculator will return a 1 for a true statement and a 0 for a false statement. In Exercises 5–8, what possible values for X, *in the interval* $[0, 2\pi)$, *will produce the accompanying graphing calculator screen?*

5.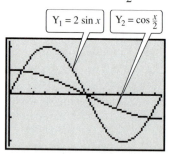

6.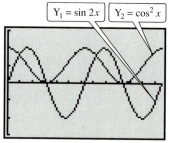

7.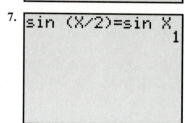

8.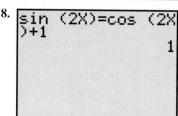

Solve each equation for solutions in the interval $[0, 2\pi)$ *by first solving for the trigonometric function. See Examples 1–3.*

9. $2 \cot x + 1 = -1$

10. $\sin x + 2 = 3$

11. $2 \sin x + 3 = 4$

12. $2 \sec x + 1 = \sec x + 3$

13. $\tan^2 x + 3 = 0$

14. $\sec^2 x + 2 = -1$

15. $(\cot x - 1)(\sqrt{3} \cot x + 1) = 0$

16. $(\csc x + 2)(\csc x - \sqrt{2}) = 0$

17. $\cos^2 x + 2 \cos x + 1 = 0$

18. $2 \cos^2 x - \sqrt{3} \cos x = 0$

19. $-2 \sin^2 x = 3 \sin x + 1$

20. $4(1 + \sin x)(1 - \sin x) = 3$

Determine all solutions of the following equations in radians.

21. $2 \sin^2 x - \sin x - 1 = 0$

22. $2 \cos^2 x + \cos x = 1$

23. $4 \cos^2 x - 1 = 0$

24. $\cos^2 x + \cos x - 6 = 0$

Solve each equation for solutions in the interval $[0°, 360°)$. *See Examples 1–3.*

25. $(\cot \theta - \sqrt{3})(2 \sin \theta + \sqrt{3}) = 0$

26. $(\tan \theta - 1)(\cos \theta - 1) = 0$

27. $2 \sin \theta - 1 = \csc \theta$

28. $\tan \theta + 1 = \sqrt{3} + \sqrt{3} \cot \theta$

29. $\tan \theta - \cot \theta = 0$

30. $\cos^2 \theta = \sin^2 \theta + 1$

31. $\sin^2 \theta \cos \theta = \cos \theta$

32. $2 \tan^2 \theta \sin \theta - \tan^2 \theta = 0$

33. $\sin^2 \theta \cos^2 \theta = 0$

34. $\sec^2 \theta \tan \theta = 2 \tan \theta$

35. $\sin \theta + \cos \theta = 1$

36. $\sec \theta - \tan \theta = 1$

Solve each equation for solutions in the interval $[0°, 360°)$. *Use a calculator and express approximate solutions to the nearest tenth of a degree. In Exercises 43–48, you will need to use the quadratic formula. See Examples 4 and 5.*

37. $3 \sin^2 x - \sin x = 2$

38. $\dfrac{2 \tan x}{3 - \tan^2 x} = 1$

39. $\sec^2 \theta = 2 \tan \theta + 4$

40. $5 \sec^2 \theta = 6 \sec \theta$

41. $3 \cot^2 \theta = \cot \theta$

42. $8 \cos \theta = \cot \theta$

43. $9 \sin^2 x - 6 \sin x = 1$

44. $4 \cos^2 x + 4 \cos x = 1$

45. $\tan^2 x + 4 \tan x + 2 = 0$

46. $3 \cot^2 x - 3 \cot x - 1 = 0$

47. $\sin^2 x - 2 \sin x + 3 = 0$

48. $2 \cos^2 x + 2 \cos x - 1 = 0$

Solve each equation for solutions in the interval $[0, 2\pi)$. *See Examples 6–8.*

49. $\cos 2x = \dfrac{\sqrt{3}}{2}$

50. $\cos 2x = -\dfrac{1}{2}$

51. $\sqrt{2} \cos 2x = -1$

52. $2\sqrt{3} \sin 2x = \sqrt{3}$

53. $\sin \dfrac{x}{2} = \sqrt{2} - \sin \dfrac{x}{2}$

54. $\sin x = \sin 2x$

55. $\tan 4x = 0$

56. $\cos 2x - \cos x = 0$

57. $\sin \dfrac{x}{2} = \cos \dfrac{x}{2}$

58. $\sec \dfrac{x}{2} = \cos \dfrac{x}{2}$

59. $\cos 2x + \cos x = 0$

60. $\sin x \cos x = \dfrac{1}{4}$

Solve each equation for solutions in the interval $[0°, 360°)$. *Use a calculator as necessary to find solutions to the nearest tenth of a degree. See Examples 1–4.*

61. $\sqrt{2} \sin 3\theta - 1 = 0$

62. $-2 \cos 2\theta = \sqrt{3}$

63. $2\sqrt{3} \sin \dfrac{\theta}{2} = 3$

64. $2\sqrt{3} \cos \dfrac{\theta}{2} = -3$

65. $2 \sin \theta = 2 \cos 2\theta$

66. $\cos \theta - 1 = \cos 2\theta$

67. $1 - \sin \theta = \cos 2\theta$

68. $\sin 2\theta = 2 \cos^2 \theta$

69. $\csc^2 \dfrac{\theta}{2} = 2 \sec \theta$

70. $\cos \theta = \sin^2 \dfrac{\theta}{2}$

71. A coil of wire rotating in a magnetic field induces a voltage given by

$$e = 20 \sin\left(\frac{\pi t}{4} - \frac{\pi}{2}\right),$$

where t is time in seconds. Find the smallest positive time to produce the following voltages.
(a) 0   **(b)** $10\sqrt{3}$

72. The equation

$$.342D \cos \theta + h \cos^2 \theta = \frac{16D^2}{V_0^2}$$

is used in reconstructing accidents in which a vehicle vaults into the air after hitting an obstruction. V_0 is the velocity in feet per second of the vehicle when it hits, D is the distance (in feet) from the obstruction to the landing point, and h is the difference in height (in feet) between the landing point and the takeoff point. Angle θ is the takeoff angle, the angle between the horizontal and the path of the vehicle. Find θ to the nearest degree if $V_0 = 60$, $D = 80$, and $h = 2$.

73. If $0 < k < 1$, how many solutions does the equation $\sin^2 \theta = k$ have in the interval $[0°, 360°)$?

74. The seasonal variation in the length of daylight can be represented by a sine function. For example, the daily number of hours of daylight in New Orleans is given by

$$h = \frac{35}{3} + \frac{7}{3} \sin \frac{2\pi x}{365},$$

where x is the number of days after March 21 (disregarding leap year). (*Source:* Bushaw, Donald et al., *A Sourcebook of Applications of School Mathematics.* Copyright © 1980 by The Mathematical Association of America. Reprinted by permission. The material was prepared with the support of National Science Foundation Grant No. SED72-01123 A05. However,

any opinions, findings, conclusions, or recommendations expressed herein are those of the authors and do not necessarily reflect the views of NSF.)
(a) On what date will there be about 14 hours of daylight?
(b) What date has the least number of hours of daylight?
(c) When will there be about 10 hours of daylight?

75. The British nautical mile is defined as the length of a minute of arc of a meridian. Since Earth is flat at its poles, the nautical mile, in feet, is given by

$$L = 6077 - 31 \cos 2\theta,$$

where θ is the latitude in degrees. (See the figure.) (*Source:* Bushaw, Donald et al., *A Sourcebook of Applications of School Mathematics.* Copyright © 1980 by The Mathematical Association of America. Reprinted by permission.)

A nautical mile is the length on any of these meridians cut by a central angle of measure 1 minute.

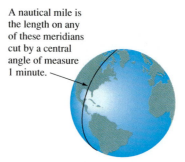

(a) Find the latitude between 0° and 90° at which the nautical mile is 6074 feet.
(b) At what latitude between 0° and 180° is the nautical mile 6108 feet?
(c) In the United States the nautical mile is defined everywhere as 6080.2 feet. At what latitude between 0° and 90° does this agree with the British nautical mile?

76. No musical instrument can generate a true pure tone. A pure tone has a unique, constant frequency and amplitude that sounds rather dull and uninteresting. The pressures caused by pure tones on the eardrum are sinusoidal. The change in pressure P in pounds per square foot on a person's eardrum from a pure tone at time t in seconds can be modeled using the equation $P = A \sin(2\pi ft + \phi)$. f is the frequency in cycles per second and ϕ is the phase angle. When P is positive there is an increase in pressure and the eardrum is pushed inward; when P is negative there is a decrease in pressure, and the eardrum is pushed outward. (*Source:* Roederer, Juan, *Introduction to the Physics and Psychophysics of Music,* The English Universities Press Ltd., London, 1973.)

(a) Middle C has a frequency of 261.63 cycles per second. Graph this tone with $A = .004$ and $\phi = \pi/7$ in the window $[0, .005]$ by $[-.005, .005]$.

(b) Determine analytically the values of t for which $P = 0$ in $[0, .005]$ and support your answers graphically.

(c) Determine graphically when $P < 0$ on $[0, .005]$.

(d) Would an eardrum hearing this tone be vibrating outward or inward when $P < 0$?

77. A piano string can vibrate at more than one frequency when it is struck. It produces a complex wave that can mathematically be modeled by a sum of several pure tones. If a piano key with a frequency of f_1 is played, then the corresponding string will not only vibrate at f_1 but it will also vibrate at the higher frequencies of $2f_1, 3f_1, 4f_1, \ldots, nf_1$. f_1 is called the **fundamental frequency** of the string and higher frequencies are called the **upper harmonics.** The human ear will hear the sum of these frequencies as one complex tone. (*Source:* Roederer, Juan, *Introduction to the Physics and Psychophysics of Music,* The English Universities Press Ltd., London, 1973.)

(a) Suppose that the A key above middle C is played. Its fundamental frequency is $f_1 = 440$ Hz and its associated pressure is expressed as $P_1 = 002 \sin 880\pi t$. The string will also vibrate at 880, 1320, 1760, . . . hertz. The corresponding pres-

sures of these upper harmonics are

$$P_2 = \frac{.002}{2} \sin 1760\pi t,$$

$$P_3 = \frac{.002}{3} \sin 2640\pi t,$$

$$P_4 = \frac{.002}{4} \sin 3520\pi t,$$

and $$P_5 = \frac{.002}{5} \sin 4400\pi t.$$

Graph each of the following expressions for P in the window $[0, .01]$ by $[-.005, .005]$.

(i) $P = P_1$
(ii) $P = P_1 + P_2$
(iii) $P = P_1 + P_2 + P_3$
(iv) $P = P_1 + P_2 + P_3 + P_4$
(v) $P = P_1 + P_2 + P_3 + P_4 + P_5$

(b) Describe the final graph of P.

(c) What is the maximum pressure of $P = P_1 + P_2 + P_3 + P_4 + P_5$? When does this maximum occur on $[0, .01]$?

78. If a string with a fundamental frequency of 110 hertz is plucked in the middle, it will vibrate at the odd harmonics of 110, 330, 550, . . . hertz but not at the even harmonics of 220, 440, 660, . . . hertz. The resulting pressure P caused by the string can be approximated using the equation

$$P = .003 \sin 220\pi t + \frac{.003}{3} \sin 660\pi t$$

$$+ \frac{.003}{5} \sin 1100\pi t + \frac{.003}{7} \sin 1540\pi t.$$

(*Sources:* Benade, Arthur, *Fundamentals of Musical Acoustics,* Oxford University Press, New York, 1976. Roederer, Juan, *Introduction to the Physics and Psychophysics of Music,* The English Universities Press Ltd., London, 1973.)

(a) Graph P in the window $[0, .03]$ by $[-.005, .005]$.

(b) Use the graph to describe the shape of the sound wave that is produced.

(c) At lower frequencies, the inner ear will hear a tone only when the eardrum is moving outward. (See Exercise 76.) Determine the times on the interval $[0, .03]$ when this will occur.

7.7 Equations with Inverse Trigonometric Functions ▼▼▼

Section 7.5 introduced the inverse trigonometric functions. Recall, for example, that $x = \sin y$, with $-\pi/2 \leq y \leq \pi/2$, means the same thing as $y = \arcsin x$ or $y = \sin^{-1} x$. Sometimes the solution of a trigonometric equation with more than one variable can be solved with inverse trigonometric functions, as shown in the following examples.

◀ **EXAMPLE 1**
Solving an equation for a variable using inverse notation

Solve $y = 3 \cos 2x$ for x.

We want $\cos 2x$ alone on one side of the equation so we can solve for $2x$ and then for x. First, divide both sides of the equation by 3.

$$y = 3 \cos 2x$$

$$\frac{y}{3} = \cos 2x$$

Now write the statement in the alternative form

$$2x = \arccos \frac{y}{3}.$$

Finally, multiply both sides by $1/2$.

$$x = \frac{1}{2} \arccos \frac{y}{3} \quad \blacktriangleright$$

The next examples show how to solve equations involving inverse trigonometric functions.

◀ **EXAMPLE 2**
Solving an equation involving an inverse trigonometric function

Solve $2 \arcsin x = \pi$.

First solve for $\arcsin x$.

$$2 \arcsin x = \pi$$

$$\arcsin x = \frac{\pi}{2} \qquad \text{Divide by 2.}$$

Use the definition of $\arcsin x$ to get

$$x = \sin \frac{\pi}{2}$$

or

$$x = 1.$$

Verify that the solution satisfies the given equation, so the solution set is $\{1\}$. $\quad \blacktriangleright$

EXAMPLE 3
Solving an equation involving
inverse trigonometric functions

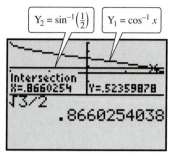

[−1, 1] by [−2, π]
Xscl = .5   Yscl = $\frac{\pi}{4}$

The exact solution found in Example 3,
$\sqrt{3}/2$, is supported graphically in this
screen. Notice that the approximation
for X corresponds to that of $\sqrt{3}/2$.

Solve $\cos^{-1} x = \sin^{-1} \dfrac{1}{2}$.

Let $\sin^{-1}(1/2) = u$. Then $\sin u = 1/2$ and the equation becomes

$$\cos^{-1} x = u,$$

for u in quadrant I. This can be written as

$$\cos u = x.$$

Sketch a triangle and label it using the facts that u is in quadrant I and $\sin u = 1/2$. See Figure 18. Since $x = \cos u$,

$$x = \frac{\sqrt{3}}{2}.$$

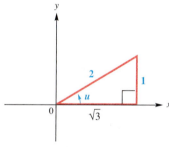

FIGURE 18

Check the solution either by substitution or graphically to see that the solution set is $\{\sqrt{3}/2\}$. ▶

Some equations with inverse trigonometric functions require the use of identities to solve.

EXAMPLE 4
Solving an inverse
trigonometric equation using
an identity

Solve $\arcsin x - \arccos x = \pi/6$.

Begin by adding $\arccos x$ to both sides of the equation so that one inverse function is alone on one side of the equation.

$$\arcsin x - \arccos x = \frac{\pi}{6}$$

$$\arcsin x = \arccos x + \frac{\pi}{6} \qquad (1)$$

Use the definition of arcsin to write this statement as

$$\sin\left(\arccos x + \frac{\pi}{6}\right) = x.$$

Let $u = \arccos x$, so $0 \le u \le \pi$ by definition. Then

$$\sin\left(u + \frac{\pi}{6}\right) = x. \tag{2}$$

Using the identity for $\sin(A + B)$,

$$\sin\left(u + \frac{\pi}{6}\right) = \sin u \cos \frac{\pi}{6} + \cos u \sin \frac{\pi}{6}.$$

Substitute this result into equation (2) to get

$$\sin u \cos \frac{\pi}{6} + \cos u \sin \frac{\pi}{6} = x. \tag{3}$$

From equation (1) and by the definition of the arcsin function,

$$-\frac{\pi}{2} \le \arccos x + \frac{\pi}{6} \le \frac{\pi}{2}.$$

Subtract $\pi/6$ from each expression to get

$$-\frac{2\pi}{3} \le \arccos x \le \frac{\pi}{3}.$$

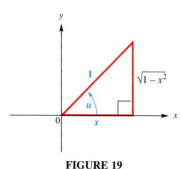

FIGURE 19

Since $0 \le \arccos x \le \pi$, it follows that here we must have $0 \le \arccos x \le \pi/3$. Thus $x > 0$, and we can sketch the triangle in Figure 19. From this triangle we find that $\sin u = \sqrt{1 - x^2}$. Now substitute into equation (3) using $\sin u = \sqrt{1 - x^2}$, $\sin \pi/6 = 1/2$, $\cos \pi/6 = \sqrt{3}/2$, and $\cos u = x$.

$$(\sqrt{1 - x^2})\frac{\sqrt{3}}{2} + x \cdot \frac{1}{2} = x$$
$$(\sqrt{1 - x^2})\sqrt{3} + x = 2x$$
$$(\sqrt{3})\sqrt{1 - x^2} = x$$

Squaring both sides gives

$$3(1 - x^2) = x^2$$
$$3 - 3x^2 = x^2$$
$$3 = 4x^2$$
$$x = \sqrt{\frac{3}{4}} \qquad \text{Choose the positive square}$$
$$\qquad\qquad \text{root because } x > 0.$$
$$= \frac{\sqrt{3}}{2}.$$

To check, replace x with $\sqrt{3}/2$ in the original equation:

$$\arcsin \frac{\sqrt{3}}{2} - \arccos \frac{\sqrt{3}}{2} = \frac{\pi}{3} - \frac{\pi}{6} = \frac{\pi}{6},$$

as required. The solution set is $\{\sqrt{3}/2\}$. ▶

7.7 Exercises ▼▼▼▼▼▼▼▼▼▼▼▼▼▼▼▼▼▼▼▼▼▼▼▼▼▼▼▼▼▼

1. The screen shows the graphs of Y_1 = arcsin x and Y_2 = arccos x in the window $[-1, 1]$ by $[-\pi/2, \pi]$, with Xscl = .5 and Yscl = $\pi/6$. What are the exact coordinates of the point of intersection of the two graphs? What is the solution of arcsin x − arccos x = 0?

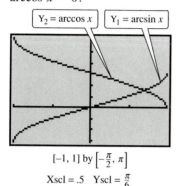

$[-1, 1]$ by $\left[-\frac{\pi}{2}, \pi\right]$
Xscl = .5   Yscl = $\frac{\pi}{6}$

2. The screen shows the graphs of Y_1 = $\cos^{-1} 2x$ and $Y_2 = \dfrac{5\pi}{3} - \cos^{-1} x$ in the window $[-1, 0]$ by $[0, 2\pi]$, with Xscl = .5 and Yscl = $\pi/3$. What are the exact coordinates of the point of intersection of the two graphs? What is the solution of $\cos^{-1} 2x + \cos^{-1} x = \dfrac{5\pi}{3}$?

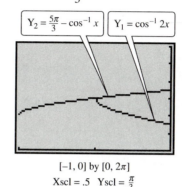

$[-1, 0]$ by $[0, 2\pi]$
Xscl = .5   Yscl = $\frac{\pi}{3}$

Recall that a graphing calculator will return a 1 for a true statement and a 0 for a false statement. In Exercises 3 and 4, what possible values for X, in the interval $[0, 1]$, will produce the accompanying graphing calculator screen?

3. `2cos⁻¹ X=π`   `1`

4. `4tan⁻¹ X=π`   `1`

Solve the equation for x. See Example 1.

5. $y = 5 \cos x$

6. $4y = \sin x$

7. $2y = \cot 3x$

8. $6y = \dfrac{1}{2} \sec x$

9. $y = 3 \tan 2x$

10. $y = 3 \sin \dfrac{x}{2}$

11. $y = 6 \cos \dfrac{x}{4}$

12. $y = -\sin \dfrac{x}{3}$

13. $y = -2 \cos 5x$

14. $y = 3 \cot 5x$

15. $y = \cos(x + 3)$

16. $y = \tan(2x - 1)$

17. $y = \sin x - 2$

18. $y = \cot x + 1$

19. $y = 2 \sin x - 4$

20. $y = 4 + 3 \cos x$

21. (Refer to Exercise 17.) A student attempting to solve this problem wrote as the first step

$$y = \sin(x - 2),$$

inserting parentheses as shown. Explain why this is incorrect.

22. Explain why the equation

$$\sin^{-1} x = \cos^{-1} 2$$

cannot have a solution. (No work needs to be shown here.)

Solve the equation. See Examples 2 and 3.

23. $\dfrac{4}{3} \cos^{-1} \dfrac{y}{4} = \pi$

24. $4\pi + 4 \tan^{-1} y = \pi$

25. $2 \arccos\left(\dfrac{y - \pi}{3}\right) = 2\pi$

26. $\arccos\left(y - \dfrac{\pi}{3}\right) = \dfrac{\pi}{6}$

27. $\arcsin x = \arctan \dfrac{3}{4}$

28. $\arctan x = \arccos \dfrac{5}{13}$

29. $\cos^{-1} x = \sin^{-1} \dfrac{3}{5}$

30. $\cot^{-1} x = \tan^{-1} \dfrac{4}{3}$

Solve the equation. See Example 4.

31. $\sin^{-1} x - \tan^{-1} 1 = -\dfrac{\pi}{4}$

32. $\sin^{-1} x + \tan^{-1} \sqrt{3} = \dfrac{2\pi}{3}$

33. $\arccos x + 2 \arcsin \dfrac{\sqrt{3}}{2} = \pi$

34. $\arccos x + 2 \arcsin \dfrac{\sqrt{3}}{2} = \dfrac{\pi}{3}$

35. $\arcsin 2x + \arccos x = \dfrac{\pi}{6}$

36. $\arcsin 2x + \arcsin x = \dfrac{\pi}{2}$

37. $\cos^{-1} x + \tan^{-1} x = \dfrac{\pi}{2}$

38. $\sin^{-1} x + \tan^{-1} x = 0$

39. Solve $d = 550 + 450 \cos \dfrac{\pi}{50} t$ for t in terms of d.

40. Solve $d = 40 + 60 \cos \dfrac{\pi}{6}(t - 2)$ for t in terms of d.

---

41. In the study of alternating electric current, instantaneous voltage is given by

$$e = E_{max} \sin 2\pi ft,$$

where f is the number of cycles per second, E_{max} is the maximum voltage, and t is time in seconds.
(a) Solve the equation for t.
(b) Find the smallest positive value of t if $E_{max} = 12$, $e = 5$, and $f = 100$. Use a calculator.

42. When a large-view camera is used to take a picture of an object that is not parallel to the film, the lens board should be tilted so that the planes containing the subject, the lens board, and the film intersect in a line (see the figure). This gives the best "depth of field." (*Source:* Bushaw, Donald et al., *A Sourcebook of Applications of School Mathematics.* Copyright © 1980 by The Mathematical Association of America. Reprinted by permission.)

(a) Write two equations, one relating α, x, and z, and the other relating β, x, y, and z.

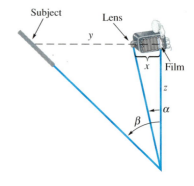

(b) Eliminate z from the equations in part (a) to get one equation relating α, β, x, and y.

(c) Solve the equation from part (b) for α.

(d) Solve the equation from part (b) for β.

43. In many computer languages, such as BASIC and FORTRAN, only the arctan function is available. To use the other inverse trigonometric functions, it is necessary to express them in terms of arctangent. This can be done as follows.

(a) Let $u = \arcsin x$. Solve the equation for x in terms of u.

(b) Use the result of part (a) to label the three sides of the triangle in the figure in terms of x.

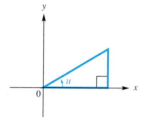

(c) Use the triangle from part (b) to write an equation for $\tan u$ in terms of x.

(d) Solve the equation from part (c) for u.

44. In the exercises for Section 6.7 we found the equation

$$y = \frac{1}{3} \sin \frac{4\pi t}{3},$$

where t is time (in seconds) and y is the angle formed by a rhythmically moving arm.

(a) Solve the equation for t.

(b) At what time(s) does the arm form an angle of .3 radian?

Chapter 7   Summary ▼▼▼▼▼▼▼▼▼▼▼▼▼▼▼▼▼▼▼▼▼▼▼▼▼▼▼▼▼▼▼▼▼▼▼▼

KEY TERMS	KEY IDEAS
7.1 Fundamental Identities	
identity	**Fundamental Identities** **Reciprocal Identities** $\cot \theta = \dfrac{1}{\tan \theta} \qquad \sec \theta = \dfrac{1}{\cos \theta} \qquad \csc \theta = \dfrac{1}{\sin \theta}$ **Quotient Identities** $\tan \theta = \dfrac{\sin \theta}{\cos \theta} \qquad \cot \theta = \dfrac{\cos \theta}{\sin \theta}$ **Pythagorean Identities** $\sin^2 \theta + \cos^2 \theta = 1 \qquad \tan^2 \theta + 1 = \sec^2 \theta \qquad 1 + \cot^2 \theta = \csc^2 \theta$ **Negative-Angle Identities** $\sin(-\theta) = -\sin \theta \qquad \cos(-\theta) = \cos \theta \qquad \tan(-\theta) = -\tan \theta$

KEY TERMS	KEY IDEAS
7.3 Sum and Difference Identities	
	Cofunction Identities
	$\cos(90° - \theta) = \sin \theta \qquad \cot(90° - \theta) = \tan \theta$
	$\sin(90° - \theta) = \cos \theta \qquad \sec(90° - \theta) = \csc \theta$
	$\tan(90° - \theta) = \cot \theta \qquad \csc(90° - \theta) = \sec \theta$
	Sum and Difference Identities
	$\cos(A - B) = \cos A \cos B + \sin A \sin B$
	$\cos(A + B) = \cos A \cos B - \sin A \sin B$
	$\sin(A + B) = \sin A \cos B + \cos A \sin B$
	$\sin(A - B) = \sin A \cos B - \cos A \sin B$
	$\tan(A + B) = \dfrac{\tan A + \tan B}{1 - \tan A \tan B}$
	$\tan(A - B) = \dfrac{\tan A - \tan B}{1 + \tan A \tan B}$
7.4 Double-Angle Identities and Half-Angle Identities	
	Double-Angle Identities
	$\cos 2A = \cos^2 A - \sin^2 A$
	$\cos 2A = 1 - 2 \sin^2 A$
	$\cos 2A = 2 \cos^2 A - 1$
	$\sin 2A = 2 \sin A \cos A$
	$\tan 2A = \dfrac{2 \tan A}{1 - \tan^2 A}$
	Half-Angle Identities
	$\sin \dfrac{A}{2} = \pm\sqrt{\dfrac{1 - \cos A}{2}} \qquad \tan \dfrac{A}{2} = \dfrac{1 - \cos A}{\sin A}$
	$\cos \dfrac{A}{2} = \pm\sqrt{\dfrac{1 + \cos A}{2}} \qquad \tan \dfrac{A}{2} = \dfrac{\sin A}{1 + \cos A}$
	$\tan \dfrac{A}{2} = \pm\sqrt{\dfrac{1 - \cos A}{1 + \cos A}}$
	(The sign is chosen based on the quadrant of $A/2$.)

KEY TERMS	KEY IDEAS

7.5 Inverse Trigonometric Functions

Inverse Trigonometric Functions

Function	Domain	Range	Quadrants of the Unit Circle Range Values Come From
$y = \sin^{-1} x$	$[-1, 1]$	$[-\pi/2, \pi/2]$	I and IV
$y = \cos^{-1} x$	$[-1, 1]$	$[0, \pi]$	I and II
$y = \tan^{-1} x$	$(-\infty, \infty)$	$(-\pi/2, \pi/2)$	I and IV
$y = \cot^{-1} x$	$(-\infty, \infty)$	$(0, \pi)$	I and II
$y = \sec^{-1} x$	$(-\infty, -1] \cup [1, \infty)$	$[0, \pi],\ y \neq \pi/2$	I and II
$y = \csc^{-1} x$	$(-\infty, -1] \cup [1, \infty)$	$[-\pi/2, \pi/2],\ y \neq 0$	I and IV

$y = \sin^{-1} x$ or arcsin x

$y = \cos^{-1} x$ or arccos x

$y = \tan^{-1} x$ or arctan x

KEY TERMS	KEY IDEAS
7.6 Trigonometric Equations	
	Solving Trigonometric Equations Analytically **1.** If functions of multiple angles are involved, use multiple-angle identities to rewrite the equation. **2.** If only one trigonometric function is present, first solve the equation for that function. **3.** If more than one trigonometric function is present, rearrange the equation so that one side equals 0. Then try to factor and set each factor equal to 0 to solve. **4.** If Step 3 does not work, try using identities to change the form of the equation. It may be helpful to square both sides of the equation first. If this is done, check for extraneous solutions. **5.** If the equation is quadratic in form, but not factorable, use the quadratic formula. Check for extraneous solutions.

Chapter 7   Review Exercises ▼▼▼▼▼▼▼▼▼▼▼▼▼▼▼▼▼▼▼▼▼▼▼▼▼▼▼▼▼▼

1. Give all the trigonometric functions that satisfy the condition $f(-x) = -f(x)$.

2. Give all the trigonometric functions that satisfy the condition $f(-x) = f(x)$.

In Exercises 3 and 4, the given graphing calculator screen is obtained for a particular stored value of X. *What will the screen display for the value of the expression in the final line of the display?*

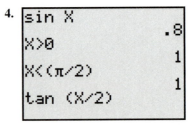

3.
```
sin X
            .6
X>(π/2)
           1
X<π
           1
sin (2X)
```

4.
```
sin X
            .8
X>0
           1
X<(π/2)
           1
tan (X/2)
```

5. Use the trigonometric identities to find the remaining five trigonometric functions of x, given that $\cos x = 3/5$ and x is in quadrant IV.

6. Find the exact values of $\sin x$, $\cos x$, and $\tan x$, for $x = \pi/12$, using
   (a) difference identities;   **(b)** half-angle identities.

For each item in Column I, give the letter of the item in Column II that completes an identity.

Column I	Column II
7. $\sin 35°$	**(a)** $\sin(-35°)$
8. $-\sin 35°$	**(b)** $\cos 55°$
9. $\cos 35°$	**(c)** $\sqrt{\dfrac{1 + \cos 150°}{2}}$
10. $\cos 75°$	**(d)** $2 \sin 150° \cos 150°$
11. $\sin 75°$	**(e)** $\cos^2 150° - \sin^2 150°$
12. $\sin 300°$	**(f)** $\sin 15° \cos 60° + \cos 15° \sin 60°$
13. $\cos 300°$	**(g)** $\cos(-35°)$
	(h) $\cot 125°$

For each item in Column I, give the letter of the item in Column II that completes an identity.

Column I	Column II
14. $\tan x$	**(a)** $\dfrac{\sin x}{\cos x}$
15. $\cot x$	**(b)** $\dfrac{1}{\cot^2 x}$
16. $\sin^2 x$	**(c)** $\dfrac{1}{\cos^2 x}$
17. $\tan^2 x + 1$	**(d)** $\dfrac{\cos x}{\sin x}$
18. $\tan^2 x$	**(e)** $\dfrac{1}{\sin^2 x}$
	(f) $1 - \cos^2 x$

Use identities to write each expression in terms of $\sin \theta$ and $\cos \theta$, and simplify.

19. $\sec^2 \theta - \tan^2 \theta$ **20.** $\tan^2 \theta (1 + \cot^2 \theta)$ **21.** $\csc \theta + \cot \theta$ **22.** $\tan \theta - \sec \theta \csc \theta$

For the following find $\sin(x + y)$, $\cos(x - y)$, $\tan(x + y)$, and the quadrant of $x + y$.

23. $\sin x = -1/4$, $\cos y = -4/5$, x and y in quadrant III

24. $\sin x = 1/10$, $\cos y = 4/5$, x in quadrant I, y in quadrant IV

Find the sine and cosine of the angle.

25. θ, given $\cos 2\theta = -3/4$, $90° < 2\theta < 180°$ **26.** B, given $\cos 2B = 1/8$, B in quadrant IV

27. $2x$, given $\tan x = 3$, $\sin x < 0$ **28.** $2y$, given $\sec y = -5/3$, $\sin y > 0$

Find each value.

29. $\cos \theta/2$, given $\cos \theta = -1/2$, with $90° < \theta < 180°$ **30.** $\sin A/2$, given $\cos A = -3/4$, with $90° < A < 180°$

31. $\tan x$, given $\tan 2x = 2$, $\pi < x < 3\pi/2$ **32.** $\sin y$, given $\cos 2y = -1/3$, $\pi/2 < y < \pi$

In Exercises 33–36, graph each expression and use the graph to conjecture an identity. Then verify your conjecture.

33. $-\dfrac{\sin 2x + \sin x}{\cos 2x - \cos x}$ **34.** $\dfrac{1 - \cos 2x}{\sin 2x}$ **35.** $\dfrac{\cos x \sin 2x}{1 + \cos 2x}$ **36.** $\dfrac{2(\sin x - \sin^3 x)}{\cos x}$

Verify that each equation is an identity.

37. $\sin^2 x - \sin^2 y = \cos^2 y - \cos^2 x$

38. $\dfrac{\sin^2 x}{2 - 2 \cos x} = \cos^2 \dfrac{x}{2}$

39. $\dfrac{\sin 2x}{\sin x} = \dfrac{2}{\sec x}$

40. $\dfrac{2 \tan B}{\sin 2B} = \sec^2 B$

41. $1 + \tan^2 \alpha = 2 \tan \alpha \csc 2\alpha$

42. $-\dfrac{\sin(A - B)}{\sin(A + B)} = \dfrac{\cot A - \cot B}{\cot A + \cot B}$

43. $\dfrac{2 \cot x}{\tan 2x} = \csc^2 x - 2$

44. $\tan \theta \sin 2\theta = 2 - 2 \cos^2 \theta$

45. $\csc A \sin 2A - \sec A = \cos 2A \sec A$

46. $2 \tan x \csc 2x - \tan^2 x = 1$

47. $\sin^3 \theta = \sin \theta - \cos^2 \theta \sin \theta$

48. $\tan \dfrac{7}{2} x = \dfrac{2 \tan \dfrac{7}{4} x}{1 - \tan^2 \dfrac{7}{4} x}$

49. $\sec^2 \alpha - 1 = \dfrac{\sec 2\alpha - 1}{\sec 2\alpha + 1}$

50. $\tan 4\theta = \dfrac{2 \tan 2\theta}{2 - \sec^2 2\theta}$

51. $2 \cos^2 \dfrac{x}{2} \tan x = \tan x + \sin x$

52. $\tan\left(\dfrac{x}{2} + \dfrac{\pi}{4}\right) = \sec x + \tan x$

53. Explain why $\cos(\arccos x)$ always equals x, but $\arccos(\cos x)$ may not equal x.

54. With a graphing calculator, graph $\cos^{-1}(\cos x)$ in the window $[0, 4\pi]$ by $[0, 8]$ and explain its shape.

55. Discuss how the accompanying screen supports the analytic work in Example 5(b) of Section 7.5. Does it matter whether the calculator is in radian mode or degree mode?

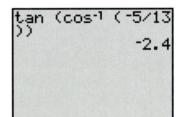

Give the exact real number value of y. Do not use a calculator.

56. $y = \sin^{-1} \dfrac{\sqrt{2}}{2}$

57. $y = \arccos\left(-\dfrac{1}{2}\right)$

58. $y = \tan^{-1}(-\sqrt{3})$

59. $y = \arcsin(-1)$

60. $y = \cos^{-1}\left(-\dfrac{\sqrt{2}}{2}\right)$

61. $y = \arctan \dfrac{\sqrt{3}}{3}$

Give the degree measure of θ. Do not use a calculator.

62. $\theta = \arccos \dfrac{1}{2}$

63. $\theta = \arcsin\left(-\dfrac{\sqrt{3}}{2}\right)$

Use a calculator to give the degree measure of θ.

64. $\theta = \arctan 1.7804675$

65. $\theta = \sin^{-1}(-.66045320)$

66. $\theta = \cos^{-1} .80396577$

67. $\theta = \cot^{-1} 4.5046388$

68. Explain why $\sin^{-1} 3$ cannot be defined.

69. $\arcsin\left(\sin \dfrac{5\pi}{6}\right) \neq \dfrac{5\pi}{6}$. Explain why this is so.

Find each value without using a calculator.

70. $\tan\left(\tan^{-1} \dfrac{2}{3}\right)$

71. $\cos(\arccos(-1))$

72. $\sin\left(\arccos \dfrac{3}{4}\right)$

73. $\cos(\arctan 3)$

74. $\sec\left(2 \sin^{-1}\left(-\dfrac{1}{3}\right)\right)$

75. $\tan\left(\arcsin \dfrac{3}{5} + \arccos \dfrac{5}{7}\right)$

Write each of the following as an expression in u.

76. $\sin(\tan^{-1} u)$

77. $\cos\left(\arctan \dfrac{u}{\sqrt{1 - u^2}}\right)$

Graph each function defined below and give the domain and range.

78. $y = \sin^{-1} x$

79. $y = \cos^{-1} x$

Solve each equation for solutions in the interval $[0, 2\pi)$. Use a calculator in Exercises 81 and 82.

80. $\sin^2 x = 1$

81. $2 \tan x - 1 = 0$

82. $3 \sin^2 x - 5 \sin x + 2 = 0$

83. $\tan x = \cot x$

84. $\tan^2 2x - 1 = 0$

85. $\cos 2x + \cos x = 0$

Solve each equation for solutions in the interval $[0°, 360°)$. Use a calculator and when appropriate, express solutions to the nearest tenth of a degree.

86. $\sin^2 \theta + 3 \sin \theta + 2 = 0$

87. $\sin 2\theta = \cos 2\theta + 1$

88. $3 \cos^2 \theta + 2 \cos \theta - 1 = 0$

89. $\cos \theta - \cos 2\theta = 2 \cos \theta$

Solve each equation for x.

90. $4y = 2 \sin x$

91. $y = 3 \cos \dfrac{x}{2}$

92. $2y = \tan(3x + 2)$

93. $\arccos x = \arcsin \dfrac{2}{7}$

94. $\arccos x + \arctan 1 = \dfrac{11\pi}{12}$

95. $\text{arccot } x = \arcsin\left(\dfrac{-\sqrt{2}}{2}\right) + \dfrac{3\pi}{4}$

96. Recall Snell's law from Exercises 83–84 of Section 6.3:

$$\frac{c_1}{c_2} = \frac{\sin \theta_1}{\sin \theta_2},$$

where c_1 is the speed of light in one medium, c_2 is the speed of light in a second medium, and θ_1 and θ_2 are the angles shown in the figure. Suppose a light is shining up through water into the air as in the figure. As θ_1 increases, θ_2 approaches 90°, at which point no light will emerge from the water. Assume the ratio c_1/c_2 in this case is .752. For what value of θ_1 does $\theta_2 = 90°$? This value of θ_1 is called the *critical angle* for water.

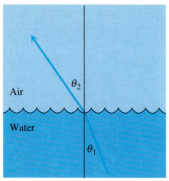

97. (Refer to Exercise 96.) What happens when θ_1 is greater than the critical angle?

 98. Musicians sometimes tune instruments by playing the same tone on two different instruments and listening for a phenomenon known as **beats.** Beats occur when two tones vary in frequency by only a few hertz. When the two instruments are in tune the beats disappear. The ear hears beats because the pressure slowly rises and falls as a result of this slight variation in the frequency. This phenomenon can be seen using a graphing calculator. (*Source:* Pierce, John, *The Science of Musical Sound,* Scientific American Books, 1992.)

(a) Consider two tones with frequencies of 220 and 223 hertz and pressures $P_1 = .005\sin 440\pi t$ and $P_2 = .005 \sin 446\pi t$, respectively. Graph the pressure $P = P_1 + P_2$ felt by an eardrum over the one-second interval [.15, 1.15]. How many beats are there in one second?

(b) Repeat part (a) with frequencies of 220 and 216.

(c) Determine a simple way to find the number of beats per second if the frequency of each tone is given.

 99. Small speakers like those found in older radios and telephones often cannot vibrate slower than 200 hertz—yet 35 keys on a piano have frequencies below 200 hertz. When a musical instrument creates a tone of 110 hertz it also creates tones at 220, 330, 440, 550, 660, . . . hertz. A small speaker cannot reproduce the 110-hertz vibration but it can reproduce the higher frequencies, which are called the upper harmonics. The low tones can still be heard because the speaker produces **difference tones** of the upper harmonics. The difference between consecutive frequencies is 110 hertz, and this difference tone will be heard by a listener. We can see this phenomenon using a graphing calculator. (*Source:* Benade, Arthur, *Fundamentals of Musical Acoustics,* Oxford University Press, New York, 1976.)

(a) Graph the upper harmonics represented by the pressure

$$P = \frac{1}{2} \sin[2\pi(220)t] + \frac{1}{3} \sin[2\pi(330)t]$$
$$+ \frac{1}{4} \sin[2\pi(440)t]$$

in the window [0, .03] by [−2, 2].

(b) Estimate all t-coordinates where P is maximum.

(c) What does a person hear in addition to the frequencies of 220, 330, and 440 hertz?

(d) Graph the pressure produced by a speaker that can vibrate at 110 Hz and above.

Chapter 7   Test ▼▼▼▼▼▼▼▼▼▼▼▼▼▼▼▼▼▼▼▼▼▼▼▼▼▼▼▼▼▼▼▼▼▼▼

1. Given $\tan x = -5/4$, where $\pi/2 < x < \pi$, use the trigonometric identities to find the other trigonometric functions of x.

2. Express $\csc^2 x + \sec^2 x$ in terms of $\sin x$ and $\cos x$ and simplify.

3. Find $\sin(x + y)$, $\cos(x - y)$, and $\tan(x + y)$ if $\sin y = -2/3$, $\cos x = -1/5$, x is in quadrant II, and y is in quadrant III.

4. Graph $\csc x - \cot x$ with a graphing calculator, and use the graph to conjecture an identity. Then verify your conjecture analytically.

Verify each identity.

5. $2 \cos A - \sec A = \cos A - \dfrac{\tan A}{\csc A}$

6. $2 \cos^2 \theta - 1 = \dfrac{1 - \tan^2 \theta}{1 + \tan^2 \theta}$

7. $-\cot \dfrac{x}{2} = \dfrac{\sin 2x + \sin x}{\cos 2x - \cos x}$

8. $\dfrac{1}{2} \cot \dfrac{x}{2} - \dfrac{1}{2} \tan \dfrac{x}{2} = \cot x$

9. With a graphing calculator, graph $\sin^{-1}(\sin x)$ in the window $[0, 4\pi]$ by $[0, 8]$ and explain its shape.

In Exercises 10 and 11, give the exact real number value of y. Do not use a calculator.

10. $y = \sec^{-1}(-2)$

11. $y = \text{arccsc} \dfrac{2\sqrt{3}}{3}$

12. If $\theta = \tan^{-1} 0$, give the degree measure of θ.

13. What is the domain of the arccot function?

Find the following without using a calculator.

14. $\tan^{-1} 1$

15. $\text{arccos}\left(\cos \dfrac{3\pi}{4}\right)$

16. $\cos(\csc^{-1}(-2))$

Solve the equations in Exercises 17–19 for solutions in the interval $[0, 2\pi)$. Give solutions that are not multiples of π to the nearest tenth.

17. $\sec\left(\dfrac{x}{2}\right) = \cos\left(\dfrac{x}{2}\right)$

18. $4 \sin x \cos x = \sqrt{3}$

19. $2 \tan^2 x = \tan x + 1$

20. Solve $\dfrac{4}{3} \arctan \dfrac{x}{2} = \pi$.

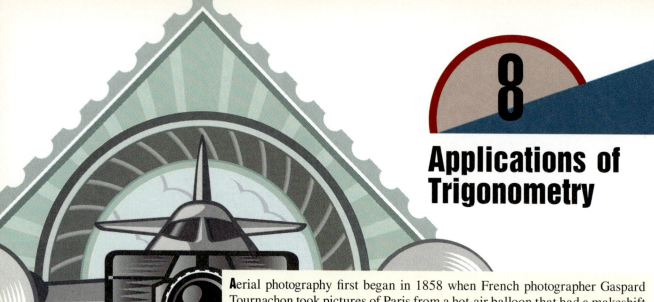

8

Applications of Trigonometry

Aerial photography first began in 1858 when French photographer Gaspard Tournachon took pictures of Paris from a hot-air balloon that had a makeshift darkroom. Since then aerial photography has been used in a variety of applications including surveying, road design, weather forecasting, military surveillance, topographic maps, and even archaeology. The first archaeological aerial photographs were taken of Stonehenge in 1906. By searching these photographs for unusual soil and marks caused by structures lying below the ground, Stonehenge Avenue was discovered. Today, hot-air balloons have been replaced by airplanes, helicopters, and satellites.

In aerial photography a series of photographs are usually taken with sufficient overlap to allow for the stereoscopic vision necessary to obtain accurate ground measurements. These photographs are often used to construct a map that gives both the coordinates and elevations of important features located on the ground. The perspective of these photographs can be affected if the airplane is not perfectly horizontal, the ground below is not level, or the camera is tilted.

Being able to determine the measurements of the sides and angles in a triangle is essential to solving applications involving aerial photography. In

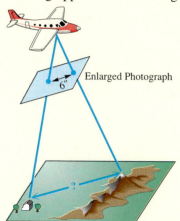

Enlarged Photograph

this chapter we will learn techniques to solve many applications using trigonometry. These trigonometric applications will allow us to interpret aerial photographs, determine the distance to the moon, calculate the area of complicated plots of land, and find the velocity of a distant star.

In this chapter, we also discuss properties of complex numbers. Historically, mathematicians felt uneasy about taking square roots of negative numbers. The famous mathematician René Descartes rejected complex numbers and coined the term "imaginary" numbers. Today, complex numbers are readily accepted and play an important role in many new and exciting fields of applied mathematics and technology. The development of complex numbers has been necessary to solve new problems like the design of airplane wings, ships, electrical circuits, noise control, and *fractals*.

At its basic level, a **fractal** is a unique and enchanting geometric figure with an endless self-similarity property. A fractal image repeats itself with ever-decreasing dimensions. Fractals have tremendous potential for applied science in the future. Through acceptance of the importance and necessity of complex numbers, a new window of understanding about modern mathematics and science has been opened.

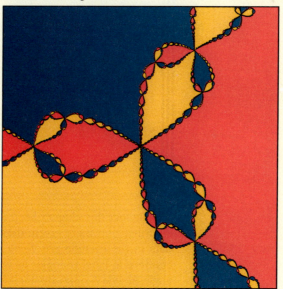

The fractal is called *Newton's basins of attraction for the cube roots of unity.*

Sources: Brooks, R. and Dieter Johannes, *Photoarchaeology,* Dioscorides Press, Portland, Oregon, 1990.

Kline, M., *MATHEMATICS The Loss of Certainty,* Oxford University Press, New York, 1980.

Lauwerier, H., *Fractals,* Princeton University Press, Princeton, New Jersey, 1991.

Moffitt, F., *Photogrammetry,* International Textbook Company, Scranton, Pennsylvania, 1967.

Figure: Kincaid, D. and W. Cheney, *Numerical Analysis,* Brooks/Cole Publishing Company, Pacific Grove, California, 1991. (Figure is from the cover.)

Until now, our applied work with trigonometry has been limited to right triangles. However, the concepts developed in earlier chapters can be extended so that our work can apply to *all* triangles. Every triangle has three sides and three angles. In this chapter we show that if any three of the six measures of a triangle (provided at least one measure is a side) are known, then the other three measures can be found. This process is called *solving a triangle.* Later in the chapter this knowledge is used to solve problems involving vectors.

8.1 Oblique Triangles and the Law of Sines ▼▼▼

The following axioms from geometry allow us to prove that two triangles are congruent (that is, their corresponding sides and angles are equal).

Congruence Axioms

Side-Angle-Side (SAS) If two sides and the included angle of one triangle are equal, respectively, to two sides and the included angle of a second triangle, then the triangles are congruent.

Angle-Side-Angle (ASA) If two angles and the included side of one triangle are equal, respectively, to two angles and the included side of a second triangle, then the triangles are congruent.

Side-Side-Side (SSS) If three sides of one triangle are equal, respectively, to three sides of a second triangle, then the triangles are congruent.

Throughout this chapter keep in mind that whenever any of the groups of data described above are given, the triangle is uniquely determined; that is, all other data in the triangle are given by one and only one set of measures. We will continue to label triangles as we did earlier with right triangles: side a opposite angle A, side b opposite angle B, and side c opposite angle C.

A triangle that is not a right triangle is called an **oblique triangle.** The measures of the three sides and the three angles of a triangle can be found if at least one side and any other two measures are known. There are four possible cases.

> ### Solving Oblique Triangles
>
> **1.** One side and two angles are known.
> **2.** Two sides and one angle not included between the two sides are known. (This case may lead to more than one triangle.)
> **3.** Two sides and the angle included between the two sides are known.
> **4.** Three sides are known.

NOTE If we know three angles of a triangle, we cannot find unique side lengths, since AAA assures us only of similarity, not congruence. For example, there are infinitely many triangles ABC with $A = 35°$, $B = 65°$, and $C = 80°$.

The first two cases require the *law of sines,* which is discussed in this section. The last two cases require the *law of cosines,* discussed in the next section.

To derive the law of sines, we start with an oblique triangle, such as the acute triangle in Figure 1(a) or the obtuse triangle in Figure 1(b). The following discussion applies to both triangles. First, construct the perpendicular from B to side AC or its extension. Let h be the length of this perpendicular. Then c is the hypotenuse of right triangle ADB, and a is the hypotenuse of right triangle BDC. By results from an earlier chapter,

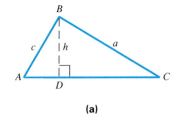

(a)

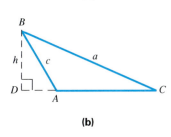

(b)

FIGURE 1

in triangle ADB, $\qquad \sin A = \dfrac{h}{c}$ or $h = c \sin A$,

in triangle BDC, $\qquad \sin C = \dfrac{h}{a}$ or $h = a \sin C$.

Since $h = c \sin A$ and $h = a \sin C$,

$$a \sin C = c \sin A,$$

or, upon dividing both sides by $\sin A \sin C$,

$$\frac{a}{\sin A} = \frac{c}{\sin C}.$$

In a similar way, by constructing the perpendiculars from other vertices, it can be shown that

$$\frac{a}{\sin A} = \frac{b}{\sin B} \quad \text{and} \quad \frac{b}{\sin B} = \frac{c}{\sin C}.$$

This discussion proves the following theorem, called the **law of sines.**

Law of Sines

In any triangle ABC, with sides a, b, and c,

$$\frac{a}{\sin A} = \frac{b}{\sin B} = \frac{c}{\sin C} \qquad \text{or} \qquad \frac{\sin A}{a} = \frac{\sin B}{b} = \frac{\sin C}{c}.$$

In some cases, the second form is easier to use.

If two angles and the side opposite one of the angles are known, the law of sines can be used directly to solve for the side opposite the other known angle. The triangle can then be solved completely, as shown in the first example.

◀EXAMPLE 1
Using the law of sines to solve a triangle

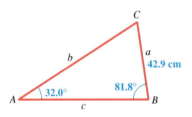

FIGURE 2

Solve triangle ABC if $A = 32.0°$, $B = 81.8°$, and $a = 42.9$ centimeters. See Figure 2.

Start by drawing a triangle, roughly to scale, and labeling the given parts as in Figure 2. Since the values of A, B, and a are known, use the part of the law of sines that involves these variables.

$$\frac{a}{\sin A} = \frac{b}{\sin B}$$

Substituting the known values gives

$$\frac{42.9}{\sin 32.0°} = \frac{b}{\sin 81.8°}.$$

Multiply both sides of the equation by $\sin 81.8°$.

$$b = \frac{42.9 \sin 81.8°}{\sin 32.0°}$$

When using a calculator to find b, keep intermediate answers in the calculator until the final result is found. Then round to the proper number of significant digits. In this case, find $\sin 81.8°$, and then multiply that number by 42.9. Keep the result in the calculator while you find $\sin 32.0°$, and then divide. Since the given information is accurate to three significant digits, round the value of b to get

$$b = 80.1 \text{ centimeters.}$$

Find C from the fact that the sum of the angles of any triangle is 180°.

$$A + B + C = 180°$$
$$C = 180° - A - B$$
$$C = 180° - 32.0° - 81.8°$$
$$= 66.2°$$

Now use the law of sines again to find c. (Why does the Pythagorean theorem not apply?)

$$\frac{a}{\sin A} = \frac{c}{\sin C}$$

$$\frac{\mathbf{42.9}}{\sin \mathbf{32.0°}} = \frac{c}{\sin \mathbf{66.2°}}$$

$$c = \frac{42.9 \sin 66.2°}{\sin 32.0°}$$

$$c = \mathbf{74.1} \text{ centimeters} \quad \blacktriangleright$$

CAUTION In applications of oblique triangles, such as the one in Example 2, a correctly labeled sketch is essential in order to set up the correct equation.

◀ **EXAMPLE 2**
Using the law of sines in an application

Tri Nguyen wishes to measure the distance across the Big Muddy River. See Figure 3. He finds that $C = 112.90°, A = 31.10°$, and $b = 347.6$ feet. Find the required distance.

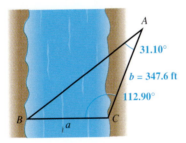

FIGURE 3

```
1ST ANGLE
?112.90
2ND ANGLE
?31.10
COMMON SIDE
?347.6
```

```
OTHER ANGLE
          36.0
OTHER SIDES
          544.8
          305.5
          Done
```

The triangle in Example 2 is completely solved above, using a program. Programs such as this one and the ones illustrated later in this chapter are available from user's groups and the manufacturer.

To use the law of sines, one side and the angle opposite it must be known. Since the only side whose length is given is b, angle B must be found before the law of sines can be used.

$$B = 180° - A - C$$
$$= 180° - \mathbf{31.10°} - \mathbf{112.90°} = 36.00°$$

Now use the form of the law of sines involving A, B, and b to find a.

$$\frac{a}{\sin A} = \frac{b}{\sin B}$$

$$\frac{a}{\sin \mathbf{31.10°}} = \frac{\mathbf{347.6}}{\sin \mathbf{36.00°}} \qquad \text{Substitute.}$$

$$a = \frac{347.6 \sin 31.10°}{\sin 36.00°} \qquad \text{Multiply by } \sin 31.10°.$$

$$a = \mathbf{305.5} \text{ feet} \qquad \text{Use a calculator.} \quad \blacktriangleright$$

If we are given the lengths of two sides and the angle opposite one of them, it is possible that 0, 1, or 2 such triangles exist. (Recall that there is no "SSA" congruence theorem.) To illustrate, suppose that the measure of acute angle A of triangle ABC, the length of side a, and the length of side b are given. Draw angle A having a terminal side of length b. Now draw a side of length a opposite angle A. The following chart shows that there might be more than one possible outcome. This situation is called the **ambiguous case of the law of sines.**

Number of Possible Triangles	Sketch	Condition Necessary for Case to Hold
0	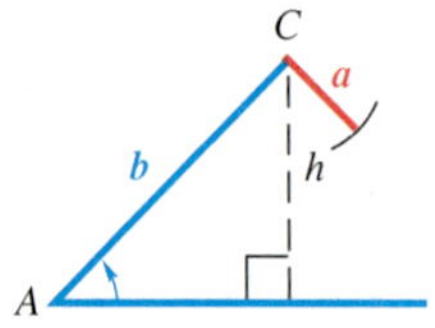	$a < h$ ($h = b \sin A$)
1	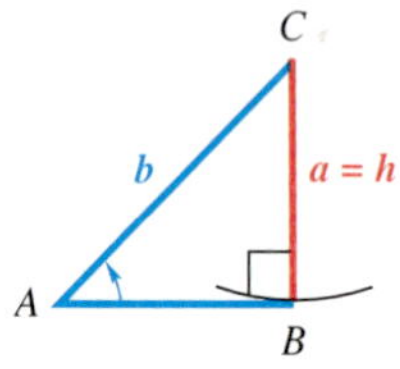	$a = h$
1	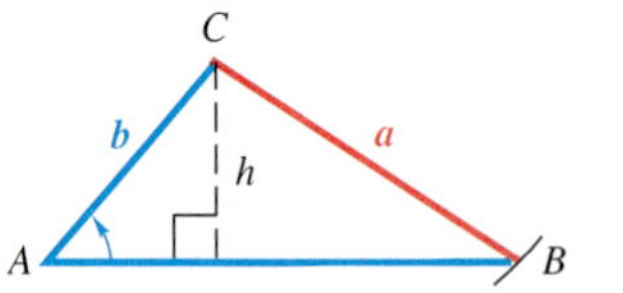	$a > b$
2	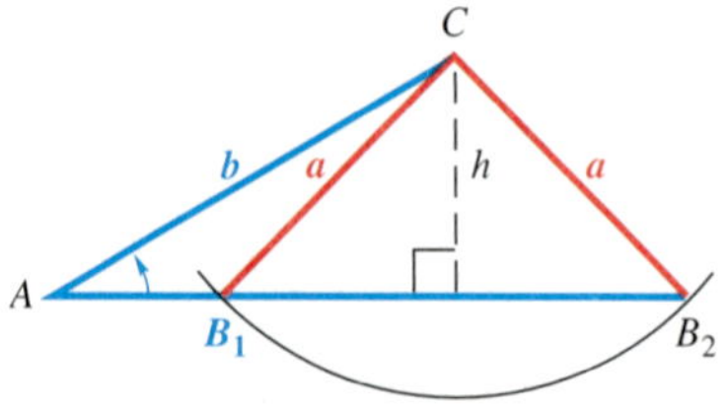	$b > a > h$

If angle A is obtuse, there are two possible outcomes, as shown in the next chart.

Number of Possible Triangles	Sketch	Condition Necessary for Case to Hold
0	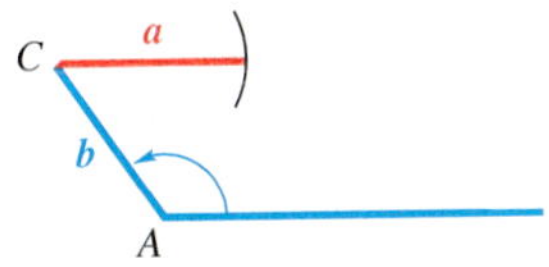	$a \leq b$
1	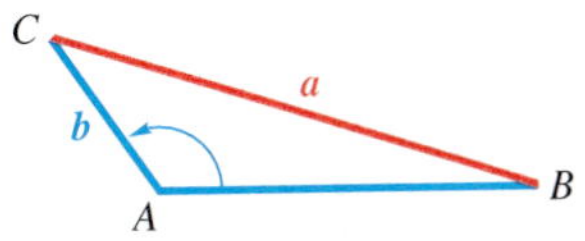	$a > b$

Applying the law of sines to the values of a, b, and A, with some basic properties of geometry and trigonometry, will allow us to determine which case applies. The following basic facts should be kept in mind.

1. For any angle θ of a triangle, $0 < \sin \theta \leq 1$. If $\sin \theta = 1$, then $\theta = 90°$ and the triangle is a right triangle.
2. $\sin \theta = \sin(180° - \theta)$ (That is, supplementary angles have the same sine value.)
3. The smallest angle is opposite the shortest side, the largest angle is opposite the longest side, and the middle-valued angle is opposite the medium side.

The law of sines is a good example of a formula that can be programmed into a graphing calculator. See your owner's manual for guidelines to programming. Remember to provide for the ambiguous case in your program.

◀EXAMPLE 3
Solving a triangle using the
law of sines (no such triangle)

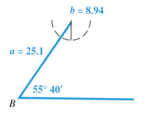

FIGURE 4

Solve triangle ABC if $B = 55° 40'$, $b = 8.94$ meters, and $a = 25.1$ meters.
   Since we are given B, b, and a, use the law of sines to find A, the larger of the unknown angles.

$$\frac{\sin A}{a} = \frac{\sin B}{b}$$

Substitute the given values.

$$\frac{\sin A}{25.1} = \frac{\sin 55° 40'}{8.94}$$

$$\sin A = \frac{25.1 \sin 55° 40'}{8.94}$$

$$\sin A = 2.3184379$$

Since $\sin A$ cannot be greater than 1, there can be no such angle A and thus no triangle with the given information. An attempt to sketch such a triangle leads to the situation seen in Figure 4. ◗

◀EXAMPLE 4
Solving a triangle using the
law of sines (two triangles)

Solve triangle ABC if $A = 55.3°$, $a = 22.8$ feet, and $b = 24.9$ feet.
   To begin, use the law of sines to find angle B.

$$\frac{a}{\sin A} = \frac{b}{\sin B}$$

$$\frac{22.8}{\sin 55.3°} = \frac{24.9}{\sin B}$$

$$\sin B = \frac{24.9 \sin 55.3°}{22.8}$$

$$\sin B = .8978678$$

Since $\sin B = .8978678$, to the nearest tenth we have one value of B as

$$B = 63.9°.$$

Supplementary angles have the same sine value, so another *possible* value of B is

$$B = 180° - 63.9° = 116.1°.$$

To see if $B = 116.1°$ is a valid possibility, simply add $116.1°$ to the measure of the given value of A, $55.3°$. Since $116.1° + 55.3° = 171.4°$, and this sum is less than $180°$ (the sum of the angles of a triangle), we know that it is a valid angle measure for this triangle.
   To keep track of these two different values of B, let

$$B_1 = 116.1° \quad \text{and} \quad B_2 = 63.9°.$$

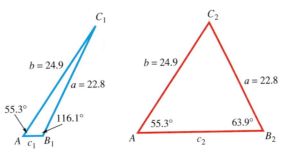

FIGURE 5

Now separately solve triangles AB_1C_1 and AB_2C_2 shown in Figure 5. Let us begin with AB_1C_1. Find C_1 first.

$$C_1 = 180° - A - B_1 = 8.6°.$$

Now, use the law of sines to find c_1.

$$\frac{a}{\sin A} = \frac{c_1}{\sin C_1}$$

$$\frac{22.8}{\sin 55.3°} = \frac{c_1}{\sin 8.6°}$$

$$c_1 = \frac{22.8 \sin 8.6°}{\sin 55.3°}$$

$$c_1 = 4.15 \text{ feet}$$

To solve triangle AB_2C_2, first find C_2.

$$C_2 = 180° - A - B_2 = 60.8°$$

By the law of sines,

$$\frac{22.8}{\sin 55.3°} = \frac{c_2}{\sin 60.8°}$$

$$c_2 = \frac{22.8 \sin 60.8°}{\sin 55.3°}$$

$$c_2 = 24.2 \text{ feet.} \quad \blacktriangleright$$

CAUTION   When solving a triangle using the type of data given in Example 4, do not forget to find the possible obtuse angle. The inverse sine function of the calculator will not give it directly. As we shall see in the next example, it is possible that the obtuse angle will not be a valid measure.

EXAMPLE 5
Solving a triangle using the
law of sines (one triangle)

Solve triangle ABC given $A = 43.5°$, $a = 10.7$ inches, and $b = 7.2$ inches.

To find angle B use the law of sines.

$$\frac{\sin B}{7.2} = \frac{\sin 43.5°}{10.7}$$

$$\sin B = \frac{7.2 \sin 43.5°}{10.7} = .46319186$$

The inverse sine function of the calculator gives us

$$B = 27.6°$$

as the acute angle. The other possible value of B is $180° - 27.6° = 152.4°$. However, when we add this possible obtuse angle to the given angle $A = 43.5°$, we get $152.4° + 43.5° = 195.9°$, which is greater than $180°$. So there can be only one triangle. (Notice that this is the third case listed in the chart earlier in this section.) Then angle $C = 180° - 27.6° - 43.5° = 108.9°$, and side c can be found with the law of sines.

$$\frac{c}{\sin 108.9°} = \frac{10.7}{\sin 43.5°}$$

$$c = \frac{10.7 \sin 108.9°}{\sin 43.5°}$$

$$c = 14.7 \text{ inches} \quad \blacktriangleright$$

EXAMPLE 6
Analyzing data involving an
obtuse angle

Without using the law of sines, explain why the data

$$A = 104°, \quad a = 26.8 \text{ meters}, \quad b = 31.3 \text{ meters}$$

cannot be valid for a triangle ABC.

Since A is an obtuse angle, the largest side of the triangle must be a, the side opposite A. However, we are given $b > a$, which is impossible if A is obtuse. Therefore, no such triangle ABC exists. $\quad \blacktriangleright$

CONNECTIONS As mentioned in the Chapter Opener, aerial photography has become important in many situations. Sometimes it is helpful to use coordinates of ordered pairs to determine distances on the ground. Suppose we assign coordinates as shown in the figure. If an object's photographic coordinates are (x, y), then its ground coordinates (X, Y) in feet can be computed using the following formulas.

$$X = \frac{(a - h)x}{f \sec \theta - y \sin \theta}, \qquad Y = \frac{(a - h)y \cos \theta}{f \sec \theta - y \sin \theta}$$

Here, f is the focal length of the camera in inches, a is the altitude in feet of the airplane, and h is the elevation in feet of the object. Suppose that a house

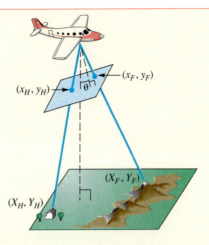

has photographic coordinates of $(x_H, y_H) = (.9, 3.5)$ with an elevation of 150 feet, while a nearby forest fire has photographic coordinates $(x_F, y_F) = (2.1, -2.4)$ and is at an elevation of 690 feet. If the photograph was taken at 7400 feet by a camera with a focal length of 6 inches and a tilt angle $\theta = 4.1°$, we can use these formulas to find the distance on the ground (in feet) between the house and the fire.*

FOR DISCUSSION OR WRITING

1. Use the formulas to find the ground coordinates of the house and the fire.
2. Use the distance formula given in Section 3.1 to find the required distance on the ground to the nearest tenth of a foot.

AREA The method used to derive the law of sines can also be used to derive a useful formula to find the area of a triangle. A familiar formula for the area of a triangle is $K = (1/2)bh$, where K represents the area, b the base, and h the height. This formula cannot always be used easily, since in practice h is often unknown. To find a more useful formula, refer to acute triangle ABC in Figure 6(a) or obtuse triangle ABC in Figure 6(b).

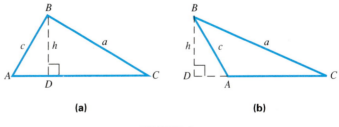

(a)   (b)

FIGURE 6

---

* Moffitt, F., *Photogrammetry*, International Textbook Company, Scranton, Pennsylvania, 1967.

A perpendicular has been drawn from B to the base of the triangle (or the extension of the base). This perpendicular forms two right triangles. Using triangle ABD,

$$\sin A = \frac{h}{c},$$

or

$$h = c \sin A.$$

Substituting into the formula $K = (1/2)bh$,

$$K = \frac{1}{2}b(c \sin A)$$

or

$$K = \frac{1}{2}bc \sin A.$$

Any other pair of sides and the angle between them could have been used, as stated in the next theorem.

Area of a Triangle

In any triangle ABC, the area K is given by any of the following formulas:

$$K = \frac{1}{2}bc \sin A, \qquad K = \frac{1}{2}ab \sin C, \qquad K = \frac{1}{2}ac \sin B.$$

That is, the area is given by half the product of the lengths of two sides and the sine of the angle included between them.

◖**EXAMPLE 7**
Finding the area of a triangle

Find the area of triangle MNP if $m = 29.7$ meters, $n = 53.9$ meters, and $P = 28.7°$.

By the last result, the area of the triangle is

$$\frac{1}{2}(29.7)(53.9) \sin 28.7° = 384 \text{ square meters.} \quad ▶$$

8.1 Exercises ▼▼▼▼▼▼▼▼▼▼▼▼▼▼▼▼▼▼▼▼▼▼▼▼▼▼▼▼▼▼▼▼▼▼▼▼▼▼▼

In Exercises 1 and 2, solve the equation for x.

1. $\dfrac{6}{\sin 30°} = \dfrac{x}{\sin 45°}$

2. $\dfrac{\sqrt{2}}{\sin 45°} = \dfrac{x}{\sin 60°}$

In Exercises 3 and 4, find the length of side a. Do not use a calculator.

3.

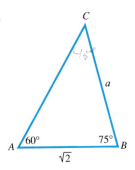

4.

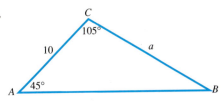

In Exercises 5–16, determine the remaining sides and angles. See Example 1.

5.

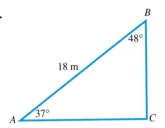

6.

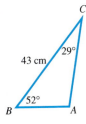

7.

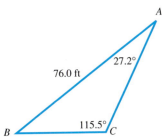

8.

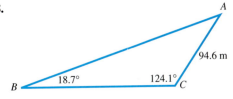

9. $A = 87.2°$, $b = 75.9$ yd, $C = 74.3°$

10. $B = 38° 40'$, $a = 19.7$ cm, $C = 91° 40'$

11. $B = 20° 50'$, $C = 103° 10'$, $AC = 132$ ft

12. $A = 35.3°$, $B = 52.8°$, $AC = 675$ ft

13. $A = 39.70°$, $C = 30.35°$, $b = 39.74$ m

14. $C = 71.83°$, $B = 42.57°$, $a = 2.614$ cm

15. $B = 42.88°$, $C = 102.40°$, $b = 3974$ ft

16. $A = 18.75°$, $B = 51.53°$, $c = 2798$ yd

17. Explain why the law of sines cannot be used to solve a triangle if we are given the lengths of the three sides of a triangle.

18. In Example 1, we ask the question, "Why does the Pythagorean theorem not apply?" Answer this question.

19. State the law of sines in your own words.

20. Kathleen Burk, a perceptive trigonometry student, makes the statement "If we know *any* two angles and one side of a triangle, then the triangle is uniquely determined." Is this a valid statement? Explain, referring to the congruence axioms given in this section.

21. Can one portion of the law of sines be written as $\dfrac{a}{b} = \dfrac{\sin A}{\sin B}$?

22. If a is twice as long as b, is A twice as large as B?

In Exercises 23 and 24, solve the equation for θ, where θ is the measure of an angle in a triangle. Give the possible values of θ.

23. $\dfrac{\sqrt{2}}{\sin 30°} = \dfrac{2}{\sin \theta°}$

24. $\dfrac{\sqrt{6}}{\sin 45°} = \dfrac{3}{\sin \theta°}$

25. In the left figure below, a line of length *h* is to be drawn from the point (3, 4) to the positive *x*-axis in order to form a triangle. For what value(s) of *h* can you draw the following?
(a) Two triangles   **(b)** Exactly one triangle   **(c)** No triangle

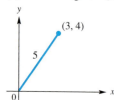 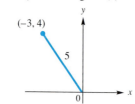

26. In the right figure above, a line of length *h* is to be drawn from the point (−3, 4) to the positive *x*-axis in order to form a triangle. For what value(s) of *h* can you draw the following?
(a) Two triangles   **(b)** Exactly one triangle   **(c)** No triangle

Determine the number of triangles possible with the given parts.

27. $a = 50, b = 26, A = 95°$

28. $b = 60, a = 82, B = 100°$

29. $a = 31, b = 26, B = 48°$

30. $a = 35, b = 30, A = 40°$

In Exercises 31 and 32, find angle B. Do not use a calculator.

31.

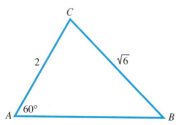

32.

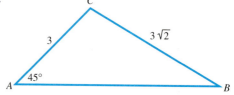

Find the unknown angles in triangle ABC for each triangle that exists. See Examples 3–6.

33. $A = 29.7°, b = 41.5$ ft, $a = 27.2$ ft

34. $B = 48.2°, a = 890$ cm, $b = 697$ cm

35. $C = 41° 20', b = 25.9$ m, $c = 38.4$ m

36. $B = 48° 50', a = 3850$ in, $b = 4730$ in

37. $B = 74.3°, a = 859$ m, $b = 783$ m

38. $C = 82.2°, a = 10.9$ km, $c = 7.62$ km

Solve each triangle that exists. See Examples 3–6.

39. $A = 42.5°, a = 15.6$ ft, $b = 8.14$ ft

40. $C = 52.3°, a = 32.5$ yd, $c = 59.8$ yd

41. $B = 72.2°, b = 78.3$ m, $c = 145$ m

42. $C = 68.5°, c = 258$ cm, $b = 386$ cm

43. $A = 38° 40', a = 9.72$ km, $b = 11.8$ km

44. $C = 29° 50', a = 8.61$ m, $c = 5.21$ m

45. $B = 39.68°, a = 29.81$ m, $b = 23.76$ m

46. $A = 51.20°, c = 7986$ cm, $a = 7208$ cm

47. Apply the law of sines to the following: $a = \sqrt{5}$, $c = 2\sqrt{5}$, $A = 30°$. What is the value of sin C? What is the measure of C? Based on its angle measures, what kind of triangle is triangle ABC?

48. Apply the law of sines to the data given in Example 6. Describe in your own words what happens when you try to find the measure of angle B using a calculator.

49. A surveyor reported the following data about a piece of property: "The property is triangular in shape, with dimensions as shown in the figure." Use the law of sines to see whether such a piece of property could exist.

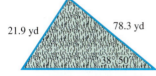

21.9 yd   78.3 yd

38° 50′

Can such a triangle exist?

50. The surveyor tries again: "A second triangular piece of property has dimensions as shown." This time it turns out that the surveyor did not consider every possible case. Use the law of sines to show why.

21.2 yd   26.5 yd

28° 10′

Solve each problem. See Example 2.

51. Three gears are arranged as shown in the figure. Find angle θ.

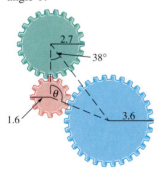

2.7

38°

θ

1.6   3.6

52. Three atoms with atomic radii of 2.0, 3.0, and 4.5 are arranged as in the figure. Find the distance between the centers of atoms A and C.

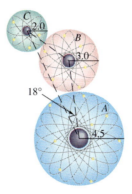

C
2.0
B
3.0
18°
A
4.5

53. The bearing of a lighthouse from a ship was found to be N 37° E. After the ship sailed 2.5 miles due south, the new bearing was N 25° E. Find the distance between the ship and the lighthouse at each location.

54. A balloonist is directly above a straight road 1.5 miles long that joins two villages. She finds that the town closer to her is at an angle of depression of 35° and the farther town is at an angle of depression of 31°. How high above the ground is the balloon? See the figure.

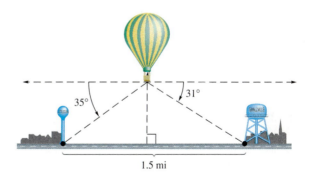

35°   31°

1.5 mi

Find the area of the triangle using the formula $K = (1/2)bh$ and then verify that the formula $K = (1/2)ab \sin C$ gives the same result.

55.

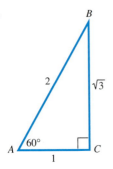

56.

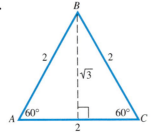

57.

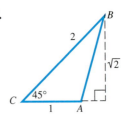

Find the area of each triangle. See Example 7.

58. $A = 42.5°$, $b = 13.6$ m, $c = 10.1$ m

59. $C = 72.2°$, $b = 43.8$ ft, $a = 35.1$ ft

60. $B = 124.5°$, $a = 30.4$ cm, $c = 28.4$ cm

61. $C = 142.7°$, $a = 21.9$ km, $b = 24.6$ km

62. $A = 56.80°$, $b = 32.67$ in, $c = 52.89$ in

63. $A = 34.97°$, $b = 35.29$ m, $c = 28.67$ m

64. A painter is going to apply a special coating to a triangular metal plate on a new building. Two sides measure 16.1 m and 15.2 m. She knows that the angle between these sides is 125°. What is the area of the surface she plans to cover with the coating?

65. A real estate agent wants to find the area of a triangular lot. A surveyor takes measurements and finds that two sides are 52.1 m and 21.3 m, and the angle between them is 42.2°. What is the area of the lot?

▼▼▼▼▼▼▼▼▼▼▼▼ **DISCOVERING CONNECTIONS** (Exercises 66–70) ▼▼▼▼▼▼▼▼▼▼▼▼

In any triangle, the longest side is opposite the largest angle. This result from geometry can be proven with trigonometry. Let us first prove it for acute triangles. (The result will be proven for obtuse triangles in 8.2 Exercises.)

66. Is the graph of the function $y = \sin x$ increasing or decreasing over the interval $0 \le x \le \pi/2$?

67. Suppose angle A is the largest angle of an acute triangle ABC and let B be an angle smaller than A. Explain why $\dfrac{\sin B}{\sin A} < 1$.

68. Solve for b in the law of sines.

69. Use the result of Exercise 68 to show that $b < a$.

70. Use the results of Exercises 66–69 to explain why no triangle ABC satisfies $A = 83°$, $B = 56°$, $a = 14$, $b = 20$.

71. Since the moon is a relatively close celestial object, its distance can be measured directly using trigonometry. To find this distance, two different photographs of the moon are taken at precisely the same time in two different locations with a known distance between them. The moon will have a different angle of elevation at each location. On April 29, 1976, at 11:35 A.M., the lunar angles of elevation during a partial solar eclipse at Bochum in upper Germany and at Donaueschingen in lower Germany were measured as 52.6997° and 52.7430°, respectively. The two cities are 398 kilometers apart. Calculate the distance to the moon from Bochum on this day and compare it with the actual value of 406,000 km. Disregard the curvature of Earth in this calculation. (*Source:* Scholosser, W., T. Schmidt-Kaler, and E. Milone, *Challenges of Astronomy,* Springer-Verlag, New York, 1991.)

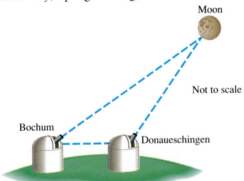

Not to scale

Moon

Bochum

Donaueschingen

72. The distance covered by an aerial photograph is determined by both the focal length of the camera and the tilt of the camera from the perpendicular to the ground. Although the tilt is usually small, both archaeological and Canadian photographs often use larger tilts. A camera lens with a 12-inch focal length will have an angular coverage of 60°. If an aerial photograph is taken with this camera tilted $\theta = 35°$ at an altitude of 5000 feet, calculate the ground distance d in miles that will be shown in this photograph.

(*Sources:* Brooks, R. and Dieter Johannes, *Photoarchaeology,* Dioscorides Press, Portland, Oregon, 1990; Moffitt, F., *Photogrammetry,* International Textbook Company, Scranton, Pennsylvania, 1967.)

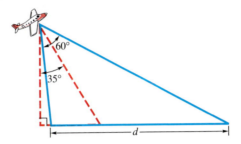

60°

35°

d

73. (Refer to the previous exercise.) A camera lens with a 6-inch focal length has an angular coverage of 86°. Suppose an aerial photograph is taken vertically with no tilt at an altitude of 3500 feet over ground with an increasing slope of 5° as shown in the figure. Calculate the ground distance *CB* that would appear in the resulting photograph. (*Source:* Moffitt, F., *Photogrammetry,* International Textbook Company, Scranton, Pennsylvania, 1967.)

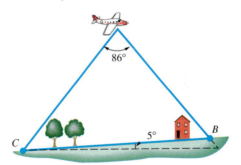

86°

C

5°

B

74. Repeat the previous exercise if the camera lens has an 8.25-inch focal length with an angular coverage of 72°. Why do cameras used in aerial photography usually have shorter focal lengths?

8.2 The Law of Cosines ▼▼▼

As mentioned in Section 8.1, if we are given two sides and the included angle or three sides of a triangle, a unique triangle is formed. These are the SAS and SSS cases, respectively. In these cases, however, we cannot begin the solution of the triangle by using the law of sines because we are not given a side and the angle opposite it. Both of these cases require the use of the law of cosines, introduced in this section.

It will be helpful to remember the following property of triangles when applying the law of cosines.

> In any triangle, the sum of the lengths of any two sides must be greater than the length of the remaining side.

For example, it would be impossible to construct a triangle with sides of lengths 3, 4, and 10. See Figure 7.

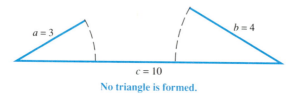

No triangle is formed.

FIGURE 7

To derive the law of cosines, let *ABC* be any oblique triangle. Choose a coordinate system so that vertex *B* is at the origin and side *BC* is along the positive *x*-axis. See Figure 8.

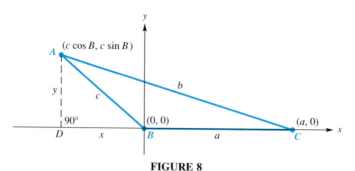

FIGURE 8

Let (x, y) be the coordinates of vertex *A* of the triangle. Verify that for angle *B*, whether obtuse or acute,

$$\sin B = \frac{y}{c} \quad \text{and} \quad \cos B = \frac{x}{c}.$$

(Here we assume that x is negative if B is obtuse.) From these results

$$y = c \sin B \quad \text{and} \quad x = c \cos B,$$

so that the coordinates of point A become

$$(c \cos B, c \sin B).$$

Point C has coordinates $(a, 0)$, and AC has length b. By the distance formula,

$$b = \sqrt{(c \cos B - a)^2 + (c \sin B)^2}.$$

Squaring both sides and simplifying gives

$$
\begin{aligned}
b^2 &= (c \cos B - a)^2 + (c \sin B)^2 \\
&= c^2 \cos^2 B - 2ac \cos B + a^2 + c^2 \sin^2 B \\
&= a^2 + c^2(\cos^2 B + \sin^2 B) - 2ac \cos B \\
&= a^2 + c^2(1) - 2ac \cos B \\
&= a^2 + c^2 - 2ac \cos B.
\end{aligned}
$$

This result is one form of the law of cosines. In the work above, we could just as easily have placed A or C at the origin. This would have given the same result, but with the variables rearranged. These various forms of the law of cosines are summarized in the following theorem.

Law of Cosines

In any triangle ABC, with sides a, b, and c,

$$a^2 = b^2 + c^2 - 2bc \cos A$$
$$b^2 = a^2 + c^2 - 2ac \cos B$$
$$c^2 = a^2 + b^2 - 2ab \cos C.$$

The law of cosines says that the square of a side of a triangle is equal to the sum of the squares of the other two sides, minus twice the product of those two sides and the cosine of the angle included between them.

NOTE   If we let $C = 90°$ in the third form of the law of cosines given above, we have $\cos C = \cos 90° = 0$, and the formula becomes

$$c^2 = a^2 + b^2,$$

the familiar equation of the Pythagorean theorem. Thus, the Pythagorean theorem is a special case of the law of cosines.

The first example shows how the law of cosines can be used to solve an applied problem.

EXAMPLE 1
Using the law of cosines in an application

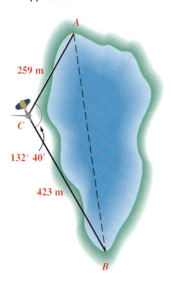

259 m

C

132° 40′

423 m

B

FIGURE 9

A surveyor wishes to find the distance between two inaccessible points *A* and *B* on opposite sides of a lake. While standing at point *C*, she finds that *AC* = 259 meters, *BC* = 423 meters, and angle *ACB* measures 132° 40′. Find the distance *AB*. See Figure 9.

The law of cosines can be used here, since we know the lengths of two sides of the triangle and the measure of the included angle.

$$AB^2 = 259^2 + 423^2 - 2(259)(423) \cos 132° \, 40'$$

$$AB^2 = 394,510.6 \qquad \text{Use a calculator.}$$

$$AB \approx 628 \qquad \text{Take the square root and round to 3 significant digits.}$$

The distance between the points is approximately 628 meters. ▸

The law of cosines is another useful formula to program into a graphing calculator. For the SSS case, only one of the three forms need be programmed. Be sure to consider the SAS case also.

EXAMPLE 2
Using the law of cosines to solve a triangle (SAS)

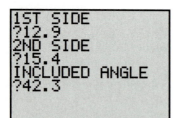

```
1ST SIDE
?12.9
2ND SIDE
?15.4
INCLUDED ANGLE
?42.3
```

```
OTHER ANGLES
      81.71255972
      55.98744028
OTHER SIDE
      10.47376596
            Done
```

A program can be used to solve the triangle in Example 2. (There are slight discrepancies due to roundoff error.)

Solve triangle *ABC* if *A* = 42.3°, *b* = 12.9 meters, and *c* = 15.4 meters. See Figure 10.

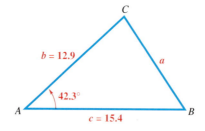

C

b = 12.9

a

42.3°

A

c = 15.4

B

FIGURE 10

Start by finding *a* with the law of cosines.

$$a^2 = b^2 + c^2 - 2bc \cos A$$

$$a^2 = 12.9^2 + 15.4^2 - 2(12.9)(15.4) \cos 42.3°$$

$$a^2 = 109.7$$

$$a = 10.5 \text{ meters}$$

We now must find the measures of angles *B* and *C*. There are several approaches that can be used at this point. Let us use the law of sines to find one of these angles. Of the two remaining angles, *B* must be the smaller since it is opposite the shorter of the two sides b and c. Therefore, it cannot be obtuse, and we will avoid any ambiguity when we find its sine.

$$\frac{\sin \mathbf{42.3°}}{\mathbf{10.5}} = \frac{\sin B}{\mathbf{12.9}}$$

$$\sin B = \frac{12.9 \sin 42.3°}{10.5}$$

$$B = 55.8° \qquad \text{Use the inverse sine function}$$
$$\text{of a calculator.}$$

The easiest way to find C is to subtract the sum of A and B from $180°$.

$$C = 180° - (A + B) = 81.9°. \quad \blacktriangleright$$

CAUTION   Had we chosen to use the law of sines to find C rather than B in Example 2, we would not have known whether C equals $81.9°$ or its supplement, $98.1°$.

EXAMPLE 3
Using the law of cosines to solve a triangle (SSS)

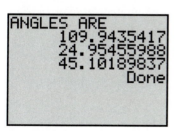

A program can be used to solve the triangle in Example 3.

Solve triangle ABC if $a = 9.47$ feet, $b = 15.9$ feet, and $c = 21.1$ feet.
   We are given the lengths of three sides of the triangle, so we may use the law of cosines to solve for any angle of the triangle. Let us solve for C, the largest angle, using the law of cosines. We will be able to tell if C is obtuse if $\cos C < 0$. Use the form of the law of cosines that involves C.

$$c^2 = a^2 + b^2 - 2ab \cos C,$$

or
$$\cos C = \frac{a^2 + b^2 - c^2}{2ab}.$$

Inserting the given values leads to

$$\cos C = \frac{(9.47)^2 + (15.9)^2 - (21.1)^2}{2(9.47)(15.9)}$$

$$\cos C = -.34109402. \qquad \text{Use a calculator.}$$

Using the inverse cosine function of the calculator, we get the obtuse angle C.

$$C = 109.9°$$

We can use either the law of sines or the law of cosines to find $B = 45.1°$. (Verify this.) Since $A = 180° - B - C$,

$$A = 25.0°. \quad \blacktriangleright$$

As shown in this section and the previous one, four possible cases can occur when solving an oblique triangle. These cases are summarized in the chart that follows, along with a suggested procedure for solving in each case. There are other procedures that work, but we give the one that is most efficient. In all four cases, it is assumed that the given information actually produces a triangle.

Case	*Suggested Procedure for Solving*
One side and two angles are known. (SAA or ASA)	**1.** Find the remaining angle using the angle sum formula ($A + B + C = 180°$). **2.** Find the remaining sides using the law of sines.
Two sides and one angle (not included between the two sides) are known. (SSA)	*Be aware of the ambiguous case; there may be two triangles.* **1.** Find an angle using the law of sines. **2.** Find the remaining angle using the angle sum formula. **3.** Find the remaining side using the law of sines. *If two triangles exist, repeat Steps 1, 2, and 3.*
Two sides and the included angle are known. (SAS)	**1.** Find the third side using the law of cosines. **2.** Find the smaller of the two remaining angles using the law of sines. **3.** Find the remaining angle using the angle sum formula.
Three sides are known. (SSS)	**1.** Find the largest angle using the law of cosines. **2.** Find either remaining angle using the law of sines. **3.** Find the remaining angle using the angle sum formula.

AREA The law of cosines can be used to derive a formula for the area of a triangle when only the lengths of the three sides are known. This formula is known as Heron's formula, named after the Greek mathematician Heron of Alexandria, who lived around A.D. 75. It is found in his work *Metrica*.

> ### Heron's Area Formula
>
> If a triangle has sides of lengths a, b, and c, and if the **semiperimeter** is
>
> $$s = \frac{1}{2}(a + b + c),$$
>
> then the area of the triangle is
>
> $$K = \sqrt{s(s - a)(s - b)(s - c)}.$$

A proof of Heron's formula is suggested in Exercises 55–60.

EXAMPLE 4
Finding an area using Heron's formula

```
ENTER SIDES
A
?29.7
B
?42.3
C
?38.4
```

```
AREA =
    552.3179085
         Done
```

A program for Heron's formula can be used to find the area of a triangle. The screens above support the result of Example 4.

Find the area of the triangle having sides of lengths $a = 29.7$ feet, $b = 42.3$ feet, and $c = 38.4$ feet.

To use Heron's area formula, first find s.

$$s = \frac{1}{2}(a + b + c)$$

$$s = \frac{1}{2}(29.7 + 42.3 + 38.4)$$

$$= 55.2$$

The area is

$$K = \sqrt{s(s - a)(s - b)(s - c)}$$
$$K = \sqrt{55.2(55.2 - 29.7)(55.2 - 42.3)(55.2 - 38.4)}$$
$$= \sqrt{55.2(25.5)(12.9)(16.8)}$$
$$= 552 \text{ square feet.} \quad \blacktriangleright$$

CONNECTIONS We have introduced two new formulas for the area of a triangle in this chapter. You should now be able to find the area K of a triangle using one of three formulas:

(a) $K = \frac{1}{2}bh$

(b) $K = \frac{1}{2}ab \sin C$ (or $K = \frac{1}{2}ac \sin B$ or $K = \frac{1}{2}bc \sin A$)

(c) $K = \sqrt{s(s - a)(s - b)(s - c)}$.

Another area formula can be used when the coordinates of the vertices of the triangle are given. If the vertices are the ordered pairs (x_1, y_1), (x_2, y_2), and (x_3, y_3), then

$$K = \left| \frac{1}{2}(x_2y_3 - x_3y_2 - x_1y_3 + x_3y_1 + x_1y_2 - x_2y_1) \right|.$$

FOR DISCUSSION OR WRITING
Consider triangle PQR with vertices $P(2, 5)$, $Q(-1, 3)$, and $R(4, 0)$.

1. Find the area of the triangle using the new formula just introduced.
2. Find the area of the triangle using (c) above. Use the distance formula to find the lengths of the three sides.
3. Find the area of the triangle using (b) above. First use the law of cosines to find the measure of an angle.

8.2 Exercises ▼▼▼▼▼▼▼▼▼▼▼▼▼▼▼▼▼▼▼▼▼▼▼▼▼▼▼▼▼▼▼▼▼▼▼▼

1. In your own words, describe the types of problems that require the law of sines and those that require the law of cosines.

2. Can the law of cosines be used to solve any triangle for which two angles and a side are known? Explain your answer.

In Exercises 3 and 4, find the length of the remaining side. Do not use a calculator.

3.
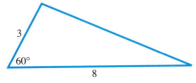

4.
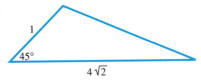

In Exercises 5 and 6, find the measure of θ. Do not use a calculator.

5.

6.
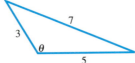

Solve each triangle. See Examples 2 and 3.

7. $C = 28.3°$, $b = 5.71$ in, $a = 4.21$ in

8. $A = 41.4°$, $b = 2.78$ yd, $c = 3.92$ yd

9. $C = 45.6°$, $b = 8.94$ m, $a = 7.23$ m

10. $A = 67.3°$, $b = 37.9$ km, $c = 40.8$ km

11. $A = 80°\ 40'$, $b = 143$ cm, $c = 89.6$ cm

12. $C = 72°\ 40'$, $a = 327$ ft, $b = 251$ ft

13. $B = 74.80°$, $a = 8.919$ in, $c = 6.427$ in

14. $C = 59.70°$, $a = 3.725$ mi, $b = 4.698$ mi

15. $A = 112.8°$, $b = 6.28$ m, $c = 12.2$ m

16. $B = 168.2°$, $a = 15.1$ cm, $c = 19.2$ cm

Find all of the angles in each triangle. See Example 3.

17. $a = 3.0$ ft, $b = 5.0$ ft, $c = 6.0$ ft

18. $a = 4.0$ ft, $b = 5.0$ ft, $c = 8.0$ ft

19. $a = 9.3$ cm, $b = 5.7$ cm, $c = 8.2$ cm

20. $a = 28$ ft, $b = 47$ ft, $c = 58$ ft

21. $a = 42.9$ m, $b = 37.6$ m, $c = 62.7$ m

22. $a = 187$ yd, $b = 214$ yd, $c = 325$ yd

23. $AB = 1240$ ft, $AC = 876$ ft, $BC = 918$ ft

24. $AB = 298$ m, $AC = 421$ m, $BC = 324$ m

Find the measure of the angle θ to two decimal places.

25.
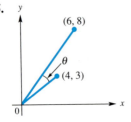

26.
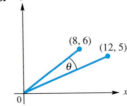

Find the area of the triangle using the formula K = (1/2)bh and then verify that Heron's formula gives the same result.

27.

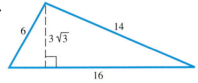

28.

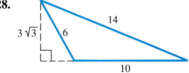

Find the area of each triangle. See Example 4.

29. $a = 12$ m, $b = 16$ m, $c = 25$ m

30. $a = 22$ in, $b = 45$ in, $c = 31$ in

31. $a = 154$ cm, $b = 179$ cm, $c = 183$ cm

32. $a = 25.4$ yd, $b = 38.2$ yd, $c = 19.8$ yd

33. $a = 76.3$ ft, $b = 109$ ft, $c = 98.8$ ft

34. $a = 15.89$ in, $b = 21.74$ in, $c = 10.92$ in

Solve the problem.

35. A painter needs to cover a triangular region 75 m by 68 m by 85 m. A can of paint covers 75 sq m of area. How many cans (to the next higher number of cans) will be needed?

36. How many cans of paint would be needed in Exercise 35 if the region were 8.2 m by 9.4 m by 3.8 m?

37. Find the area of the Bermuda Triangle, if the sides of the triangle have the approximate lengths 850 miles, 925 miles, and 1300 miles.

38. Find the area of a triangle in a rectangular coordinate plane whose vertices are $(0, 0)$, $(3, 4)$, and $(-8, 6)$ using Heron's area formula.

Solve the problem, using the law of sines or the law of cosines. See Example 1.

39. Points A and B are on opposite sides of Lake Yankee. From a third point, C, the angle between the lines of sight to A and B is $46.3°$. If AC is 350 m long and BC is 286 m long, find AB.

40. The sides of a parallelogram are 4.0 cm and 6.0 cm. One angle is $58°$ while another is $122°$. Find the lengths of the diagonals of the parallelogram.

41. Airports A and B are 450 km apart, on an east-west line. Tom flies in a northeast direction from A to airport C. From C he flies 359 km on a bearing of $128° 40'$ to B. How far is C from A?

42. Two ships leave a harbor together, traveling on courses that have an angle of $135° 40'$ between them. If they each travel 402 mi, how far apart are they?

43. The layout for a child's playhouse in her backyard shows the dimensions given in the figure. Find x.

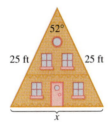

44. A hill slopes at an angle of $12.47°$ with the horizontal. From the base of the hill, the angle of elevation of a 459.0-ft tower at the top of the hill is $35.98°$. How much rope would be required to reach from the top of the tower to the bottom of the hill?

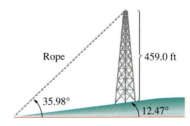

45. A crane with a counterweight is shown in the figure. Find the horizontal distance between points *A* and *B*.

46. A weight is supported by cables attached to both ends of a balance beam, as shown in the figure. What angles are formed between the beam and the cables?

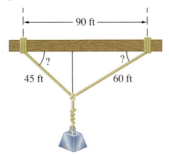

47. A satellite traveling in a circular orbit 1600 km above Earth is due to pass directly over a tracking station at noon. See the figure. Assume that the satellite takes 2 hr to make an orbit and that the radius of Earth is 6400 km. Find the distance between the satellite and the tracking station at 12:03 P.M. (*Source:* Kastner, Bernice, Ph.D., *Spacemathematics*, National Aeronautics and Space Administration (NASA), 1985.)

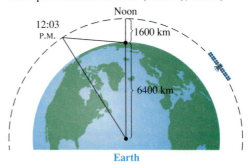

48. Two factories blow their whistles at exactly 5:00. A man hears the two blasts at 3 seconds and 6 seconds after 5:00, respectively. The angle between his lines of sight to the two factories is 42.2°. If sound travels 344 m per sec, how far apart are the factories?

49. A parallelogram has sides of lengths 25.9 cm and 32.5 cm. The longer diagonal has a length of 57.8 cm. Find the angle opposite the diagonal.

50. A person in a plane flying a straight course observes a mountain at a bearing 24.1° to the right of its course. At that time the plane is 7.92 km from the mountain. A short time later, the bearing to the mountain becomes 32.7°. How far is the airplane from the mountain when the second bearing is taken?

To help predict eruptions from the volcano Mauna Loa on the island of Hawaii, scientists keep track of the volcano's movement by using a "super triangle" with vertices on the three volcanoes shown on the map at right. (For example, in a recent year, Mauna Loa moved 6 inches, a result of increasing internal pressure.) Refer to the map to work Exercises 51 and 52.

51. *AB* = 22.47928 mi, *AC* = 28.14276 mi,
   A = 58.56989°;   find *BC*

52. *AB* = 22.47928 mi, *BC* = 25.24983 mi,
   A = 58.56989°;   find *B*

53. Refer to Figure 7. If you attempt to find any angle of a triangle using the values $a = 3$, $b = 4$, and $c = 10$ with the law of cosines, what happens?

54. A familiar saying is "The shortest distance between two points is a straight line." Explain how this relates to the geometric property that states that the sum of the lengths of any two sides of a triangle must be greater than the remaining side.

Use the fact that $\cos A = \dfrac{b^2 + c^2 - a^2}{2bc}$ *to show that each of the following is true.*

55. $1 + \cos A = \dfrac{(b + c + a)(b + c - a)}{2bc}$

56. $1 - \cos A = \dfrac{(a - b + c)(a + b - c)}{2bc}$

57. $\cos \dfrac{A}{2} = \sqrt{\dfrac{s(s - a)}{bc}}$ $\left(Hint: \cos \dfrac{A}{2} = \sqrt{\dfrac{1 + \cos A}{2}}\right)$

58. $\sin \dfrac{A}{2} = \sqrt{\dfrac{(s - b)(s - c)}{bc}}$ $\left(Hint: \sin \dfrac{A}{2} = \sqrt{\dfrac{1 - \cos A}{2}}\right)$

59. The area of a triangle having sides b and c and angle A is given by $(1/2)bc \sin A$. Show that this result can be written as

$$\sqrt{\frac{1}{2}bc(1 + \cos A) \cdot \frac{1}{2}bc(1 - \cos A)}.$$

60. Use the results of Exercises 55–59 to prove Heron's area formula.

▼▼▼▼▼▼▼▼▼▼▼▼▼ **DISCOVERING CONNECTIONS** (Exercises 61–64) ▼▼▼▼▼▼▼▼▼▼▼▼▼

In any triangle, the longest side is opposite the largest angle. This result from geometry was proven for acute triangles in 8.1 Exercises. Let us now prove it for obtuse triangles.

61. Suppose angle A is the largest angle of an obtuse triangle. Explain why $\cos A$ is negative.

62. Consider the law of cosines expression for a and show that $a^2 > b^2 + c^2$.

63. Use the result of Exercise 62 to show that $a > b$ and $a > c$.

64. Use the results of Exercises 61–63 to explain why no triangle ABC satisfies $A = 103°$, $a = 25$, $c = 30$.

8.3 Vectors and Their Applications ▼▼▼

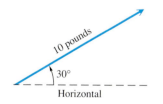

FIGURE 11

We have seen that the measures of all six parts of a triangle can be found, given at least one side and any two other measures. In this section we discuss applications of this work to *vectors.*

Many quantities in mathematics involve magnitudes, such as 45 pounds or 60 miles per hour. These quantities are called **scalars.** Other quantities, called **vector quantities,** involve both magnitude and direction. Typical vector quantities are velocity, acceleration, and force.

A vector quantity is often represented with a directed line segment, which is called a **vector.** The length of the vector represents the **magnitude** of the vector quantity. The direction of the vector is indicated with an arrowhead. For example, the vector in Figure 11 represents a force of 10 pounds applied at an angle of 30° from the horizontal.

The symbol for a vector is often printed in boldface type. When writing vectors by hand, it is customary to use an arrow over the letter or letters. Thus **OP** and $\overrightarrow{OP}$ both represent vector OP. Vectors may be named with either one lowercase or uppercase letter, or two uppercase letters. When two letters are used, the first indicates the **initial point** and the second indicates the **terminal point** of the vector. Knowing these points gives the direction of the vector. For example, vectors **OP** and **PO** in Figure 12 are not the same vector. They have the same magnitude, but they have opposite directions. The magnitude of vector **OP** is written $|\mathbf{OP}|$.

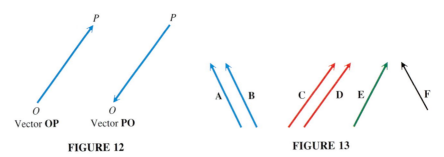

FIGURE 12   FIGURE 13

Two vectors are *equal* if and only if they both have the same direction and the same magnitude. In Figure 13 vectors **A** and **B** are equal, as are vectors **C** and **D.** As Figure 13 shows, equal vectors need not coincide, but they must be parallel. Vectors **A** and **E** are unequal because they do not have the same direction, while **A** $\neq$ **F** because they have different magnitudes, as indicated by their different lengths.

To find the *sum* of two vectors **A** and **B,** written **A** + **B,** we place the initial point of vector **B** at the terminal point of vector **A,** as shown in Figure 14. The vector with the same initial point as **A** and the same terminal point as **B** is the sum **A** + **B**. The sum of two vectors is also a vector.

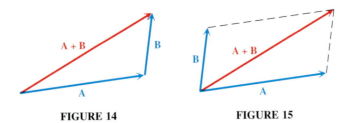

FIGURE 14   FIGURE 15

Another way to find the sum of two vectors is to use the **parallelogram rule.** Place vectors **A** and **B** so that their initial points coincide. Then complete a parallelogram that has **A** and **B** as two adjacent sides. The diagonal of the parallelogram with the same initial point as **A** and **B** is the same vector sum **A** + **B** found by the definition. See Figure 15.

The vector sum **A** + **B** is the **resultant** of vectors **A** and **B**. Each of the vectors **A** and **B** is a **component** of vector **A** + **B**. In many practical applications, such as surveying, it is necessary to break a vector into its **vertical** and **horizontal components.** These components are two vectors, one vertical and one horizontal, whose resultant is the original vector. As shown in Figure 16, vector **OR** is the vertical component and vector **OS** is the horizontal component of **OP**.

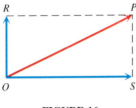

FIGURE 16

For every vector **v** there is a vector −**v** with the same magnitude as **v** but opposite direction. Vector −**v** is the **opposite** of **v**. See Figure 17. The sum of **v** and −**v** has magnitude 0 and is a **zero vector.** As with real numbers, to *subtract* vector **B** from vector **A**, we find the vector sum **A** + (−**B**). See Figure 18.

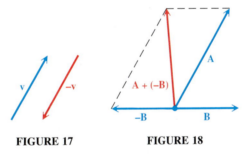

FIGURE 17   **FIGURE 18**

The **scalar product** of a real number (or scalar) k and a vector **u** is the vector $k\mathbf{u}$, which has magnitude $|k|$ times the magnitude of **u**. As shown in Figure 19, $k\mathbf{u}$ has the same direction as **u** if $k > 0$, and the opposite direction if $k < 0$.

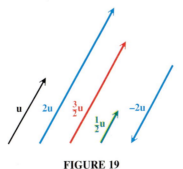

FIGURE 19

EXAMPLE 1
Finding magnitudes of vertical and horizontal components

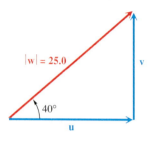

$|\mathbf{w}| = 25.0$

$\mathbf{v}$

$40°$

$\mathbf{u}$

FIGURE 20

Vector $\mathbf{w}$ has magnitude 25.0 and is inclined at an angle of 40° from the horizontal. Find the magnitudes of the horizontal and vertical components of the vector.

In Figure 20, the vertical component is labeled $\mathbf{v}$ and the horizontal component is labeled $\mathbf{u}$. Vectors $\mathbf{u}$, $\mathbf{v}$, and $\mathbf{w}$ form a right triangle. In this right triangle,

$$\sin 40° = \frac{|\mathbf{v}|}{|\mathbf{w}|} = \frac{|\mathbf{v}|}{25.0},$$

and

$$|\mathbf{v}| = 25.0 \sin 40° = 16.1.$$

In the same way,

$$\cos 40° = \frac{|\mathbf{u}|}{25.0},$$

with

$$|\mathbf{u}| = 25.0 \cos 40° = 19.2. \quad \blacktriangleright$$

It is helpful to review some of the properties of parallelograms when studying vectors. A parallelogram is a quadrilateral whose opposite sides are parallel. The opposite sides and opposite angles of a parallelogram are equal, and consecutive angles of a parallelogram are supplementary. The diagonals of a parallelogram bisect each other, but do not necessarily bisect the angles of the parallelogram.

Some of these properties are used in the following example.

EXAMPLE 2
Finding the magnitude of the resultant of two vectors in an application

Two forces of 15 newtons and 22 newtons (a newton is a unit of force used in physics) act at a point in the plane. If the angle between the forces is 100°, find the magnitude of the resultant force.

As shown in Figure 21, a parallelogram that has the forces as adjacent sides can be formed. The angles of the parallelogram adjacent to angle P each measure 80°, since adjacent angles of a parallelogram are supplementary. (Angle SPQ measures 100°.) Opposite sides of the parallelogram are equal in length. The resultant force divides the parallelogram into two triangles. Use the law of cosines to get

$$|\mathbf{v}|^2 = 15^2 + 22^2 - 2(15)(22) \cos 80°$$
$$|\mathbf{v}| = 24. \quad \blacktriangleright$$

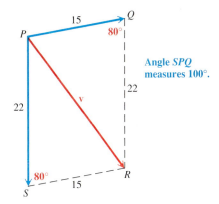

Angle *SPQ* measures 100°.

FIGURE 21

CONNECTIONS It is often useful to place a vector in the coordinate plane with its initial point at the origin and its endpoint at the point (a, b). This orientation gives us a one-to-one correspondence between points in the plane and the set of all vectors. If its endpoint is at (a, b), vector **u** is also denoted vector $\langle a, b \rangle$. The angle θ from the positive *x*-axis to the vector is called the **direction angle** of **u**. See the figure. From earlier results, if **u** $= \langle a, b \rangle$, we get the following relationships.

(a) $|\mathbf{u}| = \sqrt{a^2 + b^2}$

(b) $a = |\mathbf{u}| \cos \theta$

(c) $b = |\mathbf{u}| \sin \theta$

(d) $\tan \theta = b/a$

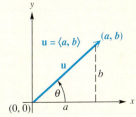

FOR DISCUSSION OR WRITING

1. Find the magnitude and direction angle rounded to the nearest tenth of a degree for **u** $= \langle 3, -2 \rangle$.
2. Use the definitions of a and b given above to find **u** $= \langle a, b \rangle$ if **u** has magnitude 5 and direction angle 60°.
3. For real numbers a, b, c, d, and k, $\langle a, b \rangle + \langle c, d \rangle = \langle a + b, c + d \rangle$ and $k\langle a, b \rangle = \langle ka, kb \rangle$. Let **u** $= \langle -2, 1 \rangle$ and **v** $= \langle 4, 3 \rangle$. Find the following.

   (a) **u** + **v**   (b) $-2\mathbf{u}$ (c) $4\mathbf{u} - 3\mathbf{v}$

In many applications it is necessary to find a vector that will counterbalance the resultant. This opposite vector is called the **equilibrant**; the equilibrant of vector **u** is the vector $-\mathbf{u}$.

EXAMPLE 3
Finding the magnitude and direction of an equilibrant

Find the magnitude of the equilibrant of forces of 48 newtons and 60 newtons acting on a point A, if the angle between the forces is 50°. Then find the angle between the equilibrant and the 48-newton force.

In Figure 22, the equilibrant is $-\mathbf{v}$. The magnitude of $\mathbf{v}$, and hence of $-\mathbf{v}$, is found by using triangle ABC and the law of cosines:

$$|\mathbf{v}|^2 = \mathbf{48}^2 + \mathbf{60}^2 - 2(\mathbf{48})(\mathbf{60})\cos\mathbf{130}°$$
$$= 9606.5,$$

and

$$|\mathbf{v}| = 98 \text{ newtons}$$

to two significant digits.

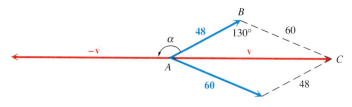

FIGURE 22

The required angle, labeled α in Figure 22, can be found by subtracting angle CAB from 180°. Use the law of sines to find angle CAB.

$$\frac{98}{\sin 130°} = \frac{60}{\sin CAB}$$
$$\sin CAB = .46900680$$
$$CAB = 28°$$

Finally,

$$\alpha = 180° - 28°$$
$$= 152°. \quad \blacktriangleright$$

EXAMPLE 4
Pulling a weight up an incline

Use vectors to solve the following incline problems.

(a) Find the force required to pull a 50-pound weight up a ramp inclined at 20° to the horizontal.

In Figure 23, the vertical 50-pound force $\mathbf{BA}$ represents the force of gravity. Its components are $\mathbf{BC}$ and $\mathbf{AC}$. The component $\mathbf{BC}$ represents the force with which the body pushes against the ramp. The vector $\mathbf{BF}$ represents a force that would pull the body up the ramp. Since the vectors $\mathbf{BF}$ and $\mathbf{AC}$ are equal, $|\mathbf{AC}|$ gives the magnitude of the required force.

Vectors $\mathbf{BF}$ and $\mathbf{AC}$ are parallel, so angle EBD equals angle A. Since angle BDE and angle C are right angles, triangles CBA and DEB have two corresponding angles equal and so are similar triangles. Therefore, angle *ABC* equals angle *E*, which is 20°. From right triangle *ABC*,

$$\sin 20° = \frac{|\mathbf{AC}|}{50}$$

$$|\mathbf{AC}| = 50 \sin 20°$$

$$|\mathbf{AC}| = 17.$$

To the nearest pound, a 17-pound force will be required to pull the weight up the ramp.

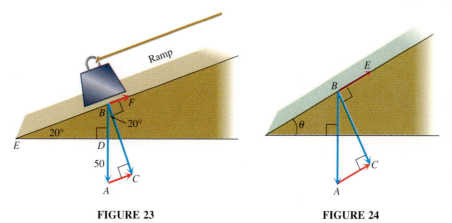

FIGURE 23   FIGURE 24

(b) A force of 16 pounds is required to hold a 40-pound lawn mower on an incline. What angle does the incline make with the horizontal?

Figure 24 illustrates the situation. Consider right triangle *ABC*. Angle *B* = angle θ, the magnitude of vector **BA** represents the weight of the mower, and vector **AC** equals vector **BE**, which represents the force required to push the mower up the incline. From the figure,

$$\sin B = \frac{16}{40}$$

$$= .4$$

$$B \approx 23.5782°.$$

Therefore, the hill makes an angle of about 24° with the horizontal. ▶

Problems involving bearing (defined in Section 6.4) can also be worked with vectors, as shown in the next example.

◖EXAMPLE 5
Applying vectors to a
navigation problem

A ship leaves port on a bearing of 28° and travels 8.2 miles. The ship then turns due east and travels 4.3 miles. How far is the ship from port? What is its bearing from port?

In Figure 25, vectors **PA** and **AE** represent the ship's path. The magnitude and bearing of the resultant **PE** can be found as follows. Triangle *PNA* is a right triangle, so angle *NAP* = 90° − 28° = 62°. Then angle *PAE* = 180° − 62° = 118°. Use the law of cosines to find $|\mathbf{PE}|$, the magnitude of vector **PE**.

$$|\mathbf{PE}|^2 = 8.2^2 + 4.3^2 - 2(8.2)(4.3) \cos 118°$$
$$|\mathbf{PE}|^2 = 118.84$$

Therefore, $\quad |\mathbf{PE}| = 10.9,$

or 11 miles, rounded to two significant digits.

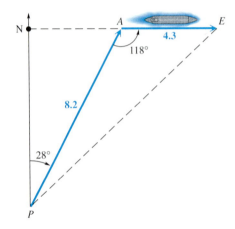

FIGURE 25

To find the bearing of the ship from port, first find angle *APE*. Use the law of sines, along with the value of $|\mathbf{PE}|$ before rounding.

$$\frac{\sin APE}{4.3} = \frac{\sin 118°}{10.9}$$

$$\sin APE = \frac{4.3 \sin 118°}{10.9}$$

$$\text{angle } APE = 20.4°$$

After rounding, angle *APE* is 20°, and the ship is 11 miles from port on a bearing of 28° + 20° = 48°. ◗

In air navigation, the **airspeed** of a plane is its speed relative to the air, while the **groundspeed** is its speed relative to the ground. Because of wind, these two speeds are usually different. The groundspeed of the plane is represented by the vector sum of the airspeed and windspeed vectors. See Figure 26.

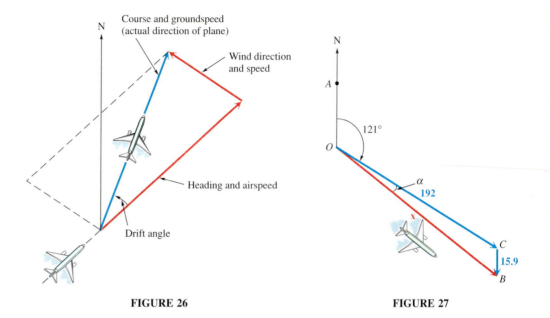

FIGURE 26   FIGURE 27

EXAMPLE 6
Applying vectors to a
navigation problem

A plane with an airspeed of 192 miles per hour is headed on a bearing of 121°.
A north wind is blowing (from north to south) at 15.9 miles per hour. Find the
groundspeed and the actual bearing of the plane.

In Figure 27 the groundspeed is represented by $|\mathbf{x}|$. We must find angle α
to determine the bearing, which will be $121° + \alpha$. From Figure 27, angle BCO
equals angle AOC, which equals 121°. Find $|\mathbf{x}|$ by the law of cosines.

$$|\mathbf{x}|^2 = 192^2 + 15.9^2 - 2(192)(15.9) \cos 121°$$
$$|\mathbf{x}|^2 = 40{,}261$$

Therefore, $|\mathbf{x}| \approx 200.7,$

or 201 miles per hour. Now find α by using the law of sines. As before, use the
value of $|\mathbf{x}|$ before rounding.

$$\frac{\sin \alpha}{15.9} = \frac{\sin 121°}{200.7}$$
$$\sin \alpha \approx .06790713$$
$$\alpha \approx 3.89°$$

After rounding, α is 3.9°. The groundspeed is about 201 miles per hour, on a
bearing of 125°, to three significant digits. ▶

8.3 Exercises ▼▼▼▼▼▼▼▼▼▼▼▼▼▼▼▼▼▼▼▼▼▼▼▼▼▼▼▼▼▼▼▼▼▼▼▼▼

1. In your own words, write a few sentences describing how a vector differs from a scalar.

2. Is a scalar product a vector or a scalar? Explain.

Exercises 3–6 refer to the vectors in the figure to the right.

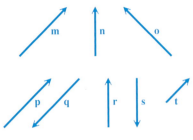

3. Name all pairs of vectors that appear to be equal.

4. Name all pairs of vectors that are opposites.

5. Name all pairs of vectors where the first is a scalar multiple of the other, with the scalar positive.

6. Name all pairs of vectors where the first is a scalar multiple of the other, with the scalar negative.

Exercises 7–18 refer to the vectors pictured below.

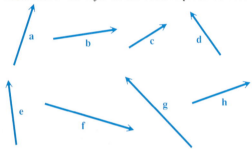

Draw a sketch to represent each vector. For example, find **a** + **e** by placing **a** and **e** so that their initial points coincide. Then use the parallelogram rule to find the resultant, shown in the figure to the left.

7. −**b**	8. −**g**	9. 3**a**	10. 2**h**
11. **a** + **b**	12. **h** + **g**	13. **a** − **c**	14. **d** − **e**
15. **a** + (**b** + **c**)	16. (**a** + **b**) + **c**	17. **c** + **d**	18. **d** + **c**

19. From the results of Exercises 15 and 16, do you think vector addition is associative?

20. From the results of Exercises 17 and 18, do you think vector addition is commutative?

For each pair of vectors **u** *and* **w** *with angle* θ *between them, sketch the resultant.*

21. $|\mathbf{u}| = 12, |\mathbf{w}| = 20, \theta = 27°$

22. $|\mathbf{u}| = 8, |\mathbf{w}| = 12, \theta = 20°$

23. $|\mathbf{u}| = 20, |\mathbf{w}| = 30, \theta = 30°$

24. $|\mathbf{u}| = 50, |\mathbf{w}| = 70, \theta = 40°$

For each of the following, vector **v** *has the given magnitude and direction. Find the magnitudes of the horizontal and vertical components of* **v,** *if* α *is the inclination angle from the horizontal. See Example 1.*

25. $\alpha = 38°, |\mathbf{v}| = 12$

26. $\alpha = 70°, |\mathbf{v}| = 150$

27. $\alpha = 35° \, 50', |\mathbf{v}| = 47.8$

28. $\alpha = 27° \, 30', |\mathbf{v}| = 15.4$

29. $\alpha = 128.5°, |\mathbf{v}| = 198$

30. $\alpha = 146.3°, |\mathbf{v}| = 238$

State a condition on **a** *and* **b** *that implies each of the following equations or inequalities.*

31. $|a + b| = 0$

32. $|a + b| = |a| + |b|$

33. $|a + b| = |a - b|$

34. $|a + b| > |a - b|$

35. Explain why the sum of two nonzero vectors can be a zero vector.

In each of the following, two forces act at a point in the plane. The angle between the two forces is given. Find the magnitude of the resultant force. See Example 2.

36. Forces of 250 and 450 newtons, forming an angle of 85°

37. Forces of 19 and 32 newtons, forming an angle of 118°

38. Forces of 17.9 and 25.8 lb, forming an angle of 105.5°

39. Forces of 37.8 and 53.7 lb, forming an angle of 68.5°

Find the magnitude and direction angle for **u** *with measures rounded to the nearest tenth. See the Connections box.*

40. $u = \langle 5, 12 \rangle$

41. $u = \langle 6, -8 \rangle$

42. $u = \langle -3, 4 \rangle$

43. $u = \langle -4, 5 \rangle$

Write **u** *in the form* $\langle a, b \rangle$. *See the Connections box.*

44. $|u| = 6$, direction angle of $u = 30°$

45. $|u| = 8$, direction angle of $u = 45°$

46. $|u| = 4$, direction angle of $u = 120°$

47. $|u| = 10$, direction angle of $u = 150°$

Solve the problem. See Examples 3–6.

48. A force of 25 lb is required to push an 80-lb crate up a hill. What angle does the hill make with the horizontal?

49. Find the force required to keep a 3000-lb car parked on a hill that makes an angle of 15° with the horizontal?

50. To build the pyramids in Egypt, it is believed that giant causeways were built to transport the building materials to the site. One such causeway is said to have been 3000 ft long, with a slope of about 2.3°. How much force would be required to pull a 60-ton monolith along this causeway?

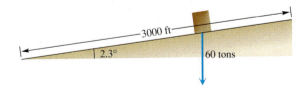

51. A force of 500 lb is required to pull a boat up a ramp inclined at 18° with the horizontal. How much does the boat weigh?

52. Two tugboats are pulling a disabled speedboat into port with forces of 1240 lb and 1480 lb. The angle between these forces is 28.2°. Find the direction and magnitude of the equilibrant.

53. Two people are carrying a box. One person exerts a force of 150 lb at an angle of 62.4° with the horizontal. The other person exerts a force of 114 lb at an angle of 54.9°. Find the weight of the box.

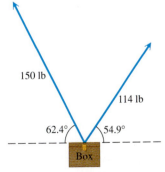

54. A crate is supported by two ropes. One rope makes an angle of 46° 20′ with the horizontal and has a tension of 89.6 lb on it. The other rope is horizontal. Find the weight of the crate and the tension in the horizontal rope.

55. Three forces acting at a point are in equilibrium. The forces are 980 lb, 760 lb, and 1220 lb. Find the angles between the directions of the forces. (*Hint:* Arrange the forces to form the sides of a triangle.)

56. A force of 176 lb makes an angle of 78° 50′ with a second force. The resultant of the two forces makes an angle of 41° 10′ with the first force. Find the magnitudes of the second force and of the resultant.

57. A force of 28.7 lb makes an angle of 42° 10′ with a second force. The resultant of the two forces makes an angle of 32° 40′ with the first force. Find the magnitudes of the second force and of the resultant.

58. A plane flies 650 mph on a bearing of 175.3°. A 25-mph wind, from a direction of 266.6°, blows against the plane. Find the resulting bearing of the plane.

59. A pilot wants to fly on a bearing of 74.9°. By flying due east, he finds that a 42-mph wind, blowing from the south, puts him on course. Find the airspeed and the groundspeed.

60. Starting at point A, a ship sails 18.5 km on a bearing of 189°, then turns and sails 47.8 km on a bearing of 317°. Find the distance of the ship from point A.

61. Two towns 21 mi apart are separated by a dense forest. (See the figure.) To travel from A to town B, a person must go 17 mi on a bearing of 325°, then turn

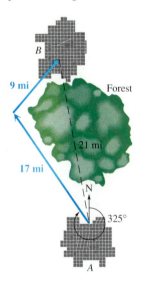

and continue for 9 mi to reach town B. Find the bearing of B from A.

62. An airline route from San Francisco to Honolulu is on a bearing of 233°. A jet flying at 450 mph on that bearing flies into a wind blowing at 39 mph from a direction of 114°. Find the resulting bearing and groundspeed of the plane.

63. A pilot is flying at 168 mph. She wants her flight path to be on a bearing of 57° 40′. A wind is blowing from the south at 27.1 mph. Find the bearing the pilot should fly, and find the plane's groundspeed.

64. What bearing and airspeed are required for a plane to fly 400 mi due north in 2.5 hr if the wind is blowing from a direction of 328° at 11 mph?

65. A plane is headed due south with an airspeed of 192 mph. A wind from a direction of 78° is blowing at 23 mph. Find the groundspeed and resulting bearing of the plane.

66. An airplane is headed on a bearing of 174° at an airspeed of 240 kph. A 30 kph wind is blowing from a direction of 245°. Find the groundspeed and resulting bearing of the plane.

67. A ship sailing due east in the North Atlantic has been warned to change course to avoid a group of icebergs. The captain turns and sails on a bearing of 62° for a while, then changes course again to a bearing of 115° until the ship reaches its original course. (See the figure.) How much farther did the ship travel to avoid the icebergs?

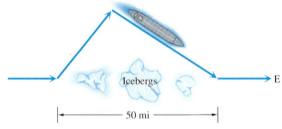

68. The aircraft carrier *Tallahassee* is traveling at sea on a steady course with a bearing of 30° at 32 mph. Patrol planes on the carrier have enough fuel for 2.6 hr of flight when traveling at a speed of 520 mph. One of the pilots takes off on a bearing of 338° and then turns and heads in a straight line, so as to be able to catch the carrier and land on the deck at the exact instant that his fuel runs out. If the pilot left at 2 P.M., at what time did he turn to head for the carrier?

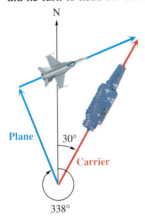

69. A car going around a banked curve is subject to the forces shown in the figure. If the radius of the curve is

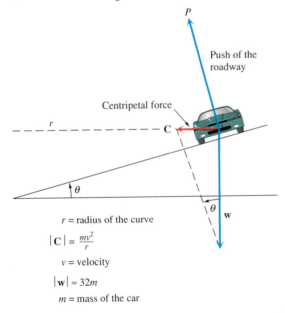

r = radius of the curve

$$|C| = \frac{mv^2}{r}$$

v = velocity

$|w| \approx 32m$

m = mass of the car

100 ft, what value of θ to the nearest degree would allow an automobile to travel around the curve at a speed of 40 ft per sec without depending on friction?

70. The space velocity **v** of a star relative to the sun can be expressed as the resultant vector of two perpendicular vectors—the radial velocity v_r and the tangential velocity v_t where $v = v_r + v_t$. Refer to the figure. If a star is located near the sun and its space velocity is large, then its motion across the sky will also be large. Barnard's Star is a relatively close star with a distance of 35 trillion miles from the sun. It moves across the sky through an angle 10.34″ per year, which is the largest motion of any known star. Its radial velocity is $v_r = 67$ mi/sec toward the sun. (*Sources:* Zelik, M., S. Gregory, and E. Smith, *Introductory Astronomy and Astrophysics,* Saunders College Publishing, 1992. Acker, A. and C. Jaschek, *Astronomical Methods and Calculations,* John Wiley & Sons, 1986.)

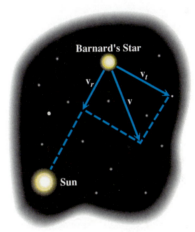

Not to scale

(a) Approximate the tangential velocity v_t of Barnard's Star. (*Hint:* Use the Length of Arc Formula: $s = r\theta$.)

(b) Compute the magnitude of **v.**

8.4 Products and Quotients of Complex Numbers in Trigonometric Form ▼▼▼

Unlike real numbers, complex numbers cannot be ordered. One way to organize them is with a graph. To graph a complex number such as $2 - 3i$, the familiar coordinate system must be modified. One way to do this is by calling the horizontal axis the **real axis** and the vertical axis the **imaginary axis.** Then complex numbers can be graphed in this **complex plane,** as shown in Figure 28 for the complex number $2 - 3i$.

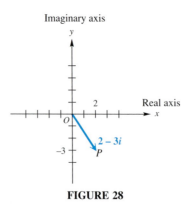

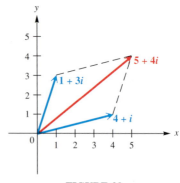

FIGURE 28   **FIGURE 29**

VECTOR REPRESENTATION   Each nonzero complex number graphed in this way determines a unique directed line segment, the segment from the origin to the point representing the complex number. Recall that such directed line segments (like **OP** of Figure 28) are called vectors.

By definition, the sum of the two complex numbers $4 + i$ and $1 + 3i$ is

$$(4 + i) + (1 + 3i) = 5 + 4i.$$

Graphically, the sum of two complex numbers is represented by the vector that is the resultant of the vectors corresponding to the two numbers. The vectors representing the complex numbers $4 + i$ and $1 + 3i$ and the resultant vector that represents their sum, $5 + 4i$, are shown in Figure 29.

◖**EXAMPLE 1**
Expressing the sum of
complex numbers graphically

Find the sum of $6 - 2i$ and $-4 - 3i$. Graph both complex numbers and their resultant.

The sum is found by adding the two numbers.

$$(6 - 2i) + (-4 - 3i) = 2 - 5i$$

The graphs are shown in Figure 30. ◗

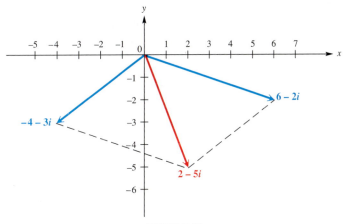

FIGURE 30

CONNECTIONS In Section 8.3 we saw that the vector **u** with its initial point at the origin and its endpoint at (a, b) could be designated $\langle a, b \rangle$. We then showed how to add and subtract vectors using this new notation. Now we see that the complex number $a + bi$ corresponds to the vector **u** described above. Thus, we have

$$\mathbf{u} = \langle a, b \rangle = a + bi$$

as three ways to designate a complex number or vector.

We can use addition of vectors in the form $\langle a, b \rangle$ to find the sum in Example 1.

$$(6 - 2i) + (-4 - 3i) = \langle 6, -2 \rangle + \langle -4, -3 \rangle$$
$$= \langle 2, -5 \rangle$$
$$= 2 - 5i$$

FOR DISCUSSION OR WRITING

1. Find $(6 - 2i) - (-4 - 3i)$ using vectors in the form $\langle a, b \rangle$. Then graph the given complex numbers and the difference.
2. Describe a general method for finding the difference of two vectors.

Figure 31 shows the complex number $x + yi$ that corresponds to a vector **OP** with direction angle θ and magnitude r. The following relationships among r, θ, x, and y can be verified from Figure 31.

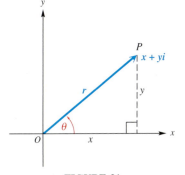

FIGURE 31

Relationships Among x, y, r, and θ
$x = r \cos \theta$ $\qquad$ $r = \sqrt{x^2 + y^2}$
$y = r \sin \theta$ $\qquad$ $\tan \theta = \dfrac{y}{x}$, if $x \neq 0$

TRIGONOMETRIC FORM Substituting $x = r \cos \theta$ and $y = r \sin \theta$ from the relationships given above into $x + yi$ gives

$$x + yi = r \cos \theta + (r \sin \theta)i$$
$$= r(\cos \theta + i \sin \theta).$$

Trigonometric or Polar Form of a Complex Number

The expression

$$r(\cos \theta + i \sin \theta)$$

is called the **trigonometric form** or **polar form** of the complex number $x + yi$. The expression $\cos \theta + i \sin \theta$ is sometimes abbreviated cis θ. Using this notation,

$$r(\cos \theta + i \sin \theta) \text{ is written as } r \text{ cis } \theta.$$

The number r is called the **modulus** or **absolute value** of $x + yi$, while θ is the **argument** of $x + yi$. In this section we choose the value of θ in the interval $[0°, 360°)$. However, angles coterminal with such angles are also possible; that is, the argument for a particular complex number is not unique.

◖ EXAMPLE 2
Converting from trigonometric form to standard form

Express $2(\cos 300° + i \sin 300°)$ in standard form.
Since $\cos 300° = 1/2$ and $\sin 300° = -\sqrt{3}/2$,

$$2(\cos 300° + i \sin 300°) = 2\left(\frac{1}{2} - i\frac{\sqrt{3}}{2}\right) = 1 - i\sqrt{3}. \quad \text{◗}$$

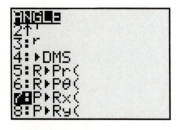

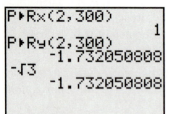

Choices 7 and 8 in the top screen show how a graphing calculator can convert from trigonometric (*polar*) form to rectangular (*x-y*) form. The screen at the bottom supports the result of Example 2. The calculator is in degree mode.

NOTE In the examples of this section, we will write arguments using degree measure. Arguments may also be written with radian measure.

In order to convert from standard form to trigonometric form, the following procedure is used.

Steps for Converting From Standard to Trigonometric Form

1. Sketch a graph of the number in the complex plane.
2. Find r by using the equation $r = \sqrt{x^2 + y^2}$.
3. Find θ by using the equation $\tan \theta = y/x$, $x \neq 0$, choosing the quadrant indicated in Step 1.

CAUTION Errors often occur in Step 3 described above. Be sure that the correct quadrant for θ is chosen by referring to the graph sketched in Step 1.

EXAMPLE 3
Converting from standard form
to trigonometric form

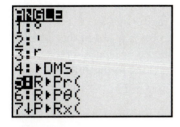

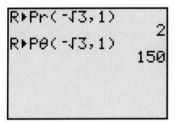

Choices 5 and 6 in the top screen
show how a graphing calculator can
convert from rectangular (*x*-*y*) form to
trigonometric (*polar*) form. The
screen at the bottom supports the
result of Example 3(a). The calculator
is in degree mode.

Write the following complex numbers in trigonometric form.

(a) $-\sqrt{3} + i$

Start by sketching the graph of $-\sqrt{3} + i$ in the complex plane, as shown in Figure 32. Next, find r. Since $x = -\sqrt{3}$ and $y = 1$,

$$r = \sqrt{x^2 + y^2} = \sqrt{(-\sqrt{3})^2 + 1^2} = \sqrt{3 + 1} = 2.$$

Then find θ.

$$\tan \theta = \frac{y}{x} = \frac{1}{-\sqrt{3}} = -\frac{\sqrt{3}}{3}$$

Since $\tan \theta = -\sqrt{3}/3$, the reference angle for θ is $30°$. From the sketch we see that θ is in quadrant II, so $\theta = 180° - 30° = 150°$. Therefore, in trigonometric form,

$$-\sqrt{3} + i = 2(\cos 150° + i \sin 150°)$$
$$= 2 \text{ cis } 150°.$$

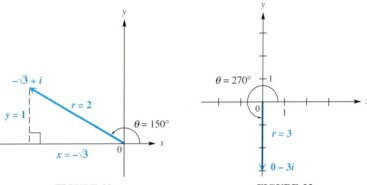

FIGURE 32   **FIGURE 33**

(b) $-3i$

The sketch of $-3i$ is shown in Figure 33. Since $-3i = 0 - 3i$, we have $x = 0$ and $y = -3$. Find r as follows.

$$r = \sqrt{0^2 + (-3)^2} = \sqrt{0 + 9} = \sqrt{9} = 3$$

We cannot find θ by using $\tan \theta = y/x$, since $x = 0$. In a case like this, refer to the graph and determine the argument directly from the sketch. A value for θ here is $270°$. In trigonometric form,

$$-3i = 3(\cos 270° + i \sin 270°)$$
$$= 3 \text{ cis } 270°. \quad \blacktriangleright$$

NOTE   In Examples 2 and 3 we gave answers in both forms: $r(\cos \theta + i \sin \theta)$ and r cis θ. These forms will be used interchangeably throughout the rest of this chapter.

The next example discusses a fractal, a relatively new application of complex numbers.

EXAMPLE 4
Deciding whether a complex number is in the Julia set

The fractal called the **Julia set** is shown in Figure 34. It can be created by graphing a special set of complex numbers.* To determine if a complex number $z = a + bi$ is in this Julia set, perform the following sequence of calculations. Repeatedly compute the values of $z^2 - 1$, $(z^2 - 1)^2 - 1, [(z^2 - 1)^2 - 1]^2 - 1, \ldots$. If the moduli of any of the resulting complex numbers exceeds 2, then the complex number z is not in the Julia set. Otherwise z is part of this set and the point (a, b) should be shaded in the graph. Determine whether or not the following numbers belong to the Julia set.

FIGURE 34

(a) $z = 0 + 0i$

Since $z = 0 + 0i = 0$, $z^2 - 1 = 0^2 - 1 = -1$,
$$(z^2 - 1)^2 - 1 = (-1)^2 - 1 = 0,$$
$$[(z^2 - 1)^2 - 1]^2 - 1 = 0^2 - 1 = -1,$$
and so on. We see that the calculations repeat as $0, -1, 0, -1$, and so on. The moduli are either 0 or 1, which do not exceed 2, so $0 + 0i$ is in the Julia set and the point $(0, 0)$ is part of the graph.

(b) $z = 1 + 1i$

We have $z^2 - 1 = (1 + i)^2 - 1 = (1 + 2i + i^2) - 1 = -1 + 2i$. The modulus is $\sqrt{(-1)^2 + 2^2} = \sqrt{5}$. Since $\sqrt{5}$ is greater than 2, $1 + 1i$ is not in the Julia set and $(1, 1)$ is not part of the graph. ▶

PRODUCTS OF COMPLEX NUMBERS The product of the two complex numbers $1 + i\sqrt{3}$ and $-2\sqrt{3} + 2i$ can be found by the FOIL method shown earlier.

$$(1 + i\sqrt{3})(-2\sqrt{3} + 2i) = -2\sqrt{3} + 2i - 2i(3) + 2i^2\sqrt{3}$$
$$= -2\sqrt{3} + 2i - 6i - 2\sqrt{3}$$
$$= -4\sqrt{3} - 4i$$

---

*Source: Crownover, R., *Introduction to Fractals and Chaos*, Jones and Bartlett Publishers, Boston, 1995.

Figure: Crownover, R., *Introduction to Fractals and Chaos*, Jones and Bartlett Publishers, Boston, 1995. Figure 8-1, p. 199.

This same product also can be found by first converting the complex numbers $1 + i\sqrt{3}$ and $-2\sqrt{3} + 2i$ to trigonometric form.

$$1 + i\sqrt{3} = 2(\cos 60° + i \sin 60°)$$
$$-2\sqrt{3} + 2i = 4(\cos 150° + i \sin 150°)$$

If the trigonometric forms are now multiplied together and if the trigonometric identities for the cosine and the sine of the sum of two angles are used, the result is

$$[2(\cos 60° + i \sin 60°)][4(\cos 150° + i \sin 150°)]$$
$$= 2 \cdot 4(\cos 60° \cdot \cos 150° + i \sin 60° \cdot \cos 150°$$
$$+ i \cos 60° \cdot \sin 150° + i^2 \sin 60° \cdot \sin 150°)$$
$$= 8[(\cos 60° \cdot \cos 150° - \sin 60° \cdot \sin 150°)$$
$$+ i(\sin 60° \cdot \cos 150° + \cos 60° \cdot \sin 150°)]$$
$$= 8[\cos(60° + 150°) + i \sin(60° + 150°)]$$
$$= 8(\cos 210° + i \sin 210°).$$

The modulus of the product, 8, is equal to the product of the moduli of the factors, $2 \cdot 4$, while the argument of the product, 210°, is the sum of the arguments of the factors, $60° + 150°$.

As we would expect, the product obtained upon multiplying by the first method is the standard form of the product obtained upon multiplying by the second method.

$$8(\cos 210° + i \sin 210°) = 8\left(-\frac{\sqrt{3}}{2} - \frac{1}{2}i\right)$$
$$= -4\sqrt{3} - 4i$$

Generalizing, the product of the two complex numbers, $r_1(\cos \theta_1 + i \sin \theta_1)$ and $r_2(\cos \theta_2 + i \sin \theta_2)$, is

$$[r_1(\cos \theta_1 + i \sin \theta_1)] \cdot [r_2(\cos \theta_2 + i \sin \theta_2)]$$
$$= r_1 r_2(\cos \theta_1 \cos \theta_2 + i \sin \theta_1 \cos \theta_2 + i \cos \theta_1 \sin \theta_2 + i^2 \sin \theta_1 \sin \theta_2)$$
$$= r_1 r_2[(\cos \theta_1 \cos \theta_2 - \sin \theta_1 \sin \theta_2) + i(\sin \theta_1 \cos \theta_2 + \cos \theta_1 \sin \theta_2)]$$
$$= r_1 r_2[\cos(\theta_1 + \theta_2) + i \sin(\theta_1 + \theta_2)].$$

This work is summarized in the following *product theorem*.

Product Theorem

If $r_1(\cos \theta_1 + i \sin \theta_1)$ and $r_2(\cos \theta_2 + i \sin \theta_2)$ are any two complex numbers, then

$$[r_1(\cos \theta_1 + i \sin \theta_1)] \cdot [r_2(\cos \theta_2 + i \sin \theta_2)]$$
$$= r_1 r_2[\cos(\theta_1 + \theta_2) + i \sin(\theta_1 + \theta_2)].$$

In compact form, this is written

$$(r_1 \text{ cis } \theta_1)(r_2 \text{ cis } \theta_2) = r_1 r_2 \text{ cis}(\theta_1 + \theta_2).$$

EXAMPLE 5
Using the product theorem

Find the product of $3(\cos 45° + i \sin 45°)$ and $2(\cos 135° + i \sin 135°)$. Using the product theorem,

$$[3(\cos 45° + i \sin 45°)][2(\cos 135° + i \sin 135°)]$$
$$= 3 \cdot 2[\cos (45° + 135°) + i \sin(45° + 135°)]$$
$$= 6(\cos 180° + i \sin 180°),$$

which can be expressed as $6(-1 + i \cdot 0) = 6(-1) = -6$. The two complex numbers in this example are complex factors of -6. ▶

QUOTIENTS OF COMPLEX NUMBERS In standard form the quotient of the complex numbers $1 + i\sqrt{3}$ and $-2\sqrt{3} + 2i$ is

$$\frac{1 + i\sqrt{3}}{-2\sqrt{3} + 2i} = \frac{(1 + i\sqrt{3})(-2\sqrt{3} - 2i)}{(-2\sqrt{3} + 2i)(-2\sqrt{3} - 2i)}$$
$$= \frac{-2\sqrt{3} - 2i - 6i - 2i^2\sqrt{3}}{12 - 4i^2}$$
$$= \frac{-8i}{16} = -\frac{1}{2}i.$$

Writing $1 + i\sqrt{3}$, $-2\sqrt{3} + 2i$, and $-(1/2)i$ in trigonometric form gives

$$1 + i\sqrt{3} = 2(\cos 60° + i \sin 60°)$$
$$-2\sqrt{3} + 2i = 4(\cos 150° + i \sin 150°)$$
$$-\frac{1}{2}i = \frac{1}{2}[\cos(-90°) + i \sin(-90°)].$$

The modulus of the quotient, $1/2$, is the quotient of the two moduli, 2 and 4. The argument of the quotient, $-90°$, is the difference of the two arguments, $60° - 150° = -90°$. It would be easier to find the quotient of these two complex numbers in trigonometric form than in standard form. Generalizing from this example leads to another theorem. The proof is similar to the proof of the product theorem, after the numerator and denominator are multiplied by the conjugate of the denominator.

Quotient Theorem

If $r_1(\cos \theta_1 + i \sin \theta_1)$ and $r_2(\cos \theta_2 + i \sin \theta_2)$ are complex numbers, where $r_2(\cos \theta_2 + i \sin \theta_2) \neq 0$, then

$$\frac{r_1(\cos \theta_1 + i \sin \theta_1)}{r_2(\cos \theta_2 + i \sin \theta_2)} = \frac{r_1}{r_2}[\cos(\theta_1 - \theta_2) + i \sin(\theta_1 - \theta_2)].$$

In compact form, this is written

$$\frac{r_1 \text{ cis } \theta_1}{r_2 \text{ cis } \theta_2} = \frac{r_1}{r_2} \text{cis}(\theta_1 - \theta_2).$$

◖EXAMPLE 6
Using the quotient theorem

Find the quotient

$$\frac{10 \text{ cis}(-60°)}{5 \text{ cis } 150°}.$$

Write the result in standard form.
By the quotient theorem,

$$\frac{10 \text{ cis}(-60°)}{5 \text{ cis } 150°} = \frac{10}{5} \text{cis}(\mathbf{-60° - 150°})$$ Quotient theorem

$$= 2 \text{ cis}(-210°)$$ Subtract.

$$= 2[\cos(-210°) + i \sin(-210°)]$$

$$= 2\left[-\frac{\sqrt{3}}{2} + i\left(\frac{1}{2}\right) \right]$$ $\cos(-210°) = -\frac{\sqrt{3}}{2}$; $\sin(-210°) = \frac{1}{2}$

$$= -\sqrt{3} + i.$$ Standard form ▶

8.4 Exercises ▼▼▼▼▼▼▼▼▼▼▼▼▼▼▼▼▼▼▼▼▼▼▼▼▼▼▼▼▼▼▼▼▼▼

1. The modulus of a complex number represents the _____ of the vector representing it in the complex plane.

2. Describe geometrically the argument of a complex number.

Graph each complex number. See Example 1.

3. $-2 + 3i$ 4. $8 - 5i$ 5. $-4i$ 6. $3i$

7. What must be true in order for a complex number to also be a real number?

8. If a real number is graphed in the complex plane, on what axis does the vector lie?

Find the resultant of each pair of complex numbers. See Example 1.

9. $4 - 3i$, $-1 + 2i$ 10. $2 + 3i$, $-4 - i$ 11. -3, $3i$

12. 6, $-2i$ 13. $2 + 6i$, $-2i$ 14. $4 - 2i$, 5

Write the given complex number in standard form. See Example 2.

15. $10(\cos 90° + i \sin 90°)$ 16. $8(\cos 270° + i \sin 270°)$

17. $4(\cos 240° + i \sin 240°)$ 18. $2(\cos 330° + i \sin 330°)$

19. cis 30°   20. 5 cis 300°

21. 6 cis 135°   22. $\sqrt{2}$ cis 180°

Write each complex number in trigonometric form $r(\cos \theta + i \sin \theta)$, with θ in the interval $[0°, 360°)$. See Example 3.

23. $3 - 3i$ 24. $-2 + 2i\sqrt{3}$ 25. $-3 - 3i\sqrt{3}$ 26. $1 + i\sqrt{3}$

27. $-5 - 5i$ 28. $-\sqrt{2} + i\sqrt{2}$ 29. -4 30. $-2i$

Perform the following conversions, using a calculator as necessary.

Standard Form	Trigonometric Form
31. $2 + 3i$	_____
32. _____	$(\cos 35° + i \sin 35°)$
33. _____	$3(\cos 250° + i \sin 250°)$
34. $-4 + i$	_____
35. $3 + 5i$	_____
36. _____	cis $110.5°$

The complex number z, where $z = x + yi$, can be graphed in the plane as (x, y). Describe the graphs of all complex numbers z satisfying the conditions in Exercises 37–40.

37. The modulus of z is 1. **38.** The real and imaginary parts of z are equal.

39. The real part of z is 1. **40.** The imaginary part of z is 1.

41. Refer to Example 4. Is $z = -.2i$ in the Julia set?

42. Refer to Example 4. The graph of the Julia set in Figure 34 appears to be symmetric with respect to both the *x*-axis and *y*-axis. Complete the following to show that this is true.
  (a) Show that complex conjugates have the same modulus.
  (b) Compute $z_1^2 - 1$ and $z_2^2 - 1$ where $z_1 = a + bi$ and $z_2 = a - bi$.
  (c) Discuss why if (a, b) is in the Julia set then so is $(a, -b)$.
  (d) Conclude that the graph of the Julia set must be symmetric with respect to the *x*-axis.
  (e) Using a similar argument, show that the Julia set must also be symmetric with respect to the *y*-axis.

43. Figure 29 shows a geometric method for finding the sum of two complex numbers. Suppose you are given a similar figure containing two blue arrows corresponding to complex numbers (without the actual complex numbers given) and are asked to draw the red arrow corresponding to their product. Explain how you would proceed if you had a protractor and a ruler.

44. Answer the problem in Exercise 43 for the quotient of two complex numbers.

Find the product. Write each product in standard form. See Example 5.

45. $[3(\cos 60° + i \sin 60°)][2(\cos 90° + i \sin 90°)]$ **46.** $[4(\cos 30° + i \sin 30°)][5(\cos 120° + i \sin 120°)]$

47. $[2(\cos 45° + i \sin 45°)][2(\cos 225° + i \sin 225°)]$ **48.** $[8(\cos 210° + i \sin 210°)][2(\cos 330° + i \sin 330°)]$

49. $[5 \text{ cis } 90°][3 \text{ cis } 45°]$ **50.** $[6 \text{ cis } 120°][5 \text{ cis}(-30°)]$

51. $[\sqrt{3} \text{ cis } 45°][\sqrt{3} \text{ cis } 225°]$ **52.** $[\sqrt{2} \text{ cis } 300°][\sqrt{2} \text{ cis } 270°]$

Find the quotient. Write each quotient in standard form. See Example 6.

53. $\dfrac{4(\cos 120° + i \sin 120°)}{2(\cos 150° + i \sin 150°)}$ **54.** $\dfrac{24(\cos 150° + i \sin 150°)}{2(\cos 30° + i \sin 30°)}$

55. $\dfrac{3 \text{ cis } 305°}{9 \text{ cis } 65°}$ **56.** $\dfrac{12 \text{ cis } 293°}{6 \text{ cis } 23°}$

57. $\dfrac{8}{\sqrt{3} + i}$ **58.** $\dfrac{2i}{-1 - i\sqrt{3}}$

59. $\dfrac{-i}{1 + i}$ **60.** $\dfrac{1}{2 - 2i}$

Use a calculator to perform the indicated operations. Give answers in standard form.

61. $[2.5(\cos 35° + i \sin 35°)][3.0(\cos 50° + i \sin 50°)]$

62. $(4 \text{ cis } 19.25°)(7 \text{ cis } 41.75°)$

63. $\dfrac{45(\cos 127° + i \sin 127°)}{22.5(\cos 43° + i \sin 43°)}$

64. $\dfrac{30(\cos 130° + i \sin 130°)}{10(\cos 21° + i \sin 21°)}$

65. $\left[2 \text{ cis } \dfrac{5\pi}{9} \right]^2$

66. $\left[24.3\left(\cos \dfrac{7\pi}{12} + i \sin \dfrac{7\pi}{12} \right) \right]^2$

▼▼▼▼▼▼▼▼▼▼▼▼▼ **DISCOVERING CONNECTIONS** (Exercises 67–73) ▼▼▼▼▼▼▼▼▼▼▼▼▼

Consider the complex numbers

$$w = -1 + i \qquad and \qquad z = -1 - i.$$

67. Multiply w and z using their standard forms and the "FOIL" method. Leave the product in standard form.

68. Find the trigonometric forms of w and z.

69. Multiply w and z using their trigonometric forms and the method described in this section.

70. Use the result of Exercise 69 to find the standard form of wz. How does this compare to your result in Exercise 67?

71. Find the quotient w/z using their standard forms and multiplying both the numerator and the denominator by the conjugate of the denominator. Leave the quotient in standard form.

72. Use the trigonometric forms of w and z, found in Exercise 68, to divide w by z using the method described in this section.

73. Use the result of Exercise 72 to find the standard form of w/z. How does this compare to your result in Exercise 71?

---

74. Notice that

$$(r \text{ cis } \theta)^2 = (r \text{ cis } \theta)(r \text{ cis } \theta)$$
$$= r^2 \text{ cis}(\theta + \theta)$$
$$= r^2 \text{ cis } 2\theta.$$

State in your own words how we can square a complex number in trigonometric form. (In the next section, we will develop this idea more fully.)

75. The alternating current in an electric inductor is

$$I = \frac{E}{Z}$$

amperes, where E is the voltage and $Z = R + X_L i$ is the impedance. If $E = 8(\cos 20° + i \sin 20°)$, $R = 6$, and $X_L = 3$, find the current. Give the answer in standard form.

76. The current I in a circuit with voltage E, resistance R, capacitive reactance X_c, and inductive reactance X_L is

$$I = \frac{E}{R + (X_L - X_c)i}.$$

Find I if $E = 12(\cos 25° + i \sin 25°)$, $R = 3$, $X_L = 4$, and $X_c = 6$. Give the answer in standard form.

77. In the parallel electrical circuit shown in the figure below, the impedance Z can be calculated using the equation $Z = \dfrac{1}{\dfrac{1}{Z_1} + \dfrac{1}{Z_2}}$ where Z_1 and Z_2 are the impedances for each branch of the circuit. If $Z_1 = 50 + 25i$ and $Z_2 = 60 + 20i$, calculate Z.

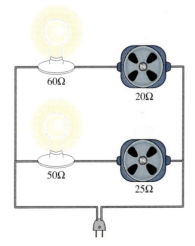

60Ω
20Ω
50Ω
25Ω

8.5 Powers and Roots of Complex Numbers ▼▼▼

In the previous section we studied the product and quotient theorems for complex numbers in trigonometric form. Because raising a number to a positive integer power is a repeated application of the product rule, it would seem likely that a theorem for finding powers of complex numbers exists. This is indeed the case. For example, the square of the complex number $r(\cos \theta + i \sin \theta)$ is

$$[r(\cos \theta + i \sin \theta)]^2 = [r(\cos \theta + i \sin \theta)][r(\cos \theta + i \sin \theta)]$$
$$= r \cdot r[\cos(\theta + \theta) + i \sin(\theta + \theta)]$$
$$= r^2(\cos 2\theta + i \sin 2\theta).$$

In the same way, $\qquad [r(\cos \theta + i \sin \theta)]^3 = r^3(\cos 3\theta + i \sin 3\theta).$

These results suggest the plausibility of the following theorem for positive integer values of n. Although the following theorem is stated and can be proved for all n, we will use it only for positive integer values of n and their reciprocals.

De Moivre's Theorem

If $r(\cos \theta + i \sin \theta)$ is a complex number and if n is any real number, then

$$[r(\cos \theta + i \sin \theta)]^n = r^n(\cos n\theta + i \sin n\theta).$$

In compact form, this is written

$$[r \operatorname{cis} \theta]^n = r^n(\operatorname{cis} n\theta).$$

This theorem is named after the French expatriate friend of Isaac Newton, Abraham De Moivre (1667–1754), although he never explicity stated it.

◀ **EXAMPLE 1**
Applying De Moivre's theorem (finding a power of a complex number)

Find $(1 + i\sqrt{3})^8$ and express the result in standard form.

To use De Moivre's theorem, first convert $1 + i\sqrt{3}$ into trigonometric form.

$$1 + i\sqrt{3} = 2(\cos 60° + i \sin 60°)$$

Now apply De Moivre's theorem.

$$(1 + i\sqrt{3})^8 = [2(\cos 60° + i \sin 60°)]^8$$
$$= 2^8[\cos(8 \cdot 60°) + i \sin(8 \cdot 60°)]$$
$$= 256(\cos 480° + i \sin 480°)$$
$$= 256(\cos 120° + i \sin 120°) \qquad \text{480° and 120° are coterminal.}$$
$$= 256\left(-\frac{1}{2} + i\frac{\sqrt{3}}{2}\right) \qquad \cos 120° = -\tfrac{1}{2}; \sin 120° = \tfrac{\sqrt{3}}{2}$$
$$= -128 + 128i\sqrt{3} \qquad \text{Standard form} \quad \blacktriangleright$$

In algebra it is shown that every nonzero complex number has exactly n distinct complex nth roots. De Moivre's theorem can be extended to find all nth roots of a complex number. An nth root of a complex number is defined as follows.

---

nth Root

For a positive integer n, the complex number $a + bi$ is an **nth root** of the complex number $x + yi$ if

$$(a + bi)^n = x + yi.$$

To find the cube roots of the complex number $8(\cos 135° + i \sin 135°)$, for example, look for a complex number, say $r(\cos \alpha + i \sin \alpha)$, that will satisfy

$$[r(\cos \alpha + i \sin \alpha)]^3 = 8(\cos 135° + i \sin 135°).$$

By De Moivre's theorem, this equation becomes

$$r^3(\cos 3\alpha + i \sin 3\alpha) = 8(\cos 135° + i \sin 135°).$$

One way to satisfy this equation is to set $r^3 = 8$ and also $\cos 3\alpha + i \sin 3\alpha = \cos 135° + i \sin 135°$. The first of these conditions implies that $r = 2$, and the second implies that

$$\cos 3\alpha = \cos 135° \qquad \text{and} \qquad \sin 3\alpha = \sin 135°.$$

For these equations to be satisfied, 3α must represent an angle that is coterminal with $135°$. Therefore, we must have

$$3\alpha = 135° + 360° \cdot k, \quad k \text{ any integer,}$$

or

$$\alpha = \frac{135° + 360° \cdot k}{3}, \quad k \text{ any integer.}$$

Now let k take on the integer values 0, 1, and 2.

If $k = 0$, $\qquad \alpha = \dfrac{135° + 0°}{3} = 45°.$

If $k = 1$, $\qquad \alpha = \dfrac{135° + 360°}{3} = \dfrac{495°}{3} = 165°.$

If $k = 2$, $\qquad \alpha = \dfrac{135° + 720°}{3} = \dfrac{855°}{3} = 285°.$

In the same way, $\alpha = 405°$ when $k = 3$. But note that $\sin 405° = \sin 45°$ and $\cos 405° = \cos 45°$. If $k = 4$, $\alpha = 525°$ which has the same sine and cosine values as $165°$. To continue with larger values of k would just be repeating solutions already found. Therefore, all of the cube roots (three of them) can be found by letting $k = 0$, 1, and 2.

When $k = 0$, the root is

$$2(\cos \mathbf{45°} + i \sin \mathbf{45°}).$$

When $k = 1$, the root is

$$2(\cos \mathbf{165°} + i \sin \mathbf{165°}).$$

When $k = 2$, the root is

$$2(\cos \mathbf{285°} + i \sin \mathbf{285°}).$$

In summary,

$$2(\cos 45° + i \sin 45°), \quad 2(\cos 165° + i \sin 165°),$$
$$\text{and} \quad 2(\cos 285° + i \sin 285°)$$

are the three cube roots of $8(\cos 135° + i \sin 135°)$.

Notice that the formula for α in the discussion above can be written in an alternative form as

$$\alpha = \frac{135°}{3} + \frac{360° \cdot k}{3} = 45° + 120° \cdot k,$$

for $k = 0, 1,$ and 2, which is easier to use.

Generalizing the work above leads to the following theorem.

---

nth Root Theorem

If n is any positive integer and r is a positive real number, then the complex number $r(\cos \theta + i \sin \theta)$ has exactly n distinct nth roots, given by

$$\sqrt[n]{r}(\cos \boldsymbol{\alpha} + i \sin \boldsymbol{\alpha}) \quad \text{or} \quad \sqrt[n]{r} \text{ cis } \boldsymbol{\alpha},$$

where

$$\alpha = \frac{\theta + 360° \cdot k}{n} \quad \text{or} \quad \alpha = \frac{\theta}{n} + \frac{360° \cdot k}{n},$$

$k = 0, 1, 2, \ldots, n - 1.$

EXAMPLE 2
Finding roots of a complex number

Find all fourth roots of $-8 + 8i\sqrt{3}$. Write the roots in standard form.

First write $-8 + 8i\sqrt{3}$ in trigonometric form as

$$-8 + 8i\sqrt{3} = 16 \text{ cis } 120°.$$

Here $r = 16$ and $\theta = 120°$. The fourth roots of this number have modulus $\sqrt[4]{16} = 2$ and arguments given as follows. Using the alternative formula for α,

$$\alpha = \frac{120°}{4} + \frac{360° \cdot k}{4} = 30° + 90° \cdot k.$$

If $k = 0$, $\qquad \alpha = 30° + 90° \cdot \mathbf{0} = 30°.$

If $k = 1$, $\qquad \alpha = 30° + 90° \cdot \mathbf{1} = 120°.$

If $k = 2$, $\qquad \alpha = 30° + 90° \cdot \mathbf{2} = 210°.$

If $k = 3$, $\qquad \alpha = 30° + 90° \cdot \mathbf{3} = 300°.$

Using these angles, the fourth roots are

$$2 \text{ cis } 30°,$$
$$2 \text{ cis } 120°,$$
$$2 \text{ cis } 210°,$$

and

$$2 \text{ cis } 300°.$$

These four roots can be written in standard form as $\sqrt{3} + i$, $-1 + i\sqrt{3}$, $-\sqrt{3} - i$, and $1 - i\sqrt{3}$. The graphs of these roots are all on a circle that has center at the origin and radius 2, as shown in Figure 35. Notice that the roots are equally spaced about the circle 90° apart. ▶

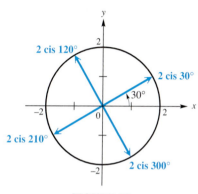

FIGURE 35

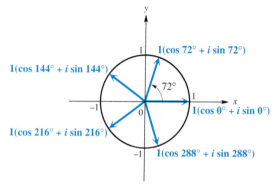

FIGURE 36

▌**EXAMPLE 3**
Solving an equation by finding complex roots

Find all complex number solutions of $x^5 - 1 = 0$.
Write the equation as

$$x^5 - 1 = 0$$
$$x^5 = 1.$$

While there is only one real number solution, 1, there are five complex number solutions. To find these solutions, first write 1 in trigonometric form as

$$1 = 1 + 0i = 1(\cos 0° + i \sin 0°).$$

The modulus of the fifth roots is $\sqrt[5]{1} = 1$, and the arguments are given by

$$0° + 72° \cdot k, \qquad k = 0, 1, 2, 3, \text{ or } 4.$$

By using these arguments, the fifth roots are

$$1(\cos 0° + i \sin 0°), \qquad k = 0$$
$$1(\cos 72° + i \sin 72°), \qquad k = 1$$
$$1(\cos 144° + i \sin 144°), \qquad k = 2$$
$$1(\cos 216° + i \sin 216°), \qquad k = 3$$

and

$$1(\cos 288° + i \sin 288°). \qquad k = 4$$

The first of these roots equals 1; the others cannot easily be expressed in standard form. The five fifth roots all lie on a unit circle and are equally spaced around it every 72°, as shown in Figure 36. ▶

8.5 Exercises ▼▼▼▼▼▼▼▼▼▼▼▼▼▼▼▼▼▼▼▼▼▼▼▼▼▼▼▼▼▼▼▼

Find each power. Write each answer in standard form. See Example 1.

1. $[3(\cos 30° + i \sin 30°)]^3$ **2.** $[2(\cos 135° + i \sin 135°)]^4$ **3.** $(\cos 45° + i \sin 45°)^8$

4. $[2(\cos 120° + i \sin 120°)]^3$ **5.** $[3 \operatorname{cis} 100°]^3$ **6.** $[3 \operatorname{cis} 40°]^3$

7. $(\sqrt{3} + i)^5$ **8.** $(2\sqrt{2} - 2i\sqrt{2})^6$ **9.** $(2 - 2i\sqrt{3})^4$

10. $\left(\dfrac{\sqrt{2}}{2} - \dfrac{\sqrt{2}}{2}i\right)^8$ **11.** $(-2 - 2i)^5$ **12.** $(-1 + i)^7$

Find and graph all cube roots of each complex number. Leave answers in trigonometric form. See Example 2.

13. $(\cos 0° + i \sin 0°)$ **14.** $(\cos 90° + i \sin 90°)$ **15.** $8 \operatorname{cis} 60°$ **16.** $27 \operatorname{cis} 300°$

17. $-8i$ **18.** $27i$ **19.** -64 **20.** 27

21. $1 + i\sqrt{3}$ **22.** $2 - 2i\sqrt{3}$ **23.** $-2\sqrt{3} + 2i$ **24.** $\sqrt{3} - i$

Find and graph the roots of 1.

25. Second (square)   **26.** Fourth   **27.** Sixth   **28.** Eighth

Find and graph the roots of i.

29. Second (square)   **30.** Fourth

31. Explain why a real number must have a real nth root if n is odd.

32. How many complex 64th roots does 1 have? How many are real? How many are not?

33. True or false: Every real number must have two real square roots.

34. True or false: Some real numbers have three real cube roots.

35. Explain why a real number can have only one real cube root.

36. Explain why the n nth roots of 1 are equally spaced around the unit circle.

37. Refer to Figure 36. A regular pentagon can be created by joining the tips of the arrows. Explain how you can use this principle to create a regular octagon.

38. Show that if z is an nth root of 1, then so is $1/z$.

Find all solutions of each equation. Leave answers in trigonometric form. See Example 3.

39. $x^3 - 1 = 0$ **40.** $x^3 + 1 = 0$ **41.** $x^3 + i = 0$ **42.** $x^4 + i = 0$

43. $x^3 - 8 = 0$ **44.** $x^3 + 27 = 0$ **45.** $x^4 + 1 = 0$ **46.** $x^4 + 16 = 0$

47. $x^4 - i = 0$ **48.** $x^5 - i = 0$ **49.** $x^3 - (4 + 4i\sqrt{3}) = 0$ **50.** $x^4 - (8 + 8i\sqrt{3}) = 0$

Use a calculator to find all solutions of each equation in standard form.

51. $x^3 + 4 - 5i = 0$ **52.** $x^5 + 2 + 3i = 0$

53. Solve the equation $x^3 - 1 = 0$ by factoring the left side as the difference of two cubes and setting each factor equal to zero. Apply the quadratic formula as needed. Then compare your solutions to those of Exercise 39.

54. Solve the equation $x^3 + 27 = 0$ by factoring the left side as the sum of two cubes and setting each factor equal to zero. Apply the quadratic formula as needed. Then compare your solutions to those of Exercise 44.

55. One of the three cube roots of a complex number is $2 + 2\sqrt{3}i$. Determine the standard form of its other two cube roots.

▼▼▼▼▼▼▼▼▼▼▼▼ **DISCOVERING CONNECTIONS** (Exercises 56–58) ▼▼▼▼▼▼▼▼▼▼▼▼

56. De Moivre's theorem states that $(\cos \theta + i \sin \theta)^2 = $ _____ .

57. Expand the left side of the equation in Exercise 56 as a binomial and collect terms to write the left side in the form $a + bi$.

58. Use the result of Exercise 57 to obtain the double-angle formulas for the cosine and sine.

59. Set your graphing calculator to degree and parametric modes, and set the window and functions as shown here. (*Note:* 72 is 360/5.) Graph to see the pentagon in the graphing calculator screen whose corners are the five 5th roots of one in the complex plane. Trace to find the coordinates of these points and thereby determine the five 5th roots of one. Compare your answers with those found in Example 3.

```
Tmin=0
Tmax=360
Tstep=72
Xmin=-1.8
Xmax=1.8
Xscl=1
Ymin=-1.2
Ymax=1.2
Yscl=1
```

$X_{1T}=\cos T$
$Y_{1T}=\sin T$

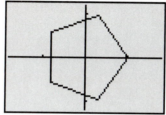

[−1.8, 1.8] by [−1.2, 1.2]

60. Use the method of Exercise 59 to find the first three of the ten 10th roots of 1.

61. The fractal called the **Mandelbrot set** is shown in the figure. To determine if a complex number $z = a + bi$ is in this set, perform the following sequence of calculations. Repeatedly compute z, $z^2 + z$, $(z^2 + z)^2 + z$, $[(z^2 + z)^2 + z]^2 + z$, ... In a manner analogous to the Julia set, the complex number z does not belong to the Mandelbrot set if any of the resulting moduli exceeds 2. Otherwise z is in the set and the point (a, b) should be shaded in the graph. Determine whether or not the following numbers belong to the Mandelbrot set. (*Sources:* Lauwerier, H., *Fractals,* Princeton University Press, Princeton, New Jersey, 1991; *Figure:* Crownover, R., *Introduction to Fractals and Chaos,* Jones and Bartlett Publishers, Boston, 1995.)

(a) $z = 0 + 0i$

(b) $z = 1 - 1i$

(c) $z = -.5i$

62. The fractal shown in the figure is the solution to Cayley's problem of determining the basins of attraction for the cube roots of unity. The three cube roots of unity are $w_1 = 1$, $w_2 = -1/2 + (\sqrt{3}/2)i$, and $w_3 = -1/2 - (\sqrt{3}/2)i$. This fractal can be generated by repeatedly evaluating the function $f(z) = \dfrac{2z^3 + 1}{3z^2}$, where z is a complex number. Begin by picking $z_1 = a + bi$ and then successively computing $z_2 = f(z_1)$, $z_3 = f(z_2)$, $z_4 = f(z_3)$, ... If the resulting values of $f(z)$ approach w_1, color the pixel at (a, b) red. If it approaches w_2, color it blue and if it approaches w_3, color it yellow. If this process continues for a large number of different z_1, the fractal in the figure will appear. Determine the appropriate color of the pixel for each value of z_1. (*Sources:* Crownover, R., *Introduction to Fractals and Chaos,* Jones and Bartlett Publishers, Boston, 1995; *Figure:* Kincaid, D. and W. Cheney, *Numerical Analysis,* Brooks/Cole Publishing Company, Pacific Grove, California, 1991.)

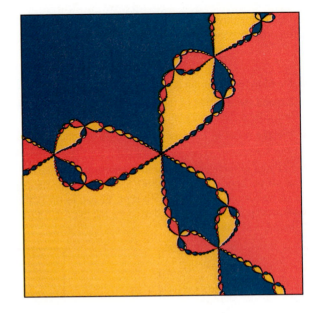

(a) $z_1 = i$

(b) $z_1 = 2 + i$

(c) $z_1 = -1 - i$

8.6 Polar Equations ▼▼▼

Throughout this text we have been using the Cartesian coordinate system to graph equations. Another coordinate system that is particularly useful for graphing many relations is the **polar coordinate system.** The system is based on a point, called the **pole,** and a ray, called the **polar axis.** The polar axis is usually drawn in the direction of the positive x-axis, as shown in Figure 37.

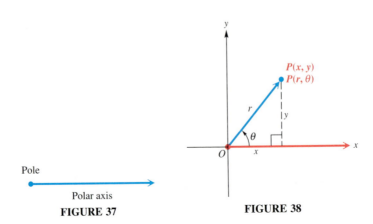

Pole

Polar axis

FIGURE 37

FIGURE 38

In Figure 38 the pole has been placed at the origin of a Cartesian coordinate system, so that the polar axis coincides with the positive x-axis. Point P has coordinates (x, y) in the Cartesian coordinate system. Point P can also be located by giving the directed angle θ from the positive x-axis to ray OP and the directed distance r from the pole to point P. The ordered pair (r, θ) gives the **polar coordinates** of point P.

EXAMPLE 1
Graphing points with polar coordinates

Plot each point, given its polar coordinates.

(a) $P(2, 30°)$

In this case $r = 2$ and $\theta = 30°$, so the point P is located 2 units from the origin in the positive direction on a ray making a 30° angle with the polar axis, as shown in Figure 39.

(b) $Q(-4, 120°)$

Since r is negative, Q is 4 units in the negative direction from the pole on an extension of the 120° ray. See Figure 40.

(c) $R(5, -45°)$

Point R is shown in Figure 41. Since θ is negative, the angle is measured in the clockwise direction. ▶

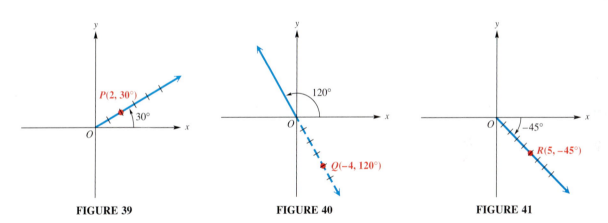

FIGURE 39   FIGURE 40   FIGURE 41

One important difference between Cartesian coordinates and polar coordinates is that while a given point in the plane can have only one pair of Cartesian coordinates, this same point can have an infinite number of pairs of polar coordinates. For example, $(2, 30°)$ locates the same point as $(2, 390°)$ or $(2, -330°)$ or $(-2, 210°)$.

EXAMPLE 2
Giving alternate forms of a pair of polar coordinates

Give three other pairs of polar coordinates for the point $P(3, 140°)$.

Three pairs that could be used for the point are $(3, -220°)$, $(-3, 320°)$, and $(-3, -40°)$. See Figure 42. ▶

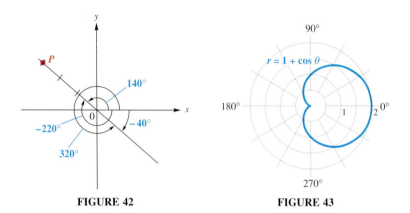

FIGURE 42   FIGURE 43

An equation like $r = 3 \sin \theta$ where r and θ are the variables, is a **polar equation.** (Equations in x and y are called **rectangular** or **Cartesian equations.**) The simplest equation for many curves turns out to be a polar equation.

Graphing a polar equation is much the same as graphing a Cartesian equation. Find some representative ordered pairs, (r, θ), satisfying the equation, and then sketch the graph.

EXAMPLE 3
Graphing a polar equation (cardioid)

Graph $r = 1 + \cos \theta$.

To graph this equation, find some ordered pairs (as in the table) and then connect the points in order—from $(2, 0°)$ to $(1.9, 30°)$ to $(1.7, 45°)$ and so on. The graph is shown in Figure 43. This curve is called a **cardioid** because of its heart shape.

θ	0°	30°	45°	60°	90°	120°	135°	150°	180°	270°	315°
$\cos \theta$	1	.9	.7	.5	0	−.5	−.7	−.9	−1	0	.7
$r = 1 + \cos \theta$	2	1.9	1.7	1.5	1	.5	.3	.1	0	1	1.7

Once the pattern of values of r becomes clear, it is not necessary to find more ordered pairs. That is why we stopped with the ordered pair $(1.7, 315°)$ in the table above. From the pattern, the pair $(1.9, 330°)$ also would satisfy the relation. ▶

▦ A graphing calculator can be used to graph an equation in the form $r = f(\theta)$. Refer to your owner's manual to see how your model handles polar graphs. As always it is necessary to set the window appropriately and choose the correct angle mode (radians or degrees). You will need to decide on maximum and minimum values of θ. Keep in mind the periods of the functions, so that the entire set of function values are generated. For example, to graph $r = 1 + \cos \theta$, which we graphed by hand in Example 3, we use degree mode. We might choose the intervals $[0, 360°]$, $[-3, 3]$, and $[-3, 3]$ for θ, x, and y, respectively, using the results of our work in Example 3 to determine these choices. The calculator-generated graph and the corresponding window are shown below. (The second window is a continuation of the first one.)

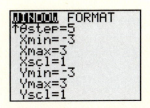

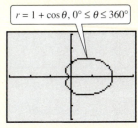

$r = 1 + \cos \theta, 0° \le \theta \le 360°$

$[-3, 3]$ by $[-3, 3]$

You can experiment with your calculator by graphing the curves shown in other examples.

EXAMPLE 4
Graphing a polar equation
(lemniscate)

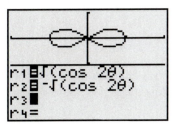

$[-2, 2]$ by $[-1, 1]$

To graph $r^2 = \cos 2\theta$ with a graphing calculator, define r_1 as $\sqrt{\cos 2\theta}$ and r_2 as $-\sqrt{\cos 2\theta}$. Compare to the graph in Figure 44.

Graph $r^2 = \cos 2\theta$.

First complete a table of ordered pairs as shown, and then sketch the graph, as in Figure 44. The point $(-1, 0°)$, with r negative, may be plotted as $(1, 180°)$. Also, $(-.7, 30°)$ may be plotted as $(.7, 210°)$, and so on. This curve is called a **lemniscate.**

θ	0°	30°	45°	135°	150°	180°
2θ	0°	60°	90°	270°	300°	360°
$\cos 2\theta$	1	.5	0	0	.5	1
$r = \pm\sqrt{\cos 2\theta}$	± 1	$\pm .7$	0	0	$\pm .7$	± 1

Values of θ for $45° < \theta < 135°$ are not included in the table because the corresponding values of $\cos 2\theta$ are negative (quadrants II and III) and so do not have real square roots. Values of θ larger than $180°$ give 2θ larger than $360°$, and would repeat the points already found. ▶

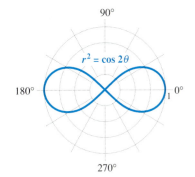

FIGURE 44

EXAMPLE 5
Graphing a polar equation
(rose)

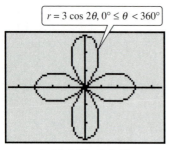

$r = 3 \cos 2\theta, 0° \le \theta < 360°$

[−4.6, 4.6] by [−3, 3]

Compare this calculator-generated graph to the one in Figure 45.

Graph $r = 3 \cos 2\theta$.

Because of the 2θ, the graph requires a large number of points. A few ordered pairs are given below. You should complete the table similarly through the first 360°.

θ	0°	15°	30°	45°	60°	75°	90°
2θ	0°	30°	60°	90°	120°	150°	180°
$\cos 2\theta$	1	.9	.5	0	−.5	−.9	−1
r	3	2.7	1.5	0	−1.5	−2.7	−3

Plotting these points in order gives the graph, called a **four-leaved rose.** Notice in Figure 45 how the graph is developed with a continuous curve, beginning with the upper half of the right horizontal leaf and ending with the lower half of that leaf. As the graph is traced, the curve goes through the pole four times.

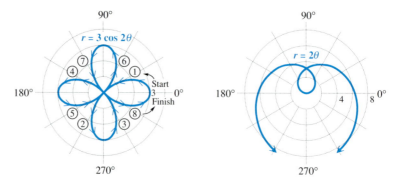

FIGURE 45   FIGURE 46

The graph in Figure 45 is one of a family of curves called **roses.** The graphs of $r = \sin n\theta$ and $r = \cos n\theta$ are roses, with n petals if n is odd, and $2n$ petals if n is even.

EXAMPLE 6
Graphing a polar equation
(spiral of Archimedes)

Graph $r = 2\theta$ (θ measured in radians).

Some ordered pairs are shown below. Since $r = 2\theta$, rather than a trigonometric function of θ, it is also necessary to consider negative values of θ. The radian measures have been rounded for simplicity.

θ (degrees)	−180	−90	−45	0	30	60	90	180	270	360
θ (radians)	−3.1	−1.6	−.8	0	.5	1	1.6	3.1	4.7	6.3
$r = 2\theta$	−6.2	−3.2	−1.6	0	1	2	3.2	6.2	9.4	12.6

Figure 46 shows this graph, called a **spiral of Archimedes.**

It is quite tedious to graph the spiral of Archimedes using traditional methods. Much more of the graph is shown in the calculator-generated graph below.

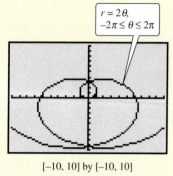

$r = 2\theta,$
$-2\pi \le \theta \le 2\pi$

[−10, 10] by [−10, 10]

Sometimes an equation given in polar form is easier to graph in Cartesian form. To convert a polar equation to a Cartesian equation, use the following relationships, which were introduced in Section 8.4. See triangle *POQ* in Figure 47.

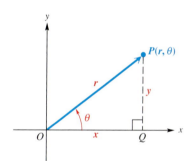

FIGURE 47

Converting Between Polar and Rectangular Coordinates

$$x = r \cos \theta \qquad r = \sqrt{x^2 + y^2}$$

$$y = r \sin \theta \qquad \tan \theta = \frac{y}{x}, \text{ if } x \ne 0$$

◖**EXAMPLE 7**
Converting a polar equation to a Cartesian equation

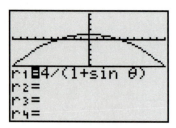

[−10, 10] by [−10, 10]
Xscl = 1   Yscl = 2

This screen shows a calculator-generated graph of the equation in Example 7, using the polar graphing mode. See Exercise 35 for more on this graph and its relationship to that of $x^2 = -8(y - 2)$.

Convert the equation

$$r = \frac{4}{1 + \sin \theta}$$

to Cartesian coordinates, and graph.
    Multiply both sides of the equation by the denominator on the right, to clear the fraction.

$$r = \frac{4}{1 + \sin \theta}$$

$$r + r \sin \theta = 4$$

Now substitute $\sqrt{x^2 + y^2}$ for r and y for $r \sin \theta$.

$$\sqrt{x^2 + y^2} + y = 4$$

$$\sqrt{x^2 + y^2} = 4 - y$$

Square both sides to eliminate the radical.

$$x^2 + y^2 = (4 - y)^2$$

$$x^2 + y^2 = 16 - 8y + y^2$$

$$x^2 = -8y + 16$$

$$x^2 = -8(y - 2)$$

The final equation represents a parabola and can be graphed using rectangular coordinates. See Figure 48. ▶

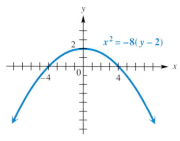

FIGURE 48

◖**EXAMPLE 8**
Converting a Cartesian equation to a polar equation

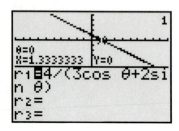

Convert the equation $3x + 2y = 4$ to a polar equation.
    Use $x = r \cos \theta$ and $y = r \sin \theta$ to get

$$3x + 2y = 4$$

$$3r \cos \theta + 2r \sin \theta = 4.$$

Now solve for r. First factor out r on the left.

$$r(3 \cos \theta + 2 \sin \theta) = 4$$

$$r = \frac{4}{3 \cos \theta + 2 \sin \theta}$$

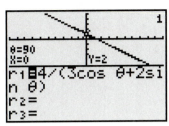

[−10, 10] by [−10, 10]
Xscl = 1   Yscl = 2

The screens here indicate that the
x- and *y*- intercepts of the line are
$\frac{4}{3}$ and 2, respectively, further
supporting the result of Example 8.
The rectangular form is $3x + 2y = 4$.

The polar equation of the line $3x + 2y = 4$ is

$$r = \frac{4}{3 \cos \theta + 2 \sin \theta}.$$ ▸

8.6 Exercises ▼▼▼▼▼▼▼▼▼▼▼▼▼▼▼▼▼▼▼▼▼▼▼▼▼▼▼▼▼▼▼▼▼▼▼▼

1. Explain why the points (r, θ) and $(r, \theta + 360°)$ have the same graph.

2. Explain why, if $r > 0$, the points (r, θ) and $(-r, \theta + 180°)$ have the same graph.

Plot each point, given its polar coordinates. Give two other pairs of polar coordinates for each point. See Examples 1 and 2.

3. $(1, 45°)$ 4. $(3, 120°)$ 5. $(-2, 135°)$ 6. $(-4, 27°)$ 7. $(5, -60°)$

8. $(2, -45°)$ 9. $(-3, -210°)$ 10. $(-1, -120°)$ 11. $(3, 300°)$ 12. $(4, 270°)$

13. If a point lies on an axis in the Cartesian plane, then what kind of angle must θ be if (r, θ) represents the point in polar coordinates?

14. What will the graph of $r = k$ be, for $k > 0$?

Graph the equation for θ in $[0°, 360°)$. See Examples 3–6.

15. $r = 2 + 2 \cos \theta$ (cardioid)   16. $r = 2(4 + 3 \cos \theta)$ (cardioid)   17. $r = 3 + \cos \theta$ (limaçon)

18. $r = 2 - \cos \theta$ (limaçon)   19. $r = 4 \cos 2\theta$ (four-leaved rose)   20. $r = 3 \cos 5\theta$ (five-leaved rose)

21. $r^2 = 4 \cos 2\theta$ (lemniscate)   22. $r^2 = 4 \sin 2\theta$ (lemniscate)   23. $r = 4(1 - \cos \theta)$ (cardioid)

24. $r = 3(2 - \cos \theta)$ (cardioid)   25. $r = 2 \sin \theta \tan \theta$ (cissoid)   26. $r = \dfrac{\cos 2\theta}{\cos \theta}$ (cissoid with a loop)

In Exercises 27–30 identify the geometric symmetry (A, B, or C) that the graph will possess.

   A. symmetry with respect to the origin
   B. symmetry with respect to the *y*-axis
   C. symmetry with respect to the *x*-axis

27. Whenever (r, θ) is on the graph, then so is $(-r, -\theta)$.

28. Whenever (r, θ) is on the graph, then so is $(-r, \theta)$.

29. Whenever (r, θ) is on the graph, then so is $(r, -\theta)$.

30. Whenever (r, θ) is on the graph, then so is $(r, \pi - \theta)$.

31. Graph the equations $r = 4 \sin \theta$ and $r = 4 \cos \theta$. How are they the same and how do they differ? Describe the graphs of $r = a \sin \theta$ and $r = a \cos \theta$ for positive a.

32. Graph the equations $r = 4(1 + \cos \theta)$, $r = 4(1 - \cos \theta)$, $r = 4(1 + \sin \theta)$, and $r = 4(1 - \sin \theta)$. How are they the same and how do they differ? Describe the graphs of $r = a(1 + \cos \theta)$, $r = a(1 - \cos \theta)$, $r = a(1 + \sin \theta)$, and $r = a(1 - \sin \theta)$ for positive a.

In Exercises 33 and 34, find the greatest value of $|r|$ of any point on the graph. Also, find all values of θ for which $r = 0$.

33. $r = 4 \cos 2\theta$, $0° \leq \theta < 360°$

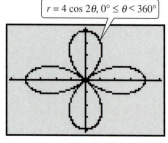

$r = 4 \cos 2\theta, 0° \leq \theta < 360°$

[−5, 5] by [−5, 5]

34. $r = 5 \sin 3\theta$, $0° \leq \theta < 180°$

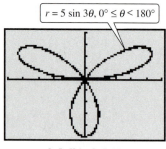

$r = 5 \sin 3\theta, 0° \leq \theta < 180°$

[−5, 5] by [−5, 5]

35. The screen displays below indicate the same point. Verify analytically that the polar coordinates shown in the left screen and the rectangular coordinates shown in the right screen are equivalent.

$$R = \frac{4}{1 + \sin \theta}$$

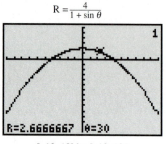

R=2.6666667 θ=30

[−10, 10] by [−10, 10]

Xscl = 1   Yscl = 1

When $\theta = 30°$, $R = \frac{8}{3}$.

$$Y = 2 - \tfrac{1}{8}X^2$$

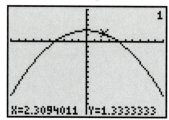

X=2.3094011 Y=1.3333333

[−10, 10] by [−10, 10]

Xscl = 1   Yscl = 1

When $X = \frac{4\sqrt{3}}{3}$, $Y = \frac{4}{3}$.

Find the polar coordinates of the points of intersection of the given curves for the specified interval of θ.

36. $r = 4 \sin \theta$, $r = 1 + 2 \sin \theta$; $0 \leq \theta < 2\pi$

37. $r = 2 + \sin \theta$, $r = 2 + \cos \theta$; $0 \leq \theta < 2\pi$

38. $r = \sin 2\theta$, $r = \sqrt{2} \cos \theta$; $0 \leq \theta < \pi$

39. Explain the method used to plot a point (r, θ) in polar coordinates, if $r < 0$.

40. Refer to Example 8. Would you find it easier to graph the equation using the Cartesian form or the polar form? Why?

For each equation, find an equivalent equation in Cartesian coordinates, and sketch the graph. See Example 7.

41. $r = 2 \sin \theta$

42. $r = 2 \cos \theta$

43. $r = \dfrac{2}{1 - \cos \theta}$

44. $r = \dfrac{3}{1 - \sin \theta}$

45. $r + 2 \cos \theta = -2 \sin \theta$

46. $r = \dfrac{3}{4 \cos \theta - \sin \theta}$

47. $r = 2 \sec \theta$

48. $r = -5 \csc \theta$

49. $r(\cos \theta + \sin \theta) = 2$

50. $r(2 \cos \theta + \sin \theta) = 2$

For each equation, find an equivalent equation in polar coordinates. See Example 8.

51. $x + y = 4$

52. $2x - y = 5$

53. $x^2 + y^2 = 16$

54. $x^2 + y^2 = 9$

55. $y = 2$

56. $x = 4$

57. Graph $r = \theta$, a spiral of Archimedes. (See Example 6.)

58. Show that the distance between (r_1, θ_1) and (r_2, θ_2) is $\sqrt{r_1^2 + r_2^2 - 2r_1 r_2 \cos(\theta_1 - \theta_2)}$.

In Exercises 59 and 60, write a polar equation of the line through the given points.

59. $(1, 0°), (2, 90°)$

60. $(2, 30°), (1, 90°)$

61. The polar equation $r = \dfrac{a(1 - e^2)}{1 + e \cos \theta}$ can be used to graph the orbits of the planets, where a is the average distance in astronomical units from the sun and e is a constant called the eccentricity. The sun will be located at the pole. The table lists a and e for the planets.

(a) Graph the orbits of the four closest planets on the same polar axis. Choose a viewing window that results in a graph with nearly circular orbits.

(b) Plot the orbits of Earth, Jupiter, Uranus, and Pluto on the same polar axis. How does Earth's distance from the sun compare to these planets?

(c) Use graphing to determine whether or not Pluto is always the farthest planet from the sun.

Planet	a	e
Mercury	.39	.206
Venus	.78	.007
Earth	1.00	.017
Mars	1.52	.093
Jupiter	5.20	.048
Saturn	9.54	.056
Uranus	19.2	.047
Neptune	30.1	.009
Pluto	39.4	.249

(*Sources:* Karttunen, H., P. Kröger, H. Oja, M. Putannen, and K. Donners (Editors), *Fundamental Astronomy*, Springer-Verlag, 1994;

Zeilik, M., S. Gregory, and E. Smith, *Introductory Astronomy and Astrophysics*, Saunders College Publishers, 1992.)

8.7 Parametric Equations ▼▼▼

Throughout this text, we have graphed sets of ordered pairs of real numbers that corresponded to a function of the form $y = f(x)$ or $r = g(\theta)$. Another way to determine a set of ordered pairs involves two functions f and g defined by $x = f(t)$ and $y = g(t)$, where t is a real number in some interval I. Each value of t leads to a corresponding x-value and a corresponding y-value, and thus to an ordered pair (x, y).

Parametric Equations of a Plane Curve

A **plane curve** is a set of points (x, y) such that $x = f(t)$, $y = g(t)$, and f and g are both defined on an interval I. The equations $x = f(t)$ and $y = g(t)$ are **parametric equations** with **parameter t.**

◀ **EXAMPLE 1**
Graphing a plane curve defined parametrically

Let $x = t^2$ and $y = 2t + 3$ for t in $[-3, 3]$. Graph the set of ordered pairs (x, y).
    Begin by making a table of values.

t	-3	-2	-1	0	1	2	3
x	9	4	1	0	1	4	9
y	-3	-1	1	3	5	7	9

Now graph the points (x, y) from the table of values and connect them with a smooth curve as in Figure 49. Since the domain of t is a closed interval, the graph has endpoints at $(9, -3)$ and $(9, 9)$. ▶

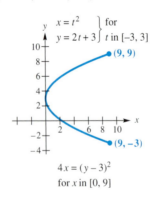

$4x = (y - 3)^2$
for x in $[0, 9]$

FIGURE 49

⊡ In addition to graphing rectangular and polar equations, graphing calculators are capable of graphing plane curves defined by parametric equations. The calculator must be set in parametric mode, and the window requires intervals for the parameter T, as well as for X and Y. The window and graph below are for the graph of $X = T^2$ and $Y = 2T + 3$, with T in $[-3, 3]$, as in Example 1.

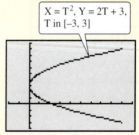

X = T², Y = 2T + 3, T in [–3, 3]

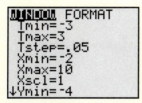

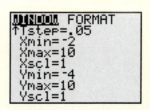

Sometimes it is possible to eliminate the parameter from a pair of parametric equations to get a **rectangular equation,** an equation relating x and y.

EXAMPLE 2
Finding an equivalent rectangular equation

Find a rectangular equation for the plane curve defined as follows, and graph the curve.

$$x = t^2, y = 2t + 3, \quad \text{for } t \text{ in } [-3, 3]$$

This is the curve of Example 1. To eliminate the parameter t, solve either equation for t. Here, only the second equation, $y = 2t + 3$, leads to a unique solution for t, so choose it.

$$y = 2t + 3$$
$$2t = y - 3$$
$$t = \frac{y - 3}{2}$$

Now substitute the result in the first equation to get

$$x = t^2 = \left(\frac{y - 3}{2}\right)^2 = \frac{(y - 3)^2}{4}$$

or $$4x = (y - 3)^2.$$

This is the equation of a horizontal parabola opening to the right, which agrees with the graph given in Figure 49. Because t is in $[-3, 3]$, x is in $[0, 9]$ and y is in $[-3, 9]$. The rectangular equation must be given with its restricted domain as

$$4x = (y - 3)^2, \quad \text{for } x \text{ in } [0, 9]. \quad \blacktriangleright$$

Trigonometric functions are often used to define a plane curve parametrically.

EXAMPLE 3
Defining a plane curve parametrically with trigonometric functions

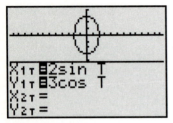

[−12.4, 12.4] by [−4, 4]

Compare this graph to the one in Figure 50. It is an ellipse with x-intercepts ±2 and y-intercepts ±3.

Graph the plane curve defined by $x = 2 \sin t$, $y = 3 \cos t$, for t in $[0, 2\pi]$.

It is awkward to solve either equation for t. Instead, use the fact that $\sin^2 t + \cos^2 t = 1$ to suggest another approach. Square each equation; solve one for $\sin^2 t$, the other for $\cos^2 t$.

$$x = 2 \sin t \qquad\qquad y = 3 \cos t$$
$$x^2 = 4 \sin^2 t \qquad\qquad y^2 = 9 \cos^2 t$$
$$\frac{x^2}{4} = \sin^2 t \qquad\qquad \frac{y^2}{9} = \cos^2 t$$

Now add corresponding sides of the two equations to get

$$\frac{x^2}{4} + \frac{y^2}{9} = \sin^2 t + \cos^2 t$$
$$\frac{x^2}{4} + \frac{y^2}{9} = 1,$$

the equation of an ellipse with vertical major axis as shown in Figure 50. ▶

$x = 2 \sin t$ ⎫ for
$y = 3 \cos t$ ⎭ t in $[0, 2\pi]$

$$\frac{x^2}{4} + \frac{y^2}{9} = 1$$

FIGURE 50

Parametric representations of a curve are not unique. In fact, there are infinitely many parametric representations of a given curve. If the curve can be described by a rectangular equation $y = f(x)$, with domain X, then one simple parametric representation is

$$x = t, \, y = f(t), \quad \text{for } t \text{ in } X.$$

EXAMPLE 4
Finding alternative parametric equation forms

Give three parametric representations for the parabola

$$y = (x - 2)^2 + 1.$$

The simplest choice is to let

$$x = t, \, y = (t - 2)^2 + 1, \quad \text{for } t \text{ in } (-\infty, \infty).$$

Another choice that leads to a simpler equation for y is

$$x = t + 2, \ y = t^2 + 1, \quad \text{for } t \text{ in } (-\infty, \infty).$$

Sometimes trigonometric functions are desirable; one choice here might be

$$x = 2 + \tan t, \ y = \sec^2 t, \quad \text{for } t \text{ in } \left(-\frac{\pi}{2}, \frac{\pi}{2}\right). \ \blacktriangleright$$

An important application of parametric equations is to determine the path of a moving object whose position is given by the functions $x = f(t), \ y = g(t)$, where t represents time. The parametric equations give the position of the object at any time t.

EXAMPLE 5
Examining parametric equations defining the position of an object in motion

The motion of a projectile (neglecting air resistance) is given by

$$x = (v_0 \cos \theta)t, \ y = (v_0 \sin \theta)t - 16t^2, \quad \text{for } t \text{ in } [0, k],$$

where t is time in seconds, v_0 is the initial speed of the projectile in the direction θ with the horizontal, x and y are in feet, and k is a positive real number. See Figure 51.

Solving the first equation for t and substituting the result into the second equation gives (after simplification)

$$y = (\tan \theta)x - \frac{16}{v_0^2 \cos^2 \theta}x^2,$$

the equation of a vertical parabola opening downward, as shown in Figure 51. $\blacktriangleright$

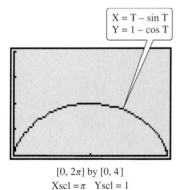

$$X = T - \sin T$$
$$Y = 1 - \cos T$$

$[0, 2\pi]$ by $[0, 4]$
Xscl $= \pi$ Yscl $= 1$

This screen shows a cycloid with $a = 1$.

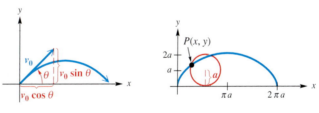

FIGURE 51

FIGURE 52

The path traced by a fixed point on the circumference of a circle rolling along a line is called a *cycloid*. See Figure 52. The **cycloid** is defined by

$$x = at - a \sin t, \ y = a - a \cos t, \quad \text{for } t \text{ in } (-\infty, \infty).$$

EXAMPLE 6
Graphing a cycloid

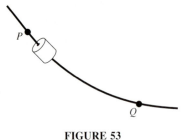

FIGURE 53

Graph the cycloid for t in $[0, 2\pi]$.

There is no simple way to find a rectangular equation for the cycloid from its parametric equations. Instead, begin with a table of values.

t	0	$\dfrac{\pi}{4}$	$\dfrac{\pi}{2}$	π	$\dfrac{3\pi}{2}$	2π
x	0	$.08a$	$.6a$	πa	$5.7a$	$2\pi a$
y	0	$.3a$	a	$2a$	a	0

Plotting the ordered pairs (x, y) from the table of values leads to the portion of the graph in Figure 52 on page 691 from 0 to $2\pi a$. ▶

The cycloid has an interesting physical property. If a flexible cord or wire goes through points P and Q as in Figure 53 and a bead is allowed to slide without friction along this path from P to Q, the path that requires the shortest time takes the shape of the graph of an inverted cycloid.

8.7 Exercises ▼▼▼▼▼▼▼▼▼▼▼▼▼▼▼▼▼▼▼▼▼▼▼▼▼▼▼▼▼▼▼▼▼▼▼▼▼▼

1. The graphing calculator screen at the right shows the graph of the parametric equations $X = -2 + 6T$, $Y = 1 + 2T$, T in $[0, 1]$. What is the first point graphed (corresponding to $T = 0$) and the last point graphed (corresponding to $T = 1$) as the curve is traced out? What point is plotted when $T = .6$?

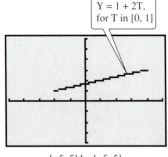

[–5, 5] by [–5, 5]

2. The parametric equations $X = 4 - 6T$, $Y = 3 - 2T$, T in $[0, 1]$ have the same graph as in Exercise 1. What is the first point graphed (corresponding to $T = 0$) and the last point graphed (corresponding to $T = 1$) as the curve is traced out? What point is plotted when $T = .6$?

3. The parametric equations $x = \cos t$, $y = \sin t$, t in $[0, 2\pi]$ and the parametric equations $x = \cos t$, $y = -\sin t$, t in $[0, 2\pi]$ both have the unit circle as their graph. However, in one case the circle is traced out clockwise (as t moves from 0 to 2π) and in the other case the circle is traced out counterclockwise. For which equations is the circle traced out in the clockwise direction?

4. Consider the parametric equations $x = f(t)$, $y = g(t)$, t in $[a, b]$. How is the graph affected if the equation $x = f(t)$ is replaced by $x = c + f(t)$? How is the graph affected if the equation $y = g(t)$ is replaced by $y = d + g(t)$? (Assume that c and d are positive.)

Use a table of values to graph each plane curve defined by the parametric equations. Find a rectangular equation for each curve. See Examples 1 and 2.

5. $x = 2t, y = t + 1$, for t in $[-2, 3]$

6. $x = t + 2, y = t^2$, for t in $[-1, 1]$

7. $x = \sqrt{t}, y = 3t - 4$, for t in $[0, 4]$

8. $x = t^2, y = \sqrt{t}$, for t in $[0, 4]$

9. $x = t^3 + 1, y = t^3 - 1$, for t in $(-\infty, \infty)$

10. $x = 2t - 1, y = t^2 + 2$, for t in $(-\infty, \infty)$

11. $x = 2^t, y = \sqrt{3t - 1}$, for t in $[1, \infty)$

12. $x = \ln(t - 1), y = 2t - 2$, for t in $(1, \infty)$

13. $x = 2 \sin \theta, y = 2 \cos \theta$, for θ in $[0, 2\pi]$

14. $x = \sqrt{5} \sin \theta, y = \sqrt{3} \cos \theta$, for θ in $[0, 2\pi]$

15. $x = 3 \tan \theta, y = 2 \sec \theta$, for θ in $\left(-\dfrac{\pi}{2}, \dfrac{\pi}{2}\right)$

16. $x = \cot \theta, y = \csc \theta$, for θ in $(0, \pi)$

Find a rectangular equation for each curve defined and graph each curve. See Examples 1 and 2.

17. $x = \sin \theta, y = \csc \theta$, for θ in $(0, \pi)$

18. $x = \tan \theta, y = \cot \theta$, for θ in $\left(0, \dfrac{\pi}{2}\right)$

19. $x = t, y = \sqrt{t^2 + 2}$, for t in $(-\infty, \infty)$

20. $x = \sqrt{t}, y = t^2 - 1$, for t in $[0, \infty)$

21. $x = e^t, y = e^{-t}$, for t in $(-\infty, \infty)$

22. $x = e^{2t}, y = e^t$, for t in $(-\infty, \infty)$

23. $x = 2 + \sin \theta, y = 1 + \cos \theta$, for θ in $[0, 2\pi]$

24. $x = 1 + 2 \sin \theta, y = 2 + 3 \cos \theta$, for θ in $[0, 2\pi]$

25. $x = t + 2, y = \dfrac{1}{t + 2}, t \neq -2$

26. $x = t - 3, y = \dfrac{2}{t - 3}, t \neq 3$

27. $x = t^2, y = 2 \ln t, t$ in $(0, \infty)$

28. $x = \ln t, y = 3 \ln t$, for t in $(0, \infty)$

Graph each curve defined in Exercises 29 and 30. Assume the interval for t is all real numbers for which $x = f(t)$ and $y = g(t)$ are both defined. See Examples 2 and 3.

29. (a) $x = \sin t, y = \cos t$ **(b)** $x = t, y = \dfrac{\sqrt{4 - 4t^2}}{2}$

30. (a) $x = t + 2, y = t - 4$ **(b)** $x = t^2 + 2, y = t^2 - 4$

Graph each cycloid defined in Exercises 31 and 32 for θ in the given interval. See Example 6.

31. $x = \theta - \sin \theta, y = 1 - \cos \theta, \theta$ in $[0, 4\pi]$

32. $x = 2\theta - 2 \sin \theta, y = 2 - 2 \cos \theta, \theta$ in $[0, 8\pi]$

33. A projectile is fired with an initial velocity of 400 ft per sec at an angle of 45° with the horizontal. Find each of the following. See Example 5. **(a)** the time when it strikes the ground **(b)** the range (horizontal distance covered) **(c)** the maximum altitude

34. Repeat Exercise 33 if the projectile is fired at 800 ft per sec at an angle of 30° with the horizontal.

35. Show that the rectangular equation for the curve describing the motion of a projectile defined by

$$x = (v_0 \cos \theta)t, \quad y = (v_0 \sin \theta)t - 16t^2, \quad \text{for } t \text{ in } [0, k],$$

is

$$y = (\tan \theta)x - \frac{16}{v_0^2 \cos^2 \theta}x^2.$$

36. Find the vertex of the parabola given by the rectangular equation of Exercise 35.

37. Give two parametric representations of the line through the point (x_1, y_1) with slope m.

38. Give two parametric representations of the parabola $y = a(x - h)^2 + k$.

39. Give a parametric representation of the hyperbola $(x^2/a^2) - (y^2/b^2) = 1$.

40. Give a parametric representation of the ellipse $(x^2/a^2) + (y^2/b^2) = 1$.

41. The spiral of Archimedes has polar equation $r = a\theta$, where $r^2 = x^2 + y^2$. Show that a parametric representation of the spiral of Archimedes is

$$x = a\theta \cos \theta, \; y = a\theta \sin \theta, \quad \text{for } \theta \text{ in } (-\infty, \infty).$$

42. Show that the hyperbolic spiral $r\theta = a$, where $r^2 = x^2 + y^2$, is given parametrically by

$$x = \frac{a \cos \theta}{\theta}, y = \frac{a \sin \theta}{\theta}, \quad \text{for } \theta \text{ in } (-\infty, 0) \cup (0, \infty).$$

Chapter 8   Summary ▼▼▼▼▼▼▼▼▼▼▼▼▼▼▼▼▼▼▼▼▼▼▼▼▼▼▼▼▼▼

KEY TERMS	KEY IDEAS
8.1 Oblique Triangles and the Law of Sines	
oblique triangle ambiguous case	**Law of Sines** In any triangle ABC, with sides a, b, and c, $$\frac{a}{\sin A} = \frac{b}{\sin B}, \quad \frac{a}{\sin A} = \frac{c}{\sin C}, \quad \frac{b}{\sin B} = \frac{c}{\sin C}.$$ **Area of a Triangle** The area of a triangle is given by half the product of the lengths of two sides and the sine of the angle between the two sides. $$K = \frac{1}{2}bc \sin A, \quad K = \frac{1}{2}ab \sin C, \quad K = \frac{1}{2}ac \sin B$$
8.2 The Law of Cosines	
semiperimeter	**Law of Cosines** In any triangle ABC, with sides a, b, and c, $$a^2 = b^2 + c^2 - 2bc \cos A$$ $$b^2 = a^2 + c^2 - 2ac \cos B$$ $$c^2 = a^2 + b^2 - 2ab \cos C.$$ **Heron's Area Formula** If a triangle has sides of lengths a, b, and c, and if the semiperimeter is $$s = \frac{1}{2}(a + b + c),$$ then the area of the triangle is $$K = \sqrt{s(s - a)(s - b)(s - c)}.$$

KEY TERMS	KEY IDEAS

8.3 Vectors and Their Applications

scalar
vector
magnitude of a vector
initial point
terminal point
parallelogram rule
resultant
component (vertical, horizontal)
zero vector
scalar product
equilibrant
airspeed
groundspeed

Sum of Two Vectors

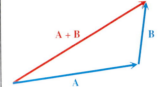 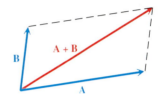

Opposite Vectors

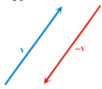

8.4 Products and Quotients of Complex Numbers in Trigonometric Form

real axis
imaginary axis
complex plane
modulus (absolute value)
argument

Trigonometric Form of Complex Numbers
If the complex number $x + yi$ corresponds to the vector with direction
angle θ and magnitude r, then

$$x = r \cos \theta \qquad r = \sqrt{x^2 + y^2}$$

$$y = r \sin \theta \qquad \tan \theta = \frac{y}{x}, \text{ if } x \neq 0$$

and

$$r(\cos \theta + i \sin \theta) \qquad \text{or} \qquad r \operatorname{cis} \theta$$

is the trigonometric form (or polar form) of $x + yi$.

Product and Quotient Theorems
For any two complex numbers $r_1(\cos \theta_1 + i \sin \theta_1)$ and
$r_2(\cos \theta_2 + i \sin \theta_2)$,

$$[r_1(\cos \theta_1 + i \sin \theta_1)] \cdot [r_2(\cos \theta_2 + i \sin \theta_2)]$$
$$= r_1 r_2[\cos(\theta_1 + \theta_2) + i \sin (\theta_1 + \theta_2)]$$

and

$$\frac{r_1(\cos \theta_1 + i \sin \theta_1)}{r_2(\cos \theta_2 + i \sin \theta_2)} = \frac{r_1}{r_2} [\cos(\theta_1 - \theta_2) + i \sin (\theta_1 - \theta_2)],$$

where $r_2 \operatorname{cis} \theta_2 \neq 0$.

KEY TERMS	KEY IDEAS
8.5 Powers and Roots of Complex Numbers	
	De Moivre's Theorem $$[r(\cos \theta + i \sin \theta)]^n = r^n(\cos n\theta + i \sin n\theta)$$ **nth Root Theorem** If n is any positive integer and r is a positive real number, then the complex number $r(\cos \theta + i \sin \theta)$ has exactly n distinct nth roots, given by $$\sqrt[n]{r}(\cos \alpha + i \sin \alpha),$$ where $$\alpha = \frac{\theta + 360°k}{n} \quad \text{or} \quad \alpha = \frac{\theta}{n} + \frac{360°k}{n},$$ $k = 0, 1, 2, \ldots, n - 1.$
8.6 Polar Equations	
polar coordinate system pole polar axis polar coordinates polar equation cardioid lemniscate n-leaved rose spiral of Archimedes	Polar coordinates determine a point by locating it θ degrees from the polar axis (the positive x-axis) and r units from the origin. Polar equations are graphed in the same way as Cartesian equations, by point plotting or with a graphing calculator.
8.7 Parametric Equations	
plane curve parametric equations parameter rectangular equation cycloid	A plane curve is a set of points (x, y) such that $x = f(t)$, $y = g(t)$, and f and g are both defined on an interval I. The equations $x = f(t)$ and $y = g(t)$ are parametric equations with parameter t.

Chapter 8   Review Exercises ▼▼▼▼▼▼▼▼▼▼▼▼▼▼▼▼▼▼▼▼▼▼▼▼▼▼▼▼▼▼

Use the law of sines to find the indicated part of each triangle ABC.

1. $C = 74.2°$, $c = 96.3$ m, $B = 39.5°$;   find b

2. $A = 129.7°$, $a = 127$ ft, $b = 69.8$ ft;   find B

3. $C = 51.3°$, $c = 68.3$ m, $b = 58.2$ m;   find B

4. $a = 165$ m, $A = 100.2°$, $B = 25.0°$;   find b

5. $B = 39° 50'$, $b = 268$ m, $a = 340$ m;   find A

6. $C = 79° 20'$, $c = 97.4$ mm, $a = 75.3$ mm;   find A

7. If we are given a, A, and C in a triangle ABC, does the possibility of the ambiguous case exist? If not, explain why.

8. Can triangle *ABC* exist if *a* = 4.7, *b* = 2.3, and *c* = 7.0? If not, explain why. Answer this question without using trigonometry.

9. Given *a* = 10 and *B* = 30°, determine the values of *b* for which *A* has
   (a) Exactly one value   **(b)** Two values   **(c)** No value.

Use the law of cosines to find the indicated part of each triangle ABC.

10. *a* = 86.14 in, *b* = 253.2 in, *c* = 241.9 in;   find *A*

11. *B* = 120.7°, *a* = 127 ft, *c* = 69.8 ft;   find *b*

12. *A* = 51° 20′, *c* = 68.3 m, *b* = 58.2 m;   find *a*

13. *a* = 14.8 m, *b* = 19.7 m, *c* = 31.8 m;   find *B*

14. *A* = 46° 10′, *b* = 184 cm, *c* = 192 cm;   find *a*

15. *a* = 7.5 ft, *b* = 12.0 ft, *c* = 6.9 ft;   find *C*

Solve each triangle ABC having the given information.

16. *A* = 61.7°, *a* = 78.9 m, *b* = 86.4 m

17. *a* = 27.6 cm, *b* = 19.8 cm, *C* = 42° 30′

18. *a* = 94.6 yd, *b* = 123 yd, *c* = 109 yd

Find the area of each triangle ABC with the given information.

19. *b* = 840.6 m, *c* = 715.9 m, *A* = 149.3°

20. *a* = 6.90 ft, *b* = 10.2 ft, *C* = 35° 10′

21. *a* = .913 km, *b* = .816 km, *c* = .582 km

22. *a* = 43 m, *b* = 32 m, *c* = 51 m

Solve the problem.

23. Raoul plans to paint a triangular wall in his A-frame cabin. Two sides measure 7 m each, and the third side measures 6 m. How much paint will he need to buy if a can of paint covers 7.5 sq m?

24. A lot has the shape of a quadrilateral. (See the figure.) What is its area?

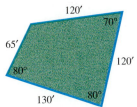

25. A tree leans at an angle of 8.0° from the vertical. (See the figure.) From a point 7.0 m from the bottom of the tree, the angle of elevation to the top of the tree is 68°. How tall is the tree?

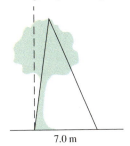

7.0 m

26. A hill makes an angle of 14.3° with the horizontal. From the base of the hill, the angle of elevation to the top of a tree on top of the hill is 27.2°. The distance along the hill from the base to the tree is 212 ft. Find the height of the tree.

27. A ship is sailing east. At one point, the bearing of a submerged rock is 45° 20′. After sailing 15.2 mi, the bearing of the rock has become 308° 40′. Find the distance of the ship from the rock at the latter point.

28. From an airplane flying over the ocean, the angle of depression to a submarine lying just under the surface is 24° 10′. At the same moment the angle of depression from the airplane to a battleship is 17° 30′. (See the figure.) The distance from the airplane to the battleship is 5120 ft. Find the distance between the battleship and the submarine. (Assume the airplane, submarine, and battleship are in a vertical plane.)

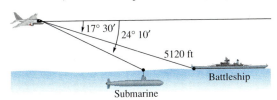

29. To measure the distance through a mountain for a proposed tunnel, a point C is chosen that can be

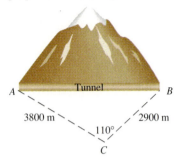

reached from each end of the tunnel. See the figure. If $AC = 3800$ m, $BC = 2900$ m, and angle $C = 110°$, find the length of the tunnel.

30. The Vietnam Veterans' Memorial in Washington, D.C., is in the shape of an unenclosed isosceles triangle (that is, V-shaped) with equal sides of length 246.75 feet and the angle between these sides measuring 125° 12′. Find the distance between the ends of the two equal sides.

31. If angle C of a triangle ABC measures 90°, what does the law of cosines $c^2 = a^2 + b^2 - 2ab \cos C$ become?

In Exercises 32–34, use the vectors pictured here. Find each of the following.

32. **a + b**

33. **a − b**

34. **a + 3c**

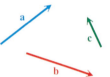

35. True or false: Opposite angles of a parallelogram are congruent.

36. True or false: A diagonal of a parallelogram must bisect two angles of the parallelogram.

Find the horizontal and vertical components of each vector, where α is the inclination of the vector from the horizontal.

37. $\alpha = 45°$, magnitude 50

38. $\alpha = 75°$, magnitude 69.2

Given two forces and the angle between them, find the magnitude of the resultant force.

39. Forces of 142 and 215 newtons, forming an angle of 112°

40. Forces of 85.2 and 69.4 newtons, forming an angle of 58° 20′

Find the magnitude and direction angle for **u** *rounded to the nearest tenth. See the Connections box in Section 8.3.*

41. $\mathbf{u} = \langle 21, -20 \rangle$

42. $\mathbf{u} = \langle -9, 12 \rangle$

Write **u** *in the form* $\langle a, b \rangle$. *See the Connections box in Section 8.3.*

43. $|\mathbf{u}| = 6$, direction angle of $\mathbf{u} = 60°$

44. $|\mathbf{u}| = 8$, direction angle of $\mathbf{u} = 120°$

Solve the problem.

45. One rope pulls a barge directly east with a force of 100 newtons. Another rope pulls the barge to the northeast with a force of 200 newtons. Find the resultant force acting on the barge, and the angle between the resultant and the first rope.

46. Paula and Steve are pulling their daughter Jessie on a sled. Steve pulls with a force of 18 lb at an angle of 10°. Paula pulls with a force of 12 lb at an angle of 15°. What is the weight of Jessie and the sled? See the figure. (*Hint:* Find the resultant force.)

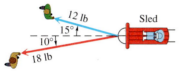

47. A 186-lb force just keeps a 2800-lb car from rolling down a hill. What angle does the hill make with the horizontal?

48. A plane has an airspeed of 520 mph. The pilot wishes to fly on a bearing of 310°. A wind of 37 mph is blowing from a bearing of 212°. What direction should the pilot fly, and what will be her actual speed?

49. In order to obtain accurate aerial photographs, ground control must determine the coordinates of **control points** located on the ground that can be identified in the photographs. Using these known control points, the orientation and scale of each photograph can be determined. Then, unknown positions and distances can easily be determined. Before an aerial photograph is taken for highway design, horizontal control points must be found and the distances between them calcu-

lated. The figure shows three consecutive control points *A*, *B*, and *C*. A surveyor measures a baseline distance of 92.13 feet from *B* to an arbitrary point *P*. Angles *BAP* and *BCP* are found to be 2° 22′ 47″ and 5° 13′ 11″, respectively. Then, angles *APB* and *CPB* are determined to be 63° 4′ 25″ and 74° 19′ 49″, respectively. Determine the distances between the control points *A* and *B* and between *B* and *C*. (*Source:* Moffitt, F., *Photogrammetry,* International Textbook Company, Scranton, Pennsylvania, 1967.)

50. In order to find the coordinates of control points for aerial photographs, ground control must first locate basic control monuments established by the U. S. Coast and Geodetic Survey and the U. S. Geological Survey. These monuments have published *x*- and *y*-coordinates called **state plane coordinates.** Using these monuments and common surveying techniques the coordinates of the control points can be determined. Two basic control monuments *A* and *B* have coordinates in feet of $x_A = 2,101,345.1$, $y_A = 998,764.3$ and $x_B = 2,131,667.8$, $y_B = 923,541.7$. The location of an unknown control point *P* is to be determined. If angles *PAB* and *PBA* are measured as 37° 41′ 37″ and 57° 52′ 04″, respectively, discuss the steps you would take to determine the state plane coordinates of control point *P*. (*Source:* Moffitt, F., *Photogrammetry,* International Textbook Company, Scranton, Pennsylvania, 1967.)

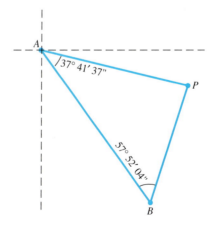

Perform the given operations. Write answers in standard form.

51. $[5(\cos 90° + i \sin 90°)][6(\cos 180° + i \sin 180°)]$

52. $[3 \text{ cis } 135°][2 \text{ cis } 105°]$

53. $\dfrac{2(\cos 60° + i \sin 60°)}{8(\cos 300° + i \sin 300°)}$

54. $\dfrac{4 \text{ cis } 270°}{2 \text{ cis } 90°}$

55. $(\sqrt{3} + i)^3$

56. $(\cos 100° + i \sin 100°)^6$

57. The vector representing a real number will lie on the _____-axis in the complex plane.

58. Explain the geometric similarity between the absolute value of a real number and the absolute value (or modulus) of a complex number. (*Hint:* Think in terms of distance.)

Graph each complex number.

59. $-4 + 2i$

60. $3 - 3i\sqrt{3}$

Find and graph the resultant of each pair of complex numbers.

61. $7 + 3i$ and $-2 + i$

62. $2 - 4i$ and $5 + i$

Complete the chart in Exercises 63–68.

Standard form	Trigonometric form
63. $-2 + 2i$	_____
64. _____	$3(\cos 90° + i \sin 90°)$
65. _____	$2 \text{ cis } 225°$
66. $1 - i$	_____
67. $-4i$	_____
68. _____	$2 \text{ cis } 180°$

The complex number z, where $z = x + yi$, can be graphed in the plane as (x, y). Describe the graphs of all complex numbers z satisfying the conditions in Exercises 69 and 70.

69. The modulus of z is 2.

70. The imaginary part of z is the negative of the real part of z.

71. Find the fifth roots of $-2 + 2i$.

72. Find the cube roots of $1 - i$.

73. How many real fifth roots does -32 have?

74. How many real sixth roots does -64 have?

Solve the equation. Express solutions in polar form.

75. $x^3 + 125 = 0$

76. $x^4 + 16 = 0$

Graph each polar equation for θ in $[0°, 360°)$.

77. $r = 4 \cos \theta$

78. $r = -1 + \cos \theta$

79. $r = 1 - \cos \theta$

80. $r = 2 \sin 4\theta$

81. $r = 3 \cos 3\theta$

Find an equivalent equation in rectangular coordinates.

82. $r = \dfrac{3}{1 + \cos \theta}$

83. $r = \dfrac{4}{2 \sin \theta - \cos \theta}$

84. $r = \sin \theta + \cos \theta$

85. $r = 2$

Find an equivalent equation in polar coordinates.

86. $y = x$

87. $y = x^2$

88. $x = y^2$

In Exercises 89–91, find a polar equation having the given graph. (Note: The values of Xscl and Yscl are 1.)

89.

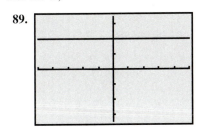

[−5, 5] by [−3.3, 3.3]

90.

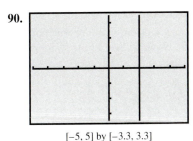

[−5, 5] by [−3.3, 3.3]

91.

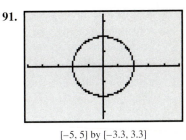

[−5, 5] by [−3.3, 3.3]

Find a rectangular equation for each plane curve with the following parametric equations.

92. $x = 3t + 2$, $y = t − 1$, for t in $[−5, 5]$

93. $x = t^2 + 1$, $y = 2t^2 − 1$, for t in $[−2, 3]$

94. $x = \sqrt{t − 1}$, $y = \sqrt{t}$, for t in $[1, ∞)$

95. $x = t^2 + 5$, $y = \dfrac{1}{t^2 + 1}$, for t in $(−∞, ∞)$

96. $x = 5 \tan t$, $y = 3 \sec t$, for t in $(−π/2, π/2)$

97. $x = \cos 2t$, $y = \sin t$, for t in $(−π, π)$

98. Show that the graph of $x = a + r \cos t$, $y = b + r \sin t$, t in $[0, 2π]$, is a circle with center (a, b) and radius r.

99. Refer to Exercise 98. Find a pair of parametric equations whose graph is the circle with center $(3, 4)$ and containing the origin.

100. Follow the steps in Exercise 42 of Section 8.4 to show that the graph of the Mandelbrot set in Exercise 61 of Section 8.5 is symmetric with respect to the *x*-axis.

Chapter 8   Test ▼▼▼▼▼▼▼▼▼▼▼▼▼▼▼▼▼▼▼▼▼▼▼▼▼▼▼▼▼▼

Find the indicated part of each triangle.

1. $A = 25.2°$, $a = 6.92$ yd, $b = 4.82$ yd;   find C

2. $C = 118°$, $b = 132$ km, $a = 75.1$ km;   find c

3. $a = 17.3$ ft, $b = 22.6$ ft, $c = 29.8$ ft;   find B

4. Given $a = 10$ and $B = 150°$ in triangle *ABC*, determine the values of b for which A has **(a)** exactly one value **(b)** two values **(c)** no values.

5. What conditions determine whether or not three positive numbers can represent the sides of a triangle?

Solve each applied problem.

6. Two boats leave a dock together. Each travels in a straight line. The angle between their courses measures 54.2°. One boat travels 36.2 km per hr, and the other travels 45.6 km per hr. How far apart will they be after 3 hr?

7. A baseball diamond is a square, 90 ft on a side, with home plate and the three bases at the vertices. The pitcher's rubber is located 60.5 ft from home plate. Find the distance from the pitcher's rubber to each of the bases.

8. Find the horizontal and vertical components of the vector with magnitude 964 lb that is inclined 154° 20′ from the horizontal.

9. Find the magnitude of the resultant of forces of 475 lb and 586 lb that form an angle of 78.2°.

10. A long-distance swimmer starts swimming a steady 3.2 mph due north. A 5.1-mph current is flowing on a bearing of 12°. What is the swimmer's resulting bearing and speed?

11. Find the resultant of $-5 + 8i$ and $6 - 4i$.

12. Write $-4 + 4i\sqrt{3}$ in trigonometric form.

13. Write 4 cis 240° in standard form.

Perform the indicated operation. Give the answer in standard form.

14. $3(\cos 30° + i \sin 30°) \cdot 5(\cos 90° + i \sin 90°)$

15. $\dfrac{2 \text{ cis } 315°}{4 \text{ cis } 45°}$ 16. $(2 - 2i)^5$

17. Find the fourth roots of $\sqrt{3} + i$. Leave the answers in trigonometric form.

18. Graph $r = 5 \cos \theta$.

19. Find an equivalent equation in polar coordinates for $x^2 - 3x + y^2 - 2y = 0$.

20. Find an equivalent equation in rectangular coordinates for

$$r = \frac{4}{2 \sin \theta - \cos \theta}.$$

21. Find a rectangular equation of the plane curve determined by $x = 4t - 3$, $y = t^2$, for t in $[-3, 4]$.

22. Sketch the graph of the plane curve given by $x = t + \ln t$, $y = t + e^t$, for t in $(0, 2]$.

Systems of Equations and Inequalities

One of the great achievements of mathematics has been its ability to create mathematical models that can accurately describe a wide range of phenomena and predict future results. Models have been used successfully to address environmental concerns such as predicting animal populations. The following example uses mathematics to predict the antelope population in Wyoming and requires knowledge of systems of equations involving more than one variable.

The Bureau of Land Management has been studying the antelope population found in the Thunder Basin of Wyoming for several years. This study has tried to identify variables that affect the newborn antelope population each spring. The goal of the Bureau of Land Management is to be able to predict the new fawn count. Several variables might affect the antelope population. The variables considered were the size of the adult population, the total precipitation for the past year, and the severity of the winter. The severity of the winter was scaled between 1 and 5 with 1 being mild and 5 severe. Although other factors might have influenced the antelope population,

Sources: Al-Khafaji, A. and J. Tooley, *Numerical Methods in Engineering and Practice,* Holt, Rinehart, and Winston, Inc., 1986.

Boeing Aircraft Company

Brase, C. and C. Brase, *Understandable Statistics,* D. C. Heath and Company, Lexington, Massachusetts, 1995.

Bureau of Land Management

Tucker, A., A. Bernat, W. Bradley, R. Cupper, G. Scragg, *Fundamentals of Computing I Logic, Problem Solving, Programs, and Computers,* McGraw-Hill, 1995.

they were not included. The following table lists the results of four different years.

Fawns	Adults	Precip.	Winter
239	871	11.5	3
234	847	12.2	2
192	685	10.6	5
343	969	14.2	1
?	960	12.6	3

In the final row of the table the adult antelope population is 960, the precipitation is 12.6 inches, and the winter has a severity of 3. How can we predict the new spring fawn count? This question will be answered later in the chapter.

In order to make predictions and forecasts about the future, professionals in many fields attempt to determine relationships between different factors. These relationships often result in equations containing more than one variable. When quantities are interrelated, *systems of equations* in several variables are used to describe their relationship. As early as 4000 B.C. in Mesopotamia, people were able to solve up to 10 equations having 10 variables. In 1940 John Atanasoff, a physicist from Iowa State University, needed to solve a system of equations containing 29 equations and 29 variables. This need to solve a large *linear system* led him to invent the first fully electronic digital computer. Modern supercomputers are capable of performing billions of calculations in a single second and solving more than 600,000 equations simultaneously. This capability has enabled people to predict the weather, design modern aircraft, and develop faster computer chips.

In this chapter we will see how systems of equations are used to make predictions and forecasts in areas like the environment and business. Whether one is predicting the number of newborn antelope in Wyoming, forecasting gross sales for a business, or analyzing data related to the greenhouse effect, systems of equations are involved. Many of the techniques presented in this chapter are currently used by professionals throughout the world to solve important applications.

9.1 Linear Systems of Equations ▼▼▼

A set of equations is called a **system of equations.** The **solutions** of a system of equations must satisfy every equation in the system. The definition of a linear equation given earlier can be extended to more variables: any equation of the form

$$a_1x_1 + a_2x_2 + \cdots + a_nx_n = b$$

for real numbers $a_1, a_2, \ldots, a_n$ (not all of which are 0), and b, is a **linear equation.** If all the equations in a system are linear, the system is a **system of linear equations,** or a **linear system.**

In this section we discuss linear systems of two or three equations and variables. In a later section we develop a systematic method for solving systems of linear equations having any number of equations and variables.

A solution of the linear equation

$$ax + by = c$$

is an ordered pair of numbers (r, s) such that

$$ar + bs = c.$$

In earlier courses, you have solved systems of two linear equations. The methods you studied earlier are substitution and elimination (addition or subtraction). The following four examples illustrate these methods.

◀ **EXAMPLE 1**
Solving a system by substitution

Solve the system

$$5x + 3y = 95 \qquad \text{(1)}$$
$$2x - 7y = -3. \qquad \text{(2)}$$

Any solution of this system of two equations with two variables will be an ordered pair of numbers (x, y) that satisfies both equations. In this example we will use the **substitution method** to solve the system.

First solve either equation for one variable. Let us solve equation (2) for x.

$$2x - 7y = -3$$
$$2x = 7y - 3$$
$$x = \frac{7y - 3}{2} \qquad \text{(3)}$$

Now substitute the result for x in equation (1).

$$5x + 3y = 95$$
$$5\left(\frac{7y - 3}{2}\right) + 3y = 95$$

To solve for y, first multiply both sides of the equation by 2 to eliminate the denominator.

$$5(7y - 3) + 6y = 190$$
$$35y - 15 + 6y = 190$$
$$41y = 205$$
$$y = 5$$

Find x by substituting 5 for y in equation (3).

$$x = \frac{7y - 3}{2} = \frac{7(5) - 3}{2} = 16$$

Check that the solution set is $\{(16, 5)\}$ by substitution in the original system. ▶

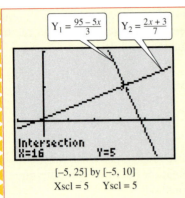

$Y_1 = \dfrac{95 - 5x}{3}$ $Y_2 = \dfrac{2x + 3}{7}$

Intersection
X=16   Y=5

[−5, 25] by [−5, 10]
Xscl = 5   Yscl = 5

A system of equations with two variables can be solved with a graphing calculator by graphing the equations in the same window, and then using the capability of the calculator to find the coordinates of the point (or points) of intersection. Before graphing, each equation must be solved for one of the variables. In Example 1, for instance, solving both equations for y gives

$$Y_1 = \frac{95 - 5x}{3} \quad \text{and} \quad Y_2 = \frac{2x + 3}{7}.$$

The graph is shown in the figure, with the coordinates of the point of intersection at the bottom of the window.

Another way to solve a system of two equations, called the **elimination method,** uses multiplication and addition to eliminate a variable from both equations. The elimination can be done if the coefficients of that variable in the two equations are additive inverses of each other. To achieve this, properties of algebra are used to change the system to obtain an **equivalent system,** one with the same solution set as the given system. Three transformations may be applied to a system to get an equivalent system. These are listed below.

> ### Transformation of a Linear System
>
> 1. Any two equations of the system may be interchanged.
> 2. Both sides of any equation of the system may be multiplied by any nonzero real number.
> 3. Any equation of the system may be replaced by the sum of that equation and a multiple of another equation in the system.

The next example illustrates the use of these transformations in the elimination method to solve a linear system.

EXAMPLE 2
Solving a system by elimination

Solve the system

$$3x - 4y = 1 \tag{1}$$
$$2x + 3y = 12. \tag{2}$$

The goal is to use the transformations to change one or both equations so the coefficients of one variable in the two equations are additive inverses of each other. Then addition of the two equations will eliminate that variable. One way to eliminate a variable in this example is to use the second transformation and multiply both sides of equation (2) by -3, giving the equivalent system

$$3x - 4y = 1$$
$$-6x - 9y = -36. \tag{3}$$

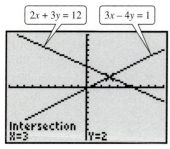

[−10, 10] by [−10, 10]

The screen supports the result of Example 2.

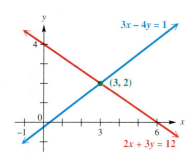

FIGURE 1

Now we multiply both sides of equation (1) by 2, and use the third transformation to add the result to equation (3), eliminating x.

$$
\begin{array}{r}
6x - 8y = 2 \\
-6x - 9y = -36 \\
\hline
-17y = -34
\end{array}
$$

The result is the system

$$3x - 4y = 1$$
$$-17y = -34. \qquad (4)$$

Multiplying both sides of equation (4) by $-1/17$ gives the equivalent system

$$3x - 4y = 1$$
$$y = 2.$$

Now we substitute 2 for y in equation (1).

$$3x - 4(2) = 1$$
$$3x - 8 = 1$$
$$3x = 9$$
$$x = 3$$

The solution set of the original system is $\{(3, 2)\}$. The graphs of the equations of the system in Figure 1 confirm that $(3, 2)$ satisfies both equations of the system. ▶

EXAMPLE 3
Solving an inconsistent system

Solve the system

$$3x - 2y = 4 \qquad (1)$$
$$-6x + 4y = 7. \qquad (2)$$

Multiply both sides of equation (1) by 2 and add it to equation (2).

$$
\begin{array}{r}
6x - 4y = 8 \\
-6x + 4y = 7 \\
\hline
0 = 15
\end{array}
$$

The new equivalent system is

$$3x - 2y = 4$$
$$0 = 15.$$

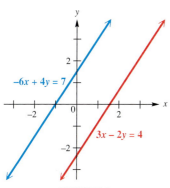

FIGURE 2

Since $0 = 15$ is never true, the system has no solution. As suggested by Figure 2, this means that the graphs of the equations of the system never intersect (the lines are parallel). The system has no solution and the solution set is $\emptyset$, the empty set. ▶

A system of equations with no solution is **inconsistent,** and the graphs of an inconsistent linear system of two equations in two variables are parallel lines. If the two equations of a system of two variables are multiples of each other, the equations are said to be **dependent equations,** and their graphs will coincide as in Figure 3. In this case the system has infinitely many solutions.

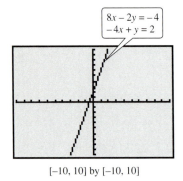

[−10, 10] by [−10, 10]

EXAMPLE 4
Solving a system with dependent equations

Solve the system

$$8x - 2y = -4 \qquad (1)$$
$$-4x + y = 2 \qquad (2)$$

(graphed in Figure 3) algebraically.

Multiply both sides of equation (1) by $1/2$, and add the result to equation (2), to get the equivalent system

$$8x - 2y = -4$$
$$0 = 0.$$

The second equation, $0 = 0$, is always true, which indicates that the equations of the original system are equivalent. (With this system, the second transformation can be used to change either equation into the other.) Any ordered pair (x, y) that satisfies either equation will satisfy the system. From equation (2),

$$-4x + y = 2,$$

or

$$y = 2 + 4x.$$

The graphs of $8x - 2y = -4$ and $-4x + y = 2$ coincide, as seen in the screen at the top. The table indicates that $Y_1 = Y_2$ for selected values of X, providing another means of support that the equations of the system lead to the same straight-line graph. (What is the expression for Y_2 in the table?)

The solution of the system can be written in the form of a set of ordered pairs $(x, 2 + 4x)$, for any real number x. Typical ordered pairs in the solution set are $(0, 2 + 4 \cdot 0) = (0, 2)$, $(-4, 2 + 4(-4)) = (-4, -14)$, $(3, 14)$, and $(7, 30)$. As shown in Figure 3, both equations of the original system lead to the same straight line graph. The solution set is written $\{(x, 2 + 4x)\}$. ▶

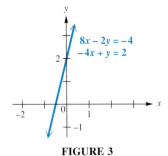

FIGURE 3

NOTE In Example 4 we wrote the solution set in a form with the variable x arbitrary. However, it would be acceptable to write the ordered pair with y arbitrary. In this case, the solution set would be written

$$\left\{ \left(\frac{y - 2}{4}, y \right) \right\}.$$

By selecting values for y and solving for x in the ordered pair above, individual solutions can be found. Verify that $(-1, -2)$ is a solution.

■ **PROBLEM SOLVING** Many applied problems involve more than one un-known quantity. Although some problems with two unknowns can be solved using just one variable, many times it is easier to use two variables. To solve a problem using a system, determine the unknown quantities you are asked to find, and let different variables represent each of these quantities. Then write a system of equations, and solve it using one of the methods of this section. Be sure that you answer the question(s) posed in the problem, and check to see that your answer is reasonable. ■

EXAMPLE 5
Solving an application using a system

Usually, as the price of an item goes up, demand for the item goes down and supply of the item goes up. Changes in gasoline prices illustrate this situation. The price where supply and demand are equal is called the *equilibrium price,* and the resulting supply or demand is called the *equilibrium supply* or *equilibrium demand.*

(a) Suppose the supply of a product is related to its price by the equation

$$p = \frac{2}{3}q,$$

where p is price in dollars and q is supply in appropriate units. (Here, q stands for quantity.) Find the price for the supply levels $q = 9$ and $q = 18$. When $q = 9$,

$$p = \frac{2}{3}q = \frac{2}{3}(9) = 6.$$

When $q = 18$,

$$p = \frac{2}{3}q = \frac{2}{3}(18) = 12.$$

(b) Suppose demand and price for the same product are related by

$$p = -\frac{1}{3}q + 18,$$

where p is price and q is demand. Find the price for the demand levels $q = 6$ and $q = 18$.
When $q = 6$,

$$p = -\frac{1}{3}q + 18 = -\frac{1}{3}(6) + 18 = 16,$$

and when $q = 18$,

$$p = -\frac{1}{3}q + 18 = -\frac{1}{3}(18) + 18 = 12.$$

(c) Graph both functions on the same axes.
Use the ordered pairs found in parts (a) and (b) and the p-intercepts to get the graphs shown in Figure 4.

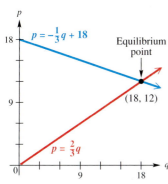

FIGURE 4

(d) Find the equilibrium price, supply, and demand.
Solve the system

$$p = \frac{2}{3}q \tag{1}$$

$$p = -\frac{1}{3}q + 18 \tag{2}$$

using substitution.

Substitute $(2/3)q$ from equation (1) for p into equation (2), and solve for q.

$$\frac{2}{3}q = -\frac{1}{3}q + 18$$

$$q = 18$$

This gives 18 units as the equilibrium supply or demand. Find the equilibrium price by substituting 18 for q in either equation. Using $p = (2/3)q$ gives

$$p = \frac{2}{3}(18) = 12$$

or \$12, the equilibrium price. The point (18, 12) that gives the equilibrium values is shown in Figure 4 on page 709. ◗

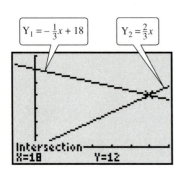

$Y_1 = -\frac{1}{3}x + 18$ $Y_2 = \frac{2}{3}x$

Intersection
X=18   Y=12

[−3, 21] by [−3, 21]
Xscl = 3   Yscl = 3

The screen supports the result of Example 5.

Transformations can also be used to solve a system of three linear equations in three variables. In that case, we first eliminate a variable from any two of the equations. Then we eliminate the *same variable* from a different pair of equations. We eliminate a second variable using the resulting two equations in two variables to get an equation with just one variable whose value can now be determined. We find the values of the remaining variables by substitution.

◖**EXAMPLE 6**
Solving a system of three equations with three variables

Solve the system

$$3x + 9y + 6z = 3 \tag{1}$$
$$2x + \ \ y - \ \ z = 2 \tag{2}$$
$$x + \ \ y + \ \ z = 2. \tag{3}$$

To begin, eliminate z by simply adding equations (2) and (3) to get

$$3x + 2y = 4. \tag{4}$$

To eliminate z from another pair of equations, multiply both sides of equation (2) by 6 and add the result to equation (1).

$$\begin{array}{r} 3x + \ \ 9y + 6z = \ \ 3 \\ \underline{12x + \ \ 6y - 6z = 12} \\ 15x + 15y \qquad = 15 \end{array} \tag{5}$$

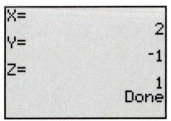

Modern graphing calculators usually have built-in system solving capability or can be programmed to solve systems. The screens here show a sample of how a typical program can be used to solve the system in Example 6.

To eliminate x from equations (4) and (5), multiply both sides of equation (4) by -5 and add the result to equation (5). Solve the new equation for y.

$$
\begin{aligned}
-15x - 10y &= -20 \\
15x + 15y &= 15 \\
\hline
5y &= -5 \\
y &= -1
\end{aligned}
$$

Using $y = -1$, find x from equation (4) by substitution.

$$3x + 2(-1) = 4$$
$$x = 2$$

Substitute 2 for x and -1 for y in equation (3) to find z.

$$2 + (-1) + z = 2$$
$$z = 1$$

Verify that the **ordered triple** $(2, -1, 1)$ satisfies all three equations. ▶

CONNECTIONS In earlier chapters, we solved linear equations in one variable. Now, in this section we solve linear equations in two variables and in three variables. Let's consider the similarities and differences of an equation like $2x + 1 = 7$ in each of these three settings. Just as we know whether $(-3, 5)$ is an ordered pair or an interval on the number line, we know whether $2x + 1 = 7$ is a linear equation in one, two, or three variables by the context of the problem. The solution $x = 3$ is written in set notation differently for each situation and the corresponding graphs differ.

One variable	Two variables	Three variables
$\{3\}$	$\{(x, y) \mid x = 3\}$	$\{(x, y, z) \mid x = 3\}$
a point	a line	a plane

FOR DISCUSSION OR WRITING
1. Give the solution sets in each setting for the equation $5x + 3 = 13$.
2. The equation $2x + 5y = 10$ can be considered a linear equation in either two or three variables. Give its solution set and describe the graph in each case.
3. Write the solution set of the equation in three variables $2x + 5y + z = 25$. Compare this solution set with those for a three-variable context, but less than three variables.

◀ EXAMPLE 7
Solving a system of two
equations with three variables

Solve the system

$$x + 2y + z = 4 \qquad (6)$$
$$3x - y - 4z = -9. \qquad (7)$$

Geometrically, the solution is the intersection of the two planes given by equations (6) and (7). The intersection of two different nonparallel planes is a line. Thus there will be an infinite number of ordered triples in the solution set, representing the points on the line of intersection. To describe these ordered triples, proceed as follows.

To eliminate x, multiply both sides of equation (6) by -3 and add to equation (7). (Either y or z could have been eliminated instead.)

$$\begin{array}{r} -3x - 6y - 3z = -12 \\ 3x - y - 4z = -9 \\ \hline -7y - 7z = -21 \end{array} \qquad (8)$$

Now solve equation (8) for z.

$$-7y - 7z = -21$$
$$-7z = 7y - 21$$
$$z = -y + 3$$

This gives z in terms of y. Express x also in terms of y by solving equation (6) for x and substituting $-y + 3$ for z in the result.

$$x + 2y + z = 4$$
$$x = -2y - z + 4$$
$$x = -2y - (-y + 3) + 4$$
$$x = -y + 1$$

The system has an infinite number of solutions. For any value of y, the value of z is given by $-y + 3$ and x equals $-y + 1$. For example, if $y = 1$, then $x = -1 + 1 = 0$ and $z = -1 + 3 = 2$, giving the solution $(0, 1, 2)$. Verify that another solution is $(-1, 2, 1)$.

With y arbitrary, the solution set is of the form $\{(-y + 1, y, -y + 3)\}$. Had equation (8) been solved for y instead of z, the solution would have had a different form but would have led to the same set of solutions. In that case we would have z arbitrary, and the solution set would be of the form $\{(-2 + z, 3 - z, z)\}$. By choosing $z = 2$, one solution would be $(0, 1, 2)$, which was verified above. ▶

Applications with three unknowns often require solving systems of three linear equations. As an example, the equation of the parabola $y = ax^2 + bx + c$ that passes through three given points can be found by solving a system of three equations with three unknowns.

EXAMPLE 8
Solving a geometric
application using a system

Find the equation of the parabola $y = ax^2 + bx + c$ that passes through $(2, 4)$, $(-1, 1)$, and $(-2, 5)$.

Since the three points lie on the graph of the equation $y = ax^2 + bx + c$, they must satisfy the equation. Substituting each ordered pair into the equation gives three equations with three variables.

$$4 = a(2)^2 + b(2) + c \qquad \text{or} \qquad 4 = 4a + 2b + c \qquad (9)$$
$$1 = a(-1)^2 + b(-1) + c \qquad \text{or} \qquad 1 = a - b + c \qquad (10)$$
$$5 = a(-2)^2 + b(-2) + c \qquad \text{or} \qquad 5 = 4a - 2b + c \qquad (11)$$

This system can be solved by the addition method. First eliminate c using equations (9) and (10).

$$
\begin{array}{rl}
4 = & 4a + 2b + c \\
-1 = & -a + b - c \qquad \text{-1 times equation (10)} \\
\hline
3 = & 3a + 3b \qquad\qquad\qquad\qquad\qquad (12)
\end{array}
$$

Now, use equations (10) and (11) to also eliminate c.

$$
\begin{array}{rl}
1 = & a - b + c \\
-5 = & -4a + 2b - c \qquad \text{-1 times equation (11)} \\
\hline
-4 = & -3a + b \qquad\qquad\qquad\qquad\qquad (13)
\end{array}
$$

Solve the system of equations (12) and (13) in two variables by eliminating a.

$$
\begin{array}{rl}
3 = & 3a + 3b \\
-4 = & -3a + b \\
\hline
-1 = & 4b
\end{array}
$$

$$-\frac{1}{4} = b$$

Find a by substituting $-1/4$ for b in equation (12), which is equivalent to $1 = a + b$.

$$1 = a + b \qquad \text{Equation (12) divided by 3}$$
$$1 = a - \frac{1}{4} \qquad \text{Let } b = -\frac{1}{4}.$$
$$\frac{5}{4} = a$$

Finally, find c by substituting $a = 5/4$ and $b = -1/4$ in equation (10).

$$1 = a - b + c$$
$$1 = \frac{5}{4} - \left(-\frac{1}{4}\right) + c \qquad a = \frac{5}{4}, b = -\frac{1}{4}$$
$$1 = \frac{6}{4} + c$$
$$-\frac{1}{2} = c$$

An equation of the parabola is $y = \dfrac{5}{4}x^2 - \dfrac{1}{4}x - \dfrac{1}{2}$. ▶

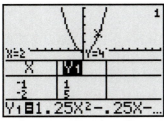

[−10, 10] by [−2, 10]
Xscl = 2   Yscl = 2

The screen supports the result of
Example 8: the points $(2, 4)$, $(-1, 1)$
and $(-2, 5)$ lie on the graph of
$y = 1.25x^2 - .25x - .5$.

■ **PROBLEM SOLVING** To solve applied problems requiring three unknowns we must write a system of three equations, as shown in the next example. ■

EXAMPLE 9
Solving an application using a system of three equations

An animal feed is made from three ingredients: corn, soybeans, and cottonseed. One unit of each ingredient provides units of protein, fat, and fiber as shown in the table below. How many units of each ingredient should be used to make a feed that contains 22 units of protein, 28 units of fat, and 18 units of fiber?

	Corn	Soybeans	Cottonseed	Total
Protein	.25	.4	.2	22
Fat	.4	.2	.3	28
Fiber	.3	.2	.1	18

Let x represent the number of units of corn, y, the number of units of soybeans, and z, the number of units of cottonseed that are required. Since the total amount of protein is to be 22 units,

$$.25x + .4y + .2z = 22.$$

Also, for the 28 units of fat,

$$.4x + .2y + .3z = 28,$$

and, for the 18 units of fiber,

$$.3x + .2y + .1z = 18.$$

Multiply the first equation on both sides by 100, and the second and third equations by 10 to get the system

$$25x + 40y + 20z = 2200$$
$$4x + 2y + 3z = 280$$
$$3x + 2y + z = 180.$$

Using the methods described earlier in this section, we can show that $x = 40$, $y = 15$, and $z = 30$. The feed should contain 40 units of corn, 15 units of soybeans, and 30 units of cottonseed to fulfill the given requirements. ▶

NOTE The table shown in Example 9 is useful in setting up the equations of the system, since the coefficients in each equation can be read from left to right. This idea is extended in Section 9.3, where we introduce solution of systems by matrices.

7·15 ODD

9.1 Exercises ▼▼▼▼▼▼▼▼▼▼▼▼▼▼▼▼▼▼▼▼▼▼▼▼▼▼▼▼▼▼▼▼▼▼▼▼▼

A solution of a system of two equations in two variables is represented by a point of intersection of the graphs of the two equations. While the graphs shown in the accompanying figure are not straight lines, the concept of finding a solution graphically is the same as it is for two linear equations. Refer to the figure to answer Exercises 1–6. Give answers with numbers rounded to the nearest unit.

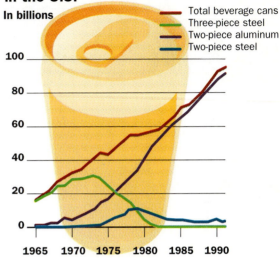

Annual Beverage Can Production in the U.S.

In billions

— Total beverage cans
— Three-piece steel
— Two-piece aluminum
— Two-piece steel

Source: Scientific American, September 1994

1. In what year did two-piece aluminum can production equal three-piece steel can production?

2. Refer to each type mentioned in Exercise 1. In the year in which production of two-piece aluminum cans equaled production of three-piece steel cans, how many of each type were produced?

3. Express as an ordered pair the solution of the system containing the graphs of the two-piece steel production and the three-piece steel production.

4. Use the terms *increasing* and *decreasing* to describe the trends for three-piece steel can production between 1965 and 1980.

5. If an equation were determined that modeled the production of total beverage cans produced, then x would represent the _____ and y would represent the _____ .

6. Explain why each graph shown is that of a function.

Solve each system by substitution. See Example 1.

7. $x - 5y = 8$
   $x = 6y$

8. $8x - 10y = -22$
   $3x + y = 6$

9. $6x - y = 5$
   $y = 11x$

10. $4x - 5y = -11$
    $2x + y = 5$

11. $7x - y = -10$
    $3y - x = 10$

12. $4x + 5y = 7$
    $9y = 31 + 2x$

13. $-2x = 6y + 18$
    $-29 = 5y - 3x$

14. $3x - 7y = 15$
    $3x + 7y = 15$

15. $3y = 5x + 6$
    $x + y = 2$

16. Only one of the following screens gives the correct graphical solution of the system in Exercise 10. Which one is it? (*Hint:* Solve for y first in each equation and use the slope-intercept forms to help you answer the question.)

(a)

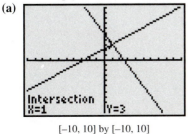

[−10, 10] by [−10, 10]

(b)

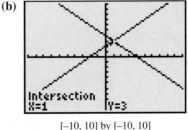

[−10, 10] by [−10, 10]

(c)

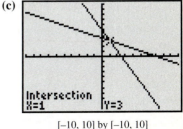

[−10, 10] by [−10, 10]

Solve each system by elimination. In Exercises 25–30, first clear denominators. See Example 2.

17. $4x + 2y = 6$
$5x - 2y = 12$

18. $2x - y = 8$
$3x + 2y = 5$

19. $3x - y = -4$
$x + 3y = 12$

20. $2x - 3y = -7$
$5x + 4y = 17$

21. $4x + 3y = -1$
$2x + 5y = 3$

22. $5x + 7y = 6$
$10x - 3y = 46$

23. $12x - 5y = 9$
$3x - 8y = -18$

24. $6x + 7y = -2$
$7x - 6y = 26$

25. $\dfrac{x}{2} + \dfrac{y}{3} = 4$
$\dfrac{3x}{2} + \dfrac{3y}{2} = 15$

26. $\dfrac{x}{3} + \dfrac{2y}{5} = 4$
$x + y = 11$

27. $9x - 2y = 30$
$\dfrac{5x}{2} + \dfrac{2y}{3} = 12$

28. $\dfrac{3x}{2} + \dfrac{y}{2} = -2$
$\dfrac{x}{2} + \dfrac{y}{2} = 0$

29. $\dfrac{2x - 1}{3} + \dfrac{y + 2}{4} = 4$
$\dfrac{x + 3}{2} - \dfrac{x - y}{3} = 3$

30. $\dfrac{x + 6}{5} + \dfrac{2y - x}{10} = 1$
$\dfrac{x + 2}{4} + \dfrac{3y + 2}{5} = -3$

Use a graphing calculator to solve each system. Express solutions with approximations to the nearest thousandth.

31. $\sqrt{3}\,x - y = 5$
$100x + y = 9$

32. $\dfrac{11}{3}x + y = .5$
$.6x - y = 3$

33. $.2x + \sqrt{2}y = 1$
$\sqrt{5}x + .7y = 1$

34. $\sqrt{7}x + \sqrt{2}y - 3 = 0$
$\sqrt{6}x - y - \sqrt{3} = 0$

Each system has either no solution or infinitely many solutions. Use either substitution or elimination to solve the system. State whether the system is inconsistent or has dependent equations. See Examples 3 and 4.

35. $9x - 5y = 1$
$-18x + 10y = 1$

36. $3x + 2y = 5$
$6x + 4y = 8$

37. $4x - y = 9$
$-8x + 2y = -18$

38. $3x + 5y + 2 = 0$
$9x + 15y + 6 = 0$

39. Refer to Example 2 in this section. If we began solving the system by eliminating *y*, by what numbers might we have multiplied equations (1) and (2)?

40. Explain how one can determine whether a system is inconsistent or has dependent equations when using the substitution or elimination method.

Solve each system of equations in three variables. See Example 6.

41. $x + y + z = 2$
$2x + y - z = 5$
$x - y + z = -2$

42. $2x + y + z = 9$
$-x - y + z = 1$
$3x - y + z = 9$

43. $x + 3y + 4z = 14$
$2x - 3y + 2z = 10$
$3x - y + z = 9$

44. $4x - y + 3z = -2$
$3x + 5y - z = 15$
$-2x + y + 4z = 14$

45. $x + 4y - z = 6$
$2x - y + z = 3$
$3x + 2y + 3z = 16$

46. $4x - 3y + z = 9$
$3x + 2y - 2z = 4$
$x - y + 3z = 5$

47. $5x + y - 3z = -6$
$2x + 3y + z = 5$
$-3x - 2y + 4z = 3$

48. $2x - 5y + 4z = -35$
$5x + 3y - z = 1$
$x + y + z = 1$

49. $x - 3y - 2z = -3$
$3x + 2y - z = 12$
$-x - y + 4z = 3$

50. $x + y + z = 3$
$3x - 3y - 4z = -1$
$x + y + 3z = 11$

51. $2x + 6y - z = 6$
$4x - 3y + 5z = -5$
$6x + 9y - 2z = 11$

52. $8x - 3y + 6z = -2$
$4x + 9y + 4z = 18$
$12x - 3y + 8z = -2$

53. Consider the linear equation in three variables $x + y + z = 4$. Find a pair of linear equations that, when considered together with the given equation, will form a system having the following.
 (a) Exactly one solution   **(b)** No solution   **(c)** Infinitely many solutions

54. (a) Using your immediate surroundings, give an example of three planes that intersect in a single point.
 (b) Using your immediate surroundings, give an example of three planes that intersect in a line.

Each system has either no solution or infinitely many solutions. Use either elimination or substitution or a combination of both to solve the system. State whether the system is inconsistent or has dependent equations. See Examples 3, 4, 6, and 7.

55. $3x + 5y - z = -2$
$4x - y + 2z = 1$
$-6x - 10y + 2z = 0$

56. $3x + y + 3z = 1$
$x + 2y - z = 2$
$2x - y + 4z = 4$

57. $5x - 3y + z = 1$
$2x + y - z = 4$

58. $x - 8y + z = 4$
$3x - y + 2z = -1$

59. $x + y + z = 6$
$2x - y - z = 3$

60. $x - y + 2z + w = 4$
$y + z = 3$
$z - w = 2$

Use a system of equations to solve each problem. See Example 8.

61. Find the equation of the line that passes through the points $(-2, 1)$ and $(-1, -2)$. Give it in the form $y = ax + b$.

62. Repeat Exercise 61, using the points $(2, 5)$ and $(-1, 4)$.

63. Find the equation of the parabola $y = ax^2 + bx + c$ that passes through the points $(2, 3)$, $(-1, 0)$, and $(-2, 2)$.

64. Repeat Exercise 63, using the points $(2, 9)$, $(-2, 1)$, and $(-3, 4)$.

65. Find the equation of the parabola shown. Three views of the same curve are given.

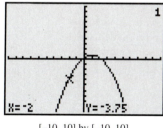

[−10, 10] by [−10, 10]

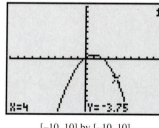

[−10, 10] by [−10, 10]

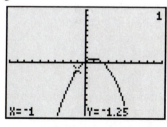

[−10, 10] by [−10, 10]

66. The table shown was generated using a function defined by $ax^2 + bx + c$. Use any three points from the table to find the equation that defines the function.

X	Y1
-3	5.48
-2	2.9
-1	1.26
0	.56
1	.8
2	1.98
3	4.1

X= -3

Given three noncollinear points, there is one and only one circle that passes through them. Knowing that the equation of a circle may be written in the form

$$x^2 + y^2 + ax + by + c = 0,$$

find the equation of the circle described or graphed in Exercises 67 and 68.

67. Passing through the points $(2, 1)$, $(-1, 0)$, and $(3, 3)$

68.

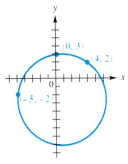

Solve each problem. See Example 5.

69. Suppose the demand and price for a certain model of electric can opener are related by

$$p = 16 - \frac{5}{4}q,$$

where p is price, in dollars, and q is demand, in appropriate units. Find the price when the demand is at the following levels.
(a) 0 units   **(b)** 4 units   **(c)** 8 units
Find the demand for the electric can opener at the following prices.
(d) $6   **(e)** $11   **(f)** $16
(g) Graph $p = 16 - (5/4)q$.
Suppose the price and supply of the item above are related by

$$p = \frac{3}{4}q,$$

where q represents the supply and p the price. Find the supply at the following prices.
(h) $0   **(i)** $10   **(j)** $20
(k) Graph $p = (3/4)q$ on the same axes used for part (g).
(l) Find the equilibrium supply.
(m) Find the equilibrium price.

70. Let the supply and demand equations for banana smoothies be

$$\text{supply: } p = \frac{3}{2}q \quad \text{and} \quad \text{demand: } p = 81 - \frac{3}{4}q.$$

(a) Graph these on the same axes.
(b) Find the equilibrium demand.
(c) Find the equilibrium price.

71. Let the supply and demand equations for chocolate frozen yogurt be given by

$$\text{supply: } p = \frac{2}{5}q \quad \text{and} \quad \text{demand: } p = 100 - \frac{2}{5}q.$$

(a) Graph these on the same axes.
(b) Find the equilibrium demand.
(c) Find the equilibrium price.

72. Let the supply and demand equations for onions be given by

$$\text{supply: } p = 1.4q - .6 \quad \text{and}$$
$$\text{demand: } p = -2q + 3.2.$$

(a) Graph these on the same axes.
(b) Find the equilibrium demand.
(c) Find the equilibrium price.

Use a system of equations to solve the problem. See Example 9.

73. A sparkling-water distributor wants to make up 300 gallons of sparkling water to sell for $6.00 per gallon. She wishes to mix three grades of water selling for $9.00, $3.00, and $4.50 per gallon, respectively. She must use twice as much of the $4.50 water as the $3.00 water. How many gallons of each should she use?

74. A glue company needs to make some glue that it can sell for $120 per barrel. It wants to use 150 barrels of glue worth $100 per barrel, along with some glue worth $150 per barrel, and glue worth $190 per barrel. It must use the same number of barrels of $150 and $190 glue. How much of the $150 and $190 glue will be needed? How many barrels of $120 glue will be produced?

75. The perimeter of a triangle is 59 inches. The longest side is 11 inches longer than the medium side, and the medium side is 3 inches more than the shortest side. Find the length of each side of the triangle.

76. The sum of the measures of the angles of any triangle is 180°. In a certain triangle, the largest angle measures 55° less than twice the medium angle, and the smallest angle measures 25° less than the medium angle. Find the measures of each of the three angles.

77. Sam Abo-zahrah wins $100,000 in the Louisiana state lottery. He invests part of the money in real estate with an annual return of 5% and another part in a money market account at 4.5% interest. He invests the rest, which amounts to $20,000 less than the sum of the other two parts, in certificates of deposit that pay 3.75%. If the total annual interest on the money is $4450, how much was invested at each rate?

78. Mary Ann O'Brien invests $10,000 received in an inheritance in three parts. With one part she buys mutual funds that offer a return of 4% per year. The second part, which amounts to twice the first, is used to buy government bonds paying 4.5% per year. She puts the rest of the money into a savings account that pays 2.5% annual interest. During the first year, the total interest is $415. How much did she invest at each rate?

Use the method of Example 8 to work Exercises 79 and 80.

79. Determining the amount of carbon dioxide in the atmosphere is important because carbon dioxide is known to be a greenhouse gas. Carbon dioxide concentrations (in parts per million) have been measured at Mauna Loa, Hawaii, over the past 30 years. This concentration has increased quadratically. The table lists readings for three years.

Year	CO_2
1958	315
1973	325
1988	352

(*Source:* Nilsson, A., *Greenhouse Earth*, John Wiley & Sons, New York, 1992.)

(a) If the quadratic relationship between the carbon dioxide concentration C and the year t is expressed as $C = at^2 + bt + c$ where $t = 0$ corresponds to 1958, use a linear system of equations to determine the constants a, b, and c.

(b) Predict the year when the amount of carbon dioxide in the atmosphere will double from its 1958 level.

80. For certain aircraft there exists a quadratic relationship between an airplane's maximum speed S (in knots) and its ceiling C or highest altitude possible (in thousands of feet). The table lists three airplanes that conform to this relationship.

Airplane	Max Speed	Ceiling
Hawkeye	320	33
Corsair	600	40
Tomcat	1283	50

(*Source:* Sanders, D., *Statistics: A First Course*, Fifth Edition, McGraw-Hill, Inc., 1995.)

(a) If the quadratic relationship between C and S is written as $C = aS^2 + bS + c$, use a linear system of equations to determine the constants a, b, and c.

(b) A new aircraft of this type has a ceiling of 45,000 feet. Predict its top speed.

▼▼▼▼▼▼▼▼▼▼▼▼ **DISCOVERING CONNECTIONS** (Exercises 81–86) ▼▼▼▼▼▼▼▼▼▼▼▼

The system

$$\frac{5}{x} + \frac{15}{y} = 16$$

$$\frac{5}{x} + \frac{4}{y} = 5$$

is not a linear system, because the variables appear in the denominator. However, it can be solved in a manner similar to the method for solving a linear system by using a substitution-of-variable technique. Let $t = 1/x$ and let $u = 1/y$.

81. Write a system of equations in t and u by making the appropriate substitutions.

82. Solve the system in Exercise 81 for t and u.

83. Solve the given system for x and y by using the equations relating t and x, and u and y.

84. Refer to the first equation in the given system, and solve for y in terms of x to obtain a rational function.

85. Repeat Exercise 84 for the second equation in the given system.

86. Using a viewing window of $[0, 10]$ by $[0, 2]$, show that the point of intersection of the graphs of the functions in Exercises 84 and 85 has the same x and y values as found in Exercise 83.

Use the substitution-of-variable technique to solve the systems in Exercises 87–90.

87. $\dfrac{2}{x} + \dfrac{1}{y} = \dfrac{3}{2}$

$\dfrac{3}{x} - \dfrac{1}{y} = 1$

88. $\dfrac{2}{x} + \dfrac{1}{y} = 11$

$\dfrac{3}{x} - \dfrac{5}{y} = 10$

89. $\dfrac{1}{x} + \dfrac{1}{y} - \dfrac{1}{z} = \dfrac{1}{4}$

$\dfrac{2}{x} - \dfrac{1}{y} + \dfrac{3}{z} = \dfrac{9}{4}$

$-\dfrac{1}{x} - \dfrac{2}{y} + \dfrac{4}{z} = 1$

90. $2x^{-1} - 2y^{-1} + z^{-1} = -1$

$4x^{-1} + y^{-1} - 2z^{-1} = -9$

$x^{-1} + y^{-1} - 3z^{-1} = -9$

9.2   Nonlinear Systems of Equations ▼▼▼

A system of equations in which at least one equation is *not* linear is called a **nonlinear system.** The substitution method works well for solving many such systems, particularly when one of the equations is linear, as in the next example.

EXAMPLE 1
Solving a nonlinear system by substitution

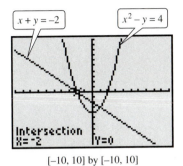

[−10, 10] by [−10, 10]

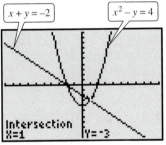

[−10, 10] by [−10, 10]

These screens support the result of Example 1.

Solve the system

$$x^2 - y = 4 \qquad \text{(1)}$$
$$x + y = -2. \qquad \text{(2)}$$

When one of the equations in a nonlinear system is linear, it is usually best to begin by solving the linear equation for either variable. With this system, we begin by solving equation (2) for y, giving

$$y = -2 - x.$$

Now substitute this result for y in equation (1) to get

$$x^2 - (-2 - x) = 4$$
$$x^2 + 2 + x = 4$$
$$x^2 + x - 2 = 0$$
$$(x + 2)(x - 1) = 0$$
$$x = -2 \qquad \text{or} \qquad x = 1.$$

Substituting -2 for x in equation (2) gives $y = 0$. Also, if $x = 1$, then $y = -3$. The solution set of the given system is $\{(-2, 0), (1, -3)\}$. A graph of the system is shown in Figure 5. ▶

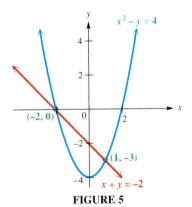

FIGURE 5

CAUTION   If we had solved for x in equation (2) to begin the solution, we would find $y = 0$ or $y = -3$. Substituting $y = 0$ into equation (1) gives $x^2 = 4$, so $x = 2$ or $x = -2$, leading to the ordered pairs $(2, 0)$ and $(-2, 0)$. The ordered pair $(2, 0)$ does not satisfy equation (2), however. This shows the *necessity* of checking by substituting all potential solutions into each equation of the system.

Nonlinear systems where both variables are squared in both equations are best solved by elimination, as shown in the next example.

EXAMPLE 2
Solving a nonlinear system by elimination

Solve the system

$$x^2 + y^2 = 4 \qquad \text{(1)}$$
$$2x^2 - y^2 = 8. \qquad \text{(2)}$$

Adding equation (1) to equation (2) (to eliminate y) gives the new system

$$x^2 + y^2 = 4$$
$$3x^2 \quad\;\; = 12. \qquad \text{(3)}$$

Solve equation (3) for x.

$$x^2 = 4$$
$$x = 2 \qquad \text{or} \qquad x = -2$$

Find y by substituting back into equation (1). If $x = 2$, then $y = 0$, and if $x = -2$, then $y = 0$. The solutions of the given system are $(2, 0)$ and $(-2, 0)$, so the solution set is $\{(2, 0), (-2, 0)\}$. ▶

To graph the system in Example 2, we must split both equations into two functions, so four functions will have to be graphed. For example, solving each equation for y gives the following four functions.

$$x^2 + y^2 = 4$$
$$y^2 = 4 - x^2$$
$$y = \sqrt{4 - x^2} \qquad \text{or} \qquad y = -\sqrt{4 - x^2}$$
$$2x^2 - y^2 = 8$$
$$-y^2 = 8 - 2x^2$$
$$y^2 = -8 + 2x^2$$
$$y = \sqrt{-8 + 2x^2} \qquad \text{or} \qquad y = -\sqrt{-8 + 2x^2}$$

The graphs of these four functions are shown in the figure. The first two functions give the top and bottom of the circle. The other two equations represent the two branches of the hyperbola (Section 10.3). The graph shows the two points of intersection at $(2, 0)$ and $(-2, 0)$.

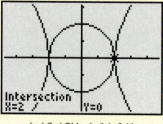
[−4.7, 4.7] by [−3.1, 3.1]

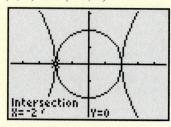
[−4.7, 4.7] by [−3.1, 3.1]

NOTE   The elimination method works with the system in Example 2 since the system can be thought of as a system of linear equations where the variables are x^2 and y^2. In other words, the system is **linear in x^2 and y^2**. To see this, substitute u for x^2 and v for y^2. The resulting system is linear in u and v.

Sometimes a combination of the elimination method and the substitution method is effective in solving a system, as illustrated in Example 3.

EXAMPLE 3
Solving a nonlinear system by a combination of methods

Solve the system

$$x^2 + 3xy + y^2 = 22 \qquad (6)$$
$$x^2 - xy + y^2 = 6. \qquad (7)$$

Multiply both sides of equation (7) by -1, and then add the result to equation (6), as follows.

$$\begin{array}{r} x^2 + 3xy + y^2 = 22 \\ -x^2 + xy - y^2 = -6 \\ \hline 4xy = 16 \end{array} \qquad (8)$$

Now solve equation (8) for either x or y and substitute the result into one of the given equations. Solving for y gives

$$y = \frac{4}{x}, \quad x \neq 0. \qquad (9)$$

(The restriction $x \neq 0$ is included since if $x = 0$ there is no value of y that satisfies the system.) Substituting for y in equation (7) (equation (6) could have been used) and simplifying gives

$$x^2 - x\left(\frac{4}{x}\right) + \left(\frac{4}{x}\right)^2 = 6.$$

Now solve for x.

$$x^2 - 4 + \frac{16}{x^2} = 6$$
$$x^4 - 4x^2 + 16 = 6x^2$$
$$x^4 - 10x^2 + 16 = 0$$
$$(x^2 - 2)(x^2 - 8) = 0$$
$$x^2 = 2 \quad \text{or} \quad x^2 = 8$$
$$x = \sqrt{2} \quad \text{or} \quad x = -\sqrt{2} \quad \text{or} \quad x = 2\sqrt{2} \quad \text{or} \quad x = -2\sqrt{2}$$

Substitute these x values into equation (9) to find corresponding values of y.

$$\text{If } x = \sqrt{2}, y = \frac{4}{\sqrt{2}} = 2\sqrt{2}.$$

$$\text{If } x = -\sqrt{2}, y = \frac{4}{-\sqrt{2}} = -2\sqrt{2}.$$

$$\text{If } x = 2\sqrt{2}, y = \frac{4}{2\sqrt{2}} = \sqrt{2}.$$

$$\text{If } x = -2\sqrt{2}, y = \frac{4}{-2\sqrt{2}} = -\sqrt{2}.$$

The solution set of the system is

$$\{(\sqrt{2}, 2\sqrt{2}), (-\sqrt{2}, -2\sqrt{2}), (2\sqrt{2}, \sqrt{2}), (-2\sqrt{2}, -\sqrt{2})\}.$$

Verify these solutions by substitution in the original system. ▶

EXAMPLE 4
Solving a nonlinear system with an absolute value equation

Solve the system

$$x^2 + y^2 = 16 \tag{10}$$
$$|x| + y = 4. \tag{11}$$

The substitution method is required here. Equation (11) can be rewritten as $|x| = 4 - y$; then the definition of absolute value can be used to get

$$x = 4 - y \quad \text{or} \quad x = -(4 - y) = y - 4. \tag{12}$$

(Since $|x| \geq 0$ for all real x, $4 - y \geq 0$, or $4 \geq y$.) Substituting from either part of equation (12) into equation (10) gives the same result.

$$(4 - y)^2 + y^2 = 16 \quad \text{or} \quad (y - 4)^2 + y^2 = 16$$

Since $(4 - y)^2 = (y - 4)^2 = 16 - 8y + y^2$, either equation becomes

$$(16 - 8y + y^2) + y^2 = 16$$
$$2y^2 - 8y = 0$$
$$2y(y - 4) = 0$$
$$y = 0 \quad \text{or} \quad y = 4.$$

From equation (12),

$$\text{If } y = 0, \text{ then } x = 4 - 0 \qquad \text{or} \qquad x = 0 - 4.$$
$$x = 4 \qquad\qquad\qquad x = -4$$
$$\text{If } y = 4, \text{ then } x = 4 - 4 = 0.$$

The solution set, $\{(4, 0), (-4, 0), (0, 4)\}$, includes the points of intersection shown in Figure 6. Be sure to check the solutions in the original system. ▶

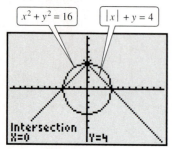

[−12.5̄3̄, 12.5̄3̄] by [−8.2̄6̄, 8.2̄6̄]

This screen indicates that $(0, 4)$ is a solution of the system in Example 4. The ordered pairs $(-4, 0)$ and $(4, 0)$ are also solutions.

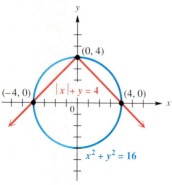

FIGURE 6

Nonlinear systems sometimes lead to solutions that are imaginary numbers.

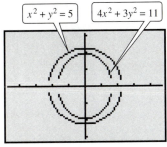

EXAMPLE 5
Solving a nonlinear system with imaginary numbers in its solutions

$[-4.7, 4.7]$ by $[-3.1, 3.1]$

The system in Example 5 has no solutions with ordered pairs consisting of two real numbers. The graphs do not intersect.

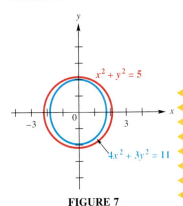

FIGURE 7

Solve the system

$$x^2 + y^2 = 5 \qquad \text{(13)}$$
$$4x^2 + 3y^2 = 11. \qquad \text{(14)}$$

Multiplying equation (13) on both sides by -3 and adding the result to equation (14) gives

$$\begin{aligned} -3x^2 - 3y^2 &= -15 \\ 4x^2 + 3y^2 &= 11 \\ \hline x^2 &= -4. \end{aligned}$$

By the square root property,

$$x = \pm\sqrt{-4}$$
$$x = 2i \qquad \text{or} \qquad x = -2i.$$

Find y by substitution. Using equation (13) gives

$$-4 + y^2 = 5$$
$$y^2 = 9$$
$$y = 3 \qquad \text{or} \qquad y = -3,$$

for either
$$x = 2i \qquad \text{or} \qquad x = -2i.$$

Checking the solutions in the given system shows that the solution set is $\{(2i, 3), (2i, -3), (-2i, 3), (-2i, -3)\}$. As the graph in Figure 7 suggests, imaginary solutions may occur when the graphs of the equations do not intersect. ▶

CONNECTIONS It is often helpful to visualize the types of graphs involved in a nonlinear system to get an idea of the possible numbers of ordered pairs of real numbers that may be in the solution set of the system. (Graphs of some nonlinear equations were discussed in Chapter 3.) For example, a line and a parabola may have 0, 1, or 2 points of intersection, as shown in the figure.

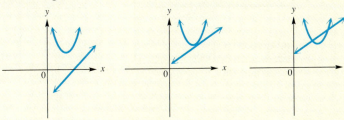

No points of intersection   One point of intersection   Two points of intersection

FOR DISCUSSION OR WRITING
What are the possible number of points of intersection of the following figures?

1. A circle and a line
2. A circle and a parabola
3. Two circles

9.2 Exercises ▼▼▼▼▼▼▼▼▼▼▼▼▼▼▼▼▼▼▼▼▼▼▼▼▼▼▼▼▼▼▼▼▼▼▼▼▼▼

In Exercises 1–6 a nonlinear system is given, along with the graphs of both equations in the system. Verify that the points of intersection specified on the graph are solutions of the system by substituting directly into both equations.

1. $x^2 = y - 1$
$\quad y = 3x + 5$

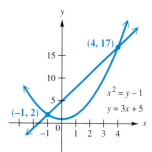

2. $2x^2 = 3y + 23$
$\quad\quad y = 2x - 5$

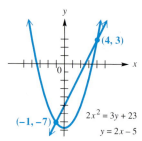

3. $\quad x^2 + y^2 = 5$
$\quad -3x + 4y = 2$

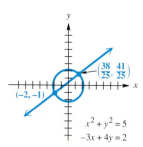

4. $x^2 + y^2 = 45$
$\quad x + y = -3$

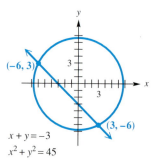

5. $y = x^2$
$\quad x + y = 2$

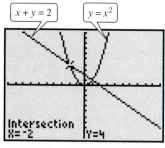

$[-10, 10]$ by $[-10, 10]$ $\quad\quad\quad\quad$ $[-10, 10]$ by $[-10, 10]$

6. $y = x^2 + 6x + 9$
$\quad x + 2y = -2$

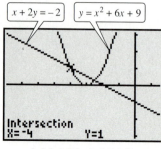
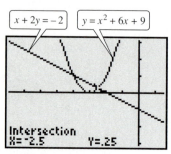

$[-7.9, 1.5]$ by $[-3.1, 3.1]$ $\quad\quad\quad$ $[-7.9, 1.5]$ by $[-3.1, 3.1]$

7. In Example 1 we solved the system

$$x^2 - y = 4$$
$$x + y = -2.$$

Explain how we can tell before doing any work that this system cannot have more than two solutions.

8. In Example 5, there are four solutions to the system, but there are no points of intersection of the graphs. When will the number of solutions of a system be less than the number of points of intersection of the graphs of the equations of the system?

Give all solutions of the following nonlinear systems of equations, including those with imaginary values. See Examples 1–5.

9. $y = x^2$
$x + y = 2$

10. $y = -x^2 + 2$
$x - y = 0$

11. $y = (x - 1)^2$
$x - 3y = -1$

12. $y = (x + 3)^2$
$x + 2y = -2$

13. $y = x^2 + 4x$
$2x - y = -8$

14. $y = 6x + x^2$
$3x - 2y = 10$

15. $3x^2 + 2y^2 = 5$
$x - y = -2$

16. $x^2 + y^2 = 5$
$-3x + 4y = 2$

17. $x^2 + y^2 = 8$
$x^2 - y^2 = 0$

18. $x^2 + y^2 = 10$
$2x^2 - y^2 = 17$

19. $5x^2 - y^2 = 0$
$3x^2 + 4y^2 = 0$

20. $x^2 + y^2 = 4$
$2x^2 - 3y^2 = -12$

21. $3x^2 + y^2 = 3$
$4x^2 + 5y^2 = 26$

22. $x^2 + 2y^2 = 9$
$3x^2 - 4y^2 = 27$

23. $2x^2 + 3y^2 = 5$
$3x^2 - 4y^2 = -1$

24. $3x^2 + 5y^2 = 17$
$2x^2 - 3y^2 = 5$

25. $2x^2 + 2y^2 = 20$
$4x^2 + 4y^2 = 30$

26. $x^2 + y^2 = 4$
$5x^2 + 5y^2 = 28$

27. $2x^2 - 3y^2 = 8$
$6x^2 + 5y^2 = 24$

28. $xy = -15$
$4x + 3y = 3$

29. $xy = 8$
$3x + 2y = -16$

30. $2xy + 1 = 0$
$x + 16y = 2$

31. $-5xy + 2 = 0$
$x - 15y = 5$

32. $x^2 + 4y^2 = 25$
$xy = 6$

33. $5x^2 - 2y^2 = 6$
$xy = 2$

34. $x^2 + 2xy - y^2 = 14$
$x^2 - y^2 = -16$

35. $3x^2 + xy + 3y^2 = 7$
$x^2 + y^2 = 2$

36. $x^2 - xy + y^2 = 5$
$2x^2 + xy - y^2 = 10$

37. $3x^2 + 2xy - y^2 = 9$
$x^2 - xy + y^2 = 9$

38. $x = |y|$
$x^2 + y^2 = 18$

39. $2x + |y| = 4$
$x^2 + y^2 = 5$

40. $2x^2 - y^2 = 4$
$|x| = |y|$

▼▼▼▼▼▼▼▼▼▼▼▼▼ **DISCOVERING CONNECTIONS** (Exercises 41–50) ▼▼▼▼▼▼▼▼▼▼▼▼▼

Consider the nonlinear system

$$y = |x - 1|$$
$$y = x^2 - 4$$

and work Exercises 41–50 in order to see how concepts from previous chapters relate to the graphs and the solutions of this system.

41. How is the graph of $y = |x - 1|$ obtained by transforming the graph of $y = |x|$?

42. How is the graph of $y = x^2 - 4$ obtained by transforming the graph of $y = x^2$?

43. Use the definition of absolute value to write $y = |x - 1|$ as a function defined piecewise.

44. Write two quadratic equations that will be used to solve the system. (*Hint:* Set both parts of the piecewise-defined function in Exercise 43 equal to $x^2 - 4$.)

45. Use the quadratic formula to solve the equation from Exercise 44 that involves the expression $x - 1$. Pay close attention to the restriction on x.

46. Use the value of x found in Exercise 45 to find one solution of the system.

47. Use the quadratic formula to solve the equation from Exercise 44 that involves the expression $1 - x$. Pay close attention to the restriction on x.

48. Use the value of x found in Exercise 47 to find another solution of the system.

49. What is the solution set of the original system?

50. Explain how the two calculator-generated screens here support your answer in Exercise 49.

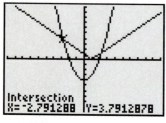
[−10, 10] by [−10, 10]

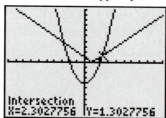
[−10, 10] by [−10, 10]

Many nonlinear systems cannot be solved analytically and as a result, graphical analysis is the only way to determine the solutions of such systems. Use a graphing calculator to solve each nonlinear system. Give x- and y-coordinates to the nearest hundredth.

51. $y = \log(x + 5)$
$y = x^2$

52. $y = 5^x$
$xy = 1$

53. $y = e^{x+1}$
$2x + y = 3$

54. $x^2 - y^2 = 4$
$y = \log x$

55. $y = \sqrt[3]{x - 4}$
$x^2 + y^2 = 6$

56. $y = \ln(2x + 3)$
$x^2 + 2y^2 = 4$

Solve the problem using a system of equations in two variables.

57. Find two numbers whose sum is 17 and whose product is 42.

58. Find two numbers whose sum is 10 and whose squares differ by 20.

59. Find two numbers whose squares have a sum of 100 and a difference of 28.

60. The longest side of a right triangle is 13 m in length. One of the other sides is 7 m longer than the shortest side. Find the length of the two shorter sides of the triangle.

61. Find two numbers whose ratio is 9 to 2 and whose product is 162.

62. Find two numbers whose ratio is 4 to 3 such that the sum of their squares is 100.

63. Does the straight line $3x - 2y = 9$ intersect the circle $x^2 + y^2 = 25$? (To find out, solve the system made up of these two equations.)

64. Find the equation of the line passing through the points of intersection of the graphs of $y = x^2$ and $x^2 + y^2 = 90$.

65. For what value of b will the line $x + 2y = b$ touch the circle $x^2 + y^2 = 9$ in only one point?

66. Suppose you are given the equations of two circles that are known to intersect in exactly two points. Explain how you would find the equation of the only chord common to these circles.

67. The emissions of carbon into the atmosphere from 1950 to 1990 are modeled in the graph for both Western Europe and Eastern Europe together with the former USSR. This carbon combines with oxygen to form carbon dioxide, which is believed to contribute to the greenhouse effect.

(a) Interpret this graph. How are emissions changing with time?

Annual Carbon Emissions

Millions of metric tons

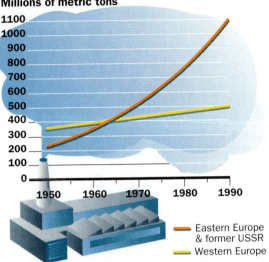

Source: Rosenberg, N. (ed.), *Greenhouse Warming: Abatement and Adaptation*, Resources for the Future. Washington, D.C. 1995.

(b) Use the graph to estimate the year and the amount when the carbon emissions were equal.

(c) The equation $W = 375(1.008)^{(t-1950)}$ models the emissions in Western Europe while the equation $E = 260(1.038)^{(t-1950)}$ models the emissions from Eastern Europe and the former USSR. Use these equations to determine the year and emission levels when $W = E$.

68. In electronics, circuit gain is given by

$$G = \frac{Bt}{R + R_t}$$

where R is the value of a resistor, t is temperature, and B is a constant. The sensitivity of the circuit to temperature is given by

$$S = \frac{BR}{(R + R_t)^2}.$$

If $B = 3.7$ and t is 90 K (Kelvin), find the values of R and R_t that will make $G = .4$ and $S = .001$.

9.3 Matrix Solution of Linear Systems ▼▼▼

Since systems of linear equations occur in so many practical situations, computer methods have been developed for efficiently solving linear systems. Computer solutions of linear systems depend on the idea of a **matrix** (plural **matrices**), a rectangular array of numbers enclosed in brackets. For example,

$$\begin{bmatrix} 2 & 3 & 7 \\ 5 & -1 & 10 \end{bmatrix}$$

is a matrix. Each number is called an **element** of the matrix.

Matrices in general are discussed in more detail later in this chapter. In this section, a method is developed for solving linear systems using matrices. As an example, start with the system

$$x + 3y + 2z = 1$$
$$2x + y - z = 2$$
$$x + y + z = 2,$$

and write the coefficients of the variables and the constants as a matrix, called the **augmented matrix** of the system.

$$\text{Rows} \begin{bmatrix} 1 & 3 & 2 & | & 1 \\ 2 & 1 & -1 & | & 2 \\ 1 & 1 & 1 & | & 2 \end{bmatrix}$$

Columns

The vertical line, which is optional, is used only to separate the coefficients from the constants. This matrix has 3 rows (horizontal) and 4 columns (vertical). To refer to a number in the matrix, use its row and column numbers. For example, the number 3 is in the first row and second column.

The rows of this matrix can be treated just like the equations of a system of linear equations. Since the augmented matrix is nothing more than a short form of the system, any tranformation of the matrix that results in an equivalent system of equations can be performed. Operations that produce such transformations are given below.

Matrix Row Transformations

For any augmented matrix of a system of linear equations, the following row transformations will result in the matrix of an equivalent system.

1. Any two rows may be interchanged.
2. The elements of any row may be multiplied by a nonzero real number.
3. Any row may be changed by adding to its elements a multiple of the elements of another row.

These transformations are just a restatement in matrix form of the transformations of systems discussed earlier. From now on, when referring to the third transformation, "a multiple of the elements of a row" will be abbreviated as "a multiple of a row."

Before using matrices to solve a linear system, the system must be arranged in the proper form, with variable terms on the left side of the equation and constant terms on the right. The variable terms must be in the same order in each of the equations.

The **Gauss-Jordan method** is a systematic technique for applying matrix row transformations in an attempt to reduce a matrix to a form such as

$$\left[\begin{array}{cc|c} 1 & 0 & a \\ 0 & 1 & b \end{array}\right] \quad \text{or} \quad \left[\begin{array}{ccc|c} 1 & 0 & 0 & a \\ 0 & 1 & 0 & b \\ 0 & 0 & 1 & c \end{array}\right]$$

from which the solutions are easily obtained. Because 1s are found along the diagonals from upper left to lower right, these matrices are in *diagonal form*.

Using the Gauss-Jordan Method to Put a Matrix into Diagonal Form

1. Obtain 1 as the first element of the first column.
2. Use the first row to transform the remaining entries in the first column to 0.
3. Obtain 1 as the second entry in the second column.
4. Use the second row to transform the remaining entries in the second column to 0.
5. Continue in this manner as far as possible.

NOTE The Gauss-Jordan method proceeds *column by column,* from left to right. When you are working with a particular column, no row operation should undo the form of a preceding column.

The following example illustrates the Gauss-Jordan method.

◀EXAMPLE 1
Using the Gauss-Jordan method

Solve the linear system

$$3x - 4y = 1$$
$$5x + 2y = 19.$$

The equations should all be in the same form, with the variable terms in the same order on the left, and the constant term on the right. Begin by writing the augmented matrix.

$$\begin{bmatrix} 3 & -4 & 1 \\ 5 & 2 & 19 \end{bmatrix}$$

The goal is to transform this augmented matrix into one in which the value of the variables will be easy to see. That is, since each column in the matrix represents the coefficients of one variable, the augmented matrix should be transformed so that it is of the form

$$\begin{bmatrix} 1 & 0 & k \\ 0 & 1 & j \end{bmatrix}$$

for real numbers k and j. Once the augmented matrix is in this form, the matrix can be rewritten as a linear system to get

$$x = k$$
$$y = j.$$

The necessary transformations are performed as follows. It is best to work in columns beginning in each column with the element that is to become 1. In the augmented matrix

$$\begin{bmatrix} 3 & -4 & 1 \\ 5 & 2 & 19 \end{bmatrix},$$

there is a 3 in the first row, first column position. Use transformation 2, multiplying each entry in the first row by 1/3 to get a 1 in this position. (This step is abbreviated as (1/3)R1.)

$$\begin{bmatrix} 1 & -4/3 & 1/3 \\ 5 & 2 & 19 \end{bmatrix} \quad \tfrac{1}{3}R1$$

Get 0 in the second row, first column by multiplying each element of the first row by -5 and adding the result to the corresponding element in the second row, using transformation 3.

$$\begin{bmatrix} 1 & -4/3 & 1/3 \\ 0 & 26/3 & 52/3 \end{bmatrix} \quad -5R1 + R2$$

Get 1 in the second row, second column by multiplying each element of the second row by 3/26, using transformation 2.

$$\begin{bmatrix} 1 & -4/3 & | & 1/3 \\ 0 & 1 & | & 2 \end{bmatrix} \quad \tfrac{3}{26} R2$$

Finally, get 0 in the first row, second column by multiplying each element of the second row by 4/3 and adding the result to the corresponding element in the first row.

$$\begin{bmatrix} 1 & 0 & | & 3 \\ 0 & 1 & | & 2 \end{bmatrix} \quad \tfrac{4}{3} R2 + R1$$

This last matrix corresponds to the system

$$x = 3$$
$$y = 2,$$

that has the solution set $\{(3, 2)\}$. This solution could have been read directly from the third column of the final matrix. ▶

A linear system with three equations is solved in a similar way. Row transformations are used to get 1s down the diagonal from left to right and 0s above and below each 1.

EXAMPLE 2
Using the Gauss-Jordan method

Use the Gauss-Jordan method to solve the system

$$x - y + 5z = -6$$
$$3x + 3y - z = 10$$
$$x + 3y + 2z = 5.$$

Since the system is in proper form, begin by writing the augmented matrix of the linear system.

$$\begin{bmatrix} 1 & -1 & 5 & | & -6 \\ 3 & 3 & -1 & | & 10 \\ 1 & 3 & 2 & | & 5 \end{bmatrix}$$

The final matrix is to be of the form

$$\begin{bmatrix} 1 & 0 & 0 & | & m \\ 0 & 1 & 0 & | & n \\ 0 & 0 & 1 & | & p \end{bmatrix},$$

where m, n, and p are real numbers. This final form of the matrix gives the system $x = m$, $y = n$, and $z = p$, so the solution set is $\{(m, n, p)\}$.

There is already a 1 in the first row, first column. Get a 0 in the second row of the first column by multiplying each element in the first row by -3 and adding the result to the corresponding element in the second row, using transformation 3.

$$\left[\begin{array}{ccc|c} 1 & -1 & 5 & -6 \\ 0 & 6 & -16 & 28 \\ 1 & 3 & 2 & 5 \end{array}\right] \quad -3R1 + R2$$

Now, to change the last element in the first column to 0, use transformation 3 and multiply each element of the first row by -1, then add the results to the corresponding elements of the third row.

$$\left[\begin{array}{ccc|c} 1 & -1 & 5 & -6 \\ 0 & 6 & -16 & 28 \\ 0 & 4 & -3 & 11 \end{array}\right] \quad -1R1 + R3$$

The same procedure is used to transform the second and third columns. For both of these columns perform the additional step of getting 1 in the appropriate position of each column. Do this by multiplying the elements of the row by the reciprocal of the number in that position.

$$\left[\begin{array}{ccc|c} 1 & -1 & 5 & -6 \\ 0 & 1 & -8/3 & 14/3 \\ 0 & 4 & -3 & 11 \end{array}\right] \quad \tfrac{1}{6}R2$$

$$\left[\begin{array}{ccc|c} 1 & 0 & 7/3 & -4/3 \\ 0 & 1 & -8/3 & 14/3 \\ 0 & 4 & -3 & 11 \end{array}\right] \quad R2 + R1$$

$$\left[\begin{array}{ccc|c} 1 & 0 & 7/3 & -4/3 \\ 0 & 1 & -8/3 & 14/3 \\ 0 & 0 & 23/3 & -23/3 \end{array}\right] \quad -4R2 + R3$$

$$\left[\begin{array}{ccc|c} 1 & 0 & 7/3 & -4/3 \\ 0 & 1 & -8/3 & 14/3 \\ 0 & 0 & 1 & -1 \end{array}\right] \quad \tfrac{3}{23}R3$$

$$\left[\begin{array}{ccc|c} 1 & 0 & 0 & 1 \\ 0 & 1 & -8/3 & 14/3 \\ 0 & 0 & 1 & -1 \end{array}\right] \quad -\tfrac{7}{3}R3 + R1$$

$$\left[\begin{array}{ccc|c} 1 & 0 & 0 & 1 \\ 0 & 1 & 0 & 2 \\ 0 & 0 & 1 & -1 \end{array}\right] \quad \tfrac{8}{3}R3 + R2$$

The linear system associated with this final matrix is

$$x = 1$$
$$y = 2$$
$$z = -1,$$

and the solution set is $\{(1, 2, -1)\}$. ▶

⊡ Row operations can be performed with graphing calculators that have ▦ matrix capability. The matrix size (or dimension) and elements are entered. Then using the row operations of the calculator, the matrix is transformed until it is in a form where the solutions can be read. The windows below show typical entries for the matrices in the second and third steps of Example 2.

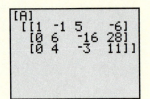

This typical menu shows various options for matrix row transformations in choices 8, 9, 0, and A.

This is the matrix that results when row 2 is multiplied by 1/6, as explained in Example 2.

If your calculator performs matrix algebra, check the manual for specific instructions. As you will probably agree, a calculator or computer is a necessity for solving larger systems.

The next two examples show how to recognize inconsistent systems or systems with dependent equations when solving such systems using row transformations.

◀**EXAMPLE 3**
Recognizing an inconsistent system

Use the Gauss-Jordan method to solve the system

$$x + y = 2$$
$$2x + 2y = 5.$$

Write the augmented matrix

$$\begin{bmatrix} 1 & 1 & | & 2 \\ 2 & 2 & | & 5 \end{bmatrix}.$$

Multiply the elements in the first row by -2 and add the result to the corresponding elements in the second row.

$$\begin{bmatrix} 1 & 1 & | & 2 \\ 0 & 0 & | & 1 \end{bmatrix} \quad -2R1 + R2$$

The next step would be to get a 1 in the second row, second column. Because of the zero there, it is impossible to go further. Since the second row corresponds to the equation

$$0x + 0y = 1,$$

which has no solution, the system is inconsistent, and the solution set is ∅. ▶

EXAMPLE 4
Solving a system with
dependent equations

Use the Gauss-Jordan method to solve the system

$$2x - 5y + 3z = 1$$
$$x - 2y - 2z = 8.$$

    Recall from Section 9.1 that a system with two equations and three variables usually has an infinite number of solutions. The Gauss-Jordan method can be used to give the solution with z arbitrary. Start with the augmented matrix

$$\begin{bmatrix} 2 & -5 & 3 & | & 1 \\ 1 & -2 & -2 & | & 8 \end{bmatrix}.$$

Exchange rows to get a 1 in the first row, first column position.

$$\begin{bmatrix} 1 & -2 & -2 & | & 8 \\ 2 & -5 & 3 & | & 1 \end{bmatrix}$$

Now multiply each element in the first row by -2 and add to the corresponding element in the second row.

$$\begin{bmatrix} 1 & -2 & -2 & | & 8 \\ 0 & -1 & 7 & | & -15 \end{bmatrix} \quad -2R1 + R2$$

Multiply each element in the second row by -1.

$$\begin{bmatrix} 1 & -2 & -2 & | & 8 \\ 0 & 1 & -7 & | & 15 \end{bmatrix} \quad -1R2$$

Multiply each element in the second row by 2 and add to the corresponding element in the first row.

$$\begin{bmatrix} 1 & 0 & -16 & | & 38 \\ 0 & 1 & -7 & | & 15 \end{bmatrix} \quad 2R2 + R1$$

It is not possible to go further with the Gauss-Jordan method. The equations that correspond to the final matrix are

$$x - 16z = 38 \quad \text{and} \quad y - 7z = 15.$$

Solve these equations for x and y, respectively.

$$\begin{aligned} x - 16z &= 38 & y - 7z &= 15 \\ x &= 16z + 38 & y &= 7z + 15 \end{aligned}$$

The solution can now be written with z arbitrary, as

$$\{(16z + 38, 7z + 15, z)\}. \quad \blacktriangleright$$

The cases that might occur when matrix methods are used to solve a system of linear equations are summarized below.

Summary of Possible Cases

When matrix methods are used to solve a system of linear equations and the resulting matrix is written in diagonal form:

1. If the number of rows with nonzero elements to the left of the vertical line is equal to the number of variables in the system, then the system has a single solution. See Example 2.
2. If one of the rows has the form $[0\ 0\ \cdots\ 0 \mid a]$ with $a \neq 0$, then the system has no solution. See Example 3.
3. If there are fewer rows in the matrix containing nonzero elements than the number of variables, then there are infinitely many solutions for the system. These solutions should be given in terms of an arbitrary variable. See Example 4.

CONNECTIONS The Gaussian reduction method (which is similar to the Gauss-Jordan method) is named after the mathematician Carl F. Gauss (1777–1855). In 1811, Gauss published a paper showing how he determined the orbit of the asteroid Pallas. Over the years he had kept data on his numerous observations. Each observation produced a linear equation in six unknowns. A typical equation was

$$.79363x + 143.66y + .39493z + .95929u - .18856v + .17387w$$
$$= 183.93.$$

Eventually he had twelve equations of this type. Certainly a method was needed to simplify the computation of a simultaneous solution. The method he developed was the reduction of the system from a rectangular to a triangular one, hence the Gaussian reduction. However, Gauss did not use matrices to carry out the computation.

When computers are programmed to solve the large linear systems involved in applications like designing aircraft or large electrical circuits, they frequently use an algorithm that is similar to the Gauss-Jordan method presented here. Solving a linear system with n equations and n variables requires the computer to perform a total of $T(n) = (2/3)n^3 + (3/2)n^2 - (7/6)n$ arithmetic calculations (additions, subtractions, multiplications, and divisions).*

*Source: Burden, R. and J. Faires, *Numerical Analysis,* Fifth Edition, PWS-KENT Publishing Company, Boston, 1993.

FOR DISCUSSION OR WRITING

1. Compute T for n = 3, 6, 10, 29, 100, 200, 400, 1000, 10,000, 100,000 and write the results in a table.
2. John Atanasoff wanted to solve a 29×29 linear system of equations. How many arithmetic operations would this have required? Is this too many to do by hand?
3. If the number of equations and variables is doubled, does the number of arithmetic operations double?
4. A Cray-2 supercomputer can execute up to 1.8 billion arithmetic operations per second. How many hours would be required to solve a linear system with 100,000 variables?
5. Discuss why it is difficult to solve large linear systems.

9.3 Exercises ▼▼▼▼▼▼▼▼▼▼▼▼▼▼▼▼▼▼▼▼▼▼▼▼▼▼▼▼▼▼▼▼▼▼▼

Use the third row transformation to change the matrix as indicated. See Example 1.

1. $\begin{bmatrix} 2 & 4 \\ 4 & 7 \end{bmatrix}$; -2 times row 1 added to row 2

2. $\begin{bmatrix} -1 & 4 \\ 7 & 0 \end{bmatrix}$; 7 times row 1 added to row 2

3. $\begin{bmatrix} 1 & 5 & 6 \\ -2 & 3 & -1 \\ 4 & 7 & 0 \end{bmatrix}$; 2 times row 1 added to row 2

4. $\begin{bmatrix} 2 & 5 & 6 \\ 4 & -1 & 2 \\ 3 & 7 & 1 \end{bmatrix}$; -6 times row 3 added to row 1

Write the augmented matrix for the system. Do not solve the system.

5. $2x + 3y = 11$
   $x + 2y = 8$

6. $3x + 5y = -13$
   $2x + 3y = -9$

7. $2x + y + z = 3$
   $3x - 4y + 2z = -7$
   $x + y + z = 2$

8. $4x - 2y + 3z = 4$
   $3x + 5y + z = 7$
   $5x - y + 4z = 7$

Write the system of equations associated with each augmented matrix. Do not try to solve.

9. $\begin{bmatrix} 3 & 2 & 1 & | & 1 \\ 0 & 2 & 4 & | & 22 \\ -1 & -2 & 3 & | & 15 \end{bmatrix}$

10. $\begin{bmatrix} 2 & 1 & 3 & | & 12 \\ 4 & -3 & 0 & | & 10 \\ 5 & 0 & -4 & | & -11 \end{bmatrix}$

11. $\begin{bmatrix} 1 & 0 & 0 & | & 2 \\ 0 & 1 & 0 & | & 3 \\ 0 & 0 & 1 & | & -2 \end{bmatrix}$

12. $\begin{bmatrix} 1 & 0 & 0 & | & 4 \\ 0 & 1 & 0 & | & 2 \\ 0 & 0 & 1 & | & 3 \end{bmatrix}$

13.
```
[A]
   [[1  1  0   3 ]
    [0  2  1  -4]
    [1  0 -1   5 ]]
```

14.
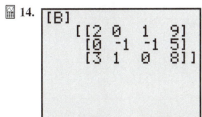
```
[B]
   [[2  0  1   9]
    [0 -1 -1   5]
    [3  1  0   8]]
```

Use the Gauss-Jordan method to solve the system of equations. For systems with dependent equations in 3 variables, give the solution with z arbitrary; for any such equations with 4 variables, let w be the arbitrary variable. See Examples 1–4.

15. $x + y = 5$
$x - y = -1$

16. $x + 2y = 5$
$2x + y = -2$

17. $x + y = -3$
$2x - 5y = -6$

18. $2x - 3y = 10$
$2x + 2y = 5$

19. $6x + y - 5 = 0$
$5x + y - 3 = 0$

20. $2x - 5y - 10 = 0$
$3x + y - 15 = 0$

21. $4x - y - 3 = 0$
$-2x + 3y - 1 = 0$

22. $-x + 2y + 6z = 2$
$3x + 2y + 6z = 6$
$x + 4y - 3z = 1$

23. $x + y - z = 6$
$2x - y + z = -9$
$x - 2y + 3z = 1$

24. $x + 3y - 6z = 7$
$2x - y + z = 1$
$x + 2y + 2z = -1$

25. $x - z = -3$
$y + z = 9$
$x + z = 7$

26. $-x + y = -1$
$y - z = 6$
$x + z = -1$

27. $y = -2x - 2z + 1$
$x = -2y - z + 2$
$z = x - y$

28. $x + y = 1$
$2x - z = 0$
$y + 2z = -2$

29. $2x - y + 3z = 0$
$x + 2y - z = 5$
$2y + z = 1$

30. $4x + 2y - 3z = 6$
$x - 4y + z = -4$
$-x + 2z = 2$

31. $3x + 5y - z + 2 = 0$
$4x - y + 2z - 1 = 0$
$-6x - 10y + 2z = 0$

32. $3x + y + 3z = 1$
$x + 2y - z = 2$
$2x - y + 4z = 4$

33. $x - 8y + z = 4$
$3x - y + 2z = -1$

34. $5x - 3y + z = 1$
$2x + y - z = 4$

35. $x - y + 2z + w = 4$
$y + z = 3$
$z - w = 2$

36. $x + 3y - 2z - w = 9$
$4x + y + z + 2w = 2$
$-3x - y + z - w = -5$
$x - y - 3z - 2w = 2$

37. Compare the use of an augmented matrix as a shorthand way of writing a system of linear equations and the use of synthetic division as a shorthand way to divide polynomials.

Solve each system using a graphing calculator capable of performing row operations. Give solutions with values correct to the nearest thousandth.

38. $.937x + .429y = 1.458$
$-.146x + .839y = -4.066$

39. $.268x + y = 9.814$
$x - .329y = 1.414$

40. $5.1x + \sqrt{3}y = \sqrt{2}$
$9x - 2.2y = 5$

41. $.3x + 2.7y - \sqrt{2}z = 3$
$\sqrt{7}x - 20y + 12z = -2$
$4x + \sqrt{3}y - 1.2z = \dfrac{3}{4}$

42. $\sqrt{5}x - 1.2y + z = -3$
$\dfrac{1}{2}x - 3y + 4z = \dfrac{4}{3}$
$4x + 7y - 9z = \sqrt{2}$

For each equation, determine the constants A and B that make the equation an identity. (Hint: Combine terms on the right and set coefficients of corresponding terms in the numerators equal.)

43. $\dfrac{1}{(x-1)(x+1)} = \dfrac{A}{x-1} + \dfrac{B}{x+1}$

44. $\dfrac{x+4}{x^2} = \dfrac{A}{x} + \dfrac{B}{x^2}$

45. $\dfrac{x}{(x-a)(x+a)} = \dfrac{A}{x-a} + \dfrac{B}{x+a}$

46. $\dfrac{2x}{(x+2)(x-1)} = \dfrac{A}{x+2} + \dfrac{B}{x-1}$

Solve each problem using matrices.

47. A working couple earned a total of $4352. The wife earned $64 per day; the husband earned $8 less per day. Find the number of days each worked if the total number of days worked by both was 72.

48. Midtown Manufacturing Company makes two products—plastic plates and plastic cups. Both require time on two machines: a unit of plates requires 1 hr on machine *A* and 2 hr on machine *B*, and a unit of cups requires 3 hr on machine *A* and 1 hr on machine *B*. Both machines operate 15 hr a day. How many units of each product can be produced in a day under these conditions?

49. A chemist has two prepared acid solutions, one of which is 2% acid by volume, another 7% acid. How many cubic centimeters of each should the chemist mix together to obtain 40 cm³ of a 3.2% acid solution?

50. To get the necessary funds for a planned expansion, a small company took out three loans totaling $25,000. The company was able to borrow some of the money at 8% interest. It borrowed $2000 more than one-half the amount of the 8% loan at 10%, and the rest at 9%. The total annual interest was $2220. How much did the company borrow at each rate?

51. In Exercise 50, suppose we drop the condition that the amount borrowed at 10% is $2000 more than one-

half the amount borrowed at 8%. How is the solution changed?

52. Suppose the company in Exercise 50 can borrow only $6000 at 9%. Is a solution possible that still meets the given conditions? Explain.

53. A hospital dietitian is planning a special diet for a certain patient. The total amount per meal of food groups A, B, and C must equal 400 g. The diet should include one-third as much of group A as of group B, and the sum of the amounts of group A and group C should equal twice the amount of group B. How many grams of each food group should be included?

54. In Exercise 53, suppose that, in addition to the conditions given there, foods A and B cost 2¢ per gram and food C costs 3¢ per gram, and that a meal must cost $8. Is a solution possible? Explain.

55. The relationship between a professional basketball player's height *H* (in inches) and weight *W* (in pounds) was modeled using two different samples of players. The resulting equations that modeled each sample were $W = 7.46H - 374$ and $W = 7.93H - 405$.

(a) Use both equations to predict the weight of a 6′ 11″ professional basketball player.

(b) According to each model, what change in weight is associated with a 1-inch increase in height?

(c) Determine the weight and height where the two models agree.

56. At rush hours, substantial traffic congestion is encountered at the traffic intersections shown in the figure. (All streets are one-way.) The city wishes to improve the signals at these corners to speed the flow of traffic. The traffic engineers first gather data. As the figure shows, 700 cars per hour come down M Street to intersection A, and 300 cars per hour come to intersection A on 10th Street. A total of x_1 of these cars leave A on M Street, while x_4 cars leave A on 10th Street. The number of cars entering A must equal the number leaving, so that

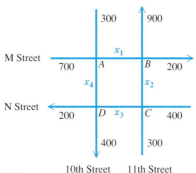

$$x_1 + x_4 = 700 + 300$$

or

$$x_1 + x_4 = 1000.$$

For intersection B, x_1 cars enter B on M Street, and x_2 cars enter B on 11th Street. The figure shows that 900 cars leave B on 11th, while 200 leave on M. We have

$$x_1 + x_2 = 900 + 200$$

$$x_1 + x_2 = 1100.$$

At intersection C, 400 cars enter on N Street and 300 on 11th Street, while x_2 leave on 11th Street and x_3 leave on N Street. This gives

$$x_2 + x_3 = 400 + 300$$

$$x_2 + x_3 = 700.$$

Finally, intersection D has x_3 cars entering on N and x_4 entering on 10th. There are 400 cars leaving D on 10th and 200 leaving on N.

(a) Set up an equation for intersection D.

(b) Use the four equations to set up an augmented matrix, and then use the Gaussian method to reduce it to triangular form.

(c) Since you got a row of all zeros, the system of equations does not have a unique solution. Write three equations, corresponding to the three nonzero rows of the matrix. Solve each of the equations for x_4.

(d) One of your equations should have been $x_4 = 1000 - x_1$. What is the largest possible value of x_1 so that x_4 is not negative?

(e) Another equation should have been $x_4 = x_2 - 100$. Find the smallest possible value of x_2 so that x_4 is not negative.

(f) Find the largest possible values of x_3 and x_4 so that neither variable is negative.

(g) Use the results of (a) − (f) to give a solution for the problem in which all the equations are satisfied and all variables are nonnegative. Is the solution unique?

▼▼▼▼▼▼▼▼▼▼▼▼▼ **DISCOVERING CONNECTIONS** (Exercises 57–60) ▼▼▼▼▼▼▼▼▼▼▼▼▼

To see how matrices can be used to solve a system of equations that arises from an environmental study, work Exercises 57–60 in order. To model the spring fawn count F from the adult antelope population A, the precipitation P, and the severity of the winter W, environmentalists have found that the equation $F = a + bA + cP + dW$ can be used, where the coefficients a, b, c, and d are constants that must be determined before one can use the equation. (Sources: Brase, C. and C. Brase, Understandable Statistics, Lexington, MA: D.C. Heath and Company, 1995; Bureau of Land Management.)

57. Substitute the values for *F*, *W*, *A*, and *P* from the table given in the Chapter Opener on page 704 for each of the four years into the equation $F = a + bA + cP + dW$ and obtain four linear equations involving *a*, *b*, *c*, and *d*.

58. Write an augmented matrix representing the system in Exercise 57, and solve for *a*, *b*, *c*, and *d*.

59. Write the equation for *F* using these values for the coefficients.

60. If a winter has a severity of 3, the adult antelope population is 960, and the precipitation is 12.6 inches, predict the spring fawn count. (Compare this with the actual count of 320.)

9.4 Properties of Matrices ▼▼▼

CONNECTIONS The word *matrix* for a rectangular array of numbers was first used in 1850 by the English mathematician James Joseph Sylvester (1814–1897). His friend and colleague Arthur Cayley (1821–1895) developed the theory further and in 1855 wrote about multiplying and finding inverses of square matrices. At the time, he was looking for an efficient way of computing the result of substituting one linear system into another. Several of the important results in the theory of matrices can be found in the correspondence of Cayley and Sylvester. Georg Frobenius (1848–1917) showed that matrices could be used to describe other mathematical systems. By 1924, matrices with complex numbers as entries were found to be the easiest way to describe atomic systems. Today, matrices are used in nearly every branch of mathematics.

FOR DISCUSSION OR WRITING
Discuss some reasons why matrices are useful in many applications.

The use of matrix notation in solving a system of linear equations was shown in the previous section. In this section, the algebraic properties of matrices are discussed.

It is customary to use capital letters to name matrices. Also, subscript notation is often used to name the elements of a matrix, as in the following matrix *A*.

$$A = \begin{bmatrix} a_{11} & a_{12} & a_{13} & \cdots & a_{1n} \\ a_{21} & a_{22} & a_{23} & \cdots & a_{2n} \\ a_{31} & a_{32} & a_{33} & \cdots & a_{3n} \\ \vdots & \vdots & \vdots & & \vdots \\ a_{m1} & a_{m2} & a_{m3} & \cdots & a_{mn} \end{bmatrix}$$

With this notation, the first-row, first-column element is a_{11} (read "a-sub-one-one"); the second-row, third-column element is a_{23}; and in general, the *i*th-row, *j*th-column element is a_{ij}.

Matrices are classified by their size or dimension—that is, by the number of rows and columns they contain, For example, the matrix

$$\begin{bmatrix} 2 & 7 & -5 \\ 3 & -6 & 0 \end{bmatrix}$$

has two rows (horizontal) and three columns (vertical), and is called a 2 × 3 matrix (read "two-by-three matrix"). A matrix with m rows and n columns is an $m \times n$ matrix. The number of rows is always given first.

Certain matrices have special names: an $n \times n$ matrix is a **square matrix** of order n. Also, a matrix with just one row is a **row matrix**, and a matrix with just one column is a **column matrix.**

Two matrices are **equal** if they are the same size and if corresponding elements, position by position, are equal. Using this definition, the matrices

$$\begin{bmatrix} 2 & 1 \\ 3 & -5 \end{bmatrix} \quad \text{and} \quad \begin{bmatrix} 1 & 2 \\ -5 & 3 \end{bmatrix}$$

are *not* equal (even though they contain the same elements and are the same size), since the corresponding elements differ.

EXAMPLE 1
Deciding whether two matrices are equal

(a) From the definition of equality given above, the only way that the statement

$$\begin{bmatrix} 2 & 1 \\ p & q \end{bmatrix} = \begin{bmatrix} x & y \\ -1 & 0 \end{bmatrix}$$

can be true is if $2 = x$, $1 = y$, $p = -1$, and $q = 0$.

(b) The statement

$$\begin{bmatrix} x \\ y \end{bmatrix} = \begin{bmatrix} 1 \\ 4 \\ 0 \end{bmatrix}$$

can never be true, since the two matrices are different sizes. (One is 2 × 1 and the other is 3 × 1.) ▶

ADDING MATRICES   Addition of matrices is defined as follows.

Definition of Addition of Matrices

To **add** two matrices of the same size, add corresponding elements. Only matrices of the same size can be added.

It can be shown that matrix addition satisfies the commutative, associative, closure, identity, and inverse properties. (See Exercises 71–75.)

EXAMPLE 2
Adding matrices

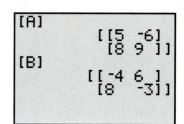

By defining matrices *A* and *B* as in Example 2(a), a graphing calculator gives the correct sum *A* + *B*.

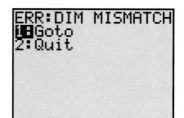

By entering the matrices in Example 2(b) directly on the home screen, their sum is displayed.

ERR:DIM MISMATCH
1▮Goto
2:Quit

A graphing calculator will return an ERROR message if it is directed to perform an operation on matrices that is not possible due to incompatible dimensions (sizes). See Example 2(c).

Find each of the following sums.

(a) $\begin{bmatrix} 5 & -6 \\ 8 & 9 \end{bmatrix} + \begin{bmatrix} -4 & 6 \\ 8 & -3 \end{bmatrix} = \begin{bmatrix} 5 + (-4) & -6 + 6 \\ 8 + 8 & 9 + (-3) \end{bmatrix} = \begin{bmatrix} 1 & 0 \\ 16 & 6 \end{bmatrix}$

(b) $\begin{bmatrix} 2 \\ 5 \\ 8 \end{bmatrix} + \begin{bmatrix} -6 \\ 3 \\ 12 \end{bmatrix} = \begin{bmatrix} -4 \\ 8 \\ 20 \end{bmatrix}$

(c) The matrices $A = \begin{bmatrix} 5 & 8 \\ 6 & 2 \end{bmatrix}$ and $B = \begin{bmatrix} 3 & 9 & 1 \\ 4 & 2 & 5 \end{bmatrix}$ are of different sizes, so the sum $A + B$ does not exist. ◗

A matrix containing only zero elements is called a **zero matrix.** For example, $O = \begin{bmatrix} 0 & 0 & 0 \end{bmatrix}$ is the 1×3 zero matrix, while

$$O = \begin{bmatrix} 0 & 0 & 0 \\ 0 & 0 & 0 \end{bmatrix}$$

is the 2×3 zero matrix.

By the additive inverse property in Chapter 1, each real number has an additive inverse: if a is a real number, there is a real number $-a$ such that

$$a + (-a) = 0 \qquad \text{and} \qquad -a + a = 0.$$

What about matrices? Given the matrix

$$A = \begin{bmatrix} -5 & 2 & -1 \\ 3 & 4 & -6 \end{bmatrix},$$

is there a matrix $-A$ such that

$$A + (-A) = O$$

where O is the 2×3 zero matrix? The answer is yes: the matrix $-A$ has as elements the additive inverses of the elements of A. (Remember, each element of A is a real number and therefore has an additive inverse.)

$$-A = \begin{bmatrix} 5 & -2 & 1 \\ -3 & -4 & 6 \end{bmatrix}$$

To check, test that $A + (-A)$ equals the zero matrix, O.

$$A + (-A) = \begin{bmatrix} -5 & 2 & -1 \\ 3 & 4 & -6 \end{bmatrix} + \begin{bmatrix} 5 & -2 & 1 \\ -3 & -4 & 6 \end{bmatrix} = \begin{bmatrix} 0 & 0 & 0 \\ 0 & 0 & 0 \end{bmatrix} = O$$

Matrix $-A$ is called the **additive inverse,** or **negative,** of matrix A. Every matrix has an additive inverse.

SUBTRACTING MATRICES The real number b is subtracted from the real number a, written $a - b$, by adding a and the additive inverse of b. That is,

$$a - b = a + (-b).$$

The same definition works for **subtraction** of matrices.

> **Definition of Subtraction of Matrices**
>
> If A and B are two matrices of the same size, then
>
> $$A - B = A + (-B).$$

In practice, the difference of two matrices of the same size is found by subtracting corresponding elements.

EXAMPLE 3
Subtracting matrices

Find each of the following differences.

(a) $\begin{bmatrix} -5 & 6 \\ 2 & 4 \end{bmatrix} - \begin{bmatrix} -3 & 2 \\ 5 & -8 \end{bmatrix} = \begin{bmatrix} -5 - (-3) & 6 - 2 \\ 2 - 5 & 4 - (-8) \end{bmatrix} = \begin{bmatrix} -2 & 4 \\ -3 & 12 \end{bmatrix}$

(b) $\begin{bmatrix} 8 & 6 & -4 \end{bmatrix} - \begin{bmatrix} 3 & 5 & -8 \end{bmatrix} = \begin{bmatrix} 5 & 1 & 4 \end{bmatrix}$

(c) The matrices

$$\begin{bmatrix} -2 & 5 \\ 0 & 1 \end{bmatrix} \quad \text{and} \quad \begin{bmatrix} 3 \\ 5 \end{bmatrix}$$

are of different sizes and cannot be subtracted. ▶

```
[A]
        [[2 -3],
         [0 4 ]]
5[A]
        [[10 -15],
         [0  20 ]]
```

This screen supports the result in Example 4(a).

MULTIPLYING MATRICES In work with matrices, a real number is called a **scalar** to distinguish it from a matrix. The product of a scalar k and a matrix X is the matrix kX, each of whose elements is k times the corresponding element of X.

EXAMPLE 4
Multiplying a matrix by a scalar

(a) $5\begin{bmatrix} 2 & -3 \\ 0 & 4 \end{bmatrix} = \begin{bmatrix} 10 & -15 \\ 0 & 20 \end{bmatrix}$

(b) $\dfrac{3}{4}\begin{bmatrix} 20 & 36 \\ 12 & -16 \end{bmatrix} = \begin{bmatrix} 15 & 27 \\ 9 & -12 \end{bmatrix}$ ▶

```
[B]
        [[20 36 ],
         [12 -16]]
(3/4)[B]
        [[15 27 ],
         [9  -12]]
```

This screen supports the result in Example 4(b). Note the careful use of parentheses around the scalar 3/4.

The proofs of the following properties of scalar multiplication are left for Exercises 79–82.

Properties of Scalar Multiplication

If A and B are matrices of the same size and c and d are real numbers:

$$(c + d)A = cA + dA$$
$$c(A + B) = cA + cB$$
$$c(A)d = cd(A)$$
$$(cd)A = c(dA).$$

We have seen how to multiply a real number (scalar) and a matrix. Now we define the product of two matrices. The procedure developed below for finding the product of two matrices may seem artificial, but it is useful in applications. The method will be illustrated before a formal rule is given. To find the product of

$$A = \begin{bmatrix} -3 & 4 & 2 \\ 5 & 0 & 4 \end{bmatrix} \quad \text{and} \quad B = \begin{bmatrix} -6 & 4 \\ 2 & 3 \\ 3 & -2 \end{bmatrix},$$

first locate *row* 1 of A and *column* 1 of B, shown shaded below.

$$A = \begin{bmatrix} -3 & 4 & 2 \\ 5 & 0 & 4 \end{bmatrix} \quad B = \begin{bmatrix} -6 & 4 \\ 2 & 3 \\ 3 & -2 \end{bmatrix}$$

Multiply corresponding elements, and find the sum of the products.

$$(-3)(-6) + (4)(2) + (2)(3) = 32$$

This result is the element for row 1, column 1 of the product matrix.

Now use *row* 1 of A and *column* 2 of B (shown shaded below) to determine the element in row 1 and column 2 of the product matrix.

$$A = \begin{bmatrix} -3 & 4 & 2 \\ 5 & 0 & 4 \end{bmatrix} \quad B = \begin{bmatrix} -6 & 4 \\ 2 & 3 \\ 3 & -2 \end{bmatrix}$$

Multiply corresponding elements, and add the products:

$$(-3)(4) + (4)(3) + (2)(-2) = -4,$$

which is the row 1, column 2 element of the product matrix.

Next, use *row* 2 of *A* and *column* 1 of *B*; this will give the row 2, column 1 entry of the product matrix.

$$\begin{bmatrix} -3 & 4 & 2 \\ 5 & 0 & 4 \end{bmatrix} \begin{bmatrix} -6 & 4 \\ 2 & 3 \\ 3 & -2 \end{bmatrix} \qquad 5(-6) + 0(2) + 4(3) = -18$$

Finally, use *row* 2 of *A* and *column* 2 of *B* to find the entry for row 2, column 2 of the product matrix.

$$\begin{bmatrix} -3 & 4 & 2 \\ 5 & 0 & 4 \end{bmatrix} \begin{bmatrix} -6 & 4 \\ 2 & 3 \\ 3 & -2 \end{bmatrix} \qquad 5(4) + 0(3) + 4(-2) = 12$$

The product matrix can now be written.

$$\begin{bmatrix} -3 & 4 & 2 \\ 5 & 0 & 4 \end{bmatrix} \begin{bmatrix} -6 & 4 \\ 2 & 3 \\ 3 & -2 \end{bmatrix} = \begin{bmatrix} 32 & -4 \\ -18 & 12 \end{bmatrix}$$

As seen in this example, the product of a 2×3 matrix and a 3×2 matrix is a 2×2 matrix.

A graphing calculator with matrix capability will perform the matrix operations of addition, subtraction, and multiplication. Since calculators vary considerably in how matrices are entered and how these operations are performed, consult your owner's manual for details. The three screens below show matrix multiplication using a popular calculator. Compare to the product above.

```
[A]
      [[-3 4 2]
       [5  0 4]]
```

```
[B]
      [[-6  4 ]
       [2   3 ]
       [3  -2]]
```

```
[A][B]
      [[32  -4]
       [-18 12]]
```

By definition, the **product** *AB* of an $m \times n$ matrix *A* and an $n \times p$ matrix *B* is found as follows. Multiply each element of the first row of *A* by the corresponding element of the first column of *B*. The sum of these *n* products is the first-row, first-column element of *AB*.

Also, the sum of the products found by multiplying the elements of the first row of *A* times the corresponding elements of the second column of *B* gives the first-row, second-column element of *AB*, and so on.

To find the *i*th-row, *j*th-column element of *AB*, multiply each element in the *i*th row of *A* by the corresponding element in the *j*th column of *B* (note the colored areas in the matrices below). The sum of these products will give the element of row *i*, column *j* of *AB*.

$$A = \begin{bmatrix} a_{11} & a_{12} & a_{13} & \cdots & a_{1n} \\ a_{21} & a_{22} & a_{23} & \cdots & a_{2n} \\ \vdots & & & & \\ a_{i1} & a_{i2} & a_{i3} & \cdots & a_{in} \\ \vdots & & & & \\ a_{m1} & a_{m2} & a_{m3} & \cdots & a_{mn} \end{bmatrix} \qquad B = \begin{bmatrix} b_{11} & b_{12} & \cdots & b_{1j} & \cdots & b_{1p} \\ b_{21} & b_{22} & \cdots & b_{2j} & \cdots & b_{2p} \\ \vdots & & & & & \\ b_{n1} & b_{n2} & \cdots & b_{nj} & \cdots & b_{np} \end{bmatrix}$$

> ## Definition of Matrix Multiplication
>
> If the number of columns of matrix *A* is the same as the number of rows of matrix *B*, then entry c_{ij} of the product matrix $C = AB$ is found as follows:
>
> $$c_{ij} = a_{i1}b_{1j} + a_{i2}b_{2j} + \cdots + a_{in}b_{nj}.$$
>
> The final product will have as many rows as *A* and as many columns as *B*.

◀ EXAMPLE 5
Deciding whether two matrices can be multiplied

Suppose matrix *A* is 3 × 2, while matrix *B* is 2 × 4. Can the product *AB* be calculated? What is the size of the product? Can the product *BA* be calculated? What is the size of *BA*?

The following diagram helps answer the questions about the product *AB*.

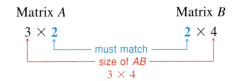

The product *AB* exists, since the number of columns of *A* equals the number of rows of *B* (both are 2). The product is a 3 × 4 matrix. Make a similar diagram for *BA*.

The product *BA* is not defined, since *B* has 4 columns and *A* has only 3 rows. **▶**

EXAMPLE 6
Multiplying two matrices

Find AB and BA, if possible, where

$$A = \begin{bmatrix} 1 & -3 \\ 7 & 2 \end{bmatrix} \quad \text{and} \quad B = \begin{bmatrix} 1 & 0 & -1 & 2 \\ 3 & 1 & 4 & -1 \end{bmatrix}.$$

First decide whether AB can be found. Since A is 2×2 and B is 2×4, the product can be found and will be a 2×4 matrix. Now use the definition of matrix multiplication.

$$AB = \begin{bmatrix} 1 & -3 \\ 7 & 2 \end{bmatrix}\begin{bmatrix} 1 & 0 & -1 & 2 \\ 3 & 1 & 4 & -1 \end{bmatrix}$$

$$= \begin{bmatrix} 1(1) + (-3)(3) & 1(0) + (-3)1 & 1(-1) + (-3)4 & 1(2) + (-3)(-1) \\ 7(1) + 2(3) & 7(0) + 2(1) & 7(-1) + 2(4) & 7(2) + 2(-1) \end{bmatrix}$$

$$= \begin{bmatrix} -8 & -3 & -13 & 5 \\ 13 & 2 & 1 & 12 \end{bmatrix}$$

Since B is a 2×4 matrix, and A is a 2×2 matrix, the product BA cannot be found. ▶

EXAMPLE 7
Multiplying square matrices in different orders

If $A = \begin{bmatrix} 1 & 3 \\ -2 & 5 \end{bmatrix}$ and $B = \begin{bmatrix} -2 & 7 \\ 0 & 2 \end{bmatrix}$, then the definition of matrix multiplication can be used to show that

$$AB = \begin{bmatrix} -2 & 13 \\ 4 & -4 \end{bmatrix} \quad \text{and} \quad BA = \begin{bmatrix} -16 & 29 \\ -4 & 10 \end{bmatrix}. \quad ▶$$

```
[A]
        [[1   3]
         [-2  5]]
[B]
        [[-2  7]
         [0   2]]
```

```
[A][B]
        [[-2  13]
         [4   -4]]
[B][A]
        [[-16 29]
         [-4  10]]
```

The products AB and BA are not equal, as seen in Example 7. Therefore, matrix multiplication is not in general a *commutative* operation.

CAUTION Examples 5 and 6 showed that the order in which two matrices are to be multiplied may determine whether their product can be found. Example 7 showed that even when both AB and BA can be found, they may not be equal. In general, for matrices A and B, $AB \neq BA$, so *matrix multiplication is not commutative.*

Matrix multiplication does, however, satisfy the associative and distributive properties.

Properties of Matrix Multiplication

If A, B, and C are matrices such that all the following products and sums exist, then

$$(AB)C = A(BC)$$
$$A(B + C) = AB + AC$$
$$(B + C)A = BA + CA.$$

For proofs of these results for the special cases when A, B, and C are square matrices, see Exercises 76 and 77. The identity and inverse properties for matrix multiplication are discussed in a later section of this chapter.

The final example shows how matrices can be used in an application.

EXAMPLE 8
Applying matrix multiplication

A contractor builds three kinds of houses, models A, B, and C, with a choice of two styles, colonial or ranch. Matrix P below shows the number of each kind of house the contractor is planning to build for a new 100-home subdivision. The amounts for each of the main materials used depend on the style of the house. These amounts are shown in matrix Q below, while matrix R gives the cost in dollars for each kind of material. Concrete is measured here in cubic yards, lumber in 1000 board feet, brick in 1000's, and shingles in 100 square feet.

$$
\begin{array}{c}
 \\
\text{Model A} \\
\text{Model B} \\
\text{Model C}
\end{array}
\begin{array}{cc}
\text{Colonial} & \text{Ranch} \\
\left[\begin{array}{cc}
0 & 30 \\
10 & 20 \\
20 & 20
\end{array}\right] & = P
\end{array}
$$

$$
\begin{array}{cc}
\begin{array}{c}
\text{Colonial} \\
\text{Ranch}
\end{array}
&
\begin{array}{cccc}
\text{Concrete} & \text{Lumber} & \text{Brick} & \text{Shingles} \\
\left[\begin{array}{cccc}
10 & 2 & 0 & 2 \\
50 & 1 & 20 & 2
\end{array}\right] & & & = Q
\end{array}
\end{array}
\qquad
\begin{array}{c}
\text{Cost per} \\
\text{unit}
\end{array}
\begin{array}{cc}
\begin{array}{c}
\text{Concrete} \\
\text{Lumber} \\
\text{Brick} \\
\text{Shingles}
\end{array}
&
\left[\begin{array}{c}
20 \\
180 \\
60 \\
25
\end{array}\right] = R
\end{array}
$$

(a) What is the total cost of materials for all houses of each model?

To find the materials cost for each model, first find matrix PQ, which will show the total amount of each material needed for all houses of each model.

$$
PQ = \left[\begin{array}{cc}
0 & 30 \\
10 & 20 \\
20 & 20
\end{array}\right]
\left[\begin{array}{cccc}
10 & 2 & 0 & 2 \\
50 & 1 & 20 & 2
\end{array}\right]
=
\begin{array}{c}
\begin{array}{cccc}
\text{Concrete} & \text{Lumber} & \text{Brick} & \text{Shingles}
\end{array} \\
\left[\begin{array}{cccc}
1500 & 30 & 600 & 60 \\
1100 & 40 & 400 & 60 \\
1200 & 60 & 400 & 80
\end{array}\right]
\begin{array}{c}
\text{Model A} \\
\text{Model B} \\
\text{Model C}
\end{array}
\end{array}
$$

Multiplying PQ and the cost matrix R gives the total cost of materials for each model.

$$
(PQ)R = \left[\begin{array}{cccc}
1500 & 30 & 600 & 60 \\
1100 & 40 & 400 & 60 \\
1200 & 60 & 400 & 80
\end{array}\right]
\left[\begin{array}{c}
20 \\
180 \\
60 \\
25
\end{array}\right]
=
\begin{array}{c}
\text{Cost} \\
\left[\begin{array}{c}
72{,}900 \\
54{,}700 \\
60{,}800
\end{array}\right]
\begin{array}{c}
\text{Model A} \\
\text{Model B} \\
\text{Model C}
\end{array}
\end{array}
$$

(b) How much of each of the four kinds of material must be ordered?

The totals of the columns of matrix PQ will give a matrix whose elements represent the total amounts of each material needed for the subdivision. Call this matrix T and write it as a row matrix.

$$T = [\begin{array}{cccc} 3800 & 130 & 1400 & 200 \end{array}]$$

(c) What is the total cost of the materials?

The total cost of all the materials is given by the product of matrix R, the cost matrix, and matrix T, the total amounts matrix. To multiply these and get a 1×1 matrix, representing the total cost, requires multiplying a 1×4 matrix and a 4×1 matrix. This is why in (b) a row matrix was written rather than a column matrix. The total materials cost is given by TR, so

$$TR = [3800 \quad 130 \quad 1400 \quad 200] \begin{bmatrix} 20 \\ 180 \\ 60 \\ 25 \end{bmatrix} = [188,400].$$

The total cost of the materials is $188,400. ▶

To help keep track of the quantities a matrix represents, let matrix P, from Example 8, represent models/styles, matrix Q represent styles/materials, and matrix R represent materials/cost. In each case the meaning of the rows is written first and that of the columns second. When the product PQ was found in Example 8, the rows of the matrix represented models and the columns represented materials. Therefore, the matrix product PQ represents models/materials. The common quantity, styles, in both P and Q was eliminated in the product PQ. Do you see that the product $(PQ)R$ represents models/cost?

In practical problems this notation helps to identify the order in which two matrices should be multiplied so that the results are meaningful. In Example 8(c), either product RT or product TR could have been found. However, since T represents subdivisions/materials and R represents materials/cost, only TR gave the required matrix representing subdivisions/cost.

9.4 Exercises ▼▼▼▼▼▼▼▼▼▼▼▼▼▼▼▼▼▼▼▼▼▼▼▼▼▼▼▼▼▼▼▼▼▼▼▼

Find the values of the variables. See Example 1.

1. $\begin{bmatrix} w & x \\ y & z \end{bmatrix} = \begin{bmatrix} 3 & 2 \\ -1 & 4 \end{bmatrix}$

2. $\begin{bmatrix} 0 & 5 & x \\ -1 & 3 & y+2 \\ 4 & 1 & z \end{bmatrix} = \begin{bmatrix} 0 & w+3 & 6 \\ -1 & 3 & 0 \\ 4 & 1 & 8 \end{bmatrix}$

3. $\begin{bmatrix} 2 & 5 & 6 \\ 1 & m & n \end{bmatrix} = \begin{bmatrix} z & y & w \\ 1 & 8 & -2 \end{bmatrix}$

4. $\begin{bmatrix} -7+z & 4r & 8s \\ 6p & 2 & 5 \end{bmatrix} + \begin{bmatrix} -9 & 8r & 3 \\ 2 & 5 & 4 \end{bmatrix} = \begin{bmatrix} 2 & 36 & 27 \\ 20 & 7 & 12a \end{bmatrix}$

5. $\begin{bmatrix} a+2 & 3z+1 & 5m \\ 8k & 0 & 3 \end{bmatrix} + \begin{bmatrix} 3a & 2z & 5m \\ 2k & 5 & 6 \end{bmatrix} = \begin{bmatrix} 10 & -14 & 80 \\ 10 & 5 & 9 \end{bmatrix}$

6. A 3×8 matrix has _____ columns and _____ rows.

Find the size of the matrix. Identify any square, column, or row matrices.

7. $\begin{bmatrix} -4 & 8 \\ 2 & 3 \end{bmatrix}$

8. $\begin{bmatrix} -9 & 6 & 2 \\ 4 & 1 & 8 \end{bmatrix}$

9. $\begin{bmatrix} -6 & 8 & 0 & 0 \\ 4 & 1 & 9 & 2 \\ 3 & -5 & 7 & 1 \end{bmatrix}$

10. $[8 \quad -2 \quad 4 \quad 6 \quad 3]$ **11.** $\begin{bmatrix} 2 \\ 4 \end{bmatrix}$ **12.** $[-9]$

13. Your friend missed the lecture on adding matrices. In your own words, explain to him how to add two matrices.

14. Explain to a friend in your own words how to subtract two matrices.

Perform each operation in Exercises 15–22, whenever possible. See Examples 2 and 3.

15. $\begin{bmatrix} 6 & -9 & 2 \\ 4 & 1 & 3 \end{bmatrix} + \begin{bmatrix} -8 & 2 & 5 \\ 6 & -3 & 4 \end{bmatrix}$ **16.** $\begin{bmatrix} 9 & 4 \\ -8 & 2 \end{bmatrix} + \begin{bmatrix} -3 & 2 \\ -4 & 7 \end{bmatrix}$

17. $\begin{bmatrix} -6 & 8 \\ 0 & 0 \end{bmatrix} - \begin{bmatrix} 0 & 0 \\ -4 & -2 \end{bmatrix}$ **18.** $\begin{bmatrix} 1 & -4 \\ 2 & -3 \\ -8 & 4 \end{bmatrix} - \begin{bmatrix} -6 & 9 \\ -2 & 5 \\ -7 & -12 \end{bmatrix}$

19. $\begin{bmatrix} 3x + y & x - 2y & 2x \\ 5x & 3y & x + y \end{bmatrix} + \begin{bmatrix} 2x & 3y & 5x + y \\ 3x + 2y & x & 2x \end{bmatrix}$

20. $\begin{bmatrix} 4k - 8y \\ 6z - 3x \\ 2k + 5a \\ -4m + 2n \end{bmatrix} - \begin{bmatrix} 5k + 6y \\ 2z + 5x \\ 4k + 6a \\ 4m - 2n \end{bmatrix}$

21. $\begin{bmatrix} 3 \\ 2 \end{bmatrix} + [2 \quad 3]$ **22.** $\begin{bmatrix} 0 \\ 0 \end{bmatrix} - [0 \quad 0 \quad 0]$

Let $A = \begin{bmatrix} -2 & 4 \\ 0 & 3 \end{bmatrix}$ *and* $B = \begin{bmatrix} -6 & 2 \\ 4 & 0 \end{bmatrix}$. *Find each of the following. See Example 4.*

23. $2A$ **24.** $-3B$ **25.** $2A - B$

26. $-2A + 4B$ **27.** $-A + \dfrac{1}{2}B$ **28.** $\dfrac{3}{4}A - B$

Find each matrix product, whenever possible. See Examples 5–7.

29. $\begin{bmatrix} 1 & 2 \\ 3 & 4 \end{bmatrix}\begin{bmatrix} -1 \\ 7 \end{bmatrix}$ **30.** $\begin{bmatrix} -1 & 5 \\ 7 & 0 \end{bmatrix}\begin{bmatrix} 6 \\ 2 \end{bmatrix}$ **31.** $\begin{bmatrix} 3 & -4 & 1 \\ 5 & 0 & 2 \end{bmatrix}\begin{bmatrix} -1 \\ 4 \\ 2 \end{bmatrix}$

32. $\begin{bmatrix} -6 & 3 & 5 \\ 2 & 9 & 1 \end{bmatrix}\begin{bmatrix} -2 \\ 0 \\ 3 \end{bmatrix}$ **33.** $\begin{bmatrix} 5 & 2 \\ -1 & 4 \end{bmatrix}\begin{bmatrix} 3 & -2 \\ 1 & 0 \end{bmatrix}$ **34.** $\begin{bmatrix} -4 & 0 \\ 1 & 3 \end{bmatrix}\begin{bmatrix} -2 & 4 \\ 0 & 1 \end{bmatrix}$

35. $\begin{bmatrix} 2 & 2 & -1 \\ 3 & 0 & 1 \end{bmatrix}\begin{bmatrix} 0 & 2 \\ -1 & 4 \\ 0 & 2 \end{bmatrix}$ **36.** $\begin{bmatrix} -9 & 2 & 1 \\ 3 & 0 & 0 \end{bmatrix}\begin{bmatrix} 2 \\ -1 \\ 4 \end{bmatrix}$ **37.** $\begin{bmatrix} -1 & 2 & 0 \\ 0 & 3 & 2 \\ 0 & 1 & 4 \end{bmatrix}\begin{bmatrix} 2 & -1 & 2 \\ 0 & 2 & 1 \\ 3 & 0 & -1 \end{bmatrix}$

38. $\begin{bmatrix} -2 & -3 & -4 \\ 2 & -1 & 0 \\ 4 & -2 & 3 \end{bmatrix}\begin{bmatrix} 0 & 1 & 4 \\ 1 & 2 & -1 \\ 3 & 2 & -2 \end{bmatrix}$ **39.** $[-2 \quad 4 \quad 1]\begin{bmatrix} 3 & -2 & 4 \\ 2 & 1 & 0 \\ 0 & -1 & 4 \end{bmatrix}$ **40.** $[0 \quad 3 \quad -4]\begin{bmatrix} -2 & 6 & 3 \\ 0 & 4 & 2 \\ -1 & 1 & 4 \end{bmatrix}$

41. $\begin{bmatrix} -3 & 0 & 2 & 1 \\ 4 & 0 & 2 & 6 \end{bmatrix}\begin{bmatrix} -4 & 2 \\ 0 & 1 \end{bmatrix}$ **42.** $\begin{bmatrix} -1 & 2 & 4 & 1 \\ 0 & 2 & -3 & 5 \end{bmatrix}\begin{bmatrix} 1 & 2 & 4 \\ -2 & 5 & 1 \end{bmatrix}$

Three screens depicting matrices A, B, and C on a graphing calculator are shown here. Use the matrix capabilities of a graphing calculator to find the result of each operation (if possible) as depicted on the screen.

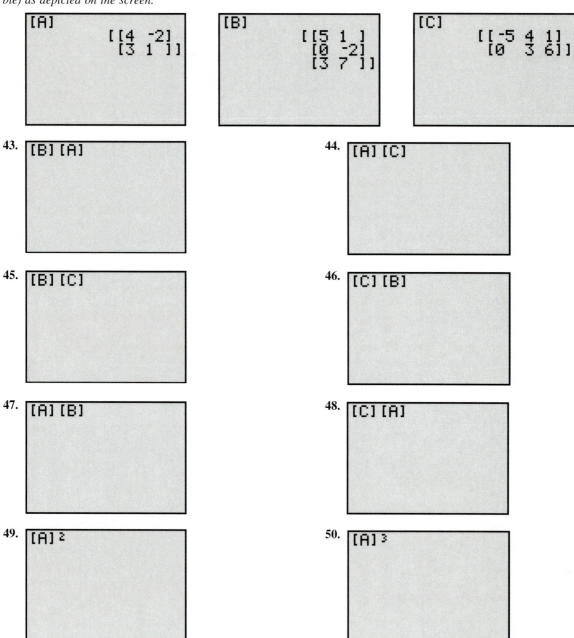

[A]
$$\begin{bmatrix} 4 & -2 \\ 3 & 1 \end{bmatrix}$$

[B]
$$\begin{bmatrix} 5 & 1 \\ 0 & -2 \\ 3 & 7 \end{bmatrix}$$

[C]
$$\begin{bmatrix} -5 & 4 & 1 \\ 0 & 3 & 6 \end{bmatrix}$$

43. [B] [A]

44. [A] [C]

45. [B] [C]

46. [C] [B]

47. [A] [B]

48. [C] [A]

49. [A]^2

50. [A]^3

51. Based on the screen shown here, what is matrix A?

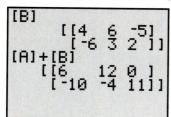

```
[B]
      [[4  6 -5]
       [-6  3  2 ]]
[A]+[B]
      [[6   12  0 ]
       [-10 -4 11]]
```

52. Based on the screen shown here, what is matrix B?

```
[A]
      [[3  6  5]
       [-2  1  4]]
[A]-[B]
      [[9  0 -5]
       [-4  6 -3]]
```

Let $A = \begin{bmatrix} -2 & 4 \\ 1 & 3 \end{bmatrix}$, $B = \begin{bmatrix} -2 & 1 \\ 3 & 6 \end{bmatrix}$, and $C = \begin{bmatrix} 5 & -2 & 1 \\ 0 & 3 & 7 \end{bmatrix}$. Find each product. See Examples 5–7.

53. AB **54.** BA **55.** AC **56.** CA

57. Did you get the same answer in Exercises 53 and 54? What about Exercises 55 and 56? Do you think matrix multiplication is commutative?

58. For any matrices P and Q, what must be true for both PQ and QP to exist?

Solve each problem. See Example 8.

59. A hardware chain does an inventory of a particular size of screw and finds that its Adelphi store has 100 flat-head and 150 round-head screws, its Beltsville store has 125 flat and 50 round, and its College Park store has 175 flat and 200 round. Write this information first as a 3 × 2 matrix and then as a 2 × 3 matrix.

60. At the grocery store, Miguel bought 4 quarts of milk, 2 loaves of bread, 4 potatoes, and an apple. Mary bought 2 quarts of milk, a loaf of bread, 5 potatoes, and 4 apples. Write this information first as a 2 × 4 matrix and then as a 4 × 2 matrix.

61. Yummy Yogurt sells three types of yogurt: nonfat, regular, and super creamy at three locations. Location I sells 50 gallons of nonfat, 100 gallons of regular, and 30 gallons of super creamy each day. Location II sells 10 gallons of nonfat and Location III sells 60 gallons of nonfat each day. Daily sales of regular yogurt are 90 gallons at Location II and 120 gallons at Location III. At Location II, 50 gallons of super creamy are sold each day, and 40 gallons of super creamy are sold each day at Location III.

(a) Write a 3 × 3 matrix that shows the sales figures for the three locations.

(b) The income per gallon for nonfat, regular, and super creamy is $12, $10, and $15, respectively. Write a 1 × 3 or 3 × 1 matrix displaying the income.

(c) Find a matrix product that gives the daily income at each of the three locations.

(d) What is Yummy Yogurt's total daily income from the three locations?

62. The Bread Box, a small neighborhood bakery, sells four main items: sweet rolls, bread, cakes, and pies. The amount of each ingredient (in cups, except for eggs) required for these items is given by matrix A.

	Eggs	Flour	Sugar	Shortening	Milk
Rolls (doz)	1	4	1/4	1/4	1
Bread (loaves)	0	3	0	1/4	0
Cake	4	3	2	1	1
Pies (crust)	0	1	0	1/3	0

$= A$

The cost (in cents) for each ingredient when purchased in large lots or small lots is given in matrix B.

	Cost Large lot	Small lot
Eggs	5	5
Flour	8	10
Sugar	10	12
Shortening	12	15
Milk	5	6

$= B$

(a) Use matrix multiplication to find a matrix giving the comparative cost per item for the two purchase options.

(b) Suppose a day's orders consist of 20 dozen sweet rolls, 200 loaves of bread, 50 cakes, and 60 pies. Write the orders as a 1×4 matrix and,

using matrix multiplication, write as a matrix the amount of each ingredient needed to fill the day's orders.

(c) Use matrix multiplication to find a matrix giving the costs under the two purchase options to fill the day's orders.

---

*Classified messages are often written in code so that unauthorized people cannot read them. Matrices are sometimes used to code these messages. One cryptographic technique that uses matrices is called the **polygraphic system.** In this system each letter in the alphabet is associated with a number between 1 and 26 with $A = 1, B = 2, C = 3, \ldots, Z = 26$. For example, to code the word MATRIX , first write it as* 13 1 20 18 9 24. *Then, these numbers are entered by columns into a matrix* $B = \begin{bmatrix} 13 & 20 & 9 \\ 1 & 18 & 24 \end{bmatrix}$ *having two rows. To code the letters a 2×2 matrix* $A = \begin{bmatrix} 2 & 1 \\ -5 & -2 \end{bmatrix}$ *is multiplied times B to form the product AB.*

$$AB = \begin{bmatrix} 2 & 1 \\ -5 & -2 \end{bmatrix} \begin{bmatrix} 13 & 20 & 9 \\ 1 & 18 & 24 \end{bmatrix} = \begin{bmatrix} 27 & 58 & 42 \\ -67 & -136 & -93 \end{bmatrix}$$

Since the resulting elements of AB are often less than 1 or greater than 26, they are scaled between 1 and 26 by adding or subtracting multiples of 26 to the number.

$$27 - (1)26 = 1, \quad 58 - 2(26) = 6, \quad 42 - 1(26) = 16,$$
$$-67 + 3(26) = 11, \quad -136 + 6(26) = 20, \quad -93 + 4(26) = 11.$$

The new coded matrix is then written as $C = \begin{bmatrix} 1 & 6 & 16 \\ 11 & 20 & 11 \end{bmatrix}$. *The entries of C are written as*

1 11 6 20 16 11 *which represent AKFTPK. Thus, the word MATRIX is coded as the word AKFTPK. The code is varied by changing the matrix A. If there is an odd number of letters, a dummy letter like Z may be added at the end of the word. (Source: Sinkov, A.,* Elementary Cryptanalysis: A Mathematical Approach, *Random House, Inc., New York, 1968.)*

63. Use the polygraphic system with the same matrix A to code each of the following words.
(a) HELP **(b)** LETTER

64. Extend the polygraphic system to code the word EQUATIONS using the 3×3 coding matrix

$$A = \begin{bmatrix} -1 & 2 & 1 \\ -2 & -2 & 1 \\ -1 & 1 & 1 \end{bmatrix}.$$

(Hint: Let the matrix B have three rows and then calculate AB.)

📱 A **stochastic matrix** *is a square matrix for which the sum of the entries in each row is* 1. *An*
example is $A = \begin{bmatrix} .8 & .2 \\ .4 & .6 \end{bmatrix}$. *Use a graphing calculator in Exercises 65–67.*

65. Show that A^2 is also a stochastic matrix. (*Note:* $A^2 = AA$.)

66. Compute A^3, A^4, A^5, and A^6 by successively multiplying by A. (*Note:* The rows of these powers
should get closer and closer to $[2/3 \quad 1/3]$.)

67. Repeat Exercises 65 and 66 for the matrix $\begin{bmatrix} .1 & .6 & .3 \\ .3 & .4 & .3 \\ .7 & .2 & .1 \end{bmatrix}$. What matrix do the rows get closer and
closer to?

*The **transpose**,* A^T, *of a matrix A is found by exchanging the rows and columns of A. That is, if*
$A = \begin{bmatrix} a & b \\ c & d \end{bmatrix}$, *then* $A^T = \begin{bmatrix} a & c \\ b & d \end{bmatrix}$. *Show that the following equations are true for matrices A and*
B, where $B = \begin{bmatrix} m & n \\ p & q \end{bmatrix}$.

68. $(A^T)^T = A$ **69.** $(A + B)^T = A^T + B^T$ **70.** $(AB)^T = B^T A^T$

For Exercises 71–84, let

$$A = \begin{bmatrix} a_{11} & a_{12} \\ a_{21} & a_{22} \end{bmatrix}, \quad B = \begin{bmatrix} b_{11} & b_{12} \\ b_{21} & b_{22} \end{bmatrix}, \quad \text{and} \quad C = \begin{bmatrix} c_{11} & c_{12} \\ c_{21} & c_{22} \end{bmatrix}$$

where all the elements are real numbers. Decide which of the following statements are true for
these three matrices. If a statement is true, prove that it is true. If it is false, give a numerical
example to show it is false.

71. $A + B = B + A$ (commutative property)

72. $A + (B + C) = (A + B) + C$ (associative property)

73. $A + B$ is a 2×2 matrix.   (closure property)

74. There exists a matrix O such that $A + O = A$ and $O + A = A$.   (identity property)

75. There exists a matrix $-A$ such that $A + (-A) = O$ and $-A + A = O$.   (inverse property)

76. $(AB)C = A(BC)$ (associative property)

77. $A(B + C) = AB + AC$ (distributive property)

78. AB is a 2×2 matrix.   (closure property)

79. $c(A + B) = cA + cB$ for any real number c.

80. $(c + d)A = cA + dA$ for any real numbers c and d.

81. $c(A)d = cd(A)$

82. $(cd)A = c(dA)$

83. $(A + B)(A - B) = A^2 - B^2$ (where $A^2 = AA$)

84. If $AB = O$, then $A = O$ or $B = O$.

9.5 Determinants ▼▼▼

Earlier in this chapter we saw that not every system of linear equations has a single solution. Sometimes a system of equations has no solution or an infinite number of solutions. In this section we introduce the *determinant* of a matrix. In the next section, determinants are used to determine whether a system of equations has a single solution, and if so, to solve the system.

Every square matrix A is associated with a real number called the **determinant** of A, written $|A|$. The determinant of a 2×2 matrix is defined as follows.

Definition of Determinant of a 2 × 2 Matrix

$$\text{If } A = \begin{bmatrix} a_{11} & a_{12} \\ a_{21} & a_{22} \end{bmatrix}, \text{ then } |A| = \begin{vmatrix} a_{11} & a_{12} \\ a_{21} & a_{22} \end{vmatrix} = a_{11}a_{22} - a_{21}a_{12}.$$

NOTE Notice that matrices are enclosed with square brackets, while determinants are denoted with vertical bars.

◀ **EXAMPLE 1**
Evaluating a 2×2 determinant

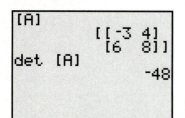

In this screen, the symbol

$$\det[A]$$

represents the determinant of matrix A. With a graphing calculator, we can define a matrix and then use the capability of the calculator to find the determinant of the matrix. Compare the result here to the one in Example 1.

Let $A = \begin{bmatrix} -3 & 4 \\ 6 & 8 \end{bmatrix}$. Find $|A|$.

Use the definition above.

$$|A| = \begin{vmatrix} -3 & 4 \\ 6 & 8 \end{vmatrix} = -3(8) - 6(4) = -48 \quad \blacktriangleright$$

The determinant of a 3×3 matrix A is defined as follows.

Definition of Determinant of a 3 × 3 Matrix

$$\text{If} \quad A = \begin{bmatrix} a_{11} & a_{12} & a_{13} \\ a_{21} & a_{22} & a_{23} \\ a_{31} & a_{32} & a_{33} \end{bmatrix}, \quad \text{then}$$

$$|A| = \begin{vmatrix} a_{11} & a_{12} & a_{13} \\ a_{21} & a_{22} & a_{23} \\ a_{31} & a_{32} & a_{33} \end{vmatrix} = (a_{11}a_{22}a_{33} + a_{12}a_{23}a_{31} + a_{13}a_{21}a_{32}) - (a_{31}a_{22}a_{13} + a_{32}a_{23}a_{11} + a_{33}a_{21}a_{12}).$$

The terms on the right side of the equation in the definition of $|A|$ can be rearranged to get

$$\begin{vmatrix} a_{11} & a_{12} & a_{13} \\ a_{21} & a_{22} & a_{23} \\ a_{31} & a_{32} & a_{33} \end{vmatrix} = a_{11}(a_{22}a_{33} - a_{32}a_{23}) - a_{21}(a_{12}a_{33} - a_{32}a_{13}) + a_{31}(a_{12}a_{23} - a_{22}a_{13}).$$

Each of the quantities in parentheses represents the determinant of a 2×2 matrix that is the part of the 3×3 matrix remaining when the row and column of the multiplier are eliminated, as shown below.

$$a_{11}(a_{22}a_{33} - a_{32}a_{23}) \quad \begin{bmatrix} a_{11} & a_{12} & a_{13} \\ a_{21} & a_{22} & a_{23} \\ a_{31} & a_{32} & a_{33} \end{bmatrix}$$

$$a_{21}(a_{12}a_{33} - a_{32}a_{13}) \quad \begin{bmatrix} a_{11} & a_{12} & a_{13} \\ a_{21} & a_{22} & a_{23} \\ a_{31} & a_{32} & a_{33} \end{bmatrix}$$

$$a_{31}(a_{12}a_{23} - a_{22}a_{13}) \quad \begin{bmatrix} a_{11} & a_{12} & a_{13} \\ a_{21} & a_{22} & a_{23} \\ a_{31} & a_{32} & a_{33} \end{bmatrix}$$

These determinants of the 2×2 matrices are called **minors** of an element in the 3×3 matrix. The symbol M_{ij} represents the determinant of the matrix that results when row i and column j are eliminated. The following list gives some of the minors from the matrix above.

Element	Minor	Element	Minor
a_{11}	$M_{11} = \begin{vmatrix} a_{22} & a_{23} \\ a_{32} & a_{33} \end{vmatrix}$	a_{22}	$M_{22} = \begin{vmatrix} a_{11} & a_{13} \\ a_{31} & a_{33} \end{vmatrix}$
a_{21}	$M_{21} = \begin{vmatrix} a_{12} & a_{13} \\ a_{32} & a_{33} \end{vmatrix}$	a_{23}	$M_{23} = \begin{vmatrix} a_{11} & a_{12} \\ a_{31} & a_{32} \end{vmatrix}$
a_{31}	$M_{31} = \begin{vmatrix} a_{12} & a_{13} \\ a_{22} & a_{23} \end{vmatrix}$	a_{33}	$M_{33} = \begin{vmatrix} a_{11} & a_{12} \\ a_{21} & a_{22} \end{vmatrix}$

In a 4×4 matrix, the minors are determinants of 3×3 matrices, and an $n \times n$ matrix has minors that are determinants of $(n - 1) \times (n - 1)$ matrices.

To find the determinant of a 3×3 or larger matrix, first choose any row or column. Then the minor of each element in that row or column must be multiplied by $+1$ or -1, depending on whether the sum of the row number and column number is even or odd. The product of a minor and the number $+1$ or -1 is called a *cofactor*.

Definition of Cofactor

Let M_{ij} be the minor for element a_{ij} in an $n \times n$ matrix. The **cofactor** of a_{ij}, written A_{ij}, is

$$A_{ij} = (-1)^{i+j} \cdot M_{ij}.$$

EXAMPLE 2
Finding the cofactor of
an element

For the matrix

$$\begin{bmatrix} 6 & 2 & 4 \\ 8 & 9 & 3 \\ 1 & 2 & 0 \end{bmatrix},$$

find the cofactor of each of the following elements.

(a) 6

Since 6 is in the first row and first column of the matrix, $i = 1$ and $j = 1$.

$$M_{11} = \begin{vmatrix} 9 & 3 \\ 2 & 0 \end{vmatrix} = -6$$

The cofactor is $(-1)^{1+1} \cdot -6 = 1 \cdot -6 = -6$.

(b) 3

Here $i = 2$ and $j = 3$.

$$M_{23} = \begin{vmatrix} 6 & 2 \\ 1 & 2 \end{vmatrix} = 10$$

The cofactor is $(-1)^{2+3} \cdot 10 = -1 \cdot 10 = -10$.

(c) 8

We have $i = 2$ and $j = 1$.

$$M_{21} = \begin{vmatrix} 2 & 4 \\ 2 & 0 \end{vmatrix} = -8$$

The cofactor is $(-1)^{2+1} \cdot -8 = -1 \cdot -8 = 8$. ▶

Finally, the determinant of a 3×3 or larger matrix is found as follows.

Finding the Determinant of a Matrix

Multiply each element in any row or column of the matrix by its cofactor. The sum of these products gives the value of the determinant.

The process of forming this sum of products is called **expansion by a given row or column.** (See Exercises 73 and 74.)

EXAMPLE 3
Evaluating a 3×3
determinant

Evaluate $\begin{vmatrix} 2 & -3 & -2 \\ -1 & -4 & -3 \\ -1 & 0 & 2 \end{vmatrix}$. Expand by the second column.

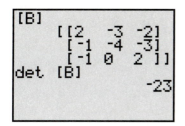

The result in Example 3 is supported with a graphing calculator.

To find this determinant, first get the minors of each element in the second column.

$$M_{12} = \begin{vmatrix} -1 & -3 \\ -1 & 2 \end{vmatrix} = -1(2) - (-1)(-3) = -5$$

$$M_{22} = \begin{vmatrix} 2 & -2 \\ -1 & 2 \end{vmatrix} = 2(2) - (-1)(-2) = 2$$

$$M_{32} = \begin{vmatrix} 2 & -2 \\ -1 & -3 \end{vmatrix} = 2(-3) - (-1)(-2) = -8$$

Now find the cofactor of each of these minors.

$$A_{12} = (-1)^{1+2} \cdot M_{12} = (-1)^3 \cdot (-5) = (-1)(-5) = 5$$
$$A_{22} = (-1)^{2+2} \cdot M_{22} = (-1)^4 \cdot (2) = 1 \cdot 2 = 2$$
$$A_{32} = (-1)^{3+2} \cdot M_{32} = (-1)^5 \cdot (-8) = (-1)(-8) = 8$$

The determinant is found by multiplying each cofactor by its corresponding element in the matrix and finding the sum of these products.

$$\begin{vmatrix} 2 & -3 & -2 \\ -1 & -4 & -3 \\ -1 & 0 & 2 \end{vmatrix} = a_{12} \cdot A_{12} + a_{22} \cdot A_{22} + a_{32} \cdot A_{32}$$

$$= -3(5) + (-4)(2) + (0)(8)$$
$$= -15 + (-8) + 0 = -23$$

Exactly the same answer would be found using any row or column of the matrix. One reason column 2 was used here is that it contains a 0 element, so that it was not really necessary to calculate M_{32} and A_{32} above. One learns quickly that 0s are friends in work with determinants.

Instead of calculating $(-1)^{i+j}$ for a given element, the following sign checkerboards can be used.

Array of Signs

For 3 × 3 matrices	For 4 × 4 matrices
+ − +	+ − + −
− + −	− + − +
+ − +	+ − + −
	− + − +

The signs alternate for each row and column, beginning with + in the first row, first column position. Thus, these arrays of signs can be reproduced as needed. If we expand a 3 × 3 matrix about row 3, for example, the first minor would have a + sign associated with it, the second minor a − sign, and the third minor a + sign. These arrays of signs can be extended in this way for determinants of 5 × 5, 6 × 6, and larger matrices.

EXAMPLE 4
Evaluating a 4 × 4
determinant

Evaluate
$$\begin{vmatrix} -1 & -2 & 3 & 2 \\ 0 & 1 & 4 & -2 \\ 3 & -1 & 4 & 0 \\ 2 & 1 & 0 & 3 \end{vmatrix}.$$

Expand about the fourth row, and do the arithmetic that has been left out.

$$-2 \begin{vmatrix} -2 & 3 & 2 \\ 1 & 4 & -2 \\ -1 & 4 & 0 \end{vmatrix} + 1 \begin{vmatrix} -1 & 3 & 2 \\ 0 & 4 & -2 \\ 3 & 4 & 0 \end{vmatrix} - 0 \begin{vmatrix} -1 & -2 & 2 \\ 0 & 1 & -2 \\ 3 & -1 & 0 \end{vmatrix} + 3 \begin{vmatrix} -1 & -2 & 3 \\ 0 & 1 & 4 \\ 3 & -1 & 4 \end{vmatrix}$$

$$= -2(6) + 1(-50) - 0 + 3(-41) = -185 \quad \blacktriangleright$$

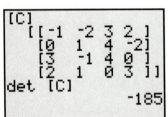

[C]
[[-1 -2 3 2]
 [0 1 4 -2]
 [3 -1 4 0]
 [2 1 0 3]]
det [C]
        -185

The result in Example 4 is supported
with a graphing calculator.

CONNECTIONS If a triangle has vertices (x_1, y_1), (x_2, y_2), and (x_3, y_3), then its area is equal to the absolute value of D where

$$D = \frac{1}{2} \begin{vmatrix} x_1 & y_1 & 1 \\ x_2 & y_2 & 1 \\ x_3 & y_3 & 1 \end{vmatrix}.$$

FOR DISCUSSION OR WRITING

1. Find the area of the triangle with vertices at **(a)** (0, 1), (2, 0), and (1, 3); **(b)** (0, 0), (0, 2), and (1, 1).
2. Compare this method of finding the area of a triangle with the familiar formula $A = (1/2)bh$. When would you choose to use one rather than the other?

There are several theorems that make it easier to calculate determinants. The theorems are true for square matrices of any order, but they are proved here only for determinants of 3 × 3 matrices.

Determinant Theorem 1

If every element in a row (or column) of matrix A is 0, then $|A| = 0$.

To prove the theorem, expand the given matrix by the row (or column) of zeros. Each term of this expansion will have a zero factor, making the final determinant 0. For example,

$$\begin{vmatrix} a_{11} & a_{12} & a_{13} \\ 0 & 0 & 0 \\ a_{31} & a_{32} & a_{33} \end{vmatrix} = -0 \cdot \begin{vmatrix} a_{12} & a_{13} \\ a_{32} & a_{33} \end{vmatrix} + 0 \cdot \begin{vmatrix} a_{11} & a_{13} \\ a_{31} & a_{33} \end{vmatrix} - 0 \cdot \begin{vmatrix} a_{11} & a_{12} \\ a_{31} & a_{32} \end{vmatrix} = 0.$$

EXAMPLE 5
Applying Theorem 1

$$\begin{vmatrix} -3 & 7 & 0 \\ 4 & 9 & 0 \\ -6 & 8 & 0 \end{vmatrix} = 0 \; \blacktriangleright$$

> ### Determinant Theorem 2
> If the rows of matrix A are the corresponding columns of matrix B, then $|B| = |A|$.

To prove the theorem, let

$$A = \begin{bmatrix} a_{11} & a_{12} & a_{13} \\ a_{21} & a_{22} & a_{23} \\ a_{31} & a_{32} & a_{33} \end{bmatrix} \quad \text{and} \quad B = \begin{bmatrix} a_{11} & a_{21} & a_{31} \\ a_{12} & a_{22} & a_{32} \\ a_{13} & a_{23} & a_{33} \end{bmatrix},$$

where B was obtained by interchanging the corresponding rows and columns of A. Find $|A|$ by expansion about row 1. Then find $|B|$ by expansion about column 1. You should find that $|B| = |A|$.

EXAMPLE 6
Applying Theorem 2

(a) Let $A = \begin{bmatrix} 2 & 1 \\ 3 & 4 \end{bmatrix}$. Interchange the rows and columns of A to get matrix B.

$$B = \begin{bmatrix} 2 & 3 \\ 1 & 4 \end{bmatrix}$$

Check that $|A| = 5$ and $|B| = 5$, so that $|B| = |A|$.

(b) By Determinant Theorem 2, $\begin{vmatrix} 2 & 1 & 6 \\ 3 & 0 & 5 \\ -4 & 6 & 9 \end{vmatrix} = \begin{vmatrix} 2 & 3 & -4 \\ 1 & 0 & 6 \\ 6 & 5 & 9 \end{vmatrix}$. $\blacktriangleright$

We now consider how each of the matrix row transformations (and corresponding column transformations) affect the determinant of a matrix.

> ### Determinant Theorem 3
> If any two rows (or columns) of matrix A are interchanged to form matrix B, then $|B| = -|A|$.

This theorem is proved by steps very similar to those used to prove the previous theorem. (See Exercise 75.)

EXAMPLE 7
Applying Theorem 3

(a) Let $A = \begin{bmatrix} 2 & 5 \\ 3 & 4 \end{bmatrix}$. Exchange the two columns of A to get the matrix $B = \begin{bmatrix} 5 & 2 \\ 4 & 3 \end{bmatrix}$. Check that $|A| = -7$ and $|B| = 7$, so that $|B| = -|A|$.

(b) By Determinant Theorem 3,
$$\begin{vmatrix} 2 & 1 & 6 \\ 3 & 0 & 5 \\ -4 & 6 & 9 \end{vmatrix} = - \begin{vmatrix} -4 & 6 & 9 \\ 3 & 0 & 5 \\ 2 & 1 & 6 \end{vmatrix}.$$ ▶

Determinant Theorem 4

Suppose matrix B is formed by multiplying every element of a row (or column) of matrix A by the real number k. Then $|B| = k \cdot |A|$.

The proof of this theorem is left for Exercise 76.

EXAMPLE 8
Applying Theorem 4

Let $A = \begin{bmatrix} 2 & -3 \\ 4 & 1 \end{bmatrix}$. Form the new matrix B by multiplying each element of the second row of A by -5.

$$B = \begin{bmatrix} 2 & -3 \\ 4(-5) & 1(-5) \end{bmatrix} = \begin{bmatrix} 2 & -3 \\ -20 & -5 \end{bmatrix}$$

Check that $|B| = -5 \cdot |A|$. ▶

Determinant Theorem 5

If two rows (or columns) of a matrix A are identical, then $|A| = 0$.

To prove this theorem, note that if two rows or columns of a matrix A are interchanged to form matrix B, then $|A| = -|B|$; while if two rows of matrix A are identical and are interchanged, we still have matrix A. But then $|A| = -|A|$, which can only happen if $|A| = 0$.

EXAMPLE 9
Applying Theorem 5

Since two rows are identical, $\begin{vmatrix} -4 & 2 & 3 \\ 0 & 1 & 6 \\ -4 & 2 & 3 \end{vmatrix} = 0$. ▶

The last theorem of this section is perhaps the most useful of all.

Determinant Theorem 6

Changing a row (or column) of a matrix by adding to it a constant times another row (or column) does not change the determinant of the matrix.

This theorem is proved in much the same way as the others in this section. (See Exercises 77 and 78.) It provides a powerful method for simplifying the work of finding the determinant of a 3×3 or larger matrix, as shown in the next example.

EXAMPLE 10
Applying Theorem 6

Let $A = \begin{bmatrix} -2 & 4 & 1 \\ 2 & 1 & 5 \\ 4 & 0 & 2 \end{bmatrix}$. Find $|A|$.

First obtain a new matrix B (using Determinant Theorem 6) by adding row 1 to row 2 and then (using Theorem 6 again) adding 2 times row 1 to row 3.

$$B = \begin{bmatrix} -2 & 4 & 1 \\ 0 & 5 & 6 \\ 0 & 8 & 4 \end{bmatrix} \qquad \begin{matrix} \\ R_1 + R_2 \\ 2R_1 + R_3 \end{matrix}$$

Now find $|B|$ by expanding about the first column.

$$|B| = -2 \begin{vmatrix} 5 & 6 \\ 8 & 4 \end{vmatrix} = -2(20 - 48) = 56$$

By the theorem above, $|B| = |A|$, so $|A| = 56$. ▶

The following examples show how the properties of determinants are used to simplify the calculation of determinants.

EXAMPLE 11
Simplifying calculation of determinants

Without expanding, show that the value of the following determinant is 0.

$$\begin{vmatrix} 2 & 5 & -1 \\ 1 & -15 & 3 \\ -2 & 10 & -2 \end{vmatrix}$$

Examining the columns of the array shows that each element in the second column is -5 times the corresponding element in the third column. By Determinant Theorem 6, add to the elements of the second column the results of multiplying the elements of the third column by 5 (abbreviated below as $5C_3 + C_2$), to get the determinant

$$\begin{vmatrix} 2 & 0 & -1 \\ 1 & 0 & 3 \\ -2 & 0 & -2 \end{vmatrix}. \qquad 5C_3 + C_2$$

The value of this determinant is 0, by Theorem 1. ▶

EXAMPLE 12
Simplifying calculation of determinants

Find $|A| = \begin{vmatrix} 4 & 2 & 1 & 0 \\ -2 & 4 & -1 & 7 \\ -5 & 2 & 3 & 1 \\ 6 & 4 & -3 & 2 \end{vmatrix}$.

The goal is to change row 1 of the matrix (any row or column could be selected) to a row in which every element but one is 0. To begin, multiply the elements of column 2 of the matrix by -2 and add the results to the elements of column 1.

$$\begin{vmatrix} 0 & 2 & 1 & 0 \\ -10 & 4 & -1 & 7 \\ -9 & 2 & 3 & 1 \\ -2 & 4 & -3 & 2 \end{vmatrix} \quad -2C_2 + C_1$$

Add to the elements of column 2 of the matrix the results of multiplying the elements of column 3 by -2.

$$\begin{vmatrix} 0 & 0 & 1 & 0 \\ -10 & 6 & -1 & 7 \\ -9 & -4 & 3 & 1 \\ -2 & 10 & -3 & 2 \end{vmatrix} \quad -2C_3 + C_2$$

Row 1 of the matrix has only one nonzero number, so expand about the first row.

$$|A| = +1 \begin{vmatrix} -10 & 6 & 7 \\ -9 & -4 & 1 \\ -2 & 10 & 2 \end{vmatrix}$$

Now change column 3 of the matrix to a column with two zeros.

$$\begin{vmatrix} 53 & 34 & 0 \\ -9 & -4 & 1 \\ -2 & 10 & 2 \end{vmatrix} \quad -7R_2 + R_1$$

$$\begin{vmatrix} 53 & 34 & 0 \\ -9 & -4 & 1 \\ 16 & 18 & 0 \end{vmatrix} \quad -2R_2 + R_3$$

Finally, expand about column 3 of the matrix to find the value of $|A|$.

$$|A| = -1 \begin{vmatrix} 53 & 34 \\ 16 & 18 \end{vmatrix} = -1(954 - 544) = -410 \quad \blacktriangleright$$

NOTE In Example 12, working with *rows* of the matrix led to a *column* with only one nonzero number and working with *columns* of the matrix led to a *row* with one nonzero number.

9.5 Exercises ▼▼▼▼▼▼▼▼▼▼▼▼▼▼▼▼▼▼▼▼▼▼▼▼▼▼▼▼▼▼▼▼▼▼▼

Find the value of each determinant. All variables represent real numbers. See Example 1.

1. $\begin{vmatrix} 2 & 5 \\ 4 & -7 \end{vmatrix}$ **2.** $\begin{vmatrix} 3 & 4 \\ 5 & -2 \end{vmatrix}$ **3.** $\begin{vmatrix} -9 & 7 \\ 2 & 6 \end{vmatrix}$ **4.** $\begin{vmatrix} 0 & 4 \\ 4 & 0 \end{vmatrix}$ **5.** $\begin{vmatrix} y & 3 \\ -2 & x \end{vmatrix}$ **6.** $\begin{vmatrix} y & 2 \\ 8 & y \end{vmatrix}$

Find the cofactor of each element in the second row for each determinant. See Example 2.

7. $\begin{vmatrix} -2 & 0 & 1 \\ 1 & 2 & 0 \\ 4 & 2 & 1 \end{vmatrix}$ **8.** $\begin{vmatrix} 1 & -1 & 2 \\ 1 & 0 & 2 \\ 0 & -3 & 1 \end{vmatrix}$ **9.** $\begin{vmatrix} 1 & 2 & -1 \\ 2 & 3 & -2 \\ -1 & 4 & 1 \end{vmatrix}$ **10.** $\begin{vmatrix} 2 & -1 & 4 \\ 3 & 0 & 1 \\ -2 & 1 & 4 \end{vmatrix}$

Find the value of each determinant. All variables represent real numbers. See Examples 3 and 4.

11. $\begin{vmatrix} 1 & 0 & 0 \\ 0 & 1 & 0 \\ 0 & 0 & 1 \end{vmatrix}$ **12.** $\begin{vmatrix} 1 & 0 & 0 \\ 0 & -1 & 0 \\ 1 & 0 & 1 \end{vmatrix}$ **13.** $\begin{vmatrix} -2 & 0 & 1 \\ 0 & 1 & 0 \\ 0 & 0 & -1 \end{vmatrix}$ **14.** $\begin{vmatrix} 1 & -2 & 3 \\ 0 & 0 & 0 \\ 1 & 10 & -12 \end{vmatrix}$

15. $\begin{vmatrix} 0 & 5 & 2 \\ 0 & 3 & -1 \\ 0 & -4 & 7 \end{vmatrix}$ **16.** $\begin{vmatrix} 3 & 3 & -1 \\ 2 & 6 & 0 \\ -6 & -6 & 2 \end{vmatrix}$ **17.** $\begin{vmatrix} 0 & 3 & y \\ 0 & 4 & 2 \\ 1 & 0 & 1 \end{vmatrix}$ **18.** $\begin{vmatrix} 3 & 2 & 0 \\ 0 & 1 & x \\ 2 & 0 & 0 \end{vmatrix}$

19. $\begin{vmatrix} .4 & -.8 & .6 \\ .3 & .9 & .7 \\ 3.1 & 4.1 & -2.8 \end{vmatrix}$ **20.** $\begin{vmatrix} -.3 & -.1 & .9 \\ 2.5 & 4.9 & -3.2 \\ -.1 & .4 & .8 \end{vmatrix}$ **21.** $\begin{vmatrix} 2 & 0 & 0 & 1 \\ -2 & 0 & 6 & 0 \\ 2 & 4 & 0 & 1 \\ 2 & 4 & 1 & 2 \end{vmatrix}$ **22.** $\begin{vmatrix} 2 & 6 & 1 & -3 \\ 4 & 0 & -1 & 2 \\ 3 & -5 & 1 & 4 \\ -1 & 2 & -3 & 0 \end{vmatrix}$

▼▼▼▼▼▼▼▼▼▼▼▼▼ **DISCOVERING CONNECTIONS** (Exercises 23–26) ▼▼▼▼▼▼▼▼▼▼▼▼▼

The equation

$$\begin{vmatrix} x & 2 & 1 \\ -1 & x & 4 \\ -2 & 0 & 5 \end{vmatrix} = 45$$

is an example of a determinant equation. It can be solved by finding an expression in x for the determinant and then solving the resulting equation. Work Exercises 23–26 in order, to see how to solve this equation.

23. Use one of the methods described in this section to write the determinant as a polynomial in *x*.

24. Replace the determinant with the expression you found in Exercise 23. What kind of equation is this (based on the degree of the polynomial)?

25. Solve the equation found in Exercise 24.

26. Verify that when the solutions are substituted for *x* in the original determinant, the equation is satisfied.

Solve the equation for x.

27. $\begin{vmatrix} -2 & 0 & 1 \\ -1 & 3 & x \\ 5 & -2 & 0 \end{vmatrix} = 3$

28. $\begin{vmatrix} 4 & 3 & 0 \\ 2 & 0 & 1 \\ -3 & x & -1 \end{vmatrix} = 5$

29. $\begin{vmatrix} 5 & 3x & -3 \\ 0 & 2 & -1 \\ 4 & -1 & x \end{vmatrix} = -7$

30. $\begin{vmatrix} 2x & 1 & -1 \\ 0 & 4 & x \\ 3 & 0 & 2 \end{vmatrix} = x$

Tell why each determinant has a value of 0. All variables represent real numbers. See Examples 5–10.

31. $\begin{vmatrix} 2 & 3 \\ 2 & 3 \end{vmatrix}$

32. $\begin{vmatrix} -5 & -5 \\ 6 & 6 \end{vmatrix}$

33. $\begin{vmatrix} 2 & 0 \\ 3 & 0 \end{vmatrix}$

34. $\begin{vmatrix} 6 & -8 \\ -3 & 4 \end{vmatrix}$

35. $\begin{vmatrix} 1 & 0 & 0 \\ 1 & 0 & 1 \\ 3 & 0 & 0 \end{vmatrix}$

36. $\begin{vmatrix} -1 & 2 & 4 \\ 4 & -8 & -16 \\ 3 & 0 & 5 \end{vmatrix}$

37. $\begin{vmatrix} 7z & 8x & 2y \\ z & x & y \\ 7z & 7x & 7y \end{vmatrix}$

38. $\begin{vmatrix} m & 2 & 2m \\ 3n & 1 & 6n \\ 5p & 6 & 10p \end{vmatrix}$

Use the appropriate theorems from this section to tell why each statement is true. Do not evaluate the determinants. All variables represent real numbers. See Examples 5–10.

39. $\begin{vmatrix} 4 & -2 \\ 3 & 8 \end{vmatrix} = \begin{vmatrix} 4 & 3 \\ -2 & 8 \end{vmatrix}$

40. $\begin{vmatrix} 2 & 1 & 6 \\ 3 & 0 & 2 \\ 4 & 1 & 8 \end{vmatrix} = \begin{vmatrix} 2 & 3 & 4 \\ 1 & 0 & 1 \\ 6 & 2 & 8 \end{vmatrix}$

41. $\begin{vmatrix} -1 & 8 & 9 \\ 0 & 2 & 1 \\ 3 & 2 & 0 \end{vmatrix} = -\begin{vmatrix} 8 & -1 & 9 \\ 2 & 0 & 1 \\ 2 & 3 & 0 \end{vmatrix}$

42. $\begin{vmatrix} 2 & 6 \\ 3 & 5 \end{vmatrix} = -\begin{vmatrix} 3 & 5 \\ 2 & 6 \end{vmatrix}$

43. $-\dfrac{1}{2} \begin{vmatrix} 5 & -8 & 2 \\ 3 & -6 & 9 \\ 2 & 4 & 4 \end{vmatrix} = \begin{vmatrix} 5 & 4 & 2 \\ 3 & 3 & 9 \\ 2 & -2 & 4 \end{vmatrix}$

44. $3 \begin{vmatrix} 6 & 0 & 2 \\ 4 & 1 & 3 \\ 2 & 8 & 6 \end{vmatrix} = \begin{vmatrix} 6 & 0 & 2 \\ 4 & 3 & 3 \\ 2 & 24 & 6 \end{vmatrix}$

45. $\begin{vmatrix} 3 & -4 \\ 2 & 5 \end{vmatrix} = \begin{vmatrix} 3 & -4 \\ 5 & 1 \end{vmatrix}$

46. $\begin{vmatrix} -1 & 6 \\ 3 & -5 \end{vmatrix} = \begin{vmatrix} -1 & 5 \\ 3 & -2 \end{vmatrix}$

47. $\begin{vmatrix} -4 & 2 & 1 \\ 3 & 0 & 5 \\ -1 & 4 & -2 \end{vmatrix} = \begin{vmatrix} -4 & 2 & 1 + (-4)k \\ 3 & 0 & 5 + 3k \\ -1 & 4 & -2 + (-1)k \end{vmatrix}$

48. $2 \begin{vmatrix} 4 & 2 & -1 \\ m & 2n & 3p \\ 5 & 1 & 0 \end{vmatrix} = \begin{vmatrix} 4 & 2 & -1 \\ 2m & 4n & 6p \\ 5 & 1 & 0 \end{vmatrix}$

Use the method of Examples 11 and 12 to find the value of each determinant.

49. $\begin{vmatrix} -5 & 10 \\ 6 & -12 \end{vmatrix}$

50. $\begin{vmatrix} 2 & 4 \\ 3 & 6 \end{vmatrix}$

51. $\begin{vmatrix} 6 & 8 & -12 \\ -1 & 0 & 2 \\ 4 & 0 & -8 \end{vmatrix}$

52. $\begin{vmatrix} 4 & 8 & 0 \\ -1 & -2 & 1 \\ 2 & 4 & 3 \end{vmatrix}$

53. $\begin{vmatrix} -4 & 1 & 4 \\ 2 & 0 & 1 \\ 0 & 2 & 4 \end{vmatrix}$

54. $\begin{vmatrix} 6 & 3 & 2 \\ 1 & 0 & 2 \\ 5 & 7 & 3 \end{vmatrix}$

55. $\begin{vmatrix} 2 & -1 & 1 & 0 \\ 1 & 1 & 0 & 1 \\ 0 & -1 & 1 & 1 \\ 1 & 2 & 1 & 2 \end{vmatrix}$

56. $\begin{vmatrix} 1 & 0 & 2 & 2 \\ 2 & 4 & 1 & -1 \\ 1 & -3 & 1 & 0 \\ 1 & 1 & 0 & 1 \end{vmatrix}$

Use the method described in the Connections box in this section to find the area of triangle PQR having vertices with the coordinates given.

57. $P(0, 0)$, $Q(0, 2)$, $R(1, 4)$

58. $P(0, 1)$, $Q(2, 0)$, $R(1, 5)$

59. $P(2, 5)$, $Q(-1, 3)$, $R(4, 0)$

60. $P(2, -2)$, $Q(0, 0)$, $R(-3, -4)$

61. $P(4, 7)$, $Q(5, -2)$, $R(1, 1)$

62. Find the area of a triangular lot whose vertices have coordinates in feet of $(101.3, 52.7)$, $(117.2, 253.9)$, and $(313.1, 301.6)$. (*Source:* Al-Khafaji, A., and J. Tooley, *Numerical Methods in Engineering Practice,* Holt, Rinehart, and Winston, Inc., 1986.)

Use a graphing calculator with matrix capability to find each determinant. Give as many decimal places as the calculator shows.

63. $\begin{vmatrix} \pi & \sqrt{2} \\ e & \sqrt{3} \end{vmatrix}$

64. $\begin{vmatrix} -\pi & -\sqrt{3} \\ 1/e & -\sqrt{2} \end{vmatrix}$

65. $\begin{vmatrix} .29 & .36 & -.51 \\ -.16 & 1.24 & 3.26 \\ 2.43 & 3.84 & -6.15 \end{vmatrix}$

66. $\begin{vmatrix} \sqrt[3]{6} & \sqrt[4]{2} & -\sqrt{7} \\ -\sqrt[3]{9} & \sqrt[5]{6} & -\sqrt[3]{19} \\ \sqrt[4]{8} & \sqrt[9]{2} & -\sqrt[8]{6} \end{vmatrix}$

67. Prove that the straight line through the distinct points (x_1, y_1) and (x_2, y_2) has equation

$$\begin{vmatrix} x & y & 1 \\ x_1 & y_1 & 1 \\ x_2 & y_2 & 1 \end{vmatrix} = 0.$$

68. Use the result of Exercise 67 to show that three distinct points (x_1, y_1), (x_2, y_2), and (x_3, y_3) lie on a straight line if

$$\begin{vmatrix} x_1 & y_1 & 1 \\ x_2 & y_2 & 1 \\ x_3 & y_3 & 1 \end{vmatrix} = 0.$$

69. Show that the lines $a_1x + b_1y = c_1$ and $a_2x + b_2y = c_2$, when $c_1 \neq c_2$, are parallel if

$$\begin{vmatrix} a_1 & b_1 \\ a_2 & b_2 \end{vmatrix} = 0.$$

70. Prove that $\begin{vmatrix} 1 & 1 & 1 \\ a & b & c \\ a^2 & b^2 & c^2 \end{vmatrix} = (a - b)(b - c)(c - a).$

71. Find $\begin{vmatrix} 1 & 1 & 1 \\ 1 + x & 1 + y & 1 \\ 1 & 1 & 1 \end{vmatrix}$.

72. Prove that $\begin{vmatrix} 1 & 1 & 1 \\ a & b & c \\ bc & ca & ab \end{vmatrix} = (a - b)(a - c)(c - b).$

Let $A = \begin{bmatrix} a_{11} & a_{12} & a_{13} \\ a_{21} & a_{22} & a_{23} \\ a_{31} & a_{32} & a_{33} \end{bmatrix}$ *for Exercises 73–80.*

73. Find $|A|$ by expansion about row 3 of the matrix. Show that your result is really equal to $|A|$ as given in the definition of the determinant of a 3×3 matrix.

74. Repeat Exercise 73 for column 3.

75. Obtain matrix B by interchanging columns 1 and 3 of matrix A. Show that $|B| = -|A|$.

76. Obtain matrix B by multiplying each element of row 3 of matrix A by the real number k. Show that $|B| = k \cdot |A|$.

77. Obtain matrix B by adding to column 1 of matrix A the result of multiplying each element of column 2 of A by the real number k. Show that $|B| = |A|$.

78. Obtain matrix B by adding to row 1 of matrix A the result of multiplying each element of row 3 of A by the real number k. Show that $|B| = |A|$.

79. Let A and B be any 2×2 matrices. Show that $|AB| = |A| \cdot |B|$, where $|AB|$ is the determinant of matrix AB.

80. Show that
$$\begin{vmatrix} a_{11} + a & a_{12} & a_{13} \\ a_{21} + b & a_{22} & a_{23} \\ a_{31} + c & a_{32} & a_{33} \end{vmatrix} = \begin{vmatrix} a_{11} & a_{12} & a_{13} \\ a_{21} & a_{22} & a_{23} \\ a_{31} & a_{32} & a_{33} \end{vmatrix} + \begin{vmatrix} a & a_{12} & a_{13} \\ b & a_{22} & a_{23} \\ c & a_{32} & a_{33} \end{vmatrix}.$$

Use this fact and Determinant Theorems 4 and 5 to prove Determinant Theorem 6.

9.6 Cramer's Rule ▼▼▼

We have now seen how to solve a system of n linear equations with n variables using the following methods: elimination, substitution, and row transformations of matrices. Most of these systems can also be solved with determinants, as shown here. An advantage of this method is that by finding the determinant of the matrix of coefficients, we can decide whether a single solution exists before actually solving the system.

To see how determinants arise in solving a system, we solve the linear system

$$a_1 x + b_1 y = c_1 \tag{1}$$
$$a_2 x + b_2 y = c_2 \tag{2}$$

by elimination as follows. Eliminate y and solve for x by first multiplying both sides of equation (1) by b_2 and both sides of equation (2) by $-b_1$. Then add these results and solve for x.

$$\begin{array}{rcl} a_1 b_2 x + b_1 b_2 y &=& c_1 b_2 \\ -a_2 b_1 x - b_1 b_2 y &=& -c_2 b_1 \\ \hline (a_1 b_2 - a_2 b_1)x &=& c_1 b_2 - c_2 b_1 \end{array}$$

$$x = \frac{c_1 b_2 - c_2 b_1}{a_1 b_2 - a_2 b_1}, \quad \text{if } a_1 b_2 - a_2 b_1 \neq 0$$

Solve for y by multiplying both sides of equation (1) by $-a_2$ and equation (2) by a_1 and then adding the two equations.

$$\begin{array}{rcl} -a_1 a_2 x - a_2 b_1 y &=& -a_2 c_1 \\ a_1 a_2 x + a_1 b_2 y &=& a_1 c_2 \\ \hline (a_1 b_2 - a_2 b_1)y &=& a_1 c_2 - a_2 c_1 \end{array}$$

$$y = \frac{a_1 c_2 - a_2 c_1}{a_1 b_2 - a_2 b_1}, \quad \text{if } a_1 b_2 - a_2 b_1 \neq 0$$

Both numerators and the common denominator of these values for x and y can be written as determinants, since

$$c_1 b_2 - c_2 b_1 = \begin{vmatrix} c_1 & b_1 \\ c_2 & b_2 \end{vmatrix}, \qquad a_1 c_2 - a_2 c_1 = \begin{vmatrix} a_1 & c_1 \\ a_2 & c_2 \end{vmatrix},$$

$$\text{and} \qquad a_1 b_2 - a_2 b_1 = \begin{vmatrix} a_1 & b_1 \\ a_2 & b_2 \end{vmatrix}.$$

Using these determinants, the solutions for x and y become

$$x = \frac{\begin{vmatrix} c_1 & b_1 \\ c_2 & b_2 \end{vmatrix}}{\begin{vmatrix} a_1 & b_1 \\ a_2 & b_2 \end{vmatrix}} \quad \text{and} \quad y = \frac{\begin{vmatrix} a_1 & c_1 \\ a_2 & c_2 \end{vmatrix}}{\begin{vmatrix} a_1 & b_1 \\ a_2 & b_2 \end{vmatrix}}, \quad \text{if } \begin{vmatrix} a_1 & b_1 \\ a_2 & b_2 \end{vmatrix} \neq 0.$$

For convenience, denote the three determinants in the solution as

$$\begin{vmatrix} a_1 & b_1 \\ a_2 & b_2 \end{vmatrix} = D, \qquad \begin{vmatrix} c_1 & b_1 \\ c_2 & b_2 \end{vmatrix} = D_x, \quad \text{and} \quad \begin{vmatrix} a_1 & c_1 \\ a_2 & c_2 \end{vmatrix} = D_y.$$

NOTE The elements of D are the four coefficients of the variables in the given system, the elements of D_x are obtained by replacing the coefficients of x in D by the respective constants, and the elements of D_y are obtained by replacing the coefficients of y in D by the respective constants.

These results are summarized as **Cramer's rule.**

> ### Cramer's Rule for 2 Equations in 2 Variables
>
> Given the system
>
> $$a_1 x + b_1 y = c_1$$
> $$a_2 x + b_2 y = c_2,$$
>
> if $D \neq 0$, the system has the unique solution
>
> $$x = \frac{D_x}{D} \quad \text{and} \quad y = \frac{D_y}{D}.$$

Cramer was looking for a method to determine the equation of a curve when he knew several points on the curve. In 1750 he wrote down the general equation for a curve and then substituted each point for which he had two coordinates into the equation. For this system of equations he gave "a rule very convenient and general to solve any number of equations and unknowns which are of no more than first degree." This is the rule that now bears his name.

CAUTION As indicated above, Cramer's rule does not apply if $D = 0$. When $D = 0$, the system is inconsistent or has dependent equations. For this reason, it is a good idea to evaluate D first.

EXAMPLE 1
Applying Cramer's rule to a
2 × 2 system

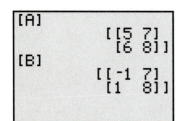

By defining matrices and using a graphing calculator to compute their determinants, Cramer's rule can be applied to solve a system of equations. These two screens show how the system in Example 1 can be solved with a graphing calculator.

Use Cramer's rule to solve the system

$$5x + 7y = -1$$
$$6x + 8y = 1.$$

By Cramer's rule, $x = D_x/D$ and $y = D_y/D$. As mentioned above, it is a good idea to find D first, since if $D = 0$, Cramer's rule does not apply. If $D \neq 0$, then find D_x and D_y.

$$D = \begin{vmatrix} 5 & 7 \\ 6 & 8 \end{vmatrix} = 5(8) - 6(7) = -2$$

$$D_x = \begin{vmatrix} -1 & 7 \\ 1 & 8 \end{vmatrix} = (-1)(8) - (1)(7) = -15$$

$$D_y = \begin{vmatrix} 5 & -1 \\ 6 & 1 \end{vmatrix} = 5(1) - (6)(-1) = 11$$

From Cramer's rule,

$$x = \frac{D_x}{D} = \frac{-15}{-2} = \frac{15}{2}$$

and

$$y = \frac{D_y}{D} = \frac{11}{-2} = -\frac{11}{2}.$$

The solution set is $\{(15/2, -11/2)\}$, as can be verified by substituting in the given system. ▶

By much the same method as used above, Cramer's rule can be generalized to a system of n linear equations with n variables.

General Form of Cramer's Rule

Let an $n \times n$ system have linear equations of the form

$$a_{11}x_1 + a_{12}x_2 + a_{13}x_3 + \cdots + a_{1n}x_n = b_1.$$

Define D as the determinant of the $n \times n$ matrix of all coefficients of the variables. Define D_{x1} as the determinant obtained from D by replacing the entries in column 1 of D with the constants of the system. Define D_{xi} as the determinant obtained from D by replacing the entries in column i with the constants of the system. If $D \neq 0$, the unique solution of the system is

$$x_1 = \frac{D_{x1}}{D}, x_2 = \frac{D_{x2}}{D}, x_3 = \frac{D_{x3}}{D}, \ldots, x_n = \frac{D_{xn}}{D}.$$

EXAMPLE 2
Applying Cramer's rule to a
3 × 3 system

Use Cramer's rule to solve the system

$$x + y - z + 2 = 0$$
$$2x - y + z + 5 = 0$$
$$x - 2y + 3z - 4 = 0.$$

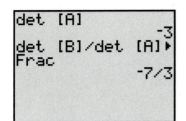

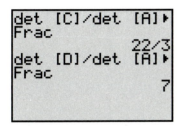

When using Cramer's rule, be sure to check that the determinant that appears in the denominator is not zero, as shown in the first display (it is -3). Refer to Example 2. The values of x, y, and z are shown in their fraction forms: $-7/3$, $22/3$, and 7.

For Cramer's rule, the system must be rewritten in the form

$$x + y - z = -2$$
$$2x - y + z = -5$$
$$x - 2y + 3z = 4.$$

Verify that the required determinants are as follows:

$$D = \begin{vmatrix} 1 & 1 & -1 \\ 2 & -1 & 1 \\ 1 & -2 & 3 \end{vmatrix} = -3, \qquad D_x = \begin{vmatrix} -2 & 1 & -1 \\ -5 & -1 & 1 \\ 4 & -2 & 3 \end{vmatrix} = 7,$$

$$D_y = \begin{vmatrix} 1 & -2 & -1 \\ 2 & -5 & 1 \\ 1 & 4 & 3 \end{vmatrix} = -22, \qquad D_z = \begin{vmatrix} 1 & 1 & -2 \\ 2 & -1 & -5 \\ 1 & -2 & 4 \end{vmatrix} = -21.$$

Thus

$$x = \frac{D_x}{D} = \frac{7}{-3} = -\frac{7}{3}, \qquad y = \frac{D_y}{D} = \frac{-22}{-3} = \frac{22}{3},$$

and

$$z = \frac{D_z}{D} = \frac{-21}{-3} = 7,$$

so the solution set is $\{(-7/3, 22/3, 7)\}$. ▶

CAUTION   As shown in Example 2, each equation in the system must be written in the form $ax + by + cz + \cdots = k$ before using Cramer's rule.

◀ **EXAMPLE 3**
Applying Cramer's rule when $D = 0$

Use Cramer's rule to solve the system

$$2x - 3y + 4z = 10$$
$$6x - 9y + 12z = 24$$
$$x + 2y - 3z = 5.$$

Verify that $D = 0$, so Cramer's rule does not apply. Use another method to determine that this system is inconsistent and thus has no solution. ▶

Several different methods for solving systems of equations have now been shown. In general, if a small system of linear equations must be solved by pencil and paper, substitution is the best method if the various variables can easily be found in terms of each other. This happens rarely. The next choice, perhaps the best choice of all, is the elimination method. Some people like the Gauss-Jordan method, which is really just a systematic way of doing the elimination method. The Gauss-Jordan method is probably superior where four or more equations are involved. Cramer's rule is seldom the method of choice simply because it involves more calculations than any other method.

9.6 Exercises ▼▼▼▼▼▼▼▼▼▼▼▼▼▼▼▼▼▼▼▼▼▼▼▼▼▼▼▼▼▼▼▼▼▼▼▼▼

Use Cramer's rule to solve each of the following systems of equations. If $D = 0$, use another method to determine the solution. See Example 1.

1. $x + y = 4$
$\quad 2x - y = 2$

2. $3x + 2y = -4$
$\quad 2x - y = -5$

3. $4x + 3y = -7$
$\quad 2x + 3y = -11$

4. $4x - y = 0$
$\quad 2x + 3y = 14$

5. $5x + 4y = 10$
$\quad 3x - 7y = 6$

6. $3x + 2y = -4$
$\quad 5x - y = 2$

7. $2x - 3y = -5$
$\quad x + 5y = 17$

8. $x + 9y = -15$
$\quad 3x + 2y = 5$

9. $1.5x + 3y = 5$
$\quad 2x + 4y = 3$

10. $12x + 8y = 3$
$\quad 15x + 10y = 9$

11. $3x + 2y = 4$
$\quad 6x + 4y = 8$

12. $4x + 3y = 9$
$\quad 12x + 9y = 27$

Use Cramer's rule to solve each of the following systems of equations. If $D = 0$, use another method to complete the solution. See Examples 2 and 3.

13. $4x - y + 3z = -3$
$\quad 3x + y + z = 0$
$\quad 2x - y + 4z = 0$

14. $5x + 2y + z = 15$
$\quad 2x - y + z = 9$
$\quad 4x + 3y + 2z = 13$

15. $2x - y + 4z = -2$
$\quad 3x + 2y - z = -3$
$\quad x + 4y + 2z = 17$

16. $x + y + z = 4$
$\quad 2x - y + 3z = 4$
$\quad 4x + 2y - z = -15$

17. $4x - 3y + z = -1$
$\quad 5x + 7y + 2z = -2$
$\quad 3x - 5y - z = 1$

18. $2x - 3y + z = 8$
$\quad -x - 5y + z = -4$
$\quad 3x - 5y + 2z = 12$

19. $x + 2y + 3z - 4 = 0$
$\quad 4x + 3y + 2z - 1 = 0$
$\quad -x - 2y - 3z = 0$

20. $2x - y + 3z - 1 = 0$
$\quad -2x + y - 3z - 2 = 0$
$\quad 5x - y + z - 2 = 0$

21. $-2x - 2y + 3z = 4$
$\quad 5x + 7y - z = 2$
$\quad 2x + 2y - 3z = -4$

22. $3x - 2y + 4z = 1$
$\quad 4x + y - 5z = 2$
$\quad -6x + 4y - 8z = -2$

23. $2x + 3y = 13$
$\quad 2y - z = 5$
$\quad x + 2z = 4$

24. $3x - z = -10$
$\quad y + 4z = 8$
$\quad x + 2z = -1$

25. $5x - y = -4$
$\quad 3x + 2z = 4$
$\quad 4y + 3z = 22$

26. $3x + 5y = -7$
$\quad 2x + 7z = 2$
$\quad 4y + 3z = -8$

27. $x + 2y = 10$
$\quad 3x + 4z = 7$
$\quad -y - z = 1$

28. $5x - 2y = 3$
$\quad 4y + z = 8$
$\quad x + 2z = 4$

29. In your own words, explain what it means in applying Cramer's rule if $D = 0$.

30. Describe D_x, D_y, and D_z in terms of the coefficients and constants in the given system of equations.

Use Cramer's rule to solve the system.

31. $x + 3y - 2z - w = 9$
$\quad 4x + y + z + 2w = 2$
$\quad -3x - y + z - w = -5$
$\quad x - y - 3z - 2w = 2$

32. $3x + 2y - w = 0$
$\quad 2x + z + 2w = 5$
$\quad x + 2y - z = -2$
$\quad 2x - y + z + w = 2$

33. $5x + 3y - 2z + w = 9$
$\quad -3x + y - 6z + 2w = -33$
$\quad 2x + 2y - z + 3w = 5$
$\quad 4x + 3y - z + 8w = 12$

34. $x + 2y - z + w = 8$
$\quad 2x - y + 2w = 8$
$\quad y + 3z = 5$
$\quad x - z = 4$

▼▼▼▼▼▼▼▼▼▼▼▼ **DISCOVERING CONNECTIONS** (Exercises 35–40) ▼▼▼▼▼▼▼▼▼▼▼▼

The system shown here has the property that when the coefficients and constants are read from left to right, starting at upper left and ending at lower right, they are consecutive integers:

$$2x + 3y = 4$$
$$5x + 6y = 7.$$

A system of two equations in two variables such as this has an interesting solution, which we will examine in Exercises 35–40. Work these exercises in order.

35. Use any method you wish to solve the system. What is the solution set?

36. Replace the coefficients and constants 2, 3, 4, 5, 6, 7 by another group of six consecutive integers. Solve the resulting system. What is the solution set?

37. Compare the solution sets in Exercises 35 and 36. How do they compare?

38. The system

$$nx + (n + 1)y = n + 2$$
$$(n + 3)x + (n + 4)y = n + 5$$

is a generalized form of the type of system examined in Exercises 35 and 36. Solve this system by Cramer's rule by first finding D. What is the value of D?

39. Find the values of D_x and D_y for the system in Exercise 38.

40. Use the results in Exercises 38 and 39 to find x and y for the system in Exercise 38. What conclusion can you make?

Use Cramer's rule to solve the systems in Exercises 41 and 42.

41. Linear systems occur in the design of roof trusses for new homes and buildings. The simplest type of roof truss is a triangle. The truss shown in the figure is used to frame roofs of small buildings. If a 100-pound force is applied at the peak of the truss, then the forces or weights W_1 and W_2 exerted parallel to each rafter of the truss are determined by the following linear system of equations.

$$\frac{\sqrt{3}}{2}(W_1 + W_2) = 100$$
$$W_1 - W_2 = 0$$

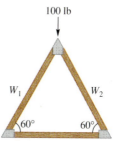

Solve the system to find W_1 and W_2. (*Source:* Hibbeler, R., *Structural Analysis,* Prentice Hall, Englewood Cliffs, 1995.)

42. (Refer to Exercise 41.) Use the following system of equations to determine the forces or weights W_1 and W_2 exerted on each rafter for the truss shown in the figure.

$$W_1 + \sqrt{2}W_2 = 300$$
$$\sqrt{3}W_1 - \sqrt{2}W_2 = 0$$

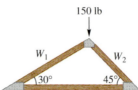

Solve each system for x and y using Cramer's rule. Assume a and b are nonzero constants.

43. $bx + y = a^2$
$ax + y = b^2$

44. $ax + by = \dfrac{b}{a}$
$x + y = \dfrac{1}{b}$

45. $b^2x + a^2y = b^2$
$ax + by = a$

46. $x + \dfrac{1}{b}y = b$
$\dfrac{1}{a}x + y = a$

9.7   Matrix Inverses ▼▼▼

In this section multiplicative identity elements and multiplicative inverses are introduced and used to solve matrix equations. This leads to another method for solving systems of equations.

IDENTITY MATRICES   The identity property for real numbers says that $a \cdot 1 = a$ and $1 \cdot a = a$ for any real number a. If there is to be a multiplicative identity matrix I, such that

$$AI = A \qquad \text{and} \qquad IA = A,$$

for any matrix A, then A and I must be square matrices of the same size. Otherwise it would not be possible to find both products.

---

2 × 2 Identity

If I_2 represents the 2 × 2 identity, then

$$I_2 = \begin{bmatrix} 1 & 0 \\ 0 & 1 \end{bmatrix}.$$

To verify that I_2 is the 2 × 2 identity matrix, show that $AI = A$ and $IA = A$ for any 2 × 2 matrix. Let

$$A = \begin{bmatrix} x & y \\ z & w \end{bmatrix}.$$

Then

$$AI = \begin{bmatrix} x & y \\ z & w \end{bmatrix}\begin{bmatrix} 1 & 0 \\ 0 & 1 \end{bmatrix} = \begin{bmatrix} x \cdot 1 + y \cdot 0 & x \cdot 0 + y \cdot 1 \\ z \cdot 1 + w \cdot 0 & z \cdot 0 + w \cdot 1 \end{bmatrix} = \begin{bmatrix} x & y \\ z & w \end{bmatrix} = A.$$

and

$$IA = \begin{bmatrix} 1 & 0 \\ 0 & 1 \end{bmatrix} \begin{bmatrix} x & y \\ z & w \end{bmatrix} = \begin{bmatrix} 1 \cdot x + 0 \cdot z & 1 \cdot y + 0 \cdot w \\ 0 \cdot x + 1 \cdot z & 0 \cdot y + 1 \cdot w \end{bmatrix} = \begin{bmatrix} x & y \\ z & w \end{bmatrix} = A.$$

Generalizing from this example, there is an $n \times n$ identity matrix having 1s on the main diagonal and 0s elsewhere.

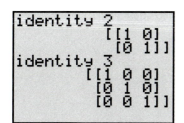

This screen shows identity matrices for $n = 2$ and $n = 3$.

$n \times n$ Identity Matrix

The $n \times n$ identity matrix is given by I_n where

$$I_n = \begin{bmatrix} 1 & 0 & \cdots & 0 \\ 0 & 1 & \cdots & 0 \\ \vdots & \vdots & a_{ij} & \vdots \\ \vdots & \vdots & & \vdots \\ 0 & 0 & \cdots & 1 \end{bmatrix}.$$

The element $a_{ij} = 1$ when $i = j$ (the diagonal elements) and $a_{ij} = 0$ otherwise.

◀ **EXAMPLE 1**
Stating and verifying the 3×3 identity matrix

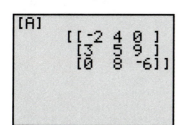

These screens support the result in Example 1.

Let $A = \begin{bmatrix} -2 & 4 & 0 \\ 3 & 5 & 9 \\ 0 & 8 & -6 \end{bmatrix}$. Give the 3×3 identity matrix I and show that $AI = A$.

The 3×3 identity matrix is

$$I = \begin{bmatrix} 1 & 0 & 0 \\ 0 & 1 & 0 \\ 0 & 0 & 1 \end{bmatrix}.$$

By the definition of matrix multiplication,

$$AI = \begin{bmatrix} -2 & 4 & 0 \\ 3 & 5 & 9 \\ 0 & 8 & -6 \end{bmatrix} \begin{bmatrix} 1 & 0 & 0 \\ 0 & 1 & 0 \\ 0 & 0 & 1 \end{bmatrix} = \begin{bmatrix} -2 & 4 & 0 \\ 3 & 5 & 9 \\ 0 & 8 & -6 \end{bmatrix} = A. \; \blacktriangleright$$

MULTIPLICATIVE INVERSES   For every nonzero real number a, there is a multiplicative inverse $1/a$ such that

$$a \cdot \frac{1}{a} = 1 \qquad \text{and} \qquad \frac{1}{a} \cdot a = 1.$$

(Recall: $1/a$ is also written a^{-1}.) In a similar way, if A is an $n \times n$ matrix, then its **multiplicative inverse,** written A^{-1}, must satisfy both

$$AA^{-1} = I_n \qquad \text{and} \qquad A^{-1}A = I_n.$$

This means that only a square matrix can have a multiplicative inverse.

CAUTION Although $a^{-1} = 1/a$ for any nonzero real number a, if A is a matrix,

$$A^{-1} \neq \frac{1}{A}.$$

In fact, $1/A$ has no meaning, since 1 is a *number* and A is a *matrix*.

The matrix A^{-1} can be found by using the row transformations introduced earlier in this chapter. As an example, let us find the inverse of

$$A = \begin{bmatrix} 2 & 4 \\ 1 & -1 \end{bmatrix}.$$

Let the unknown inverse matrix be

$$A^{-1} = \begin{bmatrix} x & y \\ z & w \end{bmatrix}.$$

By the definition of matrix inverse, $AA^{-1} = I_2$, or

$$AA^{-1} = \begin{bmatrix} 2 & 4 \\ 1 & -1 \end{bmatrix}\begin{bmatrix} x & y \\ z & w \end{bmatrix} = \begin{bmatrix} 1 & 0 \\ 0 & 1 \end{bmatrix}.$$

By matrix multiplication,

$$\begin{bmatrix} 2x + 4z & 2y + 4w \\ x - z & y - w \end{bmatrix} = \begin{bmatrix} 1 & 0 \\ 0 & 1 \end{bmatrix}.$$

Setting corresponding elements equal gives the system of equations

$$2x + 4z = 1 \qquad\qquad (1)$$
$$2y + 4w = 0 \qquad\qquad (2)$$
$$x - z = 0 \qquad\qquad (3)$$
$$y - w = 1. \qquad\qquad (4)$$

Since equations (1) and (3) involve only x and z, while equations (2) and (4) involve only y and w, these four equations lead to two systems of equations,

$$2x + 4z = 1 \qquad\qquad 2y + 4w = 0$$
$$\qquad\qquad\qquad \text{and}$$
$$x - z = 0 \qquad\qquad y - w = 1.$$

Writing the two systems as augmented matrices gives

$$\begin{bmatrix} 2 & 4 & | & 1 \\ 1 & -1 & | & 0 \end{bmatrix} \qquad \text{and} \qquad \begin{bmatrix} 2 & 4 & | & 0 \\ 1 & -1 & | & 1 \end{bmatrix}.$$

Each of these systems can be solved by the Gaussian method. However, since the elements to the left of the vertical bar are identical, the two systems can be combined into one matrix,

$$\begin{bmatrix} 2 & 4 & | & 1 & 0 \\ 1 & -1 & | & 0 & 1 \end{bmatrix},$$

and solved simultaneously using matrix row transformations. We need to change the numbers on the left of the vertical bar to the 2 × 2 identity matrix.

Exchange the two rows to get a 1 in the upper left corner.

$$\begin{bmatrix} \mathbf{1} & \mathbf{-1} & | & \mathbf{0} & \mathbf{1} \\ 2 & 4 & | & 1 & 0 \end{bmatrix}$$

Multiply the first row by −2 and add the results to the second row to get

$$\begin{bmatrix} 1 & -1 & | & 0 & 1 \\ \mathbf{0} & \mathbf{6} & | & \mathbf{1} & \mathbf{-2} \end{bmatrix}. \qquad -2R_1 + R_2$$

Now, to get a 1 in the second-row, second-column position, multiply the second row by 1/6.

$$\begin{bmatrix} 1 & -1 & | & 0 & 1 \\ \mathbf{0} & \mathbf{1} & | & \mathbf{1/6} & \mathbf{-1/3} \end{bmatrix}. \qquad \tfrac{1}{6}R_2$$

Finally, add the second row to the first row to get a 0 in the second column above the 1.

$$\begin{bmatrix} \mathbf{1} & \mathbf{0} & | & \mathbf{1/6} & \mathbf{2/3} \\ 0 & 1 & | & 1/6 & -1/3 \end{bmatrix} \qquad R_2 + R_1$$

The numbers in the first column to the right of the vertical bar give the values of x and z. The second column gives the values of y and w. That is,

$$\begin{bmatrix} 1 & 0 & | & x & y \\ 0 & 1 & | & z & w \end{bmatrix} = \begin{bmatrix} 1 & 0 & | & \mathbf{1/6} & \mathbf{2/3} \\ 0 & 1 & | & \mathbf{1/6} & \mathbf{-1/3} \end{bmatrix}$$

so that

$$A^{-1} = \begin{bmatrix} x & y \\ z & w \end{bmatrix} = \begin{bmatrix} 1/6 & 2/3 \\ 1/6 & -1/3 \end{bmatrix}.$$

To check, multiply A by A^{-1}. The result should be I_2.

$$AA^{-1} = \begin{bmatrix} 2 & 4 \\ 1 & -1 \end{bmatrix}\begin{bmatrix} 1/6 & 2/3 \\ 1/6 & -1/3 \end{bmatrix} = \begin{bmatrix} 1/3 + 2/3 & 4/3 - 4/3 \\ 1/6 - 1/6 & 2/3 + 1/3 \end{bmatrix}$$

$$= \begin{bmatrix} 1 & 0 \\ 0 & 1 \end{bmatrix} = I_2$$

Finally,

$$A^{-1} = \begin{bmatrix} 1/6 & 2/3 \\ 1/6 & -1/3 \end{bmatrix}.$$

The process for finding the multiplicative inverse A^{-1} for any $n \times n$ matrix A that has an inverse is summarized below.

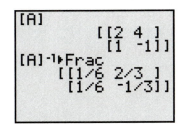

A graphing calculator can find the inverse of a matrix. In this screen the elements of the inverse are expressed as fractions.

Finding an Inverse Matrix

To obtain A^{-1} for any $n \times n$ matrix A for which A^{-1} exists, follow these steps.

1. Form the augmented matrix $[A \mid I_n]$, where I_n is the $n \times n$ identity matrix.
2. Perform row transformations on $[A \mid I_n]$ to get a matrix of the form $[I_n \mid B]$.
3. Matrix B is A^{-1}.

 Consult your owner's manual to see how to find the inverse of a matrix with a graphing calculator.

NOTE To confirm that two $n \times n$ matrices A and B are inverses of each other, it is sufficient to show that $AB = I_n$. It is not necessary to show also that $BA = I_n$.

EXAMPLE 2
Finding the inverse of a 3×3 matrix

Find A^{-1} if $A = \begin{bmatrix} 1 & 0 & 1 \\ 2 & -2 & -1 \\ 3 & 0 & 0 \end{bmatrix}$.

Use row transformations as follows.

Step 1 Write the augmented matrix $[A \mid I_3]$.

$$\begin{bmatrix} 1 & 0 & 1 & \mid & 1 & 0 & 0 \\ 2 & -2 & -1 & \mid & 0 & 1 & 0 \\ 3 & 0 & 0 & \mid & 0 & 0 & 1 \end{bmatrix}$$

Step 2 Since 1 is already in the upper left-hand corner as desired, begin by using the row transformation that will result in a 0 for the first element in the second row. Multiply the elements of the first row by -2, and add the results to the second row.

$$\begin{bmatrix} 1 & 0 & 1 & \mid & 1 & 0 & 0 \\ 0 & -2 & -3 & \mid & -2 & 1 & 0 \\ 3 & 0 & 0 & \mid & 0 & 0 & 1 \end{bmatrix} \quad -2R_1 + R_2$$

Step 3 To get 0 for the first element in the third row, multiply the elements of the first row by -3 and add to the third row.

$$\left[\begin{array}{ccc|ccc} 1 & 0 & 1 & 1 & 0 & 0 \\ 0 & -2 & -3 & -2 & 1 & 0 \\ 0 & 0 & -3 & -3 & 0 & 1 \end{array}\right] \quad -3R_1 + R_3$$

Step 4 To get 1 for the second element in the second row, multiply the elements of the second row by $-1/2$.

$$\left[\begin{array}{ccc|ccc} 1 & 0 & 1 & 1 & 0 & 0 \\ 0 & 1 & 3/2 & 1 & -1/2 & 0 \\ 0 & 0 & -3 & -3 & 0 & 1 \end{array}\right] \quad -\tfrac{1}{2}R_2$$

Step 5 To get 1 for the third element in the third row, multiply the elements of the third row by $-1/3$.

$$\left[\begin{array}{ccc|ccc} 1 & 0 & 1 & 1 & 0 & 0 \\ 0 & 1 & 3/2 & 1 & -1/2 & 0 \\ 0 & 0 & 1 & 1 & 0 & -1/3 \end{array}\right] \quad -\tfrac{1}{3}R_3$$

Step 6 To get 0 for the third element in the first row, multiply the elements of the third row by -1 and add to the first row.

$$\left[\begin{array}{ccc|ccc} 1 & 0 & 0 & 0 & 0 & 1/3 \\ 0 & 1 & 3/2 & 1 & -1/2 & 0 \\ 0 & 0 & 1 & 1 & 0 & -1/3 \end{array}\right] \quad -1R_3 + R_1$$

Step 7 To get 0 for the third element in the second row, multiply the elements of the third row by $-3/2$ and add to the second row.

$$\left[\begin{array}{ccc|ccc} 1 & 0 & 0 & 0 & 0 & 1/3 \\ 0 & 1 & 0 & -1/2 & -1/2 & 1/2 \\ 0 & 0 & 1 & 1 & 0 & -1/3 \end{array}\right] \quad -\tfrac{3}{2}R_3 + R_2$$

The last transformation shows that the inverse is

$$A^{-1} = \left[\begin{array}{ccc} 0 & 0 & 1/3 \\ -1/2 & -1/2 & 1/2 \\ 1 & 0 & -1/3 \end{array}\right].$$

Confirm this by forming the product $A^{-1}A$ or AA^{-1}, each of which should equal the matrix I_3. ◗

```
[A]
     [[1  0   1 ]
      [2 -2  -1],
      [3  0   0 ]]
```

```
[A]⁻¹
[[0      0   .3333...
 [-.5  -.5  .5    ...
 [1     0   -.333...
```

These screens support the result of Example 2. The elements of the inverse are expressed in decimal form.

As illustrated by the examples, the most efficient order of steps is to make the changes column by column from left to right, so that for each column the required 1 is the result of the first change. Next, perform the steps that obtain the zeros in that column. Then proceed to another column.

◖EXAMPLE 3
Identifying a matrix with no inverse

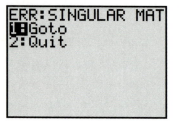

If the inverse of a matrix does not exist, the matrix is called *singular*. This occurs when the determinant of the matrix is 0. (See Exercises 27–32.)

Find A^{-1} given $A = \begin{bmatrix} 2 & -4 \\ 1 & -2 \end{bmatrix}$.

Using row transformations to change the first column of the augmented matrix

$$\begin{bmatrix} 2 & -4 & | & 1 & 0 \\ 1 & -2 & | & 0 & 1 \end{bmatrix}$$

results in the following matrices:

$$\begin{bmatrix} 1 & -2 & | & 1/2 & 0 \\ 1 & -2 & | & 0 & 1 \end{bmatrix} \quad \text{and} \quad \begin{bmatrix} 1 & -2 & | & 1/2 & 0 \\ 0 & 0 & | & -1/2 & 1 \end{bmatrix}.$$

(We multiplied the elements in row one by $1/2$ in the first step.) At this point, the matrix should be changed so that the second-row, second-column element will be 1. Since that element is now 0, there is no way to complete the desired transformation, so A^{-1} does not exist for this matrix A. What is wrong? Just as there is no multiplicative inverse for the real number 0, not every matrix has a multiplicative inverse. Matrix A is an example of such a matrix. ◗

If the inverse of a matrix exists, it is unique. That is, any given square matrix has no more than one inverse. The proof of this is left as Exercise 57 of this section.

SOLVING SYSTEMS BY INVERSES Matrix inverses can be used to solve square linear systems of equations. (A square system has the same number of equations as variables.) For example, given the linear system

$$a_{11}x + a_{12}y + a_{13}z = b_1$$
$$a_{21}x + a_{22}y + a_{23}z = b_2$$
$$a_{31}x + a_{32}y + a_{33}z = b_3,$$

the definition of matrix multiplication can be used to rewrite the system as

$$\begin{bmatrix} a_{11} & a_{12} & a_{13} \\ a_{21} & a_{22} & a_{23} \\ a_{31} & a_{32} & a_{33} \end{bmatrix} \cdot \begin{bmatrix} x \\ y \\ z \end{bmatrix} = \begin{bmatrix} b_1 \\ b_2 \\ b_3 \end{bmatrix}. \tag{1}$$

(To see this, multiply the matrices on the left.)

$$\text{If} \quad A = \begin{bmatrix} a_{11} & a_{12} & a_{13} \\ a_{21} & a_{22} & a_{23} \\ a_{31} & a_{32} & a_{33} \end{bmatrix}, \quad X = \begin{bmatrix} x \\ y \\ z \end{bmatrix}, \quad \text{and} \quad B = \begin{bmatrix} b_1 \\ b_2 \\ b_3 \end{bmatrix},$$

the system given in (1) becomes

$$AX = B.$$

If A^{-1} exists, then both sides of $AX = B$ can be multiplied on the left to get

$$A^{-1}(AX) = A^{-1}B$$
$$(A^{-1}A)X = A^{-1}B \qquad \text{Associative property}$$
$$I_3 X = A^{-1}B \qquad \text{Inverse property}$$
$$X = A^{-1}B. \qquad \text{Identity property}$$

Matrix $A^{-1}B$ gives the solution of the system.

Solution of the Matrix Equation $AX = B$

If A is an $n \times n$ matrix with inverse A^{-1}, X is an $n \times 1$ matrix of variables, and B is an $n \times 1$ matrix, then the matrix equation

$$AX = B$$

has the solution

$$X = A^{-1}B.$$

This method of using matrix inverses to solve systems of equations is useful when the inverse is already known or when many systems of the form $AX = B$ must be solved and only B changes.

EXAMPLE 4
Solving a system of equations using a matrix inverse

Use the inverse of the coefficient matrix to solve the following systems.

(a) $2x - 3y = 4$
$\qquad x + 5y = 2$

To represent the system as a matrix equation, use one matrix for the coefficients, one for the variables, and one for the constants, as follows.

$$A = \begin{bmatrix} 2 & -3 \\ 1 & 5 \end{bmatrix}, \qquad X = \begin{bmatrix} x \\ y \end{bmatrix}, \qquad \text{and} \qquad B = \begin{bmatrix} 4 \\ 2 \end{bmatrix}$$

The system can then be written in matrix form as the equation $AX = B$, since

$$AX = \begin{bmatrix} 2 & -3 \\ 1 & 5 \end{bmatrix}\begin{bmatrix} x \\ y \end{bmatrix} = \begin{bmatrix} 2x - 3y \\ x + 5y \end{bmatrix} = \begin{bmatrix} 4 \\ 2 \end{bmatrix} = B.$$

To solve the system, first find A^{-1}.

$$A^{-1} = \begin{bmatrix} 5/13 & 3/13 \\ -1/13 & 2/13 \end{bmatrix}$$

Next, find the product $A^{-1}B$.

$$A^{-1}B = \begin{bmatrix} 5/13 & 3/13 \\ -1/13 & 2/13 \end{bmatrix}\begin{bmatrix} 4 \\ 2 \end{bmatrix} = \begin{bmatrix} 2 \\ 0 \end{bmatrix}$$

[A]
$$\begin{bmatrix} 2 & -3 \\ 1 & 5 \end{bmatrix}$$
[B]
$$\begin{bmatrix} 1 \\ 20 \end{bmatrix}$$

[A]⁻¹*[B]
$$\begin{bmatrix} 5 \\ 3 \end{bmatrix}$$

These screens support the result in Example 4(b).

Since $X = A^{-1}B$,

$$X = \begin{bmatrix} x \\ y \end{bmatrix} = \begin{bmatrix} 2 \\ 0 \end{bmatrix},$$

The final matrix shows that the solution of the system is (2, 0).

(b) $2x - 3y = 1$

$\quad\quad x + 5y = 20$

This system has the same matrix of coefficients. Only matrix B is different. Use A^{-1} from part (a) and multiply by B to get

$$X = A^{-1}B = \begin{bmatrix} 5/13 & 3/13 \\ -1/13 & 2/13 \end{bmatrix}\begin{bmatrix} 1 \\ 20 \end{bmatrix} = \begin{bmatrix} 5 \\ 3 \end{bmatrix},$$

giving the solution (5, 3). ▶

Despite the incredible power of technology, we must always *understand the concepts* before relying on technology, such as a graphing calculator. For example, if we carry out the solution of the system shown in Example 4(a) with a popular model, the following screen results:

[A]
$$\begin{bmatrix} 2 & -3 \\ 1 & 5 \end{bmatrix}$$
[B]
$$\begin{bmatrix} 4 \\ 2 \end{bmatrix}$$

[A]⁻¹[B]
$$\begin{bmatrix} 2 \\ -1E-14 \end{bmatrix}$$

If we read exactly what is shown on the screen, we would interpret the y-value of the solution to be -1×10^{-14} (represented as $-1E-14$), or $-.00000000000001$. However, we know from our analytic work that the y-value is 0. The product is actually

$$\begin{bmatrix} 2 \\ 0 \end{bmatrix},$$

and thus the solution of the system is (2, 0).

As seen above, the user of the graphing calculator must be aware that reading *exactly* what is seen on the screen can lead to incorrect conclusions. Technology does have its limitations. For an excellent treatment of the perils and pitfalls associated with graphing calculators, see "The Graphics Calculator: A Tool for Critical Thinking" by Gloria Dion in the October 1990 issue of *Mathematics Teacher*.

9.7 Exercises ▼▼▼▼▼▼▼▼▼▼▼▼▼▼▼▼▼▼▼▼▼▼▼▼▼▼▼▼▼▼▼▼▼▼▼

Decide whether or not the given matrices are inverses of each other. (Check to see if their product is the identity matrix I_n.)

1. $\begin{bmatrix} 5 & 7 \\ 2 & 3 \end{bmatrix}$ and $\begin{bmatrix} 3 & -7 \\ -2 & 5 \end{bmatrix}$

2. $\begin{bmatrix} 2 & 3 \\ 1 & 1 \end{bmatrix}$ and $\begin{bmatrix} -1 & 3 \\ 1 & -2 \end{bmatrix}$

3. $\begin{bmatrix} -1 & 2 \\ 3 & -5 \end{bmatrix}$ and $\begin{bmatrix} -5 & -2 \\ -3 & -1 \end{bmatrix}$

4. $\begin{bmatrix} 2 & 1 \\ 3 & 2 \end{bmatrix}$ and $\begin{bmatrix} 2 & 1 \\ -3 & 2 \end{bmatrix}$

5. $\begin{bmatrix} 0 & 1 & 0 \\ 0 & 0 & -2 \\ 1 & -1 & 0 \end{bmatrix}$ and $\begin{bmatrix} 1 & 0 & 1 \\ 1 & 0 & 0 \\ 0 & -1 & 0 \end{bmatrix}$

6. $\begin{bmatrix} 1 & 2 & 0 \\ 0 & 1 & 0 \\ 0 & 1 & 0 \end{bmatrix}$ and $\begin{bmatrix} 1 & -2 & 0 \\ 0 & 1 & 0 \\ 0 & -1 & 1 \end{bmatrix}$

7. $\begin{bmatrix} -1 & -1 & -1 \\ 4 & 5 & 0 \\ 0 & 1 & -3 \end{bmatrix}$ and $\begin{bmatrix} 15 & 4 & -5 \\ -12 & -3 & 4 \\ -4 & -1 & 1 \end{bmatrix}$

8. $\begin{bmatrix} 1 & 3 & 3 \\ 1 & 4 & 3 \\ 1 & 3 & 4 \end{bmatrix}$ and $\begin{bmatrix} 7 & -3 & -3 \\ -1 & 1 & 0 \\ -1 & 0 & 1 \end{bmatrix}$

Find the inverse, if it exists, for each matrix. See Examples 2 and 3.

9. $\begin{bmatrix} -1 & 2 \\ -2 & -1 \end{bmatrix}$

10. $\begin{bmatrix} 1 & -1 \\ 2 & 0 \end{bmatrix}$

11. $\begin{bmatrix} -1 & -2 \\ 3 & 4 \end{bmatrix}$

12. $\begin{bmatrix} 3 & -1 \\ -5 & 2 \end{bmatrix}$

13. $\begin{bmatrix} 5 & 10 \\ -3 & -6 \end{bmatrix}$

14. $\begin{bmatrix} -6 & 4 \\ -3 & 2 \end{bmatrix}$

15. $\begin{bmatrix} 1 & 0 & 1 \\ 0 & -1 & 0 \\ 2 & 1 & 1 \end{bmatrix}$

16. $\begin{bmatrix} 1 & 0 & 0 \\ 0 & -1 & 0 \\ 1 & 0 & 1 \end{bmatrix}$

17. $\begin{bmatrix} 1 & 3 & 3 \\ 1 & 4 & 3 \\ 1 & 3 & 4 \end{bmatrix}$

18. $\begin{bmatrix} -2 & 2 & 4 \\ -3 & 4 & 5 \\ 1 & 0 & 2 \end{bmatrix}$

19. $\begin{bmatrix} 2 & 2 & -4 \\ 2 & 6 & 0 \\ -3 & -3 & 5 \end{bmatrix}$

20. $\begin{bmatrix} 2 & 4 & 6 \\ -1 & -4 & -3 \\ 0 & 1 & -1 \end{bmatrix}$

21. $\begin{bmatrix} 1 & 1 & 0 & 2 \\ 2 & -1 & 1 & -1 \\ 3 & 3 & 2 & -2 \\ 1 & 2 & 1 & 0 \end{bmatrix}$

22. $\begin{bmatrix} 1 & -2 & 3 & 0 \\ 0 & 1 & -1 & 1 \\ -2 & 2 & -2 & 4 \\ 0 & 2 & -3 & 1 \end{bmatrix}$

🖩 *A graphing calculator screen shows A^{-1} for some matrix A. What is matrix A? (Hint: $(A^{-1})^{-1} = A$.)*

23.

```
[A]-1
        [[5  -9]
         [-1  2 ]]
```

24.

```
[A]-1▶Frac
        [[3/20  1/4]
         [-1/20 1/4]]
```

25.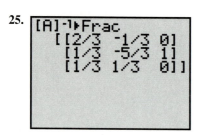

26.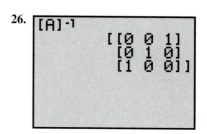

▼▼▼▼▼▼▼▼▼▼▼▼▼ **DISCOVERING CONNECTIONS** (Exercises 27–32) ▼▼▼▼▼▼▼▼▼▼▼▼▼

It can be shown that the inverse of matrix $A = \begin{bmatrix} a & b \\ c & d \end{bmatrix}$ *is*

$$A^{-1} = \begin{bmatrix} \dfrac{d}{ad - bc} & \dfrac{-b}{ad - bc} \\ \dfrac{-c}{ad - bc} & \dfrac{a}{ad - bc} \end{bmatrix}.$$

Work Exercises 27–32 in order, so that you can discover connections between the material in this section and a topic studied earlier in this chapter.

27. With respect to the matrix $A = \begin{bmatrix} a & b \\ c & d \end{bmatrix}$, what do we call $ad - bc$?

28. Refer to A^{-1} as given above, and write it using determinant notation.

29. Write A^{-1} using scalar multiplication, where the scalar is $\dfrac{1}{|A|}$.

30. Explain in your own words how the inverse of matrix A can be found using a determinant.

31. Use the method described here to find the inverse of $A = \begin{bmatrix} 4 & 2 \\ 7 & 3 \end{bmatrix}$.

32. Complete the following statement: The inverse of a 2×2 matrix A does not exist if the determinant of A has value _____ . (*Hint:* Look at the denominators in A^{-1} as given above.)

Solve the system by using the inverse of the coefficient matrix. See Example 4.

33. $-x + y = 1$
    $2x - y = 1$

34. $x + y = 5$
    $x - y = -1$

35. $2x - y = -8$
    $3x + y = -2$

36. $x + 3y = -12$
    $2x - y = 11$

37. $2x + 3y = -10$
    $3x + 4y = -12$

38. $2x - 3y = 10$
    $2x + 2y = 5$

Solve each system of equations by using the inverse of the coefficient matrix. The inverses were found in Exercises 17–20. See Example 4.

39. $x + 3y + 3z = 1$
    $x + 4y + 3z = 0$
    $x + 3y + 4z = -1$

40. $-2x + 2y + 4z = 3$
    $-3x + 4y + 5z = 1$
    $x + 2z = 2$

41. $2x + 2y - 4z = 12$
    $2x + 6y = 16$
    $-3x - 3y + 5z = -20$

42. $2x + 4y + 6z = 4$
    $-x - 4y - 3z = 8$
    $y - z = -4$

Solve each system of equations by using the inverse of the coefficient matrix. The inverses were found in Exercises 21 and 22.

43. $\begin{aligned} x + y + 2w &= 3 \\ 2x - y + z - w &= 3 \\ 3x + 3y + 2z - 2w &= 5 \\ x + 2y + z &= 3 \end{aligned}$

44. $\begin{aligned} x - 2y + 3z &= 1 \\ y - z + w &= -1 \\ -2x + 2y - 2z + 4w &= 2 \\ 2y - 3z + w &= -3 \end{aligned}$

45. The amount of plate-glass sales S (in millions of dollars) can be affected by the number of new building contracts B issued (in millions) and automobiles A produced (in millions). A plate-glass company in California wants to forecast future sales by using the past three years of sales. The totals for three years are given in the table.

S	A	B
602.7	5.543	37.14
656.7	6.933	41.30
778.5	7.638	45.62

In order to describe the relationship between these variables, the equation $S = a + bA + cB$ was used where the coefficients a, b, and c are constants that must be determined. (*Source:* Makridakis, S. and S. Wheelwright, *Forecasting Methods for Management*, John Wiley & Sons, Inc., 1989.)

(a) Substitute the values for S, A, and B for each year from the table into the equation $S = a + bA + cB$ and obtain three linear equations involving a, b, and c.

(b) Use a graphing calculator to solve this linear system for a, b, and c. Use matrix inverse methods.

(c) Write the equation for S using these values for the coefficients.

(d) For the next year it is estimated that $A = 7.752$ and $B = 47.38$. Predict S. (The actual value for S was 877.6.)

(e) It is predicted that in six years $A = 8.9$ and $B = 66.25$. Find the equation for S in this situation and discuss its validity.

46. The number of automobile tire sales is dependent on several variables. In one study the relationship between annual tire sales S (in thousands), automobile registrations R (in millions), and personal disposable income I (in millions of dollars) was investigated. The results for three years are given in the table.

S	R	I
10,170	112.9	307.5
15,305	132.9	621.63
21,289	155.2	1937.13

In order to describe the relationship between these variables, mathematicians often use the equation $S = a + bR + cI$ where the coefficients a, b, and c are constants that must be determined before the equation can be used. (*Source:* Jarrett, J., *Business Forecasting Methods*, Basil Blackwell, Ltd., Cambridge, MA, 1991.)

(a) Substitute the values for S, R, and I for each year from the table into the equation $S = a + bR + cI$ and obtain three linear equations involving a, b, and c.

(b) Use a graphing calculator to solve this linear system for a, b, and c. Use matrix inverse methods.

(c) Write the equation for S using these values for the coefficients.

(d) If $R = 117.6$ and $I = 310.73$ predict S. (The actual value for S was 11,314.)

(e) If $R = 143.8$ and $I = 829.06$ predict S. (The actual value for S was 18,481.)

(Refer to the discussion of cryptography preceding Exercises 63 and 64 in Section 9.4.) It is important in cryptography to be able to decode a message easily. To do this in the polygraphic system one uses the inverse of the coding matrix A. For example, if the message is AKFTPK then it can be represented by 1 11 6 20 16 11. *Writing these numbers in a matrix with two rows produces the matrix* $C = \begin{bmatrix} 1 & 6 & 16 \\ 11 & 20 & 11 \end{bmatrix}$. *If* $A = \begin{bmatrix} 2 & 1 \\ -5 & -2 \end{bmatrix}$ *then* $A^{-1} = \begin{bmatrix} -2 & -1 \\ 5 & 2 \end{bmatrix}$. *To reverse the process we must multiply C by the inverse of A.*

$$A^{-1}C = \begin{bmatrix} -2 & -1 \\ 5 & 2 \end{bmatrix} \begin{bmatrix} 1 & 6 & 16 \\ 11 & 20 & 11 \end{bmatrix} = \begin{bmatrix} -13 & -32 & -43 \\ 27 & 70 & 102 \end{bmatrix}$$

Scaling the matrix elements between 1 and 26 results in the matrix B.

$$-13 + (1)26 = 13, \quad -32 + 2(26) = 20, \quad -43 + 2(26) = 9,$$
$$27 - 1(26) = 1, \quad 70 - 2(26) = 18, \quad 102 - 3(26) = 24.$$
$$B = \begin{bmatrix} 13 & 20 & 9 \\ 1 & 18 & 24 \end{bmatrix}$$

From B the numbers are 13 1 20 18 9 24, *which decode to the word MATRIX.*

47. Use the same matrix A^{-1} to decode each of the following words.
   (a) UBNL   **(b)** QNABMV

48. (Refer to Exercise 47.) If it is known that the coding matrix is $A = \begin{bmatrix} -1 & 2 & 1 \\ -2 & -2 & 1 \\ -1 & 1 & 1 \end{bmatrix}$, then decode the message XCGVSBFMR.

⊞ *Use a graphing calculator to find the inverse of each of the following matrices. Give as many decimal places as the calculator shows.*

49. $\begin{bmatrix} \sqrt{2} & .5 \\ -17 & 1/2 \end{bmatrix}$

50. $\begin{bmatrix} 2/3 & .7 \\ 22 & \sqrt{3} \end{bmatrix}$

⊞ *Use a graphing calculator and the method of matrix inverses to solve each of the following systems. Give as many decimal places as the calculator shows.*

51.
$$x - \sqrt{2}y = 2.6$$
$$.75x + \quad y = -7$$

52.
$$2.1x + \quad y = \sqrt{5}$$
$$\sqrt{2}x - 2y = 5$$

53.
$$\pi x + ey + \sqrt{2}z = 1$$
$$ex + \pi y + \sqrt{2}z = 2$$
$$\sqrt{2}x + ey + \quad \pi z = 3$$

54.
$$(\log 2)x + (\ln 3)y + (\ln 4)z = 1$$
$$(\ln 3)x + (\log 2)y + (\ln 8)z = 5$$
$$(\log 12)x + (\ln 4)y + (\ln 8)z = 9$$

Let $A = \begin{bmatrix} a & b \\ c & d \end{bmatrix}$ and let O be the 2 × 2 matrix of all zeros.

Show that the statements in Exercises 55 and 56 are true.

55. $A \cdot O = O \cdot A = O$

56. For square matrices A and B of the same order, if $AB = O$ and if A^{-1} exists, then $B = O$.

57. Prove that any square matrix has no more than one inverse.

58. Give an example of two matrices A and B, where $(AB)^{-1} \neq A^{-1}B^{-1}$.

59. Suppose A and B are matrices where A^{-1}, B^{-1}, and AB all exist. Show that $(AB)^{-1} = B^{-1}A^{-1}$.

60. Let $A = \begin{bmatrix} a & 0 & 0 \\ 0 & b & 0 \\ 0 & 0 & c \end{bmatrix}$, where a, b, and c are nonzero real numbers. Find A^{-1}.

61. Let $A = \begin{bmatrix} 1 & 0 & 0 \\ 0 & 0 & -1 \\ 0 & 1 & -1 \end{bmatrix}$. Show that $A^3 = I$ and use this result to find the inverse of A.

62. What are the inverses of I, $-A$ (in terms of A), and kA (k a scalar)?

63. Give two ways to use matrices to solve a system of linear equations. Will they both work in all situations? In which situations does each method excel?

64. Discuss the similarities and differences between solving the linear equation $ax = b$ and solving the matrix equation $AX = B$.

9.8 Systems of Inequalities; Linear Programming ▼▼▼

Many mathematical descriptions of real situations are best expressed as inequalities rather than equations. For example, a firm might be able to use a machine *no more* than 12 hours a day, while production of *at least* 500 cases of a certain product might be required to meet a contract. The simplest way to see the solution of an inequality in two variables is to draw its graph.

A line divides a plane into three sets of points: the points of the line itself and the points belonging to the two regions determined by the line. Each of these two regions is called a **half-plane.** In Figure 8 line r divides the plane into three different sets of points: line r, half-plane P, and half-plane Q. The points on r belong neither to P nor to Q. Line r is the **boundary** of each half-plane.

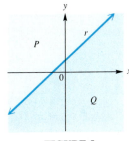

FIGURE 8

A **linear inequality in two variables** is an inequality of the form

$$Ax + By \leq C,$$

where A, B, and C are real numbers, with A and B not both equal to 0. (The symbol $\leq$ could be replaced with $\geq$, $<$, or $>$.) The graph of a linear inequality is a half-plane, perhaps with its boundary. For example, to graph the linear inequality $3x - 2y \leq 6$, first graph the boundary, $3x - 2y = 6$, as shown in Figure 9.

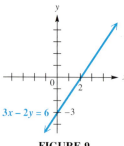

FIGURE 9   **FIGURE 10**

Since the points of the line $3x - 2y = 6$ satisfy $3x - 2y \leq 6$, this line is part of the solution. To decide which half-plane (the one above the line $3x - 2y = 6$ or the one below the line) is part of the solution, solve the original inequality for y.

$$3x - 2y \leq 6$$
$$-2y \leq -3x + 6$$
$$y \geq \frac{3}{2}x - 3 \qquad \text{Multiply by } -\tfrac{1}{2}; \text{ change } \leq \text{ to } \geq.$$

For a particular value of x, the inequality will be satisfied by all values of y that are *greater than* or equal to $(3/2)x - 3$. This means that the solution contains the half-plane *above* the line, as shown in Figure 10.

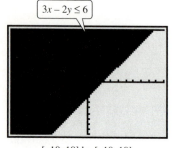

[−10, 10] by [−10, 10]

This is a calculator-generated graph, using a standard viewing window, of the inequality illustrated in Figure 10.

◖EXAMPLE 1
Graphing a linear inequality

Graph $x + 4y > 4$.

The boundary here is the straight line $x + 4y = 4$. Since the points on this line do not satisfy $x + 4y > 4$, it is customary to make the line dashed, as in Figure 11. To decide which half-plane represents the solution, solve for y.

$$x + 4y > 4$$
$$4y > -x + 4$$
$$y > -\frac{1}{4}x + 1$$

Since y is *greater than* $(-1/4)x + 1$, the graph of the solution is the half-plane above the boundary, as shown in Figure 11.

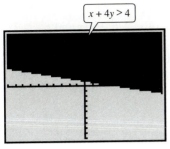

[−10, 10] by [−10, 10]

This is a calculator-generated graph, using a standard viewing window, of the inequality in Example 1 and illustrated in Figure 11.

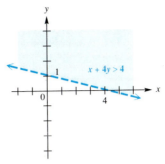

FIGURE 11

Alternatively, or as a check, choose a point not on the boundary line and substitute into the inequality. The point $(0, 0)$ is a good choice if it does not lie on the boundary, since the substitution is easily done. Here, substitution of $(0, 0)$ into the original inequality gives

$$x + 4y > 4$$
$$0 + 4(0) > 4$$
$$0 > 4,$$

a false statement. Since the point $(0, 0)$ is below the line, the points that satisfy the inequality must be above the line, which agrees with the result above. ▶

The methods used to graph linear inequalities can be used for other inequalities of the form $y \leq f(x)$ as summarized here. (Similar statements can be made for $<$, $>$, and $\geq$.)

Graphing Inequalities

I. For a function f, the graph of $y < f(x)$ consists of all the points that are *below* the graph of $y = f(x)$; the graph of $y > f(x)$ consists of all the points that are *above* the graph of $y = f(x)$.

II. If the inequality is not or cannot be solved for y, choose a test point not on the boundary. If the test point satisfies the inequality, the graph includes all points on the same side of the boundary as the test point. Otherwise, the graph includes all points on the other side of the boundary.

The solution set of a **system of inequalities,** such as

$$x > 6 - 2y$$
$$x^2 < 2y,$$

is the intersection of the solution sets of its members. This intersection is found by graphing the solution sets of both inequalities on the same coordinate axes and identifying, by shading, the region common to all graphs.

EXAMPLE 2
Graphing a system of inequalities

Graph the solution set of the system above.

In Figure 12, parts (a) and (b) show the graphs of $x > 6 - 2y$ and $x^2 < 2y$. The methods of an earlier section can be used to show that the boundaries intersect at the points $(2, 2)$ and $(-3, 9/2)$. The solution set of the system is shown in part (c). The points on the boundaries of $x > 6 - 2y$ and $x^2 < 2y$ do not belong to the graph of the solution. For this reason, the boundaries are dashed lines. ▶

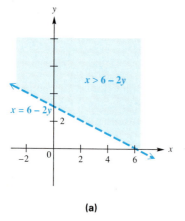

(a)

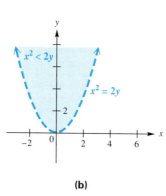

(b)

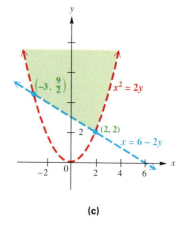

(c)

FIGURE 12

In the study of calculus it is often necessary to consider a region of the plane that lies *above* one graph and *below* another. Graphing calculators can be used to shade such regions. For example, if we are interested in viewing the solution set of the system

$$y > x^2 - 5$$
$$y < x$$

which consists of the points above the first graph and below the second, we direct the calculator to shade that region, using the commands necessary for the particular model. Using the standard window, the region looks like this:

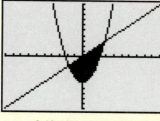

[−10, 10] by [−10, 10]

For this system, the boundaries of the graphs are not included, but this cannot be distinguished by looking at the screen. Once again, we see the need to understand the concepts before utilizing the technology.

EXAMPLE 3
Graphing a system of
inequalities

Graph the solution set of the system

$$|x| \leq 3$$
$$y \leq 0$$
$$y \geq |x| + 1.$$

Writing $|x| \leq 3$ as $-3 \leq x \leq 3$ shows that this inequality is satisfied by points in the region between $x = -3$ and $x = 3$. See part (a) of Figure 13. The set of points that satisfies $y \leq 0$ includes the points below or on the x-axis. See Figure 13(b). Graph $y = |x| + 1$ and use a test point to see that the solutions of $y \geq |x| + 1$ are above or on the boundary. See Figure 13(c). Parts (b) and (c) of Figure 13 show that the solution sets of $y \leq 0$ and $y \geq |x| + 1$ have no points in common; therefore, the solution set for the system is $\emptyset$. ▶

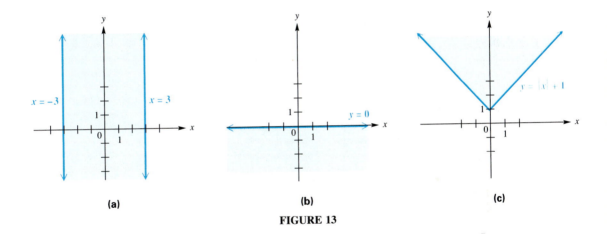

(a)   (b)   (c)

FIGURE 13

NOTE   While we illustrated three graphs in the solutions of Examples 1 and 2, in practice it is customary to give only the final graph. The three individual inequalities were shown simply to illustrate the procedure.

Linear programming   An important application of mathematics to business and social science is called linear programming. **Linear programming** is used to find such things as minimum cost and maximum profit. It was developed to solve problems in allocating supplies for the U.S. Air Force during World War II. The basic ideas of this technique will be explained with an example.

EXAMPLE 4
Finding maximum profit

The Smith Company makes two products—tape decks and amplifiers. Each tape deck gives a profit of $30, while each amplifier produces $70. The company must manufacture at least ten tape decks per day to satisfy one of its customers, but no more than fifty because of production problems. Also, the number of amplifiers produced cannot exceed sixty per day. As a further requirement, the number of tape decks cannot exceed the number of amplifiers. How many of each should the company manufacture in order to obtain the maximum profit?

To begin, translate the statement of the problem into symbols.

Let x = number of tape decks to be produced daily,

y = number of amplifiers to be produced daily.

According to the statement of the problem given above, the company must produce at least ten tape decks (ten or more), so that

$$x \geq 10.$$

The requirement that no more than 50 tape decks may be produced means that

$$x \leq 50.$$

Since no more than 60 amplifiers may be made in one day,

$$y \leq 60.$$

The fact that the number of tape decks may not exceed the number of amplifiers translates as

$$x \leq y.$$

The number of tape decks and of amplifiers cannot be negative, so

$$x \geq 0 \quad \text{and} \quad y \geq 0.$$

Listing all the restrictions, or **constraints,** that are placed on production gives

$$x \geq 10, \quad x \leq 50, \quad y \leq 60, \quad x \leq y, \quad x \geq 0, \quad y \geq 0.$$

To find the maximum possible profit that the company can make, subject to these constraints, begin by sketching the graph of each constraint. The only feasible values of x and y are those that satisfy all constraints—that is, the values that lie in the intersection of the graphs of the constraints. The intersection is shown in Figure 14. Any point lying inside the shaded region or on the boundary in Figure 14 satisfies the restrictions as to the number of tape decks and amplifiers that may be produced. (For practical purposes, however, only points with integer coefficients are useful.) This region is called the **region of feasible solutions.**

Since each tape deck gives a profit of $30, the daily profit from the production of x tape decks is $30x$ dollars. Also, the profit from the production of y amplifiers will be $70y$ dollars per day. The total daily profit is thus

$$\text{Profit} = 30x + 70y.$$

The problem of the Smith Company may now be stated as follows: Find values of x and y in the shaded region of Figure 14 that will produce the

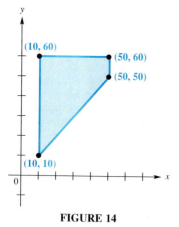

(10, 60)   (50, 60)
            (50, 50)

(10, 10)

FIGURE 14

maximum possible value of $30x + 70y$. To locate the point (x, y) that gives the maximum profit, add to the graph of Figure 14 lines corresponding to profits of $0, $1000, $3000, and $7000.

$$30x + 70y = 0$$
$$30x + 70y = 1000$$
$$30x + 70y = 3000$$
$$30x + 70y = 7000$$

For instance, each point on the line $30x + 70y = 3000$ corresponds to production values that yield a profit of $3000. Figure 15 shows the region of feasible solutions together with these lines. The lines are parallel, and the higher the line, the higher the profit. The line $30x + 70y = 7000$ has the highest profit but does not contain any points of the region of feasible solutions. To find the feasible solution of greatest profit, lower the line $30x + 70y = 7000$ until it contains a feasible solution—that is, until it just touches the region of feasible solutions. This occurs at point *A*, a **vertex** (or corner point) of the region. See Figure 16. Since the coordinates of this point are $(50, 60)$, the maximum profit is obtained when fifty tape decks and sixty amplifiers are produced each day. The maximum profit will be $30(50) + 70(60) = 5700$ dollars per day. ▶

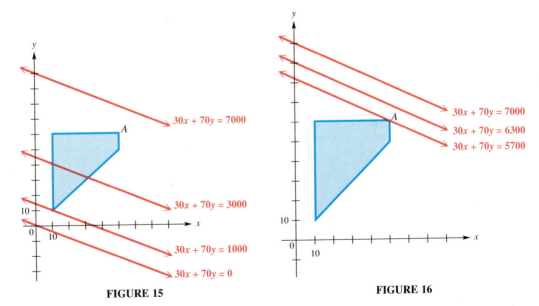

FIGURE 15   FIGURE 16

The result observed in Figure 16 holds for *every* linear programming problem.

Fundamental Theorem of Linear Programming

The optimal value for a linear programming problem occurs at a vertex of the region of feasible solutions.

EXAMPLE 5
Finding minimum cost

Robin, who is ill, takes vitamin pills. Each day, she must have at least 16 units of Vitamin A, at least 5 units of Vitamin B_1, and at least 20 units of Vitamin C. She can choose between red pills, costing 10¢ each, which contain 8 units of A, 1 of B_1, and 2 of C; and blue pills, costing 20¢ each, which contain 2 units of A, 1 of B_1, and 7 of C. How many of each pill should she buy in order to minimize her cost and yet fulfill her daily requirements?

Let x represent the number of red pills to buy, and let y represent the number of blue pills to buy. Then the cost in pennies per day is given by

$$\text{Cost} = 10x + 20y.$$

Since Robin buys x of the 10¢ pills and y of the 20¢ pills, she gets Vitamin A as follows: 8 units from each red pill and 2 units from each blue pill. Altogether, she gets $8x + 2y$ units of A per day. Since she must get at least 16 units,

$$8x + 2y \geq 16.$$

Each red pill and each blue pill supplies 1 unit of Vitamin B_1. Robin needs at least 5 units per day, so that

$$x + y \geq 5.$$

For Vitamin C the inequality is

$$2x + 7y \geq 20.$$

Also, $x \geq 0$ and $y \geq 0$, since Robin cannot buy negative numbers of the pills.

Again, total cost of the pills is minimized by the solution of the system of inequalities formed by the constraints. (See Figure 17.) The solution to this minimizing problem will also occur at a vertex point. By substituting the coordinates of the vertex points in the cost function, the lowest cost can be found.

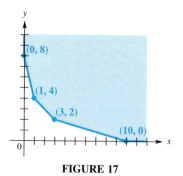

FIGURE 17

Point	Cost $= 10x + 20y$	
(10, 0)	$10(10) + 20(0) = 100$	
(3, 2)	$10(3) + 20(2) = 70$	← Minimum
(1, 4)	$10(1) + 20(4) = 90$	
(0, 8)	$10(0) + 20(8) = 160$	

Robin's best bet is to buy 3 red pills and 2 blue ones, for a total cost of 70¢ per day. She receives just the minimum amounts of Vitamins B_1 and C, but an excess of Vitamin A. Even though she has an excess of A, this is still the best buy. ▶

CONNECTIONS Linear programming was used during the Berlin Airlift after World War II to determine which combination of goods to pack on each plane. Many different constraints were involved, including volume and weight.

For example, suppose we want to ship food and clothing to hurricane victims in Mexico. Commercial carriers have volunteered to transport the packages, provided they fit in the available cargo space. Each 20-cubic-foot box of food weighs 40 pounds and each 30-cubic-foot box of clothing weighs 10 pounds. The total weight cannot exceed 16,000 pounds and the total volume must be less than 18,000 cubic feet. Each carton of food will feed 10 people, while each carton of clothing will help 8 people.

Let

$$F = \text{the number of cartons of food to send, and}$$
$$C = \text{the number of cartons of clothing to send.}$$

These constraints lead to the following inequalities:

$$40F + 10C \leq 16{,}000$$
$$20F + 30C \leq 18{,}000.$$

We want to maximize the number of people we can help, $10F + 8C$. With C on the vertical axis, the corners of the feasible region are $(0, 600)$, $(300, 400)$, and $(400, 0)$.

$$10(0) + 8(600) = 4800$$
$$\mathbf{10(300) + 8(400) = 6200} \qquad \leftarrow \text{Maximum}$$
$$10(400) + 8(0) = 4000$$

We should send 300 cartons of food and 400 cartons of clothes.

FOR DISCUSSION OR WRITING
1. Earthquake victims in China need medical supplies and bottled water. Each medical kit measures 1 cubic foot and weighs 10 pounds. Each container of water is also 1 cubic foot but weighs 20 pounds. The plane can only carry 80,000 pounds with a total volume of 6000 cubic feet. Each medical kit will aid 4 people, while each container of water will serve 10 people. How many of each should be sent?
2. If each medical kit could aid 6 people instead of 4, how would the results above change?

9.8 Exercises ▼▼▼▼▼▼▼▼▼▼▼▼▼▼▼▼▼▼▼▼▼▼▼▼▼▼▼▼▼▼▼▼▼▼▼▼▼▼

Graph the inequalities. See Example 1.

1. $x \le 3$

2. $y \le -2$

3. $x + 2y \le 6$

4. $x - y \ge 2$

5. $2x + 3y \ge 4$

6. $4y - 3x < 5$

7. $3x - 5y > 6$

8. $x < 3 + 2y$

9. $5x \le 4y - 2$

10. $2x > 3 - 4y$

11. $y < 3x^2 + 2$

12. $y \le x^2 - 4$

13. $y > (x - 1)^2 + 2$

14. $y > 2(x + 3)^2 - 1$

15. $x^2 + (y + 3)^2 \le 16$

16. $(x - 4)^2 + (y + 3)^2 \le 9$

17. In your own words, explain how to determine whether the boundary of an inequality is solid or dashed.

18. When graphing $y \le 3x - 6$, would you shade above or below the line $y = 3x - 6$? Explain your answer.

19. For $Ax + By \ge C$, if $B > 0$, would you shade above or below the line?

20. For $Ax + By \ge C$, if $B < 0$, would you shade above or below the line?

21. Which one of the following is a description of the graph of the inequality

$$(x - 5)^2 + (y - 2)^2 < 4?$$

   (a) the region inside a circle with center $(-5, -2)$ and radius 2
   (b) the region inside a circle with center $(5, 2)$ and radius 2
   (c) the region inside a circle with center $(-5, -2)$ and radius 4
   (d) the region outside a circle with center $(5, 2)$ and radius 4

22. Without graphing, write a description of the graph of the inequality $y > 2(x - 3)^2 + 2$.

▦ *In Exercises 23–26, match the inequality with the appropriate calculator-generated graph. You should not use your calculator; instead, use your knowledge of the concepts involved in graphing inequalities.*

23. $y \le 3x - 6$

24. $y \ge 3x - 6$

25. $y \le -3x - 6$

26. $y \ge -3x - 6$

A.

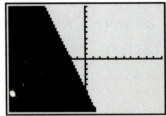

[−10, 10] by [−10, 10]

B.

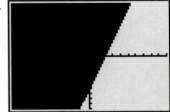

[−10, 10] by [−10, 10]

C.

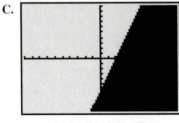

[−10, 10] by [−10, 10]

D.

[−10, 10] by [−10, 10]

Graph the solution set of each system of inequalities. See Examples 2 and 3.

27. $x + y \geq 0$
$2x - y \geq 3$

28. $x + y \leq 4$
$x - 2y \geq 6$

29. $2x + y > 2$
$x - 3y < 6$

30. $4x + 3y < 12$
$y + 4x > -4$

31. $3x + 5y \leq 15$
$x - 3y \geq 9$

32. $y \leq x$
$x^2 + y^2 < 1$

33. $4x - 3y \leq 12$
$y \leq x^2$

34. $y \leq -x^2$
$y \geq x^2 - 6$

35. $x + y \leq 9$
$x \leq -y^2$

36. $x + 2y \leq 4$
$y \geq x^2 - 1$

37. $y \leq (x + 2)^2$
$y \geq -2x^2$

38. $x - y < 1$
$-1 < y < 1$

39. $x + y \leq 36$
$-4 \leq x \leq 4$

40. $y \geq x^2 + 4x + 4$
$y < -x^2$

41. $y \geq (x - 2)^2 + 3$
$y \leq -(x - 1)^2 + 6$

42. $x \geq 0$
$x + y \leq 4$
$2x + y \leq 5$

43. $3x - 2y \geq 6$
$x + y \leq -5$
$y \leq 4$

44. $-2 < x < 3$
$-1 \leq y \leq 5$
$2x + y < 6$

45. $-2 < x < 2$
$y > 1$
$x - y > 0$

46. $x + y \leq 4$
$x - y \leq 5$
$4x + y \leq -4$

47. $x \leq 4$
$x \geq 0$
$y \geq 0$
$x + 2y \geq 2$

48. $2y + x \geq -5$
$y \leq 3 + x$
$x \leq 0$
$y \leq 0$

49. $2x + 3y \leq 12$
$2x + 3y > -6$
$3x + y < 4$
$x \geq 0$
$y \geq 0$

50. $y \geq 3^x$
$y \geq 2$

51. $y \leq \left(\dfrac{1}{2}\right)^x$
$y \geq 4$

52. $\ln x - y \geq 1$
$x^2 - 2x - y \leq 1$

53. $y \leq \log x$
$y \geq |x - 2|$

54. $e^{-x} - y \leq 1$
$x - 2y \geq 4$

In Exercises 55–58, match the system of inequalities with the appropriate calculator-generated graph. You should not use your calculator; instead, use your knowledge of the concepts involved in graphing systems of inequalities.

55. $y \geq x$
$y \leq 2x - 3$

56. $y \geq x^2$
$y < 5$

57. $x^2 + y^2 \leq 16$
$y \geq 0$

58. $y \leq x$
$y \geq 2x - 3$

A.
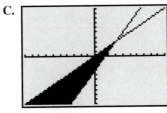
[−10, 10] by [−10, 10]

B.
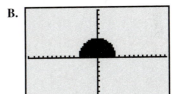
[−15, 15] by [−10, 10]

C.
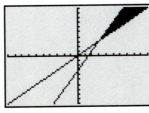
[−10, 10] by [−10, 10]

D.
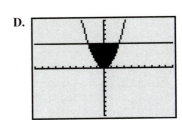
[−10, 10] by [−10, 10]

Use the shading capabilities of your graphing calculator to graph each inequality or system of inequalities.

59. $3x + 2y \geq 6$

60. $y \leq x^2 + 5$

61. $x + y \geq 2$
 $x + y \leq 6$

62. $y \geq |x + 2|$
 $y \leq 6$

63. $y \geq 2^x$
 $y \leq 8$

64. $y \leq x^3 + x^2 - 4x - 4$

65. Find a system of linear inequalities for which the graph is the region in the first quadrant between and inclusive of the pair of lines $x + 2y - 8 = 0$ and $x + 2y = 12$.

66. Find a linear inequality in two variables whose graph does not intersect the graph of $y \geq 3x + 5$.

The graphs in Exercises 67 and 68 show regions of feasible solutions. Find the maximum and minimum values of the given expressions. See Examples 4 and 5.

67. $3x + 5y$

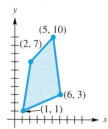

68. $6x + y$

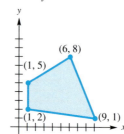

For Exercises 69–72, find the maximum and minimum values of the given expressions over the region of feasible solutions shown here. See Examples 4 and 5.

69. $3x + 5y$

70. $5x + 5y$

71. $10y$

72. $3x - y$

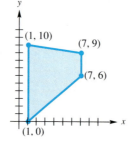

Write a system of inequalities for the problem and then graph the region of feasible solutions of the system. See Examples 4 and 5.

73. Ms. Oliveras was given the following advice. She should supplement her daily diet with at least 6000 USP units of Vitamin A, at least 195 mg of Vitamin C, and at least 600 USP units of Vitamin D. Ms. Oliveras finds that Mason's Pharmacy carries Brand X and Brand Y vitamins. Each Brand X pill contains 3000 USP units of A, 45 mg of C, and 75 USP units of D, while the Brand Y pills contain 1000 USP units of A, 50 mg of C, and 200 USP units of D.

74. The California Almond Growers have 2400 boxes of almonds to be shipped from their plant in Sacramento to Des Moines and San Antonio. The Des Moines market needs at least 1000 boxes, while the San Antonio market must have at least 800 boxes.

Solve the linear programming problem. See Examples 4 and 5.

75. Farmer Jones raises only pigs and geese. She wants to raise no more than 16 animals with no more than 12 geese. She spends $50 to raise a pig and $20 to raise a goose. She has $500 available for this purpose. Find the maximum profit she can make if she makes a profit of $80 per goose and $40 per pig.

76. A wholesaler of party goods wishes to display her products at a convention of social secretaries in such a way that she gets the maximum number of inquiries about her whistles and hats. Her booth at the convention has 12 sq m of floor space to be used for display purposes. A display unit for hats requires 2 sq m, and for whistles, 4 sq m. Experience tells the wholesaler that she should never have more than a total of 5 units of whistles and hats on display at one time. If she receives three inquiries for each unit of hats and two inquiries for each unit of whistles on display, how many of each should she display in order to get the maximum number of inquiries?

77. An office manager wants to buy some filing cabinets. He knows that cabinet #1 costs $10 each, requires 6 sq ft of floor space, and holds 8 cu ft of files. Cabinet #2 costs $20 each, requires 8 sq ft of floor space, and holds 12 cu ft. He can spend no more than $140 due to budget limitations, while his office has room for no more than 72 sq ft of cabinets. He wants the maximum storage capacity within the limits imposed by funds and space. How many of each type of cabinet should he buy?

78. Theo, who is dieting, requires two food supplements, I and II. He can get these supplements from two different products, A and B, as shown in the following table.

Supplement (g/serving)	I	II
Product A	3	2
Product B	2	4

Theo's physician has recommended that he include at least 15 g of each supplement in his daily diet. If product A costs 25¢ per serving and product B costs 40¢ per serving, how can he satisfy his requirements most economically?

79. The manufacturing process requires that oil refineries manufacture at least 2 gal of gasoline for each gal of fuel oil. To meet the winter demand for fuel oil, at least 3 million gal a day must be produced. The demand for gasoline is no more than 6.4 million gal per day. If the price of gasoline is $1.90 per gal and the price of fuel oil is $1.50 per gal, how much of each should be produced to maximize revenue?

80. Seall Manufacturing Company makes color television sets. It produces a bargain set that sells for $100 profit and a deluxe set that sells for $150 profit. On the assembly line the bargain set requires 3 hr, while the deluxe set takes 5 hr. The cabinet shop spends 1 hr on the cabinet for the bargain set and 3 hr on the cabinet for the deluxe set. Both sets require 2 hr of time for testing and packing. On a particular production run the Seall Company has available 3900 work hours on the assembly line, 2100 work hours in the cabinet shop, and 2200 work hours in the testing and packing department. How many sets of each type should it produce to make the maximum profit? What is the maximum profit?

Chapter 9   Summary ▼▼▼▼▼▼▼▼▼▼▼▼▼▼▼▼▼▼▼▼▼▼▼▼▼▼▼▼▼▼▼▼

KEY TERMS	KEY IDEAS
9.1 Linear Systems of Equations	
linear equation equivalent system linear system inconsistent system substitution method dependent equations elimination method ordered triple	**Transformations of a Linear System** 1. Any two equations of the system may be interchanged. 2. Both sides of any equation of the system may be multiplied by any nonzero real number. 3. Any equation of the system may be replaced by the sum of that equation and a multiple of another equation in the system.
9.3 Matrix Solution of Linear Systems	
matrix (matrices) element augmented matrix Gauss-Jordan method	**Matrix Row Transformations** For any augmented matrix of a system of linear equations, the following row transformations will result in the matrix of an equivalent system. 1. Any two rows may be interchanged. 2. The elements of any row may be multiplied by a nonzero real number. 3. Any row may be changed by adding to its elements a multiple of the elements of another row.

KEY TERMS	KEY IDEAS

9.4 Properties of Matrices

square matrix row matrix column matrix zero matrix scalar	**Addition of Matrices** To add two matrices of the same size, add corresponding elements. Only matrices of the same size can be added. **Subtraction of Matrices** If A and B are two matrices of the same size, then $A - B = A + (-B)$. **Multiplication of Matrices** The product AB of an $m \times n$ matrix A and an $n \times k$ matrix B is found as follows. To get the ith row, jth column element of AB, multiply each element in the ith row of A by the corresponding element in the jth column of B. The sum of these products will give the element of row i, column j of AB.

9.5 Determinants

determinant minor cofactor expansion by a row or column	**Determinant of a 2 × 2 Matrix** If $A = \begin{bmatrix} a_{11} & a_{12} \\ a_{21} & a_{22} \end{bmatrix}$, then $|A| = \begin{vmatrix} a_{11} & a_{12} \\ a_{21} & a_{22} \end{vmatrix} = a_{11}a_{22} - a_{21}a_{12}$. **Determinant of a 3 × 3 Matrix** If $A = \begin{bmatrix} a_{11} & a_{12} & a_{13} \\ a_{21} & a_{22} & a_{23} \\ a_{31} & a_{32} & a_{33} \end{bmatrix}$, then $|A| = \begin{vmatrix} a_{11} & a_{12} & a_{13} \\ a_{21} & a_{22} & a_{23} \\ a_{31} & a_{32} & a_{33} \end{vmatrix} = (a_{11}a_{22}a_{33} + a_{12}a_{23}a_{31} + a_{13}a_{21}a_{32}) - (a_{31}a_{22}a_{13} + a_{32}a_{23}a_{11} + a_{33}a_{21}a_{12})$. **Determinant Theorems** **1.** If every element in a row (or column) of matrix A is 0, then $|A| = 0$. **2.** If the rows of matrix A are the corresponding columns of matrix B, then $|B| = |A|$. **3.** If any two rows (or columns) of matrix A are interchanged to form matrix B, then $|B| = -|A|$. **4.** Suppose matrix B is formed by multiplying every element of a row (or column) of matrix A by the real number k. Then $|B| = k \cdot |A|$. **5.** If two rows (or columns) of a matrix A are identical, then $|A| = 0$. **6.** Changing a row (or column) of a matrix by adding to it a constant times another row (or column) does not change the determinant of the matrix.

KEY TERMS	KEY IDEAS

9.6 Cramer's Rule

Cramer's Rule for 2 Equations in 2 Variables

Given the system

$$a_1 x + b_1 y = c_1$$
$$a_2 x + b_2 y = c_2.$$

If $D \neq 0$, the system has the unique solution

$$x = \frac{D_x}{D} \quad \text{and} \quad y = \frac{D_y}{D}.$$

General Form of Cramer's Rule

Let an $n \times n$ system have linear equations of the form

$$a_{11}x_1 + a_{12}x_2 + a_{13}x_3 + \cdots + a_{1n}x_n = b_1.$$

Define D as the determinant of the $n \times n$ matrix of coefficients of the variables. Define D_{x1} as the determinant obtained from D by replacing the entries in column 1 of D with the constants of the system. Define D_{xi} as the determinant obtained from D by replacing the entries in column i with the constants of the system. If $D \neq 0$, the unique solution of the system is

$$x_1 = \frac{D_{x1}}{D}, \ x_2 = \frac{D_{x2}}{D}, \ x_3 = \frac{D_{x3}}{D}, \ldots, \ x_n = \frac{D_{xn}}{D}.$$

9.7 Matrix Inverses

identity matrix
multiplicative inverse

Finding an Inverse Matrix

To obtain A^{-1} for any $n \times n$ matrix A for which A^{-1} exists, follow these steps.

1. Form the augmented matrix $[A \,|\, I_n]$, where I_n is the $n \times n$ identity matrix.
2. Perform row transformations on $[A \,|\, I_n]$ to get a matrix of the form $[I_n \,|\, B]$.
3. Matrix B is A^{-1}.

9.8 Systems of Inequalities; Linear Programming

system of inequalities
half-plane
boundary
linear inequality in 2 variables
linear programming
constraints
region of feasible solutions
vertex (corner point)

Graphing Inequalities

I. For a function f, the graph of $y < f(x)$ consists of all the points that are below the graph of $y = f(x)$; the graph of $y > f(x)$ consists of all the points that are above the graph of $y = f(x)$.

II. If the inequality is not or cannot be solved for y, choose a test point not on the boundary. If the test point satisfies the inequality, the graph includes all points on the same side of the boundary as the test point. Otherwise, the graph includes all points on the other side of the boundary.

Fundamental Theorem of Linear Programming

The optimal value for a linear programming problem occurs at a vertex of the region of feasible solutions.

Chapter 9 Review Exercises ▼▼▼▼▼▼▼▼▼▼▼▼▼▼▼▼▼▼▼▼▼▼▼▼▼▼▼▼▼▼

Use the elimination or substitution method to solve the linear system. Identify any inconsistent systems or systems with dependent equations.

1. $3x - 5y = 7$
$2x + 3y = 30$

2. $-x + 4y = 3$
$x + 2y = 9$

3. $6x - 2y = 4$
$4x + 5y = 9$

4. $\frac{1}{6}x + \frac{1}{3}y = 8$
$\frac{1}{4}x + \frac{1}{2}y = 12$

5. $.2x + .5y = 6$
$.4x + y = 9$

6. $3x - 2y = 0$
$9x + 8y = 7$

7. $2x - 5y + 3z = -1$
$x + 4y - 2z = 9$
$-x + 2y + 4z = 5$

8. $5x - y = 26$
$4y + 3z = -4$
$3x + 3z = 15$

Write a system of linear equations, and then use the system to solve the problem.

9. A cup of uncooked rice contains 15 g of protein and 810 cal. A cup of uncooked soybeans contains 22.5 g of protein and 270 cal. How many cups of each should be used for a meal containing 9.5 g of protein and 324 cal?

10. A company sells $3\frac{1}{2}''$ diskettes for 40¢ each and sells $5\frac{1}{4}''$ diskettes for 30¢ each. The company receives $38 for an order of 100 diskettes. However, the customer neglected to specify how many of each size to send. Determine the number of each size of diskette that should be sent.

11. The Waputi Indians make woven blankets, rugs, and skirts. Each blanket requires 24 hr for spinning the yarn, 4 hr for dyeing the yarn, and 15 hr for weaving. Rugs require 30, 5, and 18 hr and skirts 12, 3, and 9 hr, respectively. If there are 306, 59, and 201 hr available for spinning, dyeing, and weaving, respectively, how many of each item can be made? (*Hint:* Simplify the equations you write, if possible, before solving the system.)

12. Find the equation of the parabola $y = ax^2 + bx + c$ that passes through the points $(-1, -3.25)$, $(1, -2.25)$, and $(2, -4)$.

13. Find the equation of the vertical parabola that passes through the points shown in this table.

X	Y1	
1	-2.3	
2	-1.3	
3	4.5	
4	15.1	
5	30.5	
6	50.7	
7	75.7	

X=1

Find solutions for the system with the specified arbitrary variable.

14. $2x - 6y + 4z = 5$
$5x + y - 3z = 1$; z

15. $3x - 4y + z = 2$
$2x + y = 1$; x

Solve each system in Exercises 16–19.

16. $x^2 = 2y - 3$
$x + y = 3$

17. $2x^2 + 3y^2 = 30$
$x^2 + y^2 = 13$

18. $xy = -2$
$y - x = 3$

19. $x^2 + 2xy + y^2 = 4$
$x - 3y = -2$

20. Find all values of b so that the straight line $3x - y = b$ touches the circle $x^2 + y^2 = 25$ at only one point.

21. Do the circle $x^2 + y^2 = 144$ and the line $x + 2y = 8$ have any points in common? If so, what are they?

Use the Gauss-Jordan method to solve the system.

22. $2x + 3y = 10$
$-3x + y = 18$

23. $5x + 2y = -10$
$3x - 5y = -6$

24. $3x + y = -7$
$x - y = -5$

25. $x - z = -3$
$y + z = 6$
$2x - 3z = -9$

26. $2x - y + 4z = -1$
$-3x + 5y - z = 5$
$2x + 3y + 2z = 3$

Solve each problem by writing a system of equations and then solving it using the Gauss-Jordan method.

27. Three kinds of tea worth $4.60, $5.75, and $6.50 per pound are to be mixed to get 20 pounds of tea worth $5.25 per pound. The amount of $4.60 tea used is to be equal to the total amount of the other two kinds together. How many pounds of each tea should be used?

28. A 5% solution of a drug is to be mixed with some 15% solution and some 10% solution to get 20 ml of 8%

solution. The amount of 5% solution used must be 2 ml more than the sum of the other two solutions. How many ml of each solution should be used?

29. The cashier at an amusement park has a total of $2480, made up of fives, tens, and twenties. The total number of bills is 290, and the value of the tens is $60 more than the value of the twenties. How many of each type of bill does the cashier have?

30. Can a system consisting of two equations in three variables have a unique solution? Explain.

Find the values of all variables.

31. $\begin{bmatrix} 5 & x + 2 \\ -6y & z \end{bmatrix} = \begin{bmatrix} a & 3x - 1 \\ 5y & 9 \end{bmatrix}$

32. $\begin{bmatrix} -6 + k & 2 & a + 3 \\ -2 + m & 3p & 2r \end{bmatrix} + \begin{bmatrix} 3 - 2k & 5 & 7 \\ 5 & 8p & 5r \end{bmatrix} = \begin{bmatrix} 5 & y & 6a \\ 2m & 11 & -35 \end{bmatrix}$

Perform the operation whenever possible.

33. $\begin{bmatrix} 3 \\ 2 \\ 5 \end{bmatrix} - \begin{bmatrix} 8 \\ -4 \\ 6 \end{bmatrix} + \begin{bmatrix} 1 \\ 0 \\ 2 \end{bmatrix}$

34. $4\begin{bmatrix} 3 & -4 & 2 \\ 5 & -1 & 6 \end{bmatrix} + \begin{bmatrix} -3 & 2 & 5 \\ 1 & 0 & 4 \end{bmatrix}$

35. $\begin{bmatrix} -3 & 4 \\ 2 & 8 \end{bmatrix}\begin{bmatrix} -1 & 0 \\ 2 & 5 \end{bmatrix}$

36. $\begin{bmatrix} 2 & 5 & 8 \\ 1 & 9 & 2 \end{bmatrix} - \begin{bmatrix} 3 & 4 \\ 7 & 1 \end{bmatrix}$

37. $\begin{bmatrix} 1 & -2 & 4 & 2 \\ 0 & 1 & -1 & 8 \end{bmatrix}\begin{bmatrix} -1 \\ 2 \\ 0 \\ 1 \end{bmatrix}$

38. $\begin{bmatrix} 3 & 2 & -1 \\ 4 & 0 & 6 \end{bmatrix}\begin{bmatrix} -2 & 0 \\ 0 & 2 \\ 3 & 1 \end{bmatrix}$

In Exercises 68–70 of Section 9.4, we examined some properties of the transpose A^T of a matrix A. Answer the questions in Exercises 39 and 40 concerning the transpose.

39. The transpose A^T of a matrix A is shown on a graphing calculator screen. What is matrix A?

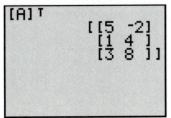

```
[A]ᵀ
      [[5 -2]
       [1 4 ]
       [3 8 ]]
```

40. What matrix would you expect the screen to show when the indicated sum is found?

```
[A]
      [[-9 2  10]
       [4  -4 5 ]]
[B]
      [[-5 6 8]
       [-3 6 4]]
[A]ᵀ+[B]ᵀ
```

Evaluate the determinant.

41. $\begin{vmatrix} -2 & 4 \\ 0 & 3 \end{vmatrix}$

42. $\begin{vmatrix} -1 & 8 \\ 2 & 9 \end{vmatrix}$

43. $\begin{vmatrix} -2 & 4 & 1 \\ 3 & 0 & 2 \\ -1 & 0 & 3 \end{vmatrix}$

44. $\begin{vmatrix} -1 & 2 & 3 \\ 4 & 0 & 3 \\ 5 & -1 & 2 \end{vmatrix}$

45. $\begin{vmatrix} -1 & 0 & 2 & -3 \\ 0 & 4 & 4 & -1 \\ -6 & 0 & 3 & -5 \\ 0 & -2 & 1 & 0 \end{vmatrix}$

Explain why the statement is true.

46. $\begin{vmatrix} 4 & 6 \\ 3 & 5 \end{vmatrix} = \begin{vmatrix} 4 & 3 \\ 6 & 5 \end{vmatrix}$

47. $\begin{vmatrix} 8 & 9 & 2 \\ 0 & 0 & 0 \\ 3 & 1 & 4 \end{vmatrix} = 0$

48. $\begin{vmatrix} 4 & 6 & 2 \\ -3 & 8 & -5 \\ 4 & 6 & 2 \end{vmatrix} = 0$

49. $\begin{vmatrix} 8 & 2 \\ 4 & 3 \end{vmatrix} = 2\begin{vmatrix} 4 & 1 \\ 4 & 3 \end{vmatrix}$

50. $\begin{vmatrix} 8 & 2 & -5 \\ -3 & 1 & 4 \\ 2 & 0 & 5 \end{vmatrix} = -\begin{vmatrix} 8 & -5 & 2 \\ -3 & 4 & 1 \\ 2 & 5 & 0 \end{vmatrix}$

51. $\begin{vmatrix} 5 & -1 & 2 \\ 3 & -2 & 0 \\ -4 & 1 & 2 \end{vmatrix} = \begin{vmatrix} 5 & -1 & 2 \\ 8 & -3 & 2 \\ -4 & 1 & 2 \end{vmatrix}$

52. Solve for x: $\begin{vmatrix} 3x & 7 \\ -x & 4 \end{vmatrix} = 8$.

53. Solve for t: $\begin{vmatrix} 6t & 2 & 0 \\ 1 & 5 & 3 \\ t & 2 & -1 \end{vmatrix} = 2t$.

Solve the system by Cramer's rule if possible. Use another method if this is not possible. Identify any dependent equations or inconsistent systems.

54. $\begin{aligned} 3x + 7y &= 2 \\ 5x - y &= -22 \end{aligned}$

55. $\begin{aligned} 3x + y &= -1 \\ 5x + 4y &= 10 \end{aligned}$

56. $\begin{aligned} 5x - 2y - z &= 8 \\ -5x + 2y + z &= -8 \\ x - 4y - 2z &= 0 \end{aligned}$

57. $\begin{aligned} 3x + 2y + z &= 2 \\ 4x - y + 3z &= -16 \\ x + 3y - z &= 12 \end{aligned}$

Find the inverse of each matrix that has an inverse.

58. $\begin{bmatrix} -4 & 2 \\ 0 & 3 \end{bmatrix}$

59. $\begin{bmatrix} 2 & 1 \\ 5 & 3 \end{bmatrix}$

60. $\begin{bmatrix} 2 & 3 & 5 \\ -2 & -3 & -5 \\ 1 & 4 & 2 \end{bmatrix}$

61. $\begin{bmatrix} 2 & -1 & 0 \\ 1 & 0 & 1 \\ 1 & -2 & 0 \end{bmatrix}$

Use the method of matrix inverses to solve each system.

62. $\begin{aligned} 2x + y &= 5 \\ 3x - 2y &= 4 \end{aligned}$

63. $\begin{aligned} x + y + z &= 1 \\ 2x - y &= -2 \\ 3y + z &= 2 \end{aligned}$

64. $\begin{aligned} x &= -3 \\ y + z &= 6 \\ 2x - 3z &= -9 \end{aligned}$

Graph the solution of the system of inequalities.

65. $\begin{aligned} x + y &\leq 6 \\ 2x - y &\geq 3 \end{aligned}$

66. $\begin{aligned} y &\leq \frac{1}{3}x - 2 \\ y^2 &\leq 16 - x^2 \end{aligned}$

67. $\begin{aligned} x^2 + y^2 &\leq 144 \\ x^2 + y^2 &\geq 16 \end{aligned}$

68. Find $x \geq 0$ and $y \geq 0$ such that

$$3x + 2y \leq 12$$
$$5x + y \geq 5$$

and $2x + 4y$ is maximized.

69. Find $x \geq 0$ and $y \geq 0$ such that

$$y \leq -3x + 5$$
$$y \geq 4x - 3$$

and $7x + 14y$ is maximized.

Solve the linear programming problem.

70. A bakery makes both cakes and cookies. Each batch of cakes requires 2 hr in the oven and 3 hr in the decorating room. Each batch of cookies needs 1 1/2 hr in the oven and 2/3 hr in the decorating room. The oven is available no more than 16 hr a day, while the decorating room can be used no more than 12 hr per day. A batch of cookies produces a profit of $20; the profit on a batch of cakes is $30. Find the number of batches of each item that will maximize profit.

71. A candy company has 100 kg of chocolate-covered nuts and 125 kg of chocolate-covered raisins to be sold as two different mixtures. One mix will contain 1/2 nuts and 1/2 raisins, while the other mix will contain 1/3 nuts and 2/3 raisins. How much of each mixture should be made to maximize revenue if the first mix sells for $6.00 per kilogram and the second mix sells for $4.80 per kilogram?

Chapter 9   Test ▼▼▼▼▼▼▼▼▼▼▼▼▼▼▼▼▼▼▼▼▼▼▼▼▼▼▼▼▼▼▼▼▼▼▼▼▼

Use substitution or elimination to solve each system. Identify any system that is inconsistent or has dependent equations. If a system has dependent equations, express the solution with y arbitrary.

1. $3x - y = 9$
$x + 2y = 10$

2. $6x + 9y = -21$
$4x + 6y = -14$

3. $\dfrac{1}{4}x - \dfrac{1}{3}y = -\dfrac{5}{12}$
$\dfrac{1}{10}x + \dfrac{1}{5}y = \dfrac{1}{2}$

4. $x - 2y = 4$
$-2x + 4y = 6$

5. $2x + y + z = 3$
$x + 2y - z = 3$
$3x - y + z = 5$

Solve the nonlinear system of equations.

6. $2x^2 + y^2 = 6$
$x^2 - 4y^2 = -15$

7. $x^2 + y^2 = 25$
$x + y = 7$

8. If a system of two nonlinear equations contains one equation whose graph is a circle and another equation whose graph is a line, can the system have exactly one solution? If so, draw a sketch to indicate this situation.

9. Find two numbers such that their sum is -1 and the sum of their squares is 61.

Use the Gauss-Jordan method to solve the system.

10. $3a - 2b = 13$
$4a - b = 19$

11. $3a - 4b + 2c = 15$
$2a - b + c = 13$
$a + 2b - c = 5$

12. Find the equation that defines the parabola shown on the screen, using the information given at the bottom of the screen and in the table.

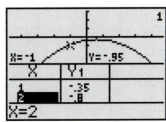

[−4.7, 4.7] by [−3.1, 3.1]

13. The sum of three numbers is 2. The first number is equal to the sum of the other two, and the third number is the result of subtracting the first from the second. Find the numbers by solving a system of equations.

14. Find the values of all variables in the equation $\begin{bmatrix} 5 & x+6 \\ 0 & 4 \end{bmatrix} = \begin{bmatrix} y-2 & 4-x \\ 0 & w+7 \end{bmatrix}$.

Perform the operations whenever possible.

15. $3\begin{bmatrix} 2 & 3 \\ 1 & -4 \\ 5 & 9 \end{bmatrix} - \begin{bmatrix} -2 & 6 \\ 3 & -1 \\ 0 & 8 \end{bmatrix}$

16. $\begin{bmatrix} 1 \\ 2 \end{bmatrix} + \begin{bmatrix} 4 \\ -6 \end{bmatrix} + \begin{bmatrix} 2 & 8 \\ -7 & 5 \end{bmatrix}$

Find the matrix product whenever possible.

17. $\begin{bmatrix} 2 & 1 & -3 \\ 4 & 0 & 5 \end{bmatrix} \begin{bmatrix} 1 & 3 \\ 2 & 4 \\ 3 & -2 \end{bmatrix}$

18. $\begin{bmatrix} 2 & -4 \\ 3 & 5 \end{bmatrix} \begin{bmatrix} 4 \\ 2 \\ 7 \end{bmatrix}$

19. Which of the following properties does not apply to multiplication of matrices?
(a) commutative   (b) associative   (c) distributive   (d) identity

Evaluate the determinant.

20. $\begin{vmatrix} 6 & 8 \\ 2 & -7 \end{vmatrix}$

21. $\begin{vmatrix} 2 & 0 & 8 \\ -1 & 7 & 9 \\ 12 & 5 & -3 \end{vmatrix}$

Give the theorem (or theorems) that justifies the statement.

22. $\begin{vmatrix} 6 & 7 \\ -5 & 2 \end{vmatrix} = -\begin{vmatrix} -5 & 2 \\ 6 & 7 \end{vmatrix}$

23. $\begin{vmatrix} 7 & 2 & -1 \\ 5 & -4 & 3 \\ 6 & -2 & 1 \end{vmatrix} = \begin{vmatrix} 7 & 2 & -1 \\ -7 & 0 & 1 \\ 13 & 0 & 0 \end{vmatrix}$

Solve the system by Cramer's rule.

24. $2x - 3y = -33$
$\quad 4x + 5y = 11$

25. $\quad x + y - z = -4$
$\quad 2x - 3y - z = 5$
$\quad x + 2y + 2z = 3$

Find the inverse, if it exists, for the matrix.

26. $\begin{bmatrix} -8 & 5 \\ 3 & -2 \end{bmatrix}$

27. $\begin{bmatrix} 4 & 12 \\ 2 & 6 \end{bmatrix}$

28. $\begin{bmatrix} 1 & 3 & 4 \\ 2 & 7 & 8 \\ -2 & -5 & -7 \end{bmatrix}$

Use matrix inverses to solve each of the following.

29. $2x + y = -6$
$\quad 3x - y = -29$

30. $\quad x + y = 5$
$\quad y - 2z = 23$
$\quad x + 3z = -27$

31. Graph the solution set of

$$x - 3y \geq 6$$
$$y^2 \leq 16 - x^2.$$

32. Find $x \geq 0$ and $y \geq 0$ such that

$$x + 2y \leq 24$$
$$3x + 4y \leq 60$$

and $2x + 3y$ is maximized.

33. The J. J. Gravois Company designs and sells two types of rings: the VIP and the SST. The company can produce up to 24 rings each day using up to 60 total hours of labor. It takes 3 hours to make one VIP ring, and 2 hours to make one SST ring. How many of each type of ring should be made daily in order to maximize the company's profit, if the profit on a VIP ring is $30 and the profit on the SST ring is $40?

10

Analytic Geometry

Since the beginning of civilization, people have been fascinated by and compelled to understand the universe they live in. In 1887 T. H. Huxley wrote:

> The known is finite, the unknown infinite; intellectually we stand on an islet in the midst of an illimitable ocean of inexplicability. Our business in every generation is to reclaim a little more land.

The Greeks together with the early Christian astronomers believed that Earth was the center of the universe and the sun and planets traveled in circular orbits around Earth. The circle was regarded as the perfect geometric shape. Later, in the sixteenth century, the greatest observational astronomer of the age, Tycho Brahe, recorded precise data on planetary movement in the sky. With Brahe's data, Johannes Kepler in 1619 empirically determined that the planets do not move in circular orbits, but rather in elliptical orbits around the sun. In 1686 Newton used Kepler's work to show analytically that elliptical orbits were a result of his famous theory of gravitation. Edmund Halley determined that comets also followed elliptical orbits around the sun and accurately predicted the return of Halley's Comet. People now know that both celestial objects and satellites move through space in one of three types of paths: elliptical, parabolic, or hyperbolic. These curves are called *conic sections* and were discovered in 200 B.C. by the Greek geometer Apollonius. Today scientists are searching the sky for information about the beginning of the universe and for signs of life elsewhere. Enormous radio telescopes with parabolic dishes—much like television satellite dishes—search the sky continuously for new information.

Source: Mars, J. and H. Liebowitz, *Structure Technology for Large Radio and Radar Telescopes Systems*, The MIT Press, Cambridge, MA, 1969.

The search to understand the universe has been directed toward both the infinite and the infinitesimal. In 1911 Ernest Rutherford determined the structure of the atom. Much like our solar system, it was composed of mostly empty space. It consisted of electrons orbiting the small, dense nucleus. Small subatomic particles were also capable of traveling in trajectories described by conic sections. Since Einstein's introduction of the photon, scores of subatomic particles have been discovered and a complete understanding of the structure of the atom appears distant.

Throughout history, parabolas, ellipses, and hyperbolas have played a central role in our understanding of the universe. How can we determine the shape of a satellite's orbit? Is Pluto the farthest planet from the sun? Where should the receiver be placed on a 210-foot radio telescope? Why is the antenna on a radio telescope a parabolic dish? What type of orbit is necessary for a spaceship to escape Earth's gravitational field? Does a projectile travel in a similar path on both Mars and the moon? The answers to these questions require knowledge about conic sections. In this chapter we will learn about these age-old curves that have had such a profound influence on our understanding of who we are and the cosmos we live in.*

Sources: Boorse, H., L. Motz, and J. Weaver, *The Atomic Scientists,* John Wiley & Sons, Inc., 1989.
National Council of Teachers of Mathematics, *Historical Topics for the Mathematics Classroom,* Thirty-first Yearbook, Washington, D. C., 1969.
Sagan, C., *Cosmos,* Random House, New York, 1980.

Figure 1 shows how conic sections are formed by the intersection of a plane and a cone. Two examples of conic sections—circles and parabolas—were studied in Chapters 3 and 4. In this chapter, we look at parabolas in more detail and study two other types of conic sections, called *ellipses* and *hyperbolas*. In the last section of this chapter, we discuss the special characteristics of the equations and graphs of the four types of conic sections.

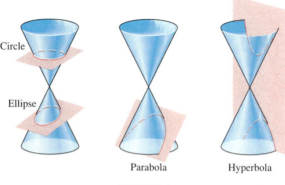

FIGURE 1

10.1   Parabolas ▼▼▼

From Chapter 4 we know that the graph of the equation $y = a(x - k)^2 + h$ is a parabola with vertex at (k, h) and the vertical line $x = k$ as axis. This equation can also be written as

$$y - h = a(x - k)^2.$$

The equation

$$x - h = a(y - k)^2$$

also has a parabola as its graph. The graph of this new equation is a reflection of the graph of $y - h = a(x - k)^2$ about the line $y = x$ since the new equation can be obtained by interchanging x and y in the original equation. Since the graph of $y - h = a(x - k)^2$ has a vertical axis and vertex at (k, h), the graph of $x - h = a(y - k)^2$ has a horizontal axis and vertex at (h, k).

Parabola with Horizontal Axis

The parabola with vertex at (h, k) and the horizontal line $y = k$ as axis has an equation of the form

$$x - h = a(y - k)^2.$$

The parabola opens to the right if $a > 0$ and to the left if $a < 0$.

EXAMPLE 1
Graphing a parabola with horizontal axis

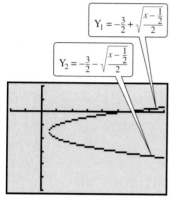

$Y_1 = -\frac{3}{2} + \sqrt{\dfrac{x - \frac{1}{2}}{2}}$

$Y_2 = -\frac{3}{2} - \sqrt{\dfrac{x - \frac{1}{2}}{2}}$

[−2, 8] by [−6, 2]

To graph a horizontal parabola, it is necessary to solve for two functions, Y_1 and Y_2.

Graph $x = 2y^2 + 6y + 5$.

Write this equation in the form $x - h = a(y - k)^2$ by completing the square on y as follows.

$$x = 2(y^2 + 3y \quad\quad) + 5$$

$$= 2\left(y^2 + 3y + \frac{9}{4} - \frac{9}{4}\right) + 5$$

$$= 2\left(y^2 + 3y + \frac{9}{4}\right) + 2\left(-\frac{9}{4}\right) + 5$$

$$= 2\left(y + \frac{3}{2}\right)^2 + \frac{1}{2}$$

$$x - \frac{1}{2} = 2\left(y - \left(-\frac{3}{2}\right)\right)^2$$

As this result shows, the vertex of the parabola is the point $(1/2, -3/2)$. The axis is the horizontal line $y = k$, or $y = -3/2$. Using the vertex and the axis and plotting a few additional points gives the graph in Figure 2. ◗

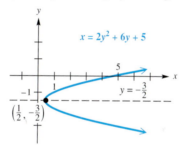

FIGURE 2

📈 The graph shows that $x - h = a(y - k)^2$ does not define a function. To graph this relation with a graphing calculator we must express it as the union of two functions. For instance, when $a > 0$ and $x \geq h$, the graph can be obtained by graphing both

$$Y_1 = k + \sqrt{\frac{x - h}{a}} \quad \text{and} \quad Y_2 = k - \sqrt{\frac{x - h}{a}}.$$

See the graphing calculator-generated figure in the margin.

The equation of a parabola can be developed from the geometric definition of a parabola as a set of points.

Definition of Parabola

A **parabola** is the set of points in a plane equidistant from a fixed point and a fixed line. The fixed point is called the **focus** and the fixed line is called the **directrix** of the parabola.

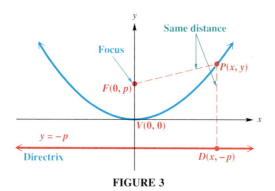

FIGURE 3

We get an equation of a parabola from the definition as follows. Let the directrix be the line $y = -p$ and the focus be the point F with coordinates $(0, p)$, as shown in Figure 3. To get the equation of the set of points that are the same distance from the line $y = -p$ and the point $(0, p)$, choose one such point P and give it coordinates (x, y). Since $d(P, F)$ and $d(P, D)$ must be the same, using the distance formula gives

$$d(P, F) = d(P, D)$$
$$\sqrt{(x - 0)^2 + (y - p)^2} = \sqrt{(x - x)^2 + (y - (-p))^2}$$
$$\sqrt{x^2 + (y - p)^2} = \sqrt{(y + p)^2}$$
$$x^2 + y^2 - 2yp + p^2 = y^2 + 2yp + p^2$$
$$x^2 = 4py.$$

This discussion is summarized below.

Parabola with a Vertical Axis and Vertex (0, 0)

The parabola with focus at $(0, p)$ and directrix $y = -p$ has equation

$$x^2 = 4py.$$

The parabola has a vertical axis, opens upward if $p > 0$, and opens downward if $p < 0$.

The definition of a parabola given above has led to another form of the equation $y = ax^2$, discussed in Chapter 4, with $a = 1/(4p)$.

If the directrix is the line $x = -p$ and the focus is at $(p, 0)$, using the definition of a parabola and the distance formula leads to the equation of a parabola with a horizontal axis. (See Exercise 58.)

> ## Parabola with a Horizontal Axis and Vertex $(0, 0)$
>
> The parabola with focus at $(p, 0)$ and directrix $x = -p$ has equation
>
> $$y^2 = 4px.$$
>
> The parabola opens to the right if $p > 0$ or to the left if $p < 0$, and it has a horizontal axis.

EXAMPLE 2
Determining information about a parabola from its equation

Find the focus, directrix, vertex, and axis of the following parabolas.

(a) $x^2 = 8y$

The equation has the form $x^2 = 4py$, so set $4p = 8$, from which $p = 2$. Since the x-term is squared, the parabola is vertical, with focus at $(0, p) = (0, 2)$ and directrix $y = -2$. The vertex is $(0, 0)$, and the axis of the parabola is the y-axis. See Figure 4.

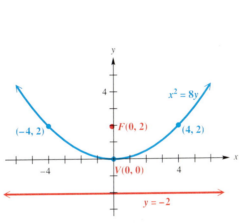

FIGURE 4

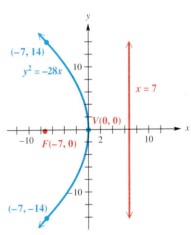

FIGURE 5

(b) $y^2 = -28x$

This equation has the form $y^2 = 4px$, with $4p = -28$, so $p = -7$. The parabola is horizontal, with focus $(-7, 0)$, directrix $x = 7$, vertex $(0, 0)$, and x-axis as axis of the parabola. Since p is negative, the graph opens to the left, as shown in Figure 5. ◗

EXAMPLE 3
Writing an equation of a parabola

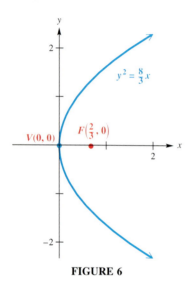

FIGURE 6

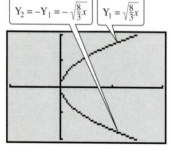

[-1, 2] by [-2, 2]
Compare to Figure 6.

EXAMPLE 4
Writing an equation of a parabola

Write an equation for each of the following parabolas.

(a) Focus $(2/3, 0)$ and vertex at the origin

Since the focus $(2/3, 0)$ is on the x-axis, the parabola is horizontal and opens to the right because $p = 2/3$ is positive. See Figure 6. The equation, which will have the form $y^2 = 4px$, is

$$y^2 = 4\left(\frac{2}{3}\right)x \qquad \text{or} \qquad y^2 = \frac{8}{3}x.$$

(b) Vertical axis, vertex at the origin, through the point $(-2, 12)$

The parabola will have an equation of the form $x^2 = 4py$ because the axis is vertical. Since the point $(-2, 12)$ is on the graph, it must satisfy the equation. Substitute $x = -2$ and $y = 12$ into $x^2 = 4py$ to get

$$(-2)^2 = 4p(12)$$
$$4 = 48p$$
$$p = \frac{1}{12},$$

which gives

$$x^2 = \frac{1}{3}y \qquad \text{or} \qquad y = 3x^2$$

as an equation of the parabola.

The equations $x^2 = 4py$ and $y^2 = 4px$ can be extended to parabolas having vertex at (h, k) by replacing x and y by $x - h$ and $y - k$.

The parabola with vertex (h, k) has an equation of the form

$$(x - h)^2 = 4p(y - k) \qquad \text{Vertical axis}$$
or
$$(y - k)^2 = 4p(x - h), \qquad \text{Horizontal axis}$$

where the focus is distance p or $-p$ from the vertex.

Write an equation for a parabola with vertex at $(1, 3)$ and focus at $(-1, 3)$.

Since the focus is to the left of the vertex, the axis is horizontal and the parabola opens to the left. See Figure 7. The distance between the vertex and the focus is $1 - (-1)$ or 2, so $p = -2$ and the equation of the parabola is

$$(y - 3)^2 = 4(-2)(x - 1)$$
or
$$(y - 3)^2 = -8(x - 1).$$

The negative sign was chosen because the parabola opens to the left.

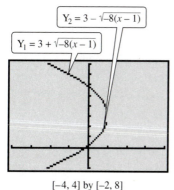

$Y_2 = 3 - \sqrt{-8(x-1)}$

$Y_1 = 3 + \sqrt{-8(x-1)}$

[-4, 4] by [-2, 8]
Compare to Figure 7.

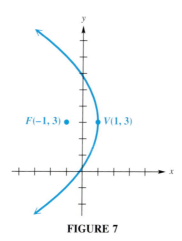

$F(-1, 3)$ •   • $V(1, 3)$

FIGURE 7

CONNECTIONS Parabolas have a special property, called the *reflecting property,* that makes them useful in the design of telescopes, radar equipment, auto headlights, and solar furnaces. When a ray of light or a sound wave traveling parallel to the axis of a parabolic shape bounces off the parabola, it passes through the focus. For example, in the solar furnace shown in the figure, a parabolic mirror collects light at the focus and thereby generates intense heat at that point. The reflecting property can be used in reverse. If a light source is placed at the focus, then the reflected light rays will be directed straight ahead. This is the reason why the reflector in a car headlight is parabolic.

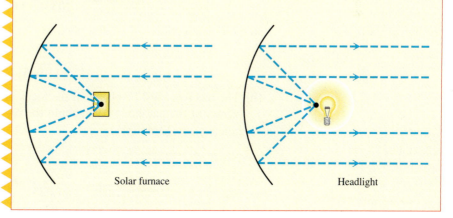

Solar furnace   Headlight

EXAMPLE 5
Applying the reflective
property of parabolas

The Parkes radio telescope has a parabolic dish shape with a diameter of 210 feet and a depth of 32 feet. Because of this parabolic shape, distant rays hitting the dish will be reflected directly toward the focus. A cross-section of the dish is shown in Figure 8. (*Source:* Mar, J., and H. Liebowitz, *Structure Technology for Large Radio and Radar Telescope Systems,* The MIT Press, Massachusetts Institute of Technology, Cambridge, Massachusetts, 1969.)

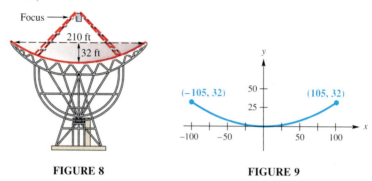

FIGURE 8   FIGURE 9

(a) Determine an equation describing this cross-section by placing the vertex at the origin with the parabola opening upward.

Locate the vertex at the origin as shown in Figure 9. Then, the form of the parabola will be $y = ax^2$. The parabola must pass through the point $(210/2, 32) = (105, 32)$. Thus,

$$32 = a(105)^2$$

$$a = \frac{32}{105^2} = \frac{32}{11,025},$$

so the cross-section can be described by

$$y = \frac{32}{11,025}x^2.$$

(b) The receiver must be placed at the focus of the parabola. How far from the vertex of the parabolic dish should the receiver be located?

Since $y = (32/11,025)x^2$,

$$4p = \frac{1}{a}$$

$$= \frac{11,025}{32}$$

$$p = \frac{11,025}{128} \approx 86.1.$$

The receiver should be located at $(0, 86.1)$ or 86.1 feet above the vertex. ▶

10.1 Exercises ▼▼

1. An equation of the form $y = ax^2 + bx + c$ or $x = ay^2 + by + c$ has a parabola as its graph. Explain in your own words how you can tell before actually graphing whether the parabola opens up, down, left, or right.

2. Explain why the vertex of a parabola is the point closest to the focus.

Graph each horizontal parabola. See Example 1.

3. $x = -y^2$

4. $x = y^2 + 2$

5. $x = (y - 3)^2$

6. $x = (y + 1)^2$

7. $x = (y - 4)^2 + 2$

8. $x = (y + 2)^2 - 1$

9. $x = -3(y - 1)^2 + 2$

10. $x = -2(y + 3)^2$

11. $x = \frac{1}{2}(y - 1)^2 + 4$

12. $x = -\frac{1}{3}(y - 3)^2 + 3$

13. $x = y^2 + 4y + 2$

14. $x = 2y^2 - 4y + 6$

15. $x = -4y^2 - 4y + 3$

16. $x = -2y^2 + 2y - 3$

17. $2x = y^2 - 4y + 6$

18. $x + 3y^2 + 18y + 22 = 0$

Give the focus, directrix, and axis for each parabola. See Example 2.

19. $x^2 = 24y$

20. $y = 8x^2$

21. $y = -4x^2$

22. $9y = x^2$

23. $x = -32y^2$

24. $x = 16y^2$

25. $x = -\frac{1}{4}y^2$

26. $x = -\frac{1}{16}y^2$

27. $(y - 3)^2 = 12(x - 1)$

28. $(x + 2)^2 = 20y$

29. $(x - 7)^2 = 16(y + 5)$

30. $(y - 2)^2 = 24(x - 3)$

Write an equation for each parabola with vertex at the origin. See Example 3.

31. Focus $(5, 0)$

32. Focus $(-1/2, 0)$

33. Focus $(0, 1/4)$

34. Focus $(0, -1/3)$

35. Through $(\sqrt{3}, 3)$, opening upward

36. Through $(2, -2\sqrt{2})$, opening to the right

37. Through $(3, 2)$, symmetric with respect to the x-axis

38. Through $(2, -4)$, symmetric with respect to the y-axis

Write an equation for each parabola. See Example 4.

39. Vertex $(4, 3)$, focus $(4, 5)$

40. Vertex $(-2, 1)$, focus $(-2, -3)$

41. Vertex $(-5, 6)$, focus $(2, 6)$

42. Vertex $(1, 2)$, focus $(4, 2)$

Use a graphing calculator to graph each parabola. You may need to rewrite the equation in the form of two functions, as described in the explanation immediately following Example 1.

43. $x = 3y^2 + 6y - 4$

44. $x = -2y^2 + 4y + 3$

45. $x + 2 = -(y + 1)^2$

46. $x - 5 = 2(y - 2)^2$

Solve each problem. See Example 5.

47. When an object moves under the influence of a constant force (without air resistance), its path is parabolic. This would occur if a ball is thrown near the surface of a planet or other celestial object. The graphing calculator can be used to simulate something that would be impossible to view in real life. Suppose two balls are simultaneously thrown upward at a 45° angle on two different planets. If their initial velocities are both 30 miles per hour, then their xy-coordinates in feet at time x in seconds can be expressed by the equation $y = x - (g/1922)x^2$, where g is the acceleration due

to gravity. The value of g will vary depending on the mass and size of the planet. (*Source:* Zeilik, M., S. Gregory, and E. Smith, *Introductory Astronomy and Astrophysics,* Saunders College Publishers, 1992.)

(a) For Earth $g = 32.2$ while on Mars $g = 12.6$. Find the two equations, and graph on the same coordinate axes the paths of the two balls thrown on Earth and Mars. Use the interval $[0, 180]$ for x. (*Hint:* If possible, set the mode on your graphing calculator to simultaneous.)

(b) Determine the difference in the horizontal distances traveled by the two balls.

48. 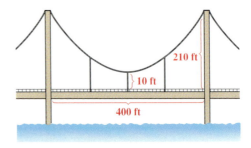 (Refer to Exercise 47.) Suppose the two balls are now thrown upward at a 60° angle on Mars and the moon. If their initial velocity is 60 miles per hour, then their xy-coordinates in feet at time x in seconds can be expressed by the equation

$$y = \frac{19}{11}x - \frac{g}{3872}x^2.$$

(*Source:* Zeilik, M., S. Gregory, and E. Smith, *Introductory Astronomy and Astrophysics,* Saunders College Publishers, 1992.)

(a) Graph on the same coordinate axes the paths of the balls if $g = 5.2$ for the moon.

(b) Determine the maximum height of each ball to the nearest foot.

49. The U.S. Naval Research Laboratory designed a giant radio telescope weighing 3450 tons. Its parabolic dish had a diameter of 300 feet with a focal length (the distance from the focus to the parabolic surface) of 128.5 feet. Determine the maximum depth of the 300-foot dish. (*Source:* Mar, J., and H. Liebowitz, *Structure Technology for Large Radio and Radar Telescope Systems,* The MIT Press, Cambridge, Massachusetts, 1969.)

50. When an alpha particle is moving in a horizontal path along the positive x-axis and passes between charged plates it is deflected in a parabolic path. If the plate is charged with 2000 volts and is .4 meter long, an alpha particle's path can be described by the equation $y = (-k/(2v_0))x^2$ where $k = 5 \times 10^{-9}$ is constant and v_0 is the initial velocity of the alpha particle. If $v_0 = 10^7$ meters/sec, what is the deflection of the alpha particle's path in the y-direction when $x = .4$ meter? (*Source:* Semat, H., and J. Albright, *Introduction to Atomic and Nuclear Physics,* Holt, Rinehart, and Winston, Inc., 1972.)

51. The cable in the center portion of a bridge is supported as shown in the figure to form a parabola. The center vertical cable is 10 ft high, the supports are 210 ft high, and the distance between the two supports is 400 ft. Find the height of the remaining vertical cables, if the vertical cables are evenly spaced. (Ignore the width of the supports and cables.)

52. An arch in the shape of a parabola has the dimensions shown in the figure. How wide is the arch 9 ft up?

12 ft

12 ft

▼▼▼▼▼▼▼▼▼▼▼▼ **DISCOVERING CONNECTIONS** (Exercises 53–56) ▼▼▼▼▼▼▼▼▼▼▼▼

Given three noncollinear points, the equation of a horizontal parabola joining them can be found. The parabola will have an equation of the form $x = ay^2 + by + c$, and it can be found by solving a system of equations. Work Exercises 53–56 in order, so that the equation of the horizontal parabola containing $(-5, 1)$, $(-14, -2)$, and $(-10, 2)$ can be found.

53. Write three equations in a, b, and c, by substituting the given values of x and y into the equation $x = ay^2 + by + c$.

54. Solve the system of three equations determined in Exercise 53.

55. Does the horizontal parabola open to the left or to the right? Why?

56. Write the equation of the horizontal parabola.

57. Find the equation of the parabola that has vertex $(1, 2)$, has its axis parallel to the x-axis, and passes through the point $(13, 4)$.

58. Prove that the parabola with focus $(p, 0)$ and directrix $x = -p$ has equation $y^2 = 4px$.

10.2 Ellipses ▼▼▼

Like the parabola, the ellipse is defined as a set of points.

> ### Definition of Ellipse
>
> An **ellipse** is the set of all points in a plane the sum of whose distances from two fixed points is constant. Each fixed point is called a **focus** (plural, **foci**) of the ellipse.

For example, the ellipse in Figure 10 has foci at points F and F'. By the definition, the ellipse consists of all points P such that sum $d(P, F) + d(P, F')$ is constant. The ellipse in Figure 10 has its center at the origin. As the vertical line test shows, the graph of Figure 10 is not the graph of a function, since one value of x can lead to two values of y.

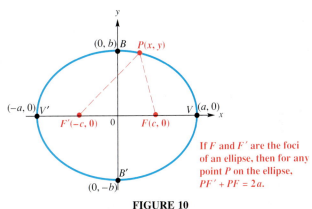

If F and F' are the foci of an ellipse, then for any point P on the ellipse, $PF' + PF = 2a$.

FIGURE 10

To obtain an equation for an ellipse centered at the origin, let the two foci have coordinates $(-c, 0)$ and $(c, 0)$, respectively. Let the sum of the distances from any point $P(x, y)$ on the ellipse to the two foci be $2a$. By the distance formula, segment PF has length

$$d(P, F) = \sqrt{(x - c)^2 + y^2},$$

while segment PF' has length

$$d(P, F') = \sqrt{[x - (-c)]^2 + y^2} = \sqrt{(x + c)^2 + y^2}.$$

The sum of the lengths $d(P, F)$ and $d(P, F')$ must be $2a$.

$$\sqrt{(x - c)^2 + y^2} + \sqrt{(x + c)^2 + y^2} = 2a$$

$$\sqrt{(x - c)^2 + y^2} = 2a - \sqrt{(x + c)^2 + y^2} \qquad \text{Isolate } \sqrt{(x - c)^2 + y^2}.$$

$$(x - c)^2 + y^2 = 4a^2 - 4a\sqrt{(x + c)^2 + y^2} + (x + c)^2 + y^2$$

Square both sides.

$$x^2 - 2cx + c^2 + y^2 = 4a^2 - 4a\sqrt{(x + c)^2 + y^2} + x^2 + 2cx + c^2 + y^2$$

$$4a\sqrt{(x + c)^2 + y^2} = 4a^2 + 4cx \qquad \text{Isolate } 4a\sqrt{(x + c)^2 + y^2}.$$

$$a\sqrt{(x + c)^2 + y^2} = a^2 + cx \qquad \text{Divide both sides by 4.}$$

$$a^2[x^2 + 2cx + c^2 + y^2] = a^4 + 2ca^2x + c^2x^2 \qquad \text{Square both sides.}$$

$$a^2x^2 + 2ca^2x + a^2c^2 + a^2y^2 = a^4 + 2ca^2x + c^2x^2 \qquad \text{Multiply.}$$

$$a^2x^2 + a^2c^2 + a^2y^2 = a^4 + c^2x^2 \qquad \text{Subtract } 2ca^2x \text{ from both sides.}$$

$$a^2x^2 - c^2x^2 + a^2y^2 = a^4 - a^2c^2 \qquad \text{Rearrange terms.}$$

$$(a^2 - c^2)x^2 + a^2y^2 = a^2(a^2 - c^2) \qquad \text{Factor.}$$

$$\frac{x^2}{a^2} + \frac{y^2}{a^2 - c^2} = 1 \qquad (*) \qquad \text{Divide both sides by } a^2(a^2 - c^2).$$

Since $B(0, b)$ is on the ellipse in Figure 10, we have

$$d(B, F) + d(B, F') = 2a$$

$$\sqrt{(-c)^2 + b^2} + \sqrt{c^2 + b^2} = 2a$$

$$2\sqrt{c^2 + b^2} = 2a$$

$$\sqrt{c^2 + b^2} = a$$

$$c^2 + b^2 = a^2$$

$$b^2 = a^2 - c^2.$$

Replacing $a^2 - c^2$ with b^2 in equation (*) gives

$$\frac{x^2}{a^2} + \frac{y^2}{b^2} = 1,$$

the **standard form** of the equation of an ellipse centered at the origin with foci on the x-axis.

Letting $y = 0$ in the standard form gives

$$\frac{x^2}{a^2} + \frac{0^2}{b^2} = 1$$

$$\frac{x^2}{a^2} = 1$$

$$x^2 = a^2$$

$$x = \pm a$$

as the x-intercepts of the ellipse. The points $V'(-a, 0)$ and $V(a, 0)$ are the **vertices** of the ellipse; the segment VV' is the **major axis.** In a similar manner, letting $x = 0$ shows that the y-intercepts are $\pm b$; the segment connecting $(0, b)$ and $(0, -b)$ is the **minor axis.** We assumed throughout the work above that the foci were on the x-axis. If the foci were on the y-axis, an almost identical proof could be used to get the standard form

$$\frac{x^2}{b^2} + \frac{y^2}{a^2} = 1.$$

CAUTION   Do not be confused by the two standard forms—in one case a^2 is associated with x^2; in the other case a^2 is associated with y^2. However, in practice it is necessary only to find the intercepts of the graph—if the positive x-intercept is larger than the positive y-intercept, the major axis is horizontal; otherwise, it is vertical. When using the relationship $a^2 - c^2 = b^2$, or $a^2 - b^2 = c^2$, choose a^2 and b^2 so that $a^2 > b^2$.

A summary of this work with ellipses follows.

Standard Equations for Ellipses

The ellipse with center at the origin and equation

$$\frac{x^2}{a^2} + \frac{y^2}{b^2} = 1 \quad (a > b)$$

has vertices $(\pm a, 0)$, endpoints of the minor axis $(0, \pm b)$, and foci $(\pm c, 0)$, where $c^2 = a^2 - b^2$.

The ellipse with center at the origin and equation

$$\frac{x^2}{b^2} + \frac{y^2}{a^2} = 1 \quad (a > b)$$

has vertices $(0, \pm a)$, endpoints of the minor axis $(\pm b, 0)$, and foci $(0, \pm c)$, where $c^2 = a^2 - b^2$.

An ellipse is symmetric with respect to its major axis, its minor axis, and its center.

EXAMPLE 1
Graphing an ellipse centered at the origin

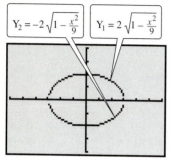

[−6, 6] by [−4, 4]

To graph the ellipse $\frac{x^2}{9} + \frac{y^2}{4} = 1$, we find two functions Y_1 and Y_2 whose union is the graph of the ellipse. Compare to Figure 11.

Graph $4x^2 + 9y^2 = 36$ and find the coordinates of the foci.

To obtain the standard form for the equation of an ellipse, divide each side by 36 to get

$$\frac{x^2}{9} + \frac{y^2}{4} = 1.$$

The x-intercepts of this ellipse are ± 3, and the y-intercepts ± 2. Additional ordered pairs satisfying the equation of the ellipse may be found if desired by choosing x-values and using the equation to find the corresponding y-values. The graph of the ellipse is shown in Figure 11. Since $9 > 4$, find the foci by letting $c^2 = 9 - 4 = 5$ so that $c = \sqrt{5}$. The major axis is along the x-axis, so the foci are at $(-\sqrt{5}, 0)$ and $(\sqrt{5}, 0)$. ▶

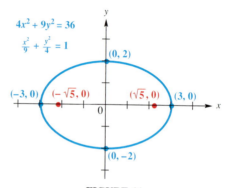

FIGURE 11

EXAMPLE 2
Finding the equation of an ellipse

Find the equation of the ellipse having center at the origin, foci at $(0, 3)$ and $(0, -3)$, and major axis of length 8 units.

Since the major axis is 8 units long,

$$2a = 8$$

or

$$a = 4.$$

To find b^2, use the relationship $a^2 - b^2 = c^2$. Here $a = 4$ and $c = 3$. Substituting for a and c gives

$$a^2 - b^2 = c^2$$
$$4^2 - b^2 = 3^2$$
$$16 - b^2 = 9$$
$$b^2 = 7.$$

Since the foci are on the y-axis, the larger intercept, a, is used to find the denominator for y^2, giving the equation in standard form as

$$\frac{x^2}{7} + \frac{y^2}{16} = 1.$$

A graph of this ellipse is shown in Figure 12. ▶

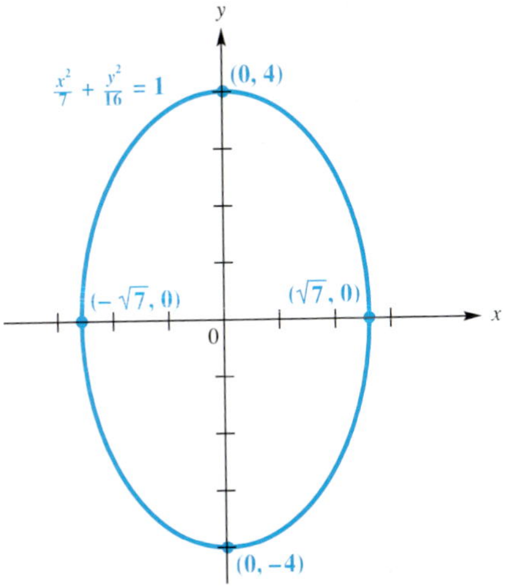

$$\frac{x^2}{7} + \frac{y^2}{16} = 1$$

(0, 4)

$(-\sqrt{7}, 0)$ $(\sqrt{7}, 0)$

0

(0, −4)

FIGURE 12

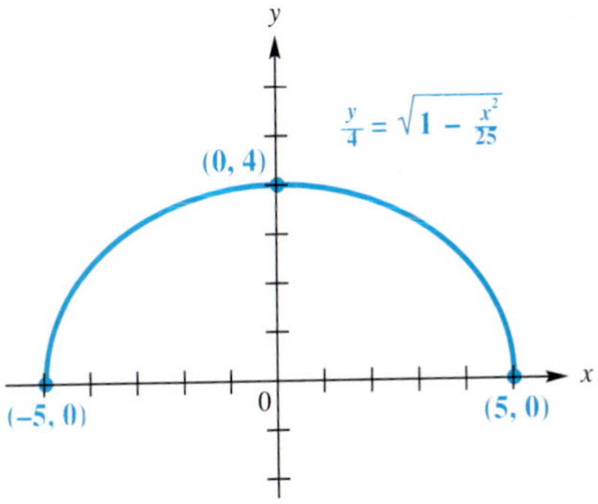

$$\frac{y}{4} = \sqrt{1 - \frac{x^2}{25}}$$

(0, 4)

(−5, 0)   0   (5, 0)

FIGURE 13

EXAMPLE 3
Graphing a half-ellipse

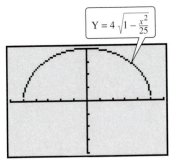

$Y = 4\sqrt{1 - \frac{x^2}{25}}$

[−6, 6] by [−4, 4]

Compare this graph with the one shown in Figure 13.

Graph $\dfrac{y}{4} = \sqrt{1 - \dfrac{x^2}{25}}$.

Square both sides to get

$$\frac{y^2}{16} = 1 - \frac{x^2}{25} \qquad \text{or} \qquad \frac{x^2}{25} + \frac{y^2}{16} = 1$$

as the equation of an ellipse with x-intercepts ± 5 and y-intercepts ± 4. Since $\sqrt{1 - (x^2/25)} \geq 0$, the only possible values of y are those making $y/4 \geq 0$, giving the half-ellipse shown in Figure 13. While the graph of the ellipse $x^2/25 + y^2/16 = 1$ is not the graph of a function, the half-ellipse in Figure 13 is the graph of a function. The domain of this function *is* the interval $[-5, 5]$, and the range is $[0, 4]$. ▶

Just as a circle need not have its center at the origin, an ellipse may also have its center translated away from the origin.

Ellipse Centered at (h, k)

An ellipse centered at (h, k) with horizontal major axis of length $2a$ has equation

$$\frac{(x - h)^2}{a^2} + \frac{(y - k)^2}{b^2} = 1.$$

There is a similar result for ellipses having a vertical major axis.

This result can be proven from the definition of an ellipse.

EXAMPLE 4

Graphing an ellipse translated away from the origin

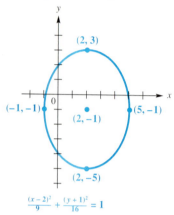

FIGURE 14

Graph $\dfrac{(x-2)^2}{9} + \dfrac{(y+1)^2}{16} = 1$.

The graph of this equation is an ellipse centered at $(2, -1)$. As mentioned earlier, ellipses always have $a > b$. For this ellipse, then, $a = 4$ and $b = 3$. Since $a = 4$ is associated with y^2, the vertices of the ellipse are on the vertical line through $(2, -1)$. Find the vertices by locating two points on the vertical line through $(2, -1)$, one 4 units up from $(2, -1)$ and one 4 units down. The vertices are $(2, 3)$ and $(2, -5)$. Locate two other points on the ellipse by locating points on a horizontal line through $(2, -1)$, one 3 units to the right and one 3 units to the left. Find additional points as needed. The final graph is shown in Figure 14. ▶

The **eccentricity**, e, of an ellipse is defined by

$$e = \frac{\sqrt{a^2 - b^2}}{a} = \frac{c}{a}.$$

Since $0 < c < a$, we have $0 < e < 1$. Nearly circular ellipses have e near 0, and flat ellipses have e near 1. See Figure 15.

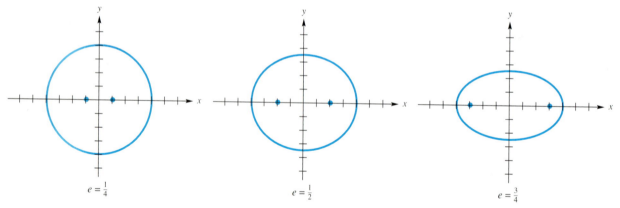

$e = \frac{1}{4}$ $e = \frac{1}{2}$ $e = \frac{3}{4}$

FIGURE 15

EXAMPLE 5

Applying an ellipse to the orbit of a planet

The orbit of the planet Mars is an ellipse with the sun at one focus. The eccentricity of the ellipse is .0935 and the closest distance that Mars comes to the sun is 128.5 million miles. (*Source: The World Almanac and Book of Facts, 1995.*) Find the maximum distance of Mars from the sun.

Figure 16 shows the orbit of Mars with the origin at the center of the ellipse and the sun at one focus. Mars is closest to the sun when Mars is at the right endpoint of the major axis and farthest from the sun when Mars is at the left endpoint. Therefore, the smallest distance is $a - c$ and the greatest distance is $a + c$.

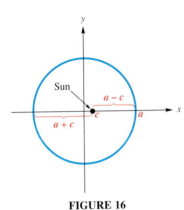

FIGURE 16

Since $a - c = 128.5$, $c = a - 128.5$. Therefore,

$$e = \frac{c}{a} = \frac{a - 128.5}{a} = .0935$$

$$a - 128.5 = .0935a$$

$$.9065a = 128.5$$

$$a \approx 141.8$$

$$c = 141.8 - 128.5 = 13.3$$

$$a + c = 141.8 + 13.3 = 155.1.$$

Thus, the maximum distance of Mars from the sun is about 155.1 million miles. ▶

CONNECTIONS When a ray of light or sound emanating from one focus of an ellipse bounces off the ellipse, it passes through the other focus. See the figure. This reflecting property is responsible for "whispering galleries," rooms with ellipsoidal ceilings in which a person whispering at one focus can be heard clearly at the other focus. In one medical application of the reflecting property, the lithotripter, patients with kidney stones are placed in a water bath in a tub with an elliptical cross section. Several hundred spark discharges are produced at one focus of the ellipse, with the kidney stone at the other focus. The discharges go through the water to the kidney stone, causing the stone to break up into small pieces that can readily be excreted from the body.

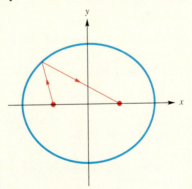

FOR DISCUSSION OR WRITING

1. If a lithotripter is based on the ellipse $\dfrac{x^2}{36} + \dfrac{y^2}{27} = 1$, determine how many units the kidney stone and the wave source must be placed from the center of the ellipse.

2. A patient is placed 12 units away from the source of shock waves. The lithotripter is based on an ellipse with a minor axis of 16 total units. Find an equation of an ellipse that would satisfy this situation.

10.2 Exercises ▼▼▼▼▼▼▼▼▼▼▼▼▼▼▼▼▼▼▼▼▼▼▼▼▼▼▼▼▼▼▼▼▼▼▼▼

1. Compare the definition of a circle given in Section 3.1 to the definition of an ellipse. How can a circle be considered a "special case" of an ellipse?

2. What are the x-intercepts of the ellipse $\dfrac{x^2}{a^2} + \dfrac{y^2}{b^2} = 1$ $(a^2 \neq b^2)$? What are its y-intercepts? What are the domain and the range of this relation?

Sketch the graph of each ellipse. Give the center, vertices, endpoints of the minor axis, and the foci for each figure. See Examples 1 and 4.

3. $\dfrac{x^2}{25} + \dfrac{y^2}{9} = 1$

4. $\dfrac{x^2}{16} + \dfrac{y^2}{25} = 1$

5. $\dfrac{x^2}{9} + y^2 = 1$

6. $\dfrac{x^2}{36} + \dfrac{y^2}{16} = 1$

7. $9x^2 + y^2 = 81$

8. $4x^2 + 16y^2 = 64$

9. $4x^2 + 25y^2 = 100$

10. $4x^2 + y^2 = 16$

11. $\dfrac{(x-2)^2}{25} + \dfrac{(y-1)^2}{4} = 1$

12. $\dfrac{(x+2)^2}{16} + \dfrac{(y+1)^2}{9} = 1$

13. $\dfrac{(x+3)^2}{16} + \dfrac{(y-2)^2}{36} = 1$

14. $\dfrac{(x-1)^2}{9} + \dfrac{(y+3)^2}{25} = 1$

Find an equation for each ellipse. See Example 2.

15. x-intercepts ±5; foci at $(-3, 0)$, $(3, 0)$

16. y-intercepts ±4; foci at $(0, -1)$, $(0, 1)$

17. Major axis with length 6; foci at $(0, 2)$, $(0, -2)$

18. Minor axis with length 4; foci at $(-5, 0)$, $(5, 0)$

19. Center at $(5, 2)$; minor axis vertical, with length 8; $c = 3$

20. Center at $(-3, 6)$; major axis vertical, with length 10; $c = 2$

21. Vertices at $(4, 9)$, $(4, 1)$; minor axis with length 6

22. Foci at $(-3, -3)$, $(7, -3)$; $(2, 1)$ on ellipse

23. Foci at $(0, -3)$, $(0, 3)$; $(8, 3)$ on ellipse

24. Foci at $(-4, 0)$, $(4, 0)$; sum of distances from foci to point on ellipse is 9 (*Hint:* Consider one of the vertices.)

25. Foci at $(0, 4)$, $(0, -4)$; sum of distances from foci to point on ellipse is 10

26. Eccentricity $\dfrac{1}{2}$; vertices at $(-4, 0)$, $(4, 0)$

27. Eccentricity $\dfrac{3}{4}$; foci at $(0, -2)$, $(0, 2)$

28. Eccentricity $\dfrac{2}{3}$; foci at $(0, -9)$, $(0, 9)$

Sketch the graph of each of the following. Identify any that are the graphs of functions. See Example 3.

29. $\dfrac{y}{2} = \sqrt{1 - \dfrac{x^2}{25}}$

30. $\dfrac{x}{4} = \sqrt{1 - \dfrac{y^2}{9}}$

31. $x = -\sqrt{1 - \dfrac{y^2}{64}}$

32. $y = -\sqrt{1 - \dfrac{x^2}{100}}$

📟 *Use a graphing calculator to graph each of the following ellipses.*

33. $\dfrac{x^2}{16} + \dfrac{y^2}{4} = 1$

34. $\dfrac{x^2}{4} + \dfrac{y^2}{25} = 1$

35. $\dfrac{(x-3)^2}{25} + \dfrac{y^2}{9} = 1$

36. $\dfrac{x^2}{36} + \dfrac{(y+4)^2}{4} = 1$

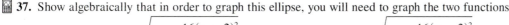

▼▼▼▼▼▼▼▼▼▼▼▼ **DISCOVERING CONNECTIONS** (Exercises 37–44) ▼▼▼▼▼▼▼▼▼▼▼▼

In Example 4 we show how to graph the ellipse

$$\frac{(x-2)^2}{9} + \frac{(y+1)^2}{16} = 1$$

in a traditional manner. In Exercises 37–44 we will investigate how to graph it using a graphing calculator, and in so doing see how its domain relates to another type of conic section.

37. Show algebraically that in order to graph this ellipse, you will need to graph the two functions

$$Y_1 = -1 + \sqrt{16 - \frac{16(x-2)^2}{9}} \quad \text{and} \quad Y_2 = -1 - \sqrt{16 - \frac{16(x-2)^2}{9}}.$$

Then graph them using a graphing calculator.

38. What is the domain of the ellipse in Example 4? See Figure 14.

39. The relations Y_1 and Y_2 in Exercise 37 are defined when their radicand is greater than or equal to 0. Write the inequality that would need to be solved in order to find the domain analytically.

40. Let y represent the expression in x found in the radicand. What conic section is the graph of this function?

41. Graph the function defined by the radicand with a graphing calculator in the window $[-10, 10]$ by $[-10, 20]$.

42. Use the graph in Exercise 41 to solve the inequality of Exercise 39.

43. Explain how the solution set from Exercise 42 confirms what was found earlier using the graph of the original ellipse.

44. Solve the inequality of Exercise 39 analytically, using a sign graph (as explained in Section 2.7).

---

45. Draftspeople often use the method shown in the sketch to draw an ellipse. Explain why the method works.

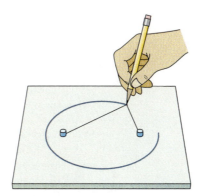

46. Explain how the method of Exercise 45 can be modified to draw a circle.

Solve each problem. See Example 5.

47. The coordinates in miles for the orbit of the artificial satellite Explorer VII can be described by the equation $(x^2/a^2) + (y^2/b^2) = 1$ where $a = 4465$ and $b = 4462$. Earth's center is located at one focus of its elliptical orbit. (*Sources:* Loh, W., *Dynamics and Thermodynamics of Planetary Entry,* Prentice Hall, Inc., Englewood Cliffs, New Jersey, 1963; Thomson, W., *Introduction to Space Dynamics,* John Wiley & Sons, Inc., New York, 1961.)

(a) Graph both the orbit of Explorer VII and of Earth on the same coordinate axes if the average radius of Earth is 3960 miles. Use the window $[-6750, 6750]$ by $[-4500, 4500]$.

(b) Determine the maximum and minimum heights of the satellite above Earth's surface.

48. Neptune and Pluto both have elliptical orbits with the sun at one focus. Neptune's orbit has a semimajor axis of $a = 30.1$ astronomical units (AU) with an eccentricity of $e = .009$ whereas Pluto's orbit has $a = 39.4$ and $e = .249$. (1 AU is equal to the average distance from Earth to the sun and is approximately 149,600,000 kilometers.) (*Source:* Zeilik, M., S. Gregory, and E. Smith, *Introductory Astronomy and Astrophysics,* Saunders College Publishers, 1992.)

(a) Position the sun at the origin and determine equations for each orbit.

(b) Graph both equations on the same coordinate axes. Use the window $[-60, 60]$ by $[-40, 40]$.

49. The maximum and minimum velocities in kilometers per second of a planet moving in an elliptical orbit can be calculated with the equations

$$v_{max} = \frac{2\pi a}{P}\sqrt{\frac{1+e}{1-e}}$$

and

$$v_{min} = \frac{2\pi a}{P}\sqrt{\frac{1-e}{1+e}}$$

where a is the semimajor axis in kilometers, P is its orbital period in seconds, and e is the eccentricity of the orbit. (*Source:* Zeilik, M., S. Gregory, and E. Smith, *Introductory Astronomy and Astrophysics,* Saunders College Publishers, 1992.)

(a) Calculate v_{max} and v_{min} for Earth if $a = 1.496 \times 10^8$ kilometers and $e = .0167$.

(b) If a planet has a circular orbit, what can be said about its orbital velocity?

(c) Kepler showed that the sun is located at a focus of a planet's elliptical orbit. He also showed that a planet's minimum velocity occurs when its distance from the sun is maximum and a planet's maximum velocity occurs when its distance from the sun is minimum. Where do the maximum and minimum velocities occur in an elliptical orbit?

50. A one-way road passes under an overpass in the form of half of an ellipse, 15 ft high at the center and 20 ft wide. Assuming a truck is 12 ft wide, what is the tallest truck that can pass under the overpass?

51. The Roman Coliseum is an ellipse with major axis 620 ft and minor axis 513 ft. Find the distance between the foci of this ellipse.

52. A formula for the approximate circumference of an ellipse is

$$C \approx 2\pi\sqrt{\frac{a^2 + b^2}{2}},$$

where a and b are the lengths as shown in the figure. Use this formula to find the approximate circumference of the Roman Coliseum (see Exercise 51).

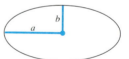

53. Halley's Comet has an elliptical orbit of eccentricity .9673 with the sun at one of the foci. The greatest distance of the comet from the sun is 3281 million miles. (*Source: The World Almanac and Book of Facts,* 1995.) Find the shortest distance between Halley's Comet and the sun.

54. The orbit of planet Earth is an ellipse with the sun at one focus. The distance between Earth and the sun ranges from 91.4 to 94.6 million miles. (*Source: The World Almanac and Book of Facts,* 1995.) Find the eccentricity of Earth's orbit.

55. Graph the ellipse $(x^2/16) + (y^2/12) = 1$ with a graphing calculator. The ellipse has foci $(-2, 0)$ and $(2, 0)$. Use the tracing feature to find the coordinates of several points on the ellipse. For each of these points P, verify that

$$[\text{Distance of } P \text{ from } (-2, 0)] + [\text{Distance of } P \text{ from } (2, 0)] = 8.$$

10.3 Hyperbolas ▼▼▼

An ellipse was defined as the set of all points in a plane the sum of whose distances from two fixed points is a constant. A *hyperbola* is defined similarly.

Definition of Hyperbola

A **hyperbola** is the set of all points in a plane the *difference* of whose distances from two fixed points is a constant. The two fixed points are called the **foci** of the hyperbola.

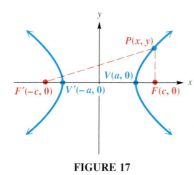

FIGURE 17

Suppose a hyperbola has center at the origin and foci at $F'(-c, 0)$ and $F(c, 0)$. The midpoint of the segment $F'F$ is the **center** of the hyperbola and the points $V'(-a, 0)$ and $V(a, 0)$ are the **vertices** of the hyperbola. The line segment $V'V$ is the **transverse axis** of the hyperbola. See Figure 17.

As with the ellipse,

$$d(V, F') - d(V, F) = (c + a) - (c - a) = 2a,$$

so the constant in the definition is $2a$, and

$$d(P, F') - d(P, F) = 2a.$$

The distance formula and algebraic manipulation similar to that used for finding an equation for an ellipse (see Exercise 55) produce the result

$$\frac{x^2}{a^2} - \frac{y^2}{c^2 - a^2} = 1.$$

Letting $b^2 = c^2 - a^2$ gives

$$\frac{x^2}{a^2} - \frac{y^2}{b^2} = 1$$

as an equation of the hyperbola in Figure 17.

Letting $y = 0$ shows that the x-intercepts are $\pm a$. If $x = 0$ the equation becomes

$$\frac{0^2}{a^2} - \frac{y^2}{b^2} = 1$$

$$-\frac{y^2}{b^2} = 1$$

$$y^2 = -b^2,$$

which has no real number solutions, showing that this hyperbola has no y-intercepts.

◖EXAMPLE 1
Graphing a hyperbola centered at the origin

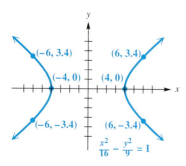

FIGURE 18

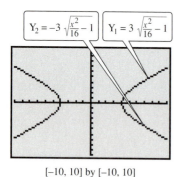

[–10, 10] by [–10, 10]

The graph of $\frac{x^2}{16} - \frac{y^2}{9} = 1$ is the union of the graphs of the two functions Y_1 and Y_2 shown above. Compare to Figure 18.

Graph $\dfrac{x^2}{16} - \dfrac{y^2}{9} = 1.$

This hyperbola has x-intercepts 4 and -4 and no y-intercepts. To sketch the graph, we can find other points that lie on the graph. For example, letting $x = 6$ gives

$$\frac{6^2}{16} - \frac{y^2}{9} = 1$$

$$-\frac{y^2}{9} = 1 - \frac{6^2}{16}$$

$$\frac{y^2}{9} = \frac{20}{16}$$

$$y^2 = \frac{180}{16} = \frac{45}{4}$$

$$y \approx \pm 3.4.$$

The graph includes the points $(6, 3.4)$ and $(6, -3.4)$. Also, letting $x = -6$ would still give $y \approx \pm 3.4$, with the points $(-6, 3.4)$ and $(-6, -3.4)$ also on the graph. These points, along with other points on the graph, were used to help sketch the final graph shown in Figure 18. ◗

Starting with the equation for a hyperbola $(x^2/a^2) - (y^2/b^2) = 1$ and solving for y, we have

$$\frac{x^2}{a^2} - 1 = \frac{y^2}{b^2}$$

$$\frac{x^2 - a^2}{a^2} = \frac{y^2}{b^2}$$

or

$$y = \pm \frac{b}{a}\sqrt{x^2 - a^2}. \qquad (*)$$

If x^2 is very large in comparison to a^2, the difference $x^2 - a^2$ would be very close to x^2. If this happens, then the points satisfying equation (*) above would be very close to one of the lines

$$y = \pm \frac{b}{a}x.$$

Thus, as $|x|$ gets larger and larger, the points of the hyperbola $(x^2/a^2) - (y^2/b^2) = 1$ come closer and closer to the lines $y = (\pm b/a)x$. These lines, called the **asymptotes** of the hyperbola, are very helpful for graphing the hyperbola. They make it possible to avoid having to plot points.

EXAMPLE 2
Using asymptotes to graph a hyperbola

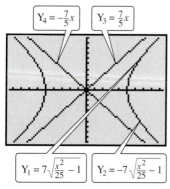

$Y_4 = -\frac{7}{5}x$ $Y_3 = \frac{7}{5}x$

$Y_1 = 7\sqrt{\frac{x^2}{25} - 1}$ $Y_2 = -7\sqrt{\frac{x^2}{25} - 1}$

[–9.4, 9.4] by [–10, 10]

The graphs of the hyperbola $\frac{x^2}{25} - \frac{y^2}{49} = 1$ and its asymptotes $Y = \pm\frac{7}{5}x$ are shown here. By tracing you can see how the branches approach the asymptotes.

Graph $\dfrac{x^2}{25} - \dfrac{y^2}{49} = 1$ and find the coordinates of the foci.

For this hyperbola, $a = 5$ and $b = 7$. With these values, $y = (\pm b/a)x$ becomes $y = (\pm 7/5)x$. If we choose $x = 5$, then $y = (\pm 7/5)(5) = \pm 7$. Choosing $x = -5$ also gives $y = \pm 7$. These four points, $(5, 7)$, $(5, -7)$, $(-5, 7)$, and $(-5, -7)$, are the corners of the rectangle shown in Figure 19. The extended diagonals of this rectangle are the asymptotes of the hyperbola. The hyperbola crosses the x-axis at 5 and -5, as shown in Figure 19. Find the foci by letting $c^2 = a^2 + b^2 = 25 + 49 = 74$ so that $c = \sqrt{74}$. Therefore, the foci are $(\sqrt{74}, 0)$ and $(-\sqrt{74}, 0)$. ▶

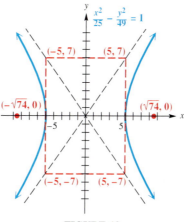

FIGURE 19

The rectangle used to graph the hyperbola in Example 2 is called the **fundamental rectangle.**

While $a > b$ for an ellipse, the examples above show that for hyperbolas, it is possible that $a > b$ or $a < b$; other examples would show that a might equal b, also. If the foci of a hyperbola are on the y-axis, the equation of the hyperbola is of the form

$$\frac{y^2}{a^2} - \frac{x^2}{b^2} = 1, \quad \text{with asymptotes} \quad y = \pm\frac{a}{b}x.$$

CAUTION   If the foci of the hyperbola are on the x-axis, we found that the asymptotes have equations $y = \pm(b/a)x$, while foci on the y-axis lead to asymptotes $y = \pm(a/b)x$. There is an obvious chance for confusion here; to avoid mistakes write the equation of the hyperbola in either the form

$$\frac{x^2}{a^2} - \frac{y^2}{b^2} = 1 \qquad \text{or} \qquad \frac{y^2}{a^2} - \frac{x^2}{b^2} = 1,$$

and replace 1 with 0. Solving the resulting equation for y produces the proper equations for the asymptotes. (The reason why this process works is explained in more advanced courses.)

The basic information on hyperbolas is summarized as follows.

Standard Equations for Hyperbolas

The hyperbola with center at the origin and equation

$$\frac{x^2}{a^2} - \frac{y^2}{b^2} = 1$$

has vertices $(\pm a, 0)$, asymptotes $y = \pm (b/a)x$, and foci $(\pm c, 0)$, where $c^2 = a^2 + b^2$.

The hyperbola with center at the origin and equation

$$\frac{y^2}{a^2} - \frac{x^2}{b^2} = 1$$

has vertices $(0, \pm a)$, asymptotes $y = \pm (a/b)x$, and foci $(0, \pm c)$, where $c^2 = a^2 + b^2$.

◀ **EXAMPLE 3**
Using the fundamental rectangle to graph a hyperbola

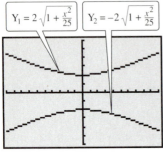

[−10, 10] by [−6, 6]

The graph of $25y^2 - 4x^2 = 100$ is the union of the graphs of the two functions Y_1 and Y_2 shown above. Compare to Figure 20.

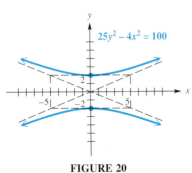

FIGURE 20

Graph $25y^2 - 4x^2 = 100$.

Divide each side by 100 to get

$$\frac{y^2}{4} - \frac{x^2}{25} = 1.$$

This hyperbola is centered at the origin, has foci on the y-axis, and has y-intercepts 2 and -2. To find the equations of the asymptotes, replace 1 with 0.

$$\frac{y^2}{4} - \frac{x^2}{25} = 0$$

$$\frac{y^2}{4} = \frac{x^2}{25}$$

$$y^2 = \frac{4x^2}{25}$$

$$y = \pm \frac{2}{5}x$$

To graph the asymptotes, use the points $(5, 2)$, $(5, -2)$, $(-5, 2)$, and $(-5, -2)$ that determine the fundamental rectangle shown in Figure 20. The diagonals of this rectangle are the asymptotes for the graph, as shown in Figure 20. ▶

In each of the graphs of hyperbolas considered so far, the center is the origin and the asymptotes pass through the origin. This feature holds in general; the asymptotes of *any* hyperbola pass through the center of the hyperbola. The next example involves a hyperbola with its center translated away from the origin.

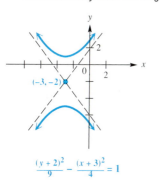

$$\frac{(y + 2)^2}{9} - \frac{(x + 3)^2}{4} = 1$$

FIGURE 21

EXAMPLE 4
Graphing a hyperbola
translated away from the origin

Graph $\dfrac{(y + 2)^2}{9} - \dfrac{(x + 3)^2}{4} = 1.$

This equation represents a hyperbola centered at $(-3, -2)$. For this vertical hyperbola, $a = 3$ and $b = 2$. Locate the y-values of the vertices by taking the y-value of the center, -2, and adding and subtracting 3. The x-values of the vertices are -3. Thus, the vertices are at $(-3, 1)$ and $(-3, -5)$. The asymptotes have slopes $\pm 3/2$ and pass through the center $(-3, -2)$. The equations of the asymptotes, $y + 2 = \pm(3/2)(x + 3)$, can be found either by using the point-slope form of the equation of a line or by replacing 1 by 0 in the equation of the hyperbola as was done in Example 3. The completed graph appears in Figure 21. ◗

Our next example shows an equation whose graph is only half of a hyperbola.

EXAMPLE 5
Graphing a half-hyperbola

Graph $x = -\sqrt{1 + 4y^2}.$
  Squaring both sides gives

$$x^2 = 1 + 4y^2$$

or $$x^2 - 4y^2 = 1.$$

To find the asymptotes, rewrite 4 as $1/(1/4)$ to change this equation into

$$x^2 - \frac{y^2}{1/4} = 0$$

or $$\frac{1}{4}x^2 = y^2,$$

giving $$y = \pm\frac{1}{2}x.$$

Since the given equation $x = -\sqrt{1 + 4y^2}$ restricts x to negative values, the graph is the left branch of the hyperbola, as shown in Figure 22. ◗

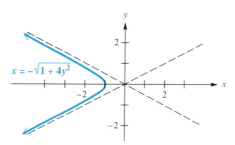

$$x = -\sqrt{1 + 4y^2}$$

FIGURE 22

The **eccentricity,** e, of a hyperbola is defined by

$$e = \frac{\sqrt{a^2 + b^2}}{a} = \frac{c}{a}.$$

Since $c > a$, we have $e > 1$. Narrow hyperbolas have e near 1 and wide hyperbolas have large e. See Figure 23.

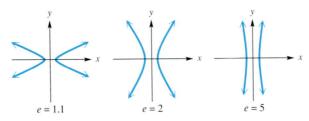

$e = 1.1$ $e = 2$ $e = 5$

FIGURE 23

EXAMPLE 6
Finding the equation of a hyperbola

Find the equation of the hyperbola with eccentricity 2 and foci at $(-9, 5)$ and $(-3, 5)$.

Since the foci have the same y-coordinate, the line through them, and therefore the hyperbola, is horizontal. The center of the hyperbola is halfway between the two foci at $(-6, 5)$. The distance from each focus to the center is $c = 3$. Since $e = c/a$,

$$a = \frac{c}{e} = \frac{3}{2} \quad \text{and} \quad a^2 = \frac{9}{4}$$

$$b^2 = c^2 - a^2 = 9 - \frac{9}{4} = \frac{27}{4}.$$

Therefore, the equation of the hyperbola is

$$\frac{x^2}{9/4} - \frac{y^2}{27/4} = 1$$

or

$$\frac{4x^2}{9} - \frac{4y^2}{27} = 1. \quad \blacktriangleright$$

CONNECTIONS Ships and planes often use a location-finding system called LORAN. With this system, a radio transmitter at M in the figure sends out a series of pulses. When each pulse is received at transmitter S, it then sends out a pulse. A ship at P receives pulses from both M and S. A receiver on the ship measures the difference in the arrival times of the pulses. The navigator then consults a special map showing hyperbolas that correspond to the differences in arrival times (which give the distances d_1 and d_2 in the figure). In this way the ship can be located as lying on a branch of a particular hyperbola.

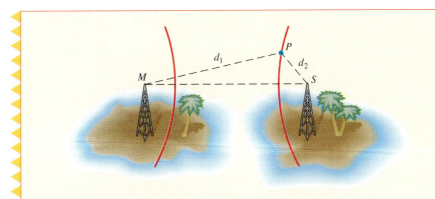

FOR DISCUSSION OR WRITING

Suppose in the figure $d_1 = 80$ miles, $d_2 = 30$ miles, and the distance between the transmitters is 100 miles. Use the definition of the hyperbola to find an equation of the hyperbola that the ship is located on.

10.3 Exercises ▼▼▼▼▼▼▼▼▼▼▼▼▼▼▼▼▼▼▼▼▼▼▼▼▼▼▼▼▼▼▼▼

Based on the material in this section and the previous one, match the equation with the correct graph.

1. $\dfrac{x^2}{25} + \dfrac{y^2}{9} = 1$

2. $\dfrac{x^2}{9} + \dfrac{y^2}{25} = 1$

3. $\dfrac{x^2}{9} - \dfrac{y^2}{25} = 1$

4. $\dfrac{x^2}{25} - \dfrac{y^2}{9} = 1$

A.

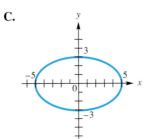

B.

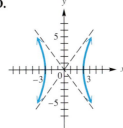

C.

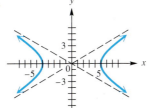

D.

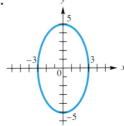

Sketch the graph of the hyperbola. Give the center, vertices, foci, and equations of the asymptotes for each figure. See Examples 1–4.

5. $\dfrac{x^2}{16} - \dfrac{y^2}{9} = 1$
   6. $\dfrac{x^2}{25} - \dfrac{y^2}{144} = 1$
   7. $\dfrac{y^2}{25} - \dfrac{x^2}{49} = 1$
   8. $\dfrac{y^2}{64} - \dfrac{x^2}{4} = 1$

9. $x^2 - y^2 = 9$
   10. $x^2 - 4y^2 = 64$
   11. $9x^2 - 25y^2 = 225$
   12. $25x^2 - 4y^2 = -100$

13. $4x^2 - y^2 = -16$
   14. $\dfrac{x^2}{4} - y^2 = 4$
   15. $9x^2 - 4y^2 = 1$
   16. $25y^2 - 9x^2 = 1$

17. $\dfrac{(y - 7)^2}{36} - \dfrac{(x - 4)^2}{64} = 1$
   18. $\dfrac{(x + 6)^2}{144} - \dfrac{(y + 4)^2}{81} = 1$
   19. $\dfrac{(x + 3)^2}{16} - \dfrac{(y - 2)^2}{9} = 1$

20. $\dfrac{(y + 5)^2}{4} - \dfrac{(x - 1)^2}{16} = 1$
   21. $16(x + 5)^2 - (y - 3)^2 = 1$
   22. $4(x + 9)^2 - 25(y + 6)^2 = 100$

Sketch the graph of each of the following. Identify any that are the graphs of functions. See Example 5.

23. $\dfrac{y}{3} = \sqrt{1 + \dfrac{x^2}{16}}$
   24. $\dfrac{x}{3} = -\sqrt{1 + \dfrac{y^2}{25}}$
   25. $5x = -\sqrt{1 + 4y^2}$
   26. $3y = \sqrt{4x^2 - 16}$

Find equations for each of the following hyperbolas. See Example 6.

27. x-intercepts ± 4; foci at $(-5, 0)$, $(5, 0)$

28. y-intercepts ± 9; foci at $(0, -15)$, $(0, 15)$

29. Vertices at $(0, 6)$, $(0, -6)$; asymptotes $y = \pm(1/2)x$

30. Vertices at $(-10, 0)$, $(10, 0)$; asymptotes $y = \pm 5x$

31. Vertices at $(-3, 0)$, $(3, 0)$; passing through $(6, 1)$

32. Vertices at $(0, 5)$, $(0, -5)$; passing through $(3, 10)$

33. Foci at $(0, \sqrt{13})$, $(0, -\sqrt{13})$; asymptotes $y = \pm 5x$

34. Foci at $(-\sqrt{45}, 0)$, $(\sqrt{45}, 0)$; asymptotes $y = \pm 2x$

35. Vertices at $(4, 5)$, $(4, 1)$; asymptotes $y - 3 = \pm 7(x - 4)$

36. Vertices at $(5, -2)$, $(1, -2)$; asymptotes $y + 2 = \pm(3/2)(x - 3)$

37. Center at $(1, -2)$; focus at $(4, -2)$; vertex at $(3, -2)$

38. Center at $(9, -7)$; focus at $(9, 3)$; vertex at $(9, -1)$

39. Eccentricity 3; center at $(0, 0)$; vertex at $(0, 7)$

40. Center at $(8, 7)$; focus at $(13, 7)$; eccentricity $5/3$

41. Vertices at $(-2, 10)$, $(-2, 2)$; eccentricity $5/4$

42. Foci at $(9, 2)$, $(-11, 2)$; eccentricity $25/9$

🖩 *Use a graphing calculator to graph each hyperbola.*

43. $\dfrac{x^2}{4} - \dfrac{y^2}{16} = 1$
   44. $\dfrac{x^2}{25} - \dfrac{y^2}{49} = 1$

45. $4y^2 - 36x^2 = 144$
   46. $y^2 - 9x^2 = 9$

▼▼▼▼▼▼▼▼▼▼▼▼▼ **DISCOVERING CONNECTIONS** (Exercises 47–52) ▼▼▼▼▼▼▼▼▼▼▼▼▼

From the discussion in this section, we know that the graph of $(x^2/4) - y^2 = 1$ is a hyperbola. We know that the graph of this hyperbola approaches its asymptotes as x gets larger and larger. Work Exercises 47–52 in order to see the relationship between the hyperbola and one of its asymptotes.

47. Solve $\dfrac{x^2}{4} - y^2 = 1$ for y, and choose the positive square root.

48. Find the equation of the asymptote with positive slope.

49. Use a calculator to evaluate the y-coordinate of the point where $x = 50$ on the graph of the portion of the hyperbola represented by the equation obtained in Exercise 47. Round your answer to the nearest hundredth.

50. Find the y-coordinate of the point where $x = 50$ on the graph of the asymptote found in Exercise 48.

51. Compare your results in Exercises 49 and 50. How do they support the following statement?

When $x = 50$, the graph of the function defined by the equation found in Exercise 47 lies *below* the graph of the asymptote found in Exercise 48.

52. What do you think will happen if we choose x-values larger than 50?

———————————————————

Solve the following problems.

53. 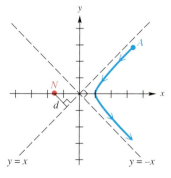 In 1911 Ernest Rutherford discovered the basic structure of the atom by "shooting" positively charged alpha particles with a speed of 10^7 m/sec at a piece of gold foil 6×10^{-7} m thick. Only a small percentage of the alpha particles struck a gold nucleus head-on and were deflected directly back toward their source. The rest of the particles often followed a hyperbolic trajectory because they were repelled by positively charged gold nuclei. As a result of this famous experiment, Rutherford proposed that the atom was composed of mostly empty space with a small and dense nucleus. The figure shows an alpha particle A initially approaching a gold nucleus N and being deflected at an angle $\theta = 90°$. N is located at a focus of the hyperbola and the trajectory of A passes through a vertex of the hyperbola. (*Source:* Semat, H. and J. Albright, *Introduction to Atomic and Nuclear Physics,* Holt, Rinehart, and Winston, Inc., 1972.)

(a) Determine the equation of the trajectory of the alpha particle if $d = 5 \times 10^{-14}$ m.

(b) What was the minimum distance between the centers of the alpha particle and the gold nucleus?

54. Microphones are placed at points $(-c, 0)$ and $(c, 0)$. An explosion occurs at point $P(x, y)$ having positive x-coordinate. See the figure. The sound is detected at the closer microphone t sec before being detected at the farther microphone. Assume that sound travels at a speed of 330 m per sec, and show that P must be on the hyperbola

$$\frac{x^2}{330^2 t^2} - \frac{y^2}{4c^2 - 330^2 t^2} = \frac{1}{4}.$$

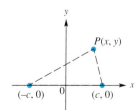

55. Suppose a hyperbola has center at the origin, foci at $F'(-c, 0)$ and $F(c, 0)$, and the value $d(P, F') - d(P, F) = 2a$. Let $b^2 = c^2 - a^2$, and show that an equation of the hyperbola is

$$\frac{x^2}{a^2} - \frac{y^2}{b^2} = 1.$$

56. Graph the hyperbola $\dfrac{x^2}{4} - \dfrac{y^2}{12} = 1$ with a graphing calculator. The hyperbola has foci $(-4, 0)$ and $(4, 0)$. Use the tracing feature to find the coordinates of several points on the right half of the hyperbola. For each

of these points P, verify that

[Distance of P from $(-4, 0)$]
$- $ [Distance of P from $(4, 0)$] $= 4$.

Find the coordinates of several points on the left half of the hyperbola. For each of these points P, verify that

[Distance of P from $(4, 0)$]
$- $ [Distance of P from $(-4, 0)$] $= 4$.

57. Duplicate the graph in Figure 21 on a graphing calculator.

58. Duplicate the graph in Figure 22 on a graphing calculator.

10.4 Conic Sections ▼▼▼

The graphs of parabolas, circles, hyperbolas, and ellipses are called **conic sections** since each graph can be obtained by cutting a cone with a plane as suggested by Figure 1 at the beginning of the chapter. It turns out that all conic sections of the types presented in this chapter have equations of the form

$$Ax^2 + Cy^2 + Dx + Ey + F = 0,$$

where either A or C must be nonzero. The graphs of the conic sections are summarized in the following chart. Ellipses and hyperbolas having centers not at the origin can be shown in much the same way as we show circles and parabolas. Following the chart, the special characteristics of the equations of each of the conic sections are summarized.

Equation	Graph	Description	Identification
$y - k = a(x - h)^2$		Opens upward if $a > 0$, downward if $a < 0$. Vertex is at (h, k).	x^2 term y is not squared.
$x - h = a(y - k)^2$		Opens to right if $a > 0$, to left if $a < 0$. Vertex is at (h, k).	y^2 term x is not squared.

Equation	Graph	Description	Identification
$(x - h)^2 + (y - k)^2 = r^2$		Center is at (h, k), radius is r.	x^2 and y^2 terms have the same positive coefficient.
$\dfrac{x^2}{a^2} + \dfrac{y^2}{b^2} = 1 \quad (a > b)$		x-intercepts are a and $-a$. y-intercepts are b and $-b$.	x^2 and y^2 terms have different positive coefficients.
$\dfrac{x^2}{b^2} + \dfrac{y^2}{a^2} = 1 \quad (a > b)$		x-intercepts are b and $-b$. y-intercepts are a and $-a$.	x^2 and y^2 terms have different positive coefficients.
$\dfrac{x^2}{a^2} - \dfrac{y^2}{b^2} = 1$		x-intercepts are a and $-a$. Asymptotes found from (a, b), $(a, -b)$, $(-a, -b)$, and $(-a, b)$.	x^2 has a positive coefficient. y^2 has a negative coefficient.
$\dfrac{y^2}{a^2} - \dfrac{x^2}{b^2} = 1$		y-intercepts are a and $-a$. Asymptotes found from (b, a), $(b, -a)$, $(-b, a)$, and $(-b, -a)$.	y^2 has a positive coefficient. x^2 has a negative coefficient.

Equations of Conic Sections

Conic Section	Characteristic	Example
Parabola	Either $A = 0$ or $C = 0$, but not both.	$x^2 = y + 4$ $(y - 2)^2 = -(x + 3)$
Circle	$A = C \neq 0$	$x^2 + y^2 = 16$
Ellipse	$A \neq C$, $AC > 0$	$\dfrac{x^2}{16} + \dfrac{y^2}{25} = 1$
Hyperbola	$AC < 0$	$x^2 - y^2 = 1$

In order to recognize the type of graph that a given conic section has, it is sometimes necessary to transform the equation into a more familiar form, as shown in the next examples.

EXAMPLE 1
Determining the type of a conic section from its equation

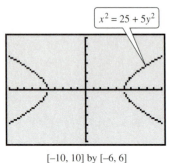

[−10, 10] by [−6, 6]

Compare to Figure 24.

Decide on the type of conic section represented by each of the following equations, and sketch each graph.

(a) $x^2 = 25 + 5y^2$

Rewriting the equation as

$$x^2 - 5y^2 = 25$$

or

$$\frac{x^2}{25} - \frac{y^2}{5} = 1$$

shows that the equation represents a hyperbola centered at the origin, with asymptotes

$$\frac{x^2}{25} - \frac{y^2}{5} = 0,$$

or

$$y = \frac{\pm\sqrt{5}}{5}x.$$

The x-intercepts are ± 5; the graph is shown in Figure 24.

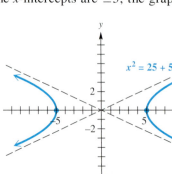

FIGURE 24

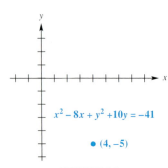

FIGURE 25

(b) $x^2 - 8x + y^2 + 10y = -41$

Complete the square on both x and y, as follows:

$$(x^2 - 8x + 16 - 16) + (y^2 + 10y + 25 - 25) = -41$$
$$(x - 4)^2 + (y + 5)^2 = 16 + 25 - 41$$
$$(x - 4)^2 + (y + 5)^2 = 0.$$

The result shows that the equation is that of a circle of radius 0; that is the point $(4, -5)$. See Figure 25. Had a negative number been obtained on the right (instead of 0), the equation would have no solution at all, and there would be no graph.

(c) $4x^2 - 16x + 9y^2 + 54y = -61$

Since the coefficients of the x^2 and y^2 terms are unequal and both positive, this equation might represent an ellipse. (It might also represent a single point or no points at all.) To find out, complete the square on x and y.

$$4(x^2 - 4x \quad) + 9(y^2 + 6y \quad) = -61$$
$$4(x^2 - 4x + 4 - 4) + 9(y^2 + 6y + 9 - 9) = -61$$
$$4(x^2 - 4x + 4) - 16 + 9(y^2 + 6y + 9) - 81 = -61$$
$$4(x - 2)^2 + 9(y + 3)^2 = 36$$
$$\frac{(x - 2)^2}{9} + \frac{(y + 3)^2}{4} = 1$$

This equation represents an ellipse having center at $(2, -3)$ and graph as shown in Figure 26.

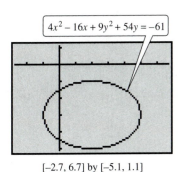

[–2.7, 6.7] by [–5.1, 1.1]

Compare to Figure 26.

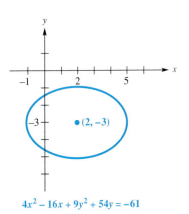

$4x^2 - 16x + 9y^2 + 54y = -61$

FIGURE 26

(d) $x^2 - 6x + 8y - 7 = 0$

Since only one variable is squared (x, and not y), the equation represents a parabola. Rearrange the terms to get the term with y (the variable that is not squared) alone on one side. Then complete the square on the other side of the equation.

$$8y = -x^2 + 6x + 7$$
$$8y = -(x^2 - 6x \quad) + 7$$
$$8y = -(x^2 - 6x + 9) + 7 + 9$$
$$8y = -(x - 3)^2 + 16$$
$$y = -\frac{1}{8}(x - 3)^2 + 2 \qquad \text{Multiply both sides by } \tfrac{1}{8}.$$
$$y - 2 = -\frac{1}{8}(x - 3)^2 \qquad \text{Subtract 2 from both sides.}$$

The parabola has vertex at (3, 2), and opens downward, as shown in Figure 27. ▶

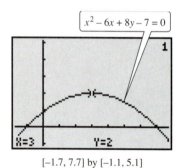

[−1.7, 7.7] by [−1.1, 5.1]

Compare to Figure 27.

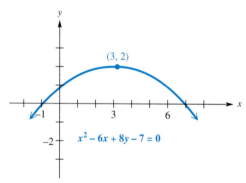

FIGURE 27

CAUTION   The next example is designed to serve as a warning about a very common error.

EXAMPLE 2
Writing the equation of a hyperbola in standard form

Graph $4y^2 - 16y - 9x^2 + 18x = -43$.

Complete the square on x and on y.

$$4(y^2 - 4y \quad) - 9(x^2 - 2x \quad) = -43$$
$$4(y^2 - 4y + 4) - 9(x^2 - 2x + 1) = -43 + 16 - 9$$
$$4(y - 2)^2 - 9(x - 1)^2 = -36$$

Because of the -36, it is very tempting to say that this equation does not have a graph. However, the minus sign in the middle on the left shows that the graph is that of a hyperbola. Dividing through by -36 and rearranging terms gives

$$\frac{(x-1)^2}{4} - \frac{(y-2)^2}{9} = 1,$$

the hyperbola centered at $(1, 2)$, with graph as shown in Figure 28. ▶

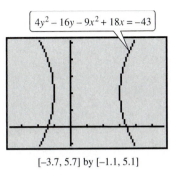

$4y^2 - 16y - 9x^2 + 18x = -43$

$[-3.7, 5.7]$ by $[-1.1, 5.1]$

Compare to Figure 28.

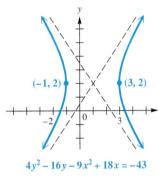

$4y^2 - 16y - 9x^2 + 18x = -43$

FIGURE 28

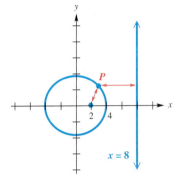

$x = 8$

FIGURE 29

In Section 10.1, the parabola was defined as the set of points in a plane whose distance from a fixed point (focus) equals their distance from a fixed line (directrix). Actually, this definition can be generalized to apply also to the ellipse and the hyperbola. Figure 29 shows an ellipse with $a = 4$, $c = 2$, and $e = 1/2$. The line $x = 8$ is shown also. For any point P on the ellipse,

$$[\text{Distance of } P \text{ from the focus}] = \frac{1}{2}[\text{Distance of } P \text{ from the line}].$$

Figure 30 shows a hyperbola with $a = 2$, $c = 4$, and $e = 2$, along with the line $x = 1$. For any point P on the hyperbola,

$$[\text{Distance of } P \text{ from the focus}] = 2[\text{Distance of } P \text{ from the line}].$$

A parabola is said to have eccentricity 1. The following geometric characterization applies to all conic sections.

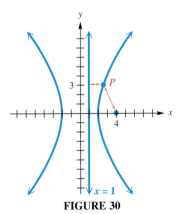

$x = 1$

FIGURE 30

Geometric Characterization of Conic Sections

Given a fixed point F, a fixed line L, and a positive number e, the set of all points P in the plane such that

$$[\text{Distance of } P \text{ from } F] = e \cdot [\text{Distance of } P \text{ from } L]$$

is a conic section of eccentricity e. The conic section is a parabola when $e = 1$, an ellipse when $e < 1$, and a hyperbola when $e > 1$.

The circle, covered in Section 3.1, has eccentricity $e = 0$. (We can think of a circle as an ellipse with $a = b$.)

10.4 Exercises ▼▼▼▼▼▼▼▼▼▼▼▼▼▼▼▼▼▼▼▼▼▼▼▼▼▼▼▼▼▼▼▼▼▼▼

In Exercises 1–12, the equation of a conic section is given in a familiar form. Identify the type of graph that the equation has, without actually graphing.

1. $x^2 + y^2 = 144$

2. $(x - 2)^2 + (y + 3)^2 = 25$

3. $y = 2x^2 + 3x - 4$

4. $x = 3y^2 + 5y - 6$

5. $x - 1 = -3(y - 4)^2$

6. $\dfrac{x^2}{25} + \dfrac{y^2}{36} = 1$

7. $\dfrac{x^2}{49} + \dfrac{y^2}{100} = 1$

8. $x^2 - y^2 = 1$

9. $\dfrac{x^2}{4} - \dfrac{y^2}{16} = 1$

10. $\dfrac{(x + 2)^2}{9} + \dfrac{(y - 4)^2}{16} = 1$

11. $\dfrac{x^2}{25} - \dfrac{y^2}{25} = 1$

12. $y + 7 = 4(x + 3)^2$

For each equation that has a graph, identify the type of graph. It may be necessary to transform the equation. See Examples 1 and 2.

13. $\dfrac{x^2}{4} = 1 - \dfrac{y^2}{9}$

14. $\dfrac{x^2}{4} = 1 + \dfrac{y^2}{9}$

15. $\dfrac{x^2}{4} + \dfrac{y^2}{4} = 1$

16. $\dfrac{x^2}{4} + \dfrac{y^2}{4} = -1$

17. $x^2 = 25 + y^2$

18. $x^2 = 25 - y^2$

19. $9x^2 + 36y^2 = 36$

20. $x^2 = 4y - 8$

21. $\dfrac{(x + 3)^2}{16} + \dfrac{(y - 2)^2}{16} = 1$

22. $\dfrac{(x - 4)^2}{8} + \dfrac{(y + 1)^2}{2} = 0$

23. $y^2 - 4y = x + 4$

24. $11 - 3x = 2y^2 - 8y$

25. $(x + 7)^2 + (y - 5)^2 + 4 = 0$

26. $4(x - 3)^2 + 3(y + 4)^2 = 0$

27. $3x^2 + 6x + 3y^2 - 12y = 12$

28. $2x^2 - 8x + 2y^2 + 20y = 12$

29. $x^2 - 6x + y = 0$

30. $x - 4y^2 - 8y = 0$

31. $4x^2 - 8x - y^2 - 6y = 6$

32. $x^2 + 2x = -4y$

33. $4x^2 - 8x + 9y^2 + 54y = -84$

34. $3x^2 + 12x + 3y^2 = -11$

35. $6x^2 - 12x + 6y^2 - 18y + 25 = 0$

36. $4x^2 - 24x + 5y^2 + 10y + 41 = 0$

37. Identify the type of conic section consisting of the set of all points in the plane for which the sum of the distances from the points $(5, 0)$ and $(-5, 0)$ is 14.

38. Identify the type of conic section consisting of the set of all points in the plane for which the absolute value of the difference of the distances from the points $(3, 0)$ and $(-3, 0)$ is 2.

39. Identify the type of conic section consisting of the set of all points in the plane for which the distance from the point $(3, 0)$ is one and one-half times the distance from the line $x = 4/3$.

40. Identify the type of conic section consisting of the set of all points in the plane for which the distance from the point $(2, 0)$ is one-third of the distance from the line $x = 10$.

In Exercises 41–46, find the eccentricity of the conic section. The point shown on the x-axis is a focus and the line shown is a directrix.

41.

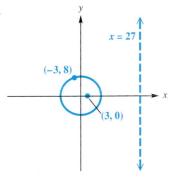

42.

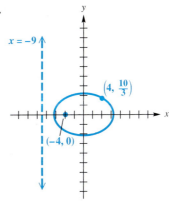

43.

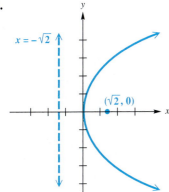

44.

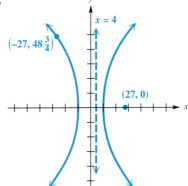

45.

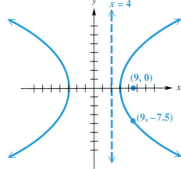

46.

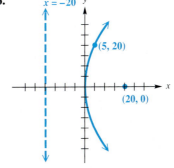

47. Match each equation with its calculator-generated graph. Do this first wihout actually using your calculator. Then check your answer by generating a calculator graph of your own. (Every window has Xscl = Yscl = 1.)

(a) $y = x^2$

(b) $x = y^2$

(c) $x = 2(y + 3)^2 - 4$

(d) $y = 2(x + 3)^2 - 4$

(e) $y = -\dfrac{1}{3}x^2$

(f) $x = -\dfrac{1}{3}y^2$

(g) $x^2 + y^2 = 25$

(h) $(x - 3)^2 + (y + 4)^2 = 25$

(i) $(x + 3)^2 + (y - 4)^2 = 25$

(j) $x^2 + y^2 = -4$

A.

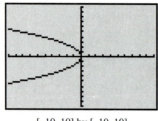

[-10, 10] by [-10, 10]

B.

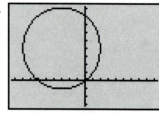

[-9.4, 9.4] by [-3.1, 9.1]

C.

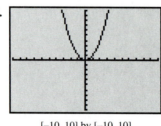

[-10, 10] by [-10, 10]

D.

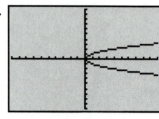

[-10, 10] by [-10, 10]

E.

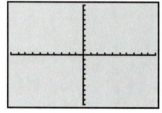

[-10, 10] by [-10, 10]

F.

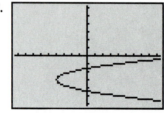

[-9.4, 9.4] by [-6.2, 6.2]

G.

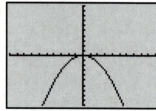

[-10, 10] by [-10, 10]

H.

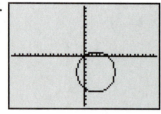

[-18.4, 18.4] by [-12.2, 12.2]

I.

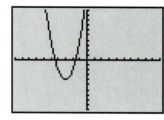

[-10, 10] by [-10, 10]

J.

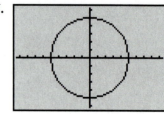

[-9.4, 9.4] by [-6.2, 6.2]

48. Match each equation with its calculator-generated graph. Do this first without actually using your calculator. Then check your answer by generating a calculator graph of your own. (Every window has Xscl = Yscl = 1.)

(a) $\dfrac{y^2}{16} + \dfrac{x^2}{4} = 1$

(b) $\dfrac{x^2}{16} + \dfrac{y^2}{4} = 1$

(c) $\dfrac{x^2}{64} - \dfrac{y^2}{16} = 1$

(d) $\dfrac{y^2}{4} - \dfrac{x^2}{16} = 1$

(e) $\dfrac{(y-4)^2}{25} + \dfrac{(x+2)^2}{9} = 1$

(f) $\dfrac{(y+4)^2}{25} + \dfrac{(x-2)^2}{9} = 1$

(g) $\dfrac{(x+2)^2}{9} - \dfrac{(y-4)^2}{25} = 1$

(h) $\dfrac{(x-2)^2}{9} - \dfrac{(y+4)^2}{25} = 1$

(i) $36x^2 + 4y^2 = 144$

(j) $9x^2 - 4y^2 = 36$

A.

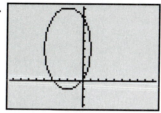

[−9.4, 9.4] by [−3.1, 9.4]

B.

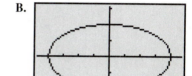

[−4.7, 4.7] by [−3.1, 3.1]

C.

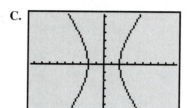

[−9.4, 9.4] by [−6.2, 6.2]

D.

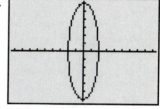

[−9.4, 9.4] by [−6.2, 6.2]

E.

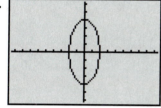

[−9.4, 9.4] by [−6.2, 6.2]

F.

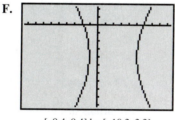

[−9.4, 9.4] by [−10.2, 2.2]

G.

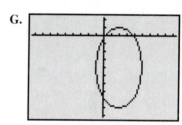

[−9.4, 9.4] by [−10.2, 2.2]

H.

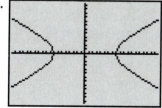

[−18.8, 18.8] by [−12.4, 12.4]

I.

[−9.4, 9.4] by [−2.2, 10.2]

J.

[−9.4, 9.4] by [−6.2, 6.2]

When a satellite is near Earth, its orbital trajectory may trace out a hyperbola, parabola, or ellipse. The type of trajectory depends on the satellite's velocity V in meters per second. It will be hyperbolic if $V > k/\sqrt{D}$, parabolic if $V = k/\sqrt{D}$, and elliptic if $V < k/\sqrt{D}$ where $k = 2.82 \times 10^7$ is a constant and D is the distance in meters from the satellite to the center of Earth. (Sources: Loh, W., Dynamics and Thermodynamics of Planetary Entry, Prentice Hall, Inc., Englewood Cliffs, New Jersey, 1963; Thomson, W., Introduction to Space Dynamics, John Wiley & Sons, Inc., New York, 1961.)

49. When the artificial satellite Explorer IV was at a maximum distance D of 42.5×10^6 m from Earth's center, it had a velocity V of 2090 m/sec. Determine the shape of its trajectory.

50. If a satellite is scheduled to leave Earth's gravitational influence, its velocity must be increased so that its trajectory changes from elliptic to hyperbolic. Determine the minimum increase in velocity necessary for Explorer IV to escape Earth's gravitational influence when $D = 42.5 \times 10^6$ m.

51. Explain why it is easier to change a satellite's trajectory from an ellipse to a hyperbola when D is maximum rather than minimum.

52. If $Ax^2 + Cy^2 + Dx + Ey + F = 0$ is the general equation of an ellipse, find its center point by completing the square.

53. Graph the ellipse $\dfrac{x^2}{16} + \dfrac{y^2}{12} = 1$ with a graphing calculator. Use tracing to find the coordinates of several points on the ellipse. For each of these points P, verify that

[Distance of P from (2, 0)]

$$= \frac{1}{2} \text{[Distance of } P \text{ from the line } x = 8].$$

54. Graph the hyperbola $\dfrac{x^2}{4} - \dfrac{y^2}{12} = 1$ with a graphing calculator. Use tracing to find the coordinates of several points on the hyperbola. For each of these points P, verify that

[Distance of P from (4, 0)]

$$= 2\text{[Distance of } P \text{ from the line } x = 1].$$

Chapter 10   Summary ▼▼▼▼▼▼▼▼▼▼▼▼▼▼▼▼▼▼▼▼▼▼▼▼▼▼▼▼▼▼▼▼▼

KEY TERMS	KEY IDEAS
10.1 Parabolas	
conic sections focus directrix	**Equations of Parabolas** Vertical $\quad y - k = a(x - h)^2$ $\qquad\qquad (x - h)^2 = 4p(y - k) \quad \left(a = \dfrac{1}{4p}\right)$ Horizontal $\quad x - h = a(y - k)^2$ $\qquad\qquad (y - k)^2 = 4p(x - h) \quad \left(a = \dfrac{1}{4p}\right)$
10.2 Ellipses	
ellipse foci vertices major axis minor axis eccentricity	**Equations of Ellipses** Horizontal $\qquad\qquad\qquad\qquad$ Vertical $\dfrac{(x - h)^2}{a^2} + \dfrac{(y - k)^2}{b^2} = 1 \quad$ or $\quad \dfrac{(y - k)^2}{a^2} + \dfrac{(x - h)^2}{b^2} = 1$

KEY TERMS	KEY IDEAS

10.3 Hyperbolas

hyperbola vertices transverse axis asymptotes fundamental rectangle	**Equations of Hyperbolas** Horizontal$\qquad\qquad\qquad$Vertical $$\frac{(x-h)^2}{a^2}-\frac{(y-k)^2}{b^2}=1 \quad \text{or} \quad \frac{(y-k)^2}{a^2}-\frac{(x-h)^2}{b^2}=1$$

10.4 Conic Sections

For $Ax^2 + Cy^2 + Dx + Ey + F = 0$, where A or C is nonzero:

Conic Section	Characteristic	Example
Parabola	Either $A = 0$ or $C = 0$, but not both.	$y = x^2$ $x = 3y^2 + 2y - 4$
Circle	$A = C \neq 0$	$x^2 + y^2 = 16$
Ellipse	$A \neq C, AC > 0$	$\dfrac{x^2}{16} + \dfrac{y^2}{25} = 1$
Hyperbola	$AC < 0$	$x^2 - y^2 = 1$

Conic Section	Eccentricity
Parabola	$e = 1$
Ellipse	$e < 1$
Hyperbola	$e > 1$

Chapter 10   Review Exercises ▼▼▼▼▼▼▼▼▼▼▼▼▼▼▼▼▼▼▼▼▼▼▼▼▼▼▼▼▼

Graph each equation. Give the vertex and axis of each figure.

1. $x = 4(y - 5)^2 + 2$ $\qquad$ **2.** $x = -(y + 1)^2 - 7$ $\qquad$ **3.** $x = 5y^2 - 5y + 3$ $\qquad$ **4.** $x = 2y^2 - 4y + 1$

Graph each equation. Give the focus, directrix, and axis of each figure.

5. $y^2 = -\dfrac{2}{3}x$ $\qquad\qquad$ **6.** $y^2 = 2x$ $\qquad\qquad$ **7.** $3x^2 = y$ $\qquad\qquad$ **8.** $x^2 + 2y = 0$

Write an equation for each parabola with vertex at the origin.

9. Focus $(4, 0)$ $\qquad\qquad\qquad\qquad\qquad$ **10.** Focus $(0, -3)$

11. Through $(-3, 4)$, opening upward $\qquad$ **12.** Through $(2, 5)$, opening to the right

In Exercises 13–20, an equation of a conic section is given. Identify the type of conic section. It may be necessary to transform the equation into a more familiar form.

13. $y^2 + 9x^2 = 9$ $\qquad\qquad$ **14.** $9x^2 - 16y^2 = 144$ $\qquad\qquad$ **15.** $3y^2 - 5x^2 = 30$

16. $y^2 + x = 4$ $\qquad\qquad\qquad$ **17.** $4x^2 - y = 0$ $\qquad\qquad\qquad\qquad$ **18.** $x^2 + y^2 = 25$

19. $4x^2 - 8x + 9y^2 + 36y = -4$ $\qquad\qquad$ **20.** $9x^2 - 18x - 4y^2 - 16y - 43 = 0$

In Exercises 21–26, match the equation with its calculator-generated graph. Then use your calculator to support your answers.

21. $4x^2 + y^2 = 36$

22. $x = 2y^2 + 3$

23. $(x - 2)^2 + (y + 3)^2 = 36$

24. $\dfrac{x^2}{36} + \dfrac{y^2}{9} = 1$

25. $(y - 1)^2 - (x - 2)^2 = 36$

26. $y^2 = 36 + 4x^2$

A.

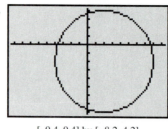

[−9.4, 9.4] by [−8.2, 4.2]

B.

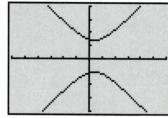

[−30, 30] by [−20, 20]
Xscl = 5   Yscl = 5

C.

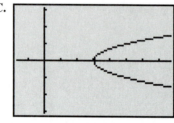

[−1.7, 7.7] by [−3.1, 3.1]

D.

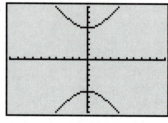

[−10, 10] by [−10, 10]

E.

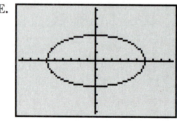

[−9.4, 9.4] by [−6.2, 6.2]

F.

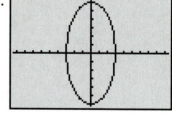

[−9.4, 9.4] by [−6.2, 6.2]

Graph each relation and identify each graph. Give the coordinates of the vertices for each ellipse
or hyperbola, and give the equations of the asymptotes for each hyperbola.

27. $\dfrac{x^2}{4} + \dfrac{y^2}{9} = 1$

28. $\dfrac{x^2}{16} + \dfrac{y^2}{4} = 1$

29. $\dfrac{x^2}{64} - \dfrac{y^2}{36} = 1$

30. $\dfrac{y^2}{25} - \dfrac{x^2}{9} = 1$

31. $\dfrac{(x + 1)^2}{16} + \dfrac{(y - 1)^2}{16} = 1$

32. $(x - 3)^2 + (y + 2)^2 = 9$

33. $4x^2 + 9y^2 = 36$

34. $x^2 = 16 + y^2$

35. $\dfrac{(x - 3)^2}{4} + (y + 1)^2 = 1$

36. $\dfrac{(x - 2)^2}{9} + \dfrac{(y + 3)^2}{4} = 1$

37. $\dfrac{(y + 2)^2}{4} - \dfrac{(x + 3)^2}{9} = 1$

38. $\dfrac{(x + 1)^2}{16} - \dfrac{(y - 2)^2}{4} = 1$

Graph each relation. State whether the relation is a function.

39. $\dfrac{x}{3} = -\sqrt{1 - \dfrac{y^2}{16}}$

40. $x = -\sqrt{1 - \dfrac{y^2}{36}}$

41. $y = -\sqrt{1 + x^2}$

42. $y = -\sqrt{1 - \dfrac{x^2}{25}}$

Write an equation for each conic section (centers at the origin).

43. Ellipse; vertex at $(0, 4)$, focus at $(0, 2)$

44. Ellipse; x-intercept 6, focus at $(-2, 0)$

45. Hyperbola; focus at $(0, -5)$, transverse axis of length 8

46. Hyperbola; y-intercept -2, passing through $(2, 3)$

Write an equation for each conic section satisfying the given conditions.

47. Parabola with focus at $(3, 2)$ and directrix $x = -3$

48. Parabola with vertex at $(-3, 2)$ and y-intercepts 5 and -1

49. Ellipse with foci at $(-2, 0)$ and $(2, 0)$ and major axis of length 10

50. Ellipse with foci at $(0, 3)$ and $(0, -3)$ and vertex at $(0, 7)$

51. Hyperbola with x-intercepts ± 3; foci at $(-5, 0)$, $(5, 0)$

52. Hyperbola with foci at $(0, 12)$, $(0, -12)$; asymptotes $y = \pm x$

53. Find the equation of the ellipse consisting of all points in the plane the sum of whose distances from $(0, 0)$ and $(4, 0)$ is 8.

54. Find the equation of the hyperbola consisting of all points in the plane for which the difference of the distances from $(0, 0)$ and $(0, 4)$ is 2.

55. Calculator-generated graphs are shown in Figures A–D. Arrange the figures in order so that the first in the list has the smallest eccentricity and the rest have eccentricities in increasing order.

A.

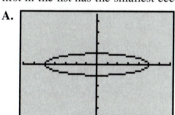

[−7.05, 7.05] by [−4.65, 4.65]

B.

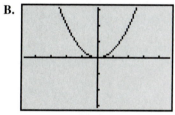

[−4.7, 4.7] by [−3.1, 3.1]

C.

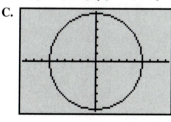

[−9.4, 9.4] by [−6.1, 6.1]

D.

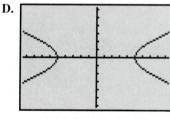

[−9.4, 9.4] by [−6.2, 6.2]

56. The orbit of Venus is an ellipse with the sun at one of the foci. The eccentricity of the orbit is $e = .006775$ and the major axis has length 134.5 million miles. (*Source: The World Almanac and Book of Facts,* 1995.) Find the smallest and greatest distances of Venus from the sun.

57. Comet Swift-Tuttle has an elliptical orbit of eccentricity $e = .964$, with the sun at one of the foci. Find the equation of the comet given that the closest it comes to the sun is 89 million miles.

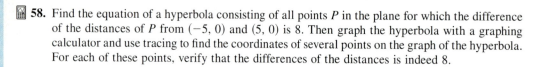 **58.** Find the equation of a hyperbola consisting of all points P in the plane for which the difference of the distances of P from $(-5, 0)$ and $(5, 0)$ is 8. Then graph the hyperbola with a graphing calculator and use tracing to find the coordinates of several points on the graph of the hyperbola. For each of these points, verify that the differences of the distances is indeed 8.

Chapter 10   Test ▼▼▼▼▼▼▼▼▼▼▼▼▼▼▼▼▼▼▼▼▼▼▼▼▼▼▼▼▼▼▼▼▼▼

Graph each parabola. Give the coordinates of the vertex and the equation of the axis.

1. $y = -x^2 + 6x$ **2.** $x = 4y^2 + 8y$

3. Give the coordinates of the focus and the equation of the directrix for the parabola with equation $x = 8y^2$.

4. Write an equation for the parabola with vertex $(2, 3)$, passing through the point $(-18, 1)$, and opening to the left.

5. Explain how to determine just by looking at the equation whether a parabola has a vertical or a horizontal axis, and whether it opens upward, downward, to the left, or to the right.

Graph each ellipse.

6. $\dfrac{(x-8)^2}{100} + \dfrac{(y-5)^2}{49} = 1$ **7.** $16x^2 + 4y^2 = 64$

8. Graph $y = -\sqrt{1 - \dfrac{x^2}{36}}$. Tell whether the graph is that of a function.

9. Write an equation for the ellipse centered at the origin having horizontal major axis with length 6 and minor axis with length 4.

10. An arch of a bridge has the shape of the top half of an ellipse. The arch is 40 ft wide and 12 ft high at the center. Find the equation of the complete ellipse. Find the height of the arch 10 ft from the center of the bottom.

Graph each hyperbola. Give the equations of the asymptotes.

11. $\dfrac{x^2}{4} - \dfrac{y^2}{4} = 1$ **12.** $9x^2 - 4y^2 = 36$

13. Find the equation of the hyperbola with x-intercepts ± 5 and foci at $(-6, 0)$ and $(6, 0)$.

Identify the type of graph, if any, defined by the equation.

14. $x^2 + 8x + y^2 - 4y + 2 = 0$ **15.** $5x^2 + 10x - 2y^2 - 12y - 23 = 0$

16. $3x^2 + 10y^2 - 30 = 0$ **17.** $x^2 - 4y = 0$

18. $(x + 9)^2 + (y - 3)^2 = 0$ **19.** $x^2 + 4x + y^2 - 6y + 30 = 0$

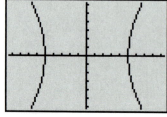 **20.** The screen shown here gives the graph of $\dfrac{x^2}{25} - \dfrac{y^2}{49} = 1$ as generated by a graphing calculator. What two functions Y_1 and Y_2 were used to obtain the graph?

$[-9.4, 9.4]$ by $[-6.2, 6.2]$

$\dfrac{x^2}{25} - \dfrac{y^2}{49} = 1$

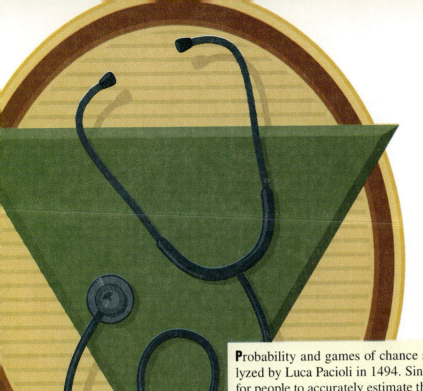

11

Further Topics in Algebra

Probability and games of chance such as gambling were first formally analyzed by Luca Pacioli in 1494. Since then, mathematics has made it possible for people to accurately estimate the probability of an event occurring. In the study of genetics and the spread of disease, the concepts of counting and probability both play a central role in predicting outcomes. Genetic engineering has already influenced society with the food people eat and the medical drugs they take. It will continue to have a dramatic effect on many aspects of society.

Largely due to the frequent use of antibiotics in society, many strains of bacteria are becoming resistant to one or more of these drugs. This phenomenon is of interest to both the medical community and genetic engineers. Many types of bacteria contain genetic material called plasmids. Plasmids are inherited during cell division. The genetic material of bacteria can sometimes contain two different plasmids causing drug resistance to both the antibiotics ampicillin and tetracycline. When the bacteria reproduce, the type of plasmids passed on to each new cell is random. Genetic engineers want to predict what will happen to these bacteria after many generations. Will the bacteria remain resistant to both antibiotics indefinitely?*

In order to answer this and other questions related to the spread of disease, further topics in algebra are needed. Bacteria growth can be modeled and simulated using sequences. The behavior of the bacteria after a long period of time often involves an element of chance. To analyze the phenomenon of chance, the study of probability is required. Solving real-world

---

Sources: Hoppensteadt, F., and C. Peskin, *Mathematics in Medicine and the Life Sciences,* Springer-Verlag, 1992.

National Council of Teachers of Mathematics, *Historical Topics for the Mathematics Classroom,* Thirty-first Yearbook, Washington, D. C., 1969.

applications usually requires knowledge from several areas of mathematics. The solution to this genetic problem will require not only the new concepts of sequences and probability, but also previously learned concepts like matrices. As our mathematical ability has increased during this mathematics course, so has the complexity of the applications that we can solve.

In this chapter we explore topics related to sums of n terms, where n is a positive integer. First, we look at *sequences* (lists of numbers) and *series* (sums of sequences), paying special attention to two very useful types of sequences, *arithmetic* and *geometric*. Next, we see how *mathematical induction* is used to prove theorems about sequences and series. Then we revisit the binomial theorem, a formula for writing out the terms of the series that is equivalent to $(x + y)^n$, which was introduced in Chapter 1. The chapter ends with two related sections on counting theory and probability theory.

11.1 Sequences and Series ▼▼▼

SEQUENCES Defined informally, a sequence is a list of numbers. We are most interested in lists of numbers that satisfy some pattern. For example,

$$2, 4, 6, 8, 10, \ldots$$

is a list of the natural-number multiples of 2. This can be writen $2n$, where n is a natural number, so a sequence may be defined (as is a function) by a variable expression. More formally, a sequence is defined as follows.

> **Sequence**
>
> A **sequence** is a function that has a set of natural numbers as its domain.

Instead of using $f(x)$ notation to indicate a sequence, it is customary to use a_n, where n represents an element in the domain of the sequence. Thus, $a_n = f(n)$. The letter n is used instead of x as a reminder that n represents a *natural number*. The elements in the range of a sequence, called the **terms** of the sequence, are $a_1, a_2, a_3, \ldots$. The elements of both the domain and the range of a sequence are *ordered*. The first term (range element) is found by letting $n = 1$, the second term is found by letting $n = 2$, and so on. The **general term,** or **nth term,** of the sequence is a_n.

Figure 1 shows graphs of $f(x) = 2x$ and $a_n = 2n$. Notice that $f(x)$ defines a "continuous" function, while a_n is "discontinuous."

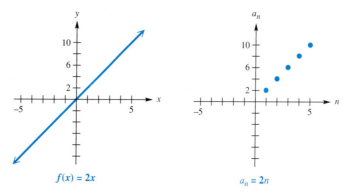

$f(x) = 2x$ $a_n = 2n$

FIGURE 1

A graphing calculator provides a convenient way to list the terms in a sequence. Methods may vary, so you should refer to your manual. Using the sequence mode to list the terms of the sequence with general term $n + (1/n)$ produces the result shown in the figure on the left. Additional terms of the sequence can be seen by scrolling to the right. Sequences can also be graphed by using the sequence mode. In the figure on the right, we show a calculator screen with the graph of $a_n = n + (1/n)$. Notice that for $n = 5$, the term is $5 + 1/5 = 5.2$.

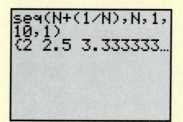

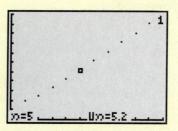

[0, 11] by [0, 11]
Xscl = 1   Yscl = 1

The fifth term is 5.2.

EXAMPLE 1
Finding terms of a sequence
from the general term

Write the first five terms for each of the following sequences.

(a) $a_n = \dfrac{n+1}{n+2}$

Replacing n, in turn, with 1, 2, 3, 4, and 5 gives

$$\frac{2}{3}, \frac{3}{4}, \frac{4}{5}, \frac{5}{6}, \frac{6}{7}.$$

(b) $a_n = (-1)^n \cdot n$

Replace n with 1, 2, 3, 4, and 5 to get

$$n = 1: a_1 = (-1)^1 \cdot 1 = -1$$
$$n = 2: a_2 = (-1)^2 \cdot 2 = 2$$
$$n = 3: a_3 = (-1)^3 \cdot 3 = -3$$
$$n = 4: a_4 = (-1)^4 \cdot 4 = 4$$
$$n = 5: a_5 = (-1)^5 \cdot 5 = -5.$$

(c) $b_n = \dfrac{(-1)^n}{2^n}$

Here, we have $b_1 = -1/2$, $b_2 = 1/4$, $b_3 = -1/8$, $b_4 = 1/16$, and $b_5 = -1/32$. ▶

A sequence is a **finite sequence** if the domain is the set $\{1, 2, 3, 4, \ldots, n\}$, where n is a natural number. An **infinite sequence** has the set of all natural numbers as its domain.

EXAMPLE 2
Distinguishing between finite
and infinite sequences

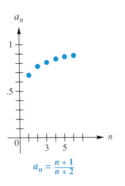

$a_n = \dfrac{n+1}{n+2}$

FIGURE 2

The sequence of natural-number multiples of 2,

$$2, 4, 6, 8, 10, 12, 14, \ldots,$$

is infinite, but the sequence of days in June is finite:

$$1, 2, 3, 4, \ldots, 29, 30. \quad ▶$$

If the terms of an infinite sequence get closer and closer to some real number, the sequence is said to be **convergent** and to **converge** to that real number. Graphs of sequences illustrate this property. The sequence in Example 1(a) is graphed in Figure 2. What number do you think this sequence converges to? A sequence that does not converge to some number is **divergent.**

Some sequences are defined by a **recursive definition,** one in which each term is defined as an expression involving the previous term. On the other hand, the sequences in Example 1 were defined *explicitly,* with a formula for a_n that does not depend on a previous term.

EXAMPLE 3
Using a recursion formula

Find the first four terms for the sequences defined as follows.

(a) $a_1 = 4$; for $n > 1$, $a_n = 2 \cdot a_{n-1} + 1$

This is an example of a recursive definition. We know $a_1 = 4$. Since $a_n = 2 \cdot a_{n-1} + 1$,

$$a_2 = 2 \cdot a_1 + 1 = 2 \cdot 4 + 1 = 9$$
$$a_3 = 2 \cdot a_2 + 1 = 2 \cdot 9 + 1 = 19$$
$$a_4 = 2 \cdot a_3 + 1 = 2 \cdot 19 + 1 = 39.$$

(b) $a_1 = 2$; for $n > 1$, $a_n = a_{n-1} + n - 1$

$$a_1 = 2$$
$$a_2 = a_1 + 2 - 1 = 2 + 1 = 3$$
$$a_3 = a_2 + 3 - 1 = 3 + 2 = 5$$
$$a_4 = a_3 + 4 - 1 = 5 + 3 = 8$$

CONNECTIONS One of the most famous sequences in mathematics is the **Fibonacci sequence:**

$$1, 1, 2, 3, 5, 8, 13, 21, 34, 55, \ldots.$$

This sequence is named for the Italian mathematician Leonardo of Pisa (1170–1250), who was also known as Fibonacci. The Fibonacci sequence is found in numerous places in nature. For example, male honeybees hatch from eggs that have not been fertilized, so a male bee has only one parent, a female. On the other hand, female honeybees hatch from fertilized eggs, so a female has two parents, one male and one female. The number of ancestors in consecutive generations of bees follows the Fibonacci sequence. Successive terms in the sequence also appear in plants: in the daisy head, the pineapple, and the pine cone, for instance.

FOR DISCUSSION OR WRITING
1. See if you can discover the pattern in the Fibonacci sequence. (*Hint:* Can you explain how to find the next term, given the preceding two terms?)
2. Draw a tree showing the number of ancestors of a male bee in each generation following the description given above.

SERIES Suppose a sequence has terms $a_1, a_2, a_3, \ldots$. Then S_n is defined as the sum of the first n terms. That is,

$$S_n = a_1 + a_2 + a_3 + \cdots + a_n.$$

The sum of the first n terms of a sequence is called a **series**. Special notation is used to represent a series. The symbol Σ, the Greek capital letter *sigma*, is used to indicate a sum.

Series

A **finite series** is an expression of the form

$$S_n = a_1 + a_2 + a_3 + \cdots + a_n = \sum_{i=1}^{n} a_i,$$

and an **infinite series** is an expression of the form

$$S_n = a_1 + a_2 + a_3 + \cdots + a_n + \cdots = \sum_{i=1}^{\infty} a_i.$$

The letter i is called the **index of summation.**

CAUTION Do not confuse this use of i with the use of i to represent an imaginary number. Other letters may be used for the index of summation.

◀ EXAMPLE 4
Using summation notation

Evaluate the series $\displaystyle\sum_{k=1}^{6} (2^k + 1)$.

Write out each of the 6 terms, then evaluate the sum.

$$\sum_{k=1}^{6} (2^k + 1) = (2^1 + 1) + (2^2 + 1) + (2^3 + 1) + (2^4 + 1)$$
$$+ (2^5 + 1) + (2^6 + 1)$$
$$= (2 + 1) + (4 + 1) + (8 + 1) + (16 + 1)$$
$$+ (32 + 1) + (64 + 1)$$
$$= 3 + 5 + 9 + 17 + 33 + 65 = 132 \quad \blacktriangleright$$

◀ EXAMPLE 5
Using summation notation with subscripts

Write out the terms for each of the following series. Evaluate each sum if possible.

(a) $\displaystyle\sum_{j=3}^{6} a_j = a_3 + a_4 + a_5 + a_6$

(b) $\displaystyle\sum_{i=1}^{3} (6x_i - 2)$ if $x_1 = 2$, $x_2 = 4$, $x_3 = 6$

Let $i = 1$, 2, and 3 respectively to get

$$\sum_{i=1}^{3} (6x_i - 2) = (6x_1 - 2) + (6x_2 - 2) + (6x_3 - 2).$$

Now substitute the given values for x_1, x_2, and x_3.

$$\sum_{i=1}^{3} (6x_i - 2) = (6 \cdot 2 - 2) + (6 \cdot 4 - 2) + (6 \cdot 6 - 2)$$

$$= 10 + 22 + 34 = 66$$

(c) $\displaystyle\sum_{i=1}^{4} f(x_i)\, \Delta x$ if $f(x) = x^2$, $x_1 = 0$, $x_2 = 2$, $x_3 = 4$, $x_4 = 6$, and $\Delta x = 2$

$$\sum_{i=1}^{4} f(x_i)\, \Delta x = f(x_1)\, \Delta x + f(x_2)\, \Delta x + f(x_3)\, \Delta x + f(x_4)\, \Delta x$$

$$= x_1^2\, \Delta x + x_2^2\, \Delta x + x_3^2\, \Delta x + x_4^2\, \Delta x$$
$$= 0^2(2) + 2^2(2) + 4^2(2) + 6^2(2)$$
$$= 0 + 8 + 32 + 72 = 112 \quad \blacktriangleright$$

The list feature of a graphing calculator can be used to find the sum of the terms of a finite series. First, we save the definition of the sequence in a list. Then we can use the capability of the calculator to get the sum of the list. Typical screens for the series $\sum_{i=0}^{3} [1000(1.06)^i]$ are shown in the figure.

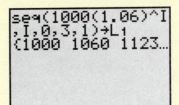

A given series can be represented by summation notation in more than one way, as shown in the next example.

EXAMPLE 6
Representing a series with different summations

Use summation notation to rewrite each series with the index of summation starting at the indicated number.

(a) $\displaystyle\sum_{i=1}^{8} (3i - 4); 0$

Let the new index be j. Since the new index is to start at 0, which is $1 - 1$, $j = i - 1$, or $i = j + 1$. Substitute $j + 1$ for i in the summation.

$$\sum_{i=1}^{8} (3i - 4) = \sum_{j+1=1}^{j+1=8} [3(j + 1) - 4]$$

$$= \sum_{j=0}^{j=7} (3j - 1) \quad \text{or} \quad \sum_{j=0}^{7} (3j - 1)$$

(b) $\displaystyle\sum_{i=2}^{10} i^2; -1$

Here, if the new index is j, then $i = j + 3$ and

$$\sum_{i=2}^{10} i^2 = \sum_{j+3=2}^{j+3=10} (j + 3)^2$$

$$= \sum_{j=-1}^{j=7} (j + 3)^2 \quad \text{or} \quad \sum_{j=-1}^{7} (j + 3)^2. \quad \blacktriangleright$$

Polynomial functions, defined by expressions of the form

$$f(x) = a_n x^n + a_{n-1}x^{n-1} + \cdots + a_1 x + a_0,$$

can be written in compact form, using summation notation, as

$$f(x) = \sum_{i=0}^{n} a_n x^n.$$

Several properties of summation are given below. These provide useful shortcuts for evaluating series.

Summation Properties

If $a_1, a_2, a_3, \ldots, a_n$ and $b_1, b_2, b_3, \ldots, b_n$ are two sequences, and c is a constant, then for every positive integer n,

(a) $\displaystyle\sum_{i=1}^{n} c = nc$

(b) $\displaystyle\sum_{i=1}^{n} ca_i = c \sum_{i=1}^{n} a_i$

(c) $\displaystyle\sum_{i=1}^{n} (a_i + b_i) = \sum_{i=1}^{n} a_i + \sum_{i=1}^{n} b_i$

(d) $\displaystyle\sum_{i=1}^{n} (a_i - b_i) = \sum_{i=1}^{n} a_i - \sum_{i=1}^{n} b_i.$

To prove property (a), expand the series to get

$$c + c + c + c + \cdots + c,$$

where there are n terms of c, so the sum is nc.

Property (c) also can be proved by first expanding the series:

$$\sum_{i=1}^{n} (a_i + b_i) = (a_1 + b_1) + (a_2 + b_2) + \cdots + (a_n + b_n).$$

Now use the commutative and associative properties to rearrange the terms.

$$\sum_{i=1}^{n} (a_i + b_i) = (a_1 + a_2 + \cdots + a_n) + (b_1 + b_2 + \cdots + b_n)$$

$$= \sum_{i=1}^{n} a_i + \sum_{i=1}^{n} b_i$$

Proofs of the other two properties are similar.

The following results are proved in the text and exercises of Section 5 of this chapter.

$$\sum_{i=1}^{n} i^2 = 1^2 + 2^2 + \cdots + n^2 = \frac{n(n + 1)(2n + 1)}{6}$$

and

$$\sum_{i=1}^{n} i = 1 + 2 + \cdots + n = \frac{n(n + 1)}{2}$$

These summations are used in the next example.

◀EXAMPLE 7
Using the summation properties

Use the properties of series to evaluate $\displaystyle\sum_{i=1}^{6} (i^2 + 3i + 5)$.

$$\sum_{i=1}^{6} (i^2 + 3i + 5) = \sum_{i=1}^{6} i^2 + \sum_{i=1}^{6} 3i + \sum_{i=1}^{6} 5 \qquad \text{Property (c)}$$

$$= \sum_{i=1}^{6} i^2 + 3 \sum_{i=1}^{6} i + \sum_{i=1}^{6} 5 \qquad \text{Property (b)}$$

$$= \sum_{i=1}^{6} i^2 + 3 \sum_{i=1}^{6} i + 6(5) \qquad \text{Property (a)}$$

By substituting the results given just before this example, we get

$$= \frac{6(6 + 1)(2 \cdot 6 + 1)}{6} + 3\left[\frac{6(6 + 1)}{2}\right] + 6(5)$$

$$= 91 + 3(21) + 6(5)$$

$$= 184. \quad \blacktriangleright$$

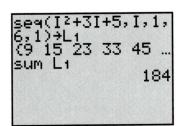

```
seq(I²+3I+5,I,1,
6,1)→L₁
{9 15 23 33 45 ...
sum L₁
              184
```

The result of Example 7 is supported in this screen. The calculator stores a sequence into a list, and then computes the sum of the terms.

11.1 Exercises ▼▼▼▼▼▼▼▼▼▼▼▼▼▼▼▼▼▼▼▼▼▼▼▼▼▼▼▼▼▼▼▼▼▼

Write the first five terms of each sequence. See Example 1.

1. $a_n = 4n + 10$ **2.** $a_n = 6n - 3$ **3.** $a_n = 2^{n-1}$ **4.** $a_n = -3^n$

5. $a_n = \left(\dfrac{1}{3}\right)^n (n - 1)$ **6.** $a_n = (-2)^n(n)$ **7.** $a_n = (-1)^n(2n)$ **8.** $a_n = (-1)^{n-1}(n + 1)$

9. $a_n = \dfrac{4n - 1}{n^2 + 2}$ **10.** $a_n = \dfrac{n^2 - 1}{n^2 + 1}$

11. Your friend does not understand what is meant by the *n*th term or general term of a sequence. How would you explain this idea?

12. How are sequences related to functions? Discuss some similarities and some differences.

Decide whether the given sequence is finite or infinite. See Example 2.

13. the sequence of days of the week   **14.** the sequence of dates in the month of November

15. 1, 2, 3, 4   **16.** $-1, -2, -3, -4$

17. 1, 2, 3, 4, . . .   **18.** $-1, -2, -3, -4, \ldots$

19. $a_1 = 3$;   for $2 < n < 10$, $a_n = 3 \cdot a_{n-1}$ **20.** $a_1 = 1$; $a_2 = 3$;   for $n \geq 3$, $a_n = a_{n-1} + a_{n-2}$

Find the first four terms for each sequence. See Example 3.

21. $a_1 = -2$, $a_n = a_{n-1} + 3$, for $n > 1$

22. $a_1 = -1$, $a_n = a_{n-1} - 4$, for $n > 1$

23. $a_1 = 1$, $a_2 = 1$, $a_n = a_{n-1} + a_{n-2}$, for $n \geq 3$ (the Fibonacci sequence)

24. $a_1 = 2$, $a_n = n \cdot a_{n-1}$, for $n > 1$

Evaluate each series. See Example 4.

25. $\displaystyle\sum_{i=1}^{5} (2i + 1)$ **26.** $\displaystyle\sum_{i=1}^{6} (3i - 2)$ **27.** $\displaystyle\sum_{j=1}^{4} \dfrac{1}{j}$ **28.** $\displaystyle\sum_{i=1}^{5} (i + 1)^{-1}$

29. $\displaystyle\sum_{i=1}^{4} i^i$ **30.** $\displaystyle\sum_{k=1}^{4} (k + 1)^2$ **31.** $\displaystyle\sum_{k=1}^{6} (-1)^k \cdot k$ **32.** $\displaystyle\sum_{i=1}^{7} (-1)^{i+1} \cdot i^2$

Evaluate the terms for each sum where $x_1 = -2$, $x_2 = -1$, $x_3 = 0$, $x_4 = 1$, and $x_5 = 2$. See Examples 5(a) and 5(b).

33. $\displaystyle\sum_{i=1}^{5} (2x_i + 3)$ **34.** $\displaystyle\sum_{i=1}^{4} x_i^2$ **35.** $\displaystyle\sum_{i=1}^{3} (3x_i - x_i^2)$

36. $\displaystyle\sum_{i=1}^{3} (x_i^2 + 1)$ **37.** $\displaystyle\sum_{i=2}^{5} \dfrac{x_i + 1}{x_i + 2}$ **38.** $\displaystyle\sum_{i=1}^{5} \dfrac{x_i}{x_i + 3}$

Evaluate the terms of $\sum_{i=1}^{4} f(x_i)\, \Delta x$ with $x_1 = 0$, $x_2 = 2$, $x_3 = 4$, $x_4 = 6$, and $\Delta x = .5$ for the functions defined. See Example 5(c).

39. $f(x) = 4x - 7$ **40.** $f(x) = 6 + 2x$ **41.** $f(x) = 2x^2$

42. $f(x) = x^2 - 1$ **43.** $f(x) = \dfrac{-2}{x + 1}$ **44.** $f(x) = \dfrac{5}{2x - 1}$

Use summation notation to rewrite each series with the index of summation starting at the indicated number. See Example 6.

45. $\displaystyle\sum_{i=1}^{5} (6 - 3i);$ 3

46. $\displaystyle\sum_{i=1}^{7} (5i + 2);$ -2

47. $\displaystyle\sum_{i=1}^{10} 2(3)^i;$ 0

48. $\displaystyle\sum_{i=-1}^{6} 5(2)^i;$ 3

49. $\displaystyle\sum_{i=-1}^{9} (i^2 - 2i);$ 0

50. $\displaystyle\sum_{i=3}^{11} (2i^2 + 1);$ 0

Use the summation properties to evaluate each series. See Example 7. The following sums may be needed.

$$\sum_{i=1}^{n} i = \frac{n(n + 1)}{2}$$

$$\sum_{i=1}^{n} i^2 = \frac{n(n + 1)(2n + 1)}{6}$$

$$\sum_{i=1}^{n} i^3 = \frac{n^2(n + 1)^2}{4}$$

51. $\displaystyle\sum_{i=1}^{5} (5i + 3)$

52. $\displaystyle\sum_{i=1}^{5} (8i - 1)$

53. $\displaystyle\sum_{i=1}^{5} (4i^2 - 2i + 6)$

54. $\displaystyle\sum_{i=1}^{6} (2 + i - i^2)$

55. $\displaystyle\sum_{i=1}^{4} (3i^3 + 2i - 4)$

56. $\displaystyle\sum_{i=1}^{6} (i^2 + 2i^3)$

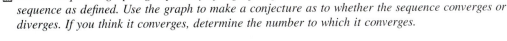 *Use the sequence graphing capability of a graphing calculator to graph the first ten terms of each sequence as defined. Use the graph to make a conjecture as to whether the sequence converges or diverges. If you think it converges, determine the number to which it converges.*

57. $a_n = \dfrac{n + 4}{2n}$

58. $a_n = \dfrac{1 + 4n}{2n}$

59. $a_n = 2e^n$

60. $a_n = n(n + 2)$

61. $a_n = \left(1 + \dfrac{1}{n}\right)^n$

62. $a_n = (1 + n)^{1/n}$

Solve the problems involving sequences.

63. Certain strains of bacteria cannot produce an amino acid called histidine. This amino acid is necessary for them to produce proteins and reproduce. If these bacteria are cultured in a medium with sufficient histidine, they will double in size and then divide every 40 minutes. Let N_1 be the initial number of bacteria cells, N_2 the number after 40 minutes, N_3 the number after 80 minutes, and N_j the number after $40(j - 1)$ minutes. (*Source:* Hoppensteadt, F., and C. Peskin, *Mathematics in Medicine and the Life Sciences,* Springer-Verlag, 1992.)
(a) Write N_{j+1} in terms of N_j for $j \geq 1$.
(b) Determine the number of bacteria after two hours if $N_1 = 230$.
(c) Graph the sequence N_j for $j = 1, 2, 3, \ldots, 7$. Use the window $[0, 10]$ by $[0, 15{,}000]$.
(d) Describe the growth of these bacteria when there are unlimited nutrients.

64. Refer to Exercise 63. If the bacteria are not cultured in a medium with sufficient nutrients, competition will ensue and the growth will slow. According to Verhulst's Model, the number of bacteria N_j at time $40(j - 1)$ in minutes can be determined by the sequence $N_{j+1} = \left[\dfrac{2}{1 + (N_j/K)} \right] N_j$ where K is a constant and $j \geq 1$. (*Source:* Hoppensteadt, F., and C. Peskin, *Mathematics in Medicine and the Life Sciences,* Springer-Verlag, 1992.)

(a) If $N_1 = 230$ and $K = 5000$, make a table of N_j for $j = 1, 2, 3, \ldots, 20$. Round values in the table to the nearest integer.

(b) Graph the sequence N_j for $j = 1, 2, 3, \ldots, 20$. Use the window $[0, 20]$ by $[0, 6000]$.

(c) Describe the growth of these bacteria when there are limited nutrients.

(d) Make a conjecture as to why K is called the *saturation constant.* Test your conjecture.

11.2 Arithmetic Sequences and Series ▼▼▼

A sequence in which each term after the first is obtained by adding a fixed number to the previous term is an **arithmetic sequence** (or **arithmetic progression**). The fixed number that is added is the **common difference.** The sequence

$$5, 9, 13, 17, 21, \ldots$$

is an arithmetic sequence since each term after the first is obtained by adding 4 to the previous term. That is,

$$9 = 5 + 4$$
$$13 = 9 + 4$$
$$17 = 13 + 4$$
$$21 = 17 + 4,$$

and so on. The common difference is 4.

If the common difference of an arithmetic sequence is d, then by the definition of an arithmetic sequence,

$$d = a_{n+1} - a_n$$

for every positive integer n in its domain.

◀EXAMPLE 1
Finding the common difference

Find the common difference, d, for the arithmetic sequence

$$-9, -7, -5, -3, -1, \ldots.$$

Since this sequence is arithmetic, d can be found by choosing any two adjacent terms and subtracting the first from the second. Choosing -7 and -5 gives

$$d = -5 - (-7) = 2.$$

Choosing -9 and -7 would give $d = -7 - (-9) = 2$, the same result. ▶

If a_1 and d are known, then all the terms of an arithmetic sequence can be found.

EXAMPLE 2
Finding the terms given a_1 and d

Write the first five terms for each of the following arithmetic sequences.

(a) The first term is 7, and the common difference is -3.
Here

$$a_1 = 7 \quad \text{and} \quad d = -3.$$
$$a_2 = 7 + (-3) = 4,$$
$$a_3 = 4 + (-3) = 1,$$
$$a_4 = 1 + (-3) = -2,$$
$$a_5 = -2 + (-3) = -5.$$

(b) $a_1 = -12, d = 5$
Starting with a_1, add d to each term to get the next term.

$$a_1 = -12$$
$$a_2 = -12 + d = -12 + 5 = -7$$
$$a_3 = -7 + d = -7 + 5 = -2$$
$$a_4 = -2 + d = -2 + 5 = 3$$
$$a_5 = 3 + d = 3 + 5 = 8 \quad \blacktriangleright$$

If a_1 is the first term of an arithmetic sequence and d is the common difference, then the terms of the sequence are given by

$$a_1 = a_1$$
$$a_2 = a_1 + d$$
$$a_3 = a_2 + d = a_1 + d + d = a_1 + 2d$$
$$a_4 = a_3 + d = a_1 + 2d + d = a_1 + 3d$$
$$a_5 = a_1 + 4d$$
$$a_6 = a_1 + 5d,$$

and, by this pattern $a_n = a_1 + (n - 1)d$. This result can be proven by mathematical induction (see Section 5 of this chapter); a summary is given below.

---

nth Term of an Arithmetic Sequence

In an arithmetic sequence with first term a_1 and common difference d, the nth term, a_n, is given by

$$a_n = a_1 + (n - 1)d.$$

EXAMPLE 3
Using the formula for the nth term

Find a_{13} and a_n for the arithmetic sequence

$$-3, 1, 5, 9, \ldots .$$

Here $a_1 = -3$ and $d = 1 - (-3) = 4$. To find a_{13}, substitute 13 for n in the preceding formula.

$$a_{13} = a_1 + (\mathbf{13} - 1)d$$
$$a_{13} = -3 + (12)4$$
$$a_{13} = -3 + 48$$
$$a_{13} = 45$$

Find a_n by substituting values for a_1 and d in the formula for a_n.

$$a_n = \mathbf{-3} + (n - 1) \cdot \mathbf{4}$$
$$a_n = -3 + 4n - 4 \qquad \text{Distributive property}$$
$$a_n = 4n - 7 \quad \blacktriangleright$$

EXAMPLE 4
Using the formula for the nth term

Find a_{18} and a_n for the arithmetic sequence having $a_2 = 9$ and $a_3 = 15$.
Find d first; $d = a_3 - a_2 = 15 - 9 = 6$.

Since
$$a_2 = a_1 + d,$$
$$\mathbf{9} = a_1 + \mathbf{6} \qquad \text{Let } a_2 = 9, d = 6.$$
$$a_1 = 3.$$

Then
$$a_{18} = 3 + (\mathbf{18} - 1) \cdot 6 \qquad \text{Formula for } a_n; \quad n = 18$$
$$a_{18} = 105,$$

and
$$a_n = 3 + (n - 1) \cdot 6$$
$$a_n = 3 + 6n - 6 \qquad \text{Distributive property}$$
$$a_n = 6n - 3. \quad \blacktriangleright$$

EXAMPLE 5
Using the formula for the nth term

Suppose that an arithmetic sequence has $a_8 = -16$ and $a_{16} = -40$. Find a_1.
We must find d first. Since $a_8 = a_1 + (8 - 1)d$, replacing a_8 with -16 gives $-16 = a_1 + 7d$ or $a_1 = -16 - 7d$. Similarly, $-40 = a_1 + 15d$ or $a_1 = -40 - 15d$. From these two equations, using the substitution method from Chapter 9,

$$-16 - 7d = -40 - 15d,$$

so $d = -3$. To find a_1, substitute -3 for d in $-16 = a_1 + 7d$:

$$-16 = a_1 + 7d$$
$$-16 = a_1 + 7(\mathbf{-3}) \qquad \text{Let } d = -3.$$
$$a_1 = 5. \quad \blacktriangleright$$

SUM OF THE FIRST n TERMS It is often necessary to add the terms of an arithmetic sequence. For example, suppose that a person borrows $3000 and agrees to pay $100 per month plus interest of 1% per month on the unpaid balance until the loan is paid off. The first month $100 is paid to reduce the loan, plus interest of $(.01)3000 = 30$ dollars. The second month another \$100 is paid toward the loan and $(.01)2900 = 29$ dollars is paid for interest. Since the loan is reduced by \$100 each month, interest payments decrease by $(.01)100 = 1$ dollar each month, forming the arithmetic sequence

$$30, 29, 28, \ldots, 3, 2, 1.$$

The total amount of interest paid is given by the sum S_n of the terms of this sequence. A formula will be developed here to find this sum without adding all thirty numbers directly.

Since the sequence is arithmetic, the sum of the first n terms can be written as follows.

$$S_n = a_1 + [a_1 + d] + [a_1 + 2d] + \cdots + [a_1 + (n - 1)d]$$

The formula for the general term was used in the last expression. Now write the same sum in reverse order, beginning with a_n and *subtracting d.*

$$S_n = a_n + [a_n - d] + [a_n - 2d] + \cdots + [a_n - (n - 1)d]$$

Adding respective sides of these two equations term by term gives

$$S_n + S_n = (a_1 + a_n) + (a_1 + a_n) + \cdots + (a_1 + a_n)$$

or

$$2S_n = n(a_1 + a_n),$$

since there are n terms of $a_1 + a_n$ on the right. Now solve for S_n to get

$$S_n = \frac{n}{2}(a_1 + a_n).$$

Using the formula $a_n = a_1 + (n - 1)d$, this result for S_n can also be written as

$$S_n = \frac{n}{2}[a_1 + a_1 + (n - 1)d]$$

or

$$S_n = \frac{n}{2}[2a_1 + (n - 1)d],$$

an alternative formula for the sum of the first n terms of an arithmetic sequence. A summary of this work with sums of arithmetic sequences follows.

Sum of the First n Terms of an Arithmetic Sequence

If an arithmetic sequence has first term a_1 and common difference d, then the sum of the first n terms is given by

$$S_n = \frac{n}{2}(a_1 + a_n)$$

or

$$S_n = \frac{n}{2}[2a_1 + (n - 1)d].$$

The first formula is used when the first and last terms are known; otherwise the second formula is used.

Either one of these formulas can be used to find the total interest on the $3000 loan discussed above. In the sequence of interest payments, $a_1 = 30$, $d = -1$, $n = 30$, and $a_n = 1$. Choosing the first formula,

$$S_n = \frac{n}{2}(a_1 + a_n),$$

gives $S_{30} = \frac{30}{2}(30 + 1) = 15(31) = 465,$

so a total of $465 interest will be paid over the 30 months.

EXAMPLE 6
Using the sum formulas

(a) Evaluate S_{12} for the arithmetic sequence

$$-9, -5, -1, 3, 7, \ldots.$$

We want the sum of the first twelve terms. Using $a_1 = -9$, $n = 12$, and $d = 4$ in the second formula,

$$S_n = \frac{n}{2}[2a_1 + (n - 1)d]$$

$$S_{12} = \frac{12}{2}[2(-9) + 11(4)]$$

$$= 6(-18 + 44) = 156.$$

(b) Use the formula for S_n to evaluate the sum of the first 60 positive integers.
In this example, $n = 60$, $a_1 = 1$, and $a_{60} = 60$, so it is convenient to use the first of the two formulas:

$$S_n = \frac{n}{2}(a_1 + a_n)$$

$$S_{60} = \frac{60}{2}(1 + 60)$$

$$= 30 \cdot 61 = 1830. \quad \blacktriangleright$$

EXAMPLE 7
Using the sum formulas

The sum of the first 17 terms of an arithmetic sequence is 187. If $a_{17} = -13$, find a_1 and d.
Use the first formula for S_n, with $n = 17$, to find a_1.

$$S_{17} = \frac{17}{2}(a_1 + a_{17}) \qquad \text{Let } n = 17.$$

$$187 = \frac{17}{2}(a_1 - 13) \qquad \text{Let } S_{17} = 187, a_{17} = -13.$$

$$22 = a_1 - 13 \qquad \text{Multiply by } \tfrac{2}{17}.$$

$$a_1 = 35$$

Since $a_{17} = a_1 + (17 - 1)d$,

$$-13 = 35 + 16d \qquad \text{Let } a_{17} = -13, a_1 = 35.$$
$$-48 = 16d$$
$$d = -3. \quad \text{⬤}$$

Any sum of the form

$$\sum_{i=1}^{n} (mi + p),$$

where m and p are real numbers, represents the sum of the terms of an arithmetic sequence having first term

$$a_1 = m(1) + p = m + p$$

and common difference $d = m$. These sums can be evaluated by the formulas in this section, as shown by the next example.

⬤**EXAMPLE 8**
Using summation notation

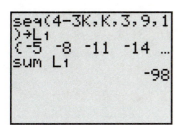

```
sum (seq(4I+8,I,
1,10,1)
                300
```

This screen supports the result in Example 8(a).

```
seq(4-3K,K,3,9,1
)→L1
{-5 -8 -11 -14 …
sum L1
                -98
```

This screen supports the result in Example 8(b).

Evaluate the following sums.

(a) $\displaystyle\sum_{i=1}^{10} (4i + 8)$

This sum represents the sum of the first ten terms of the arithmetic sequence having

$$a_1 = 4 \cdot 1 + 8 = 12,$$
$$n = 10,$$
$$\text{and} \qquad a_n = a_{10} = 4 \cdot 10 + 8 = 48.$$

Thus $\displaystyle\sum_{i=1}^{10} (4i + 8) = S_{10} = \frac{10}{2}(12 + 48) = 5(60) = 300.$

(b) $\displaystyle\sum_{k=3}^{9} (4 - 3k)$

The first few terms are

$$[4 - 3(3)] + [4 - 3(4)] + [4 - 3(5)] + \cdots$$
$$= -5 + (-8) + (-11) + \cdots.$$

Thus, $a_1 = -5$ and $d = -3$. If the sequence started with $k = 1$, there would be 9 terms. Since it starts at 3, 2 of those terms are missing, so there are 7 terms and $n = 7$.

$$\sum_{k=3}^{9} (4 - 3k) = \frac{7}{2}[2(-5) + (6)(-3)] = -98 \quad \text{⬤}$$

11.2 Exercises ▼▼▼▼▼▼▼▼▼▼▼▼▼▼▼▼▼▼▼▼▼▼▼▼▼▼▼▼▼▼▼▼▼▼▼▼▼▼▼

Find the common difference d for each arithmetic sequence. See Example 1.

1. 2, 5, 8, 11, . . .   **2.** 4, 10, 16, 22, . . .

3. 3, −2, −7, −12, . . .   **4.** −8, −12, −16, −20, . . .

5. $x + 3y, 2x + 5y, 3x + 7y, \ldots$ **6.** $t^2 + q, -4t^2 + 2q, -9t^2 + 3q, \ldots$

Write the first five terms for each arithmetic sequence. See Example 2.

7. The first term is 8, and the common difference is 6.

8. The first term is −2, and the common difference is 12.

9. $a_1 = 5, d = -2$ **10.** $a_1 = 4, d = 3$

11. $a_3 = 10, d = -2$ **12.** $a_1 = 3 - \sqrt{2}, a_2 = 3$

Find a_8 and a_n for each arithmetic sequence. See Examples 3 and 4.

13. $a_1 = 5, d = 2$ **14.** $a_1 = -3, d = -4$

15. $a_3 = 2, d = 1$ **16.** $a_4 = 5, d = -2$

17. $a_1 = 8, a_2 = 6$ **18.** $a_1 = 6, a_2 = 3$

19. $a_{10} = 6, a_{12} = 15$ **20.** $a_{15} = 8, a_{17} = 2$

21. $a_1 = x, a_2 = x + 3$ **22.** $a_2 = y + 1, d = -3$

Find a_1 for each arithmetic sequence. See Example 5.

23. $a_5 = 27, a_{15} = 87$ **24.** $a_{12} = 60, a_{20} = 84$

25. $S_{16} = -160, a_{16} = -25$ **26.** $S_{28} = 2926, a_{28} = 199$

Find the sum of the first ten terms for each arithmetic sequence. See Example 6.

27. $a_1 = 8, d = 3$ **28.** $a_1 = -9, d = 4$ **29.** $a_3 = 5, a_4 = 8$

30. $a_2 = 9, a_4 = 13$ **31.** 5, 9, 13, . . .   **32.** 8, 6, 4, . . .

33. $a_1 = 10, a_{10} = 5.5$ **34.** $a_1 = -8, a_{10} = -1.25$

Find a_1 and d for each arithmetic sequence. See Example 7.

35. $S_{20} = 1090, a_{20} = 102$ **36.** $S_{31} = 5580, a_{31} = 360$

37. $S_{12} = -108, a_{12} = -19$ **38.** $S_{25} = 650, a_{25} = 62$

Evaluate each sum. See Example 8.

39. $\displaystyle\sum_{i=1}^{3} (i + 4)$ **40.** $\displaystyle\sum_{i=1}^{5} (i - 8)$ **41.** $\displaystyle\sum_{j=1}^{10} (2j + 3)$ **42.** $\displaystyle\sum_{j=1}^{15} (5j - 9)$

43. $\displaystyle\sum_{i=1}^{12} (-5 - 8i)$ **44.** $\displaystyle\sum_{k=1}^{19} (-3 - 4k)$ **45.** $\displaystyle\sum_{i=1}^{1000} i$ **46.** $\displaystyle\sum_{k=1}^{2000} k$

▼▼▼▼▼▼▼▼▼▼▼▼ **DISCOVERING CONNECTIONS** (Exercises 47–50) ▼▼▼▼▼▼▼▼▼▼▼▼

Let $f(x) = mx + b$.

47. Find $f(1), f(2)$, and $f(3)$.

48. Consider the sequence $f(1), f(2), f(3), \ldots$. Is it an arithmetic sequence?

49. If the sequence is arithmetic, what is the common difference?

50. What is a_n for the sequence described in Exercise 48?

Use the sequence feature of a graphing calculator to evaluate the sum of the first ten terms of the arithmetic sequence. In Exercises 53 and 54, round to the nearest thousandth.

51. $a_n = 4.2n + 9.73$

52. $a_n = 8.42n + 36.18$

53. $a_n = \sqrt{8}n + \sqrt{3}$

54. $a_n = -\sqrt[3]{4}n + \sqrt{7}$

Solve each problem involving arithmetic sequences.

55. Find the sum of all the integers from 51 to 71.

56. Find the sum of the integers from -8 to 30.

57. If a clock strikes the proper number of chimes each hour on the hour, how many times will it chime in a month of 30 days?

58. A stack of telephone poles has 30 in the bottom row, 29 in the next, and so on, with one pole in the top row. How many poles are in the stack?

---

59. The population of a city was 49,000 five years ago. Each year the zoning commission permits an increase of 580 in the population. What will the maximum population be five years from now?

60. A super slide of uniform slope is to be built on a level piece of land. There are to be twenty equally spaced supports, with the longest support 15 m long and the shortest 2 m long. Find the total length of all the supports.

61. How much material would be needed for the rungs of a ladder of 31 rungs, if the rungs taper uniformly from 18 in to 28 in?

62. The normal growth pattern for children aged 3–11 follows that of an arithmetic sequence. An increase in height of about 6 cm per year is expected. Thus, 6 would be the common difference of the sequence.

A child who measures 96 cm at age 3 would have his expected height in subsequent years represented by the sequence 102, 108, 114, 120, 126, 132, 138, 144. Each term differs from the adjacent terms by the common difference, 6.

(a) If a child measures 98.2 cm at age 3 and 109.8 cm at age 5, what would be the common difference of the arithmetic sequence describing her yearly height?

(b) What would we expect her height to be at age 8?

63. Find all arithmetic sequences $a_1, a_2, a_3, \ldots$, such that $a_1^2, a_2^2, a_3^2, \ldots$, is also an arithmetic sequence.

64. Suppose that $a_1, a_2, a_3, \ldots$ and $b_1, b_2, b_3, \ldots$ are both arithmetic sequences. Let $d_n = a_n + c \cdot b_n$, for any real number c and every positive integer n. Show that $d_1, d_2, d_3, \ldots$ is an arithmetic sequence.

65. Suppose that $a_1, a_2, a_3, a_4, a_5, \ldots$ is an arithmetic sequence. Is $a_1, a_3, a_5, \ldots$ an arithmetic sequence?

66. Explain why the sequence log 2, log 4, log 8, log 16, . . . is arithmetic.

11.3 Geometric Sequences and Series ▼▼▼

Suppose you agreed to work for 1¢ the first day, 2¢ the second day, 4¢ the third day, 8¢ the fourth day, and so on, with your wages doubling each day. How much will you earn on day 20, after working 5 days a week for a month? How much will you have earned altogether in 20 days? These questions will be answered in this section.

A **geometric sequence** (or **geometric progression**) is a sequence in which each term after the first is obtained by multiplying the preceding term by a constant nonzero real number, called the **common ratio.** The sequence discussed above,

$$1, 2, 4, 8, 16, \ldots$$

is an example of a geometric sequence in which the first term is 1 and the common ratio is 2.

If the common ratio of a geometric sequence is r, then by the definition of a geometric sequence,

$$r = \frac{a_{n+1}}{a_n}$$

for every positive integer n. Therefore, the common ratio can be found by choosing any term except the first and dividing it by the preceding term.

In the geometric sequence

$$2, 8, 32, 128, \ldots$$

$r = 4$. Notice that

$$8 = \mathbf{2 \cdot 4}$$
$$32 = 8 \cdot 4 = (2 \cdot 4) \cdot 4 = \mathbf{2 \cdot 4^2}$$
$$128 = 32 \cdot 4 = (2 \cdot 4^2) \cdot 4 = \mathbf{2 \cdot 4^3}.$$

To generalize this, assume that a geometric sequence has first term a_1 and common ratio r. The second term can be written as $a_2 = a_1 r$, the third as $a_3 = a_2 r = (a_1 r)r = a_1 r^2$, and so on. Following this pattern, the nth term is $a_n = a_1 r^{n-1}$. Again, this result is proven by mathematical induction. (See Section 5 of this chapter.)

---

nth Term of a Geometric Sequence

In the geometric sequence with first term a_1 and common ratio r, the nth term is

$$a_n = a_1 r^{n-1}.$$

EXAMPLE 1
Finding the nth term of a geometric sequence

The formula for the nth term of a geometric sequence can be used to answer the first question posed at the beginning of this section. How much will be earned on day 20 if daily wages follow the sequence

$$1, 2, 4, 8, 16, \ldots$$

with $a_1 = 1$ and $r = 2$?
To answer the question, find a_{20}.

$$a_{20} = a_1 r^{19} = 1(2)^{19} = 524,288 \text{ cents, or } \$5242.88 \quad \blacktriangleright$$

EXAMPLE 2
Using the formula for the nth term

Find a_5 and a_n for the following geometric sequence.

$$4, 12, 36, 108, \ldots$$

The first term, a_1, is 4. Find r by choosing any term except the first and dividing it by the preceding term. For example,

$$r = \frac{36}{12} = 3.$$

Since $a_4 = 108$, $a_5 = 3 \cdot 108 = 324$. The fifth term also could be found using the formula for a_n, $a_n = a_1 r^{n-1}$, and replacing n with 5, r with 3, and a_1 with 4.

$$a_5 = 4 \cdot (3)^{5-1} = 4 \cdot 3^4 = 324$$

By the formula, $a_n = 4 \cdot 3^{n-1}.$ $\blacktriangleright$

EXAMPLE 3
Using the formula for the nth term

Find a_1 and r for the geometric sequence with third term 20 and sixth term 160.
Use the formula for the nth term of a geometric sequence.

$$\text{For } n = 3, a_3 = a_1 r^2 = 20;$$
$$\text{for } n = 6, a_6 = a_1 r^5 = 160.$$

We have $a_1 r^2 = 20$, so $a_1 = 20/r^2$. Substituting this in the second equation gives

$$a_1 r^5 = 160$$
$$\left(\frac{20}{r^2}\right) r^5 = 160$$
$$20 r^3 = 160$$
$$r^3 = 8$$
$$r = 2.$$

Since $a_1 r^2 = 20$ and $r = 2$, we know $a_1 = 5$. $\blacktriangleright$

EXAMPLE 4
Solving a geometric growth problem

A population of fruit flies is growing in such a way that each generation is 1.5 times as large as the last generation. Suppose there were 100 insects in the first generation. How many would there be in the fourth generation?

The population of each generation can be written as a geometric sequence with a_1 as the first-generation population, a_2 the second-generation population, and so on. Then the fourth-generation population will be a_4. Using the formula for a_n, with $n = 4$, $r = 1.5$, and $a_1 = 100$, gives

$$a_4 = a_1 r^3 = 100(1.5)^3 = 100(3.375) = 337.5.$$

In the fourth generation, the population will number about 338 insects. ▶

SUM OF FIRST n TERMS In applications of geometric sequences, it is often necessary to know the sum of the first n terms for the sequence. For example, a scientist might want to know the total number of insects in four generations of the population discussed in Example 4.

To find a formula for the sum of the first n terms of a geometric sequence, S_n, first write the sum as

$$S_n = a_1 + a_2 + a_3 + \cdots + a_n$$

or
$$S_n = a_1 + a_1 r + a_1 r^2 + \cdots + a_1 r^{n-1}. \qquad (1)$$

If $r = 1$, $S_n = na_1$, which is a correct formula for this case. If $r \neq 1$, multiply both sides of equation (1) by r, obtaining

$$rS_n = a_1 r + a_1 r^2 + a_1 r^3 + \cdots + a_1 r^n. \qquad (2)$$

If (2) is subtracted from (1),

$$
\begin{array}{lll}
S_n = a_1 + a_1 r + a_1 r^2 + \cdots + a_1 r^{n-1} & & \\
rS_n = \quad\quad a_1 r + a_1 r^2 + \cdots + a_1 r^{n-1} + a_1 r^n & & \\
\hline
S_n - rS_n = a_1 \quad\quad\quad\quad\quad\quad\quad\quad\quad\quad - a_1 r^n & & \text{Subtract.}
\end{array}
$$

or $\quad S_n(1 - r) = a_1(1 - r^n),$ ⠀⠀⠀⠀⠀⠀⠀⠀⠀⠀ Factor.

which finally gives

$$S_n = \frac{a_1(1 - r^n)}{1 - r}, \quad \text{where } r \neq 1. \qquad \text{Divide by } 1 - r.$$

This discussion is summarized below.

Sum of the First n Terms of a Geometric Sequence

If a geometric sequence has first term a_1 and common ratio r, then the sum of the first n terms is given by

$$S_n = \frac{a_1(1 - r^n)}{1 - r}, \quad \text{where } r \neq 1.$$

This formula can be used to find the total fruit fly population in Example 4 over the four-generation period. With $n = 4$, $a_1 = 100$, and $r = 1.5$,

$$S_4 = \frac{100(1 - 1.5^4)}{1 - 1.5} = \frac{100(1 - 5.0625)}{-.5} = 812.5,$$

so the total population for the four generations will amount to about 813 insects.

EXAMPLE 5
Applying the sum of the first n terms

```
sum seq(2^I,I,0,
19,1)
           1048575
```

This screen supports the result of Example 5.

To answer the second question posed at the beginning of this section, we must find the total amount earned in 20 days with daily wages of

$$1, 2, 4, 8, \ldots$$

cents. Since $a_1 = 1$ and $r = 2$,

$$S_{20} = \frac{1(1 - 2^{20})}{1 - 2}$$

$$= \frac{1 - 1,048,576}{-1}$$

$$= 1,048,575 \text{ cents,}$$

or \$10,485.75. Not bad for 20 days of work! ▶

EXAMPLE 6
Finding the sum of the first n terms

```
sum seq(2*3^I,I,
1,6,1)
              2184
```

This screen supports the result of Example 6.

Find $\displaystyle\sum_{i=1}^{6} 2 \cdot 3^i$.

This sum is the sum of the first six terms of a geometric sequence having $a_1 = 2 \cdot 3^1 = 6$ and $r = 3$. From the formula for S_n,

$$\sum_{i=1}^{6} 2 \cdot 3^i = S_6 = \frac{6(1 - 3^6)}{1 - 3}$$

$$= \frac{6(1 - 729)}{-2} = \frac{6(-728)}{-2} = 2184. \quad ▶$$

INFINITE GEOMETRIC SERIES   Now the discussion of sums of sequences will be extended to include infinite geometric sequences such as the infinite sequence

$$2, 1, \frac{1}{2}, \frac{1}{4}, \frac{1}{8}, \frac{1}{16}, \ldots$$

with first term 2 and common ratio 1/2. Using the formula above gives the following sequence.

$$S_1 = 2, S_2 = 3, S_3 = \frac{7}{2}, S_4 = \frac{15}{4}, S_5 = \frac{31}{8}, S_6 = \frac{63}{16}$$

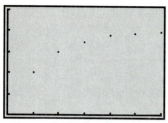

[0, 6] by [0, 5]

As n gets larger, S_n approaches 4.

FIGURE 3

As Figure 3 suggests, these sums seem to be getting closer and closer to the number 4. For no value of n is $S_n = 4$. However, if n is large enough, then S_n is as close to 4 as desired.* As mentioned earlier, we say the sequence converges to 4. This is expressed as

$$\lim_{n \to \infty} S_n = 4.$$

(Read: "the limit of S_n as n increases without bound is 4.") Since

$$\lim_{n \to \infty} S_n = 4,$$

the number 4 is called the **sum** of the infinite geometric sequence

$$2, 1, \frac{1}{2}, \frac{1}{4}, \dots$$

and

$$2 + 1 + \frac{1}{2} + \frac{1}{4} + \dots = 4.$$

EXAMPLE 7
Finding the sum of the terms of an infinite geometric sequence

Find $1 + \frac{1}{3} + \frac{1}{9} + \frac{1}{27} + \dots$.

Use the formula for the first n terms of a geometric sequence to get

$$S_1 = 1, \qquad S_2 = \frac{4}{3}, \qquad S_3 = \frac{13}{9}, \qquad S_4 = \frac{40}{27},$$

and in general

$$S_n = \frac{1\left[1 - \left(\frac{1}{3}\right)^n\right]}{1 - \frac{1}{3}}. \qquad \text{Let } a_1 = 1, r = \frac{1}{3}.$$

The chart below shows the value of $(1/3)^n$ for larger and larger values of n.

n	1	10	100	200
$\left(\frac{1}{3}\right)^n$	$\frac{1}{3}$	1.69×10^{-5}	1.94×10^{-48}	3.76×10^{-96}

As n gets larger and larger, $(1/3)^n$ gets closer and closer to 0. That is,

$$\lim_{n \to \infty} \left(\frac{1}{3}\right)^n = 0,$$

making it reasonable that

$$\lim_{n \to \infty} S_n = \lim_{n \to \infty} \frac{1\left[1 - \left(\frac{1}{3}\right)^n\right]}{1 - \frac{1}{3}} = \frac{1(1 - 0)}{1 - \frac{1}{3}} = \frac{1}{\frac{2}{3}} = \frac{3}{2}.$$

Hence,

$$1 + \frac{1}{3} + \frac{1}{9} + \frac{1}{27} + \dots = \frac{3}{2}. \quad \blacktriangleright$$

---

* The phrases "large enough" and "as close as desired" are not nearly precise enough for mathematicians; much of a standard calculus course is devoted to making them more precise.

If a geometric sequence has a first term a_1 and a common ratio r, then

$$S_n = \frac{a_1(1 - r^n)}{1 - r} \quad (r \neq 1)$$

for every positive integer n. If $-1 < r < 1$, then $\lim\limits_{n \to \infty} r^n = 0$, and

$$\lim_{n \to \infty} S_n = \frac{a_1(1 - 0)}{1 - r} = \frac{a_1}{1 - r}.$$

This quotient, $a_1/(1 - r)$, is called the **sum of an infinite geometric sequence.** The limit $\lim\limits_{n \to \infty} S_n$ is often expressed as S_∞ or $\sum_{i=1}^{\infty} a_i$. These results lead to the following definition.

> **Sum of an Infinite Geometric Sequence**
>
> The sum of an infinite geometric sequence with first term a_1 and common ratio r, where $-1 < r < 1$, is given by
>
> $$S_\infty = \frac{a_1}{1 - r}.$$

If $|r| > 1$, the terms get larger and larger in absolute value, so there is no limit as $n \to \infty$. Hence the sequence will not have a sum.

EXAMPLE 8
Finding the sum of the terms of an infinite geometric sequence

Find each sum.

(a) $\displaystyle\sum_{i=1}^{\infty} \left(-\frac{3}{4}\right)\left(-\frac{1}{2}\right)^{i-1}$

Here, $a_1 = -3/4$ and $r = -1/2$. Since $-1 < r < 1$, the formula above applies, and

$$S_\infty = \frac{a_1}{1 - r} = \frac{-\dfrac{3}{4}}{1 - \left(-\dfrac{1}{2}\right)} = -\frac{1}{2}.$$

(b) $\displaystyle\sum_{i=1}^{\infty} \left(\frac{3}{5}\right)^i = \sum_{i=1}^{\infty} \left(\frac{3}{5}\right)\left(\frac{3}{5}\right)^{i-1} = \frac{\dfrac{3}{5}}{1 - \dfrac{3}{5}} = \frac{3}{2}$ ◗

CONNECTIONS Geometric sequences and series are very important in the mathematics of finance. An example is a sequence of equal payments made at equal periods of time, such as car payments or house payments, called an *annuity*. If the payments are accumulated in an account (with no withdrawals) the sum of the payments and interest on the payments is called the *future value* of the annuity.

To save money for a trip to Europe, Meg Holden deposited $1000 at the *end* of each year for four years in an account paying 6% interest compounded annually. To find the future value of this annuity, we use the formula for compound interest, $A = P(1 + r)^t$. The first payment earns interest for three years, the second payment for two years, and the third payment for one year. The last payment earns no interest. The total amount is

$$1000(1.06)^3 + 1000(1.06)^2 + 1000(1.06) + 1000.$$

This is the sum of a geometric sequence with first term (starting at the end of the sum as written above) $a_1 = 1000$ and common ratio $r = 1.06$. Using the formula for S_4, the sum of four terms, gives

$$S_4 = \frac{1000[1 - (1.06)^4]}{1 - 1.06}$$

$$\approx 4374.62.$$

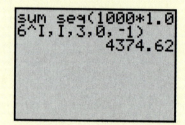

The screen supports the result of the computation in the example in the Connections.

The future value of the annuity is $4374.62.

FOR DISCUSSION OR WRITING

A loan is *amortized* or paid off if both the principal and interest are paid by a sequence of equal periodic payments. The formula

$$R = \frac{Pi}{1 - (1 + i)^{-n}}$$

gives the amount of each payment R required to pay off a loan of P dollars at an interest rate i per period for n periods. David Horwitz takes out a loan of $6000 to buy a car. He will make payments at the end of each month for four years (48 payments). He pays interest of 1% compounded monthly.

1. Find the amount of each payment.
2. What is the total amount he will pay over the four years?
3. How much of the total is interest?

11.3 Exercises ▼▼▼▼▼▼▼▼▼▼▼▼▼▼▼▼▼▼▼▼▼▼▼▼▼▼▼▼▼▼▼▼▼▼

Write out the terms of the geometric sequence that satisfies the given conditions.

1. $a_1 = \dfrac{5}{3}, r = 3, n = 4$

2. $a_1 = -\dfrac{3}{4}, r = \dfrac{2}{3}, n = 4$

3. $a_4 = 5, a_5 = 10, n = 5$

4. $a_3 = 16, a_4 = 8, n = 5$

Find a_5 and a_n for each geometric sequence. See Example 2.

5. $a_1 = 5, r = -2$

6. $a_1 = 8, r = -5$

7. $a_2 = -4, r = 3$

8. $a_3 = -2, r = 4$

9. $a_4 = 243, r = -3$

10. $a_4 = 18, r = 2$

11. $-4, -12, -36, -108, \ldots$

12. $-2, 6, -18, 54, \ldots$

13. $\dfrac{4}{5}, 2, 5, \dfrac{25}{2}, \ldots$

14. $\dfrac{1}{2}, \dfrac{2}{3}, \dfrac{8}{9}, \dfrac{32}{27}, \ldots$

▼▼▼▼▼▼▼▼▼▼▼▼▼▼ **DISCOVERING CONNECTIONS** (Exercises 15–18) ▼▼▼▼▼▼▼▼▼▼▼▼▼

Using the definition of difference *for an arithmetic sequence and* ratio *for a geometric sequence, we can find the appropriate middle term in a group of three terms so that the resulting sequence is the type desired. For example, consider the three terms*

$$5, x, .6,$$

where x is to be determined. Work Exercises 15–18 in order.

15. For these terms to form an arithmetic sequence, the difference $x - 5$ must be the same as the difference $.6 - x$. Write an equation that makes this statement.

16. Solve the equation of Exercise 15, and write the three terms of the arithmetic sequence.

17. For these terms to form a geometric sequence, the ratio $x/5$ must be the same as the ratio $.6/x$. Write an equation that makes this statement.

18. Solve the equation of Exercise 17 for its positive solution, and write the three terms of the geometric sequence.

Find a_1 and r for each geometric sequence. See Example 3.

19. $a_3 = 5, a_8 = \dfrac{1}{625}$

20. $a_2 = -6, a_7 = -192$

21. $a_4 = -\dfrac{1}{4}, a_9 = -\dfrac{1}{128}$

22. $a_3 = 50, a_7 = .005$

Use the formula for S_n to find the sum of the first five terms for each geometric sequence. See Example 5.

23. $2, 8, 32, 128, \ldots$

24. $4, 16, 64, 256, \ldots$

25. $18, -9, \dfrac{9}{2}, -\dfrac{9}{4}, \ldots$

26. $12, -4, \dfrac{4}{3}, -\dfrac{4}{9}, \ldots$

27. $a_1 = 8.423, r = 2.859$

28. $a_1 = -3.772, r = -1.553$

Find each sum. See Example 6.

29. $\displaystyle\sum_{i=1}^{5} 3^i$

30. $\displaystyle\sum_{i=1}^{4} (-2)^i$

31. $\displaystyle\sum_{j=1}^{6} 48\left(\frac{1}{2}\right)^j$

32. $\displaystyle\sum_{j=1}^{5} 243\left(\frac{2}{3}\right)^j$

33. $\displaystyle\sum_{k=4}^{10} 2^k$

34. $\displaystyle\sum_{k=3}^{9} (-3)^k$

35. Under what conditions does the sum of an infinite geometric series exist?

36. The number .999 . . . can be written as the sum of the terms of an infinite geometric sequence: .9 + .09 + .009 + ... Here we have $a_1 = .9$ and $r = .1$. Use the formula for S_∞ to find this sum. Does your intuition indicate that your answer is correct?

Find r for each infinite geometric sequence. Identify any whose sum would not converge.

37. 12, 24, 48, 96, . . .

38. 625, 125, 25, 5, . . .

39. −48, −24, −12, −6, . . .

40. 2, −10, 50, −250, . . .

Find each sum that converges by using the formula from this section where it applies. See Example 8.

41. $16 + 2 + \dfrac{1}{4} + \dfrac{1}{32} + \cdots$

42. $18 + 6 + 2 + \dfrac{2}{3} + \cdots$

43. $100 + 10 + 1 + \cdots$

44. $128 + 64 + 32 + \cdots$

45. $\dfrac{4}{3} + \dfrac{2}{3} + \dfrac{1}{3} + \cdots$

46. $\dfrac{1}{4} - \dfrac{1}{6} + \dfrac{1}{9} - \dfrac{2}{27} + \cdots$

47. $\displaystyle\sum_{i=1}^{\infty} 3\left(\frac{1}{4}\right)^{i-1}$

48. $\displaystyle\sum_{i=1}^{\infty} 5\left(-\frac{1}{4}\right)^{i-1}$

49. $\displaystyle\sum_{k=1}^{\infty} (.3)^k$

50. $\displaystyle\sum_{k=1}^{\infty} 10^{-k}$

▼▼▼▼▼▼▼▼▼▼▼▼▼ **DISCOVERING CONNECTIONS** (Exercises 51–54) ▼▼▼▼▼▼▼▼▼▼▼▼▼

Let $g(x) = ab^x$.

51. Find $g(1)$, $g(2)$, and $g(3)$.

52. Consider the sequence $g(1)$, $g(2)$, $g(3)$, ... Is it a geometric sequence? If so, what is the common ratio?

53. What is the general term of the sequence in Exercise 52?

54. Explain how geometric sequences are related to exponential functions.

🖩 *Use the sequence feature of a graphing calculator to evaluate each of the following sums. Round to the nearest thousandth.*

55. $\displaystyle\sum_{i=1}^{10} (1.4)^i$

56. $\displaystyle\sum_{j=1}^{6} -(3.6)^j$

57. $\displaystyle\sum_{j=3}^{8} 2(.4)^j$

58. $\displaystyle\sum_{i=4}^{9} 3(.25)^i$

Solve each problem involving geometric sequences. See Examples 1, 4, and 5.

59. The strain of bacteria described in Exercise 63 in Section 11.1 will double in size and then divide every 40 minutes. Let a_1 be the initial number of bacteria cells, a_2 the number after 40 minutes, and a_n the number after $40(n - 1)$ minutes. (*Source:* Hoppensteadt, F., and C. Peskin, *Mathematics in Medicine and the Life Sciences,* Springer-Verlag, 1992.)

(a) Write a formula for the nth term a_n of the geometric sequence $a_1, a_2, a_3, \ldots, a_n, \ldots$.

(b) Determine the first n where $a_n > 1,000,000$ when $a_1 = 100$.

(c) How long does it take for the number of bacteria to exceed one million?

60. The final step in processing a black-and-white photographic print is to immerse the print in a chemical called "fixer." The print is then washed in running water. Under certain conditions, 98% of the fixer in a print will be removed with 15 min of washing. How much of the original fixer would be left after 1 hr of washing?

61. A scientist has a vat containing 100 L of a pure chemical. Twenty liters is drained and replaced with water. After complete mixing, 20 L of the mixture is drained and replaced with water. What will be the strength of the mixture after 9 such drainings?

62. The half-life of a radioactive substance is the time it takes for half the substance to decay. Suppose the half-life of a substance is 3 yr, and 10^{15} molecules of the substance are present initially. How many molecules will be present after 15 yr?

63. Each year a machine loses 20% of the value it had at the beginning of the year. Find the value of the machine at the end of 6 yr if it cost $100,000 new.

64. A sugar factory receives an order for 1000 units of sugar. The production manager thus orders production of 1000 units of sugar. He forgets, however, that the production of sugar requires some sugar (to prime the machines, for example), and so he ends up with only 900 units of sugar. He then orders an additional 100 units, and receives only 90 units. A further order for 10 units produces 9 units. Finally seeing he is wrong, the manager decides to try mathematics. He views the production process as an infinite geometric progression with $a_1 = 1000$ and $r = .1$. Using this, find the number of units of sugar that he should have ordered originally.

65. A pendulum bob swings through an arc 40 cm long on its first swing. Each swing thereafter, it swings only 80% as far as on the previous swing. How far will it swing altogether before coming to a complete stop?

66. Mitzi drops a ball from a height of 10 m and notices that on each bounce the ball returns to about 3/4 of its previous height. About how far will the ball travel before it comes to rest? (*Hint:* Consider the sum of two sequences.)

67. Each person has two parents, four grandparents, eight great-grandparents, and so on. What is the total number of ancestors a person has, going back five generations? Ten generations?

68. Certain medical conditions are treated with a fixed dose of a drug administered at regular intervals. Suppose a person is given 2 mg of a drug each day and that during each 24-hr period, the body utilizes 40% of the amount of drug that was present at the beginning of the period.

(a) Show that the amount of the drug present in the body at the end of n days is $\sum_{i=1}^{n} 2(.6)^i$.

(b) What will be the approximate quantity of the drug in the body at the end of each day after the treatment has been administered for a long period of time?

69. A sequence of equilateral triangles is constructed. The first triangle has sides 2 m in length. To get the second triangle, midpoints of the sides of the original triangle are connected. What is the length of the side of the eighth such triangle? See the figure below.

70. In Exercise 69, if the process could be continued indefinitely, what would be the total perimeter of all the triangles? What would be the total area of all the triangles, disregarding the overlapping?

Find the future value of each annuity. See the Connections feature in this section.

71. Payments of $1000 at the end of each year for 9 years at 8% interest compounded annually

72. Payments of $800 at the end of each year for 12 years at 7% interest compounded annually

73. Payments of $2430 at the end of each year for 10 years at 6% interest compounded annually

74. Payments of $1500 at the end of each year for 6 years at 5% interest compounded annually

75. Let $a_1, a_2, a_3, \ldots$ and $b_1, b_2, b_3, \ldots$ be geometric sequences. Let $d_n = c \cdot a_n \cdot b_n$ for any real number c and every positive integer n. Show that $d_1, d_2, d_3, \ldots$ is a geometric sequence.

76. Explain why the sequence log 6, log 36, log 1296, log 1,679,616, . . . is geometric.

11.4 The Binomial Theorem Revisited ▼▼▼

In Chapter 1 we observed a parallel between the numbers in Pascal's triangle, shown below, and the coefficients of the terms in expansions of powers of binomials of the form $(x + y)^n$. As we saw earlier, the *n*th row of the triangle gives the coefficients of the terms of $(x + y)^n$. Also the *variables* in the expansion have the pattern

$$x^n, x^{n-1}y, x^{n-2}y^2, x^{n-3}y^3, \ldots, xy^{n-1}, y^n.$$

As the rows of Pascal's triangle show, there are $n + 1$ terms in the expansion of $(x + y)^n$. For example, in the fifth row, there are 6 coefficients, and therefore 6 terms, in the expansion of $(x + y)^5$.

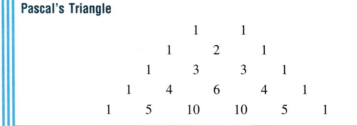

Pascal's Triangle

```
              1     1
           1     2     1
        1     3     3     1
     1     4     6     4     1
  1     5     10    10    5     1
```

CONNECTIONS Over the years, many interesting patterns have been discovered in Pascal's triangle. In the figure at the top of the next page, the triangular array is written in a different form. The indicated sums along the diagonals shown are the terms of the *Fibonacci sequence,* mentioned at the beginning of the chapter. The presence of this sequence in the triangle apparently was not recognized by Pascal.

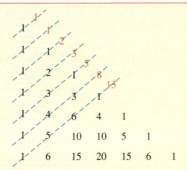

```
         1
       1   1
     1   2   1
   1   3   3   1
 1   4   6   4   1
1   5  10  10   5   1
1   6  15  20  15   6   1
```

FOR DISCUSSION OR WRITING

1. Predict the next two numbers in the sequence of sums of the diagonals of Pascal's triangle.

2. Look up and report on other occurrences of this sequence.

Although it is possible to use Pascal's triangle to find the coefficients of $(x + y)^n$ for any positive integer value of n, this becomes impractical for large values of n because of the need to write out all the preceding rows. A more efficient way of finding these coefficients uses factorial notation. The number $n!$ (read "n-factorial") is defined as follows.

Definition of n-factorial

For any positive integer n,

$$n! = n(n - 1)(n - 2) \cdots (3)(2)(1)$$

and $$0! = 1.$$

For example, $5! = 5 \cdot 4 \cdot 3 \cdot 2 \cdot 1 = 120$, $7! = 7 \cdot 6 \cdot 5 \cdot 4 \cdot 3 \cdot 2 \cdot 1 = 5040$, $2! = 2 \cdot 1 = 2$, and so on.

Many calculators have the capability of finding $n!$. A calculator with a 10-digit display will give the exact value of $n!$ for $n \leq 13$ and approximate values of $n!$ for $14 \leq n \leq 69$. The figure shows the display for 13!, 25!, and 69!.

```
13!
           6227020800
25!
    1.551121004E25
69!
    1.711224524E98
```

Now look at the coefficients of the expression

$$(x + y)^5 = x^5 + 5x^4y + 10x^3y^2 + 10x^2y^3 + 5xy^4 + y^5.$$

The coefficient of the second term, $5x^4y$, is 5, and the exponents on the variables are 4 and 1. Note that

$$5 = \frac{5!}{4! \ 1!}.$$

The coefficient of the third term is 10, with exponents of 3 and 2, and

$$10 = \frac{5!}{3! \ 2!}.$$

The last term (the sixth term) can be written as $y^5 = 1x^0y^5$, with coefficient 1, and exponents of 0 and 5. Since $0! = 1$, check that

$$1 = \frac{5!}{0! \ 5!}.$$

Generalizing from these examples, the coefficient for the term of the expansion of $(x + y)^n$ in which the variable part is x^ry^{n-r} (where $r \le n$) will be

$$\frac{n!}{r! \ (n - r)!}.$$

This number, called a **binomial coefficient**, is often symbolized $\binom{n}{r}$ (read "n choose r").

Definition of Binomial Coefficient

For nonnegative integers n and r, with $r \le n$, the symbol $\binom{n}{r}$ is defined as

$$\binom{n}{r} = \frac{n!}{r! \ (n - r)!}.$$

These binomial coefficients are just numbers from Pascal's triangle. For example, $\binom{3}{0}$ is the first number in the third row, and $\binom{7}{4}$ is the fifth number in the seventh row. Another common notation for the binomial coefficient is $_nC_r$. Many calculators have a key to use for finding binomial coefficients. Others can be programmed to calculate them.

Most graphing calculators have the ability to calculate binomial coefficients. This function is often found in the math mode. The figure shows a graphing calculator display for $_{24}C_6$ and $_{16}C_5$.

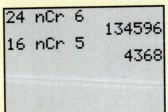

EXAMPLE 1
Evaluating binomial coefficients

(a) $\dbinom{6}{2} = \dfrac{6!}{2!\,(6-2)!} = \dfrac{6!}{2!\,4!} = \dfrac{6 \cdot 5 \cdot 4 \cdot 3 \cdot 2 \cdot 1}{2 \cdot 1 \cdot 4 \cdot 3 \cdot 2 \cdot 1} = 15$

(b) $\dbinom{8}{0} = \dfrac{8!}{0!\,(8-0)!} = \dfrac{8!}{0!\,8!} = \dfrac{8!}{1 \cdot 8!} = 1$

(c) $\dbinom{10}{10} = \dfrac{10!}{10!\,(10-10)!} = \dfrac{10!}{10!\,0!} = 1$ ▶

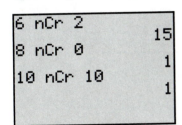

This screen supports the results in Example 1.

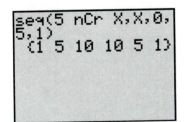

The two screens above illustrate how the SEQUENCE and TABLE capabilities of a graphing calculator can generate rows of Pascal's triangle.

Refer again to Pascal's triangle. Notice the symmetry in each row. This suggests that the binomial coefficients should have the same property. That is,

$$\binom{n}{r} = \binom{n}{n-r}.$$

This is true, since

$$\binom{n}{r} = \frac{n!}{r!\,(n-r)!} \quad \text{and} \quad \binom{n}{n-r} = \frac{n!}{(n-r)!\,r!}.$$

Our conjectures about the expansion of $(x + y)^n$ may be summarized as follows.

1. There are $n + 1$ terms in the expansion.
2. The first term is x^n and the last term is y^n.
3. The exponent on x decreases by 1, and the exponent on y increases by 1 in each succeeding term.
4. The sum of the exponents on x and y in any term is n.
5. The coefficient of the term with $x^r y^{n-r}$ or $x^{n-r} y^r$ is $\binom{n}{r}$.

These observations about the expansion of $(x + y)^n$ for any positive integer value of n suggest the **binomial theorem.**

Binomial Theorem

For any positive integer n and any complex numbers x and y,

$$(x + y)^n = x^n + \binom{n}{1}x^{n-1}y + \binom{n}{2}x^{n-2}y^2 + \binom{n}{3}x^{n-3}y^3 + \cdots$$

$$+ \binom{n}{r}x^{n-r}y^r + \cdots + \binom{n}{n-1}xy^{n-1} + y^n.$$

As stated above, the binomial theorem is a conjecture, determined inductively by looking at $(x + y)^n$ for several values of n. A proof of the binomial theorem using *mathematical induction* is given in Section 5 of this chapter.

NOTE   The binomial theorem also looks much more manageable written in summation notation. The theorem can be summarized as follows:

$$(x + y)^n = \sum_{r=0}^{n} \binom{n}{r}x^{n-r}y^r.$$

◀ EXAMPLE 2
Applying the binomial theorem

Write out the binomial expansion of $(x + y)^9$.
    Using the binomial theorem,

$$(x + y)^9 = x^9 + \binom{9}{1}x^8y + \binom{9}{2}x^7y^2 + \binom{9}{3}x^6y^3 + \binom{9}{4}x^5y^4 + \binom{9}{5}x^4y^5$$

$$+ \binom{9}{6}x^3y^6 + \binom{9}{7}x^2y^7 + \binom{9}{8}xy^8 + y^9.$$

Now evaluate each of the binomial coefficients.

$$(x + y)^9 = x^9 + \frac{9!}{1!\ 8!}x^8y + \frac{9!}{2!\ 7!}x^7y^2 + \frac{9!}{3!\ 6!}x^6y^3 + \frac{9!}{4!\ 5!}x^5y^4$$

$$+ \frac{9!}{5!\ 4!}x^4y^5 + \frac{9!}{6!\ 3!}x^3y^6 + \frac{9!}{7!\ 2!}x^2y^7 + \frac{9!}{8!\ 1!}xy^8 + y^9$$

$$= x^9 + 9x^8y + 36x^7y^2 + 84x^6y^3 + 126x^5y^4 + 126x^4y^5$$

$$+ 84x^3y^6 + 36x^2y^7 + 9xy^8 + y^9 \quad \blacktriangleright$$

◀ EXAMPLE 3
Applying the binomial theorem

Expand $\left(a - \dfrac{b}{2}\right)^5$.
    Write the binomial as follows.

$$\left(a - \frac{b}{2}\right)^5 = \left(a + \left(-\frac{b}{2}\right)\right)^5$$

Now use the binomial theorem with $x = a$, $y = -b/2$, and $n = 5$ to get

$$\left(a - \frac{b}{2}\right)^5 = a^5 + \binom{5}{1}a^4\left(-\frac{b}{2}\right) + \binom{5}{2}a^3\left(-\frac{b}{2}\right)^2 + \binom{5}{3}a^2\left(-\frac{b}{2}\right)^3 + \binom{5}{4}a\left(-\frac{b}{2}\right)^4 + \left(-\frac{b}{2}\right)^5$$

$$= a^5 + 5a^4\left(-\frac{b}{2}\right) + 10a^3\left(-\frac{b}{2}\right)^2 + 10a^2\left(-\frac{b}{2}\right)^3 + 5a\left(-\frac{b}{2}\right)^4 + \left(-\frac{b}{2}\right)^5$$

$$= a^5 - \frac{5}{2}a^4 b + \frac{5}{2}a^3 b^2 - \frac{5}{4}a^2 b^3 + \frac{5}{16}ab^4 - \frac{1}{32}b^5. \quad \blacktriangleright$$

NOTE As Example 3 illustrates, any expansion of the *difference* of two terms has alternating signs.

EXAMPLE 4
Applying the binomial theorem

Expand $\left(\dfrac{3}{m^2} - 2\sqrt{m}\right)^4$. (Assume $m > 0$.)

By the binomial theorem,

$$\left(\frac{3}{m^2} - 2\sqrt{m}\right)^4 = \left(\frac{3}{m^2}\right)^4 + \binom{4}{1}\left(\frac{3}{m^2}\right)^3 (-2\sqrt{m})^1 + \binom{4}{2}\left(\frac{3}{m^2}\right)^2 (-2\sqrt{m})^2$$

$$+ \binom{4}{3}\left(\frac{3}{m^2}\right)^1 (-2\sqrt{m})^3 + (-2\sqrt{m})^4$$

$$= \frac{81}{m^8} + 4\left(\frac{27}{m^6}\right)(-2m^{1/2}) + 6\left(\frac{9}{m^4}\right)(4m)$$

$$+ 4\left(\frac{3}{m^2}\right)(-8m^{3/2}) + 16m^2.$$

Here, we used the fact that $\sqrt{m} = m^{1/2}$. Finally,

$$\left(\frac{3}{m^2} - 2\sqrt{m}\right)^4 = \frac{81}{m^8} - \frac{216}{m^{11/2}} + \frac{216}{m^3} - \frac{96}{m^{1/2}} + 16m^2. \quad \blacktriangleright$$

Earlier in this section, we wrote the binomial theorem in summation notation as $\sum_{r=0}^{n} \binom{n}{r} x^{n-r}y^r$, which gives the form of each term. We can use this form to write any particular term of a binomial expansion without writing out the entire expansion. For example, to find the tenth term of $(x + y)^n$, where $n \geq 9$, first notice that in the tenth term y is raised to the ninth power (since y has the power 1 in the second term, the power 2 in the third term, and so on). Because the exponents on x and y in any term must have a sum of n, the exponent on x in the tenth term is $n - 9$. Thus, the tenth term of the expansion is

$$\binom{n}{9}x^{n-9}y^9 = \frac{n!}{9!\,(n-9)!}\,x^{n-9}y^9.$$

This same idea can be used to obtain the result given in the following theorem.

---

*k*th Term of the Binomial Expansion

The *k*th term of the binomial expansion of $(x + y)^n$, where $n \geq k - 1$, is

$$\binom{n}{k-1}x^{n-(k-1)}y^{k-1}.$$

To find the *k*th term of a binomial expansion, use the following steps.

1. Find $k - 1$. This is the exponent on the second term of the binomial.
2. Subtract the exponent on the second term from n to get the exponent on the first term of the binomial.
3. Determine the coefficient by using the exponents found in the first two steps and n.

◀ **EXAMPLE 5**
Finding a particular term of a binomial expansion

Find the seventh term of $(a + 2b)^{10}$.

In the seventh term $2b$ has an exponent of 6, while a has an exponent of $10 - 6$, or 4. The seventh term is

$$\binom{10}{6}a^4(2b)^6 = 210a^4(64b^6)$$

$$= 13{,}440a^4b^6. \quad ▶$$

11.4 Exercises ▼▼▼▼▼▼▼▼▼▼▼▼▼▼▼▼▼▼▼▼▼▼▼▼▼▼▼▼▼▼▼▼▼▼

Evaluate the following. See Example 1.

1. $\dfrac{6!}{3!\,3!}$

2. $\dfrac{5!}{2!\,3!}$

3. $\dfrac{7!}{3!\,4!}$

4. $\dfrac{8!}{5!\,3!}$

5. $\binom{8}{3}$

6. $\binom{7}{4}$

7. $\binom{10}{8}$

8. $\binom{9}{6}$

9. $\binom{13}{13}$

10. $\binom{12}{12}$

11. $\binom{n}{n-1}$

12. $\binom{n}{n-2}$

13. Describe in your own words how you would determine the binomial coefficient for the fifth term in the expansion of $(x + y)^8$.

14. How many terms are there in the expansion of $(x + y)^{10}$?

Write out the binomial expansion for the following. See Examples 2–4.

15. $(x + y)^6$

16. $(m + n)^4$

17. $(p - q)^5$

18. $(a - b)^7$

19. $(r^2 + s)^5$

20. $(m + n^2)^4$

21. $(p + 2q)^4$

22. $(3r - s)^6$

23. $(7p + 2q)^4$

24. $(4a - 5b)^5$

25. $(3x - 2y)^6$

26. $(7k - 9j)^4$

27. $\left(\dfrac{m}{2} - 1\right)^6$

28. $\left(3 + \dfrac{y}{3}\right)^5$

29. $\left(\sqrt{2r} + \dfrac{1}{m}\right)^4$

30. $\left(\dfrac{1}{k} - \sqrt{3p}\right)^3$

Write the indicated term of the binomial expansion. See Example 5.

31. Sixth term of $(4h - j)^8$

32. Eighth term of $(2c - 3d)^{14}$

33. Fifteenth term of $(a^2 + b)^{22}$

34. Twelfth term of $(2x + y^2)^{16}$

35. Fifteenth term of $(x - y^3)^{20}$

36. Tenth term of $(a^3 + 3b)^{11}$

Use the concepts of this section to work Exercises 37–40.

37. Find the middle term of $(3x^7 + 2y^3)^8$.

38. Find the two middle terms of $(-2m^{-1} + 3n^{-2})^{11}$.

39. Find the value of n for which the coefficients of the fifth and eighth terms in the expansion of $(x + y)^n$ are the same.

40. Find the term in the expansion of $(3 + \sqrt{x})^{11}$ that contains x^4.

In later courses, it is shown that

$$(1 + x)^n = 1 + nx + \frac{n(n - 1)}{2!}x^2 + \frac{n(n - 1)(n - 2)}{3!}x^3 + \cdots$$

for any real number n (not just positive integer values) and any real number x where $|x| < 1$. This result, a generalized binomial theorem, may be used to find approximate values of powers and roots. For example,

$$(1.008)^{1/4} = (1 + .008)^{1/4}$$

$$= 1 + \frac{1}{4}(.008) + \frac{(1/4)(-3/4)}{2!}(.008)^2 + \frac{(1/4)(-3/4)(-7/4)}{3!}(.008)^3 + \cdots$$

$$\approx 1.002.$$

Use this result to approximate the quantities in Exercises 41–44 to the nearest thousandth.

41. $(1.02)^{-3}$ **42.** $\dfrac{1}{1.04^5}$ **43.** $(1.01)^{3/2}$ **44.** $(1.03)^2$

45. Let $n = -1$ and expand $(1 + x)^{-1}$.

46. Use polynomial division to find the first four terms when $1 + x$ is divided into 1. Compare the result with the result of Exercise 45. What do you find? Explain.

47. Find the sum of the first four terms in the expansion of $(1 + 3)^{1/2}$ using $x = 3$ and $n = 1/2$ in the formula above. Is the result close to $(1 + 3)^{1/2} = 4^{1/2} = 2$? Why not? Explain.

48. Use the result above to show that for small values of x, $\sqrt{1 + x} \approx 1 + (1/2)x$.

▼▼▼▼▼▼▼▼▼▼▼▼▼ **DISCOVERING CONNECTIONS** (Exercises 49–52) ▼▼▼▼▼▼▼▼▼▼▼▼▼

In this section we saw how the factorial of a positive integer n can be computed as a product:
$n! = 1 \cdot 2 \cdot 3 \cdot \ldots \cdot n$. *Calculators and computers are capable of evaluating factorials very quickly. Before the days of technology, mathematicians developed a formula for approximating large factorials. Interestingly enough, the formula involves the irrational numbers π and e. It is called **Stirling's formula:***

$$n! \approx \sqrt{2\pi n} \cdot n^n \cdot e^{-n}.$$

As an example for a small value of n, we observe that the exact value of 5! is 120, while Stirling's formula gives the approximation as 118.019160 using a graphing calculator. This is "off" by less than 2, an error of only 1.65%.

Work Exercises 49–52 in order.

49. Use a calculator to find the exact value of 10! and the approximation using Stirling's formula.

50. Subtract the larger value from the smaller value in Exercise 49. Divide it by 10! and convert to a percent. What is the percent error?

51. Repeat Exercises 49 and 50 for $n = 12$.

52. Repeat Exercises 49 and 50 for $n = 13$. What seems to happen as n gets larger?

---

53. When $(4x - 5)^7$ is written in the form $a_7 x^7 + a_6 x^6 + \cdots + a_1 x + a_0$, what is the sum of the numbers $a_7, a_6, \ldots, a_1, a_0$? (*Hint:* This question can be answered without determining the values of the coefficients.)

54. Show that $\binom{n}{0} + \binom{n}{1} + \binom{n}{2} + \cdots + \binom{n}{n} = 2^n$. (*Hint:* Set $x = 1$ in the binomial expansion of $(1 + x)^n$.)

11.5   Mathematical Induction ▼▼▼

Many results in mathematics are claimed to be true for every positive integer. Any of these results could be checked for $n = 1$, $n = 2$, $n = 3$, and so on, but since the set of positive integers is infinite it would be impossible to check every possible case. For example let S_n represent the statement that the sum of the first n positive integers is $n(n + 1)/2$.

$$S_n: \quad 1 + 2 + 3 + \cdots + n = \frac{n(n + 1)}{2}$$

The truth of this statement is easily verified for the first few values of n:

If $n = 1$, then S_1 is	$1 = \dfrac{1(1 + 1)}{2}$,	which is true.
If $n = 2$, then S_2 is	$1 + 2 = \dfrac{2(2 + 1)}{2}$,	which is true.
If $n = 3$, then S_3 is	$1 + 2 + 3 = \dfrac{3(3 + 1)}{2}$,	which is true.
If $n = 4$, then S_4 is	$1 + 2 + 3 + 4 = \dfrac{4(4 + 1)}{2}$,	which is true.

Continuing in this way for any amount of time would still not prove that S_n is true for *every* positive integer value of n. To prove that such statements are true for every positive integer value of n, the following principle is often used.

Principle of Mathematical Induction

Let S_n be a statement concerning the positive integer n. Suppose that

1. S_1 is true;
2. for any positive integer k, $k \leq n$, if S_k is true, then S_{k+1} is also true.

Then S_n is true for every positive integer value of n.

A proof by mathematical induction can be explained as follows. By assumption (1) above, the statement is true when $n = 1$. By (2) above, the fact that the statement is true for $n = 1$ implies that it is true for $n = 1 + 1 = 2$. Using (2) again, the statement is thus true for $2 + 1 = 3$, for $3 + 1 = 4$, for $4 + 1 = 5$, and so on. Continuing in this way shows that the statement must be true for *every* positive integer, no matter how large.

The situation is similar to that of a number of dominoes lined up as shown in Figure 4. If the first domino is pushed over, it pushes the next, which pushes the next, and so on until all are down.

FIGURE 4

Another example of the principle of mathematical induction might be an infinite ladder. Suppose the rungs are spaced so that, whenever you are on a rung, you know you can move to the next rung. Then *if* you can get to the first rung, you can go as high up the ladder as you wish.

As these comments show, two separate steps are required for a proof by mathematical induction.

Proof by Mathematical Induction

Step 1   Prove that the statement is true for $n = 1$.
Step 2   Show that, for any positive integer k, $k \leq n$, if S_k is true, then S_{k+1} is also true.

Mathematical induction is used in the next example to prove the statement S_n mentioned at the beginning of this section.

EXAMPLE 1
Proving a statement by
mathematical induction

Let S_n represent the statement

$$1 + 2 + 3 + \cdots + n = \frac{n(n + 1)}{2}.$$

Prove that S_n is true for every positive integer n.

Proof   The proof by mathematical induction is as follows.

Step 1   Show that the statement is true when $n = 1$. If $n = 1$, S_1 becomes

$$1 = \frac{1(1 + 1)}{2},$$

which is true.

Step 2   Show that S_k implies S_{k+1}, where S_k is the statement

$$1 + 2 + 3 + \cdots + k = \frac{k(k + 1)}{2},$$

and S_{k+1} is the statement

$$1 + 2 + 3 + \cdots + k + (k + 1) = \frac{(k + 1)[(k + 1) + 1]}{2}.$$

Start with S_k.

$$1 + 2 + 3 + \cdots + k = \frac{k(k + 1)}{2}$$

How can S_k be changed algebraically to match S_{k+1}? Adding $k + 1$ to both sides of S_k gives

$$1 + 2 + 3 + \cdots + k + (k + 1) = \frac{k(k + 1)}{2} + (k + 1)$$

Now factor out the common factor $k + 1$ on the right to get

$$= (k + 1)\left(\frac{k}{2} + 1\right)$$

$$= (k + 1)\left(\frac{k + 2}{2}\right)$$

$$1 + 2 + 3 + \cdots + k + (k + 1) = \frac{(k + 1)[(k + 1) + 1]}{2}. \quad \blacktriangleright$$

This final result is the statement for $n = k + 1$; it has been shown that S_k implies S_{k+1}. The two steps required for a proof by mathematical induction have now been completed, so the statement S_n is true for every positive integer value of n.

CAUTION   Notice that the left side of the statement always includes *all* the terms up to the nth term, as well as the nth term.

CONNECTIONS Notice in Example 1 that the sum on the left in the statement of S_n can be written as a sum of functions of the form $f(n) = n$:

$$S_n = f(1) + f(2) + f(3) + \cdots + f(n) = g(n).$$

With this notation, the step where we add the term $k + 1$ to both sides of S_k becomes

$$f(1) + f(2) + f(3) + \cdots + f(k) = g(k)$$

$$f(1) + f(2) + f(3) + \cdots + f(k) + f(k + 1) = g(k) + f(k + 1),$$

so that we must prove that $g(k) + f(k + 1) = g(k + 1)$.

FOR DISCUSSION OR WRITING

1. Let $f(n) = 3n + 1$ and find $f(1), f(2), f(3)$.
2. If the statement

$$4 + 7 + 10 + \cdots + (3n + 1) = \frac{n(3n + 5)}{2}$$

is written using function notation, as shown above, what is $g(k)$?
3. Show that $g(k) + f(k + 1) = g(k + 1)$.

EXAMPLE 2
Proving a statement by mathematical induction

Prove: If x is a real number between 0 and 1, then for every positive integer n,

$$0 < x^n < 1.$$

Proof   Here S_1 is the statement

if $0 < x < 1$, then $0 < x^1 < 1$,

which is true. S_k is the statement

if $0 < x < 1$, then $0 < x^k < 1$.

To show that this implies that S_{k+1} is true, multiply all expressions of $0 < x^k < 1$ by x to get

$$x \cdot 0 < x \cdot x^k < x \cdot 1.$$

(Here the fact that $0 < x$ is used.) Simplify to get

$$0 < x^{k+1} < x.$$

Since $x < 1$,

$$x^{k+1} < x < 1$$

and $$0 < x^{k+1} < 1.$$

This work shows that S_k implies S_{k+1}, and since S_1 is true, the given statement is true for every positive integer n. ▶

Some statements S_n are not true for the first few values of n, but are true for all values of n that are at least equal to some fixed integer j. The following slightly generalized form of the principle of mathematical induction takes care of these cases.

Generalized Principle of Mathematical Induction

Let S_n be a statement concerning the positive integer n. Let j be a fixed positive integer. Suppose that

(a) S_j is true;
(b) for any positive integer k, $k \geq j$, S_k implies S_{k+1}.

Then S_n is true for all positive integers n, where $n \geq j$.

◀ EXAMPLE 3
Using the generalized principle

Let S_n represent the statement $2^n > 2n + 1$. Show that S_n is true for all values of n such that $n \geq 3$.

(Check that S_n is false for $n = 1$ and $n = 2$.) As before, the proof requires two steps.

Step 1 Show that S_n is true for $n = 3$. If $n = 3$, S_n is

$$2^3 > 2 \cdot 3 + 1,$$

or

$$8 > 7,$$

which is true.

Step 2 Now show that S_k implies S_{k+1}, where $k \geq 3$ and

$$S_k \quad \text{is} \quad 2^k > 2k + 1,$$
$$S_{k+1} \text{ is } 2^{k+1} > 2(k + 1) + 1.$$

Multiply both sides of $2^k > 2k + 1$ by 2, obtaining

$$2 \cdot 2^k > 2(2k + 1),$$

or

$$2^{k+1} > 4k + 2.$$

Rewrite $4k + 2$ as $2(k + 1) + 2k$, giving

$$2^{k+1} > 2(k + 1) + 2k. \tag{1}$$

Since k is a positive integer greater than 3,

$$2k > 1. \tag{2}$$

Adding $2(k + 1)$ to both sides of inequality (2) gives

$$2(k + 1) + 2k > 2(k + 1) + 1. \tag{3}$$

From inequalities (1) and (3),

$$2^{k+1} > 2(k + 1) + 2k > 2(k + 1) + 1,$$

or

$$2^{k+1} > 2(k + 1) + 1,$$

as required. Thus, S_k implies S_{k+1}, and this, together with the fact that S_3 is true, shows that S_n is true for every positive integer value of n greater than or equal to 3. ▶

EXAMPLE 4
Proving the binomial theorem

The binomial theorem can be proved by mathematical induction. That is, for any positive integer n and any complex numbers x and y,

$$(x + y)^n = x^n + \binom{n}{1}x^{n-1}y + \binom{n}{2}x^{n-2}y^2 + \binom{n}{3}x^{n-3}y^3$$

$$+ \cdots + \binom{n}{r}x^{n-r}y^r + \cdots + \binom{n}{n-1}xy^{n-1} + y^n. \quad \text{(4)}$$

Proof Let S_n be statement (4) above. Begin by verifying S_n for $n = 1$,

$$S_1: \quad (x + y)^1 = x^1 + y^1,$$

which is true.

Now assume that S_n is true for the positive integer k. Statement S_k becomes (using the definition of the binomial coeffcient)

$$S_k: \quad (x + y)^k = x^k + \frac{k!}{1! \, (k - 1)!}x^{k-1}y + \frac{k!}{2! \, (k - 2)!}x^{k-2}y^2$$

$$+ \cdots + \frac{k!}{(k - 1)! \, 1!}xy^{k-1} + y^k. \quad \text{(5)}$$

Multiply both sides of equation (5) by $x + y$.

$$(x + y)^k \cdot (x + y)$$
$$= x(x + y)^k + y(x + y)^k$$
$$= \left[x \cdot x^k + \frac{k!}{1! \, (k - 1)!}x^k y + \frac{k!}{2! \, (k - 2)!}x^{k-1}y^2 + \cdots + \frac{k!}{(k - 1)! \, 1!}x^2 y^{k-1} + xy^k \right]$$
$$+ \left[x^k \cdot y + \frac{k!}{1! \, (k - 1)!}x^{k-1}y^2 + \cdots + \frac{k!}{(k - 1)! \, 1!}xy^k + y \cdot y^k \right]$$

Rearrange terms to get

$$(x + y)^{k+1} = x^{k+1} + \left[\frac{k!}{1! \, (k - 1)!} + 1 \right]x^k y + \left[\frac{k!}{2! \, (k - 2)!} + \frac{k!}{1! \, (k - 1)!} \right]x^{k-1}y^2$$

$$+ \cdots + \left[1 + \frac{k!}{(k - 1)! \, 1!} \right]xy^k + y^{k+1}. \quad \text{(6)}$$

The first expression in brackets in equation (6) simplifies to $\binom{k+1}{1}$. To see this, note that

$$\binom{k + 1}{1} = \frac{(k + 1)(k)(k - 1)(k - 2) \cdots 1}{1 \cdot (k)(k - 1)(k - 2) \cdots 1} = k + 1.$$

Also,

$$\frac{k!}{1! \, (k - 1)!} + 1 = \frac{k(k - 1)!}{1(k - 1)!} + 1 = k + 1.$$

The second expression becomes $\binom{k+1}{2}$, the last $\binom{k+1}{k}$, and so on. The result of equation (6) is just equation (5) with every k replaced by $k + 1$. Thus, the truth of S_n when $n = k$ implies the truth of S_n for $n = k + 1$, which completes the proof of the theorem by mathematical induction. ▶

11.5 Exercises ▼▼▼▼▼▼▼▼▼▼▼▼▼▼▼▼▼▼▼▼▼▼▼▼▼▼▼▼▼▼▼▼▼▼▼▼▼

Write out in full and verify the statements S_1, S_2, S_3, S_4, and S_5 for the following. Then use mathematical induction to prove that each statement is true for every positive integer n. See Example 1.

1. $2 + 4 + 6 + \cdots + 2n = n(n + 1)$

2. $1 + 3 + 5 + \cdots + (2n - 1) = n^2$

Use the method of mathematical induction to prove the statement. Assume that n is a positive integer. See Example 1.

3. $3 + 6 + 9 + \cdots + 3n = \dfrac{3n(n + 1)}{2}$

4. $5 + 10 + 15 + \cdots + 5n = \dfrac{5n(n + 1)}{2}$

5. $2 + 4 + 8 + \cdots + 2^n = 2^{n+1} - 2$

6. $3 + 3^2 + 3^3 + \cdots + 3^n = \dfrac{3(3^n - 1)}{2}$

7. $1^2 + 2^2 + 3^2 + \cdots + n^2 = \dfrac{n(n + 1)(2n + 1)}{6}$

8. $1^3 + 2^3 + 3^3 + \cdots + n^3 = \dfrac{n^2(n + 1)^2}{4}$

9. $5 \cdot 6 + 5 \cdot 6^2 + 5 \cdot 6^3 + \cdots + 5 \cdot 6^n = 6(6^n - 1)$

10. $7 \cdot 8 + 7 \cdot 8^2 + 7 \cdot 8^3 + \cdots + 7 \cdot 8^n = 8(8^n - 1)$

11. $\dfrac{1}{1 \cdot 2} + \dfrac{1}{2 \cdot 3} + \dfrac{1}{3 \cdot 4} + \cdots + \dfrac{1}{n(n + 1)} = \dfrac{n}{n + 1}$

12. $\dfrac{1}{1 \cdot 4} + \dfrac{1}{4 \cdot 7} + \dfrac{1}{7 \cdot 10} + \cdots + \dfrac{1}{(3n - 2)(3n + 1)} = \dfrac{n}{3n + 1}$

13. $\dfrac{1}{2} + \dfrac{1}{2^2} + \dfrac{1}{2^3} + \cdots + \dfrac{1}{2^n} = 1 - \dfrac{1}{2^n}$

14. $\dfrac{4}{5} + \dfrac{4}{5^2} + \dfrac{4}{5^3} + \cdots + \dfrac{4}{5^n} = 1 - \dfrac{1}{5^n}$

In Exercises 15–18, find all natural number values for n for which the given statement is not true.

15. $2^n > 2n$

16. $3^n > 2n + 1$

17. $2^n > n^2$

18. $n! > 2n$

Prove each statement by mathematical induction. See Examples 2 and 3.

19. $(a^m)^n = a^{mn}$ (Assume a and m are constant.)

20. $(ab)^n = a^n b^n$ (Assume a and b are constant.)

21. $2^n > 2n$, if $n \geq 3$

22. $3^n > 2n + 1$, if $n \geq 2$

23. If $a > 1$, then $a^n > 1$

24. If $a > 1$, then $a^n > a^{n-1}$

25. If $0 < a < 1$, then $a^n < a^{n-1}$

26. $2^n > n^2$, for $n > 4$

27. If $n \geq 4$, then $n! > 2^n$, where $n! = n(n - 1)(n - 2) \cdots (3)(2)(1)$.

28. $4^n > n^4$, for $n \geq 5$

29. Suppose that n straight lines (with $n \geq 2$) are drawn in a plane, where no two lines are parallel and no three lines pass through the same point. Show that the number of points of intersection of the lines is $(n^2 - n)/2$.

30. The series of sketches below starts with an equilateral triangle having sides of length 1. In the following steps, equilateral triangles are constructed on each side of the preceding figure. The lengths of the sides of these new triangles is $1/3$ the length of the sides of preceding triangles. Develop a formula for the number of sides of the nth figure. Use mathematical induction to prove your answer.

31. Find the perimeter of the nth figure in Exercise 30.

32. Show that the area of the nth figure in Exercise 30 is

$$\sqrt{3}\left[\frac{2}{5} - \frac{3}{20}\left(\frac{4}{9}\right)^{n-1}\right].$$

33. A pile of n rings, each ring smaller than the one below it, is on a peg. Two other pegs are attached to a board with this peg. In the game called the *Tower of Hanoi* puzzle, all the rings must be moved to a different peg, with only one ring moved at a time, and with no ring ever placed on top of a smaller ring. Find the least number of moves that would be required. Prove your result with mathematical induction.

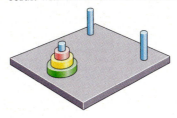

11.6 Counting Theory ▼▼▼

If there are 3 roads from Albany to Baker and 2 roads from Baker to Creswich, in how many ways can one travel from Albany to Creswich by way of Baker? For each of the 3 roads from Albany to Baker, there are 2 different roads from Baker to Creswich. Hence, there are $3 \cdot 2 = 6$ different ways to make the trip, as shown in the **tree diagram** in Figure 5.

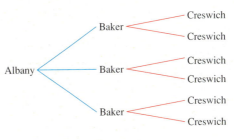

FIGURE 5

Two events are **independent events** if neither influences the outcome of the other. The opening example illustrates the fundamental principle of counting with independent events.

Fundamental Principle of Counting

If n independent events occur, with

$$m_1 \text{ ways for event 1 to occur,}$$
$$m_2 \text{ ways for event 2 to occur,}$$
$$\vdots$$

and $\quad\quad\quad\quad\quad m_n$ ways for event n to occur,

then there are
$$m_1 \cdot m_2 \cdot \ldots \cdot m_n$$

different ways for all n events to occur.

EXAMPLE 1
Using the fundamental principle of counting

A restaurant offers a choice of 3 salads, 5 main dishes, and 2 desserts. Use the fundamental principle of counting to find the number of different 3-course meals that can be selected.

Three events are involved: selecting a salad, selecting a main dish, and selecting a dessert. The first event can occur in 3 ways, the second event can occur in 5 ways, and the third event can occur in 2 ways; thus there are

$$3 \cdot 5 \cdot 2 = 30 \text{ possible meals.} \quad \blacktriangleright$$

EXAMPLE 2
Using the fundamental principle of counting

A teacher has 5 different books that he wishes to arrange in a row. How many different arrangements are possible?

Five events are involved: selecting a book for the first spot, selecting a book for the second spot, and so on. For the first spot the teacher has 5 choices. After a choice has been made, the teacher has 4 choices for the second spot. Continuing in this manner, there are 3 choices for the third spot, 2 for the fourth spot, and 1 for the fifth spot. By the fundamental principle of counting, there are

$$5 \cdot 4 \cdot 3 \cdot 2 \cdot 1 \text{ or } 120 \text{ different arrangements.} \quad \blacktriangleright$$

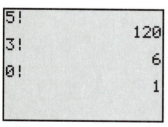

The discussion following Example 2 is supported in this screen.

In using the fundamental principle of counting, products such as $5 \cdot 4 \cdot 3 \cdot 2 \cdot 1$ occur often. For convenience in writing these products, we use the symbol $n!$ (read "n factorial"), which was defined earlier for any counting number n, as follows.

$$n! = n(n-1)(n-2)(n-3) \ldots (2)(1)$$

Thus, $5 \cdot 4 \cdot 3 \cdot 2 \cdot 1$ is written as $5!$. Also, $3! = 3 \cdot 2 \cdot 1 = 6$. By the definition of $n!$, $n[(n-1)!] = n!$ for all natural numbers $n \geq 2$. It is convenient to have this relation hold also for $n = 1$, so, by definition,

$$0! = 1.$$

EXAMPLE 3
Arranging r of n items $(r < n)$

Suppose the teacher in Example 2 wishes to place only 3 of the 5 books in a row. How many arrangements of 3 books are possible?

The teacher still has 5 ways to fill the first spot, 4 ways to fill the second spot, and 3 ways to fill the third. Since only 3 books will be used, there are only 3 spots to be filled (3 events) instead of 5, with

$$5 \cdot 4 \cdot 3 = 60 \text{ arrangements.} \quad \blacktriangleright$$

PERMUTATIONS   Since each ordering of three books is considered a different *arrangement*, the number 60 in the preceding example is called the number of permutations of 5 things taken 3 at a time, written $P(5, 3) = 60$. The number of ways of arranging 5 elements from a set of 5 elements, written $P(5, 5) = 120$, was found in Example 2. A **permutation** of n elements taken r at a time is one of the *arrangements* of r elements from a set of n elements. Generalizing from the examples above, the number of permutations of n elements, taken r at a time, denoted by $P(n, r)$, is

$$P(n, r) = n(n - 1)(n - 2) \cdots (n - r + 1)$$

$$= \frac{n(n - 1)(n - 2) \cdots (n - r + 1)(n - r)(n - r - 1) \cdots (2)(1)}{(n - r)(n - r - 1) \cdots (2)(1)}$$

$$= \frac{n!}{(n - r)!}.$$

In summary, we have the following result.

Permutations of n Elements Taken r at a Time

If $P(n, r)$ denotes the number of permutations of n elements taken r at a time, with $r \le n$, then

$$P(n, r) = \frac{n!}{(n - r)!}.$$

Alternative notations for $P(n, r)$ are P_r^n and $_nP_r$.

Many scientific calculators have a key for finding permutations. Although most graphing calculators do not have a dedicated key for permutations, a permutations function is usually found in the math menu, along with $_nC_r$. The screen here shows how the calculator evaluates $P(5, 2)$, $P(7, 0)$, and $P(4, 4)$.

```
5 nPr 2
                20
7 nPr 0
                 1
4 nPr 4
                24
```

This screen supports the following permutations calculations:

$$P(5, 2) = \frac{5!}{(5 - 2)!} = \frac{5!}{3!} = 4 \cdot 5 = 20$$

$$P(7, 0) = \frac{7!}{(7 - 0)!} = \frac{7!}{7!} = 1$$

$$P(4, 4) = \frac{4!}{(4 - 4)!} = \frac{4!}{0!} = 24$$

Using the permutations
formula

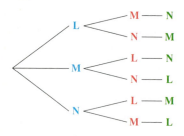

FIGURE 6

The results in Examples 4 and 5 are
supported here.

Using the permutations
formula

Find the following.

(a) The number of permutations of the letters L, M, and N
By the formula for $P(n, r)$, with $n = 3$ and $r = 3$,

$$P(3, 3) = \frac{3!}{(3 - 3)!}$$

$$= \frac{3!}{0!} = \frac{3!}{1} = 3 \cdot 2 \cdot 1 = 6.$$

As shown in the tree diagram in Figure 6, the 6 permutations are

LMN, LNM, MLN, MNL, NLM, NML.

(b) The number of permutations of 2 of the 3 letters M, N, and L
Find $P(3, 2)$.

$$P(3, 2) = \frac{3!}{(3 - 2)!}$$

$$= \frac{3!}{1!} = \frac{3!}{1} = 6$$

This result is the same as the answer in part (a) because after the first 2 choices
are made, the third is already determined, since only one letter is left.

EXAMPLE 5
Using the permutations
formula

Suppose 8 people enter an event in a swim meet. In how many ways could the
gold, silver, and bronze prizes be awarded?
    Using the fundamental principle of counting, there are 3 choices to be made,
giving $8 \cdot 7 \cdot 6 = 336$. However, the formula for $P(n, r)$ also can be used to get
the same result.

$$P(8, 3) = \frac{8!}{5!} = \frac{8 \cdot 7 \cdot 6 \cdot 5 \cdot 4 \cdot 3 \cdot 2 \cdot 1}{5 \cdot 4 \cdot 3 \cdot 2 \cdot 1}$$

$$= 8 \cdot 7 \cdot 6 = 336$$

EXAMPLE 6
Using the permutations
formula

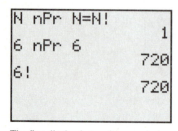

The first display is true for *any* whole
number value of N. The calculations in
Example 6 are supported in the second
and third displays.

In how many ways can 6 students be seated in a row of 6 desks?
    Use $P(n, n)$ with $n = 6$ to get

$$P(6, 6) = 6! = 6 \cdot 5 \cdot 4 \cdot 3 \cdot 2 \cdot 1 = 720.$$

COMBINATIONS   Earlier, we saw that there are 60 ways that a teacher can
arrange 3 of 5 different books in a row. That is, there are 60 permutations of
5 things taken 3 at a time. Suppose now that the teacher does not wish to arrange
the books in a row, but rather wishes to choose, without regard to order, any 3
of the 5 books to donate to a book sale to raise money for the school. In how
many ways can this be done?

At first glance, we might say 60 again, but this is incorrect. The number 60 counts all possible *arrangements* of 3 books chosen from 5. The following 6 arrangements, however, would all lead to the same set of 3 books being given to the book sale.

mystery-biography-textbook	biography-textbook-mystery
mystery-textbook-biography	textbook-biography-mystery
biography-mystery-textbook	textbook-mystery-biography

The list shows 6 different *arrangements* of 3 books but only one *set* of 3 books. A subset of items selected *without regard to order* is called a **combination.** The number of combinations of 5 things taken 3 at a time is written $\binom{5}{3}$, or $C(5, 3)$. In this book, we will use the more common notation $\binom{5}{3}$.

NOTE This combinations notation also represents the binomial coefficient defined in Section 4 of this chapter. That is, binomial coefficients are the combinations of n elements chosen r at a time.

To evaluate $\binom{5}{3}$, start with the $5 \cdot 4 \cdot 3$ *permutations* of 5 things taken 3 at a time. Since order doesn't matter, and each subset of 3 items from the set of 5 items can have its elements rearranged in $3 \cdot 2 \cdot 1 = 3!$ ways, $\binom{5}{3}$ can be found by dividing the number of permutations by $3!$, or

$$\binom{5}{3} = \frac{5 \cdot 4 \cdot 3}{3!} = \frac{5 \cdot 4 \cdot 3}{3 \cdot 2 \cdot 1} = 10.$$

There are 10 ways that the teacher can choose 3 books for the book sale.

Generalizing this discussion gives the following formula for the number of combinations of n elements taken r at a time:

$$\binom{n}{r} = \frac{P(n, r)}{r!}.$$

A more useful version of this formula is found as follows.

$$\binom{n}{r} = \frac{P(n, r)}{r!}$$

$$= \frac{n!}{(n - r)!} \cdot \frac{1}{r!}$$

$$= \frac{n!}{(n - r)! \, r!}$$

This last version is the most useful for calculation and is the one we used earlier to calculate binomial coefficients. This discussion is summarized in the following result.

> **Combinations of n Elements Taken r at a Time**
>
> If $\binom{n}{r}$ represents the number of combinations of n things taken r at a time, with $r \leq n$, then
>
> $$\binom{n}{r} = \frac{n!}{(n-r)!\,r!}.$$

EXAMPLE 7
Using the combinations formula

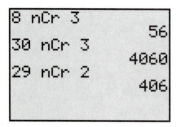

The results in Examples 7 and 8 are supported here.

How many different committees of 3 people can be chosen from a group of 8 people?

Since a committee is an unordered set, use combinations to get

$$\binom{8}{3} = \frac{8!}{5!\,3!} = \frac{8 \cdot 7 \cdot 6 \cdot 5 \cdot 4 \cdot 3 \cdot 2 \cdot 1}{5 \cdot 4 \cdot 3 \cdot 2 \cdot 1 \cdot 3 \cdot 2 \cdot 1} = 56. \;\blacktriangleright$$

EXAMPLE 8
Using the combinations formula

Three lawyers are to be selected from a group of 30 to work on a special project.

(a) In how many different ways can the lawyers be selected?

Here we wish to know the number of 3-element combinations that can be formed from a set of 30 elements. (We want combinations, not permutations, since order within the group of 3 does not matter.)

$$\binom{30}{3} = \frac{30!}{27!\,3!} = \frac{30 \cdot 29 \cdot 28 \cdot 27!}{27! \cdot 3 \cdot 2 \cdot 1}$$

$$= \frac{30 \cdot 29 \cdot 28}{3 \cdot 2 \cdot 1}$$

$$= 4060$$

There are 4060 ways to select the project group.

(b) In how many ways can the group of 3 be selected if a certain lawyer must work on the project?

Since 1 lawyer already has been selected for the project, the problem is reduced to selecting 2 more from the remaining 29 lawyers.

$$\binom{29}{2} = \frac{29!}{27!\,2!} = \frac{29 \cdot 28 \cdot 27!}{27! \cdot 2 \cdot 1} = \frac{29 \cdot 28}{2 \cdot 1} = 29 \cdot 14 = 406$$

In this case, the project group can be selected in 406 ways. $\;\blacktriangleright$

The formulas for permutations and combinations given in this section will be very useful in solving probability problems in the next section. Any difficulty in using these formulas usually comes from being unable to differentiate between them. Both permutations and combinations give the number of ways to choose r objects from a set of n objects. The differences between permutations and combinations are outlined in the following chart.

Permutations	Combinations
Different orderings or arrangements of the r objects are different permutations. $$P(n, r) = \frac{n!}{(n - r)!}$$ Clue words: arrangement, schedule, order	Each choice or subset of r objects gives one combination. Order within the group of r objects does not matter. $$\binom{n}{r} = \frac{n!}{(n - r)! \, r!}$$ Clue words: group, committee, sample, selection

In the next example, concentrate on recognizing which formula should be applied.

◀EXAMPLE 9
Distinguishing between combinations and permutations

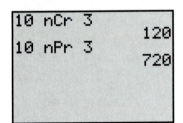

This screen supports the results in Example 9.

A sales representative has 10 accounts in a certain city.

(a) In how many ways can 3 accounts be selected to call on?

Within a selection of 3 accounts, the arrangement of the visits is not important, so there are

$$\binom{10}{3} = \frac{10!}{7! \, 3!} = \frac{10 \cdot 9 \cdot 8}{3 \cdot 2 \cdot 1} = 120$$

ways to select 3 accounts.

(b) In how many ways can calls be scheduled for 3 of the 10 accounts?

To schedule calls, the sales representative must *order* each selection of 3 accounts. Use permutations here, since order is important.

$$P(10, 3) = \frac{10!}{(10 - 3)!} = \frac{10!}{7!} = 10 \cdot 9 \cdot 8 = 720$$

There are 720 different orders in which to call on 3 of the accounts. ▶

EXAMPLE 10
Distinguishing between permutations and combinations

To illustrate the differences between permutations and combinations in another way, suppose 2 cans of soup are to be selected from 4 cans on a shelf: noodle (N), bean (B), mushroom (M), and tomato (T). As shown in Figure 7(a), there are 12 ways to select 2 cans from the 4 cans if the order matters (if noodle first and bean second is considered different from bean, then noodle, for example). On the other hand, if order is unimportant, then there are 6 ways to choose 2 cans of soup from the 4, as illustrated in Figure 7(b). ▶

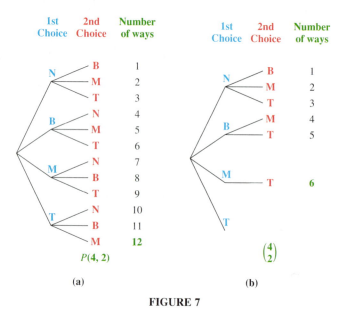

FIGURE 7

CAUTION It should be stressed that not all counting problems lend themselves to either permutations or combinations. Whenever a tree diagram or the multiplication principle can be used directly, as in Example 10, then use it.

11.6 Exercises ▼▼▼▼▼▼▼▼▼▼▼▼▼▼▼▼▼▼▼▼▼▼▼▼▼▼▼▼▼▼▼

Evaluate the following. See Examples 4–9.

1. $P(12, 8)$ **2.** $P(5, 5)$ **3.** $P(9, 2)$ **4.** $P(10, 9)$

5. $P(5, 1)$ **6.** $P(6, 0)$ **7.** $\binom{4}{2}$ **8.** $\binom{9}{3}$

9. $\binom{6}{0}$ **10.** $\binom{8}{1}$ **11.** $\binom{12}{4}$ **12.** $\binom{16}{3}$

13. Decide whether the situation described involves a permutation or a combination of objects. See Examples 9 and 10.
    (a) a telephone number
    (b) a social security number
    (c) a hand of cards in poker
    (d) a committee of politicians
    (e) the ''combination'' on a combination lock
    (f) a lottery choice of six numbers where the order does not matter
    (g) an automobile license plate

14. Explain the difference between a permutation and a combination. What should you look for in a problem to decide which of these is an appropriate method of solution?

Use the fundamental principle of counting to solve each problem. See Examples 1–3.

15. How many different types of homes are available if a builder offers a choice of 5 basic plans, 3 roof styles, and 2 exterior finishes?

16. An auto manufacturer produces 7 models, each available in 6 different colors, with 4 different upholstery fabrics, and 5 interior colors. How many varieties of the auto are available?

17. How many different 4-letter radio-station call letters can be made
    (a) if the first letter must be K or W and no letter may be repeated;
    (b) if repeats are allowed (but the first letter is K or W)?
    (c) How many of the 4-letter call letters (starting with K or W) with no repeats end in R?

18. A menu offers a choice of 3 salads, 8 main dishes, and 5 desserts. How many different 3-course meals (salad, main dish, dessert) are possible?

19. A couple has narrowed down the choice of a name for their new baby to 3 first names and 5 middle names. How many different first- and middle-name arrangements are possible?

20. A concert to raise money for an economics prize is to consist of 5 works: 2 overtures, 2 sonatas, and a piano concerto. In how many ways can a program with these 5 works be arranged?

21. For many years, the state of California used three letters followed by three digits on its automobile license plates.

    (a) How many different license plates are possible with this arrangement?
    (b) When the state ran out of new plates, the order was reversed to three digits followed by three letters. How many additional plates were then possible?
    (c) Several years ago, the plates described in (b) were also used up. The state then issued plates with one letter followed by three digits and then three letters. How many plates does this scheme provide?

22. How many 7-digit telephone numbers are possible if the first digit cannot be zero and
    (a) only odd digits may be used;
    (b) the telephone number must be a multiple of 10 (that is, it must end in zero);
    (c) the telephone number must be a multiple of 100;
    (d) the first 3 digits are 481;
    (e) no repetitions are allowed?

Solve the following problems involving permutations. See Examples 4–6.

23. In an experiment on social interaction, 6 people will sit in 6 seats in a row. In how many ways can this be done?

24. In how many ways can 7 of 10 monkeys be arranged in a row for a genetics experiment?

25. A business school offers courses in typing, shorthand, transcription, business English, technical writing, and accounting. In how many ways can a student arrange a schedule if 3 courses are taken?

26. If your college offers 400 courses, 20 of which are in mathematics, and your counselor arranges your schedule of 4 courses by random selection, how many schedules are possible that do not include a math course?

27. In a club wih 15 members, how many ways can a slate of 3 officers consisting of president, vice-president, and secretary/treasurer be chosen?

28. A baseball team has 20 players. How many 9-player batting orders are possible?

29. In how many ways can 5 players be assigned to the 5 positions on a basketball team, assuming that any player can play any position? In how many ways can 10 players be assigned to the 5 positions?

30. How many ways can all the letters of the word TOUGH be arranged?

Solve the following problems involving combinations. See Examples 7 and 8.

31. A club has 30 members. If a committee of 4 is selected at random, how many committees are possible?

32. How many different samples of 3 apples can be drawn from a crate of 25 apples?

33. Hal's Hamburger Heaven sells hamburgers with cheese, relish, lettuce, tomato, mustard, or ketchup. How many different hamburgers can be made using any 3 of the extras?

34. Three students are to be selected from a group of 12 students to participate in a special class. In how many ways can this be done? In how many ways can the group that will not participate be selected?

35. Five cards are marked with the numbers 1, 2, 3, 4, and 5, shuffled, and 2 cards are then drawn. How many different 2-card combinations are possible?

36. If a bag contains 15 marbles, how many samples of 2 marbles can be drawn from it? How many samples of 4 marbles?

37. In Exercise 36, if the bag contains 3 yellow, 4 white, and 8 blue marbles, how many samples of 2 can be drawn in which both marbles are blue?

38. In Exercise 32, if it is known that there are 5 rotten apples in the crate:
   (a) How many samples of 3 could be drawn in which all 3 are rotten?
   (b) How many samples of 3 could be drawn in which there are 1 rotten apple and 2 good apples?

39. A city council is composed of 5 liberals and 4 conservatives. Three members are to be selected randomly as delegates to a convention.
   (a) How many delegations are possible?
   (b) How many delegations could have all liberals?
   (c) How many delegations could have 2 liberals and 1 conservative?
   (d) If 1 member of the council serves as mayor, how many delegations are possible that include the mayor?

40. Seven workers decide to send a delegation of 2 to their supervisor to discuss their grievances.
   (a) How many different delegations are possible?
   (b) If it is decided that a certain employee must be in the delegation, how many different delegations are possible?

   (c) If there are 2 women and 5 men in the group, how many delegations would include at least one woman?

Use any or all of the methods described in this section to solve the following problems. See Examples 1–10.

41. If Matthew has 8 courses to choose from, how many ways can he arrange his schedule if he must pick 4 of them?

42. How many samples of 3 pineapples can be drawn from a crate of 12?

43. Velma specializes in making different vegetable soups with carrots, celery, beans, peas, mushrooms, and potatoes. How many different soups can she make using any 4 ingredients?

44. From a pool of 7 secretaries, 3 are selected to be assigned to 3 managers, 1 secretary to each manager. In how many ways can this be done?

45. In a game of musical chairs, 12 children will sit in 11 chairs (1 will be left out). How many seatings are possible?

46. In an experiment on plant hardiness, a researcher gathers 6 wheat plants, 3 barley plants, and 2 rye plants. She wishes to select 4 plants at random.
   (a) In how many ways can this be done?
   (b) In how many ways can this be done if exactly 2 wheat plants must be included?

47. In a club with 8 men and 11 women members, how many 5-member committees can be chosen that have the following?
   (a) all men
   (b) all women
   (c) 3 men and 2 women
   (d) no more than 3 women

48. From 10 names on a ballot, 4 will be elected to a political party committee. In how many ways can the committee of 4 be formed if each person will have a different responsibility?

49. In how many ways can 5 out of 9 plants be arranged in a row on a window sill?

50. In how many ways can all the letters of CHAMBER-POT be arranged?

Prove each statement for positive integers n and r, with r ≤ n.

51. $P(n, n - 1) = P(n, n)$ **52.** $P(n, 1) = n$ **53.** $P(n, 0) = 1$ **54.** $\binom{n}{n} = 1$

55. $\binom{n}{0} = 1$ **56.** $\binom{n}{n - 1} = n$ **57.** $\binom{n}{n - r} = \binom{n}{r}$

58. Explain why the restriction $r \leq n$ is needed in the formula for $P(n, r)$.

59. Series are often used in mathematics and science to make approximations. Large values of factorials often occur in counting theory. The value of $n!$ can quickly become too large for most calculators to evaluate. To estimate $n!$ for large values of n one can use the property of logarithms that

$$\log n! = \log(1 \times 2 \times 3 \times \cdots \times n)$$
$$= \log 1 + \log 2 + \log 3 + \cdots + \log n.$$

Using a sum and sequence utility on a calculator, one can then determine an r such that $n! \approx 10^r$ since $r = \log n!$. For example, the screen shown here illustrates that a calculator will give the same approximation for $30!$ using the factorial function and the formula just discussed.

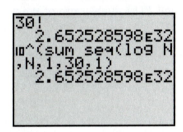

Use this technique to approximate each of the factorials. Then, try to compute the value directly on your calculator.

(a) 50!   (b) 60!   (c) 65!

60. Use the technique described in the previous exercise to approximate each permutation $P(n, r)$.

(a) $P(47, 13)$ (b) $P(50, 4)$ (c) $P(29, 21)$

11.7 Basics of Probability ▼▼▼

The study of probability has become increasingly popular because it has a wide range of practical applications. The basic ideas of probability are introduced in this section.

Consider an experiment that has one or more possible **outcomes,** each of which is equally likely to occur. For example, the experiment of tossing a fair coin has two equally likely possible outcomes: landing heads up (H) or landing tails up (T). Also, the experiment of rolling a fair die has 6 equally likely outcomes: landing so the face that is up shows 1, 2, 3, 4, 5, or 6 points.

The set S of all possible outcomes of a given experiment is called the **sample space** for the experiment. (In this text all sample spaces are finite.) One sample space for the experiment of tossing a coin could consist of the outcomes H

and T. This sample space can be written in set notation as

$$S = \{H, T\}.$$

Similarly, a sample space for the experiment of rolling a single die might be

$$S = \{1, 2, 3, 4, 5, 6\}.$$

Any subset of the sample space is called an **event.** In the experiment with the die, for example, "the number showing is a three" is an event, say E_1, such that $E_1 = \{3\}$. "The number showing is greater than three" is also an event, say E_2, such that $E_2 = \{4, 5, 6\}$. To represent the number of outcomes that belong to event E, the notation $n(E)$ is used. Then $n(E_1) = 1$ and $n(E_2) = 3$.

The notation $P(E)$ is used for the *probability* of an event E. If the outcomes in the sample space for an experiment are equally likely, then the probability of event E occurring is found as follows.

Definition of Probability of Event E

In a sample space with equally likely outcomes, the **probability** of an event E, written $P(E)$, is the ratio of the number of outcomes in sample space S that belong to event E, $n(E)$, to the total number of outcomes in sample space S, $n(S)$. That is,

$$P(E) = \frac{n(E)}{n(S)}.$$

To use this definition to find the probability of the event E_1 given above, start with the sample space for the experiment, $S = \{1, 2, 3, 4, 5, 6\}$, and the desired event, $E_1 = \{3\}$. Since $n(E_1) = 1$ and since there are 6 outcomes in the sample space,

$$P(E_1) = \frac{n(E_1)}{n(S)} = \frac{1}{6}.$$

◀ EXAMPLE 1
Finding probabilities of events

A single die is rolled. Write the following events in set notation and give the probability for each event.

(a) E_3: the number showing is even
Use the definition above. Since $E_3 = \{2, 4, 6\}$, $n(E_3) = 3$. As shown above, $n(S) = 6$, so

$$P(E_3) = \frac{3}{6} = \frac{1}{2}.$$

(b) E_4: the number showing is greater than 4
Again $n(S) = 6$. Event $E_4 = \{5, 6\}$, with $n(E_4) = 2$. By the definition,

$$P(E_4) = \frac{2}{6} = \frac{1}{3}.$$

(c) E_5: the number showing is less than 7

$$E_5 = \{1, 2, 3, 4, 5, 6\} \quad \text{and} \quad P(E_5) = \frac{6}{6} = 1$$

(d) E_6: the number showing is 7

$$E_6 = \emptyset \quad \text{and} \quad P(E_6) = \frac{0}{6} = 0 \quad \blacktriangleright$$

In Example 1(c), $E_5 = S$. Therefore, the event E_5 is certain to occur every time the experiment is performed. An event that is certain to occur, such as E_5, always has a probability of 1. On the other hand, $E_6 = \emptyset$ and $P(E_6)$ is 0. The probability of an impossible event, such as E_6, is always 0, since none of the outcomes in the sample space satisfy the event. For any event E, $P(E)$ is between 0 and 1 inclusive.

A standard deck of 52 cards has four suits: hearts, clubs, diamonds, and spades, with thirteen cards of each suit. Each suit has an ace, king, queen, jack, and cards numbered from 2 to 10. The hearts and diamonds are red and the spades and clubs are black. We will refer to this standard deck of cards in this section.

The set of all outcomes in the sample space that do *not* belong to event E is called the **complement** of E, written E'. For example, in the experiment of drawing a single card from a standard deck of 52 cards, let E be the event "the card is an ace." Then E' is the event "the card is not an ace." From the definition of E', for an event E,

$$E \cup E' = S \quad \text{and} \quad E \cap E' = \emptyset.*$$

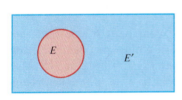

FIGURE 8

Probability concepts can be illustrated using **Venn diagrams,** as shown in Figure 8. The rectangle in Figure 8 represents the sample space in an experiment. The area inside the circle represents event E, while the area inside the rectangle, but outside the circle, represents event E'.

EXAMPLE 2
Using the complement in a probability problem

In the experiment of drawing a card from a well-shuffled deck, find the probability of event E, the card is an ace, and event E'.

Since there are four aces in the deck of 52 cards, $n(E) = 4$ and $n(S) = 52$. Therefore,

$$P(E) = \frac{n(E)}{n(S)} = \frac{4}{52} = \frac{1}{13}.$$

Of the 52 cards, 48 are not aces, so

$$P(E') = \frac{n(E')}{n(S)} = \frac{48}{52} = \frac{12}{13}. \quad \blacktriangleright$$

---

*The **union** of two sets A and B is the set $A \cup B$ made up of all the elements from either A or B, or both. The **intersection** of sets A and B, written $A \cap B$, is made up of all the elements that belong to both sets.

In Example 2, $P(E) + P(E') = (1/13) + (12/13) = 1$. This is always true for any event E and its complement E'. That is,

$$P(E) + P(E') = 1.$$

This can be restated as

$$P(E) = 1 - P(E') \qquad \text{or} \qquad P(E') = 1 - P(E).$$

These two equations suggest an alternative way to compute the probability of an event. For example, if it is known that $P(E) = 1/10$, then

$$P(E') = 1 - \frac{1}{10} = \frac{9}{10}.$$

ODDS Sometimes probability statements are expressed in terms of odds, a comparison of $P(E)$ with $P(E')$. The **odds** in favor of an event E are expressed as the ratio of $P(E)$ to $P(E')$ or as the fraction $P(E)/P(E')$. For example, if the probability of rain can be established as $1/3$, the odds that it will rain are

$$P(\text{rain}) \text{ to } P(\text{no rain}) = \frac{1}{3} \text{ to } \frac{2}{3} = \frac{1/3}{2/3} = \frac{1}{2} \qquad \text{or} \qquad 1 \text{ to } 2.$$

On the other hand, the odds that it will not rain are 2 to 1 (or $2/3$ to $1/3$). If the odds in favor of an event are, say, 3 to 5, then the probability of the event is $3/8$, while the probability of the complement of the event is $5/8$. If the odds favoring event E are m to n, then

$$P(E) = \frac{m}{m + n} \qquad \text{and} \qquad P(E') = \frac{n}{m + n}.$$

EXAMPLE 3
Finding odds in favor of an event

A shirt is selected at random from a dark closet containing 6 blue shirts and 4 shirts that are not blue. Find the odds in favor of a blue shirt being selected.

Let E represent "a blue shirt is selected." Then $P(E) = 6/10$ or $3/5$. Also, $P(E') = 1 - (3/5) = 2/5$. Therefore, the odds in favor of a blue shirt being selected are

$$P(E) \text{ to } P(E') = \frac{3}{5} \text{ to } \frac{2}{5} = \frac{3/5}{2/5} = \frac{3}{2} \qquad \text{or} \qquad 3 \text{ to } 2. \quad \blacktriangleright$$

THE UNION OF TWO EVENTS We now extend the rules for probability to more complex events. Since events are sets, we can use set operations to find the union of two events. (The *union* of sets A and B includes all elements of set A in addition to all elements of set B.)

Suppose a fair die is tossed. Let H be the event "the result is a 3," and K the event "the result is an even number." From the results earlier in this section,

$$H = \{3\} \qquad\qquad P(H) = \frac{1}{6}$$

$$K = \{2, 4, 6\} \qquad\qquad P(K) = \frac{3}{6} = \frac{1}{2}$$

$$H \cup K = \{2, 3, 4, 6\} \qquad P(H \cup K) = \frac{4}{6} = \frac{2}{3}.$$

Notice that $P(H) + P(K) = P(H \cup K)$.

Before assuming that this relationship is true in general, consider another event for this experiment, "the result is a 2," event G.

$$G = \{2\} \qquad\qquad P(G) = \frac{1}{6}$$

$$K = \{2, 4, 6\} \qquad\qquad P(K) = \frac{3}{6} = \frac{1}{2}$$

$$K \cup G = \{2, 4, 6\} \qquad P(K \cup G) = \frac{3}{6} = \frac{1}{2}$$

In this case $P(K) + P(G) \neq P(K \cup G)$. See Figure 9.

As Figure 9 suggests, the difference in the two examples above comes from the fact that events H and K cannot occur simultaneously. Such events are called **mutually exclusive events.** In fact, $H \cap K = \emptyset$, which is true for any two mutually exclusive events. Events K and G, however, can occur simultaneously. Both are satisfied if the result of the roll is a 2, the element in their intersection $(K \cap G = \{2\})$. This example suggests the following property.

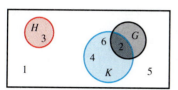

FIGURE 9

> ### Probability of the Union of Two Events
>
> For any events E and F:
> $$P(E \text{ or } F) = P(E \cup F) = P(E) + P(F) - P(E \cap F).$$

EXAMPLE 4
Finding the probability of a union

One card is drawn from a well-shuffled deck of 52 cards. What is the probability of the following outcomes?

(a) The card is an ace or a spade.

The events "drawing an ace" and "drawing a spade" are not mutually exclusive since it is possible to draw the ace of spades, an outcome satisfying both events. The probability is

$$P(\text{ace or spade}) = P(\text{ace}) + P(\text{spade}) - P(\text{ace and spade})$$

$$= \frac{4}{52} + \frac{13}{52} - \frac{1}{52} = \frac{16}{52} = \frac{4}{13}.$$

(b) The card is a three or a king.

"Drawing a 3" and "drawing a king" are mutually exclusive events because it is impossible to draw one card that is both a 3 and a king. Using the rule given above,

$$P(3 \text{ or } K) = P(3) + P(K) - P(3 \text{ and } K)$$

$$= \frac{4}{52} + \frac{4}{52} - 0 = \frac{8}{52} = \frac{2}{13}. \quad \blacktriangleright$$

```
4/52+13/52-1/52▶
Frac
                4/13
6/36+11/36-2/36▶
Frac
                5/12
```

The arithmetic in Examples 4(a) and
5(a) is easily accomplished with a
calculator.

Suppose two fair dice are rolled. Find each of the following probabilities.

(a) The first die shows a 2, or the sum of the two dice is 6 or 7.

Think of the two dice as being distinguishable, one red and one green for example. (Actually, the sample space is the same even if they are not apparently distinguishable.) A sample space with equally likely outcomes is shown in Figure 10, where (1, 1) represents the event "the first (red) die shows a 1 and the second die (green) shows a 1," (1, 2) represents "the first die shows a 1 and the second die shows a 2," and so on. Let A represent the event "the first die shows a 2," and B represent the event "the sum of the results is 6 or 7." These events are indicated in Figure 10. From the diagram, event A has 6 elements, B has 11 elements, and the sample space has 36 elements. Thus,

$$P(A) = \frac{6}{36}, \quad P(B) = \frac{11}{36}, \quad \text{and} \quad P(A \cap B) = \frac{2}{36}.$$

By the union rule,

$$P(A \cup B) = P(A) + P(B) - P(A \cap B)$$

$$P(A \cup B) = \frac{6}{36} + \frac{11}{36} - \frac{2}{36} = \frac{15}{36} = \frac{5}{12}.$$

(b) The sum of the points showing is at most 4.

"At most 4" can be written as "2 or 3 or 4." (A sum of 1 is meaningless here.) Then

$$P(\textbf{at most 4}) = P(\textbf{2 or 3 or 4})$$
$$= P(2) + P(3) + P(4), \qquad (*)$$

since the events represented by "2," "3," and "4" are mutually exclusive.

The sample space for this experiment includes the 36 possible pairs of numbers shown in Figure 10. The pair (1, 1) is the only one with a sum of 2, so

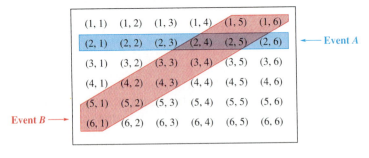

FIGURE 10

$P(2) = 1/36$. Also $P(3) = 2/36$ since both $(1, 2)$ and $(2, 1)$ give a sum of 3. The pairs, $(1, 3)$, $(2, 2)$, and $(3, 1)$ have a sum of 4, so $P(4) = 3/36$. Substituting into equation (*) above gives

$$P(\text{at most } 4) = \frac{1}{36} + \frac{2}{36} + \frac{3}{36}$$

$$= \frac{6}{36} = \frac{1}{6}. \quad \blacktriangleright$$

The properties of probability discussed in this section are summarized as follows.

Properties of Probability

For any events E and F:

1. $0 \le P(E) \le 1$
2. $P(\text{a certain event}) = 1$
3. $P(\text{an impossible event}) = 0$
4. $P(E') = 1 - P(E)$
5. $P(E \text{ or } F) = P(E \cup F) = P(E) + P(F) - P(E \cap F)$.

CAUTION   When finding the probability of a union, don't forget to subtract the probability of the intersection from the sum of the probabilities of the individual events.

BINOMIAL PROBABILITY   If an experiment consists of repeated independent trials with only two outcomes in each trial, success or failure, it is called a **binomial experiment.** Let the probability of success in one trial be p. Then the probability of failure is $1 - p$, and the probability of r successes in n trials is given by

$$\binom{n}{r} p^r (1 - p)^{n-r}.$$

Notice that this expression is equivalent to the general term of the binomial expansion given in Section 4 of this chapter. Thus the terms of the binomial expansion give the probabilities of r successes in n trials, for $0 \le r \le n$, in a binomial experiment.

◀ **EXAMPLE 6**
Finding probabilities using a
binomial experiment

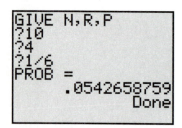

An experiment consists of rolling a die 10 times. Find the following probabilities.

(a) The probability that exactly 4 of the tosses result in a three.

The probability of a three on one roll is $p = 1/6$. The required probability is

$$\binom{10}{4}\left(\frac{1}{6}\right)^4\left(1 - \frac{1}{6}\right)^{10-4} = 210\left(\frac{1}{6}\right)^4\left(\frac{5}{6}\right)^6$$
$$\approx .054.$$

(b) The probability that in 9 of the 10 tosses the result is not a three.
This probability is

$$\binom{10}{1}\left(\frac{1}{6}\right)^1\left(\frac{5}{6}\right)^9 \approx .323. \quad ▶$$

A short program for a graphing
calculator can be written to compute
binomial probabilities. The two
screens above support the results in
Example 6.

11.7 Exercises ▼▼▼▼▼▼▼▼▼▼▼▼▼▼▼▼▼▼▼▼▼▼▼▼▼▼▼▼▼▼▼▼▼▼▼▼

Write a sample space with equally likely outcomes for each experiment.

1. A two-headed coin is tossed once.

2. Two ordinary coins are tossed.

3. Three ordinary coins are tossed.

4. Slips of paper marked with the numbers 1, 2, 3, 4, and 5 are placed in a box. After mixing well, two slips are drawn.

5. An unprepared student takes a three-question true/false quiz in which he guesses the answers to all three questions.

6. A die is rolled and then a coin is tossed.

Write the event in set notation and give the probability of the event. See Example 1.

7. In Exercise 1:
   (a) the result of the toss is heads;
   (b) the result of the toss is tails.

8. In Exercise 2:
   (a) both coins show the same face;
   (b) at least one coin turns up heads.

9. In Exercise 5:
   (a) the student gets all three answers correct;
   (b) he gets all three answers wrong;
   (c) he gets exactly two answers correct;
   (d) he gets at least one answer correct.

10. In Exercise 4:
    (a) both slips are marked with even numbers;
    (b) both slips are marked with odd numbers;
    (c) both slips are marked with the same number;
    (d) one slip is marked with an odd number, the other with an even number.

11. A student gives the answer to a probability problem as 6/5. Explain why this answer must be incorrect.

12. If the probability of an event is .857, what is the probability that the event will not occur?

Work each problem. See Examples 1–5.

13. A marble is drawn at random from a box containing 3 yellow, 4 white, and 8 blue marbles. Find the probabilities in (a)–(c).
    (a) A yellow marble is drawn.
    (b) A black marble is drawn.
    (c) The marble is yellow or white.
    (d) What are the odds in favor of drawing a yellow marble?
    (e) What are the odds against drawing a blue marble?

14. A baseball player with a batting average of .300 comes to bat. What are the odds in favor of his getting a hit?

15. In Exercise 4, what are the odds that the sum of the numbers on the two slips of paper is 5?

16. If the odds that it will rain are 4 to 5, what is the probability of rain?

17. If the odds that a candidate will win an election are 3 to 2, what is the probability that the candidate will lose?

18. A card is drawn from a well-shuffled deck of 52 cards. Find the probability that the card is
    (a) a 9;   (b) black;   (c) a black 9;
    (d) a heart;
    (e) a face card (K, Q, J of any suit);
    (f) red or a 3;
    (g) less than a 4 (consider aces as 1s).

19. Mrs. Elliott invites 10 relatives to a party: her mother, two uncles, three brothers, and four cousins. If the chances of any one guest arriving first are equally likely, find the following probabilities.
    (a) The first guest is an uncle or a brother.
    (b) The first guest is a brother or cousin.
    (c) The first guest is a brother or her mother.

20. Two dice are rolled. Find the probability of the following events.
    (a) The sum of the points is at least 10.
    (b) The sum of the points is either 7 or at least 10.
    (c) The sum of the points is 2 or the dice both show the same number.

21. The table shows the probability that a customer of a department store will make a purchase in the indicated price range.

Cost	Probability
Below $5	.25
$5–$19.99	.37
$20–$39.99	.11
$40–$69.99	.09
$70–$99.99	.07
$100–$149.99	.08
$150 or more	.03

Find the probability that a customer makes a purchase that is
(a) less than $20;   (b) $40 or more;
(c) more than $99.99;   (d) less than $100.

22. One game in a state lottery requires you to pick one heart, one club, one diamond, and one spade, in that order, from the thirteen cards in each suit. What is the probability of getting all four picks correct and winning $5000?

23. If three of the four selections in Exercise 22 are correct, the player wins $200. Find the probability of this outcome.

24. The law firm of Alam, Bartolini, Chinn, Dickinson, and Ellsberg has two senior partners, Alam and Bartolini. Two of the attorneys are to be selected to attend a conference. Assuming that all are equally likely to be selected, find the following probabilities.
(a) Chinn is selected.
(b) Alam and Dickinson are selected.
(c) At least one senior partner is selected.

25. The management of a firm wishes to survey the opinions of its workers, classified as follows for the purpose of an interview:

30% have worked for the company more than 5 years,
28% are female,
65% contribute to a voluntary retirement plan, and
1/2 of the female workers contribute to the retirement plan.
Find the following probabilities.
(a) A male worker is selected.
(b) A worker is selected who has worked for the company less than 5 years.
(c) A worker is selected who contributes to the retirement plan or is female.

26. Explain in your own words why the probability of an event must be a number between and inclusive of 0 and 1.

Suppose a family has 5 children. Also, suppose that the probability of having a girl is 1/2. Find the probability that the family has the following children. See Example 6.

27. Exactly 2 girls and 3 boys

30. No boys

28. Exactly 3 girls and 2 boys

31. At least 3 boys

29. No girls

32. No more than 4 girls

A die is rolled 12 times. Find the probability of rolling the following. See Example 6.

33. Exactly 12 ones

35. No more than 3 ones

34. Exactly 6 ones

36. No more than 1 one

The screens shown here illustrate how the TABLE feature of a graphing calculator can be used to find the probabilities of having 0, 1, 2, 3, or 4 girls in a family of 4 children. (Note that 0 appears for X = 5 and X = 6. Why is this so?) Use this approach to work Exercises 37 and 38.

37. Find the probabilities of having 0, 1, 2, or 3 boys in a family of 3 children.

38. Find the probabilities of having 0, 1, 2, 3, 4, 5, or 6 girls in a family of 6 children.

39. What will happen when an infectious disease is introduced into a family? Suppose a family has I infected members and S members who are not infected but are susceptible to contracting the disease. The probability P of k people not contracting the disease during a one-week period can be calculated by the formula $P = \binom{S}{k} q^k (1 - q)^{S-k}$ where $q = (1 - p)^I$, and p is the probability that a susceptible person contracts the disease from an infectious person. For example, if $p = .5$ then there is a 50% chance that a susceptible person exposed to one infectious person for one week will contract the disease. (*Source:* Hoppensteadt, F., and C. Peskin, *Mathematics in Medicine and the Life Sciences,* Springer-Verlag, 1992.)

(a) Compute the probability P of 3 family members not becoming infected within one week if there are currently 2 infected and 4 susceptible members. Assume that $p = .1$.

(b) A highly infectious disease can have $p = .5$. Repeat part (a) with this value of p.

(c) Determine the probability that everyone would become sick in a large family if initially, $I = 1$, $S = 9$, and $p = .5$. Discuss the results.

40. (Refer to Exercise 39.) Suppose that in a family $I = 2$ and $S = 4$. If the probability P is .25 of there being $k = 2$ uninfected members after one week, estimate graphically the possible values of p. (*Hint:* Write P as a function of p.)

41. As mentioned in the chapter opener, some bacteria, called **haploid organisms,** contain genetic material called **plasmids.** Plasmids can cause bacteria to become resistant to certain types of antibiotic drugs. Suppose the bacteria carry two plasmids that cause drug resistance. Plasmid R_1 is resistant to the antibiotic ampicillin whereas plasmid R_2 is resistant to the antibiotic tetracycline. If a bacterium has both plasmids it will be resistant to both antibiotics. R_1 and R_2 are passed in cell division to daughter cells. The type of plasmids inherited by a daughter cell is random. It can have 0, 1, or 2 plasmids of type R_1. The probability $P_{i,j}$ that a mother cell with i plasmids of type R_1 produces a daughter cell with j plasmids of type R_1 can be calculated by the formula

$$P_{i,j} = \frac{\binom{2i}{j}\binom{4-2i}{2-j}}{\binom{4}{2}}.$$

(*Source:* Hoppensteadt, F., and C. Peskin, *Mathematics in Medicine and the Life Sciences,* Springer-Verlag, 1992.)

(a) Compute the nine values of $P_{i,j}$ for $0 \le i, j \le 2$. Assume that $\binom{0}{0} = 1$ and $\binom{i}{j} = 0$ whenever $i < j$.

(b) Write your results from part (a) in the matrix

$$P = \begin{bmatrix} P_{00} & P_{01} & P_{02} \\ P_{10} & P_{11} & P_{12} \\ P_{20} & P_{21} & P_{22} \end{bmatrix}.$$

(c) Describe the matrix. Where are the probabilities the greatest?

42. (Continuation of Exercise 41) The genetic makeup of future generations of the haploid bacteria can be modeled using matrices. Let $A = \begin{bmatrix} a_1 & a_2 & a_3 \end{bmatrix}$ be a 1×3 matrix containing three probabilities. a_1 is the probability that a cell has two R_1 plasmids, a_2 is the probability that it has one R_1 plasmid and one R_2 plasmid, and a_3 is the probability that a cell has two R_2 plasmids. If the current generation of bacteria only have one plasmid of each type, then $A_1 = \begin{bmatrix} 0 & 1 & 0 \end{bmatrix}$. The probabilities A_{i+1} for plasmids R_1 and R_2 in each future generation can be calculated from the equation $A_{i+1} = A_i P$ where $i \ge 1$ and the 3×3 matrix P was determined in the previous exercise. The phenomenon that results from this sequence of calculations was not well understood until fairly recently. It is now used in the genetic engineering of plasmids.

(a) If all bacteria initially have both plasmids, make a conjecture as to the types of plasmids future generations of bacteria will have.

(b) Test your conjecture by repeatedly computing with your calculator the matrix product $A_{i+1} = A_i P$ with $A_1 = \begin{bmatrix} 0 & 1 & 0 \end{bmatrix}$ for $i = 1, 2, 3, \ldots, 12$. Interpret the result. It may surprise you.

Chapter 11 Summary ▼▼▼▼▼▼▼▼▼▼▼▼▼▼▼▼▼▼▼▼▼▼▼▼▼▼▼▼▼▼

KEY TERMS	KEY IDEAS
11.1 Sequences and Series	
sequence terms of a sequence general or nth term finite sequence infinite sequence convergent sequence divergent sequence recursive definition series index of summation	**Summation Properties** If $a_1, a_2, a_3, \ldots, a_n$ and $b_1, b_2, b_3, \ldots, b_n$ are sequences and c is a constant, then for every positive integer n, (a) $\displaystyle\sum_{i=1}^{n} c = nc$ (b) $\displaystyle\sum_{i=1}^{n} ca_i = c\sum_{i=1}^{n} a_i$ (c) $\displaystyle\sum_{i=1}^{n} (a_i \pm b_i) = \sum_{i=1}^{n} a_i \pm \sum_{i=1}^{n} b_i.$
11.2 Arithmetic Sequences and Series	
arithmetic sequence common difference	**nth Term of an Arithmetic Sequence** In an arithmetic sequence with first term a_1 and common difference d, the nth term, a_n, is given by $$a_n = a_1 + (n - 1)d.$$ **Sum of the First n Terms of an Arithmetic Sequence** If an arithmetic sequence has first term a_1 and common difference d, then the sum of the first n terms is given by $$S_n = \frac{n}{2}(a_1 + a_n)$$ or $\quad S_n = \dfrac{n}{2}[2a_1 + (n - 1)d].$
11.3 Geometric Sequences and Series	
geometric sequence common ratio	**nth Term of a Geometric Sequence** In a geometric sequence with first term a_1 and common ratio r, the nth term is $$a_n = a_1 r^{n-1}.$$ **Sum of the First n Terms of a Geometric Sequence** If a geometric sequence has first term a_1 and common ratio r, then the sum of the first n terms is given by $$S_n = \frac{a_1(1 - r^n)}{1 - r}, \quad \text{where } r \neq 1.$$ **Sum of an Infinite Geometric Sequence** The sum of an infinite geometric sequence with first term a_1 and common ratio r, where $-1 < r < 1$, is given by $$S_\infty = \frac{a_1}{1 - r}.$$

KEY TERMS	KEY IDEAS
11.4 The Binomial Theorem Revisited	
n-factorial binomial coefficient	**Binomial Coefficient** $$\binom{n}{r} = \frac{n!}{(n-r)!\,r!}$$ **Binomial Theorem** For any positive integer n: $$(x+y)^n = x^n + \binom{n}{1}x^{n-1}y + \binom{n}{2}x^{n-2}y^2$$ $$+ \binom{n}{3}x^{n-3}y^3 + \cdots + \binom{n}{r}x^{n-r}y^r$$ $$+ \cdots + \binom{n}{n-1}xy^{n-1} + y^n.$$
11.5 Mathematical Induction	
	Principle of Mathematical Induction Let S_n be a statement concerning the positive integer n. Suppose that **1.** S_1 is true; **2.** for any positive integer k, $k \leq n$, if S_k is true, then S_{k+1} is also true. Then S_n is true for every positive integer value of n. **Generalized Principle of Mathematical Induction** Let S_n be a statement about the positive integer n. Let j be a fixed positive integer. If **(a)** S_j is true; **(b)** for any positive integer k, $k \geq j$, S_k implies S_{k+1}; then S_n is true for all positive integers n, where $n \geq j$.

KEY TERMS	KEY IDEAS
11.6 Counting Theory	

tree diagram
independent events
permutation
combination
	Fundamental Principle of Counting
	If n independent events occur, with
	
	$\qquad m_1$ ways for event 1 to occur,
	
	$\qquad m_2$ ways for event 2 to occur,
	
	$\qquad\qquad \vdots$
	
	and $\qquad m_n$ ways for event n to occur,
	
	then there are
	$$m_1 \cdot m_2 \cdot \ldots \cdot m_n$$
	different ways for all n events to occur.
	
	Permutations of n Elements, r at a Time
	If $P(n, r)$ denotes the number of permutations of n elements taken r at a time, $r \leq n$, then
	$$P(n, r) = \frac{n!}{(n-r)!}.$$
	
	Combinations of n Elements, r at a Time
	If $\binom{n}{r}$ represents the number of combinations of n elements taken r at a time, $r \leq n$, then
	$$\binom{n}{r} = \frac{n!}{(n-r)!\, r!}.$$

11.7 Basics of Probability	
outcome sample space event complement Venn diagram odds mutually exclusive events binomial experiment	**Probability of Event E** In a sample space S with equally likely outcomes, the probability of an event E is $$P(E) = \frac{n(E)}{n(S)}.$$ **Probability of the Union of Two Events** $$P(E \text{ or } F) = P(E \cup F) = P(E) + P(F) - P(E \cap F)$$ **Binomial Probability** If the probability of success in a binomial experiment is p, the probability of r successes in n trials is $$\binom{n}{r} p^r (1-p)^{n-r}.$$

Chapter 11   Review Exercises ▼▼▼▼▼▼▼▼▼▼▼▼▼▼▼▼▼▼▼▼▼▼▼▼▼▼▼▼▼

Write the first five terms for each sequence. State whether the sequence is arithmetic, geometric, or neither.

1. $a_n = \dfrac{n}{n+1}$

2. $a_n = (-2)^n$

3. $a_n = 2(n+3)$

4. $a_n = n(n+1)$

5. $a_1 = 5$;   for $n \geq 2$, $a_n = a_{n-1} - 3$

In Exercises 6–9, write the first five terms of the sequence described.

6. Arithmetic, $a_2 = 10$, $d = -2$

7. Arithmetic, $a_3 = \pi$, $a_4 = 1$

8. Geometric, $a_1 = 6$, $r = 2$

9. Geometric, $a_1 = -5$, $a_2 = -1$

10. An arithmetic sequence has $a_5 = -3$ and $a_{15} = 17$. Find a_1 and a_n.

11. A geometric sequence has $a_1 = -8$ and $a_7 = -1/8$. Find a_4 and a_n.

Find a_8 for each arithmetic sequence.

12. $a_1 = 6$, $d = 2$

13. $a_1 = 6x - 9$, $a_2 = 5x + 1$

Find S_{12} for each arithmetic sequence.

14. $a_1 = 2$, $d = 3$

15. $a_2 = 6$, $d = 10$

Find a_5 for each geometric sequence.

16. $a_1 = -2$, $r = 3$

17. $a_3 = 4$, $r = \dfrac{1}{5}$

Find S_4 for each geometric sequence.

18. $a_1 = 3$, $r = 2$

19. $a_1 = -1$, $r = 3$

20. $\dfrac{3}{4}, -\dfrac{1}{2}, \dfrac{1}{3}, \cdots$

Evaluate the sums that exist.

21. $\displaystyle\sum_{i=1}^{7} (-1)^{i-1}$

22. $\displaystyle\sum_{i=1}^{5} (i^2 + i)$

23. $\displaystyle\sum_{i=1}^{4} \dfrac{i+1}{i}$

24. $\displaystyle\sum_{j=1}^{10} (3j - 4)$

25. $\displaystyle\sum_{j=1}^{2500} j$

26. $\displaystyle\sum_{i=1}^{5} 4 \cdot 2^i$

27. $\displaystyle\sum_{i=1}^{\infty} \left(\dfrac{4}{7}\right)^i$

28. $\displaystyle\sum_{i=1}^{\infty} -2\left(\dfrac{6}{5}\right)^i$

Evaluate the sums that converge. If the series diverges, say so.

29. $24 + 8 + \dfrac{8}{3} + \dfrac{8}{9} + \cdots$

30. $-\dfrac{3}{4} + \dfrac{1}{2} - \dfrac{1}{3} + \dfrac{2}{9} - \cdots$

31. $\dfrac{1}{12} + \dfrac{1}{6} + \dfrac{1}{3} + \dfrac{2}{3} + \cdots$

32. $.9 + .09 + .009 + .0009 + \cdots$

Evaluate the sum, where $x_1 = 0$, $x_2 = 1$, $x_3 = 2$, $x_4 = 3$, $x_5 = 4$, and $x_6 = 5$.

33. $\displaystyle\sum_{i=1}^{4} (x_i^2 - 6)$

34. $\displaystyle\sum_{i=1}^{6} f(x_i)\,\Delta x$; $f(x) = (x - 2)^3$, $\Delta x = .1$

Write each sum using summation notation.

35. $4 - 1 - 6 - \cdots - 66$

36. $10 + 14 + 18 + \cdots + 86$

37. $4 + 12 + 36 + \cdots + 972$

38. $\dfrac{5}{6} + \dfrac{6}{7} + \dfrac{7}{8} + \cdots + \dfrac{12}{13}$

Use the binomial theorem to expand the following.

39. $(x + 2y)^4$

40. $(3z - 5w)^3$

41. $\left(3\sqrt{x} - \dfrac{1}{\sqrt{x}}\right)^5$

42. $(m^3 - m^{-2})^4$

Find the indicated term or terms for each expansion.

43. Sixth term of $(4x - y)^8$

44. Seventh term of $(m - 3n)^{14}$

45. First four terms of $(x + 2)^{12}$

46. Last three terms of $(2a + 5b)^{16}$

47. Describe a proof by mathematical induction.

48. What kinds of statements are proved by mathematical induction? Give examples.

Use mathematical induction to prove that each statement is true for every positive integer n.

49. $1 + 3 + 5 + 7 + \cdots + (2n - 1) = n^2$

50. $2 + 6 + 10 + 14 + \cdots + (4n - 2) = 2n^2$

51. $2 + 2^2 + 2^3 + \cdots + 2^n = 2(2^n - 1)$

52. $1^3 + 3^3 + 5^3 + \cdots + (2n - 1)^3 = n^2(2n^2 - 1)$

53. How do permutations and combinations differ? How are they alike?

Find the value of each expression.

54. $P(9, 2)$

55. $P(6, 0)$

56. $\dbinom{8}{3}$

57. $9!$

58. $C(10, 5)$

Solve each problem.

59. Two people are planning their wedding. They can select from 2 different chapels, 4 soloists, 3 organists, and 2 ministers. How many different wedding arrangements are possible?

60. Bob Schiffer, who is furnishing his apartment, wants to buy a new couch. He can select from 5 different styles, each available in 3 different fabrics, with 6 color choices. How many different couches are available?

61. Four students are to be assigned to 4 different summer jobs. Each student is qualified for all 4 jobs. In how many ways can the jobs be assigned?

62. A student body council consists of a president, vice-president, secretary/treasurer, and 3 representatives at large. Three members are to be selected to attend a conference.
(a) How many different such delegations are possible?
(b) How many are possible if the president must attend?

63. Nine football teams are competing for first-, second-, and third-place titles in a statewide tournament. In how many ways can the winners be determined?

64. How many different license plates can be formed with a letter followed by 3 digits and then 3 letters? How many such license plates have no repeats?

65. A marble is drawn at random from a box containing 4 green, 5 black, and 6 white marbles. Find the following probabilities.
  (a) A green marble is drawn.
  (b) A marble that is not black is drawn.
  (c) A blue marble is drawn.

66. Refer to Exercise 65 and answer each question.
  (a) What are the odds in favor of drawing a green marble?
  (b) What are the odds against drawing a white marble?
  (c) What are the odds in favor of drawing a marble that is not white?

A card is drawn from a standard deck of 52 cards. Find the probability that the following is drawn.

67. A black king

68. A face card or an ace

69. An ace or a diamond

70. A card that is not a diamond

A sample shipment of 5 swimming pool filters is chosen. The probability of exactly 0, 1, 2, 3, 4, or 5 filters being defective is given in the following table.

Number defective	0	1	2	3	4	5
Probability	.31	.25	.18	.12	.08	.06

Find the probability that the given number of filters are defective.

71. No more than 3   **72.** At least 2   **73.** More than 5

74. A die is rolled 12 times. Find the probability that exactly 2 of the rolls result in a five.

75. A coin is tossed 10 times. Find the probability that exactly 4 of the tosses result in a tail.

Chapter 11   Test ▼▼▼▼▼▼▼▼▼▼▼▼▼▼▼▼▼▼▼▼▼▼▼▼▼▼▼▼▼▼▼▼▼▼▼

Write the first five terms for each sequence. State whether the sequence is arithmetic, geometric, or neither.

1. $a_n = (-1)^n(n^2 + 2)$

2. $a_n = -3 \cdot \left(\dfrac{1}{2}\right)^n$

3. $a_1 = 2, a_2 = 3, a_n = a_{n-1} + 2a_{n-2},$ for $n \geq 3$

4. A certain arithmetic sequence has $a_1 = 1$ and $a_3 = 25$. Find a_5.

5. A certain geometric sequence has $a_1 = 81$ and $r = -2/3$. Find a_6.

Find the sum of the first ten terms of the sequence described.

6. Arithmetic, with $a_1 = -43$ and $d = 12$

7. Geometric, with $a_1 = 5$ and $r = -2$

Evaluate each sum that exists.

8. $\displaystyle\sum_{i=1}^{30} (5i + 2)$

9. $\displaystyle\sum_{i=1}^{5} (-3 \cdot 2^i)$

10. $\displaystyle\sum_{i=1}^{\infty} (2^i) \cdot 4$

11. $\displaystyle\sum_{i=1}^{\infty} 54\left(\dfrac{2}{9}\right)^i$

Use the binomial theorem to expand each of the following.

12. $(x + y)^6$

13. $(2x - 3y)^4$

14. Find the third term in the expansion of $(w - 2y)^6$.

Evaluate each of the following.

15. $C(10, 2)$

16. $\binom{7}{3}$

17. $P(11, 3)$

18. $8!$

19. Use mathematical induction to prove that for all positive integers n,

$$8 + 14 + 20 + 26 + \cdots + (6n + 2) = 3n^2 + 5n.$$

Solve each problem involving counting theory.

20. A sports-shoe manufacturer makes athletic shoes in four different styles. Each style comes in three different colors, and each color comes in two different shades. How many different types of shoes can be made?

21. A club with 20 members plans to elect a president, a secretary, and a treasurer from its membership. If a member can hold at most one office, in how many ways can the three offices be filled?

22. Refer to the problem in Exercise 21. If there are 8 men and 12 women in the club, in how many different ways can 2 men and 3 women be chosen to attend a conference?

23. Write a few sentences to a friend explaining how to determine when to use permutations and when to use combinations in an applied problem.

A card is drawn from a standard deck of 52 cards. Find the probability that each of the following is drawn.

24. A red three

25. A card that is not a face card

26. A king or a spade

27. In the card-drawing experiment above, what are the odds in favor of drawing a face card?

28. A sample of 4 transistors is chosen. The probability of exactly 0, 1, 2, 3, or 4 transistors being defective is given in the following table.

Number defective	0	1	2	3	4
Probability	.19	.43	.30	.07	.01

Find the probability that at most two are defective.

An experiment consists of rolling a die 8 times. Find the probability of the event described.

29. Exactly 3 rolls result in a four.

30. All 8 rolls result in a six.

Answers to Selected Exercises

TO THE STUDENT

If you need further help with algebra, you may want to obtain a copy of the *Student's Solution Manual* that goes with this book. It contains solutions to all the odd-numbered exercises and all the chapter test exercises. You also may want the *Student's Study Guide*. It has extra examples and exercises to complete, corresponding to each section of the book. In addition, there is a practice test for each chapter. Your college bookstore either has the *Manual* or *Guide* or can order them for you.

In this section we provide the answers that we think most students will obtain when they work the exercises using the methods explained in the text. If your answer does not look exactly like the one given here, it is not necessarily wrong. In many cases there are equivalent forms of the answer. For example, if the answer section shows 3/4 and your answer is .75, you have obtained the correct answer but written it in a different (yet equivalent) form. Unless the directions specify otherwise, .75 is just as valid an answer as 3/4. In general, if your answer does not agree with the one given in the text, see whether it can be transformed into the other form. If it can, then it is the correct answer. If you still have doubts, talk with your instructor.

CHAPTER 1 Algebraic Expressions ▼▼▼

1.1 EXERCISES (page 11) **1.** 1, 3 **3.** $-6, -\dfrac{12}{4}$ (or -3), 0, 1, 3 **5.** $-\sqrt{3}, 2\pi, \sqrt{12}$ **7.** natural, whole, integer, rational, real **9.** rational, real **11.** irrational, real **15.** -243 **17.** 81 **19.** -243 **21.** negative, positive **23.** 79 **25.** -6
27. -60 **29.** -12 **31.** $-\dfrac{25}{36}$ **33.** $-\dfrac{6}{7}$ **35.** 23 **37.** 36 **39.** $-\dfrac{1}{2}$ **41.** $-\dfrac{23}{20}$ **43.** $M = 5$ **45.** distributive
47. inverse **49.** identity **53.** $(8 - 14)p = -6p$ **55.** $6(3y + 1)$ **57.** $-3z + 3y$ **59.** $ar + as - at$ **61.** $20z$
63. $15p$ **65.** $-5m - 2y + 8z$ **67.** 67.75 **69.** approximately 6.36 miles per hour **71.** 2400 **73. (a)** 31 feet **(b)** No. The chart shows that 5-foot waves occur with a wind speed of 17.3 mph and a duration of 20 hours or more. It may also be possible for two different wind speeds and durations to produce the same wave height. **(c)** Stronger winds produce higher waves. The duration of the wind will increase the wave height as the wind becomes stronger. At 11.5 mph the wind duration does not affect the wave height whereas at 57.5 mph the wind duration has a dramatic effect on the wave height.

1.2 EXERCISES (page 21) **1.** true **3.** true **5.** true **7.** $-|9|, -|-6|, |-8|$
9. $-5, -4, -2, -\sqrt{3}, \sqrt{6}, \sqrt{8}, 3$ **11.** $\dfrac{3}{4}, \dfrac{7}{5}, \sqrt{2}, \dfrac{22}{15}, \dfrac{8}{5}$ **15.** $-6 < 15$ **17.** $1 \leq 2$ **19.** $x < 3$ **21.** $k < -7$ **23.** 8
25. 6 **27.** 4 **29.** -5 **31.** $.58\overline{3}$ or $\dfrac{7}{12}$ **33.** $\pi - 3$ **35.** $x - 4$ **37.** $8 - 2k$ **39.** $8 + 4m$ **41.** $y - x$ **43.** $3 + x^2$
45. The absolute value of any number is greater than or equal to zero. **47.** addition property of order **49.** transitive property of order **51.** addition property of order **53.** triangle inequality, $|a + b| \leq |a| + |b|$ **55.** property of absolute value, $|a| \geq 0$
57. $P_d = 9$ **59.** $42°F$ **61.** $36°F$ **63.** 56 billion dollars; in the black **65.** 170 billion dollars; in the red

CONNECTIONS (page 30)   **1.** 9999   **2.** 3591   **3.** 10,404   **4.** 5041   **5.** Answers will vary.
(page 32)   **1.** The next four terms are 8, 13, 21, 34. Starting with the third term, a new term is found by adding the preceding two terms.   **2.** Answers will vary.

1.3 EXERCISES (page 34)   **1.** false   **3.** true   **5.** true   **7.** 2^{10} **9.** $2^3 x^{15} y^{12}$ **11.** $-\dfrac{p^8}{q^2}$ **13.** polynomial;   degree 11;   monomial   **15.** polynomial;   degree 6;   binomial   **17.** polynomial;   degree 6;   binomial   **19.** polynomial;   degree 6;   trinomial   **21.** not a polynomial   **23.** $x^2 - x + 3$ **25.** $9y^2 - 4y + 4$ **27.** $6m^4 - 2m^3 - 7m^2 - 4m$ **29.** $28r^2 + r - 2$ **31.** $15x^2 - \dfrac{7}{3}x - \dfrac{2}{9}$ **33.** $12x^5 + 8x^4 - 20x^3 + 4x^2$ **35.** $-2z^3 + 7z^2 - 11z + 4$
37. $m^2 + mn - 2n^2 - 2km + 5kn - 3k^2$ **39.** $(x + y)^2 = x^2 + 2xy + y^2$ is true for all replacements of x and y.
41. (a) $(x + y)^2$ **(b)** $x^2 + 2xy + y^2$ **(c)** Since they both represent the area of the largest square, and it can have only one area, the two expressions must be equal.   **(d)** It reinforces the special product for squaring a binomial.   **43.** $4m^2 - 9$
45. $16m^2 + 16mn + 4n^2$ **47.** $25r^2 + 30rt^2 + 9t^4$ **49.** $4p^2 - 12p + 9 + 4pq - 6q + q^2$ **51.** $9q^2 + 30q + 25 - p^2$
53. $9a^2 + 6ab + b^2 - 6a - 2b + 1$ **55.** $p^3 - 7p^2 - p - 7$ **57.** $49m^2 - 4n^2$ **59.** $-14q^2 + 11q - 14$ **61.** $4p^2 - 16$
63. $11y^3 - 18y^2 + 4y$ **65.** $k^{2m} - 4$ **67.** $3p^{2x} - 5p^x - 2$ **69.** $q^{2p} - 10q^p p^q + 25p^{2q}$
71. m **73.** $m + n$ **75.** $x^6 + 6x^5 y + 15x^4 y^2 + 20x^3 y^3 + 15x^2 y^4 + 6xy^5 + y^6$
77. $p^5 - 5p^4 q + 10p^3 q^2 - 10p^2 q^3 + 5pq^4 - q^5$ **79.** $r^{10} + 5r^8 s + 10r^6 s^2 + 10r^4 s^3 + 5r^2 s^4 + s^5$
81. $729r^6 - 1458r^5 s + 1215r^4 s^2 - 540r^3 s^3 + 135r^2 s^4 - 18rs^5 + s^6$
83. $1024a^5 - 6400a^4 b + 16{,}000a^3 b^2 - 20{,}000a^2 b^3 + 12{,}500ab^4 - 3125b^5$ **85. (a)** approximately 60,501,067 cubic feet

(b) The shape becomes a rectangular box with a square base, with volume $b^2 h$.   **(c)** If we let $a = b$ then $\dfrac{1}{3}h(a^2 + ab + b^2)$

becomes $\dfrac{1}{3}h(b^2 + bb + b^2)$ which simplifies to hb^2. Yes, the Egyptian formula gives the same result.   **87.** 1;   1   **88.** $x + 1$;   11
89. $x^2 + 2x + 1$;   121   **90.** $x^3 + 3x^2 + 3x + 1$;   1331   **91.** In each case, the coefficients in the polynomial correspond to the digits in the power of 11.   **92.** $x^4 + 4x^3 + 6x^2 + 4x + 1$;   A logical prediction is 14,641, which is the value of 11^4.

93. 5.6;   .1 off   **95.** 4.7;   1.0 off   **97.** $2x^5 + 7x^4 - 5x^2 + 7$ **99.** $-5x^2 + 8 + \dfrac{2}{x^2}$ **101.** $x^2 + \dfrac{7}{2}x - \dfrac{9}{4} + \dfrac{31/4}{2x - 1}$

103. $x^2 + \dfrac{1}{3}x + \dfrac{7}{9} + \dfrac{(61/9)x - 1}{3x^2 - x}$

1.4 EXERCISES (page 44)   **1.** Since the polynomial was not *completely* factored, the teacher was justified. The correct answer is $4x^2 y^4(xy - 2)$.   **3.** $4k^2 m^3(1 + 2k^2 - 3m)$ **5.** $2(a + b)(1 + 2m)$ **7.** $(r + 3)(3r - 5)$ **9.** $(m - 1)(2m^2 - 7m + 7)$
11. $(2s + 3)(3t - 5)$ **13.** $(m^4 + 3)(2 - a)$ **15.** $(5z - 2x)(4z - 9x)$ **17.** $6(a - 10)(a + 2)$ **19.** $3m(m + 1)(m + 3)$
21. $(3k - 2p)(2k + 3p)$ **23.** $(5a + 3b)(a - 2b)$ **25.** $x^2(3 - x)^2$ **27.** $2a^2(4a - b)(3a + 2b)$ **29.** $(3m - 2)^2$
31. $2(4a - 3b)^2$ **33.** $(2xy + 7)^2$ **35.** $(a - 3b - 3)^2$ **37.** $(3a + 4)(3a - 4)$ **39.** $(5s^2 + 3t)(5s^2 - 3t)$
41. $(a + b + 4)(a + b - 4)$ **43.** $(p^2 + 25)(p + 5)(p - 5)$ **45. (b)**   **47.** $(2 - a)(4 + 2a + a^2)$
49. $(5x - 3)(25x^2 + 15x + 9)$ **51.** $(3y^3 + 5z^2)(9y^6 - 15y^3 z^2 + 25z^4)$ **53.** $r(r^2 + 18r + 108)$
55. $(3 - m - 2n)(9 + 3m + 6n + m^2 + 4mn + 4n^2)$ **57.** $(x - 1)(x^2 + x + 1)(x + 1)(x^2 - x + 1)$
58. $(x - 1)(x + 1)(x^4 + x^2 + 1)$ **59.** $(x^2 - x + 1)(x^2 + x + 1)$ **60.** additive inverse property (0 in the form $x^2 - x^2$ was added on the right.);   associative property of addition;   factoring a perfect square trinomial;   factoring the difference of two squares;   commutative property of addition   **61.** They are the same.   **62.** $(x^4 - x^2 + 1)(x^2 + x + 1)(x^2 - x + 1)$
63. $(m^2 - 5)(m^2 + 2)$ **65.** $9(7k - 3)(k + 1)$ **67.** $(3a - 7)^2$ **69.** $(2b + c + 4)(2b + c - 4)$ **71.** $(x + y)(x - 5)$
73. $(m - 2n)(p^4 + q)$ **75.** $(2z + 7)^2$ **77.** $(10x + 7y)(100x^2 - 70xy + 49y^2)$ **79.** $(5m^2 - 6)(25m^4 + 30m^2 + 36)$
81. $(6m - 7n)(2m + 5n)$ **83.** $(4p - 1)(p + 1)$ **85.** prime   **87.** $4xy$ **89.** $(r + 3s^q)(r - 2s^q)$ **91.** $(3a^{2k} + b^{4k})(3a^{2k} - b^{4k})$
93. $(2y^a - 3)^2$ **97.** ± 36 **99.** 9

CONNECTIONS (page 47)   **1.** 0 miles   **2.** 2300 miles   **3.** -2.74; y increases without bound.

1.5 EXERCISES (page 53)   **1.** $x \neq -6$ **3.** $x \neq \dfrac{3}{5}$ **5.** no restrictions   **7. (a)**   **9.** $k = -3$ **11.** $\dfrac{8}{9}$ **13.** $\dfrac{3}{t - 3}$
15. $\dfrac{2x + 4}{x}$ **17.** $\dfrac{m - 2}{m + 3}$ **19.** $\dfrac{2m + 3}{4m + 3}$ **21.** $\dfrac{25p^2}{9}$ **23.** $\dfrac{2}{9}$ **25.** $\dfrac{5x}{y}$ **27.** $\dfrac{2(a + 4)}{a - 3}$ **29.** 1   **31.** $\dfrac{m + 6}{m + 3}$ **33.** $\dfrac{m - 3}{2m - 3}$
35. $\dfrac{x^2 - xy + y^2}{x^2 + xy + y^2}$ **37. (b) and (c)**   **39.** $\dfrac{19}{6k}$ **41.** 1   **43.** $\dfrac{6 + p}{2p}$ **45.** $\dfrac{137}{30m}$ **47.** $\dfrac{a - b}{a^2}$ **49.** $\dfrac{2x}{(x + z)(x - z)}$

51. $\dfrac{4}{a-2}$ or $\dfrac{-4}{2-a}$ **53.** $\dfrac{3x+y}{2x-y}$ or $\dfrac{-3x-y}{y-2x}$ **55.** $\dfrac{x-11}{(x+4)(x-4)(x-3)}$ **57. (a)** 20.1 (thousand dollars)

(b) 127.3 (thousand dollars) **(c)** 439.97 (thousand dollars) **59. (a)** \$65.5 tens of millions, or \$655,000,000

(b) \$64 tens of millions, or \$640,000,000 **(c)** \$60 tens of millions, or \$600,000,000 **(d)** \$40 tens of millions, or

\$400,000,000 **(e)** \$0 **61.** $\dfrac{x+1}{x-1}$ **63.** $\dfrac{-1}{x+1}$ **65.** $\dfrac{(2-b)(1+b)}{b(1-b)}$ **67.** $\dfrac{m^3-4m-1}{m-2}$ **69.** $\dfrac{p+5}{p(p+1)}$ **71.** $\dfrac{-1}{x(x+h)}$

73. approximately .02 cm

CONNECTIONS (page 61) **1.** They can expect the storm to last .56 hour. **2.** The storm will last approximately 1.3 hours, not quite long enough to meet their need.

1.6 EXERCISES (page 63) **1.** $-\dfrac{1}{64}$ **3.** 8 **5.** -2 **7.** 4 **9.** $\dfrac{1}{9}$ **11.** not defined **13.** $\dfrac{256}{81}$ **15.** 512

17. $\dfrac{1}{32}$ **19.** $\dfrac{243}{32}$ **21.** $\dfrac{1}{100,000}$ **25.** (d) **27.** $\dfrac{1}{27^3}$ **29.** 1 **31.** $m^{7/3}$ **33.** $(1+n)^{5/4}$ **35.** $\dfrac{6z^{2/3}}{y^{5/4}}$ **37.** $2^6a^{1/4}b^{37/2}$

39. $\dfrac{r^6}{s^{15}}$ **41.** $-\dfrac{1}{ab^3}$ **43.** $12^{9/4}y$ **45.** $\dfrac{1}{2p^2}$ **47.** $\dfrac{m^3p}{n}$ **49.** $-4a^{5/3}$ **51.** $\dfrac{1}{(k+5)^{1/2}}$ **53.** \$64 **55.** \$977.78

57. \$64,000,000 **59.** \$10,000,000 **61.** \$386 million **62.** \$455 million **63.** \$537 million **64.** \$634 million

65. \$748 million **66. (a)** \$36 million **(b)** greater than **67. (a)** \$45 million **(b)** less than

68. (a) \$63 million **(b)** less than **69. (a)** \$34 million **(b)** greater than **70. (a)** \$2 million **(b)** less than **71.** 60 species

73. 177 species **75.** r^{6+p} **77.** $m^{3/2}$ **79.** $x^{(2n^2-1)/n}$ **81.** $p^{(m+n+m^2)/(mn)}$ **83.** $y-10y^2$ **85.** $-4k^{10/3}+24k^{4/3}$ **87.** x^2-x

89. $r-2+r^{-1}$ or $r-2+\dfrac{1}{r}$ **91.** $k^{-2}(4k+1)$ or $\dfrac{4k+1}{k^2}$ **93.** $z^{-1/2}(9+2z)$ or $\dfrac{9+2z}{z^{1/2}}$

95. $p^{-7/4}(p-2)$ or $\dfrac{p-2}{p^{7/4}}$ **97.** $(p+4)^{-3/2}(p^2+9p+21)$ or $\dfrac{p^2+9p+21}{(p+4)^{3/2}}$ **99.** $b+a$ **101.** -1 **103.** $\dfrac{y(xy-9)}{x^2y^2-9}$

CONNECTIONS (page 74) **2.** 6; for $x=6$, $\dfrac{1}{\sqrt{x}+\sqrt{6}}=\dfrac{1}{2\sqrt{6}}$

1.7 EXERCISES (page 75) **1.** (f) **3.** (h) **5.** (g) **7.** (c) **9.** $\sqrt[3]{(-m)^2}$ or $(\sqrt[3]{-m})^2$

11. $\sqrt[3]{(2m+p)^2}$ or $(\sqrt[3]{2m+p})^2$ **13.** $k^{2/5}$ **15.** $-3\cdot 5^{1/2}p^{3/2}$ **21.** 5 **23.** -5 **25.** $5\sqrt{2}$ **27.** $3\sqrt[3]{3}$ **29.** $-2\sqrt[4]{2}$

31. $-\dfrac{3\sqrt{5}}{5}$ **33.** $-\dfrac{\sqrt[3]{100}}{5}$ **35.** $32\sqrt[3]{2}$ **37.** $2x^2z^4\sqrt{2x}$ **39.** $2zx^2y\sqrt[3]{2z^2x^2y}$ **41.** $np^2\sqrt[4]{m^2n^3}$ **43.** cannot simplify further

45. $\dfrac{\sqrt{6x}}{3x}$ **47.** $\dfrac{x^2y\sqrt{xy}}{z}$ **49.** $\dfrac{2\sqrt[3]{x}}{x}$ **51.** $\dfrac{h\sqrt[4]{9g^3hr^2}}{3r^2}$ **53.** $\dfrac{m\sqrt[3]{n^2}}{n}$ **55.** $2\sqrt[4]{x^3y^3}$ **57.** $\sqrt[3]{2}$ **59.** true **61.** $23\sqrt{2k}$

63. $3\sqrt[3]{4}$ **65.** $\dfrac{11\sqrt{2}}{8}$ **67.** $\dfrac{-25\sqrt[3]{9}}{18}$ **69.** 3 **71.** 34 **73.** $3-2\sqrt{2}$ **75.** $58+5\sqrt{5}$ **77.** $\dfrac{\sqrt{15}-3}{2}$

79. $\dfrac{3\sqrt{5}+3\sqrt{15}-2\sqrt{3}-6}{33}$ **81.** $\dfrac{p(\sqrt{p}-2)}{p-4}$ **83.** $\dfrac{a(\sqrt{a+b}+1)}{a+b-1}$ **85. (a)** $-60.9°\text{F}$ **(b)** $-64.4°\text{F}$

86. (a) $-63.4°\text{F}$ **(b)** $-46.7°\text{F}$ **87. (a)** $-63°\text{F}$ **(b)** $-47°\text{F}$ **88.** The formula in Exercise 86 provides a better model.

89. $|m+n|$ **91.** $|z-3x|$ **93.** $\dfrac{-1}{2(1-\sqrt{2})}$ **95.** $\dfrac{x}{\sqrt{x}+x}$ **97.** $\dfrac{-1}{2x-2\sqrt{x(x+1)}+1}$

99. (a) approximately 5.10×10^8 square kilometers **(b)** approximately 3.62×10^8 square kilometers **(c)** approximately 2.76×10^6 cubic kilometers **(d)** approximately 7.6 meters; This is approximately 25 feet and very close to the published value of 7.5 meters. **(e)** An increase in the sea level of 7.6 meters would displace millions of people and cause enormous damage to property. Since the elevations of Boston, New Orleans, and San Diego are all less than 7.6 meters, or about 25 feet, they would be below sea level and underwater without a dike system. **(f)** If inland seas and fresh-water lakes had been accounted for, the calculated area for the oceans would have been less. Therefore, the increase in sea level would have been greater, but not significantly. The two largest inland bodies of water are the Caspian Sea and Lake Superior which account for only .12% of the total surface area covered by water.

1.8 EXERCISES (page 84) **3.** real **5.** imaginary **7.** imaginary **9.** $10i$ **11.** $-20i$ **13.** $-i\sqrt{39}$ **15.** $5 + 2i$
17. $9 - 5i\sqrt{2}$ **19.** -5 **21.** 2 **23.** $7 - i$ **25.** 2 **27.** $1 - 10i$ **29.** $-14 + 2i$ **31.** $5 - 12i$ **33.** 13 **35.** 7
37. $25i$ **39.** i **41.** i **43.** 1 **45.** $-i$ **47.** 1 **49.** i **53.** i **55.** $\dfrac{7}{25} - \dfrac{24}{25}i$ **57.** $\dfrac{26}{29} + \dfrac{7}{29}i$ **59.** $-2 + i$ **61.** $-2i$
65. $4 + 6i$ **67.** A rule for exponents says that $a^3 = a^2 \cdot a$. **68.** $3 + 4i$ **69.** $2 + 11i$; Yes, it agrees.
70. $(1 + i)^6 = 1^6 + 6(1)^5i + 15(1)^4i^2 + 20(1)^3i^3 + 15(1)^2i^4 + 6(1)i^5 + i^6$; $-8i$

CHAPTER 1 REVIEW EXERCISES (page 88) **1.** $-12, -6, -\sqrt{4}$ (or -2), $0, 6$ **3.** $-\sqrt{7}, \dfrac{\pi}{4}, \sqrt{11}$ **5.** integer, rational,
real **7.** irrational, real **9.** 9 **11.** 31 **13.** $-\dfrac{37}{20}$ **15.** $-\dfrac{19}{42}$ **17.** -32 **21.** commutative **23.** associative
25. identity **27.** $kr + ks - kt$ **29.** $-|3 - (-2)|, -|-2|, |6 - 4|, |8 + 1|$ **31.** -3 **33.** $3 - \sqrt{8}$
35. $m - 3$ **37.** $4 - \pi$ **39.** $7q^3 - 9q^2 - 8q + 9$ **41.** $16y^2 + 42y - 49$ **43.** $9k^2 - 30km + 25m^2$ **45. (a)** 3.0 million
(b) 2.99 million **47. (a)** 6.5 million **(b)** 6.46 million **49.** $x^4 + 8x^3y + 24x^2y^2 + 32xy^3 + 16y^4$ **51.** $9r + 4$
53. $5m - 7 + \dfrac{10m}{m^2 - 2}$ **55.** $z(7z - 9z^2 + 1)$ **57.** $(r + 7p)(r - 6p)$ **59.** $(3m + 1)(2m - 5)$ **61.** $(13y^2 + 1)(13y^2 - 1)$
63. $8(y - 5z^2)(y^2 + 5yz^2 + 25z^4)$ **65.** $(r - 3s)(a + 5b)$ **67.** $(4m - 7 + 5a)(4m - 7 - 5a)$ **69. (a)** **71.** $\dfrac{3}{8r}$
73. $(3m + n)(3m - n)$ **75.** $\dfrac{37}{20y}$ **77.** $\dfrac{x + 9}{(x - 3)(x - 1)(x + 1)}$ **79.** $\dfrac{3m^2 + 2m - 12}{5(m + 2)}$ **81.** $\dfrac{1}{64}$ **83.** $\dfrac{16}{25}$ **85.** $-10z^8$ **87.** 1
89. $-8y^{11}p$ **91.** $\dfrac{1}{(p + q)^5}$ **93.** $-14r^{17/12}$ **95.** $y^{1/2}$ **97.** $10z^{7/3} - 4z^{1/3}$ **99.** $3p^2 + 3p^{3/2} - 5p - 5p^{1/2}$ **101.** $10\sqrt{2}$
103. $5\sqrt[4]{2}$ **105.** $-\dfrac{\sqrt[3]{50p}}{5p}$ **107.** $\sqrt[12]{m}$ **109.** 66 **111.** $-9m\sqrt{2m} + 5m\sqrt{m}$ or $m(-9\sqrt{2m} + 5\sqrt{m})$
113. $\dfrac{6(3 + \sqrt{2})}{7}$ **115.** $x - 1 - \sqrt{x(x - 2)}$ **117.** $7i$ **119.** $10 - 3i$ **121.** $-8 + 13i$ **123.** $19 + 17i$ **125.** 146
127. $7 - 24i$ **129.** $\dfrac{5}{2} + \dfrac{7}{2}i$ **131.** real **133.** $-i$ **135.** i

CHAPTER 1 TEST (page 92) **1. (a)** $-13, -\dfrac{12}{4}$ (or -3), $0, \sqrt{49}$ (or 7) **(b)** $-13, -\dfrac{12}{4}$ (or -3), $0, \dfrac{3}{5}, 5.9, \sqrt{49}$ (or 7)
(c) All are real numbers. **2.** 4 **3. (a)** associative **(b)** commutative **(c)** distributive **(d)** inverse
4. $11x^2 - x + 2$ **5.** $36r^2 - 60r + 25$ **6.** $3t^3 + 5t^2 + 2t + 8$ **7.** $2x^2 - x - 5 + \dfrac{3}{x - 5}$ **8.** approximately \$5298
10. $16x^4 - 96x^3y + 216x^2y^2 - 216xy^3 + 81y^4$ **11.** $(x^2 + 4)(x + 2)(x - 2)$ **12.** $2m(4m + 3)(3m - 4)$
13. $(x - 2)(x^2 + 2x + 4)(y + 3)(y - 3)$ **14.** $\dfrac{x^4(x + 1)}{3(x^2 + 1)}$ **15.** $\dfrac{x(4x + 1)}{(x + 2)(x + 1)(2x - 3)}$ **16.** $\dfrac{2a}{2a - 3}$ or $\dfrac{-2a}{3 - 2a}$
17. $\dfrac{y}{y + 2}$ **18.** $\dfrac{yz}{x}$ **19.** $3x^2y^4\sqrt{2x}$ **20.** $2\sqrt{2x}$ **21.** $x - y$ **22.** approximately 2.1 seconds
23. $5 - 8i$ **24.** $-29 - 3i$ **25.** $2 + i$ **26.** i

CHAPTER 2 Equations and Inequalities ▼▼▼

CONNECTIONS (page 97) **1.** The relation $<$ satisfies only the transitive axiom. **2.** Not necessarily—for example, if A defeats
B and B defeats C, it is not a "sure thing" that A will defeat C. **3.** Some examples are "likes," "is next to (in a line)," and "studies
with."

2.1 EXERCISES (page 101) **1.** true **3.** false **7.** identity; {all real numbers} **9.** conditional; {7}
11. contradiction; ∅ **13.** equivalent **15.** not equivalent **17. (b)** **19.** {12} **21.** $\left\{-\dfrac{2}{7}\right\}$ **23.** $\left\{-\dfrac{7}{8}\right\}$ **25.** {−1}
27. {3} **29.** $\left\{\dfrac{3}{4}\right\}$ **31.** $\left\{-\dfrac{12}{5}\right\}$ **33.** ∅ **35.** $\left\{\dfrac{27}{7}\right\}$ **37.** $\left\{-\dfrac{59}{6}\right\}$ **39.** ∅ **41.** {0} **43.** {50} **45.** −4
47. 35 **48.** −55 **49.** −2 **50.** −27 **51.** 68°F **53.** 15°C **55.** 37.8°C **57.** 13% **59.** \$432 **61.** \$205.41
63. \$66.50 **65.** approximately 40 liters per second **67. (a)** year 6, or 1988 **(b)** 1988; They correspond very closely.

(c) 39.33%   **69.** {4}   **71.** {1.6}   **73.** {6}   **75.** {−.5}   **77.** {−1.2}   **79.** {$1.\overline{3}$}   **81.** {6}   **83.** {$−3.\overline{6}$}   **85.** {−3}
87. {16.07}   **89.** {−1.46}   **91.** {−3.92}

2.2 EXERCISES (page 113)   **1.** 20 miles   **3.** $8   **5.** $l = \dfrac{V}{wh}$ **7.** $c = P - a - b$ **9.** $B = \dfrac{2A - hb}{h}$ or $B = \dfrac{2A}{h} - b$

11. $h = \dfrac{S - 2\pi r^2}{2\pi r}$ or $h = \dfrac{S}{2\pi r} - r$ **13.** $F = \dfrac{9C}{5} + 32$ **15.** $f = \dfrac{un(n + 1)}{k(k + 1)}$ **17.** It is not solved for x, since x appears on

the left side of the equation as well.   **19.** (d)   **21.** 10 cm   **23.** 328.6 yd   **25.** 4 in   **27.** (b) and (c)   **29.** 84   **31.** 4 liters of

92-octane and 8 liters of 98-octane   **33.** 125   **35.** 2.7 mi   **37.** about 840 mi   **39.** 15 min   **41.** 78 hr   **43.** $\dfrac{40}{3}$ hr

45. 2 liters   **47.** 2.4 liters   **49.** Short-term note is for $60,000; long-term note is for $65,000.

51. $15,000 at 7%; $60,000 at 11%   **53.** $20,000 at 6.5%; $14,560 at 6.25%   **55. (a)** $V = 900x$ **(b)** $\dfrac{3}{50}x$

(c) $A = 2.4$ ach   **(d)** It should be increased by $3\frac{1}{3}$ times. (Smoking areas require more than triple the ventilation.)
57. (a) approximately .000021 for each individual   **(b)** $C = .000021x$ **(c)** approximately 2.1 cases   **(d)** approximately 413,000
deaths per year   **59. (a)** $17.96 **(b)** $20.13 **(c)** $21.21   **(d)** Congress regulated rates for cable television in 1992 and called for
more competition. We cannot use the model reliably because of influences such as this.

CONNECTIONS (page 125)   **1. (a)** $\left\{ -\dfrac{1}{3}, 2 \right\}$ **(b)** $\left\{ \dfrac{3 \pm \sqrt{17}}{4} \right\}$ **(c)** $\left\{ \dfrac{-2 \pm i\sqrt{2}}{3} \right\}$ **3.** Answers will vary.

2.3 EXERCISES (page 128)   **1.** (d); $\left\{ -\dfrac{1}{3}, 7 \right\}$ **3.** (c); {−4, 3}   **5.** {±4}   **7.** {$±3\sqrt{3}$}   **9.** {±4i}   **11.** {$±3i\sqrt{2}$}

13. $\left\{ \dfrac{1 \pm 2\sqrt{3}}{3} \right\}$ **15.** {2, 3}   **17.** $\left\{ \dfrac{3}{5} \pm \dfrac{\sqrt{3}}{5}i \right\}$ **19.** {3, 5}   **21.** {$1 \pm \sqrt{5}$}   **23.** $\left\{ -\dfrac{1}{2} \pm \dfrac{1}{2}i \right\}$ **25.** He is incorrect because

$c = 0$.   **27.** $\left\{ \dfrac{1 \pm \sqrt{5}}{2} \right\}$ **29.** {$3 \pm \sqrt{2}$}   **31.** $\left\{ \dfrac{3}{2} \pm \dfrac{\sqrt{2}}{2}i \right\}$ **33.** $\left\{ \dfrac{-1 \pm \sqrt{97}}{4} \right\}$ **35.** $\left\{ \dfrac{-3 \pm \sqrt{41}}{8} \right\}$ **37.** {$2, -1 \pm i\sqrt{3}$}

39. $\left\{ 3, -\dfrac{3}{2} \pm \dfrac{3\sqrt{3}}{2}i \right\}$ **41.** {$-4, 2 \pm 2i\sqrt{3}$}   **43.** $\left\{ -\dfrac{5}{4}, \dfrac{5}{4} \pm \dfrac{5\sqrt{3}}{4}i \right\}$ **45.** {$3 \pm \sqrt{5}$}   **47.** $\left\{ -\dfrac{1}{2} \pm \dfrac{\sqrt{3}}{2}i \right\}$

49. $\left\{ -\dfrac{1}{3} \pm \dfrac{\sqrt{7}}{3}i \right\}$ **51.** {1. 931851653, −.5176380902}   **53.** {.7071067812, 1.414213562}

55. {−1.618033989, −.6180339887}   **57.** {$−.6123724357 \pm 1.767766953i$}   **59.** {$−1.020620726 \pm .4993273296i$}

61. {$.5 \pm .8380817098i$}   **63.** {−.5, 7}   **65.** {−.4509456768, 1.267442258}   **67.** $t = \dfrac{\pm\sqrt{2sg}}{g}$ **69.** $v = \dfrac{\pm\sqrt[4]{Frk^3 M^3}}{kM}$

71. $R = \dfrac{E^2 - 2Pr \pm E\sqrt{E^2 - 4Pr}}{2P}$ **73. (a)** $x = \dfrac{y \pm \sqrt{8 - 11y^2}}{4}$ **(b)** $y = \dfrac{x \pm \sqrt{6 - 11x^2}}{3}$ **75.** 0; one rational solution

77. 1; two different rational solutions   **79.** 84; two different irrational solutions   **81.** −23; two different imaginary solutions

83. 2304; two different rational solutions   **85.** 0   **86.** $121 - 4k$ **87.** $121 - 4k = 0$ **88.** $\left\{ \dfrac{121}{4} \right\}$

89. $x^2 + 11x + \dfrac{121}{4} = 0$ **90.** $\left\{ -\dfrac{11}{2} \right\}$ **91.** $k = 1$; $\left\{ \dfrac{1}{5} \right\}$ **92.** $k = -40$ or $k = 40$; When $k = -40$, the solution set is $\left\{ \dfrac{5}{4} \right\}$;

when $k = 40$, the solution set is $\left\{ -\dfrac{5}{4} \right\}$.   **93.** $x \neq 7, x \neq -\dfrac{2}{3}$ **95.** $y \neq -\dfrac{1}{4}$ **97.** no restrictions

CONNECTIONS (page 135)   **2.** 5, 12, 13   **3.** No, because r must be larger than s.

2.4 EXERCISES (page 136)   **1. (a)**   **3.** (d)   **5.** 100 yd by 400 yd   **7.** 10 in by 13 in   **9.** 2 cm   **11.** Felipe: 36 hr; Felix:
45 hr   **13.** 4 hr   **15.** 50 mph   **17.** 2.5 hr (The solution 16 is not realistic.)   **19.** 40 ft   **21.** 61 min   **23.** .68 sec, 7.32 sec
25. .19 sec, 10.92 sec; 11.32 sec   **27.** approximately 19.2 hours   **29.** 10.3 hr   **31.** When $x \approx 3.6$; Since 3.6 occurs during year
3, which is 1988, and year 4 corresponds to 1989, we would expect that in 1989, the number was about 7500 million.
33. $80 - x$ **34.** $300 + 20x$ **35.** $(80 - x)(300 + 20x)$ or $24,000 + 1300x - 20x^2$ **36.** $20x^2 - 1300x + 11,000 = 0$
37. {10, 55}; Because of the restriction, only 10 is valid here, and the number of units rented is 70.   **38.** 80

🖩 **(page 143)** The graph is not a straight line because the expression includes a linear variable under a radical. See Section 3.4.

2.5 EXERCISES (page 145) **1.** (d) **3.** The number -1 is not a solution. A check will indicate this. **5.** $\{\pm\sqrt{3}, \pm i\sqrt{5}\}$

7. $\left\{\pm 1, \pm\dfrac{\sqrt{10}}{2}\right\}$ **9.** $\{4, 6\}$ **11.** $\left\{\dfrac{-5 \pm \sqrt{21}}{2}\right\}$ **13.** $\left\{\dfrac{-6 \pm 2\sqrt{3}}{3}, \dfrac{-4 \pm \sqrt{2}}{2}\right\}$ **15.** $\left\{-\dfrac{1}{3}, \dfrac{7}{2}\right\}$ **17.** $\{-63, 28\}$

19. $\{-1\}$ **21.** $\{5\}$ **23.** $\{9\}$ **25.** When squaring the right side, the binomial $3z + 5$ must be squared to get

$(3z + 5)^2 = 9z^2 + 30z + 25.$ **27.** $\{9\}$ **29.** $\emptyset$ **31.** $\{8\}$ **33.** $\{-2\}$ **35.** $\{-2\}$ **37.** $\{2\}$ **39.** $\left\{\dfrac{3}{2}\right\}$ **41.** $\{-27, 3\}$

43. $\left\{\dfrac{1}{4}, 1\right\}$ **45.** $\{16\}$; $u = -3$ does not lead to a solution of the equation. **46.** $\{16\}$; 9 does not satisfy the equation.

47. Answers will vary. **48.** $\{4\}$ **49.** $\{4\}$ **51.** $\{-.4542187292\}$ **53.** Dividing both sides by a variable expression (in this case, x^2) causes us to "lose" the solution 0. **55.** $y = (a^{2/3} - x^{2/3})^{3/2}$

2.6 EXERCISES (page 151) **1.** width; 5 **3.** directly; square; radius; π **5.** directly; radius; 2π

7. inversely; height; 24 **9.** $\dfrac{220}{7}$ **11.** $\dfrac{32}{15}$ **13.** $\dfrac{18}{125}$ **15.** $\dfrac{300}{x} = \dfrac{5}{400}$ **16.** $400x$ **17.** $120{,}000 = 5x$

18. $\{24{,}000\}$; about 24,000 **19.** increases; decreases **21.** y is half as large as it was before. **23.** y is one-third as large as it

was before. **25.** p is $\dfrac{1}{32}$ as large as it was before. **27.** 16 in **29.** $\dfrac{875}{72}$ candela **31.** $\dfrac{450}{11}$ km **33.** 799.5 cu cm **35.** $\dfrac{1024}{9}$ kg

37. 39.3 km per hr **39.** 4.94 **41.** 7.4 km **43.** $F = 4$ **45.** approximately 12,500

🖩 **(page 161)** **1.** (a) $\{-5/2, 2\}$ (b) $(-\infty, -5/2) \cup (2, \infty)$ (c) the open interval $(-5/2, 2)$
2. For $x = -5/2$ or $x = 2$, Y_1 is on the x-axis; in $(-\infty, -5/2) \cup (2, \infty)$, Y_1 is above the x-axis; in $(-5/2, 2)$, Y_1 is below the x-axis. **3.** At the x-values where the graph of $Y_1 = 2x^2 + x - 10$ is on the x-axis, the expression equals zero; when it is below the x-axis, the expression is less than zero (negative); and when it is above the x-axis, the expression is greater than zero (positive).

2.7 EXERCISES (page 165) **1.** F **3.** A **5.** I **7.** B **9.** E **13.** $[-1, \infty)$

15. $(-\infty, 6]$ **17.** $(-\infty, 4)$

19. $\left[-\dfrac{11}{5}, \infty\right)$ **21.** $[1, 4]$

23. $(-6, -4)$ **25.** $(-16, 19]$

27. $x \approx 2$, so it happened first in 1987. **29.** $[500, \infty)$ **31.** $[45, \infty)$ **33.** $[-3, 3]$

35. $(-\infty, -3] \cup [-1, \infty)$ **37.** $[-2, 3]$

39. $\left(-\infty, \dfrac{1}{2}\right) \cup (4, \infty)$ **41.** $(-\infty, 0) \cup (0, \infty)$

43. $\left(\dfrac{-5 - \sqrt{33}}{2}, \dfrac{-5 + \sqrt{33}}{2}\right)$ **45.** $[1 - \sqrt{2}, 1 + \sqrt{2}]$

47. $(-5, 3]$ **49.** $(-\infty, -2)$ **51.** $(-\infty, 6) \cup \left[\dfrac{15}{2}, \infty\right)$ **53.** $(-\infty, 1) \cup \left(\dfrac{9}{5}, \infty\right)$ **55.** $\left(-\infty, -\dfrac{3}{2}\right) \cup \left[-\dfrac{1}{2}, \infty\right)$ **57.** $(-2, \infty)$

59. $\left(0, \dfrac{4}{11}\right) \cup \left(\dfrac{1}{2}, \infty\right)$ **61.** $(-\infty, -2] \cup (1, 2)$ **63.** $(-\infty, 5)$ **65.** $\left\{\dfrac{4}{3}, -2, -6\right\}$ **66.**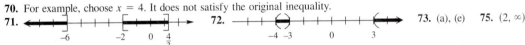

67. For example, choose $x = -10$. It satisfies the original inequality.

68. For example, choose $x = -4$. It does not satisfy the original inequality.
69. For example, choose $x = 0$. It satisfies the original inequality.

70. For example, choose $x = 4$. It does not satisfy the original inequality.
71.   **72.**   **73.** (a), (e)   **75.** $(2, \infty)$

77. $[-1, 5]$ **79.** $(-\infty, -6] \cup (-2, \infty)$ **83.** (a) $2.08 \times 10^{-5} \le \dfrac{R}{72} \le 8.33 \times 10^{-5}$ (b) between 5400 and 21,700

CONNECTIONS (page 173)   **1.** $-.05 \le x \le .05$ **2.** Some possible answers are quality control on a production line or in a laboratory, or in any process where tolerance is limited.

2.8 EXERCISES (page 174)   **1.** F   **3.** D   **5.** G   **7.** C   **9.** $\left\{-\dfrac{1}{3}, 1\right\}$ **11.** $\left\{\dfrac{2}{3}, \dfrac{8}{3}\right\}$ **13.** $\{-6, 14\}$ **15.** $\left\{\dfrac{5}{2}, \dfrac{7}{2}\right\}$

17. $\left\{-3, \dfrac{3}{2}\right\}$ **19.** $\left\{-\dfrac{3}{2}\right\}$ **21.** $\left\{-\dfrac{4}{3}, \dfrac{2}{9}\right\}$ **23.** $\left\{-\dfrac{7}{3}, -\dfrac{1}{7}\right\}$ **25.** $\{1\}$ **29.** $\{1, 3\}$ **31.** $\{-6, -1\}$ **33.** $\left\{\dfrac{2}{11}, 6\right\}$

35. $(-\infty, -1) \cup (1, \infty)$ **37.** $(-4, -1)$ **39.** $(-\infty, 0) \cup (6, \infty)$ **41.** $\left(-\infty, -\dfrac{8}{3}\right] \cup [2, \infty)$ **43.** $\left[-1, -\dfrac{1}{2}\right]$

45. $\left(-\dfrac{3}{2}, \dfrac{13}{10}\right)$ **47.** $(-\infty, \infty)$ **49.** $(-\infty, -12) \cup (-12, \infty)$ **51.** $|p - q| = 5$ (or $|q - p| = 5$)   **55.** $|m - 9| \le 8$

57. $|p - 9| \ge 5$ **59.** $|r - 3| = 5$ **61.** -6 or 6 **62.** $x^2 - x = 6$ **63.** $\{-2, 3\}$ **64.** $x^2 - x = -6$

65. $\left\{\dfrac{1}{2} \pm \dfrac{\sqrt{23}}{2}i\right\}$ **66.** $\left\{-2, 3, \dfrac{1}{2} \pm \dfrac{\sqrt{23}}{2}i\right\}$ **67.** $25.33 \le R_L \le 28.17$; $36.58 \le R_E \le 40.92$ **69.** There are many possible explanations. Students may work harder during an exam or there may be more stress, so students may breathe more frequently.
71. $[6.5, 9.5]$ **73.** $|x - 123| \le 25$; $|x - 21| \le 5$

CHAPTER 2 REVIEW EXERCISES (page 178)   **1.** $\{6\}$ **3.** $\left\{-\dfrac{11}{3}\right\}$ **5.** $\emptyset$ **7.** $\left\{\dfrac{1}{60}\right\}$ **9.** $x = -6b - a - 6$

11. $f = \dfrac{AB(p + 1)}{24}$ **13.** $m = \dfrac{Pi}{A - P}$ **15.** earned runs allowed: 88   **17.** E.R.A.: 3.37   **19.** innings pitched: 241

21. 20 lb of hearts and 10 lb of kisses   **23.** 3 3/7 liters   **25.** $800   **27.** 15 mph   **29.** (a) $36.525x$ (b) 2629.8 mg

31. $\{-7 \pm \sqrt{5}\}$ **33.** $\left\{\dfrac{5}{2}, -3\right\}$ **35.** $\left\{-\dfrac{3}{2}, 7\right\}$ **37.** $\left\{-\dfrac{1}{2}, 3\right\}$ **39.** $\{\sqrt{2} \pm 1\}$ **41.** (b) and (c)

43. -188;   two different imaginary solutions   **45.** 484;   two different rational solutions   **47.** 4 sec and 9.75 sec

49. 50 m by 225 m   or   112.5 m by 100 m   **51.** 5 in, 12 in, 13 in   **53.** $\dfrac{27}{2}$ **55.** 150 kg per sq m

57. 33,750 units   **59.** $\left\{\pm\dfrac{1}{2}, \pm i\right\}$ **61.** $\left\{-15, \dfrac{5}{2}\right\}$ **63.** $\{3\}$ **65.** $\{-2, -1\}$ **67.** $\emptyset$ **69.** $\{-1\}$ **71.** $\left(-\dfrac{7}{13}, \infty\right)$

73. $(-\infty, 1]$ **75.** $(1, \infty)$ **77.** $[4, 5]$ **79.** $[-4, 1]$ **81.** $\left(-\dfrac{2}{3}, \dfrac{5}{2}\right)$ **83.** $\{3\}$ **85.** $\left(-\dfrac{1}{3}, 0\right)$ **87.** $(-2, 4) \cup (16, \infty)$

91. The amount of ozone remaining after filtration is 79.8 ppb. Since $79.8 > 50$, it did not remove enough of the ozone.

93. $[300, \infty)$ **95.** $\{-11, 3\}$ **97.** $\left\{\dfrac{11}{27}, \dfrac{25}{27}\right\}$ **99.** $\left\{-\dfrac{2}{7}, \dfrac{4}{3}\right\}$ **101.** $[-7, 7]$ **103.** $(-\infty, -3) \cup (3, \infty)$ **105.** $[-6, -3]$

107. $\left(-\dfrac{2}{7}, \dfrac{8}{7}\right)$ **109.** $(-\infty, -4) \cup \left(-\dfrac{2}{3}, \infty\right)$

CHAPTER 2 TEST (page 183)   **1.** {0}   **2.** ∅   **3.** $W = \dfrac{S - 2LH}{2H + 2L}$ **4.** $C \approx 7.029$ pCi/L Since the level is above 4 pCi/L, it is

unsafe.   **5.** One possible action would be to seal any cracks in the basement walls and increase ventilation.   **6.** $13\frac{1}{3}$ qt

7. 225 mi   **8.** $\left\{\frac{2}{3}, 1\right\}$ **9.** $\left\{\dfrac{3 \pm \sqrt{17}}{5}\right\}$ **10.** $\left\{1 \pm \dfrac{\sqrt{6}}{6}i\right\}$ **11.** $-\dfrac{25}{16}$ **12.** (b)   **13.** (a) 6.6 million   (b) 5.725 million

(c) 4.388 million   **14.** (a) 1 second and 5 seconds   (b) 6 seconds   **15.** {2}   **16.** {±2, ±$i\sqrt{10}$}   **17.** {−2}   **18.** $\left\{\dfrac{4}{3}, \dfrac{9}{4}\right\}$

19. 1.1 mm   **20.** $26\frac{2}{3}$ days   **21.** $(-3, \infty)$ **22.** $[-10, 2]$ **23.** $(-\infty, 1] \cup \left[\dfrac{3}{2}, \infty\right)$ **24.** $(-\infty, 3) \cup (4, \infty)$ **25.** $\left\{-3, -\dfrac{1}{3}\right\}$
26. $(-2, 7)$ **27.** $(-\infty, -6] \cup [5, \infty)$

CHAPTER 3   Relations and Functions ▼▼▼

CONNECTIONS (page 189)   **1.** Each solution would require three numbers and could be written as an ordered *triple*.   **2.** Three
number lines would be required.   **3.** Answers will vary.

3.1 EXERCISES (page 196)   **1.** true   **3.** true   **5.** $(-4, 6), (3, 2), (5, 7)$;   domain: {−4, 3, 5};   range: {6, 2, 7}
In Exercises 7–13, there are other possible answers for the three points.
7. $(1, 6), (2, 15), (-1, -12)$;   domain: $(-\infty, \infty)$;   range: $(-\infty, \infty)$ **9.** $(0, 0), (1, -1), (4, -2)$;   domain: $[0, \infty)$;   range: $(-\infty, 0]$
11. $(0, 2), (1, 3), (-1, 1)$;   domain: $(-\infty, \infty)$;   range: $[0, \infty)$ **13.** $(0, 3), (1, 5.1), (2, 7.2)$;   domain: {0, 1, 2, 3, 4, 5, 6};

range: {3, 5.1, 7.2, 9.3, 11.4, 13.5, 15.6}   **15.** (a) $8\sqrt{2}$ (b) $(-9, -3)$ **17.** (a) $\sqrt{34}$ (b) $\left(\dfrac{11}{2}, \dfrac{7}{2}\right)$ **19.** (a) $\sqrt{133}$

(b) $\left(2\sqrt{2}, \dfrac{3\sqrt{5}}{2}\right)$ **21.** (a) $2\sqrt{10}$ (b) $(-1, 4)$ **23.** (a) $3495   (b) If the data are related linearly the midpoint formula will be

exact.   (c) No. The midpoint of (1970, 3968) and (1980, 8414) is (1975, 6191). The actual cutoff for 1975 is $5500, not $6191.
25. yes   **27.** no   **29.** yes   **31.** no   **33.** $(-3, 6)$ **35.** $(5, -4)$ **39.** $x^2 + y^2 = 36$;   domain: $[-6, 6]$;   range: $[-6, 6]$

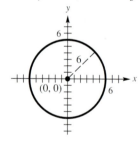

$x^2 + y^2 = 36$

41. $(x - 2)^2 + y^2 = 36$;   domain: $[-4, 8]$;   range: $[-6, 6]$ **43.** $(x + 2)^2 + (y - 5)^2 = 16$;   domain: $[-6, 2]$;   range: $[1, 9]$

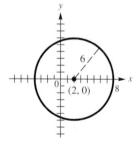

$(x - 2)^2 + y^2 = 36$

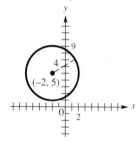

$(x + 2)^2 + (y - 5)^2 = 16$

45. $(x - 5)^2 + (y + 4)^2 = 49$;   domain: $[-2, 12]$;   range: $[-11, 3]$ **47.** $(x - 3)^2 + (y - 2)^2 = 4$

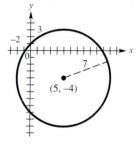

$(x - 5)^2 + (y + 4)^2 = 49$

49. $Y_1 = 2 + \sqrt{25 - (x + 4)^2}$; $Y_2 = 2 - \sqrt{25 - (x + 4)^2}$ **51.** $(-3, -4)$; $r = 4$ **53.** $(2, -6)$; $r = 6$

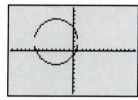

Square viewing window

55. $(0, 1)$; $r = 7$ **57.** It is the point $(3, 3)$.   **61.** $(2, -3)$ **62.** $3\sqrt{5}$ **63.** $3\sqrt{5}$ **64.** $3\sqrt{5}$ **65.** $(x - 2)^2 + (y + 3)^2 = 45$
66. $(x + 2)^2 + (y + 1)^2 = 41$ **67.** Show that the point $(-3, 4)$ lies on each of the following circles: $(x - 1)^2 + (y - 4)^2 = 16$,
$(x + 6)^2 + y^2 = 25$, $(x - 5)^2 + (y + 2)^2 = 100$.   **71.** $(x + \sqrt{2})^2 + (y + \sqrt{2})^2 = 2$ **73.** $(2 + \sqrt{7}, 2 + \sqrt{7})$,
$(2 - \sqrt{7}, 2 - \sqrt{7})$ **75.** $(2, 3)$ or $(4, 1)$ **77.** $9 + \sqrt{119}, 9 - \sqrt{119}$

CONNECTIONS (page 201)   **1.** 3000 gallons   **2.** at 25 hours   **3.** for 25 hours;   for 25 hours   **4.** 2000 gallons

3.2 EXERCISES (page 208)   **1.** yes   **3.** no   **5.** yes   **7.** yes   **9.** no   **11.** no   **13.** yes   **15.** no   **17.** -14
19. 67   **21.** $4k - 2$ **23.** 78   **25.** $x^2 + 2xk + k^2 + 3$ **27.** -4 **29. (a)** 0 **(b)** 4 **(c)** 2 **(d)** 4   **31. (a)** -3 **(b)** -2
(c) 0 **(d)** 2   **33.** -14 **35.** 1.5   **37.** 2   **39. (a)** approximately 89% **(b)** approximately 47%   **41.** approximately 12 seconds
In Exercises 43–61, we give the domain first and the range second.   **43.** $[-5, 4]$; $[-2, 6]$ **45.** $(-\infty, \infty)$; $(-\infty, 12]$
47. $[-3, 4]$; $[-6, 8]$ **49.** $(-\infty, \infty)$; $[-4, \infty)$ **51.** $(-\infty, \infty)$; $(-\infty, \infty)$ **53.** $(-\infty, \infty)$; $[0, \infty)$ **55.** $[-9, \infty)$; $[0, \infty)$
57. $[-2, 2]$; $[-2, 0]$ **59.** $(-\infty, -7) \cup (-7, \infty)$; $(-\infty, 0) \cup (0, \infty)$ **61.** $(-\infty, \infty)$; $(-\infty, \infty)$ **63. (a)** $[4, \infty)$
(b) $(-\infty, -1]$ **(c)** $[-1, 4]$ **65. (a)** $(-\infty, 4]$ **(b)** $[4, \infty)$ **(c)** none   **67. (a)** none **(b)** $(-\infty, -2]$; $[3, \infty)$ **(c)** $(-2, 3)$
69. (a) $[0, 25]$ **(b)** $[50, 75]$ **(c)** $[25, 50]$; $[75, 100]$ **71. (a)** $C(x) = 10x + 500$ **(b)** $R(x) = 35x$ **(c)** $P(x) = 25x - 500$
(d) 20 units; do not produce   **73. (a)** $C(x) = 150x + 2700$ **(b)** $R(x) = 280x$ **(c)** $P(x) = 130x - 2700$ **(d)** 20.77 or 21 units;
produce   **75. (a)** 25 units **(b)**

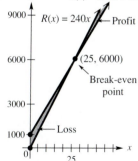

(c) $6000

3.3 EXERCISES (page 219)   **1.** true   **3.** true   **5.** false   In Exercises 7–15, we give the domain first and then the range.
7. $(-\infty, \infty)$; $(-\infty, \infty)$ **9.** $(-\infty, \infty)$; $(-\infty, \infty)$ **11.** $(-\infty, \infty)$; $(-\infty, \infty)$ **13.** $(-\infty, \infty)$; $\{-4\}$

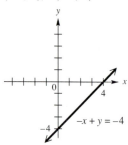

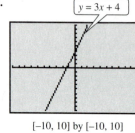

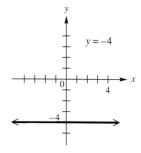

15. $(-\infty, \infty)$; $(-\infty, \infty)$ **17.** A   **19.** D   **21.**   **23.** $\left(\text{Use } y = -\dfrac{3}{4}x + \dfrac{3}{2}.\right)$

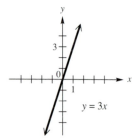

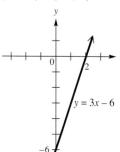

[−10, 10] by [−10, 10]

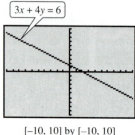

[−10, 10] by [−10, 10]

25.   **27.** $\dfrac{2}{5}$ **29.** 0   **31.** 0   **33.** undefined   **35.** -3.5 **39.**

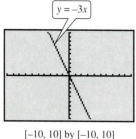

[−10, 10] by [−10, 10]

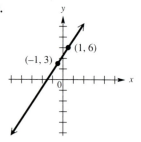

41.

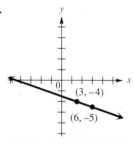

43.

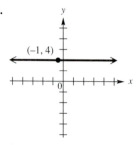

45. (a) The slope of $-.0221$ indicates that on the average from 1912 to 1992 the 5000-meter run is being run .0221 second faster every year. It is negative because the times are generally decreasing as time progresses.   **(b)** World War II (1939–1945) included the years 1940 and 1944.   **(c)** Yes. If it continues in a linear pattern, eventually the winning time is 0, which is unrealistic.   **47.** 3   **48.** 3   **49.** equal   **50.** $\sqrt{10}$ **51.** $2\sqrt{10}$ **52.** $3\sqrt{10}$ **53.** The sum is $3\sqrt{10}$, which is equal to the answer in Exercise 52.   **54.** B; C; A; C (The order of the last two may be reversed.)

55. The midpoint is (3, 3), which is the same as the middle entry in the table. **56.** 7.5 **57. (a)** $6x + 6h + 2$ **(b)** $6h$ **(c)** 6
59. (a) $-2x - 2h + 5$ **(b)** $-2h$ **(c)** -2 **61. (a)** $x^2 + 2xh + h^2 - 4$ **(b)** $2xh + h^2$ **(c)** $2x + h$

3.4 EXERCISES (page 227) **1.** true **3.** false **5.** $2x + y = 5$ **7.** $3x + 2y = -7$ **9.** $x = -8$ **11.** $y = \frac{1}{4}x + \frac{13}{4}$

13. $y = \frac{2}{3}x - 2$ **15.** $x = -6$ (cannot be written in slope-intercept form) **17.** -2; does not; undefined; $\frac{1}{2}$;

does not; zero **19. (a)** B **(b)** D **(c)** A **(d)** C **21.** slope: 3; y-intercept: -1 **23.** slope: 4; y-intercept: -7

25. slope: $-\frac{3}{4}$; y-intercept: 0 **27.** $x + 3y = 11$ **29.** $5x - 3y = -13$ **31.** $x = -5$ **33. (a)** $-\frac{1}{2}$ **(b)** $-\frac{7}{2}$

35. the Pythagorean theorem and its converse **36.** $\sqrt{x_1{}^2 + m_1{}^2 x_1{}^2}$ **37.** $\sqrt{x_2{}^2 + m_2{}^2 x_2{}^2}$ **38.** $\sqrt{(x_2 - x_1)^2 + (m_2 x_2 - m_1 x_1)^2}$
40. $-2x_1 x_2(m_1 m_2 + 1) = 0$ **41.** Since $x_1 \neq 0$, $x_2 \neq 0$, $m_2 m_1 + 1 = 0$, implying $m_1 m_2 = -1$. **42.** The product of the slopes
is -1. **43. (a)** There appears to be a linear relationship. **(b)** $y = 76.9x$

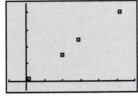

[−100, 600] by [−5000, 45,000]
Xscl = 100   Yscl = 10,000

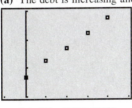

[−100, 600] by [−5000, 45,000]
Xscl = 100   Yscl = 10,000

(c) approximately 780 megaparsecs (about 1.5×10^{22} miles) **(d)** approximately 12.4 billion years **(e)** $A(50) \approx 1.9 \times 10^{10}$,
$A(100) \approx 9.5 \times 10^9$; between 9.5 billion and 19 billion years **45. (a)** The debt is increasing and the data appear to be linear.

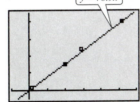

[−1, 5] by [1500, 3000]
Xscl = 1   Yscl = 500

(b) $f(x) = 263.25x + 1828$. The slope of 263.25 indicates that the federal debt is increasing at a rate of 263.25 billion dollars per year.

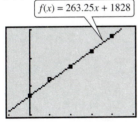

[−1, 5] by [1500, 3000]
Xscl = 1   Yscl = 500

(c) 1984: 1564.75 billion dollars; 1990: 3144.25 billion dollars; In both cases, the results are less than the true values. **(d)** 1980:
511.75 billion dollars; 1994: 4197.25 billion dollars; In both cases, the results are less than the true values. **(e)** A linear
approximation is more accurate over a smaller time interval. As the time interval gets larger, the approximation becomes less

accurate. **47. (a)** $f(x) = \frac{4}{75}x - \frac{13}{300}$ or $f(x) = .05\overline{3}x - .04\overline{3}$ **(b)** The slope of $.05\overline{3}$ tells us that the percent of HIV patients

who get AIDS will increase by $5.\overline{3}\%$ a year. **49.** $f(x) = -\frac{64}{3}x + 717$ **51.** $y = 4x + 120$; 4 ft **55.** It is not true if one line

is vertical and the other is horizontal.

3.5 EXERCISES (page 238)   1. H   3. F   5. A   7. J   9. B   11. J

In Exercises 13–31, we give the domain first and the range second.

13. $(-\infty, \infty)$; $(-\infty, \infty)$

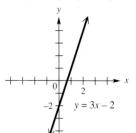

15. $[0, \infty)$; $(-\infty, \infty)$

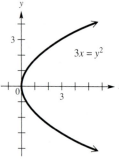

17. $(-\infty, \infty)$; $(-\infty, 0]$

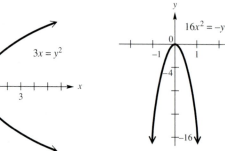

19. $(-\infty, \infty)$; $[4, \infty)$

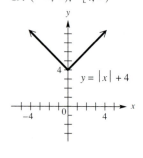

21. $[1, \infty)$; $(-\infty, \infty)$

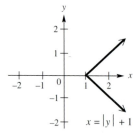

23. $(-\infty, \infty)$; $(-\infty, 0]$

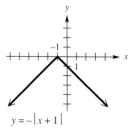

25. $[-2, \infty)$; $[0, \infty)$

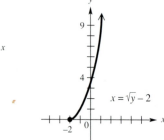

27. $(-\infty, 0]$; $[2, \infty)$

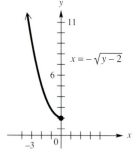

29. $[-2, \infty)$; $[0, \infty)$

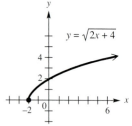

31. $[0, \infty)$; $(-\infty, 0]$

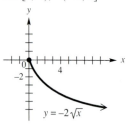

33. the relations in Exercises 13, 14, 17, 18, 19, 20, 23, 24, 29, 30, 31, and 32   **35. (a)** -10 **(b)** -2 **(c)** -1 **(d)** 2

37.

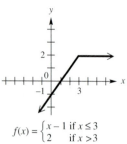

$$f(x) = \begin{cases} x-1 & \text{if } x \le 3 \\ 2 & \text{if } x > 3 \end{cases}$$

39.

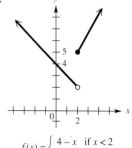

$$f(x) = \begin{cases} 4-x & \text{if } x < 2 \\ 1+2x & \text{if } x \ge 2 \end{cases}$$

41.

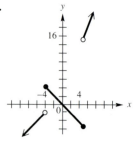

$$f(x) = \begin{cases} 2+x & \text{if } x < -4 \\ -x & \text{if } -4 \le x \le 5 \\ 3x & \text{if } x > 5 \end{cases}$$

43. $f(x) = \begin{cases} -1 & \text{if } x \leq 0 \\ 1 & \text{if } x > 0 \end{cases}$; $(-\infty, \infty)$; $\{-1, 1\}$ **45.** $f(x) = \begin{cases} 2 & \text{if } x \leq 0 \\ -1 & \text{if } x > 1 \end{cases}$; $(-\infty, 0] \cup (1, \infty)$; $\{-1, 2\}$

47. domain: $(-\infty, \infty)$;
range: $\{\ldots, -2, -1, 0, 1, 2, \ldots\}$

49. domain: $(-\infty, \infty)$;
range: $\{\ldots, -2, -1, 0, 1, 2, \ldots\}$

51.

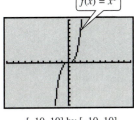

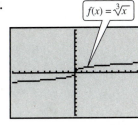

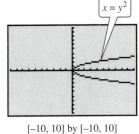

[−10, 10] by [−10, 10]

53.

55.

57. $Y_1 = \sqrt{x}$, $Y_2 = -\sqrt{x}$

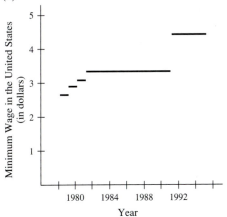

[−10, 10] by [−10, 10]   [−10, 10] by [−10, 10]   [−10, 10] by [−10, 10]

59. (h)

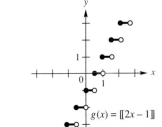

61. (a) 140 **(b)** 220 **(c)** 220 **(d)** 220 **(e)** 220 **(f)** 60 **(g)** 60

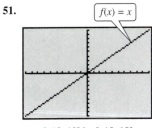

$i(t) = \begin{cases} 40t + 100 & \text{if } 0 \leq t \leq 3 \\ 220 & \text{if } 3 < t \leq 8 \\ -80t + 860 & \text{if } 8 < t \leq 10 \\ 60 & \text{if } 10 < t \leq 24 \end{cases}$

63. B   **65.** D

3.6 EXERCISES (page 252)   **1.** true   **3.** true   **5.** false   **7.**

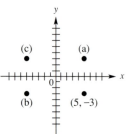

9.

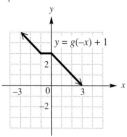

11. (a) The graph of $g(x)$ is reflected about the y-axis and translated up 1 unit.

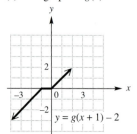

(b) The graph of $g(x)$ is translated to the right 2 units.   **(c)** The graph of $g(x)$ is translated to the left 1 unit and down 2 units.

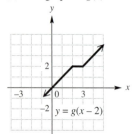

(d) The graph of $g(x)$ is reflected about the x-axis and translated up 2 units.

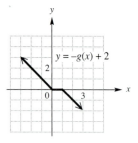

13. y-axis   **15.** x-axis, y-axis, origin   **17.** origin   **19.** none of these

21.   **23.**   **25.**   **27.**

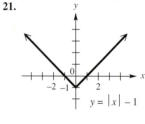

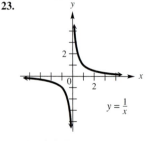

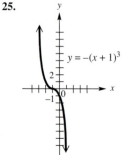

29. $f(-3) = -6$ **31.** $f(9) = 6$ **33.** $f(-3) = -6$ **35.** $g(x) = 2x + 13$

37. (a)

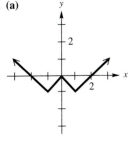

(b)

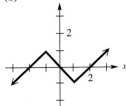

41. Many answers are possible.   **43.** F   **45.** D   **47.** B

49.

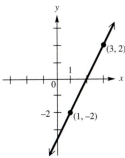

50. 2   **51.** $y_1 = 2x - 4$ **52.** (1, 4), (3, 8)   **53.** 2   **54.** $y_2 = 2x + 2$

55. The graph of Y_2 is obtained by shifting the graph of Y_1 up 6 units. The constant, 6, comes from the 6 added in Exercise 52.

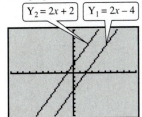

[−10, 10] by [−10, 10]

56. c; c; the same as; c; upward   **57.** The general trend in these carbon monoxide levels has been to gradually decrease from 9.125 to 7.454 parts per million.

CONNECTIONS (page 260)   We used composite functions when function notation was introduced: $f(x + h)$, for example, and in Section 3.6, we used composite functions to indicate horizontal translations.

3.7 EXERCISES (page 261)   **1.** true   **3.** true   **5.** $5x - 1$; $x + 9$; $6x^2 - 7x - 20$; $(3x + 4)/(2x - 5)$; all domains are $(-\infty, \infty)$ except for that of f/g, which is $(-\infty, 5/2) \cup (5/2, \infty)$ **7.** $3x^2 - 4x + 3$; $x^2 - 2x - 3$; $2x^4 - 5x^3 + 9x^2 - 9x$; $(2x^2 - 3x)/(x^2 - x + 3)$; all domains are $(-\infty, \infty)$ **9.** $\sqrt{4x - 1} + \sqrt{x + 3}$; $\sqrt{4x - 1} - \sqrt{x + 3}$; $\sqrt{(4x - 1)(x + 3)}$; $\sqrt{(4x - 1)/(x + 3)}$; all domains are $[1/4, \infty)$ **11.** 61   **13.** 2016   **15.** -3.5 or $-\dfrac{7}{2}$ **17.** $5m^2 - 8m - 4$ **19.** 1248   **21.** 100   **23.** 5   **25.** 0   **27.** 3   **29.** 2   **31.** 1   **33.** 9   **35.** 1   **37.** $g(1) = 9$, and $f(9)$ cannot be determined from the table given.   **39.** $-30x - 33$; $-30x + 52$ **41.** $4x^2 + 42x + 118$; $4x^2 + 2x + 13$ **43.** $\dfrac{2}{(2 - x)^4}$; $2 - \dfrac{2}{x^4}$ **45.** $36x + 72 - 22\sqrt{x + 2}$; $2\sqrt{9x^2 - 11x + 2}$

51.

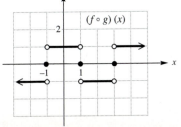

In Exercises 57–61, we give only one of the many possible ways.
57. $g(x) = 6x - 2$, $f(x) = x^2$ **59.** $g(x) = x^2 - 1$, $f(x) = \sqrt{x}$ **61.** $g(x) = 6x$, $f(x) = \sqrt{x} + 12$ **63. (a)** -5 **(b)** -3
(c) 3 **(d)** 13 **65.** $(f \circ g)(x) = 63{,}360x$ computes the number of inches in x miles. **67. (a)** $A(2x) = \sqrt{3}x^2$ **(b)** $64\sqrt{3}$ square
units **69. (a)** $(A \circ r)(t) = 16\pi t^2$ **(b)** It defines the area of the leak in terms of the time t, in minutes. **(c)** 144π sq ft
71. (a) $N(x) = 100 - x$ **(b)** $G(x) = 2 + .2x$ **(c)** $C(x) = (100 - x)(2 + .2x)$ **(d)** $560

CHAPTER 3 REVIEW EXERCISES (page 267) **1.** domain: $\{-3, -1, 8\}$; range: $\{6, 4, 5\}$ **3.** $\sqrt{85}$; $\left(-\dfrac{1}{2}, 2\right)$

5. 5; $\left(-6, \dfrac{11}{2}\right)$ **7.** -7; -1; 8; 23 **9.** $(x + 2)^2 + (y - 3)^2 = 225$ **11.** $(x + 8)^2 + (y - 1)^2 = 289$

13. $(2, -3)$; $r = 1$ **15.** $\left(-\dfrac{7}{2}, -\dfrac{3}{2}\right)$; $r = \dfrac{3\sqrt{6}}{2}$ **17.** $3 + 2\sqrt{5}$; $3 - 2\sqrt{5}$ **19.** $\left(\dfrac{-5 + \sqrt{71}}{2}, \dfrac{5 - \sqrt{71}}{2}\right)$;
$\left(\dfrac{-5 - \sqrt{71}}{2}, \dfrac{5 + \sqrt{71}}{2}\right)$ **21.** no; $(-\infty, \infty)$; $[0, \infty)$ **23.** yes; $(-\infty, -2] \cup [2, \infty)$; $[0, \infty)$ **25.** yes; $(-\infty, \infty)$; $(-\infty, \infty)$
27. not a function **29.** function **31.** $(-\infty, \infty)$ **33.** $(-\infty, \infty)$ **35. (a)** $[2, \infty)$ **(b)** $(-\infty, -2]$ **37.** 2 **39. (a)** no
(b) 1986; 1990 **(c)** about 7; about $27.50 **(d)** down **(e)** a constant, stable price **41.** $\dfrac{6}{5}$ **43.** undefined slope **45.** $\dfrac{9}{4}$
47. $\dfrac{1}{5}$ **49.** -3 **51.** $(-\infty, \infty)$; $(-\infty, \infty)$ **53.** $(-\infty, \infty)$; $(-\infty, \infty)$ **55.** $\{-5\}$; $(-\infty, \infty)$ **57.** $x + 3y = 10$

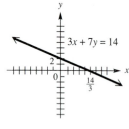

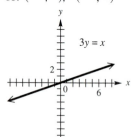

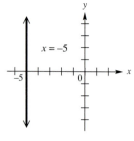

59. $5x - 3y = -15$ **61.** $5x - 8y = -40$ **63.** $y = -5$ **65.** 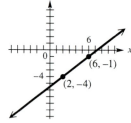 **67.** $y = f(x) = 2.065x - .0456$

69. $f(4.9) \approx 10.1$ million passengers; this agrees favorably with the FAA prediction of 10.3 million.
71. 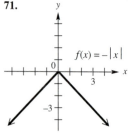 **73.** 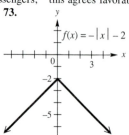 **75.** 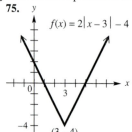 **77.**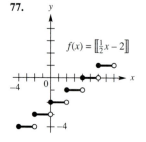

79.
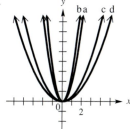

$$f(x) = \begin{cases} 3x + 1 \text{ if } x < 2 \\ -x + 4 \text{ if } x \geq 2 \end{cases}$$

81. $f(x) = \begin{cases} \dfrac{3}{250}x + \dfrac{5}{100} & \text{if } 0 \leq x \leq 15 \\ .23 & \text{if } 15 < x \leq 20 \end{cases}$ **83.** x-axis, y-axis, origin   **85.** none of these symmetries

87. x-axis

89. x-axis, y-axis, origin   **91.** Translate the graph of $f(x) = |x|$ down 2 units.   **93.** $y = -3x + 4$ **95.** $y = 3x + 4$
97. $4x^2 - 3x - 8$ **99.** 44   **101.** $16k^2 - 6k - 8$ **103.** undefined   **105.** $(-\infty, -1) \cup (-1, 4) \cup (4, \infty)$ **107.** $\sqrt{x^2 - 2}$
109. $\sqrt{34}$ **111.** 1   **113.** $(P \circ f)(a) = 18a^2 + 24a + 9$ **115.** $P = 2x + x + 2x + x$; $P(x) = 6x$;   linear function

CHAPTER 3 TEST (page 272)   **1.** positive   **2.** $\dfrac{3}{5}$ **3.** $\sqrt{34}$ **4.** $\left(\dfrac{1}{2}, \dfrac{5}{2}\right)$ **5.** $3x - 5y = -11$ **6.** $f(x) = \dfrac{3}{5}x + \dfrac{11}{5}$

7. (a) $x = 5$ **(b)** $y = -3$ **8. (a)** $y = -3x + 9$ **(b)** $y = \dfrac{1}{3}x + \dfrac{7}{3}$ **9.** $y = -4x + 3$ **10.** not a function;   domain:

$[0, 4]$;   range: $[-4, 4]$ **11.** function;   domain: $(-\infty, -1) \cup (-1, \infty)$;   range: $(-\infty, 0) \cup (0, \infty)$;   decreasing on $(-\infty, -1)$ and
$(-1, \infty)$ **12.**   **13.**   **14.**

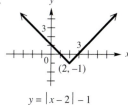

$y = |x - 2| - 1$

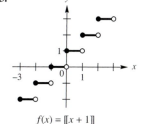

$f(x) = [\![x + 1]\!]$

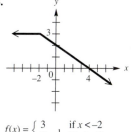

$f(x) = \begin{cases} 3 & \text{if } x < -2 \\ 2 - \dfrac{1}{2}x \text{ if } x \geq -2 \end{cases}$

15. Translate the graph of $y = \sqrt{x}$ 2 units to the left, stretch by a factor of 2, reflect across the x-axis, and translate 3 units down.
16. (a) yes   **(b)** yes   **(c)** yes   **17.** 29   **18.** $4x + 2h - 3$ **19.** $8x^2 - 2x + 1$ **20. (a)** $y = 295.25x + 1147$
(b) The predicted debt based on the model was 3509 billion dollars. This is slightly higher than the actual debt.

CHAPTER 4   Polynomial and Rational Functions ▼▼▼

4.1 EXERCISES (page 282)   **1. (a)** domain: $(-\infty, \infty)$;   range: $[-4, \infty)$ **(b)** $(-3, -4)$ **(c)** $x = -3$ **(d)** 5   **(e)** $-5, -1$
3. (a) domain: $(-\infty, \infty)$;   range: $(-\infty, 2]$ **(b)** $(-3, 2)$ **(c)** $x = -3$ **(d)** -16 **(e)** $-4, -2$ **5.** B   **7.** D
9.   **11.**   **13. (a)** III   **(b)** II   **(c)** IV   **(d)** I

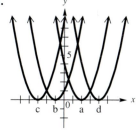

15. vertex: $(2, 0)$;   axis: $x = 2$;   domain: $(-\infty, \infty)$;   range: $[0, \infty)$

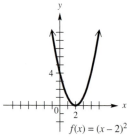

$f(x) = (x - 2)^2$

17. vertex: $(-3, -4)$;   axis: $x = -3$;   domain: $(-\infty, \infty)$;   range: $[-4, \infty)$

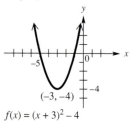

$f(x) = (x + 3)^2 - 4$

19. vertex: $(-3, 2)$;   axis: $x = -3$;   domain: $(-\infty, \infty)$;   range: $(-\infty, 2]$

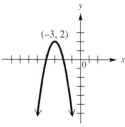

$f(x) = -2(x + 3)^2 + 2$

21. vertex: $(-1, -3)$;   axis: $x = -1$;   domain: $(-\infty, \infty)$;   range: $(-\infty, -3]$

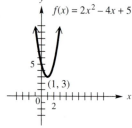

$f(x) = -\frac{1}{2}(x + 1)^2 - 3$

23. vertex: $(1, 2)$;   axis: $x = 1$;   domain: $(-\infty, \infty)$;   range: $[2, \infty)$

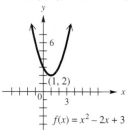

$f(x) = x^2 - 2x + 3$

25. vertex: $(1, 3)$;   axis: $x = 1$;   domain: $(-\infty, \infty)$;   range: $[3, \infty)$

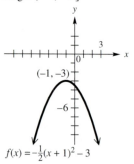

$f(x) = 2x^2 - 4x + 5$

27. 3   **29.** none   **31.** No, they are not equivalent.   **33.** E   **35.** D   **37.** C   **39.** The x-intercepts are -4 and 2.

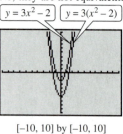

$y = 3x^2 - 2$ $y = 3(x^2 - 2)$

$[-10, 10]$ by $[-10, 10]$

40. the open interval $(-4, 2)$

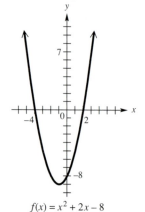

$f(x) = x^2 + 2x - 8$

41. The graph of g is obtained by reflecting the graph of f across the x-axis. **42.** the open interval $(-4, 2)$ **43.** They are the same.

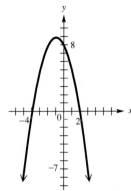

$g(x) = -f(x) = -x^2 - 2x + 8$

45. $f(x) = -2(x - 1)^2 + 4$ or $f(x) = -2x^2 + 4x + 2$ **47.** approximately 8723 billion dollars **49.** 1986

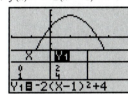

$[-2, 4]$ by $[-2, 5]$

51. (a)

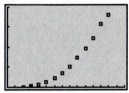

$[0, 15]$ by $[-5000, 400,000]$
Xscl = 1 Yscl = 100,000

(b) One can see that the data are not linear but instead resemble the right half of a parabola opening upward. A quadratic function will model the data better than a linear function.
(c) $f(x) = 2974.76(x - 2)^2 + 1563$ (Other choices will lead to other models.)
(d) There is a relatively good fit.

$f(x) = 2974.76(x - 2)^2 + 1563$

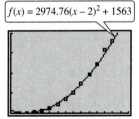

$[0, 15]$ by $[-5000, 400,000]$
Xscl = 1   Yscl = 100,000

(e) 1999: 861,269; 2000: 965,385; In the year 2000, nearly 1 million people will have been diagnosed with AIDS since 1981.
(f) approximately 104,116
53. (a) 3.5 ft **(b)** approximately .2 sec and 2.3 sec **(c)** 1.25 ft **(d)** approximately 3.78 ft **55.** 80 ft wide, 160 ft long
57. (a) 2.5 sec **(b)** 200 ft **59. (a)** 6.25 sec, 675 ft **(b)** between 1.4 and 11.1 sec **(c)** approximately 12.75 sec **61.** $c = 25$

63. $b = -6$ or $b = 6$ **65.** $f(x) = -2x^2 - 2x + 4$ **67.** $y = 3$ **(a)** $\sqrt{3}$ **(b)** $\dfrac{1}{3}$

CONNECTIONS (page 293) **1.** -18 **2.** yes, yes **3.** One example is $f(x) = x^2 + 1$.

4.2 EXERCISES (page 293)   **1.** true   **3.** false   **5.** $x^2 - 2x + 7$ **7.** $4x^2 - 4x + 1 + \dfrac{-3}{x + 1}$ **9.** $x^3 - 4x$

11. $x^4 + x^3 + 2x - 1 + \dfrac{3}{x + 2}$ **13.** $f(x) = (x + 1)(2x^2 - x + 2) - 10$ **15.** $f(x) = (x + 2)(-x^2 + 4x - 8) + 20$

17. $f(x) = (x - 3)(4x^3 + 9x^2 + 7x + 20) + 60$ **19.** 2   **21.** -1 **23.** -6 **25.** 0   **27.** 11   **29.** $-6 - i$ **31.** no

33. yes   **35.** no   **37.** no   **39.** 1   **40.** It is equal to the real number, because 1 is the identity element for multiplication.

41. Add the coefficients of f.   **42.** $1 + (-4) + 9 + (-6) = 0$; the answers agree.

43. $f(-x) = -x^3 - 4x^2 - 9x - 6$; $f(-1) = -20$ **44.** Both are -20. To find $f(-1)$, add the coefficients of $f(-x)$.

CONNECTIONS (page 296)   **1.** $\pm 1, \pm \dfrac{1}{3}, \pm \dfrac{1}{5}, \pm \dfrac{1}{15}$ **2.** $\pm 1, \pm 2, \pm 3, \pm 6, \pm 9, \pm 18$

4.3 EXERCISES (page 302)   **1.** true   **3.** false   **5.** no   **7.** yes   **9.** yes   **11.** $-1 \pm i$ **13.** $-1 \pm \dfrac{\sqrt{2}}{2}$ **15.** $i, \pm 2i$

17. $f(x) = -3x^3 + 6x^2 + 33x - 36$ **19.** $f(x) = -\dfrac{1}{2}x^3 - \dfrac{1}{2}x^2 + x$ **21.** $f(x) = -\dfrac{1}{3}x^3 + \dfrac{5}{3}x^2 - \dfrac{1}{3}x + \dfrac{5}{3}$

23. $g(x) = x^2 - 4x - 5$ **24.** The function g is a quadratic function. The x-intercepts of g are also x-intercepts of f.

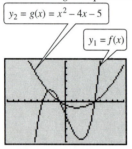

$y_2 = g(x) = x^2 - 4x - 5$

$y_1 = f(x)$

$[-10, 10]$ by $[-60, 60]$
Xscl = 1   Yscl = 10

25. $h(x) = x + 1$ **26.** The function h is a linear function. The x-intercept of h is also an x-intercept of g.

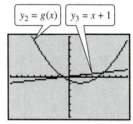

$y_2 = g(x)$ $y_3 = x + 1$

$[-10, 10]$ by $[-60, 60]$
Xscl = 1   Yscl = 10

27. $f(x) = (x - 2)(2x - 5)(x + 3)$ **29.** $f(x) = (x + 3)(3x - 1)(2x - 1)$ **31.** $0, \pm \dfrac{\sqrt{7}}{7}i$ **33.** $2, -3, 1, -1$

35. -2 (multiplicity 5), 1 (multiplicity 5), $1 - \sqrt{3}$ (multiplicity 2)   **37.** $x^2 - 6x + 10$ **39.** $x^3 - 5x^2 + 5x + 3$

41. $x^4 + 4x^3 - 4x^2 - 36x - 45$ **43.** $x^3 - 2x^2 + 9x - 18$ **45.** $x^4 - 6x^3 + 17x^2 - 28x + 20$ **47.** $-1, 3$;

$f(x) = (x + 2)^2(x + 1)(x - 3)$ **53.** $-.88, 2.12, 4.86$ **55.** $.44, 1.81$ **57.** 1.40

4.4 EXERCISES (page 315)   **1.** A   **3.** one   **5.** B and D   **7.** one   **9.**

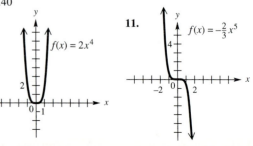

$f(x) = 2x^4$

11.

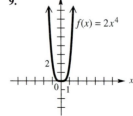

$f(x) = -\dfrac{2}{3}x^5$

13.

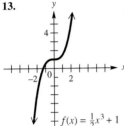

$f(x) = \frac{1}{2}x^3 + 1$

15.

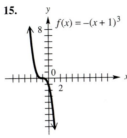

$f(x) = -(x + 1)^3$

17.

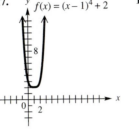

$f(x) = (x - 1)^4 + 2$

19. (c)

21.

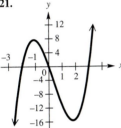

$f(x) = 2x(x - 3)(x + 2)$

23.

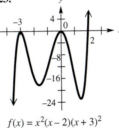

$f(x) = x^2(x - 2)(x + 3)^2$

25.

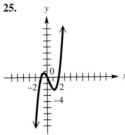

$f(x) = x^3 - x^2 - 2x$

27.

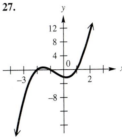

$f(x) = (x + 2)(x - 1)(x + 1)$

29.

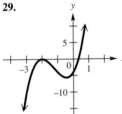

$f(x) = (3x - 1)(x + 2)^2$

31.

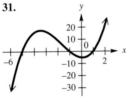

$f(x) = x^3 + 5x^2 - x - 5$

33.

$f(x) = 2x(x - 3)(x + 2)$

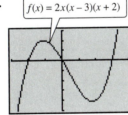

[−3, 4] by [−20, 12]
Xscl = 1   Yscl = 4

35.

$f(x) = (3x - 1)(x + 2)^2$

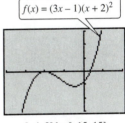

[−4, 2] by [−15, 15]
Xscl = 1   Yscl = 5

37. $f(2) = -2 < 0$, while $f(3) = 1 > 0$.   **39.** $f(0) = 7 > 0$, while $f(1) = -1 < 0$.   **41.** $f(1) = -6 < 0$, while $f(2) = 16 > 0$.
43. 2.7807764064   **45.** 2.19332495204   **47.** $f(x) = .5(x + 6)(x - 2)(x - 5)$ **53.** $-3.0, -1.4, 1.4$ **55.** $-1.1, 1.2$

59.

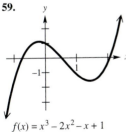

$f(x) = x^3 - 2x^2 - x + 1$

61.

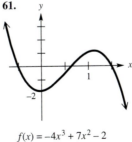

$f(x) = -4x^3 + 7x^2 - 2$

63.

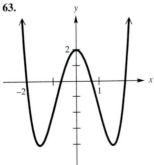

$f(x) = x^4 - 5x^2 + 2$

65. (a)

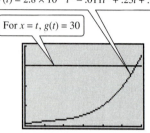

For $x = t$,
$f(t) = 2.8 \times 10^{-4}t^3 - .011t^2 + .23t + .93$

For $x = t$, $g(t) = 30$

[0, 60] by [0, 40]
Xscl = 10   Yscl = 10

(b) The graphs intersect at $x = t \approx 56.9$. Since $t = 0$ corresponds to 1930, this would be during 1986.   **(c)** An increasing percentage of females have smoked during this time period. Smoking has been shown to increase the likelihood of lung cancer.

67. (a)

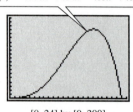

$g(x) = -.006x^4 + .14x^3 - .05x^2 + .02x$

[0, 24] by [0, 200]
Xscl = 1   Yscl = 10

(b) 17.3 hr   **(c)** from 11.4 hr to 21.2 hr

69. (a)

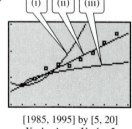

(i)   (ii)   (iii)

[1985, 1995] by [5, 20]
Xscl = 1   Yscl = 5

(b) All three approximate the data near 1986, but only the linear function (ii) $f(x) = 1.088(x - 1986) + 8.6$ approximates the data near 1994.

71. (a) 49%   **(b)** approximately 10.2 yr   **73. (a)** If the length of the pendulum increases, so does the period of oscillation T.
(b) There are a number of ways. One way is to realize that $k = \dfrac{L}{T^n}$ for some integer n. The ratio should be the constant k for each data point when the correct n is found.   **(c)** $k \approx .81$; $n = 2$ **(d)** 2.48 sec   **(e)** T increases by a factor of $\sqrt{2} \approx 1.414$.
75. (a) $0 < x < 10$ **(b)** $A(x) = x(20 - 2x)$ or $A(x) = -2x^2 + 20x$ **(c)** $x = 5$;   maximum cross section area: 50 square in
(d) between 0 and 2.76   or   between 7.24 and 10
77. (a) $x - 1$ **(b)** $\sqrt{x^2 - (x - 1)^2}$ **(c)** $2x^3 - 5x^2 + 4x - 28,225 = 0$ **(d)** hypotenuse: 25 in;   legs: 24 in and 7 in
79. $y = 13,333.\overline{3}x^3 - 87,000x^2 + 122,666.\overline{6}x + 118,000$ **(a)** 167,000   **(b)** There is quite a large discrepancy.

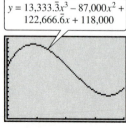

$y = 13,333.\overline{3}x^3 - 87,000x^2 +$
$122,666.\overline{6}x + 118,000$

[0, 4] by [0, 180,000]
Xscl = 1   Yscl = 10,000

4.5 EXERCISES (page 332) 1. A, B, C   3. A   5. A   7. A, C, D

9.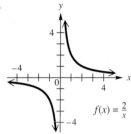
$f(x) = \frac{2}{x}$

11. $f(x) = \frac{1}{x+2}$
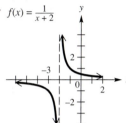
$x = -2$

13.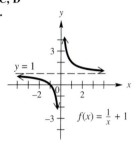
$y = 1$
$f(x) = \frac{1}{x} + 1$

In Exercises 15–21, V. A. stands for *vertical asymptote,* H. A. stands for *horizontal asymptote,* and O. A. stands for *oblique asymptote.*   **15.** V. A.: $x = 5$;   H. A.: $y = 0$ **17.** V. A.: $x = -\frac{1}{2}$;   H. A.: $y = -\frac{3}{2}$ **19.** V. A.: $x = -3$;   O. A.: $y = x - 3$

21. V. A.: $x = -2, x = \frac{5}{2}$;   H. A.: $y = \frac{1}{2}$ **23.** (a)   **25.** (a) C   (b) A   (c) B   (d) D   **27.**

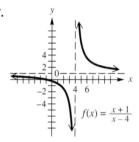
$f(x) = \frac{x+1}{x-4}$

29.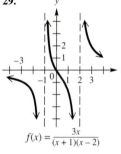
$f(x) = \frac{3x}{(x+1)(x-2)}$

31.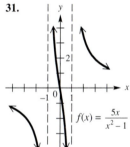
$f(x) = \frac{5x}{x^2 - 1}$

33.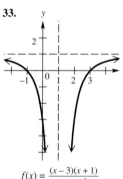
$f(x) = \frac{(x-3)(x+1)}{(x-1)^2}$

35.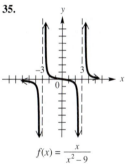
$f(x) = \frac{x}{x^2 - 9}$

37.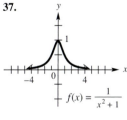
$f(x) = \frac{1}{x^2 + 1}$

39.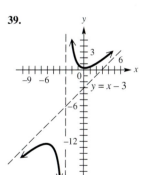
$y = x - 3$
$f(x) = \frac{x^2 + 1}{x + 3}$

41.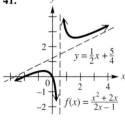
$y = \frac{1}{2}x + \frac{5}{4}$
$f(x) = \frac{x^2 + 2x}{2x - 1}$

43.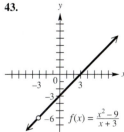
$f(x) = \frac{x^2 - 9}{x + 3}$

45. $f(x) = \dfrac{(x-3)(x+2)}{(x-2)(x+2)}$ or $f(x) = \dfrac{x^2 - x - 6}{x^2 - 4}$ **47.** $f(x) = \dfrac{x-2}{x(x-4)}$ or $f(x) = \dfrac{x-2}{x^2 - 4x}$

49.

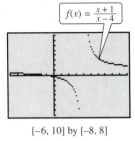

$f(x) = \dfrac{x+1}{x-4}$

51.

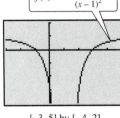

$f(x) = \dfrac{(x-3)(x+1)}{(x-1)^2}$

55. (a) 1   **57. (a)** approximately 52.1 mi per hour

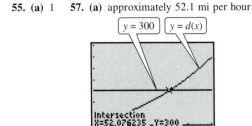

$y = 300$ $y = d(x)$

Intersection
X=52.076235 Y=300

[−6, 10] by [−8, 8]
Dot mode

[−3, 5] by [−4, 2]
Connected mode

[20, 70] by [0, 700]
Xscl = 10   Yscl = 100

(b)

x	20	25	30	35	40	45	50	55	60	65	70
d(x)	34	56	85	121	164	215	273	340	415	499	591

(c) From the table we can see that if the speed doubles, the stopping distance more than doubles.   **(d)** If the stopping distance doubled when the speed doubled, there would be a linear relationship between speed and distance. The graph would be linear and not curved like the graph of d.

59. (a) choice (iii), $r_3(x)$ **(b)** choice (ii), $r_2(x)$

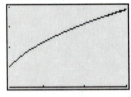

[1, 15] by [0, 4]
Xscl = 5 Yscl = 1

$f_1(x) \approx r_3(x)$

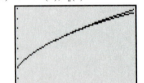

[1, 15] by [0, 8]
Xscl = 5 Yscl = 1

$f_2(x) \approx r_2(x)$

(c) choice (iv), $r_4(x)$ **(d)** choice (i), $r_1(x)$

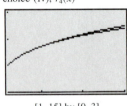

[1, 15] by [0, 3]
Xscl = 5 Yscl = 1

$f_3(x) \approx r_4(x)$

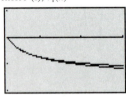

[1, 15] by [−1, .5]
Xscl = 5 Yscl = .5

$f_4(x) \approx r_1(x)$

The graphs of the function and rational approximation are given on the same screen. From the graphs we can see that all the rational approximations give excellent results on the interval [1, 10].

CHAPTER 4 REVIEW EXERCISES (page 338)

1. vertex: $(-4, -5)$
axis: $x = -4$
x-intercepts: $\dfrac{-12 \pm \sqrt{15}}{3}$
y-intercept: 43
$f(x) = 3(x + 4)^2 - 5$

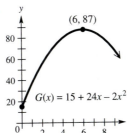

3. vertex: $(-2, 11)$
axis: $x = -2$
x-intercepts: $\dfrac{-6 \pm \sqrt{33}}{3}$
y-intercept: -1
$f(x) = -3x^2 - 12x - 1$

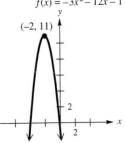

5. k **7.** $ah^2 + k$ **9.** $c - \dfrac{b^2}{4a}$ **11. (a)**

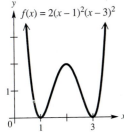

$G(x) = 15 + 24x - 2x^2$

(b) 87 particles on June 6

13. Because the discriminant is 67.3033, a positive number, there are two x-intercepts.   **15. (a)** the open interval $(-.52, 2.59)$
(b) $(-\infty, -.52) \cup (2.59, \infty)$ **19.** $q(x) = 3x^2 + 2x + 1$; $r = 8$ **21.** 6   **23.** 40
In Exercises 25 and 27, other answers are possible.
25. $f(x) = x^3 - 10x^2 + 17x + 28$ **27.** $f(x) = x^4 - 5x^3 + 3x^2 + 15x - 18$ **29.** no
31. $f(x) = 2x^4 - 4x^3 + 10x^2 - 68x + 60$
33. Any polynomial that can be factored into $a(x - b)^2(x - c)^2$ works. One example is $f(x) = 2(x - 1)^2(x - 3)^2$.

$f(x) = 2(x - 1)^2(x - 3)^2$

35. $1 - i, 1 + i, 4, -3$ **37.** -9 **39.** $4 \times 4 \times 4$ in   **41.** $f(-1) = -10 < 0$, while $f(0) = 2 > 0$; $f(2) = -4 < 0$, while
$f(3) = 14 > 0$.   **45.** $4.58039972384, -2.25883787095$ $f(x) = x^4 - 4x^3 - 5x^2 + 14x - 15$

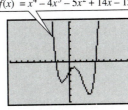

[−10, 10] by [−60, 60]
Xscl = 1   Yscl = 10

47. $f(x) = (x + 1)(x - 1)(x + \sqrt{2})(x - \sqrt{2})$ **49.**

$f(x) = x^4 - 3x^2 + 2$
$\quad = (x + 1)(x - 1)(x + \sqrt{2})(x - \sqrt{2})$

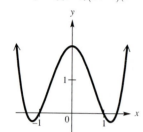

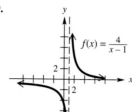

$f(x) = \dfrac{4}{x - 1}$

51.

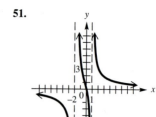

$f(x) = \dfrac{6x}{(x - 1)(x + 2)}$

53.

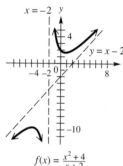

$f(x) = \dfrac{x^2 + 4}{x + 2}$

55.

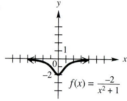

$f(x) = \dfrac{-2}{x^2 + 1}$

57. (a)

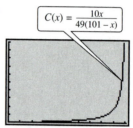

$C(x) = \dfrac{10x}{49(101 - x)}$

[0, 101] by [0, 10]
Xscl = 10   Yscl = 1

(b) approximately 3.23 thousand dollars   **59. (a)**

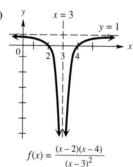

$f(x) = \dfrac{(x - 2)(x - 4)}{(x - 3)^2}$

(b) One possibility is $f(x) = \dfrac{(x - 2)(x - 4)}{(x - 3)^2}$.

CHAPTER 4 TEST (page 341)   **1.** x-intercepts: $\dfrac{-3 \pm \sqrt{3}}{-2}$ $\left(\text{or } \dfrac{3 \pm \sqrt{3}}{2}\right)$

y-intercept: -3

vertex: $\left(\dfrac{3}{2}, \dfrac{3}{2}\right)$

axis: $x = \dfrac{3}{2}$

domain: $(-\infty, \infty)$

range: $\left(-\infty, \dfrac{3}{2}\right]$

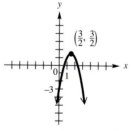

$f(x) = -2x^2 + 6x - 3$

2. (a) 599,814   **(b)** during 1987;   592,642   **3.** $q(x) = 3x^2 - 2x - 5$; $r = 16$ **4.** $q(x) = 2x^2 - x - 5$; $r = 3$ **5.** 53

6. It is a factor. The other factor is $6x^3 + 7x^2 - 14x - 8$.   **7.** $-2, -\dfrac{1}{2}, 3$ **8.** $f(x) = 2x^4 - 2x^3 - 2x^2 - 2x - 4$

10. (a) $f(1) = 5 > 0$, while $f(2) = -1 < 0$.   **(b)** 4.09376345695, 1.83703814322, $-.930801600173$

11. To obtain the graph of f_2, shift the graph of f_1 5 units to the left, stretch by a factor of 2, reflect across the x-axis, and shift 3 units up.

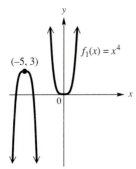

$f_2(x) = -2(x + 5)^4 + 3$

12. C   **13.**   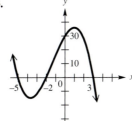   **14.**   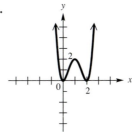   **15.** $f(x) = 2(x - 2)^2(x + 3)$

$f(x) = (3 - x)(x + 2)(x + 5)$ $f(x) = 2x^4 - 8x^3 + 8x^2$

16. (a) 270.08   **(b)** increasing from $t = 0$ to $t = 5.9$ and $t = 9.5$ to $t = 15$;   decreasing from $t = 5.9$ to $t = 9.5$

17.   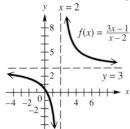   **18.**   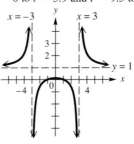   **19.**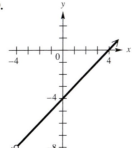

$f(x) = \dfrac{3x - 1}{x - 2}$ $f(x) = \dfrac{x^2 - 1}{x^2 - 9}$ $f(x) = \dfrac{x^2 - 16}{x + 4}$

20. (a) $y = 2x + 3$ **(b)** $-2, \dfrac{3}{2}$ **(c)** 6   **(d)** $x = 1$ **(e)**

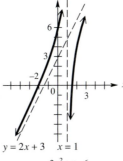

$y = 2x + 3$ $x = 1$

$f(x) = \dfrac{2x^2 + x - 6}{x - 1}$

CHAPTER 5 Exponential and Logarithmic Functions ▼▼▼

CONNECTIONS (page 349) **1.** $f^{-1}(x) = \dfrac{x + 2}{3}$; the message reads MIGUEL HAS ARRIVED.
2. 6858   124   2743   63   511   124   1727   4095; $f^{-1}(x) = \sqrt[3]{x + 1}$

5.1 EXERCISES (page 352)
1. one-to-one   **3.** x; $(g \circ f)(x)$ **5.** (b, a) **7.** $y = x$ **9.** one-to-one   **11.** one-to-one   **13.** not one-to-one   **15.** not one-to-one   **17.** not one-to-one   **19.** one-to-one   **21.** one-to-one   **23.** Yes, they are inverses, because, for the ordered pairs shown, the values of x and y in the second function are the reverse of the values of x and y in the first function.   **24.** x; x;   inverse   **25. (a)** 6;   subtraction   **(b)** 4;   division   **26.** identity   **27.** untying your shoelaces   **28.** stopping a car   **29.** leaving a room   **30.** descending the stairs   **31.** landing in an airplane   **32.** emptying a cup   **33.** inverses   **35.** not inverses   **37.** inverses   **39.** not inverses   **41.** inverses   **43.** not inverses
45.

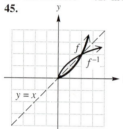

47.

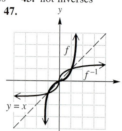

49.

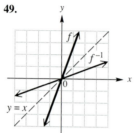

51. 4   **53.** 2   **55.** -2

57. $f^{-1}(x) = \dfrac{x + 4}{3}$

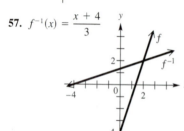

59. $f^{-1}(x) = \sqrt[3]{x - 1}$

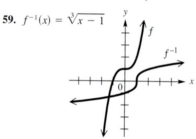

61. not one-to-one

63. $f^{-1}(x) = \dfrac{1}{x}$

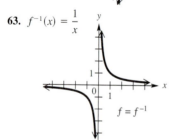

65. $f^{-1}(x) = x^2 - 6, x \geq 0$

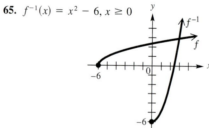

67. 1.14   **69.** the number of dollars required to build 1000 cars
71. $d = \dfrac{k}{B}$ **73.** $\dfrac{1}{a}$ **75.** not one-to-one
77. one-to-one; $f^{-1}(x) = \dfrac{-5 - 3x}{x - 1}$

CONNECTIONS (page 359)   **1.** One possibility is $h(x) = 2^x$ and $g(x) = x + 3$.   **2.** $h(g(x)) = 2^{-x^2}$ **3.** One possibility is $f(x) = kx^n$, $g(x) = 1 + x$.

CONNECTIONS (page 365)   **1.** 2.717   **2.** .9512   **3.** $\dfrac{x^6}{6 \cdot 5 \cdot 4 \cdot 3 \cdot 2 \cdot 1}$

5.2 EXERCISES (page 367)   **1.** yes;   an inverse function   **2.**

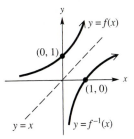

3. $x = a^y$ **4.** $x = 10^y$ **5.** $x = e^y$

7. (a) 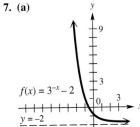   **(b)** 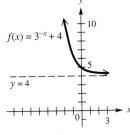   **(c)** 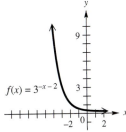   **(d)**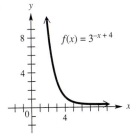

15. 2.3   **17.** .75   **19.** .31

9. 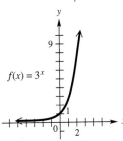   **11.** 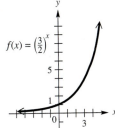   **13.**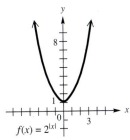

21. If $a > 1$, the function value increases. If $0 < a < 1$, the function value decreases.   **25.** $f(x) = 2^x$ **27.** $f(t) = 27 \cdot 9^t$

31. $\left\{\dfrac{1}{3}\right\}$ **33.** $\{-2\}$ **35.** $\left\{\dfrac{1}{2}\right\}$ **37.** $\{-3, 3\}$ **39.** $\{4\}$ **41.** $\left\{\dfrac{4}{9}\right\}$ **43.** $\left\{-\dfrac{3}{5}\right\}$

45. 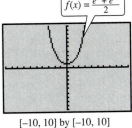   **47.** 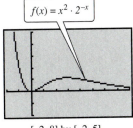   **49.** $76,855.95   **51.** $41,845.63   **53.** 8.0%

[−10, 10] by [−10, 10]   [−2, 8] by [−2, 5]

55. (a) about 63,000 **(b)** about 42,000 **(c)** about 21,000   **57. (a)** 440 g **(b)** 387 g **(c)** 264 g **(d)**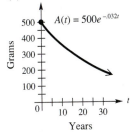

59. (a) linear **(b)** $T(R) = 1.03R$ **(c)** $5.15°F$ **61.** $\emptyset$ **63.** $\{0, .73\}$ **65.** $f(x)$ approaches the line $y = 2.71828$.

5.3 EXERCISES (page 378) **1.** $x = a^y$ **3.** 3; 5; 125 **5.** $\log_3 81 = 4$ **7.** $\log_{2/3}\left(\dfrac{27}{8}\right) = -3$ **9.** $6^2 = 36$

11. $(\sqrt{3})^8 = 81$ **15.** 2 **17.** -3 **19.** $-\dfrac{1}{6}$ **21.** 9 **23.** $\{5\}$ **25.** $\left\{\dfrac{1}{5}\right\}$

29. (a) y

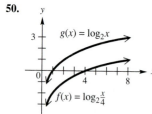

$f(x) = (\log_{1/2} x) - 2$

(b) y

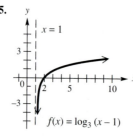

$f(x) = \log_{1/2}(x - 2)$

$x = 2$

(c) y

$f(x) = |\log_{1/2}(x - 2)|$

$x = 2$

31. y

$f(x) = \log_3 x$

33.

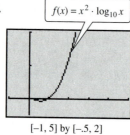

$x = 1$

$f(x) = \log_{1/2}(1 - x)$

35. y

$x = 1$

$f(x) = \log_3(x - 1)$

37. They are not, because the domains are different.

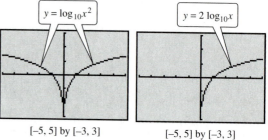

$y = \log_{10}x^2$

$y = 2\log_{10}x$

$[-5, 5]$ by $[-3, 3]$ $[-5, 5]$ by $[-3, 3]$

39. D **41.** C **43.** A **45.**

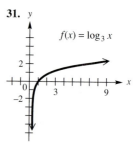

$f(x) = x^2 \cdot \log_{10} x$

$[-1, 5]$ by $[-.5, 2]$

47. $\{1.87\}$ **48.** $\log_a x - \log_a y$ **49.** Since $\log_2\left(\dfrac{x}{4}\right) = \log_2 x - \log_2 4$

by the quotient rule, the graph of $y = \log_2\left(\dfrac{x}{4}\right)$ can be obtained by shifting the graph of $y = \log_2 x$ down $\log_2 4 = 2$ units.

50. y

$g(x) = \log_2 x$

$f(x) = \log_2\frac{x}{4}$

51. 0; 2; 2; 0; By the quotient rule, $\log_2\left(\dfrac{x}{4}\right) = \log_2 x - \log_2 4$. Both sides should equal 0. Since $2 - 2 = 0$, they do.

53. $\log_3 4 + \log_3 p - \log_3 q$ **55.** $1 + \left(\dfrac{1}{2}\right)\log_2 3 - \log_2 5$ **57.** cannot be simplified

59. $\left(\dfrac{1}{3}\right)(5\log_p m + 4\log_p n - 2\log_p t)$ **61.** $\log_b\left(\dfrac{k}{ma}\right)$ **63.** $\log_y(p^{-7/6})$ **65.** $\log_b\left(\dfrac{2y + 5}{\sqrt{y + 3}}\right)$ **67.** 1.0791

69. $-.1303$ **71.** 6

5.4 EXERCISES (page 386)   **1.** 1.6335   **3.** -1.8539 **5.** 6.3630   **7.** $-.3567$ **9.** $\log 8 \approx .90308999$ **11.** $\log_3 4$

12. 2   **13.** 3   **14.** It lies between 2 and 3. Because the function defined by $y = \log_3 x$ is increasing and $9 < 16 < 27$, we have

$\log_3 9 < \log_3 16 < \log_3 27$.   **15.** By the change-of-base-theorem, $\log_3 16 = \dfrac{\log 16}{\log 3} = \dfrac{\ln 16}{\ln 3} \approx 2.523719014$.   **16.** -1; 0

17. It lies between -1 and 0. $\dfrac{1}{5} = .2 < .68 < 1$, so $\log_5 .2 < \log_5 .68 < \log_5 1$; $\log_5 .68 = \dfrac{\log .68}{\log 5} = \dfrac{\ln .68}{\ln 5} \approx -.2396255723$

19. 1.13   **21.** -1.58 **23.** .97   **25.** 1.45   **27.** The function is undefined for $X \geq 4$ because the domain of $Y = \log_a X$ is $X > 0$. This means $4 - X > 0$ here, so $X < 4$ is the domain.

29. The vertical line simulates an asymptote at $x = 1$. The base must be greater than 0 and not equal to 1.

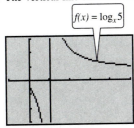

$[-1, 5]$ by $[-3, 3]$

Connected mode

31. 3.2   **33.** 1.8   **35.** 2.0×10^{-3} **37.** 1.6×10^{-5} **39.** (a) 20 (b) 30 (c) 50 (d) 60   **41.** (a) 3 (b) 6 (c) 8

43. (a) about $200{,}000{,}000 I_0$ (b) about $13{,}000{,}000 I_0$ (c) The 1906 earthquake had a magnitude more than 15 times greater than the 1989 earthquake.   **45.** (a) 2 (b) 2 (c) 2 (d) 1   **47.** 1   **49.** (a) 3 (b) $5^2 = 25$ (c) $1/e$ **51.** (a) 5 (b) $\ln 3$

(c) $2 \ln 3$ or $\ln 9$ **53.** about 66 million;   We must assume that the rate of increase continues to be logarithmic.

55. between 7°F and 11°F

57. (a) Let $x = \ln D$ and $y = \ln P$ for each planet. From the graph, the data appear to be linear.

Planet	ln D	ln P
Mercury	$-.94$	-1.43
Venus	$-.33$	$-.48$
Earth	0	0
Mars	.42	.64
Jupiter	1.65	2.48
Saturn	2.26	3.38
Uranus	2.95	4.43
Neptune	3.40	5.10

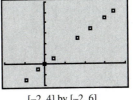

$[-2, 4]$ by $[-2, 6]$

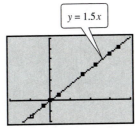

$[-2, 4]$ by $[-2, 6]$

(b) The points $(0, 0)$ and $(3.40, 5.10)$ determine the line $y = 1.5x$ or $\ln P = 1.5 \ln D$. (Answers will vary.) (c) $P \approx 248.3$ years

CONNECTIONS (page 392)   **2.** 0 is extraneous because $e^0 = 1$, so the denominator of the original equation becomes 0. The solution set is $\{\ln 3\}$.   **3.** First, solve for e^x;   second, solve for x.

5.5 EXERCISES (page 396)   **5.** $\{1.631\}$ **7.** $\{-.080\}$ **9.** $\{2.386\}$ **11.** $\{-.123\}$ **13.** Ø   **15.** $\{2\}$ **17.** $\{17.475\}$ **19.** $\{11\}$

21. $\{10\}$ **23.** $\{4\}$ **25.** $\{11\}$ **27.** Ø   **29.** $\{8\}$ **31.** $\{-2, 2\}$ **33.** $\{1, 10\}$ **37.** $t = -\dfrac{2}{R} \ln\left(1 - \dfrac{RI}{E}\right)$ **39.** $x = e^{k/(p-a)}$

42. $(e^x - 1)(e^x - 3) = 0$ **43.** $\{0, \ln 3\}$ **44.** The graph intersects the x-axis at 0 and $1.099 \approx \ln 3$.

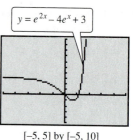

$y = e^{2x} - 4e^x + 3$

$[-5, 5]$ by $[-5, 10]$

45. $(-\infty, 0) \cup (\ln 3, \infty)$ **46.** $(0, \ln 3)$ **47.** $\{1.52\}$ **49.** $\{0\}$ **51.** $\{2.45, 5.66\}$ **53.** $f^{-1}(x) = \dfrac{1}{3}(\ln(x) - 1); (0, \infty); (-\infty, \infty)$

55. $(27, \infty)$ **57.** 89 decibels is about twice as loud as 86 decibels, for a 100% increase.   **59.** 1.25 yr   **61.** 4.27%   **63.** 1999
65. (a) $P(T) = 1 - e^{-.0034 - .0053T}$ **(b)**

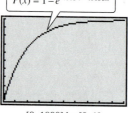

For $t = x$,

$P(x) = 1 - e^{-.0034 - .0053x}$

$[0, 1000]$ by $[0, 1]$
Xscl = 100 Yscl = .1

(c) $P(60) \approx .275$ or 27.5%. The reduction in carbon emissions from a tax of $60 per ton of carbon is 27.5%.   **(d)** $T = \$130.14$

5.6 EXERCISES (page 405)   **1. (a)** 440 g   **(b)** 387 g   **(c)** 264 g   **(d)** 21.66 yr   **3.** 1611.97 yr   **7. (a)** 11%   **(b)** 36%
(c) 84%   **9.** 9000 yr   **11.** about 16,000 yr   **13.** 30 min   **15. (a)** about 46.2 yr   **(b)** about 46.0 yr   **17. (a)** about 27.81 yr
(b) about 27.73 yr   **19.** about 4.3 yr   **21.** When $x = 25$, $Y_1 \approx 35.533417$ and $Y_2 \approx 36.94528$. When $x = 25$, $Y_1 \approx 36.70088$.
23. about 1503 thousand, or 1.503 million   **25.**

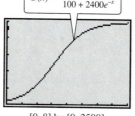

For $x = t$,

$G(x) = \dfrac{250,000}{100 + 2400e^{-x}}$

$[0, 8]$ by $[0, 2500]$
Xscl = 1 Yscl = 500

27. 2.8; 2.7726   **29.** about 2349 million, or 2.349 billion

31. 1997   **33.** about 6.9 yr   **35.** about 11.6 yr   **37. (a)** 11   **(b)** 12.6   **(c)** 18.0   **(d)**

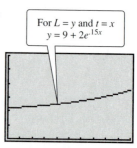

For $L = y$ and $t = x$
$y = 9 + 2e^{.15x}$

$[0, 10]$ by $[0, 30]$
Xscl = 1 Yscl = 5

(e) Living standards are increasing, but at a slow rate.   **39.** $r = 10^{-n/7600}$ **(a)** .74   **(b)** .47   **41.** 5.31%; Social Security taxes rose at a faster rate than consumer prices.   **43.** 2016

CHAPTER 5 REVIEW EXERCISES (page 410)   **1.** $f^{-1}(x) = \sqrt[3]{x+3}$ **3.** the number of years after 1992 required for the investment to reach \$50,000   **5.** yes   **7.** decreasing   **9.** A   **11.** D   **13.** $\log_{100} 10 = \dfrac{1}{2}$ **15.** $\log_{3/4}\left(\dfrac{4}{3}\right) = -1$ **17.** $\log_e 12 = 2.4849$ or $\ln 12 = 2.4849$ **19.** $a < 1$ **21.** $(0, \infty)$ **23.**

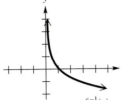

$y = f^{-1}(x)$

25. $10^{-3} = .001$ **27.** $10^{.537819} = 3.45$

29. $9^{3/2} = 27$ **31.** 3   **33.** $\log_3 m + \log_3 n - \log_3 5 - \log_3 r$ **35.** $2 \log_5 x + 4 \log_5 y + \left(\dfrac{1}{5}\right)(3 \log_5 m + \log_5 p)$ **37.** The correct statement is $\log_5 125 - \log_5 25 = \log_5\left(\dfrac{125}{25}\right) = \log_5 5 = 1.$ **39.** -1.386 **41.** 11.878   **43.** 1.159   **45.** by a factor of $2^{1/2} \approx 1.4$

47. $\left\{\dfrac{5}{3}\right\}$ **49.** {2.115}   **51.** {4}   **53.** {3}   **55.** {−3}   **57.** 4.0 yr   **59.** \$25,149.59   **61. (a)** 2.2×10^{12} tons   **(b)** 1987

63. $(-.7666647, .58777476)$ **65.** The x-intercept is .5, which agrees with the x-value found in Exercise 64.

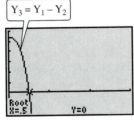

[0, 3] by [−20, 60]
Xscl = 1   Yscl = 10

67. $Y_1 = \dfrac{\log x}{\log 2} + \dfrac{\log (x + 2)}{\log 2} - 3$ **69. (a)** There appears to be a linear relationship.   **(b)** By taking logarithms on both sides, the function can be written as $\ln P = kx + \ln C$, which is a linear function in the form $\ln P = ax + b$. Here $a = k$ and $b = \ln C$, with $y = \ln P$.

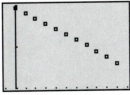

[−1000, 11,000] by [5, 7]
Xscl = 1000   Yscl = 1

(b) Function (iii) best describes the data because it increases at a faster rate as x increases.

(c) $f(x) = 2300e^{.3241x}$ is one answer. Answers may vary somewhat.

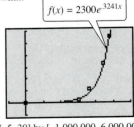

$f(x) = 2300e^{.3241x}$

[−5, 30] by [−1,000,000, 6,000,000]
Xscl = 5   Yscl = 1,000,000

71. (a)

[−5, 30] by [−1,000,000, 6,000,000]
Xscl = 5   Yscl = 1,000,000

(d) about 28,000,000

CHAPTER 5 TEST (page 414)

1. {5} **2.** $\log_a b = 2$ **3.** $\ln 4.82 = c$ **4.** $3^{3/2} = \sqrt{27}$ **5.** $e^a = 5$ **6.** $(a, 1)$ and $(1, 0)$

7.

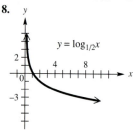

8.

$y = \log_{1/2} x$

9. $2 \log_7 x + \dfrac{1}{4} \log_7 y - 3 \log_7 z$ **10.** 7.741 **11.** 4.581 **12.** $(-\infty, 3/2]$ **14.** {3.107} **15.** {7}

16. {20.125} **17. (a)** 329.3 g **(b)** 13.9 days **18.** 10 sec **19.** 18.9 yr **20. (a)** New York: $y = 18.2(1.001)^x$;
Florida: $y = 14.0(1.017)^x$ **(b)** The graphs intersect at (16.5, 18.5), so Florida's population will exceed New York's during January 2011, based on this model.

CHAPTER 6 Trigonometric Functions ▼▼▼

6.1 EXERCISES (page 421) **3.** 45° **5.** counterclockwise **7.** 70°; 110° **9.** 55°; 35° **11.** 80°; 100°
13. $90 - x$ degrees **15.** $x - 360$ degrees **17.** 83° 59′ **19.** 119° 27′ **21.** 38° 32′ **23.** 17° 1′ 49″ **25.** 20.900°
27. 91.598° **29.** 274.316° **31.** 31° 25′ 47″ **33.** 89° 54′ 1″ **35.** 178° 35′ 58″ **39.** 320° **41.** 235° **43.** 179°
45. 130° **47.** $30° + n \cdot 360°$ **49.** $60° + n \cdot 360°$ **51.** $135° + n \cdot 360°$ **53.** $-90° + n \cdot 360°$ **57.** 320°
Angles other than those given are possible in Exercises 59–70. **59.** 435°; −285°; quadrant I

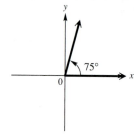

61. 482°; −238°; quadrant II

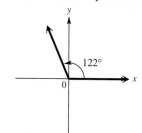

63. 594°; −126°; quadrant III

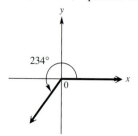

65. 660°; −60°; quadrant IV

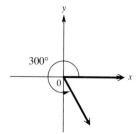

67. 264°; −96°; quadrant III

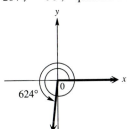

69. 299°; −421°; quadrant IV

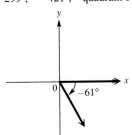

71. $3\sqrt{2}$

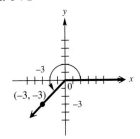

73. $\sqrt{34}$

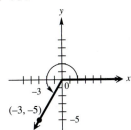

75. 4

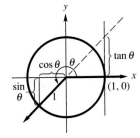

77. 1.5   **79.** 1800°   **81.** 12.5 rotations per hour
83. 4 seconds

CONNECTIONS (page 427)   **1.** $\sin\theta = \dfrac{y}{1} = y = PQ;$ $\cos\theta = \dfrac{x}{1} = x = OQ;$ $\tan\theta = \dfrac{y}{x} = \dfrac{PQ}{OQ} = \dfrac{BA}{1},$ so $BA = \tan\theta$

θ in Quadrant III

6.2 EXERCISES (page 436)   **1.**

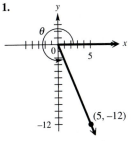

In Exercises 3–7, we give, in order, sine, cosine, tangent, cotangent, secant, and cosecant.

3. $\dfrac{4}{5};$ $-\dfrac{3}{5};$ $-\dfrac{4}{3};$ $-\dfrac{3}{4};$ $-\dfrac{5}{3};$ $\dfrac{5}{4}$ **5.** 1; 0; undefined; 0; undefined; 1   **7.** $\dfrac{\sqrt{3}}{2};$ $\dfrac{1}{2};$ $\sqrt{3};$ $\dfrac{\sqrt{3}}{3};$ 2; $\dfrac{2\sqrt{3}}{3}$

11. negative   **13.** negative

In Exercises 15 and 17, we give, in order, sine, cosine, tangent, cotangent, secant, and cosecant.

15. $-\dfrac{2\sqrt{5}}{5}$; $\dfrac{\sqrt{5}}{5}$; -2; $-\dfrac{1}{2}$; $\sqrt{5}$; $-\dfrac{\sqrt{5}}{2}$

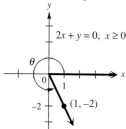

$2x + y = 0,\ x \geq 0$

$(1, -2)$

17. $\dfrac{6\sqrt{37}}{37}$; $-\dfrac{\sqrt{37}}{37}$; -6; $-\dfrac{1}{6}$; $-\sqrt{37}$; $\dfrac{\sqrt{37}}{6}$

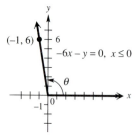

$(-1, 6)$

$-6x - y = 0,\ x \leq 0$

21. -7 **23.** 3 **25.** 1 **27.** 0 **29.** They are negatives of each other. **33.** about .940, about. 342 **35.** $35°$
37. decrease, increase **39.** 1; $\theta = 90°$ **41.** $-.4$ **43.** $2\sqrt{2}$ **45.** .70069071 **49.** $-\dfrac{1}{3}$ **51.** $2°$ **53.** II **55.** I or III

In Exercises 57–61, we give, in order, sine and cosecant, cosine and secant, and tangent and cotangent.

57. $-$; $-$; $+$ **59.** $-$; $+$; $-$ **61.** $-$; $+$; $-$ **63.** $\tan 30°$ **65.** impossible **67.** possible **69.** impossible

71. possible **73.** The graph is the horizontal line $y = 1$. **75.** $-2\sqrt{2}$ **77.** $-2\sqrt{2}$ **79.** $-\dfrac{\sqrt{15}}{4}$

81. .38443820 **83.** .36 **85.** yes

In Exercises 87 and 89, we give, in order, sine, cosine, tangent, cotangent, secant, and cosecant.

87. $-\dfrac{4}{5}$; $-\dfrac{3}{5}$; $\dfrac{4}{3}$; $\dfrac{3}{4}$; $-\dfrac{5}{3}$; $-\dfrac{5}{4}$ **89.** $\dfrac{\sqrt{5}}{7}$; $\dfrac{2\sqrt{11}}{7}$; $\dfrac{\sqrt{55}}{22}$; $\dfrac{2\sqrt{55}}{5}$; $\dfrac{7\sqrt{11}}{22}$; $\dfrac{7\sqrt{5}}{5}$

91. .164215; $-.986425$; $-.166475$; -6.00691; -1.01376; 6.08958 **95.** false; $\sin \theta \leq 1$ for any θ.

6.3 EXERCISES (page 447) **1.** $\sin A = \dfrac{n}{p}$; $\cos A = \dfrac{m}{p}$; $\tan A = \dfrac{n}{m}$; $\cot A = \dfrac{m}{n}$; $\sec A = \dfrac{p}{m}$; $\csc A = \dfrac{p}{n}$

In Exercises 3 and 5, we give, in order, the unknown side, sine, cosine, tangent, cotangent, secant, and cosecant.

3. $c = 13$; $\dfrac{12}{13}$; $\dfrac{5}{13}$; $\dfrac{12}{5}$; $\dfrac{5}{12}$; $\dfrac{13}{5}$; $\dfrac{13}{12}$ **5.** $b = \sqrt{13}$; $\dfrac{\sqrt{13}}{7}$; $\dfrac{6}{7}$; $\dfrac{\sqrt{13}}{6}$; $\dfrac{6\sqrt{13}}{13}$; $\dfrac{7}{6}$; $\dfrac{7\sqrt{13}}{13}$ **7.** $\dfrac{\sqrt{3}}{3}$ **9.** $\sqrt{2}$

15. $\dfrac{\sqrt{3}}{3}$; $\sqrt{3}$ **17.** $\dfrac{\sqrt{3}}{2}$; $\dfrac{\sqrt{3}}{3}$; $\dfrac{2\sqrt{3}}{3}$ **19.** -1; -1 **21.** $-\dfrac{\sqrt{3}}{2}$; $-\dfrac{2\sqrt{3}}{3}$

In Exercises 23–31, we give, in order, sine, cosine, tangent, cotangent, secant, and cosecant.

23. $-\dfrac{\sqrt{3}}{2}$; $\dfrac{1}{2}$; $-\sqrt{3}$; $-\dfrac{\sqrt{3}}{3}$; 2; $-\dfrac{2\sqrt{3}}{3}$ **25.** $-\dfrac{1}{2}$; $\dfrac{\sqrt{3}}{2}$; $-\dfrac{\sqrt{3}}{3}$; $-\sqrt{3}$; $\dfrac{2\sqrt{3}}{3}$; -2

27. $\dfrac{\sqrt{2}}{2}$; $\dfrac{\sqrt{2}}{2}$; 1; 1; $\sqrt{2}$; $\sqrt{2}$ **29.** $\dfrac{1}{2}$; $\dfrac{\sqrt{3}}{2}$; $\dfrac{\sqrt{3}}{3}$; $\sqrt{3}$; $\dfrac{2\sqrt{3}}{3}$; 2

31. $-\dfrac{\sqrt{3}}{2}$; $-\dfrac{1}{2}$; $\sqrt{3}$; $\dfrac{\sqrt{3}}{3}$; -2; $-\dfrac{2\sqrt{3}}{3}$ **33.** $\sin x$, $\tan x$ **35.** $60°$ **37.** $\left(\dfrac{\sqrt{2}}{2}, \dfrac{\sqrt{2}}{2}\right)$; $45°$ **39.** $y = \dfrac{\sqrt{3}}{3}x$

41. (a) 45° **(b)** $k\sqrt{2}$ **(c)** $\sqrt{2}$ **43.** $x = \dfrac{9\sqrt{3}}{2}$; $y = \dfrac{9}{2}$; $z = \dfrac{3\sqrt{3}}{2}$; $w = 3\sqrt{3}$

45. $p = 15$; $r = 15\sqrt{2}$; $q = 5\sqrt{6}$; $t = 10\sqrt{6}$ **49.** .5657728   **51.** 1.1342773   **53.** 1.0273488   **55.** 1.7768146

57. .4771588   **59.** -5.7297416 **61.** .9668234   **63.** .6494076   **65.** false; $\dfrac{1 + \sqrt{3}}{2} \neq 1$ **67.** true

69. false; $.7660444 \neq 1.8793852$ **71.** true   **73.** 57.997172°   **75.** 30.502748°   **77.** 68.673241°   **79.** 56
81. (a) ≈ 155 ft **(b)** ≈ 194 ft   **83.** 2×10^8 m per sec   **85.** 19°   **87.** 48.7°   **89.** ≈ -100.5 lb   **91.** ≈ 2771 lb
93. the 2200-pound car   **95. (a)** ≈ 550 ft **(b)** ≈ 369 ft

CONNECTIONS (page 458)   **1.** $y = (\tan 36.7°)(x - 50)$ **2.** (110.49675, 45.092889);   yes

6.4 EXERCISES (page 461)   **3.** .05   **5.** Both 2 and 65 are exact measurements.
7. $B = 53° \, 40'$; $a = 571$ m; $b = 777$ m   **9.** $M = 38.8°$; $n = 154$ m; $p = 198$ m
17. $B = 62.00°$; $a = 8.17$ ft; $b = 15.4$ ft   **19.** $A = 17.00°$; $a = 39.1$ in; $c = 134$ in
21. $c = 85.9$ yd; $A = 62° \, 50'$; $B = 27° \, 10'$ **23.** 9.35 m   **25.** 33.4 m   **27.** 134.7 cm   **29.** 1.95 mi   **31.** 84.7 m

37. 13.3 ft   **39.** 37.58°   **41.** 42,600 ft   **43.** 270°; N 90° W   **45.** 315°; N 45° W   **47.** $y = \dfrac{\sqrt{3}}{3}x, \; x \leq 0$

49. 220 mi   **51.** 5856 m   **53.** 2.01 mi   **55.** $x = \dfrac{b}{a - c}$ **57.** 433 ft   **59.** 114 ft   **61. (a)** $d = \dfrac{b}{2}\left(\cot\dfrac{\alpha}{2} + \cot\dfrac{\beta}{2}\right)$

(b) 345.3951 m   **63. (a)** ≈ 23.4 ft   **(b)** ≈ 48.3 ft   **(c)** The faster the speed, the more land needs to be cleared on the inside of the curve.

6.5 EXERCISES (page 475)   **1.** 1   **3.** 3   **5.** $\dfrac{\pi}{3}$ **7.** $\dfrac{5\pi}{6}$ **9.** $\dfrac{7\pi}{4}$ **15.** 60°   **17.** 315°   **19.** 288°   **21.** 48°   **23.** .68

25. 2.43   **27.** 1.48   **29.** 114.6°   **31.** 99.7°   **33.** 564.2276°   **35.** $\dfrac{\sqrt{3}}{2}$ **37.** $\sqrt{2}$ **39.** -1 **41.** $-\dfrac{\sqrt{3}}{3}$
43. We begin the answers with the blank next to 30°, and then proceed counterclockwise from there:
$\dfrac{\pi}{6}$; 45; $\dfrac{\pi}{3}$; 120; 135; $\dfrac{5\pi}{6}$; π; $\dfrac{7\pi}{6}$; $\dfrac{5\pi}{4}$; 240; 300; $\dfrac{7\pi}{4}$; $\dfrac{11\pi}{6}$.   **45.** radian   **47.** 2π **49.** 8   **51.** 25.8 cm
53. 318 m   **55.** 5.5 m   **57.** The length is doubled.   **59.** 3500 km   **61.** 5900 km   **63.** 44° N
65. (a) 11.6 in **(b)** 37° 5′   **67.** 38.5°   **69.** 146 in   **71.** .06 km   **73.** 6π **75.** 1.5   **77.** 1300 cm²

79. 114 cm²   **81.** 3.6   **83.** The area of a circle of radius r is πr^2.   **85.** $\dfrac{\pi}{3}$ **87. (a)** 13.85° **(b)** 76 m²

89. (a) ≈ 500 meters **(b)** ≈ 1800 meters

6.6 EXERCISES (page 488)   **3.** $-\dfrac{1}{2}$ **5.** -1 **7.** -2 **9.** $-\dfrac{1}{2}$ **13.** 1.2800079   **15.** 1.2723944   **17.** .42442278

19. $-.99668945$ **21.** 1.4429646   **23.** .38370341   **25.** 1.3631380   **27.** $\dfrac{5\pi}{6}$ **29.** $\dfrac{4\pi}{3}$ **31.** $\dfrac{7\pi}{4}$ **33.** $\approx .9846$

35. $(-.80114362, .59847214)$ **37.** $(.43854733, -.89870810)$ **39.** I   **41.** II   **43. (a)** 30° **(b)** 60° **(c)** 75° **(d)** 86°
(e) 86° **(f)** 60°   **45.** $\left\{\left(\dfrac{\pi}{2}\right) - 1, \left(\dfrac{3\pi}{2}\right) - 1\right\}$ **47.** 2π sec   **51.** 2π radians   **53.** $\dfrac{6}{5}$ min   **55.** .180311 radian per sec

57. 8π meters per sec   **59.** 1.83333 radians per sec   **61.** 18π cm   **63.** 12 sec   **67.** $\dfrac{\pi}{6}$ radian per hr   **69.** $\dfrac{14\pi}{15}$ mm per sec

71. 1500π m per min   **73. (a)** $\dfrac{2\pi}{365}$ radian **(b)** $\dfrac{\pi}{4380}$ radian per hr **(c)** about 66,700 mi per hr

75. larger pulley: $\dfrac{25\pi}{18}$ radians per sec;   smaller pulley: $\dfrac{125\pi}{48}$ radians per sec   **77. (a)** 2π radians per day; $\dfrac{\pi}{12}$ radian per hr
(b) 0 **(c)** $12,800\pi$ km per day or about 533π km per hr **(d)** about 9050π km per day or about 377π km per hr
79. ≈ 523.6 radians/sec

🖩 **(page 505)**   **1.** The graphs coincide.   **2.** One example is $y = -.5 \sin\left(x + \dfrac{\pi}{2}\right)$.

6.7 EXERCISES (page 507)   **5.** 1   **7.** $\dfrac{3\pi}{2}$ **9.** 3   **11.** $\dfrac{3}{4}$

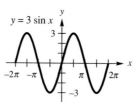

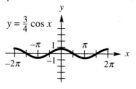

13. 1

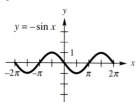

15. 3

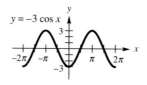

17. 3π, 1, none

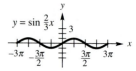

19. $\dfrac{2\pi}{3}$, 1, none

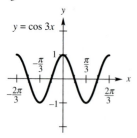

21. $\dfrac{\pi}{2}$, 1, none

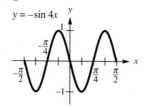

23. 6π, 3, none

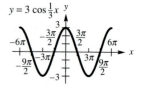

25. $\dfrac{4\pi}{3}$, 2, none

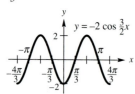

27. 4π, $\dfrac{2}{3}$, none

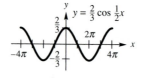

29. 2π, 1, up 1

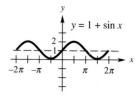

31. 2π, 3, up 2

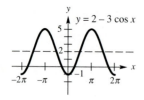

33. 1, 2π, none, $\dfrac{\pi}{2}$ to the left

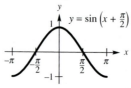

35. 2, 2π, none, $\dfrac{\pi}{2}$ to the left

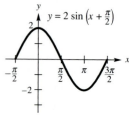

37. 2, 2π, none, $\dfrac{\pi}{3}$ to the right   **39.** 2, $\dfrac{\pi}{2}$, none, $\dfrac{\pi}{3}$ to the left   **41.** $\dfrac{1}{3}$, 8π, none, $\dfrac{4\pi}{3}$ to the right

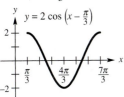

$y = 2 \cos\left(x - \dfrac{\pi}{3}\right)$

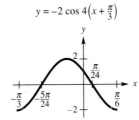

$y = -2 \cos 4\left(x + \dfrac{\pi}{3}\right)$

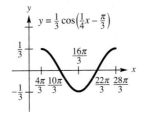

$y = \dfrac{1}{3} \cos\left(\dfrac{1}{4}x - \dfrac{\pi}{3}\right)$

43. 3, 2π, 4 up, π to the left   **45.** D   **47.** B   **49.** F   **51.** E   **55. (a)** about 2 hr   **(b)** 1 year   **57.** 1; $240°$ or $\dfrac{4\pi}{3}$

$y = 4 - 3 \cos (x + \pi)$

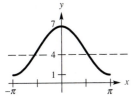

59. (a) 5; $\dfrac{1}{60}$ **(b)** 60   **(c)** 5;   1.545; -4.045; -4.045;   1.545   **(d)**

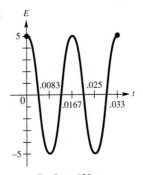

$E = 5 \cos 120\pi t$

61. (a)

$L(x) = .022x^2 + .55x + 316 + 3.5 \sin (2\pi x)$

[15, 35] by [325, 365]
Xscl = 5   Yscl = 5

(b) maximums: $x = \dfrac{1}{4}, \dfrac{5}{4}, \dfrac{9}{4}, \ldots$;   minimums: $x = \dfrac{3}{4}, \dfrac{7}{4}, \dfrac{11}{4}, \ldots$ **63.** $a = 1, \alpha = \dfrac{\pi}{3}$

65. $y = \pi \sin \pi(x - .5)$ (There are other correct answers.)   **67.** $y = \pi \cos \pi(x + 1)$ **69.** $-1 + 2 \sin\left(\dfrac{2}{3}\left(x - \dfrac{\pi}{2}\right)\right)$

71. (a) 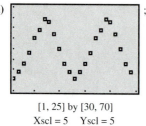 ; yes **(b)** 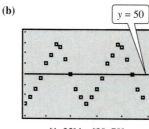 $y = 50$; average yearly temperature **(c)** 14; 12; $\dfrac{\pi}{6}$

[1, 25] by [30, 70]
Xscl = 5   Yscl = 5

[1, 25] by [30, 70]
Xscl = 5   Yscl = 5

(d) $f(x) = 14 \sin\left[\dfrac{\pi}{6}(x - 4.2)\right] + 50$ **(e)** 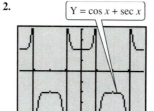 $f(x) = 14 \sin\left[\dfrac{\pi}{6}(x-4.2)\right]+50$; The function gives an excellent model for the data.

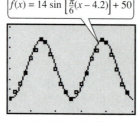

[1, 25] by [30, 70]
Xscl = 5   Yscl = 5

(page 523) **1.** 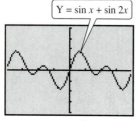 $Y = \sin x + \sin 2x$ **2.** 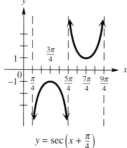 $Y = \cos x + \sec x$

Trig Window

Trig Window

6.8 EXERCISES (page 524) **3.** false; $3 \csc x = \dfrac{3}{\sin x}$ **5.** true **7.** B **9.** E **11.** D

13.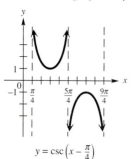

$y = \csc\left(x - \dfrac{\pi}{4}\right)$

15.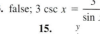

$y = \sec\left(x + \dfrac{\pi}{4}\right)$

17.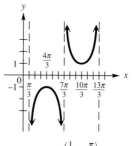

$y = \sec\left(\dfrac{1}{2}x + \dfrac{\pi}{3}\right)$

19.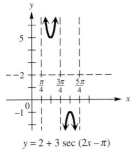

$y = 2 + 3 \sec (2x - \pi)$

21.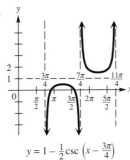

$y = 1 - \frac{1}{2} \csc \left(x - \frac{3\pi}{4}\right)$

23.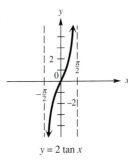

$y = 2 \tan x$

25.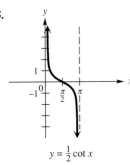

$y = \frac{1}{2} \cot x$

27.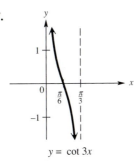

$y = \cot 3x$

29.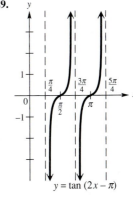

$y = \tan (2x - \pi)$

31.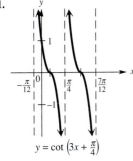

$y = \cot \left(3x + \frac{\pi}{4}\right)$

33.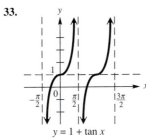

$y = 1 + \tan x$

35.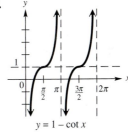

$y = 1 - \cot x$

37.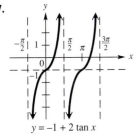

$y = -1 + 2 \tan x$

39.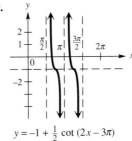

$y = -1 + \frac{1}{2} \cot (2x - 3\pi)$

41.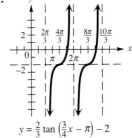

$y = \frac{2}{3} \tan \left(\frac{3}{4}x - \pi\right) - 2$

43. $y = .5 \csc(2x)$ **45.** **(a)** 0 m **(b)** -2.9 m **(c)** -12.3 m **(d)** 12.3 m **(e)** It leads to $\tan \frac{\pi}{2}$, which is undefined.

CHAPTER 6 REVIEW EXERCISES (page 531)   **1.** $309°$ **3.** $1280°$

In Exercises 5–9, we give, in order, sine, cosine, tangent, cotangent, secant, and cosecant.

5. $-\dfrac{\sqrt{2}}{2}$; $-\dfrac{\sqrt{2}}{2}$; 1; 1; $-\sqrt{2}$; $-\sqrt{2}$ **7.** 0; -1; 0; undefined; -1; undefined

9. $-\dfrac{4}{5}$; $\dfrac{3}{5}$; $-\dfrac{4}{3}$; $-\dfrac{3}{4}$; $\dfrac{5}{3}$; $-\dfrac{5}{4}$ **11.** tangent and secant   **13.** $\dfrac{5\sqrt{26}}{26}$; $-\dfrac{\sqrt{26}}{26}$ **15.** IV;   negative

In Exercises 17–21, we give answers in the order sine, cosine, tangent, cotangent, secant, and cosecant.

17. $\dfrac{20}{29}$; $\dfrac{21}{29}$; $\dfrac{20}{21}$; $\dfrac{21}{20}$; $\dfrac{29}{21}$; $\dfrac{29}{20}$ **19.** $\dfrac{\sqrt{3}}{2}$; $-\dfrac{1}{2}$; $-\sqrt{3}$; $-\dfrac{\sqrt{3}}{3}$; -2; $\dfrac{2\sqrt{3}}{3}$

21. $-\dfrac{1}{2}$; $\dfrac{\sqrt{3}}{2}$; $-\dfrac{\sqrt{3}}{3}$; $-\sqrt{3}$; $\dfrac{2\sqrt{3}}{3}$; -2 **23.** $.95371695$ **25.** $.01737737$ **27.** (d)   **29.** II

31. $A = 42.12°$, $C = 90°$, $c = 402.5$ m, $a = 270.0$ m   **33.** 20.4 m   **35.** 140 mi   **37.** OC **39.** (a) $x_Q = x_P + d \cos \theta$,

$y_Q = y_P + d \sin \theta$ (b) $(308.69, 395.67)$ **43.** $\dfrac{\pi}{4}$ **45.** $480°$ **47.** $\sqrt{3}$ **49.** $-\sqrt{2}$ **51.** 2263 in^2 **53.** (a) $\dfrac{\pi}{3}$ radians

(b) 2π in   **55.** -11.426605 **57.** $\dfrac{7\pi}{6}$ **59.** $35°$; $\dfrac{7\pi}{36}$ radian   **61.** $\dfrac{15}{32}$ sec   **63.** 4500 km   **65.** $\sin x$, $\csc x$, $\cos x$, $\sec x$

67. 2; 2π; none; none   **69.** $\dfrac{1}{2}$; $\dfrac{2\pi}{3}$; none; none   **71.** 2; 8π; 1; none   **73.** 3; 2π; none; $\dfrac{\pi}{2}$ units to the left

75. tangent   **77.** cosine   **79.**

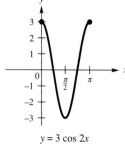

$y = 3 \cos 2x$

81.

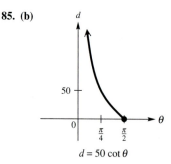

$y = \tan\left(x - \dfrac{\pi}{2}\right)$

83.

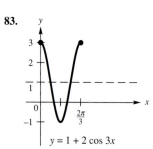

$y = 1 + 2 \cos 3x$

85. (b)

$d = 50 \cot \theta$

87. (a) about 20 years   (b) maximum: about 150,000;   minimum: about 5000

CHAPTER 6 TEST (page 538)   **1.** $203°$ **2.** $2700°$ **3.** $-\dfrac{5\sqrt{29}}{29}$; $\dfrac{2\sqrt{29}}{29}$; $-\dfrac{5}{2}$ **4.** IV   **5.** $\sin\theta = -\dfrac{3}{5}$, $\tan\theta = -\dfrac{3}{4}$,

$\cot\theta = -\dfrac{4}{3}$, $\sec\theta = \dfrac{5}{4}$, $\csc\theta = -\dfrac{5}{3}$ **8.** $-\sqrt{3}$ **9. (a)** $.97939940$ **(b)** $.20834446$ **(c)** 1.9362132

10. $B = 31°\,30'$, $a = 638$, $b = 391$ **11.** 18.75 cm   **12.** 110 km   **13.** $\dfrac{4}{3}$ **14.** $\dfrac{2\pi}{3}$ **15.** $162°$ **16.** $55.673870°$

17. (a) 240π radians per sec   **(b)** 2160π in per sec   **18.** 2, 8π, up 1, none

19.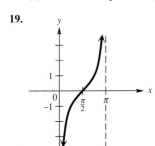

$y = \tan\left(x - \dfrac{\pi}{2}\right)$

20.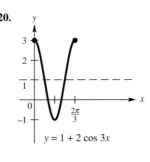

$y = 1 + 2\cos 3x$

CHAPTER 7   Trigonometric Identities and Equations ▼▼▼

CONNECTIONS (page 543)   The tangent function is odd. Yes, $\tan(-x) = -\tan x$, because tangent is an odd function.

7.1 EXERCISES (page 547)   **1.** -2.6 **3.** 2.5 **5.** $\dfrac{\sqrt{7}}{4}$ **7.** $-\dfrac{2\sqrt{5}}{5}$ **9.** $-\dfrac{\sqrt{105}}{11}$ **13.** $\dfrac{\sqrt{21}}{2}$

15. $\cos\theta = -\dfrac{\sqrt{5}}{3}$; $\tan\theta = -\dfrac{2\sqrt{5}}{5}$; $\cot\theta = -\dfrac{\sqrt{5}}{2}$; $\sec\theta = -\dfrac{3\sqrt{5}}{5}$; $\csc\theta = \dfrac{3}{2}$

17. $\sin\theta = -\dfrac{\sqrt{17}}{17}$; $\cos\theta = \dfrac{4\sqrt{17}}{17}$; $\cot\theta = -4$; $\sec\theta = \dfrac{\sqrt{17}}{4}$; $\csc\theta = -\sqrt{17}$

19. $\sin\theta = \dfrac{3}{5}$; $\cos\theta = \dfrac{4}{5}$; $\tan\theta = \dfrac{3}{4}$; $\sec\theta = \dfrac{5}{4}$; $\csc\theta = \dfrac{5}{3}$

21. $\sin\theta = -\dfrac{\sqrt{7}}{4}$; $\cos\theta = \dfrac{3}{4}$; $\tan\theta = -\dfrac{\sqrt{7}}{3}$; $\cot\theta = -\dfrac{3\sqrt{7}}{7}$; $\csc\theta = -\dfrac{4\sqrt{7}}{7}$ **23.** (b)   **25.** (e)   **27.** (a)

29. (a)   **31.** (d)   **35.** $\sin\theta = \dfrac{\pm\sqrt{2x+1}}{x+1}$ **37.** $\cos\theta$ **39.** $\cot\theta$ **41.** $\cos^2\theta$ **43.** $\sec\theta - \cos\theta$ **45.** $\cot\theta - \tan\theta$

47. $\sec\theta\csc\theta$ **49.** $\cos^2\theta$ **51.** $\sec^2\theta$ **53.** $\dfrac{25\sqrt{6}-60}{12}$; $\dfrac{-25\sqrt{6}-60}{12}$ **55.** $\cot x$ and $\csc x$ are odd; $\sec x$ is even

57. $y = -\sin(2x)$ **58.** It is the negative of $y = \sin(2x)$.   **59.** $y = \cos(4x)$ **60.** It is the same function.   **61. (a)** $y = -\sin(4x)$
(b) $y = \cos(2x)$ **(c)** $y = 5\sin(3x)$ **63.** The graph of $y = \csc(-x)$ is a reflection across the x-axis as compared to the graph of
$y = \csc x$.   **65.** The graph of $y = \cot(-x)$ is a reflection across the x-axis as compared to the graph of $y = \cot x$.

CONNECTIONS (page 556) $\sqrt{(1-x^2)^3} = \sin^3\theta$

7.2 EXERCISES (page 557)   **1.** $\dfrac{1}{\sin\theta\cos\theta}$ or $\csc\theta\sec\theta$ **3.** $1 + \cos s$ **5.** 1 **7.** 1 **9.** $2 + 2\sin t$

11. $\dfrac{-2\cos x}{\sin^2 x}$ or $-2\cot x\csc x$ **13.** $(\sin\gamma + 1)(\sin\gamma - 1)$ **15.** $4\sin x$ **17.** $(2\sin x + 1)(\sin x + 1)$

19. $(\cos^2 x + 1)^2$ **21.** $(\sin x - \cos x)(1 + \sin x \cos x)$ **23.** $\sin \theta$ **25.** 1   **27.** $\tan^2 \beta$ **29.** $\tan^2 x$ **31.** $\sec^2 x$

71. $(\sec \theta + \tan \theta)(1 - \sin \theta) = \cos \theta$ **73.** $\dfrac{\cos \theta + 1}{\sin \theta + \tan \theta} = \cot \theta$ **75.** identity   **77.** not an identity   **79.** not an identity

81. not an identity   **83.** identity

7.3 EXERCISES (page 568)

5. $\dfrac{\sqrt{6} + \sqrt{2}}{4}$ **7.** $2 - \sqrt{3}$ **9.** $\dfrac{-\sqrt{6} - \sqrt{2}}{4}$ **11.** $\dfrac{\sqrt{6} - \sqrt{2}}{4}$ **13.** $2 - \sqrt{3}$

15. $\dfrac{\sqrt{2}}{2}$ **17.** -1 **19.** 0   **21.** $\cot 3°$ **23.** $\tan\left(-\dfrac{2\pi}{5}\right)$ **25.** $\tan$ **27.** $\cos$ **29.** $\dfrac{\sqrt{2}(\cos \theta + \sin \theta)}{2}$

31. $\dfrac{\sqrt{2}(\cos x + \sin x)}{2}$ **33.** $\sin \theta$ **37.** $\dfrac{63}{65}; \dfrac{33}{65}; \dfrac{63}{16}; \dfrac{33}{56};$ I; I

39. $\dfrac{4\sqrt{2} + \sqrt{5}}{9}; \dfrac{4\sqrt{2} - \sqrt{5}}{9}; \dfrac{-8\sqrt{5} - 5\sqrt{2}}{20 - 2\sqrt{10}}$ or $\dfrac{4\sqrt{2} + \sqrt{5}}{2 - 2\sqrt{10}}; \dfrac{-8\sqrt{5} + 5\sqrt{2}}{20 + 2\sqrt{10}}$ or $\dfrac{-4\sqrt{2} + \sqrt{5}}{2\sqrt{10} + 2};$ II; II

41. $\dfrac{77}{85}; \dfrac{13}{85}; -\dfrac{77}{36}; \dfrac{13}{84};$ II; I   **43.** $\sin\left(\dfrac{3\pi}{2} + x\right) = -\cos x$ **45.** $\dfrac{1 + \tan x}{1 - \tan x} = \tan\left(\dfrac{\pi}{4} + x\right)$ **57.** $110°, 430°, 470°,$ etc.

59. $-\dfrac{\sqrt{6} + \sqrt{2}}{4}$ **60.** $-\dfrac{\sqrt{6} + \sqrt{2}}{4}$ **61. (a)** $\dfrac{\sqrt{2} - \sqrt{6}}{4}$ **(b)** $\dfrac{\sqrt{6} - \sqrt{2}}{4}$ **64.** $180° - \beta$ **65.** $\theta = \beta - \alpha$

66. $\tan \theta = \dfrac{\tan \beta - \tan \alpha}{1 + \tan \alpha \tan \beta}$ **68.** $18.4°$ **69.** $80.8°$ **71.** $163; -163;$ no

73. (a)
For $x = t$,
$V = V_1 + V_2 = 30 \sin 120\pi t + 40 \cos 120\pi t$

[0, .05] by [−60, 60]
Xscl = .01   Yscl = 10

(b) $a = 50, \phi = -5.353$

7.4 EXERCISES (page 580)

1. .2   **3.** .1270166538   **5.** $-$ **7.** $+$

9. $\cos \theta = \dfrac{2\sqrt{5}}{5}; \sin \theta = \dfrac{\sqrt{5}}{5}; \tan \theta = \dfrac{1}{2}; \sec \theta = \dfrac{\sqrt{5}}{2}; \csc \theta = \sqrt{5}; \cot \theta = 2$

11. $\tan 2x = -\dfrac{4}{3}; \sec 2x = -\dfrac{5}{3}; \cos 2x = -\dfrac{3}{5}; \cot 2x = -\dfrac{3}{4}; \sin 2x = \dfrac{4}{5}; \csc 2x = \dfrac{5}{4}$ **13.** $\dfrac{\sqrt{3}}{2}$ **15.** $\dfrac{\sqrt{3}}{3}$

17. $\dfrac{\sqrt{3}}{2}$ **19.** $\dfrac{1}{2} \tan 102°$ **21.** $\sin 10x$ **23.** 1   **25.** $\csc^2 3r$ **27.** $\dfrac{\sqrt{2 + \sqrt{2}}}{2}$ **29.** $\dfrac{-\sqrt{2 + \sqrt{3}}}{2}$ **31.** $\dfrac{-\sqrt{2 + \sqrt{3}}}{2}$

35. $\dfrac{\sqrt{50 - 10\sqrt{5}}}{10}$ **37.** $-\sqrt{7}$ **39.** $\dfrac{\sqrt{5}}{5}$ **41.** $\tan 73.5°$ **43.** $\cos 9x$ **45.** $\tan 79.1°$

47. $\dfrac{4 \tan x \cos^2 x - 2 \tan x}{1 - \tan^2 x} = \sin 2x$ **49.** $1 - 8 \sin^2 \dfrac{x}{2} \cos^2 \dfrac{x}{2} = \cos 2x$ **57.** $\dfrac{1}{2}(\sin 70° - \sin 20°)$ **59.** $\dfrac{3}{2}(\cos 8x + \cos 2x)$

61. $\dfrac{1}{2}[\cos 2\theta - \cos(-4\theta)] = \dfrac{1}{2}(\cos 2\theta - \cos 4\theta)$ **63.** $-4[\cos 9y + \cos(-y)] = -4(\cos 9y + \cos y)$ **67.** $84°$ **69.** 3.9

71. (a)

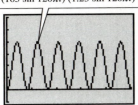

For $x = t$, $W = VI =$
$(163 \sin 120\pi t)(1.23 \sin 120\pi t)$

[0, .05] by [−50, 300]
Xscl = .01   Yscl = 50

(b) maximum: 200.49 watts;   minimum: 0 watts
(c) $a = -100.245$, $\omega = 240\pi$, $c = 100.245$
(e) 100.245 watts

73. They are both radii of the circle.   **74.** It is the supplement of a 30° angle.   **75.** Their sum is $180 - 150 = 30$ degrees and they are equal.   **76.** $2 + \sqrt{3}$ **78.** $\dfrac{\sqrt{6} + \sqrt{2}}{4}$ **79.** $\dfrac{\sqrt{6} - \sqrt{2}}{4}$ **80.** $2 - \sqrt{3}$

7.5 EXERCISES (page 593)   **1.** $-\dfrac{\pi}{6}$ **3.** $\cos^{-1}\left(\dfrac{1}{a}\right)$ **5.** $-\dfrac{\pi}{6}$ **7.** $\dfrac{\pi}{4}$ **9.** π **11.** $-\dfrac{\pi}{3}$ **13.** 0   **15.** $\dfrac{\pi}{2}$ **17.** $\dfrac{\pi}{4}$

19. $\dfrac{5\pi}{6}$ **21.** $\dfrac{3\pi}{4}$ **23.** $-\dfrac{\pi}{6}$ **25.** $\dfrac{\pi}{6}$ **27.** $\dfrac{\pi}{3}$ **29.** $-45°$ **31.** $-60°$ **33.** 120°   **35.** $-30°$ **37.** .83798122   **39.** 2.3154725

41. 1.1900238   **43.** $-7.6713835°$ **45.** 113.500970°   **47.** 30.987961°   **49.** $(-\infty, \infty)$; $(0, \pi)$

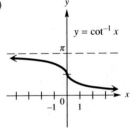

$y = \cot^{-1} x$

51. $(-\infty, -1] \cup [1, \infty)$; $\left[0, \dfrac{\pi}{2}\right) \cup \left(\dfrac{\pi}{2}, \pi\right]$

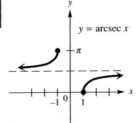

$y = \text{arcsec } x$

57. $\dfrac{\sqrt{7}}{3}$ **59.** $\dfrac{\sqrt{5}}{5}$ **61.** $-\dfrac{\sqrt{5}}{2}$ **63.** 2   **65.** $\dfrac{\pi}{4}$

67. $\dfrac{\pi}{3}$ **69.** $\dfrac{120}{169}$ **71.** $-\dfrac{7}{25}$ **73.** $\dfrac{4\sqrt{6}}{25}$ **75.** $-\dfrac{24}{7}$ **77.** $\dfrac{\sqrt{10} - 3\sqrt{30}}{20}$ **79.** $-\dfrac{16}{65}$ **81.** .89442719   **83.** .12343998

85. $\sqrt{1 - u^2}$ **87.** $\dfrac{\sqrt{1 - u^2}}{u}$ **89.** $\dfrac{\sqrt{u^2 - 4}}{|u|}$ **91.** $\dfrac{u\sqrt{2}}{2}$ **95.** **(a)** 113°   **(b)** 84°   **(c)** 60°   **(d)** 47°

7.6 EXERCISES (page 604)   **1.** $-1 < b < 1$; $b = \pm 1$; $b > 1$ or $b < -1$ **3.** .5, 3.1, 5.8   **5.** $\dfrac{\pi}{3}, \dfrac{5\pi}{3}$

7. $0, \dfrac{2\pi}{3}$ **9.** $\left\{\dfrac{3\pi}{4}, \dfrac{7\pi}{4}\right\}$ **11.** $\left\{\dfrac{\pi}{6}, \dfrac{5\pi}{6}\right\}$ **13.** $\emptyset$ **15.** $\left\{\dfrac{\pi}{4}, \dfrac{2\pi}{3}, \dfrac{5\pi}{4}, \dfrac{5\pi}{3}\right\}$ **17.** $\{\pi\}$ **19.** $\left\{\dfrac{7\pi}{6}, \dfrac{3\pi}{2}, \dfrac{11\pi}{6}\right\}$

21. $\left\{\dfrac{\pi}{2} + 2n\pi, \dfrac{7\pi}{6} + 2n\pi, \dfrac{11\pi}{6} + 2n\pi, \text{ where } n \text{ is an integer}\right\}$

23. $\left\{\dfrac{\pi}{3} + 2n\pi, \dfrac{2\pi}{3} + 2n\pi, \dfrac{4\pi}{3} + 2n\pi, \dfrac{5\pi}{3} + 2n\pi, \text{ where } n \text{ is an integer}\right\}$ **25.** $\{30°, 210°, 240°, 300°\}$ **27.** $\{90°, 210°, 330°\}$

29. $\{45°, 135°, 225°, 315°\}$ **31.** $\{90°, 270°\}$ **33.** $\{0°, 90°, 180°, 270°\}$ **35.** $\{0°, 90°\}$ **37.** $\{90°, 221.8°, 318.2°\}$
39. $\{135°, 315°, 71.6°, 251.6°\}$ **41.** $\{71.6°, 90°, 251.6°, 270°\}$ **43.** $\{53.6°, 126.4°, 187.9°, 352.1°\}$

45. $\{149.6°, 329.6°, 106.3°, 286.3°\}$ **47.** $\emptyset$ **49.** $\left\{\dfrac{\pi}{12}, \dfrac{11\pi}{12}, \dfrac{13\pi}{12}, \dfrac{23\pi}{12}\right\}$ **51.** $\left\{\dfrac{3\pi}{8}, \dfrac{5\pi}{8}, \dfrac{11\pi}{8}, \dfrac{13\pi}{8}\right\}$ **53.** $\left\{\dfrac{\pi}{2}, \dfrac{3\pi}{2}\right\}$

55. $\left\{0, \dfrac{\pi}{4}, \dfrac{\pi}{2}, \dfrac{3\pi}{4}, \pi, \dfrac{5\pi}{4}, \dfrac{3\pi}{2}, \dfrac{7\pi}{4}\right\}$ **57.** $\left\{\dfrac{\pi}{2}\right\}$ **59.** $\left\{\dfrac{\pi}{3}, \pi, \dfrac{5\pi}{3}\right\}$ **61.** $\{15°, 45°, 135°, 165°, 255°, 285°\}$ **63.** $\{120°, 240°\}$

65. $\{30°, 150°, 270°\}$ **67.** $\{0°, 30°, 150°, 180°\}$ **69.** $\{60°, 300°\}$ **71. (a)** 2 sec   **(b)** $\dfrac{10}{3}$ sec   **73.** 4

75. (a) 42.2°   **(b)** 90°   **(c)** 48.0°   **77. (a)**

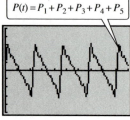

For $x = t$,
$P(t) = P_1 + P_2 + P_3 + P_4 + P_5$

(b) The graph approximates a sawtooth shape.
(c) The maximum value of P is approximately .00317 and occurs when $x \approx$ .000188, .00246, .00474, .00701, and .00928.

[0, .01] by [−.005, .005]
Xscl = .001   Yscl = .001

7.7 EXERCISES (page 611)   **1.** $\left(\dfrac{\sqrt{2}}{2}, \dfrac{\pi}{4}\right)$; $\dfrac{\sqrt{2}}{2}$ **3.** 0   **5.** $x = \arccos \dfrac{y}{5}$ **7.** $x = \dfrac{1}{3} \text{arccot } 2y$ **9.** $x = \dfrac{1}{2} \arctan \dfrac{y}{3}$

11. $x = 4 \arccos \dfrac{y}{6}$ **13.** $x = \dfrac{1}{5} \arccos\left(-\dfrac{y}{2}\right)$ **15.** $x = -3 + \arccos y$ **17.** $x = \arcsin(y + 2)$ **19.** $x = \arcsin\left(\dfrac{y + 4}{2}\right)$

23. $\{-2\sqrt{2}\}$ **25.** $\{\pi - 3\}$ **27.** $\left\{\dfrac{3}{5}\right\}$ **29.** $\left\{\dfrac{4}{5}\right\}$ **31.** $\{0\}$ **33.** $\left\{\dfrac{1}{2}\right\}$ **35.** $\left\{-\dfrac{1}{2}\right\}$ **37.** $\{0\}$ **39.** $t = \dfrac{50}{\pi} \arccos\left(\dfrac{d - 550}{450}\right)$

41. (a) $t = \dfrac{1}{2\pi f} \arcsin \dfrac{e}{E_{\max}}$ **(b)** .00068 sec

43. (a) $x = \sin u$, $-\dfrac{\pi}{2} \le u \le \dfrac{\pi}{2}$ **(b)**

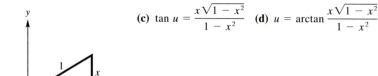

(c) $\tan u = \dfrac{x\sqrt{1 - x^2}}{1 - x^2}$ **(d)** $u = \arctan \dfrac{x\sqrt{1 - x^2}}{1 - x^2}$

CHAPTER 7 REVIEW EXERCISES (page 616)

1. sine, tangent, cotangent, cosecant   **3.** $-.96$

5. $\sin x = -\dfrac{4}{5}$; $\tan x = -\dfrac{4}{3}$; $\sec x = \dfrac{5}{3}$; $\csc x = -\dfrac{5}{4}$; $\cot x = -\dfrac{3}{4}$ **7.** (b)   **9.** (g)   **11.** (f)   **13.** (e)   **15.** (d)

17. (c)   **19.** 1   **21.** $\dfrac{1 + \cos \theta}{\sin \theta}$ **23.** $\dfrac{4 + 3\sqrt{15}}{20}$; $\dfrac{4\sqrt{15} + 3}{20}$; $\dfrac{192 + 25\sqrt{15}}{231}$;   I   **25.** $\sin \theta = \dfrac{\sqrt{14}}{4}$; $\cos \theta = \dfrac{\sqrt{2}}{4}$

27. $\sin 2x = \dfrac{3}{5}$; $\cos 2x = -\dfrac{4}{5}$ **29.** $\dfrac{1}{2}$ **31.** $\dfrac{\sqrt{5} - 1}{2}$ **33.** $-\dfrac{\sin 2x + \sin x}{\cos 2x - \cos x} = \cot \dfrac{x}{2}$ **35.** $\dfrac{\cos x \sin 2x}{1 + \cos 2x} = \sin x$

57. $\dfrac{2\pi}{3}$ **59.** $-\dfrac{\pi}{2}$ **61.** $\dfrac{\pi}{6}$ **63.** $-60°$ **65.** $-41.334446°$ **67.** $12.516313°$ **71.** -1 **73.** $\dfrac{\sqrt{10}}{10}$

75. $\dfrac{294 + 125\sqrt{6}}{92}$ **77.** $\sqrt{1 - u^2}$ **79.** $[-1, 1]$; $[0, \pi]$ **81.** $\{.46364761, 3.6052403\}$

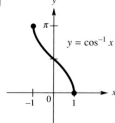

$y = \cos^{-1} x$

83. $\left\{ \dfrac{\pi}{4}, \dfrac{3\pi}{4}, \dfrac{5\pi}{4}, \dfrac{7\pi}{4} \right\}$ **85.** $\left\{ \dfrac{\pi}{3}, \pi, \dfrac{5\pi}{3} \right\}$ **87.** $\{45°, 90°, 225°, 270°\}$ **89.** $\{60°, 180°, 300°\}$ **91.** $x = 2 \arccos \dfrac{y}{3}$ **93.** $\left\{ \dfrac{3\sqrt{5}}{7} \right\}$

95. $\{0\}$ **97.** The light beam is completely under water.

99. (a)

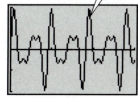

For x = t,
$P(t) = \dfrac{1}{2} \sin\left[2\pi(220)t\right] + \dfrac{1}{3} \sin\left[2\pi(330)t\right] + \dfrac{1}{4} \sin\left[2\pi(440)t\right]$

[0, .03] by [−2, 2]
Xscl = .01   Yscl = 1

(b) .0007576, .009847, .01894, .02803   **(c)** 110 Hz   **(d)**

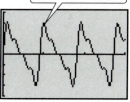

For x = t,
$P(t) = \sin\left[2\pi(110)t\right] + \dfrac{1}{2} \sin\left[2\pi(220)t\right] + \dfrac{1}{3} \sin\left[2\pi(330)t\right] + \dfrac{1}{4} \sin\left[2\pi(440)t\right]$

[0, .03] by [−2, 2]
Xscl = .01   Yscl = .5

CHAPTER 7 TEST (page 621)

1. $\sec x = -\dfrac{\sqrt{41}}{4}$; $\cos x = -\dfrac{4\sqrt{41}}{41}$; $\cot x = -\dfrac{4}{5}$; $\sin x = \dfrac{5\sqrt{41}}{41}$; $\csc x = \dfrac{\sqrt{41}}{5}$

2. $\dfrac{1}{\sin^2 x \cos^2 x}$ **3.** $\dfrac{2 - 2\sqrt{30}}{15}$; $\dfrac{\sqrt{5} - 4\sqrt{6}}{15}$; $\dfrac{-10\sqrt{6} + 2\sqrt{5}}{5 + 4\sqrt{30}}$ or $\dfrac{-50\sqrt{5} + 18\sqrt{6}}{91}$ **4.** $\csc x - \cot x = \tan \dfrac{x}{2}$

10. $\dfrac{2\pi}{3}$ **11.** $\dfrac{\pi}{3}$ **12.** $0°$ **13.** $(-\infty, \infty)$ **14.** $\dfrac{\pi}{4}$ **15.** $\dfrac{3\pi}{4}$ **16.** $\dfrac{\sqrt{3}}{2}$ **17.** $\{0, \pi\}$ **18.** $\left\{ \dfrac{\pi}{6}, \dfrac{\pi}{3}, \dfrac{7\pi}{6}, \dfrac{4\pi}{3} \right\}$

19. $\left\{ \dfrac{\pi}{4}, 2.7, \dfrac{5\pi}{4}, 5.8 \right\}$ **20.** $\varnothing$

CHAPTER 8 Applications of Trigonometry ▼▼▼

CONNECTIONS (page 633)   **1.** House: $X_H = 1131.8$ ft, $Y_H = 4390.2$ ft;   Fire: $X_F = 2277.5$ ft, $Y_F = -2596.2$ ft   **2.** 7079.7 ft

8.1 EXERCISES (page 634)   **1.** $\{6\sqrt{2}\}$ **3.** $\sqrt{3}$ **5.** $C = 95°$, $b = 13$ m, $a = 11$ m   **7.** $B = 37.3°$, $a = 38.5$ ft, $b = 51.0$ ft   **9.** $B = 18.5°$, $a = 239$ yd, $c = 230$ yd   **11.** $A = 56° \ 00'$, $AB = 361$ ft, $BC = 308$ ft   **13.** $B = 109.95°$, $a = 27.01$ m, $c = 21.36$ m   **15.** $A = 34.72°$, $a = 3326$ ft, $c = 5704$ ft   **21.** yes   **23.** $\{45, 135\}$ **25. (a)** $4 < h < 5$ **(b)** $h = 4$ and $h > 5$ **(c)** $h < 4$ **27.** 1   **29.** 2   **31.** 45°   **33.** $B_1 = 49.1°$, $C_1 = 101.2°$, $B_2 = 130.9°$, $C_2 = 19.4°$ **35.** $B = 26° \ 30'$, $A = 112° \ 10'$ **37.** no such triangle   **39.** $B = 20.6°$, $C = 116.9°$, $c = 20.6$ ft   **41.** no such triangle **43.** $B_1 = 49° \ 20'$, $C_1 = 92° \ 00'$, $c_1 = 15.5$ km, $B_2 = 130° \ 40'$, $C_2 = 10° \ 40'$, $c_2 = 2.88$ km   **45.** $A_1 = 53.23°$, $C_1 = 87.09°$, $c_1 = 37.16$ m, $A_2 = 126.77°$, $C_2 = 13.55°$, $c_2 = 8.719$ m   **47.** 1;   90°;   a right triangle   **49.** The triangle does not exist. **51.** 111°   **53.** first location: 5.1 mi;   second location: 7.2 mi   **55.** $\dfrac{\sqrt{3}}{2}$ **57.** $\dfrac{\sqrt{2}}{2}$ **59.** 732 ft²   **61.** 163 km²

63. 289.9 m²   **65.** 373 m²   **66.** increasing   **67.** If $0 \le A \le \dfrac{\pi}{2}$ and $0 \le B \le \dfrac{\pi}{2}$ and $A > B$, then sin A is greater than sin B.

68. $b = \dfrac{a \sin B}{\sin A}$ **69.** From Exercises 67 and 68, $b = a \dfrac{\sin B}{\sin A} < a \cdot 1 < a$.   **71.** $\approx 419{,}000$ km   **73.** 6596.4 ft

CONNECTIONS (page 645)   All three formulas give the area as 9.5 square units.

8.2 EXERCISES (page 646)   **3.** 7   **5.** 30°   **7.** $c = 2.83$ in, $A = 44.9°$, $B = 106.8°$ **9.** $c = 6.46$ m, $A = 53.1°$, $B = 81.3°$ **11.** $a = 156$ cm, $B = 64° \ 50'$, $C = 34° \ 30'$ **13.** $b = 9.529$ in, $A = 64.59°$, $C = 40.61°$ **15.** $a = 15.7$ m, $B = 21.6°$, $C = 45.6°$ **17.** $A = 30°$, $B = 56°$, $C = 94°$ **19.** $A = 82°$, $B = 37°$, $C = 61°$ **21.** $A = 42.0°$, $B = 35.9°$, $C = 102.1°$ **23.** $A = 47.7°$, $B = 44.9°$, $C = 87.4°$ **25.** 16.26°   **27.** $24\sqrt{3}$ or ≈ 41.57 **29.** 78 m²   **31.** 12,600 cm²   **33.** 3650 ft² **35.** 33 cans   **37.** 392,000 mi²   **39.** 257 m   **41.** 281 km   **43.** 22 ft   **45.** 18 ft   **47.** 2000 km   **49.** 163.5° **51.** 25.24983 mi   **61.** An obtuse triangle has an angle that measures between 90° and 180°. Since A is the largest angle, $90° < A < 180°$. The cosine of a quadrant II angle is negative.   **62.** By the law of cosines, $a^2 = b^2 + c^2 - 2bc \cos A$. From Exercise 61, we know that cos A is negative, so $-2bc \cos A$ is positive. Thus, $a^2 = b^2 + c^2 + $ (a positive quantity) so $a^2 > b^2 + c^2$.   **63.** Here, we know that a, b, and c are all positive. Because $a^2 > b^2 + c^2$, $a^2 > b^2$ and $a^2 > c^2$, we have $a > b$ and $a > c$.   **64.** Since A is 103°, the triangle is obtuse and A is the obtuse angle. Thus side a should be the longest side.

CONNECTIONS (page 653)   **1.** $|\mathbf{u}| = \sqrt{13}$, $\theta = 326.3°$ **2.** $\mathbf{u} = \left\langle \dfrac{5}{2}, \dfrac{5\sqrt{3}}{2} \right\rangle$ **3. (a)** $\langle 2, 4 \rangle$ **(b)** $\langle 4, -2 \rangle$ **(c)** $\langle -20, -5 \rangle$

8.3 EXERCISES (page 658)   **3.** **m** and **p**; **n** and **r**   **5.** **m** and **p** equal 2**t**, or **t** is one half **m** or **p**;   also, **m** = 1**p** and **n** = 1**r**   **7.**   **9.**   **11.**   **13.**   **15.**

−**b**

3**a**

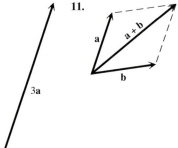

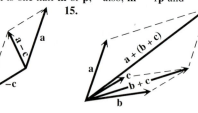

17.   **19.** Yes, vector addition is associative.   **21.** 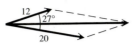   **23.**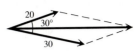

25. 9.5, 7.4   **27.** 38.8, 28.0   **29.** 123, 155   **31.** $\mathbf{a} = -\mathbf{b}$ **33.** $\mathbf{a}$ and $\mathbf{b}$ must be perpendicular.   **37.** 29 newtons   **39.** 53.1 lb
41. $|\mathbf{u}| = 10$, $\theta = 306.9°$ **43.** $|\mathbf{u}| = \sqrt{41}$, $\theta = 128.7°$ **45.** $\langle 4\sqrt{2}, 4\sqrt{2} \rangle$ **47.** $\langle -5\sqrt{3}, 5 \rangle$ **49.** 780 lb   **51.** 1600 lb
53. 226 lb   **55.** 126.6° between the 760-lb and the 1220-lb forces; 91.9° between the 760-lb and the 980-lb forces; 141.5° between the 1220-lb and the 980-lb forces   **57.** magnitude of resultant is 117 lb;   other force is 93.9 lb   **59.** groundspeed: 161 mph;
airspeed: 156 mph   **61.** 350°   **63.** 65° 30′;   181 mph   **65.** groundspeed: 198 mph;   bearing: 186.5°   **67.** The ship traveled
55.9 mi on its modified course, for an additional distance of $55.9 - 50 = 5.9$ mi.   **69.** 27°

CONNECTIONS (page 663)   **1.** $10 + i$;

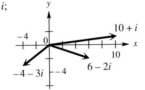

8.4 EXERCISES (page 669)   **1.** magnitude (length)   **3.** 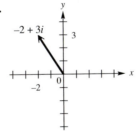   **5.**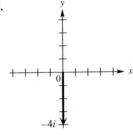

7. In $a + bi$ form, we must have $b = 0$.   **9.** $3 - i$ **11.** $-3 + 3i$ **13.** $2 + 4i$ **15.** $10i$ **17.** $-2 - 2i\sqrt{3}$ **19.** $\dfrac{\sqrt{3}}{2} + \dfrac{1}{2}i$
21. $-3\sqrt{2} + 3i\sqrt{2}$ **23.** $3\sqrt{2}(\cos 315° + i \sin 315°)$ **25.** $6(\cos 240° + i \sin 240°)$ **27.** $5\sqrt{2}(\cos 225° + i \sin 225°)$
29. $4(\cos 180° + i \sin 180°)$ **31.** $\sqrt{13}(\cos 56.31° + i \sin 56.31°)$ **33.** $-1.0260604 - 2.8190779i$
35. $\sqrt{34}(\cos 59.04° + i \sin 59.04°)$ **37.** the circle of radius one centered at the origin   **39.** the vertical line $x = 1$ **41.** yes
45. $-3\sqrt{3} + 3i$ **47.** $-4i$ **49.** $-\dfrac{15\sqrt{2}}{2} + \dfrac{15\sqrt{2}}{2}i$ **51.** $-3i$ **53.** $\sqrt{3} - i$ **55.** $-\dfrac{1}{6} - \dfrac{\sqrt{3}}{6}i$ **57.** $2\sqrt{3} - 2i$

59. $-\dfrac{1}{2} - \dfrac{1}{2}i$ **61.** $.65366807 + 7.4714602i$ **63.** $.20905693 + 1.9890438i$ **65.** $-3.7587705 - 1.3680806i$ **67.** 2
68. $w = \sqrt{2}$ cis 135°; $z = \sqrt{2}$ cis 225°   **69.** 2 cis 0°   **70.** 2;   It is the same.   **71.** $-i$ **72.** cis$(-90°)$ **73.** $-i$;   It is the
same.   **75.** $1.2 - .14i$ **77.** $\approx 27.43 + 11.5i$

8.5 EXERCISES (page 676)   **1.** $27i$ **3.** 1   **5.** $\dfrac{27}{2} - \dfrac{27\sqrt{3}}{2}i$ **7.** $-16\sqrt{3} + 16i$ **9.** $-128 + 128i\sqrt{3}$ **11.** $128 + 128i$

13. $(\cos 0° + i \sin 0°)$, $(\cos 120° + i \sin 120°)$, $(\cos 240° + i \sin 240°)$

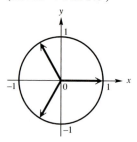

15. 2 cis 20°, 2 cis 140°, 2 cis 260°

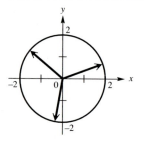

17. $2(\cos 90° + i \sin 90°)$, $2(\cos 210° + i \sin 210°)$, $2(\cos 330° + i \sin 330°)$

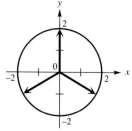

19. $4(\cos 60° + i \sin 60°)$, $4(\cos 180° + i \sin 180°)$, $4(\cos 300° + i \sin 300°)$

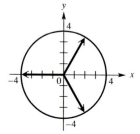

21. $\sqrt[3]{2}(\cos 20° + i \sin 20°)$, $\sqrt[3]{2}(\cos 140° + i \sin 140°)$, $\sqrt[3]{2}(\cos 260° + i \sin 260°)$

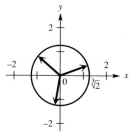

23. $\sqrt[3]{4}(\cos 50° + i \sin 50°)$, $\sqrt[3]{4}(\cos 170° + i \sin 170°)$, $\sqrt[3]{4}(\cos 290° + i \sin 290°)$

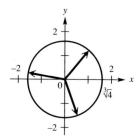

25. $(\cos 0° + i \sin 0°)$, $(\cos 180° + i \sin 180°)$

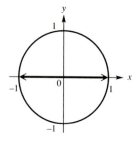

27. $(\cos 0° + i \sin 0°)$, $(\cos 60° + i \sin 60°)$, $(\cos 120° + i \sin 120°)$, $(\cos 180° + i \sin 180°)$, $(\cos 240° + i \sin 240°)$, $(\cos 300° + i \sin 300°)$

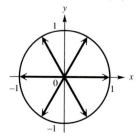

29. $(\cos 45° + i \sin 45°)$, $(\cos 225° + i \sin 225°)$ **33.** false

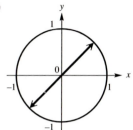

39. $\{(\cos 0° + i \sin 0°), (\cos 120° + i \sin 120°), (\cos 240° + i \sin 240°)\}$
41. $\{(\cos 90° + i \sin 90°), (\cos 210° + i \sin 210°), (\cos 330° + i \sin 330°)\}$
43. $\{2(\cos 0° + i \sin 0°), 2(\cos 120° + i \sin 120°), 2(\cos 240° + i \sin 240°)\}$
45. $\{(\cos 45° + i \sin 45°), (\cos 135° + i \sin 135°), (\cos 225° + i \sin 225°), (\cos 315° + i \sin 315°)\}$
47. $\{(\cos 22.5° + i \sin 22.5°), (\cos 112.5° + i \sin 112.5°), (\cos 202.5° + i \sin 202.5°), (\cos 292.5° + i \sin 292.5°)\}$
49. $\{2(\cos 20° + i \sin 20°), 2(\cos 140° + i \sin 140°), 2(\cos 260° + i \sin 260°)\}$
51. $\{1.3606 + 1.2637i, -1.7747 + .5464i, .4141 - 1.8102i\}$ **53.** $\left\{1, -\dfrac{1}{2} + \dfrac{\sqrt{3}}{2}i, -\dfrac{1}{2} - \dfrac{\sqrt{3}}{2}i\right\}$ **55.** $-4, 2 - 2\sqrt{3}i$
56. $\cos 2\theta + i \sin 2\theta$ **57.** $(\cos^2 \theta - \sin^2 \theta) + i(2\cos \theta \sin \theta) = \cos 2\theta + i \sin 2\theta$ **58.** $\cos 2\theta = \cos^2 \theta - \sin^2 \theta$; $\sin 2\theta = 2\cos \theta \sin \theta$ **59.** $1, .30901699 + .95105652i, -.809017 + .58778525i, -.809017 - .5877853i, .30901699 - .9510565i$
61. (a) yes   **(b)** no   **(c)** yes

8.6 EXERCISES (page 685)   Answers may vary in Exercises 1–11.
3. $(1, 450°), (-1, 225°)$ **5.** $(-2, 495°), (2, 315°)$ **7.** $(5, 300°), (-5, 120°)$

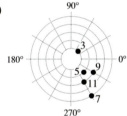

9. $(-3, 150°), (3, -30°)$ **11.** $(3, 660°), (-3, 120°)$ **13.** quadrantal
15.

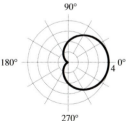

$r = 2 + 2 \cos \theta$

17.

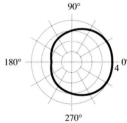

$r = 3 + \cos \theta$

19.

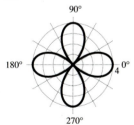

$r = 4 \cos 2\theta$

21.

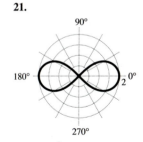

$r^2 = 4 \cos 2\theta$

23.

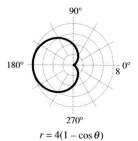

$r = 4(1 - \cos \theta)$

25.

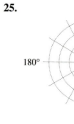

$r = 2 \sin \theta \tan \theta$

27. B   **29.** C   **33.** 4; 45°, 135°, 225°, 315°

37. $\left(\dfrac{4 + \sqrt{2}}{2}, \dfrac{\pi}{4} \right), \left(\dfrac{4 - \sqrt{2}}{2}, \dfrac{5\pi}{4} \right)$

41. $x^2 + (y - 1)^2 = 1$

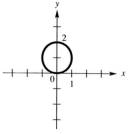

$r = 2 \sin \theta$
$x^2 + (y - 1)^2 = 1$

43. $y^2 = 4(x + 1)$

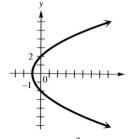

$r = \dfrac{2}{1 - \cos \theta}$
$y^2 = 4(x + 1)$

45. $(x + 1)^2 + (y + 1)^2 = 2$

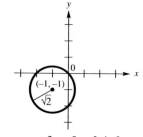

$r + 2 \cos \theta = -2 \sin \theta$
$(x + 1)^2 + (y + 1)^2 = 2$

47. $x = 2$

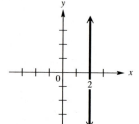

$r = 2 \sec \theta$
$x = 2$

49. $x + y = 2$

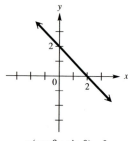

$r (\cos \theta + \sin \theta) = 2$
$x + y = 2$

51. $r(\cos \theta + \sin \theta) = 4$ **53.** $r = 4$

55. $r = 2 \csc \theta$ or $r \sin \theta = 2$ **57.**

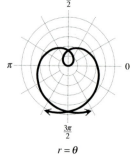

$r = \theta$

59. $r = \dfrac{2}{2 \cos \theta + \sin \theta}$

61. (a) 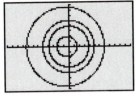   **(b)** Earth is closest to the sun of these four planets.

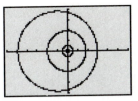

$[-2.4, 2.4]$ by $[-1.6, 1.6]$
Xscl = .2   Yscl = .2

$[-60, 60]$ by $[-40, 40]$
Xscl = 10   Yscl = 10

(c) Pluto is not always the farthest planet from the sun.

8.7 EXERCISES (page 692)   **1.** $(-2, 1);$ $(4, 3);$ $(1.6, 2.2)$ **3.** the second set of equations

5. $y = \dfrac{1}{2}x + 1$, for x in $[-4, 6]$ **7.** $y = 3x^2 - 4$, for x in $[0, 2]$ **9.** $y = x - 2$, for x in $(-\infty, \infty)$

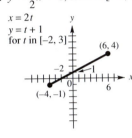

$x = 2t$
$y = t + 1$
for t in $[-2, 3]$

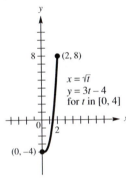

$x = \sqrt{t}$
$y = 3t - 4$
for t in $[0, 4]$

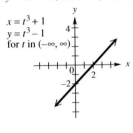

$x = t^3 + 1$
$y = t^3 - 1$
for t in $(-\infty, \infty)$

11. $x = 2^{(y^2+1)/3}$, for y in $[\sqrt{2}, \infty);$ or $y^2 = \dfrac{3 \ln x}{\ln 2} - 1$, for x in $[2, \infty)$ **13.** $x^2 + y^2 = 4$, for x in $[-2, 2]$

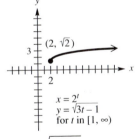

$x = 2^t$
$y = \sqrt{3t - 1}$
for t in $[1, \infty)$

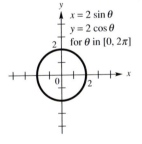

$x = 2 \sin \theta$
$y = 2 \cos \theta$
for θ in $[0, 2\pi]$

15. $y = 2\sqrt{1 + \dfrac{x^2}{9}}$, for x in $(-\infty, \infty)$ **17.**

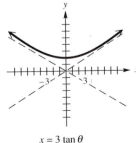

$x = 3 \tan \theta$
$y = 2 \sec \theta$
for θ in $\left(-\dfrac{\pi}{2}, \dfrac{\pi}{2}\right)$

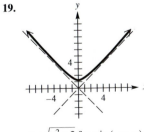

$xy = 1$
for x in $(0, 1]$

19.

$y = \sqrt{x^2 + 2}$ for x in $(-\infty, \infty)$

21.

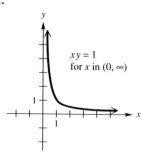

$xy = 1$
for x in $(0, \infty)$

23.

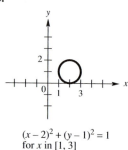

$(x - 2)^2 + (y - 1)^2 = 1$
for x in $[1, 3]$

25.

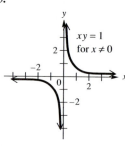

$xy = 1$
for $x \neq 0$

27.

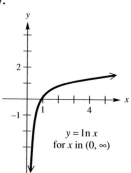

$y = \ln x$
for x in $(0, \infty)$

29. (a)

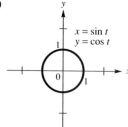

$x = \sin t$
$y = \cos t$

(b)

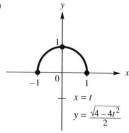

$x = t$
$y = \dfrac{\sqrt{4 - 4t^2}}{2}$

31.

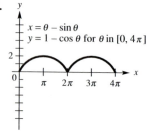

$x = \theta - \sin \theta$
$y = 1 - \cos \theta$ for θ in $[0, 4\pi]$

33. (a) 17.7 sec   **(b)** 5000 ft   **(c)** 1250 ft   **37.** Many answers are possible, two of which are $x = t$, $y = m(t - x_1) + y_1$ and $x = t^2$, $y = m(t^2 - x_1) + y_1$.   **39.** Many answers are possible; for example, $x = a \sec \theta$, $y = b \tan \theta$; $x = t$, $y^2 = \dfrac{b^2}{a^2}(t^2 - a^2)$.

CHAPTER 8 REVIEW EXERCISES (page 696)   **1.** 63.7 m   **3.** 41.7°   **5.** 54° 20′ or 125° 40′   **9. (a)** $b = 5$, $b \geq 10$
(b) $5 < b < 10$ **(c)** $b < 5$ **11.** 173 ft   **13.** 26.5° or 26° 30′   **15.** 32°   **17.** $c = 18.7$ cm, $A = 91° 40′$, $B = 45° 50′$
19. 153,600 m²   **21.** .234 km²   **23.** He needs about 2.5 cans, so he must buy 3 cans.   **25.** 13 m   **27.** 10.8 mi   **29.** 5500 m
31. It becomes the Pythagorean theorem.   **33.**   **35.** true   **37.** $25\sqrt{2}$, $25\sqrt{2}$ (about 35 for each)

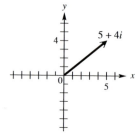

39. 209 newtons   **41.** 29, 316.4°   **43.** $\langle 3, 3\sqrt{3} \rangle$ **45.** 280 newtons, 30.4°
47. 3° 50′   **49.** $AB = 1978.28$ ft; $BC = 975.05$ ft   **51.** $-30i$
53. $-\dfrac{1}{8} + \dfrac{\sqrt{3}}{8}i$ **55.** $8i$ **57.** x **59.**

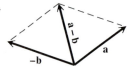

$-4 + 2i$

61. $5 + 4i$

$5 + 4i$

63. $2\sqrt{2}(\cos 135° + i \sin 135°)$ **65.** $-\sqrt{2} - i\sqrt{2}$ **67.** $4(\cos 270° + i \sin 270°)$ **69.** a circle of radius 2 with the origin as center   **71.** $\sqrt[10]{8}(\cos 27° + i \sin 27°)$, $\sqrt[10]{8}(\cos 99° + i \sin 99°)$, $\sqrt[10]{8}(\cos 171° + i \sin 171°)$, $\sqrt[10]{8}(\cos 243° + i \sin 243°)$, $\sqrt[10]{8}(\cos 315° + i \sin 315°)$ **73.** one   **75.** $\{5(\cos 60° + i \sin 60°), 5(\cos 180° + i \sin 180°), 5(\cos 300° + i \sin 300°)\}$

77.

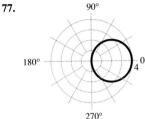

$r = 4 \cos \theta$

79.

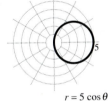

$r = 1 - \cos \theta$

81.

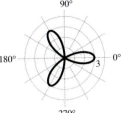

$r = 3 \cos 3\theta$

83. $2y - x = 4$ **85.** $x^2 + y^2 = 4$ **87.** $r = \tan \theta \sec \theta$ or $r = \dfrac{\tan \theta}{\cos \theta}$ **89.** $r = 2 \csc \theta$ **91.** $r = 2$

93. $y = 2x - 3$, for x in $[5, 10]$ **95.** $y = \dfrac{1}{x - 4}$, for x in $[5, \infty)$ **97.** $y^2 = -\dfrac{1}{2}(x - 1)$ or $2y^2 + x - 1 = 0$, for x in $[-1, 1]$

99. $x = 3 + 5 \cos t, y = 4 + 5 \sin t, t$ in $[0, 2\pi]$

CHAPTER 8 TEST (page 701)   **1.** $137.5°$ **2.** 180 km   **3.** $49.0°$ **4. (a)** $b > 10$ **(b)** not possible   **(c)** $b \le 10$
6. 115 km   **7.** distance to both first and third bases: 63.7 ft;   distance to second base: 66.8 ft
8. horizontal: 869 lb;   vertical: 418 lb   **9.** 826 lb   **10.** bearing: $7.4°$;   speed: 8.3 mph   **11.** $1 + 4i$

12. $8(\cos 120° + i \sin 120°)$ or 8 cis $120°$ **13.** $-2 - 2i\sqrt{3}$ **14.** $-\dfrac{15}{2} + \dfrac{15\sqrt{3}}{2}i$ **15.** $-\dfrac{1}{2}i$ **16.** $-128 + 128i$

17. $\sqrt[4]{2}$ cis $7.5°$, $\sqrt[4]{2}$ cis $97.5°$, $\sqrt[4]{2}$ cis $187.5°$, $\sqrt[4]{2}$ cis $277.5°$ **18.**

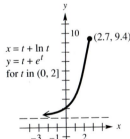

$r = 5 \cos \theta$

19. $r = 2 \sin \theta + 3 \cos \theta$

20. $-x + 2y = 4$ **21.** $y = \left(\dfrac{x + 3}{4}\right)^2$, for x in $[-15, 13]$ **22.**

$x = t + \ln t$
$y = t + e^t$
for t in $(0, 2]$

$(2.7, 9.4)$

CHAPTER 9 Systems of Equations and Inequalities ▼▼▼

CONNECTIONS (page 711) **1.** $\{2\}, \{(x, y) \mid x = 2\}; \{(x, y, z) \mid x = 2\}$ **2.** $\{(x, y) \mid 2x + 5y = 10\}$ or $\left\{\left(x, \dfrac{10 - 2x}{5}\right)\right\}$, a line; $\{(x, y, z) \mid 2x + 5y = 10\}$, a plane **3.** $\{(x, y, z) \mid 2x + 5y + z = 25\}$

9.1 EXERCISES (page 715) **1.** approximately 1976 **3.** (1979, 10 billion) **5.** year; number of cans produced (in billions)
7. $\{(48, 8)\}$ **9.** $\{(-1, -11)\}$ **11.** $\{(-1, 3)\}$ **13.** $\{(3, -4)\}$ **15.** $\{(0, 2)\}$ **17.** $\{(2, -1)\}$ **19.** $\{(0, 4)\}$ **21.** $\{(-1, 1)\}$
23. $\{(2, 3)\}$ **25.** $\{(4, 6)\}$ **27.** $\{(4, 3)\}$ **29.** $\{(5, 2)\}$ **31.** $\{(.138, -4.762)\}$ **33.** $\{(.236, .674)\}$ **35.** $\emptyset$; inconsistent system
37. $\left\{\left(\dfrac{y + 9}{4}, y\right)\right\}$; dependent equations **39.** Multiply (1) by 3 and (2) by 4. **41.** $\{(1, 2, -1)\}$ **43.** $\{(2, 0, 3)\}$ **45.** $\{(1, 2, 3)\}$
47. $\{(-1, 2, 1)\}$ **49.** $\{(4, 1, 2)\}$ **51.** $\left\{\left(\dfrac{1}{2}, \dfrac{2}{3}, -1\right)\right\}$
53. (a) for example: $x + 2y + z = 5, 2x - y + 3z = 4$ (There are others.) **(b)** for example: $x + y + z = 5, 2x - y + 3z = 4$
(There are others.) **(c)** for example: $2x + 2y + 2z = 8, 2x - y + 3z = 4$ (There are others.) **55.** $\emptyset$; inconsistent system
57. $\left\{\left(\dfrac{13}{11} + \dfrac{2}{11}z, \dfrac{7}{11}z + \dfrac{18}{11}, z\right)\right\}$; dependent equations **59.** $\{(3, 3 - z, z)\}$; dependent equations **61.** $y = -3x - 5$
63. $y = \dfrac{3}{4}x^2 + \dfrac{1}{4}x - \dfrac{1}{2}$ **65.** $y = -\dfrac{1}{2}x^2 + x + \dfrac{1}{4}$ **67.** $x^2 + y^2 + x - 7y = 0$ **69. (a)** 16 **(b)** 11 **(c)** 6 **(d)** 8 **(e)** 4
(f) 0 **(g), (k)**   **(h)** 0 **(i)** $\dfrac{40}{3}$ **(j)** $\dfrac{80}{3}$ **(l)** 8 **(m)** \$6 **71. (a)**

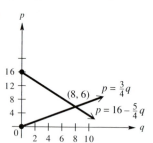

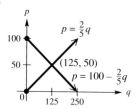

(b) 125 **(c)** \$50 **73.** 120 gallons of \$9.00; 60 gallons of \$3.00; 120 gallons of \$4.50 **75.** 28 inches, 17 inches, 14 inches
77. \$50,000 at 5%; \$10,000 at 4.5%; \$40,000 at 3.75% **79. (a)** $C = \dfrac{17}{450}t^2 + \dfrac{1}{10}t + 315$ **(b)** 2048 **81.** $5t + 15u = 16$
$\qquad\qquad 5t + 4u = 5$
82. $t = \dfrac{1}{5}, u = 1$ **83.** $x = 5, y = 1$ **84.** $y = \dfrac{-15x}{5 - 16x}$ **85.** $y = \dfrac{-4x}{5 - 5x}$ **86.**

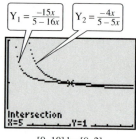

$Y_1 = \dfrac{-15x}{5 - 16x}$ $Y_2 = \dfrac{-4x}{5 - 5x}$

[0, 10] by [0, 2]
Dot mode

87. $\{(2, 2)\}$ **89.** $\{(2, 4, 2)\}$

CONNECTIONS (page 725) **1.** 0, 1, or 2 **2.** 0, 1, 2, 3, or 4 **3.** 0, 1, 2, or an infinite number

9.2 EXERCISES (page 726) **7.** A line and a parabola cannot intersect in more than two points. **9.** $\{(1, 1), (-2, 4)\}$
11. $\left\{(2, 1), \left(\dfrac{1}{3}, \dfrac{4}{9}\right)\right\}$ **13.** $\{(2, 12), (-4, 0)\}$ **15.** $\left\{\left(-\dfrac{3}{5}, \dfrac{7}{5}\right), (-1, 1)\right\}$ **17.** $\{(2, 2), (2, -2), (-2, 2), (-2, -2)\}$
19. $\{(0, 0)\}$ **21.** $\{(i, \sqrt{6}), (-i, \sqrt{6}), (i, -\sqrt{6}), (-i, -\sqrt{6})\}$ **23.** $\{(1, -1), (-1, 1), (1, 1), (-1, -1)\}$ **25.** $\emptyset$

27. $\{(2, 0), (-2, 0)\}$ **29.** $\left\{(-4, -2), \left(-\dfrac{4}{3}, -6\right)\right\}$ **31.** $\left\{\left(6, \dfrac{1}{15}\right), \left(-1, -\dfrac{2}{5}\right)\right\}$

33. $\left\{\left(\dfrac{2\sqrt{5}}{5}i, -i\sqrt{5}\right), \left(-\dfrac{2\sqrt{5}}{5}i, i\sqrt{5}\right), (\sqrt{2}, \sqrt{2}), (-\sqrt{2}, -\sqrt{2})\right\}$ **35.** $\{(1, 1), (-1, -1)\}$

37. $\left\{(\sqrt{3}, 2\sqrt{3}), (-\sqrt{3}, -2\sqrt{3}), \left(\dfrac{6\sqrt{7}}{7}, -\dfrac{3\sqrt{7}}{7}\right), \left(-\dfrac{6\sqrt{7}}{7}, \dfrac{3\sqrt{7}}{7}\right)\right\}$ **39.** $\{(1, 2), (1, -2)\}$ **41.** Shift the graph of $y = |x|$ one

unit to the right.   **42.** Shift the graph of $y = x^2$ four units down.   **43.** $y = \begin{cases} x - 1 & \text{if } x \ge 1 \\ 1 - x & \text{if } x < 1 \end{cases}$

44. $x^2 - 4 = x - 1$ $(x \ge 1)$; $x^2 - 4 = 1 - x$ $(x < 1)$

45. $\dfrac{1 + \sqrt{13}}{2}$ **46.** $\left(\dfrac{1 + \sqrt{13}}{2}, \dfrac{-1 + \sqrt{13}}{2}\right)$ **47.** $\dfrac{-1 - \sqrt{21}}{2}$ **48.** $\left(\dfrac{-1 - \sqrt{21}}{2}, \dfrac{3 + \sqrt{21}}{2}\right)$

49. $\left\{\left(\dfrac{1 + \sqrt{13}}{2}, \dfrac{-1 + \sqrt{13}}{2}\right), \left(\dfrac{-1 - \sqrt{21}}{2}, \dfrac{3 + \sqrt{21}}{2}\right)\right\}$ **50.** The displays at the bottoms of the screens correspond to

approximations of the x- and y-values in the ordered pair solutions.   **51.** $\{(-.79, .62), (.88, .77)\}$ **53.** $\{(.06, 2.88)\}$ **55.** $\{(-1.68,$
$-1.78), (2.12, -1.24)\}$ **57.** 14 and 3   **59.** 8 and 6, 8 and -6, -8 and 6, -8 and -6 **61.** 27 and 6, -27 and -6 **63.** yes
65. $\pm 3\sqrt{5}$ **67. (a)** The emission of carbon is increasing with time. The carbon emissions from the former USSR and Eastern
Europe have surpassed the emissions of Western Europe.   **(b)** They were equal in 1962 or 1963 when the levels were approximately
400 million metric tons.   **(c)** year: 1962;   emission level: approximately 414 million metric tons

CONNECTIONS (page 736)   **1.**

n	T
3	28
6	191
10	805
29	17,487
100	681,550
200	5,393,100
400	42,906,200
1000	668,165,500
5000	8.3×10^{10}
10,000	6.7×10^{11}
100,000	6.7×10^{14}

2. 17,487;   yes   **3.** No, it increases by almost a factor of 8.   **4.** 103 hr

9.3 EXERCISES (page 737)   **1.** $\begin{bmatrix} 2 & 4 \\ 0 & -1 \end{bmatrix}$ **3.** $\begin{bmatrix} 1 & 5 & 6 \\ 0 & 13 & 11 \\ 4 & 7 & 0 \end{bmatrix}$ **5.** $\begin{bmatrix} 2 & 3 & | & 11 \\ 1 & 2 & | & 8 \end{bmatrix}$ **7.** $\begin{bmatrix} 2 & 1 & 1 & | & 3 \\ 3 & -4 & 2 & | & -7 \\ 1 & 1 & 1 & | & 2 \end{bmatrix}$

9. $\begin{aligned} 3x + 2y + z &= 1 \\ 2y + 4z &= 22 \\ -x - 2y + 3z &= 15 \end{aligned}$ **11.** $\begin{aligned} x &= 2 \\ y &= 3 \\ z &= -2 \end{aligned}$ **13.** $\begin{aligned} x + y &= 3 \\ 2y + z &= -4 \\ x - z &= 5 \end{aligned}$ **15.** $\{(2, 3)\}$ **17.** $\{(-3, 0)\}$ **19.** $\{(2, -7)\}$ **21.** $\{(1, 1)\}$

23. $\{(-1, 23, 16)\}$ **25.** $\{(2, 4, 5)\}$ **27.** $\left\{\left(\dfrac{1}{2}, 1, -\dfrac{1}{2}\right)\right\}$ **29.** $\{(2, 1, -1)\}$ **31.** $\emptyset$ **33.** $\left\{\left(-\dfrac{15}{23}z - \dfrac{12}{23}, \dfrac{1}{23}z - \dfrac{13}{23}, z\right)\right\}$

35. $\{(1 - 4w, 1 - w, 2 + w, w)\}$ **39.** $\{(4.267, 8.671)\}$ **41.** $\{(.571, 7.041, 11.442)\}$ **43.** $A = \dfrac{1}{2}, B = -\dfrac{1}{2}$ **45.** $A = \dfrac{1}{2}, B = \dfrac{1}{2}$

47. wife: 40 days;   husband: 32 days   **49.** 9.6 cm^3 of 7%; 30.4 cm^3 of 2%   **53.** 44.4 g of A;   133.3 g of B;   222.2 g of C
55. (a) using the first equation, approximately 245 pounds; using the second, approximately 253 pounds   **(b)** for the first, 7.46
pounds, and for the second, 7.93 pounds   **(c)** at approximately 66 inches and 118 pounds

57. $a + 871b + 11.5c + 3d = 239$
$a + 847b + 12.2c + 2d = 234$
$a + 685b + 10.6c + 5d = 192$
$a + 969b + 14.2c + 1d = 343$

58. $\begin{bmatrix} 1 & 871 & 11.5 & 3 & | & 239 \\ 1 & 847 & 12.2 & 2 & | & 234 \\ 1 & 685 & 10.6 & 5 & | & 192 \\ 1 & 969 & 14.2 & 1 & | & 343 \end{bmatrix}$
The solution is $a \approx -715.457$, $b \approx .34756$, $c \approx 48.6585$, and $d \approx 30.71951$.

59. $F = -715.457 + .34756A + 48.6585P + 30.71951W$ **60.** approximately 323

9.4 EXERCISES (page 750)

1. $w = 3, x = 2, y = -1, z = 4$ **3.** $m = 8, n = -2, z = 2, y = 5, w = 6$
5. $a = 2, z = -3, m = 8, k = 1$ **7.** 2×2; square   **9.** 3×4 **11.** 2×1; column

15. $\begin{bmatrix} -2 & -7 & 7 \\ 10 & -2 & 7 \end{bmatrix}$ **17.** $\begin{bmatrix} -6 & 8 \\ 4 & 2 \end{bmatrix}$ **19.** $\begin{bmatrix} 5x + y & x + y & 7x + y \\ 8x + 2y & x + 3y & 3x + y \end{bmatrix}$ **21.** cannot be added   **23.** $\begin{bmatrix} -4 & 8 \\ 0 & 6 \end{bmatrix}$

25. $\begin{bmatrix} 2 & 6 \\ -4 & 6 \end{bmatrix}$ **27.** $\begin{bmatrix} -1 & -3 \\ 2 & -3 \end{bmatrix}$ **29.** $\begin{bmatrix} 13 \\ 25 \end{bmatrix}$ **31.** $\begin{bmatrix} -17 \\ -1 \end{bmatrix}$ **33.** $\begin{bmatrix} 17 & -10 \\ 1 & 2 \end{bmatrix}$ **35.** $\begin{bmatrix} -2 & 10 \\ 0 & 8 \end{bmatrix}$ **37.** $\begin{bmatrix} -2 & 5 & 0 \\ 6 & 6 & 1 \\ 12 & 2 & -3 \end{bmatrix}$

39. $[2 \quad 7 \quad -4]$ **41.** not possible   **43.** $\begin{bmatrix} 23 & -9 \\ -6 & -2 \\ 33 & 1 \end{bmatrix}$ **45.** $\begin{bmatrix} -25 & 23 & 11 \\ 0 & -6 & -12 \\ -15 & 33 & 45 \end{bmatrix}$ **47.** not possible   **49.** $\begin{bmatrix} 10 & -10 \\ 15 & -5 \end{bmatrix}$

51. $\begin{bmatrix} 2 & 6 & 5 \\ -4 & -7 & 9 \end{bmatrix}$ **53.** $\begin{bmatrix} 16 & 22 \\ 7 & 19 \end{bmatrix}$ **55.** $\begin{bmatrix} -10 & 16 & 26 \\ 5 & 7 & 22 \end{bmatrix}$ **57.** no; the answers are different (CA is not possible); no

59. $\begin{bmatrix} 100 & 150 \\ 125 & 50 \\ 175 & 200 \end{bmatrix}$; $\begin{bmatrix} 100 & 125 & 175 \\ 150 & 50 & 200 \end{bmatrix}$ **61. (a)** $\begin{bmatrix} 50 & 100 & 30 \\ 10 & 90 & 50 \\ 60 & 120 & 40 \end{bmatrix}$ **(b)** $\begin{bmatrix} 12 \\ 10 \\ 15 \end{bmatrix}$ (If the rows and columns are interchanged in

part (a), this should be a 1×3 matrix.)   **(c)** $\begin{bmatrix} 2050 \\ 1770 \\ 2520 \end{bmatrix}$ (This may be a 1×3 matrix.)   **(d)** $6340

63. (a) UBNL   **(b)** CHHPBQ   **65.** $A^2 = \begin{bmatrix} .72 & .28 \\ .56 & .44 \end{bmatrix}$ The sum of the entries in each row is 1.   **67.** $\begin{bmatrix} 1 & 5 & 1 \\ 3 & 12 & 4 \end{bmatrix}$

71. true   **73.** true   **75.** true   **77.** true   **79.** true   **81.** true   **83.** false; Let $A = \begin{bmatrix} 1 & 3 \\ 4 & 5 \end{bmatrix}$ and $B = \begin{bmatrix} 2 & 3 \\ 1 & 5 \end{bmatrix}$;

then $(A + B)(A - B) = \begin{bmatrix} 15 & 0 \\ 25 & 0 \end{bmatrix}$ while $A^2 - B^2 = \begin{bmatrix} 6 & -3 \\ 17 & 9 \end{bmatrix}$.

CONNECTIONS (page 760)

1. (a) $\frac{5}{2}$ **(b)** 1   **2.** The determinant method requires expressing the vertices of the triangle as ordered pairs, while the formula requires the lengths of the base and the height of the triangle. Which one is used depends on the given information about the triangle.

9.5 EXERCISES (page 765)

1. -34 **3.** -68 **5.** $yx + 6$ **7.** $2, -6, 4$ **9.** $-6, 0, -6$ **11.** 1   **13.** 2   **15.** 0
17. $6 - 4y$ **19.** -5.5 **21.** -40 **23.** $5x^2 + 2x - 6$ **24.** $5x^2 + 2x - 6 = 45$; quadratic   **25.** $\{3, -3.4\}$

26. $\begin{vmatrix} 3 & 2 & 1 \\ -1 & 3 & 4 \\ -2 & 0 & 5 \end{vmatrix} = 45$ and $\begin{vmatrix} -3.4 & 2 & 1 \\ -1 & -3.4 & 4 \\ -2 & 0 & 5 \end{vmatrix} = 45$ are both true.   **27.** $\{-4\}$ **29.** $\{13\}$ **31.** Two rows are identical.

33. One column contains all zeros.   **35.** One column contains all zeros.   **37.** If each element of the second row is multiplied by 7, then two rows are identical.   **39.** Rows and columns are interchanged.   **41.** Two columns are interchanged.   **43.** Each element of the second column is multiplied by $-\frac{1}{2}$.   **45.** Elements of the first row are multiplied by 1; products are added to the elements of the second row.   **47.** Elements of the first column are multiplied by the constant k; products are added to the third column.   **49.** 0
51. 0   **53.** 16   **55.** -6 **57.** 1   **59.** 9.5   **61.** 16.5   **63.** 1.597167065   **65.** -1.494192 **71.** 0

9.6 EXERCISES (page 772)   **1.** $\{(2, 2)\}$ **3.** $\{(2, -5)\}$ **5.** $\{(2, 0)\}$ **7.** $\{(2, 3)\}$ **9.** can't use Cramer's rule, $D = 0$; $\emptyset$

11. can't use Cramer's rule, $D = 0$; $\left\{\left(\dfrac{4 - 2y}{3}, y\right)\right\}$ **13.** $\{(-1, 2, 1)\}$ **15.** $\{(-3, 4, 2)\}$ **17.** $\{(0, 0, -1)\}$

19. can't use Cramer's rule, $D = 0$; $\emptyset$ **21.** can't use Cramer's rule, $D = 0$; $\left\{\left(\dfrac{-32 + 19z}{4}, \dfrac{24 - 13z}{4}, z\right)\right\}$ **23.** $\{(2, 3, 1)\}$

25. $\{(0, 4, 2)\}$ **27.** $\left\{\left(\dfrac{31}{5}, \dfrac{19}{10}, -\dfrac{29}{10}\right)\right\}$ **31.** $\{(0, 2, -2, 1)\}$ **33.** $\{(2, 3, 5, 0)\}$ **35.** $\{(-1, 2)\}$ **36.** $\{(-1, 2)\}$ **37.** They are

the same.   **38.** -3 **39.** $D_x = 3$; $D_y = -6$ **40.** $x = \dfrac{D_x}{D} = -1$; $y = \dfrac{D_y}{D} = 2$;   The solution set is always $\{(-1, 2)\}$ for such a

system.   **41.** $W_1 = W_2 = \dfrac{100\sqrt{3}}{3} \approx 58$ lb   **43.** $\{(-a - b, a^2 + ab + b^2)\}$ **45.** $\{(1, 0)\}$

9.7 EXERCISES (page 783)   **1.** yes   **3.** no   **5.** no   **7.** yes   **9.** $\begin{bmatrix} -1/5 & -2/5 \\ 2/5 & -1/5 \end{bmatrix}$ **11.** $\begin{bmatrix} 2 & 1 \\ -3/2 & -1/2 \end{bmatrix}$

13. Inverse does not exist.   **15.** $\begin{bmatrix} -1 & 1 & 1 \\ 0 & -1 & 0 \\ 2 & -1 & -1 \end{bmatrix}$ **17.** $\begin{bmatrix} 7 & -3 & -3 \\ -1 & 1 & 0 \\ -1 & 0 & 1 \end{bmatrix}$ **19.** $\begin{bmatrix} -15/4 & -1/4 & -3 \\ 5/4 & 1/4 & 1 \\ -3/2 & 0 & -1 \end{bmatrix}$

21. $\begin{bmatrix} 1/2 & 0 & 1/2 & -1 \\ 1/10 & -2/5 & 3/10 & -1/5 \\ -7/10 & 4/5 & -11/10 & 12/5 \\ 1/5 & 1/5 & -2/5 & 3/5 \end{bmatrix}$ **23.** $\begin{bmatrix} 2 & 9 \\ 1 & 5 \end{bmatrix}$ **25.** $\begin{bmatrix} 1 & 0 & 1 \\ -1 & 0 & 2 \\ -2 & 1 & 3 \end{bmatrix}$ **27.** It is the determinant of A.

28. $A^{-1} = \begin{bmatrix} d/|A| & -b/|A| \\ -c/|A| & a/|A| \end{bmatrix}$ **29.** $A^{-1} = \dfrac{1}{|A|}\begin{bmatrix} d & -b \\ -c & a \end{bmatrix}$ **31.** $A^{-1} = \begin{bmatrix} -3/2 & 1 \\ 7/2 & -2 \end{bmatrix}$ **32.** zero   **33.** $\{(2, 3)\}$

35. $\{(-2, 4)\}$ **37.** $\{(4, -6)\}$ **39.** $\{(10, -1, -2)\}$ **41.** $\{(11, -1, 2)\}$ **43.** $\{(1, 0, 2, 1)\}$

45. **(a)** $602.7 = a + 5.543b + 37.14c$; $656.7 = a + 6.933b + 41.30c$; $778.5 = a + 7.638b + 45.62c$ **(b)** $a \approx -490.547$,

$b \approx -89, c \approx 42.71875$ **(c)** $S = -490.547 - 89A + 42.71875B$ **(d)** approximately 843.5   **(e)** $S \approx 1547.5$

Using only three consecutive years to forecast six years into the future is probably not very accurate.   **47.** **(a)** HELP

(b) DIVIDE   **49.** $\begin{bmatrix} .0543058761 & -.0543058761 \\ 1.846399787 & .153600213 \end{bmatrix}$ **51.** $\{(-3.542308934, -4.343268299)\}$

53. $\{(-.9704156959, 1.391914631, .1874077432)\}$ **61.** $A^{-1} = A^2 = \begin{bmatrix} 1 & 0 & 0 \\ 0 & -1 & 1 \\ 0 & -1 & 0 \end{bmatrix}$

CONNECTIONS (page 795)   **1.** Ship no medical kits and 4000 containers of water.   **2.** Ship 4000 medical kits and 2000 containers of water.

9.8 EXERCISES (page 796)   **1.**   **3.**   **5.**

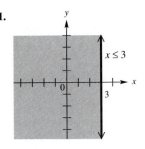

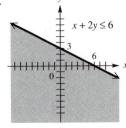

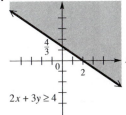

7.

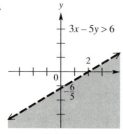

$3x - 5y > 6$

9.

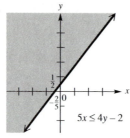

$5x \leq 4y - 2$

11.

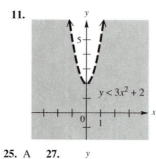

$y < 3x^2 + 2$

13.

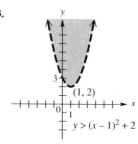

$(1, 2)$

$y > (x - 1)^2 + 2$

15.

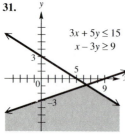

$(0, -3)$

$x^2 + (y + 3)^2 \leq 16$

19. above **21.** (b) **23.** C **25.** A **27.**

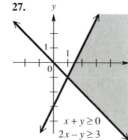

$x + y \geq 0$
$2x - y \geq 3$

29.

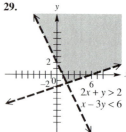

$2x + y > 2$
$x - 3y < 6$

31.

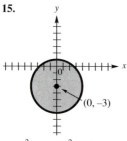

$3x + 5y \leq 15$
$x - 3y \geq 9$

33.

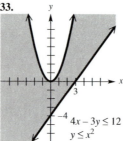

$4x - 3y \leq 12$
$y \leq x^2$

35.

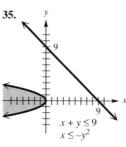

$x + y \leq 9$
$x \leq -y^2$

37.

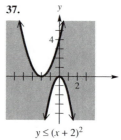

$y \leq (x + 2)^2$
$y \geq -2x^2$

39.

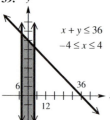

$x + y \leq 36$
$-4 \leq x \leq 4$

41.

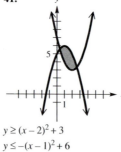

$y \geq (x - 2)^2 + 3$
$y \leq -(x - 1)^2 + 6$

43. $3x - 2y \geq 6$
$x + y \leq -5$
$y \leq 4$

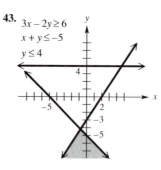

45.

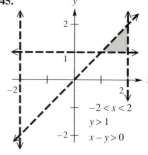

$-2 < x < 2$
$y > 1$
$x - y > 0$

47.

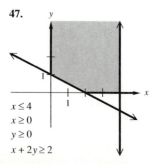

$x \leq 4$
$x \geq 0$
$y \geq 0$
$x + 2y \geq 2$

49.

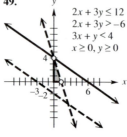

$2x + 3y \leq 12$
$2x + 3y > -6$
$3x + y < 4$
$x \geq 0, y \geq 0$

51.

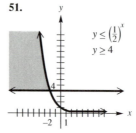

$y \leq \left(\frac{1}{2}\right)^x$
$y \geq 4$

53.

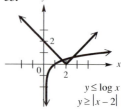

$y \leq \log x$
$y \geq |x - 2|$

55. A   **57.** B   **59.**

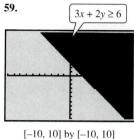

$3x + 2y \geq 6$

[−10, 10] by [−10, 10]

61.

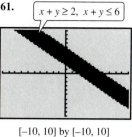

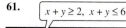

$x + y \geq 2, \; x + y \leq 6$

[−10, 10] by [−10, 10]

63.

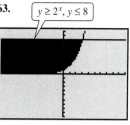

$y \geq 2^x, \; y \leq 8$

[−10, 10] by [−10, 10]

65. $x + 2y - 8 \geq 0, x + 2y \leq 12, x \geq 0, y \geq 0$ **67.** maximum of 65 at (5, 10);   minimum of 8 at (1, 1)   **69.** maximum of 66 at (7, 9);   minimum of 3 at (1, 0)   **71.** maximum of 100 at (1, 10);   minimum of 0 at (1, 0)
73. Let x = number of Brand X pills and y = number of Brand Y pills.
Then $3000x + 1000y \geq 6000, 45x + 50y \geq 195, 75x + 200y \geq 600, x \geq 0, y \geq 0$.

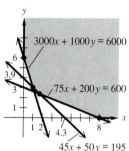

75. \$1120, with 4 pigs and 12 geese   **77.** 8 of #1 and 3 of #2, for 100 cubic feet of storage   **79.** 6.4 million gallons of gasoline and 3.2 million gallons of fuel oil, for a maximum revenue of \$16,960,000

CHAPTER 9 REVIEW EXERCISES (page 802)   **1.** {(9, 4)}   **3.** {(1, 1)}   **5.** ∅;   inconsistent system   **7.** {(3, 2, 1)}
9. $\frac{1}{3}$ cup of rice; $\frac{1}{5}$ cup of soybeans   **11.** 5 blankets, 3 rugs, 8 skirts   **13.** $y = 2.4x^2 - 6.2x + 1.5$
15. $\{(x, 1 - 2x, 6 - 11x)\}$ **17.** {(3, 2), (3, −2), (−3, 2), (−3, −2)}   **19.** {(−2, 0), (1, 1)}
21. yes; $\left(\dfrac{8 - 8\sqrt{41}}{5}, \dfrac{16 + 4\sqrt{41}}{5}\right), \left(\dfrac{8 + 8\sqrt{41}}{5}, \dfrac{16 - 4\sqrt{41}}{5}\right)$ **23.** {(−2, 0)}   **25.** {(0, 3, 3)}
27. 10 lb of \$4.60 tea;   8 lb of \$5.75 tea;   2 lb of \$6.50 tea   **29.** 164 fives;   86 tens;   40 twenties
31. $a = 5, x = \dfrac{3}{2}, y = 0, z = 9$ **33.** $\begin{bmatrix} -4 \\ 6 \\ 1 \end{bmatrix}$ **35.** $\begin{bmatrix} 11 & 20 \\ 14 & 40 \end{bmatrix}$ **37.** $\begin{bmatrix} -3 \\ 10 \end{bmatrix}$ **39.** $\begin{bmatrix} 5 & 1 & 3 \\ -2 & 4 & 8 \end{bmatrix}$ **41.** −6   **43.** −44
45. 138   **47.** One row is all zeros.   **49.** The elements of the first row are multiplied by 2.
51. The second row on the right is the sum of the first two rows on the left.   **53.** $\left\{\dfrac{1}{31}\right\}$ **55.** {(−2, 5)}   **57.** {(−4, 6, 2)}
59. $\begin{bmatrix} 3 & -1 \\ -5 & 2 \end{bmatrix}$ **61.** $\begin{bmatrix} 2/3 & 0 & -1/3 \\ 1/3 & 0 & -2/3 \\ -2/3 & 1 & 1/3 \end{bmatrix}$ **63.** {(−1, 0, 2)}   **65.**

$x + y \leq 6$
$2x - y \geq 3$

67.

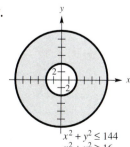

$x^2 + y^2 \leq 144$
$x^2 + y^2 \geq 16$

69. maximum of 70 at (0, 5)   **71.** 150 kg of half and half and 75 kg of the other, for a maximum revenue of \$1260

CHAPTER 9 TEST (page 805)   **1.** $\{(4, 3)\}$ **2.** $\left\{\left(\dfrac{-3y - 7}{2}, y\right)\right\}$;   dependent equations   **3.** $\{(1, 2)\}$
4. $\emptyset$;   inconsistent system   **5.** $\{(2, 0, -1)\}$ **6.** $\{(1, 2), (-1, 2), (1, -2), (-1, -2)\}$ **7.** $\{(3, 4), (4, 3)\}$ **8.** yes

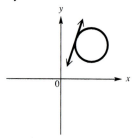

9. 5 and -6 **10.** $\{(5, 1)\}$ **11.** $\{(5, 3, 6)\}$ **12.** $y = -.25x^2 + .3x - .4$ **13.** 1, 1, 0   **14.** $w = -3$; $x = -1$; $y = 7$

15. $\begin{bmatrix} 8 & 3 \\ 0 & -11 \\ 15 & 19 \end{bmatrix}$ **16.** not possible   **17.** $\begin{bmatrix} -5 & 16 \\ 19 & 2 \end{bmatrix}$ **18.** not possible   **19.** (a)   **20.** -58 **21.** -844

22. Two rows are interchanged.   **23.** Row 3 is multiplied by -2 and added to row 2, and row 1 is added to row 3.   **24.** $\{(-6, 7)\}$

25. $\{(1, -2, 3)\}$ **26.** $\begin{bmatrix} -2 & -5 \\ -3 & -8 \end{bmatrix}$ **27.** Inverse does not exist.   **28.** $\begin{bmatrix} -9 & 1 & -4 \\ -2 & 1 & 0 \\ 4 & -1 & 1 \end{bmatrix}$ **29.** $\{(-7, 8)\}$ **30.** $\{(0, 5, -9)\}$

31.   **32.** maximum of 42 at $(12, 6)$ **33.** 0 VIP rings and 24 SST rings, for a maximum profit of $960

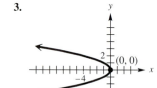

$x - 3y \geq 6$
$y^2 \leq 16 - x^2$

CHAPTER 10   Analytic Geometry ▼▼▼

10.1 EXERCISES (page 817)

3.

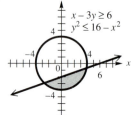

$x = -y^2$

5.

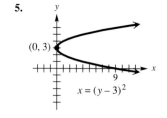

$x = (y - 3)^2$

7.

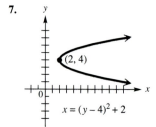

$x = (y - 4)^2 + 2$

9.

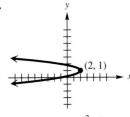

$x = -3(y-1)^2 + 2$

11.

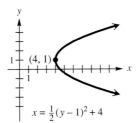

$x = \frac{1}{2}(y-1)^2 + 4$

13.

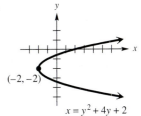

$x = y^2 + 4y + 2$

15.

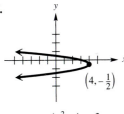

$x = -4y^2 - 4y + 3$

17.

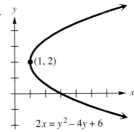

$2x = y^2 - 4y + 6$

19. $(0, 6), y = -6$, y-axis

21. $\left(0, -\dfrac{1}{16}\right), y = \dfrac{1}{16}$, y-axis   **23.** $\left(-\dfrac{1}{128}, 0\right), x = \dfrac{1}{128}$, x-axis   **25.** $(-1, 0), x = 1$, x-axis   **27.** $(4, 3), x = -2, y = 3$

29. $(7, -1), y = -9, x = 7$ **31.** $y^2 = 20x$ **33.** $x^2 = y$ **35.** $x^2 = y$ **37.** $y^2 = \dfrac{4}{3}x$

39. $(x-4)^2 = 8(y-3)$ **41.** $(y-6)^2 = 28(x+5)$ **43.**

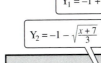

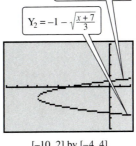

$[-10, 2]$ by $[-4, 4]$

45.

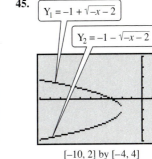

$[-10, 2]$ by $[-4, 4]$

47. (a) Earth: $y = x - \dfrac{16.1}{961}x^2$;   Mars: $y = x - \dfrac{6.3}{961}x^2$ **(b)** approximately 93 feet

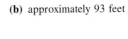

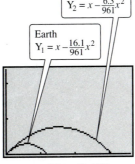

$[0, 180]$ by $[0, 120]$
Xscl = 50   Yscl = 50

49. approximately 43.8 feet   **51.** 60 feet   **53.** $a + b + c = -5$; $4a - 2b + c = -14$; $4a + 2b + c = -10$

54. $\{(-2, 1, -4)\}$ **55.** It opens to the left, because $a = -2 < 0$.   **56.** $x = -2y^2 + y - 4$ **57.** $(y - 2)^2 = \dfrac{1}{3}(x - 1)$

CONNECTIONS (page 825)   **1.** The kidney stone and the wave source must each be placed 3 units from the center of the ellipse.
2. One such ellipse would have an equation of $\dfrac{x^2}{100} + \dfrac{y^2}{64} = 1$.

10.2 EXERCISES (page 826)   **1.** If the two foci of an ellipse are allowed to coincide, the resulting figure is a circle.
3. $(0, 0)$; $(-5, 0), (5, 0)$; $(0, -3), (0, 3)$; $(-4, 0), (4, 0)$

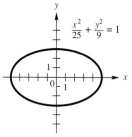

5. $(0, 0)$; $(-3, 0), (3, 0)$; $(0, -1), (0, 1)$; $(-2\sqrt{2}, 0), (2\sqrt{2}, 0)$

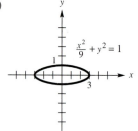

7. $(0, 0)$; $(0, -9), (0, 9)$; $(-3, 0), (3, 0)$; $(0, -6\sqrt{2}), (0, 6\sqrt{2})$

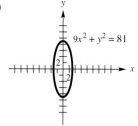

9. $(0, 0)$; $(-5, 0), (5, 0)$; $(0, -2), (0, 2)$; $(-\sqrt{21}, 0), (\sqrt{21}, 0)$

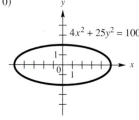

11. $(2, 1)$; $(-3, 1), (7, 1)$; $(2, -1), (2, 3)$; $(2 - \sqrt{21}, 1), (2 + \sqrt{21}, 1)$

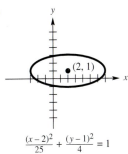

$$\frac{(x-2)^2}{25} + \frac{(y-1)^2}{4} = 1$$

13. $(-3, 2)$; $(-3, -4), (-3, 8)$; $(-7, 2), (1, 2)$; $(-3, 2 - 2\sqrt{5}), (-3, 2 + 2\sqrt{5})$

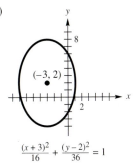

$$\frac{(x+3)^2}{16} + \frac{(y-2)^2}{36} = 1$$

15. $\dfrac{x^2}{25} + \dfrac{y^2}{16} = 1$ **17.** $\dfrac{x^2}{5} + \dfrac{y^2}{9} = 1$ **19.** $\dfrac{(x-5)^2}{25} + \dfrac{(y-2)^2}{16} = 1$ **21.** $\dfrac{(x-4)^2}{9} + \dfrac{(y-5)^2}{16} = 1$ **23.** $\dfrac{x^2}{72} + \dfrac{y^2}{81} = 1$

25. $\dfrac{x^2}{9} + \dfrac{y^2}{25} = 1$ **27.** $\dfrac{9x^2}{28} + \dfrac{9y^2}{64} = 1$

29. function

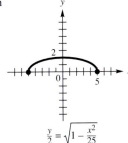

$$\frac{y}{2} = \sqrt{1 - \frac{x^2}{25}}$$

31.

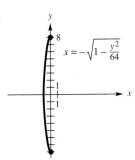

$$x = -\sqrt{1 - \frac{y^2}{64}}$$

33.

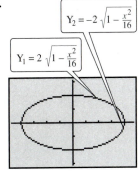

$$Y_1 = 2\sqrt{1 - \frac{x^2}{16}}$$

$$Y_2 = -2\sqrt{1 - \frac{x^2}{16}}$$

$[-4.7, 4.7]$ by $[-3.1, 3.1]$

35.

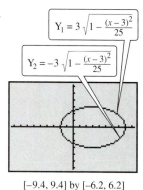

$Y_1 = 3\sqrt{1 - \frac{(x-3)^2}{25}}$

$Y_2 = -3\sqrt{1 - \frac{(x-3)^2}{25}}$

[−9.4, 9.4] by [−6.2, 6.2]

37. Solve the original equation for y.

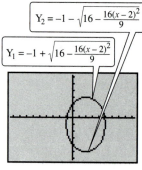

$Y_2 = -1 - \sqrt{16 - \frac{16(x-2)^2}{9}}$

$Y_1 = -1 + \sqrt{16 - \frac{16(x-2)^2}{9}}$

[−9.4, 9.4] by [−6.2, 6.2]

38. $[-1, 5]$ **39.** $16 - \frac{16(x-2)^2}{9} \geq 0$ **40.** Its graph is a parabola.   **41.**

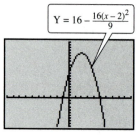

$Y = 16 - \frac{16(x-2)^2}{9}$

[−10, 10] by [−10, 20]

42. The graph of $Y = 16 - \frac{16(x-2)^2}{9}$ lies above or on the x-axis in the interval $[-1, 5]$.

43. In Figure 14, we see that the domain is $[-1, 5]$. This corresponds to the solution set found graphically in Exercise 42.
44. $[-1, 5]$
47. (a)

$Y_3 = \sqrt{3960^2 - (x - 163.6)^2}$

$Y_1 = 4462\sqrt{1 - \frac{x^2}{4465^2}}$

$Y_4 = -Y_3$ $Y_2 = -Y_1$

[−6750, 6750] by [−4500, 4500]
Xscl = 1000   Yscl = 1000

(b) minimum: approximately 341 miles;   maximum: approximately 669 miles

49. (a) $v_{max} \approx 30.3$ km/sec; $v_{min} \approx 29.3$ km/sec   **(b)** For a circle $e = 0$, so $v_{max} = v_{min} = \frac{2\pi}{P}$. The minimum and maximum velocities are equal. Therefore, the planet's velocity is constant.   **(c)** A planet is at its maximum and minimum distance from a focus when it is located at the vertices of the ellipse. Thus, the minimum and maximum velocities of a planet will occur at the vertices of the elliptical orbit.   **51.** 348.2 ft   **53.** approximately 55 million miles   **55.** Answers will vary.

CONNECTIONS (page 835) $\dfrac{x^2}{625} - \dfrac{y^2}{1875} = 1$

10.3 EXERCISES (page 835)   **1.** C   **3.** D   **5.** (0, 0); (−4, 0), (4, 0);

(−5, 0), (5, 0); $y = \pm\dfrac{3}{4}x$

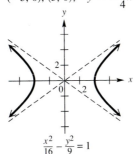

$\dfrac{x^2}{16} - \dfrac{y^2}{9} = 1$

7. (0, 0); (0, −5), (0, 5);

(0, −$\sqrt{74}$), (0, $\sqrt{74}$);

$y = \pm\dfrac{5}{7}x$

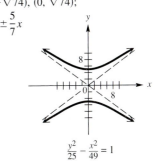

$\dfrac{y^2}{25} - \dfrac{x^2}{49} = 1$

9. (0, 0); (−3, 0), (3, 0);

(−3$\sqrt{2}$, 0), (3$\sqrt{2}$, 0); $y = \pm x$

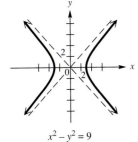

$x^2 - y^2 = 9$

11. (0, 0); (−5, 0), (5, 0);

(−$\sqrt{34}$, 0), ($\sqrt{34}$, 0);

$y = \pm\dfrac{3}{5}x$

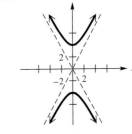

$9x^2 - 25y^2 = 225$

13. (0, 0); (0, −4), (0, 4);

(0, −2$\sqrt{5}$), (0, 2$\sqrt{5}$);

$y = \pm 2x$

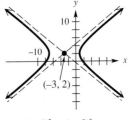

$4x^2 - y^2 = -16$

15. (0, 0); $\left(-\dfrac{1}{3}, 0\right), \left(\dfrac{1}{3}, 0\right)$;

$\left(-\dfrac{\sqrt{13}}{6}, 0\right), \left(\dfrac{\sqrt{13}}{6}, 0\right)$;

$y = \pm\dfrac{3}{2}x$

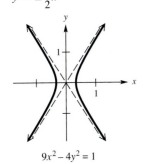

$9x^2 - 4y^2 = 1$

17. (4, 7); (4, 1), (4, 13); (4, −3), (4, 17);

$y = 7 \pm\dfrac{3}{4}(x - 4)$

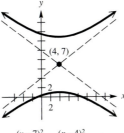

$\dfrac{(y-7)^2}{36} - \dfrac{(x-4)^2}{64} = 1$

19. (−3, 2); (−7, 2), (1, 2);

(−8, 2), (2, 2); $y = 2 \pm\dfrac{3}{4}(x + 3)$

$\dfrac{(x+3)^2}{16} - \dfrac{(y-2)^2}{9} = 1$

21. $(-5, 3)$; $\left(-\dfrac{21}{4}, 3\right), \left(-\dfrac{19}{4}, 3\right)$;

$\left(-5 - \dfrac{\sqrt{17}}{4}, 3\right), \left(-5 + \dfrac{\sqrt{17}}{4}, 3\right)$;

$y = 3 \pm 4(x + 5)$

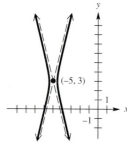

$16(x + 5)^2 - (y - 3)^2 = 1$

23. function

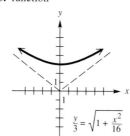

$\dfrac{y}{3} = \sqrt{1 + \dfrac{x^2}{16}}$

25.

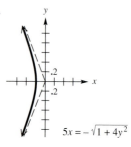

$5x = -\sqrt{1 + 4y^2}$

27. $\dfrac{x^2}{16} - \dfrac{y^2}{9} = 1$ **29.** $\dfrac{y^2}{36} - \dfrac{x^2}{144} = 1$ **31.** $\dfrac{x^2}{9} - 3y^2 = 1$ **33.** $\dfrac{2y^2}{25} - 2x^2 = 1$ **35.** $\dfrac{(y - 3)^2}{4} - \dfrac{49(x - 4)^2}{4} = 1$

37. $\dfrac{(x - 1)^2}{4} - \dfrac{(y + 2)^2}{5} = 1$ **39.** $\dfrac{y^2}{49} - \dfrac{x^2}{392} = 1$ **41.** $\dfrac{(y - 6)^2}{16} - \dfrac{(x + 2)^2}{9} = 1$ **43.**

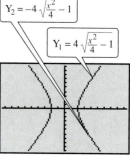

$Y_2 = -4\sqrt{\dfrac{x^2}{4} - 1}$

$Y_1 = 4\sqrt{\dfrac{x^2}{4} - 1}$

$[-9.4, 9.4]$ by $[-10, 10]$

45.

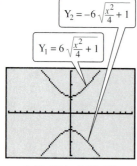

$Y_2 = -6\sqrt{\dfrac{x^2}{4} + 1}$

$Y_1 = 6\sqrt{\dfrac{x^2}{4} + 1}$

$[-10, 10]$ by $[-15, 15]$
Xscl $= 1$ Yscl $= 5$

47. $y = \sqrt{\dfrac{x^2}{4} - 1}$ **48.** $y = \dfrac{1}{2}x$ **49.** $y \approx 24.98$ **50.** $y = 25$

51. Because $24.98 < 25$, the graph of $y = \sqrt{\dfrac{x^2}{4} - 1}$ lies below the graph of $y = \dfrac{1}{2}x$ when $x = 50$.

52. The y-values on the hyperbola will approach the y-values on the asymptote.   **53. (a)** $x = \sqrt{y^2 + 2.5 \times 10^{-27}}$

(b) approximately 1.2×10^{-13} m   **57.**

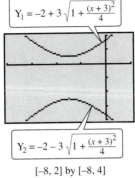

$$Y_1 = -2 + 3\sqrt{1 + \frac{(x+3)^2}{4}}$$

$$Y_2 = -2 - 3\sqrt{1 + \frac{(x+3)^2}{4}}$$

$[-8, 2]$ by $[-8, 4]$
Xscl = 2   Yscl = 2

10.4 EXERCISES (page 844)   **1.** circle   **3.** parabola   **5.** parabola   **7.** ellipse   **9.** hyperbola   **11.** hyperbola
13. ellipse   **15.** circle   **17.** hyperbola   **19.** ellipse   **21.** circle   **23.** parabola   **25.** no graph   **27.** circle   **29.** parabola

31. hyperbola   **33.** ellipse   **35.** no graph   **37.** ellipse   **39.** hyperbola   **41.** $\dfrac{1}{3}$ **43.** 1   **45.** $\dfrac{3}{2}$

47. (a) C   **(b)** D   **(c)** F   **(d)** I   **(e)** G   **(f)** A   **(g)** J   **(h)** H   **(i)** B   **(j)** E   **49.** elliptic   **53.** Answers will vary.

CHAPTER 10 REVIEW EXERCISES (page 849)   **1.** $(2, 5)$; $y = 5$ **3.** $\left(\dfrac{7}{4}, \dfrac{1}{2}\right)$; $y = \dfrac{1}{2}$

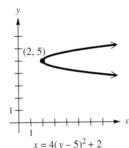

$x = 4(y - 5)^2 + 2$

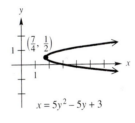

$x = 5y^2 - 5y + 3$

5. $\left(-\dfrac{1}{6}, 0\right)$; $x = \dfrac{1}{6}$; x-axis   **7.** $\left(0, \dfrac{1}{12}\right)$; $y = -\dfrac{1}{12}$; y-axis   **9.** $x = \dfrac{1}{16}y^2$ **11.** $y = \dfrac{4}{9}x^2$ **13.** ellipse

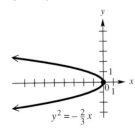

$y^2 = -\dfrac{2}{3}x$

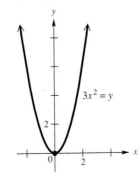

$3x^2 = y$

15. hyperbola   **17.** parabola   **19.** ellipse   **21.** F   **23.** A   **25.** B   **27.** ellipse;   (0, −3), (0, 3)

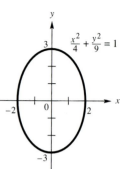
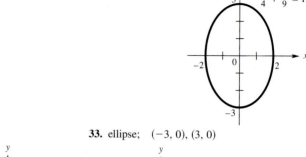

29. hyperbola;   (−8, 0), (8, 0); $y = \pm\dfrac{3}{4}x$ **31.** circle   **33.** ellipse;   (−3, 0), (3, 0)

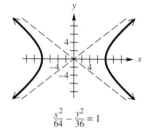

$$\frac{x^2}{64} - \frac{y^2}{36} = 1$$

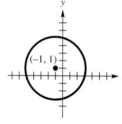

$$\frac{(x+1)^2}{16} + \frac{(y-1)^2}{16} = 1$$

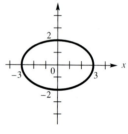

$$4x^2 + 9y^2 = 36$$

35. ellipse;   (1, −1), (5, −1)   **37.** hyperbola;   (−3, −4), (−3, 0); $y = \pm\dfrac{2}{3}(x + 3) - 2$

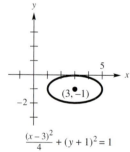

$$\frac{(x-3)^2}{4} + (y+1)^2 = 1$$

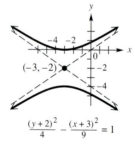

$$\frac{(y+2)^2}{4} - \frac{(x+3)^2}{9} = 1$$

39.

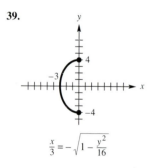

$$\frac{x}{3} = -\sqrt{1 - \frac{y^2}{16}}$$

41. function

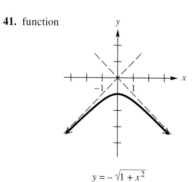

$$y = -\sqrt{1 + x^2}$$

43. $\dfrac{x^2}{12} + \dfrac{y^2}{16} = 1$ **45.** $\dfrac{y^2}{16} - \dfrac{x^2}{9} = 1$ **47.** $x = \dfrac{1}{12}(y - 2)^2$ **49.** $\dfrac{x^2}{25} + \dfrac{y^2}{21} = 1$ **51.** $\dfrac{x^2}{9} - \dfrac{y^2}{16} = 1$

53. $\dfrac{(x - 2)^2}{16} + \dfrac{y^2}{12} = 1$ **55.** C, A, B, D   **57.** $\dfrac{x^2}{6{,}111{,}883} + \dfrac{y^2}{432{,}135} = 1$

CHAPTER 10 TEST (page 852)   **1.** $(3, 9)$; $x = 3$ **2.** $(-4, -1)$; $y = -1$ **3.** $\left(\dfrac{1}{32}, 0\right)$; $x = -\dfrac{1}{32}$

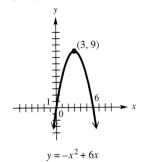

$y = -x^2 + 6x$

$x = 4y^2 + 8y$

4. $(y - 3)^2 = -\dfrac{1}{5}(x - 2)$ **6.**

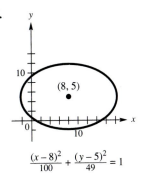

$\dfrac{(x - 8)^2}{100} + \dfrac{(y - 5)^2}{49} = 1$

7.

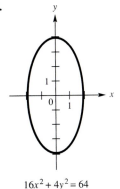

$16x^2 + 4y^2 = 64$

8. function

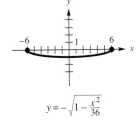

$y = -\sqrt{1 - \dfrac{x^2}{36}}$

9. $\dfrac{x^2}{9} + \dfrac{y^2}{4} = 1$ **10.** $\dfrac{x^2}{400} + \dfrac{y^2}{144} = 1$;   approximately 10.39 ft

11. asymptotes: $y = \pm x$ **12.** asymptotes: $y = \pm \dfrac{3}{2}x$ **13.** $\dfrac{x^2}{25} - \dfrac{y^2}{11} = 1$ **14.** circle   **15.** hyperbola   **16.** ellipse

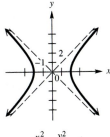

$\dfrac{x^2}{4} - \dfrac{y^2}{4} = 1$

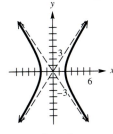

$9x^2 - 4y^2 = 36$

17. parabola   **18.** point   **19.** no graph   **20.** $Y_1 = 7\sqrt{\dfrac{x^2}{25} - 1}$; $Y_2 = -7\sqrt{\dfrac{x^2}{25} - 1}$

CHAPTER 11   Further Topics in Algebra ▼▼▼

CONNECTIONS (page 857)   **1.** After the first term, each term is the sum of the two preceding terms.
2. We show only the first four generations in this sketch.

11.1 EXERCISES (page 862)   **1.** 14, 18, 22, 26, 30   **3.** 1, 2, 4, 8, 16   **5.** $0, \frac{1}{9}, \frac{2}{27}, \frac{1}{27}, \frac{4}{243}$ **7.** $-2, 4, -6, 8, -10$

9. $1, \frac{7}{6}, 1, \frac{5}{6}, \frac{19}{27}$ **13.** finite   **15.** finite   **17.** infinite   **19.** finite   **21.** $-2, 1, 4, 7$ **23.** 1, 1, 2, 3   **25.** 35   **27.** $\frac{25}{12}$

29. 288   **31.** 3   **33.** $-1 + 1 + 3 + 5 + 7$ **35.** $-10 - 4 + 0$ **37.** $0 + \frac{1}{2} + \frac{2}{3} + \frac{3}{4}$ **39.** $-3.5 + .5 + 4.5 + 8.5$

41. $0 + 4 + 16 + 36$ **43.** $-1 - \frac{1}{3} - \frac{1}{5} - \frac{1}{7}$ **45.** $\sum_{j=3}^{7} (12 - 3j)$ **47.** $\sum_{j=0}^{9} 2(3)^{j+1}$

49. $\sum_{j=0}^{10} [(j-1)^2 - 2(j-1)]$ or $\sum_{j=0}^{10} (j^2 - 4j + 3)$ **51.** 90   **53.** 220   **55.** 304   **57.** converges to $\frac{1}{2}$ **59.** diverges
61. converges to $e \approx 2.71828$ **63.** (a) $N_{j+1} = 2N_j$ for $j \geq 1$ (b) 14,720   (c)

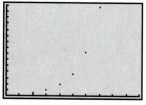

[0, 10] by [0, 15,000]
Xscl = 1   Yscl = 1000

(d) The growth is very rapid. Since there is a doubling of the bacteria at equal intervals, their growth is exponential.

11.2 EXERCISES (page 870)   **1.** 3   **3.** -5 **5.** $x + 2y$ **7.** 8, 14, 20, 26, 32   **9.** 5, 3, 1, -1, -3 **11.** 14, 12, 10, 8, 6

13. $a_8 = 19$; $a_n = 3 + 2n$ **15.** $a_8 = 7$; $a_n = n - 1$ **17.** $a_8 = -6$; $a_n = 10 - 2n$ **19.** $a_8 = -3$; $a_n = -39 + \frac{9n}{2}$

21. $a_8 = x + 21$; $a_n = x + 3n - 3$ **23.** 3   **25.** 5   **27.** 215   **29.** 125   **31.** 230   **33.** 77.5   **35.** $a_1 = 7, d = 5$

37. $a_1 = 1, d = -\frac{20}{11}$ **39.** 18   **41.** 140   **43.** -684 **45.** 500,500   **47.** $f(1) = m + b$; $f(2) = 2m + b$; $f(3) = 3m + b$
48. yes   **49.** m **50.** $a_n = mn + b$ **51.** 328.3   **53.** 172.884   **55.** 1281   **57.** 4680   **59.** 54,800   **61.** 713 in
63. All terms are the same constant.   **65.** yes

CONNECTIONS (page 878)   **1.** $158.00   **2.** $7584.00   **3.** $1584.00

11.3 EXERCISES (page 879)   **1.** $\frac{5}{3}$, 5, 15, 45   **3.** $\frac{5}{8}, \frac{5}{4}, \frac{5}{2}$, 5, 10   **5.** 80; $5(-2)^{n-1}$

7. -108; $\left(-\frac{4}{3}\right)(3)^{n-1}$ or $(-4)(3)^{n-2}$ **9.** -729; $(-9)(-3)^{n-1}$ or $-(-3)^{n+1}$ **11.** -324; $-4(3)^{n-1}$

13. $\frac{125}{4}$; $\left(\frac{4}{5}\right)\left(\frac{5}{2}\right)^{n-1}$ or $\frac{5^{n-2}}{2^{n-3}}$ **15.** $x - 5 = .6 - x$ **16.** The solution is 2.8;   5, 2.8, .6   **17.** $\frac{x}{5} = \frac{.6}{x}$

18. The solution is $\sqrt{3}$; 5, $\sqrt{3}$, .6   **19.** 125; $\dfrac{1}{5}$ **21.** -2; $\dfrac{1}{2}$ **23.** 682   **25.** $\dfrac{99}{8}$ **27.** 860.95   **29.** 363   **31.** $\dfrac{189}{4}$

33. 2032   **35.** The sum exists if $|r| < 1$.   **37.** 2; does not converge   **39.** $\dfrac{1}{2}$ **41.** $\dfrac{128}{7}$ **43.** $\dfrac{1000}{9}$ **45.** $\dfrac{8}{3}$ **47.** 4

49. $\dfrac{3}{7}$ **51.** $g(1) = ab$; $g(2) = ab^2$; $g(3) = ab^3$ **52.** yes; the common ratio is b.   **53.** $a_n = ab^n$ **55.** 97.739

57. .212   **59. (a)** $a_n = a_1 \cdot 2^{n-1}$ **(b)** 15 (rounded from 14.28)   **(c)** 560 minutes, or 9 hours, 20 minutes   **61.** $\approx 13.4\%$

63. \$26,214.40   **65.** 200 cm   **67.** 62; 2046   **69.** $\dfrac{1}{64}$ m   **71.** \$12,487.56   **73.** \$32,029.33

CONNECTIONS (page 883)   **1.** 21, 34, and 55

11.4 EXERCISES (page 888)   **1.** 20   **3.** 35   **5.** 56   **7.** 45   **9.** 1   **11.** n
15. $x^6 + 6x^5y + 15x^4y^2 + 20x^3y^3 + 15x^2y^4 + 6xy^5 + y^6$ **17.** $p^5 - 5p^4q + 10p^3q^2 - 10p^2q^3 + 5pq^4 - q^5$
19. $r^{10} + 5r^8s + 10r^6s^2 + 10r^4s^3 + 5r^2s^4 + s^5$ **21.** $p^4 + 8p^3q + 24p^2q^2 + 32pq^3 + 16q^4$
23. $2401p^4 + 2744p^3q + 1176p^2q^2 + 224pq^3 + 16q^4$
25. $729x^6 - 2916x^5y + 4860x^4y^2 - 4320x^3y^3 + 2160x^2y^4 - 576xy^5 + 64y^6$
27. $\dfrac{m^6}{64} - \dfrac{3m^5}{16} + \dfrac{15m^4}{16} - \dfrac{5m^3}{2} + \dfrac{15m^2}{4} - 3m + 1$ **29.** $4r^4 + \dfrac{8\sqrt{2}r^3}{m} + \dfrac{12r^2}{m^2} + \dfrac{4\sqrt{2}r}{m^3} + \dfrac{1}{m^4}$ **31.** $-3584h^3j^5$
33. $319{,}770a^{16}b^{14}$ **35.** $38{,}760x^6y^{42}$ **37.** $90{,}720x^{28}y^{12}$ **39.** 11   **41.** .942   **43.** 1.015
45. $1 - x + x^2 - x^3 + x^4 - \ldots$ **49.** exact: 3,628,800; approximate: 3,598,695.619   **50.** $\approx .830\%$
51. exact: 479,001,600; approximate: 475,687,486.5; $\approx .692\%$ **52.** exact: 6,227,020,800; approximate: 6,187,239,475;
$\approx .639\%$; As n gets larger, the percent error decreases.   **53.** -1

CONNECTIONS (page 893)   **1.** $f(1) = 3(1) + 1 = 4$; $f(2) = 3(2) + 1 = 7$; $f(3) = 3(3) + 1 = 10$

2. $g(k) = \dfrac{k(3k + 5)}{2}$ **3.** $g(k) + f(k + 1) = \dfrac{k(3k + 5)}{2} + [3(k + 1) + 1]$

$$= \dfrac{3k^2 + 5k}{2} + 3k + 4$$

$$= \dfrac{3k^2 + 5k + 2(3k + 4)}{2}$$

$$= \dfrac{3k^2 + 11k + 8}{2}$$

$$= \dfrac{(k + 1)(3k + 8)}{2}$$

$$= \dfrac{(k + 1)[3(k + 1) + 5]}{2}$$

$$= g(k + 1)$$

11.5 EXERCISES (page 896)   **1.** S_1: $2 = 1(1 + 1)$; S_2: $2 + 4 = 2(2 + 1)$; S_3: $2 + 4 + 6 = 3(3 + 1)$;
S_4: $2 + 4 + 6 + 8 = 4(4 + 1)$; S_5: $2 + 4 + 6 + 8 + 10 = 5(5 + 1)$ **15.** $n = 1$ or 2   **17.** $n = 2, 3,$ or 4

31. $\dfrac{4^{n-1}}{3^{n-2}}$ or $3\left(\dfrac{4}{3}\right)^{n-1}$

11.6 EXERCISES (page 904)   **1.** 19,958,400   **3.** 72   **5.** 5   **7.** 6   **9.** 1   **11.** 495
13. (a) permutation   **(b)** permutation   **(c)** combination   **(d)** combination   **(e)** permutation   **(f)** combination   **(g)** permutation
15. 30   **17. (a)** 27,600   **(b)** 35,152   **(c)** 1104   **19.** 15   **21. (a)** 17,576,000   **(b)** 17,576,000   **(c)** 456,976,000   **23.** 720
25. 120   **27.** 2730   **29.** 120; 30,240   **31.** 27,405   **33.** 20   **35.** 10   **37.** 28   **39. (a)** 84   **(b)** 10   **(c)** 40   **(d)** 28
41. 1680   **43.** 15   **45.** 479,001,600   **47. (a)** 56   **(b)** 462   **(c)** 3080   **(d)** 8526   **49.** 15,120
59. (a) $3.04140932 \times 10^{64}$ **(b)** $8.320987113 \times 10^{81}$ **(c)** $8.247650592 \times 10^{90}$

11.7 EXERCISES (page 914)

1. Let h = heads, t = tails. $S = \{h\}$

3. $S = \{(h, h, h), (h, h, t), (h, t, h), (t, h, h), (h, t, t), (t, h, t), (t, t, h), (t, t, t)\}$

5. Let c = correct, w = wrong. $S = \{(c, c, c), (c, c, w), (c, w, c), (w, c, c), (w, w, c), (w, c, w), (c, w, w), (w, w, w)\}$

7. (a) $\{h\}$;   1   **(b)** $\emptyset$;   0   **9. (a)** $\{(c, c, c)\}$; $\dfrac{1}{8}$ **(b)** $\{(w, w, w)\}$; $\dfrac{1}{8}$ **(c)** $\{(c, c, w), (c, w, c), (w, c, c)\}$; $\dfrac{3}{8}$

(d) $\{(c, w, w), (w, c, w), (w, w, c), (c, c, w), (c, w, c), (w, c, c), (c, c, c)\}$; $\dfrac{7}{8}$ **13. (a)** $\dfrac{1}{5}$ **(b)** 0   **(c)** $\dfrac{7}{15}$ **(d)** 1 to 4   **(e)** 7 to 8

15. 1 to 4   **17.** $\dfrac{2}{5}$ **19. (a)** $\dfrac{1}{2}$ **(b)** $\dfrac{7}{10}$ **(c)** $\dfrac{2}{5}$ **21. (a)** .62   **(b)** .27   **(c)** .11   **(d)** .89   **23.** $\dfrac{48}{28{,}561} \approx .001681$

25. (a) .72   **(b)** .70   **(c)** .79   **27.** $\approx .313$ **29.** $\approx .031$ **31.** .5   **33.** $\approx 4.6 \times 10^{-10}$ **35.** $\approx .875$

37. The probabilities, in order, are .125, .375, .375, and .125.   **39. (a)** $\approx 40.4\%$ **(b)** $\approx 4.7\%$ **(c)** $\approx .2\%$;   This means that in a large family or group of people, it is highly unlikely that everyone will become sick even though the disease is highly infectious.

41. (a) $P_{00} = 1$; $P_{01} = 0$; $P_{02} = 0$; $P_{10} = \dfrac{1}{6}$; $P_{11} = \dfrac{2}{3}$; $P_{12} = \dfrac{1}{6}$; $P_{20} = 0$; $P_{21} = 0$; $P_{22} = 1$

(b) $P = \begin{bmatrix} P_{00} & P_{01} & P_{02} \\ P_{10} & P_{11} & P_{12} \\ P_{20} & P_{21} & P_{22} \end{bmatrix} = \begin{bmatrix} 1 & 0 & 0 \\ \dfrac{1}{6} & \dfrac{2}{3} & \dfrac{1}{6} \\ 0 & 0 & 1 \end{bmatrix}$

(c) The matrix is symmetric. The sum of the probabilities in each row is equal to 1. The greatest probabilities lie along the diagonal. This means that a mother cell is most likely to produce a daughter cell like itself.

CHAPTER 11 REVIEW EXERCISES (page 921)

1. $\dfrac{1}{2}, \dfrac{2}{3}, \dfrac{3}{4}, \dfrac{4}{5}, \dfrac{5}{6}$;   neither   **3.** 8, 10, 12, 14, 16;   arithmetic

5. 5, 2, -1, -4, -7;   arithmetic   **7.** $3\pi - 2, 2\pi - 1, \pi, 1, -\pi + 2$ **9.** $-5, -1, -\dfrac{1}{5}, -\dfrac{1}{25}, -\dfrac{1}{125}$

11. -1; $-8\left(\dfrac{1}{2}\right)^{n-1} = -\left(\dfrac{1}{2}\right)^{n-4}$ or 1; $-8\left(-\dfrac{1}{2}\right)^{n-1} = \left(-\dfrac{1}{2}\right)^{n-4}$ **13.** $-x + 61$ **15.** 612   **17.** $\dfrac{4}{25}$ **19.** -40

21. 1   **23.** $\dfrac{73}{12}$ **25.** 3,126,250   **27.** $\dfrac{4}{3}$ **29.** 36   **31.** diverges   **33.** -10 **35.** $\displaystyle\sum_{i=1}^{15} (-5i + 9)$ **37.** $\displaystyle\sum_{i=1}^{6} 4(3)^{i-1}$

39. $x^4 + 8x^3y + 24x^2y^2 + 32xy^3 + 16y^4$ **41.** $243x^{5/2} - 405x^{3/2} + 270x^{1/2} - 90x^{-1/2} + 15x^{-3/2} - x^{-5/2}$ **43.** $-3584x^3y^5$

45. $x^{12} + 24x^{11} + 264x^{10} + 1760x^9$ **55.** 1   **57.** 362,880   **59.** 48   **61.** 24   **63.** 504

65. (a) $\dfrac{4}{15}$ **(b)** $\dfrac{2}{3}$ **(c)** 0   **67.** $\dfrac{1}{26}$ **69.** $\dfrac{4}{13}$ **71.** .86   **73.** 0   **75.** $\approx .205$

CHAPTER 11 TEST (page 923)

1. $-3, 6, -11, 18, -27$;   neither   **2.** $-\dfrac{3}{2}, -\dfrac{3}{4}, -\dfrac{3}{8}, -\dfrac{3}{16}, -\dfrac{3}{32}$;   geometric

3. 2, 3, 7, 13, 27;   neither   **4.** 49   **5.** $-\dfrac{32}{3}$ **6.** 110   **7.** -1705 **8.** 2385   **9.** -186 **10.** does not exist

11. $\dfrac{108}{7}$ **12.** $x^6 + 6x^5y + 15x^4y^2 + 20x^3y^3 + 15x^2y^4 + 6xy^5 + y^6$ **13.** $16x^4 - 96x^3y + 216x^2y^2 - 216xy^3 + 81y^4$

14. $60w^4y^2$ **15.** 45   **16.** 35   **17.** 990   **18.** 40,320   **20.** 24   **21.** 6840   **22.** 6160   **24.** $\dfrac{1}{26}$ **25.** $\dfrac{10}{13}$ **26.** $\dfrac{4}{13}$

27. 3 to 10   **28.** .92   **29.** $\approx .104$ **30.** $\approx .000000595$

Index